BRUCE A. KAPLAN

 **W9-CPO-150**

Regression Graphics

Regression Graphics

Ideas for Studying Regressions through Graphics

R. DENNIS COOK

The University of Minnesota
St. Paul, Minnesota

A Wiley-Interscience Publication

JOHN WILEY & SONS, INC.

New York · Chichester · Weinheim · Brisbane · Singapore · Toronto

Library of Congress Cataloging-in-Publication Data:

Cook, R. Dennis.
    Regression graphics: ideas for studying regressions through graphics/R. Dennis Cook.
      p.   cm. – (Wiley series in probability and statistics.
  Probability and statistics section)
    "A Wiley-Interscience publication."
    Includes bibliographical references and index.
    ISBN 0-471-19365-8 (cloth : alk. paper)
    1. Regression analysis—Graphic methods. I. Title. II. Series: Wiley series in probability and statistics. Probability and statistics.

QA278.2.C6647   1998
519.5'36–dc21
                                                   98-3628
                                                    CIP

Printed in the United States of America

10 9 8 7 6 5 4 3 2 1

To Jami

Contents

Preface

Humans are good, she knew, at discerning subtle patterns that are really there, but equally so at imagining them when they are altogether absent.

Carl Sagan (1985), *Contact*

Simple graphs have always played a useful role in the analysis and presentation of data. Until about 12 years ago, my personal view of statistical graphics was mostly confined to relatively simple displays that could be produced on a teletype or CRT terminal. Many displays were based on six-line plots. Today statistical graphics aren't so simple. Advances in computing have stimulated ideas that go far beyond the historically dominant graphics, and that have the potential to substantially expand the role of visualization in statistical analyses. I understand that much of modern computer graphics can be traced back to the pioneering work on *PRIM-9* (Fisherkeller, Friedman, and Tukey 1974) and to Peter Huber's visions for *PRIM-ETH* and *PRIM-H* (Cleveland and McGill 1988). David Andrew's Macintosh program *McCloud* provided my first exposure to three-dimensional plots, although computer graphics didn't really become a concrete tool for me until after Luke Tierney began work on *XLISP-STAT*, a programming language that allows the user to implement graphical ideas with relatively little difficulty (Tierney 1990).

This book is about ideas for the graphical analysis of regression data. The original motivation came from wondering how far computer graphics could be pushed in a regression analysis. In the extreme, is it possible to conduct a regression analysis by using just graphics? The answer depends on the semantics of the question, but under certain rather weak restrictions it seems that the possibility does exist. And in some regressions such an analysis may even be desirable.

This book is not about how to integrate graphical and nongraphical methodology in pursuit of a comprehensive analysis. The discussion is single-minded, focusing on graphics unless nongraphical methods seem essential for progress. This should not be taken to imply that nongraphical methods are somehow less appropriate or less desirable. I hope that this framework will facilitate an understanding of the potential roles for graphics in regression.

CONTEXT

In practice, regression graphics, like most of statistics, requires both an application context and a statistical context. Statistics exists as a discipline because statistical contexts apply across diverse areas and, together with an application context, provide a foundation for scientific inquiries subject to random variation. Much of this book is devoted to a relatively new statistical context for regression and regression graphics. This new context is intended to blend with rather than replace more traditional paradigms for regression analysis. (See, for example, Box 1980.) It imposes few scope-limiting restrictions on the nature of the regression and for this reason it may be particularly useful at the beginning of an analysis for guiding the choice of a first model, or during the model-checking phase when the response is replaced by a residual. Basing an entire regression analysis on graphics is also a possibility that is discussed.

OUTLINE

Chapter 1 is a light introduction to selected graphical issues that arise in a familiar context. Notational conventions are described at the end of this chapter. Chapter 2 consists of a number of miscellaneous topics to set the stage for later developments, including two-dimensional scatterplots and scatterplot matrices, smoothing, response transformations in regressions with a single predictor, plotting exchangeable pairs, and a little history. In the same spirit, some background on constructing an illusion of a rotating three-dimensional scatterplot on a two-dimensional computer screen is given in Chapter 3. The main theme of this book begins in Chapter 4.

Much of the book revolves around the idea of reducing the dimension of the predictor vector through the use of central dimension-reduction subspaces and sufficient summary plots. These and other central ideas are introduced in Chapter 4 where I make extensive use of three-dimensional scatterplots for graphical analyses of regression problems with two predictors and a many-valued response. In the same vein, graphics for regressions with a binary response are introduced in Chapter 5. The development of ideas stemming from central dimension-reduction subspaces is continued in Chapters 6, 7, and 8 by allowing for many predictors. Practical relevance of these ideas is explored in Chapter 9 through a number of examples.

Starting in Chapter 4, steps in the development of various ideas for regression graphics are expressed as propositions with justifications separated from the main text. This formal style is not intended to imply a high degree of mathematical formalism, however. Rather, I found it convenient to keep track of results and to separate justifications to facilitate reading. Generally, knowledge of mathematical statistics and finite dimensional vector spaces is required for the justifications.

The graphical foundations are expanded in Chapter 10 by incorporating inverse regressions. Numerical methods for estimating a central dimension-reduction subspace via inverse regressions are discussed in Chapters 11 and 12.

Traditional models start to play a more central role in Chapter 13, which is devoted to ideas for studying the roles of individual predictors. Chapter 14 is on graphical methods for visualizing predictor transformations in linear models. Graphical methods for model assessment are studied in Chapter 15. Finally, each chapter ends with a few problems for those who might like to explore the ideas and methodology further.

Residuals are an important part of graphics for regression analyses, and they play key roles in this book. But they are not singled out for special study. Rather, they occur throughout the book in different roles depending on the context.

No color is used in the plots of this book. Nevertheless, color can facilitate the interpretation of graphical displays. Color and three-dimensional versions of selected plots, data sets, links to recent developments, and other supple-mental information will be available via *http://www.stat.umn.edu/RegGraph/*. A few data sets are included in the book.

ACKNOWLEDGMENTS

Earlier versions of this book were used over the past six years as lecture notes for a one-quarter course at the University of Minnesota. Most students who attended the course had passed the Ph.D. preliminary examination, which in part requires a year of mathematical statistics and two quarters of linear models. The students in these courses contributed to the ideas and flavor of the book. In particular, I would like to thank Efstathia Bura, Francesca Chiaromonte, Rodney Croos–Dabrera, and Hakbae Lee, who each worked through an entire manuscript and furnished help that went far beyond the limits of the course. Dave Nelson was responsible for naming "central subspaces." Bret Musser helped with computing and the design of the Web page.

Drafts of this book formed the basis for various short courses sponsored by Los Alamos National Laboratory, the Brazilian Statistical Association, South-ern California Chapter of the American Statistical Association, University of Birmingham (U.K.), University of Waikato (New Zealand), International Bio-metric Society, Seoul National University (Korea), Universidad Carlos III de Madrid, the Winter Hemavan Conference (Sweden), the University of Hong Kong, and the American Statistical Association.

I would like to thank many friends and colleagues who were generous with their help and encouragement during this project, including Richard Atkinson,

Frank Critchley, Doug Hawkins, Ker-Chau Li, Bret Musser, Chris Nachtsheim, Rob Weiss, Nate Wetzel, and Joe Whittaker. I am grateful to Joe Eaton for helping me reason through justifications that I found a bit tricky, and to Harold Henderson for his careful reading of the penultimate manuscript. Sandy Weisberg deserves special recognition for his willingness to engage new ideas.

Some of the material in this book was discussed in the recent text *An Introduction to Regression Graphics* by Cook and Weisberg (1994a). That text comes with a computer program, the R-code, that can be used to implement all of the ideas in this book, many of which have been incorporated in the second generation of the R-code. All plots in this book were generated with the R-code, but the development is not otherwise dependent on this particular computer program.

I was supported by the National Science Foundation's Division of Mathematical Sciences from the initial phases of this work in 1991 through its completion.

R. DENNIS COOK

St. Paul, Minnesota
January 1998

CHAPTER 1

Introduction

The focus of this book is fairly narrow relative to what could be included under the umbrella of statistical graphics. We will concentrate almost exclusively on regression problems in which the goal is to extract information from the data about the statistical dependence of a response variable y on a $p \times 1$ vector of predictors $x = (x_j)$, $j = 1, \ldots, p$. The intent is to study existing graphical methods and to develop new methods that can facilitate understanding how the conditional distribution of $y \mid x$ changes as a function of the value of x, often concentrating on the *regression function* $E(y \mid x)$ and on the *variance function* $Var(y \mid x)$. If the conditional distribution of $y \mid x$ was completely known, the regression problem would be solved definitively, although further work may be necessary to translate this knowledge into actions. Just how to choose an effective graphical construction for extracting information from the data on the distribution of $y \mid x$ depends on a variety of factors, including the specific goals of the analysis, the nature of the response and the regressors themselves, and available prior information. The graphical techniques of choice may be quite different depending on whether the distribution of $y \mid x$ is essentially arbitrary or is fully specified up to a few unknown parameters, for example.

Familiarity with standard graphical and diagnostic methods based on a linear regression model is assumed (see, for example, Cook and Weisberg 1982). These methods will be reviewed briefly as they are needed to illustrate specific ideas within the general development, but we will not usually be studying them in isolation. Similar remarks hold for smoothing two-dimensional scatterplots.

Regardless of the specific attributes of the regression problem at hand, it is useful to recognize three aspects of using graphical methodology to gain insights about the distribution of $y \mid x$: *construction*, *characterization*, and *inference* (Cook 1994a).

1.1. C C & I

1.1.1. Construction

Construction refers to everything involved in the production of the graphical display, including questions of what to plot and how to plot. Deciding what

1

to plot is not always easy and again depends on what we want to accomplish. In the initial phases of an analysis, two-dimensional displays of the response against each of the p predictors are obvious choices for gaining insights about the data, choices that are often recommended in the introductory regression literature. Displays of residuals from an initial exploratory fit are frequently used as well.

Recent developments in computer graphics have greatly increased the flexibility that we have in deciding how to plot. Some relatively new techniques include scatterplot rotation, touring, scatterplot matrices, linking, identification, brushing, slicing, and animation. Studying how to plot is important, but is not a focal point of this book. As for the graphical techniques themselves, we will rely on scatterplot displays along with various graphical enhancements. Two- and three-dimensional scatterplots and scatterplot matrices will be used most frequently, but higher-dimensional scatterplots will be encountered as well. The graphical enhancements include brushing, linking, rotation, slicing, and smoothing. While familiarity with basic construction methods is assumed, some of the more central ideas for this book are introduced briefly in Chapter 2. Methods for constructing a rotating three-dimensional plot on a two-dimensional computer screen are discussed in Chapter 3.

In striving for versatile displays, the issue of what to plot has been relatively neglected. In linear regression, for example, would a rotating three-dimensional scatterplot of case indices, leverages, and residuals be a useful display? More generally, how can three-dimensional plots be used effectively in regression problems with many predictors? Huber's (1987) account of his experiences with three-dimensional scatterplots is interesting reading as a reference point in the development of statistical graphics. His emphasis was more on the construction of displays and how the user could interact with them, and less on how the displays could be used to advance data analysis. For the most part, three-dimensional scatterplots were seen as effective tools for viewing spatial objects (e.g., galaxies) and for finding " ... outliers, clusters and other remarkable structures." Three-dimensional plotting of linear combinations of variables was discouraged.

In this book, quantities to plot will generally be dictated by the development starting in Chapter 4.

1.1.2. Characterization

Characterization refers to what we see in the plot itself. What aspects of the plot are meaningful or relevant? Is there a linear or nonlinear trend? clusters? outliers? heteroscedasticity? a combination thereof?

There are several standard characterizations for two-dimensional scatterplots of residuals versus fitted values from an ordinary least squares (OLS) fit of a linear regression model, including a curvilinear trend, a fan-shaped pattern, isolated points, or no apparent systematic tendency. For further reading on these characterizations, see Anscombe (1973), Anscombe and Tukey

(1963), or Cook and Weisberg (1982, p. 37). Characterizing a two-dimensional scatterplot is relatively easy, particularly with the full range of recently developed graphical enhancements at hand. However, standard patterns to watch for in three-dimensional plots are not as well understood as they are in many two-dimensional plots. We can certainly look for very general characteristics like curvature in three-dimensional plots, but it may not be clear how or if the curvature itself should be characterized. It is also possible to obtain useful insights into higher-dimensional scatterplots, but for the most part their interpretation must rely on lower-dimensional constructions. Similar statements apply to scatterplot matrices and various linked plots. Scatterplot matrices are discussed briefly in Section 2.5.

Beginning with Chapter 4, one central theme of this book is that characterizations of various two- and three-dimensional scatterplots can provide useful information about the distribution of $y \mid x$, even when the number of predictors p is appreciable.

1.1.3. Inference

It would do little good to construct and characterize a display if we don't then know what to do with the information.

Imagine inspecting a scatterplot matrix of data on (y, x^T). Such a display consists of a square array of $p(p + 1)$ scatterplots, one for each ordered pair of distinct univariate variables from (y, x^T). The plots involving just pairs of predictors may be useful for diagnosing the presence of collinearity and for spotting high-leverage cases. The marginal plots of the response against each of the individual predictors allow visualization of aspects of the marginal regressions problems as represented by $y \mid x_j$, $j = 1, \ldots, p$, particularly the marginal regression functions $\mathrm{E}(y \mid x_j)$ and the corresponding variance functions. There is surely a wealth of information in a scatterplot matrix. But in the absence of a *context* that establishes a connection with the conditional distribution of $y \mid x$, its all about sidelights having little to do with the fundamental problem of regression. For example, the presence of collinearity is generally of interest because the distributions of $y \mid (x = a)$ and $y \mid (x = b)$ are relatively difficult to distinguish when a and b are close. The presence of collinearity may tell us something about the relatively difficulty of the analysis, but it says nothing about the distribution of $y \mid x$ *per se*.

Similar remarks apply to other modern graphical displays like rotating 3D displays. What can a three-dimensional plot of the response and two predictors tell us about the distribution of $y \mid x$ when $p > 2$? when $p = 2$? Again, without a context that establishes a connection between the graphic and the full regression $y \mid x$, the characterization of a display may be of limited use. In a commentary on graphics in the decades to come, Tukey (1990) stressed that " ... we badly need a detailed understanding of purpose," particularly for deciding when a display reflects relevant phenomena. A similar theme can be found in an overview of graphical methods by Cox (1978). The construction of imponderable displays seems much too easy with all the modern graphical machinery. Even the distinction between ponderable and imponderable displays may be lost without a context. The work of Tufte (1983) provides an important lesson that such distinctions are important at all levels.

Ideally, the analysis of a plot would be concluded by forming a well-grounded inference about the nature of the distribution of $y \mid x$, about the data itself, about how to carry on the analysis, or about the possibility of unexpected phenomena. Inferences need not depend solely on the characterization, although this is often the case. An inference might also depend on other aspects of the analysis or on prior information, for example.

Detecting a fan-shaped pattern in the usual plot of residuals versus fitted values from the OLS regression of y on x leads to the data-analytic observation that the residual variance increases with the fitted values. This is a characterization of a common regression scatterplot. The inference comes when this characterization is used to infer heteroscedastic errors for the true but unknown regression relationship and to justify a weighted regression or transformation. As a second example, Cook and Weisberg (1989) suggested that a saddle-shaped pattern in certain detrended added variable plots can be used to infer the need for interaction terms. While the regression literature is replete with this sort of advice, bridging the gap between the data-analytic characterization of a plot and the subsequent inference often requires a leap of faith regarding properties of $y \mid x$.

The characterization of a plot, which is a data-analytic task, and the subsequent inference have often been merged in the literature, resulting in some confusion over crucial aspects of regression graphics. Nevertheless, it still seems useful to have a collective term: We will use *interpretation* to indicate the combined characterization-inference phase of a graphical analysis.

1.2. ILLUSTRATIONS

1.2.1. Residuals versus fitted values

Figure 1.1 gives a plot of the OLS residuals versus fitted values from fitting a plane with 100 observations on two predictors $x = (x_1, x_2)^T$. The predictors were sampled from a Pearson type II distribution on the unit disk in $\mathbb{R}^2$ (Johnson 1987, p. 111), where $\mathbb{R}^t$ denotes t-dimensional Euclidean space. This plot exhibits the classical fan-shaped pattern that characterizes heteroscedasticity. While the construction and characterization of this plot are straightforward, any inferences must depend strongly on context.

Residual plots like the one in Figure 1.1 are often interpreted in the context of a linear regression model, say

$$y \mid x = \beta_0 + \beta^T x + \sigma(\beta^T x)\varepsilon \tag{1.1}$$

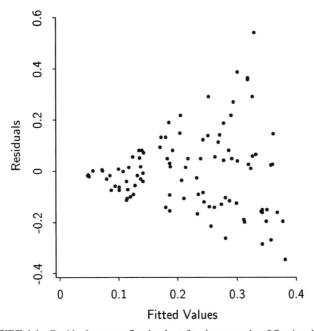

FIGURE 1.1 Residuals versus fitted values for the example of Section 1.2.1.

where ε is independent of x, $\mathrm{E}(\varepsilon) = 0$, and $\mathrm{Var}(\varepsilon) = 1$. The OLS estimate of β is unbiased under this model, even if the standard deviation function is nonconstant. Thus, the residual plot provides useful information on $\sigma(\beta^T x)$, allowing the viewer to assess whether it is constant or not. In a related context, nonconstant variance in residual plots can indicate that a homoscedastic linear model would be appropriate after transformation of the response. Because of such results, a variance-stabilizing transformation of the response or a weighted regression is often recommended as the next step when faced with a residual plot like the one in Figure 1.1. While these ideas are surely useful, tying graphical methods to a narrow target model like (1.1) severely restricts their usefulness. Graphics are at their best in contexts with few scope-limiting constraints, allowing the user an unfettered view of the data.

How should we interpret Figure 1.1 if the context is expanded to allow essentially arbitrary regression and variance functions,

$$y \mid x = \mathrm{E}(y \mid x) + \sigma(x)\varepsilon$$

where ε is as defined previously? Now the heteroscedastic pattern can be a manifestation of a nonlinear regression function or a nonconstant variance function, or both, and we can conclude only that the homoscedastic linear regression model does not reasonably describe the data.

The response for Figure 1.1 was generated as

$$y \mid x = \frac{|x_1|}{2 + (1.5 + x_2)^2} + \varepsilon$$

with homoscedastic error ε that is independent of x. A rotating three-dimensional plot with y on the vertical axis and the predictors on the horizontal axes resembles a round tilted sheet of paper that is folded down the middle, the fold increasing in sharpness from one side to the other. The plot in Figure 1.1 is essentially a projection of the sheet onto the plane determined by the vertical axis and the direction of the fold, so that the paper appears as a triangle. Interpreting Figure 1.1 in the context of a linear regression model could not yield the best possible explanation with the available data. A similar example was given by Cook (1994a).

1.2.2. Residuals versus the predictors

Another common interpretation arises in connection with a scatterplot of the OLS residuals from a linear regression versus a selected predictor x_j. In reference to such a plot Atkinson (1985, p. 3), for example, provides the standard interpretation, "The presence of a curvilinear relationship suggests that a higher-order term, perhaps a quadratic in the explanatory variable, should be added to the model." In other words, if a plot of the residuals versus x_j is characterized by a curvilinear trend, then infer that a higher-order term in x_j is needed in the model.

For example, consider another constructed regression problem with 100 observations on a univariate response y and 3×1 predictor vector $x = (x_1, x_2, x_3)^T$ generated as a multivariate normal random variable with mean 0 and nonsingular covariance matrix. Figure 1.2 gives a scatterplot matrix of the predictors and the OLS residuals $\hat{e}$ from the linear model $y \mid x = \beta_0 + \beta^T x + \sigma\varepsilon$. The residuals here estimate $\sigma\varepsilon$.

The bottom row of plots in Figure 1.2 shows the residuals plotted against each of the three predictors. The plot of $\hat{e}$ versus x_1 and the plot of $\hat{e}$ versus x_3 both exhibit a nonlinear trend (characterization), an indication that the distribution of the residuals depends on x_1 and x_3, and thus that the linear model is deficient (inference). Restricting the context to the class of homoscedastic regression models $y \mid x = E(y \mid x) + \sigma\varepsilon$, where the distribution of ε does not depend on x, what additional inferences can be obtained from Figure 1.2? Since the plot of $\hat{e}$ versus x_2 shows no clear systematic tendencies, would restricting attention to the manner in which x_1 and x_3 enter the model necessarily point us in the right direction?

The response for the data of Figure 1.2 was generated as

$$y \mid x = |x_2 + x_3| + 0.5\varepsilon$$

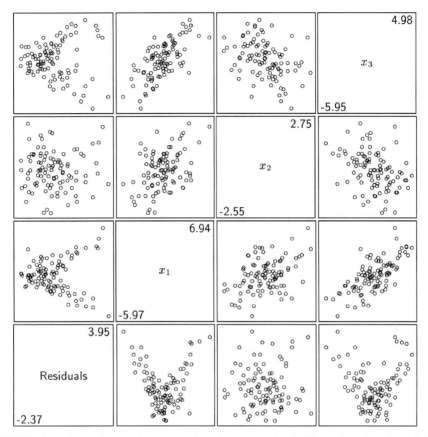

FIGURE 1.2 Scatterplot matrix of residuals and predictors for the illustration of Section 1.2.2.

where ε is a standard normal random variable. Clearly revisions of the linear model that restrict attention to x_1 and x_3 cannot yield the best explanation. Plots of residuals versus individual predictors cannot generally be regarded as sufficient diagnostics of model failure. They can fail to indicate model deficiency when there is a nonlinearity in the predictor in question (x_2), they can indicate model failure in terms of a predictor that is not needed in the regression (x_1), and they can correctly indicate model failure for a directly relevant predictor (x_3).

Even when the context is correct, popular interpretations of standard graphical constructions can result in misdirection.

1.2.3. Residuals versus the response

Draper and Smith (1966) played a notable role in the evolution of regression methodology. They stressed the importance of graphics and promoted a

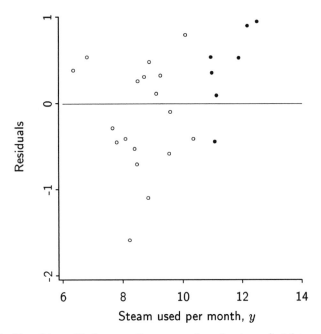

FIGURE 1.3 Plot of the residuals versus the response from the steam plant data used by Draper and Smith (1966). The 7 observations with the largest responses are highlighted.

number of specific graphical methods for assessing model adequacy. The first edition of their text stood for a number of years as a point of introduction to regression. Many regression texts of today offer relatively little graphical methodology beyond that originally available in Draper and Smith, who relied primarily on unadorned two-dimensional scatterplots, particularly scatterplots of residuals versus fitted values or selected predictors.

However, Draper and Smith (1966, p. 122) also used a scatterplot of the OLS residuals versus the response in an example relating the pounds of steam used monthly y at a large industrial concern to the number of operating days per month x_1 and the average monthly atmospheric temperature x_2. Their plot is reconstructed here as Figure 1.3. They *characterized* the plot by noting that " ... six out of the seven largest values of Y have positive residuals." From this they *inferred* that " ... the model should be amended to provide better prediction at higher [response] values." We now know that this inference was not well founded. While the characterization is valid, the response and residuals from an OLS fit are always positively correlated, so the finding that "six out of the seven largest values of Y have positive residuals" is not necessarily noteworthy. In their second edition, Draper and Smith (1981, p. 147) describe the pitfalls of plotting the residuals versus the responses. The point of this illustration is not to criticize Draper and Smith. Rather, it is to reenforce the general idea that there is much more to regression graphics than construction and characterization. Many different types of graphical displays

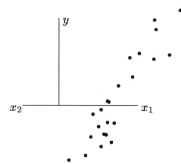

FIGURE 1.4 Two-dimensional view of the three-dimensional scatterplot of the steam data; steam used per month y, operating days per month x_1, monthly atmospheric temperature x_2.

have been developed since the first edition of Draper and Smith (1966), most of which are included in widely available software. And there has probably been a corresponding increase in the number of ways in which we can be misled.

A three-dimensional plot of the steam data from Draper and Smith is quite informative and seems easy to interpret relative to some three-dimensional plots. It shows that the two months with the smallest operating days do not conform to the linear regression established by the remaining data. Figure 1.4 gives one two-dimensional view of this three-dimensional plot. This view was obtained by rotating the three-dimensional plot to the two-dimensional projection that provides the best visual fit while mentally ignoring the two outlying months. The idea of visual fitting is developed in Chapter 4 for regressions with $p = 2$ predictors. In later chapters these ideas are adapted to allow visual fitting when $p > 2$.

1.3. ON THINGS TO COME

Progress on the types of issues raised in this chapter may be possible if we can establish a context allowing the development of connections between graphical methodology and the object of interest, principally the distribution of $y \mid x$. Beginning in Chapter 4, a central theme of this book is based on finding a simplified version of $y \mid x$ by reducing the algebraic dimension of the predictor vector

- without losing information on the response, and
- avoiding, as far as possible, the introduction of assumptions on the nature of the conditional distribution of $y \mid x$.

Let η denote a $p \times k$ matrix, $k \leq p$, so that $y \mid x$ and $y \mid \eta^T x$ are identically distributed. The *dimension-reduction subspace* $\mathcal{S}(\eta)$ spanned by the columns of η can be used as a superparameter to index the distribution of $y \mid x$ and thus taken as the target for a graphical inquiry. A plot of y versus the k-dimensional vector $\eta^T x$ is called a *sufficient summary plot*. In the end, an estimate $\hat{\eta}$ of η can be used to form a plot of y versus $\hat{\eta}^T x$ that serves as a summary of the data. This paradigm may have practical merit because very often it seems that good summaries can be obtained with $k = 1$ or 2. In developing this paradigm, all plots are judged on their ability to provide graphical information about $\mathcal{S}(\eta)$, and the three aspects of regression graphics—construction, characterization, and inference—come packaged together.

Dimension-reduction subspaces, summary plots, and other central notions are introduced in Chapter 4, which makes extensive use of three-dimensional scatterplots for graphical analyses of regression problems with two predictors. The basic ideas and methodology developed in that chapter will also play a central role in regression with many predictors. The fundamental material in Chapter 4 does not depend on the nature of the response, but graphical implementation does. Basic constructions are adapted for regressions with a binary response in Chapter 5.

The development of ideas stemming from dimension-reduction subspaces is continued in Chapters 6, 7, and 8 by permitting many predictors and refining the foundations. Practical relevance of these ideas is explored in Chapter 9 through a number of examples.

The graphical foundations are expanded in Chapter 10 by incorporating inverse regressions, where x plays the role of the response and y becomes the predictor. Numerical methods for estimating a dimension-reduction subspace via inverse regressions are discussed in Chapters 11 and 12.

Traditional models start to play a more central role in Chapter 13, which is devoted to ideas for studying the roles of individual predictors. Chapter 14 is on graphical methods for visualizing predictor transformations in linear models. Finally, graphical methods for model assessment are studied in Chapter 15.

1.4. NOTATIONAL CONVENTIONS

Abbreviations

- OLS, ordinary least squares
- d, the structural dimension for the regression under consideration
- DRS, dimension-reduction subspace
- $\mathcal{S}_{y|x}$ central DRS for the regression of y on x

Data and random variables

When discussing generic regression problems, the response will generally be denoted by y and the $p \times 1$ vector of predictors will be denoted by x. Typically, $x \in \mathbb{R}^p$ is a random vector, where $\mathbb{R}^p$ denotes p-dimensional Euclidean space, and $\text{Var}(x) = \Sigma$. Random variables are not represented solely by uppercase letters and are not normally distinguished from their realizations. Once the difference between random variables and their realizations is understood, it seems somewhat pedantic to carry on careful notational distinctions.

Notation of the form $y \mid x$ refers to the (distribution of the) response given x. More complete notation, $y \mid (x = x_0)$ or $y \mid (x \in H)$, is occasionally used when it seems useful to emphasize the conditioning value.

The subscript "i" will be reserved to indicate instances of y and x so that the raw data are of the form (y_i, x_i), $i = 1, \ldots, n$. Alphabetic subscripts other than i indicate coordinates of x, that is, individual predictors. Thus, for example, x_j and x_k are the jth and kth coordinates of x, respectively. Numerical subscripts attached to x often indicate a partition: $x^T = (x_1^T, x_2^T)$, for example. The data (y_i, x_i), $i = 1, \ldots, n$, are almost always assumed to be independent and identically distributed observations from the joint distribution of (y, x).

Distribution functions
Following a common convention, all cumulative distribution functions (CDFs) will be denoted by $F(\cdot)$ with subscripts indicating the random variables involved, except when a specific family is assumed or clarity of exposition seems to require different notation. Thus, for example, $F_{y|x}(\cdot)$ is the distribution function of y given x, $F_x(\cdot)$ is the marginal distribution function of x, and $F_{y|a^T x}(\cdot)$ is the distribution function of y given the indicated linear combinations of x. Typically, CDFs are used only as a mechanism for referencing distributions; the argument of a CDF will rarely be needed. All densities are with respect to Lebesgue measure and will be denoted similarly by f.

Expectations
$E_{u|z}(v(u))$ is the expectation of $v(u)$, a function of u, with respect to the conditional distribution of $u \mid z$. It is more conventional to write conditional expectations as $E(v(u) \mid z)$, but such expressions can be difficult to read with multiple conditioning when v is a lengthy expression.

Independence
Let u, v, and z be random vectors. The notation $u \perp\!\!\!\perp v$ will be used to indicate that u and v are independent. The notation $u \perp\!\!\!\perp v \mid z$ indicates conditional independence: u and v are independent given any value for z. Similarly, $u \perp\!\!\!\perp v \mid (z = z_0)$ indicates that u and v are independent when $z = z_0$, without necessarily implying anything about the independence of u and v at other values of z.

Matrices
No special notational conventions are used for matrices or vectors.

Models

A model will be represented by a statement of the form

$$y \mid x = g(x, \varepsilon)$$

which means that the distribution of $y \mid x$ is the same as the distribution of $g(x, \varepsilon) \mid x$ for all values of x. The function g may be known or unknown and x may be random or fixed depending on context. The error ε is assumed to be independent of the predictor unless explicitly stated otherwise. For example, the usual homoscedastic linear model for $y \mid x$ can be represented as

$$y \mid x = \beta_0 + \beta^T x + \sigma \varepsilon$$

with ε independent of x,

$$E(\varepsilon \mid x) = E(\varepsilon) = 0$$

and

$$Var(\varepsilon \mid x) = Var(\varepsilon) = 1.$$

Models will also be represented directly in terms of the conditional distribution function. For example, letting $a \in \mathbb{R}^p$ be a fixed vector, the equality $F_{y|x} = F_{y|a^T x}$ represents a model. Such expressions are restricted to versions of $F_{y|x}$ and $F_{y|a^T x}$ so that $F_{y|x} = F_{y|a^T x}$ for all values of x in the marginal sample space.

Projection operators

Let B denote a symmetric positive definite matrix. The notation $P_{A(B)}$ denotes the perpendicular projection operator which projects onto $S(A)$ relative to the inner product $(a, b)_B = a^T B b$, $P_{A(B)} = A(A^T B A)^{-1} A^T B$. Also, $Q_{A(B)} = I - P_{A(B)}$, $P_A = P_{A(I)}$, and $Q_A = I - P_A$.

Residuals

A generic residual will be denoted by r. Other designations may be used for residuals from specific situations. Sample residuals from an ordinary least squares fit of a linear model will usually be denoted by $\hat{e}$, with e used to denote the population version.

Scatterplots

When referring to a *plot* of y_i versus x_i, we shall mean a $(p + 1)$-dimensional scatterplot of the points (y_i, x_i) in Cartesian coordinates with y on the "vertical" axis and the coordinates of x allocated to the remaining p axes in any convenient way. A plot of a_i versus b_i will often be denoted by $\{a_i, b_i\}$ or more simply by $\{a, b\}$ with the understanding that the first argument, which will always be a scalar, is assigned to the vertical axis. The coordinates of the second argument b_i, which may be a vector, can be assigned to the "horizontal" axes in any convenient way unless indicated otherwise. The plot $\{a_i, b_i\}$ will occasionally be denoted by $\{A, B\}$ where A denotes the vector with elements a_i, and B is the matrix with rows b_i^T.

When describing three-dimensional scatterplots the letters H, V, and O will be used to designate the horizontal, vertical, and out-of-page axes.

Subsets
The notation $A \subset B$ means that every point of A is contained in B; that is, either $A = B$ or A is a proper subset of B.

Subspaces, S
A subspace will always be denoted by S. When used with a $t \times w$ matrix argument, say A, $S(A)$ will have one of two meanings. If A is defined, then $S(A)$ is the subspace of $\mathbb{R}^t$ spanned by the columns of A. If the subspace S is defined, then $S(A)$ indicates that subspace with the columns of A implicitly defined to form a basis for S.

PROBLEMS

1.1. Consider a regression problem in which (y, x_1, x_2) follows a trivariate normal distribution with a nonsingular covariance matrix. Describe how to construct examples so that

- The plot $\{y, x_1\}$ shows a clear linear relationship, and yet x_1 is not needed in the regression of y on (x_1, x_2). Equivalently, $y \mid x_1$ depends on the value of x_1, but $y \perp\!\!\!\perp x_1 \mid x_2$.
- The plot $\{y, x_1\}$ shows no dependence and yet x_1 is needed in the full regression. Equivalently, $y \mid x_1$ does not depend on the value of x_1, but $y \mid x$ does depend on the value of x_1.

1.2. Let (y, x_1, x_2) be distributed as a trivariate normal random vector. Is it possible to construct an example so that $y \perp\!\!\!\perp x_1 \mid x_2$ and $y \perp\!\!\!\perp x_2 \mid x_1$, and yet y is dependent on (x_1, x_2)?

CHAPTER 2

Introduction to 2D Scatterplots

Two-dimensional (2D) scatterplots can be effective for obtaining an initial impression of regression data, particularly with the aid of various graphical enhancements such as scatterplot smoothers. We begin the study of 2D scatterplots in the context of a simple regression problem with univariate response y and univariate predictor x, assuming throughout that the data (y_i, x_i) are independent and identically distributed realizations on the bivariate random variable (y, x). A variety of miscellaneous topics is introduced in this chapter, including smoothing, transforming the response for linearity, visualizing exchangeable data, and scatterplot matrices. To provide a little historical perspective, the work of Mordecai Ezekiel and Louis Bean on graphically fitting generalized additive models is discussed near the end of this chapter.

When V and H are scalars, the notation $\{V, H\}$ indicates a 2D scatterplot with V on the vertical axis and H on the horizontal axis.

2.1. RESPONSE PLOTS IN SIMPLE REGRESSION

There are three primary reasons for inspecting the 2D *response plot* $\{y, x\}$ in simple regression problems:

- to study the bivariate distribution of (y, x),
- to study how properties of $y \mid (x = x_o)$ vary with x_o, and
- to identify outliers and other anomalies in the data.

Inverse regression, the study of the conditional distribution of $x \mid y$, may be relevant in some problems as well. In this chapter we focus mostly on properties of the *forward regression* $y \mid x$.

Initial impressions of a 2D response plot are often sufficient to decide if the data contradict the possibility that $y \perp\!\!\!\perp x$. When the distribution of $y \mid x$ depends on the value of x, interest usually centers on the *regression function* $\mathrm{E}(y \mid x)$ and on the *variance function* $\mathrm{Var}(y \mid x)$. Studying higher-order moments graphically seems difficult without fairly large sample sizes, although skewness can often be detected visually.

14

Smooths of 2D response plots can serve as useful graphical enhancements for inferring about qualitative characteristics of regression or variance functions. They can be particularly useful for detecting departures from linearity, and for selecting predictor transformations to linearize a regression function when appropriate. Although smoothing is used from time to time in this book, it is not a topic for study *per se*. Knowledge of the basic ideas underlying smoothing is assumed.

For an introduction to the literature on smoothing, see Altman (1992), Wand and Jones (1995), or Simonoff (1996). Background on kernel smoothers is available in Härdle (1990); Green and Silverman (1994) is a useful source of information on smoothing via splines. We will rely mostly on the *LOWESS* smoother (Cleveland 1979) as implemented in the *R-code* (Cook and Weisberg 1994a); *LOWESS* is an acronym for locally weighted scatterplot smoothing. A more recent version of the LOWESS smoother, called *loess*, was developed by Cleveland, Devlin, and Grosse (1988). While there is considerable literature on how to choose an "optimal" value for the tuning parameter that controls the amount of smoothing, we will be using smooths primarily as visual enhancements and will not be striving for optimal procedures. Most smooths in this book were determined visually by interactively manipulating the tuning parameter while observing the changing smooth superimposed on a scatterplot of the data. Marron and Tsybakov (1995) discussed distinctions between visual fitting and other methods, and described possibilities for quantifying visual impressions when smoothing scatterplots.

Cleveland and McGill (1984) gave an informative discussion on various aspects of scatterplot construction that are not revisited here, including smoothing for the conditional distribution of $y \mid x$ and the joint distribution of (y, x), point-cloud sizing, construction of point symbols, and category codes. The aspect ratio, shape parameter, and other aspects of two-dimensional plotting were studied by Cleveland, McGill, and McGill (1988). See also Cook and Weisberg (1994a).

2.2. NEW ZEALAND HORSE MUSSELS

A sample of 201 horse mussels was collected at 5 sites in the Marlborough Sounds at the Northeast of New Zealand's South Island (Camden 1989). The response variable is muscle mass M, the edible portion of the mussel, in grams. The quantitative predictors all relate characteristics of mussel shells: shell width W, shell height H, shell length L, each in *mm*, and shell mass S in grams. Indicator predictors for site may be relevant when variation in muscle mass from site to site is of interest.

The actual sampling plan used to generate these data is unknown. Nevertheless, to ease interpretation we will assume that the data are independent and identically distributed observations from the total combined population over

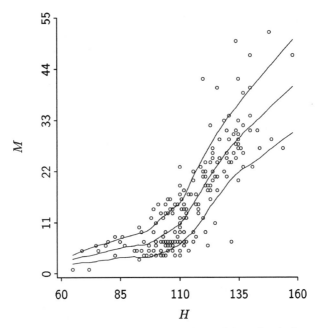

FIGURE 2.1 Scatterplot of muscle mass M versus shell height H for the horse mussel data. The curves on the plot are formed by LOWESS $\pm$ the LOWESS smooth of the standard deviation.

the five sites. This assumption may be appropriate if the sites were sampled with probability proportional to size.

We begin the study of the mussel data by using the 2D plot $\{M, H\}$, as shown in Figure 2.1, to investigate the behavior of the conditional distribution of $M \mid H$ as the value of H changes. For now, ignore the curves superimposed on the plot; they will be discussed shortly. Figure 2.1 displays observations from a series of conditional distributions, one for each value of H in the data. There is not really enough data to allow the distribution of $M \mid H$ to be studied by conditioning on various values of H. However, if the first two moments of $M \mid H$ are sufficiently smooth functions, then they can be approximated by *slicing* (Tukey and Tukey 1983) the plot parallel to the vertical axis around a selected value H_o for H. Figure 2.2 shows three such slices for $H_o = 105$, 120 and 140. In effect, slicing is used to approximate observations on $M \mid (H = H_o)$ with observations on $M \mid (h_L \leq H \leq h_U)$, where H_o is at the midpoint of the *slice window*, $[h_L, h_U]$. The *window width*, $\lambda = h_U - h_L$, may be chosen visually so that the magnitudes of the changes in the distribution of $M \mid (H = H_o)$, as H_o varies in the slice window, are small relative to those as H_o varies in the observed range of H. The window widths of the slices in Figure 2.2 are about 8 *mm*. In the smoothing literature, slicing is sometimes referred to as *binning*.

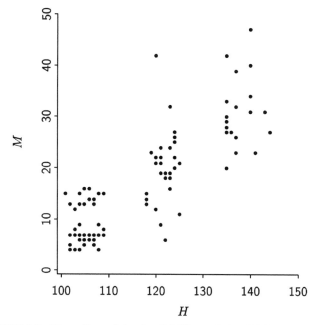

FIGURE 2.2 Three slices of the plot $\{M,H\}$ rescaled to fill the plotting area.

The middle curve superimposed on the plot of Figure 2.1 is a LOWESS smooth (Cleveland 1979); LOWESS uses the fitted value at H_i from a local linear fit to estimate $E(M \mid H_i)$. Instead of using fixed width slices as described earlier, LOWESS bases each local fit on the nearest neighbors of H_i. The fraction of points taken as the nearest neighbors is the tuning parameter for a LOWESS smooth. The smooth of Figure 2.1 suggests that the regression function consists roughly of two linear phases with a transition around 105 *mm*.

The upper and lower curves of Figure 2.1 were constructed as

$$\hat{m}(H_i) \pm \hat{s}(H_i), \qquad i = 1,\ldots,201$$

where $\hat{m}$ represents the LOWESS smooth of the regression function and $\hat{s}$ represents the corresponding LOWESS estimate of the conditional standard deviation function $s(H) = \sqrt{\text{Var}(M \mid H)}$. The standard deviation was estimated by taking the square root of the LOWESS smooth of $(M_i - \hat{m}(H_i))^2$ against H_i, with tuning parameter the same as that used for the regression function. Using the same value of the tuning parameter for both the regression and variance functions facilitates interactive analysis because only a single parameter needs to be manipulated while observing the smooths. Additionally, this restriction is in line with Stone's (1977) discussion on smooth estimation of conditional second moments.

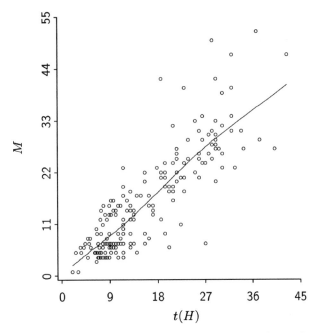

FIGURE 2.3   Scatterplot of M versus t with a LOWESS smooth.

The literature on nonparametric estimation of the variance function is rather sparse, at least relative to that for the regression function. However, Ruppert, Wand, Holst, and Hössjer (1997) recently reported a study of local polynomial smoothers for estimation of the variance function. They show it is possible to estimate the variance function with no asymptotic loss of information due to lack of knowledge of the regression function. This paper provides a useful entry point to the literature on nonparametric estimation of the variance function.

We see from Figures 2.1 and 2.2 that there is ample evidence to contradict the possibility that $M \perp\!\!\!\perp H$: The regression function $E(M \mid H)$ increases nonlinearly with the value of H and the conditional variance function $\text{Var}(M \mid H)$ seems to increase as well. The two small clusters in the slice at $H = 105$ in Figure 2.2 may indicate a mixture of distinct regressions, perhaps reflecting site differences.

Let $t_i = \hat{m}(H_i)$ denote shell height transformed according to the LOWESS smooth of the regression function in Figure 2.1. If the smooth is a good characterization of the regression function then $E(M \mid t)$ should be linear in t. To check this conjecture we can extract the transformed values t_i, $i = 1, \ldots, 201$, from Figure 2.1 and then inspect the plot of muscle mass versus the transformed values of shell height, as shown in Figure 2.3. Characterizing $E(M \mid t)$ as an approximately linear function of t may be reasonable in view of the LOWESS smooth superimposed on the scatterplot.

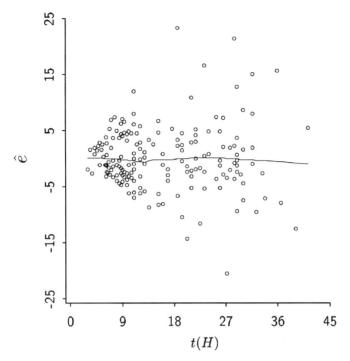

FIGURE 2.4 Scatterplot of the OLS residuals $\hat{e}$ versus t from the mussel data.

Turning to heteroscedasticity, Figure 2.4 shows the plot $\{\hat{e}, t\}$, where $\hat{e}$ denotes a typical residual from the OLS linear regression of M on t. The presence of heteroscedasticity is indicated by the fan-shaped pattern in the plot. The rather sharp boundary on the lower left of the scatter in Figure 2.4 is due to the fact that there are several values of M near the minimum value for muscle mass. More carefully, the residuals are

$$\hat{e}_i = M_i - \hat{b}_0 - \hat{b}_1 t_i$$

where $\hat{b}_0$ and $\hat{b}_1$ are the OLS coefficients from the linear regression of M on t. If we select the points in the plot $\{\hat{e}, t\}$ that correspond to nearly equal values M_0 of M, then among the selected points we will have

$$\hat{e}_i = M_0 - \hat{b}_0 - \hat{b}_1 t_i.$$

Thus, the selected points will fall close to a line with slope $-\hat{b}_1$ and intercept $M_0 - \hat{b}_0$. The lower-left boundary in Figure 2.4 corresponds to choosing M_0 near the minimum muscle mass. For further discussion of this type of phenomenon in residual plots see Searle (1988) and Nelder (1989).

2.3. TRANSFORMING y VIA INVERSE RESPONSE PLOTS

In addition to predictor transformations described briefly in the last section, monotonic transformations of the response y can also facilitate analysis of simple regression problems. The transformation methodology pioneered by Box and Cox (1964) is perhaps the most common.

To review the basic ideas behind the Box–Cox methodology, assume that $y > 0$ and that there is a scaled power transformation

$$y^{(\lambda)} = \frac{y^\lambda - 1}{\lambda} \tag{2.1}$$

such that

$$y^{(\lambda)} \mid x = \beta_0 + \beta x + \sigma\varepsilon \tag{2.2}$$

where ε is an approximately normal random variable that is independent of x, $E(\varepsilon) = 0$, and $\text{Var}(\varepsilon) = 1$. The transformation is monotonic, and $y^{(\lambda)} \to \log(y)$ as $\lambda \to 0$. For a fixed value of λ, let $\hat{\beta}_0(\lambda)$ and $\hat{\beta}(\lambda)$ be the estimates of β_0 and β, respectively, computed from the ordinary least squares regression of $y^{(\lambda)}$ on x. Inference for λ can be based on the profile log likelihood $L(\lambda)$ that stems from model (2.2),

$$L(\lambda) = -\frac{n}{2} \log \left[\sum_{i=1}^{n} \{y_i^{(\lambda)} - \hat{\beta}_0(\lambda) - \hat{\beta}(\lambda)x_i\}^2 \right] + \log[J(\lambda)]$$

where (x_i, y_i) is the ith realization of (x, y), $i = 1, \ldots, n$, and $J(\lambda)$ is the Jacobian of the transformation. For additional review, see Atkinson (1985), Carroll and Ruppert (1988), and Cook and Weisberg (1982). The Box–Cox method is used widely. And yet it provides no direct way of visualizing the transformation against the data.

A graphical method for determining a response transformation to linearize the regression function is developed in this section. The method was proposed by Cook and Weisberg (1994b) and is based on the *inverse response plot* $\{x, y\}$. The goal of choosing a response transformation to linearize the regression function is different from that for the Box–Cox method, which tries to achieve simultaneously a linear regression function, a constant variance function, and normally distributed residuals.

2.3.1. Response transformations

Assume that there is a strictly monotonic transformation $t(y)$ so that a simple linear regression model holds in the transformed scale,

$$t(y) \mid x = \beta_0 + \beta x + \varepsilon \tag{2.3}$$

where $\varepsilon \perp\!\!\!\perp x$, $E(\varepsilon) = 0$, and $\text{Var}(\varepsilon) = \sigma^2$. If $\beta = 0$ then $y \perp\!\!\!\perp x$ and the regression function is linear for all transformations. We avoid this simple case by requiring $\beta \neq 0$. Model (2.3) is similar to the model (2.2) underlying Box–Cox transformations, but there are two important differences: The errors are not assumed to be normally distributed and a parametric family of transformations is not required. The transformation is not unique since if $t(y)$ satisfies (2.3) so will any linear function of $t(y)$. This nonuniqueness will not hinder the development that follows, however. Any strictly monotonic response transformation that satisfies (2.3) will be denoted by $t(y)$.

If an appropriate transformation were known then it could be visualized directly in the plot $\{t(y), y\}$. From (2.3), $t(y)$ is known up to a linear transformation of x plus an error, and this suggests that the inverse response plot $\{x, y\}$ might be used to visualize t. For this suggestion to be most useful, we should have the *inverse regression function* $E(x \mid y) = t(y)$, or approximately so. Equivalently, we require that

$$E(w \mid y) = t(y) \tag{2.4}$$

where $w = \beta_0 + \beta x$. Working in terms of w is equivalent to working in terms of x because if (2.4) holds then it also holds for any linear transformation of w.

To see situations in which (2.4) may hold to a reasonable approximation, consider the following:

$$E(w \mid y) = E(w \mid t(y)) \tag{2.5}$$

$$= E(w + \varepsilon - \varepsilon \mid t(y))$$

$$= t(y) - E(\varepsilon \mid w + \varepsilon) \tag{2.6}$$

where (2.5) follows because $t(y)$ is a strictly monotonic function of y. Thus, a sufficient condition for $E(w \mid y)$ to provide an appropriate transformation is that $E(\varepsilon \mid w + \varepsilon)$ is linear in $t(y) = w + \varepsilon$, which constrains the joint distribution of w and ε. If (w, ε) follows an elliptically contoured distribution[1] then $E(\varepsilon \mid t(y))$ is linear in $t(y)$ and (2.4) is guaranteed. Because we have assumed that $\varepsilon \perp\!\!\!\perp w$, the only possible elliptically contoured distribution is the bivariate normal. Normality is stronger than is really needed since linearity may hold in a number of other cases as well. If ε and w have the same marginal distribution, for example, then (2.4) again holds.

Equation (2.4) may be a practically useful approximation when $E(w \mid y)$ is not an exact linear function of $t(y)$, provided the nonlinearity is not too strong. The population correlation ρ between $E(w \mid t(y))$ and $t(y)$ can be used

---

[1] Elliptically contoured distributions are reviewed in Section 7.3.2.

to characterize the degree of linearity:

$$\rho = \frac{\text{Cov}(\text{E}(w \mid t), t)}{[\text{Var}(\text{E}(w \mid t))\text{Var}(t)]^{1/2}} \tag{2.7}$$

where $t(y)$ has been written as t for notational convenience. We will need the following to get this correlation into a more informative form:

$$\text{Var}(t) = \text{Var}(w) + \text{Var}(\varepsilon)$$

because ε and w are independent, and

$$\text{Cov}(\text{E}(w \mid t), t) = \text{Cov}(t - \text{E}(\varepsilon \mid t), t)$$

$$= \text{Var}(t) - \text{E}(t \times \varepsilon)$$

$$= \text{Var}(w).$$

Substituting these results into (2.7) yields

$$\rho = \left[\left(1 + \frac{\text{Var}(\varepsilon)}{\text{Var}(w)} \right) \left(\frac{\text{Var}(\text{E}(w \mid t))}{\text{Var}(w)} \right) \right]^{-1/2}. \tag{2.8}$$

Next, the term $\text{Var}(\text{E}(w \mid t))$ can be rewritten by decomposing $\text{E}(w \mid t)$ as the sum of the fitted values and residuals r from the population OLS linear regression on t,

$$\text{E}(w \mid t) = \text{E}(w) + \gamma(t - \text{E}(t)) + r$$

where $r = \text{E}(w \mid t) - \text{E}(w) - \gamma(t - \text{E}(t))$ and

$$\gamma = \frac{\text{Cov}(\text{E}(w \mid t), t)}{\text{Var}(t)} = \frac{\text{Var}(w)}{\text{Var}(w) + \text{Var}(\varepsilon)}.$$

Because $\gamma(t - \text{E}(t))$ and r are uncorrelated,

$$\text{Var}(\text{E}(w \mid t)) = \gamma^2 \text{Var}(t) + \text{Var}(r).$$

Substituting this result into (2.8) yields one desired form for the correlation,

$$\rho = \left[1 + \frac{\text{Var}(r)}{\text{Var}(w)} \left(1 + \frac{\text{Var}(\varepsilon)}{\text{Var}(w)} \right) \right]^{-1/2}. \tag{2.9}$$

If $\text{E}(w \mid t)$ is linear in t then $\text{Var}(r) = 0$ and consequently $\rho = 1$. Otherwise, equation (2.9) shows that ρ depends on a noise-to-signal ratio $\text{Var}(\varepsilon)/\text{Var}(w)$

and a nonlinearity ratio $\text{Var}(r)/\text{Var}(w)$. Any deviations of $E(w \mid t)$ from linearity are represented by the ratio $\text{Var}(r)/\text{Var}(w)$. If the residual variance is small relative to the variance in w, then the inverse response plot should still be useful for visually assessing a response transformation. If $\text{Var}(\varepsilon)$ is small, then the inverse response plot should again be useful because $\text{Var}(\varepsilon) = 0$ implies $\text{Var}(r) = 0$.

There is another form for the correlation that may be more helpful in practice. Before considering that form we need a brief aside to construct a metaphor for visually assessing strength of relationship in a generic 2D plot $\{y, x\}$. When the regression function $E(y \mid x)$ appears to be visually well-determined, the variance ratio

$$R_{y|x} = \frac{E[\text{Var}(y \mid x)]}{\text{Var}[E(y \mid x)]} \tag{2.10}$$

should be *small*. This ratio is reminiscent of a one-way analysis of variance, with the numerator representing the average within group (slice) variance and the denominator representing the variance between group (slice) means. Equivalently, $R_{y|x}$ can be viewed as the average residual variance divided by the variance of population fitted values.

Two forms of (2.10) are needed to re-express ρ. The first reflects the strength of relationship on the transformed scale,

$$R_{t|w} = \frac{E[\text{Var}(t(y) \mid w)]}{\text{Var}[E(t(y) \mid w)]} = \frac{\text{Var}(\varepsilon)}{\text{Var}(w)} \tag{2.11}$$

and the second reflects the strength of relationship in the inverse response plot

$$R_{w|y} = \frac{E[\text{Var}(w \mid y)]}{\text{Var}[E(w \mid y)]} = \frac{E[\text{Var}(w \mid t)]}{\text{Var}[E(w \mid t)]}. \tag{2.12}$$

Using these definitions in combination with (2.8) we have

$$\rho = \left(\frac{1 + R_{w|y}}{1 + R_{t|w}} \right)^{1/2}. \tag{2.13}$$

If $\text{Var}(r) = 0$, then $R_{w|y} = R_{t|w}$ and $\rho = 1$ as we saw in connection with (2.9). Generally, $R_{w|y} \leq R_{t|w}$ because $\rho \leq 1$. This means that the strength of relationship on the transformed scale, as it appears in the plot $\{t(y), w\}$, cannot exceed that in the inverse response plot $\{w, y\}$.

These results suggest the following graphical procedure for determining a response transformation to linearize the regression function in simple regression problems with response y and predictor x.

- Inspect the inverse response plot $\{x,y\}$. If it seems visually clear that the regression function is nonlinear and monotonic then a transformation of the response may be appropriate. An empirical transformation $\hat{t}(y)$ can be determined by smoothing the inverse response plot, either parametrically or nonparametrically.
- Inspect the transformed data $\{\hat{t}(y),x\}$. If the strength of relationship does not seem noticeably worse than that in the inverse response plot then it is reasonable to conclude that $R_{w|y}$ is close to $R_{t|w}$ and thus that ρ is large. In such cases, nonlinearity in $E(x\,|\,t(y))$ may not be a serious issue.

This graphical method for selecting a response transformation can be adapted for use with multiple predictors $x \in \mathbb{R}^p$: Assume that there is a strictly monotonic transformation $t(y)$ so that a linear regression holds on the transformed scale

$$t(y)\,|\,x = \beta_0 + \beta^T x + \sigma\varepsilon$$

where ε is as defined in (2.3). If we can find a consistent estimate b of a basis for $S(\beta)$ without requiring knowledge of t then the 2D inverse response plot $\{b^T x, y\}$ can be used to select a linearizing transformation as described here. Methods for constructing such consistent estimates are described in later chapters, particularly Section 8.1.

2.3.2. Response transformations: Mussel data

We now use the regression of muscle mass M on shell width W to illustrate the use of an inverse response plot to determine a response transformation. The inverse response plot $\{W,M\}$ is shown in Figure 2.5a. The curve on the plot was determined visually from the family of transformations

$$t(M;\lambda) = b_0 + b_1 M^{(\lambda)}$$

where $M^{(\lambda)}$ is the scaled power transformation given in (2.1). For each value of λ, the constants b_0 and b_1 were determined from the OLS linear regression of W on $M^{(\lambda)}$, and then the resulting curve was superimposed on the plot $\{W,M\}$. Manipulating λ interactively to obtain the best visual fit led to the cube root transformation, $\lambda = 1/3$ shown on Figure 2.5a. A *parametric smooth* was used because it seems to provide a sufficiently flexible family of transformations in this example. A monotonic nonparametric smooth could have been used as well.

A plot of the transformed data $\{M^{1/3}, W\}$ is shown in Figure 2.5b along with a *lowess* smooth and the OLS fitted values. The cube root transformation does seem to linearize the response function quite well. Additionally, the strength of relationship in Figure 2.5a seems similar to that in Figure 2.5b, indicating

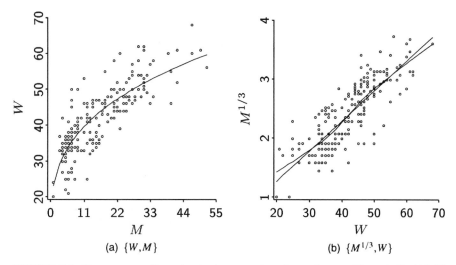

FIGURE 2.5 Response transformation in the regression of muscle mass M on shell width W. (a) Inverse response plot with a cube root transformation superimposed. (b) Response plot for the regression of $M^{1/3}$ on W with a LOWESS smooth and the OLS fitted values superimposed.

that $R_{W|M}$ and $R_{M^{1/3}|W}$ are close. Thus, nonlinearity in $E(W \mid t(M))$ may not be worrisome in this example.

From subject matter considerations, we expect M to be proportional to S, and S to be proportional to the product LWH. Thus, the cube root transformation of M seems reasonable in light of basic allometric considerations.

The next section addresses an entirely different issue.

2.4. DANISH TWINS

The Danish Twin Registry was established in 1954 as a means of collecting and maintaining data on Danish twins. It covers all twins born in Denmark between 1870 and 1910, and all same-sex twins born between 1911 and 1930. The Registry is well-suited for studying genetic influences on aging since it contains accurate information on zygosity and death date. The lifetimes of adult twins from the Registry were studied by Hougaard, Harvald, and Holm (1992). Anderson, Louis, Holm, and Harvald (1992) used Registry data in their development of time-dependent measures of association for bivariate survival distributions. Both articles contain background on the Registry, including descriptions of how the data were collected and verified. Here we consider how an initial graphical inspection of Registry data might be used to gain information about genetic influences on aging. Discussion is restricted to female monozygotic and dizygotic (identical and nonidentical) twins born between 1870 and 1880, and surviving to an age of at least 40 years. Most of the discussion will concern monozygotic twins. Contrasting results for dizygotic twins will be presented a bit later.

Given the death age of a twin, we would like a graphic that can provide information about the distribution of the death age for the other member of the twin pair, without assuming that either twin survives the other. Let D_{at} denote the death age for *a twin* and let D_{ot} denote the death age for the *other twin*. The goal is to construct a plot that provides information on the conditional density functions $f_{D_{ot}|D_{at}}$.

The first three pairs of death ages in the data file for monozygotic twins are $(71.79, 77.80)$, $(67.16, 78.21)$, and $(73.16, 82.74)$. One way to gain graphical information on $f_{D_{ot}|D_{at}}$ is to simply plot the first death age listed in the file against the second death age and then use the ideas discussed previously for 2D plots. In doing so we would run into an annoying problem, however, because such a plot is not unique. Each death age listed in the data file is a realization of D_{at} and of D_{ot} so the data could be permuted randomly within pairs and there would be no way to distinguish the new permuted file from the original file. Using any particular permutation could yield misleading results. In the three pairs listed above, for example, the first item in each pair is smaller than the second.

In most simple regression problems we plot *ordered* pairs (y, x), but in the twin data the underlying random variables D_{ot} and D_{at} are *exchangeable*, so the realized pairs of death ages are not ordered. One consequence of the exchangeability is that the marginal distributions of D_{ot} and D_{at} are the same. It would be good if we could preserve this property in the empirical marginal distributions for any plot.

Suppose we could define ordered death age random variables T_1 and T_2 that underlie the construction of the data file. Then

$$f_{D_{ot}|D_{at}} = \tfrac{1}{2}f_{T_1|T_2} + \tfrac{1}{2}f_{T_2|T_1}.$$

This suggests that both orderings of each pair of death ages be plotted, resulting in a 2D plot with the number of points being double the number of pairs in the data file. For example, the first three twin pairs in the data file for monozygotic twins would be represented by six points $(71.79, 77.80)$, $(77.80, 71.79)$, $(67.16, 78.21)$, $(78.21, 67.16)$, $(73.16, 82.74)$, and $(82.74, 73.16)$. The empirical marginal distributions would be identical in such a doubled plot.

The doubled plot consisting of 222 points for the 111 monozygotic Danish twins born between 1870 and 1880 is shown in Figure 2.6. The horizontal line in this figure is at the average death age, about 76 years. The contrast between the LOWESS smooth and the average death line indicates that there may be genetic factors that influence aging. The same plot for 184 dizygotic twins is shown in Figure 2.7.

A simple random effects model might be useful for the Danish twin data. Let D_{ij} denote the death age of twin $j = 1, 2$ in family $i = 1, \ldots, n$. Then rep-

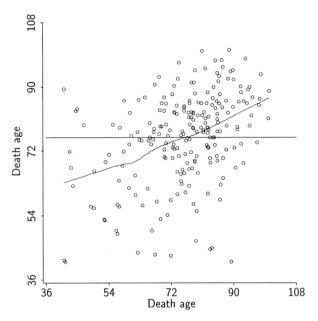

FIGURE 2.6 Doubled plot of death ages for 111 monozygotic Danish twins born between 1870 and 1880.

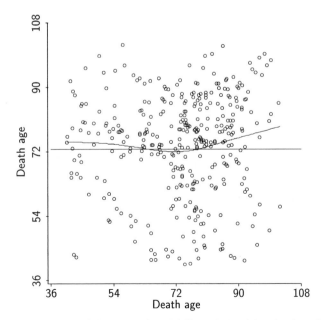

FIGURE 2.7 Doubled plot of death ages for 184 dizygotic Danish twins born between 1870 and 1880.

resenting D in terms of random effects ϕ for family and τ for twin,

$$D_{ij} = \mu + \phi_i + \tau_{ij} \qquad (2.14)$$

where the ϕ_i and τ_{ij} are mutually independent random effects with $\mathrm{E}(\phi_i) = \mathrm{E}(\tau_{ij}) = 0$, $\mathrm{Var}(\phi_i) = \sigma_\phi^2$, and $\mathrm{Var}(\tau_{ij}) = \sigma^2$. Under this model, $\mathrm{E}(D_{ij}) = \mu$, $\mathrm{Var}(D_{ij}) = \sigma_\phi^2 + \sigma^2$, $\mathrm{Cov}(D_{i1}, D_{i2}) = \sigma_\phi^2$, and

$$\mathrm{E}(D_{i1} \mid D_{i2} = d) = \mu + \rho_I(d - \mu) \qquad (2.15)$$

where $\rho_I = \mathrm{corr}(D_{i1}, D_{i2}) = \sigma_\phi^2/(\sigma_\phi^2 + \sigma^2)$ is the *intraclass correlation coefficient*, the fraction of the total variance in death age that can be attributed to variation between families. Relatively high values of the intraclass correlation may indicate a genetic effect on death age.

It may seem that the regression function in (2.15) could be estimated by using the usual OLS methods, but there is still no way to assign death ages uniquely. However, if we add the condition that the random effects in model (2.14) are normally distributed, then the maximum likelihood estimate $\hat{\rho}_I$ of ρ_I is the slope of the OLS regression function computed from the doubled data as in Figures 2.6 and 2.7. Thus the plot of the doubled data provides a way of visualizing the maximum likelihood estimate of ρ_I in terms of OLS. For example, intuition about outliers and influential observations in OLS estimation can be applied to assess visually the effects of individual observations on $\hat{\rho}_I$ when the data are displayed as in Figures 2.6 and 2.7. The only essential change is that the influence of an individual point (D_{i1}, D_{i2}) on $\hat{\rho}_I$ corresponds to the influence of a pair of points (D_{i1}, D_{i2}) and (D_{i2}, D_{i1}) on the OLS slope in a plot of the doubled data.

Perhaps a more common method of estimating ρ_I is to use the moment estimate $\tilde{\rho}_I$ based on the expected mean squares from a one-way analysis of variance,

$$\tilde{\rho}_I = \frac{\mathrm{msb} - \mathrm{msw}}{\mathrm{msb} + \mathrm{msw}} \qquad (2.16)$$

where msb and msw are the usual mean squares between and within families. It is straightforward to verify that

$$|\hat{\rho}_I - \tilde{\rho}_I| = \frac{1}{n}|\hat{\rho}_I - 1|\frac{\mathrm{msb}}{\mathrm{msb} + \mathrm{msw}}. \qquad (2.17)$$

Thus, for reasonably large values of n, $\hat{\rho}_I$ and $\tilde{\rho}_I$ should be close, and the doubled plot can serve to visualize either estimate.

A disadvantage of basing visualization on doubled data is that the plots give the impression that there is more data than is really present. The plots suggested by Ernst, Guerra, and Schucany (1996) might be adapted to deal with this issue.

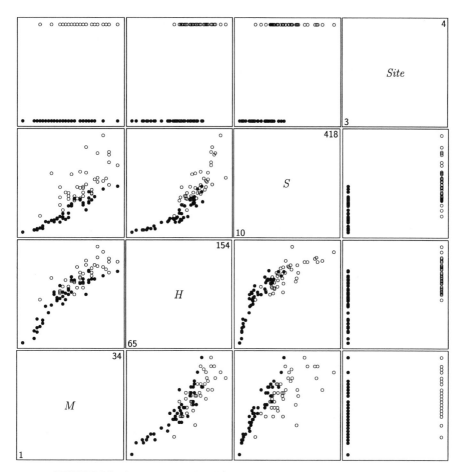

FIGURE 2.8   Scatterplot matrix of four variables from the horse mussel data.

2.5. SCATTERPLOT MATRICES

2.5.1. Construction

A scatterplot matrix is a two-dimensional array of 2D plots as illustrated in Figure 2.8 for the response and three predictors from the horse mussel data. Data on only two of the sites is shown to avoid overplotting in later discussion. Also, ignore the highlighted points for now; they will be discussed a bit later.

Except for the diagonal, each frame in the scatterplot matrix is a 2D scatterplot. The names on the diagonal of the matrix serve to name the axes of each of the 12 2D scatterplots in Figure 2.8. For example, the three scatterplots along the bottom row of the matrix are, from left to right, $\{M,H\}$, $\{M,S\}$, and $\{M,site\}$. The plots above the diagonal are the "transposes" of the plots below the diagonal: The plot in the lower-right frame of Figure 2.8 is $\{M,site\}$

while the plot in the upper left is $\{site, M\}$. The scatterplot matrices produced by some computer packages are symmetric about the other diagonal.

The frames of a scatterplot matrix fall into three categories that reflect distinct roles in the graphical regression methods to be developed in this book. Those with the response on the vertical axis correspond to simple regression problems with each predictor taken in turn. For ease of later reference, such plots will be called *2D marginal response plots*. The frames with the response on the horizontal axis are inverse response plots corresponding to *inverse regression* problems of the form $x \mid y$. The remaining frames of a scatterplot matrix display marginal relationships among the predictors.

In Figure 2.8 the response is M and the three predictors are S, H, and *site*. The bottom row of the scatterplot matrix contains the three marginal response plots. The plot for shell height was discussed previously. The marginal response plot for shell mass suggests that $E(M \mid S)$ is a linear function of the value of S and that $Var(M \mid S)$ is an increasing function of S. The marginal response plot for *site* shows how the conditional distribution of $M \mid site$ varies with site. The predictor plots for shell mass and shell height exhibit both curved regression functions and nonconstant variance functions. Finally, the scatterplot matrix provides a visualization of the usual sample covariance matrix for the variables in the plot.

Depending on the implementation, the frames in a scatterplot matrix may be *linked*, so that when actions such as selecting or deleting points are applied to one frame they are automatically applied to all others. For example, when the points for site 3 were selected in the lower-right frame of Figure 2.8, the corresponding points in all other frames were simultaneously highlighted. The highlighted points in the marginal response plot for shell height allow a visualization of the joint conditional distribution of $(M, H) \mid (site = 3)$ and a visualization of the family of conditional distributions $M \mid (H, site = 3)$ for varying values of $H \mid (site = 3)$. Generally, the highlighted points in Figure 2.8 can be viewed as described in Section 2.1 with the data restricted to the subpopulation in which $site = 3$. Moving a selection rectangle from site to site while observing the other frames allows for contrasting information on the various subpopulations. This type of procedure is called *brushing*.

Slicing a continuous predictor may also be used to select data for relevant subpopulations. For example, slicing around a value for H and observing the marginal response plot for S allows visualizing data on the joint conditional distribution $(M, S) \mid (h_L \leq H \leq h_U)$, which in turn allows visualizing data on $M \mid (S, h_L \leq H \leq h_U)$ as the value of $S \mid (h_L \leq H \leq h_U)$ varies. For further background on scatterplot matrices see Becker and Cleveland (1987), Becker, Cleveland, and Wilks (1987), and Cleveland (1993). Scatterplot matrix techniques to deal with the problem of overplotting when the number of observations is large were discussed by Carr, Littlefield, Nicholson, and Littlefield (1987).

The marginal information provided by scatterplot matrices can be helpful at the preliminary stages of a regression analysis. Nevertheless, it may not

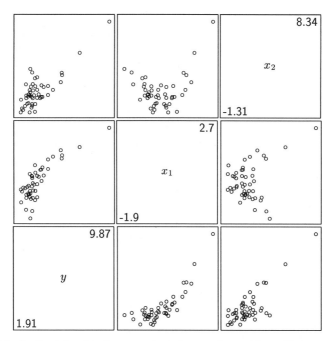

FIGURE 2.9 Scatterplot matrix for a constructed regression problem with two predictor, and 50 observations.

be clear if a scatterplot matrix provides any usable joint information on the regression of the response on all of the predictors. Does Figure 2.8 provide any clear information on the behavior of $M \mid (H,S,site)$ as the values of the three predictors vary? Since $E(M \mid H)$ is apparently a nonlinear function of the value of H, does it necessarily follow that $E(M \mid H,S,site)$ is a nonlinear function of H? The development of a context to help answer such questions is one of the tasks of this book. The next section contains an example to help define the issues.

2.5.2. Example

Figure 2.9 shows a scatterplot matrix for a constructed regression problem with two predictors $x = (x_1, x_2)^T$ and 50 observations. The marginal response plot for x_1 exhibits curvature, and there may be a suggestion of heteroscedasticity as well, although this is questionable because of the uneven distribution of points along the horizontal axis. The marginal response plot for x_2 doesn't seem to show curvature, but the two cases with the largest values of x_2 may turn out to be outliers or influential observations in the fit of a linear model.

Assume now that it is reasonable to describe the distribution of $y \mid x$ in terms of the model

$$y \mid x = g(x_1, x_2) + \sigma \varepsilon$$

where $\varepsilon \perp\!\!\!\perp x$, $E(\varepsilon) = 0$, and $Var(\varepsilon) = 1$. What does the scatterplot matrix tell us about g? Are we justified inferring that g is nonlinear, perhaps containing a quadratic in x_1, because the marginal response plot for x_1 is curved? More generally, does the scatterplot matrix contain any useful information that can be applied to $y \mid x$?

The individual response plots in a scatterplot matrix do provide a lower bound on the fraction of the variation in the response that can be explained jointly by using all the predictors. Otherwise, the study of marginal plots may provide little relevant information about a higher-dimensional regression without further knowledge or constraints. The model used to generate the scatterplot matrix in Figure 2.9 is $y \mid x = 3 + x_1 + x_2/2$—without error.

One reason that the marginal response plot for x_1 may give a misleading impression has to do with the empirical distribution of the predictors represented in the predictor frames of Figure 2.9. There is a curvilinear relationship between the predictors, and this is what forces the curved pattern in the marginal response plot, although the true regression function $E(y \mid x)$ is linear in both predictors. In particular,

$$E(y \mid x_1) = 3 + x_1 + E(x_2 \mid x_1)/2.$$

The frame $\{x_2, x_1\}$ in Figure 2.9 shows that $E(x_2 \mid x_1)$ is a nonlinear function of x_1, and thus $E(y \mid x_1)$ is a nonlinear function of x_1. If we are to have a chance of using marginal plots to infer about higher-dimensional regressions, then the behavior of the predictors must be taken into account.

Returning briefly to the scatterplot matrix for the horse mussel data, we see that there is curvature in some of the predictor plots. This means that the curvature in the marginal response plot for shell height H may be due to curvature among the predictors rather than curvature in H of the regression function $E(M \mid H, S, site)$.

2.6. REGRESSION GRAPHICS IN THE 1920s

In this section we review some of the contributions by Mordecai Ezekiel and Louis H. Bean to the development of regression graphics during the 1920s. Many of the graphical ideas in modern statistical computing packages can be found in the writings of Bean and Ezekiel.

2.6.1. Ezekiel's successive approximations

Around the turn of the century, *curvilinear regression* involved the fitting of additive regressions of the form

$$E(y \mid x) = \alpha + \sum_{j=1}^{p} g_j(x_j) \tag{2.18}$$

where the functions g_j are unknown. It was apparently standard at the time to use power terms like $g_j = \alpha_j x_j^2$ or $g_j = \alpha_j \sqrt{x_j}$ to obtain adaptable regressions. Based on graduate work at the University of Minnesota in 1923, Ezekiel (1924) went beyond parametric functions and suggested a smoothing algorithm for estimating the g's nonparametrically. He expanded and refined his algorithm in his 1930 book *Methods of Correlation Analysis*, and in his subsequent second edition (Ezekiel 1930a; Ezekiel 1941). Ezekiel's treatment of regression anticipated many of the techniques that are popular today, including *component-plus-residual* plots (Wood 1973), which are the same as *partial residual plots* (Larsen and McClearly 1972), and additive modeling (Hastie and Tibshirani 1990). In developing his algorithm, he made extensive use of residual plots, smoothing, and slicing.

The following is a sketch of Ezekiel's algorithm for fitting (2.18):

- Initialize

$$g_j^{(0)}(x_j) = b_j x_j, \qquad j = 1 \ldots p$$

and

$$\hat{y}^{(0)} = b_0 + \sum_{j=1}^{p} b_j x_j$$

where the b_j's are the regression coefficients from the OLS linear regression of y on x.

- For each $\ell = 1, 2, \ldots$, set

$$g_k^{(\ell)} = \text{Smooth of } \left\{ \hat{y}^{(\ell-1)} - \sum_{j \neq k}^{p} g_j^{(\ell-1)}(x_j), x_k \right\}, \qquad j = 1 \ldots p$$

$$\alpha^{(\ell)} = \text{Average} \left[y - \sum_{j=1}^{p} g_j^{(\ell-1)}(x_j) \right]$$

$$(2.19)$$

and

$$\hat{y}^{(\ell)} = \alpha^{(\ell)} + \sum_{j=1}^{p} g_j^{(\ell)}(x_j).$$

Continue until there are no further noteworthy changes in any of the estimates $g_k^{(\ell)}$.

The backfitting algorithm described by Hastie and Tibshirani (1990) is quite similar to Ezekiel's algorithm, which would not be very difficult to implement today. However, in Ezekiel's time, implementation required considerable effort. The smoothing of the plots in (2.19) was carried out "freehand." As an aid to freehand smoothing, Ezekiel suggested plotting points for the averages of slices along the horizontal axes, and somewhat later (Ezekiel 1930b; Ezekiel 1940) reported on an investigation to develop a method for determining standard errors of a freehand curve.

Ezekiel's idea was apparently well-received, prompting Bean (1929) to suggest a simpler graphical method for fitting additive models, becoming known as "the graphic method," "the short-cut graphic method," or "the Bean method" (Wellman 1941). Writing on Bean's graphic method, Waite's (1941) opening two sentences read as follows

> The controversy that has raged over the graphic method of multiple correlation analysis is well known. Its violence attests to the great divergence of opinions regarding the place and the effectiveness of the method.

2.6.2. Bean's graphic method

Following his initial paper in 1929, Bean (1930) later discussed various refinements and presented examples to further illustrate his method. To preserve some historical flavor, the following discussion of the graphic method is set in the context of Bean's first example involving the regression of the yearly yield Y of wheat in an eastern state on three weather variables,

- the rainfall R in inches during February, March, and April,
- snow days S, the number of days on which the average snow depth was one inch or more, and
- the average temperature T in March and April.

Consider now the problem of fitting the regression function

$$\mathrm{E}(Y \mid R,S,T) = \alpha + g_R(R) + g_S(S) + g_T(T).$$

Bean's graphical analysis is based on the following idea: If we can find a subset of the data in which S and T are relatively constant, then a plot of Y versus R for the points in that subset may give useful information on g_R. Bean recognized that such subsets could be quite small and so stated "The steps involved ... center around one important consideration, namely, that the nature of the net relation between the dependent variable [Y] and any of the independents, say [R], can be detected in a few observations at a time in which the relation between [Y] and each of the other variables, [S] and [T], appears to be constant... ." Further, it seems that Bean judged the idea of approximate conditioning (slicing) as one of the most novel aspects of his method (Bean 1940; Ezekiel 1940).

Bean relied on linked scatterplots to implement his idea, where the linking was accomplished by writing unique case labels next to each point in the plots

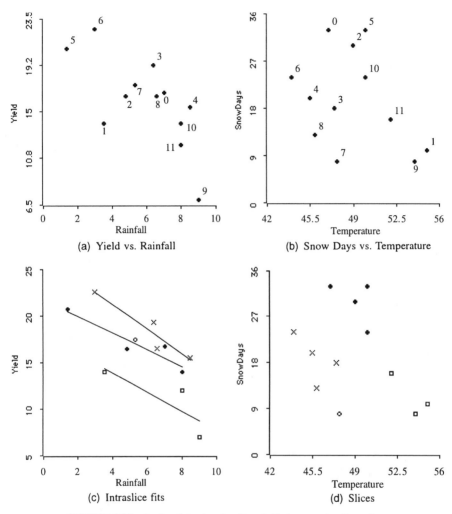

FIGURE 2.10   Scatterplots showing Bean's first approximation of g_R.

to be linked. For example, Figure 2.10a and Figure 2.10b give plots $\{Y,R\}$ and $\{S,T\}$ that are linked in the manner suggested by Bean. To construct a first approximation of g_R, find a two-dimensional slice of points in the plot $\{S,T\}$ where the values of S and T are relatively constant, and then locate the corresponding points in the linked plot $\{Y,R\}$. Letting L denote the set of case labels for the points in the slice of $\{S,T\}$, the plot of the corresponding points in $\{Y,R\}$ will be represented as $\{Y,R \mid L\}$. Next, the plot $\{Y,R \mid L\}$ is smoothed by eye and the smooth, which is sketched on the plot, is taken as a rough intraslice estimate of g_R. This procedure can be repeated over several disjoint slices resulting in several intraslice estimates of g_R displayed on the plot $\{Y,R\}$. These estimates are then averaged by eye, resulting in the first approximation g_R^1 of g_R.

For example, Figure 2.10d shows four slices identified by different plotting symbols. One slice contains only a single observation and played no role in determining a first approximation of g_R. Each slice plot $\{Y,R \mid L\}$ was judged to be linear and the intraslice estimates of g_R were determined by using OLS, rather than by eye as suggested by Bean. The three intraslice estimates of g_R shown in Figure 2.10c seem to have about the same slope, and the first approximation g_R^1 (not shown) was set equal to the fitted value from the OLS regression of Y on R using all the data.

Having determined g_R^1, a first approximation of g_S is constructed according to Bean's method by repeating the above procedure using the residuals $Y - g_R^1$ as the response, slicing on T, and constructing intraslice estimates of g_S by smoothing plots of the form $\{Y - g_R^1, S \mid L\}$, where L now represents the subsets of case indices from slicing T. For example, the upper and lower curves in Figure 2.11a are intraslice estimates of g_S formed by using LOWESS smooths of the subsets of cases corresponding to $T > 49$ and $T < 49$. The middle curve, which is the first approximation of g_S, is the pointwise average of the two intraslice estimates. The first approximation g_S^1 of g_S is distinctly different from the closest curve possible by using a LOWESS smooth of all the data in Figure 2.11a. This distinction is potentially important. Suppose for illustration that $g_R = g_R^1$ were known. Then a smooth of all the data in Figure 2.11a would be an estimate of

$$E(Y - g_R \mid S) = \alpha_0 + g_S + E(g_T \mid S).$$

The final term $E(g_T \mid S)$ represents a potential bias that is mitigated by averaging the intraslice estimates as suggested by Bean.

With the first approximations of g_S and g_R, the first approximation of g_T is constructed by smoothing the plot $\{Y - g_R^1 - g_S^1, T\}$ as shown in Figure 2.11b. Although the details of the implementation in the present example are somewhat different from Bean's (1930), the first approximation g_T^1 is quite similar.

We now have first approximations of each of the three g terms in the regression function. Second approximations can be constructed by beginning again, adjusting the response for previous approximations as the process continues. For example, Figure 2.11c gives Bean's plot for determining the second approximation of g_R with the first approximation superimposed. Ignoring the first approximation, the plot would be smoothed by eye as before to give the second approximation g_R^2. Both approximations g_R^1 and g_R^2 would now be on the same plot, allowing them to be compared. Bean recommended continuing to iterate until successive approximations are sufficiently close.

The discussion here represents little more than an introduction to the basic ideas underlying Bean's graphical method. The plotting methods used by Bean are more intricate than those used here, for example. In addition, Bean and others discussed ways to refine and expand application, including sug-

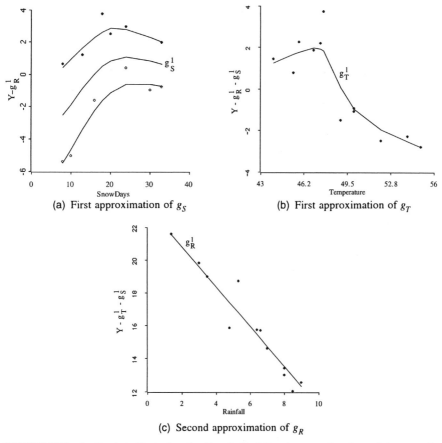

(a) First approximation of g_S

(b) First approximation of g_T

(c) Second approximation of g_R

FIGURE 2.11 Scatterplots illustrating the ideas behind Bean's approximations of the terms in a generalized additive model. The dots and crosses in plot (a) correspond to $T > 49$ and $T < 49$.

gestions for dealing with many predictors, suggestions for dealing with overplotting caused by many observations, how to transfer residuals from one plot to another without doing arithmetic, how to use the intraslice smooths to detect departures from additivity, and how to use plotting symbols to represent the levels of a third variable in a 2D plot. For additional examples of Bean's graphic method, see Ezekiel (1930a, 1941), Malenbaum and Black (1937), and the text *Practical Business Statistics* by Croxton and Cowden (1935, p. 435).

2.7. DISCUSSION

Two-dimensional response plots are relatively easy to characterize, particularly when interactive smoothing is available. Scatterplot matrices for higher-dimensional regressions provide a useful first look at the data. They may show outliers and influential observations, or other relevant anomalies. The interpretation of individual frames in a scatterplot matrix is relatively easy when they are viewed in isolation as simple regression problems. But the interpretation of a scatterplot matrix can be a difficult and perplexing task when viewed in the context of the full regression $y \mid x$. Brushing can help but may serve to complicate matters. We will return to scatterplot matrices in later chapters after the development of results that can facilitate interpretation.

A three-dimensional (3D) plot is another widely available graphical tool that might be used in regression analyses. The construction of 3D plots, which is somewhat more involved than the construction of two-dimensional plots or scatterplot matrices, is considered in the next chapter.

PROBLEMS

2.1. With a single predictor x, assume that the distribution of $y \mid x$ can be described by the simple linear regression model $y \mid x = \beta_0 + \beta x + \sigma \varepsilon$ where $\varepsilon \perp\!\!\!\perp x$, $E(\varepsilon) = 0$, and $Var(\varepsilon) = 1$. Find an expression for $R_{y|x}$ in terms of β, σ^2, and $Var(x)$.

2.2. Verify (2.9), (2.13), and that $R_{w|y} = R_{t|w}$ when $E(w \mid t)$ is linear in t.

2.3. Investigate the usefulness of inverse response plots for determining a response transformation $t(y)$ when $E(t(y) \mid w) = w$ and $E(w \mid t) = b_0 + b_1 t + b_2 t^2$.

2.4. Construct an example in which the Box–Cox method gives a transformation different from the graphical method based on the inverse response plot. When will these methods give essentially the same transformation?

2.5. Suppose model (2.3) is replaced with

$$t(y) \mid x = \beta_0 + \beta x + \sigma(x)\delta \qquad (2.20)$$

where $E(\delta) = 0$, $Var(\delta) = 1$, δ is independent of x, and $\sigma(x) > 0$. This model is like (2.3) except that the error variance is no longer assumed to be constant. Is the graphical procedure for determining t based on the inverse response plot still valid under this more general model? If not, can it be modified appropriately? Are there any important differences in interpretation of the inverse response plot?

2.6. The following problems relate to the discussion of Section 2.4.

    2.6.1. Verify the claim that $\hat{\rho}_I$ is the OLS slope in the plot of the doubled data.

2.6.2. Does the plot of the doubled data in Figure 2.6 provide any useful visual information on $\text{Var}(\hat{\rho}_I)$?

2.6.3. Verify (2.17).

2.7. Suggest a plot for visualizing the intraclass correlation coefficient in a problem where each of n "families" has t "twins." Provide support for your suggestion on plotting this type of exchangeable data.

2.8. In reference to the mussel data, is the data consistent with the conjecture that $M \perp\!\!\!\perp site \mid H$? If not, describe how the regression of $M \mid site$ on $H \mid site$ changes with the site.

CHAPTER 3

Constructing 3D Scatterplots

In this chapter we discuss the construction of three-dimensional (3D) scatter-plots and various associated tuning parameters. Interpretations of 3D plots are considered in the next chapter.

3.1. GETTING AN IMPRESSION OF 3D

A 3D scatterplot is simply a plot of triplets $\ell_i = (H_i, V_i, O_i)^T$, $i = 1, \ldots, n$, relative to the Cartesian coordinate system illustrated in Figure 3.1, where H, V, and O refer to the variables plotted on the horizontal, vertical, and out-of-page axes, respectively. Since it is not possible to construct a true 3D scatterplot on a flat computer screen, we must be content with *kinetic* displays that give an illusion of three dimensions. A kinetic graphic is any graphical display that uses motion to convey information. We assume that the data have been centered at an appropriate location, often midrange of the axis variables, and

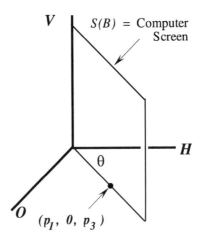

FIGURE 3.1 Cartesian coordinate system for a three-dimensional plot.

that the origin of the coordinate system is at the origin of the centered variables. Centering is used to insure that the origin of the coordinate system falls near the center of the data.

Rotation is the basic operation that allows an impression of three dimensions. This discussion will be restricted to rotation about the vertical axis since this is the primary method for viewing regression data. Imagine a point cloud superimposed on the coordinate system in Figure 3.1 and consider rotating about the vertical axis. Let $p = (p_1, 0, p_3)^T$, as illustrated in Figure 3.1. The subspace $\mathcal{S}(B)$ spanned by the columns of

$$B = \begin{pmatrix} 0 & p_1/\|p\| \\ 1 & 0 \\ 0 & p_3/\|p\| \end{pmatrix}$$

corresponds to the two-dimensional computer screen. At any instant during rotation about the vertical axis, the user sees the point cloud projected onto the subspace $\mathcal{S}(B)$, that is, the point cloud projected onto the surface of the computer screen. By incrementing the angle θ shown in Figure 3.1 and rapidly updating the 2D projection onto $\mathcal{S}(B)$, the display takes on the appearance of a rotating 3D point cloud.

When the 3D plot is not rotating, a standard 2D plot is displayed. For example, prior to rotation ($\theta = 0$) the plot on the computer screen is simply $\{V, H\}$. The out-of-page variable O is not visible because it is orthogonal to the computer screen and so we see just the projection of the data onto the subspace spanned by the horizontal and vertical axes.

For the actual construction, consider projecting a typical point $\ell = (H, V, O)^T$ onto the subspace $\mathcal{S}(B)$ and let P_B denote the orthogonal projection operator for $\mathcal{S}(B)$ with the usual inner product. The projected point has the form

$$P_B\ell = \begin{pmatrix} 0 \\ 1 \\ 0 \end{pmatrix} \times V + \begin{pmatrix} p_1/\|p\| \\ 0 \\ p_3/\|p\| \end{pmatrix} \times \frac{p_1 H + p_3 O}{\|p\|}$$

which is still three dimensional and thus cannot be plotted directly in two dimensions. The point on the computer screen corresponding to $P_B\ell$ consists of the coordinates of $P_B\ell$ relative to the orthonormal basis given by the columns of B. The 2D plot obtained after projecting all points onto $\mathcal{S}(B)$ has the form $\{V, H_p\}$ where

$$H_p = \frac{p_1 H + p_3 O}{\sqrt{p_1^2 + p_3^2}}.$$

Equivalently, since the cosine of the angle θ shown in Figure 3.1 is

$$\cos(\theta) = \frac{p_1}{\sqrt{p_1^2 + p_3^2}}$$

the quantity plotted on the horizontal axis can be expressed as $H(\theta) = \cos(\theta)H + \sin(\theta)O$.

In short, the 2D plot obtained after rotating about the vertical axis through an angle of θ is the plot of V versus $\cos(\theta)H + \sin(\theta)O$. These variables correspond to the coordinates of the projection of $(H, V, O)^T$ onto $S(B)$ relative to the basis B. To obtain the illusion of rotation about the vertical axis, θ is incremented in small steps and the points plotted on the computer screen are rapidly updated. An analogous method is used when rotating about any of the axes in Figure 3.1.

Rotation can be seen equivalently as reexpressing the data in terms of different bases for $\mathbb{R}^3$. The coordinates for Figure 3.1 display each point in terms of the natural orthonormal basis given by the columns of the 3×3 identity matrix. To rotate about the vertical axis we reexpress each data point ℓ in terms of a different orthonormal basis that depends on θ. Specifically, define the orthonormal matrix

$$\Gamma = \begin{pmatrix} \cos(\theta) & 0 & -\sin(\theta) \\ 0 & 1 & 0 \\ \sin(\theta) & 0 & \cos(\theta) \end{pmatrix}.$$

The coordinates of ℓ relative to the basis given by the columns of Γ are just $u = \Gamma^T \ell$ and for each θ we display the first two coordinates of u, again obtaining the projected 2D plot. The elements of u are called *screen coordinates* of the point ℓ. In most computer implementations, rotation is allowed around the screen axes only; rotation about a natural coordinate axis is generally not allowed unless the natural axis corresponds to a screen axis.

A *projective view* of a 3D plot is the 2D projection on the computer screen. The linear combinations plotted along the axes of a view will be called the *horizontal and vertical screen variables*.

3.2. DEPTH CUING

Depth cuing is a supplemental method that enhances the illusion of three dimensions. The basic idea is to let the number of pixels representing a point depend on the distance that the point is from the observer; the greater the distance the fewer the pixels.

3.3. SCALING

If we think of the computer's prototype plot as having axis ranges running from -1 to 1, then it is necessary to scale the data so that it fits within the computer's available plotting space. In most computer packages that allow 3D plotting, scales are automatically chosen so that the maximum possible range is used on each axis. Recalling that the data have been centered, this can be accomplished by scaling each variable to have a maximum absolute value of 1. The data actually plotted are then of the form $(H_i/\max|H|, V_i/\max|V|, O_i/\max|O|)$. Since each axis is scaled separately, we call this *abc-scaling*, following Cook and Weisberg (1989), to indicate that different scale factors, $a = \max|H|$, $b = \max|V|$, and $c = \max|O|$, have been applied to the three axes. The data actually plotted are then of the form $M^{-1}\ell$ where $M = \mathrm{diag}(a,b,c)$.

abc-scaling has the advantage of filling the range of each axis with data, but there are at least two related disadvantages. The first is that the relative magnitudes of the individual coordinates may no longer be apparent. For example, a comparison of two data values V_i and H_i will generally not be possible with *abc*-scaling since the values actually plotted are V_i/a and H_i/b. Second, it is generally not possible to associate angles θ with specific linear combinations of H and O. Suppose, for example, that we wish to rotate to the plot $\{V, H + O\}$. This could be accomplished in Figure 3.1 with $\theta = \pi/4$, resulting in the 2D plot $\{V, (H + O)/\sqrt{2}\}$. But with *abc*-scaling this same rotation would result in the plot $\{V, (H/a + O/c)/\sqrt{2}\}$, which is not what we want unless $a = c$. Further, to obtain the required plot with *abc*-scaling we would need to rotate through an angle of $\theta = \tan^{-1}(c/a)$. In short, while *abc*-scaling has the advantage of filling the axes with data, the characteristics of the display cannot be interpreted easily in terms of specific linear combinations of the original variables or in terms of their relative magnitudes.

Other types of scaling may be useful on occasion to compensate for the shortcomings of *abc*-scaling. If easy visual assessment of relative magnitudes and angles is important, the display could be restricted so that the same scale factor, the largest absolute element in the data $\max_i(\max\{|H_i|,|V_i|,|O_i|\})$, is applied to all three axes. This type of scaling, called *aaa-scaling*, may be particularly useful when all three variables have the same units. The disadvantage of an *aaa*-scaled plot is that the points may occupy only a small fraction of the available plotting region along one or two axes. In *aba-scaling* the same scale factor $\max_i(\max\{|H_i|,|O_i|\})$, is applied to the horizontal and out-of-page axes, while a different scale factor $\max_i|V_i|$ is applied to the vertical axis. Similarly, we can also consider *aab*- and *aba-scaling*, depending on the requirements of the problem.

3.4. ORTHOGONALIZATION

In addition to scaling, the user also has the choice of what to plot in regression problems where the relationship between a response y and two predictors x_1 and x_2 is of interest. We assume that the predictors have been centered for plotting. The response is typically assigned to the vertical axis, but the centered predictors x_1 and x_2 or some distinct linear combinations of them could be assigned to the remaining two axes. Regardless of the choice, all linear combinations of x_1 and x_2 will appear on the horizontal screen axis as the plot is rotated once around the vertical axis, so we are clearly free to choose any two linear combinations at the start. Visual impressions of the point cloud can depend on the chosen linear combinations, however. An abc-scaled plot will often be the first choice, but it may be difficult to see details of the point cloud because of collinearity between x_1 and x_2. When this happens the plot may be little more than a vertical plane of points rotating in three dimensions. In such cases a useful alternative is to select two linear combinations of x_1 and x_2, say z_1 and z_2, for plotting on the horizontal and out-of-page axes. The linear combinations should be chosen so that the data are well-spread after projecting onto any subspace $S(B)$ as represented in Figure 3.1.

Let A denote a 2×2 nonsingular matrix

$$A = \begin{pmatrix} a_1 & a_2 \\ c_1 & c_2 \end{pmatrix}$$

and consider plotting the coordinates of $z = Ax$ on the horizontal and out-of-page axes, where $x = (x_1, x_2)^T$. Under abc-scaling the variables actually plotted will be $w = M^{-1}z = M^{-1}Ax$ where $M = \text{diag}(m_1, m_2)$ is a function of all the observed values of x, x_i, $i = 1, \ldots, n$, and of A,

$$m_1 = \max_i |a_1 x_{i1} + a_2 x_{i2}|$$

and

$$m_2 = \max_i |c_1 x_{i1} + c_2 x_{i2}|.$$

Two-dimensional views then consist of plots of the general form

$$\{y, \cos(\theta)w_1 + \sin(\theta)w_2\}.$$

The problem now is to choose A to maximize the spread on the horizontal axis of this 2D plot.

Let $t(\theta) = (\cos(\theta), \sin(\theta))^T$ and $w_i = (w_{i1}, w_{i2})^T$. As a measure of spread we take the sum of squares about the origin represented by

$$\sum_{i=1}^{n} (t^T(\theta) w_i)^2 = \sum_{i=1}^{n} t^T(\theta) M^{-1} A x_i x_i^T A^T M^{-1} t(\theta)$$

$$= t^T(\theta) M^{-1} A X X^T A^T M^{-1} t(\theta) \tag{3.1}$$

where X, the $2 \times n$ matrix with columns x_i, is assumed to have full row rank. Since A is an arbitrary nonsingular matrix, we can assume $X X^T = I_2$ without loss of generality. In other words, we assume that the raw data have been transformed linearly so that the vectors of predictor values are orthogonal and have length 1. The problem is now to maximize $t^T(\theta) M^{-1} A A^T M^{-1} t(\theta)$ as a function of A. To remove the dependence on a particular marginal plot as determined by θ, we could consider selecting the optimal value $\tilde{A}$ of A by maximizing the determinant of $M^{-1} A A^T M^{-1}$

$$\tilde{A} = \arg\max_{A} |M^{-1} A A^T M^{-1}|. \tag{3.2}$$

Alternatively, $\tilde{A}$ could be chosen to maximize the spread in the 2D plot with the minimum spread,

$$\tilde{A} = \arg\max_{A} \left[\min_{\theta} t^T(\theta) M^{-1} A A^T M^{-1} t(\theta) \right]$$

$$= \arg\max_{A} \lambda_{\min}(M^{-1} A A^T M^{-1}) \tag{3.3}$$

where $\lambda_{\min}(\cdot)$ denotes the smallest eigenvalue of the symmetric matrix argument. In either case, the solution will depend on characteristics of $M^{-1} A A^T M^{-1}$, which we now investigate in additional detail.

We will need the angles α and χ that characterize z in terms of x

$$\cos(\alpha) = \frac{a_1}{\sqrt{a_1^2 + a_2^2}}$$

and

$$\cos(\chi) = \frac{c_1}{\sqrt{c_1^2 + c_2^2}}$$

and the function

$$m(\phi) = \max_{i} |\cos(\phi) x_{i1} + \sin(\phi) x_{i2}|.$$

Then

$$M^{-1}AA^T M^{-1} = \begin{pmatrix} \dfrac{1}{m^2(\alpha)} & \dfrac{\cos(\alpha - \chi)}{m(\alpha)m(\chi)} \\[2ex] \dfrac{\cos(\alpha - \chi)}{m(\alpha)m(\chi)} & \dfrac{1}{m^2(\chi)} \end{pmatrix} \qquad (3.4)$$

Although A has four entries, the essential problem for an *abc*-scaled plot depends only on the two angles α and χ. Experience has shown that good results in practice can be achieved by requiring only that $\cos(\alpha - \chi) = 0$. This requirement is met by choosing $A = I$. Since we assumed that $XX^T = I$, this means that we can choose any pair of orthonormal variables for plotting. Additional discussion of this problem is available in Cook and Weisberg (1990).

In addition to improving resolution, orthogonalization has statistical advantages that will be apparent in later chapters.

PROBLEMS

3.1. What are the disadvantages to basic spread criterion (3.1)? Are there any other criteria that may be reasonable? (Hint: Is maximizing spread really the best for visual resolution?)

3.2. Verify (3.4).

3.3. The discussion of Section 3.4 may have terminated prematurely. Do the spread criteria (3.2) and (3.3) really treat all orthonormal pairs of variables as equivalent? Further investigate the implications of (3.2) and (3.3) with emphasis on the role of orthogonality. This could be done analytically, numerically, or graphically.

3.4. Adapt the results of Section 3.4 for *aaa*-scaling.

3.5. Let (D_1, D_2, D_3) denote a trivariate normal vector with $E(D_k) = \mu$, $Var(D_k) = \sigma^2$, and $Cov(D_j, D_k) = \sigma_\phi^2$ for $j \neq k$.

3.5.1. Show that

$$E(D_1 | D_2 = d_2, D_3 = d_3) = \mu + \frac{\sigma_\phi^2}{\sigma^2 + \sigma_\phi^2}(d_2 + d_3 - 2\mu)$$

3.5.2. Suggest a 3D graphical display that can be used to visualize the maximum likelihood estimate of the intraclass correlation coefficient $\sigma_\phi^2 / (\sigma^2 + \sigma_\phi^2)$.

Is it possible to construct a 2D display for the same purpose?

CHAPTER 4

Interpreting 3D Scatterplots

The interpretation of 3D scatterplots, which is notably more intricate than the interpretation of 2D scatterplots, is considered in this chapter. The discussion centers mostly on 3D plots arising in regression problems with response y and $p = 2$ predictors $x = (x_1, x_2)^T$. The foundations introduced here do not depend on the nature of the response, but visualization methods can depend on its nature. To focus the presentation, the illustrations are restricted to regressions with continuous or many-valued responses where the 3D plots discussed in Chapter 3 are most informative. Adaptations to regression problems with binary responses are discussed in Chapter 5. Finally, the material of this chapter will eventually be applied to regression problems with more than two predictors.

We begin with an example to introduce a few basic ideas.

4.1. HAYSTACKS

Hay was sold by the stack in the Great Plains during the late 1920s, requiring estimation of stack volume to insure a fair price. The estimation of the volume of a haystack was a nontrivial task, and could require much give-and-take between buyer and seller to reach a mutually agreeable price. A study was conducted in Nebraska in 1927 and 1928 to see if a simple method could be developed to estimate the volume *Vol* of round haystacks. It was reasoned that farmers could easily use a rope to characterize the size of a haystack with two measurements: C, the circumference around the base of the stack, and *Over*, the distance from the ground on one side of the stack to the ground on the other side. The data, as reported by Ezekiel (1941, pp. 378–380), consist of measurements of the volume *Vol*, C, and *Over* on 120 round stacks. Stack volume was determined by a series of survey measurements that were not normally possible for Nebraska farmers at the time.

We now have a regression problem with response $y = Vol$ and predictors $x_1 = C$ and $x_2 = Over$. Because there are only two predictors, the full data set can be viewed in a 3D scatterplot. Four views corresponding to four angles of rotation about the vertical axis are shown in the four frames of Figure 4.1.

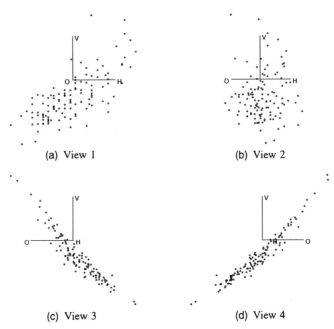

FIGURE 4.1   Four views of the 3D scatterplot for the haystack data with axes V = Volume, H = Circumference, and $O = Over$.

There is surely enough information in Figure 4.1 to contradict the possibility that $Vol \perp\!\!\!\perp (C, Over)$. After inspecting a 3D plot awhile, many analysts will stop rotation at the 2D view they find to be the most compelling or interesting, perhaps fine-tuning the projection to end with the most striking 2D view. View 4 in Figure 4.1 was selected as the most striking for the haystack data, with View 3 not far behind. As discussed in Chapter 3, each of the four views in Figure 4.1 corresponds to a 2D plot of the form $\{Vol, H(\theta)\}$, where

$$H(\theta) = \cos(\theta)C + \sin(\theta)Over$$

is the horizontal screen variable for rotation angle θ. The linear combination of the predictors on the horizontal axis of View 4 is

$$H_4 = 0.19C + 0.98Over$$

where the subscript on H_4 indicates the corresponding view in Figure 4.1. Regardless of the method used to determine the most striking view, a key question may arise once the view is selected: Is View 4 sufficient to characterize how the distribution of $Vol \mid (C, Over)$ varies with the values of C and $Over$? Stated differently, is the conjecture

$$Vol \perp\!\!\!\perp (C, Over) \mid H_4 \tag{4.1}$$

contradicted by information in the data? If not, then there may be no important loss of regression information when using View 4 as a substitute for the rotating 3D plot. Informally, we could then say that the 2D view $\{Vol, H_4\}$ is a *sufficient summary* of the 3D plot $\{Vol, (C, Over)\}$. If evidence is found indicating that (4.1) is false, then there are two possibilities: Either the statement

$$Vol \perp\!\!\!\perp (C, Over) \mid H(\theta)$$

is false for all 2D views $\{Vol, H(\theta)\}$, or it is false for the specific linear combination H_4 while it is true for some other linear combination. In the former case, no 2D view is a sufficient summary and we must study the 3D plot to understand the conditional distribution of $Vol \mid (C, Over)$. In the latter case, further analysis may bring us closer to a sufficient 2D summary view.

The ideas introduced in this example for characterizing a 3D plot are developed in the next section. The haystack data will be revisited occasionally in the main text and in the problems. See, for example, Problems 4.12 and 12.6.

4.2. STRUCTURAL DIMENSIONALITY

The *structural dimension* (d) of a regression is the smallest number of distinct linear combinations of the predictors required to characterize the conditional distribution of $y \mid x$. (Cook and Weisberg 1994a; Cook and Wetzel 1993). The structural dimension of a plot $\{y, x\}$ is the structural dimension of the regression of y, the quantity on the vertical axis, on x. Structural dimensions will be denoted by d, or by $d_{y \mid x}$ when there is more than one regression under consideration and it is necessary to distinguish between them.

4.2.1. One predictor

For a regression with a single predictor x, there are only two possible structural dimensions: Either

- $y \perp\!\!\!\perp x$, corresponding to *0-dimensional (0D) structure* because no linear combinations of x are required to characterize $y \mid x$, or
- the distribution of $y \mid x$ depends on the value of x, corresponding to *1-dimensional (1D) structure* because one linear combination of the predictor is needed.

Trivially, $y \perp\!\!\!\perp x$ if and only if the distribution of $y \mid x$ is the same as the marginal distribution of y regardless of the value of x. The graphical methods discussed in Chapter 2 can, in the first instance, be viewed as methods for deciding the structural dimension of a regression. The notion of structural dimension does not really contribute much to an understanding of regressions with a single predictor. The first nontrivial application of the idea is to regressions with 2 predictors, as illustrated previously by using the haystack data.

4.2.2. Two predictors

With two predictors $x = (x_1, x_2)^T$, there are three possible structural dimensions, each of which can be represented in terms of the single expression

$$y \perp\!\!\!\perp x \mid \eta^T x \qquad (4.2)$$

by choosing an appropriate value for the $2 \times d$ matrix η, $d \le 2$. First, if (4.2) holds with $\eta = 0$ then $y \perp\!\!\!\perp x$ and all 2D views of the 3D plot $\{y, x\}$ should appear as plots of independent variables. In that case we say that the 3D plot and the corresponding regression have *0-dimensional (0D) structure* because no linear combinations of x furnish information about the response.

Second, if y is dependent on x and if (4.2) holds for some $\eta \ne 0$ with $d = 1$ then y is independent of x given the single linear combination $\eta^T x$. In that case the regression has *1-dimensional (1D) structure* because only one linear combination of the predictors is needed to describe $F_{y|x}$ fully, where $F_{U|V}(\cdot)$ denotes the conditional CDF of $U|V$. One-dimensional structure holds if and only if $F_{y|x} = F_{y|\eta^T x}$ for all values of x in its marginal sample space. Finally, if (4.2) fails for all vectors $\eta \in \mathbb{R}^2$ then the regression has *2-dimensional (2D) structure* because two linear combinations are required to describe the regression.

When (4.2) holds, the plot $\{y, \eta^T x\}$ will be called a *sufficient summary* of the 3D plot $\{y, x\}$. The 3D plot itself is always a sufficient summary since (4.2) holds trivially when $\eta = I$. If the regression problem has 0D structure ($\eta = 0$) then a graphical representation of the marginal distribution of y, such as a histogram or smooth density estimate, is a sufficient summary. If the regression problem has 1D structure ($d = 1$) then the 2D plot $\{y, \eta^T x\}$ is a sufficient summary. There are no lower-dimensional sufficient summary plots of the form $\{y, \eta^T x\}$ for regression problems with 2D structure. In that case the 3D plot itself must be inspected to understand the distribution of $y \mid x$.

Define transformed predictors $z = A^T x$, where A is a full rank 2×2 matrix. Then $y \perp\!\!\!\perp x \mid \eta^T x$ if and only if $y \perp\!\!\!\perp z \mid \alpha^T z$, where $\alpha = A^{-1}\eta$. Because $\dim[S(\eta)] = \dim[S(\alpha)]$, where "dim" denotes dimension, the structural dimension of a regression is invariant under nonsingular linear transformations of the predictors. This means that we are free to improve resolution by orthogonalizing the predictors (Section 3.4) when assessing structural dimensionality in a 3D plot.

4.2.3. Many predictors

The ideas introduced above extend immediately to $p > 2$ predictors, the possible structural dimensions being $0, 1, \ldots, p$. However, when dealing with many predictors a broader context may be worthwhile.

Regardless of the number of predictors, if (4.2) holds then it also holds when η is replaced with any basis for the subspace $S(\eta)$ spanned by the columns of η. Thus (4.2) can be viewed in a broader context as a statement about the relationship between $y \mid x$ and the *dimension-reduction subspace* $S(\eta)$. Let $S_{y|x}$ denote a subspace of minimum dimension so that (4.2) holds when the columns of η form any basis for the subspace. The structural dimension of the regression is then just the dimension of $S_{y|x}$. The idea of a dimension-reduction subspace will play a central role when developing graphics for regressions with many predictors beginning in Chapter 6. The broader approach afforded by working with subspaces is not really necessary for the case of two predictors considered in this chapter because $S_{y|x}$ must be either the origin, $\mathbb{R}^2$, or some one-dimensional subspace of $\mathbb{R}^2$. Nevertheless, in anticipation of later developments, it may be helpful to recall this broader context from time to time.

Except for a few general results, the focus is on $p = 2$ predictors in the rest of this chapter.

4.3. ONE-DIMENSIONAL STRUCTURE

Many standard regression models have 1D structure. For example, consider the usual linear regression model with $p = 2$ predictors $x = (x_1, x_2)^T$,

$$y \mid x = \beta_0 + \beta^T x + \sigma\varepsilon \qquad (4.3)$$

where $\beta \neq 0$, $\varepsilon \perp\!\!\!\perp x$, $E(\varepsilon) = 0$, and $\text{Var}(\varepsilon) = 1$. Clearly, $y \perp\!\!\!\perp x \mid \beta^T x$ and thus the structural dimension of this model equals 1. Equation (4.2) applies in this case with $\eta = c\beta$ for any nonzero scalar c. Equivalently, (4.2) applies for any nonzero $\eta \in S_{y|x} = S(\beta)$.

A more general model with 1D structure is

$$y \mid x = g(\beta^T x) + \sigma(\beta^T x)\varepsilon$$

where $g(\beta^T x)$ and $\sigma^2(\beta^T x)$ are the regression and variance functions, respectively, each depending on the same linear combination of the predictors. Allowing for a possibly parameterized monotonic transformation $t(y)$ of the response results in a still more general model with 1D structure,

$$t(y) \mid x = g(\beta^T x) + \sigma(\beta^T x)\varepsilon. \qquad (4.4)$$

This form covers certain transform-both-sides models (Carroll and Ruppert 1988), as well as the usual Box–Cox transformations. Similar comments apply to generalized linear models. Logistic regression models with link function of the form $g(\beta^T x)$ have 1D structure, for example.

Suppose that y is a nonnegative continuous random variable representing the lifetime in some context. Models for $y \mid x$ are often stated in terms of the hazard function

$$h(t \mid x) = \frac{f_{y|x}(t)}{\Pr(y \geq t \mid x)}$$

where $f_{y|x}$ is the density of $y \mid x$. Proportional hazards models are characterized by hazard functions of the form $h(t \mid x) = h_0(t)g(x)$ for some function g. Perhaps the most common proportional hazards model (Cox 1972) is another model with 1D structure,

$$h(t \mid x) = h_0(t)\exp(\beta^T x).$$

What we might see in a 3D plot $\{y, x\}$ of data that can be described adequately by a model with 1D structure depends on the regression and variance functions, and on the distribution of the predictors. For example, suppose that linear model (4.3) holds and consider any 2D view $\{y, b^T x\}$ of the 3D plot $\{y, x\}$. The regression function for this view can be obtained by averaging the full regression function

$$E(y \mid b^T x, \beta^T x) = \beta_0 + \beta^T x$$

with respect to the conditional distribution of $\beta^T x \mid b^T x$:

$$E(y \mid b^T x) = \beta_0 + E(\beta^T x \mid b^T x).$$

If $E(\beta^T x \mid b^T x)$ is a linear function of the value of $b^T x$ for all $b \in \mathbb{R}^2$ then all 2D views will exhibit linear regression functions. The strength of the linear relationships will vary with b, but no 2D view will show curvature. If $E(\beta^T x \mid b^T x)$ is a nonlinear function of $b^T x$ for some b then the corresponding 2D view may well show curvature even though the linear regression model (4.3) holds. An example of this phenomenon was given during the discussion of scatterplot matrices in Section 2.5. Four additional views of the data for that example are shown in Figure 4.2. Three of the views show curvature, but the points fall exactly on a plane as shown in View 2.

Suppose now that the general 1D model (4.4) gives a good description of the data. Regardless of the distribution of the predictors, the view $\{y, c\beta^T x\}$ is a sufficient summary of the full 3D plot for any $c \neq 0$. Because of the scaling used in plot construction (Section 3.3), we will not normally have

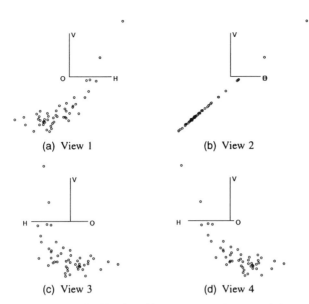

(a) View 1   (b) View 2

(c) View 3   (d) View 4

FIGURE 4.2 Four views of the 3D plot with axes $V = y$, $H = x_1$, and $O = x_2$, where $y = 3 + x_1 + x_2/2$.

$c = 1$. But this is largely irrelevant since the sufficient 2D view will still allow visualization of the regression and variance functions.

Any view of a 3D plot could be used as a 2D summary plot. But when viewing regression data with 1D structure we should strive to stop rotation at or near a sufficient summary. If this can be achieved then further analysis might be guided by the 2D summary view for which the methods of Chapter 2 are applicable. Selecting a sufficient view will hopefully correspond to the intuitive process of determining a most striking 2D view as discussed in connection with the haystack data. The visual process that leads to a most striking view is difficult to characterize analytically since it depends on subjective choices. Nevertheless, there are metaphors that may help.

While rotating a generic 3D plot about its vertical axis we observe all 2D projections of the form $\{y, H(\theta)\}$ where $H(\theta) = b^T x = \cos(\theta)x_1 + \sin(\theta)x_2$, each corresponding to a simple regression problem with response y and predictor $H(\theta)$. Determining the most striking view can be thought of as visually determining a "best" linear combination of the predictors with which to explain the variation in the response. It can also be regarded as a visual method of estimating a basis for the minimal subspace $S_{y|x}$ when $d = 1$.

Imagine mentally fitting the regression function $E(y \mid H(\theta))$ for each 2D view. The most striking view is often the one with the smallest average variation about the regression function. This leads to the following projection pursuit (Huber 1985) metaphor for visually determining the angle of rotation

$\tilde{\theta}$ associated with a "best" 2D view:

$$\tilde{\theta} = \arg\min_{\theta} E[\text{Var}(y \mid H(\theta))]$$

$$= \arg\min_{\theta} E\{E[y - E(y \mid H(\theta))]^2 \mid H(\theta)\}. \qquad (4.5)$$

Because

$$\text{Var}(y) = E[\text{Var}(y \mid H(\theta))] + \text{Var}[E(y \mid H(\theta))]$$

determining the value of θ that minimizes $E[\text{Var}(y \mid H(\theta))]$ is the same as maximizing the variance of the regression function, $\text{Var}[E(y \mid H(\theta))]$. This process is equivalent to minimizing the variance ratio $R_{y|H(\theta)}$ as previously defined in (2.10). Further, as stated in the following proposition, $\{y, H(\tilde{\theta})\}$ is a sufficient summary plot. The justification is given as Problem 4.2.

Proposition 4.1.   Assume the 1D regression model,

$$y \mid x = g(\beta^T x) + \sigma(\beta^T x)\varepsilon$$

with $\beta \neq 0$ and some nontrivial regression function g. Let

$$\tilde{\theta} = \arg\min_{\theta} R_{y|H(\theta)}$$

where $\tilde{\theta}$ is unique and $R_{y|H(\theta)}$ was defined in (2.10),

$$R_{y|H(\theta)} = \frac{E[\text{Var}(y \mid H(\theta))]}{\text{Var}[E(y \mid H(\theta))]}.$$

Then $\{y, H(\tilde{\theta})\}$ is a sufficient summary plot.

While the metaphor of Proposition 4.1 will be applicable frequently, it may not always be appropriate since clusters, outliers, and heteroscedasticity with $g = 0$ may be missed (Problem 4.1). In addition, the metaphor works best with continuous or many-valued responses which allow for visualization of the regression function $E(y \mid H(\theta))$ as the 3D plot $\{y, x\}$ is rotated. The metaphor may be of little use when the response takes on few values.

Suppose, for example, that the response is binary, $y = 0$ or 1. Each point in the 3D plot $\{y, x\}$ will fall in one of two horizontal planes which intersect the vertical axis at $y = 0$ and $y = 1$. Because of this arrangement of points, little visual information will be available on the regression function $E(y \mid H(\theta))$ as the plot is rotated. While the idea of structural dimensionality does not place constraints on the response, a different visualization method becomes necessary when the response is binary. Graphics for regressions with binary responses and at most 3 predictors are discussed in Chapter 5. The use of structural dimensionality in this chapter is mostly restricted to regressions with many-valued responses.

4.4. TWO-DIMENSIONAL STRUCTURE

A first example of a regression with 2D structure can be obtained by adding an interaction term to the linear regression model (4.3),

$$y \mid x = \beta_0 + \beta_1 x_1 + \beta_2 x_2 + \beta_{12} x_1 x_2 + \sigma \varepsilon. \tag{4.6}$$

Three-dimensional plots of data from this type of model display a saddle shape with a linear tend. Generally, adding any higher-order terms to a linear regression model (4.3) results in a model with 2D structure via the regression function g,

$$y \mid x = g(x) + \sigma(x)\varepsilon \tag{4.7}$$

where $x = (x_1, x_2)^T$. Other 2-dimensional forms are possible as well. The heteroscedastic model

$$y \mid x = g(\beta^T x) + \sigma(\alpha^T x)\varepsilon$$

has 2D structure as long as $\beta \not\propto \alpha$. Because there is no 2D view that can serve as a sufficient replacement for the 3D plot $\{y, x\}$ in regression problems with 2D structure, we must inspect the 3D plot itself to gain full information on $y \mid x$.

4.4.1. Removing linear trends

If $\sigma^2(x) = \mathrm{Var}(y \mid x)$ is sufficiently small, a useful mental image of a two-dimensional regression function might be obtained from a 3D plot, but generally this is not an easy task. Removing linear trends can help. Specifically, consider displays that may help understand the regression function g in (4.7). Let

$$\beta_{ols} = \Sigma^{-1} \mathrm{Cov}(x, g(x))$$

denote the population regression coefficients from the OLS fit of y on x, where $\Sigma = \mathrm{Var}(x)$. Then model (4.7) can be rewritten as

$$y \mid x = \beta_0 + \beta_{ols}^T x + g_r(x) + \sigma(x)\varepsilon \tag{4.8}$$

where $g_r(x) = g(x) - \beta_0 - \beta_{ols}^T x$ is the residual regression function, and β_0 is the usual intercept computed from the population. Plotting the OLS residuals

$$e = y - \beta_0 - \beta_{ols}^T x$$

in a 3D plot $\{e,x\}$ will allow visualization of the residual regression function g_r. This can be useful because strong linear trends often visually dominate other relevant features of g. For example, a strong linear trend in model (4.6) can often mask a saddle shape, making 2D structure difficult to see.

An illustration of removing linear trends is given in Section 4.5.3.

4.4.2. Identifying semiparametric regression functions

When the linear regression model (4.3) holds, the residual plot $\{e,x\}$ will have 0D structure. Otherwise, the structural dimension of the residual plot will generally be the same as the 3D plot of the raw data. There are important practical exceptions, however. Suppose that

$$y \mid x = \alpha_0 + \alpha_1^T x + g_2(\alpha_2^T x) + \sigma\varepsilon \tag{4.9}$$

where now $g(x) = \alpha_0 + \alpha_1^T x + g_2(\alpha_2^T x)$, $\alpha_2^T x$ is a scalar and $\alpha_1 \not\propto \alpha_2$ so we still have 2D structure. Will the residual plot $\{e,x\}$ have 1D or 2D structure? A guarantee of 1D structure could be useful because then we may be able to identify semiparametric forms from 3D plots.

The following calculations will be useful for exploring the structural dimension of the residual plot $\{e,x\}$ under (4.9). First, without loss of generality, assume that $E(x) = 0$. Then

$$\beta_{ols} = \Sigma^{-1}\text{Cov}(x, g(x)) = \alpha_1 + \Sigma^{-1}\text{Cov}(x, g_2(\alpha_2^T x))$$

and

$$g_r(x) = g(x) - \beta_0 - \beta_{ols}^T x$$
$$= (\alpha_0 - \beta_0) + g_2(\alpha_2^T x) - \text{Cov}(x, g_2(\alpha_2^T x))^T \Sigma^{-1} x$$

where $\Sigma = \text{Var}(x)$ is restricted to be positive definite. For the residual plot $\{e,x\}$ to have 1D structure, we must have

$$S(\Sigma^{-1}\text{Cov}(x, g_2(\alpha_2^T x))) = S(\alpha_2) \tag{4.10}$$

where

$$\text{Cov}(x, g_2(\alpha_2^T x)) = E[g_2(\alpha_2^T x)E(x \mid \alpha_2^T x)]$$

because then the residual regression function g_r will depend only on the single linear combination $\alpha_2^T x$. Clearly, there is no guarantee that (4.10) will hold and thus no general conclusion about the structural dimension of the plot $\{e, x\}$ is possible. We can make some progress, however, by constraining the conditional expectation $E(x \mid \alpha_2^T x)$ to be a linear function of $\alpha_2^T x$ via a 2×1 vector A:

$$E(x \mid \alpha_2^T x) = A \alpha_2^T x.$$

The general result needed is stated as the following proposition which will be used in subsequent chapters. The justification is an exercise at the end of this chapter.

Proposition 4.2 (Linear Conditional Expectations). Let x be a $p \times 1$ random vector with $E(x) = 0$ and positive definite covariance matrix Σ, and let α be a full-rank $p \times q$, $q \leq p$, matrix. Assume that $E(x \mid \alpha^T x = u)$ is a linear function of u: $E(x \mid \alpha^T x = u) = Mu$ for some fixed $p \times q$ matrix M. Then

1. $M = \Sigma \alpha (\alpha^T \Sigma \alpha)^{-1}$.
2. M^T is a generalized inverse of α.
3. αM^T is the orthogonal projection operator for $S(\alpha)$ relative to the inner product $(v_1, v_2)_\Sigma = v_1^T \Sigma v_2$.

Returning to the structural dimension of the residual plot $\{e, x\}$ under model (4.9) and assuming that $E(x \mid \alpha_2^T x) = A \alpha_2^T x$,

$$\Sigma^{-1} \text{Cov}(x, g_2(\alpha_2^T x)) = [E(g_2(\alpha_2^T x)\alpha_2^T x)]\Sigma^{-1}A.$$

It now follows from Proposition 4.2 that $\Sigma^{-1}\text{Cov}(x, g_2(\alpha_2^T x)) \in S(\alpha_2)$ and thus that the residual plot $\{e, x\}$ has 1D structure.

As a result of this line of reasoning we can conclude the following. If model (4.9) holds and if $E(x \mid \alpha_2^T x)$ is linear in $\alpha_2^T x$, then the residual plot $\{e, x\}$ must have 1D structure. In practice, if $\{y, x\}$ has 2D structure, if an inspection of the predictor plot $\{x_1, x_2\}$ indicates that $E(x \mid \alpha_2^T x)$ is linear, and if $\{e, x\}$ has 1D structure, then models of the form (4.9) would be natural to consider next.

Since α_2 is unknown, we may want to insure that $E(x \mid b^T x)$ is linear for all b when inspecting the predictor plot $\{x_1, x_2\}$. Eaton (1986) has shown that a random vector V is elliptically contoured if and only if $E(V \mid b^T V)$ is linear for all b. Thus, if the plot $\{x_1, x_2\}$ is consistent with an elliptically contoured distribution for x, assuming the linearity of $E(x \mid \alpha_2^T x)$ is probably quite reasonable. The discussion of 3D plots under model (4.9) is continued in Problem 4.6. Elliptically contoured distributions are discussed in Section 7.3.2.

We have seen several examples in this and previous chapters where marginal response plots and views of 3D plots can be quite misleading about the nature of $y \mid x$. Because we cannot obtain accurate visual impressions in high dimensions, we will eventually be forced to use 2D and 3D plots for graphical analysis in regression problems with more than 2 predictors. Restrictions are needed on the marginal distribution of the predictors for this to be successful. Proposition 4.2 represents a first step in the development of such restrictions.

4.5. ASSESSING STRUCTURAL DIMENSIONALITY

The discussion of structural dimensionality d so far was mostly under the assumption that d is known. Graphical methods for assessing the structural dimensionality of a regression problem with response y and $p = 2$ predictors $x = (x_1, x_2)^T$ are discussed in this section. Since structural dimensionality is preserved under nonsingular linear transformations of x, orthogonalization can be used freely to improve resolution. As in previous sections, the basic rationale is developed in the population context. Consistent estimates can be substituted where needed for use in practice.

Determining if $d = 0$ is perhaps the first step in characterizing the 3D plot $\{y, x\}$. It is very often easy to rule out the possibility of 0D structure, as in the haystack data, but on occasion this decision may not be obvious. If any 2D view $\{y, b^T x\}$ shows that the distribution of $y \mid b^T x$ depends on $b^T x$ then the structural dimension must be at least 1. On the other hand, suppose that no 2D view indicates dependence so that $y \perp\!\!\!\perp b^T x$ for all b. Is that sufficient information to infer that $y \perp\!\!\!\perp x$? How carefully do we need to study a rotating point cloud to infer 0D structure? An answer is in the following proposition.

Proposition 4.3. Let y be a $q \times 1$ random vector and let x be a $p \times 1$ be random vector. Then $y \perp\!\!\!\perp x$ if and only if $y \perp\!\!\!\perp b^T x$ for all $b \in \mathbb{R}^p$.

Justification. The conclusion follows immediately by use of characteristic functions: y and x are independent if and only if their joint characteristic function

$$\varphi(t_1, t_2) = \mathrm{E}(e^{it_1^T y + it_2^T x})$$

is of the form

$$\varphi(t_1, t_2) = \varphi_y(t_1)\varphi_x(t_2).$$

Because $y \perp\!\!\!\perp b^T x$ for all $b \in \mathbb{R}^p$ the factorization is immediate. □

This proposition indicates that we do not have to look deeply into a rotating point cloud to conclude that $y \perp\!\!\!\perp x$ because the absence of dependence in every 2D view implies 0D structure. The methods of Chapter 2 can help decide if $y \perp\!\!\!\perp b^T x$ in any particular view.

If the variation is sufficiently small, it may be easy to see from a rotating 3D plot that $d = 2$. The choice between $d = 1$ and $d = 2$ is usually the most difficult, however, and in Sections 4.5.2 and 4.6 we discuss various graphical methods to aid that choice. First, we discuss a visual metaphor that may help on occasion.

4.5.1. A visual metaphor for structural dimension

Begin with a sheet of lined paper from a writing tablet. Most writing tablets have two kinds of lines: horizontal lines running across the sheet to be used as writing guides and a vertical line or two that marks the left margin. We will refer to the horizontal lines as "blue lines" since they are often this color; similarly, the vertical margin lines will be called "red lines." This sheet will be used to represent the regression function $E(y \mid x)$ in the additive error model $y \mid x = E(y \mid x) + \varepsilon$ with two predictors $x = (x_1, x_2)^T$ and $\varepsilon \perp\!\!\!\perp x$.

Holding the sheet of paper parallel to the floor, the (x_1, x_2) plane, gives a representation of 0D structure.

Next, remembering that the paper represents a function, consider tilting and/or bending it so that all of the blue lines remain parallel to the floor. For example, the sheet may be bent to resemble a trigonometric function with a linear trend. As long as all the blue lines remain parallel to the floor, there is 1D structure, $y \perp\!\!\!\perp x \mid \eta^T x$. The sheet may be creased, so the surface isn't everywhere differentiable, but there is still 1D structure as long as all the blue lines remain parallel to the floor. Projecting the red line on the floor gives a 1D affine subspace that is parallel to $S(\eta)$. We can maintain 1D structure while allowing $\mathrm{Var}(\varepsilon \mid x)$ to depend on x provided $\mathrm{Var}(\varepsilon \mid x)$ is constant along the blue lines.

Still maintaining 1D structure, rotate the sheet so it is viewed from an edge and the blue lines are no longer visible. The sheet might now look like $\vee$ or $\sim$, for example. This view, which results from the visual fitting metaphor (4.5), is then the sufficient summary plot. The horizontal screen axis for this view corresponds to $\eta^T x$.

Bending the sheet so that some blue lines are parallel to the floor while others are not results in a regression with 2D structure.

This metaphor may be useful when viewing a 3D plot in practice, provided the variability in the data doesn't completely obscure the regression function.

4.5.2. A first method for deciding $d = 1$ or 2

To decide if $d = 1$ or 2 in the haystack data, we begin with the conjecture stated in (4.1),

$$Vol \perp\!\!\!\perp (C, Over) \mid H_4 \tag{4.11}$$

where $H_4 = (0.19C + 0.98Over)$ is the linear combination of the predictors on the horizontal screen axis of View 4 in Figure 4.1. This conjecture cannot be assessed directly because there are too few observations for the formation of adequately sized subsets in which H_4 is fixed. However, progress is possible by

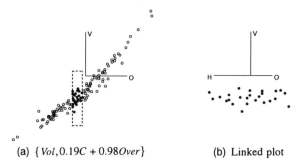

(a) $\{Vol, 0.19C + 0.98Over\}$ (b) Linked plot

FIGURE 4.3 Two views of the 3D plot $\{Vol, (C, Over)\}$ for the haystack data. (a) The view giving the "best" visual fit. (b) Linked view showing only the slice points for visual clarity.

slicing to condition approximately on a value of H_4, as illustrated in Figure 4.3a for a single slice. Let $\{J_s\}$ denote a partition of the range of H_4 into equal-width slices J_s, $s = 1 \ldots h$. The restriction to slices of equal width is used only to facilitate discussion and is not a necessary constraint. The slices are analogous to the blue lines discussed in Section 4.5.1.

Slicing is used to evaluate the condition

$$Vol \perp\!\!\!\perp (C, Over) \mid (H_4 \in J_s) \qquad (4.12)$$

which means that Vol is conditionally independent of $(C, Over)$ in every slice of the partition $\{J_s\}$, following our usual notational convention. Condition (4.12) is sufficient to imply condition (4.11). (See Problem 4.5 for justification.) On the other hand, if (4.11) is a good approximation and the window width is sufficiently narrow then Vol should appear independent of $(C, Over)$ within each slice. In other words, if the window width is sufficiently small, (4.11) and (4.12) are approximately equivalent, and we can concentrate on assessing whether the data support (4.12). To perform such an assessment, rotate the plot $\{Vol, (C, Over)\}$ with the points in a slice highlighted. They should appear as a rotating horizontal band if (4.12) holds. Any other pattern is evidence against (4.12) and therefore is also evidence against (4.11).

Beginning with the view in Figure 4.3a, the plot in Figure 4.3b was constructed by rotating and focusing on the points within the slice for visual clarity. Since the points in Figure 4.3b seem to form a horizontal band, there is no evidence to contradict (4.11) within the slice selected in Figure 4.3a. Repeating this operation over several slices that cover the range of H_4 yields similar results in each case, so there is no clear evidence that (4.11) fails. The general conclusion is that the empirical distribution of $Vol \mid (C, Over)$ is not noticeably different from the empirical distribution of $Vol \mid H_4$ and thus we might proceed by treating the 2D plot $\{Vol, H_4\}$ as a sufficient summary of the 3D plot $\{Vol, (C, Over)\}$.

The two frames of Figure 4.4 illustrate what can happen when visual fitting of the "best" linear combination does not yield a sufficient summary plot.

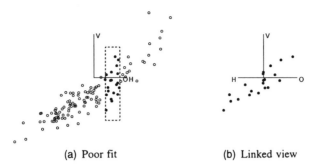

(a) Poor fit   (b) Linked view

FIGURE 4.4 Two views of the 3D plot $\{V,(C,Over)\}$ for the haystack data. (a) Relatively poor fit. (b) Linked view showing slice points only.

They were constructed like the frames in Figure 4.3, except that the view in Figure 4.4a has been rotated away from the view in Figure 4.3a. Because there is an increasing trend in the points of the linked view, Figure 4.4b, we have evidence that the view shown in Figure 4.4a is an insufficient summary plot.

This first method of assessing the structural dimensionality of a 3D plot can be summarized as follows:

- Observe the rotating 3D plot $\{y,x\}$. If any 2D view $\{y,b^T x\}$ shows that the distribution of $y \mid b^T x$ depends on the value of $b^T x$ then the structural dimension must be at least 1. If it is concluded that $y \perp\!\!\!\perp b^T x$ in all 2D views $\{y,b^T x\}$ then, from Proposition 4.3, $y \perp\!\!\!\perp x$ and the structural dimension is 0.
- To decide between 1D and 2D structure, rotate the 3D plot $\{y,x\}$ to the "best" visual fit, say $\{y,\hat{b}^T x\}$. While the summary plot $\{y,\hat{b}^T x\}$ remains on the computer screen, create another copy of $\{y,x\}$ and start rotating it about the vertical axis, as suggested by McCulloch (1993). Finally, brush the summary view while observing the rotating copy. Lack of dependence in the highlighted points produced while brushing provides support for 1D structure, for using $\{y,\hat{b}^T x\}$ as a sufficient summary plot, and for using $\hat{b}$ as an estimated basis for the minimal subspace $S_{y|x}$.
- If dependence is seen in $\{y,x\}$ while brushing the present summary view then there are two possibilities. Either the structural dimension is 2, or the present summary plot $\{y,\hat{b}^T x\}$ is noticeably far from a sufficient summary plot. In the former case, the full 3D plot is the sufficient summary. In the latter case, rotating to a different summary view may help.

A refinement of this first method is discussed in Section 4.6.4.

4.5.3. Natural rubber

Rice (1988, p. 506) reports the results of an experiment to study the modulus of natural rubber M as a function of the amount of decomposed dicumyl

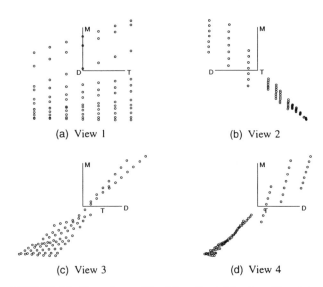

FIGURE 4.5   Four views of the 3D scatterplot for the rubber data.

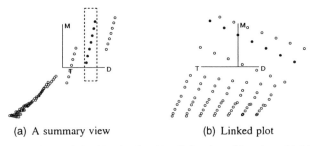

FIGURE 4.6   Two views of the 3D plot $\{M,(T,D)\}$ for the rubber data. (a) Mostly arbitrary summary view. (b) Linked 2D view.

peroxide D and its temperature T in degrees Celsius. Figure 4.5 gives four views of the 3D plot $\{M,(T,D)\}$. A general impression is that the response surface resembles a plane that has been distorted by twisting and stretching. The stretching can be seen in the upper-right corner of the scatter in View 1, while the twisting may be evident in the other three views. View 3 was selected to provide the best visual fit for the points in the upper-right corner of the plot, while View 4 was selected to provide the best fit of the points in the lower left. Since two different views are needed, the structural dimension must be 2. This can be confirmed by brushing any summary view and observing the corresponding highlighted points in a rotating copy of the 3D plot $\{M,(T,D)\}$, as illustrated in Figure 4.6.

Figure 4.7 shows the four views of the 3D residual plot $\{\hat{e},(T,D)\}$ that correspond to the four views of Figure 4.5. The impression of a saddle is greatly

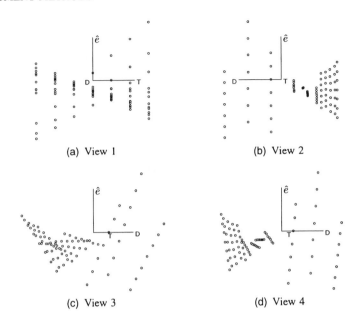

FIGURE 4.7   Four views of the 3D residual plot $\{\hat{e}, (D,T)\}$ for the rubber data.

enhanced in the residual plot, suggesting the possibility of an interaction between the predictors. It is also evident that the saddle is not symmetric, so a
model of the form (4.6) is not likely to be adequate.

4.6. ASSESSMENT METHODS

In this section we discuss several procedures for assessing the structural dimensionality d of a 3D plot. The procedures come with varying levels of
difficulty and varying requirements. One such procedure was illustrated in
Section 4.5.2, but there are several others as well. The methods discussed in
Sections 4.6.1 and 4.6.2 are often the most useful in practice. Sections 4.6.4–
4.6.6 can be skipped at first reading without loss of essential background for
subsequent chapters.

The discussion here and in later chapters will be facilitated by the following
three propositions on conditional independence. For background on these
propositions and conditional independence in general see Basu and Pereira
(1983), Cox and Wermuth (1996), Dawid (1979a, 1979b, 1980), Mouchart
and Rolin (1984), and Whittaker (1990, Section 2.2). In each proposition, U,
V, W, and Z are random vectors.

Proposition 4.4. $U \perp\!\!\!\perp V \mid W$ if and only if $U \perp\!\!\!\perp (V, W) \mid W$.

Proposition 4.5. If $U \perp\!\!\!\perp V \mid W$ and if U^* is a function of U then (a) $U^* \perp\!\!\!\perp V \mid W$ and (b) $U \perp\!\!\!\perp V \mid (W, U^*)$.

Proposition 4.6 (Conditional Independence). The following pair of conditions a_1 and a_2 is equivalent to the pair of conditions b_1 and b_2 which is equivalent to condition c.

$$
\begin{array}{ll}
(a_1) \quad U \perp\!\!\!\perp W \mid (Z,V) & (a_2) \quad U \perp\!\!\!\perp V \mid Z \\
(b_1) \quad U \perp\!\!\!\perp V \mid (Z,W) & (b_2) \quad U \perp\!\!\!\perp W \mid Z \\
(c) \quad U \perp\!\!\!\perp (V,W) \mid Z
\end{array}
$$

Conditions (a_1, a_2) and (b_1, b_2) of the Conditional Independence Proposition are really redundant, since each is implied by condition c. These conditions are stated separately to facilitate later use of the proposition and to emphasize the symmetric roles of W and V.

Consider now a regression problem with response y and $p = 2$ predictors $x = (x_1, x_2)^T$. Assume that graphical analysis of the 3D plot $\{y, x\}$ has led to the conjecture that the 2D plot $\{y, b^T x\}$ may serve as a sufficient summary. To assess this conjecture we need to check the condition

$$
y \perp\!\!\!\perp x \mid b^T x. \tag{4.13}
$$

This requirement for a sufficient 2D summary plot will provide a reference point in the methods that follow.

4.6.1. Using independence

Slicing may be unnecessary to assess condition (4.13) when there is a second linear combination $b_o^T x$ such that $b_o^T x \perp\!\!\!\perp b^T x$. Instead of observing a 3D plot for each slice as discussed in Section 4.5.2, a single 3D plot would be enough to check the sufficiency of $\{y, b^T x\}$. To develop the plot, first select a nonzero vector b_o so that $b_o^T x$ and $b^T x$ are uncorrelated,

$$
b_o^T \mathrm{Var}(x) b = 0 \tag{4.14}
$$

Next, rewrite (4.13) as

$$
y \perp\!\!\!\perp (b_o^T x, b^T x) \mid b^T x.
$$

From Proposition 4.4, $y \perp\!\!\!\perp x \mid b^T x$ if and only if

$$
y \perp\!\!\!\perp b_o^T x \mid b^T x. \tag{4.15}
$$

If $b_o^T x \perp\!\!\!\perp b^T x$ then, by setting $U = b_o^T x$, $V = y$, and $W = b^T x$, the Conditional Independence Proposition implies that (4.15) holds if and only if

$$b_o^T x \perp\!\!\!\perp (y, b^T x).$$

The implication of this result is summarized in the following proposition.

Proposition 4.7. If $b_o^T x \perp\!\!\!\perp b^T x$ then $\{y, b^T x\}$ is a sufficient summary plot if and only if the 3D plot $\{b_o^T x, (y, b^T x)\}$ has 0-dimensional structure.

If x is normally distributed then with b_o constructed according to (4.14) the condition $b_o^T x \perp\!\!\!\perp b^T x$ holds. Extreme deviations from normality can often be seen in the 2D plot $\{x_1, x_2\}$ at the start of the analysis.

A streamlined procedure for deciding between 1D and 2D structure can be implemented as follows.

- Rotate the 3D plot $\{y, x\}$ to the desired 2D view $\{y, b^T x\}$ and construct b_o according to a sample version of (4.14).
- Construct the 2D plot $\{b_o^T x, b^T x\}$. If the plot is consistent with the requirement $b_o^T x \perp\!\!\!\perp b^T x$ then the sufficiency of $\{y, b^T x\}$ can be assessed with a single 3D plot as in the next point. Otherwise, a different method may be necessary.
- Construct the 3D plot $\{b_o^T x, (y, b^T x)\}$ and consider its structural dimension. Resolution can be improved by orthogonalization (Section 3.4) when y and $b^T x$ are highly correlated, as may often be the case. The conclusion that the structural dimension of this plot is 0 supports using $\{y, b^T x\}$ as a sufficient summary.

This paradigm is easier than the first method discussed in Section 4.5.2 because it requires inspecting only one 2D and one 3D plot, rather than one 3D plot for each slice. Also, characterizing the checking plot $\{b_o^T x, (y, b^T x)\}$ will generally be easier than characterizing the original plot $\{y, x\}$ because deciding between $d = 0$ and $d > 0$ is generally easier than deciding between $d = 1$ and $d = 2$.

The two plots $\{y, x\}$ and $\{b_o^T x, (y, b^T x)\}$ can be linked dynamically with sufficient computing power. As $\{y, x\}$ is rotated about the vertical axis, the horizontal screen variable $H(\theta)$ and the orthogonal variable $H_o(\theta)$ calculated from a sample version of (4.14) are updated in real time and sent to the rotating plot $\{H_o(\theta), (y, H(\theta))\}$. As $\{y, x\}$ is analyzed the checking plot is then instantly available.

4.6.2. Using uncorrelated 2D views

Assessment procedures that require inspecting 3D plots can be a little time consuming. This could be a problem because it may be necessary to determine structural dimensionality for many 3D plots in regression problems with many predictors, as described in later chapters. Thus there may be some advantages to methods that require only 2D plots because characterizing a 2D plot is generally easier than characterizing a 3D plot.

An assessment of $y \perp\!\!\!\perp x \mid b^T x$ can be confined to a 2D construction by slicing on $b^T x$. Let

$$J_s = \{b^T x \mid \ell_s < b^T x < u_s\}$$

denote a typical slice of $b^T x$ with fixed width $w = u_s - \ell_s$, as illustrated in Figures 4.3a and 4.4a. As in Section 4.5.2, the fixed window width is used only to facilitate the discussion, and is not a necessary requirement. The slice constraint $b^T x \in J_s$ will be indicated by using just J_s for notational convenience. Statements such as $u \perp\!\!\!\perp v \mid J_s$ mean that u is independent of v, given $b^T x \in J_s$ for any slice J_s of width w. The expanded plot notation $\{u, v \mid J_s\}$ indicates plots only of those points that correspond to $b^T x \in J_s$. This notation is similar to that used during the discussion of Bean's graphic method in Section 2.6.2.

The following equivalences stem from the Conditional Independence Proposition and form the basis for using 2D constructions to evaluate the summary plot $\{y, b^T x\}$:

(a_1) $\quad b_o^T x \perp\!\!\!\perp b^T x \mid (y, J_s)$ $\qquad$ (a_2) $\quad b_o^T x \perp\!\!\!\perp y \mid J_s$

(b_1) $\quad b_o^T x \perp\!\!\!\perp y \mid (b^T x, J_s)$ $\qquad$ (b_2) $\quad b_o^T x \perp\!\!\!\perp b^T x \mid J_s$

(c) $\quad b_o^T x \perp\!\!\!\perp (y, b^T x) \mid J_s$

where b_o is still constructed according to (4.14). Condition (b_1) is the same as $y \perp\!\!\!\perp b_o^T x \mid b^T x$, which, from the discussion around (4.15), holds if and only if the summary plot is sufficient. The goal of this discussion is to find ways of using other conditions, principally (a_2), to check condition (b_1) indirectly.

Assuming that $b_o^T x \perp\!\!\!\perp b^T x$ as in the previous section, condition (b_2) holds for all window widths, including the case where there is only a single slice $J_s = \mathbb{R}^1$. (See Proposition 4.5.) Conditions (b_1), (b_2), and (c) with $J_s = \mathbb{R}^1$ reduce to the case considered in Proposition 4.7. On the other hand, if there is dependence between $b_o^T x$ and $b^T x$, condition (b_2) might be forced to an adequate approximation by choosing the window width to be sufficiently small.

The validity of condition (a_1) also depends on the window width w, and becomes trivially true as $w \to 0$. Condition (a_2) involves the response and will not necessarily be forced as $w \to 0$. However, it can be checked straightforwardly by inspecting subsets of the 2D plot $\{y, b_o^T x\}$ that correspond to slices J_s of $\{y, b^T x\}$ formed by brushing along $b^T x$ with a window width of w. An inference about the validity of (a_2) can be used to infer about the key condition (b_1) by using one of the following three logical implications:

1. If (a_1) and (b_2) both hold then (b_1) is true if and only if (a_2) is true.
2. If (a_2) is true then (a_1) implies (b_1).
3. If (a_2) is false then (b_2) implies that (b_1) is false.

The first implication will be useful in practice when the window width is sufficiently small to force (a_1) and (b_2) because then (b_1) is equivalent to (a_2). The second and third implications may be useful when there is some doubt about (a_1) and (b_2). These implications are not exhaustive. For example, if (a_2) is true and (a_1) is false, we have no way of inferring about (b_1) without further information.

The following procedure for assessing a summary plot is an expanded version of that used in Cook and Weisberg (1994a).

- Rotate the 3D plot $\{y,x\}$ to the desired summary view $\{y,b^Tx\}$ and construct b_o according to a sample version of (4.14).
- Construct the checking plot $\{y,b_o^Tx\}$ and link it to the summary view. The checking plot is called the *uncorrelated 2D view* by Cook and Weisberg (1994a).
- Brush the summary view parallel to the vertical axis and observe the corresponding highlighted points in the uncorrelated 2D view. The conclusion $y \perp\!\!\!\perp b_o^Tx \mid J_s$ supports the use of $\{y,b^Tx\}$ as a sufficient summary plot, assuming that the slices are sufficiently narrow to force (a_1) and (b_2). The steps to be taken when there is doubt depend on the conclusion regarding (a_2):
  - If no information is found to contradict condition (a_2), then check condition (a_1) using the same window width. If both (a_1) and (a_2) are inferred to be true, then condition (b_1) follows, again supporting the use of $\{y,b^Tx\}$ as a sufficient summary plot.
  - If information is found to contradict condition (a_2), then check condition (b_2) using the same window width. If condition (b_2) is then inferred to be true, it follows that (b_1) must be false and thus that $\{y,b^Tx\}$ is an insufficient summary plot.

A scatterplot matrix of y, b^Tx, and b_o^Tx is often a useful environment in which to implement this procedure because then the various conditions can be assessed with relative ease. Condition (a_1), for example, could be checked first by using a small rectangular brush to condition on (y,J_s) while observing the corresponding highlighted points in $\{b_o^Tx,b^Tx\}$.

4.6.3. Uncorrelated 2D views: Haystack data

We return to the haystack data described in Section 4.1 to illustrate the assessment paradigms introduced in Sections 4.6.1 and 4.6.2.

The predictor plot $\{Over,C\}$ shown in Figure 4.8 seems reasonably consistent with the assumption of bivariate normality, save perhaps two or three outliers. One of the advantages of graphical analyses is that outlying points can be neglected mentally or emphasized, as appropriate. Except for a little more curvature in the LOWESS smooth, the characterization of the trans-

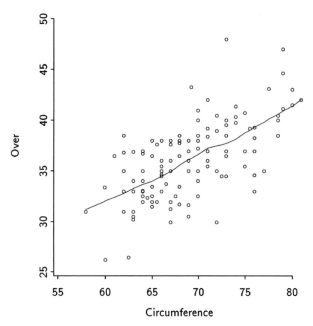

FIGURE 4.8 Predictor plot $\{Over, C\}$ for the haystack data with a LOWESS smooth.

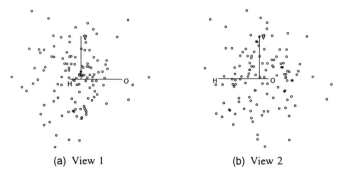

(a) View 1	(b) View 2

FIGURE 4.9 Two views of the 3D plot $\{b_o^T x, (y, b^T x)\}$ for the haystack data: $b_o^T x$ is on the V axis, $b^T x$ is on the H axis, and the O axis contains the residuals from the OLS fit of Vol on $b^T x$.

posed plot $\{C, Over\}$ is similar to that for Figure 4.8. Thus, both $E(C \mid Over)$ and $E(Over \mid C)$ are approximately linear and treating $(Over, C)$ as bivariate normal may not be unreasonable for the purpose of checking the structural dimension.

Following the paradigm of Section 4.6.1, Figure 4.9 gives two views of the 3D checking plot $\{b_o^T x, (y, b^T x)\}$ where $x = (C, Over)^T$, $b = (0.19, 0.98)^T$, and b_o was constructed from a sample version of (4.14). As indicated in the legend of the figure, y and $b^T x$ were orthogonalized to improve resolution. Neither

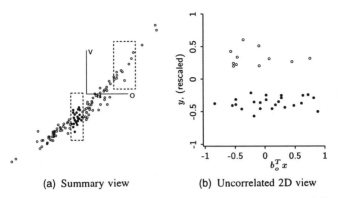

(a) Summary view	(b) Uncorrelated 2D view

FIGURE 4.10   Slicing the summary view of the haystack data to assess structural dimensionality. (a) Sliced summary view $\{y, b^T x\}$ as determined from the 3D plot $\{y, x\}$. (b) Uncorrelated 2D view $\{y, b_o^T x\}$ showing slice points only.

view indicates clear dependence, nor does any other view encountered during rotation. Because bivariate normality of the predictors implies that $b_o^T x \perp\!\!\!\perp b^T x$, we have evidence supporting the use of $\{y, b^T x\}$ as a sufficient summary plot.

Figure 4.10a is the summary view $\{Vol, 0.19C + 0.98Over\}$ as in Figure 4.1d but now shows two slices. The corresponding points in the uncorrelated 2D view $\{y, b_o^T x\}$ are shown in the plot of Figure 4.10b. Neither slice exhibits clear dependence, nor does any other slice encountered during brushing. From the paradigm of Section 4.6.2 this again supports the use of $\{y, b^T x\}$ as a sufficient summary plot.

4.6.4. Intraslice residuals

The paradigms outlined in Sections 4.6.1 and 4.6.2 are different from that illustrated in Section 4.5.2, which begins with an operational version of (4.13):

$$y \perp\!\!\!\perp x \mid J_s \qquad\qquad (4.16)$$

This is equivalent to

$$y \perp\!\!\!\perp (b_o^T x, b^T x) \mid J_s \qquad\qquad (4.17)$$

where b_o is chosen according to (4.14). From Proposition 4.5, (4.17) implies condition (4.13), $y \perp\!\!\!\perp x \mid b^T x$. Thus, failing to find evidence to contradict the sufficiency of the summary plot $\{y, b^T x\}$ by using the operational condition (4.17) provides support for 1D structure, as illustrated in Section 4.5.2. It looks as if (4.17) may serve as a practically useful substitute for the condition $y \perp\!\!\!\perp x \mid b^T x$.

However, a potential problem arises because (4.17) also implies

$$y \perp\!\!\!\perp b^T x \mid J_s. \qquad\qquad (4.18)$$

If it were possible to condition exactly on a value of $b^T x$, (4.18) would be true trivially. Otherwise, (4.18) requires that the window width be small enough so that there is no apparent intraslice dependence of y on $b^T x$. If the window width is not sufficiently small, (4.18) and thus (4.16) could be noticeably false, while it could still be true that $y \perp\!\!\!\perp x \mid b^T x$. In other words, relatively large window widths that could be useful with relatively small samples may leave a *remnant intraslice regression*, leading to the possibly wrong conclusion that the condition $y \perp\!\!\!\perp x \mid b^T x$ is false. This is a possibility for the rubber data in Figure 4.6a where a relatively large window width was used to capture all points with a common value of D. The remnant intraslice relationship may contribute to the linear trend in the highlighted points of Figure 4.6b.

To mitigate the effects of a remnant intraslice regression, we make use of intraslice residuals. Assume that the window widths are sufficiently narrow so that any intraslice dependence of $y \mid b^T x$ on $b^T x$ can be described adequately by the *intraslice models*

$$y \mid b^T x = \alpha_s + \beta_s(b^T x) + \sigma_s \varepsilon, \qquad b^T x \in J_s \qquad (4.19)$$

where $E(\varepsilon) = 0$, $Var(\varepsilon) = 1$, and $\varepsilon \perp\!\!\!\perp b^T x$. Let

$$e_s = [y - \alpha_s - \beta_s(b^T x)]/\sigma_s$$

denote a typical standardized population *intraslice residual* from an intraslice simple linear regression of y on $b^T x$. Passing from the response to the residuals is intended to remove the remnant intraslice relationship as mentioned previously. Requirement (4.17) is now replaced with

$$e_s \perp\!\!\!\perp (b_o^T x, b^T x) \mid J_s. \qquad (4.20)$$

This can be assessed by using 3D plots, as illustrated in Section 4.5.2 with the haystack data. Nevertheless, an assessment paradigm that relies only on 2D plots may facilitate routine application.

The following three equivalent sets of conditions stem from (4.20) and the Conditional Independence Proposition (Proposition 4.6):

(a_1) $\quad e_s \perp\!\!\!\perp b^T x \mid (b_o^T x, J_s) \qquad (a_2)$ $\quad e_s \perp\!\!\!\perp b_o^T x \mid J_s$

(b_1) $\quad e_s \perp\!\!\!\perp b_o^T x \mid (b^T x, J_s) \qquad (b_2)$ $\quad e_s \perp\!\!\!\perp b^T x \mid J_s$

(c) $\quad e_s \perp\!\!\!\perp (b_o^T x, b^T x) \mid J_s.$

Condition (c) is the same as (4.20) and is equivalent to the pairs of conditions (a_1, a_2) and (b_1, b_2). Condition (b_2) follows from the intraslice model (4.19). This condition does not seem hard to satisfy in most applications, but it might be relaxed by using a quadratic intraslice model, for example. Condition (a_2) can be assessed easily with 2D plots. We would like (a_2) to imply (b_1), but this requires condition (a_1). Thus, (a_2) and (b_1) are equivalent in cases where (a_1) holds to a good approximation.

This argument leads to another method of assessing a summary plot based on window widths that are small enough to force condition (a_1):

- Rotate the 3D plot $\{y,x\}$ to the desired 2D view $\{y,b^Tx\}$ and construct the uncorrelated direction b_o according to a sample version of (4.14).
- Slice the plot $\{y,b^Tx\}$ around a value for b^Tx, construct the intraslice residuals e_s, and send the residuals to the linked plot $\{e_s,b_o^Tx \mid J_s\}$.
- The conclusion $e_s \perp\!\!\!\perp b_o^Tx \mid J_s$ for all slices supports the sufficiency of the summary plot provided condition (a_1) is a reasonable approximation.
- If there is information to contradict $e_s \perp\!\!\!\perp b_o^Tx \mid J_s$ then it follows that the summary plot is not sufficient. Recall that condition (b_2) is assumed to follow from the intraslice model.

This paradigm is similar to that in Section 4.6.2 based on uncorrelated 2D views, but intraslice residuals are used instead of the response on the vertical axis of each uncorrelated view.

4.6.5. Intraslice orthogonalization

The methods of Sections 4.6.1 and 4.6.2 are based in part on the requirement that $b^Tx \perp\!\!\!\perp b_o^Tx \mid J_s$, with $J_s = \mathbb{R}^1$ for the single 3D checking plot in Section 4.6.1. In some cases it may not be possible to approximate these requirements adequately because sufficiently small window widths result in slices with too few observations. But it may be possible to extend applicability of the methods by using linear combinations of predictors that are approximately independent within slices.

Construct the slice-specific vector b_s so that

$$b_s^T \mathrm{Var}(x \mid b^Tx \in J_s)b = 0 \qquad (4.21)$$

and consider the equivalences

$$(a_1) \quad b_s^Tx \perp\!\!\!\perp b^Tx \mid (y,J_s) \qquad (a_2) \quad b_s^Tx \perp\!\!\!\perp y \mid J_s$$

$$(b_1) \quad b_s^Tx \perp\!\!\!\perp y \mid (b^Tx,J_s) \qquad (b_2) \quad b_s^Tx \perp\!\!\!\perp b^Tx \mid J_s$$

$$(c) \quad b_s^Tx \perp\!\!\!\perp (y,b^Tx) \mid J_s.$$

As in Section 4.6.2, condition (b_1) corresponds to the key requirement $y \perp\!\!\!\perp x \mid b^Tx$ for 1D structure. If the window width is sufficiently small then it should be reasonable to take (4.21) a step further and assume condition (b_2). The intraslice structure is now similar to that encountered in Sections 4.6.1 and 4.6.2. Like the previous paradigms, this permits two checking procedures, one based on the 3D plots implied by condition (c) and one based on 2D plots implied by condition (a_2).

3D paradigm

The paradigm based on 3D plots via condition (c) is roughly as follows:

- Rotate the 3D plot $\{y,x\}$ to the desired 2D view $\{y,b^T x\}$.
- Slice the plot $\{y,b^T x\}$ around a value for $b^T x$ and construct the intraslice uncorrelated direction b_s according to a sample version of (4.21).
- Send the intraslice information to the linked rotating 3D plot $\{b_s^T x, (y, b^T x) \mid J_s\}$. The conclusion of 0D structure for all of the 3D plots constructed in that way supports the sufficiency of $\{y, b^T x\}$.

This paradigm is only partially dependent on the window width. The conclusion of 0D structure for all 3D plots $\{b_s^T x, (y, b^T x) \mid J_s\}$ supports the conjecture that $\{y, b^T x\}$ is a sufficient summary regardless of the window width. But the conclusion that $d > 0$ for some 3D plots does not necessarily imply that the summary plot is insufficient unless the window width is small enough to insure that condition (b_2) is a good approximation. Finally, this procedure can be automated in sufficiently fast computing environments so that brushing $\{y, b^T x\}$ parallel to the vertical axis causes the linked plot $\{b_s^T x, (y, b^T x) \mid J_s\}$ to be updated in real time.

2D paradigm

A paradigm based on 2D plots via condition (a_2) is outlined as follows:

- Rotate the 3D plot $\{y,x\}$ to the desired 2D view $\{y,b^T x\}$.
- Slice the plot $\{y,b^T x\}$ around a value for $b^T x$ and construct the intraslice uncorrelated direction b_s according to a sample version of (4.21).
- Send the intraslice information to the 2D plot $\{y, b_s^T x \mid J_s\}$. With sufficiently small window width, the conclusion that $b_s^T x \perp\!\!\!\perp y \mid J_s$ in each of the 2D plots supports the sufficiency of $\{y, b^T x\}$.

This paradigm is fully dependent on the window width. Unless conditions (a_1) and (b_2) hold, the presence of dependence in every 2D plot $\{y, b_s^T x \mid J_s\}$ does not necessarily imply that $\{y, b^T x\}$ is insufficient, and the absence of dependence in any 2D plot does not necessarily imply that the summary plot is sufficient.

4.6.6. Mussels again

Return to the mussel data and consider a regression problem with mussel mass M as the response and predictors shell width W and shell mass S, $x = (S, W)^T$.

Two views of the 3D plot $\{M, (S, W)\}$ are shown in Figure 4.11. The first view is the selected summary plot. There is noticeably more variation in this

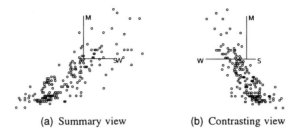

(a) Summary view   (b) Contrasting view

FIGURE 4.11 Two views of the 3D plot $\{M,(S,W)\}$ from the mussel data. (a) Summary view, $\{M,b^Tx\}$. (b) Contrasting view.

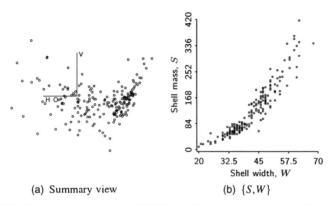

(a) Summary view   (b) $\{S,W\}$

FIGURE 4.12 Mussel regression $\{M,(S,W)\}$. (a) Summary view of the 3D checking plot $\{b_o^Tx,(M,b^Tx)\}$. (b) Predictor plot $\{S,W\}$.

problem than in the haystack data, making precise determination of the summary view somewhat more difficult. The second view was selected for contrast.

Curvature is apparent in the plot of Figure 4.12a, the summary view of the 3D checking plot $\{b_o^Tx,(M,b^Tx)\}$ constructed according to the paradigm of Section 4.6.1. Does this curvature provide evidence to conclude that the summary view of Figure 4.11a is insufficient? Not necessarily. The difficulty rests with the predictor plot shown in Figure 4.12b. Because the predictor distribution is apparently not normal, b_o^Tx and b^Tx are not guaranteed to be independent, and we cannot conclude that the summary plot is insufficient from this evidence alone.

Application of the methods in Sections 4.6.4 and 4.6.5 does provide support for using the plot of Figure 4.11a as a sufficient summary.

4.6.7. Discussion

Any of the procedures outlined in this section can be implemented for routine application. Those based on 3D plots are better than those based on 2D plots because there are fewer requirements. Those based on 2D plots have an edge, however, because characterization is generally easier and this may be important when it is necessary to determine the structural dimensionality of many 3D plots. Flexible systems providing access to both 2D and 3D assessment methods will be able to make use of their respective advantages.

The idea of structural dimensionality is based on reducing the dimension of the predictor vector by using linear combinations of the predictors. When a regression with $p = 2$ predictors has 2D structure, there could still be a lower-dimensional sufficient summary plot that requires a nonlinear transformation of x. Suppose, for example, that $y \perp\!\!\!\perp x \mid \|x\|$. Then the 2D plot $\{y, \|x\|\}$ is a sufficient summary but no 2D plot of the form $\{y, \eta^T x\}$ will be sufficient. The approach in this book is restricted to sufficient summary plots that can be constructed from linear combinations of the predictors.

PROBLEMS

4.1. Consider a rotating 3D plot in which $E(y \mid H(\theta))$ is constant for all θ, while $Var(y \mid H(\theta))$ clearly depends on θ. Suggest a procedure for selecting a 2D summary view of the 3D plot.

4.2. Justify Proposition 4.1.

4.3. Consider a 2D view $\{y, b^T x\}$ of the 3D plot $\{y, x\}$ for which the linear regression model (4.3) holds. Characterize the behavior of the variance function $Var(y \mid b^T x)$. Is it possible for $Var(y \mid b^T x)$ to depend nontrivially on $b^T x$ even though model (4.3) has constant variance?

4.4. Assume the 1D regression model (4.4),

$$t(y) \mid x = g(\beta^T x) + \sigma(\beta^T x)\varepsilon.$$

Then there exists a sufficient 2D view $\{y, \beta^T x\}$ of the 3D plot $\{y, x\}$. Is this sufficient view unique?

4.5. Using Proposition 4.5, show that (4.12) implies (4.11).

4.6. Suppose that $p = 2$, that the regression of y on x has 2D structure, and that

$$y \mid x = g(x) + \sigma\varepsilon$$

with $x \perp\!\!\!\perp \varepsilon$, $E(\varepsilon) = 0$, and $Var(\varepsilon) = 1$. Let e denote the residuals from the population OLS fit of y on x. If $E(x \mid a^T x)$ is linear in $a^T x$ for all vectors

$a \in \mathbb{R}^2$ and if the regression of e on x has 1D structure, does it necessarily follow that the response function must be of the semiparametric form given in (4.9)?

4.7. Justify Proposition 4.2.

4.8. Let U, V, and W be random vectors. Using the propositions on conditional independence, show that $U \perp\!\!\!\perp V \mid W$ if and only if

$$(U,W) \perp\!\!\!\perp (V,W) \mid W.$$

4.9. Let U, V, and W be random vectors, and let W^* be any (measurable) function of W. Using the propositions on conditional independence, show that $U \perp\!\!\!\perp V \mid W$ and $U \perp\!\!\!\perp W \mid W^*$ imply $U \perp\!\!\!\perp V \mid W^*$.

4.10. Consider a regression problem with response y and $p = 2$ predictors. Suppose that analysis of the 3D plot $\{y,x\}$ leads to the conjecture that the 2D summary plot $\{y,b^T x\}$ is sufficient. Suppose further that an analysis of the simple regression of y on $b^T x$ produces a set of residuals $r = k(y,b^T x)$ for some function k so that the transformation from $(y,b^T x)$ to $(r,b^T x)$ is one-to-one. Finally, suppose that inspection of various 2D plots leads to the conclusions that $r \perp\!\!\!\perp b^T x$, $r \perp\!\!\!\perp b_o^T x$, and $b_o^T x \perp\!\!\!\perp b^T x$, where b_o is constructed so that $b_o^T \Sigma b = 0$.

With this structure show that $y \perp\!\!\!\perp x \mid b^T x$ if and only if one of the following equivalent conditions holds:
1. $r \perp\!\!\!\perp b_o^T x \mid b^T x$
2. $r \perp\!\!\!\perp (b^T x, b_o^T x)$
3. $b^T x \perp\!\!\!\perp (r, b_o^T x)$
4. $b_o^T x \perp\!\!\!\perp (r, b^T x)$

One implication of this result is that with certain marginal independence, 0D structure in any of the three 3D plots 2–4 is enough to guarantee that $\{y,b^T x\}$ is sufficient.

4.11. Let b_s, b, e_s, and J_s be as defined in Sections 4.6.4 and 4.6.5, and assume that $b^T x \perp\!\!\!\perp b_s^T x \mid J_s$. Show that $b^T x \perp\!\!\!\perp e_s \mid (J_s, b_s^T x)$ if and only if $b^T x \perp\!\!\!\perp (b_s^T x, e_s) \mid J_s$. How does this relate to the graphical procedures in Section 4.6.4?

4.12. Depending on prior knowledge, predictor transformations might be used in an effort to reduce the structural dimension of a regression. For example, the following multiplicative model for the haystack data could yield a reasonable approximation to the distribution of $Vol \mid (C, Over)$,

$$Vol = \beta_0 C^{\beta_1} Over^{\beta_2} \varepsilon$$

where the positive error $\varepsilon \perp\!\!\!\perp (C, Over)$. As long as the βs are nonzero, this model has 2D structure. However, taking logarithms of both sides yields a model with 1D structure in the transformed variables,

$$\log Vol = \log \beta_0 + \beta_1 \log C + \beta_2 \log Over + \log \varepsilon.$$

Study the 3D plot of the log-transformed predictors for the haystack data and infer about the structural dimension of the corresponding regression $\log Vol \mid (\log C, \log Over)$. Assuming 1D structure for this regression, comment on the structural dimensions for the regressions, (a) $Vol \mid (\log C, \log Over)$, (b) $Vol \mid (C, Over)$, and (c) $\log Vol \mid (C, Over)$. Is your conclusion regarding $Vol \mid (C, Over)$ consistent the discussion of the haystack data in this chapter?

4.13. Visually estimate the structural dimension of the regression of C on $(Vol, Over)$ in haystack data.

4.14. Croxton and Cowden (1935, pp. 435–445) used the data of Table 4.1 to illustrate Bean's graphical method of fitting a generalized additive model as discussed in Section 2.6.2. The response variable is the sales of a "nationally known toilet preparation" per 1,000 persons by state including the District of Columbia. The two predictors are the percent minority M by state in 1931, and the McCann–Erickson index of per capita buying power as a percent of the U.S. average. The first approximation by Croxton and Cowden produced a generalized additive model of the form discussed in Section 4.4.2:

$$E[Sales \mid (M, BP)] = \alpha_0 + \alpha_1^T x + g(\alpha_2^T x) \qquad (4.22)$$

where $x = (M, BP)^T$.

Use the graphical methods discussed in this chapter to investigate the structural dimension of the sales data in Table 4.1, and to construct a summary plot for (a) the regression of *Sales* on M and BP, and (b) the regression of *Sales* on $\log(M)$ and $\log(BP)$. For each regression discuss the appropriateness of model (4.22). Using the predictor logarithms may help with the requirement of linear conditional expectations introduced in Section 4.4.2.

TABLE 4.1. Sales Data from Croxton and Cowden (1935, pp. 435–445)

State	Sales	M	BP	State	Sales	M	BP
DC	69.4	27.1	151.8	FL	56.9	29.4	104.9
LA	47.5	36.9	64.6	MS	47.5	50.2	47.8
SC	37.2	45.6	49.0	DE	30.6	13.7	111.2
MD	43.1	16.9	99.5	GA	37.5	36.8	54.8
AL	41.9	35.7	53.6	VA	30.6	26.8	62.8
NC	37.5	29.0	55.6	AR	36.9	25.8	54.4
TX	42.8	14.7	74.9	TN	43.1	18.3	61.3
CA	46.9	1.4	174.0	NY	36.9	3.3	144.7
NJ	20.6	5.2	121.8	IL	25.6	4.3	121.6
OH	22.9	4.7	108.7	MO	27.5	6.2	89.6
PA	18.7	4.5	105.2	MI	18.6	3.5	112.0
KY	14.4	8.6	63.3	CT	21.9	1.8	130.1
WV	12.2	6.6	76.0	OK	27.5	7.2	70.7
MA	14.7	1.2	132.6	IN	13.8	3.5	97.4
RI	15.6	1.4	117.7	KS	15.9	3.5	83.8
NV	17.5	.6	123.8	AZ	30	2.5	88.5
OR	21	.2	122.8	WY	11.2	.6	114.4
CO	24.7	1.1	101.7	WA	20.9	.4	112.8
NH	10.6	.2	104.5	MN	12.8	.4	99.4
UT	24.7	.2	101.0	IA	11	.7	89.6
NE	12.5	1.0	85.1	VT	9.8	.2	97.8
ME	10.5	.1	97.3	WI	10.5	.4	92.2
MT	12.8	.2	88.4	ID	9.8	.2	81.7
NM	9.8	.7	61.2	SD	6.9	.1	70.2
ND	4.4	.1	67.9				

State is identified by the U.S. postal code. *Sales* is per 1,000 persons in dollars. *BP* is the McCann–Erickson index of per capita buying power as a percent of the U.S. average. *M* is the percent for a minority population in 1930.

CHAPTER 5

Binary Response Variables

The idea of structural dimensionality introduced in Chapter 4 does not require constraints on the support of the response variable, but visualization techniques for assessing structural dimensionality do depend on the nature of the response. In this chapter, which follows the development in Cook (1996b), we address the problem of visually assessing structural dimensionality in regressions with a binary response taking values $y = 0$ and $y = 1$. Much of the material can be extended to few-valued responses with little difficulty.

5.1. ONE PREDICTOR

For a regression with a single predictor x there are only two possible structural dimensions: Either $y \perp\!\!\!\perp x$, corresponding to 0D structure and $\mathcal{S}_{y|x} = \mathcal{S}(\text{origin})$, or y is dependent on x, corresponding to 1D structure and $\mathcal{S}_{y|x} = \mathbb{R}^1$, where $\mathcal{S}_{y|x}$ is the subspace defined in Section 4.2. Trivially, $y \perp\!\!\!\perp x$ if and only if $x \mid (y = 0)$ and $x \mid (y = 1)$ have the same distribution. The problem of assessing structural dimensionality can therefore be reduced to a comparison of the empirical distribution of the predictor for $y = 0$ with the empirical distribution of the predictor for $y = 1$.

A scatterplot $\{y, x\}$ may not be very useful for deciding between 0D and 1D structure because it will consist of two lines of points perpendicular to the vertical axis at $y = 0$ and at $y = 1$. Jittering the plotted points to provide separation and thus to facilitate visualization of relative density may help some. Visualization might also be improved by plotting all points at the same location on the vertical axis prior to jittering and using different plotting symbols or colors to mark the states of y. There are, of course, many standard graphics for the two-sample problem that could be applied. For example, back-to-back stem-and-leaf displays, histograms with kernel density estimates (Scott 1992) superimposed, or side-by-side boxplots may furnish sufficient information to decide between 0D and 1D structure. For background on boxplots see McGill, Tukey, and Larsen (1978), Kafadar (1985), or Cleveland (1993). There are also many nonparametric tests available for comparing two samples.

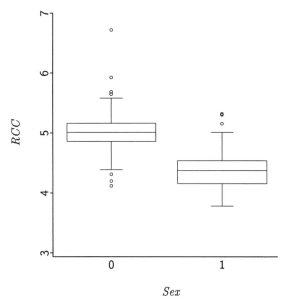

FIGURE 5.1 Side-by-side boxplots of the red blood cell count *RCC* grouped by values of the response *y* = *Sex* (0 = males, 1 = females) from the Australian Institute of Sport data.

Figure 5.1 gives side-by-side boxplots of the red blood cell count *RCC* grouped by the response *y* = *Sex* from a data set on 202 athletes at the Australian Institute of Sport (Cook and Weisberg 1994a). The visual evidence suggests that *RCC* | (*Sex* = 0) and *RCC* | (*Sex* = 1) are distributed differently, and thus that the regression of *Sex* on *RCC* has 1D structure.

5.2. TWO PREDICTORS

As in Chapter 4, assessing structural dimensionality with $p = 2$ predictors $x = (x_1, x_2)^T$ involves characterizing the regression in terms of the subspace $\mathcal{S}_{y|x}(\eta)$ based on statements of the form $y \perp\!\!\!\perp x \mid \eta^T x$ where η is a $2 \times d$ matrix, $d \leq 2$ (cf. (4.2)). If $\eta = 0$ then the structural dimension is 0, and if $d = 1$ then there is 1D structure. Otherwise, the regression has 2D structure.

To illustrate graphical issues that come with binary responses, 200 binary observations were generated independently according to a logistic regression model with probability of "success"

$$\Pr(y = 1 \mid x) = \frac{\exp[\gamma(x)]}{1 + \exp[\gamma(x)]} \tag{5.1}$$

where $\gamma(x) = 2.5(x_1 + x_2)$. The predictors (x_1, x_2) were constructed as independent standard normal random variables. According to (5.1), $y \perp\!\!\!\perp x \mid (x_1 + x_2)$

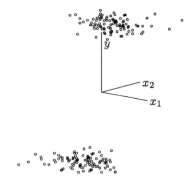

FIGURE 5.2   Three-dimensional scatterplot $\{y,x\}$ with binary response on the vertical axis.

and thus $\eta = (1,1)^T$ and $\{y, x_1 + x_2\}$ is a sufficient summary plot. We should be able to come close to this plot by using appropriate graphical methods. Figure 5.2 gives the usual 3D scatterplot of y versus x. The plot, which was pitched and rotated slightly to enhance the visual impression, suggests that the distributions of $x \mid (y = 0)$ and $x \mid (y = 1)$ are not equal and thus that the structural dimension is greater than 0. The plot is not very useful otherwise. The problem is that all points fall on one of two planes that are perpendicular to the vertical axis at $y = 1$ and $y = 0$.

The visualization of binary response data might be facilitated by passing to a different plot that uses symbols or colors to indicate the two values of the response. Beginning with Figure 5.2, the points corresponding to $y = 0$ and $y = 1$ were marked with distinct symbols. Visualization can be improved considerably by using different colors, but for this discussion different symbols must suffice. Next, the 3D plot was rotated so that the plot in the plane of the computer screen was the 2D plot $\{x_1, x_2\}$ with the different symbols standing for the values of the response, as shown in Figure 5.3. The *binary response plot* in Figure 5.3 could be viewed as a static 3D plot since it shows the relative values of three variables simultaneously without motion. Nevertheless, the plot will be termed a *2D binary response plot* to indicate the dimension of the plot itself and the number of predictors involved. The binary response plot was constructed beginning with the 3D plot in Figure 5.2 to allow certain rotations a bit later.

A visual analysis of structural dimension using binary response plots such as that in Figure 5.3 can be guided by the following simple idea. Imagine viewing Figure 5.3 as a small rectangular window

$$\Delta = \{x \mid \ell_j \le x_j \le u_j, \ j = 1,2\}$$

is moved around the plot and, at each window location, counting the number of symbols of each type. The ratio of the counts can be related to the condition

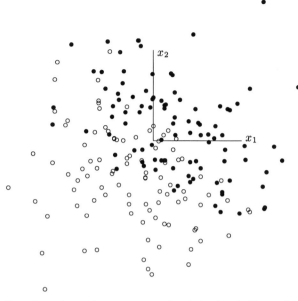

FIGURE 5.3 Two-dimensional binary response plot of the data in Figure 5.2. An open point denotes $y = 0$ and a filled point denotes $y = 1$.

$y \perp\!\!\!\perp x \mid \eta^T x$ as follows:

$$\frac{\Pr(x \in \Delta, \, y = 1)}{\Pr(x \in \Delta, \, y = 0)} = \frac{\Pr(y = 1 \mid x \in \Delta)\Pr(x \in \Delta)}{[1 - \Pr(y = 1 \mid x \in \Delta)]\Pr(x \in \Delta)}$$

$$= \frac{E(y \mid x \in \Delta)}{1 - E(y \mid x \in \Delta)}.$$

Letting $\Delta_\eta = \{\eta^T x \mid x \in \Delta\}$,

$$E(y \mid x \in \Delta) = E(y \mid x \in \Delta, \, \eta^T x \in \Delta_\eta).$$

If $y \perp\!\!\!\perp x \mid \eta^T x$ then

$$E(y \mid x \in \Delta, \, \eta^T x \in \Delta_\eta) = E(y \mid \eta^T x \in \Delta_\eta)$$

and thus

$$\frac{\Pr(x \in \Delta, \, y = 1)}{\Pr(x \in \Delta, \, y = 0)} = \frac{E(y \mid \eta^T x \in \Delta_\eta)}{1 - E(y \mid \eta^T x \in \Delta_\eta)}. \qquad (5.2)$$

This result indicates that if Δ is moved so that $\eta^T x$ is effectively constant, then the ratio of symbol counts should be constant up to the implicit binomial variation regardless of the particular value of $\eta^T x$. Further, provided the window Δ is sufficiently small, the implication can be inverted: Consider moving Δ so that $\eta^T x$ is fixed, or approximately so. If the ratio of symbol counts is constant (again, up to the implicit binomial variation) regardless of the value for $\eta^T x$, then $y \perp\!\!\!\perp x \mid \eta^T x$.

Alternatively, (5.2) can be reexpressed in terms of the fraction of one symbol in Δ:

$$\frac{\Pr(x \in \Delta, \ y = 1)}{\Pr(x \in \Delta)} = \mathrm{E}(y \mid \eta^T x \in \Delta_\eta). \tag{5.3}$$

This second expression indicates that if $y \perp\!\!\!\perp x \mid \eta^T x$ then the fraction of each type of symbol should be roughly constant when moving Δ so that $\eta^T x$ is effectively constant. Visually assessing the composition of Δ in terms of relative counts as in (5.2) or the count of one symbol relative to the total as in (5.3) is mostly a matter of taste since either can be used to gain visual information on $S_{y|x}(\eta)$. With a little practice, it seems possible to assess relative symbol density visually without actually counting.

Assuming that x has a density f and letting f_1 denote the conditional density of $x \mid (y = 1)$, (5.3) can be reduced to

$$\frac{f_1(x)}{f(x)} \propto \mathrm{E}(y \mid \eta^T x). \tag{5.4}$$

Like (5.3), this expression connects relative symbol density, as represented by the ratio f_1/f, to η via the regression function. We are now in a position to use these ideas for a visual assessment of structural dimension via binary response plots. The different methods of visualizing density will be referred to collectively with the phrase "relative symbol density."

5.2.1. Checking 0D structure

According to the previous discussion, dependence of $y \mid x$ on the value of x is reflected by changes in the relative symbol density across a binary response plot. The condition $y \perp\!\!\!\perp x$ would be supported by an approximately constant density, for example. In Figure 5.3, the relative density is quite different in the lower left of the plot than it is in the upper right, so the possibility that $y \perp\!\!\!\perp x$ is refuted easily.

5.2.2. Checking 1D structure

Consider using the plot in Figure 5.3 to assess an instance of 1D structure:

$$y \perp\!\!\!\perp (x_1, x_2) \mid x_1 \tag{5.5}$$

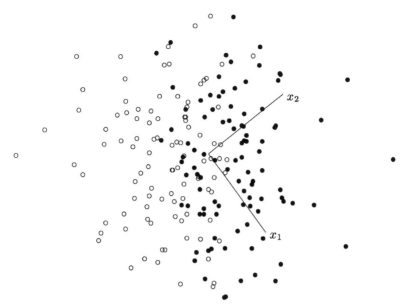

FIGURE 5.4 Rotated version of the 2D binary response plot in Figure 5.3.

or equivalently that $y \perp\!\!\!\perp x_2 \mid x_1$. Conditioning on x_1 can be approximated in Figure 5.3 by slicing the plot parallel to the vertical screen axis. If (5.5) were true then, from (5.3), relative symbol density should be constant throughout any slice, and conversely. The relative intraslice density can change from slice to slice without violating (5.5), but must be constant within any slice. Mentally slice the plot of Figure 5.3 just to the right of the x_2-axis. The density of filled points is much higher at the top of such a slice than at the bottom, so (5.5) is not sustained.

As may now be clear, the regression of y on x has 1D structure when there is a linear combination of the predictors $\eta^T x = \eta_1 x_1 + \eta_2 x_2$ so that $y \perp\!\!\!\perp x \mid \eta^T x$. Further, it follows from (5.3) that this is the case if and only if relative symbol density is constant when conditioning on $\eta^T x$. The possibility of 1D structure can be checked by fitting visually to achieve constant intraslice symbol density: Consider rotating the plot of Figure 5.3 about the stationary out-of-screen axis. As the plot is fully rotated once about the out-of-screen axis, the stationary *horizontal screen axis* will correspond to all possible linear combinations of the two predictors. The goal of a visual fit is to stop rotation at a point where the relative intraslice density is constant in any slice parallel to the stationary *vertical screen axis*.

One possibility is shown in Figure 5.4. The horizontal screen axis in this plot corresponds to the linear combination of the predictors

$$b^T x = 0.49 x_1 + 0.47 x_2.$$

The subspace $S[(0.49, 0.47)^T]$ is nearly the same as the defining subspace $S_{y|x} = S[(1, 1)^T]$. While visual fitting has come close to recovering the linear combination of the predictors used to generate the data, such population information will be unavailable in practice.

To check the conjecture

$$y \perp\!\!\!\perp x \mid b^T x \tag{5.6}$$

we can view Figure 5.4 as we did Figure 5.3, slicing the plot parallel to the vertical screen axis and looking for nonuniform intraslice density. Slices at relatively low or high values of $b^T x$ contain only one type of symbol so 1D structure is sustained at the extremes. There is also no clear visual evidence to suggest variable intraslice density at the middle values of $b^T x$ so again 1D structure is sustained. In short, the plot in Figure 5.4 seems visually consistent with 1D structure, implying that $\{y, b^T x\}$ might be used as a sufficient summary plot for further analysis.

If the intraslice density is clearly variable in all views obtained by rotating about the out-of-page axis then the structural dimension is 2, $S_{y|x} = \mathbb{R}^2$, and we must view the full 2D binary response plot to understand how the distribution of $y \mid x$ depends on the value of x.

5.2.3. Comparison with previous checking methods

The basic paradigm discussed in Sections 5.2.1 and 5.2.2 for assessing the structural dimension of a 2D binary response plot may seem distinctly different from the methods for many-valued responses in Chapter 4, but it is essentially the same as those methods.

Application of the first method for many-valued responses described in Section 4.5.2 begins by rotating the 3D plot $\{y, x\}$ to a potentially sufficient 2D view $\{y, b^T x\}$. Next, the 2D view is sliced around a selected value v for $b^T x$ and the slice is rotated to assess the condition $y \perp\!\!\!\perp x \mid (b^T x = v)$. Failure to find dependence in any such slice supports 1D structure. The paradigm is the same for binary responses, except rotation is no longer necessary to assess the condition $y \perp\!\!\!\perp x \mid (b^T x = v)$. Instead, this condition is checked by inspecting vertical slices of 2D binary response plots such as that in Figure 5.4.

The predictors in our example are independent standard normal variables. Hence, Proposition 4.7 indicates that conjecture (5.6) can be checked by inspecting the 3D plot $\{b_o^T x, (y, b^T x)\}$, where $b = (0.49, 0.47)^T$ and the sample covariance between $b_o^T x$ and $b^T x$ is zero. The horizontal screen axis of Figure 5.4 is $b^T x$, the vertical screen axis is $b_o^T x$, and the different symbols correspond to the two values of y. Application of Proposition 4.7 requires only reinterpreting Figure 5.4 in terms of a regression problem with response $b_o^T x$ and predictors y and $b^T x$. In particular, the plot can be rotated about the vertical screen axis as necessary to check for 0D structure according to Proposition 4.7. This operation supports the previous conclusion of 1D structure for the regression of y on x in the present illustration. More generally, application of Proposition 4.7 can be facilitated by orthogonalizing the predictors at the outset so that $\text{Var}(x) \propto I$. As a result, $b_o^T x$ will always be the vertical screen axis of plots analogous to Figure 5.4.

Describing the connections between 2D binary response plots and uncorrelated 2D views (Section 4.6.2) is one of the problems of this chapter.

Visually recognizing changes in relative density, as required for the characterization of a binary response plot, is not as easy as recognizing trends in scatterplots with a many-valued response. Nevertheless, with some practice using simulated data the ideas discussed here can be useful when there is sufficient variation in the regression function $E(y \mid x)$ across the observed values of the predictors. When there is little variation in $E(y \mid x)$, recognizing changes in relative density becomes quite difficult. Of course, the same type of situation occurs with many-valued responses when the magnitude of the trend is small relative to background variation. Additionally, seeing changes in relative density can be difficult when there are so many points that overplotting becomes a problem. Overplotting may also be a problem when y is binomial (m, p) and $m > 1$.

The ideas in the next section rely less on visualizing changes in relative density.

5.2.4. Exploiting the binary response

The direct method based on slicing for assessing the structural dimension of a 2D binary response plot does not require constraints on the predictors. The methods based on Proposition 4.7 do require conditions on the marginal distribution of x, however. In this section we consider indirect methods that exploit the binary nature of the response. These methods impose the same constraints on the distribution of x as do the methods based on Proposition 4.7. To facilitate later use, the main ideas are developed in the more general context of pD binary response plots.

Let b denote a full rank, $p \times d$ matrix, $d \le p$, and suppose that we wish a graphical assessment of the conjecture

$$y \perp\!\!\!\perp x \mid b^T x \tag{5.7}$$

where x is a $p \times 1$ predictor vector. Construct a full rank, $p \times (p-d)$ matrix b_o so that

$$b_o^T \Sigma b = 0 \tag{5.8}$$

where Σ is the covariance matrix of the predictors. The matrix $B = (b, b_o)$ forms a basis for $\mathbb{R}^p$, and $b^T x$ and $b_o^T x$ are uncorrelated. In practice, the usual sample covariance matrix could be used as an estimate of Σ. The next proposition gives conditions for checking (5.7) by looking for deviations from 0D structure. The justification is set as a problem at the end of this chapter.

Proposition 5.1. Assume that $b^T x \perp\!\!\!\perp b_o^T x$. Then $y \perp\!\!\!\perp x \mid b^T x$ if and only if (a) $b^T x \perp\!\!\!\perp b_o^T x \mid y$ and (b) $y \perp\!\!\!\perp b_o^T x$.

The checking conditions (a) and (b) of Proposition 5.1 are sufficient to imply both $y \perp\!\!\!\perp x \mid b^T x$ and the independence condition $b^T x \perp\!\!\!\perp b_o^T x$. However, $y \perp\!\!\!\perp x \mid b^T x$ need not imply (a) and (b) unless the independence condition holds.

When y is binary, condition (a) of Proposition 5.1 is equivalent to the pair of subconditions $b^T x \perp\!\!\!\perp b_o^T x \mid (y = 0)$ and $b^T x \perp\!\!\!\perp b_o^T x \mid (y = 1)$. These subconditions require that, for each state of y, the regression of $b^T x$ on $b_o^T x$ has 0D structure. Similarly, condition (b) requires that the regression of y on $b_o^T x$ has 0D structure.

The following steps can be used to check condition (5.7) when the response is binary, $p = 2$ and $d = 1$. Begin by constructing the 2×1 vector b_o so that $b_o^T \hat{\Sigma} b = 0$, and then proceed as follows:

1. Using the methods of Chapter 2, check the condition $b^T x \perp\!\!\!\perp b_o^T x$ by examining the 2D plot $\{b^T x, b_o^T x\}$ of all the data. If the condition fails, a different checking procedure may be necessary to verify (5.7). If the condition is sustained, proceed to the next step.

2. Focusing on each state of y separately in $\{b^T x, b_o^T x\}$, check the conditions

$$b^T x \perp\!\!\!\perp b_o^T x \mid (y = 0) \qquad \text{and} \qquad b^T x \perp\!\!\!\perp b_o^T x \mid (y = 1)$$

again by using the methods of Chapter 2. Depending on the computer program, controls may be available to toggle easily between the plots. Dependence of $b^T x$ on $b_o^T x$ in the plot for $y = 0$ or in the plot for $y = 1$ is evidence against conjecture (5.7).

3. Finally, check the condition $y \perp\!\!\!\perp b_o^T x$ by using the ideas discussed in Section 5.1.

4. If nothing is found to contradict the three independence conditions in Steps 2 and 3, then conjecture (5.7) is sustained. Evidence against any of those three independence conditions can be interpreted as evidence against conjecture (5.7).

5.3. ILLUSTRATIONS

5.3.1. Australian Institute of Sport

As mentioned briefly in Section 5.1, Cook and Weisberg (1994a) describe a data set consisting of physical and hematological measurements on 202 athletes at the Australian Institute of Sport. In this section, we use selected predictors from these data for two illustrations of assessing structural dimen-

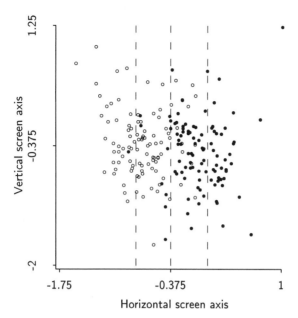

FIGURE 5.5 Two-dimensional binary response plot for the Australian Institute of Sport data with orthogonalized predictors *Ht* and *RCC*. The horizontal axis corresponds to the linear combination of the predictors from a visual fit. Highlighted points denote males.

sion with 2D binary response plots, in each case taking the response to be *Sex*, an athlete's sex.

Height and red cell count

The predictors for the first illustration are an athlete's height *Ht* and red cell count *RCC*. Orthogonalizing the predictors to improve visual resolution and then following the steps for visual fitting that led to Figure 5.4 resulted in the view given in Figure 5.5, where the highlighted symbols denote male athletes. The axes of the plot are the horizontal and vertical screen axes of the final 3D plot analogous to Figure 5.4. In particular, the horizontal axis corresponds to the linear combination of the predictors obtained from a visual fit. The vertical dashed lines on the plot are intended to facilitate visualization of intraslice density.

There is surely enough information in the plot of Figure 5.5 to rule out the possibility that $Sex \perp\!\!\!\perp (Ht, RCC)$ because the male athletes fall mostly at high values along the horizontal screen axis and the female athletes fall mostly at low values. The issue then is whether the structural dimension is 1 or 2. The density of symbols within each of the slices parallel to the vertical screen axis seems fairly constant, supporting the conclusion of 1D structure.

Apart from an additive constant, the horizontal screen axis of Figure 5.5 is

$$b^T x = 0.014 Ht + 0.47 RCC.$$

It may be informative to contrast this result from visual fitting with the corresponding result from fitting a first-order logistic regression model with canonical link and

$$\gamma(Ht, RCC) = \beta_0 + \beta_1 Ht + \beta_2 RCC$$

as defined generally in (5.1). The fitted linear combination from this logistic regression model is

$$\hat{\gamma} = 60.79 - .194Ht - 5.46RCC$$

with a deviance of 113.7 on 199 degrees of freedom. The sample correlation between $b^T x$ and $\hat{\gamma}$ is -0.995, indicating that visual fitting has produced essentially the same fitted values as logistic regression in this example.

As a final check, the second-order logistic model with

$$\gamma(Ht, RCC) = \beta_0 + \beta_1 Ht + \beta_2 RCC + \beta_{11} Ht^2 + \beta_{12} Ht \times RCC + \beta_{22} RCC^2$$

was fitted. The resulting deviance of 112.1 is only slightly smaller than that for the previous fit and none of the quadratic coefficients added significantly, again supporting the results of the visual analysis.

Because of these results, the standard scatterplot of y versus $b^T x$, as well as any of the other plots listed in Section 5.1, may serve as a sufficient summary of the regression. A close inspection of the plot in Figure 5.5 will show that there are three males surrounded by females in the left half of the plot. These points might be regarded as outliers because they occur in a region that otherwise has a very high density of females. The deviance residuals from a fit of a logistic model with canonical link supports the visual impression of these three points.

Height and lean body mass

The regression of *Sex* on height *Ht* and an athlete's lean body mass *LBM* appears to have 2D structure. The reasons for this conclusion may be seen with the aid of Figure 5.6, which was constructed in the same manner as Figure 5.5. The horizontal axis of Figure 5.6 is the linear combination of the predictors from a fit of the first-order logistic regression model in *Ht* and *LBM* with the canonical link. A portion of the highlighted points, females, overlaps the point cloud for the male points in the upper portion of the third slice of the plot. This causes the density for females to be higher at the top of the third slice than at the bottom. A similar phenomenon occurs in the second slice with the roles of males and females reversed, although this may be a little difficult to see because the highlighted points (females) partially cover several of the points for males. Using colors to distinguish males and females makes the overlap between the point clouds much easier to see. Because of the local overlap in the points for males and females, a single linear combination of the predictors cannot be found so that the intraslice symbol density is essentially

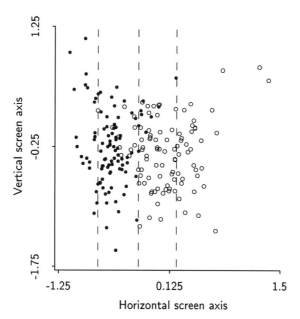

FIGURE 5.6 Two-dimensional binary response plot for the Australian Institute of Sport data with orthogonalized predictors *Ht* and *LBM*. The horizontal axis corresponds to the linear combination of the predictors from a fit of a first-order logistic model. Highlighted points correspond to females.

constant. Thus, the possibility of 1D structure is not sustained by the binary response plot and we must therefore conclude that the structural dimension is 2. Interestingly, all of the males in the data file who run 400*m* fall in or very near the overlap between males and females in the second slice.

The deviance from the fitted first-order logistic model is 103.6 with 199 degrees of freedom, while the deviance for the second-order logistic model is 82.2 with 196 degrees of freedom, and the absolute values of the estimated quadratic coefficients all exceed three times their respective standard errors. As in the previous illustration, the results from the second-order logistic model support the graphical analysis. Further, when the 7 female points at the top of the third slice are deleted, the estimated quadratic coefficients are all less than 1.5 times their respective standard errors, again supporting the idea that the local overlap in the point clouds for males and females is contributing to 2D structure.

5.3.2. Kyphosis data

Hastie and Tibshirani (1990, p. 282; see also Chambers and Hastie 1992) analyze data on spinal deformities following corrective spinal surgery. The response is the presence (*kyp* = 1) or absence (*kyp* = 0) of kyphosis *kyp*, and the

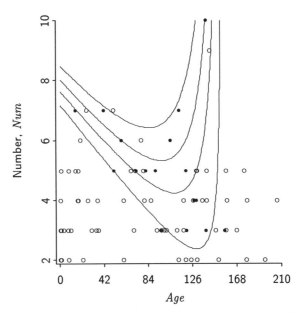

FIGURE 5.7 Two-dimensional binary response plot for the kyphosis data with predictors *Age* and *Num*. The kyphosis cases are represented by a highlighted (filled) circle. Proceeding from the lower right to the upper left, the curves correspond to the 0.2, 0.4, 0.6, and 0.8 contours of the estimated regression function.

three predictors are age in months at time of surgery *Age*, the starting vertebrae *Start* and number *Num* of vertebrae involved. Two remote points were deleted from this illustration and the forthcoming illustration in Section 5.4.2, the same two points that were deleted by Hastie and Tibshirani (1990, p. 287), leaving 81 observations with *kyp* = 1 in 17 cases. This example is restricted to the marginal regression on *Age* and *Num*. Figure 5.7 gives a 2D binary response plot with the two predictors; the contours on the plot will be discussed shortly.

A visual inspection of the pattern of points in Figure 5.7 indicates 2D structure. At *Num* = 3, 4, and 5, the probability of kyphosis evidently increases with age and then decreases, giving a suggestion that $E(y \mid x)$ is a nonmonotonic function of *Age* for every fixed value of *Num*. No kyphosis cases were observed at *Num* = 2. Because the plot can be viewed as a very coarse version of a contour plot of $E(y \mid x)$, it seems that the regression has 2D structure, perhaps with an interaction. In addition, the two cases with the largest values of *Num* are likely to be outlying or influential when attempting to model the data.

The contours on the plot are of the estimates $\hat{E}(y \mid x)$ resulting from a logistic regression on *Age*, *Num*, *Age*², and *Age* × *Num*. The coefficient estimates are all at least twice their respective standard errors. The contours nicely follow the pattern of the data and confirm the visual impressions. As anticipated, there are influential cases as well.

5.4. THREE PREDICTORS

With a binary response variable and three predictors, all the data can be handled simultaneously in a single *3D binary response plot*: Assign the three predictors $x = (x_1, x_2, x_3)^T$ to the axes of a 3D plot in any convenient way and mark the points with easily distinguishable colors for the states of y. When viewing a 3D response plot $\{y, (x_1, x_2)\}$ with two predictors, rotation is generally restricted to be about the vertical axis. However, rotation should not be restricted when viewing a 3D binary response plot.

With three predictors the possible structural dimensions d are 0, 1, 2, and 3. From (5.2), 0D structure is indicated when the relative symbol density is approximately constant throughout the 3D binary response plot. Fortunately, it is not necessary to look into the middle of the point cloud when assessing relative density. Rather, as implied by Proposition 4.3, assessing relative density in the 2D views observed while rotating is sufficient. Often the decision between $d = 0$ and $d > 0$ is relatively easy, particularly when there is substantial variation in $E(y \mid x)$. The decision between $d = 1$ and $d > 1$ is usually the most important and can be more difficult.

5.4.1. Checking 1D structure

Returning to the Australian Sport data, consider the regression of *Sex* on three of the hematological variables, *RCC*, hematocrit (Hc), and hemoglobin (Hg). An initial 3D binary response plot is not very informative because of fairly high correlations between the predictors. However, we can linearly transform to new uncorrelated predictors without loss of information on the structural dimension. Once the uncorrelated predictors are obtained, the next step is to temporarily assume 1D structure, $y \perp\!\!\!\perp x \mid \eta^T x$ with $\eta \in \mathbb{R}^3$, and then rotate so that an estimate $b^T x$ of $\eta^T x$ is on the horizontal screen axis. The ideas for obtaining $b^T x$ are the same as those discussed for two predictors. In particular, $b^T x$ could be selected visually by manipulating two rotation angles, or it could be taken from an initial numerical fit.

The view of the Australian Sport data given in Figure 5.8 was obtained by rotating the 3D binary response plot so that the fitted predictor $\hat{\gamma} = b^T x$ from a first-order logistic regression is on the horizontal axis. The plot shows good separation of the sexes with fairly uniform symbol density within vertical slices, and thus supports the assumption of 1D structure. For the conjecture $y \perp\!\!\!\perp x \mid \hat{\gamma}$ to be fully sustained, Proposition 4.3 implies that it is sufficient to have uniform intraslice densities in all views obtained by rotating about the horizontal screen axis $\hat{\gamma}$ of the plot in Figure 5.8. Little evidence to contradict uniform intraslice density was observed in any views obtained in this way, so there is no clear visual indication to contradict 1D structure.

Beginning with the fitted logistic predictor $\hat{\gamma} = b^T x$, the conjecture $y \perp\!\!\!\perp x \mid \hat{\gamma}$ could also be checked by using Proposition 5.1. Begin by constructing the

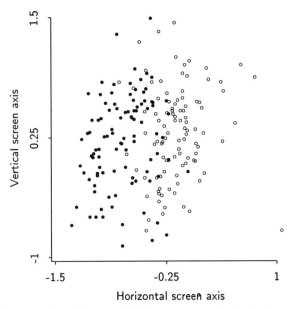

FIGURE 5.8 Two-dimensional view of a 3D binary response plot for the Australian Sport data. The horizontal axis corresponds to the linear combination of the predictors from a first-order logistic regression. Points for females are highlighted.

3×2 matrix b_o so that $b_o^T \hat{\Sigma} b = 0$ and then proceed as follows:

1. Using the methods of Chapter 4, check the necessary independence condition $b^T x \perp\!\!\!\perp b_o^T x$ by examining the 3D plot $\{b^T x, b_o^T x\}$ of all the data. If the condition fails, a different procedure might be necessary. If the condition is sustained, proceed to Step 2.

2. Examine the male and female separately in the 3D plot $\{b^T x, b_o^T x\}$ to check conditions

$$b^T x \perp\!\!\!\perp b_o^T x \mid (y = 0) \qquad \text{and} \qquad b^T x \perp\!\!\!\perp b_o^T x \mid (y = 1)$$

again by using the methods of Chapter 4.

3. Finally, check the condition $y \perp\!\!\!\perp b_o^T x$. This requires checking a single 2D binary response for deviations from 0D structure.

4. If nothing is found to contradict the three independence conditions in Steps 3 and 4, then the possibility of 1D structure is sustained. Evidence against any of the three independence conditions can be interpreted as evidence against 1D structure.

It may be important to recognize that evidence against the conjecture $y \perp\!\!\!\perp x \mid \hat{\gamma}$ is not necessarily evidence against 1D structure: There could be 1D structure $y \perp\!\!\!\perp x \mid \eta^T x$, while $\hat{\gamma} = b^T x$ is a poor estimate of $\eta^T x$; equivalently, $\mathcal{S}(b)$ is a poor estimate of $S(\eta)$. As we have seen, an alternative is to select an estimate of $\eta^T x$ visually. Circumstances when it may be reasonable to expect $S(b)$ to be a useful estimate of $S(\eta)$ are discussed in later chapters.

With a little practice it is not difficult to alternate rotation between the horizontal and out-of-screen axes to arrive at a visual estimate $b^T x$. If no such linear combination can be found that is consistent with the requirements of 1D structure then the conclusion that $d > 1$ is indicated. Nevertheless, this direct procedure for checking 1D structure can be time consuming, particularly when several 3D plots may need to be inspected. Methods that facilitate visual analysis of 3D binary response plots are described in later chapters after further developments have been presented.

5.4.2. Kyphosis data again

The 3D binary response plot for the regression of *kyp* on all three predictors, $x = (x_k) = (Age, Start, Num)^T$, indicates 2D and possibly 3D structure, suggesting that the analysis could be fairly complicated. This is in line with the marginal findings for *Age* and *Num* discussed in Section 5.3.2. Hastie and Tibshirani (1990) proceed by fitting a number of generalized additive models. In the present context, fitting a generalized additive model can be viewed as attempting to estimate predictor transformations $t(x) = (t_k(x_k))$ so that the regression of *kyp* on the transformed predictors has 1D structure. Once the transformations are estimated, 1D structure can be checked by inspecting a 3D binary response plot of the transformed predictors. Such an inspection could proceed by first rotating so that, in terms of the transformed predictors, $\hat{\gamma}$ is on the horizontal screen axis and then using the methods discussed earlier in this chapter.

One generalized additive model investigated by Hastie and Tibshirani (1990, Model iv, p. 289) is based on using a smoothing spline with 3 degrees of freedom for each of the predictors. A 3D binary response plot with the transformed predictors (not shown) seems reasonably consistent with 1D structure, although there are indications of influential cases. Influential cases might be anticipated for these data because of the relatively few kyphosis cases.

A second model investigated by Hastie and Tibshirani (1990, p. 291) is based on using logistic regression to obtain the fitted values

$$\hat{\gamma}(Age, Start) = b_1 Age + b_2 Age^2 + b_3(Start - 12)J(Start > 12)$$

$$= t_1(Age) + b_3 t_2(Start)$$

where J is the indicator function and the transformations t_1 and t_2 are defined implicitly. This model allows a quadratic effect of *Age* and distinguishes between the thoracic vertebrae, where $\hat{\gamma}$ is constant in *Start*, and the lumbar vertebrae where $\hat{\gamma}$ is linear in *Start*. A 2D binary response plot of the transformed predictors can be used to check 1D structure.

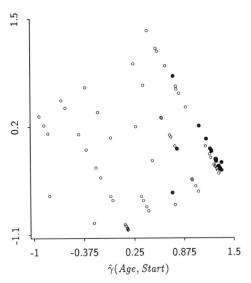

FIGURE 5.9   Two-dimensional binary response plot from the kyphosis data. The 17 kyphosis cases are highlighted. The smallest value of $\hat{\gamma}$ among the kyphosis cases is 0.74.

Figure 5.9 shows a 2D binary response plot of t_1 versus t_2 rotated so that $\hat{\gamma}$ is on the horizontal axis. Perhaps the most notable feature of this plot is that all but two of the points fall at the same value of $t_2(Start)$, leading to the conclusion that the two unique points, which correspond to kyphosis cases, may be quite influential. A small amount has been added to the vertical co-ordinates of the 17 kyphosis cases, which are represented by filled circles, to facilitate visual separation.

5.5. VISUALIZING A LOGISTIC MODEL

Binary response plots can provide visual clues about the structure of a logistic regression, in addition to providing visual information about $S_{y|x}$. This applies to the original predictors x, as well as to linear combinations $\eta^T x$ of predictors that come from visual inference about $S_{y|x}$ prior to modeling considerations, but the following discussion will be phrased in terms of x. The graphical ideas presented in this section were stimulated by the work of Kay and Little (1987).

Assume that x has a density and let f_j denote the conditional density of $x \mid (y = j)$, $j = 0, 1$. Then it follows from the discussion in Section 5.2 that

$$\log \frac{f_1(x)}{f_0(x)} = \log \frac{\Pr(y = 0)}{\Pr(y = 1)} + \log \frac{E(y \mid x)}{1 - E(y \mid x)}$$

$$= \log \frac{\Pr(y = 0)}{\Pr(y = 1)} + \gamma(x) \tag{5.9}$$

where $\gamma(x)$ is the usual "linear predictor" associated with logistic regression,

$$\Pr(y = 1 \mid x) = \frac{\exp[\gamma(x)]}{1 + \exp[\gamma(x)]}$$

as previously described in (5.1). Binary response plots provide visual information about the conditional densities f_0 and f_1. Thus, assuming we can find a visual bridge between the left and right sides of (5.9), binary response plots can provide clues about predictor transformations that may result in a relatively simple form for $\gamma(x)$. The examples of the next section are intended to be of intrinsic interest and to illustrate general principles.

5.5.1. Conditionally normal predictors

Univariate normals
Assume that there is a single predictor, $p = 1$, and that $x \mid (y = j)$ is a normal random variable with mean $\mu_j = E(x \mid y = j)$ and variance $\sigma_j^2 = \text{Var}(x \mid y = j)$. Then evaluating $\log f_1/f_0$ gives

$$2\gamma(x) = c + \log\left(\frac{\sigma_0^2}{\sigma_1^2}\right) + \frac{\mu_0^2}{\sigma_0^2} - \frac{\mu_1^2}{\sigma_1^2}$$

$$+ 2x\left(\frac{\mu_1}{\sigma_1^2} - \frac{\mu_0}{\sigma_0^2}\right)$$

$$+ x^2\left(\frac{1}{\sigma_0^2} - \frac{1}{\sigma_1^2}\right) \tag{5.10}$$

where c is a constant that depends only on the marginal distribution of y. Thus, when a binary regression problem includes only a single conditionally normal predictor, $\gamma(x)$ is generally a quadratic function of x. It reduces to a linear function of x only if the conditional variances are equal, $\sigma_0^2 = \sigma_1^2$. If both the conditional means and conditional variances are equal then $f_0 = f_1$ and $y \perp\!\!\!\perp x$ as described in Section 5.1.

The boxplots shown in Figure 5.1 for the regression of *Sex* on *RCC* from the Australian Sport data suggest that the conditional distribution of *RCC* | *Sex* is reasonably normal and, while the conditional means clearly differ, the conditional variances are approximately equal. The visual indication then is that a simple logistic model with γ set equal to a linear function of *RCC* may be sufficient. This conclusion is supported by the probability plots shown in Figure 5.10. The vertical distance between the probability plots indicates that the conditional means are different, but the similarity of the slopes suggests again that the conditional variances are similar as well.

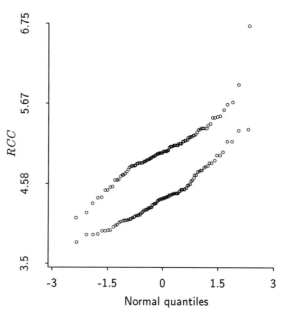

FIGURE 5.10   Separate probability plots of *RCC* for male (upper curve) and female athletes from the Australian Sport data.

Bivariate normals

When $p = 2$ and $x \mid (y = j)$ follows a bivariate normal distribution with $\mu_j = \mathrm{E}(x \mid y = j)$ and $\Sigma_j = \mathrm{Var}(x \mid y = j)$, calculations similar to those for the univariate case give

$$2\gamma(x) = c + x^T(\Sigma_0^{-1} - \Sigma_1^{-1})x + 2x^T(\Sigma_1^{-1}\mu_1 - \Sigma_0^{-1}\mu_0).$$

Thus, with $x = (x_1, x_2)^T$ there are five possible terms in γ: x_1, x_2, x_1^2, x_2^2, and $x_1 x_2$, comprising a full second-order logistic response surface model. If the conditional covariance matrices are equal, $\Sigma_0 = \Sigma_1$, then γ reduces to a linear function of x. If $x_1 \perp\!\!\!\perp x_2 \mid y$ then the cross-product term $x_1 x_2$ is not needed.

Consider, for example, the problem of distinguishing between counterfeit and genuine Swiss bank notes, as described in Problem 5.5. The response variable indicates if a note is authentic, $A = 0$ for genuine notes and $A = 1$ for counterfeit notes. Shown in Figure 5.11 is a binary response plot for the regression of A on two size measurements, the length Lg of a note at the center and the length along the bottom edge B, both in *mm*. A visual inspection of the plot suggests that

- the structural dimension of the regression is 1 with $A \perp\!\!\!\perp Lg \mid B$,
- A is marginally independent of Lg, $A \perp\!\!\!\perp Lg$,
- any dependence between the predictors within either the counterfeit or genuine notes is weak,

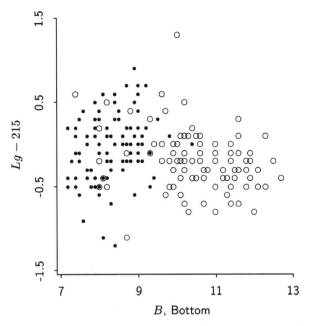

FIGURE 5.11   Binary response plot for the regression of A on length Lg and bottom B from the bank notes data (Problem 5.5). Genuine notes are marked by the highlighted (filled circle) points.

- except for two or three outlying points, the conditional densities f_0 and f_1 seem roughly consistent with the bivariate normal, although there is some evidence of skewness in a normal probability plot of B (not shown),

- the counterfeit note with the smallest value of B and the genuine note with the largest value of B may be influential cases,

- $E(B \mid A = 0) < E(B \mid A = 1)$, implying that γ should contain a term in B,

- $Var(B \mid A = 0) < Var(B \mid A = 1)$, implying that γ should contain a term in B^2, and

- overall $\gamma(B) = \beta_0 + \beta_1 B + \beta_2 B^2$ could give suitable description of the regression.

Nonlinear dependence

The potential for a complicated regression increases when the predictors are nonlinearly dependent. Again suppose that there are $p = 2$ predictors, $x = (x_1, x_2)^T$. To allow for the possibility of a nonlinear dependence write

$$\log\left(\frac{f_1(x)}{f_0(x)}\right) = \log\left(\frac{f_1(x_1 \mid x_2)}{f_0(x_1 \mid x_2)}\right) + \log\left(\frac{f_1(x_2)}{f_0(x_2)}\right) \qquad (5.11)$$

where $f_j(x_1 \mid x_2)$ indicates the conditional density of $x_1 \mid (x_2, y = j)$ and $f_j(x_2)$ indicates the conditional density of $x_2 \mid (y = j)$. Now assume that $x_2 \mid y$ and that $x_1 \mid (x_2, y)$ are both normally distributed.

The two terms on the right of (5.11) can be judged separately for their possible contributions to γ. Since $x_2 \mid y$ is normal, the term involving its marginal distribution can be interpreted as (5.10), the possible contributions being terms in x_2 and x_2^2. Further, since $x_1 \mid (x_2, y)$ is normal its potential contribution can also be judged from (5.10), but the interpretation is more complicated because the conditional means and variances may be functions of x_2. For illustration, suppose $\mu_j(x_2) = E(x_1 \mid x_2, y = j)$ depends on x_2 but $\sigma_j^2 = \mathrm{Var}(x_1 \mid x_2, y = j)$ does not. Then rewriting (5.10) to reflect this dependence,

$$
\log\left(\frac{f_1(x_1 \mid x_2)}{f_0(x_1 \mid x_2)}\right) = c + \log\left(\frac{\sigma_0^2}{\sigma_1^2}\right) + \frac{\mu_0^2(x_2)}{\sigma_0^2} - \frac{\mu_1^2(x_2)}{\sigma_1^2}
$$

$$
+ 2x_1\left(\frac{\mu_1(x_2)}{\sigma_1^2} - \frac{\mu_0(x_2)}{\sigma_0^2}\right)
$$

$$
+ x_1^2\left(\frac{1}{\sigma_0^2} - \frac{1}{\sigma_1^2}\right).
$$

Clearly, γ depends on the conditional mean functions $\mu_1(x_2)$ and $\mu_0(x_2)$. If these functions are complicated then γ may be complicated as well. Graphical inspection alone may not be very effective for deciding on a form for γ in such cases. Nevertheless, depending on the nature of the predictors, it is usually easy to get visual clues to discriminate between straightforward regression problems and regression problems where γ could be complicated.

5.5.2. Other predictor distributions

While the discussion so far has been limited to normal distributions with one or two predictors, the central ideas apply equally to regressions with many predictors and to other distributions.

Suppose that $p = 1$ and that $x \mid (y = j)$ follows a gamma distribution,

$$
f_j(x) = \frac{1}{\Gamma(\alpha_j)\beta_j^{\alpha_j}} \exp\{-x/\beta_j\} \qquad (\alpha_j > 0,\ \beta_j > 0,\ x > 0)
$$

for $j = 0, 1$. It follows immediately from the form of $\log f_j$ that γ may contain terms in x and $\log x$. Similarly, if $x \mid (y = j)$ follows a beta distribution then γ may contain terms in $\log x$ and $\log(1 - x)$.

When $p > 1$ and the predictors are independent, the structure of γ can be built up from the individual predictors because $\log f_1/f_0$ can be represented as the sum over similar terms for the individual predictors.

The structure of a logistic model can be most complicated when there are clear nonlinear dependencies among the predictors. In such cases, it is often worthwhile to use power transformations of the predictors in an attempt to bring the distributions of $x \mid (y = 0)$ and $x \mid (y = 1)$ close to multivariate normality. The transformation methodology proposed by Velilla (1993) can be adapted for this purpose.

PROBLEMS

5.1. In the case of a binary response, describe the differences between (a) using 2D binary response plots to determine structural dimension and (b) using uncorrelated 2D views to determine structural dimension.

5.2. Adapt the plotting paradigms of Sections 4.6.4 and 4.6.5 for binary responses.

5.3. Justify Proposition 5.1.

5.4. Consider a binary regression with a single predictor x. Following the discussion of Section 5.5, describe the potential structure of $\gamma(x)$ when x follows a (a) Poisson, (b) log normal $(x > 0)$, and (c) logistic distribution. How are the results for the logistic distinctly different from those for the Poisson and log normal as well as the distributions discussed in Section 5.5?

5.5. Flury and Riedwyl (1988, pp. 5–8) report six size measurements in mm on 100 counterfeit and 100 genuine Swiss bank notes. Let $x = (Lg, L, R, B, T, D)^T$ denote the 6×1 vector of size measurements where Lg is the length of a bank note, and L, R, B, T and D are measurements along the left, right, bottom, top and diagonal of a note. Also, let A denote authenticity, with $A = 0$ for the genuine notes and $A = 1$ for the counterfeit notes. Consider studying regressions of A on selected subsets of x to discover if counterfeit notes might be distinguished from genuine notes. The regression of A on x falls within the framework of this chapter, except for the sampling scheme. Although the sampling scheme is unknown, it seems reasonable to suppose that the bank notes were sampled from the conditional populations of x given A, as in case-control studies. Accordingly, the 100 genuine notes are assumed to provide independent and identically distributed observations on $x \mid (A = 0)$, while the 100 counterfeit notes are assumed to provide observations on $x \mid (A = 1)$. In contrast, straightforward application of the developments in this chapter would require that all 200 notes were obtained as independent and identically distributed observations on (A, x).

    5.5.1. Although obtained via case-control sampling, show that the data on bank notes can still be used to infer about the structural dimension of the regression of A on x. Which of the methods described in this chapter for finding sufficient summary plots are still valid under case-control sampling?

For additional background on case-control studies and possible hints on how to proceed, see Collett (1991, pp. 251–255).

5.5.2. Visually estimate the structural dimensions of the following regressions:

1. $A \mid (B, L)$
2. $A \mid (Lg, L)$
3. $A \mid (Lg, L, R)$

For regressions estimated to have 1D structure with $A \perp\!\!\!\perp x \mid \eta^T x$, include an estimate of the minimal subspace $\mathcal{S}(\eta)$, and contrast your solution with that from a first-order logistic regression. Describe the clues available about the structure γ of an adequate logistic model.

For regressions estimated to have 2D structure, include an estimate of the minimal subspace.

5.5.3. Inspect a scatterplot matrix of the predictors with the points colored to distinguish between the genuine and counterfeit notes. Based on this information, describe how you would carry out a regression of A on all 6 predictors.

CHAPTER 6

Dimension-Reduction Subspaces

So far we have used the ideas of structural dimensionality and sufficient summary plot to explore graphics for regressions with a univariate response and mostly one or two predictors. Beginning in this chapter, we use extensions and adaptations of these ideas to study regression problems with a $p \times 1$ vector of predictors x taking values in its support $\Omega_x \subset \mathbb{R}^p$. The methodology to be suggested for such studies will be based on the developments in the previous chapters. To facilitate the transition from few to many predictors, basic ideas will first be revisited in the broader context. The data (y_i, x_i^T), $i = 1, \ldots, n$, are still assumed to be independent and identically distributed observations on the random vector (y, x^T), and the goal is still to study the conditional distribution of $y \mid x$ graphically, with emphasis on the regression and variance functions.

6.1. OVERVIEW

Three-dimensional plots can be constructed easily with widely available software, as discussed in Chapter 3, and interpreted using the ideas described in Chapter 4. The high-dimensional Cartesian coordinate plots required when p is large cannot be viewed directly and methods for their interpretation must necessarily rely on plots of lower dimension. Specifically, the full $(p + 1)$-dimensional plot $\{y, x\}$ might be interpretable via $(q + 1)$-dimensional plots of the form $\{y, A^T x\}$, where A is a $p \times q$ matrix of rank $q \leq p$. Understanding when statistical characteristics of $\{y, x\}$ can be inferred reliably from $\{y, A^T x\}$ could be quite useful. Knowing that $\{y, A^T x\}$ is sufficient for $y \mid x$, for example, would surely be helpful.

One theme of this book is that, under certain conditions on the marginal distribution of x and on the structure of the regression problem, it may be possible to obtain useful insights about the distribution of $y \mid x$ from characterizations of appropriately chosen lower-dimensional plots. The regression information in the plot $\{y, A^T x\}$ can be found by studying the conditional distribution of $y \mid A^T x$. Such distributions provide information on the usefulness of lower-dimensional plots, on the relative worth of "standard" plots, on what to look for in plots (characterization) and on how to make use of what is found (inference).

When conditioning on linear combinations $A^T x$ there is no loss of generality in assuming that the columns of A are a basis for the subspace $\mathcal{S}(A)$ spanned by the columns of A because

$$F_{y|A^T x}(\cdot) = F_{y|B^T x}(\cdot) \tag{6.1}$$

for all matrices B such that $\mathcal{S}(B) = \mathcal{S}(A)$. In terms of the plots themselves, this means that if $\mathcal{S}(B) = \mathcal{S}(A)$ then $\{y, A^T x\}$ and $\{y, B^T x\}$ differ only by a change of basis for the "horizontal" axes and any statistical information contained in one plot will be contained in the other. The covariance matrix $\mathrm{Var}(A^T x)$ can influence the ease with which the plot can be characterized, however. We reasoned in Section 3.4 that visual resolution can be enhanced in 3D plots by choosing the vectors of points plotted on each horizontal axis to be orthogonal. This means that good visual resolution will be obtained in the population case when $\mathrm{Var}(A^T x) \propto I$. Thus, beginning with any basis A, visual resolution can be enhanced by plotting $\Phi A^T x$, where Φ is any nonsingular matrix such that $\Phi[A^T \Sigma A]\Phi^T = I$ and $\Sigma = \mathrm{Var}(x)$. Interpretation may be facilitated at other times by letting the columns of A correspond to linear combinations of particular interest, provided the latter are not too strongly correlated.

Plots will be described in two ways. If there are particular linearly independent plotting coordinates at hand, say $A^T x$, we will refer to a plot of y versus $A^T x$ which is denoted by $\{y, A^T x\}$. Otherwise, the only constraint on the columns of A will be that they form a basis for a relevant subspace $\mathcal{S}$. To avoid specifying a particular set of plotting coordinates, we will then refer to a plot of y *over* $\mathcal{S}$ with the understanding that the plotting coordinates do not matter, except perhaps for visual clarity as mentioned previously.

Part of the philosophy underlying modern graphical computing environments is based on the notion that experiencing data sets through many visualizations in a fairly short period of time is worthwhile. Spinning plots, grand tours (Asimov 1985; Buja and Asimov 1986), and scatterplot matrices, for example, all seem to reflect this philosophy. This approach may be useful, but it raises several key issues introduced in previous chapters: How should an interesting plot be interpreted? When can a plot be taken as a reflection of a fundamental aspect of the regression problem? And how do we determine fundamental aspects in theory or practice? In parametric regression modeling, fundamental aspects are described in terms of the parameters of the model. On the other hand, graphical analyses are at their best in exploration where an adequate parsimonious model may not yet be available. How should we think about fundamental aspects of regression problems when parameters are not available? One approach developed in this book was started in Chapter 4. In the following sections we expand that development.

6.2. DIMENSION-REDUCTION SUBSPACES

Working nonparametrically without a descriptive structure for the conditional distribution of $y \mid x$ is limiting since there is no intrinsic mechanism that allows for a parsimonious characterization of the data. Fully nonparametric approaches would permit the possibility that each distinct p-dimensional vector of predictor values indexes a different conditional distribution for $y \mid x$, but they would not allow explicitly for the possibility that less than p dimensions provide a complete or adequate description. Additionally, it is necessary to have a framework in which to assess the worth of lower-dimensional plots $\{y, A^T x\}$ as a reflection of relevant characteristics of $y \mid x$.

To allow descriptive flexibility, we expand the notion of structural dimensionality that was introduced in Chapter 4. This general type of construction has been used in related problems by Carroll and Li (1995), Duan and Li (1991), Cook (1992a, 1994a, 1994b, 1996b), Cook and Weisberg (1994a), Cook and Wetzel (1993), Li (1991, 1992) and others, and it often arises in studies of conditional independence (Dawid 1979a, 1979b, 1984). The developments of this chapter follow the results of Cook (1992a, 1994a, 1994b, 1996b).

Let $h_1, h_2, \ldots, h_k$ denote $k \le p$ linearly independent $p \times 1$ vectors so that, for each value of x in its marginal sample space and all real a,

$$F_{y|x}(a) = F_{y|h_1^T x, \ldots, h_k^T x}(a)$$

$$= F_{y|\eta^T x}(a) \tag{6.2}$$

where η denotes the $p \times k$ matrix with columns h_j, $j = 1, \ldots, k$. The equality in (6.2) means that y depends on x only through $\eta^T x$. In view of (6.2), however, $\eta^T x$ could be replaced by $\tilde{\eta}^T x$, where $\tilde{\eta}$ is any matrix such that $\mathcal{S}(\tilde{\eta}) = \mathcal{S}(\eta)$. Informally, one can then characterize (6.2) by saying that y depends on x only through the k-dimensional subspace $\mathcal{S}(\eta)$.[1]

Equation (6.2) is equivalent to x and y being independent given $\eta^T x$ (for additional discussion, see Basu and Pereira 1983):

$$y \perp\!\!\!\perp x \mid \eta^T x. \tag{6.3}$$

This is the same form of conditional independence discussed in Chapters 4 and 5 for small p. The statement implies that there would be no loss of regression information if x were replaced by $\eta^T x$, or if the full plot $\{y, x\}$ were replaced by the lower-dimensional plot $\{y, \eta^T x\}$. Phrased somewhat differently, (6.3) implies that the $(k + 1)$-dimensional plot $\{y, \eta^T x\}$ is sufficient for the regression of y on x.

---

[1]Here and throughout this discussion, sets of measure 0 can be avoided by considering only versions of CDFs so that equalities like (6.2) hold at all values of x.

If (6.3) holds, the subspace $\mathcal{S}(\eta)$ is called a *dimension-reduction subspace (DRS) for y | x*, or equivalently, for the regression of y on x. The short-hand phrase "dimension-reduction subspace" will be used when the response and predictors are clear from context.

Dimension-reduction subspaces need not result in reduced dimensionality since (6.3) holds trivially for $\eta = I_p$, the $p \times p$ identity matrix. For this same reason, a DRS is not necessarily unique. The main intent behind (6.3) is to provide a framework for reducing the dimensionality of the predictors to facilitate graphical analyses so the idea of a smallest subspace will be helpful: A subspace $\mathcal{S}$ is said to be a *minimum DRS for y | x* if $\mathcal{S}$ is a DRS and dim[$\mathcal{S}$] $\leq$ dim[$\mathcal{S}_{drs}$] for all DRSs $\mathcal{S}_{drs}$, where "dim" denotes dimension. As we will see shortly, a minimum DRS is not necessarily unique. Nevertheless, even if a regression admits several minimum DRSs, their dimensions must be the same by definition. Such unique dimension is called the *structural dimension* of the regression and denoted by d, as in Chapters 4 and 5.

Sufficient summary plots as introduced in Section 4.2 can be constructed from DRSs. If $\mathcal{S}_{drs}$ is a DRS then a plot of y over $\mathcal{S}_{drs}$ is called a *sufficient summary plot*. We carry the connection further and define a plot of y over $\mathcal{S}_{drs}$ to be a *minimal sufficient summary plot* if $\mathcal{S}_{drs}$ is a minimum DRS.

Dawid (1979a, 1984) discussed the notion of sufficient covariates in the context of comparative experiments: A collection C of covariates is sufficient if the individual experimental units contain no further information about the response given the treatment assignment and C. There is a connection between the idea of a sufficient subset of covariates and the notions of a sufficient plot and a DRS. In dimension reduction, conditional independence is applied to the measured covariates only. On the other hand, for a covariate subset to be sufficient the response must be conditionally independent of all information contained in the experimental units whether measured or not. Dimension reduction is like Dawid's covariate sufficiency relative to the distribution of the response y given the measured covariates x.

The following proposition shows what happens to a minimum DRS under full rank linear transformations of the predictors.

Proposition 6.1. Let $\mathcal{S}(\eta)$ denote a minimum dimension-reduction subspace for the regression of y on x, and let $z = A^T x$ where A is a full rank $p \times p$ matrix. Then $\mathcal{S}(A^{-1}\eta)$ is a minimum dimension-reduction subspace for the regression of y on z.

Justification. $\mathcal{S}(A^{-1}\eta)$ is a DRS for the regression of y on z because $y \perp\!\!\!\perp x \mid \eta^T x$ if and only if $y \perp\!\!\!\perp z \mid (A^{-1}\eta)^T z$. Next, suppose there is a DRS $\mathcal{S}(\alpha)$ for the regression of y on z with dim[$\mathcal{S}(\alpha)$] < dim[$\mathcal{S}(A^{-1}\eta)$]. Then $y \perp\!\!\!\perp A^T x \mid (A\alpha)^T x$ so that $\mathcal{S}(A\alpha)$ is a DRS for the regression of y on x. But this implies the contradiction dim[$\mathcal{S}(A\alpha)$] < dim[$\mathcal{S}(\eta)$] because rank($A\alpha$) = rank(α). Thus, $\mathcal{S}(A^{-1}\eta)$ is a minimum DRS for the regression of y on z. □

6.3. CENTRAL SUBSPACES

Minimum dimension-reduction subspaces (DRSs) are not generally unique. For example, let $p = 2$ and let $x = (x_1, x_2)^T$ be uniformly distributed on the unit circle, $\|x\| = 1$, and set $y \mid x = x_1^2 + \varepsilon$ where ε is an independent error. Because $x_1^2 + x_2^2 = 1$, either x_1 or x_2 provides full information about the regression of y on x. That is,

$$y \mid x = x_1^2 + \varepsilon = (1 - x_2^2) + \varepsilon.$$

And therefore $\mathcal{S}((1,0)^T)$ and $\mathcal{S}((0,1)^T)$ are both minimum DRSs.

The general goal here is to find a way of using DRSs to guide graphical analyses, particularly the search for sufficient plots of minimal dimension. It might be possible to develop graphical methods for estimation of minimum DRSs, but the fact that such subspaces are not generally unique may be annoying and could lead to elusive analyses. Avoiding such nonuniqueness by introducing more restrictive subspaces should facilitate progress. Throughout the remainder of this book we will rely on the idea of a central DRS (Cook 1994b, 1996b).

A subspace $\mathcal{S}$ is a *central dimension-reduction subspace* for the regression of y on x if $\mathcal{S}$ is a DRS and $\mathcal{S} \subset \mathcal{S}_{drs}$ for all DRSs $\mathcal{S}_{drs}$. Central DRSs will be denoted by $\mathcal{S}_{y|x}$, or by $\mathcal{S}_{y|x}(\eta)$ when a basis η is needed. A central DRS will often be referred to as simply a *central subspace*. This should not lead to confusion because the notion of centrality is relevant only within the context of dimension-reduction subspaces.

It may be clear from the definition that a central subspace exists if and only if the intersection $\cap \mathcal{S}_{drs}$ of all DRSs is itself a DRS, in which case $\mathcal{S}_{y|x} = \cap \mathcal{S}_{drs}$. The intersection $\cap \mathcal{S}_{drs}$ is always a subspace but it is not necessarily a DRS. In the previous example, $\mathcal{S}((1,0)^T)$ and $\mathcal{S}((0,1)^T)$ are both DRSs but their intersection is not a DRS. Although a central DRS might not exist, it is unique when it does exist. The idea of centrality is useful because it forces a degree of order on graphical problems that might otherwise be rather elusive. Recall that the notation $\mathcal{S}_{y|x}$ was used in Chapters 4 and 5, even though central DRSs were not yet defined and the existence/uniqueness issue was not addressed.

The following two propositions give properties of central subspaces.

Proposition 6.2. If $\mathcal{S}_{y|x}$ is the central dimension-reduction subspace for the regression of y on x then $\mathcal{S}_{y|x}$ is the unique minimum dimension-reduction subspace.

Justification. Because $\mathcal{S}_{y|x} = \cap \mathcal{S}_{drs}$, it is a minimum DRS. If $\mathcal{S}_1$ is a second minimum DRS then $\mathcal{S}_{y|x} \subset \mathcal{S}_1$. This implies $\mathcal{S}_1 = \mathcal{S}_{y|x}$ because $\dim(\mathcal{S}_{y|x}) = \dim(\mathcal{S}_1)$. Thus $\mathcal{S}_{y|x}$ is unique. $\qquad\square$

Proposition 6.3. Let $S_{y|x}(\eta)$ be the central dimension-reduction subspace for the regression of y on x, and let $S_{y|z}$ be the central subspace for the regression of y on $z = A^T x$ where A is a full rank, $p \times p$ matrix. Then $S_{y|z} = A^{-1} S_{y|x}$.

Justification. Let $S(\beta)$ be a DRS for the regression of y on $A^T x$. Then

$$y \perp\!\!\!\perp A^T x \mid (A\beta)^T x.$$

Because $S_{y|x}(\eta)$ is central, $S_{y|x}(\eta) \subset S(A\beta)$. This implies $S(A^{-1}\eta) \subset S(\beta)$. The conclusion now follows because $S(A^{-1}\eta)$ is a DRS for the regression of y on $A^T x$. □

Proposition 6.2 shows that a central subspace is a unique minimum DRS. On the other hand, unique minimum DRSs need not be central subspaces. For example, let $p = 3$ and let $x = (x_1, x_2, x_3)^T$ be uniformly distributed on the unit sphere, $\|x\| = 1$, and set $y \mid x = x_1^2 + \varepsilon$ where ε is an independent error. The subspace S_1 spanned by the single vector $(1, 0, 0)^T$ is the unique minimum DRS. The subspace S_2 spanned by the pair of vectors $(0, 1, 0)^T$ and $(0, 0, 1)^T$ is a DRS. And yet S_1 is not contained in S_2. In this example $\cap S_{drs}$ is equal to the origin and thus a central subspace does not exist. However, if we modify the regression model to be

$$y \mid x = x_1^2 + x_1 + \varepsilon$$

then $(1, 0, 0)^T$ spans the central subspace. This example illustrates that the existence of a central subspace can depend on the conditional distribution of $y \mid x$ and on the marginal distribution of x.

To explore the issues involved in forcing the existence of central subspaces, let $S_m(\alpha)$ be a minimum DRS and let $S_{drs}(\phi)$ be an arbitrary DRS. Then,

$$F_{y|x}(a) = F_{y|\alpha^T x, \phi^T x}(a) = F_{y|\alpha^T x}(a) = F_{y|\phi^T x}(a)$$

for all $a \in \mathbb{R}^1$. In the rest of this discussion, we suppress the arguments to CDFs understanding that results hold for all a. It follows that

$$F_{y|\phi^T x} = E_{\alpha^T x | \phi^T x}[F_{y|\alpha^T x, \phi^T x}]$$

$$= E_{\alpha^T x | \phi^T x}[F_{y|\alpha^T x}]$$

$$= F_{y|\alpha^T x}.$$

The key equality

$$E_{\alpha^T x | \phi^T x}[F_{y|\alpha^T x}] = F_{y|\alpha^T x} \tag{6.4}$$

says that $F_{y|\alpha^T x}$ must be constant with probability 1 relative to the conditional distribution of $\alpha^T x \mid \phi^T x$. In effect, the information that $\alpha^T x$ supplies to $F_{y|\alpha^T x}$ must also be available from $\phi^T x$. Equation (6.4) is clearly true if $S_m(\alpha)$ is central so that $S_m(\alpha) \subset S_{drs}(\phi)$, and this is what might normally be expected in practice. But (6.4) may hold in other ways as well. The presence of exact colinearities among the predictors may account for the relationship, for example.

Equation (6.4) may hold without imposing exact colinearities or centrality. To illustrate that possibility, suppose that $F_{y|x}$ is characterized by its regression function, $y \perp\!\!\!\perp x \mid E(y \mid x)$. In this case, DRSs are determined solely from the regression function (see Problem 6.2), and it can be shown that (6.4) is equivalent to

$$E_{\alpha_1^T x \mid (\phi_1^T x, \delta^T x)}[E(y \mid \alpha_1^T x, \delta^T x)] = E(y \mid \alpha_1^T x, \delta^T x) \tag{6.5}$$

where $S(\delta) = S_m(\alpha) \cap S_{drs}(\phi)$, $\alpha = (\alpha_1, \delta)$, $\phi = (\phi_1, \delta)$. The columns of the matrices ϕ_1 and α_1 extend the basis δ for the intersection to bases for the respective subspaces. Assume that α_1 is nonzero so that $S_m(\alpha)$ is not central. Because $S_m(\alpha)$ is minimal, ϕ_1 must be nonzero, and the columns of ϕ_1 and α_1 are linearly independent. For (6.5) to hold, $E(y \mid \alpha_1^T x, \delta^T x)$ must be constant with respect to the conditional distribution of $\alpha_1^T x \mid \phi^T x$. Informally, there must exist nontrivial functions g_α and g_ϕ so that $g_\alpha(\alpha_1^T x) = g_\phi(\phi_1^T x)$ and either can be used to convey the information essential to the regression function $E(y \mid x)$. The general intuition from this argument is that it could be possible to tie the regression function to the distribution of x and thereby force (6.5) without requiring centrality.

For example, consider three predictors $x = (x_1, x_2, x_3)^T$ with a joint density and the following two properties:

- The joint density of (x_1, x_2) is supported on $\mathbb{R}^2_- \cup \mathbb{R}^2_+$, where

$$\mathbb{R}^2_- = \{(a,b) \mid a < 0,\ b < 0\}$$

and

$$\mathbb{R}^2_+ = \{(a,b) \mid a > 0,\ b > 0\}$$

- $\Pr(x_j > 0) > 0$ and $\Pr(x_j < 0) > 0$, $j = 1, 2$.

Suppose that $y \perp\!\!\!\perp x \mid E(y \mid x)$, and that

$$E(y \mid x) = \beta_0 \frac{x_1}{|x_1|} + \beta_1 x_3. \tag{6.6}$$

Following the notation of (6.5), set $\alpha_1^T x = x_1$, $\phi_1^T x = x_2$ and $\delta^T x = x_3$. Because x_1 and x_2 have the same sign, $x_1/|x_1| = x_2/|x_2|$ and $S(\alpha_1, \delta)$ and $S(\phi_1, \delta)$ are both

DRSs. However, a central DRS does not exist because $\mathcal{S}(\alpha_1, \delta) \cap \mathcal{S}(\phi_1, \delta) = \mathcal{S}(\delta)$ is not a DRS for the regression.

On the other hand, defining $g_\alpha(x_1) = x_1/|x_1|$ and $g_\phi(x_2) = x_2/|x_2|$, we have

$$E(y \mid x) = g_\alpha(x_1)\beta_0 + \beta_1 x_3$$

$$= g_\phi(x_2)\beta_0 + \beta_1 x_3.$$

That is, either $g_\alpha(x_1)$ or $g_\phi(x_2)$ can be used to express the regression function. To express this in terms of (6.5), use the first of the above equalities to obtain

$$E(y \mid x_1, x_3) = E(y \mid x) = \beta_0 g_\alpha(x_1) + \beta_1 x_3.$$

Next, using the second equality,

$$E(y \mid x_1, x_3) = \beta_0 g_\phi(x_2) + \beta_1 x_3$$

and therefore

$$E_{x_1 \mid x_2, x_3}(E(y \mid x_1, x_3)) = E_{x_1 \mid x_2, x_3}(\beta_0 g_\phi(x_2) + \beta_1 x_3)$$

$$= \beta_0 g_\phi(x_2) + \beta_1 x_3$$

$$= E(y \mid x_1, x_3).$$

In the next section we consider two ways of guaranteeing the existence of a central subspace.

6.4. GUARANTEEING $\mathcal{S}_{y|x}$ BY CONSTRAINING...

6.4.1. ... the distribution of x

There are perhaps many ways of restricting the regression problem to guarantee the existence of a central subspace. The next proposition does this by requiring x to have a density with a convex support.

Proposition 6.4.   Let $\mathcal{S}(\alpha)$ and $\mathcal{S}(\phi)$ be DRSs for $y \mid x$. If x has a density $f(a) > 0$ for $a \in \Omega_x \subset \mathbb{R}^p$ and $f(a) = 0$ otherwise, and if Ω_x is a convex set, then $\mathcal{S}(\alpha) \cap \mathcal{S}(\phi)$ is a DRS.

Justification.   Let δ be a basis for $\mathcal{S}(\alpha) \cap \mathcal{S}(\phi)$ and extend δ to bases for $\mathcal{S}(\alpha)$ and $\mathcal{S}(\phi)$ so that $\alpha = (\alpha_1, \delta)$ and $\phi = (\phi_1, \delta)$. Here we assume that $\alpha_1 \neq 0$ and $\phi_1 \neq 0$. Otherwise the conclusion is trivially true. Then we need to show that $\mathcal{S}(\delta)$ is a DRS. Without loss of generality we can take Ω_x to be open in $\mathbb{R}^p$.

For notational convenience let

$$W = \begin{pmatrix} W_1 \\ W_2 \\ W_3 \end{pmatrix} = \begin{pmatrix} \alpha_1^T x \\ \phi_1^T x \\ \delta^T x \end{pmatrix}.$$

Conditioning values for W_j are denoted as w_j, $j = 1,2,3$, unless indicated otherwise. Since x has a density with convex open support and $(\alpha_1, \phi_1, \delta)$ is a full rank linear operator, W has a density with convex open support Ω_w. Similarly, the distribution of

$$(W_1, W_2) \mid (W_3 = w_3)$$

has a density with convex open support $\Omega_{12|3}(w_3)$ which depends on w_3.

Because $\mathcal{S}(\alpha)$ and $\mathcal{S}(\phi)$ are DRSs,

$$F_{y|W} = F_{y|W_1,W_3} = F_{y|W_2,W_3} \tag{6.7}$$

for all $w = (w_1, w_2, w_3)^T \in \Omega_w$, where the CDF arguments are again suppressed to simplify expressions. In order to show that the intersection $\mathcal{S}(\delta)$ is a DRS, we need to show that

$$F_{y|W} = F_{y|W_3} \qquad \text{for all} \quad (w_1, w_2, w_3)^T \in \Omega_w.$$

Since we can write $\Omega_w = \bigcup_{\Omega_3} \Omega_{12|3}(w_3)$, where Ω_3 is the support of W_3, this is equivalent to showing that, regardless of w_3,

$$F_{y|W} = F_{y|W_3} \qquad \text{for all} \quad (w_1, w_2)^T \in \Omega_{12|3}(w_3).$$

Fix any $w_3 \in \Omega_3$. Two points, $(w_1, w_2')^T$ and $(w_1', w_2)^T$, in $\Omega_{12|3}(w_3)$ are said to be *linked* if either $w_1 = w_1'$ or $w_2 = w_2'$.

As depicted in Figure 6.1, choose three points, $(w_1', w_2)^T$, $(w_1, w_2)^T$, and $(w_1, w_2')^T$ in $\Omega_{12|3}(w_3)$ and define

$$a(t) = \begin{pmatrix} tw_1 + (1-t)w_1' \\ w_2 \end{pmatrix}$$

and

$$b(s) = \begin{pmatrix} w_1 \\ sw_2 + (1-s)w_2' \end{pmatrix}.$$

Since $\Omega_{12|3}(w_3)$ is convex, $a(t), b(s) \in \Omega_{12|3}(w_3)$ for $0 \le t, s \le 1$. Thus, from (6.7),

$$F_{y|a(t),W_3} = F_{y|W_2,W_3}$$

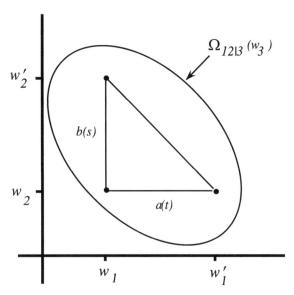

FIGURE 6.1 Illustration for Proposition 6.4.

and

$$F_{y|b(s),W_3} = F_{y|W_1,W_3}$$

where the subscript $a(t)$ in $F_{y|a(t),W_3}$ means $(W_1, W_2)^T = a(t)$. From (6.7),

$$F_{y|a(t),W_3} = F_{y|b(s),W_3}$$

for all $t, s \in [0, 1]$. In particular, evaluating at $t = s = 0$ gives

$$F_{y|W_1 = w_1', w_2, w_3} = F_{y|w_1, W_2 = w_2', w_3}. \tag{6.8}$$

Finally, because $\Omega_{12|3}(w_3)$ is convex and open, any two points in $\Omega_{12|3}(w_3)$ can be chained together by using a series of linked points, implying that (6.8) holds for two arbitrary points in $\Omega_{12|3}(w_3)$. Thus, $F_{y|W_1,W_2,W_3}$ is a constant function of $(w_1, w_2) \in \Omega_{12|3}(w_3)$ and the conclusion follows. $\qquad\square$

Proposition 6.4 implies that the central subspace exists in regressions where x has a density with a convex support. An adaptation of this result to regressions with discrete predictors is given in Problem 6.5. In the next section we insure the existence of $\mathcal{S}_{y|x}$ by requiring x to have a density and by constraining the distribution of $y \mid x$.

6.4.2. ... **the distribution of** $y \mid x$

The developments of this section are in the context of a *location regression*, a regression problem that can be characterized by its regression function:

$$y \perp\!\!\!\perp x \mid E(y \mid x). \tag{6.9}$$

Under this condition an investigation of DRSs can be confined to the regression function: If (6.9) holds then $y \perp\!\!\!\perp x \mid \alpha^T x$ if and only if $y \perp\!\!\!\perp x \mid E(y \mid \alpha^T x)$. Justification of this statement is set as Problem 6.2.

Location regressions include many generalized linear models and are considerably more general than the commonly used regression models with additive errors in which

$$y - E(y \mid x) \perp\!\!\!\perp x. \tag{6.10}$$

For example, the conditional variance Var$(y \mid x)$ or any higher-order moment of $F_{y|x}$ can depend on $E(y \mid x)$ under (6.9). We will refer to regressions satisfying (6.10) as *additive-location regressions*.

The next proposition (Cook 1996b) gives one way to guarantee a central subspace in a location regression.

Proposition 6.5. Let $S(\alpha)$ and $S(\phi)$ be dimension-reduction subspaces for a location regression, $y \perp\!\!\!\perp x \mid E(y \mid x)$. If x has a density $f > 0$ on $\Omega_x \subset \mathbb{R}^p$ and if $E(y \mid x)$ can be expressed as a convergent power series in the coordinates of $x = (x_k)$,

$$E(y \mid x) = \sum_{k_1,\ldots,k_p}^{\infty} a_{k_1,\ldots,k_p} x_1^{k_1} \cdots x_p^{k_p}$$

then $S(\alpha) \cap S(\phi)$ is a DRS.

Justification. Let α_1, ϕ_1 and δ be defined as in the justification of Proposition 6.4, and assume that $\alpha_1 \neq 0$ and $\phi_1 \neq 0$. Also, assume that $\alpha_1^T x$ and $\delta^T x$ are scalars. The extension to higher dimensions in these variables should be straightforward conceptually, albeit notationally awkward.

Because $E(y \mid x)$ can be expressed as a convergent power series in $x \in \Omega_x$ and because $E(y \mid x) = E(y \mid \alpha^T x)$, it follows that $E(y \mid \alpha^T x)$ can be expressed as a convergent power series in $\alpha_1^T x$ and $\delta^T x$,

$$E(y \mid \alpha^T x) = \sum_{k_1,k_2}^{\infty} b_{k_1,k_2} (\alpha_1^T x)^{k_1} (\delta^T x)^{k_2}.$$

Next, since $\mathcal{S}(\phi)$ is a DRS,

$$\mathrm{E}_{\alpha_1^T x | \phi^T x}[\mathrm{E}(y \mid \alpha^T x)] = \mathrm{E}(y \mid \alpha^T x)$$

so that $\mathrm{E}(y \mid \alpha^T x)$ is constant with probability 1 relative to the conditional distribution of $\alpha_1^T x \mid \phi^T x$. To emphasize that $\delta^T x$ is constant with respect to $\alpha_1^T x \mid \phi^T x$, define the coefficient function

$$c_{k_1}(\delta^T x) = \sum_{k_2}^{\infty} b_{k_1,k_2}(\delta^T x)^{k_2}.$$

Then

$$\mathrm{E}(y \mid \alpha^T x) = \sum_{k_1}^{\infty} c_{k_1}(\delta^T x)(\alpha_1^T x)^{k_1}. \tag{6.11}$$

Now for each value of $\phi^T x$, equation (6.11) is a convergent power series over $\{\alpha_1^T x \mid f(\alpha_1^T x \mid \phi^T x) > 0\}$ which must contain an open interval in $\mathbb{R}^1$. Because the convergent power series (6.11) is constant on an open interval we must have $c_{k_1} = 0$ for $k_1 > 0$. Thus,

$$\mathrm{E}(y \mid \alpha^T x) = c_0 = \sum_{k_2}^{\infty} b_{0,k_2}(\delta^T x)^{k_2}$$

and $\mathcal{S}(\delta)$ is a DRS. $\qquad\square$

Proposition 6.5 can be extended to *location-scale regressions* in which both conditional means and conditional variances are required,

$$y \perp\!\!\!\perp x \mid [\mathrm{E}(y \mid x), \mathrm{Var}(y \mid x)].$$

Additionally, central subspaces can be guaranteed in location regressions when x is discrete by restricting the regression function to be a finite degree polynomial, the maximum allowable degree depending on the distribution of x in a rather complicated way.

For further discussion on the existence of central subspaces, including results that are more general than those presented here, see Carroll and Li (1995), and Chiaromonte and Cook (1997).

6.5. IMPORTANCE OF CENTRAL SUBSPACES

The assumption of a central subspace should be reasonable in most regression problems. Accordingly, in the rest of this book *we will assume that central subspaces exist for all regression problems unless explicitly stated otherwise.* One implication of this assumption is that minimum DRSs and central subspaces are identical constructions, as stated in Proposition 6.2. Either description may be used, depending on the desired emphasis.

Central subspaces $S_{y|x}$ are really superparameters intended to serve essentially the same role in graphical analyses as the parameters do in parametric analyses. Existing graphical methods and potentially new methods can be assessed on their ability to provide useful information on $S_{y|x}$. If $S_{y|x}$ was known, further analysis could be restricted to the minimal sufficient summary plot of y over $S_{y|x}$. The hope is that the structural dimension $\dim[S_{y|x}] = d$ is small, say 1 or 2, because this would result in a relatively uncomplicated plot and a relatively simple analysis. Even with $d = 1$ the structure implied by (6.2) is quite general. All of the 1D model forms discussed in Section 4.3 for two predictors have 1D structure when considered in terms of p predictors. Similarly, the general model forms of Section 4.4 have 2D structure when there are $p > 2$ predictors. The Hougen–Watson reaction rate model (Bates and Watts 1988, p. 272)

$$y \mid x = \frac{\theta_1 \theta_3 (x_2 - x_3/1.632)}{1 + \theta_2 x_1 + \theta_3 x_2 + \theta_4 x_3} + \varepsilon \tag{6.12}$$

has $p = 3$ predictors and $d \leq 2$, with $S_{y|x}$ contained in

$$S((0, 1, -1/1.632)^T, (\theta_2, \theta_3, \theta_4)^T).$$

According to Proposition 6.4, we can guarantee that $S_{y|x}$ exists if Ω_x is convex. If the central subspace $S_{y|x}(\eta)$ exists, statements of the form

$$y \perp\!\!\!\perp x \mid \eta^T x \tag{6.13}$$

can be thought of as *dimension-reduction models*. The model designation is fair because $S_{y|x}$ is well-defined and unique. Dimension-reduction models are nearly parametric models, but they lack the necessary "link" functions for a full parametric specification. The parameter of a dimension-reduction model is the subspace $S_{y|x}$, as mentioned previously. It is important to recognize that there are no restrictive assumptions involved in (6.13) provided d is not constrained. A dimension-reduction model provides a descriptive framework that uses the central subspace as the focal point. It is then straightforward to impose additional conditions that lead to any parametric or nonparametric model. In this sense, the central subspace provides a conceptual and practically useful umbrella for many types of regression problems.

Some authors, including Carroll and Li (1995), Duan and Li (1991), and Li (1991, 1992), represented dimension-reduction in a different way:

$$y = g(h_1^T x, \ldots, h_K^T x, \varepsilon) \tag{6.14}$$

where g is an unknown function and the error $\varepsilon \perp\!\!\!\perp x$. The vectors h_j, $j = 1,\ldots,K$, and any linear combinations of them are called *effective dimension-reduction directions*, and the subspace spanned by them is the *effective dimension-reduction subspace*. These ideas are usually employed without further definition and without addressing existence or uniqueness issues, and consequently they can be elusive. As defined in this chapter, the distinction between an *arbitrary* DRS and the *central* DRS is a key to progress in later chapters.

Assuming that a central subspace exists, the dimension-reduction models (6.13) and (6.14) are technically equivalent, and we can connect them by requiring that $(h_1,\ldots,h_K)$ be a basis for $S_{y|x}$. However, when thinking in terms of (6.14), the need to conceive an independent error ε can be an obstacle. An independent error may be quite natural in some problems, particularly when (6.14) is invertible in ε so that $\varepsilon = g^{-1}(h_1^T x,\ldots,h_K^T x,y)$. The general approach to residuals by Cox and Snell (1968) is based on such invertible representations. In other problems, understanding could be impeded by requiring an independent error. For example, suppose that y is binary, taking values 0 and 1 with probability that depends on $\eta^T x$ with $\eta = (h_1,\ldots,h_K)$. This additional knowledge causes no complications in the interpretation of (6.13), while the interpretation of (6.14) becomes elusive because it is not possible to construct an independent error based on just y and $\eta^T x$ (Cox and Snell 1968). There are technical ways to get around the difficulty, but they do not seem to facilitate understanding or applicability generally.

Although (6.13) does have considerable descriptive power and is relatively easy to interpret, there are limitations nevertheless. Perhaps the main drawback is that this dimension-reduction model restricts parsimonious characterizations of $y \mid x$ to linear manifolds. As previously mentioned in Section 4.6.7, even simple nonlinear manifolds may take all of $\mathbb{R}^p$ to be characterized in terms of linear manifolds. For example, the only way to describe the model $y \perp\!\!\!\perp x \mid \|x\|$ in the context of (6.13) is with $\eta = I_p$.

We conclude in the next section with a first application showing how central subspaces might be used to guide plot construction.

6.6. h-LEVEL RESPONSE PLOTS

When the response is many-valued, the graphics of Chapter 4 allow the response and at most two predictors to be viewed simultaneously in a single plot. However, following the discussion of Section 5.4, the response and three predictors can be viewed in a single plot when the response is binary. It may be useful occasionally to consider converting a many-valued response y into a binary response $\tilde{y}$ and thereby gain the ability to view more variables in a single plot.

Specifically, let $\tilde{y}$ be the binary response formed by partitioning the range of y into two fixed, nonoverlapping slices $J_1 = (-\infty, a)$ and $J_2 = [a, \infty)$. Values of *a* selected so that $\Pr(y \in J_1) \approx 0.5$ should be reasonable, although this may not always be so. Also, for definiteness, let $\tilde{y} = s$ when $y \in J_s$, $s = 1, 2$. Considering the regression of $\tilde{y}$ on x allows for the possibility of viewing the response and three predictors in a single plot. But the regression of $\tilde{y}$ on x should also provide useful information about the regression of y on x.

Because $\tilde{y}$ is a function of y, it follows from Proposition 4.5 that $\tilde{y} \perp\!\!\!\perp x \mid \eta^T x$, where η is a basis for $S_{y|x}$. Thus, the central subspace from the regression of $\tilde{y}$ on x provides information about the central subspace from the regression of y on x:

$$S_{\tilde{y}|x} \subset S_{y|x}. \tag{6.15}$$

There remains the possibility that $S_{\tilde{y}|x}$ is a *proper* subset of $S_{y|x}$, implying that it may be possible to lose information on $S_{y|x}$ when replacing y with $\tilde{y}$. For example, suppose that $y \mid x$ is normal with constant mean 0 and variance function $\mathrm{Var}(y \mid x) = \mathrm{Var}(y \mid \eta^T x)$. With $J_1 = (-\infty, 0)$,

$$\Pr(y \in J_1 \mid x) = \Pr(y \in J_1) = 1/2$$

for all values of x. Thus, $S_{\tilde{y}|x} = S(\text{origin})$, and the regression of $\tilde{y}$ on x contains no helpful information about $S_{y|x}$. The possibility of losing information when converting to a binary response might not be worrisome for location regressions (6.9), although some loss of power should be expected. Using more than two levels when constructing $\tilde{y}$ could help avoid losing information and power.

Partition the range of y into h fixed, nonoverlapping slices J_s, $s = 1, \ldots, h$ and let $\tilde{y} = s$ when $y \in J_s$, $s = 1, \ldots, h$. An *h-level response plot* of $\tilde{y}$ versus three predictors is constructed by adapting the ideas in Chapter 5 for binary (two-level) response plots: Assign the predictors to the axes of a 3D plot and then mark the points with h distinct colors or symbols according to the values of $\tilde{y}$. Characterizing plots with more than three or four colors or symbols can be a bit tedious.

An *h*-level response plot can be interpreted by using adaptations of the ideas presented in Chapter 5. For example, Figure 6.2 contains a 2D view of a three-level response plot for the mussels data introduced in Section 2.2; the slice in this figure will be discussed shortly. The response for this example is mussel mass M, and the predictors are shell length, width, and height, $x = (L, W, H)^T$. The view in Figure 6.2 was constructed by first assigning orthogonalized predictors to the axes of a 3D plot. Next, the range of M was partitioned into three slices with approximately the same number of observations, $J_1 = [0, 8.5)$, $J_2 = [8.5, 19.5)$, and $J_3 = (19.5, \infty)$. The corresponding points in the 3D plot were then marked with an open circle, a filled circle, and an ex. Finally, the 3D plot was rotated so that the horizontal screen axis corresponds to the fitted values $\hat{M} = b_0 + b^T x$ from the OLS linear regression of M on x. If the OLS fit provides a good summary, then the data should appear as if $M \perp\!\!\!\perp x \mid b^T x$.

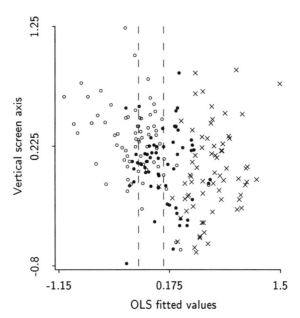

FIGURE 6.2 Two-dimensional view of a three-level response plot for the Mussels data. The horizontal axis is proportional to the fitted values from the OLS regression of M on (L,H,W).

Following the discussion of Chapter 5, imagine looking at the view in Figure 6.2 as a small rectangular window

$$\Delta = \{x \mid \ell_j \leq x_j \leq u_j, \ j = 1,2,3\}$$

is moved around the 3D plot and, at each window location, counting the number of symbols of each type. Under the condition $M \perp\!\!\!\perp x \mid b^T x$, the behavior of the ratio of the counts is indicated by the relationship

$$\frac{\Pr(x \in \Delta, \ \tilde{y} = j)}{\Pr(x \in \Delta, \ \tilde{y} = k)} = \frac{\Pr(\tilde{y} = j \mid b^T x \in \Delta_b)}{\Pr(\tilde{y} = k \mid b^T x \in \Delta_b)} \tag{6.16}$$

where $\Delta_b = \{b^T x \mid x \in \Delta\}$. This equation indicates that if Δ is moved so that $b^T x$ is effectively constant, and $M \perp\!\!\!\perp x \mid b^T x$ holds to a good approximation, then the relative symbol density of each pair of symbols should be constant up to the Bernoulli variation of the estimates of the probabilities in (6.16). In particular, the relative density of each pair of symbols should be constant throughout slices parallel to the vertical axis of the plot in Figure 6.2. In the slice shown in this figure, the density of filled circles is noticeably higher at the bottom of the slice than at the top; hence, we have visual evidence that the OLS fit misses part of the regression structure. This can mean that the structural dimension is larger than 1. It could also happen that the structural dimension equals 1, but that $S(b)$ does not provide a good approximation of the central subspace.

In the next chapter we begin considering graphical ideas for gaining information on $S_{y|x}$ by using information from various marginal plots.

PROBLEMS

6.1. Assume that the central subspace $S_{y|x}$ exists for the regression of y on x, and let $t(y)$ be a strictly monotonic transformation of y. Argue that $S_{t(y)|x} = S_{y|x}$.

6.2. Assuming a location regression as defined in (6.9), show that $y \perp\!\!\!\perp x \mid \alpha^T x$ if and only if $y \perp\!\!\!\perp x \mid E(y \mid \alpha^T x)$.

6.3. Consider a regression problem where the predictor vector x has a density that is positive everywhere in $\mathbb{R}^p$. Does a central subspace exist?

6.4. The example involving (6.6) in Section 6.3 shows that a central subspace need not exist, even in location regressions. Show how that example fails to fit the conditions of Propositions 6.4 and 6.5.

6.5. Suppose that $y \mid x$ admits two dimension-reduction subspaces S_1 and S_2 such that $S_1 \cap S_2 = S(0)$. Equivalently, in the notation of Proposition 6.4, assume that $\delta = 0$. According to the proposition, if x has a density with convex support, then $S(0)$ is a DRS; that is, $y \perp\!\!\!\perp x$.

With $\delta = 0$, the equalities in (6.7) can be expressed equivalently as the pair of independence statements,

$$y \perp\!\!\!\perp w_1 \mid w_2 \qquad \text{and} \qquad y \perp\!\!\!\perp w_2 \mid w_1. \tag{6.17}$$

The problem of forcing the existence of the central subspace is fundamentally the same as the problem of finding conditions that force these statements to imply that $y \perp\!\!\!\perp (w_1, w_2)$. Similar problems have been discussed in various probabilistic and statistical settings by Dawid (1979a, 1979b, 1984), Basu and Pereria (1983), Koehn and Thomas (1975), and Cox and Wermuth (1996, p. 37). Of course, by the Conditional Independence Proposition 4.6, $y \perp\!\!\!\perp (w_1, w_2)$ always implies (6.17).

Assume now that $(y, w_1, w_2)^T$ is a 3×1 discrete random vector supported on a finite set of points. Let Ω_w be the marginal support for the joint distribution of (w_1, w_2). The support Ω_w is said to be *connected* if any two points in Ω_w can be chained together by a series of linked points, where the definition of linked points is as given in the justification of Proposition 6.4. For example, a support with 11 points is shown in Fig-

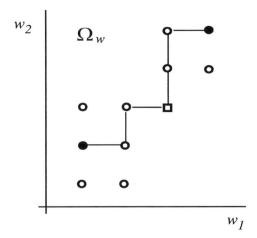

FIGURE 6.3 Graphical illustration of a connected support for a discrete random vector (w_1, w_2).

ure 6.3. The two filled points have been chained together by joining a series of linked points with lines. All pairs of points in the support can be chained together similarly and thus the support is connected. However, if the point represented by the square was removed, the support would no longer be connected.

The idea of a connected support is quite similar to the standard idea of a connected experimental design. Thinking of w_1 as defining blocks and w_2 as defining treatments, Figure 6.3 may be viewed as arising from an incomplete block design with 5 treatments and 5 blocks. Because the design is connected, all treatment effects are estimable.

6.5.1. Let $(y, w_1, w_2)^T$ be a 3×1 discrete random vector. Show that, if the marginal support Ω_w of (w_1, w_2) is connected, then (6.17) implies $y \perp\!\!\!\perp (w_1, w_2)$.

6.5.2. Still letting $(y, w_1, w_2)^T$ be a 3×1 discrete random vector, give an example showing that (6.17) does not necessarily imply $y \perp\!\!\!\perp (w_1, w_2)$ when Ω_w is not connected.

6.6. Consider a regression problem in which it is assumed that

$$y - \mathrm{E}(y \mid x) \perp\!\!\!\perp x. \tag{6.18}$$

An essential task in this setting is to find an adequate estimate of the regression function. This is often addressed by specifying a functional form for $\mathrm{E}(y \mid x)$, say $m(x \mid \theta)$, that is known up to a finite dimensional parameter vector θ. For example, m might be set to the standard linear regression function $m(x \mid \theta) = \theta_0 + \theta_1^T x$. Following estimation of θ with an estimate $\hat{\theta}$ that is consistent under m, diagnostic checks are desirable to see if there is information in the data that contradicts the fitted model $m(x \mid \hat{\theta})$.

Diagnostic plots of the sample residuals

$$\hat{r} = y - m(x \mid \hat{\theta}) \tag{6.19}$$

versus linear combinations of the predictors $a^T x$ are often useful for model checking. If the model is adequate then we would expect all plots of this type to behave as if $\hat{r} \perp\!\!\!\perp a^T x$ so that the regression of $\hat{r}$ on x has 0D structure. Any evidence to the contrary may cast doubt on the model for the regression function or on the original assumption (6.18). Finally, the ideas of Section 6.6 can be applied with the residual in place of the response to yield an *h-level residual plot* for diagnostic checking.

The substitution of estimates $\hat{\theta}$ for the unknown parameters θ may cause complications for diagnostic plots in small samples by inducing dependence between various quantities that are independent in the population. Neglect such issues in this problem. See Cook and Weisberg (1982), Cox and Snell (1968), and Davison and Snell (1991) for further discussion.

6.6.1. Construct a three-level residual plot following the discussion of the mussel data in Section 6.6. Summarize your conclusions, using the residual plot for support.

6.6.2. Construct a three-level residual plot as in Problem 6.6.1, but this time take the residuals from the OLS fit of the full second-order response surface model in L, H, and W. Summarize your conclusions on the fit of the model.

CHAPTER 7

Graphical Regression

In Chapter 6 we established the central subspace (central DRS) $S_{y|x}$ as the fundamental object for inference. If the number p of predictors is small then we can use the visualization methods of Chapters 4 and 5 to infer directly about $S_{y|x}$. Otherwise we confront the problem of using a 2D or 3D plot to infer indirectly about $S_{y|x}$. The issue then is how to use low-dimensional plots of the form $\{y, A^T x\}$ to gain information about $S_{y|x}$, where A is a selected $p \times q$, $q \leq p$, full rank matrix. Plots of the response versus selected predictors are used in the next section to introduce ideas and highlight basic issues.

7.1. INTRODUCTION TO GRAPHICAL REGRESSION

Partition $x^T = (x_1^T, x_2^T)$ where x_k is $p_k \times 1$, and $p_1 + p_2 = p$. Let $S_{y|x_k}$ denote the central subspace for the regression of y on x_k, $k = 1, 2$. Are there conditions under which $S_{y|x_k}$ is usefully related to $S_{y|x}(\eta)$, the central subspace for the regression of y on x with basis η? If so, then we may be able to use the marginal response plot $\{y, x_k\}$ as a stepping stone to inference about $S_{y|x}$. Otherwise, the marginal response plot may be of little use in this regard. To rephrase the issue, partition

$$\eta = \begin{pmatrix} \eta_1 \\ \eta_2 \end{pmatrix} \tag{7.1}$$

where the row dimension of η_1 corresponds to the dimension of x_1. Then

$$F_{y|x_1} = E_{2|1}[F_{y|\eta^T x}]$$

$$= E_{2|1}[F_{y|(\eta_1^T x_1 + \eta_2^T x_2)}] \tag{7.2}$$

where the expectation $E_{2|1}$ is with respect to the conditional distribution of $x_2 \mid x_1$. In view of (7.2), any relationship between $S_{y|x_1}$ and $S_{y|x}$ depends on the relationship between $S_{y|x_1}$ and $S(\eta_1)$. The connection between $S_{y|x_1}$ and $S(\eta_1)$, however, depends on the conditional distribution of $x_2 \mid x_1$, and useful relationships seem possible only with restrictions.

120

In full generality,

$$\mathcal{S}_{y|x} \subset \mathcal{S}(\eta_1) \oplus \mathcal{S}(\eta_2) \subset \mathbb{R}^p \qquad (7.3)$$

where $\oplus$ indicates the direct sum. A graphical analysis could be facilitated if we can find conditions which force $\mathcal{S}_{y|x_k} = \mathcal{S}(\eta_k)$ for both $k = 1, 2$, or at least for one value of k. If $\mathcal{S}_{y|x_k} = \mathcal{S}(\eta_k)$ then the marginal regression of y on x_k provides information on $\mathcal{S}_{y|x}$ via its central subspace $\mathcal{S}_{y|x_k}$. When p_k is small such information can be obtained in practice by investigating the marginal response plot $\{y, x_k\}$ with the methods described in previous chapters. Conceptually, here is how an analysis based on these ideas might be structured.

Prior to any analysis, the only known DRS for $y \mid x$ is the whole space, $\mathcal{S}_{y|x} \subset \mathbb{R}^p$. If $\mathcal{S}_{y|x_1} = \mathcal{S}(\eta_1)$, the first term of the direct sum (7.3) can be determined by investigating the marginal plot $\{y, x_1\}$. Further regression analysis can then be restricted to

$$\mathcal{S}_{y|x_1} \oplus \mathbb{R}^{p_2} \supset \mathcal{S}_{y|x}.$$

In particular, if $\dim[\mathcal{S}_{y|x_1} \oplus \mathbb{R}^{p_2}] < p$ then the dimensionality is reduced and the original predictors can be replaced with $(\gamma_1^T x_1, x_2)$, where γ_1 is a basis for $\mathcal{S}_{y|x_1}$. In the extreme, if $y \perp\!\!\!\perp x_1$ then $\gamma_1 = 0$ and x_1 can be deleted from the analysis.

Next, if $\mathcal{S}_{y|x_2} = \mathcal{S}(\eta_2)$, the second term of the direct sum (7.3) can be determined by investigating $\{y, x_2\}$ and the analysis can be further restricted to $\mathcal{S}_{y|x_1} \oplus \mathcal{S}_{y|x_2}$ which is another, potentially smaller, DRS. If

$$\dim[\mathcal{S}_{y|x_1} \oplus \mathcal{S}_{y|x_2}] < \dim[\mathcal{S}_{y|x_1} \oplus \mathbb{R}^{p_2}]$$

then the dimensionality is further reduced and the predictors can be replaced with $(\gamma_1^T x_1, \gamma_2^T x_2)$, where γ_2 is a basis for $\mathcal{S}_{y|x_2}$.

The plot $\{y, (\gamma_1^T x_1, \gamma_2^T x_2)\}$ constitutes a sufficient summary for the regression of y on x, but it is not necessarily minimal. However, if the dimension of the reduced predictor $(\gamma_1^T x_1, \gamma_2^T x_2)$ is small, then we can investigate this sufficient plot directly in an effort to reduce the dimensionality even more. Of course, we would need to use estimates of $\mathcal{S}_{y|x_1}$ and $\mathcal{S}_{y|x_2}$ in practice.

Subspaces of the form $\mathcal{S}(\eta_1)$ and $\mathcal{S}(\eta_2)$ constructed from a partition (7.1) of a basis η for $\mathcal{S}_{y|x}$ will be called *coordinate subspaces* of $\mathcal{S}_{y|x}$. By investigating coordinate subspaces we will, under certain conditions, be able to piece together an upper bound for $\mathcal{S}_{y|x}$, and hence we will be able to restrict attention to a reduced sufficient plot, as sketched above. We refer to this type of analysis as *graphical regression*.

The columns of the partition η_1 span the coordinate subspace $\mathcal{S}(\eta_1)$ by definition, but they do not necessarily form a basis for $\mathcal{S}(\eta_1)$ because there may be linear dependencies. For example, suppose there are three predictors and

$$\eta = \begin{pmatrix} 1 & 1 \\ 1 & 1 \\ 0 & 1 \end{pmatrix}.$$

The partition consisting of the first two rows of η is not of full rank.

We conclude this section with a simple illustration of basic graphical regression ideas to establish connections with Chapter 4 and to provide a little methodological context for the rest of the chapter.

One hundred observations on four predictors $x = (w_1, w_2, w_3, w_4)^T$ were generated as independent standard normal random variables. The corresponding responses were then generated as

$$y \mid x = 3(w_1 + w_2 + w_3 - w_4) + \varepsilon$$

where ε is an independent standard normal random variable. This represents an additive-location regression, $y - E(y \mid x) \perp\!\!\!\perp x$, with

$$E(y \mid x) = 3(w_1 + w_2 + w_3 - w_4).$$

Thus, the central subspace $\mathcal{S}_{y|x}$ exists and is spanned by the vector $\eta = (1, 1, 1, -1)^T$.

We first consider the 3D plot $\{y, (w_1, w_2)\}$. In our general notation, $x_1 = (w_1, w_2)^T$ and the corresponding coordinate subspace is $\mathcal{S}(\eta_1)$ with $\eta_1 = (1, 1)^T$. Because $y \mid (w_1, w_2)$ is normally distributed with

$$E(y \mid (w_1, w_2)) = 3(w_1 + w_2)$$

and constant variance, $\mathcal{S}_{y|x_1} = \mathcal{S}((1, 1)^T)$. Thus $\mathcal{S}_{y|x_1} = \mathcal{S}(\eta_1)$ and we can estimate $\mathcal{S}(\eta_1)$ visually by inspecting the 3D plot $\{y, (w_1, w_2)\}$.

Using uncorrelated 2D views as described in Chapter 4, the structural dimension of the 3D plot $\{y, (w_1, w_2)\}$ was inferred to be 1. Figure 7.1a gives the corresponding visually determined 2D summary view. Figure 7.1b gives a contrasting view. The linear combination of the predictors on the horizontal screen axis of Figure 7.1a is

$$w_{12} = 0.32w_1 + 0.25w_2.$$

Thus, $\mathcal{S}_{y|x_1}$ is estimated to be $\mathcal{S}((0.32, 0.25)^T)$. Although there is considerable variation in the plot, Figure 7.1a is quite close to the sufficient plot $\{y, w_1 + w_2\}$ and the estimate $\mathcal{S}((0.32, 0.25)^T)$ is quite close to the corresponding coordinate subspace $\mathcal{S}((1, 1)^T)$, the angle between them being about seven degrees. We now have a new regression problem with three predictors w_{12}, w_3 and w_4.

We next repeat the procedure for the 3D plot $\{y, (w_3, w_4)\}$. Now in the general notation $x_2 = (w_3, w_4)^T$ and the corresponding coordinate subspace is

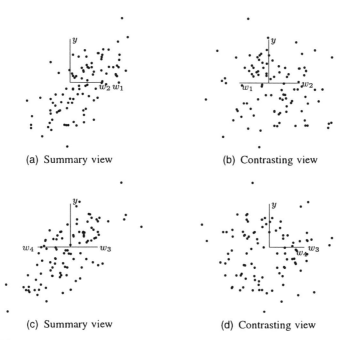

(a) Summary view   (b) Contrasting view

(c) Summary view   (d) Contrasting view

FIGURE 7.1 Four plots illustrating steps in graphical regression. The plots in (a) and (c) are visually estimated summary views; the other views are shown for contrast.

$S(\eta_2)$ with $\eta_2 = (1, -1)^T$. Figure 7.1c gives the summary view of the 3D plot $\{y, (w_3, w_4)\}$ which was also inferred to have 1D structure. The linear combination of the predictors on the horizontal axis of this figure is

$$w_{34} = 0.28w_3 - 0.34w_4.$$

The estimate of $S_{y|x_2}$ is $S((0.28, -0.34)^T)$, which is close to the corresponding coordinate subspace $S((1, -1)^T)$, the angle between the subspaces being about five degrees. The reduced regression problem now has two predictors, w_{12} and w_{34}.

Figure 7.2 gives two views of the 3D plot $\{y, (w_{12}, w_{34})\}$. Again, the first view is a visually estimated summary view and the second is provided for contrast. The dispersion about the regression function in Figure 7.2a is noticeably smaller than that around either summary view in Figure 7.1, reflecting the general fact that, on the average, the variation increases as the dimension of the projection decreases (See Problem 7.1). The linear combination of the predictors for Figure 7.2a is $w_{1234} = 0.78w_{12} + 0.74w_{34}$. Substituting for w_{12} and w_{34} from the previous analysis gives

$$w_{1234} = 0.25w_1 + 0.2w_2 + 0.21w_3 - 0.25w_4.$$

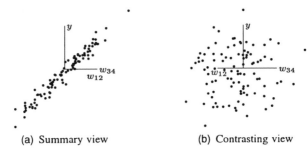

(a) Summary view   (b) Contrasting view

FIGURE 7.2 Two views of the 3D plot $\{y,(w_{12},w_{34})\}$. (a) The visually estimated summary view. (b) A contrasting view.

The angle between the vector $(0.25, 0.2, 0.21, -0.25)^T$ and the basis $\eta = (1, 1, 1, -1)^T$ for $\mathcal{S}_{y|x}$ is about six degrees, so the graphical analysis has come close to recovering the central subspace. As might be expected, the summary view in Figure 7.2a is nearly visually indistinguishable from the minimal sufficient summary plot $\{y, E(y \mid x)\}$.

This example is relatively straightforward since the predictors are independent and normally distributed, and the regression structure is not complicated. In other analyses, understanding the essential structure of a regression problem by using marginal response plots can be more difficult. In the rest of this chapter we investigate ways to apply the basic ideas illustrated in this example to more complicated regression problems.

7.2. CAPTURING $\mathcal{S}_{y|x_1}$

Again, let η be a basis for $\mathcal{S}_{y|x}$ and partition $x^T = (x_1^T, x_2^T)$ and $\eta^T = (\eta_1^T, \eta_2^T)$. In this section we describe a super-space for $\mathcal{S}_{y|x_1}$, capturing aspects of the regression problem that may influence the characterization of the marginal response plot $\{y, x_1\}$.

Let the columns of χ_1 be a basis for $\mathcal{S}_{\eta_2^T x_2|x_1}$, the central subspace for the regression of $\eta_2^T x_2$ on x_1. Similarly, let χ_2 be a basis for $\mathcal{S}_{\eta_1^T x_1|x_2}$.

Proposition 7.1.

$$\mathcal{S}_{y|x_1} \subset \mathcal{S}(\eta_1, \chi_1).$$

Justification.   The following equivalences stem from Proposition 4.6:

(a_1) $x_1 \perp\!\!\!\perp \eta_2^T x_2 \mid (\chi_1^T x_1, \eta_1^T x_1, y)$ (a_2) $x_1 \perp\!\!\!\perp y \mid (\chi_1^T x_1, \eta_1^T x_1)$

(b_1) $x_1 \perp\!\!\!\perp y \mid (\chi_1^T x_1, \eta_1^T x_1, \eta_2^T x_2)$ (b_2) $x_1 \perp\!\!\!\perp \eta_2^T x_2 \mid (\chi_1^T x_1, \eta_1^T x_1)$

(c) $x_1 \perp\!\!\!\perp (y, \eta_2^T x_2) \mid (\chi_1^T x_1, \eta_1^T x_1).$

Recall that conditions (a_1, a_2), (b_1, b_2), and (c) are equivalent. Conditions (b_1) and (b_2) follow immediately from Proposition 4.5 and the definitions of η and χ_1. This implies condition (a_2), which in turn implies the desired result because $\mathcal{S}_{y|x_1}$ is a central subspace.   □

One important implication of Proposition 7.1 is that characterizations of marginal response plots can be influenced by the distribution of the predictors, as well as by the corresponding coordinate subspace. Isolating a coordinate subspace by using marginal response plots can be difficult because dependence among the predictors can force patterns in some of the plots that detract from the main message in the data. This is essentially what happened in the examples of Section 2.5: To produce the misleading characterizations of the marginal plots it was necessary to use dependent predictors.

7.2.1. Example: Linear regression

Suppose that there are $p = 3$ predictors $x = (w_1, w_2, w_3)^T$ and that

$$y \mid x = \beta_0 + \beta_1 w_1 + \beta_2 w_2 + \beta_3 w_3 + \sigma\varepsilon$$

where ε is a standard normal random variable and $\varepsilon \perp\!\!\!\perp x$. Then $d = 1$ and $\mathcal{S}_{y|x}$ is spanned by

$$\eta = \begin{pmatrix} \beta_1 \\ \beta_2 \\ \beta_3 \end{pmatrix}.$$

Next, to correspond with the general notation, set $x_1 = (w_1, w_2)^T$ and $x_2 = w_3$ so that

$$\eta_1 = \begin{pmatrix} \beta_1 \\ \beta_2 \end{pmatrix} \quad \text{and} \quad \eta_2 = \beta_3.$$

We would like the 2D plot $\{y, \eta_1^T x_1\}$ to be a sufficient summary of the 3D plot $\{y, x_1\}$, that is, to have $\mathcal{S}_{y|x_1} \subset \mathcal{S}(\eta_1)$. According to Proposition 7.1, however, the relationship between the two subspaces depends on the regression of $\eta_2^T x_2 = \beta_3 x_2$ on x_1.

Suppose further that $x_2 \mid x_1$ is normally distributed with mean

$$E(x_2 \mid x_1) = \alpha_1 w_1 + \alpha_2 w_2 + \alpha_3^2 (w_1 - w_2)^2$$

and constant variance. Then $\mathcal{S}_{\eta_2^T x_2 | x_1}$ is spanned by the columns of

$$\chi_1 = \beta_3 \begin{pmatrix} \alpha_1 & \alpha_3 \\ \alpha_2 & -\alpha_3 \end{pmatrix}$$

and, from Proposition 7.1, $S_{y|x_1}$ is contained in the subspace spanned by

$$(\eta_1, \chi_1) = \begin{pmatrix} \beta_1 & \beta_3\alpha_1 & \beta_3\alpha_3 \\ \beta_2 & \beta_3\alpha_2 & -\beta_3\alpha_3 \end{pmatrix}$$

If all $\beta_3\alpha_k = 0$ then $\{y, \eta_1^T x_1\}$ is indeed a sufficient summary of the marginal response plot $\{y, x_1\}$. Otherwise, the structural dimension of $\{y, x_1\}$ will generally be 2.

The main point of this example can also be seen from the regression function,

$$E(y \mid x_1) = \beta_0 + (\beta_1 + \beta_3\alpha_1)w_1 + (\beta_2 + \beta_3\alpha_2)w_2 + \beta_3\alpha_3^2(w_1 - w_2)^2$$

which generally corresponds to a 2D regression problem. In particular, $S_{y|x_1}$ is spanned by the columns of

$$\begin{pmatrix} \beta_1 + \beta_3\alpha_1 & \beta_3\alpha_3 \\ \beta_2 + \beta_3\alpha_2 & -\beta_3\alpha_3 \end{pmatrix}$$

This example illustrates the possibility that $S_{y|x_1}$ may not contain much direct information about $S(\eta_1)$. The example in the next section gives a graphical illustration of this idea.

7.2.2. Example: $S_{y|x_1} = S(\eta_1)$, but $S_{y|x_2} \neq S(\eta_2)$

Consider a regression problem with two predictors $x = (x_1, x_2)^T$ and response

$$y \mid x = x_1 + .1\varepsilon$$

where ε and x_1 are independent standard normal random variables. Thus, $\eta = (\eta_1, \eta_2)^T = (1, 0)^T$. The second predictor, which is not included *per se* in the generation of y, is defined as $x_2 = |x_1| + z$ where z is also a standard normal random variable that is independent of (x_1, ε).

The relationship between the predictors in this example is nonlinear, which has the effect of making most marginal plots $\{y, b^T x\}$ show something interesting. Figure 7.3 shows a scatterplot matrix of the response and the predictors for 100 observations generated according to this model. The marginal response plot $\{y, x_1\}$ exhibits strong linearity as might be expected from the nature of the model, while heteroscedasticity is present in the plot $\{y, x_2\}$. From the structure of the model, y and x_2 are uncorrelated, but they are nevertheless dependent.

Because $\eta_2 = 0$, $\chi_1 = 0$. Thus, from Proposition 7.1, $S_{y|x_1} \subset S(\eta_1)$. In other words, the frame $\{y, x_1\}$ gives the correct impression that x_1 is required in the full regression; that is, that $S(\eta_1)$ is a nontrivial coordinate subspace.

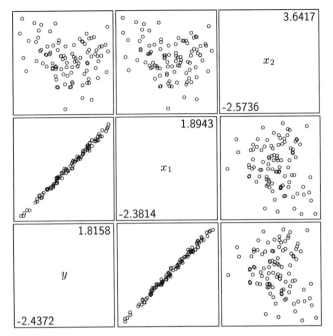

FIGURE 7.3 Scatterplot matrix of the response and two predictors for the example of Section 7.2.2.

Consider next the marginal response plot $\{y, x_2\}$. Interchanging the roles of x_1 and x_2 in Proposition 7.1, it follows that $\mathcal{S}_{y|x_2} \subset \mathcal{S}(\eta_2, \chi_2) = \mathcal{S}(\chi_2) = \mathbb{R}^1$. If we had incorrectly assumed that $\mathcal{S}_{y|x_2} \subset \mathcal{S}(\eta_2)$, an inspection of the plot $\{y, x_2\}$ would have led us to the wrong conclusion that $\mathcal{S}(\eta_2) \neq 0$.

In this example there are only two predictors so the true story would be told by the 3D plot $\{y, x\}$. In higher-dimensional problems, however, it can be difficult to understand the essential structure of the full regression from marginal response plots when the predictors are dependent, even if it seems easy to find interesting marginal projections.

In Sections 7.3 and 7.4 we concentrate on containment, $\mathcal{S}_{y|x_1} \subset \mathcal{S}(\eta_1)$. The stronger requirement $\mathcal{S}_{y|x_1} = \mathcal{S}(\eta_1)$ is deferred until Section 7.5.

7.3. FORCING $\mathcal{S}_{y|x_1} \subset \mathcal{S}(\eta_1)$

The examples in the previous two sections show that it is possible for extraneous information from the distribution of the predictors to be present in a marginal response plot. That will not be a problem, however, if $\mathcal{S}_{\eta_2^T x_2 | x_1} \subset \mathcal{S}(\eta_1)$ because then $\mathcal{S}_{y|x_1} \subset \mathcal{S}(\eta_1)$. The next two propositions address this situation.

Proposition 7.2. $S_{\eta_2^T x_2 | x_1} \subset S(\eta_1)$ if and only if $x_1 \perp\!\!\!\perp \eta_2^T x_2 \mid \eta_1^T x_1$.

Justification.   The conclusion follows immediately because $S_{\eta_2^T x_2 | x_1}$ is the central subspace for the regression of $\eta_2^T x_2$ on x_1.   □

Proposition 7.3.   If $x_1 \perp\!\!\!\perp \eta_2^T x_2 \mid \eta_1^T x_1$ then $S_{y|x_1} \subset S(\eta_1)$.

Justification.   The conclusion follows immediately from Propositions 7.1 and 7.2.   □

Proposition 7.3 relies on the requirement $x_1 \perp\!\!\!\perp \eta_2^T x_2 \mid \eta_1^T x_1$, which is of course implied by the more restrictive condition $x_1 \perp\!\!\!\perp x_2$. It will hold also when $\eta_2 = 0$ so that x_2 is not needed in the full regression. The role of this requirement is to insure that everything seen in the plot $\{y, x_1\}$ will be relevant to $S_{y|x}$ via the coordinate subspace $S(\eta_1)$.

7.3.1. Location regressions for the predictors

The condition $x_1 \perp\!\!\!\perp \eta_2^T x_2 \mid \eta_1^T x_1$ of Proposition 7.3 seems rather restrictive from a practical view. The degree of applicability may be increased by modifying the predictors to satisfy the condition while preserving $S(\eta_1)$. The next proposition shows how to modify the predictors when the regression of x_1 on x_2 follows an additive-location regression (6.10), $y - E(y \mid x) \perp\!\!\!\perp x$. First, define the population predictor residuals

$$r_{1|2} = x_1 - E(x_1 \mid x_2). \tag{7.4}$$

Proposition 7.4.   If $r_{1|2} \perp\!\!\!\perp x_2$ then

$$S_{y|r_{1|2}} \subset S(\eta_1).$$

Justification.   Writing $\eta^T x$ in terms of $r_{1|2}$,

$$\eta_1^T x_1 + \eta_2^T x_2 = \eta_1^T r_{1|2} + k(x_2)$$

where $k(x_2) = \eta_1^T E(x_1 \mid x_2) + \eta_2^T x_2$. Because $r_{1|2} \perp\!\!\!\perp k(x_2)$ the conclusion follows from Proposition 7.3 applied to the regression of y on $r_{1|2}$ and $k(x_2)$, or by noting that

$$F_{y|r_{1|2}} = E_{k(x_2)|r_{1|2}}[F_{y|(\eta_1^T r_{1|2} + k(x_2))}]$$
$$= E_{k(x_2)}[F_{y|(\eta_1^T r_{1|2} + k(x_2))}].   □$$

If the regression of x_1 on x_2 is adequately described by an additive-location regression then Proposition 7.4 indicates that all of the information in the plot $\{y, r_{1|2}\}$ is relevant to the coordinate subspace; that is, $\mathcal{S}_{y|r_{1|2}} \subset \mathcal{S}(\eta_1)$. In practice, this requires using estimates to form the residuals $r_{1|2}$ because the predictor regression function $E(x_1 \mid x_2)$ will usually be unknown. Estimates could be constructed by using smoothers or fitting polynomials, for example.

In some applications, a linear regression might be sufficient to describe the regression function $E(x_1 \mid x_2)$. Let

$$e_{1|2} = x_1 - E(x_1) - \beta_{1|2}^T(x_2 - E(x_2)) \tag{7.5}$$

where $\beta_{1|2}$ is the coefficient vector from the population OLS regression of x_1 on x_2. Partitioning $\text{Var}(x) = \Sigma = (\Sigma_{jk})$ for $j, k = 1, 2$ according to the partition of x,

$$\beta_{1|2} = \Sigma_{22}^{-1}\Sigma_{21}. \tag{7.6}$$

Sample versions of $e_{1|2}$ are relatively easy to construct in practice. Using $e_{1|2}$ in place of the more general residuals $r_{1|2}$ may be appropriate when $E(x_1 \mid x_2)$ is essentially a linear function of x_2. Coordinatewise transformations of the predictors can be used to induce linearity of the regression function and thus to extend applicability.

The following proposition summarizes the situation when x is normally distributed.

Proposition 7.5. If x is normally distributed then

$$\mathcal{S}_{y|e_{1|2}} \subset \mathcal{S}(\eta_1).$$

Justification. The conclusion is immediate because $e_{1|2} \perp\!\!\!\perp x_2$. □

A potential practical limitation of assuming an additive-location regression for the predictors ($r_{1|2} \perp\!\!\!\perp x_2$) is that the conditional variance $\text{Var}(x_1 \mid x_2)$ must be independent of x_2. To gain some insights into the effects of allowing non-constant variances, we investigate the case in which x follows an elliptically contoured distribution. As we will see shortly, $E(x_1 \mid x_2)$ is linear in this case so $r_{1|2} = e_{1|2}$, but $r_{1|2}$ is still dependent on x_2 and Proposition 7.4 does not apply, except when x is normal.

Background on elliptically contoured distributions can be found in Johnson (1987, Chapter 6), Cambanis, Huang, and Simons (1981), and Fang, Kotz, and Ng (1990). A very brief overview of elliptically contoured distributions is given in the next section before returning to the main question in Section 7.3.3.

7.3.2. Elliptically contoured distributions

An elliptically contoured random vector x can be specified by the triplet (μ, D, g) where $E(x) = \mu$, $\text{Var}(x) = \Sigma = aD$ for some positive constant a, and g

is a function that determines a particular family. Elliptically contoured densities, when they exist, are of the form

$$f_x(u) = |D|^{-1/2} g[(u - \mu)^T D^{-1}(u - \mu)].$$

The set of elliptically contoured distributions includes the nonsingular multivariate normal family for which $a = 1$ and $g(t) = \exp(-t/2)$.

Any $p \times 1$ elliptically contoured random vector x characterized by (μ, D, g) can be represented in the form

$$x = RBU + \mu$$

where $D = BB^T$, U is a random vector that is uniformly distributed on the surface of the unit sphere in $\mathbb{R}^p$, and R is an independent positive random variable (Johnson 1987, Chapter 6). The function g is determined by the distribution of R. For example, if R^2 follows a chi-squared distribution with p degrees of freedom then x is normal. It follows from this representation that

$$R^2 = (x - \mu)^T D^{-1}(x - \mu)$$

and thus that

$$R \perp\!\!\!\perp \frac{x - \mu}{[(x - \mu)^T D^{-1}(x - \mu)]^{1/2}} \tag{7.7}$$

Let ϕ denote a full rank $p \times q$ matrix, $q \leq p$, and let $P_{\phi(\Sigma)}$ denote the orthogonal projection operator onto $\mathcal{S}(\phi)$ relative to the inner product $(a, b) = a^T \Sigma b$,

$$P_{\phi(\Sigma)} = \phi(\phi^T \Sigma \phi)^{-1} \phi^T \Sigma. \tag{7.8}$$

We will also use $Q_{\phi(\Sigma)} = I - P_{\phi(\Sigma)}$. Projection operators relative to the usual inner product will be indicated without the second subscript argument on the projector, $P_{\phi} = P_{\phi(I)}$ and $Q_{\phi} = I - P_{\phi}$.

If x is a $p \times 1$ elliptically contoured random variable characterized (μ, D, g), then $\phi^T x$ is an elliptically contoured random variable characterized by $(\phi^T \mu, \phi^T D \phi, g)$. Conditioning also gives an elliptically contoured random variable of the same family.

Eaton (1986) showed that x is an elliptically contoured random variable if and only if $E(x \mid \phi^T x)$ is a linear function of $\phi^T x$ for all conforming matrices ϕ. This, in combination with Proposition 4.2 on linear conditional expectations, implies that conditional expectations $E(x \mid \phi^T x)$ follow the same form as found in the normal family (Cook 1992b):

$$E(x \mid \phi^T x) = \mu + P^T_{\phi(\Sigma)}(x - \mu). \tag{7.9}$$

Up to a proportionality function ν, conditional variances are also the same as those in the multivariate normal family (Cook 1992b):

$$
\begin{aligned}
\mathrm{Var}(x \mid \phi^T x) &= \nu(\phi^T x)[\Sigma - \Sigma\phi(\phi^T\Sigma\phi)^{-1}\phi^T\Sigma] \\
&= \nu(\phi^T x)\Sigma Q_{\phi(\Sigma)} \\
&= \nu(\phi^T x)Q_{\phi(\Sigma)}^T \Sigma Q_{\phi(\Sigma)} \\
&= \nu(\phi^T x)\Sigma^{1/2}Q_{\Sigma^{1/2}\phi}\Sigma^{1/2}.
\end{aligned}
\tag{7.10}
$$

The proportionality function $\nu(\phi^T x)$ depends only on the quadratic form $(x-\mu)^T\phi[\mathrm{Var}(\phi^T x)]^{-1}\phi^T(x-\mu)$. Furthermore, ν is constant if and only if x is normal (Kelker 1970).

7.3.3. Elliptically contoured predictors

As in the previous section, let ϕ denote a full rank $p \times q$ matrix, $q \leq p$. The following proposition gives a characterization of what we might see in the plot $\{y, \phi^T x\}$ with elliptically contoured predictors.

Proposition 7.6. Let $\mathcal{S}_{y|x}(\eta)$ denote the central dimension-reduction subspace for the regression of y on x, let $P_{\phi(\Sigma)}$ denote the projection operator as defined in (7.8), and let

$$
\omega(\phi^T x) = \{(x - \mathrm{E}(x))^T \phi[\mathrm{Var}(\phi^T x)]^{-1}\phi^T(x - \mathrm{E}(x))\}^{1/2}.
$$

If x is an elliptically contoured random variable then

$$
y \perp\!\!\!\perp \phi^T x \mid \{(P_{\phi(\Sigma)}\eta)^T x, \omega(\phi^T x)\}.
$$

Justification. Because $\mathcal{S}_{y|x}(\eta)$ is the central subspace, $y \perp\!\!\!\perp \phi^T x \mid \eta^T x$ and thus

$$
y \perp\!\!\!\perp \phi^T x \mid \{(P_{\phi(\Sigma)}\eta)^T x, (Q_{\phi(\Sigma)}\eta)^T x\}.
$$

For notational convenience, let $A = (\phi^T\Sigma\phi)^{-1}\phi^T\Sigma\eta$ and let ψ denote a basis for $\mathcal{S}(Q_{\phi(\Sigma)}\eta)$. The last conditional independence statement can now be rewritten as $y \perp\!\!\!\perp \phi^T x \mid \{A^T\phi^T x, \psi^T x\}$, which implies by Proposition 4.5 that

$$
y \perp\!\!\!\perp \phi^T x \mid \{A^T\phi^T x, \omega(\phi^T x), \psi^T x\}.
\tag{7.11}
$$

Next, because $\phi^T x$ and $\psi^T x$ are uncorrelated it follows that (Cambanis, Huang, and Simons 1981) $\phi^T x \perp\!\!\!\perp \psi^T x \mid \omega(\phi^T x)$ and thus that

$$
\phi^T x \perp\!\!\!\perp \psi^T x \mid \{A^T\phi^T x, \omega(\phi^T x)\}.
\tag{7.12}
$$

Using the Conditional Independence Proposition 4.6 with (7.11) and (7.12) gives the desired result. $\qquad\square$

Proposition 7.6 says in part that characteristics of the plot $\{y, \phi^T x\}$ depend on $P_{\phi(\Sigma)}\eta$, the projection of η onto the subspace spanned by ϕ relative to the inner product induced by Σ. The following corollary shows one way to choose ϕ so that we can again use a marginal plot to focus on the coordinate subspace $S(\eta_1)$.

Corollary 7.1. Let $S_{y|x}(\eta)$ denote the central dimension-reduction subspace for the regression of y on x. Without loss of generality, assume that $E(x) = 0$. Partition $x^T = (x_1^T, x_2^T)$ and $\eta^T = (\eta_1^T, \eta_2^T)$, and let $\phi^T = (I, -\beta_{1|2}^T)$ where $\beta_{1|2}$ is defined in (7.6). Then

$$\phi^T x = e_{1|2} = x_1 - \beta_{1|2}^T x_2$$

and, if x is elliptically contoured,

$$y \perp\!\!\!\perp e_{1|2} \mid \{\eta_1^T e_{1|2}, \omega(e_{1|2})\}$$

where

$$\omega(e_{1|2}) = \{e_{1|2}^T[\mathrm{Var}(e_{1|2})]^{-1} e_{1|2}\}^{1/2}$$

Justification. The results can be obtained by substituting for ϕ in Proposition 7.6 and evaluating $P_{\phi(\Sigma)}\eta$. $\qquad\square$

This corollary shows that, for elliptically contoured predictors,

$$S_{y|e_{1|2}} \subset S(\eta_1, \alpha_1)$$

where α_1 is determined by $\omega(e_{1|2})$. The presence of the radius $\omega(e_{1|2})$ is a potential complication. Because the conditional distribution of $y \mid e_{1|2}$ depends on $\omega(e_{1|2})$, the plot $\{y, e_{1|2}\}$ may show more than the coordinate subspace $S(\eta_1)$. This problem disappears if the predictors are normal, as shown in the next corollary, which is essentially a restatement of Proposition 7.5 using elliptical predictors as the starting point.

Corollary 7.2. In addition to the conditions and definitions of Corollary 7.1, assume that x has a nonsingular normal distribution. Then

$$y \perp\!\!\!\perp e_{1|2} \mid \eta_1^T x.$$

Justification. In the justification of Proposition 7.6, $\phi^T x$ and $\psi^T x$ are independent, and thus there is no need to condition on $w(\phi^T x)$ in (7.11) and (7.12). □

With elliptically contoured predictors, the possibility that the plot $\{y, e_{1|2}\}$ may show more than the coordinate subspace $\mathcal{S}(\eta_1)$ has not been found to be worrisome in practice, since systematic trends due to $w(e_{1|2})$ often seem to be small relative to background variation. Nevertheless, with a little playing around it may be possible to construct examples of misleading dependence.

For example, let the predictor vector $x = (w_1, w_2, w_3)^T$ follow a Pearson Type II distribution with $\mu = 0$, $D = I$, shape parameter $m = -1/2$, and support $\|x\| \leq 1$ (Johnson 1987, p. 111). Define the response to be $y \mid x = \alpha w_3 + \sigma \varepsilon$ where $\varepsilon \perp\!\!\!\perp x$ and $\mathrm{Var}(\varepsilon) = 1$. Thus, the central subspace is spanned by $\eta = (0, 0, 1)^T$. In the general notation, set $x_1 = (w_1, w_2)^T$ and $x_2 = w_3$, and consider trying to gain information on the coordinate subspace $\mathcal{S}(\eta_1) = \mathcal{S}(0)$ with the plot $\{y, e_{1|2}\}$. Since the elements of x are uncorrelated, $e_{1|2} = x_1$ and we can equivalently consider the 3D plot $\{y, x_1\}$. Because $\mathcal{S}(\eta_1) = \mathcal{S}(0)$, we may hope that $\mathcal{S}_{y|x_1} = \mathcal{S}(0)$ as well. However, while $\mathrm{E}(y \mid x_1) = 0$, the conditional variance function still depends on x_1:

$$\mathrm{Var}(y \mid x_1) = \alpha^2(1 - w_1^2 - w_2^2) + \sigma^2. \tag{7.13}$$

Depending on the values of α and σ, the plot $\{y, x_1\}$ may show an indication of heteroscedasticity as a function of $\|x_1\|$ even though $\mathcal{S}(\eta_1) = \mathcal{S}(0)$. In terms of Proposition 7.1, $\mathcal{S}_{y|x_1}$ is picking up the contribution of $\mathcal{S}_{\eta_2^T x_2 | x_1} = \mathbb{R}^2$.

The potential misleading dependence of $y \mid e_{1|2}$ on $w(e_{1|2})$ can be removed as follows. Define the normalized residuals

$$\tilde{e}_{1|2} = \frac{e_{1|2}}{[w(e_{1|2})]^{1/2}}. \tag{7.14}$$

Then

$$y \perp\!\!\!\perp \tilde{e}_{1|2} \mid [\eta_1^T \tilde{e}_{1|2}, w(e_{1|2})].$$

Because $\tilde{e}_{1|2} \perp\!\!\!\perp w(e_{1|2})$ (see (7.7)), the above translates as

$$y \perp\!\!\!\perp \tilde{e}_{1|2} \mid \eta_1^T \tilde{e}_{1|2}.$$

Thus, normalizing the residuals $e_{1|2}$ removes any contamination coming from the distribution of the predictors via $w(e_{1|2})$. It is possible that this normalization may lose information if the distribution of $y \mid x$ happens to depend on $w(e_{1|2})$, but such loss does not seem worrisome in practice. For future reference, the discussion so far is summarized in the following proposition.

Proposition 7.7. Let x be an elliptically contoured random variable and let the normalized residuals $\tilde{e}_{1|2}$ be defined as in (7.14). Then

$$\mathcal{S}_{y|\tilde{e}_{1|2}} \subset \mathcal{S}(\eta_1).$$

The practical implication of this proposition is that we could consider the plot $\{y, \tilde{e}_{1|2}\}$ involving normalized residuals rather than $\{y, e_{1|2}\}$ to help identify the coordinate subspace $\mathcal{S}(\eta_1)$ when there may be contamination from the distribution of the predictors via $\omega(e_{1|2})$. The length of each point $\tilde{e}_{1|2}$ plotted relative to the horizontal axes is constant with respect to the inner product determined by $\text{Var}(e_{1|2})$, so the points in the horizontal plane of a 3D plot will fall on an ellipse. This may make the plot seem a bit unusual, but the characterization methods of Chapter 4 still apply.

7.4. IMPROVING RESOLUTION

Four types of plots—$\{y, x_1\}$, $\{y, e_{1|2}\}$, $\{y, \tilde{e}_{1|2}\}$, and $\{y, r_{1|2}\}$—were studied in Section 7.3 for visually inferring about the coordinate subspace $\mathcal{S}(\eta_1)$, depending on the distribution of the predictors. Although not required for the general development, practical considerations force the plots to be 2D or 3D. Visual estimates of $\mathcal{S}(\eta_1)$ found using the methods of Chapters 4 and 5 can then be incorporated into the general paradigm sketched in Section 7.1. The general paradigm will be discussed in some detail in Chapter 9. For now we consider a way to increase resolution by reducing variation on the vertical axis and thus improve visual power for estimating coordinate subspaces. The general ideas apply to any of the four plots studied so far. For definiteness we focus on $\{y, e_{1|2}\}$.

Consider the original regression problem but with predictors transformed to $(e_{1|2}, x_2)$. Let $\delta = \eta_2 + \beta_{1|2}\eta_1$ where $\beta_{1|2} = \Sigma_{22}^{-1}\Sigma_{21}$ was defined in (7.6). The linear combinations $\eta^T x$ formed by using the basis η for $\mathcal{S}_{y|x}$ transform to

$$\eta^T x = \eta_1^T e_{1|2} + \delta^T x_2.$$

Thus, recognizing that $e_{1|2}$ is a function of (x_1, x_2), Proposition 4.5 gives

$$y \perp\!\!\!\perp (e_{1|2}, x_2) \mid (\eta_1^T e_{1|2} + \delta^T x_2)$$

and

$$y \perp\!\!\!\perp (e_{1|2}, x_2) \mid (\eta_1^T e_{1|2} + \delta^T x_2, x_2)$$

which in turn implies that

$$y \perp\!\!\!\perp e_{1|2} \mid (\eta_1^T e_{1|2}, x_2).$$

Assume now that $x_1 \mid x_2$ is well described by a linear additive-location model so that

$$e_{1|2} \perp\!\!\!\perp x_2 \mid \eta_1^T e_{1|2}.$$

Then, from the Conditional Independence Proposition 4.6, the last two conditional independence statements imply that

$$(y, x_2) \perp\!\!\!\perp e_{1|2} \mid \eta_1^T e_{1|2}.$$

The discussion thus far is summarized in the following proposition.

Proposition 7.8. Linearly transform the partitioned predictor vector $x = (x_1, x_2)^T$ to $(e_{1|2}, x_2)^T$, where $e_{1|2}$ is the residual from the population OLS fit of x_1 on x_2. If $e_{1|2} \perp\!\!\!\perp x_2 \mid \eta_1^T e_{1|2}$ then

$$(y, x_2) \perp\!\!\!\perp e_{1|2} \mid \eta_1^T e_{1|2}.$$

Proposition 7.8 is useful because it implies that any function of (y, x_2) is independent of $e_{1|2}$ given $\eta_1^T e_{1|2}$, and this allows the opportunity to reduce the variation on the vertical axis of the plot $\{y, e_{1|2}\}$ while preserving $S(\eta_1)$ as a DRS.

In particular, let $r_{y|2}$ denote any population residual from modeling y as a function of x_2 alone. There is no requirement that the model be correct or even adequate, although there are advantages if this is so. Then

$$S_{r_{y|2}|e_{1|2}} \subset S(\eta_1) \tag{7.15}$$

and the plot $\{r_{y|2}, e_{1|2}\}$ may be used to gain information on the coordinate subspace. Although not needed for (7.15), we will require that $r_{y|2}$ be a one-to-one function of y for each value of x_2. This prevents us from losing relevant information about $S(\eta_1)$ when passing from $\{y, e_{1|2}\}$ to $\{r_{y|2}, e_{1|2}\}$. For a very extreme case, consider defining $r_{y|2}$ to be identically zero. Equation (7.15) still holds, but all nontrivial information on $S(\eta_1)$ has been lost.

A simple example might help fix ideas. Let $x = (w_1, w_2, w_3)^T$ and

$$y \mid x = w_1^2 + w_2 + w_3 + \varepsilon$$

where $(w_1, w_2, w_3, \varepsilon)$ is multivariate normal with mean 0 and covariance matrix I. The conditional distribution of $y \mid (w_2, w_3)$ can be described as

$$y \mid (w_2, w_3) = w_2 + w_3 + \varepsilon + \chi^2$$

where χ^2 is an independent chi-squared random variable with one degree of freedom. The coordinate subspace $S((1,1)^r)$ corresponding to (w_2, w_3) may be difficult to see in the plot $\{y, (w_2, w_3)\}$ because the total error $(\varepsilon + \chi^2)$ is positively skewed. Modeling $y \mid w_1$ alone yields $y \mid w_1 = w_1^2 + \varepsilon^*$ where $\text{Var}(\varepsilon^*) = 3$. If it is successfully concluded that $E(y \mid w_1) = w_1^2$ and $r_{y|1} = y - w_1^2$ then

$$r_{y|1} \mid (w_2, w_3) = w_2 + w_3 + \varepsilon$$

which preserves the coordinate subspace for (w_2, w_3) and removes the chi-squared error.

Although the precise form of the residuals is not very important, it is possible that the wrong choice may do more harm than good. For example, if $y \perp\!\!\!\perp x_2$ and $r_{y|2} = y - b^T x_2$ for some nonzero coefficients b then $\text{Var}(r_{y|2}) > \text{Var}(y)$ so that the variation has been increased by using this particular $r_{y|2}$. Because of such potential to increase variation, the residuals $r_{y|2}$ are best used as a method of eliminating gross variation in y as a function of x_2. Toggling between the plots $\{y, e_{1|2}\}$ and $\{r_{y|2}, e_{1|2}\}$ can be useful in practice for visually assessing the variance reduction.

Let

$$e_{y|2} = y - E(y) - \beta_{y|2}^T (x_2 - E(x_2)) \tag{7.16}$$

where

$$\beta_{y|2}^T = \Sigma_{22}^{-1} \text{Cov}(x_2, y).$$

Setting $r_{y|2}$ to be the residuals $e_{y|2}$ from a linear regression of y on x_2 may be a useful default option for use in practice. The corresponding plot $\{e_{y|2}, e_{1|2}\}$ is known as an *added-variable plot* for studying the effects of x_1 on the regression after including x_2. Two-dimensional added-variable plots were studied by several authors, including Atkinson (1982, 1985), Belsley, Kuh, and Welsch (1980, p. 30), Chambers, Cleveland, Kleiner, and Tukey (1983), Cook (1986), Cook and Weisberg (1982, 1990, 1991a), Henderson and Velleman (1981), Lawrance (1986), and Mosteller and Tukey (1977, p. 343). Wang (1985) investigated added-variable plots in generalized linear models. According to Davison and Snell (1991), instances of added-variable plots date to Cox (1958, Section 4.5). The name "added-variable plot" was coined by Cook and Weisberg (1982), although other names have been used from time to time, including "adjusted variable plot," "coefficient plot," and "partial regression plot." Three-dimensional added-variable plots were suggested by Cook (1987c) and discussed by Cook and Weisberg (1989, 1994a) and Berk (1998).

Although there is a substantial literature on added-variable plots, the role suggested here is relatively new (Cook, 1994a). Added-variable plots will play a central role for estimating coordinate subspaces during the methodological discussions in Chapter 9.

7.5. FORCING $S_{y|x_1} = S(\eta_1)$

The developments of the last section centered on ways to modify the predictors to force $S_{y|x_1} \subset S(\eta_1)$. Without further conditions, however, those results do not necessarily guarantee that $S_{y|x_1} = S(\eta_1)$, even with independent predictors. The next section contains an example illustrating that with independent predictors $S_{y|x_1}$ may still be a proper subset of $S(\eta_1)$.

7.5.1. Example: x_1 independent of x_2, but $S_{y|x_1} \neq S(\eta_1)$

Let x_1 be a uniform random variable on the interval $(-1,1)$ and, given x_1, let (y,x_2) follow a bivariate normal distribution with means 0, variances 1, and $\text{Cov}(y,x_2 \mid x_1) = x_1$. Because $E(y \mid x_1,x_2) = x_1x_2$ and $\text{Var}(y \mid x_1,x_2) = 1 - x_1^2$, the central subspace $S_{y|x} = \mathbb{R}^2$. However, by writing the joint density as

$$f_{(y,x_1,x_2)} = f_{(y,x_2|x_1)} f_{x_1}$$

and then integrating over x_2 it is seen that $y \perp\!\!\!\perp x_1$. Similarly, $x_1 \perp\!\!\!\perp x_2$. The central subspace for the regression of y on x_1 is therefore $S(0)$ and the marginal plot $\{y,x_1\}$ will fail to indicate that x_1 is needed in the full regression. In reference to Proposition 7.3, the condition $x_1 \perp\!\!\!\perp x_2 \mid \eta_1^T x_1$ holds but $S_{y|x_1}$ is a proper subset of $S(\eta_1)$.

The four plots in Figure 7.4 give a visual representation of this example. Plots (a) and (b) show the marginal response plot $\{y,x_1\}$ and the predictor plot $\{x_1,x_2\}$. These plots support the conclusion that $y \perp\!\!\!\perp x_1$ and $x_2 \perp\!\!\!\perp x_1$. Plots (c) and (d) give two views of the 3D plot $\{y,(x_1,x_2)\}$ that display dependence. To aid visualization in these two plots, the data points below the 1/3 quantile of x_1 were represented by a filled disk, those above the 2/3 quantile of x_1 were represented by an open disk, and the remaining points were removed. The size of the plotting symbols was increased as well.

7.5.2. Conditions for $S_{y|x_1} = S(\eta_1)$

The following proposition gives a necessary and sufficient condition for insuring that $S_{y|x_1} = S(\eta_1)$ after imposing the condition $x_1 \perp\!\!\!\perp \eta_2^T x_2 \mid \eta_1^T x_1$ of Proposition 7.3 to force $S_{y|x_1} \subset S(\eta_1)$. First, let the columns of the matrix γ_k be a basis for $S_{y|x_k}$, the central subspace for the regression of y on x_k, $k = 1,2$.

Proposition 7.9. Assume that $x_1 \perp\!\!\!\perp \eta_2^T x_2 \mid \eta_1^T x_1$. Then $S_{y|x_1}(\gamma_1) = S(\eta_1)$ if and only if $x_1 \perp\!\!\!\perp (y,\eta_2^T x_2) \mid \gamma_1^T x_1$.

Justification. From Proposition 7.3, $S_{y|x_1} \subset S(\eta_1)$. Thus, we need show only that $S_{y|x_1} \supset S(\eta_1)$ if and only if $x_1 \perp\!\!\!\perp (y,\eta_2^T x_2) \mid \gamma_1^T x_1$.

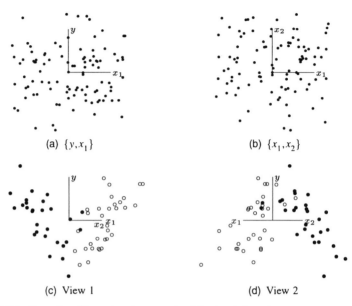

(a) $\{y, x_1\}$

(b) $\{x_1, x_2\}$

(c) View 1

(d) View 2

FIGURE 7.4   Four plots illustrating the example of Section 7.5.1. Plots (c) and (d) give views of the 3D plot $\{y,(x_1,x_2)\}$. The different symbols represent the magnitude of x_1; the points between the 1/3 and 2/3 quantile of x_1 were removed for visual clarity.

Suppose $x_1 \perp\!\!\!\perp (y, \eta_2^T x_2) \mid \gamma_1^T x_1$. Then, from the Conditional Independence Proposition 4.6,

$$x_1 \perp\!\!\!\perp y \mid (\eta_2^T x_2, \gamma_1^T x_1).$$

Trivially,

$$y \perp\!\!\!\perp (x_1, \eta_2^T x_2) \mid (\eta_2^T x_2, \gamma_1^T x_1)$$

and thus the subspace spanned by the columns of the matrix

$$A = \begin{pmatrix} \gamma_1 & 0 \\ 0 & I \end{pmatrix}$$

is a DRS for the regression of y on $(x_1, \eta_2^T x_2)$. But the central subspace for the regression of y on $(x_1, \eta_2^T x_2)$ is spanned by the columns of $(\eta_1^T, I)^T$. Consequently, there is a matrix $B^T = (B_1^T, B_2^T)$ so that

$$AB = \begin{pmatrix} \gamma_1 B_1 \\ B_2 \end{pmatrix} = \begin{pmatrix} \eta_1 \\ I \end{pmatrix}.$$

Thus, $\mathcal{S}_{y|x_1}(\gamma_1) \supset \mathcal{S}(\eta_1)$.

Suppose $\mathcal{S}_{y|x_1}(\gamma_1) \supset S(\eta_1)$ so that $\mathcal{S}_{y|x_1}(\gamma_1) = S(\eta_1)$. Consider the following equivalences from Proposition 4.6:

$$
\begin{array}{ll}
(a_1) \quad x_1 \perp\!\!\!\perp \eta_2^T x_2 \mid (\eta_1^T x_1, y) & (a_2) \quad x_1 \perp\!\!\!\perp y \mid \eta_1^T x_1 \\
(b_1) \quad x_1 \perp\!\!\!\perp y \mid (\eta_1^T x_1, \eta_2^T x_2) & (b_2) \quad x_1 \perp\!\!\!\perp \eta_2^T x_2 \mid \eta_1^T x_1 \\
(c) \quad x_1 \perp\!\!\!\perp (y, \eta_2^T x_2) \mid \eta_1^T x_1.
\end{array}
$$

Condition (b_2) holds by assumption, and conditions (a_2) and (b_1) hold by construction. Condition (c) is the desired result because we can take $\gamma_1 = \eta_1$. $\qquad\square$

The condition $x_1 \perp\!\!\!\perp (y, \eta_2^T x_2) \mid \gamma_1^T x_1$ of Proposition 7.9 may be interpreted as requiring $\mathcal{S}_{y|x_1}$ to be a DRS for the multivariate regression of $(y, \eta_2^T x_2)$ on x_1. In fact it must be a central subspace because $x_1 \perp\!\!\!\perp (y, \eta_2^T x_2) \mid \gamma_1^T x_1$ implies $x_1 \perp\!\!\!\perp y \mid \gamma_1^T x_1$. More carefully, if $S(\delta)$ is a proper subset of $\mathcal{S}_{y|x_1}$ and $x_1 \perp\!\!\!\perp (y, \eta_2^T x_2) \mid \delta^T x_1$ then $x_1 \perp\!\!\!\perp y \mid \delta^T x_1$, which would contradict the centrality of $\mathcal{S}_{y|x_1}$.

Proposition 7.9 is stated in terms of the original predictors, but the result is equally valid after transforming the predictors to insure the required initial condition of the proposition. For example, in the statement of the proposition we may set $x_1 = r_{1|2}$ where the predictor residuals $r_{1|2}$ are defined at (7.4).

7.5.3. Marginal consistency assumption

The example of Section 7.5.1 shows that it is possible for $\mathcal{S}_{y|x_1}$ to be a proper subset of $S(\eta_1)$, even with independent predictors. The example also suggests that regression problems in which that is so may be the exception rather than the rule in practice. To see why this conclusion may be reasonable, let $\mu(x) = E(y \mid x)$ and consider a location regression, $y \perp\!\!\!\perp x \mid \mu(x)$, as defined in Section 6.4.2. Assume also that the predictors have been modified as necessary to preserve the coordinate subspace $S(\eta_1)$ and insure that $\mathcal{S}_{y|x_1} \subset S(\eta_1)$. Since η is a basis for $\mathcal{S}_{y|x}$ the regression function can be written as $\mu(x) = \mu(\eta_1^T x_1 + \eta_2^T x_2)$ without loss of generality. The regression function for the marginal regression of y on x_1 becomes

$$
E(y \mid x_1) = E(y \mid \gamma_1^T x_1) = E(\mu(\eta_1^T x_1 + \eta_2^T x_2) \mid x_1)
$$

where γ_1 is still a basis for $\mathcal{S}_{y|x_1}$.

If $\mathcal{S}_{y|x_1}$ is a *proper* subset of $S(\eta_1)$ then $E(\mu(\eta_1^T x_1 + \eta_2^T x_2) \mid x_1)$ must be constant in $a^T x_1$ where a is some nonzero vector in $S(\eta_1) - \mathcal{S}_{y|x_1}$. In other words, if $E(\mu(\eta_1^T x_1 + \eta_2^T x_2) \mid x_1)$ is a nontrivial function of $c^T x$ for all nonzero vectors c in $S(\eta_1)$ then $S(\eta_1) \subset \mathcal{S}_{y|x_1}$. This same argument can be extended to the distribution function

$$
F_{y|x_1} = E_{\eta_2^T x_2 | x_1}[F_{y|(\eta_1^T x_1 + \eta_2^T x_2)}].
$$

For example, consider an interaction model in two independent predictors $x = (x_1, x_2)^T$,

$$y \mid x = \beta_0 + \beta_1 x_1 + \beta_2 x_2 + \beta_{12} x_1 x_2 + \sigma \varepsilon \qquad (7.17)$$

where $E(\varepsilon) = 0$, $\text{Var}(\varepsilon) = 1$, and ε is independent of x. For this model, $\mathcal{S}_{y|x} = \mathbb{R}^2$ as long as $\beta_{12} \neq 0$. Thus we can take $\eta_1 = (1,0)$ and $\eta_2 = (0,1)$. To force $\mathcal{S}_{y|x_1}$ to be a proper subset of $\mathcal{S}(\eta_1) = \mathbb{R}^1$, we must arrange the model so that $\mathcal{S}_{y|x_1} = \mathcal{S}(0)$ and thus $y \perp\!\!\!\perp x_1$. In particular, the marginal regression function

$$E(y \mid x_1) = \beta_0 + \beta_2 E(x_2) + (\beta_1 + \beta_{12} E(x_2)) x_1$$

must necessarily be constant, which will happen only if $\beta_1 + \beta_{12} E(x_2) = 0$. Conversely, the conditions $\beta_1 + \beta_{12} E(x_2) \neq 0$ and $\beta_2 + \beta_{12} E(x_1) \neq 0$ are sufficient to guarantee that $\mathcal{S}_{y|x_k} = \mathcal{S}(\eta_k)$, $k = 1, 2$.

We must also arrange for a constant marginal variance function to force $y \perp\!\!\!\perp x_1$:

$$\text{Var}(y \mid x_1) = E_{x_2|x_1}[\text{Var}(y \mid x)] + \text{Var}_{x_2|x_1}[E(y \mid x)]$$
$$= \sigma^2 + (\beta_2 + \beta_{12} x_1)^2 \text{Var}(x_2).$$

Clearly, $\text{Var}(y \mid x_1)$ will be constant only if $\beta_{12} = 0$. Thus, it is impossible to have $y \perp\!\!\!\perp x_1$ while maintaining $\mathcal{S}_{y|x} = \mathbb{R}^2$. Combining this with the requirement for a constant regression function, we see that it is impossible to have $y \perp\!\!\!\perp x_1$ unless $y \perp\!\!\!\perp x_1 \mid x_2$. If this is the case then the marginal response plot for x_1 gives the correct inference that x_1 is not needed in the full regression.

The situation described in the example of Section 7.5.1 corresponds to a version of (7.17) with independent predictors and a nonconstant variance function:

$$y \mid x = x_1 x_2 + (1 - x_1^2)^{1/2} \varepsilon.$$

For this model, $y \perp\!\!\!\perp x_1$, $E(y \mid x_1) = 0$, and $\text{Var}(y \mid x_1) = 1$. Encountering such carefully balanced situations in practice would seem to be the exception rather than the rule.

In the remainder of this book, *we will assume that $\mathcal{S}_{y|x_1} = \mathcal{S}(\eta_1)$ whenever conditions have been imposed to insure that $\mathcal{S}_{y|x_1} \subset \mathcal{S}(\eta_1)$, unless explicitly stated otherwise.* This will be called the *marginal consistency assumption* (Cook 1992a; Cook 1994a), indicating that the structure of a certain marginal regression contains all the information on the corresponding coordinate subspace.

7.6. VISUAL FITTING WITH h-LEVEL RESPONSE PLOTS

Ideas for visually fitting with a 2D binary response plot were illustrated in Section 5.2.2, but the discussion of 3D binary response plots in Section 5.4.1

and 3D *h*-level response plots in Section 6.6 was mostly restricted to checking for 1D structure by using results from a numerical fit. The results of this chapter provide tools that facilitate visual fitting with such plots. As an introduction to the possible applications, the following outline gives a way of visually regressing an *h*-level response $\tilde{y}$ on three predictors $x = (x_1, x_2, x_3)^T$, assuming 1D structure and approximately normal predictors with a positive-definite covariance matrix $\text{Var}(x) = \Sigma$.

- Begin by transforming to uncorrelated predictors

$$z = \hat{\Sigma}^{-1/2} x$$

  where $\hat{\Sigma}$ is a consistent estimate of Σ. The relationship between the central subspaces in the population is given by Proposition 6.3:

$$S_{\tilde{y}|x} = \Sigma^{-1/2} S_{\tilde{y}|z}.$$

  The idea here is to use a 3D *h*-level response plot to estimate $S_{\tilde{y}|z}$ visually, and then backtransform to an estimate of $S_{\tilde{y}|x}$.

- Inspect a 3D plot of z to insure visual agreement with the assumption of normal predictors. Then mark the points according to the levels of $\tilde{y}$, to obtain a 3D *h*-level response plot.

- The next step is to estimate visually the coordinate subspace corresponding to a pair of the transformed predictors, say z_1 and z_2, by using Proposition 7.3 and the marginal consistency assumption. Since $\dim[S_{\tilde{y}|z}] = 1$, adding the marginal consistency assumption forces all 2D *h*-level response plots to have either 0D or 1D structure. The coordinate subspace for (z_1, z_2) can be estimated visually as follows: First rotate so that the plot $\{z_1, z_2\}$ is in the plane of the computer screen. Next, check whether the symbol density is approximately constant across the plot. If so, then we can conclude that $\tilde{y} \perp\!\!\!\perp (z_1, z_2)$ and that the coordinate direction for z_3 is a basis for $S_{\tilde{y}|z}$. Otherwise, following the discussion of Section 5.2.2, rotate about the out-of-screen axis (z_3) to obtain the best visual fit on the horizontal screen axis. Let

$$z_{12} = b_1 z_1 + b_2 z_2$$

  denote the resulting linear combination which is now on the horizontal screen axis. The estimate of the 1D coordinate subspace is $S((b_1, b_2)^T)$.

- Rotate about the horizontal screen axis (z_{12}) until the 2D plot in the plane of the computer screen is $\{z_3, z_{12}\}$. Now, again rotate about the out-of-screen axis to obtain the best visual fit on the horizontal screen axis, say

$$z_{123} = b_{12}z_{12} + b_3 z_3$$
$$= b_{12}b_1 z_1 + b_{12}b_2 z_2 + b_3 z_3$$
$$= b^T z$$

where $b^T = (b_{12}b_1, b_{12}b_2, b_3)$.

- Finally, rotate about the horizontal screen axis (now z_{123}) to check on the fit. If the relative symbol density seems constant in vertical slices of all 2D views encountered during this rotation, then the fit, as well as the initial assumption that $\dim[S_{\tilde{y}|z}] = 1$, should be reasonable. The estimated basis for $S_{\tilde{y}|x}$ is $\hat{\Sigma}^{1/2}b$.

PROBLEMS

7.1. Consider a regression problem with univariate response y and predictor vector $x = (x_1^T, x_2^T)^T \in \mathbb{R}^p$. Show that the average variation around the regression function $E_{x_2|x_1}(\text{Var}(y \mid x_1, x_2))$ in the full $(p+1)$-dimensional plot $\{y, x\}$ cannot exceed $\text{Var}(y \mid x_1)$ from the marginal response plot $\{y, x_1\}$.

7.2. Let P denote the orthogonal projection operator for a DRS of the regression of y on x, and assume that the central subspace $S_{y|x}$ exists. Show that $S_{y|Px} = S_{y|x}$. How does this fact fit into the graphical regression procedure sketched in Section 7.1?

7.3. Let the columns of the $p \times d$ matrices η and α be bases for $S_{y|x}$ and $S_{y|A^Tx}$, respectively, where A is a full rank $p \times p$ matrix. Let η_1 and α_1 consist of the first p_1 rows of η and α. What can be said about the relationship between the coordinate subspaces $S(\eta_1)$ and $S(\alpha_1)$?

7.4. The following two problems relate to the example of misleading dependence discussed in Section 7.3.3.

    7.4.1. Verify Equation 7.13.

    7.4.2. Using the setup of the example involving (7.13), construct a simulated data set illustrating that the marginal plot $\{y, e_{1|2}\}$ may show more that the coordinate subspace $S(\eta_1)$.

7.5. What would likely happen during application of the paradigm in Section 7.6 if (a) the regression of $\tilde{y}$ on x has 2D structure? (b) the regression of $\tilde{y}$ on x has 1D structure but the predictors are far from normal?

CHAPTER 8

Getting Numerical Help

The primary approach so far has been graphical, inferring about the central subspace $S_{y|x}$ by combining visual inferences on coordinate subspaces. There is one more ingredient that will help in the exploration of graphical regression methodology in Chapter 9. Under certain conditions on the distribution of the predictors, standard regression estimation methods can be used to obtain consistent estimates of vectors in $S_{y|x}$. We first outline results for linear kernels and then turn to quadratic kernels in Section 8.2.

8.1. FITTING WITH LINEAR KERNELS

Consider summarizing the regression of the response y on the $p \times 1$ vector of predictors x by using an objective function that depends on x only through the linear kernel $a + b^T x$. Specifically, estimates $(\hat{a}, \hat{b})$ of a and b are assumed to be obtained by minimizing an objective function:

$$(\hat{a}, \hat{b}) = \arg \min_{a,b} L_n(a, b) \qquad (8.1)$$

where L_n is of the form

$$L_n(a, b) = \frac{1}{n} \sum_{i=1}^{n} L(a + b^T x_i, y_i)$$

and $L(u, v)$ is a convex function of u. OLS estimates are obtained by setting $L(a + b^T x, y) = (y - a - b^T x)^2$. Maximum likelihood estimates corresponding to a natural exponential family can be found by setting

$$L(a + b^T x, y) = -y(a + b^T x) + \phi(a + b^T x)$$

for suitable choice of the convex function ϕ (See Problem 8.1). Certain robust estimates like Huber's M-estimate fall into the class characterized by (8.1), but

143

high breakdown estimates like least median of squares are excluded because the corresponding objective functions are not convex.

This use of the objective function (8.1) is not meant to imply that any associated model is "true" or even provides an adequate fit of the data. Nevertheless, in some problems there is a useful connection between $\hat{b}$ and the central subspace $\mathcal{S}_{y|x}$.

Because (y_i, x_i), $i = 1, \ldots, n$, are independent and identically distributed, the strong law of large numbers shows that L_n converges almost surely to

$$R(a,b) = \mathrm{E}[L(a + b^T x, y)]$$

where the expectation, which is assumed to be finite, is with respect to the joint distribution of y and x. The population minimizers can then be described by

$$(\tilde{a}, \beta) = \arg\min_{a,b} R(a,b). \tag{8.2}$$

We assume that $(\tilde{a}, \beta)$ is unique. Uniqueness will be assured if $L(\cdot, \cdot)$ is strictly convex in its first argument, for example.

The following proposition shows a connection between β and an arbitrary dimension-reduction subspace (DRS) for the regression of y on x.

Proposition 8.1 (Li–Duan). Let $\mathcal{S}_{drs}(\alpha)$ be a dimension-reduction subspace for the regression of y on x, and assume that $\mathrm{E}(x \mid \alpha^T x)$ is a linear function of $\alpha^T x$ and that $\Sigma = \mathrm{Var}(x)$ is positive definite. Assume further that $L(\cdot, \cdot)$ is convex in its first argument and that β, as defined in (8.2), is unique. Then $\beta \in \mathcal{S}_{drs}(\alpha)$.

Justification. We first rewrite $R(a,b)$ making use of the fact that $\mathcal{S}_{drs}(\alpha)$ is a DRS,

$$R(a,b) = \mathrm{E}[L(a + b^T x, y)]$$

$$= \mathrm{E}_{y,\alpha^T x} \mathrm{E}_{x|y,\alpha^T x}[L(a + b^T x, y)]$$

$$= \mathrm{E}_{y,\alpha^T x} \mathrm{E}_{x|\alpha^T x}[L(a + b^T x, y)]$$

where the final step follows because $y \perp\!\!\!\perp x \mid \alpha^T x$. Because $L(\cdot, \cdot)$ is convex in its first argument, it follows from Jensen's inequality that

$$R(a,b) \geq \mathrm{E}_{y,\alpha^T x}[L(a + b^T \mathrm{E}(x \mid \alpha^T x), y)].$$

We may now apply Proposition 4.2, assuming without loss of generality that $\mathrm{E}(x) = 0$ and making use of the fact that $\mathrm{E}(x \mid \alpha^T x)$ is a linear function of $\alpha^T x$,

$$R(a,b) \geq \mathrm{E}_{y,\alpha^T x}[L(a + (P_{\alpha(\Sigma)} b)^T x, y)]$$

where $\Sigma = \text{Var}(x)$. Thus

$$R(a,b) \geq R(a, P_{\alpha(\Sigma)} b)$$

and the conclusion now follows because $P_{\alpha(\Sigma)} b \in S_{drs}(\alpha)$ and β is unique.

$\square$

Li and Duan (1989, Theorem 2.1; see also Duan and Li, 1991) gave Proposition 8.1 for the case when $\dim[S_{drs}(\alpha)] = 1$. The justification here is the same as that by Li and Duan, except for the minor modifications necessary to remove the dimension restriction. Under certain conditions, the Li–Duan Proposition states that $\hat{b}$ is a Fisher-consistent estimate of a vector $\beta \in S_{drs}(\alpha)$. (See Cox and Hinkley, 1974, p. 287, for background on Fisher consistency.) Almost sure convergence can be established as well by following Li and Duan (1989, Theorem 5.1).

The essential message of the Li–Duan Proposition is that many common estimation methods based on linear kernels $a + b^T x$ provide estimates of vectors in DRSs when the predictors have linear conditional expectations. In particular, if a central subspace $S_{y|x}(\eta)$ exists and $\text{E}(x \mid \eta^T x)$ is linear in $\eta^T x$ then we may be able to construct useful estimates of a vector in the central subspace. But such estimates may not be useful if a central subspace does not exist. For example suppose that x is an elliptically contoured random variable so that $\text{E}(x \mid \alpha^T x)$ is linear in $\alpha^T x$ for all DRSs $S_{drs}(\alpha)$. Let β be as defined in (8.2). Then the Li–Duan Proposition implies that $\beta \in \cap S_{drs}$, which may not be useful unless $\cap S_{drs}$ is itself a DRS. The first example of Section 6.3 is a case in which $\cap S_{drs}$ equals the origin, which is not a DRS because x and y are dependent.

8.1.1. Isomerization data

Bates and Watts (1988, p. 55) used the Hougen–Watson reaction rate model

$$y \mid x = \frac{\theta_1 \theta_3 (x_2 - x_3/1.632)}{1 + \theta_2 x_1 + \theta_3 x_2 + \theta_4 x_3} + \sigma\varepsilon$$

previously described in (6.12) to fit 24 observations from a data set (Carr 1960) on the reaction rate (y) of the catalytic isomerization of n-pentane to isopentane. The Hougen–Watson model has $p = 3$ predictors $x = (x_1, x_2, x_3)^T$ with $S_{y|x}$ spanned by the vectors $(0, 1, -1/1.632)^T$ and $(\theta_2, \theta_3, \theta_4)^T$. Let $\hat{S}_{y|x}$ denote the estimate of $S_{y|x}$ obtained by substituting the estimates of the θ's given by Bates and Watts (1988, p. 56), and let $\hat{b}$ denote the 3×1 vector of coefficients from the OLS regression of y on x.

Except for two mild outliers, a 3D plot of the predictors (not shown) seems consistent with an elliptically contoured distribution. According to the Li–

Duan Proposition then, we might expect $\hat{b}$ to be close to $\hat{S}_{y|x}$. It turns out that the angle between $\hat{b}$ and its projection onto $\hat{S}_{y|x}$ is only 0.046 degrees.

8.1.2. Using the Li–Duan Proposition

There are at least two basic ways in which the Li–Duan Proposition might be used to facilitate graphical regression, both of which will be employed in subsequent chapters. The first applies during the interpretation of a 3D plot $\{v, h\}$ to identify a coordinate subspace, $S_{v|h} = S(\eta_1)$, where $\{v, h\}$ may correspond to any of the 3D plots introduced in Sections 7.3 and 7.4. The required characterization of $\{v, h\}$ involves first inferring about its structural dimension, 0, 1, or 2, and then estimating $S_{v|h}$ if 1 is selected for the dimension. Let $\hat{b}_{v|h}$ denote an estimate constructed according to (8.1) for the regression of v on h and let $\beta_{v|h}$ denote the corresponding population value as defined in (8.2). If the structural dimension is 1 and if the conditions of the Li–Duan Proposition hold, then $S(\hat{b}_{v|h})$ is a consistent estimate of $S_{v|h}(\beta_{v|h})$. This suggests that the 2D view $\{v, \hat{b}_{v|h}^T h\}$ may be a useful place to start a visual analysis of $\{v, h\}$. The structural dimension should still be assessed using the methods of Chapter 4, but the analysis will be facilitated if the structural dimension is 1 and the conditions of the Li–Duan Proposition hold to a reasonable approximation. For some plots $\{v, h\}$ a structural dimension of 1 may be appropriate, while a clear visual estimate of $S_{v|h}$ could be difficult because of substantial variation in the plot. In such cases, $S(\hat{b}_{v|h})$ might be used as the estimate of $S_{v|h}$.

A second way of using the Li–Duan Proposition involves linear transforms of the predictors at the beginning of an analysis before passing to coordinate subspaces. If $\dim[S_{y|x}] = 1$ and if the conditions of the Li–Duan Proposition hold, then $S(\hat{b}_{y|x})$ may be a useful estimate of $S_{y|x}$. One way to make use of this possibility is to apply graphical regression ideas to the regression of y on the transformed predictors

$$w = (\hat{b}_{y|x}, \tilde{b}_{y|x})^T x$$

where $\tilde{b}_{y|x}$ is a $p \times (p-1)$ matrix selected to extend the basis to $\mathbb{R}^p$. If $\dim[S_{y|x}] = 1$ and if $S(\hat{b}_{y|x})$ is a reasonable estimate then the dimension of all coordinate subspaces not involving the first coordinate of w should be 0. If any such coordinate subspace is found to have dimension greater than 0 then the dimension of $S_{y|x}$ is larger than 1.

The matrix $\tilde{b}_{y|x}$ can be chosen in a number of ways. One possibility is to choose $\tilde{b}_{y|x}$ so that w consists of $\hat{b}_{y|x}^T x$ plus $p-1$ of the original predictors in x. This has the advantage of preserving the identity of all but one of the original predictors, which may be useful for maintaining a close connection with the subject matter context. Another possibility is to choose $\tilde{b}_{y|x}$ so that the linear combinations comprising w are all uncorrelated in the sample. This second possibility may improve resolution in the 3D plots used in graphical regression, but the connection to the subject matter may be elusive. Of course, the vectors of points on the horizontal axes of any 3D plot can always be orthogonalized without loss of information on the corresponding coordinate subspace. These ideas are revisited in the case study of Section 9.1.2.

Being based on linear kernels, the Li–Duan Proposition may not provide much help in practice when there is little linear trend in the regression function. For example, suppose that $\dim[\mathcal{S}_{y|x}(\eta)] = 1$, and that $y \mid x = (\eta^T x)^2 + \varepsilon$ where x is normal with mean 0 and $\text{Var}(x) = I$, and ε is an independent error. In that case, $\beta_{y|x} = 0$, which does not represent useful information on $\mathcal{S}_{y|x}$. Quadratic kernels may be useful in the presence of a weak linear trend in the regression function, as described in the next section.

8.2. QUADRATIC KERNELS

The linear kernel $a + b^T x$ discussed in the last section is just one of several common kernels that are used in practice to construct parsimonious descriptions of data. Results based on quadratic kernels are discussed in this section. Assuming that $\Sigma = \text{Var}(x)$ is positive definite, we can work in terms of the standardized predictor

$$z = \Sigma^{-1/2}(x - E(x))$$

with central subspace $\mathcal{S}_{y|z}(\gamma)$. For use in practice, the corresponding sample version $\hat{z}$ is obtained by replacing Σ and $E(x)$ with their usual moment estimates, $\hat{\Sigma}$ and $\bar{x}$. There is no loss of generality when working on the z-scale because any basis on the z-scale can be backtransformed to a basis on the x-scale, as indicated in Proposition 6.3.

Let C_k denote a real $p \times p$ symmetric matrix of rank not greater than $k \le p$, and consider summarizing the data with a fit of the quadratic kernel $a + b^T \hat{z} + \hat{z}^T C_k \hat{z}$. As in the previous section, estimates are assumed to be based on a convex objective function L,

$$(\hat{a}, \hat{b}, \hat{C}_k) = \arg \min_{a,b,C_k} L_n(a,b,C_k) \tag{8.3}$$

where

$$L_n(a,b,C_k) = \frac{1}{n} \sum_{i=1}^{n} L(a + b^T \hat{z}_i + \hat{z}_i^T C_k \hat{z}_i, y_i)$$

and $L(\cdot,\cdot)$ is again convex in its first argument. The minimization in (8.3) and the definition of L_n are over $a \in \mathbb{R}^1$, $b \in \mathbb{R}^p$ and C_k in the set of real symmetric matrices of rank not greater than k. Let

$$R(a,b,C_k) = E[L(a + b^T z + z^T C_k z, y)].$$

The corresponding population values, which are again assumed to be unique, can be described as in the last section,

$$(\tilde{a}, \beta, \Gamma_k) = \arg \min_{a,b,C_k} R(a,b,C_k). \tag{8.4}$$

The following proposition shows how the population values relate to $S_{y|z}$ when the predictors are normally distributed. Li (1992) gave a version of this proposition for OLS estimation and $k = p$. The version given here is based on Cook (1992b).

Proposition 8.2. Let $S_{y|z}(\gamma)$ denote the central subspace with basis γ, and assume that z follows a multivariate normal distribution with mean $E(z) = 0$ and covariance matrix I. Assume further that L is convex in its first argument and that β and Γ_k, as defined in (8.4), are unique. Then

$$\beta \in S_{y|z}(\gamma) \tag{8.5}$$

and

$$S(\Gamma_k) \subset S_{y|z}(\gamma). \tag{8.6}$$

Justification. Using the facts that L is convex in its first argument and that $y \perp\!\!\!\perp z \mid \gamma^T z$,

$$R(a,b,C_k) = E_{y,\gamma^T z} E_{z|\gamma^T z} [L(a + b^T z + z^T C_k z, y)]$$

$$\geq E_{y,\gamma^T z} [L(a + b^T E(z \mid \gamma^T z)$$

$$+ \operatorname{tr}[E(zz^T \mid \gamma^T z) C_k], y)] \tag{8.7}$$

where tr indicates the trace operator. Since z is assumed to be normal, it follows from Section 7.3.2 that $E(z \mid \gamma^T z) = P_\gamma z$, $\operatorname{Var}(z \mid \gamma^T z) = Q_\gamma$, and

$$E(zz^T \mid \gamma^T z) = E(z \mid \gamma^T z)E(z^T \mid \gamma^T z) + \operatorname{Var}(z \mid \gamma^T z)$$

$$= P_\gamma zz^T P_\gamma + Q_\gamma \tag{8.8}$$

so that

$$\text{tr}[E(zz^T \mid \gamma^T z)C_k] = z^T P_\gamma C_k P_\gamma z + \text{tr}[Q_\gamma C_k].$$

Substituting into (8.7) yields

$$R(a,b,C_k) \geq R(a', P_\gamma b, P_\gamma C_k P_\gamma)$$

where

$$a' = a + \text{tr}[Q_\gamma C_k].$$

The conclusion now follows because the β and Γ_k are unique. $\quad\square$

The normality assumption imposed in Proposition 8.2 was used in (8.8). If z were assumed to be elliptically contoured rather than normal, it would be necessary to replace (8.8) with

$$E(zz^T \mid \gamma^T z) = P_\gamma zz^T P_\gamma + \nu(\gamma^T z)Q_\gamma$$

and the justification would fail because of the scale factor ν as defined in (7.10) of Section 7.3.2. Nevertheless, if $\nu(\gamma^T z)$ does not deviate too far from 1, the results of Proposition 8.2 may still be a useful approximation. Almost sure convergence can be established as for the Li–Duan Proposition.

From (8.5) we see that β must be in the central subspace, just as in the Li–Duan Proposition. Further, from (8.6), any eigenvector of Γ_k that corresponds to a nonzero eigenvalue must be in $\mathcal{S}_{y|z}$. In practice we may substitute sample estimates for the population values and proceed as generally described in Li (1992). However, in this version the sample estimates are not required to be based on OLS and the matrix C of coefficients of the quadratic terms may be constrained to be less than full rank.

Setting $k = 1$, so that the maximum rank of C_k is 1, may be useful when the dimension of $\mathcal{S}_{y|z}$ is at most 2. One way to restrict $k = 1$ is by setting $C_1 = uu^T$ so that the quadratic kernel becomes $a + b^T z \pm (u^T z)^2$, where $u \in \mathbb{R}^p$. Diagnostics based on quadratic kernels of this form have been studied by St. Laurent (1987).

Another way to restrict $k = 1$ is by tying C_1 to b: $C_1 = cbb^T$ where c is a scalar. This yields the quadratic kernel

$$a + b^T z + c(b^T z)^2. \tag{8.9}$$

Kernels of this form, which are often associated with Tukey's (1949) test of additivity (St. Laurent 1990), may be useful when the dimension of $\mathcal{S}_{y|z}$ is 1 and there is not sufficient linear trend in the regression function for useful application of the Li–Duan Proposition. Illustrations are given in Chapter 9 and further comment is available in Problem 8.3.

8.3. THE PREDICTOR DISTRIBUTION

The results of the previous sections for linear and quadric kernels require constraints on the distribution of the predictors: For quadratic kernels the predictors must be normally distributed, while for linear kernels only the conditional expectation $E(x \mid \eta^T x)$ needs to be linear in $\eta^T x$, where η is any basis for $\mathcal{S}_{y|x}$. Experience indicates that minor deviations from these assumptions need not be worrisome, and Hall and Li (1993) argue that $E(x \mid \eta^T x)$ should be reasonably linear when p is large. Nevertheless, major violations can produce misleading results.

Suppose, for example, that $p = 2$ with $x = (x_1, x_2)^T$ and that the distribution of $y \mid x$ is described by

$$y \mid x = 4(1 + x_1)^2 + \varepsilon$$

where ε is an independent standard normal random variable. The central subspace of this regression is spanned by $\eta = (1,0)^T$. Suppose further that the distribution of x is such that

$$x_2 \mid x_1 = cx_1^2 + \delta$$

where x_1 and δ are independent standard normal random variables and c is a constant.

The population coefficient vector of x from the OLS regression of y on x, including a constant term, is given by

$$\beta = \Sigma^{-1} \text{Cov}(x, y)$$

where $\Sigma = \text{Var}(x)$ with $\text{Cov}(x_1, x_2) = 0$, $\text{Var}(x_2) = 2c^2 + 1$, and $\text{Var}(x_1) = 1$. Using well-known properties of the moments of a standard normal random variable, it follows that

$$\beta = \Sigma^{-1} E[(x - E(x))E(y \mid x)]$$

$$= 8 \begin{pmatrix} 1 \\ c \\ 2c^2 + 1 \end{pmatrix}. \tag{8.10}$$

If $c = 0$ then, as predicted by the Li–Duan Proposition, $\mathcal{S}(\beta) = \mathcal{S}_{y|x}$. Otherwise, $\mathcal{S}(\beta) \neq \mathcal{S}_{y|x}$, the degree of bias depending on the value of c. The maximum possible angle, about 19.5 degrees, between $\mathcal{S}(\beta)$ and $\mathcal{S}_{y|x}$ is attained when $c = 1/\sqrt{2}$.

In this example, the difference between $\mathcal{S}(\beta)$ and $\mathcal{S}_{y|x}$ is constrained, and this type of phenomenon may partially account for the empirical evidence that minor deviations from the assumptions of the Li–Duan Proposition do not matter much. Nevertheless, a plot of y over $\mathcal{S}(\beta)$ can leave a very different impression than a plot of y over $\mathcal{S}_{y|x}$. For example, the two plots in Figure 8.1

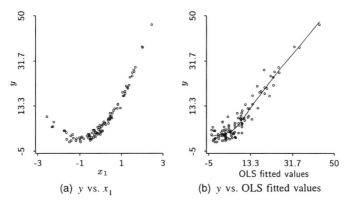

FIGURE 8.1 Illustrations of the possible consequences when $E(x \mid \eta^T x)$ is a nonlinear function. A LOWESS smooth was added to Figure b as a visual enhancement.

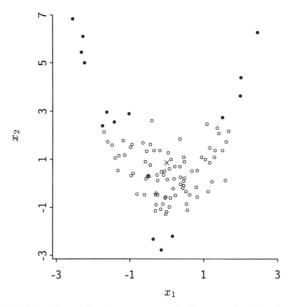

FIGURE 8.2 Scatterplot of x_1 versus x_2 for the example of Section 8.3.

were constructed according to the setup of this example with $c = 1$ and $n = 100$ observations. Figure 8.1a contains the sufficient summary plot. It shows clear curvature and relatively small variation. The OLS summary view shown in Figure 8.1b leaves quite a different impression. The dominant trend shows less curvature and the variation about it is noticeably larger than the variation about the nonlinear trend in Figure 8.1a. A scatterplot of the predictors is shown in Figure 8.2. The ex in the center of the plot corresponds to the sample means; and the filled points will be discussed shortly.

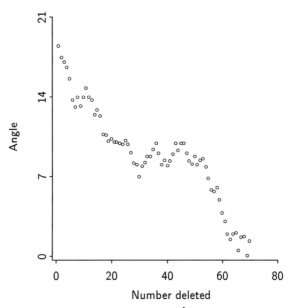

FIGURE 8.3 Plot of the angle between $S(\hat{b}_j)$ and $S_{y|x}$ for $j = 0,\ldots,70$.

In view of the difference between Figures 8.1a and 8.1b, it may be useful to consider methods for modifying the distribution of the predictors to meet the requirements of Propositions 8.1 and 8.2. One reason that the OLS summary plot in Figure 8.1b misses essential features of the sufficient summary in Figure 8.1a is because the key conditional expectation is nonlinear unless $c = 0$:

$$E(x \mid \eta^T x) = E(x \mid x_1) = \begin{pmatrix} x_1 \\ cx_1^2 \end{pmatrix}.$$

The points on the arms of the quadratic trend in Figure 8.2 surely contribute to the nonlinearity. Perhaps deleting these points will reduce the nonlinearity and produce an OLS estimate that is closer to $S_{y|x}$.

To illustrate the potential consequences of deleting remote points, the points in Figure 8.2 were ranked according to their distance

$$r = [(x - \bar{x})^T \hat{\Sigma}^{-1}(x - \bar{x})]^{1/2} \tag{8.11}$$

from the sample mean $\bar{x}$, which is marked with an ex on the plot. Let $\hat{b}_j$ denote the OLS coefficient vector of x obtained after deleting the points with the j largest values of r. Shown in Figure 8.3 is a plot of the angle between $S(\hat{b}_j)$ and $S_{y|x}$ versus $j = 0,\ldots,70$.

The largest angle is 18.5 degrees, which is near the maximum population angle of 19.5 degrees mentioned previously. The first 15 observations deleted correspond to the filled points in Figure 8.2. Sequentially deleting these 15 observations results in a systematic decrease in the angle until a plateau is reached. Although deleting the relatively remote points helped some, the remaining data still exhibit a nonlinear relationship, which probably accounts for the plateau between roughly the 15th and 50th deletion. After deleting about half the data, the nonlinear relationship begins to weaken and the angle decreases further.

There are two important indications from the results of this example. First, deleting a portion of the data, say 10–20 percent, that is relatively remote can result in improved estimates of the central subspace. The basic method used here for identifying remote points in multivariate data is only one of several possibilities (See, for example, Barnett and Lewis 1994, Beckman and Cook 1983), and may not be the best choice. However, in view of the requirements of Propositions 8.1 and 8.2, methods for multivariate normal samples may be appropriate. Cook and Nachtsheim (1994) suggest using the minimum volume ellipsoid for trimming.

Removing remote observations should be viewed as an attempt to satisfy the conditions of Propositions 8.1 and 8.2, and does not imply that the deleted data are "bad" or unrepresentative of the regression under study. In particular, all of the data would normally be displayed in summary views, even when the corresponding estimate of $S_{y|x}$ is based on reduced data. However, it is possible that deleting a large portion of the data could result in a change in the central subspace: Letting $\tilde{x}$ denote a trimmed version of x, it is possible that $S_{y|\tilde{x}} \neq S_{y|x}$.

Second, while deleting remote observations can help, notable curvature among the predictors may still remain. Continued deletion of up to 70 percent of the data, as used for illustration in Figure 8.3, would not normally be a sound procedure in practice. However, reweighting the data may offer additional gains.

8.4. REWEIGHTING FOR ELLIPTICAL CONTOURS

The reweighting method sketched in this section was proposed by Cook and Nachtsheim (1994). Additional details and discussion can be found in their article.

Let $\chi_n = \{x_1, \ldots, x_n\}$ denote an observed sample of n predictor vectors and let W_n denote the set of all probability measures on χ_n. Next, let G_{w_n} denote the distribution function for the probability measure $w_n \in W_n$. The usual empirical distribution function that places equal mass on the points of χ_n will be denoted by G_n. Finally, let G_T denote a user-specified target distribution function that satisfies the requirements of Proposition 8.1 or of Proposition 8.2. How to select G_T will be discussed shortly; for now we assume that it is available. The basic problem considered in this section is how to choose $w_n \in W_n$ so that G_{w_n} approaches G_T as n grows.

Once w_n has been determined, the analysis can proceed as indicated previously, except that G_n is replaced with G_{w_n}. Operationally, this just entails performing a weighted fit with weights $w_n(x_i)$. For example, the weighted version of the objective function (8.1) for linear kernels is

$$L_n(a,b) = \frac{1}{n} \sum_{i=1}^{n} w_n(x_i) L(a + b^T x_i, y_i).$$

The weighted version of the objective function (8.3) for quadratic kernels is constructed similarly. The case deletion results of Figure 8.3 stem from a weighted fit with weights $w_n(x_i) = 0$ or 1.

Here is the general idea behind this setup: If G_{w_n} approaches G_T and if G_T meets the requirements of Proposition 8.1 or of Proposition 8.2, then we might expect the weighted objective functions to provide better estimates of vectors in $\mathcal{S}_{y|x}$ than their unweighted counterparts, provided n is sufficiently large.

The next step is to find how to construct useful weights.

8.4.1. Voronoi weights

Let $D(x_i)$ denote the *Dirichlet cell* corresponding to x_i,

$$D(x_i) = \{x \in \mathbb{R}^p \text{ such that } \|x - x_i\| \leq \|x - x_j\|, \ j = 1,\dots,n\}.$$

The Dirichlet cell $D(x_i)$ consists of all points in $\mathbb{R}^p$ for which x_i is the closest of the sample points. Consider the distribution on χ_n given by

$$w_n^v(x_i) = \int_{D(x_i)} dG_T(x), \qquad i = 1\dots,n \tag{8.12}$$

and indicate the corresponding distribution function with G_n^v. We call $w_n^v(x_i)$ the ith *Voronoi weight* because of the relationship to Voronoi tessellation (see, for example, Okabe, Boots, and Sugihara 1992).

We would like G_n^v to approach G_T as n becomes large. This clearly cannot happen in general, but it does happen under the assumption that the support of G_T is contained within the support of G, the true distribution function for x. If it is also assumed that the support of G_T is bounded, then G_n^v converges uniformly to G_T as $n \to \infty$. The convergence of the reweighted empirical distribution function to the target G_T then has the same nature as the convergence of the empirical distribution function to the true G ($G_n \to G$ uniformly).

Thus, with two mild assumptions, the Voronoi weights (8.12) may provide better estimates of vectors in $\mathcal{S}_{y|x}$ when used in the objective functions for the linear and quadratic kernels. Unfortunately, the analytic integration of the target distribution over each Dirichlet cell is usually impossible. Instead, Cook and Nachtsheim (1994) propose estimating the weights by repeated Monte Carlo sampling from G_T: For each observation g from G_T, find the closest sample point, say x_k,

$$\|g - x_k\| \leq \|g - x_i\|, \qquad i = 1,\ldots,n$$

and then update the count for $D(x_k)$ by 1. After a series of m Monte Carlo trials, the estimated weights $\hat{w}_n^v(x_i)$ are the relative counts in the Dirichlet cells,

$$\hat{w}_n^v(x_i) = \frac{m_i}{m}, \qquad i = 1,\ldots,n$$

where m_i is the count for $D(x_i)$.

8.4.2. Target distribution

The choice for a target distribution should be guided by the conditions for Propositions 8.1 and 8.2. The Li–Duan Proposition requires that $E(x \mid \eta^T x)$ be a linear function of the value of $\eta^T x$, where η is a basis for $S_{y|x}$. Because η is unknown, this condition cannot be used directly to force narrow constraints on G_T. However, if we require that G_T be an elliptically contoured distribution as reviewed in Section 7.3.2, then the requirement of the Li–Duan Proposition will be met by G_T. Proposition 8.2 requires the predictors to follow a normal distribution, which falls in the family of elliptically contoured distributions. Thus, it seems reasonable to restrict G_T to be an elliptically contoured distribution with special emphasis on the normal.

As described in Section 7.3.2, any elliptically contoured random variable V can be represented in the form

$$V = RBU + \mathrm{E}(V) \tag{8.13}$$

where U is a random vector that is uniformly distributed on the surface of the unit sphere in $\mathbb{R}^p$, R is a positive scalar-valued random variable that is independent of U, and $\mathrm{Var}(V) \propto BB^T$. In the case of a normal distribution, R^2 has a chi-squared distribution with p degrees of freedom.

To characterize an elliptically contoured distribution, we need to specify $\mathrm{E}(V)$, $\mathrm{Var}(V)$, and the distribution for R. The mean and variance can be taken as the sample mean and variance of the full data, the data after trimming a percentage of remote points, or some robust version thereof. Setting $\mathrm{Var}(V)$ to be a fraction of a robust estimate of $\mathrm{Var}(x)$ seems to work well and may reduce the need to trim remote points at the outset.

The choice for the distribution of R is open, subject only to the support constraints described previously. That is, R must be a positive random variable such that the support of V is contained within the support of G. To obtain a normal target, set R^2 to be a chi-squared random variable as indicated previously. Otherwise, a convenient default choice that seems to work well in practice is the empirical distribution of the sample radii

$$r_i = [(x_i - \bar{x})^T \hat{\Sigma}^{-1}(x_i - \bar{x})]^{1/2}$$

8.4.3. Modifying the predictors

The ideas discussed so far are combined into the following outline of an algorithm for determining Voronoi weights.

1. Trim a portion, say 0 to 20 percent, of the data by using the minimum volume ellipsoid (Cook and Nachtsheim 1994) or an appropriate method for identifying remote multivariate points (Barnett and Lewis 1994). The rest of this outline applies to the trimmed data.

2. Standardize the trimmed data to have sample mean 0 and covariance matrix I_p. This step is not really required, but working with standardized data may facilitate thinking about the target distribution. A robust center may be worthwhile as well.

3. Select a target elliptically contoured distribution on the standardized scale. Setting G_T to be normal with mean 0 and covariance matrix $\sigma_T^2 I$, $0 < \sigma_T^2 \leq 1$, seems to work well in practice when various numerical methods for estimating $\mathcal{S}_{y|x}$ will be applied. The specific choice of σ_T depends on the distribution of the predictors. Empirical evidence indicates that $\sigma_T = 1$ is often enough to induce sufficient normality, even with no trimming, and that values less than 0.75 are rarely needed. Small values of σ_T should normally be avoided because of the associated reduction in the effective number of observations.

4. Estimate the Voronoi weights by Monte Carlo sampling from the target distribution. The characterization given in (8.3) can be used to generate the samples; details are available in Johnson (1987). The number of Monte Carlo samples depends on n and on the desired precision of the estimated weights. The variances of the estimated weights can be controlled by using standard reasoning (See Problem 8.6).

5. Use the Voronoi weights in combination with either the Li–Duan Proposition or Proposition 8.2 to estimate vectors in $\mathcal{S}_{y|x}$.

For an illustration of the potential gains from the use of Voronoi weights, we return to the example of Section 8.3, and in particular to our sample of $n = 100$ observations from the model with $c = 1$. For Step 1 no observations were trimmed, so the full data were used in all subsequent steps. Next, the data were standardized to have mean 0 and covariance matrix I, as indicated in Step 2. Notice that $\mathcal{S}_{y|z} = \mathcal{S}_{y|x} = \mathcal{S}((1,0)^T)$. The target distribution on the standardized scale was taken to be normal with mean 0 and covariance matrix

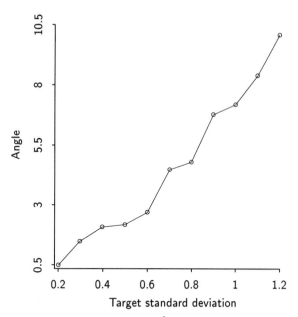

FIGURE 8.4   Scatterplot of the angle between $\mathcal{S}(\hat{b}_{\sigma_T})$ and $\mathcal{S}_{y|x}$ versus the target standard deviation σ_T.

$\sigma_T^2 I$ with values of σ_T ranging from 0.2 to 1.2 in increments of 0.1 (Step 3). For each value of σ_T, 5000 Monte Carlo replications were used to determine the Voronoi weights (Step 4). Finally, weighted least squares coefficient vectors $\hat{b}_{\sigma_T}$ were used as estimates of $\mathcal{S}_{y|x} = \mathcal{S}((1,0)^r)$.

Figure 8.4 shows a plot of the angle between $\mathcal{S}(\hat{b}_{\sigma_T})$ and $\mathcal{S}_{y|x}$ versus σ_T. The angle for $\sigma_T = 1$, about 7.2 degrees, is substantially smaller than the angle of 18.5 degrees for the OLS estimate. As σ_T decreases, the plot shows a rather dramatic drop in the angle, ending around 0.5 degrees for $\sigma_T = 0.2$. Of course, this type of behavior cannot be guaranteed in all applications, but the results do suggest that gains are possible from the use of Voronoi weights.

Plots such as that in Figure 8.4 cannot be drawn in practice because $\mathcal{S}_{y|x}$ is unknown. However, a series of summary plots $\{y, \hat{b}_{\sigma_T}^r x\}$ for various values of σ_T may help gauge the consequences of weighting. For the present example, these plots show a clear transition between the OLS summary plot in Figure 8.1b and the sufficient summary view in Figure 8.1a.

While the use of Voronoi weights can yield improved results, not all empirical distributions G_n will allow for the inducement of sufficiently elliptical contours to produce practically useful gains. But the use of Voronoi weights does seem to extend significantly the applicability of the methods based on Propositions 8.1 and 8.2. Similarly, Voronoi weighting can be used to extend methodology to be introduced in subsequent chapters.

For further discussion of the role of the predictor distribution and nonlinear confounding, see Li (1997).

PROBLEMS

8.1. As mentioned in the text, estimates corresponding to a natural exponential family can be found by setting

$$L(a + b^T x, y) = -y(a + b^T x) + \phi(a + b^T x).$$

When stemming from a natural exponential family, must the function ϕ be convex? strictly convex?

8.2. For the isomerization data (Section 8.1.1), verify that the angle between $\hat{b}$ and its projection onto $\hat{S}_{y|x}$ is about 0.046 degrees.

8.3. The quadratic kernel in (8.9) is not quite in the form of Proposition 8.2. Under the conditions of Proposition 8.2, show that $\beta \in S_{y|z}$, where β is the population value corresponding to b in (8.9).

8.4. In Proposition 8.2 the assumption that x is normally distributed implies that the central subspace $S_{y|x}$ exists. Why?

8.5. For the example in Section 8.3, verify Equation 8.10 and the corresponding claim that the maximum possible angle between $S(\beta)$ and $S_{y|x}$ is attained when $c = 1/\sqrt{2}$.

8.6. In reference to Section 8.4.3, describe a rationale based on the variances of the estimated Voronoi weights for determining the number of Monte Carlo samples.

CHAPTER 9

Graphical Regression Studies

This chapter is devoted to studies that are intended to illustrate selected practical aspects of the results developed in previous chapters, and to motivate additional ideas. The first example involves separate analyses with two different response variables and the same three predictors. Some of the operations in graphical regression are repetitive. Using examples with three predictors allows parsimonious illustration of basic ideas without the advantage of being able to display all the data in a single 3D plot.

9.1. NAPHTHALENE DATA

Franklin, Pinchbeck, and Popper (1956) report the results of a study to investigate the effects of three process variables (predictors) in the vapor phase oxidation of naphthalene. There are 80 observations in the data set which was taken from Box and Draper (1987, p. 392). The two response variables to be used in this section are the percentage mole conversion of naphthalene to naphthoquinone Y_N, and the percentage mole conversion of naphthalene to phthalic anhydride Y_P. The three predictors are

z_1 = air to naphthalene ratio (L/pg)
z_2 = contact time (seconds)
z_3 = bed temperature (deg C)

Graphical regression studies are most informative in practice when there are no strong nonlinearities in any predictor regression function of the form $E(x \mid \alpha^T x)$, as indicated in various propositions of Chapters 7 and 8. This implies that the predictor data should behave as if they were sampled from an elliptically contoured distribution that is preferably normal, or approximately so. The scatterplot matrix of the process variables $z = (z_1, z_2, z_3)^T$ given in Figure 9.1 shows enough skewness and curvature to raise questions about applicability of graphical regression methods. To compensate, simultaneous power transformations $\lambda = (\lambda_1, \lambda_2, \lambda_3)^T$ of the predictors were selected in an

159

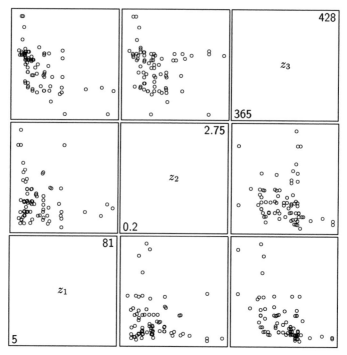

FIGURE 9.1 Scatterplot matrix of the three process variables from the naphthalene data.

effort to induce approximate multivariate normality on the transformed scale:

$$z^{(\lambda)} = (z_1^{(\lambda_1)}, z_2^{(\lambda_2)}, z_3^{(\lambda_3)})^T.$$

The methodology used to estimate λ is like that used in regression to estimate a Box–Cox power transformation of the response, except in this case the likelihood stems from the trivariate normal distribution. For further background on multivariate normalizing transformations, see Velilla (1993, 1995).

After dealing with an observation that notably influenced the estimate of λ_3, reasonable transformations to trivariate normality seem to be the negative cube root for z_1, the logarithm for z_2 and $\lambda_3 = 1$. Interestingly, Franklin et al. also transformed the predictors for their analysis, using the logarithms of z_1 and $10 \times z_2$ and just a linear transformation for z_3. They did not state a clear rationale for taking logarithms, although their reported finding that the yield surface is stabilized by these transformations may have contributed to the decision. The only difference between the transformations used by Franklin et al. and those suggested here is their use of the logarithm for z_1 rather than the negative cube root. Because these two transformations seem to produce essentially the same results for present purposes, we will use the logarithm of z_1 as well to maintain a connection with the previous analysis.

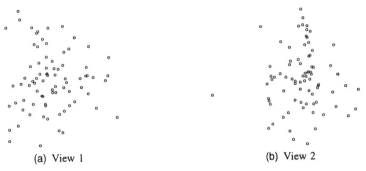

(a) View 1   (b) View 2

FIGURE 9.2 Two views of a 3D plot of the transformed predictors for the naphthalene data. Variables have been orthogonalized to remove all linear trends and axes have been removed for visual clarity.

Specifically, the predictors we will use are the same as those by Franklin et al.: $AN = \log_{10}(z_1)$, $Ctime = \log_{10}(10z_2)$, and $Btemp = 0.01(z_3 - 330)$. Two views of a 3D plot of the transformed predictors are shown in Figure 9.2. Except for a few remote points, there do not seem to be notable nonlinear trends among the transformed predictors. The graphical analyses to be illustrated here could be repeated without the outlying points to see if they cause distortions in various projections (See Problem 9.1).

We next turn to analysis with Y_N as the response.

9.1.1. Naphthoquinone, Y_N

Franklin et al. (1956) used a full second-order response surface model as the foundation of their analysis. Box and Draper (1987, p. 392) suggested an analysis based on the same model. As a base line, Table 9.1 gives the OLS coefficient estimates and t-values from a fit of the full second-order response surface model in

$$x = (AN, Ctime, Btemp)^T.$$

One conclusion from the table is that all coefficient estimates are large relative to their standard errors, suggesting that the response surface may be complicated.

Two standard diagnostic plots are shown in Figure 9.3. The plot of Cook's distances (Cook 1977) versus case numbers suggests that there may be three cases exerting relatively high influence on the estimated coefficients. The plot of residuals versus fitted values suggests that the second-order response surface model may fail to capture all of the systematic features in the data. In addition, the score test for heteroscedasticity (Cook and Weisberg 1983) based on the variance function $\text{Var}(y \mid x) \propto \exp(\alpha^T x)$ yields a chi-squared statistic for $\alpha = 0$ of 19.85 on three degrees of freedom. Evidently, an analysis of these

TABLE 9.1. Fitted Second-Order Quadratic Model with Response Y_N

Term	Coefficient	t-value
Constant	263.93	8.4
AN	−137.04	−7.1
Ctime	−155.39	−8.2
Btemp	−290.75	−6.6
AN^2	19.44	5.8
$Ctime^2$	24.02	7.4
$Btemp^2$	77.46	4.9
$AN \times Ctime$	36.28	6.7
$AN \times Btemp$	77.06	5.8
$Ctime \times Btemp$	89.39	6.9
R^2	0.83	
$\hat{\sigma}$	1.21	

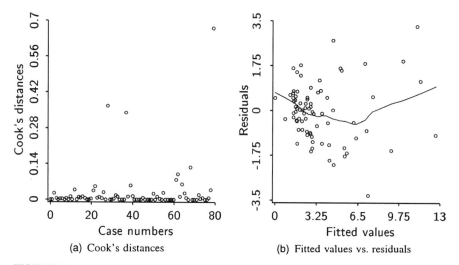

(a) Cook's distances   (b) Fitted values vs. residuals

FIGURE 9.3 Diagnostic plots from a full quadratic fit of the naphthalene data with response Y_N: (a) Cook's distances versus case numbers. (b) Residuals versus fitted values with a LOWESS smooth.

data starting with a second-order response surface model may be fairly complicated, probably requiring fits of several different models on several subsets of the data.

Using 2D added-variable plots

An inspection of a 2D added-variable plot for each predictor is often a useful way to start a graphical regression analysis. Let $\mathcal{S}_{y|x}$ denote the central subspace for the regression of $y = Y_N$ on x with basis η. Partition $\eta = (\eta_k)$ on the individual predictors so that η_k has dimension $1 \times d$, where $d = \dim[S_{y|x}]$ and $k = 1, 2, 3$. Then

$$\eta^T x = \eta_1^T AN + \eta_2^T Ctime + \eta_3^T Btemp.$$

To conform to the general notation in Chapter 7, let x_1 denote a single predictor and let x_2 denote the 2×1 vector of the remaining predictors. For initial considerations, we set $x_1 = AN$ and $x_2 = (Ctime, Btemp)^T$. According to the discussion of Section 7.4,

$$S_{e_{y|2}|e_{1|2}} \subset S(\eta_1)$$

with normally distributed predictors, where $e_{h|v}$ generally represents the population OLS residuals for the linear regression of h on v, as defined in (7.16) for y and x_2. In addition, if we invoke the marginal consistency assumption of Section 7.5.3 then $S_{e_{y|2}|e_{1|2}} = S(\eta_1)$. Thus, by inspecting a sample version of the 2D added-variable $\{e_{y|2}, e_{1|2}\}$ we may be able to estimate $S(\eta_1)$ visually. Because η_1 is a $1 \times d$ vector, the only possible choices for $S(\eta_1)$ are $S(0)$ and $\mathbb{R}^1$.

If the sample version of the 2D added-variable plot $\{e_{y|2}, e_{1|2}\}$ does not show information to contradict the possibility that $e_{y|2} \perp\!\!\!\perp e_{1|2}$ then $S(\eta_1) = S(0)$ may be appropriate, suggesting that the distribution of $Y_N \mid (AN, Ctime, Btemp)$ does not depend on the value of AN. If that were the conclusion then there would be some support for deleting AN from the analysis, but such decisions should perhaps be delayed until near the end of an analysis because variability increases as the dimension of a projection decreases (See Problem 7.1).

If the sample 2D added-variable plot does provide information to contradict the possibility that $e_{y|2} \perp\!\!\!\perp e_{1|2}$ then $S(\eta_1) = \mathbb{R}^1$ is indicated, corresponding to the conclusions that the distribution of $Y_N \mid x$ does in fact depend on the value of AN and that the structural dimension of the regression is at least 1. The added-variable plot for AN shown in Figure 9.4a indicates that $e_{y|2}$ depends on $e_{1|2}$ and thus that $S(\eta_1) = \mathbb{R}^1$. In other words, the regression of Y_N on $(AN, Ctime, Btemp)$ does depend on AN. The ideas as developed to this point provide no tools for inferring about the form of that dependence, however.

The same procedure can be applied after redefining the predictor partitions as $x_1 = Ctime$, $x_2 = (AN, Btemp)^T$ and $x_1 = Btemp$, $x_2 = (AN, Ctime)^T$. The 2D added-variable plots in Figures 9.4b and 9.4c lead us to the conclusion that the distribution of $Y_N \mid x$ depends on each of the three predictors.

In brief, once approximate normality for the predictors has been established, a collection of 2D added-variable plots allows us to investigate whether the distribution of $y \mid x$ depends on each of the predictors. This is an important first step of the analysis; nevertheless, it permits dimension reduction only through deletion of individual predictors. Approximate normality is a sufficient but not a necessary condition for progress in this context. It does seem important, however, to avoid strong nonlinearities in the predictor regression functions.

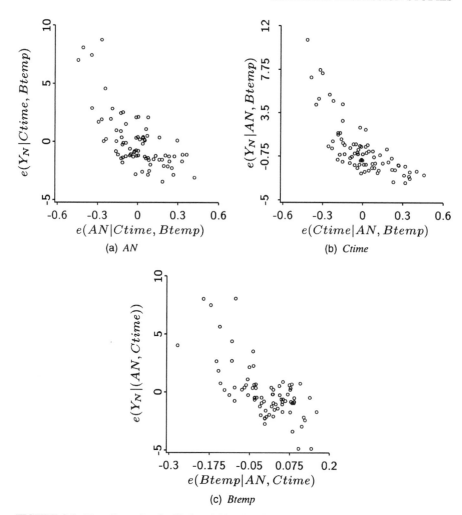

FIGURE 9.4 Two-dimensional added-variable plots for the predictors in the naphthalene data with response Y_N.

The 3D added-variable plots considered in the next section allow for dimension reduction by taking linear combinations of the predictors.

Using 3D added-variable plots

Partition $x^T = (x_1^T, x_2)$ so that $x_1 = (AN, Ctime)^T$ and $x_2 = Btemp$. Similarly, partition $\eta^T = (\eta_1^T, \eta_2^T)$. Then, as observed in Section 7.1,

$$S_{y|x} \subset S(\eta_1) \oplus S(\eta_2).$$

We concluded by using 2D added-variable plots in the last section that $S(\eta_2) = \mathbb{R}^1$. Our first task in this section is to construct a visual estimate of a basis for

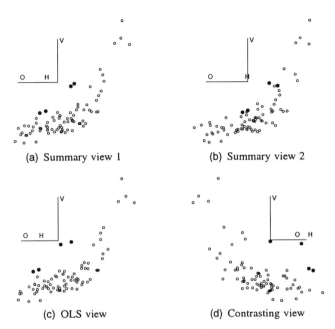

(a) Summary view 1   (b) Summary view 2

(c) OLS view   (d) Contrasting view

FIGURE 9.5   Four views of the 3D added-variable plot, $\hat{e}_{Y_N|Btemp}$ versus $(\hat{e}_{AN|Btemp}, \hat{e}_{Ctime|Btemp})$, for the naphthalene data with response Y_N. Horizontal and out-of-page axes have been orthogonalized in each view.

$S(\eta_1)$. This could be done by appealing to Proposition 7.5 plus the extensions in Section 7.4 to obtain

$$S_{e_{y|2}|e_{1|2}} \subset S(\eta_1).$$

If we adopt the marginal consistency assumption as well, then

$$S_{e_{y|2}|e_{1|2}} = S(\eta_1)$$

as in the last section. This implies that a sample version of the 3D added-variable plot $\{e_{y|2}, e_{1|2}\}$ might be used to estimate $S(\eta_1)$. Other plots could be used, as discussed in Chapter 7, but we continue with added-variable plots since they are often a useful default choice in practice.

The 3D added-variable plot $\{e_{y|2}, e_{1|2}\}$ has $e_{y|2}$, the residuals from the OLS linear regression of $y = Y_N$ on $x_2 = Btemp$, on the vertical axis. The horizontal and out-of-page axes hold the coordinates of the 2×1 vector $e_{1|2}$: the residuals from the OLS linear regression of AN on $Btemp$ and the residuals from the OLS linear regression of $Ctime$ on $Btemp$.

Figure 9.5 shows four views of a sample version of the 3D added-variable plot $\{e_{y|2}, e_{1|2}\}$. Our first task in viewing this plot is to determine its structural dimension for inference about $\dim[S(\eta_1)]$. The structural dimension is clearly at least 1. The choice between 1D and 2D structure might be facilitated by first rotating to a "best" summary view and then using the methods of Chapter 4. For these data the decision is complicated by the four highlighted points shown in each view of Figure 9.5. Figure 9.5a shows a visually determined summary view obtained by mentally neglecting the four points, while Figure 9.5b shows a corresponding view that attempts to accommodate the points. Alternatively, following the Li–Duan Proposition, the OLS linear regression of $e_{y|2}$ on $e_{1|2}$ could be used to determine a summary view. Since we have transformed the predictors to approximate normality, $e_{1|2}$ should be approximately normal, and the Li–Duan Proposition can be applied to the regression of $e_{y|2}$ on $e_{1|2}$. The view in Figure 9.5c, which is quite close to the view in Figure 9.5a, was constructed in that way after deleting the four highlighted points and then restoring them for plotting.

However the summary view is selected and checked, the decision between 1D and 2D structure is complicated by the four highlighted points. The problem is that these points give an impression of cupping in the lower portion of the 3D plot, implying 2D structure. This cupping effect might be visualized from Figure 9.5d. On the other hand, without the four points in question all summary views are about the same and there is little evidence to contradict 1D structure.

To continue, we could decide 2D structure, reasoning that the four points in question provide valuable information. Or we could adopt a parsimonious approach and reason that 1D structure captures most of the variation in the observed values of $e_{y|2}$ as a function of $e_{1|2}$. We adopt the latter view for this illustration, and select Figure 9.5a as our estimate of a sufficient summary of the 3D added-variable plot.

The linear combination of AN and $Ctime$ corresponding to the horizontal screen axis of Figure 9.5a is

$$AC = 0.666AN + 0.746Ctime$$

and thus $\mathcal{S}(\eta_1)$ is estimated to be $\mathcal{S}((0.666, 0.746)^T)$. The associated inference for $\mathcal{S}_{y|x}$ is

$$\mathcal{S}_{y|x} \subset \mathcal{S}((0.666, 0.746)^T) \oplus \mathcal{S}(\eta_2).$$

Hence, $\{Y_N, (AC, Btemp)\}$ constitutes an estimated sufficient summary for the regression of Y_N on x.

We now have a new regression problem with two predictors, AC and $Btemp$, and response $y = Y_N$ that can be viewed in a 3D plot as shown in Figure 9.6. If the 3D plot $\{Y_N, (AC, Btemp)\}$ were judged to have 2D structure then the plot itself could be taken as the estimated minimal sufficient summary of the regression. If the plot were judged to have 1D structure then the corresponding 2D plot could be taken as the estimated minimal sufficient summary.

The 3D plot $\{Y_N, (AC, Btemp)\}$ was judged to have 1D structure with summary view shown in Figure 9.6a. A contrasting view is shown in Figure 9.6b. The linear combination of the predictors on the horizontal screen axis of Fig-

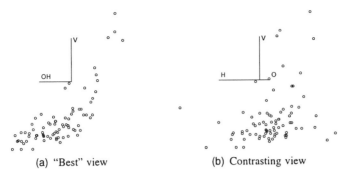

(a) "Best" view   (b) Contrasting view

FIGURE 9.6 Two views of the 3D plot $\{V,(H,O)\} = \{Y_N,(AC,Btemp)\}$ for the naphthalene data. Horizontal and out-of-page axes have been orthogonalized for visual clarity: (a) best view indicating 1D structure, (b) a view selected for contrast.

ure 9.6a is

$$ACB = 0.397AN + 0.445Ctime + 0.802Btemp$$

$$= 0.596AC + 0.802Btemp \tag{9.1}$$

and thus $S_{y|x}$ is estimated to be $S[(0.397, 0.445, 0.802)^T]$.

Using graphical methods we have been able to reduce the original regression problem to one with response Y_N and predictor ACB. The analysis could now continue using the view in Figure 9.6a for guidance, because it is estimated to contain all, or nearly all, of the regression information.

Interpreting coefficients
The values of the coefficients in the linear combination ACB shown in (9.1) depend in part on the units of the predictors, just as they do in linear regression. Similarly, coefficients can be easier to interpret if the predictors are all scaled to have unit sample standard deviation. In terms of the scaled predictors $Z = (z_j) = (AN_s, Ctime_s, temp_s)^T$ indicated by the subscript s, ACB becomes

$$ACB_s = 0.45AN_s + 0.70Ctime_s + 0.56Btemp_s.$$

Next, in view of the summary plot shown in Figure 9.6, it may be reasonable to consider a model of the form

$$Y_N = g(\eta^T Z) + \varepsilon$$

where we think of ACB_s as a graphical estimate of $\eta^T Z$. From this it follows that

$$\frac{\partial g(\eta^T Z)/\partial z_j}{\partial g(\eta^T Z)/\partial z_k} = \frac{\eta_j}{\eta_k}.$$

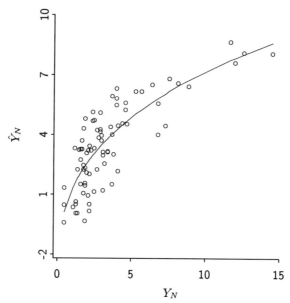

FIGURE 9.7   Inverse response plot $\{\hat{Y}_N, Y_N\}$ for the naphthalene data. The curve corresponds to the cube root transformation of Y_N.

Thus, ratios of coefficient estimates for the scaled predictors can be viewed as relative rates of change in the regression function, just as they are in linear regression.

Using the Li–Duan Proposition

We used the Li–Duan Proposition in the last section as one method for constructing a summary view of a 3D added-variable plot. But under the assumption of 1D structure, the Li–Duan Proposition could also be used at the outset to obtain the estimated sufficient summary plot $\{Y_N, \hat{Y}_N\}$, where $\hat{Y}_N$ denotes an OLS fitted value from the linear regression of Y_N on the three predictors. In this case the absolute value of the correlation between $\hat{Y}_N$ and the linear combination given in (9.1) is 0.995. The summary view shown in Figure 9.6a is nearly visually indistinguishable from the plot $\{Y_N, \hat{Y}_N\}$.

Transforming the response

The regression of Y_N on the predictor ACB from the graphical analysis, or on the OLS fitted values $\hat{Y}_N$ might be simplified by using a response transformation. Following the discussion of response transformations in Section 2.3, Figure 9.7 shows the inverse response plot $\{\hat{Y}_N, Y_N\}$. The curve superimposed on the plot corresponds to the cube root transformation

$$t(Y_N; 1/3) = b_0 + b_1 Y_N^{(1/3)}$$

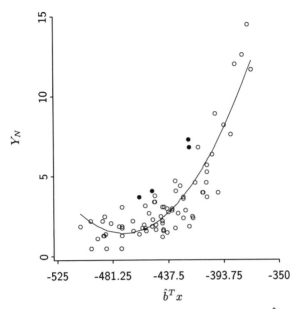

FIGURE 9.8 Plot of Y_N versus the linear combination of the predictors $\hat{b}^T x$ from the OLS fit of the quadratic kernel (9.2). The curve on the plot is the fitted quadratic.

where the constants b_0 and b_1 were determined by OLS. The curve seems to give a reasonable visual fit, so the cube root transformation of Y_N will likely result in a simplified regression function that is linear in the predictors.

In contrast, the analysis of Table 9.1 would seem to preclude the possibility of a simplifying response transformation because quadratic response surfaces are not monotonic.

Using quadratic kernels
In this section we illustrate application of Proposition 8.2 by using first the quadratic kernel given in (8.9),

$$a + b^T x + c(b^T x)^2 \tag{9.2}$$

in combination with OLS. The OLS estimate of b from this kernel is

$$\frac{\hat{b}}{\|\hat{b}\|} = -(0.350, 0.489, 0.800)^T.$$

The plot $\{Y_N, \hat{b}^T x\}$ is shown in Figure 9.8. The curve superimposed on the plot is the fitted quadratic (9.2).

The absolute value of the correlation between $\hat{b}^T x$ and the linear combination ACB from the graphical analysis is 0.984. While this correlation is large there is still notable variation around a linear trend in the plot $\{\hat{b}^T x, ACB\}$, indicating that the two solutions are not quite the same. This difference can be seen by comparing Figures 9.8 and 9.6a. It seems that the two solutions differ in their method of handling the four highlighted points in Figure 9.5. If Figure 9.5b were taken as the summary view of the 3D added variable plot, then the graphical analysis would lead to essentially the same solution as that based on the quadratic kernel (9.2), the correlation between the corresponding linear combinations being 0.9996. Finally, the fitted curve shown in Figure 9.8 suggests that the regression function $E(y \mid x)$ is noticeably different from the quadratic kernel (9.2). This accounts for the behavior of the residual plot in Figure 9.3b. In other words, the class of second-order response surfaces may not be sufficiently rich to describe the 1D regression function in this example.

Using the full second-order quadratic kernel

$$a + b^T x + x^T C_3 x \tag{9.3}$$

provides another method for constructing potentially useful summary plots. Let $\hat{C}_3$ denote the OLS estimate of C_3 and let $\hat{\Sigma}$ denote the usual estimate of $\Sigma = \mathrm{Var}(x)$. From the discussion following Proposition 8.2, the eigenvector ℓ_1 corresponding to the largest eigenvalue λ_1 of $\hat{C}_3 \hat{\Sigma}$ is

$$\ell_1 = (0.387, 0.446, 0.897)^T.$$

The elements of this vector are nearly identical to the coefficients of ACB. A characterization of the 3D plot $\{y, (\ell_1, \ell_2)\}$ is much the same as the 3D added-variable plot represented in Figure 9.5: There is a suggestion of 2D structure that depends heavily on the four points identified in Figure 9.5.

The results from the different methods considered in this example are in quite good agreement. Either Figure 9.6a or Figure 9.8 may serve as a useful parsimonious summary plot to guide further analysis, depending on the method of handling the four "outlying" points. Further, if these points are considered in the analysis, there is a mild suggestion of 2D structure that seems difficult to pin down with the present data.

Voronoi weights

Finally, portions of the previous analysis following the predictor transformations were recomputed using Voronoi weights constructed as described in the illustration of Section 8.4.3. In all cases the results from the weighted analysis were in good agreement with those above. For example, with $\sigma_T = 1$, the angle between the OLS estimate and the weighted least squares estimate was 1.8 degrees, while with $\sigma_T = 0.5$ the angle was 4.2 degrees.

9.1.2. Phthalic anhydride, Y_P

We now change the response variable to $y = Y_P$, phthalic anhydride, while keeping the three predictors discussed in the last section.

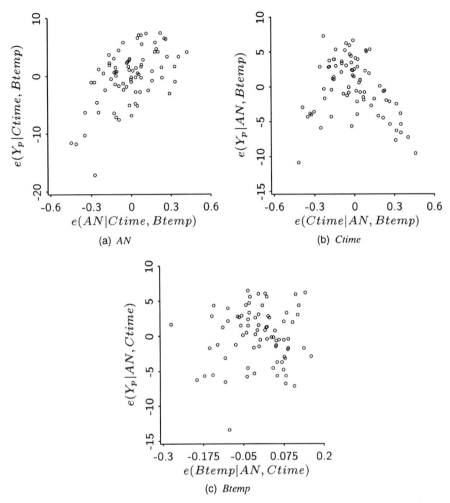

FIGURE 9.9 Two-dimensional added-variable plots for each predictor in the naphthalene data with response Y_p.

Three 2D added-variable plots for the predictors are shown in Figure 9.9. These plots give clear indications that the conditional distribution of Y_p given AN, Ctime, and Btemp depends on the values of AN and Ctime. The interpretation of the added-variable plot for Btemp does not seem as clear, although there may be a weak linear trend. Nevertheless, Btemp will not be removed at this stage. Generally, deletion of individual predictors based on 2D added-variable plots should perhaps be avoided because of the possibility that the variation encountered in low-dimensional projections may mask systematic trends. Deletion of predictors, if desirable, can always be carried out at a later stage using 3D plots.

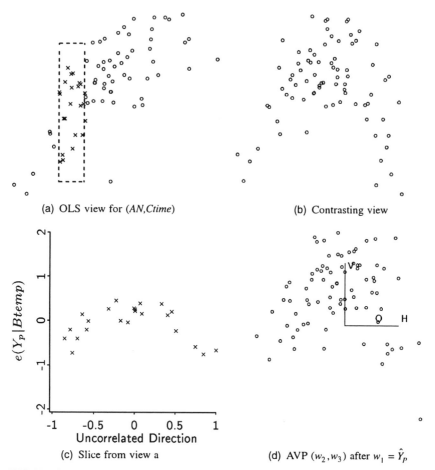

(a) OLS view for *(AN,Ctime)*

(b) Contrasting view

(c) Slice from view a

(d) AVP (w_2, w_3) after $w_1 = \hat{Y}_p$

FIGURE 9.10 Graphical analysis of the naphthalene data with response Y_p. Screen axes have been removed in selected plots for visual clarity.

When using 3D added-variable plots to reduce dimensionality, it is often helpful to start with the two predictors associated with the strongest dependence in the 2D added-variable plots. The effect of a strong predictor can visually dominate that of a weak predictor in a 3D added-variable plot, making structural characterizations relatively difficult. Accordingly, we next attempt dimension reduction by using the 3D added-variable plot for $x_1 = (AN, Ctime)^T$ after $x_2 = Btemp$.

Figure 9.10a shows the OLS view of the sample 3D added-variable plot $\{e_{y|2}, e_{1|2}\}$. An approximately orthogonal view is shown for contrast in Figure 9.10b. According to the Li–Duan Proposition, the OLS view is a useful candidate for a sufficient summary of the 3D plot. If the OLS view were sufficient, or approximately so, the points in any slice of the OLS view of

Figure 9.10a should appear as an independent and identically distributed sample in the associated uncorrelated 2D view (Section 4.6.2). Figure 9.10c gives the uncorrelated 2D view corresponding to the points in the slice shown in Figure 9.10a. Because there is noticeable dependence, the OLS view apparently misses information. A systematic tendency can be seen also in the linked slice points shown in Figure 9.10b. This suggests that the 3D added-variable plot $\{e_{y|1}, e_{2|1}\}$ has 2D structure, a suggestion that is sustained by further analysis of the plot. The corresponding inference is that $S(\eta_1) = \mathbb{R}^2$. Figure 9.10d will be discussed shortly.

The analysis so far does not help much from the point of view of the relationship

$$S_{y|x} \subset S(\eta_1) \oplus S(\eta_2).$$

By keeping *Btemp* in the analysis we have at least temporarily taken $S(\eta_2) = \mathbb{R}^1$. Hence, the conclusion that $S(\eta_1) = \mathbb{R}^2$ tells us only that $\dim[S_{y|x}] \leq 3$. On the other hand, by a purely geometrical argument $S(\eta_1) = \mathbb{R}^2$ implies $\dim[S_{y|x}] \geq 2$. It remains to decide if $\dim[S_{y|x}] = 2$ or 3.

If the projections of $S_{y|x}$ on the coordinate directions for the three predictors are all assessed to be 1D, as judged by using the three 2D added-variable plots, and if the projection of $S_{y|x}$ on the coordinate plane for *AN* and *Ctime* is assessed to be 2D, as judged by using the 3D added variable plot for (*AN*, *Ctime*) after *Btemp*, then necessarily the projections of $S_{y|x}$ on the other two coordinate planes will be 2D as well. In other words, if our judgment of the 2D added variable plot for *Btemp* was correct, it would be useless to analyze the other two 3D added-variable plots because they would present 2D structure as the first one does. Since the 2D added-variable plot for *Btemp* has an ambiguous interpretation, the other two 3D added-variable plots were considered. Unfortunately, the graphical analyses led to the conclusion that both have 2D structure. This reinforces our choice of keeping *Btemp* in the analysis, but leaves us knowing nothing beyond $\dim[S_{y|x}] \geq 2$.

Continuing our geometric line of reasoning, if $\dim[S_{y|x}] = 3$ then each population 3D added-variable plot must have 2D structure, which is consistent with our analysis of the example so far. However, it is also possible to have $\dim[S_{y|x}] = 2$ while each population 3D added-variable plot has 2D structure. The latter possibility is illustrated in the next section.

Transforming predictors linearly to estimate $S_{y|x}$
To fix ideas, consider a regression problem with 3×1 predictor vector $u = (u_j)$ and central subspace spanned by the columns of

$$\tau = \begin{pmatrix} 1 & 1 \\ 0 & 1 \\ -1 & 0 \end{pmatrix}.$$

Each partition of u leads to a τ_1 consisting of two rows of τ with $\dim[\mathcal{S}(\tau_1)] = 2$. Thus $\dim(\mathcal{S}_{y|u}) = 2$ while $\dim[\mathcal{S}(\tau_1)] = 2$ for all possible partitions leading to a 3D added-variable plot. In this simple illustration, the minimal sufficient 3D plot is $\{y, (u_1 - u_3, u_1 + u_2)\}$, but we are unable to estimate it graphically using partitions of the coordinate components of u. In other words, we cannot see that $\mathcal{S}_{y|u}$ is 2D by using projections on coordinate planes $\{u_j, u_k\}$. The situation is consistent with current knowledge about Y_p: It is possible to find that $\dim[\mathcal{S}_{y|u}] \geq 2$, but we would be unable to tell that $\dim[\mathcal{S}_{y|u}] = 2$. We may be able to make progress, however, by using linearly transformed predictors.

Consider a general regression problem with response y and $p \times 1$ predictor vector x. If $E(x \mid \eta^T x)$ is linear in $\eta^T x$ then by the Li–Duan Proposition, $\beta \in \mathcal{S}_{y|x}$, where β is as defined in the lemma. Let $B = (\beta, \tilde{\beta})$, where $\tilde{\beta}$ is any $p \times (p-1)$ matrix that extends β to a basis for $\mathbb{R}^p$, $p = \dim(x)$. Next, define the transformed predictors

$$w = B^T x = \begin{pmatrix} \beta^T x \\ \tilde{\beta}^T x \end{pmatrix}. \tag{9.4}$$

Using Proposition 6.3, a basis $\gamma = (\gamma_1, \gamma_2)$ for $\mathcal{S}_{y|w}$ can be written in the form

$$\gamma = B^{-1}\eta = \left(\begin{pmatrix} 1 \\ 0 \end{pmatrix}, \gamma_2 \right) \tag{9.5}$$

where γ_1 is a $p \times 1$ vector with a 1 in the first position and zeros elsewhere, and γ_2 is a $p \times (d-1)$ matrix. It follows from (9.5) that if $\dim[\mathcal{S}_{y|w}] = \dim[\mathcal{S}_{y|x}] = 2$ then, in the population, the collection of 3D added-variable plots defined on the coordinate components of w will lead us to the minimal sufficient 3D plot for the regression of y on w. The plot can then be backtransformed to the minimal sufficient 3D plot for the regression of y on w. The sample version of the linear transformation (9.4) was discussed in Section 8.1.2.

We apply these ideas to the naphthalene data with response Y_p in the next section.

Back to the naphthalene data

We left the naphthalene analysis with the conclusion that each 3D added-variable plot has 2D structure and hence that $d = \dim[\mathcal{S}_{y|x}] \geq 2$. We now investigate the possibility that $d = 2$ by transforming the predictors according to a sample version of (9.4). Specifically, let $\hat{b}$ be the coefficient vector from the OLS linear regression of Y_p on $x = (AN, Ctime, Btemp)^T$. Extend $\hat{b}$ to a basis $\hat{B} = (\hat{b}, \tilde{b})$ for $\mathbb{R}^3$ so that the columns of $\hat{B}$ are orthogonal relative to the inner product induced by $\hat{\Sigma}$, the sample covariance matrix for the three predictors. Finally, define the transformed predictors $w = (w_1, w_2, w_3)^T$ as in (9.4) with $\hat{B}$ in place of B. Notice that $w_1 = \hat{b}^T x$ corresponds to the fitted values $\hat{Y}_p$ from the OLS regression up to the additive constant for the intercept.

If $d = 2$ then from (9.5) we would expect the 3D added-variable plots for (w_1, w_2) after w_3 and for (w_1, w_3) after w_2 to each have 2D structure. However, the 3D added-variable plot for (w_2, w_3) after w_1 would have 1D structure. The sample version of the latter added-variable plot does indeed appear to have 1D structure; one possible summary view is shown in Figure 9.10d. Because of the strong quadratic tendency, this view might be refined by using the methods associated with Proposition 8.2.

The general conclusion is that $\dim[S_{y|x}] = 2$. An estimated basis for $S_{y|x}$ is given by the two vectors

$$\hat{b} = (0.923, -0.363, 0.131)^T$$

and

$$\hat{b}_1 = (0.386, 0.558, 0.669)^T$$

where $\hat{b}_1^T x$ is the linear combination of the predictors on the horizontal screen axis of the summary view in Figure 9.10d. Finally, $\{Y_P, (\hat{b}^T x, \hat{b}_1^T x)\}$ is the corresponding estimated minimal sufficient summary plot.

9.2. WHEAT PROTEIN

Fearn (1983) described a data set from an experiment to calibrate a near infrared reflectance (NIR) instrument for measuring the protein content of ground wheat samples. The protein content of each sample (y in percent) was measured by the standard Kjeldahl method. The predictors, $L_1, \ldots, L_6$, are measurements on log(1/reflectance) of NIR radiation by the wheat samples at six wavelengths in the range 1680–2310 nm. In all, 50 samples of ground wheat were used. The problem is to study the conditional distribution of y given the six predictors with a view toward prediction. Fearn (1983) used the first 24 samples for calibration and the last 26 samples for prediction. We will use all 50 samples in a graphical analysis.

An initial inspection of a scatterplot matrix of the six predictors and the response (not shown) indicated that there is little marginal relationship between the response and any single predictor. Additionally, the predictors were found to be highly correlated with no nonlinear trends evident. The R^2 value for the OLS linear regression of L_2 on the other predictors is 0.9993, for example. An inspection of various 3D plots of orthogonalized predictors provided support for the conclusion that there are no notable nonlinear trends, so it seems reasonable to proceed as if the predictors are elliptically contoured.

We next turn to the six 2D added-variable plots for each predictor in turn. For notational convenience, let

$$x^{(0)} = (L_1, \ldots, L_6)^T$$

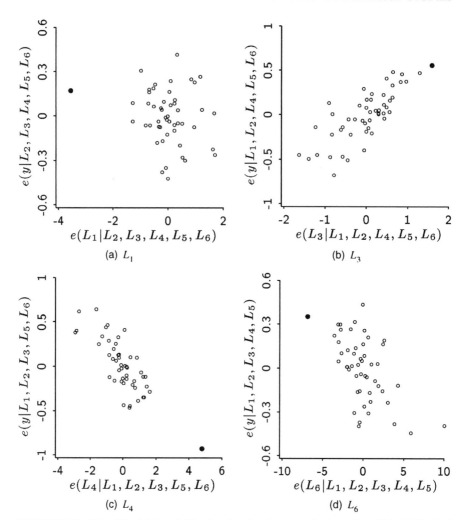

FIGURE 9.11   Four 2D added-variable plots for the wheat protein data. Case 47 is highlighted in each plot.

denote the original six predictors. The plots for L_3, L_4, and L_6 clearly indicate that the conditional distribution of $y \mid x^{(0)}$ depends on the values of these predictors. The added-variable plots for L_1, L_2, and L_5 exhibit weak dependence and do not allow for firm inference on the need for these predictors.

The 2D added-variable plots for L_1, L_3, L_4, and L_6 are shown in Figure 9.11. One observation, case 47, stands apart in the plots for L_1 and L_4, indicating that this case is relatively removed from the empirical distribution of the remaining predictor values. This case was next deleted from the analysis to allow for greater visual resolution in subsequent plots and to avoid the characterization

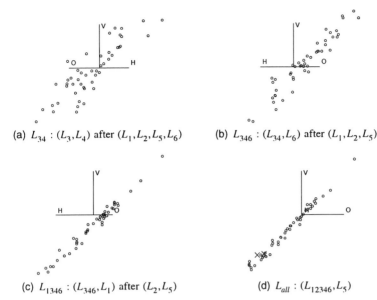

(a) $L_{34} : (L_3, L_4)$ after (L_1, L_2, L_5, L_6) (b) $L_{346} : (L_{34}, L_6)$ after (L_1, L_2, L_5)

(c) $L_{1346} : (L_{346}, L_1)$ after (L_2, L_5) (d) $L_{all} : (L_{12346}, L_5)$

FIGURE 9.12   Four OLS summary views of 3D added-variable plots encountered in the graphical analysis of the wheat protein data. Case 47 was removed from plot a. Cases 24 and 47 were removed from the analyses leading to plots c and d. They are represented as exes in plot d.

difficulties associated with remote points. The interpretation of the 2D added-variable plots is not changed by removing case 47.

We next turn to the 3D added-variable plot for $x_1^{(0)} = (L_3, L_4)^T$ after $x_2^{(0)} = (L_1, L_2, L_5, L_6)^T$ because L_3 and L_4 seem to exhibit the strongest dependence in the 2D added-variable plots. Using the methods of Chapter 4, particularly the uncorrelated 2D views, no clear information was found in the plot to contradict the possibility of 1D structure. The OLS view was taken as the summary view for further analysis.

We now have a new regression problem with response y and five predictors

$$x^{(1)} = (L_1, L_2, L_{34}, L_5, L_6)^T$$

where

$$L_{34} = 0.856L_3 - 0.517L_4$$

is the linear combination of L_3 and L_4 on the horizontal screen axis of the OLS summary view shown in Figure 9.12a.

We now begin again with $x^{(1)}$ as the predictors. Clear dependence is evident in the 2D added-variable plots for L_{34} and L_6, while dependence in the added-variable plots for the remaining predictors is not visually evident. In addition, case 24 is clearly remote in the added-variable plot for L_6. This case also reduces the resolution of the 3D added-variable plot for $x_1^{(1)} = (L_6, L_{34})^T$

after $x_2^{(1)} = (L_1, L_2, L_5)^T$. Because case 24 causes difficulty with visual characterizations, it was removed from the analysis at this point. The deletion of case 24 resulted in only negligible changes in the coefficients of the linear combination L_{34} previously determined.

After the removal of case 24, the 3D added-variable plot for $x_1^{(1)} = (L_6, L_{34})^T$ after $x_2^{(1)} = (L_1, L_2, L_5)^T$ was characterized as having 1D structure. The OLS view shown in Figure 9.12b was taken as the summary. The linear combination of the original predictors on the horizontal screen axis of Figure 9.12b is

$$L_{346} = 0.839L_3 - 0.522L_4 - 0.152L_6.$$

We again have a new regression problem with predictors

$$x^{(2)} = (L_1, L_2, L_{346}, L_5)^T.$$

The general procedure described above was next repeated three more times. Each stage consisted of the following:

- First, using 2D added-variable plots for the predictors of the previous stage to select the two predictors exhibiting the strongest dependence. There were two clear choices, say v_1 and v_2, at each stage.
- Next, using the 3D added-variable plot for (v_1, v_2) after the remaining predictors in an attempt to further reduce the dimension of the predictor vector. As it turned out, the 3D added-variable plot was inferred to have 1D structure, and hence the OLS view was taken as the summary.
- Finally, reducing the number of predictors by replacing (v_1, v_2) with the linear combination determined by the OLS view.

This general procedure produced the following sequence of linearly combined predictors:

$$L_{1346} = -0.092L_1 + 0.836L_3 - 0.519L_4 - 0.152L_6$$
$$L_{12346} = -0.091L_1 - 0.088L_2 + 0.833L_3 - 0.518L_4 - 0.151L_6$$

and finally

$$L_{all} = -0.091L_1 - 0.088L_2 + 0.833L_3 - 0.518L_4 + 0.009L_5 - 0.151L_6.$$

The OLS summary views for L_{1346} and L_{all} are shown in Figures 9.12c and 9.12d. The essential inferences resulting from this procedure are that, aside from dealing with the two deleted points, $\dim[S_{y|x^{(0)}}] = 1$ and the coefficient vector for L_{all} is an estimated basis of $S_{y|x^{(0)}}$.

The two remote points that were deleted during the analysis are plotted as exes in the final summary view of Figure 9.12d. These points seem to conform reasonably with the linear trend of the remaining data. Outlying, influential, and high-leverage cases are often easily recognized during a graphical regression analysis. Deletion followed by restoration at the end seems a useful option, as illustrated here.

The final summary view in Figure 9.12d is equivalent to the plot $\{y, \hat{y}\}$, where $\hat{y}$ denotes a fitted value from the OLS linear regression of y on the six predictors. Whenever graphical regression is based on 3D added-variable plots and each plot is summarized by its OLS view, the coefficient vector of the final combined predictor, here L_{all}, will be proportional to the coefficient vector from the OLS linear regression of y on $x^{(0)}$. The summary view shows a strong linear trend, suggesting that a linear regression function may be appropriate. The plot of residuals versus fitted values $\{\hat{e}, \hat{y}\}$ from the OLS linear regression of y on $x^{(0)}$ supports this possibility.

The graphical regression procedure used in this example can be adapted easily to allow for model checking. This involves simply replacing the response y with the residuals $\hat{e}$ and proceeding as before. If all 2D and 3D added-variable plots have 0D structure, then the model has considerable support. Any plot having more than 0D structure is evidence that the linear model may be deficient.

9.3. REACTION YIELD

Box and Draper (1987, p. 368) report 32 observations on the percentage yield y from a two-stage chemical process characterized by the temperatures ($Temp_1$ and $Temp_2$ in degrees Celsius) and times ($time_1$ and $time_2$ in hours) of reaction at the two stages, and the concentration (Con in percent) of one of the reactants. Prior experimentation using steepest ascent indicated that the maximum yield would likely be found in the region of the factor space covered by the present experiment. Accordingly, Box and Draper based their analysis on a full second-order response surface model in the five factors. The specific predictors used were

$$T_1 = \frac{(Temp_1 - 122.5)}{7.5}$$

$$Lt_1 = \frac{2[\log(time_1) - \log 5]}{\log 2} + 1$$

$$C = \frac{(Con - 70)}{10}$$

$$T_2 = \frac{Temp_2 - 32.5}{7.5}$$

$$Lt_2 = \frac{2[\log(time_2) - \log 1.25]}{\log 5} + 1.$$

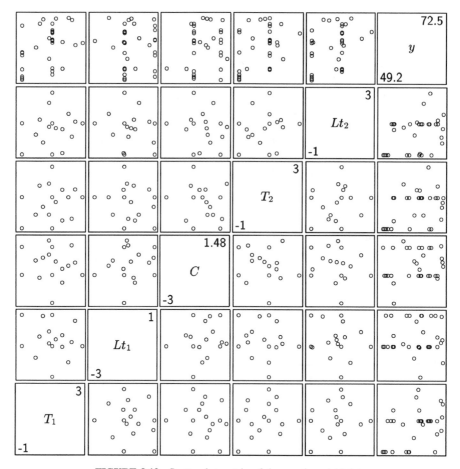

FIGURE 9.13   Scatterplot matrix of the reaction yield data.

Their choice of the logarithmic transformations for the reaction times was based on prior information. We will use these five predictors to see what might be learned about the response surface from a graphical analysis.

This data set was included because a graphical analysis seems relatively difficult, requiring more subjective judgment than the previous studies. Generally, characterizations of 3D plots based on 32 observations can be problematic unless the systematic features are quite strong.

We begin the analysis by inspecting a scatterplot matrix of the five predictors

$$x^{(0)} = (T_1, Lt_2, C, T_2, Lt_2)^T$$

and the response, as shown in Figure 9.13. The scatterplot matrix suggests that there are no strong nonlinear relationships among the predictors, a suggestion

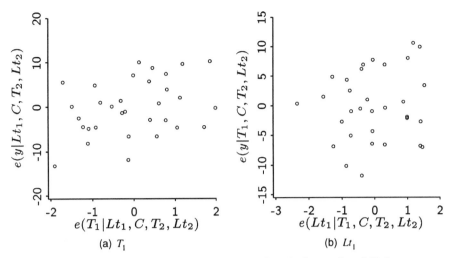

FIGURE 9.14 Added-variable plots for T_1 and Lt_1 in the reaction yield data.

that is supported by using 3D plots to further explore the predictors. Generally, there are no indications that nonlinearities among the predictors may hinder application of graphical regression ideas. The scatterplot matrix also shows that the marginal regressions of the response on the individual predictors are weak.

Like the marginal regressions in the scatterplot matrix, 2D added-variable plots show only weak relationships. For illustration, the added-variable plots for T_1 and Lt_1 are given in Figure 9.14. Because the 2D added-variable plots exhibit only weak dependence, it is difficult to select predictors based on visual assessments of dependence for dimension reduction via 3D added-variable plots.

So far, we have used the ideas discussed in Chapter 7 to explore two different sequential strategies. While the final results should not depend on the order in which the predictors are selected for dimension reduction in 3D plots, the analysis may be facilitated by using subject matter considerations.

The example under discussion involves a two-stage chemical reaction. If we conjecture that the reaction yield depends primarily on a single linear combination of T_1 and Lt_1 at stage 1 and a single linear combination of T_2 and Lt_2 at stage 2, then using 3D added-variable plots for these stage-specific combinations of variables may facilitate the analysis. Accordingly, we next consider the 3D added-variable plot for $x_1^{(0)} = (T_1, Lt_1)^T$ after $x_2^{(0)} = (C, T_2, Lt_2)^T$.

There is considerable variation in the 3D added-variable plot for $x_1^{(0)}$, as shown by the OLS view given in Figure 9.15a. This variation causes difficulty when assessing structural dimension or choosing a 2D summary view. As discussed in Section 7.4, such variation might be reduced without changing the central subspace for the plot by using residuals from modeling the marginal

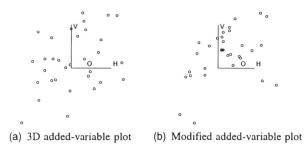

(a) 3D added-variable plot   (b) Modified added-variable plot

FIGURE 9.15 (a) OLS view of the 3D added-variable plot for (T_1, Lt_1) after (C, T_2, Lt_2) from the reaction rate data. (b) OLS view after replacing the quantity on the vertical axis in (a) with the residuals from an OLS fit of (9.6).

dependence of y on $x_2^{(0)}$ on the vertical axis. Because the underlying problem involves maximizing yield, we model the regression function $E(y \mid x_2^{(0)})$ using a full, second-order quadratic,

$$E(y \mid x_2^{(0)}) = \beta_0 + q(x_2^{(0)})^T \beta \tag{9.6}$$

where $q(x_2^{(0)})$ is the 9×1 vector-valued function containing linear, quadratic, and cross-product terms in the elements of $x_2^{(0)}$.

Shown in Figure 9.15b is the OLS view of the 3D plot obtained by replacing the quantity on the vertical axis of Figure 9.15a with the ordinary residual from an OLS fit of (9.6). The variation in Figure 9.15b is noticeably less than that in Figure 9.15a, considerably easing a visual assessment. Next, the structural dimension of the plot was inferred to be 1 and the OLS view in Figure 9.15b was taken as the summary. The associated linear combination of T_1 and Lt_1 gives the stage 1 predictor

$$S_1 = 0.861T_1 + 0.508Lt_1.$$

We now have a new regression problem with response y and predictors $x^{(1)} = (S_1, C, T_2, Lt_2)^T$. Combining the stage 2 predictors by using the methods employed to combine the stage 1 predictors again led to an inference of 1D structure with stage 2 predictor

$$S_2 = 0.979T_2 + 0.201Lt_2.$$

The graphical regression problem now contains three predictors $x^{(2)} = (S_1, C, S_2)^T$, along with the original response y. Attempts to further reduce the dimension of the predictor vector by using 3D added-variable plots were not successful because each of the three possible plots appears to have 2D structure. For further progress it was necessary to use orthogonal predictors, as dis-

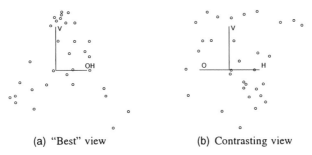

(a) "Best" view   (b) Contrasting view

FIGURE 9.16 Two views of the 3D added-variable plot for (w_2, w_3) after w_1 from the analysis of the reaction yield data: (a) best visually determined view, (b) alternate view provided for contrast.

cussed in Section 9.1.2. Specifically, let $\hat{b}$ be the coefficient vector from the OLS linear regression of y on $x^{(2)} = (S_1, C, S_2)^T$. Extend $\hat{b}$ to a basis $\hat{B} = (\hat{b}, \hat{b})$ for $\mathbb{R}^3$ so that the columns of $\hat{B}$ are orthogonal relative to the inner product induced by $\hat{\Sigma}$, the sample covariance matrix for $x^{(2)}$. Finally, define the transformed predictors $w = (w_1, w_2, w_3)^T$ as in (9.4) with $\hat{B}$ in place of B. The first transformed predictor w_1 coincides with the fitted values from the OLS regression of y on $x^{(2)}$.

The 3D added-variable plot for (w_2, w_3) after w_1 gave a clear indication of 1D structure. The corresponding best visually determined view is shown in Figure 9.16a. The view in Figure 9.16b is provided for contrast. The linear combination of x on the horizontal screen axis of Figure 9.16a is $w_{23} = b_{23}^T x$ with

$$b_{23} = (0.58, 0.34, -0.67, -0.30, -0.06)^T$$

and the linear combination corresponding to $w_1 = b_1^T x$ is

$$b_1 = (0.50, 0.30, 0.40, 0.69, 0.14)^T.$$

Finally, the 3D plot $\{y, (w_1, w_{23})\}$ was inferred to have 2D structure and thus this plot is the estimated minimal sufficient summary for the regression. Evidently, two linear combinations of the predictors must be controlled to produce maximum yield. Two views of the 3D summary plot are shown in Figure 9.17. The plot resembles a slightly distorted cone that is cut off on one side giving rise to a linear trend. The OLS view in Figure 9.17a shows this trend. The view in Figure 9.17b shows the conical shape more clearly. The general shape of the 3D summary plot is in line with the prior expectation that the maximum yield would be found by this experiment. A statistical model of the conical shape could now be used to refine the graphical results and to produce standard errors. Alternatively, the 3D summary plot may give sufficient information on how to achieve near maximum yields. We return to this example in Section 12.4.

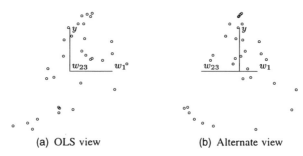

(a) OLS view	(b) Alternate view

FIGURE 9.17   Two views of the final 3D summary plot $\{y,(w_1,w_{23})\}$ for the reaction yield data.

9.4. DISCUSSION

The graphical regression ideas illustrated in this chapter are based on using 3D plots to combine pairs of predictors. In essence, each 3D plot is used to judge if it is reasonable to replace a pair of predictors with a single linear combination of them. Repeated application of this idea leads to an estimate of the central subspace and to an estimated sufficient summary plot. The order in which the predictors are combined, while unimportant in theory, can be an issue in practice. As discussed in Section 9.1.2, combining predictors based on strength of dependence in 2D added-variable plots seems to work well in practice. Subject matter considerations can be useful as well.

Sliced inverse regression (SIR) and principal Hessian directions (pHd) are numerical methods discussed in Chapters 11 and 12 for estimating portions of the central subspace. Each methods provides p linear combinations of the predictors $\hat{u}_1^T x, \ldots, \hat{u}_p^T x$ in an order that reflects numerical measures of the "likelihood" that $\hat{u}_j$ is estimating a vector in the central subspace. A graphical regression analysis could be facilitated by using the SIR or pHd predictors $\hat{u}_j^T x$ instead of the original predictors because there is a clear order based on their likely contribution to the regression.

PROBLEMS

9.1. Repeat the analysis of the naphthalene data with response Y_N as in Section 9.1.1, after deleting the eight points forming the ray in Figure 9.2a and the remote point in Figure 9.2b.

9.2. Reanalyze the naphthalene data for both responses, Y_N and Y_P, using all cases, but this time base your analysis h-level response plots. Are your results consistent with those from this chapter?

9.3. Meyers, Stollsteimer, Wims (1975) report a study based on 77 different gasolines to predict the motor octane number M from four chemical measurements: isoparaffin index (Ip), aromatic content (A, percent by volume), lead content (L, grams per gallon), and sulfur content (S, percent by weight). Use the methods discussed so far in this book to carry out a graphical regression analysis of the data, which are given in Table 9.2.

TABLE 9.2. Octane Data

M	Ip	A	L	S
87.230	0.767	19	2.160	0.050
83.692	0.787	20	1.500	0.030
85.384	0.833	26	1.150	0.030
83.076	0.931	25	0.500	0.020
85.076	0.836	23	1.170	0.050
84.153	0.860	24	0.690	0.020
84.769	0.808	25	0.950	0.020
84.923	0.866	25	0.770	0.010
86.153	0.940	21	1.470	0.020
85.230	0.948	23	0.710	0.010
84.461	0.875	21	1.490	0.020
89.538	0.875	19	2.420	0.020
85.538	0.970	23	0.560	0.010
85.538	0.850	24	0.940	0.030
93.384	1.149	29	1.910	0.010
94.307	1.209	18	2.270	0.009
91.076	1.047	24	2.190	0.010
91.384	1.347	16	1.780	0.010
86.461	1.413	11	1.150	0.050
92.307	1.325	13	1.530	0.020
90.461	1.533	8	1.020	0.050
91.692	1.234	17	1.490	0.020
92.307	1.266	27	1.360	0.009
91.384	1.120	27	1.460	0.090
91.692	1.122	27	2.020	0.010
91.384	0.958	37	2.500	0.009
92.000	1.204	17	1.870	0.010
92.769	1.309	13	1.750	0.040
84.923	0.798	20	1.450	0.030
93.538	1.170	31	2.170	0.009
83.076	1.034	26	0.016	0.010
82.307	0.983	20	0.014	0.030
84.769	1.073	25	0.023	0.009
88.461	1.242	22	0.440	0.070
86.769	1.291	17	0.430	0.009
86.000	0.803	21	2.350	0.070
85.538	0.805	25	1.760	0.070
86.153	0.737	26	1.760	0.040
86.153	0.884	24	1.030	0.030
85.384	0.910	23	1.150	0.050

TABLE 9.2. (*Continued*)

85.076	0.818	26	1.350	0.030
84.000	0.809	24	1.530	0.130
84.461	0.814	26	1.160	0.030
87.384	0.770	28	1.740	0.040
83.692	0.800	27	1.070	0.020
86.307	0.813	27	0.960	0.020
86.615	0.799	20	2.190	0.030
85.230	0.845	26	1.080	0.020
91.076	1.170	24	2.810	0.020
92.000	1.240	24	1.970	0.020
92.769	1.564	15	1.320	0.010
91.076	1.326	14	1.300	0.080
94.153	1.632	11	1.380	0.030
93.384	1.513	10	1.400	0.120
92.615	1.147	23	2.420	0.020
90.615	1.256	28	1.460	0.010
91.230	1.098	25	1.580	0.004
89.538	1.108	29	1.490	0.010
92.000	0.937	36	2.620	0.010
92.769	1.564	15	1.320	0.010
91.076	1.326	14	1.300	0.080
94.153	1.632	11	1.380	0.030
93.384	1.513	10	1.400	0.120
92.615	1.147	23	2.420	0.020
90.615	1.256	28	1.460	0.010
91.230	1.098	25	1.580	0.004
89.538	1.108	29	1.490	0.010
92.000	0.937	36	2.620	0.010
89.538	1.070	26	1.630	0.010
91.384	1.504	14	1.280	0.030
84.769	1.088	25	0.020	0.010
82.615	1.076	20	0.007	0.050
85.846	1.203	26	0.014	0.020
86.307	1.385	16	0.003	0.010
83.384	1.257	17	0.011	0.100
81.846	0.993	21	0.009	0.050
85.538	1.259	20	0.015	0.080
84.153	1.094	22	0.008	0.040
86.923	1.373	37	0.008	0.004
82.461	1.054	24	0.015	0.010
85.538	1.176	25	0.055	0.010
86.769	1.365	20	0.020	0.010
83.230	1.029	24	0.011	0.004
82.000	0.956	20	0.009	0.050
85.538	0.824	26	0.860	0.020
93.076	1.259	29	1.750	0.004

M = Motor octane number, Ip = isoparaffin index, A = Aromatic content; L = Lead content; S = sulfur content.

CHAPTER 10

Inverse Regression Graphics

The approach of the previous chapters was mostly graphical, using various visualization techniques to summarize a regression by estimating its central dimension-reduction subspace $\mathcal{S}_{y|x}$. The ideas discussed in Sections 8.1 and 8.2 provided first numerical methods for estimating vectors in the central subspace. In this chapter we continue to investigate graphical methods for inference about $\mathcal{S}_{y|x}$, still assuming that the data are independent and identically distributed observations on (y, x^T). The primary difference between this and previous approaches is that we now make use of the *inverse regression* $x \mid y$ rather than the *forward regression* $y \mid x$. The inverse regression $x \mid y$ is composed of p simple regressions, $x_j \mid y$, $j = 1, \ldots, p$, each of which can be studied in a 2D plot. Dealing with the inverse regression graphically may thus be easier than dealing directly with the forward regression.

Of course, the possibility of obtaining useful information from the inverse regression depends on developing a connection with the central subspace $\mathcal{S}_{y|x}$. For example, it was necessary to develop such a connection prior to the analysis of the bank notes data in Problem 5.5.

10.1. INVERSE REGRESSION FUNCTION

The inverse regression function $E(x \mid y)$ traces a one-parameter curve in $\mathbb{R}^p$ as the value of y varies in its marginal sample space Ω_y. Depending on the application, a complete description of this curve can be quite complicated or fairly easy.

Suppose, for example, that (y, x^T) follows a nonsingular multivariate normal distribution. Let $\Sigma_{yx} = \mathrm{Cov}(y, x)$, $\Sigma = \mathrm{Var}(x)$ and $\sigma^2 = \mathrm{Var}(y)$. Then

$$E(y \mid x) = E(y) + \Sigma_{yx}\Sigma^{-1}(x - E(x))$$

and thus $\eta = \Sigma^{-1}\Sigma_{xy}$ spans $\mathcal{S}_{y|x}$. Similarly, the inverse regression function is

$$E(x \mid y) = E(x) + \Sigma_{xy}\sigma^{-2}(y - E(y))$$

$$= E(x) + \Sigma\eta\sigma^{-2}(y - E(y)).$$

This result shows that the centered inverse regression function $E(x \mid y) - E(x)$ forms the one-dimensional subspace $S(\Sigma\eta)$ as the value of y ranges over $\mathbb{R}^1$. Clearly, the subspace $S(\Sigma\eta)$ is related to the central subspace $S_{y|x}(\eta)$ via the linear transformation Σ.

Important characteristics of this example can be preserved when multivariate normality does not hold, but certain constraints are placed on the marginal distribution of x. To see how this comes about, define the *inverse regression subspace* as

$$S_{E(x|y)} = \text{span}\{E(x \mid y) - E(x) \mid y \in \Omega_y\}. \tag{10.1}$$

The following proposition establishes a connection between $S_{y|x}$ and $S_{E(x|y)}$. A related result was given by Li (1991).

Proposition 10.1. Let η be a basis for $S_{y|x}$, and let $\Sigma = \text{Var}(x)$. Assume that $E(x \mid \eta^T x = u)$ is a linear function of u. Then

$$E(x \mid y) - E(x) = P_{\eta(\Sigma)}^T (E(x \mid y) - E(x))$$

and

$$S_{E(x|y)} \subset S(\Sigma\eta) = \Sigma S_{y|x}$$

where $P_{\eta(\Sigma)}$ is the projection operator for $S_{y|x}$ relative to the inner product induced by Σ.

Justification. Because $y \perp\!\!\!\perp x \mid \eta^T x$,

$$E(x \mid y) = E_{\eta^T x|y} E(x \mid \eta^T x, y)$$

$$= E_{\eta^T x|y} E(x \mid \eta^T x).$$

From Proposition 4.2, $E(x \mid \eta^T x) - E(x) = P_{\eta(\Sigma)}^T (x - E(x))$. Thus,

$$E(x \mid y) - E(x) = E_{\eta^T x|y} (P_{\eta(\Sigma)}^T (x - E(x)))$$

$$= P_{\eta(\Sigma)}^T (E(x \mid y) - E(x))$$

and the conclusions follow.   □

The next corollary adapts the results of Proposition 10.1 to standardized predictors.

Corollary 10.1. Beginning as in Proposition 10.1, let $z = \Sigma^{-1/2}(x - E(x))$. Then

$$S_{E(z|y)} \subset S(\Sigma^{1/2}\eta) = S_{y|z}.$$

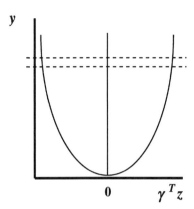

FIGURE 10.1 Stylized representation of the inverse regression function for the model $y \mid z = \beta_0 + (\gamma^T z)^2 + \sigma\varepsilon$.

Proposition 10.1 is potentially useful because it may be possible to investigate $S_{E(x|y)}$ graphically and thereby construct an estimate of $S(\Sigma\eta)$, which can then be used to infer about $S_{y|x}$. Scatterplot matrices, which were introduced in Section 2.5, allow visualization of estimates of the individual inverse regression functions $E(x_j \mid y)$, for example.

Proposition 10.1 requires that $E(x \mid \eta^T x)$ be a linear function of $\eta^T x$. This condition is the same as that used in the Li–Duan Proposition (Proposition 8.1) and is implied when x is an elliptically contoured random variable, as discussed in Section 7.3.2. Thus, Proposition 10.1 fits within the present structure and does not require a new set of constraints.

The conclusion of Proposition 10.1 allows $S_{E(x|y)}$ to be a proper subset of $S(\Sigma\eta)$, and in some situations this may be so. For example, working in terms of the standardized variable z to facilitate exposition, suppose that $\dim[S_{y|z}(\gamma)] = 1$, and that the distribution of $y \mid z$ can be described by the model $y \mid z = (\gamma^T z)^2 + \sigma\varepsilon$, where z is a standard normal random vector and ε is an independent standard normal error. Then $E(z \mid y) = 0$ for all values of y and $S_{E(z|y)} = S(0)$. A stylized representation of this example is shown in Figure 10.1, where the curve represents the quadratic $(\gamma^T z)^2$ and the dashed lines represent a slice of y. The average of $\gamma^T z$ equals 0 for each slice. While Proposition 10.1 still holds in this simple example, the result is not useful. However, if the model were changed to have a linear component,

$$y \mid z = \beta_0 + \beta_1(\gamma^T z) + (\gamma^T z)^2 + \sigma\varepsilon$$

then $E(z \mid y)$ would no longer be constant and $S_{E(z|y)}$ would equal $S_{y|z}$.

As a second example showing that $S_{E(z|y)}$ may be a proper subset of $S_{y|z}$, suppose that the four predictors $z = (z_1, \ldots, z_4)^T$ are independent and identically

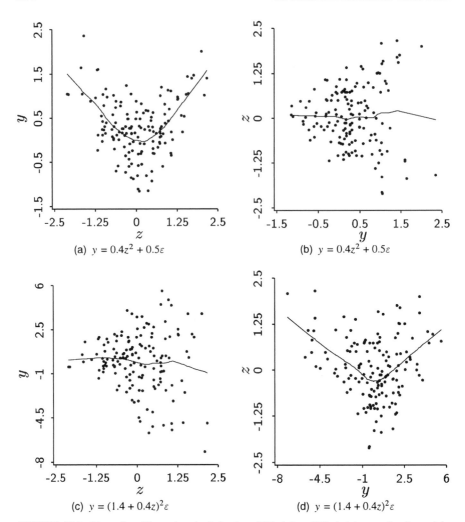

FIGURE 10.2   Four plots illustrating the behavior of $E(y \mid z)$ and $E(z \mid y)$ in two simple models. The superimposed LOWESS smooths have the same span.

distributed and that

$$y \mid z = e^{z_1 + z_2} + e^{z_3 + z_4} + \varepsilon$$

where again $\varepsilon \perp\!\!\!\perp z$. By symmetry, $E(z_j \mid y) = E(z_k \mid y)$, $j, k = 1, \ldots, 4$, and thus $\mathcal{S}_{E(z \mid y)}$ is spanned by the 4×1 vector of ones, while $\mathcal{S}_{y \mid z}$ is spanned by the pair of vectors $(1, 1, 0, 0)^T$ and $(0, 0, 1, 1)^T$.

Figures 10.2a and 10.2b give another illustration of the symmetry that may cause $\mathcal{S}_{E(z \mid y)}$ to be a proper subset of $\mathcal{S}_{y \mid z}$. The data consist of 150 observations from the model $y \mid z = 0.4z^2 + 0.5\varepsilon$, where z and ε are independent standard normal random variables. The quadratic regression function $E(y \mid z)$ that is clearly visible in Figure 10.2a results in the constant inverse regression function in Figure 10.2b. In other words, $S_{y\mid z} = \mathbb{R}^1$, while $S_{E(z\mid y)} = S(0)$.

The data in Figures 10.2c and 10.2d consist of 150 observations from the heteroscedastic model

$$y \mid z = (1.4 + 0.4z)^2 \varepsilon$$

where again z and ε are independent standard normal random variables. The forward regression function in Figure 10.2c is constant, but the inverse regression function in Figure 10.2d is nonconstant so that $S_{E(z\mid y)} = S_{y\mid z} = \mathbb{R}^1$. The point of this illustration is that the inverse regression function may well respond to nonconstant variance $\text{Var}(y \mid z)$, although as illustrated previously, it will not respond to certain types of symmetric dependence in the forward regression function $E(y \mid z)$.

Symmetry in the forward regression function can force $S_{E(z\mid y)}$ to be a proper subset of $S_{y\mid z}$. Such balance would seem to be the exception rather than the rule in practice and, when $E(z \mid \gamma^T z)$ is linear, it may be reasonable in some applications to assume that $S_{E(z\mid y)} = S_{y\mid z}$. However, it will be worthwhile to keep in mind the possibility that the inverse regression subspace can miss relevant information.

10.1.1. Mean checking condition

In this section we develop a method for checking 1D structure by visualizing the inverse regression function.

Assume that the conditions for of Proposition 10.1 hold. Then we can write

$$E(x \mid y) = E(x) + \alpha m(y)$$

where $\alpha = \Sigma\eta$ is a $p \times d$ matrix,

$$m(y) = (\eta^T \Sigma \eta)^{-1} \eta^T \{E(x \mid y) - E(x)\}$$

is a $d \times 1$ vector depending on the value of y, and $d = \dim[S_{y\mid x}]$. If we assume that $d \leq 1$, then α becomes a vector, and $m(y)$ reduces to a scalar. Further, if $\alpha = 0$ then $\dim[S_{y\mid x}] = \dim[S_{E(x\mid y)}] = 0$. If $\alpha \neq 0$ while $m(y) = 0$, then $\dim[S_{E(x\mid y)}] = 0$ and $\dim[S_{y\mid x}] = 1$, reflecting the fact that $S_{E(y\mid x)}$ is a proper subset of $S(\Sigma\eta)$.

Still assuming that $d \leq 1$, the inverse regression function can be written in terms of individual predictors x_j as

$$E(x_j \mid y) = E(x_j) + \alpha_j m(y), \qquad j = 1, \ldots, p \tag{10.2}$$

where α_j is the jth element of α. This representation, which is called the *mean checking condition* by Cook and Weisberg (1994a, p. 117), indicates that, when $d \leq 1$, the inverse regression functions for the individual predictors differ only by a location constant $E(x_j)$ and a scaling constant α_j. In practice, the regression functions $E(x_j \mid y)$ can be visualized in separate inverse response plots $\{x_j, y\}$, perhaps arranged in a scatterplot matrix of y and x. A scatterplot matrix has the advantage of allowing a check on the linear relationships among the predictors as required by Proposition 10.1. In either case, the automatic scaling used by most computer programs standardizes so that the plot produced on the computer screen is essentially

$$\text{sign}(\alpha_j) m(y) = \frac{E(x_j \mid y) - E(x_j)}{|\alpha_j|} \quad \text{vs.} \quad y$$

provided that $\alpha_j \neq 0$.

The practical implication of this reasoning is that, to be consistent with 0D or 1D structure, the inverse regression functions must satisfy one of the following two necessary conditions while allowing for any variation in the automatic scaling used by the computer program:

- $E(x_j \mid y)$ is constant. This condition is forced by either $\alpha_j = 0$ or $m(y) = 0$.
- $E(x_j \mid y)$ is the same as or the mirror image of all nonconstant inverse regression functions. When $\alpha_j \neq 0$, automatic scaling should cause $E(x_j \mid y)$ to appear to equal all other nonconstant inverse regression functions with $\text{sign}(\alpha_i) = \text{sign}(\alpha_j)$. Similarly, when $\text{sign}(\alpha_j) = -\text{sign}(\alpha_i)$, it should appear that $E(x_j \mid y) = -E(x_i \mid y)$.

These conditions do not guarantee that $d \leq 1$, since they are necessary but not sufficient. However, any other conditions can be taken as an indication of at least 2D structure.

Suppose that $\dim[\mathcal{S}_{y|x}] = \dim[\mathcal{S}_{E(x|y)}] = 2$ and let the two $p \times 1$ vectors $\alpha_1 = (\alpha_{j1})$ and $\alpha_2 = (\alpha_{j2})$ span $\mathcal{S}_{E(x|y)}$. The individual inverse regression functions must then be of the form

$$E(x_j \mid y) = E(x_j) + m_1(y)\alpha_{j1} + m_2(y)\alpha_{j2}. \tag{10.3}$$

In the case of 2D structure, the regression function for each inverse response plot will be a linear combination of two distinct functions $m_1(y)$ and $m_2(y)$. For many regressions it may be possible to tell by visual inspection that the structural dimension is greater than one because the individual inverse regression functions do not have the same shape. Deciding that the structural dimension is exactly 2 will generally be much more difficult, however, because we must be able to recognize visually linear combinations of the two functions m_1 and m_2.

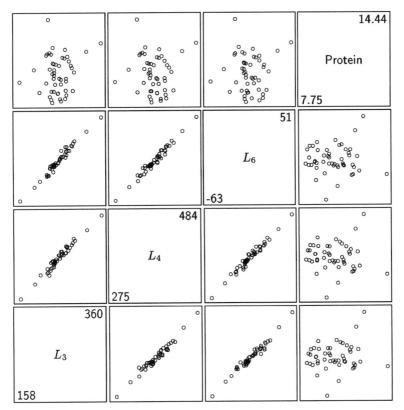

FIGURE 10.3 Scatterplot matrix of the response and three predictors from the wheat protein data.

In short, we may be able to gain useful information on the complexity of a regression problem by inspecting the inverse response plots in a scatterplot matrix. Plots from relatively straightforward regression problems with $\dim[S_{y|x}] \leq 1$ should support the mean checking condition (10.2), while this condition may fail in more complicated problems where $\dim[S_{y|x}] \geq 2$.

10.1.2. Mean checking condition: Wheat protein

The wheat protein data introduced in Section 9.2 provides a first illustration on the use of the mean checking condition. The response is still protein content y, but to facilitate the discussion only three of the original six predictors will be used. Case 24, which was discussed in Section 9.2, was deleted from all plots for visual clarity.

A scatterplot matrix of the response, L_3, L_4, and L_6 is shown in Figure 10.3. The predictors exhibit linear relationships, supporting the linearity assumption necessary for application of Proposition 10.1. An inspection of the predictors in a 3D plot provides additional support. The three inverse response plots are shown in the final column of the scatterplot matrix. These plots are nearly identical, so there is no evidence against the possibility that $d \leq 1$. Because the inverse regression functions are essentially flat, there is also no evidence against the possibility that $d = 0$. But the strong collinearity among the predictors may be hiding relevant structure.

Represent $\alpha = \Sigma \eta$ by using the spectral decomposition of Σ,

$$\Sigma \eta = \ell_1 \ell_1^T \eta \lambda_1 + \ell_2 \ell_2^T \eta \lambda_2 + \ell_3 \ell_3^T \eta \lambda_3$$

where the ℓ's are the eigenvectors and the λ's are the eigenvalues of Σ with $\lambda_1 \geq \lambda_2 \geq \lambda_3$. A 3D plot of the predictors shows that the predictor variation is substantially in a single direction, indicating that λ_1 is much larger than λ_2 and that the single linear combination $\ell_1^T x$ accounts for most of the variation in the predictors. If $\dim[S_{y|x}] = 1$ and η is close to $S[(\ell_2, \ell_3)]$, then there will be relatively little variation in the predictors to show that $\alpha_j \neq 0$. In other words, there is a possibility that the collinearity among the predictors is hiding the fact that $\eta \neq 0$. It may be reasonable in such cases to use uncorrelated predictors z (Corollary 10.1) as a means of removing the collinearity and possibly improving the power of the mean checking condition.

Shown in Figure 10.4 is a scatterplot matrix of the response and three uncorrelated predictors, $\hat{y}$, U_1, and U_2. The first uncorrelated predictor $\hat{y}$ is the fitted values from the OLS regression of y on (L_3, L_4, L_6). It was selected based on the Li–Duan Proposition (Proposition 8.1). The other two uncorrelated predictors, U_1 and U_2, were selected just to complete the basis. The response plot with the fitted values shows a clear linear trend, indicating that the structural dimension is at least one. Because the other two inverse response plots show flat regression functions, the scatterplot matrix is consistent with 1D structure and the plot $\{y, \hat{y}\}$ might be taken as an estimated sufficient summary.

See Problem 10.2 for additional comments on this example.

10.1.3. Mean checking condition: Mussel data

Shown in Figure 10.5 is a scatterplot matrix of the response M, shell length L, shell width W, and shell mass S from the mussel data introduced in Section 2.2. The inverse response plots for L and W show curved regression functions of roughly the same shape, while the inverse response plot for S seems linear. Together the inverse response plots seem to suggest at least 2D structure. This interpretation may be a bit premature, however, because the plots of S versus the other two predictors are clearly curved. This curvature raises doubts about the linearity condition needed for Proposition 10.1.

The linearity condition needed for application of Proposition 10.1 might be satisfied to a sufficient degree by using power transformations of the predictors to induce approximate multivariate normality, as previously discussed

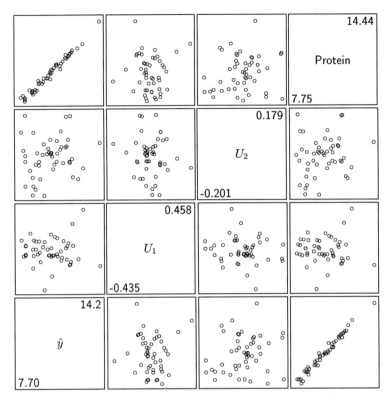

FIGURE 10.4 Scatterplot matrix of the response, the OLS fitted values, and two uncorrelated predictors U_1 and U_2 from the wheat protein data.

in Section 9.1. As a result, the shell length L was not transformed, but shell width W and shell mass S were transformed to $W^{0.36}$ and $S^{0.11}$ by using the maximum likelihood estimates. In practice, such powers might be rounded to meaningful values, recognizing the variation in the estimates. Appealing to allometry leads to consideration of the log transform for all variables. Nevertheless, lacking clear subject matter guidance, we will use the maximum likelihood estimates here and in subsequent illustrations using the regression of M on (L, W, S).

A scatterplot matrix of the response, L and the two transformed predictors is shown in Figure 10.6. The predictors now seem to support the linearity condition of Proposition 10.1, a conclusion that is sustained by a 3D plot of the transformed predictors. The inverse response plots in the scatterplot matrix of Figure 10.6 all seem to have the same shape, which is consistent with 1D structure. The plot of y versus the fitted values from the OLS regression on the transformed predictors might now be used as a minimal summary of the regression.

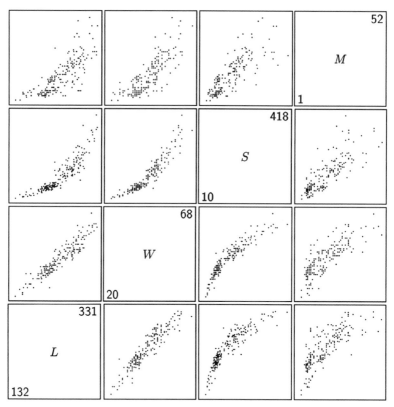

FIGURE 10.5 Scatterplot matrix of muscle mass M, shell length L, shell width W, and shell mass S from the mussel data.

10.2. INVERSE VARIANCE FUNCTION

The mean checking condition makes use of the inverse regression function $E(x \mid y)$ for inferring about the central subspace. However, as illustrated in Figures 10.2c and d, there may be information available in the inverse variance function as well. By using the inverse variance function we might be able to recover some of the information that could be overlooked by the inverse regression function because of symmetries in the forward regression function. In Figure 10.1, for example, the inverse regression function is constant, but the inverse variance does change with the value of y. An ability to extract information from the inverse variance function might be particularly useful during model criticism when residuals are used as the response.

To develop a connection between the inverse variance function and the central subspace $\mathcal{S}_{y|x}(\eta)$, we again work in terms of the standardized pre-dictors $z = \Sigma^{-1/2}(x - E(x))$ with central subspace $\mathcal{S}_{y|z}(\gamma)$, where $\Sigma = \text{Var}(x)$.

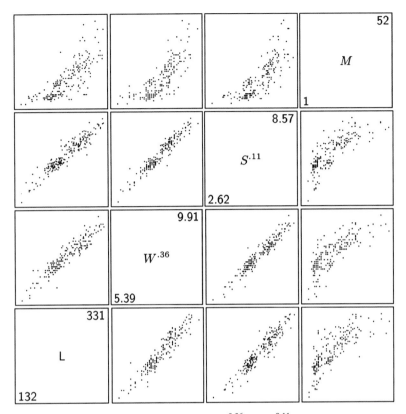

FIGURE 10.6 Scatterplot matrix of M, L, $W^{0.36}$, and $S^{0.11}$ from the mussel data.

The following proposition shows a relationship between $\Sigma_{z|y} = \text{Var}(z \mid y)$ and $\mathcal{S}_{y|z}$.

Proposition 10.2. Let the columns of γ form a basis for $\mathcal{S}_{y|z}$. Assume

1. $E(z \mid \gamma^T z) = P_\gamma z$
2. $\text{Var}(z \mid \gamma^T z) = Q_\gamma$

where P_γ is a projection operator for $\mathcal{S}_{y|z}$, $Q_\gamma = I_p - P_\gamma$, and I_k is the identity matrix of order k. Then

$$\Sigma_{z|y} = Q_\gamma + P_\gamma \Sigma_{z|y} P_\gamma \qquad (10.4)$$

and therefore

$$\mathcal{S}(I_p - \Sigma_{z|y}) \subset \mathcal{S}_{y|z}.$$

Justification.

$$\Sigma_{z|y} = E[Var(z \mid \gamma^T z, y) \mid y] + Var[E(z \mid \gamma^T z, y) \mid y]$$

$$= E[Var(z \mid \gamma^T z) \mid y] + Var[E(z \mid \gamma^T z) \mid y]$$

$$= E[Var(z \mid \gamma^T z) \mid y] + P_\gamma \Sigma_{z|y} P_\gamma$$

$$= Q_\gamma + P_\gamma \Sigma_{z|y} P_\gamma. \tag{10.5}$$

The second equality follows because $y \perp\!\!\!\perp x \mid \gamma^T x$, and the third equality follows from condition 1 of the proposition. The final equality follows from condition 2. $\square$

Proposition 10.2 shows that, with two conditions imposed on the distribution of z, the eigenvectors of $I_p - \Sigma_{z|y}$ corresponding to the nonzero eigenvalues are all in $S_{y|z}$. The first condition of the proposition is the same as that used in Proposition 10.1: If $E(z \mid \gamma^T z)$ is linear in $\gamma^T z$ then $E(z \mid \gamma^T z) = P_\gamma z$, as shown in Proposition 4.2.

With γ as defined in Proposition 10.2, let (γ, γ_o) denote a $p \times p$ matrix with columns forming an orthonormal basis for $\mathbb{R}^p$. Then the condition

$$Var(\gamma_o^T z \mid \gamma^T z) = I_{p-d} \tag{10.6}$$

is equivalent to condition 2 of Proposition 10.2,

$$Var(z \mid \gamma^T z) = Q_\gamma$$

$$= Var(Q_\gamma z + P_\gamma z \mid \gamma^T z)$$

$$= Var(Q_\gamma z \mid \gamma^T z)$$

$$= \gamma_o Var(\gamma_o^T z \mid \gamma^T z) \gamma_o^T$$

where the last equality is due to the fact that $Q_\gamma = \gamma_o \gamma_o^T$. Equation 10.6 now follows because $\gamma_o^T \gamma_o = I_{p-d}$.

Condition 2 of Proposition 10.2 holds when z is multivariate normal, as well as in other cases. For example, if the original predictors are independent and $\gamma^T z$ corresponds to a subset of the predictors, then condition 2 holds regardless of the distribution of x. Further, if $\Sigma_{z|y}$ is constant and condition 1 holds, then condition 2 again holds. To justify this conclusion, write

$$I_p = Var(z)$$

$$= E(\Sigma_{z|y}) + Var[E(z \mid y)]$$

$$= \Sigma_{z|y} + Var[E(z \mid y)] \tag{10.7}$$

because $\Sigma_{z|y}$ is constant. Further, it follows from (10.5) that

$$\Sigma_{z|y} = E[\text{Var}(z \mid \gamma^T z) \mid y] + P_\gamma \Sigma_{z|y} P_\gamma$$
$$= \text{Var}(z \mid \gamma^T z) + P_\gamma \Sigma_{z|y} P_\gamma$$

again because $\Sigma_{z|y}$ is constant. Substituting this into (10.7) and rearranging terms yields

$$\text{Var}(z \mid \gamma^T z) = I_p - P_\gamma \Sigma_{z|y} P_\gamma - \text{Var}[E(z \mid y)]$$
$$= I_p - P_\gamma \text{Var}(z) P_\gamma$$
$$= Q_\gamma$$

giving the desired conclusion because $\text{Var}(z) = I_p$.

Finally, condition 2 should hold to a practically useful approximation for many elliptically contoured distributions (see Problem 10.1).

In the next section, we use Proposition 10.2 to develop a variance checking condition for 1D structure, similar in spirit to the mean checking condition in Section 10.1.1.

10.2.1. Variance checking condition

As used for the mean checking condition near (10.2), let $\alpha = \Sigma \eta$, where $\Sigma = \text{Var}(x)$. Then from (10.4),

$$\text{Var}(x \mid y) = \Sigma - \Sigma^{1/2} \gamma \gamma^T \Sigma^{1/2} + \Sigma^{1/2} \gamma \gamma^T \Sigma^{-1/2} \text{Var}(x \mid y) \Sigma^{-1/2} \gamma \gamma^T \Sigma^{1/2}$$
$$= \Sigma - \Sigma^{1/2} P_{\Sigma^{1/2}\eta} \Sigma^{1/2} + \Sigma^{1/2} P_{\Sigma^{1/2}\eta} \Sigma^{-1/2} \text{Var}(x \mid y) \Sigma^{-1/2} P_{\Sigma^{1/2}\eta} \Sigma^{1/2}$$

where the second equality follows because γ is an orthonormal basis for $\mathcal{S}(\Sigma^{1/2}\eta)$ and thus $P_\gamma = \gamma \gamma^T = P_{\Sigma^{1/2}\eta}$. Clearly, $\text{Var}(x \mid y) = \Sigma$ if $\dim[\mathcal{S}_{y|x}(\eta)] = 0$. Assuming $\dim[\mathcal{S}_{y|x}(\eta)] = 1$,

$$\text{Var}(x \mid y) = \Sigma + \alpha \alpha^T v(y)$$

where

$$v(y) = \frac{\eta^T \text{Var}(x \mid y)\eta}{(\eta^T \Sigma \eta)^2} - \frac{1}{\eta^T \Sigma \eta}.$$

The inverse variance functions for the individual predictors x_j can now be represented as

$$\text{Var}(x_j \mid y) = \text{Var}(x_j) + \alpha_j^2 v(y) \tag{10.8}$$

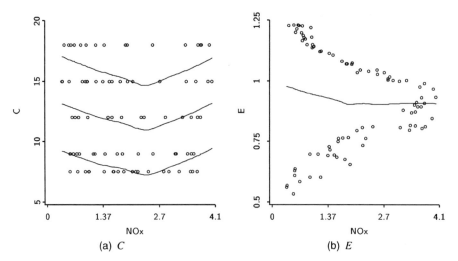

FIGURE 10.7 Inverse response plots for the ethanol data with LOWESS smooths for the mean on each plot. The plot for C contains LOWESS smooths for the standard deviation as well.

where α_j is the jth element of α, $j = 1,\ldots,p$. This expression, which is called the *variance checking condition* by Cook and Weisberg (1994a, p. 117), can be used in the same way as the mean checking condition (10.2): To be consistent with a structural dimension of at most 1, the individual inverse variance functions must be constant, corresponding to $\alpha_j = 0$ or to constant $v(y)$, or they must have the same "shape" as all other nonconstant variance functions. In addition, there is a potentially informative connection between the inverse variance functions (10.8) and the inverse regression functions (10.2) because they both depend on the same constants α_j (see Problem 10.3). The impact of collinearity on the variance checking condition is discussed in Problem 10.6.

10.2.2. Variance checking condition: Ethanol data

Brinkman (1981) describes an industrial experiment to study exhaust from an experimental one-cylinder engine using ethanol as fuel. The response variable NOx is the concentration of nitrogen oxide plus nitrogen dioxide, normalized by the work of the engine. The two predictors are E, a measure of the richness of the air and fuel mixture, and the compression ratio C.

The inverse response plots of the 88 observations in the experiment are shown in Figure 10.7, along with LOWESS smooths for the mean and standard deviation. We now investigate the structural dimension of the regression of NOx on (C,E) by using the mean and variance checking conditions, which are reproduced here for ease of reference:

$$E(x_j \mid y) = E(x_j) + \alpha_j m(y)$$

and

$$\text{Var}(x_j \mid y) = \text{Var}(x_j) + \alpha_j^2 v(y).$$

Consider first the mean checking condition, represented by the middle smooths in Figure 10.7. The mean smooth of the inverse response plot $\{C, NOx\}$ in Figure 10.7a appears to be nonconstant and V-shaped, suggesting that $\alpha_1 \neq 0$, that $m(y) \neq 0$, and thus that the structural dimension is at least 1. The mean smooth for E in Figure 10.7b is more difficult to assess visually. If it is inferred that $\text{E}(E \mid NOx)$ is constant, so that $\alpha_2 = 0$, then the mean checking condition is consistent with 1D structure. However, if it is inferred that $\text{E}(E \mid NOx)$ is a nonconstant function with a non-V-shape then $\alpha_2 \neq 0$, the mean checking condition is violated, and the structural dimension must be 2.

Next, we consider the variance checking condition for Figure 10.7b. Clearly, $\text{Var}(E \mid NOx)$ is a decreasing function of NOx, implying that $\alpha_2 \neq 0$, that $v(y)$ is a nonconstant function, and thus that the structural dimension is at least 1. In contrast, the variance function in Figure 10.7a seems constant, implying that $\alpha_1 = 0$. The variance checking condition then is visually consistent with 1D structure.

So far the mean and variance checking conditions both support 1D structure, with a possible hint of 2D structure from the mean checking condition. However, a very different picture emerges when the checking conditions are considered together. Suppose we infer that $\text{E}(E \mid NOx)$ is constant, implying that $\alpha_2 = 0$. This implication is in conflict with the inference $\alpha_2 \neq 0$ from the variance checking condition. Because the mean and variance conditions for $\{E, NOx\}$ are in conflict, 2D structure is indicated. Similarly, we concluded that $\alpha_1 = 0$ from the variance checking condition for $\{C, NOx\}$, while the mean checking condition suggested the inference that $\alpha_1 \neq 0$. Again, 2D structure is suggested.

Inspection of the 3D plot $\{NOx, (C, E)\}$ supports the conclusion of 2D structure, as discussed in detail by Cook and Weisberg (1994a, Section 7.3).

PROBLEMS

10.1. Assume that condition 1 of Proposition 10.2 holds and that the predictor vector x follows an elliptically contoured distribution. For this situation, derive a representation for $\Sigma_{z|y}$ analogous to that given in (10.4).

10.2. In Section 10.1.2 we circumvented the collinearity problem by passing to three uncorrelated predictors, $\hat{y}$, U_1, and U_2. Consider the three uncorrelated predictors constructed from the spectral decomposition of Σ: $\ell_1^T x$, $\ell_2^T x$, and $\ell_3^T x$. Is there reason to expect that use of these predictors would get around the collinearity problem as well?

10.3. The mean checking condition (10.2) and the variance checking condition (10.8) depend on the same constants α_j. Based on this connection, what could be expected of the simultaneous behavior of these checking conditions? For example, if $\alpha_j = 0$ then both $E(x_j \mid y)$ and $Var(x_j \mid y)$ must be constant.

10.4. The developments of this chapter are based on the inverse regression function $E(x \mid y)$ and the inverse variance function $Var(x \mid y)$. Investigate the possibility of gaining additional graphical information from the inverse covariance functions $Cov(x_j, x_k \mid y)$. If appropriate, include a description of a possible graphical interface.

10.5. Use the mean and variance checking conditions to investigate the regression of Y_p on the three transformed predictors in the naphthalene data (Section 9.1). Do the checking conditions suggest that the structural dimension is greater than 1?

10.6. In Section 10.1.2 we reasoned that collinearity could weaken the ability of the mean checking condition to show an appropriate structural dimension. Will the same sort of problem arise when using the variance checking condition? Let

$$\Sigma = \sum_1^p \ell_k \ell_k^T \lambda_k$$

denote the spectral decomposition of $\Sigma = Var(x)$ with eigenvalues $\lambda_1 \geq \lambda_2 \cdots \geq \lambda_p$. Let $\mathcal{S}$ denote the subspace spanned by the eigenvectors corresponding to the smallest q eigenvalues, assuming for convenience that $\lambda_{p-q} > \lambda_{p-q+1}$. Finally, assume that the variance of x is concentrated in the subspace spanned by the eigenvectors corresponding to the first $p - q$ eigenvalues, so that the last q eigenvalues are relatively small.

Argue that collinearity may be a problem for both the mean and variance checking conditions when $\mathcal{S}_{y|x} \subset \mathcal{S}$.

In Section 10.1.2 we got around the collinearity problem by using three uncorrelated predictors $\hat{y}$, U_1, and U_2. Will that idea work for the variance checking condition as well?

CHAPTER 11

Sliced Inverse Regression

The ideas discussed in Sections 8.1 and 8.2, which are based on the forward regression $y \mid x$, provided first numerical methods for estimating vectors in the central subspace $S_{y|x}(\eta)$. In this chapter we make use of the inverse regression $x \mid y$, specifically Propositions 10.1 and 10.2, to construct numerical estimates of vectors in the central subspace. We will work mostly in the scale of

$$z = \Sigma^{-1/2}(x - E(x)).$$

This involves no loss of generality since $S_{y|x} = \Sigma^{-1/2} S_{y|z}$.

11.1. INVERSE REGRESSION SUBSPACE

If $E(z \mid \eta^T z)$ is linear, then from Corollary 10.1,

$$S_{E(z|y)} \subset S_{y|z}.$$

Thus an estimate of the inverse regression subspace $S_{E(z|y)}$ would provide an estimate of at least a part of the central subspace $S_{y|z}$. The next proposition provides a portion of a rationale for estimating $S_{E(z|y)}$.

Proposition 11.1.

$$S\{\mathrm{Var}[E(z \mid y)]\} = S_{E(z|y)}.$$

Justification. The containment $S_{E(z|y)} \subset S\{\mathrm{Var}[E(z \mid y)]\}$ follows from Proposition 2.7 of Eaton (1983). The reverse containment is given as a problem at the end of this chapter. □

This proposition shows that the inverse regression subspace is spanned by the eigenvectors corresponding to the nonzero eigenvalues of $\mathrm{Var}[E(z \mid y)]$. Consequently, an estimate of $S_{E(z|y)}$ might be constructed from an estimate of $\mathrm{Var}[E(z \mid y)]$.

One way of turning these observations into methodology is based on replacing the response y with a discrete version $\tilde{y}$ formed by partitioning the range of y into h fixed, nonoverlapping slices J_s, $s = 1,\ldots,h$. Slices of the form $(-\infty,a)$ and (a,∞) are permitted. For definiteness, let $\tilde{y} = s$ when $y \in J_s$. This setup is the same as that used for the construction of h-level response plots in Section 6.6, although here the value of h could be larger than that normally used for plotting.

Clearly, $\mathcal{S}_{\tilde{y}|x} \subset \mathcal{S}_{y|x}$, as previously indicated in (6.15). Going a step further and assuming that $\mathcal{S}_{\tilde{y}|x} = \mathcal{S}_{y|x}$, as should be the case when h is sufficiently large, there is no loss of population structure when replacing y with $\tilde{y}$. Combining this with Proposition 11.1 applied to $z \mid \tilde{y}$ and assuming that $E(z \mid \eta^T z)$ is linear, we obtain

$$\mathcal{S}\{\text{Var}[E(z \mid \tilde{y})]\} = \mathcal{S}_{E(z|\tilde{y})} \subset \mathcal{S}_{\tilde{y}|z} \subset \mathcal{S}_{y|z}.$$

This string of relationships provides a basis for estimating at least a portion of the central subspace. Forcing a discrete response allows $E(z \mid \tilde{y})$ to be estimated straightforwardly by using slice means. And $\mathcal{S}\{\text{Var}[E(z \mid \tilde{y})]\}$ can then be estimated from a spectral analysis of the sample covariance matrix of the slice means. Although the starting point is different, this is essentially the idea underlying *sliced inverse regression* (SIR) proposed by Li (1991). The limitations of the inverse regression function discussed in Section 10.1 apply to SIR as well.

11.2. SIR

Let $\hat{\Sigma}$ denote the usual estimate of Σ, and define the standardized observations

$$\hat{z}_i = \hat{\Sigma}^{-1/2}(x_i - \bar{x})$$

where $\bar{x}$ is the sample mean of the predictors and $i = 1,\ldots,n$. Let n_s denote the number of observations falling in slice s. The specific algorithm for SIR proposed by Li (1991) is as follows.

- Within each slice, compute the sample mean of the $\hat{z}$'s:

$$\bar{z}_s = \frac{\sum_{y_i \in J_s} \hat{z}_i}{n_s}.$$

The slice mean $\bar{z}_s$ converges almost surely to the population mean

$$E(z \mid \tilde{y} = s) \in \mathcal{S}_{E(z|\tilde{y})} \subset \mathcal{S}_{y|z} = \Sigma^{1/2}\mathcal{S}_{y|x}.$$

In effect, the SIR algorithm works on the z-scale to first produce an estimated basis for $\mathcal{S}_{E(z|\tilde{y})}$, which is then converted to an estimate of at least a part of $\mathcal{S}_{y|x}$.

- Find the eigenvectors $\hat{\ell}_j$ and eigenvalues $\hat{\lambda}_1 \geq \hat{\lambda}_2 \geq \cdots \geq \hat{\lambda}_p$ of the weighted sample covariance matrix

$$\hat{V} = (1/n)\sum_{s=1}^{h} n_s \bar{z}_s \bar{z}_s^T. \tag{11.1}$$

The matrix $\hat{V}$ converges at root n rate to the population matrix

$$\mathrm{Var}[\mathrm{E}(z \mid \tilde{y})] = \sum_{s=1}^{h} \mathrm{Pr}(\tilde{y} = s)\mu_{z|s}\mu_{z|s}^T$$

where $\mu_{z|s} = \mathrm{E}(z \mid \tilde{y} = s)$. Consequently, the eigenvalues and eigenvectors of $\hat{V}$ converge at the same rate to the eigenvalues λ_j and eigenvectors ℓ_j of $\mathrm{Var}[\mathrm{E}(z \mid \tilde{y})]$.

- Letting $d = \dim[\mathcal{S}_{E(z|\tilde{y})}]$, the SIR estimate $\hat{\mathcal{S}}_{E(z|\tilde{y})}$ of $\mathcal{S}_{E(z|\tilde{y})}$ is

$$\hat{\mathcal{S}}_{E(z|\tilde{y})} = \mathcal{S}(\hat{\ell}_1,\ldots,\hat{\ell}_d).$$

The SIR estimate of a subspace of $\mathcal{S}_{y|x}$ is then

$$\hat{\Sigma}^{-1/2}\hat{\mathcal{S}}_{E(z|\tilde{y})} = \mathcal{S}(\hat{\Sigma}^{-1/2}\hat{\ell}_1,\ldots,\hat{\Sigma}^{-1/2}\hat{\ell}_d).$$

- Letting $\hat{u}_j = \hat{\Sigma}^{-1/2}\hat{\ell}_j$, the regression can be summarized with a plot of y versus the d *ordered SIR predictors* $\hat{u}_1^T x,\ldots,\hat{u}_d^T x$.

The SIR procedure just described could be used to facilitate a graphical analysis by using all p of the estimated *SIR predictors* $\hat{u}_1^T x,\ldots,\hat{u}_p^T x$ in graphical regression. The advantage here is that the ordering of the SIR predictors may reflect their relative contributions to explaining the variation in y, with $\hat{u}_1^T x$ having potentially the greatest contribution. Thus, a graphical regression analysis could be facilitated by bringing in the SIR predictors $\hat{u}_j^T x$ in their SIR order.

Otherwise, inference about d is still required for use in practice. If $\mathrm{Var}[\mathrm{E}(z \mid \tilde{y})]$ were known, d could be determined easily because the smallest $p - d$ eigenvalues equal 0; $\lambda_j = 0$ for $j = d + 1,\ldots,p$. The method suggested by Li (1991) for inferring about d is based on using the statistic

$$\hat{\Lambda}_m = n \sum_{j=m+1}^{p} \hat{\lambda}_j \tag{11.2}$$

to test hypotheses of the form $d = m$ versus $d > m$. Beginning with $m = 0$, compare $\hat{\Lambda}_m$ to a selected quantile of its distribution under the hypothesis $d = m$. If it is smaller, there is no information to contradict the hypothesis $d = m$. If it is larger, conclude that $d > m$, increment m by 1 and repeat the procedure. As long as $m \leq p - 1$, the inference that $d = m$ follows when $\hat{\Lambda}_{m-1}$ is relatively large, implying that $d > m - 1$, while $\hat{\Lambda}_m$ is relatively small, implying that $d = m$. The case when $m = p$ is special. Because $d \leq p$, relatively large values of $\hat{\Lambda}_{p-1}$ alone imply that $d = p$. Note also that $\hat{\Lambda}_m$ is not defined for $m = p$. For this paradigm to be possible we must find an approximation to the distribution of $\hat{\Lambda}_d$, which is n times the sum of the smallest $p - d$ eigenvalues of $\hat{V}$.

When there are only $h = 2$ slices, $\hat{\Lambda}_0$ reduces to a comparison of $\bar{x}_1$ and $\bar{x}_2$, the sample averages of the n_1 values of x in the first slice and the n_2 values of x in the second slice:

$$\hat{\Lambda}_0 = \frac{n_1 n_2}{n_1 + n_2}(\bar{x}_1 - \bar{x}_2)^T \hat{\Sigma}^{-1}(\bar{x}_1 - \bar{x}_2). \tag{11.3}$$

In general, the test statistic $\hat{\Lambda}_0$ for $d = 0$ is identical to Pillai's (1955) trace statistic applied to yield a comparison of the means of h multivariate samples (for a review, see Seber 1984).

11.3. ASYMPTOTIC DISTRIBUTION OF $\hat{\Lambda}_d$

We investigate the asymptotic distribution of $\hat{\Lambda}_d$ in this section. The most general version is discussed first, followed by a simpler form that relies on various constraints. We need the distribution of $\hat{\Lambda}_d$ only when $p > d$, which we assume throughout the discussion.

11.3.1. Overview

We first establish a little more notation. As defined previously, let n_s denote the number of observations in slice $s = 1,\ldots,h$, and let $n = \sum n_s$. The sample slice fractions $\hat{f}_s = n_s/n$ converge to the corresponding population fractions $f_s = \Pr(\tilde{y} = s)$. Let $g_s = \sqrt{f_s}$ and $\hat{g}_s = \sqrt{\hat{f}_s}$, $s = 1,\ldots,h$. The slice means of the x_i and the $\hat{z}_i$ will be represented by $\bar{x}_s$ and $\bar{z}_s$. Finally, define the $p \times h$ matrix

$$Z_n = (\hat{g}_1 \bar{z}_1,\ldots,\hat{g}_h \bar{z}_h)$$

and the corresponding population version

$$B = (g_1 \mu_{z|1},\ldots,g_h \mu_{z|h})$$

where again $\mu_{z|s} = E(z \mid \tilde{y} = s)$.

There is at least one linear dependency among the columns of B because $E(E(z \mid \tilde{y})) = 0$. The rank of B is therefore at most $\min(p, h - 1)$. We assume that h is large enough to satisfy $\min(p, h - 1) > d$. Otherwise, there would be no possibility of estimating d with the procedure in question. For example, in a problem with five predictors, three slices, and $\text{rank}(B) = 2$, we would know only that $d \geq 2$. On the other hand, if the rank of B were 1, we could conclude that $d = 1$. The constraint $\min(p, h - 1) > d$ is always satisfied if the number of slices is chosen to be $h \geq p + 1$.

The matrix $\hat{V}$ (11.1) used in the SIR algorithm is related to $Z_n : \hat{V} = Z_n Z_n^T$. We will find the asymptotic distribution of the test statistic $\hat{\Lambda}_d$ by first characterizing the joint asymptotic distribution of the smallest $\min(p - d, h - d)$ singular values $\hat{\delta}_j$ of Z_n. We will then be in a position to find the asymptotic distribution of $\hat{\Lambda}_d$ because the nonzero eigenvalues of $Z_n Z_n^T$ are the squares of the singular values of Z_n.

The asymptotic distribution of the singular values of Z_n can be obtained by using the general approach developed by Eaton and Tyler (1994): First, construct the singular value decomposition of B,

$$B = \Gamma^T \begin{pmatrix} D & 0 \\ 0 & 0 \end{pmatrix} \Psi$$

where Γ^T and Ψ are orthonormal matrices with dimension $p \times p$ and $h \times h$, and D is a $d \times d$ diagonal matrix of positive singular values. Next, partition $\Gamma^T = (\Gamma_1, \Gamma_o)$ and $\Psi^T = (\Psi_1, \Psi_o)$ where Γ_0 is $p \times (p - d)$ and Ψ_0 is $h \times (h - d)$. Then it follows from Eaton and Tyler (1994) that the asymptotic distribution of the smallest $\min(p - d, h - d)$ singular values of $\sqrt{n}(Z_n - B)$ is the same as the asymptotic distribution of the singular values of the $(p - d) \times (h - d)$ matrix

$$\sqrt{n} U_n = \sqrt{n} \Gamma_o^T (Z_n - B) \Psi_o$$

$$= \sqrt{n} \Gamma_o^T Z_n \Psi_o. \tag{11.4}$$

Thus, the asymptotic distribution of $\hat{\Lambda}_d$ is the same as that of

$$\Lambda_d = n \, \text{trace}[\Gamma_o^T Z_n \Psi_o (\Gamma_o^T Z_n \Psi_o)^T]$$

which is the sum of the squared singular values of $\sqrt{n} U_n$. Given an $r \times c$ matrix $A = (a_1, \ldots, a_c)$, let $\text{vec}(A)$ denote the $rc \times 1$ vector constructed by stacking the columns of A: $\text{vec}(A) = (a_1^T, \ldots, a_c^T)^T$. With this notation, Λ_d can be reexpressed as

$$\Lambda_d = n \, \text{vec}(U_n)^T \text{vec}(U_n).$$

The rest of the justification proceeds as follows: First, we show that the asymptotic distribution of the key variable $\sqrt{n}\,\text{vec}(U_n)$ is multivariate normal with mean 0 and $(p-d)(h-d) \times (p-d)(h-d)$ covariance matrix Ω. The derivation of a useful form for the covariance matrix is a central part of the discussion. It will then follow (see Eaton 1983, p. 112, or Guttman 1982, p. 76) that Λ_d, and thus $\hat{\Lambda}_d$, is distributed asymptotically as a linear combination of independent chi-square random variables. Finally, additional constraints will be added to force Ω to be a symmetric idempotent matrix, in which case the distribution of $\hat{\Lambda}_d$ will reduce to a central chi-square with $(p-d)(h-d-1)$ degrees of freedom.

11.3.2. The general case

Let f, $\hat{f}$, g, and $\hat{g}$ be $h \times 1$ vectors with elements f_s, $\hat{f}_s$, g_s, and $\hat{g}_s$. Additionally, let

$$M_n = (\bar{x}_1, \ldots, \bar{x}_h) : p \times h$$

$$C = (\mu_{x|1}, \ldots, \mu_{x|h}) : p \times h$$

$$G = h \times h \text{ diagonal matrix with diagonal entries } g_s = \sqrt{f_s}$$

$$Q_g = \text{projection for } S^{\perp}(g)$$

where $\mu_{x|s} = \text{E}(x \mid \tilde{y} = s)$. The estimate $\hat{G}$ is defined as G with g replaced by $\hat{g}$. The following relationships may be useful during the discussion:

$$Z_n = \hat{\Sigma}^{-1/2}M_n(I_h - \hat{f}1_h^T)\hat{G} \tag{11.5}$$

$$B = \Sigma^{-1/2}C(I_h - f1_h^T)G \tag{11.6}$$

$$\hat{G}Q_{\hat{g}} = (I_h - \hat{f}1_h^T)\hat{G} \tag{11.7}$$

$$1_h^T(I_h - \hat{f}1_h^T) = 0 \tag{11.8}$$

$$C^T\Sigma^{-1/2}\Gamma_o \in \mathcal{S}(1_h) \tag{11.9}$$

where 1_h is the $h \times 1$ vector of ones. Equalities (11.5) to (11.8) follow from straightforward algebra, while (11.9) follows from (11.6) and $\Gamma_o^T B = 0$.

Proposition 11.2.

$$\sqrt{n}U_n = \sqrt{n}\Gamma_o^T\Sigma^{-1/2}(M_n - C)GQ_g\Psi_o + o_p(n^{-1/2}).$$

Justification. For notational convenience, let $\hat{A} = \hat{\Sigma}^{-1/2}\Sigma^{1/2}$, $F = GQ_g$, and $\hat{F} = \hat{G}Q_{\hat{g}}$. Then

$$\sqrt{n}U_n = \sqrt{n}\Gamma_o^T(\hat{A} - I_p + I_p)\Sigma^{-1/2}(M_n - C + C)(\hat{F} - F + F)\Psi_o.$$

Expanding this in terms of $(\hat{A} - I_p)$, $(M_n - C)$, and $(\hat{F} - F)$, and collecting the four $o_p(n^{-1/2})$ terms leaves

$$\sqrt{n}U_n = \sqrt{n}\Gamma_o^T(\hat{A} - I_p)\Sigma^{-1/2}CF\Psi_o$$
$$+ \sqrt{n}\Gamma_o^T\Sigma^{-1/2}(M_n - C)F\Psi_o$$
$$+ \sqrt{n}\Gamma_o^T\Sigma^{-1/2}C(\hat{F} - F)\Psi_o$$
$$+ \sqrt{n}\Gamma_o^T\Sigma^{-1/2}CF\Psi_o + o_p(n^{-1/2}).$$

From (11.6), $B = \Sigma^{-1/2}CF$, and thus the first and fourth terms equal 0. The third term is 0 because of (11.8) and (11.9). That leaves the second term, which corresponds to the desired conclusion. □

The asymptotic distribution of U_n now follows from the asymptotic distribution of M_n, the only random quantity remaining in the expansion of Proposition 11.2.

Proposition 11.3. $\sqrt{n}\,\text{vec}(U_n)$ converges in distribution to a normal random vector with mean 0 and $(p - d)(h - d) \times (p - d)(h - d)$ covariance matrix

$$\Omega = (\Psi_o^T Q_g \otimes I_{(p-d)})\Delta_o(Q_g\Psi_o \otimes I_{(p-d)}) \tag{11.10}$$

where Δ_o is a $(p - d)h \times (p - d)h$ block diagonal matrix with diagonal blocks $\Gamma_o^T\text{Var}(z \mid \tilde{y} = s)\Gamma_o$, $s = 1,\ldots h$.

Justification. Writing a typical column of $\sqrt{n}(M_n - C)$ as

$$\hat{g}_s^{-1}\sqrt{n_s}(\bar{x}_s - \mu_{x|s})$$

and then applying the central limit theorem and the multivariate version of Slutsky's theorem, it follows that $\sqrt{n}\text{vec}(M_n - C)$ converges in distribution to a normal random vector with mean 0 and $ph \times ph$ block diagonal covariance matrix

$$(G^{-1} \otimes I_p)\Delta_x(G^{-1} \otimes I_p)$$

where Δ_x is a block diagonal matrix with diagonal blocks $\text{Var}(x \mid \tilde{y} = s)$, $s = 1,\ldots h$. The conclusion follows from Proposition 11.2. $\square$

The next proposition gives the asymptotic distribution of the SIR test statistic. Its justification is set as a problem at the end of this chapter.

Proposition 11.4. Let $d = \dim[S_{E(x|y)}]$ and assume that $h > d + 1$ and $p > d$. Then the asymptotic distribution of $\hat{\Lambda}_d$ is the same as the distribution of

$$C = \sum_{k=1}^{(p-d)(h-d)} \omega_k C_k$$

where the C_k's are independent chi-square random variables each with one degree of freedom, and $\omega_1 \geq \omega_2 \geq \cdots \geq \omega_{(p-d)(h-d)}$ are the eigenvalues of the covariance matrix Ω in (11.10).

This proposition allows for a general test of dimension using an estimate $\hat{\Omega}$ of Ω. Perhaps the most straightforward procedure is to simply substitute sample versions of the various quantities needed to compute Ω, including estimates of g_s and $\text{Var}(z \mid \tilde{y})$. Additionally, under a hypothesized value of d, Γ_o and Ψ_o can be estimated by using the sample versions computed from Z_n. Letting $\{\hat{\omega}_k\}$ denote the eigenvalues of $\hat{\Omega}$, the asymptotic distribution of $\hat{\Lambda}_d$ is then estimated to be the same as that of

$$\hat{C} = \sum_{k=1}^{(p-d)(h-d)} \hat{\omega}_k C_k. \qquad (11.11)$$

As in Li's original presentation of SIR, these results can be used to estimate $d = \dim(S_{E(z|y)})$: Beginning with $m = 0$, compare $\hat{\Lambda}_m$ to a selected quantile of the estimated distribution of C. If it is smaller, there is no information to contradict the hypothesis $d = m$. If it is bigger conclude that $d > m$, increment m by 1 and repeat the procedure. There is a substantial literature on computing the tail probabilities of linear combinations of chi-square random variables. See Field (1993) for an introduction.

11.3.3. Distribution of $\hat{\Lambda}_d$ with constraints

The asymptotic distribution of $\hat{\Lambda}_d$ depends on the behavior of the conditional covariance matrices

$$\text{Var}(\Gamma_o^T z \mid \tilde{y}) = \Gamma_o^T \text{Var}(z \mid \tilde{y}) \Gamma_o$$

as they appear in Ω (11.10). And this behavior depends on whether $S_{E(z|\bar{y})} = S_{\bar{y}|z}$ or not. Unless indicated otherwise, we will assume that

$$S_{E(z|\bar{y})} = S_{\bar{y}|z}(\gamma) \tag{11.12}$$

throughout this section. A treatment of the case in which $S_{E(z|\bar{y})}$ is a proper subset of $S_{\bar{y}|z}$ is described in the problems section of this chapter.

Because of (11.12), $S(\Gamma_1) = S_{\bar{y}|z}(\gamma)$ and $S(\Gamma_o) = S_{\bar{y}|z}^{\perp}$, where Γ_o and Γ_1 are as defined in the singular value decomposition of B. This means that we now have two sets of notation to describe a basis for $S_{\bar{y}|z}$: our standard notation γ for denoting bases of central subspaces when standardized predictors are used, and Γ_1. We are assuming that Γ_1 is a basis for $S_{\bar{y}|z}$, but it is not such a basis in general. To avoid carrying two sets of notation for the same thing, we now set $\gamma = \Gamma_1$ and $\gamma_o = \Gamma_o$. Then

$$\text{Var}(\gamma_o^T z) = E[\text{Var}(\gamma_o^T z \mid \gamma^T z)] + \text{Var}[E(\gamma_o^T z \mid \gamma^T z)].$$

Under the linearity requirement of Proposition 10.1, $E(z \mid \gamma^T z) = P_\gamma z$. Thus, $E(\gamma_o^T z \mid \gamma^T z) = 0$, which implies that $\text{Var}[E(\gamma_o^T z \mid \gamma^T z)] = 0$. Because $\text{Var}(z) = I$, we have that

$$E[\text{Var}(\gamma_o^T z \mid \gamma^T z)] = I_{p-d}. \tag{11.13}$$

Constancy of the conditional covariance matrix $\text{Var}(\gamma_o^T z \mid \gamma^T z)$ is one requirement for the distribution of $\hat{\Lambda}_d$ to be chi-square, as will be demonstrated shortly. Because of (11.13), requiring that $\text{Var}(\gamma_o^T z \mid \gamma^T z)$ be constant is equivalent to requiring that $\text{Var}(\gamma_o^T z \mid \gamma^T z) = I_{p-d}$, which is in turn equivalent to (cf. (10.6)):

$$\text{Var}(z \mid \gamma^T z) = Q_\gamma. \tag{11.14}$$

The following proposition gives the asymptotic distribution of $\hat{\Lambda}_d$ under (11.14), which is the same as condition 2 of Proposition 10.2, and the linearity condition of Proposition 10.1. The special case of the proposition in which z is assumed to be a normal random vector was given by Li (1991); see also Schott (1994). The justification is deferred until a bit later in this section.

Proposition 11.5. Let $d = \dim[S_{E(z|\bar{y})}]$ and let h denote the number of slices. Assume that

1. $h > d + 1$ and $p > d$
2. $S_{E(z|\bar{y})} = S_{\bar{y}|z}(\gamma)$
3. $E(z \mid \gamma^T z) = P_\gamma z$
4. $\text{Var}(z \mid \gamma^T z) = Q_\gamma$

where z is the standardized predictor. Then $\hat{\Lambda}_d$ has an asymptotic chi-square distribution with degrees of freedom $(p - d)(h - d - 1)$.

Condition 1 of Proposition 11.5 was imposed for the reasons discussed previously. Condition 2 insures that SIR will not miss a part of the central subspace based on the discrete version $\tilde{y}$ of y. If $\mathcal{S}_{E(z|\tilde{y})}$ is a proper subset of $\mathcal{S}_{\tilde{y}|z}$, then $\hat{\Lambda}_d$ may no longer have a chi-square distribution. Further discussion of this possibility is given at the end of the section. Condition 3 is the same as that required for Proposition 10.1: If $E(z \mid \gamma^T z)$ is linear in $\gamma^T z$ then $E(z \mid \gamma^T z) = P_\gamma z$ as shown in Proposition 4.2. Condition 4, which was discussed following Proposition 10.2, was imposed to force a straightforward asymptotic distribution.

Conditions 3 and 4 hold when z is normally distributed. If z is assumed to be an elliptically contoured random variable, then condition 4 forces z to be normal (see Section 7.3.2). However, normality is not a necessary requirement for these conditions.

Additionally, conditions 1–4 are trivially true when $\dim[\mathcal{S}_{\tilde{y}|z}] = 0$ so that $\tilde{y} \perp\!\!\!\perp z$. This special case could be useful for model checking by applying SIR to the regression of residuals on z. Although SIR may not respond to curvature in residuals, it may be useful for detecting heteroscedasticity.

Proposition 11.5 can be used as the basis for inference about d, in the same way as Proposition 11.4: Beginning with $m = 0$, if $\hat{\Lambda}_m$ is sufficiently large relative to the quantiles of a chi-square distribution with $(p - m)(h - m - 1)$ degrees of freedom, then infer that $d > m$. The inference that $d = m$ follows when $\hat{\Lambda}_{m-1}$ is relatively large, implying that $d > m - 1$, while $\hat{\Lambda}_m$ is relatively small, implying that $d = m$.

Justification of Proposition 11.5. The justification involves showing that Ω (11.10) is a symmetric idempotent matrix of rank $(p - d)(h - d - 1)$. The form of the terms $\gamma_o^T \text{Var}(z \mid \tilde{y}) \gamma_o$ is the key to showing the desired result. In reference to Ω, recall that for notational consistency, we have set $\gamma_o = \Gamma_o$. Now,

$$\gamma_o^T \text{Var}(z \mid \tilde{y}) \gamma_o = E[\text{Var}(\gamma_o^T z \mid \gamma^T z, \tilde{y}) \mid \tilde{y}] + \text{Var}[E(\gamma_o^T z \mid \gamma^T z, \tilde{y}) \mid \tilde{y}]$$

$$= E[\text{Var}(\gamma_o^T z \mid \gamma^T z) \mid \tilde{y}] + \text{Var}[E(\gamma_o^T z \mid \gamma^T z) \mid \tilde{y}]$$

$$= E[\text{Var}(\gamma_o^T z \mid \gamma^T z) \mid \tilde{y}]$$

$$= I_{(p-d)}. \tag{11.15}$$

The second equation follows because $\gamma_o^T z \perp\!\!\!\perp \tilde{y} \mid \gamma^T z$ by the second condition of the proposition. The third condition of the proposition forces $E(\gamma_o^T z \mid \gamma^T z) = 0$. Finally, imposing the fourth condition gives (11.15) because of (11.13). Substituting (11.15) into (11.10)

$$\Omega = \Psi_o^T Q_g \Psi_o \otimes I_{p-d}.$$

Because $Bg = 0$, $g \in \mathcal{S}(\Psi_o)$. Thus, Ω is a symmetric idempotent matrix, and $\hat{\Lambda}_d$ converges to a chi-square random variable with

$$\text{trace}(\Omega) = (p - d)(h - d - 1)$$

degrees of freedom. $\square$

Proposition 11.5 may no longer hold when condition 2 is violated so that $\mathcal{S}_{E(z|\tilde{y})}$ is a proper subset of $\mathcal{S}_{\tilde{y}|z}$ and SIR misses part of the central subspace. The main issue is that now $\mathcal{S}(\Gamma_1)$ is a proper subset of $\mathcal{S}_{\tilde{y}|x}(\gamma)$, where Γ_1 is still from the singular value decomposition of B.

Replacing γ with Γ_1, the reasoning leading to (11.15) may no longer hold. For example,

$$E(\Gamma_o^T z \mid \Gamma_1^T z, \tilde{y}) \neq E(\Gamma_o^T z \mid \Gamma_1^T z)$$

because Γ_1 no longer spans $\mathcal{S}_{\tilde{y}|z}$. However, some progress may still be possible.

Let $\gamma = (\Gamma_1, \Gamma_2)$ be an orthonormal basis for $\mathcal{S}_{\tilde{y}|z}$, where Γ_2 is the part of $\mathcal{S}_{\tilde{y}|z}$ missed by SIR. Recomputing (11.15) by conditioning on $(\Gamma_1^T z, \Gamma_2^T z)$ gives

$$\begin{aligned}
\Gamma_o^T \text{Var}(z \mid \tilde{y})\Gamma_o &= E[\text{Var}(\Gamma_o^T z \mid \Gamma_1^T z, \Gamma_2^T z, \tilde{y}) \mid \tilde{y}] \\
&\quad + \text{Var}[E(\Gamma_o^T z \mid \Gamma_1^T z, \Gamma_2^T z, \tilde{y}) \mid \tilde{y}] \\
&= E[\text{Var}(\Gamma_o^T z \mid \Gamma_1^T z, \Gamma_2^T z) \mid \tilde{y}] \\
&\quad + \text{Var}[E(\Gamma_o^T z \mid \Gamma_1^T z, \Gamma_2^T z) \mid \tilde{y}].
\end{aligned} \tag{11.16}$$

The development of this case can now proceed as in the justification of Proposition 11.5, but different conditions may be required to force an asymptotic chi-square distribution. Completion of this line of reasoning is set as a problem at the end of this chapter.

11.4. SIR: MUSSEL DATA

Table 11.1 gives results from applying SIR to the transformed mussel data as shown in Figure 10.6. Recall that the predictors were transformed to $(L, W^{0.36}, S^{0.11})$ to comply with the linearity requirement. The two halves of Table 11.1 give results for $h = 20$ and $h = 9$ slices. When there are ties in the values of the response variable, as in the present example, the output from SIR may depend on the method of breaking the ties. The *R-code* (Cook and Weisberg 1994a) was used for all computations in this example. The actual

TABLE 11.1. Results from Application of SIR to the Regression of M on L, $W^{0.36}$, and $S^{0.11}$ in the Mussel Data

m	$\hat{\Lambda}_m$	DF	p-value
$h = 20$			
0	177.6	57	0.000
1	32.99	36	0.613
2	12.13	17	0.792
$h = 9$			
0	164.6	24	0.000
1	26.60	14	0.022
2	10.18	6	0.117

slice sizes that were produced by the *R-code* algorithm for breaking response ties are

$$(19, 13, 13, 21, 14, 13, 8, 7, 7, 10, 8, 9, 11, 9, 9, 6, 9, 6, 7, 2)$$

and

$$(32, 34, 20, 22, 21, 24, 20, 20, 8)$$

for $h = 20$ and $h = 9$, respectively.

The rows of Table 11.1 summarize hypothesis tests of the form $d = m$. For example, the first row gives the statistic $\hat{\Lambda}_0 = 177.6$ with $(p - d)(h - d - 1) = 3(19) = 57$ degrees of freedom. Taken together, the three tests with $h = 20$ in Table 11.1 suggest 1D structure. The first SIR predictor is

$$\hat{u}_1^T x = 0.001L + 0.073W^{.36} + 0.997S^{.11}.$$

Figure 11.1 gives the corresponding estimated sufficient summary plot $\{M, \hat{u}_1^T x\}$. This plot is visually indistinguishable from the plot of the response versus the OLS fitted values $\hat{M}$ from the regression of M on the three transformed predictors, the correlation between $\hat{M}$ and $\hat{u}_1^T x$ being 0.998. The analysis might now be continued by investigating transformations of M to simplify the regression, as illustrated in previous chapters.

A SIR analysis can be sensitive to the number of slices used, resulting in somewhat elusive conclusions. For example, in contrast to the conclusions based on 20 slices, the results in Table 11.1 for $h = 9$ slices suggest 2D structure. The correlation between the first SIR predictors $\hat{u}_1^T x$ for 9 and 20 slices is 0.9998, and the correlation between the second SIR predictors $\hat{u}_2^T x$ is 0.97. Because of these high correlations, it seems that the two analyses are producing essentially the same estimates of the central subspace, and that the primary difference rests with the results of the SIR tests. It is possible that

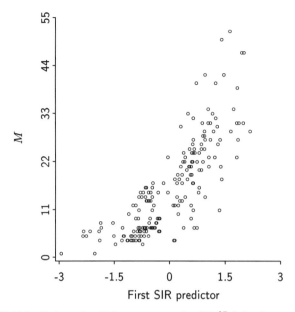

FIGURE 11.1   Estimated sufficient summary plot $\{M, \hat{u}_1^T x\}$ for the mussel data.

condition 4 of Proposition 11.5 does not hold across all slice sizes so that the nominal chi-square distribution is not an accurate approximation. In any event, when faced with apparent contradictions as in the present example, it may be useful to apply the visualization techniques discussed in Chapter 4 for determining structural dimension. Using uncorrelated 2D views to study the 3D plot $\{M, (\hat{u}_1^T x, \hat{u}_2^T x)\}$ from the analysis with nine slices does not produce compelling evidence of 2D structure.

   SIR might also be used in an attempt to diagnose deficiencies in a postulated model. To illustrate the idea, let $\hat{e}$ denote the residuals from an OLS fit of the linear model

$$M \mid x = \beta_0 + \beta^T x + \varepsilon$$

with $x = (L, W^{.36}, S^{.11})^T$ and $\varepsilon \perp\!\!\!\perp x$. Also, let e denote the population version of $\hat{e}$. If the linear model was adequate, e would be independent of x. However, it is clear from Figure 11.1 that the model has deficiencies. Should the results from applying SIR to the regression of $\hat{e}$ on x be expected to show that $\dim[\mathcal{S}_{e|x}] > 0$?

   Application of SIR to $\hat{e} \mid x$ produced a p-value of 0.081 for the test of 0D structure. The corresponding plot $\{\hat{e}, \hat{u}_1^T x\}$ for the first SIR predictor is shown in Figure 11.2a. For contrast, the plot of residuals versus OLS fitted values $\{\hat{e}, \hat{M}\}$ is shown in Figure 11.2b. Apparently, SIR failed to find curvature in the regression of $\hat{e}$ on x for the reasons discussed at the end of Section 10.1: The inverse regression function $E(x \mid e)$ is relatively flat so there is little systematic

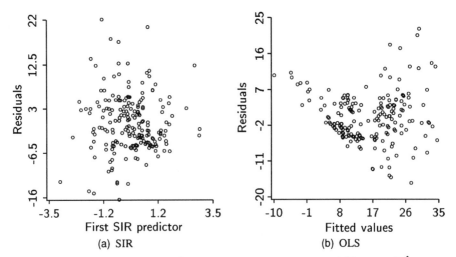

FIGURE 11.2   Regression of $\hat{e}$ on x from the mussels data: (a) $\{\hat{e}, \hat{u}_1^T x\}$, (b) $\{\hat{e}, \hat{M}\}$.

information for SIR to find. The pattern in Figure 11.2a suggests that SIR is responding somewhat to heteroscedasticity, however. Generally, SIR does much better at finding heteroscedastic patterns than it does at finding curvature when applied to regressions with residuals as the response (Cook and Weisberg 1991b). This observation is in line with the illustration of Figures 10.2a and b.

11.5. MINNEAPOLIS SCHOOLS

On November 20, 1973, the *Minneapolis Star* newspaper published various statistics on 63 Minneapolis elementary schools, including

- $p = (p_-, p_0, p_+)^T$, the percent of students who scored below average p_-, average p_0, and above average p_+ relative to established norms for a standard sixth-grade comprehension test given in 1972,
- *AFDC*, the percent of children in the school area who receive aid to families with dependent children,
- *BP*, the percentage of children in the school area who do not live with both parents, and
- *HS*, the percent of adults in the school area who completed high school.

Evidently, the *Star's* goal was to allow judgments of a school's academic performance, as reflected by p, against various social and economic variables. A unique feature of this example is that the academic performance variable p forms a trivariate response. Because the sum of the components of p is a constant 100 percent, p can be reduced to the bivariate response $y = (p_-, p_+)^T$

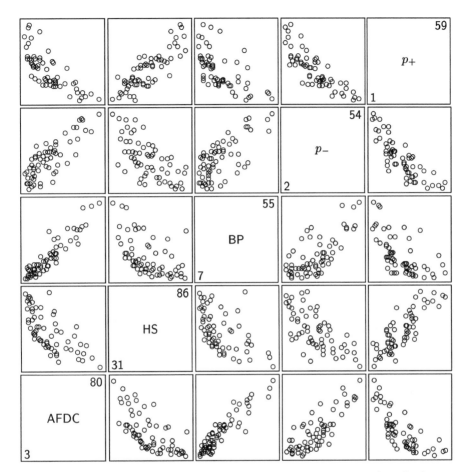

FIGURE 11.3 Scatterplot matrix of the three predictors and two responses from the data on Minneapolis elementary schools.

without loss of information. In the following, we first use SIR to aid in a graphical analysis of the regression of the academic performance variable y on the three predictors $(AFDC, BP, HS)$. And then we turn to a discussion of issues encountered when extending selected results of previous chapters to the multivariate response setting.

Shown in Figure 11.3 is a scatterplot matrix of the data. The regressions $E(HS \mid AFDC)$ and $E(HS \mid BP)$ seem curved. Although the curvature is not great, it represents a potential problem for SIR. To compensate, simultaneous power transformations of the predictors were estimated using the procedure described in Section 9.1. The 3D plot of the resulting transformed predictors

$$x = (AFCD^{1/2}, \log(HS), \log(BP))^T$$

TABLE 11.2. Test Results from Application of SIR to the Minneapolis School Data

m	$\hat{\Delta}_m$	DF	p-value
Response (p_-, p_+)			
0	95.71	57	0.001
1	43.39	36	0.185
2	18.88	17	0.335

shows no notable curvature and seems to sustain the assumption of approximate multivariate normality on the transformed scale. The transformed predictors will be used in the rest of this discussion.

Although SIR was previously described in the context of regressions with a univariate response, the same theory applies with multivariate responses. In particular, the goal of a SIR analysis is still to estimate a basis for the central subspace $\mathcal{S}_{y|x}$. For the bivariate response of this example, the methodological change entails double slicing (p_-, p_+) in the plane, rather than slicing a univariate response. In effect, the bivariate response is replaced with a discrete bivariate response $(\tilde{p}_-, \tilde{p}_+)$, assuming that $\mathcal{S}_{(\tilde{p}_-, \tilde{p}_+)|x} = \mathcal{S}_{(p_-, p_+)|x}$. Once the bivariate slices are constructed, the SIR methodology follows the steps described previously.

SIR was applied with the transformed predictors by first partitioning the 63 observations on p_+ into five slices containing about 13 cases each. The values of p_- in each slice were then partitioned into four slices of about three observations. In this way the data were partitioned into 20 slices, each containing about three observations. The SIR test results shown in Table 11.2 suggest that the bivariate regression has 1D structure so that only a single linear combination of the transformed predictors is required to characterize the regression. Let $b^T x$ denote the estimated linear combination provided by the SIR analysis. The finding of 1D structure seems to be the exception rather than the rule in regressions with a bivariate response. However, it may be reasonable in the present example if we view both responses as measuring similar aspects of a school's academic performance.

Practically, we now have a regression with a bivariate response and a single predictor $b^T x$. The nature of the dependence of the response $y = (p_-, p_+)^T$ on the value of $b^T x$ can be investigated graphically by using a number of different procedures. For example, as described in Section 2.5, we could brush a plot of $b^T x$ while observing the linked points in a scatterplot $\{p_-, p_+\}$ of the responses. This operation indicates that $E(p_+ \mid b^T x)$ is a decreasing function of the value of $b^T x$, while $E(p_- \mid b^T x)$ is an increasing function of $b^T x$. Additional discussion of brushing is available in Chapter 13.

Alternatively, we could view all three variables—p_-, p_+ and $b^T x$ – in a 3D plot. This possibility raises a new issue: How might we usefully characterize a 3D plot of a bivariate response and a univariate predictor? Our previous discussions of 3D plots were confined mostly to regressions with a univariate response and two predictors, and centered on determining structural dimension and sufficient 2D summary plots.

Assume 1D structure for the regression of y on x, and let the vector β span the central subspace. Then

$$y \perp\!\!\!\perp x \mid \beta^T x. \tag{11.17}$$

In addition, suppose that there is a nonzero vector α so that

$$\beta^T x \perp\!\!\!\perp y \mid \alpha^T y. \tag{11.18}$$

One implication of this supposition is that $\mathcal{S}(\alpha)$ is a dimension-reduction subspace for the inverse regression of $\beta^T x$ on y. Further, interchanging the roles of x and y, the pair of conditions given in equations (11.17) and (11.18) is equivalent to the pair of conditions

$$\text{(a)} \ x \perp\!\!\!\perp y \mid \alpha^T y \quad \text{and} \quad \text{(b)} \ \alpha^T y \perp\!\!\!\perp x \mid \beta^T x. \tag{11.19}$$

Condition (a) indicates that $\mathcal{S}(\alpha)$ is also a dimension-reduction subspace for the inverse regression of x on y, while (b) indicates that $\mathcal{S}(\beta)$ is a dimension-reduction subspace for the regression of $\alpha^T y$ on x.

To explore another implication of (11.19), let (α, α_o) be a basis for $\mathbb{R}^2$, let f represent a density, and consider the following relationships:

$$
\begin{aligned}
f_{\alpha^T y, \alpha_o^T y \mid x} &= f_{\alpha^T y, \alpha_o^T y \mid \beta^T x} \\
&= f_{\alpha_o^T y \mid (\beta^T x, \alpha^T y))} f_{\alpha^T y \mid \beta^T x} \\
&= f_{\alpha_o^T y \mid \alpha^T y} f_{\alpha^T y \mid \beta^T x}.
\end{aligned}
$$

The first equality follows from (11.17), the second is standard, and the third equality follows from (11.18). Taken together they imply that the fundamental structure of the regression of y on x rests with the regression of $\alpha^T y$ on $\beta^T x$ because y depends on $\beta^T x$ via $\alpha^T y$. Under such structure, $\{\alpha^T y, \beta^T x\}$ seems to be a useful summary of the regression.

If $\beta^T x$ were known, then from (11.18) we may be able to gain information on α by investigating the regression of $\beta^T x$ on y. And this suggests that we may be able to estimate α graphically by replacing $\beta^T x$ with its estimate $b^T x$ and then using the method of Chapter 4 to investigate the structural dimension of the 3D plot $\{b^T x, (p_-, p_+)\}$. This operation led to the conclusion of 1D structure, thus supporting condition (11.18), and to the 2D summary plot shown in Figure 11.4. The horizontal axis is

$$b^T x = 0.58 \log(BP) + 0.74 \log(HS) - 0.34 AFDC^{1/2}$$

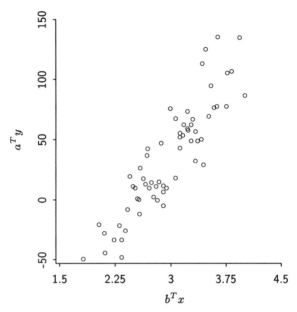

FIGURE 11.4 Summary plot for the Minneapolis school data.

and vertical axis contains the estimate $a^T y$ from a visual analysis of the 3D plot $\{b^T x, (p_-, p_+)\}$,

$$a^T y = 2.4 p_+ - p_-.$$

Reasoning from the estimates instead of the population values, an interpretation of this summary plot is as follows. First think of generating an observation on $b^T x$ from its marginal distribution. Once this value is known, the rest of x furnishes little if any information on y. Given $b^T x$, a value of $a^T y \mid b^T x$ is generated according to its conditional distribution; information on this conditional distribution is available in the summary plot of Figure 11.4. The rest of the response vector is then generated from the conditional distribution of y given $a^T y$. For example, the plots $\{p_+, a^T y\}$ and $\{p_-, a^T y\}$ shown in the two frames of Figure 11.5 depict the marginal relationships between p_+ and $a^T y$, and between p_- and $a^T y$.

11.6. DISCUSSION

The version of SIR described in this chapter is based on fixed slices: As the sample size grows the so does the number of observations per slice, but the slices themselves remain fixed. In contrast, Hsing and Carroll (1992) develop a version of SIR in which each slice is constructed to contain 2 observations.

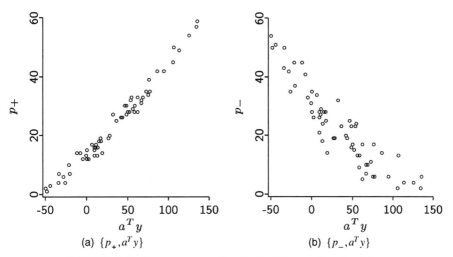

FIGURE 11.5   Response scatterplots for the Minneapolis school data.

This two-slice method, which was originally suggested by Li (1991), differs from the fixed-slice method because the number of slices grows with the sample size rather than the number of observations per slice. The results of Hsing and Carroll were extended by Zhu and Ng (1995) to allow for a fixed number c of observations per slice. The distribution of the SIR test statistic presented in Proposition 11.5 is not generally applicable under the two-slice method.

Proposition 11.1, $S\{\mathrm{Var}[\mathrm{E}(z \mid y)]\} = S_{\mathrm{E}(z\mid y)}$, forms the essential basis for SIR. Once the discrete version $\tilde{y}$ of y is constructed, the covariance matrix $\mathrm{Var}[\mathrm{E}(z \mid \tilde{y})]$ is estimated by using the intraslice predictor averages. Zhu and Fang (1996) bypass the slicing step and use kernel smoothing to estimate $\mathrm{Var}[\mathrm{E}(z \mid y)]$.

Suppose that, based on prior information or inspection of the plots $\{z_j, y\}$, $j = 1, \ldots, p$, we conclude that the inverse regression functions are sufficiently straightforward to be captured by a common family of parametric functions. Specifically, assume that

$$\mathrm{E}(z_j \mid y) = \beta_{1j} g_1(y) + \cdots + \beta_{qj} g_q(y)$$

where the g_k's are known real-valued linearly independent functions of y. This can be rewritten as

$$\mathrm{E}(z \mid y) = B \times G(y)$$

where $G(y) = (g_j(y))$ and B is the $p \times q$ matrix with elements β_{ik}. Assuming span$\{G(y)\} = \mathbb{R}^p$, the inverse regression subspace is just $S_{\mathrm{E}(z\mid y)} = S(B)$. Thus, proceeding parametrically we may be able to bypass the need to estimate

Var[E($z \mid y$)], and instead base an estimate of $\mathcal{S}_{E(z \mid y)}$ on an estimate of B. This line of reasoning leading to a parametric version of SIR was developed by Bura (1996, 1997).

In SIR the estimate of the dimension of $\mathcal{S}_{E(y \mid x)}$ is based on a series of nested tests, as described following (11.2). The procedure is designed for control of the nominal level of each test within the series, leaving the overall level open to question. A Bonferroni inequality could be used to bound the overall level. Assuming that $\mathcal{S}_{y \mid x} = \mathcal{S}_{E(y \mid x)}$, Ferré (1997, 1998) proposed a "model selection" procedure for estimating $\mathcal{S}_{E(y \mid x)}$, similar in spirit to Akaike's criterion. The basic idea is to optimize a measure of closeness between $\mathcal{S}_{E(y \mid x)}$ and its estimate.

Additional discussion of SIR is available from Cook and Weisberg (1991b), Härdle and Tsybakov (1991), Kent (1991), Kötter (1996), and Schott (1994). Sheather and McKean (1997) studied simple nonparametric methods for testing if the dimension of the SIR subspace is greater than 1.

PROBLEMS

11.1. In reference to Proposition 11.1, show that $\mathcal{S}\{\text{Var}[E(z \mid y)]\} \subset \mathcal{S}_{E(z \mid y)}$.

11.2. Show how Proposition 11.2 can be used in the completion of Proposition 11.3, and justify Proposition 11.4.

11.3. Complete the investigation of the asymptotic distribution of $\hat{\Lambda}_d$ when $\mathcal{S}_{E(z \mid y)}$ is a nontrivial proper subset of $\mathcal{S}_{z \mid y}$, as described at the end of Section 11.3. Is it possible to find conditions which force an asymptotic chi-square distribution? Do your conditions seem more or less restrictive than those used in Proposition 11.5?

11.4. Show that the pair of conditions given in (11.17) and (11.18) is equivalent to the pair of conditions given in (11.19).

11.5. The original trivariate response vector (p_+, p_0, p_-) for the Minneapolis school data is an example of compositional data consisting of vectors of proportions adding to 1. There is a long history of discussion on how to best approach the analysis of compositional data. Aitchison (1982, 1986) argued that the analyses are best conducted in terms of logarithms of ratios. Following Aitchison's advice, the bivariate response for the Minneapolis school data would be

$$Y = (Y_1, Y_2)^T = \{\log(p_+/p_0), \log(p_-/p_0)\}^T$$

11.5.1. Consider repeating the analysis of Section 11.5 using Y instead of (p_+, p_-). Describe the differences you expect. Or repeat the analysis and describe the differences you observe.

11.5.2. With either Y or (p_+, p_-) as the response vector, what are the advantages and disadvantages of using $\log(AFDC/(100 - AFDC))$ instead of $AFDC^{-1/2}$ as a predictor, with the remaining predictors as given in Section 11.5?

11.6. Apply SIR to the two univariate regressions for naphthalene data as described in Section 9.1. Specifically, use SIR to study the regression of Y_N on three transformed predictors, and the regression of Y_P on the three transformed predictors. Does SIR give results that are in reasonable agreement with the graphical regression results of Section 9.1? Suggest explanations for any notable differences.

11.7. Again in reference to the naphthalene data, use SIR to study the bivariate regressions of (a) (Y_N, Y_P) on $(AN, Btemp)$, and (b) (Y_N, Y_P) on $(AN, Ctime, Btemp)$. In each case, give a graphical summary of your results.

11.8. Without performing any calculations, give a qualitative description of the results you would expect from using SIR to study (a) the wheat protein regression discussed in Section 9.2 and (b) the reaction yield regression discussed in Section 9.3.

CHAPTER 12

Principal Hessian Directions

Methods based on the inverse regression function for estimating vectors in the central subspace may be handicapped when the forward regression function has little linear trend. Similar restrictions apply to methods based on the Li–Duan Proposition, as discussed in Section 8.1.2. In such cases, information missed by those methods might be recovered by using second moments. Graphical methods based on the inverse variance function were discussed in Section 10.2, for example. In this section we consider a second-moment paradigm for extracting information on the central subspace. The idea is based on *the method of principal Hessian directions (pHd)*, which is due to Li (1992). As in previous chapters, discussion of this idea will again be directed toward data (y_i, x_i), $i = 1, \ldots, n$, that are independent and identically distributed observations on (y, x).

Li motivated the notion of a principal Hessian direction as follows: Let $H(x)$ denote the $p \times p$ Hessian matrix of the forward regression function,

$$H(x) = \frac{\partial^2 E(y \mid x)}{\partial x \partial x^T}$$

$$= \frac{\partial^2 E(y \mid \eta^T x)}{\partial x \partial x^T}$$

$$= \eta \frac{\partial^2 E(y \mid \eta^T x)}{\partial (\eta^T x) \partial (x^T \eta)} \eta^T$$

where η is a basis for the central subspace $\mathcal{S}_{y|x}$. It follows from this representation that the Hessian matrix $H(x)$ is degenerate along directions that are orthogonal to $\mathcal{S}_{y|x}$. Li based his procedure on an estimate of $E[H(x)]$, assuming normally distributed predictors to allow application of Stein's (1981, Lemma 4) lemma on the expectation of the derivative of a function of a normal random variable. While the development in this chapter is related to Li's, it differs in several ways that generalize and refine both conceptual foundations and methodology, as described by Cook (1998). As in the development of

SIR, discussion will mostly be in terms of the standardized predictor

$$z = \Sigma^{-1/2}(x - E(x))$$

with sample version $\hat{z}$ constructed by replacing Σ and $E(x)$ with their corresponding estimates, $\hat{\Sigma}$ and $\bar{x}$.

The overarching goal in this chapter is to infer about the central subspace $S_{y|z}(\gamma)$, as it was in previous chapters. However, here we approach the problem indirectly through residuals. The possibility of applying pHd methodology with the response is discussed later in Section 12.7.

12.1. INCORPORATING RESIDUALS

Let

$$\hat{e}_i = y_i - \bar{y} - \hat{\beta}^T \hat{z}_i$$

denote the ith sample residual from the OLS linear regression of y on $\hat{z}$, and let

$$e = y - E(y) - \beta^T z$$

denote a typical population OLS residual, where $\beta = \text{Cov}(z, y)$. Additionally, let $S_{e|z}(\rho)$ denote the central subspace, with basis given by the columns of the matrix ρ, for the regression of e on z.

In broad terms, inferences about $S_{y|z}$ will be formed by combining inferences about $S_{e|z}$ with $\hat{\beta}$. But we need to be a little careful because the regression of e on z could turn out to be more complicated than the original regression of y on z: It is possible to have $\dim[S_{e|z}] > \dim[S_{y|z}]$, a situation that would not normally seem desirable. Such situations arise when $\beta \notin S_{y|z}$ so that computation of the residuals essentially adds the dimension determined by β to the central subspace. The next proposition uses linear conditional expectations to avoid this possibility and to establish a simple relationship between $S_{y|z}$, $S_{e|z}$ and $S(\beta)$. Recall that P_A is the projection matrix for $S(A)$ relative to the usual inner product and $Q_A = I - P_A$.

Proposition 12.1. Let the columns of the matrices γ and ρ be bases for the central subspaces $S_{y|z}$ and $S_{e|z}$, respectively, and let $\beta = \text{Cov}(z, y)$. If $E(z \mid \gamma^T z) = P_\gamma z$, then

$$S_{y|z} = S_{e|z} + S(\beta). \tag{12.1}$$

Justification. From the definitions of e and ρ

$$y - \beta^T z \perp\!\!\!\perp z \mid \rho^T z.$$

Making use of Propositions 4.4 and 4.5, this implies that

$$y - \beta^T z \perp\!\!\!\perp z \mid (\rho^T z, \beta^T z)$$

and

$$(y - \beta^T z, \beta^T z) \perp\!\!\!\perp z \mid (\rho^T z, \beta^T z).$$

Thus,

$$y \perp\!\!\!\perp z \mid (\rho^T z, \beta^T z)$$

which means that $\mathcal{S}(\rho, \beta)$ is a dimension-reduction subspace for the regression of y on z:

$$\mathcal{S}_{y|z} \subset \mathcal{S}_{e|z} + \mathcal{S}(\beta) = \mathcal{S}(\rho, \beta). \tag{12.2}$$

From the Li–Duan Proposition, the condition $E(z \mid \gamma^T z) = P_\gamma z$ forces $\beta \in \mathcal{S}_{y|z}$. Therefore $\mathcal{S}_{y|z}$ is a dimension-reduction subspace for the regression of the residuals e on z, and $\mathcal{S}_{e|z} \subset \mathcal{S}_{y|z}$. Using this result with (12.2) yields the desired conclusion:

$$\mathcal{S}_{y|z} \subset \mathcal{S}(\rho, \beta) \subset \mathcal{S}(\gamma, \beta) = \mathcal{S}_{y|z}. \qquad \square$$

The requirement of a linear conditional expectation $E(z \mid \gamma^T z) = P_\gamma z$ was used in Proposition 12.1 to insure that $\beta \in \mathcal{S}_{y|z}$. If $\beta \notin \mathcal{S}_{y|z}$ then the structural dimension of the residual regression will be greater than the structural dimension of the original regression, $\dim[\mathcal{S}_{e|z}] > \dim[\mathcal{S}_{y|z}]$. Regression analyses are often approached iteratively, using residual plots to indicate deficiencies and to suggest ways of improving a model. (See, for example, Cook and Weisberg 1982.) The possibility that $\dim[\mathcal{S}_{e|z}] > \dim[\mathcal{S}_{y|z}]$ casts some doubt on the usefulness of the iterative fit-residual paradigm when $E(z \mid \gamma^T z)$ is nonlinear.

For example, let z_1 be a standard normal random variable, let $z_2 \mid z_1 = z_1^2 - 1 + \varepsilon_1$, and let

$$y \mid (z_1, z_2) = 0.5(z_1^2 + 1) + \varepsilon$$

where ε_1 and ε are independent standard normal errors. Then $\beta = (0, 1)^T$ and $\dim[\mathcal{S}_{y|z}] = 1$. Although the regression of y on z has 1D structure, the regression of $e = y - z_2$ on z has 2D structure.

In cases where a linear conditional expectation is in doubt, the essential requirement that $\beta \in \mathcal{S}_{y|z}$ might be induced to an adequate approximation by using weighted least squares, with weights determined according the Voronoi weighting algorithm discussed in Section 8.4.1. Also, if it happens that the data follow a linear model

$$y \mid z = \beta_0 + \beta^T z + \varepsilon$$

then, regardless of the distribution of the predictors, $\mathcal{S}_{e|z} = 0$ and again $\beta \in \mathcal{S}_{y|z}$.

In this chapter, inferences about $\mathcal{S}_{y|z}$ will be formed by combining inferences on $\mathcal{S}(\beta)$ with inferences on $\mathcal{S}_{e|z}$ under the conditions of Proposition 12.1. The Li–Duan Proposition shows that the OLS estimate $\hat{\beta}$, or a suitable robust version, can be used as the basis for inference on $\mathcal{S}(\beta)$. pHd will be used for $\mathcal{S}_{e|x}$. However, rather than tackling $\mathcal{S}_{e|z}$ directly, pHd provides a method for inferring about subspaces of $\mathcal{S}_{e|z}$ under conditions that are similar to those of Proposition 12.1.

pHd is based on connections between the central subspace $\mathcal{S}_{e|z}$ and the subspace $\mathcal{S}_{ezz}$ spanned by the eigenvectors corresponding to the nonzero eigenvalues of the population moment matrix

$$\Sigma_{ezz} = \mathrm{E}(e \times zz^T). \tag{12.3}$$

That is, $\mathcal{S}_{ezz} = \mathcal{S}(\Sigma_{ezz})$. The subspace $\mathcal{S}_{ezz}$ can be estimated from the eigenstructure of the straightforward moment estimate of Σ_{ezz}. If a usefully simple connection can be established between $\mathcal{S}_{e|z}$ and $\mathcal{S}_{ezz}$, then inference about $\mathcal{S}_{e|z}$ can be based on an estimate of $\mathcal{S}_{ezz}$.

For later reference, let $\ell_1, \ldots, \ell_p$ denote the eigenvectors corresponding to the ordered absolute eigenvalues $|\delta_1| \geq \cdots \geq |\delta_p|$ of Σ_{ezz}, and let

$$\kappa = \mathrm{rank}(\Sigma_{ezz}) = \dim(\mathcal{S}_{ezz}).$$

Li (1992) called the eigenvectors $\ell_1, \ldots, \ell_\kappa$ *principal Hessian directions*. The apparent rationale for this name will be mentioned later in Section 12.2.3.

12.2. CONNECTING $\mathcal{S}_{e|z}$ AND $\mathcal{S}_{ezz}$ WHEN...

The conditions necessary for establishing useful connections accumulate over the following sections.

12.2.1. ... $\mathrm{E}(z \mid \rho^T z) = P_\rho z$

Recall from Proposition 10.1 that if $\mathrm{E}(z \mid \rho^T z) = P_\rho z$ then $\mathrm{E}(z \mid e) \in \mathcal{S}_{e|z}(\rho)$ and thus

$$\mathrm{E}(z \mid e) = P_\rho \mathrm{E}(z \mid e).$$

Further, from (10.5) in the justification of Proposition 10.2, the linearity condition also implies that

$$\Sigma_{z|e} = \mathrm{E}[\mathrm{Var}(z \mid \rho^T z) \mid e] + P_\rho \Sigma_{z|e} P_\rho \tag{12.4}$$

where $\Sigma_{z|e} = \text{Var}(z \mid e)$. Combining these results we obtain

$$E(zz^T \mid e) = \Sigma_{z|e} + E(z \mid e)E(z^T \mid e) \tag{12.5}$$

$$= E[\text{Var}(z \mid \rho^T z) \mid e] + P_\rho \Sigma_{z|e} P_\rho + P_\rho E(z \mid e)E(z^T \mid e)P_\rho$$

$$= E[\text{Var}(z \mid \rho^T z) \mid e] + P_\rho E(zz^T \mid e)P_\rho$$

$$= Q_\rho E[\text{Var}(z \mid \rho^T z) \mid e]Q_\rho + P_\rho E(zz^T \mid e)P_\rho. \tag{12.6}$$

Multiplying both sides of (12.6) by e and averaging gives the desired relation, which is stated in the following proposition to facilitate later reference.

Proposition 12.2. Let ρ be a basis for the central subspace $S_{e|z}$ and assume that $E(z \mid \rho^T z) = P_\rho z$. Then

$$\Sigma_{ezz} = Q_\rho E[e \times \text{Var}(z \mid \rho^T z)]Q_\rho + P_\rho \Sigma_{ezz} P_\rho. \tag{12.7}$$

This proposition is useful because it tells us that, when $E(z \mid \rho^T z)$ is linear, the eigenvectors of Σ_{ezz} corresponding to its nonzero eigenvalues will be in either $S_{e|z}(\rho)$ or $S_{e|z}^\perp(\rho)$, the orthogonal complement of $S_{e|z}(\rho)$. There is no clear way to separate the eigenvectors in $S_{e|z}(\rho)$ from those in $S_{e|z}^\perp(\rho)$ based just on Σ_{ezz}, but we may be able to use the basic graphical methods of Chapter 4 for that task once inference procedures on the eigenstructure of Σ_{ezz} have been developed. If $S_{e|z} = S(P_\rho \Sigma_{ezz} P_\rho)$ then $S_{e|z} \subset S_{ezz}$ and consequently S_{ezz} is a DRS, although it is not necessarily minimal.

The following example may help fix ideas. Partition the standardized predictor vector as

$$z = \begin{pmatrix} z_1 \\ z_2 \\ z_3 \end{pmatrix}$$

where z_1 and z_2 are scalars and the remaining predictors are collected in the $(p-2) \times 1$ vector z_3. Assume that

$$e \mid z = f(z_1) + \varepsilon$$

$$z_2 \mid z_1 = \sigma(z_1)\varepsilon_1$$

$$z_3 \perp\!\!\!\perp (z_1, z_2)$$

where ε and ε_1 are independent errors with mean 0 and variance 1, and $E(f(z_1)) = 0$. Under this structure $S_{e|z} = S(\rho)$, where $\rho = (1, 0, \ldots, 0)^T$. The condition required by Proposition 12.2 holds because $E(z \mid z_1) = z_1\rho$. In addition,

$$\text{Var}(z \mid z_1) = \begin{pmatrix} 0 & 0 & 0 \\ 0 & \sigma^2(z_1) & 0 \\ 0 & 0 & I_{p-2} \end{pmatrix}$$

The implication of these results for Σ_{ezz} can be seen from the following decomposition,

$$\Sigma_{ezz} = E[zz^T E(e \mid z)]$$

$$= E[f(z_1)E(zz^T \mid z_1)]$$

$$= E\{f(z_1)[\text{Var}(z \mid z_1) + E(z \mid z_1)E(z^T \mid z_1)]\}$$

$$= E[f(z_1)\sigma^2(z_1)]\nu\nu^T + E[f(z_1)z_1^2]\rho\rho^T$$

where $\nu = (0,1,0,\ldots,0)^T$. If $E[f(z_1)\sigma^2(z_1)] \neq 0$ and $E[f(z_1)z_1^2] \neq 0$ then $\kappa = 2$, the extra dimension represented by ν arising because of the nonconstant conditional variance $\text{Var}(z \mid z_1)$. If, in addition,

$$|E[f(z_1)\sigma^2(z_1)]| < |E[f(z_1)z_1^2]|$$

then the eigenvector ℓ_1 corresponding to the largest absolute eigenvalue $|\delta_1| = |E[f(z_1)z_1^2]|$ will equal ρ, the vector that spans $\mathcal{S}_{e|z}$. On the other hand, if

$$|E[f(z_1)\sigma^2(z_1)]| > |E[f(z_1)z_1^2]|$$

then $\ell_1 = \nu$, which is orthogonal to the central subspace. In either case, if we knew that $\kappa = \text{rank}(\Sigma_{ezz}) = 2$ then the graphical methods of Chapter 4 could be applied to the population 3D summary plot $\{e, (\ell_1^T z, \ell_2^T z)\}$ and thereby used to determine that the structural dimension is really 1, and to decide that ρ spans the central subspace. We will eventually need a method for inferring about κ for these ideas to be applicable in practice.

The present example can also be used to highlight another characteristic of the methodology under development. Namely, that there is not necessarily a useful connection between the eigenstructure of Σ_{ezz} and the central subspace $\mathcal{S}_{e|z}$. To enforce this point, suppose that z is normally distributed and that $y \mid z = z_1^3 + \varepsilon$. This implies that $e \mid z = z_1^3 - 3z_1 + \varepsilon$ so that $f(z_1) = z_1^3 - 3z_1$. Thus,

$$E[f(z_1)z_1^2] = E(z_1^5 - 3z_1^3) = 0$$

so that $\kappa = 1$ and $\ell_1 = \nu$. In this very special case, there is no useful relationship between the eigenstructure of Σ_{ezz} and the central subspace because the single unique eigenvector of Σ_{ezz} is not in $\mathcal{S}_{e|z}$. Nevertheless, the careful balance needed for such an occurrence would seem to be the exception rather than the rule in practice.

Finally, recall that the predictor constraint $E(z \mid \gamma^T z) = P_\gamma z$ was used in Proposition 12.1 to insure that the central subspace for the regression of e on z is contained in the central subspace for the regression of y on z. That constraint is similar to the constraint $E(z \mid \rho^T z) = P_\rho z$ used in Proposition 12.2 to force a connection between $\mathcal{S}_{ezz}$ and $\mathcal{S}_{e|z}$. While neither constraint implies the other, it may be likely in practice that when one holds the other will also since the two spaces "differ" only by the single vector β.

12.2.2. ... $E(z \mid \rho^T z) = P_\rho z$ and $\mathrm{Var}(z \mid \rho^T z) = Q_\rho$

The following proposition shows that adding the constant variance condition $\mathrm{Var}(z \mid \rho^T z) = Q_\rho$ forces $\mathcal{S}_{ezz} \subset \mathcal{S}_{e|z}$.

Proposition 12.3.   Let ρ be a basis for the central subspace $\mathcal{S}_{e|z}$, and let $\mathcal{S}_{ezz} = \mathcal{S}(\Sigma_{ezz})$. If

1. $E(z \mid \rho^T z) = P_\rho z$ and
2. $\mathrm{Var}(z \mid \rho^T z) = Q_\rho$

then $\Sigma_{ezz} = P_\rho \Sigma_{ezz} P_\rho$ and thus $\mathcal{S}_{ezz} \subset \mathcal{S}_{e|z}$.

Justification.   According to Proposition 12.2, the first condition of the proposition implies that

$$\Sigma_{ezz} = Q_\rho E[e \times \mathrm{Var}(z \mid \rho^T z)]Q_\rho + P_\rho \Sigma_{ezz} P_\rho.$$

Substituting $\mathrm{Var}(z \mid \rho^T z) = Q_\rho$ into this expression gives

$$\Sigma_{ezz} = P_\rho \Sigma_{ezz} P_\rho$$

from which the conclusions follow.   □

Proposition 12.3 combines the condition $\mathrm{Var}(z \mid \rho^T z) = Q_\rho$ and the linear mean requirement from Proposition 12.3. The result is that the eigenvectors of Σ_{ezz} corresponding to its nonzero eigenvalues all fall in $\mathcal{S}_{e|z}$ and there is no longer a possibility that some will fall in $\mathcal{S}_{e|z}^\perp$. As in the previous section, if we can determine the rank of Σ_{ezz} then we may be able to discover useful information about the central subspace.

Nevertheless, there is still no guarantee that $\mathcal{S}_{ezz} = \mathcal{S}_{e|z}$. For example, suppose that z is normally distributed, that the central subspace has dimension 1, and that $e \mid z = \sigma(\rho^T z)\varepsilon$, where $z \perp\!\!\!\perp \varepsilon$ and $E(\varepsilon) = 0$. Then

$$\Sigma_{ezz} = E[(\sigma(\rho^T z)\varepsilon)zz^T]$$

$$= E[\sigma(\rho^T z)zz^T]E(\varepsilon)$$

$$= 0.$$

Again, $\mathcal{S}_{ezz}$ is a proper subset of $\mathcal{S}_{e|z}$. This example suggests more generally that pHd may not be effective at finding heteroscedasticity due to linear combinations that are not otherwise reflected by Σ_{ezz}.

12.2.3. ...z is normally distributed

Finally, when z is normally distributed Li (1992) showed by using Stein's (1981, Lemma 4) lemma that the moment matrix Σ_{ezz} equals the average Hessian matrix for the regression function:

$$\Sigma_{ezz} = E\left(\frac{\partial^2 E(e \mid z)}{\partial z \partial z^T}\right) = \rho E\left(\frac{\partial^2 E(e \mid \rho^T z)}{\partial(\rho^T z)\partial z^T \rho}\right)\rho^T.$$

This result allows a relatively straightforward interpretation of Σ_{ezz}. It suggests, for example, that pHd will not be useful for finding structure that is not manifest through the residual regression function, reinforcing the idea mentioned in the last section that pHd may not respond to heteroscedasticity.

Of course, the results of the previous section hold when z is normal so we still have that $\mathcal{S}_{ezz} \subset \mathcal{S}_{e|z}$. The name *principal Hessian direction* stems from the fact that the eigenvectors of Σ_{ezz} for its nonzero eigenvalues are the same as the corresponding eigenvectors of the average Hessian matrix. The name is not strictly applicable when z is not normal because the connection with the Hessian matrix may be lost. Nevertheless, we will continue to use the name in reference to the eigenvectors corresponding to the nonzero eigenvalues of Σ_{ezz}.

12.3. ESTIMATION AND TESTING

An estimate of $\Sigma_{exx} = E\{e \times (x - E(x))(x - E(x))^T\}$ can be constructed by substituting corresponding sample quantities:

$$\hat{\Sigma}_{exx} = \frac{1}{n}\sum_{i=1}^{n}\hat{e}_i(x_i - \bar{x})(x_i - \bar{x})^T. \qquad (12.8)$$

Similarly, the corresponding sample version of Σ_{ezz} can be represented as

$$\hat{\Sigma}_{ezz} = \hat{\Sigma}^{-1/2}\hat{\Sigma}_{exx}\hat{\Sigma}^{-1/2} \tag{12.9}$$

where

$$\hat{\Sigma} = \frac{1}{n}\sum_{i=1}^{n}(x_i - \bar{x})(x_i - \bar{x})^T.$$

When $\kappa = \dim[S_{ezz}]$ is known, the pHd estimate $\hat{S}_{ezz}$ of S_{ezz} is the subspace spanned by the κ eigenvectors $\hat{\ell}_1,\ldots,\hat{\ell}_\kappa$ corresponding the κ largest of the absolute eigenvalues $|\hat{\delta}_1| \geq |\hat{\delta}_2| \geq \cdots \geq |\hat{\delta}_p|$ of $\hat{\Sigma}_{ezz}$. The eigenvectors $\hat{\ell}_j$ can then be backtransformed to vectors $\hat{u}_j$ in the original scale, $\hat{u}_j = \hat{\Sigma}^{-1/2}\hat{\ell}_j$. The resulting linear combinations $\hat{u}_j^T x$, $j = 1,\ldots,p$ will be called the pHd *predictors*.

How the estimate $\hat{S}_{ezz}$ might be used in practice to infer about $S_{e|z}$ and eventually about $S_{y|z}$ depends on the applicable conditions as described in Sections 12.2.1–12.2.3. Before discussing the possibilities, it is necessary to consider inference on κ.

We could infer about κ and simultaneously estimate S_{ezz} by using all p of the pHd predictors $\hat{u}_1^T x,\ldots,\hat{u}_p^T x$ as predictors in a graphical regression analysis. This possibility is the same as that discussed in connection with SIR in Section 11.2. The only difference is that the pHd predictors are used instead of the SIR predictors. The ordering of the pHd predictors $\hat{u}_j^T x$ reflects their relative contributions to explaining the variation in y as judged by pHd, with $\hat{u}_1^T x$ having the greatest contribution. If κ is small and if $S_{y|z} = S_{ezz} + S(\beta)$ then few graphical regression plots should be needed to settle on visual estimates of κ and $S_{y|z}$.

Additionally, an estimate of κ could be based on the test statistic $\hat{\Delta}_m$ proposed by Li (1992) for the hypothesis $\kappa = m$,

$$\hat{\Delta}_m = \frac{n\sum_{j=m+1}^{p}\hat{\delta}_j^2}{2\,\hat{\mathrm{Var}}(e)}$$

where $\hat{\mathrm{Var}}(e)$ is a consistent estimate of the marginal variance of e. After finding its asymptotic distribution, this statistic can be used to infer about κ in the same way that $\hat{\Lambda}_m$ is used in SIR.

12.3.1. Asymptotic distribution of $\hat{\Delta}_\kappa$

We need the following quantities for a description of the asymptotic distribution of $\hat{\Delta}_\kappa$. Let θ_0 be a $p \times (p - \kappa)$ matrix with columns given by the eigenvectors corresponding to the zero eigenvalues of Σ_{ezz}. Next, define the $(p - \kappa) \times 1$ vector $V = \theta_0^T z$ and let v_j denote its jth element. Finally, define the

$(p - \kappa)(p - \kappa + 1)/2 \times 1$ vector

$$W = \begin{pmatrix} \begin{pmatrix} v_1^2 - 1 \\ \sqrt{2}v_1 v_2 \\ \sqrt{2}v_1 v_3 \\ \vdots \\ \sqrt{2}v_1 v_{p-\kappa} \end{pmatrix} \\ \vdots \\ \begin{pmatrix} v_j^2 - 1 \\ \sqrt{2}v_j v_{j+1} \\ \vdots \\ \sqrt{2}v_j v_{p-\kappa} \end{pmatrix} \\ \vdots \\ \begin{pmatrix} v_{p-\kappa-1}^2 - 1 \\ \sqrt{2}v_{p-\kappa-1} v_{p-\kappa} \\ (v_{p-\kappa}^2 - 1) \end{pmatrix} \end{pmatrix} \tag{12.10}$$

The distribution of $\hat{\Delta}_\kappa$ is the subject of the next proposition.

Proposition 12.4.   Let $\kappa = \dim[\mathcal{S}_{ezz}]$. The asymptotic distribution of $\hat{\Delta}_\kappa$ is the same as the distribution of

$$C = \frac{1}{2 \operatorname{Var}(e)} \sum_{j=1}^{(p-\kappa)(p-\kappa+1)/2} \omega_j C_j \tag{12.11}$$

where the C_j are independent chi-squared random variables, each with one degree of freedom, and $\omega_1 \geq \omega_2 \geq \cdots \geq \omega_{(p-\kappa)(p-\kappa+1)/2}$ are the eigenvalues of $\operatorname{Var}(eW)$.

Justification.   Li (1992, Appendices B.2 and B.4) presented an argument for a related distribution of $\hat{\Delta}_\kappa$, requiring that the predictors be normally distributed. This development generally follows Li's argument, but we do not require normally distributed predictors. In addition, this development emphasizes portions of Li's argument that will be useful in later discussion.

To investigate the asymptotic distribution of $\hat{\Delta}_\kappa$, we begin by investigating the matrix $\tilde{\Sigma}_{ezz}$ of (12.9), assuming without loss of generality that $E(x) = 0$

and $\mathrm{Var}(x) = I$ so that $z = x$. Let

$$\xi_i = z_i z_i^T - I_p,$$

$$\bar{\xi} = \frac{1}{n} \sum_i^n \xi_i,$$

$$\phi_i = e_i \xi_i - \Sigma_{ezz} \qquad \text{and}$$

$$\bar{\phi} = \frac{1}{n} \sum_i^n \phi_i.$$

Then Li shows that

$$\hat{\Sigma}_{ezz} = \Sigma_{ezz} + B_n + o_p(n^{-1/2})$$

where

$$B_n = \bar{\phi} - \bar{z} b_{ols}^T - b_{ols} \bar{z}^T - (1/2) \bar{\xi} \Sigma_{ezz} - (1/2) \Sigma_{ezz} \bar{\xi}$$

and $b_{ols} = \mathrm{Cov}(z, e)$.

It now follows from Li, or from Eaton and Tyler (1991), that the asymptotic distribution of $\hat{\Delta}_\kappa$ is the same as the asymptotic distribution of

$$\Delta_\kappa^* = \frac{n}{2\,\mathrm{Var}(e)} \mathrm{trace}[(\theta_0^T B_n \theta_0)^2]$$

$$= \frac{n}{2\,\mathrm{Var}(e)} \sum_{i,j}^{p-\kappa} (\theta_0^T B_n \theta_0)_{ij}^2$$

where $(\theta_0^T B_n \theta_0)_{ij}$ denotes the ijth element of $\theta_0^T B_n \theta_0$. Using the facts that $\Sigma_{ezz} \theta_0 = 0$ and $b_{ols} = 0$ we have

$$\theta_0^T B_n \theta_0 = \theta_0^T[\bar{\phi} - \bar{z} b_{ols}^T - b_{ols} \bar{z}^T - (1/2)\bar{\xi}\Sigma_{ezz} - (1/2)\Sigma_{ezz}\bar{\xi}]\theta_0$$

$$= \frac{1}{n} \sum_i^n [e_i(\theta_0^T z_i z_i^T \theta_0 - I_{p-\kappa})] - \theta_0^T \bar{z} b_{ols}^T \theta_0 - \theta_0^T b_{ols} \bar{z}^T \theta_0$$

$$= \frac{1}{n} \sum_i^n [e_i(\theta_0^T z_i z_i^T \theta_0 - I_{p-\kappa})]$$

$$= \frac{1}{n} \sum_i^n [e_i(V_i V_i^T - I_{p-\kappa})] \qquad (12.12)$$

where the $V_i = \theta_0^T z_i$ are independent and identically distributed random vectors with $\mathrm{E}(V) = 0$ and $\mathrm{Var}(V) = I_{p-\kappa}$. This representation shows that $\theta_0^T B_n \theta_0$ is an average of independent and identically distributed matrices with mean 0:

$$E[e \times (VV^T - I)] = E(e \times VV^T)$$

$$= \theta_0^T E[e \times zz^T] \theta_0$$

$$= \theta_0^T \Sigma_{ezz} \theta_0$$

$$= 0.$$

Next, using (2.12) and the definition of W,

$$\Delta_\kappa^* = \frac{1}{2 \operatorname{Var}(e)} \left\| \frac{1}{\sqrt{n}} \sum_{i=1}^{n} e_i W_i \right\|^2.$$

By the multivariate central limit theorem, $1/\sqrt{n} \sum_{i=1}^{n} e_i W_i$ follows an asymptotic normal distribution with mean 0 and covariance matrix $\operatorname{Var}(eW)$. Thus, the asymptotic distribution of Δ_κ is as given in (12.11). (See, for example, Guttman 1982, p. 76.) □

The fact that $b_{ols} = \operatorname{Cov}(e, z) = 0$ plays a crucial role in the development leading to (12.12). Without this condition, the asymptotic distribution of $\hat{\Delta}_\kappa$ will not be as given in Proposition 12.4. This will be relevant when considering the possibility of using pHd with the responses in Section 12.7.

12.3.2. An algorithm for inference on κ

The asymptotic distribution of $\hat{\Delta}_\kappa$ given in Proposition 12.4 leads to the following algorithm for inference on κ.

1. Construct the sample OLS residuals $\hat{e}_i$ and the sample standardized predictors $\hat{z}_i = \hat{\Sigma}^{-1/2}(x_i - \bar{x})$, $i = 1, \ldots, n$, and then obtain the ordered absolute eigenvalues $|\hat{\delta}_1| \geq |\hat{\delta}_2| \geq \cdots \geq |\hat{\delta}_p|$ and the corresponding eigenvectors $\hat{\ell}_1, \ldots, \hat{\ell}_p$ of $\hat{\Sigma}_{ezz}$.
2. For a selected value of $\kappa = \dim[S_{ezz}]$, form
   a. the $p \times (p - \kappa)$ matrix $\hat{\theta}_0 = (\hat{\ell}_{\kappa+1}, \ldots, \hat{\ell}_p)$.
   b. the $(p - \kappa) \times 1$ vectors $\hat{V}_i = \hat{\theta}_0^T \hat{z}_i$, $i = 1, \ldots, n$. $\hat{V}_i = \hat{z}_i$ can be used when using $\kappa = 0$.
   c. the $(p - \kappa)(p - \kappa + 1) \times 1$ vectors $\hat{W}_i$ by using the elements of $\hat{V}_i$ in place of the elements of V in (12.10), $i = 1, \ldots, n$.
3. Estimate $\operatorname{Var}(eW)$ with the usual sample covariance matrix $\hat{\Sigma}_{eW}$ constructed from the vectors $\hat{e}_i \hat{W}_i$, $i = 1, \ldots, n$.

4. Construct the eigenvalues $\hat{\omega}_1 \geq \cdots \geq \hat{\omega}_{(p-\kappa)(p-\kappa+1)/2}$ of $\hat{\Sigma}_{eW}$.

5. The asymptotic distribution of $\hat{\Delta}_\kappa$ is then estimated to be the same as that of

$$\hat{C} = \frac{1}{2\,\hat{\text{Var}}(e)} \sum_{j=1}^{(p-\kappa)(p-\kappa+1)/2} \hat{\omega}_j C_j$$

where the C_j are as described in Proposition 12.4, and $\hat{\text{Var}}(e)$ can be the usual estimate of $\text{Var}(e)$. The p-value is computed as $\Pr(\hat{C} > \text{observed } \hat{\Delta}_\kappa)$. There is a substantial literature on computing tail probabilities of distributions of linear combinations of chi-squared random variables. See Farebrother (1990), Field (1993), and Wood (1989) for an introduction to the literature.

6. Once the p-values for $\kappa = 0, 1, \ldots, p-1$ are available, inference on κ can proceed as described in connection with SIR. Letting $\hat{\kappa}$ be the inferred value of κ, the estimate of $\mathcal{S}_{ezz}$ is given by $\mathcal{S}\{\hat{\ell}_1, \ldots, \hat{\ell}_{\hat{\kappa}}\}$. These eigenvectors can then be backtransformed to $\hat{u}_j = \hat{\Sigma}^{-1/2} \hat{\ell}_j$, $j = 1, \ldots, \hat{\kappa}$, for inference in the original scale.

This inference algorithm for $\mathcal{S}_{ezz}$ does not impose constraints on the marginal distribution of z. Consequently, we are free in practice to consider the relationship between $\mathcal{S}_{ezz}$ and $\mathcal{S}_{y|z}$ without being encumbered by a need to maintain conditions for inference on κ:

- If $E(z \mid \rho^T z)$ is nonlinear, we can still infer about $\mathcal{S}_{ezz}$ but such inference may be of no value because there may be no useful relationship between $\mathcal{S}_{ezz}$ and $\mathcal{S}_{e|z}$.

- If $E(z \mid \rho^T z) = P_\rho z$ then, from Proposition 12.2, the eigenvectors of Σ_{ezz} corresponding to its nonzero eigenvalues will be in either $\mathcal{S}_{e|z}(\rho)$ or $\mathcal{S}_{e|z}^\perp(\rho)$. There is no way to separate the eigenvectors in $\mathcal{S}_{e|z}$ from those in $\mathcal{S}_{e|z}^\perp$ based just on Σ_{ezz}, but we can use the graphical methods of Chapter 4 for that task when $\hat{\kappa}$ is small, as often seems to be the case in practice.

- If $E(z \mid \rho^T z) = P_\rho z$ and $\text{Var}(z \mid \rho^T z) = Q_\rho$ then, from Proposition 12.3, $\mathcal{S}_{ezz} \subset \mathcal{S}_{e|z}$ and inferences about $\mathcal{S}_{ezz}$ apply directly to $\mathcal{S}_{e|z}$.

12.3.3. Asymptotic distribution of $\hat{\Delta}_\kappa$ with constraints

The distribution of $\hat{\Delta}_\kappa$ given in Proposition 12.4 simplifies if additional structure is warranted. Let the columns of the $p \times \kappa$ matrix θ be an orthogonal basis for $\mathcal{S}_{ezz}$ so that (θ, θ_0) is an orthogonal basis for $\mathbb{R}^p$, where θ_0 is as defined in Section 12.3.1. The next lemma shows how the distribution of $\hat{\Delta}_\kappa$ simplifies when the predictors are normal and $\theta^T z$ captures all of the linear combinations of the predictors that drive the residual regression function and the residual variance function.

Proposition 12.5.   Assume the following three conditions:

1. normally distributed predictors
2. $E(e \mid z) = E(e \mid \theta^T z)$
3. $\text{Var}(e \mid z) = \text{Var}(e \mid \theta^T z)$.

Then asymptotic distribution of $\hat{\Delta}_\kappa$ is chi-squared with $(p - \kappa)(p - \kappa + 1)/2$ degrees of freedom.

Justification.   The idea here is that the three conditions allow us to factor $\text{Var}(eW) = \text{Var}(e)\text{Var}(W)$.

The assumption of normal predictors insures that $\theta^T z \perp\!\!\!\perp \theta_0^T z$ and thus using condition 2

$$E(e \mid \theta_0^T z) = E[E(e \mid z) \mid \theta_0^T z]$$
$$= E[E(e \mid \theta^T z, \theta_0^T z) \mid \theta_0^T z]$$
$$= E[E(e \mid \theta^T z) \mid \theta_0^T z]$$
$$= E[E(e \mid \theta^T z)]$$
$$= 0.$$

From this result and conditions 1, 2, and 3 of the proposition,

$$\text{Var}(e \mid \theta_0^T z) = E(e^2 \mid \theta_0^T z)$$
$$= E[\text{Var}(e \mid \theta^T z, \theta_0^T z) \mid \theta_0^T z] + \text{Var}[E(e \mid \theta^T z, \theta_0^T z) \mid \theta_0^T z]$$
$$= E[\text{Var}(e \mid z) \mid \theta_0^T z] + \text{Var}[E(e \mid z) \mid \theta_0^T z]$$
$$= E[\text{Var}(e \mid \theta^T z)] + \text{Var}[E(e \mid \theta^T z)]$$
$$= \text{Var}(e).$$

Now, returning to the end of the proof of Proposition 12.4 and recalling that $V = \theta_0^T z$,

$$\text{Var}(eW) = E(e^2 W W^T)$$
$$= E[E(e^2 \mid \theta_0^T z)W W^T]$$
$$= \text{Var}(e)\text{Var}(W)$$

because $E(W) = 0$. Because z is standard normal, $\theta_0^T z$ is standard normal and $\text{Var}(W) = 2I$, as can be shown by using straightforward calculation. Hence, the eigenvalues of $\text{Var}(eW)$ are all equal to $2\text{Var}(e)$. It follows from (12.11) that the asymptotic distribution of $\hat{\Delta}_k$ is chi-squared with $(p-\kappa)(p-\kappa+1)/2$ degrees of freedom, which is the result given by Li (1992, Theorem 4.2) under the explicit assumption that z is normal and the implicit assumption that $e \perp\!\!\!\perp V$.   □

12.3.4. Testing e independent of z

The testing procedure of Section 12.3.1 can be adapted easily to provide a test of $e \perp\!\!\!\perp z$, as indicated in the following corollary.

Corollary 12.1.   Assume that $e \perp\!\!\!\perp z$. Then the asymptotic distribution of $\hat{\Delta}_0$ is the same as the distribution of

$$D = \frac{1}{2} \sum_{j=1}^{p(p+1)/2} \omega_j C_j \qquad (12.13)$$

where the C_j are independent chi-squared random variables, each with one degree of freedom, and $\omega_1 \geq \omega_2 \geq \cdots \geq \omega_{p(p+1)/2}$ are the eigenvalues of $\text{Var}(W)$. (W is constructed with $\theta_0 = I$.)

While the test of this corollary has been found to be useful, its virtues relative to the tests on κ have not been investigated extensively.

12.4. pHd: REACTION YIELD

We use the reaction yield data introduced in Section 9.3 for a first illustration of pHd. Recall that the data consist of 32 observations on the percentage yield y from a two-stage chemical process characterized by the temperatures ($Temp_1$ and $Temp_2$ in degrees Celsius) and times ($time_1$ and $time_2$ in hours) of reaction at the two stages, and the concentration (Con in percent) of one of the reactants. The elements of the 5×1 vector of predictors $x = (T_1, Lt_1, C, T_2, Lt_2)^T$ used in the analysis are

$$T_1 = \frac{(Temp_1 - 122.5)}{7.5}$$

$$Lt_1 = \frac{2[\log(time_1) - \log 5]}{\log 2} + 1$$

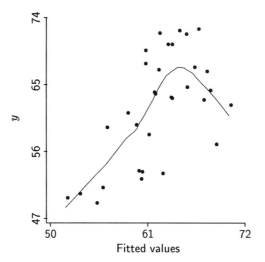

FIGURE 12.1 Scatterplot of the response versus the OLS fitted values in the reaction yield data along with a LOWESS smooth.

$$C = \frac{(Con - 70)}{10}$$

$$T_2 = \frac{Temp_2 - 32.5}{7.5}$$

$$Lt_2 = \frac{2[\log(time_2) - \log 1.25]}{\log 5} + 1.$$

The general goal of the analysis is to find predictor settings that maximize the expected response. Computations in this and subsequent examples involving pHd were carried out using the second generation of the R-code (Cook and Weisberg 1994a).

12.4.1. OLS and SIR

To get started, Figure 12.1 shows a response plot of y versus the fitted values $\hat{y}$ from the OLS regression on the five predictors. With a p-value of 0.027, the overall F-test shows that there is a significant linear trend. The curvature of the LOWESS smooth supports the prior notion that the maximum yield should be found in the region of the factor space covered by this experiment.

Figure 12.2 is a plot of y versus the first SIR predictors. With seven slices, the value of the SIR test statistic Λ_0 is 35.75 with 30 degrees of freedom, giving a p-value of about 0.22. Thus, SIR suggests that y is independent of the predictors and that the trend in Figure 12.1b can be explained by appealing to random variation. Recall, however, that SIR is not very effective for finding nonlinear trends.

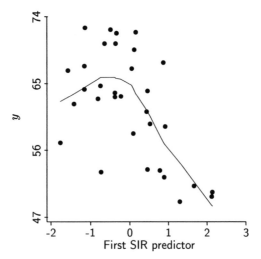

FIGURE 12.2 Scatterplot of the response versus the first SIR predictor for the reaction yield data.

TABLE 12.1. pHd Test Results from an Analysis of the Reaction Yield Data

m	$\hat{\Delta}_m$	$\hat{C}$ p-value
0	28.35	0.003
1	10.87	0.054
2	5.534	0.057
3	1.465	0.254
4	.0029	0.781

The p-values were computed according to the random variable $\hat{C}$ in Step 5 of the algorithm in Section 12.3.2.

12.4.2. pHd test results

Shown in Table 12.1 are the pHd test results from the regression of the ordinary least squares residuals $\hat{e}$ on the five predictors. The first two columns contain the hypothesized value of κ and the corresponding test statistic. The p-values in the final column were computed according to the random variable $\hat{C}$ described in Step 5 of the algorithm in Section 12.3.2.

The testing procedure given in Section 12.3.2 requires no conditions other than independent and identically distributed data with finite moments. These starting conditions are not at issue in this chapter. Because

$$\Sigma_{ezz} = E[E(e \mid z) \times zz^T] = \text{Cov}[E(e \mid z), zz^T]$$

we can always interpret the test of $\kappa = 0$ as a test of the hypothesis that the residual regression function is uncorrelated with zz^T. This is often a useful diagnostic test of the standard linear model,

$$y \mid z = \alpha_0 + \alpha^T z + \varepsilon$$

with $\kappa > 0$ suggesting a model deficiency. The p-value for $\kappa = 0$ in Table 12.1 is 0.003 so it appears that a standard linear model is not appropriate for these data, a conclusion that is consistent with the LOWESS smooth in Figure 12.1 and with the experimental objective of finding predictor settings that yield the maximum response.

Taken together, the p-values in Table 12.1 suggest the estimate $\hat{\kappa} = 3$. How we interpret this result depends on the various conditions developed in previous sections. An inspection of the scatterplot matrix of the predictors in Figure 9.13 and various 3D plots (not shown) doesn't yield clear information to contradict the linearity constraints on the conditional expectations of Propositions 12.1 and 12.2, so we assume that

$$\mathcal{S}_{y|z} = \mathcal{S}_{e|z} + \mathcal{S}(\beta) \tag{12.14}$$

where $\beta = \text{Cov}(z, y)$, as defined in the statement of Proposition 12.1. From Proposition 12.2, we assume further that some of $\hat{\ell}_j$, $j = 1, 2, 3$ are estimating vectors in $\mathcal{S}_{e|z}$, although some could be estimating vectors in $\mathcal{S}_{e|z}^{\perp}$.

From (12.14), the structural dimension of the regression of y on z depends on the location of β. If $\mathcal{S}(\beta) \in \mathcal{S}_{e|z}$ then the structural dimension is at most 3. But if $\mathcal{S}(\beta) \notin \mathcal{S}_{e|z}$ then the structural dimension of that regression could be as much as 4. The usual value of R^2 from the regression of $\hat{y}$ on the three significant pHd predictors, $\hat{u}_1^T x$, $\hat{u}_2^T x$, and $\hat{u}_3^T x$, is 0.96, suggesting that pHd has detected the curvature present in Figure 12.1, and that $\mathcal{S}(\beta) \in \mathcal{S}_{e|z}$. We conclude then that $\dim[\mathcal{S}_{y|z}]$ is at most 3.

A plot of the response versus the first pHd predictor $\hat{u}_1^T x$ is shown in Figure 12.3. The points sketch a fairly sharp peak which again is in line with the prior expectation that the maximum response would be found by this experiment. The sample correlation between the OLS fitted values $\hat{y}$ and $\hat{u}_1^T x$ is only 0.35, so $\hat{\Sigma}^{-1/2} \hat{\beta}$ and $\hat{u}_1$ are apparently estimating different vectors in $\mathcal{S}_{y|x}$, and thus it seems reasonable to infer that $\dim[\mathcal{S}_{y|x}] = \dim[\mathcal{S}_{y|z}] \geq 2$. Equivalently, $\hat{\beta}$ and $\hat{\ell}_1$, which are in the scale of z, are estimating different vectors in $\mathcal{S}_{y|z}$.

12.4.3. Subtracting β

The conclusion to this point is that $\dim[\mathcal{S}_{y|z}] = 2$ or 3. To investigate the situation a bit further, pHd was applied to the regression of $\hat{r}$, residuals from

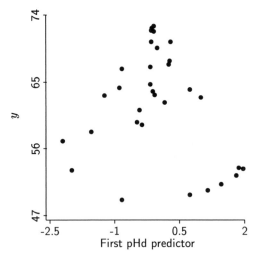

FIGURE 12.3 Scatterplot of the response versus the first pHd predictor $\hat{u}_1^T x$ for the reaction yield data.

TABLE 12.2. pHd Test Results from an Analysis of the Reaction Yield Data Using the Residuals from the LOWESS Smooth of Figure 12.1

m	$\hat{\Delta}_m$	$\hat{C}$ p-value
0	28.76	0.008
1	8.557	0.148
2	2.979	0.221
3	1.108	0.339
4	0.001	0.893

the LOWESS fit in Figure 12.1, on the five predictors. The idea here is that computation of the LOWESS residuals may remove one dimension from the central subspace; specifically, the dimension spanned by β. The results from applying the testing procedure of Section 12.3.2 are shown in Table 12.2. They suggest that $\dim[S_{r|z}] = 1$. The correlation between the first pHd predictors from the analyses of Tables 12.1 and 12.2 is 0.99, so this analysis recaptures essentially the first pHd direction from the regression of $\hat{e}$ on x. It appears that most of the systematic variation in the data can be explained by two linear combinations of the predictors, leading to $\hat{S}_{y|x} = \text{span}(\hat{\Sigma}^{-1/2}\hat{\beta}, \hat{u}_1)$ as an estimate of the central subspace. The corresponding 3D summary plot is $\{y, (\hat{y}, \hat{u}_1^T x)\}$. Figures 12.1 and 12.3 give two views of that 3D summary plot. The rotating plot resembles a cone with a section of one side removed, and tilted away from the perpendicular. The points in the semicircular arrangement along the bottom of Figure 12.3 form a little more than half of the base of the cone. The increasing trend of these basal points, which represents the cone's tilt, can be seen in Figure 12.1. The removed section of the metaphorical cone lies to the right of the points in Figure 12.1.

A practical conclusion from this graphical analysis is that the maximum yield should be reached with $63 < \hat{y} < 66$ and $\hat{u}_1^T x \approx -.14$, as can be seen from the 3D summary plot and Figures 12.1 and 12.3. A statistical model of the conical shape could be now be used to refine the graphical results and produce standard errors.

Finally, the pHd analysis in this section produced essentially the same results as the graphical regression analysis in Section 9.3, lending support to the conclusions and to the idea of basing graphical regression on the pHd predictors $\hat{u}_j^T x$.

12.4.4. Using stronger assumptions

The approach in the previous sections was based only on the condition of linear conditional expectations in the predictors. The analysis would be easier under the conditions of Proposition 12.5 because then $\mathcal{S}_{ezz} \subset \mathcal{S}_{e|z}$ and the central chi-squared distribution could be used for testing. However, when the chi-squared test of $\kappa = 0$ is rejected we must practically be prepared to question the null hypothesis in addition to all three conditions of the proposition. This often leads to a more intricate analysis than one based just on linear conditional expectations.

12.5. pHd: MUSSEL DATA

For this next example we return to the mussel data. Recall that the response is the mussel's muscle mass M. The predictors for this example are the shell length L, shell width W, shell height H, and $S^{1/3}$, the cube root of shell mass. Let $x = (L, W, H, S^{1/3})^T$, and let $\hat{e}$ represent a typical residual from the OLS regression of M on x.

The only predictor transformation used here is the cube root of S. This removes gross nonlinearity that is evident in a scatterplot matrix of the predictors, but additional transformation may be needed to remove all notable nonlinearity in the predictors. In particular, the transformations needed to achieve approximate multivariate normality (Section 9.1) are close to those used in the SIR analysis of Section 11.4.

The purpose of this example is to illustrate the importance of the predictor distribution, including outlying predictor values, in the application of pHd and to suggest that Voronoi weighting may result in improved performance in some situations. Limiting the predictor transformation to the cube root of S is intended to leave some nonlinearity that might impact the results.

TABLE 12.3. pHd Test Results for the Mussels Data

m	$\hat{\Delta}_m$	DF	χ^2 p-value	$\hat{C}$ p-value
A: Full data, M				
0	62.72	10	0.000	0.021
1	35.12	6	0.000	0.040
2	18.78	3	0.000	0.056
3	3.966	1	0.046	0.295
B: Reduced data				
0	57.37	10	0.000	0.042
1	30.99	6	0.000	0.116
2	15.99	3	0.001	0.237
3	3.961	1	0.047	0.297
C: Full data, M^*				
0	58.49	10	0.000	0.059
1	22.96	6	0.001	0.216
2	1.808	3	0.613	0.774
3	0.009	1	0.759	0.798
D: Weights, M^*				
0	41.97	10	0.000	0.001
1	3.250	6	0.777	0.767
2	0.736	3	0.865	0.878
3	0.004	1	0.842	0.854
E: Weights, M				
0	39.67	10	0.000	0.000
1	19.16	6	0.004	0.010
2	0.225	3	0.973	0.953
3	0.001	1	0.969	0.954

(A) Full data. (B) After deleting an extreme case in Figure 12.5b. (C) Full data with simulated response M^*. (D) Full data with simulated responses M^* and Voronoi weights. (E) Full data with responses M and Voronoi weights.

12.5.1. pHd test results

Shown in Part A of Table 12.3 are test results from the application of pHd to the regression of $\hat{e}$ on x. The first two columns contain the hypothesized value of κ and the corresponding test statistic. The third column contains the degrees of freedom for computing the p-value in the fourth column according to the chi-squared distribution given in Proposition 12.5. The p-values in the final column were computed using $\hat{C}$ as described in Step 5 of the algorithm in Section 12.3.2.

The p-values in Part A of Table 12.3 suggest a fairly complicated regression with a structural dimension of at least 3. This conclusion may be premature, however, because we have not yet considered the various conditions needed

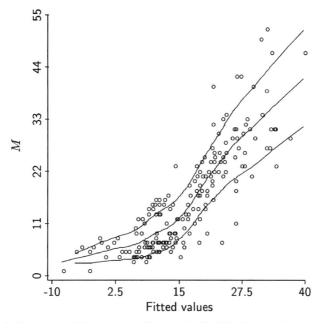

FIGURE 12.4   Scatterplot of the response M versus the OLS fitted values from the mussel data, along with LOWESS smooths for location and scale.

for such an interpretation. The small chi-square p-values may be due to $\kappa > 0$, nonnormal predictors, or failure of either of the other two conditions of Proposition 12.5. The small $\hat{C}$ p-values give a firm indication that $\kappa > 0$, but without at least the linear conditional expectation of Proposition 12.2 there is little else that can be concluded comfortably. In short, without considering the empirical distribution of the predictors about all we can conclude is $\kappa > 0$. And this rather weak conclusion could simply be indicating there is curvature in the plot shown in Figure 12.4 of M versus the fitted values from the OLS regression of M on x.

Figure 12.5 gives two views of a 3D plot of the predictors H, W, $S^{1/3}$ after orthogonalization to remove linear relationships. View 1 was selected to highlight the potential for a nonlinear relationship among the predictors, while View 2 was selected to emphasize several relatively remote points. The actual impact of the nonlinearity in View 1 on the pHd tests in Table 12.3 is unknown, but the relatively remote points in View 2 do have a substantial influence on those tests. For example, shown in Part B of Table 12.3 are the test results obtained after deleting the remote case in the lower-left corner of View 2. The chi-squared tests still result in four significant components, but the $\hat{C}$ tests now suggests only one significant component. Various other distinct test results can be obtained, depending on the particular combination of the remote points that is deleted.

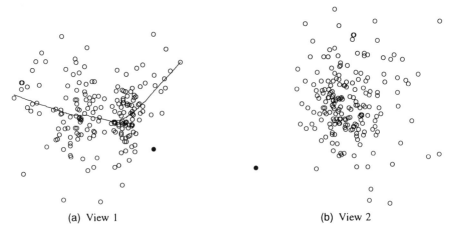

(a) View 1   (b) View 2

FIGURE 12.5 Two views of a 3D plot of the muscle mass predictors $H, W, S^{1/3}$ orthogonalized to remove the linear trends.

12.5.2. Simulating the response

To gain some further information on the potential complications that may come from the predictor distribution, simulated responses M^* were constructed as

$$M^* = m(\hat{y}) + sd(\hat{y})\varepsilon$$

where $\hat{y}$ denotes a typical fitted value from the OLS regression of M on x; m and sd are the location and scale smooths shown in Figure 12.4, and ε is a standard normal error. Letting $\hat{e}^*$ denote the residuals from the OLS regression of M^* on x, application of pHd to the regression of $\hat{e}^*$ on x should show that $\kappa = 1$, reflecting the nonlinearity in $m(\hat{y})$. The actual test results are shown in Part C of Table 12.3. The chi-square p-values suggest that $\kappa = 2$, although we know from the method of construction that $\kappa = 1$. The $\hat{C}$ p-values correctly indicate a single component.

12.5.3. Using Voronoi weights

When influential points or the potential for disruptive nonlinearity in the predictor distribution are notable issues, a weighted pHd analysis based on Voronoi weighting (Section 8.4.1) may offer some improvement by inducing necessary conditions and hopefully stabilizing the results.

Let $w_i \geq 0$, $i = 1,\ldots,n$, denote the Voronoi weights normalized to sum to 1, $\sum w_i = 1$. Briefly, a weighted pHd analysis can be performed by replacing all unweighted predictor moments by corresponding weighted moments. The weighted versions of $\bar{x}$, $\hat{\Sigma}$, and $\hat{\Sigma}_{exx}$ are

$$\bar{x}_w = \sum_{i=1}^{n} w_i x_i$$

$$\hat{\Sigma}_w = \sum_{i=1}^{n} w_i (x_i - \bar{x}_w)(x_i - \bar{x}_w)^T$$

and

$$\hat{\Sigma}_{w \cdot exx} = \sum_{i=1}^{n} w_i \hat{e}_i (x_i - \bar{x}_w)(x_i - \bar{x}_w)^T$$

where $\hat{e}$ is now constructed from the weighted least squares regression of y on x. Similarly,

$$\hat{\Sigma}_{w \cdot ezz} = \hat{\Sigma}_w^{-1/2} \hat{\Sigma}_{w \cdot exx} \hat{\Sigma}_w^{-1/2}$$

and $\text{Var}(eW)$ is estimated by using the weighted sample covariance matrix.

Part D of Table 12.3 shows the test results from application of weighted pHd to the simulated data. The weights were determined from 10,000 replications of the Voronoi algorithm with $\sigma_T = 0.85$. The chi-square and $\hat{C}$ tests are now in good agreement, each indicating 1D structure of the weighted residual regression. The correlation between the first pHd predictor and the OLS fitted values $\hat{y}$ that were used to generate the data is 0.99, so we have essentially recovered the structure in the data. Different values of σ_T between 0.75 and 1 gave results that are essentially the same as those in Table 12.3D.

Finally, Part E of Table 12.3 shows the results of applying the Voronoi weights used in Part D with the actual response M. The chi-square and $\hat{C}$ tests again show good agreement, each indicating that $\kappa = 2$. As with the simulated responses, different values of σ_T between 0.75 and 1 produced essentially the same results. For normalized weighted predictors z_w, we should have

$$\mathcal{S}_{y|z_w} = \mathcal{S}_{e|z_w} + \mathcal{S}(\beta_w)$$

and $\mathcal{S}_{ez_w z_w} \subset \mathcal{S}_{e|z_w}$, where $\beta_w = \text{Cov}(z_w, y)$. Thus, the inference so far is that $2 \leq \dim[\mathcal{S}_{y|z_w}] \leq 3$. Letting $\hat{y}_w$ denote the fitted values from the weighted regression of M on x, the R^2 value for the OLS regression of $\hat{y}_w$ on the first two pHd predictors from the final part of Table 12.3 is 0.995. Thus it seems reasonable to conclude that $\mathcal{S}(\beta_w) \subset \mathcal{S}_{e|z_w}$ and that $\dim[\mathcal{S}_{y|z_w}] = 2$. The analysis could now be continued by viewing the 3D summary plot.

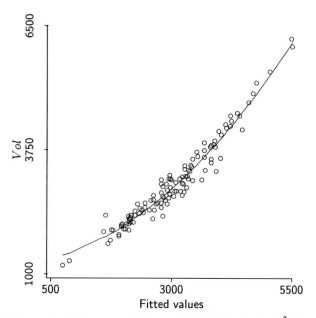

FIGURE 12.6 Scatterplot of the response *Vol* versus the OLS fitted values $\widehat{Vol}$ from the haystack data. The curve is the quadratic fit of *Vol* on $\widehat{Vol}$.

12.6. pHd: HAYSTACKS

Recall that the haystack data introduced in Section 4.1 consists of measurements on 120 haystacks. The response is stack volume *Vol* and the predictors are the circumference *C* and *Over*, the distance from the ground on one side of the stack to the ground on the other side. In Chapter 4 we concluded tentatively that the regression has 1D structure with a plot of *Vol* versus $H_4 = 0.19C + 0.98Over$ providing a summary of the regression, and that the distribution of $x = (C, Over)^T$ seems reasonably normal. Let $\widehat{Vol}$ denote the fitted values from the OLS regression of *Vol* on x and let $\hat{e} = Vol - \widehat{Vol}$.

Application of pHd to the regression of $\hat{e}$ on the two predictors indicates that $\dim[\mathcal{S}_{e|z}] = 1$, with the chi-squared and $\hat{C}$ tests in good agreement. The sample correlation between $\widehat{Vol}$ and the first pHd predictor $\hat{u}_1^T x$ is 0.997, so the plot of *Vol* versus $\widehat{Vol}$ shown in Figure 12.6 seems a good summary of the regression. The sample correlation between H_4 and $\widehat{Vol}$ is very close to 1.

The curve superimposed on the scatterplot of Figure 12.6 is the result of an OLS regression of *Vol* on $\widehat{Vol}$ and $\widehat{Vol}^2$. The curve shows a good fit, except perhaps for the two points in the lower-left corner of the plot. In any event, the plot suggests that the regression of *Vol* on a second-order response surface model in $(C, Over)$ may provide a good fit to the data. Further, after scaling the predictors to have sample standard deviation 1, $\hat{u}_1^T = (0.126, 0.992)$. The relatively small coefficient of C suggests that perhaps the quadratic term C^2 will not be estimated with sufficient precision to be useful, which agrees with indication from the t-value for C^2 in the actual fit of the second-order model. We thus conjecture that the model

$$Vol \mid x = \beta_0 + \beta_1 C + \beta_2 Over + \beta_{12} C \times Over + \beta_{22} Over^2 + \sigma\varepsilon \quad (12.15)$$

will provide a useful description of the data. Let $\hat{r}$ denote the residuals from an OLS fit of this model, and let $U = (C, Over, C \times Over, Over^2)^T$.

pHd could now be used to check the fit of model (12.15) by considering either the regression of $\hat{r}$ on x, or the regression of $\hat{r}$ on U. Because of the functional relationship between the predictors, only the test of $\kappa = 0$ or the test of Corollary 12.1 will be appropriate in the regression of $\hat{r}$ on U. Neither test suggests that model (12.15) is deficient.

12.7. DISCUSSION

12.7.1. pHd with the response

The main difference between using responses and residuals in pHd is that $Cov(e, z) = 0$ while typically $Cov(y, z) \neq 0$. Beginning with (12.12), the property $Cov(e, z) = 0$ was used throughout the development of pHd in this chapter.

To see the implications of applying pHd with the response, let z be normally distributed and assume the homoscedastic linear model

$$y \mid z = \beta z_1 + \sigma\varepsilon \quad (12.16)$$

where z_1 is the first predictor in z. Then the three conditions of Proposition 12.5 hold with e replaced with y. If, in addition, $\beta = 0$ then Proposition 12.5 will hold with the responses because we will still be able to take the critical step in the development at (12.12). However, if $\beta \neq 0$ this step is no longer possible and the distribution of $\hat{\Delta}_0$ will depend on β, although $\Sigma_{yzz} = 0$ so that $\kappa = 0$. When $\beta \neq 0$, the distribution of $\hat{\Delta}_0$ will be shifted to the right relative to the chi-squared distribution of Proposition 12.5. And this means that the pHd test has some power to detect $\beta \neq 0$, even when $\kappa = 0$.

Additionally, the eigenvector associated with the largest eigenvalue will exhibit some attraction for the central subspace, and the summary plot $\{y, \hat{\ell}_1^T z\}$ will tend to show a linear trend. This may seem surprising because $\Sigma_{yzz} = 0$. It turns out that, while there is little information in the expected eigenvalues, the eigenvalue with the largest variability will lean toward the central subspace, the degree of association depending on σ. Because estimation is based on absolute eigenvalues, variability matters.

Nevertheless, pHd is not very powerful at finding linear trends, and is substantially less powerful than other methods like OLS and SIR. Using pHd in the presence of linear trends may do little more than unnecessarily complicate the analysis and make nonlinear structure more difficult to find. It seems prudent to remove linear trends at the outset and use pHd for what it does best. See Cook (1998) for further discussion on using pHd with responses.

12.7.2. Additional developments

The development of pHd in this chapter was based on OLS residuals. Cook and Bura (1997) adapt pHd to the problem of testing the adequacy of regression functions with other types of residuals. Ferré (1998) proposed an alternative method for estimating the pHd subspace.

As a starting point for pHd and other dimension-reduction methodology, it was assumed that both x and y are random, with the conditional distribution of $y \mid x$ being of primary interest. This may seem to preclude application of dimension-reduction methods to regressions with fixed predictors, but this is not really the case. The definition of the central subspace applies whether the predictors are random or not. Other fundamental notions can be adapted for nonrandom predictors with little difficulty. In principle there seems no reason that the dimension-reduction methods discussed in this book cannot be adapted to regressions with fixed predictors, including designed experiments. Cheng and Li (1995) studied the applicability of pHd in designed experiments with special emphasis on factorial and rotatable designs. Filliben and Li (1997) showed how pHd can be used to investigate complex interaction patterns in two-level factorial designs. Ibrahimy and Cook (1995) studied designs to insure that the expectation of the OLS coefficient vector is in $S_{y|x}$ for regressions with 1D structure.

pHd is based on a linear combination of the inverse variances $\Sigma_{z|y}$, as can be seen from the discussion of Section 12.2.1. In contrast, Cook and Weisberg (1991b) suggested a dimension-reduction method, called SAVE for "sliced average variance estimation," based on slicing and the resulting inverse variances $\Sigma_{z|\tilde{y}}$ for the discrete response $\tilde{y}$ as used in SIR. Under conditions similar to those used in this chapter, it can be shown that

$$\mathcal{S}_{ezz} \subset \mathcal{S}_{save} \subset \mathcal{S}_{y|z}$$

where $\mathcal{S}_{save}$ denotes the subspace estimated under SAVE. Thus, SAVE is potentially more comprehensive than pHd because it generally captures a larger part of the central subspace. Traditional inference methods for SAVE have not yet been developed, although the SAVE predictors could be used in graphical regression in the same way as SIR or pHd predictors.

PROBLEMS

12.1. Consider a regression problem with response y and $p \times 1$ predictor vector x. Let $\hat{r}$ denote a typical residual from an OLS fit of a full second-order quadratic model in x. Would it be useful to apply pHd to the regression of $\hat{r}$ on x? Provide support for your response.

12.2. Use pHd to investigate the regressions in the naphthalene data discussed in Section 9.1.

12.3. Justify Corollary 12.1.

12.4. Find conditions under which $\Sigma_{yzz} = 0$ when

$$y \mid z = \beta_0 + \beta^T z + \sigma \varepsilon.$$

Here, $\varepsilon \perp\!\!\!\perp z$, $E(z) = 0$ and $\text{Var}(z) = I$, although z is not required to be normal. Are there situations in which pHd will identify $\mathcal{S}(\beta)$ when applied to the population regression of y on z?

12.5. Consider the following two models,

$$y \mid z = (\gamma^T z)^2 + \sigma \varepsilon$$

and

$$y \mid z = e^{z_1 + z_2} + e^{z_3 + z_4} + \sigma \varepsilon$$

where $z = (z_1, z_2, z_3, z_4)^T \in \mathbb{R}^4$ is a standard normal random variable, ε is a standard normal error, and $\gamma \in \mathbb{R}^4$. These models were used in Section 10.1 to illustrate that the inverse regression subspace may be a proper subset of the central subspace. Consequently, we expect SIR to miss the central subspace in each model.

Would you expect the pHd methodology described in this chapter to capture the central subspaces for these models? Justify your response.

12.6. Using the haystack data discussed in Sections 4.1 and 12.6, Becker, Cleveland, and Shyu (1996) develop a model for haystack volume by imagining a round haystack as the frustum of a right circular cone, the part of the cone that remains when the top is cut off with a plane parallel to the base. Let b denote the radius of the circular frustum base, let t denote the radius of the circular top, and let h denote the frustum height. Then the volume of a frustum can be written as

$$\text{frustum volume} = \frac{\pi}{3} \times h \times b^2 (1 + \alpha + \alpha^2)$$

where $\alpha = t/b$ is the ratio of the top radius to the bottom radius. Next, b and h can be related to the frustum "over" and the circumference C of the frustum base as follows,

$$b(C) = \frac{C}{2\pi}$$

and

$$h(C, Over, \alpha) = \sqrt{(Over/2 - \alpha b)^2 - b(1 - \alpha)}.$$

Substituting these into the expression for frustum volume gives frustum volume in terms of C, $Over$, and α:

frustum volume $= V(C, Over, \alpha)$

$$= \frac{\pi}{3} \times h(C, Over, \alpha) \times b^2(C)(1 + \alpha + \alpha^2).$$

Becker et al. then use this final form to model haystack volume Vol as

$$\log Vol = \mu + \log V(C, Over, \alpha) + \sigma\varepsilon \qquad (12.17)$$

where α is assumed to be constant, and $\varepsilon \perp\!\!\!\perp (C, Over)$, with $E(\varepsilon) = 0$ and $Var(\varepsilon) = 1$. This model contains three unknown parameters, μ, α, and σ. After fitting by nonlinear least squares, Becker et al., report estimates along with their standard errors: $\hat{\alpha} = 0.90 \pm 0.03$, $\hat{\mu} = 0.092 \pm 0.011$, and $\hat{\sigma} = 0.091 \pm 0.0058$.

12.6.1. Do the data provide any evidence to suggest that model (12.17) could be improved, or is deficient in some way? Letting $\hat{r}$ denote the residuals from the fit reported by Becker et al., investigate the structural dimension of the regression of $\hat{r}$ on $(C, Over)$. This could be done graphically with a 3D plot, or by using pHd or SIR, for example.

12.6.2. In Section 12.6 we reasoned that the regression of Vol on $(C, Over)$ has 1D structure with the plot of Vol versus the OLS fitted values $\hat{Vol}$ providing an estimated sufficient summary. In contrast, model (12.17) implies that the regression of Vol on $(C, Over)$ has 2D structure. Treating $\hat{Vol}$ as a component that is common to both approaches, does the remaining dimension suggested by (12.17) noticeably improve the prediction of haystack volume, beyond the satisfaction that comes from having statistical models based on physical metaphors? Stated differently, the graphical approach concludes that there is no compelling information in the data to contradict the statement $Vol \perp\!\!\!\perp (C, Over) \mid \hat{Vol}$, while model (12.17) suggests otherwise. Is there sufficient information in the data to decide the issue, or is the conclusion of 2D structure under (12.17) based largely our ability to imagine a haystack as a frustum?

CHAPTER 13

Studying Predictor Effects

The central subspace $S_{y|x}$ has been a focal point so far. In this chapter, we move away from these subspaces and consider graphics to aid in assessing the contribution of a selected predictor to a regression. The discussion follows the developments by Cook (1995).

13.1. INTRODUCTION TO NET-EFFECT PLOTS

Partition the $p \times 1$ predictor vector as $x = (x_1^T, x_2^T)^T$, where x_j contains p_j predictors, $p_1 + p_2 = p$. The issue we address in this chapter is how to construct graphical displays for the regression of the "conditional response" $y \mid (x_1 = \tilde{x}_1)$ on the "conditional predictor" $x_2 \mid (x_1 = \tilde{x}_1)$, with $\tilde{x}_1$ being the same in both variables. The conditioning is used to account for the contribution of x_1 to the regression prior to considering x_2. Because x_1 is held fixed, only x_2 is left to explain the remaining variation in y. Displays that address the "conditional regression" of $y \mid x_1$ on $x_2 \mid x_1$ can be viewed collectively as *net-effect plots* since it is the net effect of x_2 on the response that is of interest. Net-effect plots can differ depending on the particular effects of interest and on the structure of the regression problem. Added-variable plots, for example, can be instances of net-effect plots, as discussed later in this chapter.

Whether the predictors are fixed by design or random variables, a *local* net-effect plot for x_2 at a selected value for x_1 is a graphical display of y versus x_2, confined to the selected value. In the case of random predictors, the data displayed in a net-effect plot should ideally be a sample from the conditional distribution of (y, x_2) given x_1. Useful information may arise by studying a particular net-effect plot or by studying the *change* in a series of net-effect plots obtained by varying the value of x_1. Net-effect plots of this type can be considered *local* since they display the effects of x_2 at or near a selected value for x_1. Later in this chapter we consider *global* net-effect plots that apply over all values of x_1.

254

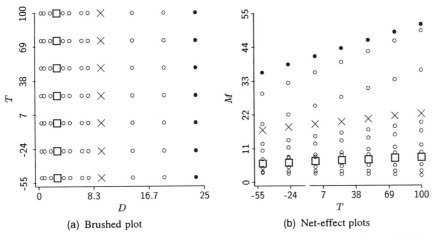

(a) Brushed plot   (b) Net-effect plots

FIGURE 13.1 Net-effect plots for temperature T in the data set on the modulus of natural rubber.

13.1.1. Natural rubber: Net-effect plots

For example, consider net-effect plots for the temperature predictor T in the data on the modulus of natural rubber described in Section 4.5.3. Figure 13.1a is the *control plot* $\{T,D\}$ to be used for conditioning on the decomposed dicumyl peroxide D; a histogram of D could also be used as a control. Figure 13.1b is a linked plot of M versus T to be used for displaying net effects. The column of filled circles in Figure 13.1a gives the seven cases in the data at the highest value of $D = 23.8$. The corresponding points in the plot $\{M,T\}$ are represented by the same symbol and form a net-effect plot for T at $D = 23.8$. These points suggest a very strong simple linear regression of M on T at the selected value of D.

Similarly, the points represented by squares ($D = 2.86$) and exes ($D = 9.52$) in Figure 13.1a yield corresponding net-effect plots for T in Figure 13.1b. These net-effect plots also suggest very strong simple linear regressions at the associated values of D. Taken together, the net-effect plots of Figure 13.1b indicate an interaction between T and D, suggesting a regression function of the form

$$E(M \mid D,T) = \beta_0(D) + \beta(D)T$$

where the slope $\beta(D)$ and intercept $\beta_0(D)$ both seem to be increasing functions of D.

The experimental design facilitated the conditioning necessary for the construction of the net-effect plots in Figure 13.1b because there are seven observations at each value of D. In addition, the variation about the regression function $E(M \mid D,T)$ is evidently quite small, making interpretation of the net-effect plots in Figure 13.1b relatively easy. Nevertheless, useful information can often be obtained by using similar graphical procedures when there is substantially more variation in the data.

13.1.2. Joint normality

Studying net effects may be more complicated when the predictors are random. To help introduce basic ideas, we begin with a relatively simple case that can be studied analytically. Suppose that $p_2 = 1$ and that $(y, x_1^T, x_2)^T$ follows a nonsingular multivariate normal distribution. For notational convenience, let $w = (y, x_2)^T$. Then the conditional distribution of $w \mid (x_1 = \tilde{x}_1)$ is bivariate normal with mean

$$E(w \mid x_1 = \tilde{x}_1) = E(w) + \text{Cov}(w, x_1)[\text{Var}(x_1)]^{-1}(\tilde{x}_1 - E(x_1))$$

and variance

$$\text{Var}(w \mid x_1 = \tilde{x}_1) = \text{Var}(w) - \text{Cov}(w, x_1)[\text{Var}(x_1)]^{-1}\text{Cov}(x_1, w).$$

The location $E(w \mid x_1)$ depends on the value of x_1, but the conditional variance $\text{Var}(w \mid x_1)$ does not. Thus the partial correlation coefficient $\rho_{y2 \cdot 1}$ between y and x_2 gives a measure of the net effect of x_2 not depending on the value of x_1. More specifically, given x_1, the population regression of y on x_2 can be written in the linear form

$$y \mid x = E(y \mid x_1) + \beta[x_2 - E(x_2 \mid x_1)] + \varepsilon$$

where the slope

$$\beta = \rho_{y2 \cdot 1} \frac{\text{Var}(y \mid x_1)}{\text{Var}(x_2 \mid x_1)}$$

does not depend on the value of x_1, and ε is normal with mean 0 and variance

$$\sigma^2 = \text{Var}(y \mid x_1)(1 - \rho_{y2 \cdot 1}^2)$$

which also does not depend on x_1. For each fixed value of x_1 the regression function is linear with normal errors. As the value of x_1 varies, the variance about the regression function and the slope remain constant, but the intercept may change.

In the case of a multivariate normal distribution then, the net effect of any predictor is characterized by a simple linear regression with normal errors, the corresponding partial correlation being a possible measure of the strength of the effect. As the value of x_1 varies, the regression structure remains constant except for the location $E(w \mid x_1)$, which changes as a linear function of x_1.

When the form of the joint distribution is unknown, graphical methods based on approximate conditioning (slicing) may be useful for studying net effects.

13.1.3. Slicing

As in Chapter 4, let J denote a generic slice, a region in the sample space of a conditioning variable where the variable is relatively constant. Then plots of the form $\{y, x_2 \mid x_1 \in J\}$ show the regression relationship between y and x_2 with x_1 confined to slice J. Brushing, interactively changing J and updating the plot $\{y, x_2 \mid x_1 \in J\}$, can be used to visualize how the regression y on x_2 changes with the slice $x_1 \in J$, as introduced during the discussion of scatterplot matrices in Section 2.5.

Trellis displays (Becker, Cleveland, and Shyu 1996) consist of several net-effect plots $\{y, x_2 \mid x_1 \in J_k\}$, $k = 1, \ldots q$, on a single page. The plots are constructed so that $\bigcup_k J_k$ covers the observed range for x_1. The advantage of such displays is that several net-effect plots can be viewed at the same time, similar to the arrangement in Figure 13.1b.

Operational versions of this general idea depend on the predictor dimensions, p_1 and p_2. With two or three predictors, scatterplot matrices or trellis displays provide a useful environment for local net-effect plots $\{y, x_2 \mid x_1 \in J\}$. With two predictors, a long narrow brush can be used to construct slices J, as illustrated a bit later in Figure 13.2, while a small square brush is required when there are $p = 3$ predictors and $p_1 = 2$. Scatterplot matrices are less useful when $p = 3$ and $p_1 = 1$ because $\{y, x_2 \mid x_1 \in J\}$ is then a 3D plot which must be linked to a separate plot of x_1 for brushing. To illustrate some of these ideas, we return to the mussel data of Section 10.1.3 where the mean checking condition was used to conjecture that the regression of muscle mass M on the transformed predictor vector $x = (L, W^{0.36}, S^{0.11})^T$ has 1D structure,

$$M \perp\!\!\!\perp x \mid \beta^T x.$$

Because the predictors were transformed to achieve approximate multivariate normality, the Li–Duan Proposition 8.1 implies that the coefficient vector from the OLS regression of M on x might provide a useful estimate of $\mathcal{S}(\beta)$. Accordingly, let $\hat{M}$ denote the OLS fitted values, and linearly transform the observed vectors of predictor values to three new predictors $(\hat{M}, U_1, U_2)$ with sample correlations of 0. Then to sustain the conjecture of 1D structure, the data should appear as if

$$M \perp\!\!\!\perp (U_1, U_2) \mid \hat{M}. \tag{13.1}$$

This condition can be checked directly by viewing the 3D net-effect plot $\{M, (U_1, U_2) \mid \hat{M} \in J\}$, using brushing to interactively change the slice J in a control plot of $\hat{M}$.

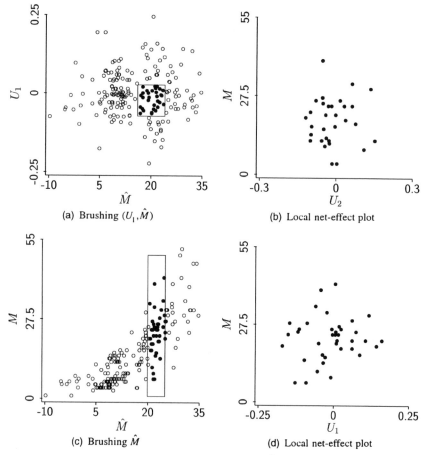

FIGURE 13.2   Four frames of a scatterplot matrix of $(M,\hat{M},U_1,U_2)$ for the mussel data to be use in checking the conditions of (13.2).

Alternatively, (13.1) might be checked by using a scatterplot matrix of $(M,\hat{M},U_1,U_2)$ to check the equivalent pair of conditions

$$\text{(a)} \quad M \perp\!\!\!\perp U_2 \mid (U_1,\hat{M}) \qquad \text{and} \qquad \text{(b)} \quad M \perp\!\!\!\perp U_1 \mid \hat{M}. \qquad (13.2)$$

The four key frames of this scatterplot matrix are shown in Figure 13.2. Figure 13.2a shows the plot $\{U_1,\hat{M}\}$ which we will slice for checking (13.2a), along with a small square of highlighted points J. Shown in Figure 13.2b is the corresponding net-effect plot

$$\{M,U_2 \mid (U_1,\hat{M}) \in J\}$$

which seems to support condition (13.2a).

The plot $\{M,\hat{M}\}$ in Figure 13.2c is the control plot for checking (13.2b). In this case, a long narrow brush was used to select the slice J of highlighted points shown on the plot. The corresponding net-effect plot $\{M,U_1 \mid \hat{M} \in J\}$ shown in Figure 13.2d seems to support condition (13.2b). If the net-effect plots illustrated in Figure 13.2b and Figure 13.2d support the respective conditions regardless of the slice J then there is no graphical evidence to contradict the original conjecture of 1D structure.

Finally, the graphical paradigm illustrated in this example is a form of model checking that does not require the computation of residuals.

13.1.4. Reducing brushing dimensions

Net-effect plots as illustrated in the previous sections can be useful for studying the net effect of x_2 on y at a selected value for x_1, and brushing, or a trellis display, can be used to visualize how the conditional distribution of $(y,x_2) \mid x_1$ changes with the value of x_1. However, direct construction of such plots is limited to regression problems in which p is small, since practical limitations are encountered otherwise. Local net-effect plots $\{y,x_2 \mid x_1 \in J\}$ are two dimensional when $p_2 = 1$ and three dimensional when $p_2 = 2$. Two-dimensional plots are the easiest to visualize. Three-dimensional plots may be useful on occasion, but visualization and interpretation are more difficult. Becker et al. (1996) gave an example of a trellis plot with 3D displays. The usefulness of such plots may be quite limited when $p_2 > 2$.

The dimension of x_1 is a second limitation on construction because slicing and brushing are practically difficult, if not impossible, when p_1 is large. The practical limitations on the dimensions p_1 and p_2 seem to limit applicability of local net-effect plots to regressions with at most four predictors.

Even if a plot of several predictors could be brushed, the sparseness usually encountered in high dimensions may make it difficult to capture subsets of the data that reasonably approximate samples on $(y,x_2) \mid x_1$. One possible way out of these difficulties is to replace x_1 with a lower-dimensional function of x_1.

13.2. DISTRIBUTIONAL INDICES

The practical difficulties in brushing a scatterplot of x_1 when $p_1 > 2$ might be overcome by replacing x_1 with a low-dimension *distributional index function*, say $\tau(x_1) \in \mathbb{R}^q$ with $q < p_1$. A distributional index function is intended to index the individual conditional distributions of $(y,x_2) \mid x_1$ just as x_1 does. The basic idea is to brush a control plot of $\tau(x_1)$ while observing the corresponding net-effect plot $\{y,x_2 \mid \tau(x_1) \in J\}$.

To avoid loss of important information, distributional indices should partition the sample space for x_1 into equivalence classes with the values in a class corresponding to identical or nearly identical distributions. Ideally, $\tau(x_1)$ will be at most two dimensional, $q \leq 2$, and have the property that

$$(y, x_2) \perp\!\!\!\perp x_1 \mid \tau(x_1). \tag{13.3}$$

Wishing that $q \leq 2$ is simply to facilitate brushing. Condition (13.3), which is trivially true when $\tau(x_1) = x_1$, insures that no information on $(y, x_2) \mid x_1$ will be lost using $\tau(x_1)$ in place of x_1. Although (13.3) may be difficult to satisfy exactly in practice, an index $\tau(x_1)$ should be a useful tool for understanding the net effect of x_2 when (13.3) holds to a reasonable approximation.

The problem of finding a low-dimensional function $\tau(x_1)$ to satisfy expression (13.3) is essentially a dimension-reduction problem for the multivariate regression of (y, x_2) on x_1. Later in this chapter we will restrict attention to linear functions $\tau(x_1) = \alpha^T x$ as we have done for past dimension-reduction problems.

A constructed example is given in the next section to help fix ideas.

13.2.1. Example

Let w_1 and w_2 be independent uniform random variables on the interval $[-1, 1]$, and let $w_3 \mid (w_1, w_2)$ be a normal random variable with mean $(w_1 + w_2)^2$ and variance 0.2. These three variables are the predictors for the example, $w = (w_1, w_2, w_3)^T$. The distribution of $y \mid w$ is described by the linear model,

$$y \mid w = 1.5(w_1 + w_2) + w_3 + 0.5\varepsilon \tag{13.4}$$

where ε is a standard normal random variable and $\varepsilon \perp\!\!\!\perp w$.

A scatterplot matrix of 150 observations generated according to this model is given in Figure 13.3. The highlighted points will be discussed shortly. Imagine inspecting this scatterplot without knowledge of the model. Several characteristics are immediately apparent: It appears that w_1 and w_2 are independent, while the predictor plots $\{w_3, w_2\}$ and $\{w_3, w_1\}$ exhibit curvature. The marginal response plots $\{y, w_1\}$ and $\{y, w_2\}$ suggest heteroscedasticity in the corresponding marginal distributions. The most curious behavior is in the plot of y versus w_3, which looks rather like a distorted "C".

Nothing in the scatterplot matrix suggests the rather simple form of model (13.4). The behaviors of the individual cells in Figure 13.3 might suggest transformations or other standard methodology. The plot $\{y, w_3\}$ might be taken to indicate a mixture of two distinct regressions. But in view of (13.4) none of this is likely to facilitate understanding.

Brushing w_1 or w_2 separately while observing the linked marginal response plot $\{y, w_3\}$ does not seem to help much, although these operations do indicate that $(y, w_3) \mid w_j$ depends on the value of w_j, $j = 1, 2$. For example, the highlighted points in Figure 13.3 represent approximate conditioning on a relatively large value of w_2. The corresponding highlighted points in the frame

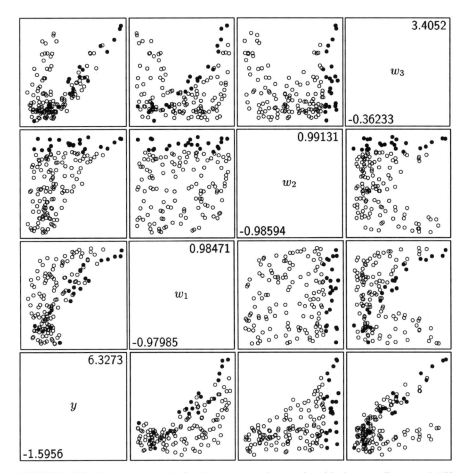

FIGURE 13.3 Scatterplot matrix for the constructed example with three predictors and 150 observations.

$\{y, w_3\}$ cover half of the "C". The results from using a small square brush to condition simultaneously on w_1 and w_2 also seem difficult to interpret.

To better understand the plot $\{y, w_3\}$ it is necessary to examine the structure of the example, particularly the distribution of $(y, w_3) \mid (w_1, w_2)$, which is bivariate normal with mean

$$\begin{pmatrix} E(y \mid w_1, w_2) \\ E(w_3 \mid w_1, w_2) \end{pmatrix} = \begin{pmatrix} 1.5(w_1 + w_2) + (w_1 + w_2)^2 \\ (w_1 + w_2)^2 \end{pmatrix} \qquad (13.5)$$

and constant covariance matrix

$$\begin{pmatrix} 0.5^2 + 0.2^2 & 0.2^2 \\ 0.2^2 & 0.2^2 \end{pmatrix}$$

This distribution represents the population net effect of w_3. Like the multivariate normal discussed previously in Section 13.1.2, the conditional covariance matrix is constant and so the partial correlation $\rho_{y3\cdot(12)} = 0.37$ may be a useful summary of the net effect of w_3. This aspect of the distribution might be visualized by using a small square brush to construct net-effect plots $\{(y, w_3) \mid (w_1, w_2) \in J\}$.

However, because $(y, w_3) \perp\!\!\!\perp (w_1, w_2) \mid (w_1 + w_2)$ it follows that,

$$\tau(w_1, w_2) = (w_1 + w_2)$$

is a valid distributional index function. This implies that the net effects of w_3 could be studied by conditioning on $(w_1 + w_2)$, rather than on (w_1, w_2). This is desirable for at least two reasons. First, one-dimensional slicing on $(w_1 + w_2)$ is generally easier than two-dimensional slicing, as required for (w_1, w_2). Second, net-effect plots obtained when slicing $(w_1 + w_2)$ will generally contain more observations than will net-effect plots from slicing (w_1, w_2), often resulting in a substantial increase of visual power.

A control plot $\{w_1 - w_2, w_1 + w_2\}$ for conditioning on $w_1 + w_2$ along with a typical brush location is shown in Figure 13.4a. The highlighted points in Figure 13.4b give the corresponding local net-effect plot $\{(y, w_3) \mid w_1 + w_2 \in J\}$ embedded in the marginal plot $\{y, w_3\}$. Figure 13.4c shows only the points in the net-effect plot. Because the variation in the net-effect plot $\{(y, w_3) \mid w_1 + w_2 \in J\}$ shown in Figure 13.4c is substantially smaller than the variation in the marginal plot $\{y, w_3\}$, there appears to be considerable covariation between (y, w_3) and $w_1 + w_2$. Consequently, the bulk of the variation in $\{y, w_3\}$ can be associated with variation in $w_1 + w_2$, suggesting that the net effect of w_3 is relatively small. The sample correlation between y and w_3 in the net-effect plot of Figure 13.4c is about 0.41, which seems a reasonable reflection of the population partial correlation 0.37.

In this example, we used the known population structure to select a distributional index. Ideas for estimating distributional indices in practice are discussed in the following subsections.

13.2.2. Location dependence

In some problems it might be reasonable to assume that the distribution of $(y, x_2) \mid x_1$ depends on x_1 only through its conditional mean

$$\mu(x_1) = \begin{pmatrix} E(y \mid x_1) \\ E(x_2 \mid x_1) \end{pmatrix}$$

so that

$$(y, x_2) \perp\!\!\!\perp x_1 \mid \mu(x_1). \tag{13.6}$$

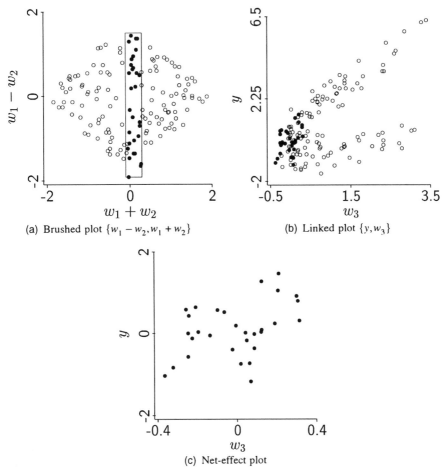

FIGURE 13.4   Net-effect plot for w_3 in the constructed example.

This is a special case of (13.3) with $\tau(x_1) = \mu(x_1)$. In practice, it will be necessary to estimate μ, which requires estimating two regression functions.

Condition (13.6) is satisfied in the example associated with (13.4). Figure 13.5 is a plot of the estimates $\hat{\mu}(w_{i1}, w_{i2})$, $i = 1, \ldots, 150$, obtained by using full second-order quadratic response surface models in $x_1 = (w_1, w_2)^T$ to construct OLS estimates of $E(y \mid w_1, w_2)$ and $E(w_3 \mid w_1, w_2)$. The pattern of points closely matches the plane curve traced by the true mean (13.5) as (w_1, w_2) varies in its sample space. Because $\hat{\mu}$ is an estimate of a distributional index, the scatterplot in Figure 13.5 could be used as a control plot for conditioning on μ. Brushing around the curve of points in Figure 13.5 is essentially equivalent to brushing $(w_1 + w_2)$, and results in the same conclusions as those previously described.

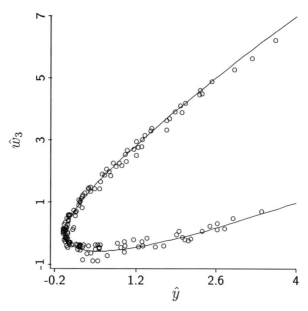

FIGURE 13.5 Estimated values for the conditional mean curve (13.5) along with the curve itself, for the constructed example.

13.2.3. Post-model net-effect plots

For some regressions we may wish to study net effects after analysis has produced a useful model for the regression of y on x. Having an estimated full model available need not necessarily mitigate interest in studying the contribution of x_2 after x_1. The following proposition will be useful in constructing net-effect plots from models. The justification is set as a problem at the end of this chapter.

Proposition 13.1. Let $\tau^T(x_1) = (\tau_1^T(x_1), \tau_2^T(x_1))$. If

   a. $y \perp\!\!\!\perp x \mid (\tau_1(x_1), x_2)$, and
   b. $x_1 \perp\!\!\!\perp x_2 \mid \tau_2(x_1)$

then $(y, x_2) \perp\!\!\!\perp x_1 \mid \tau(x_1)$.

According to this proposition, if we can replace x_1 with $\tau_1(x_1)$ in the regression of y on x without loss of information (condition (a)), and if we can similarly replace x_1 with $\tau_2(x_1)$ in the regression of x_2 on x_1 (condition (b)) then $\tau(x_1)$ is a valid distributional index function.

    To construct an index function from a model, extract from the model the lowest-dimensional function τ_1 so that condition (a) of Proposition 13.1 is satisfied. A separate analysis of the regression of x_2 on x_1 is then required for the second component τ_2. For example, suppose that $p_2 = 1$ and that the distribution of $y \mid x$ has been characterized by the homoscedastic generalized additive model

$$\hat{E}(y \mid x) = a + \sum_{k}^{p-1} g_{1k}(x_{1k}) + g_2(x_2)$$

where g_{1k} is the estimated smooth for the kth coordinate x_{1k} of x_1, and g_2 is the estimated smooth for x_2. Then we can take

$$\hat{\tau}_1(x_1) = \sum_{k}^{p-1} g_{1k}(x_{1k}).$$

Determining an estimate of a suitable τ_2 requires knowledge of the regression of x_2 on x_1. For example, assuming that the regression of x_2 on x_1 is found to have 1D structure, let b denote an estimated basis vector for $S_{x_2 \mid x_1}$. The basis could be estimated by using the Li–Duan Proposition, SIR, or any of the graphical methods discussed in previous chapters. Then $\hat{\tau}_2(x_1) = b^T x_1$ and the estimated distributional index is just

$$\hat{\tau}^T(x_1) = (\hat{\tau}_1(x_1), \hat{\tau}_2(x_1)) = \left(\sum_{k}^{p-1} g_{1k}(x_{1k}), b^T x_1 \right)$$

Finally, local net-effect plots for x_2 can be constructed by brushing a scatterplot of $\hat{\tau}$ and observing the corresponding points in a linked plot $\{y, x_2\}$.

13.2.4. Bivariate SIR

Restricting $\tau(x_1)$ to be a linear function of x_1 allows for another method of estimating a distributional index function for studying the net-effect of x_2. Specifically, letting η be a basis for the central dimension-reduction subspace $S_{(y,x_2) \mid x_1}$, it follows that

$$(y, x_2) \perp\!\!\!\perp x_1 \mid \eta^T x_1$$

and thus that $\eta^T x_1$ is a distributional index for $(y, x_2) \mid x_1$. Once an estimate $\hat{\eta}$ of η is found, we can set $\hat{\tau}(x_1) = \hat{\eta}^T x_1$ for use in practice.

When $p_2 = 1$ the bivariate SIR methodology described in Section 11.5 can be used to help find a useful distributional index. For example, SIR was applied to the bivariate regression in the example associated with (13.4) by first partitioning the 150 observations on y into 5 slices, each containing 30 observations. The 30 observations in each slice were then partitioned on w_3 into 5 slices of 6 observations each. Thus the data were partitioned into 25 slices with 6 observations each. The results from SIR strongly indicate that the dimension of $S_{(\bar{y},\bar{x}_2)|x_1}$ is 1, with the estimated basis $\hat{\eta} = (0.707, 0.708)^T$ nearly identical to the population basis $\eta = (1, 1)^T$.

13.3. GLOBAL NET-EFFECT PLOTS

The study of a collection of local net-effect plots $\{y, x_2 \mid \tau(x_1) \in J\}$ obtained by varying J should provide two kinds of information. First, any net-effect plot should correspond approximately to a sample on $(y, x_2) \mid x_1$ so that it provides visual information about the regression of y on x_2 when x_1 is restricted to J. Such net-effect plots can be interpreted as any regression scatterplot, understanding that x_1 is relatively constant. Second, the collection of plots provides information on how the conditional regressions change with the values of x_1 in the sample. In some problems, aspects of the conditional regression structure may be constant, or approximately so, allowing a single global plot based on all the data.

The problem of combining local net-effect plots translates into the problem of combining the conditional distributions of $(y, x_2) \mid x_1$ over the values of x_1. Informative combinations of these variables need not necessarily be possible since there is no general reason why $(y, x_2) \mid (x_1 = c_1)$ should be similar to $(y, x_2) \mid (x_1 = c_2)$, particularly if there are interactions. Nevertheless, a first method of constructing a global net-effect plot for x_2 is to shift the conditional distributions of $(y, x_2) \mid x_1$ so they all have the same mean.

Let $r_{y|1} = y - E(y \mid x_1)$ and $r_{2|1} = x_2 - E(x_2 \mid x_1)$. Then the expectation of $(r_{y|1}, r_{2|1}) \mid x_1$ is 0 for all values of x_1. In other words, the translation to the residuals shifts the individual conditional distributions so that their means coincide at 0, but leaves the conditional covariance structure, as well as any other nonlocation structure, unchanged. If

$$(r_{y|1}, r_{2|1}) \perp\!\!\!\perp x_1 \tag{13.7}$$

then $(r_{y|1}, r_{2|1}) \mid x_1$ and $(r_{y|1}, r_{2|1})$ have the same distribution. Using the residuals in such situations has the desired effect of removing the dependence on the value of x_1, allowing the conditional distributions to be combined over x_1 without changing the conditional regression structure, except for the predictable shift in location. The plot $\{r_{y|1}, r_{2|1}\}$ will be called a (mean) *global net-effect plot*, even when the regression functions needed for its construction are estimated. When (13.7) fails noticeably, a global net-effect plot may not be easily interpreted, or useful.

If the regression functions $E(y \mid x_1)$ and $E(x_2 \mid x_1)$ are estimated by using the fitted values from the OLS linear regression of y on x_1 and the fitted values from the OLS linear regression of x_2 on x_1 (so that $r_{\cdot|1} = e_{\cdot|1}$), then the resulting global net-effect plot is just the standard added-variable plot for x_2

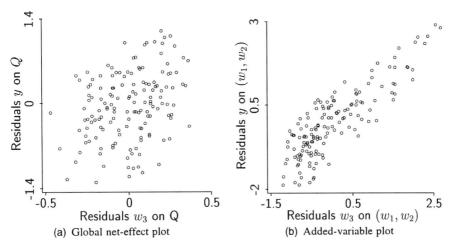

FIGURE 13.6 Constructed example: (a) Scatterplot of the OLS residuals from the regression of y on the full quadratic predictor Q in (w_1, w_2) versus the OLS residuals from the regression of w_3 on Q. (b) Added-variable plot for w_3.

after x_1. Thus an added-variable plot for x_2 is an instance of a global net-effect plot when (13.7) holds, and when $E(y \mid x_1)$ and $E(x_2 \mid x_1)$ are both linear in x_1. There is no requirement that $E(y \mid x)$ be linear in x, however.

For example, (13.7) holds in the constructed example discussed in Section 13.2.1 so a global net-effect plot is appropriate. Figure 13.6a shows the sample plot $\{\hat{r}_{y|1}, \hat{r}_{2|1}\}$ where the conditional means were estimated by fitting a full second-order response surface in (w_1, w_2). The plot is interpreted as showing the net effect of w_3; that is, the regression of y on w_3 with (w_1, w_2) fixed, the particular value influencing only the location of the point cloud. The sample correlation between the points in Figure 13.6a is about 0.36, which is quite close to the corresponding population correlation of 0.37.

The regression functions $E(y \mid w_1, w_2)$ and $E(w_3 \mid w_1, w_2)$ needed for Figure 13.6a were estimated by using quadratics in (w_1, w_2), although the full regression function $E(y \mid w_1, w_2, w_3)$ is linear. Shown in Figure 13.6b is the standard added-variable plot that results from using first-order OLS fits to estimate these regression functions. The two plots of Figure 13.6 do seem to leave distinct visual impressions about the net-effect of w_3, the added-variable plot suggesting a much stronger net effect. In this example, the added-variable plot overestimates the nonlocation net-effect of w_3 because it does not fully account for the variation in (w_1, w_2); using OLS linear regression estimates of $E(y \mid w_1, w_2)$ and $E(w_3 \mid w_1, w_2)$ does not properly account for the first-moment dependence.

Generally, added-variable plots tend to overestimate net effects unless $E(y \mid x_1)$ and $E(x_2 \mid x_1)$ are both linear in x_1.

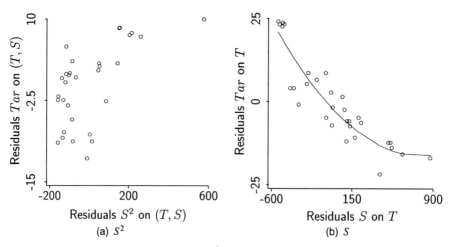

FIGURE 13.7   (a) Added-variable plot for S^2 from model (13.8). (b) Global net-effect plot for S with a quadratic fit.

13.3.1. Tar

In their discussion of added-variable plots, Chambers, Cleveland, Kleiner, and Tukey (1983, pp. 268–275, called *adjusted-variable plots*) use a data set relating tar content *Tar* of a gas to the temperature T of a chemical process and the speed S of a rotor. Because speed was expected to have a nonlinear effect on tar content, S^2 was used as an additional predictor in the initial model

$$Tar \mid (S,T) = \beta_0 + \beta_1 T + \beta_2 S + \beta_{22} S^2 + \varepsilon. \qquad (13.8)$$

To study the contribution of S^2, Chambers et al. (1983, p. 273) used an added-variable plot for S^2, as shown in Figure 13.7a.

Regardless of the adequacy of the initial model, the added-variable plot for S^2 provides visual information on the numerical calculation of the coefficient of S^2 from an OLS fit (see, for example, Draper and Smith 1966, Section 4.1, and Cook 1987b for extensions to nonlinear regression), and information on the influence of various perturbations of the data (Cook and Weisberg 1982, p. 45; Cook 1986). But the added-variable plot for S^2 cannot be interpreted as a global net-effect plot as described here because the distribution of the second coordinate of $(Tar, S^2) \mid (S,T)$ is degenerate.

However, the net-effect of S, including any nonlinear tendencies, can be investigated via $(Tar, S) \mid T$. Net-effect plots obtained by brushing on T are not very informative for these data because there are only 31 observations. But a global net-effect plot $\{r_{Tar\mid T}, r_{S\mid T}\}$ may still be useful. Inspection of the scatterplots $\{Tar, T\}$ and $\{S, T\}$ suggests that the corresponding regression functions $E(Tar \mid T)$ and $E(S \mid T)$ are reasonably linear in T. Estimating these regression functions by using the OLS fitted values from simple linear regressions results

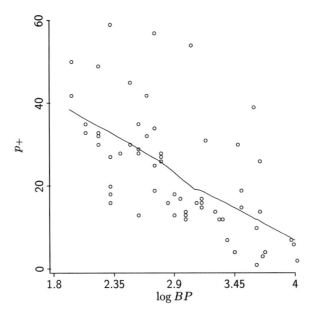

FIGURE 13.8   Scatterplot of p_+ versus $\log(BP)$ from the Minneapolis school data. The LOWESS smooth is provided as a visual enhancement.

in the global net-effect plot for S shown in Figure 13.7b. This net-effect plot provides visual information on the conditional regression structure between *Tar* and S at a fixed value for T, the particular value of T serving only to determine the location of the point cloud. The dominant conditional relationship between *Tar* and S is linear, but the quadratic fit shown on Figure 13.7b indicates that there may be curvature as well. The curvature seems largely a consequence of a few remote points.

13.3.2. Minneapolis schools again

Returning to the Minneapolis school data introduced in Section 11.5, Figure 13.8 gives a scatterplot of the response p_+, the percent of above average scores, versus $\log(BP)$, where BP is the percent of students who do not live with both parents. The plot suggests that the regression function $E(p_+ \mid \log(BP))$ is a decreasing function of $\log(BP)$. This conclusion seems to agree with a common idea that children who live with both parents would, on the average, do better academically than children who do not. In an effort to gain more information, we now consider the net-effect of $\log(BP)$ in the regression of p_+ on $\log(BP)$, $\log(HS)$, and $AFDC^{1/2}$. Recall that HS is the percentage of adults in the area who completed high school and $AFDC$ is the percentage of students receiving aid to families with dependent children.

The net-effect of $\log(BP)$ might be investigated by using a small square brush to condition on regions in the plot $\{\log(HS), AFDC^{1/2}\}$ while observing

TABLE 13.1. Results from Application of SIR to the Bivariate Regression of $(p_+, \log(BP))$ on $(\log(HS), AFDC^{1/2})$

m	$\hat{\Lambda}_m$	DF	p-value
0	81.91	38	0.000
1	22.63	18	0.205

the corresponding net-effect plots. However, with only 63 cases, such conditioning is not likely to leave many data points in the net-effect plots. In order to assess whether a 1D linear distributional index could be used, SIR was applied to the bivariate regression of $(p_+, \log(BP))$ on $x = (\log(HS), AFDC^{1/2})^T$.

The 63 observations were first partitioned into 4 slices on $\log(BP)$. The observations in each slice were then partitioned into 5 slices on p_+. Thus the data were partitioned into 20 slices with about 3 observations each. The results of SIR shown in Table 13.1 indicate 1D structure. The corresponding SIR estimate of the central subspace yields the estimated linear combination

$$b^T x = 0.76 \log(HS) - 0.65 AFDC^{1/2}.$$

Based on the results of the SIR analysis, net-effect plots

$$\{p_+, \log(BP) \mid b^T x \in J\}$$

were constructed by partitioning the observed range of $b^T x$ into five slices J of approximately equal length. Figure 13.9 shows all five local net-effect plots simultaneously, the points in a common plot being represented with a common plotting symbol. In addition, the five intraplot OLS regression lines are shown as well. In contrast to the negative slope of Figure 13.8, the OLS lines for all five net-effect plots have a positive slope, suggesting that the correlation between p_+ and $\log(BP)$ is positive in any subpopulation with HS and $AFDC$ fixed.

PROBLEMS

13.1. Justify Proposition 13.1.

13.2. Assuming that (y, x) follows a nonsingular multivariate normal distribution, describe the relationship between a global net-effect plot for a single predictor x_2 and the partial correlation coefficient $\rho_{y2 \cdot 1}$.

13.3. Construct net-effect plots for each of the predictors in the haystack data. Assume model (12.17) of Problem 12.6. Your plots can be local or

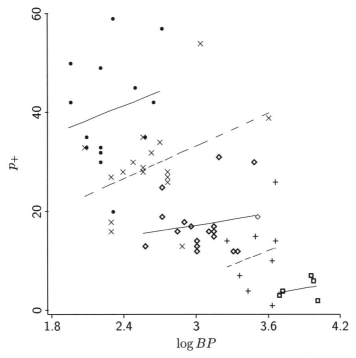

FIGURE 13.9 Five net-effect plots for $\log(BP)$ indicated by different point symbols. The lines represent OLS fitted values in each plot.

global, but you should make use of the knowledge represented by the model.

13.4. Study the net effect of the predictor $\log HS$ in the Minneapolis school regression of Section 13.3.2. Describe how to construct a net-effect plot for the pair of predictors $(\log HS, AFDC^{1/2})$.

13.5. Consider an experiment made up of four replicates of 27 treatments arranged in a completely randomized design. Suppose that the treatments are in a 3^3 design with the quantitative levels of each of the three factors coded as $-1, 0,$ and 1. Discuss the issue of net factor effects and describe how to construct net-effect plots. You may assume a model, but this is not necessary.

CHAPTER 14

Predictor Transformations

Coordinatewise transformations $g(x) = (g_j(x_j))$, $j = 1,\ldots,p$, of the vector x of predictors x_j were used in previous chapters to force linear conditional expectations in their marginal distributions. Predictor transformations may also be useful for reducing the dimension of the central subspace by choosing g so that $\dim[\mathcal{S}_{y|g(x)}] < \dim[\mathcal{S}_{y|x}]$. There is no particular reason why a dimension-reducing transformation should also be a linearizing transformation, but frequently they seem to turn out the same in practice. Nevertheless, in this chapter we consider diagnostic plots to aid in determining if predictor transformations could be useful for dimension reduction. The development is restricted to additive-error models as in Cook (1993). Cook and Croos–Dabrera (1998) studied extensions to generalized linear models.

Partition the $p \times 1$ predictor vector x into a $p_1 \times 1$ vector x_1 and a $p_2 \times 1$ vector x_2, $x = (x_1^T, x_2^T)^T$. The discussion of this chapter will mostly be in the context of the semiparametric model

$$y \mid x = \alpha_0 + \alpha_1^T x_1 + g(x_2) + \varepsilon \tag{14.1}$$

where $E(\varepsilon \mid x) = 0$, $\varepsilon \perp\!\!\!\perp x_1 \mid x_2$, α_1 is a $p_1 \times 1$ vector, and g is an unknown scalar-valued transformation of x_2. The error in (14.1) can depend on x_2, but is independent of x_1 once x_2 is fixed. For example, $\mathrm{Var}(\varepsilon \mid x_1, x_2) = \mathrm{Var}(\varepsilon \mid x_2)$, as well as any higher-order moments, can depend on x_2. The dimension of the central subspace $\mathcal{S}_{y|x}$ can be as much as $p_2 + 1$ under (14.1), but on the transformed scale $\dim[\mathcal{S}_{y|(x_1, g(x_2))}] = 1$. Thus, an appropriate transformation g could result in considerable reduction of the structural dimension.

The primary goal of this chapter is to investigate displays that allow visualization of g and that thus provide information on plausible transformations of x_2. The hope is that using the population metaphor (14.1) to guide diagnostic considerations may lead to a more useful model, or may add support to the possibility that no predictor transformations are required. Secondarily, we will watch for manifestations of the error structure as it relates to x_2.

The formal development is for regressions in which the data are independent and identically distributed observations on the multivariate random variable $(y, x_1^T, x_2^T)^T$. In some problems it may be desirable to condition on the value of the predictors rather than treat them as realizations of a random vector. All of the results can be adapted for the conditional approach, but the development tends to be a bit more complicated. In any event, the two approaches do not lead to any notable differences in practice, and the plots developed here can be used in either case. The unconditional approach is used because the development seems more straightforward.

14.1. CERES PLOTS

14.1.1. Motivation

Suppose $p_2 = 1$ or 2, as will often be the case in practice, and consider the marginal response plot of y versus x_2 as a means of visualizing g. Under (14.1) the regression function for this plot is

$$E(y \mid x_2) = \alpha_0 + \alpha_1^T E(x_1 \mid x_2) + g(x_2).$$

If the predictor regression function $E(x_1 \mid x_2)$ is not constant, the plot $\{y, x_2\}$ could give a biased visualization of g. But if $E(x_1 \mid x_2)$ is constant then

$$E(y \mid x_2) = \text{constant} + g(x_2)$$

and we can expect $\{y, x_2\}$ to allow visualization of g up to an unimportant vertical shift. This unknown shift is in part a consequence of the fact that α_0 and $E(g(x_2))$ are confounded under (14.1) and thus neither is identifiable without imposing further constraints. In view of these results, it seems that any general graphical display for g must somehow account for the predictor regression function $E(x_1 \mid x_2)$.

In rough analogy with an analysis of covariance, define the *population adjusted response*

$$u = y - \alpha_0 - \alpha_1^T x_1$$
$$= g(x_2) + \varepsilon. \tag{14.2}$$

If u were observable then a simple plot of u versus x_2 would reveal g up to the errors ε because $E(u \mid x_2) = g(x_2)$. Additionally, $\text{Var}(u \mid x_2) = \text{Var}(\varepsilon \mid x_2)$, so $\{u, x_2\}$ may also reveal relevant aspects of the error structure. Since u will not normally be observable in practice, a plot for visualizing g will be based on a *sample adjusted response* $\hat{u}$ constructed by replacing α_1 with a consistent estimate that does not depend on knowledge of g. However, as suggested above, it turns out that such an estimate of α_1 requires information about $E(x_1 \mid x_2)$. We begin the development in the next section by assuming that $E(x_1 \mid x_2)$ can be modeled parametrically. It is also possible to use smooths to estimate the components of $E(x_1 \mid x_2)$, as described near the end of the next section.

14.1.2. Estimating α_1

For notational convenience, let $m(x_2) = E(x_1 \mid x_2)$ and suppose that $m(x_2)$ can be modeled as

$$m(x_2) = A_0 + Af(x_2) \tag{14.3}$$

where A_0 is a $p_1 \times 1$ vector of known or unknown constants, A is a $p_1 \times q$ matrix of constants which may be unknown and $f(x_2) = (f_j(x_2))$ is a $q \times 1$ vector of known functions of x_2. For example, if $p_2 = 1$ and if $E(x_1 \mid x_2)$ is a linear function of x_2, then $m(x_2) = A_0 + Ax_2$ where A is now a $p_1 \times 1$ vector. Also, if $E(x_1 \mid x_2)$ is known, then simply set $A_0 = 0$, $A = I$ and $f(x_2) = E(x_1 \mid x_2)$.

Starting with n independent and identically distributed observations $\{y_i, x_i\}$, $i = 1, \ldots, n$, on (y, x), consider summarizing the data by fitting the equation

$$y = b_0 + b_1^T x_1 + b_2^T f(x_2) + \text{error} \tag{14.4}$$

via a convex objective function L,

$$\hat{b} = (\hat{b}_0, \hat{b}_1^T, \hat{b}_2^T)^T = \arg\min L_n(b_0, b_1, b_2) \tag{14.5}$$

where

$$L_n(b_0, b_1, b_2) = \frac{1}{n} \sum_{i=1}^{n} L\left(y_i - b_0 - b_1^T x_{1i} - b_2^T f(x_{2i})\right).$$

Any linear dependencies in $f(x_2)$ should be removed, leaving just linearly independent predictors and a unique estimate $\hat{b}$.

The spelling out of the final term in (14.4) is intended to emphasize that this equation is not necessarily either an adequate description of $y \mid x$, nor near the best that can be achieved. The class of objective functions L includes ordinary least squares, $L(a) = a^2$, and certain robust procedures like Huber's M-estimates. Estimation methods like least median of squares that are based on nonconvex objective functions are excluded, however.

We rely on the following calculations to motivate the use of the estimated coefficients from (14.5) in constructing sample adjusted response. The objective function L_n converges almost surely to

$$R(b_0, b_1, b_2) = E\{L(y - b_0 - b_1^T x_1 - b_2^T f(x_2))\}$$

where the expectation, which is assumed to be finite, is computed with respect to the joint distribution of y and x. The population version of (14.5) is then

$$\beta = (\beta_0, \beta_1^T, \beta_2^T)^T = \arg\min R(b_0, b_1, b_2) \tag{14.6}$$

which is assumed to be unique.

The following proposition (Cook 1993) establishes a key relationship between α_1 and β_1.

Proposition 14.1. Under models (14.1) and (14.3), and estimation procedure (14.5), $\beta_1 = \alpha_1$.

Justification. Assume without loss of generality that $E(x) = 0$ and $E(g(x_2)) = 0$. Using the conditions that $x_1 \perp\!\!\!\perp \varepsilon \mid x_2$, that L is convex, and that $E(x_1 \mid x_2) = Af(x_2)$ gives the first three relations in the following:

$$R(b_0, b_1, b_2) = E_{(\varepsilon, x_2)}\{E_{x_1 \mid x_2}[L(\alpha_0 + \alpha_1^T x_1 + g(x_2) + \varepsilon - b_0 - b_1^T x_1 - b_2^T f(x_2))]\}$$

$$\geq E[L(\alpha_0 - b_0 + (\alpha_1 - b_1)^T E(x_1 \mid x_2) + g(x_2) + \varepsilon - b_2^T f(x_2))]$$

$$= E[L(\alpha_0 - b_0 + (A^T(\alpha_1 - b_1) - b_2)^T f(x_2) + g(x_2) + \varepsilon)]$$

$$= R(b_0, \alpha_1, b_2 - A^T(\alpha_1 - b_1))$$

$$= R(b_0, \alpha_1, b_2^*)$$

where $b_2^* = b_2 - A^T(\alpha_1 - b_1)$. Thus, if $\beta_1 \neq \alpha_1$, where β_1 is the minimizing value defined in (14.6), then

$$R(\beta_0, \beta_1, \beta_2) \geq R(\beta_0, \alpha_1, \beta_2^*)$$

where $\beta_2^* = \beta_2 - A^T(\alpha_1 - \beta_1)$. Since β is assumed to be unique, the inequality must be strict, which contradicts the condition that β minimizes R. Thus it must be true that $\beta_1 = \alpha_1$. $\qquad\square$

This proposition establishes that procedure (14.5) returns α_1 in the population. Root n convergence of $\hat{b}_1$ to α_1 can be shown as well.

Let

$$\hat{u} = y - \hat{b}_0 - \hat{b}_1^T x_1 \tag{14.7}$$

denote a typical adjusted response. Then it follows from Proposition 14.1 that a plot of $\hat{u}$ versus x_2 corresponds to a population plot of

$$\alpha_0 - \beta_0 + g(x_2) + \varepsilon \quad \text{versus} \quad x_2$$

where the error is the same as that attached to the population metaphor (14.1). Thus, $\{\hat{u}, x_2\}$ might be useful for visualizing g up to an unimportant vertical shift, as well as aspects of the conditional error structure. As an alternative to parametric modeling, $E(x_1 \mid x_2)$ could be estimated by using nonparametric regression. This is fairly straightforward when $p_2 = 1$: An estimate $\hat{E}(x_{1j} \mid x_2)$ of the jth coordinate of $E(x_1 \mid x_2)$, $j = 1, \ldots, p_1$, can be obtained by extracting the fitted values from a smooth of x_{1j} against x_2. The $p_1 \times 1$ vector $\hat{E}(x_1 \mid x_2)$ of nonparametric regression estimates evaluated at the data points is then used as a replacement for f in (14.4), and the plot $\{\hat{u}, x_2\}$ constructed as previously described.

If the estimates $\hat{E}(x_{1j} \mid x_2)$ are consistent, this procedure has the same large sample properties as the parametric procedure. The parametric and nonparametric procedures could also be combined so that some elements of $E(x_1 \mid x_2)$ are represented by smooths, while others are represented parametrically.

Plots of the form $\{\hat{u}, x_2\}$ are called CERES plots for x_2 (Cook 1993). The name derives from a slightly different way of thinking about the construction. Let

$$\hat{r} = y - \hat{b}_0 - \hat{b}_1^T x_1 - \hat{b}_2^T f(x_2)$$

denote a typical residual from the fit of (14.4). Then the adjusted response can be rewritten as

$$\hat{u} = \hat{b}_2^T f(x_2) + \hat{r}. \tag{14.8}$$

In this form we see that the adjusted response is a combination of estimated conditional expectations represented by the term $\hat{b}_2^T f(x_2)$ and the residuals $\hat{r}$. The name CERES is an abbreviated acronym for "Combining Conditional Expectations and RESiduals."

14.1.3. Example

To help fix ideas, consider the following regression model with response y and 3×1 predictor vector $w = (w_1, w_2, w_3)^T$,

$$y \mid w = w_1 + w_2 + \frac{1}{1 + \exp(-w_3)}$$

$$= w_1 + w_2 + g(w_3)$$

where w_3 is a uniform random variable on the interval $(1, 26)$ and $w_1 \perp\!\!\!\perp w_2 \mid w_3$ with

$$w_1 \mid w_3 = w_3^{-1} + 0.1\varepsilon_{1|3}$$

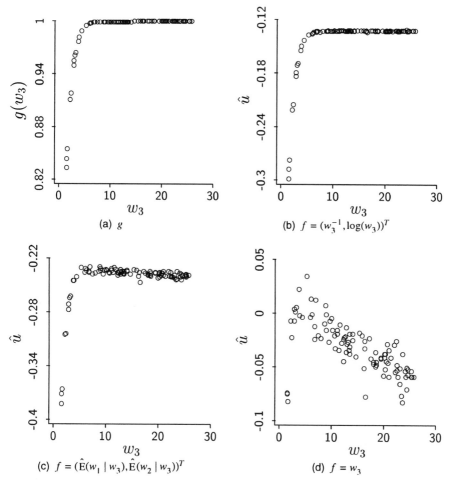

FIGURE 14.1 Plots from the constructed example of Section 14.1.3: (a) Scatterplots of g versus w_3. (b)–(d) CERES plots constructed using the indicated function $f(w_3)$.

and

$$w_2 \mid w_3 = \log(w_3) + 0.25\varepsilon_{2|3}$$

where $\varepsilon_{1|3}$ and $\varepsilon_{2|3}$ are independent standard normal random variables. The response $y \mid w$ is a deterministic function of the three predictors. This allows qualitative conclusions to be illustrated more clearly than when including an additive error. As a point of reference, a plot of g is shown in Figure 14.1a.

Using

$$f(w_3) = \begin{pmatrix} \mathrm{E}(w_1 \mid w_3) \\ \mathrm{E}(w_2 \mid w_3) \end{pmatrix} = \begin{pmatrix} w_3^{-1} \\ \log(w_3) \end{pmatrix}$$

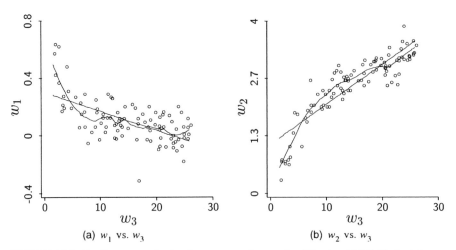

FIGURE 14.2 Predictor smooths for the constructed example of Section 14.1.3. Each plot contains a LOWESS smooth along with the linear OLS fit.

a CERES plot for w_3 can be constructed by first obtaining the residuals $\hat{r}$ from an OLS fit of the equation

$$y = b_0 + b_2 w_1 + b_2 w_2 + b_3 w_3^{-1} + b_4 \log(w_3) + \text{error}$$

and then forming the adjusted response

$$\hat{u} = \hat{b}_3 w_3^{-1} + \hat{b}_4 \log(w_3) + \hat{r}.$$

The shape of the resulting plot, which is shown in Figure 14.1b, is nearly identical to that in the plot of g. Notice that the vertical scale on the plot of Figure 14.1a is quite different from the vertical scale on the plot of Figure 14.1b, a reminder that we can identify g only up to an unimportant vertical shift.

The construction of the CERES plot of Figure 14.1b was based on the actual conditional expectations, $E(w_1 \mid w_3)$ and $E(w_2 \mid w_3)$. As discussed previously, these could be estimated with a parametric model (14.3) or smooths when they are unknown. In the present example, estimates $\hat{E}(w_j \mid w_3)$ of $E(w_j \mid w_3)$ were obtained by smoothing w_j against w_3, $j = 1, 2$, as shown in Figure 14.2. The CERES plot for w_3 shown in Figure 14.1c was then constructed from the adjusted response

$$\hat{u} = \hat{b}_3 \hat{E}(w_1 \mid w_3) + \hat{b}_4 \hat{E}(w_2 \mid w_3) + \hat{r}$$

where the residuals were taken from an OLS fit of the equation

$$y = b_0 + b_2 w_1 + b_2 w_2 + b_3 \hat{E}(w_1 \mid w_3) + b_4 \hat{E}(w_2 \mid w_3) + \text{error}.$$

Although there is some degradation in the plot relative to the previous ones, it still gives a useful visualization of $g(x_2)$.

Finally, the plot shown in Figure 14.1d was constructed by using $f(w_3) = A_0 + Aw_3$, which is not a good representation of the conditional expectation $E[(w_1, w_2)^T \mid w_3]$. This is equivalent to smoothing w_j against w_3, $j = 1, 2$, by using the linear OLS fits shown in Figure 14.2. The plot in Figure 14.1d shows little resemblance to g, and would probably be characterized as having a linear trend with two outliers in the lower-left corner, illustrating that the choice of a parametric or nonparametric representation for $m(x_2)$ is an important part of CERES plot construction. The plot would probably lead to the conclusion that no transformation is required.

As a practical aid, it can be useful to link dynamically a CERES plot to the corresponding predictor plots so that consequences of changing the predictor smooths are immediately apparent in the CERES plot. Experience with this type of dynamic linking using simulated data sets suggests that the best results are obtained when the predictor plots appear somewhat undersmoothed, as illustrated by the LOWESS smooths in the plots of Figure 14.2. This suggestion is in line with the conclusions reported by Wetzel (1996, 1997) who discusses implementation of CERES plots.

14.2. CERES PLOTS WHEN $E(x_1 \mid x_2)$ IS...

As discussed in the previous section, the construction of CERES plots in practice depends fundamentally on knowledge of the conditional expectations $E(x_1 \mid x_2)$. In this section we consider CERES plots under three important special cases for $E(x_1 \mid x_2)$.

14.2.1. ...Constant

When $E(x_1 \mid x_2)$ is constant, a CERES plot for x_2 is $\{\hat{u}, x_2\}$ with the adjusted response $\hat{u} = \hat{r}$, where $\hat{r}$ is a typical residual from a fit of

$$y = b_0 + b_1^T x_1 + \text{error}.$$

Alternatively, information on g might be obtained directly from the marginal response plot $\{y, x_2\}$ because $E(y \mid x_2) = \text{constant} + g(x_2)$. However, we would normally expect that $\text{Var}(y \mid x_2) > \text{Var}(\hat{u} \mid x_2)$, so there are still advantages to using a CERES plot.

For example, Figure 14.3 shows the response plot $\{y, z_2\}$ and the CERES plot $\{\hat{u}, z_2\}$ constructed from 100 observations that were sampled according to the regression

$$y \mid z = z_1 + |z_2| + (1/2)\varepsilon$$

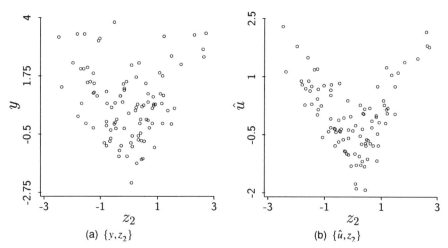

FIGURE 14.3 Two plots illustrating the potential advantage of CERES plots when $E(x_1 \mid x_2)$ is constant.

where z_1, z_2 and ε are independent standard normal random variables. The adjusted responses $\hat{u}$ were set equal to the residual from the OLS linear regression of y on z_1, including an intercept. Apart from a vertical shift, both plots in Figure 14.3 have the same population regression function $g(z_2) = |z_2|$, but the variability about the regression function in the marginal response plot is noticeably larger than in the CERES plot.

Finally, the idea that visual resolution may be increased by replacing the response with the residuals without altering the regression function is similar to the ideas for improving resolution discussed in Section 7.4.

14.2.2. ...Linear in x_2

When $E(x_1 \mid x_2)$ is a linear function of x_2, equation (14.4) is linear in x,

$$y = b_0 + b_1^T x_1 + b_2^T x_2 + \text{error}. \qquad (14.9)$$

The adjusted response is then constructed by adding the estimated conditional expectation term $\hat{b}_2^T x_2$ to the residual from a fit of (14.9) using a convex objective function L.

Plots of the form $\{\hat{b}_2^T x_2 + \hat{r}, x_2\}$ constructed from (14.9) for a single predictor ($p_2 = 1$) and using OLS estimation have a long history and, judging from their prominence in the literature, are frequently used. Their origins can be traced back to Ezekiel (1924). Starting in the early 1970s, considerable attention was devoted to their development. Larsen and McClearly (1972) called such plots *partial residual plots*, while Wood (1973) referred to them as *component plus residual plots*.

Discussions of partial residual plots can be found in many books, including Atkinson (1985), Chambers et al. (1983), Chatterjee and Hadi (1988), Cook and Weisberg (1982, 1994a), Fox (1991), Gunst and Mason (1980), and Seber (1977). Informative research articles include Berk and Booth (1995), Cook (1993), Johnson and McCulloch (1987), Mallows (1986), Mansfield and Conerly (1987), and O'Hara Hines and Carter (1993). A common thread in most of these studies is the problem of inferring about g in model (14.1) when $p_2 = 1$, $\varepsilon \perp\!\!\!\perp x$ and OLS estimation is used for (14.4), but without recognition of the role of the predictor expectation $E(x_1 \mid x_2)$.

A partial residual plot and CERES plot will be the same when $E(x_1 \mid x_2)$ is linear or inferred to be so and when the CERES plot is based on OLS estimation, but otherwise the plots differ. Because of their historical prominence, it seems important to recognize that partial residual plots may not provide a useful visualization of g unless $E(x_1 \mid x_2)$ is linear. In particular, if model (14.1) is accurate and a partial residual plot shows no curvature, we still do not have a firm indication that g is linear unless $E(x_1 \mid x_2)$ is linear. This conclusion is illustrated by the plot of Figure 14.1d, which can now be seen as a partial residual plot for w_3. For further discussion of this point, see Cook (1993), and Johnson and McCulloch (1987).

McKean and Sheather (1997) investigated using Huber and Wilcoxon estimates instead of least squares in the construction of partial residual plots. Robust estimates may be desirable in some circumstances, but they do not mitigate the impact of nonlinear relationships among the predictors, and in some cases can do more harm than good (Cook, Hawkins and Weisberg 1992).

14.2.3. ... Quadratic in x_2

When $E(x_1 \mid x_2)$ is a quadratic function of x_2, equation (14.4) is quadratic in x_2,

$$y = b_0 + b_1^T x_1 + b_2^T x_2 + x_2^T B_2 x_2 + \text{error} \qquad (14.10)$$

where B_2 is a $p_2 \times p_2$ symmetric matrix of coefficients for the quadratic terms. The adjusted response for the resulting CERES plot then takes the form

$$\hat{u} = \hat{b}_2^T x_2 + x_2^T \hat{B}_2 x_2 + \hat{r}$$

where $\hat{r}$ is a typical residual from a fit of (14.10) using a convex L. Mallows (1986) suggested plots based on this construction when $p_2 = 1$ and OLS estimation is used, but without reference to the requirement that $E(x_1 \mid x_2)$ should be a quadratic function of x_2 for good performance. Plots based on using (14.10) to form adjusted responses are sometimes called *augmented partial residual plots*.

Historically, the role of the conditional expectations $E(x_1 \mid x_2)$ was apparently not recognized and this contributed to the often elusive nature of discussions on the relative merits of partial and augmented partial residual plots. The exchange between Mallows (1988) and O'Brien (1988) typifies the nature of the historical debate.

Weisberg (1985, p. 137) described an experiment on the quantity of hydrocarbons y emitted when gasoline is pumped into a tank. There are 125 observations and four covariates: initial tank temperature TT, temperature of dispensed gasoline GT, initial vapor pressure in the tank TP, and vapor pressure of the dispensed gasoline GP. Initial application of standard Box–Cox methodology for transforming the response (Section 2.3) indicated that the square-root transformation of y may produce an approximately normal linear model in the four predictors. We use $y^{1/2}$ as the response in this example to investigate the need to transform TT.

The values of the predictors in this example were evidently selected by the experimenters rather than sampled from a population. Nevertheless, as pointed out at the beginning of this chapter, CERES plots are still applicable. The only change is in interpretation: The predictor expectations should be interpreted relative to the design points rather than relative to some larger population.

The first step is to inspect plots of TT against each of the three remaining predictors as a means of providing information on the predictor expectations $E(GT \mid TT)$, $E(TP \mid TT)$, and $E(GP \mid TT)$. A plot of GT versus TT (not shown) suggests that $E(GT \mid TT)$ is linear. However, the quadratic fits shown on the plots of TP versus TT and GP versus TT in Figure 14.4a and Figure 14.4b suggest that $E(TP \mid TT)$ and $E(GP \mid TT)$ are nonlinear. Because of the experimental design, the data fall into three clusters, which effectively limits characterizations of $E(TP \mid TT)$ and $E(GP \mid TT)$ to quadratic polynomials. Thus the adjusted response

$$\hat{u} = \hat{b}_4 TT + \hat{b}_{44} TT^2 + \hat{r}$$

was constructed from the OLS fit of

$$y^{1/2} = b_0 + b_1 GT + b_2 TP + b_3 GP + b_4 TT + b_{44} TT^2 + \text{error}.$$

The resulting CERES plot in Figure 14.4c shows clear curvature, suggesting that the usual linear model might be improved by adding a quadratic in TT. For contrast, the partial residual plot for TT is shown in Figure 14.4d. Relative to the partial residual plot, the first cluster of values in the CERES plot is moved up and the last cluster is moved down. Overall, the CERES plot shows a more pronounced effect of initial tank temperature, and stronger curvature. According to the usual t test, the coefficient of initial tank temperature is not significant in the standard linear model. But when the quadratic in tank temperature is added, the coefficients of the linear and quadratic terms in tank temperature are significant with t values of 3.1 and -4.3, respectively. The CERES plot does seem to give a more accurate visual impression of the effects of initial tank temperature than the partial residual plot.

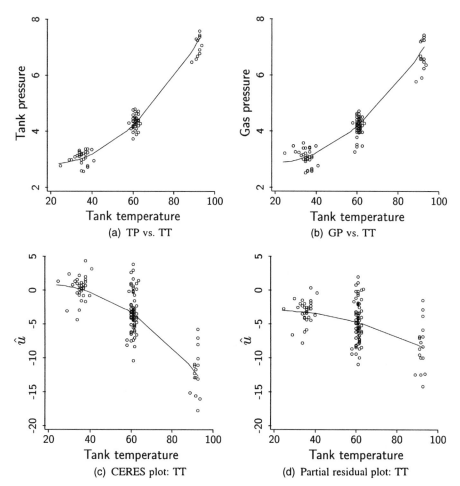

FIGURE 14.4   Four plots relating to the problem of transforming tank temperature (TT) in the sniffer data described in Section 14.2.3.

14.3. CERES PLOTS IN PRACTICE

Berk and Booth (1995) reported the results of an empirical study to contrast the relative performance of several methods for identifying $g(x_2)$ in (14.1), including visualization methods such as CERES plots and various standard residual plots, and an iterative numerical method designated AMONE based on the backfitting algorithm. They concluded that " ... the AMONE additive model plot based on smoothing just one predictor and the CERES plot seem equally good," and that both are able to adapt to any form of curve. Nevertheless, the generally good performance of these methods can be mitigated in some practical situations. In this section we consider circumstances that may limit the performance of CERES plots, as well as ways of refining application.

In addition, Berk's (1998) study of the geometry of three-dimensional CERES and added-variable plots is quite useful for gaining a deeper understanding of the operating characteristics of these plots in practice.

14.3.1. Highly dependent predictors

The justification of CERES plots in the previous sections was based on large sample considerations. In small samples it is possible for substantial dependence among the predictors to produce misleading behavior in CERES plots. To illustrate the phenomenon, assume that there are just $p = 2$ predictors and that $m(x_2) = E(x_1 \mid x_2)$ is known. Under (14.1), the adjusted response used for the vertical axis of a CERES plot for x_2 can be written as

$$\hat{u} = \delta_0 + (\alpha_1 - \hat{b}_1)x_1 + g(x_2) + \varepsilon$$

where $\hat{b}_1$ is the estimate from (14.5). As the sample size grows, $\hat{b}_1$ converges to α_1 and the adjusted response converges to the population version given previously. But in small samples observed nonzero differences $\alpha_1 - \hat{b}_1$ can be large enough to cause problems for CERES plots. In particular, $\text{Var}(\hat{b}_1)$ will be relatively large when the correlation between $m(x_2)$ and x_1 is large.

The ability of a CERES plot to give a good visualization of $g(x_2)$ depends on most aspects of the problem, including the relative magnitudes of $g(x_2)$, $m(x_2)$ and ε, and the strength of the dependence between x_1 and x_2. As an extreme case, suppose $g(x_2) = 0$ so that any nonconstant pattern in the regression function of the CERES plot for x_2 represents misleading information. The adjusted response can be written as

$$\hat{u} = \delta_0 + (\alpha_1 - \hat{b}_1)(m(x_2) + \varepsilon^*) + \varepsilon$$

where, for convenience of illustration, we assume that $x_1 \mid x_2 = m(x_2) + \varepsilon_{1|2}$ with $\varepsilon_{1|2} \perp\!\!\!\perp (x, \varepsilon)$ and $x \perp\!\!\!\perp \varepsilon$. The CERES plot might show a clear systematic pattern representing $m(x_2)$, depending on the magnitudes of the various terms in $\hat{u}$. Treating $\hat{b}_1$ as fixed, the squared correlation between $\hat{u}$ and $m(x_2)$ is

$$\rho^2\left(\hat{u}, m(x_2)\right) = \frac{(\alpha_1 - \hat{b}_1)^2 \text{Var}\left(m(x_2)\right)}{\text{Var}(\varepsilon) + (\alpha_1 - \hat{b}_1)^2 [\text{Var}\left(m(x_2)\right) + \text{Var}(\varepsilon_{1|2})]}.$$

The CERES plot for x_2 will show $m(x_2)$ when this correlation is sufficiently large. Generally, the correlation may be increased by increasing $(\alpha_1 - \hat{b}_1)^2$ or by decreasing $\text{Var}(\varepsilon)$ or $\text{Var}(\varepsilon_{1|2})$.

As a graphic illustration of these general ideas, we use data sets with $n = 199$ observations constructed according to the following setup: The values of the predictor x_2 are $-0.99, -0.98, \ldots, 0.99$. The other predictor x_1 is con-

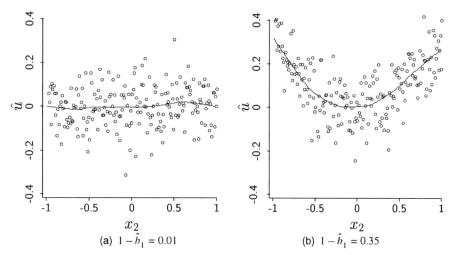

FIGURE 14.5 Two CERES plots from replications of the constructed example with highly dependent predictors in Section 14.3.1. The plots are labeled by the value of $\alpha_1 - \hat{b}_1 = 1 - \hat{b}_1$ where $\hat{b}_1$ is the estimated coefficient of x_1 from the estimative equation with linear and quadratic terms in x_2.

structed as $x_1 \mid x_2 = x_2^2 + .01\varepsilon_{1|2}$. Finally, the response is $y \mid (x_1, x_2) = x_1 + .1\varepsilon$. The terms $\varepsilon_{1|2}$ and ε are independent standard normal random variables. The predictors in this example are very highly related, the sample correlation between x_1 and x_2^2 being about 0.9995. Shown in Figure 14.5 are two CERES plots for x_2 constructed from independent replications of the data. The value of $\alpha_1 - \hat{b}_1$ is shown below each plot, where $\hat{b}_1$ is the coefficient of x_1 from the estimative equation with linear and quadratic terms in x_2. These plots illustrate the general conclusion that CERES plots can respond to $m(x_2)$ in situations where $\hat{b}_1$ is not sufficiently close to α_1.

14.3.2. Using many CERES plots

So far our treatment of CERES plots has relied on the presumption that only a few predictors may require transformation, and that we know the candidates from the components of x. The analysis of covariance is one situation when such a condition may be reasonable. For example, Cook and Weisberg (1994a, p. 146) used a 3D CERES plot ($p_2 = 2$) to provide a visualization of the relative fertility in an 8×7 array of field plots with a randomized complete block design superimposed. The x_2 predictor held the row-column position $x_2 = (row, column)^T$ of each plot, while x_1 was used to identify the additive treatments.

The usefulness of CERES plots would be increased if they could be adapted to situations in which transform of one or two *unidentified* predictors is required. It may seem reasonable in such situations to construct p CERES plots,

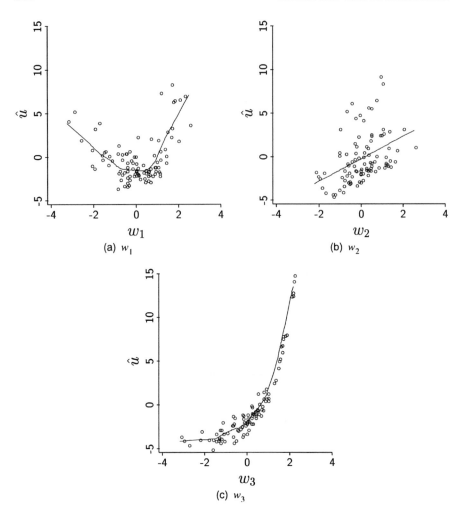

FIGURE 14.6   CERES plots for the predictors in (14.11).

one for each of the predictors, to decide which predictors to transform. Partial residual plots have often been used in the literature in this way. Such a procedure comes with complications, however.

To illustrate one possibility when viewing multiple CERES plots, consider 100 observations generated from the regression

$$y \mid w = w_1 + w_2 + 2\exp(w_3) + 0.5\varepsilon \qquad (14.11)$$

where $w_3 \mid (w_1, w_2) = w_1 + 0.5\varepsilon^*$, and w_1, w_2, ε and ε^* are independent standard normal random variables. Note for later reference that $w_2 \perp\!\!\!\perp (w_1, w_3)$.

Shown in Figure 14.6 are the three CERES plots, one for each predictor. The CERES plots in this case were constructed as partial residual plots because the relevant predictor conditional expectations are all linear. The vertical axes on the plots of Figure 14.6 are the same to facilitate later comparison. The CERES plot for w_3 gives good visualization of the exponential function $g(w_3) = 2\exp(w_3)$. However, the CERES plot for w_1 also shows curvature, while the CERES plot for w_2 seems to exhibit heteroscedasticity. Taken at face value, the three CERES plots in Figure 14.6 may suggest a regression structure that is more complicated than the regression (14.11) used to generate the data. Selecting a predictor to transform without further guidance would be problematic in this example. But notice that the variation about the smooth for w_3, the predictor that needs to be transformed, is clearly less that the variation about the smooths for the other two predictors. This type of behavior can be used to select a predictor for transformation, as developed in the remainder of this section.

To explore the issues involved in viewing multiple CERES plots, consider a situation in which x_1 or x_2 may contribute nonlinearly,

$$y \mid x = \alpha_0 + g_1(x_1) + g_2(x_2) + \varepsilon \tag{14.12}$$

where $\varepsilon \perp\!\!\!\perp (x_1, x_2)$, $E(\varepsilon) = 0$, and $\text{Var}(\varepsilon) = \sigma^2$. Without loss of generality, assume that $E(x) = 0$ and $E(g_j(x_j)) = 0$, $j = 1, 2$. Assuming that $m(x_2) = E(x_1 \mid x_2)$ is known, a CERES plot for x_2 is constructed from a fit of (14.4) with $f(x_2) = m(x_2)$. It follows that the population-adjusted response is

$$
\begin{aligned}
u &= y - \beta_0 - \beta_1^T x_1 \\
&= \delta_0 + g_1(x_1) - \beta_1^T x_1 + g_2(x_2) + \varepsilon
\end{aligned}
$$

where $\delta_0 = \alpha_0 - \beta_0$, and β_0 and β_1 are the population parameters defined at (14.6). The error at the end of this equation is the same as that in (14.12); thus, population CERES plots will still be subject to at least the variation in the original model.

The expected ordinate of a CERES plot for x_2 is

$$E(u \mid x_2) = \delta_0 + h(x_2) + g_2(x_2) \tag{14.13}$$

where $h(x_2) = E[g_1(x_1) - \beta_1^T x_1 \mid x_2]$. The following points can be deduced from (14.13).

g_1 *linear*
If $g_1(x_1) = \alpha_1^T x_1$, then $\beta_1 = \alpha_1$ as in Proposition 14.1, and $g_1(x_1) - \beta_1^T x_1 = 0$. Thus, $E(u \mid x_2) = \delta_0 + g_2(x_2)$. Additionally, the variance about the regression function is the same as that in model (14.12):

$$
\begin{aligned}
\text{Var}(u \mid x_2) &= E[(u - E(u \mid x_2))^2 \mid x_2] \\
&= \sigma^2. \tag{14.14}
\end{aligned}
$$

In reference to the example generated from (14.11), set $x_2 = w_3$ and $x_1 = (w_1, w_2)^T$. The population regression function for the CERES plot for w_3 in Figure 14.6c is $E(u \mid w_3) = 2\exp(w_3)$, and the variation about the regression function is the same as that in model (14.11).

x_1 independent of x_2

If $x_1 \perp\!\!\!\perp x_2$ then $h(x_2) = 0$, and the regression function of the CERES plots for x_2 is again as required: $E(u \mid x_2) = \delta_0 + g_2(x_2)$. Turning to the variance function of the population CERES plot,

$$\text{Var}(u \mid x_2) = E[(u - E(u \mid x_2))^2 \mid x_2]$$

$$= E[(g_1(x_1) - \beta_1^T x_1 + \varepsilon)^2 \mid x_2].$$

Since $\varepsilon \perp\!\!\!\perp (x_1, x_2)$ implies that $\varepsilon \perp\!\!\!\perp x_1 \mid x_2$, the variance function can be written

$$\text{Var}(u \mid x_2) = E[(g_1(x_1) - \beta_1^T x_1)^2 \mid x_2] + E(\varepsilon^2 \mid x_2)$$

$$= \text{Var}(g_1(x_1) - \beta_1^T x_1) + \sigma^2. \tag{14.15}$$

The second equality follows because $x_1 \perp\!\!\!\perp x_2$ and $\varepsilon \perp\!\!\!\perp x_2$. We see from (14.15) that the conditional variance $\text{Var}(u \mid x_2)$ does not depend on the value of x_2 and thus is constant. More importantly, $\text{Var}(u \mid x_2) > \sigma^2$ as long as $g_1(x_1) \neq \beta_1^T x_1$. In this case the CERES plot for x_2 will have the desired regression function regardless of g_1, but the variation around the regression function may be substantially larger than that when g_1 is linear.

This situation is illustrated by the CERES plot for w_2 in Figure 14.6b: Setting $x_2 = w_2$ and $x_1 = (w_1, w_3)^T$, it follows that $x_1 \perp\!\!\!\perp x_2$. The regression function for Figure 14.6b is linear because w_2 enters (14.11) linearly and $x_1 \perp\!\!\!\perp x_2$. However, because of the exponential function in g_1, the variation about the regression function is skewed and is noticeably larger than that in Figure 14.6c. These conclusions may be a bit clearer if we return to the original setup.

In terms of model (14.11), the population CERES plot corresponding to Figure 14.6b can be represented as

$$\{\delta_0 + w_2\} + \{w_1 + 2\exp(w_3) - (w_1, w_3)^T \beta_1 + .5\varepsilon\} \quad \text{vs.} \quad w_2.$$

Because $w_2 \perp\!\!\!\perp (w_1, w_3)$, the regression function for the plot is linear. The second term in braces represents the "conditional error" about the regression function. The conditional error distribution does not depend on the value of w_2, but the presence of the exponential function makes it quite skewed. Although the variation about the regression function in Figure 14.6b may appear nonconstant, that impression is due to the observed skewness along the regression function; $\text{Var}(u \mid w_2)$ is actually constant, as demonstrated at (14.15).

g_2 *linear*

If $g_2(x_2) = \alpha_2^T x_2$ is linear, so that the CERES plot is for a predictor that does not require transformation, then

$$E(u \mid x_2) = \delta_0 + h(x_2) + \alpha_2^T x_2$$

and

$$E(\text{Var}(u \mid x_2)) = E(u - E(u \mid x_2))^2$$

$$= E(g_1(x_1) - \beta_1^T x_1 + \varepsilon - h(x_2))^2$$

$$= E[\text{Var}(g_1(x_1) - \beta_1^T x_1 \mid x_2)] + \sigma^2. \tag{14.16}$$

Depending on the magnitude of h, the CERES plot may well show any curvature in g_1. The main problem here is that any nonlinear effect of x_1 can leak through to the CERES plot for the linear variable, in this case x_2. This *leakage effect* was mentioned by Chambers et al. (1983, p. 306) in connection with partial residual plots. This accounts for the curvature in Figure 14.6a: Although w_1 does not require transformation, the leakage effect causes the nonlinear function of w_3 to manifest in the CERES plot for w_1. The magnitude of the leakage effect depends on the size of $h(x_2)$ relative to that of $\alpha_2^T x_2$. Generally, the leakage effect may be more pronounced when x_1 and x_2 are highly dependent.

14.3.3. Transforming more than one predictor

The main points from the previous discussion that seem potentially helpful in practice are:

- Because of the leakage effect, curvature in a CERES plot is not necessarily an indication that the corresponding predictor x_2 should be transformed, unless x_2 is independent of the remaining predictors or the remaining predictors enter the model linearly.
- CERES plots for predictors that require transformation will often be distinguished by having relatively small variation about the regression function, as compared to CERES plots for predictors that do not require transformation.
- A linear regression function in a CERES plot is a good indication that the corresponding predictor does not require transformation, although sufficient variation can mask small curvature.

Applying these ideas in the example of Figure 14.6 immediately identifies w_3 as the single predictor most likely to need transformation. Nevertheless, without the prior knowledge that only one predictor needs to be transformed, there remains the possibility that w_1 should be transformed as well. There are at least two ways to proceed.

First, since only two predictors show curvature in the CERES plot of Figure 14.6, we could construct a 3D CERES plot for w_1 and w_3. If the plot has 1D structure after removing its linear trend and if the sufficient summary view is inferred to be the 2D plot $\{\hat{u}, w_3\}$, then following the discussion of Section 4.4.2 we may infer that only w_3 requires transformation. Any other structure would be an indication that the transformations required are more complicated. For example, if the plot has 1D structure after removing its linear trend and if the sufficient summary view is inferred to be a 2D plot of the form $\{\hat{u}, aw_1 + bw_3\}$ with $a \neq 0$ and $b \neq 0$, then coordinatewise transformations of the predictors may not achieve the best possible results. Application of these ideas to the example of Figure 14.6 yields results that are consistent with transforming only w_3.

Second, we could consider CERES plots for the regression of a new response

$$Y_1 = y - \hat{g}_3(w_3)$$

on the two remaining predictors w_1 and w_2. Here, the estimate $\hat{g}_3$ is the smooth shown in Figure 14.6c evaluated at the data points. These CERES plots (not shown) exhibit no notable nonlinearity and thus no further transformation is indicated. If one of them, say the CERES plot for w_1, had shown notable nonlinearity, then we could have proceeded as before, extracting a smooth $\hat{g}_1(w_1)$ of the regression function and forming another new response

$$Y_{13} = y - \hat{g}_1(w_1) - \hat{g}_3(w_3).$$

This response could then have been plotted against the remaining predictor w_2. Proceeding iteratively with this idea, we obtain a graphical fitting procedure similar to the methods proposed by Ezekiel (Section 2.6.1) and to the backfitting algorithm.

Generally, if it appears in any problem that many predictors are likely to need transformation, then iterative fitting of additive models should be considered.

14.4. BIG MAC DATA

We use economic data on 45 world cities to further illustrate the use of CERES plots in practice. The data, which were originally published by Enz (1991), are available from Cook and Weisberg (1994a). The response variable is log(*BigMac*), the logarithm of the number of minutes of labor required to purchase a Big Mac hamburger and French fries. The three predictors

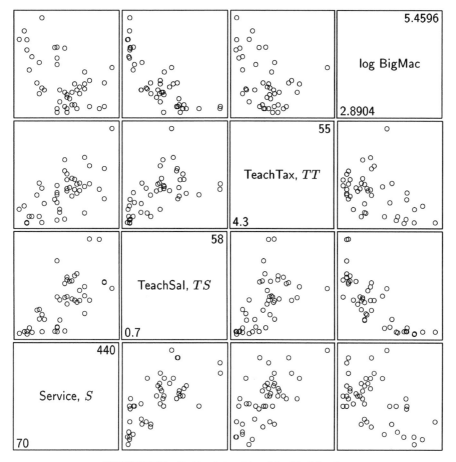

FIGURE 14.7   Scatterplot matrix of the Big Mac data.

are

- *TS*—the average annual salary of primary school teachers (Teachers' Salary), in thousands of U.S. dollars
- *TT*—the average tax rate paid by primary school teachers (Teachers' Tax)
- *S*—total annual cost of 19 services common in western countries (Services)

Shown in Figure 14.7 is a scatterplot matrix of the three predictors and the response. The predictor plots all seem reasonably linear, except perhaps for one or two relatively remote points. This is supported by a 3D plot of the predictors (not shown). Although it is a little hard to see without using smoothers, the inverse response plots for *S* and *TT* have similar regression functions, while that for *TS* seems different, suggesting that the mean checking

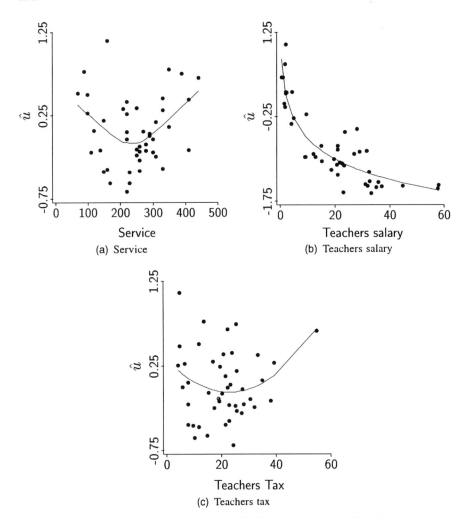

FIGURE 14.8 CERES plots for the Big Mac data, one for each predictor.

condition is not satisfied and that the regression has at least 2D structure. We now turn to CERES plots to see if a predictor transformation can reduce the structural dimension. Because the predictor plots seem reasonably linear, we assume that the conditional expectations of the predictors are all linear and construct CERES plots accordingly.

The three CERES plots, one for each predictor, are shown in Figure 14.8. As suggested by the smooths, each plot exhibits some curvature, although the curvature for teachers' tax is largely due to a single remote point. If no predictor transformations were necessary, we would expect the regression function for each plot to be linear, or approximately so. The fact that each plot shows some curvature in its regression function suggests that a predictor transforma-

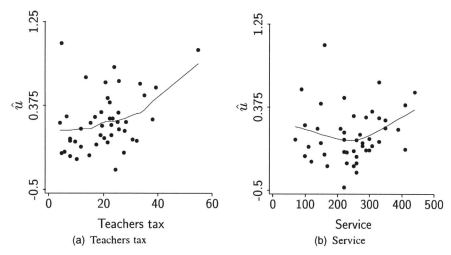

FIGURE 14.9 CERES plots with response Y_1 for the Big Mac data, one for each of the remaining two predictors.

tion may be worthwhile. Which predictor should we transform? It is possible that all predictors need to be transformed to achieve 1D structure, or that there is already 1D structure $y \mid x = g(\eta^T x) + \varepsilon$ so that transformation of any single predictor is not likely to help much. It is also possible that the curvature in each of the plots can be explained by a model like (14.1) where x_2 is just one of the predictors.

Returning to the three CERES plots shown in Figure 14.8 for the Big Mac data, it seems that the regression function in the plot for TS is the most clearly defined since it exhibits the smallest variation. Hence, we start by transforming TS. The smooth regression function shown on the CERES plot for TS corresponds to its log transformation and seems to give a good visual fit. Accordingly, we consider replacing TS with $\log(TS)$, or equivalently with the values of the smooth $\hat{g}_1(TS)$ extracted at the data points from Figure 14.8b.

To decide if a second predictor should be transformed, CERES plots can be constructed for the regression of the new response

$$Y_1 = \log(BigMac) - \hat{g}_1(TS)$$

on the two remaining predictors, TT and S. These CERES plots are shown in Figure 14.9. The nonlinearity in the CERES plot for service seems to be the most pronounced, although the curvature in both plots appears rather weak so the benefits of further transformation are not really clear. Nevertheless, a transformation of S could be incorporated by extracting the smooth $\hat{g}_2(S)$ shown in Figure 14.9a and forming the next response

$$Y_{12} = \log(BigMac) - \hat{g}_1(TS) - \hat{g}_2(S)$$

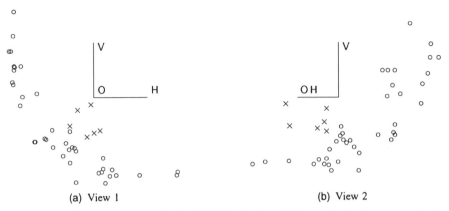

FIGURE 14.10 Two views of the 3D CERES plot for (TS,S) in the Big Mac data. $V = \hat{u}$, $H = TS$, and $O = S$.

which could then be plotted against the remaining predictor TT to assess the need to transform TT.

Additionally, a 3D CERES plot for (TS,S) could be used as a transformation aid. Shown in Figure 14.10 are two views of this 3D plot. The 2D projection $\{\hat{u},TS\}$ shown in View 1 supports the previous finding that a transformation of TS would be worthwhile. However, after removing the linear trend, the 3D plot clearly has 2D structure, a firm indication that transforming TS alone is not sufficient to describe the regression function. The 2D structure is driven by the points marked with an ex in Figure 14.10. These points, which mostly correspond to Scandinavian cities, form a distinct plane that disrupts the 1D structure formed by the remaining points.

In conclusion, depending on the specific goals of the analysis, a transformation of TS alone may account for sufficient variation. But it seems clear that some systematic variation would remain unexplained. A more complete description of the regression function would seem to require a joint transformation of the form $g(TS,S)$.

14.5. ADDED-VARIABLE PLOTS

Added-variable plots have arisen naturally in this book from time to time: In Section 7.4 we saw how added-variable plots arise in connection with visual estimation of dimension-reduction subspaces, and in Chapter 13 we found that such plots can, under certain circumstances, be used as global net-effect plots. Added-variable plots are also quite useful in studying influence (Cook 1986; Cook 1987a; Cook and Weisberg 1982).

Some have suggested that an added-variable plot is useful for visualizing curvature and predictor transformations in regression, which is the general topic of this chapter. To explore this use of added-variable plots, we again return to a version of model (14.1)

$$y \mid x = \alpha_0 + \alpha_1^T x_1 + g(x_2) + \varepsilon \qquad (14.17)$$

where $\varepsilon \perp\!\!\!\perp (x_1, x_2)$, $E(\varepsilon) = 0$, $Var(\varepsilon) = \sigma^2$, $E(g(x_2)) = 0$, and x_2 is a scalar ($p_2 = 1$). Next, an added-variable plot for x_2 is $\{\hat{e}_{y|1}, \hat{e}_{2|1}\}$, where $\hat{e}_{y|1}$ denotes a typical residual from the OLS linear regression of y on x_1, and $\hat{e}_{2|1}$ is a typical residual from the OLS linear regression of x_2 on x_1. The question for this section is whether the added-variable plot for x_2 is generally a useful method of visualizing $g(x_2)$ in (14.17).

The sample residual $\hat{e}_{y|1}$ converges to the population residual

$$e_{y|1} = g(x_2) - x_1^T [Var(x_1)]^{-1} Cov(x_1, g) + \varepsilon.$$

This expression is the population version of the ordinate of an added-variable plot under model (14.17). Similarly, the population abscissa of an added-variable plot is

$$e_{2|1} = x_2 - \beta_{2|1}^T x_1$$

where $\beta_{2|1} = [Var(x_1)]^{-1} Cov(x_1, x_2)$. To characterize the population behavior of an added-variable plot, consider the decomposition

$$e_{y|1} = E(e_{y|1} \mid e_{2|1}) + [e_{y|1} - E(e_{y|1} \mid e_{2|1})]$$

where the first term

$$\gamma(e_{2|1}) = E(e_{y|1} \mid e_{2|1}) \qquad (14.18)$$

is the regression function for the plot and the second term reflects the variance.

For an added-variable plot to work reasonably well, the regression function should equal g, $\gamma(e_{2|1}) = g(e_{2|1})$ or approximately so. Whether this holds to a useful approximation depends on the marginal distribution of the predictors and on $g(x_2)$.

Assume now that (x_1^T, x_2) is normally distributed. This implies that $x_1 \perp\!\!\!\perp e_{2|1}$ and thus that the regression function (14.18) reduces to

$$\gamma(e_{2|1}) = E[g(x_2) \mid e_{2|1}] = E[g(\beta_{2|1}^T x_1 + e_{2|1}) \mid e_{2|1}] \qquad (14.19)$$

where $x_2 = \beta_{2|1}^T x_1 + e_{2|1}$. The regression function in an added-variable plot with normal predictors is like a smoothed version of a plot of $g(e_{2|1})$ versus $e_{2|1}$, where the smoothing is with respect to the normal distribution of $\beta_{2|1}^T x_1$. This smoothing might not be a problem if the ratio

$$\frac{\text{Var}(\beta_{2|1}^T x_1)}{\text{Var}(e_{2|1})} = \frac{\rho^2}{1 - \rho^2}$$

is small, where ρ is the multiple correlation coefficient between x_2 and x_1. When ρ^2 is not small, an added-variable plot may substantially underestimate any curvature present in the model. In other words, added-variable plots are biased toward linear trends unless $\rho = 0$. As we have seen, the corresponding population CERES plot is unbiased in the sense used here.

Generally, added-variable plots should not be used as a basis for diagnosing curvature. The absence of curvature in an added-variable plot should not necessarily be taken as assurance that g is linear. For further discussion along these lines, see Cook (1996a).

14.6. ENVIRONMENTAL CONTAMINATION

In this section we consider a relatively complicated example to bring together selected techniques from the last few chapters. The data comes from a large simulation code developed at Los Alamos National Laboratory (LANL) to aid in a study of the fate of an environmental contaminant introduced into an ecosystem. Environmental contaminants have the potential for ecological and human health effects due to their toxicological properties. A good appreciation of the ecological risk associated with contamination requires an understanding of its dynamics.

The LANL code is essentially a compartmental model with eight compartments: vegetation interior, vegetation surface, terrestrial invertebrates, small herbivores, large herbivores, insectivores, predators, and litter. The litter compartment is a sink that only receives the contaminant. The model consists of a set of coupled differential equations representing the various compartments in the ecosystem. It is based on the assumption that the contaminant enters the ecosystem by dissolution to water and then moves through the food web by one organism consuming another. The concentration of the contaminant in water is assumed to be constant for any run of the code. In total, the model requires 84 inputs and, for the purposes of this study, the response y is the amount of contamination in the terrestrial invertebrates at day 5000. The general issue addressed in this example is how to gain an understanding of the conditional distribution of $y \mid x$, where $x = (x_j)$ denotes the 84×1 vector of inputs. Additionally, scientists working on the project were interested in identifying the most important inputs.

Application of any of the ideas discussed so far in this book will depend on the joint distribution of the inputs. The scientists who developed the LANL model provided ranges for each of the 84 input variables and an estimate of the nominal value. Based on this and other information, members of the LANL

Statistics Group developed beta distributions for each of the 84 inputs. No J-shaped or U-shaped beta distributions were allowed, but the distributions could be symmetric or skewed. The inputs were regarded as mutually independent so their joint distribution is the product of the individual marginal beta densities. The sample space for the inputs will be called the *operability region.*

The data for this example were generated by taking $n = 500$ Latin hypercube samples (McKay, Beckman, and Conover 1979) at the midpoints of equally probable slices across the ranges of the inputs. Although more observations could have been taken, it was felt that 500 would be sufficient to gain useful information about the inputs and to guide further sampling as necessary. All sampling was carried out by members of the LANL Statistics Group.

14.6.1. Assessing relative importance

The notion of relative importance of the input variables is perhaps easiest to quantify locally, where measures of importance might be based on rates of change in the regression function at selected points. The results of such an analysis could be elusive, however, particularly if the regression function is flat on the average but has many local peaks and valleys. The local relative importance of the inputs near the nominal value of x need not be the same as that near the edges of the operability region, for example.

Alternatively, relative importance might be approached globally, looking for inputs or functions of inputs that account for most of the variation in the response across the operability region. Such an approach might yield useful results if there are clear global trends, but it could well fail if the primary variation in the response is local. In this example we concentrate on the global approach since experience with a distinct but similar simulation code indicated that there may well be clear global trends.

Since the LANL simulation code itself is deterministic, $F_{y|x}$ places mass 1 at the value of y corresponding to the value of x, as in the example of Section 14.1.3. The basic idea in a global analysis is to find a low-dimensional function $G(x)$ that can serve as a substitute for x itself without important loss of information. In particular, $F_{y|G(x)}$ should be a good approximation of $F_{y|x}$,

$$F_{y|G(x)} \approx F_{y|x} \qquad (14.20)$$

for all values of x in the operability region. Note that, although $y \mid x$ is deterministic, $y \mid G(x)$ need not be so. If a low-dimensional function G can be found so that (14.20) holds to a useful approximation, then the problem of assessing relative importance might be eased considerably. The success of this approach depends in part on the specific types of functions allowed for G. In keeping with the overall theme of this book, we will use linear functions $G(x) = \eta^T x$ leading to dimension-reduction subspaces, and linear functions after coordinatewise transformations $g(x) = (g_j(x_j))$, so that $G(x) = \eta^T g(x)$.

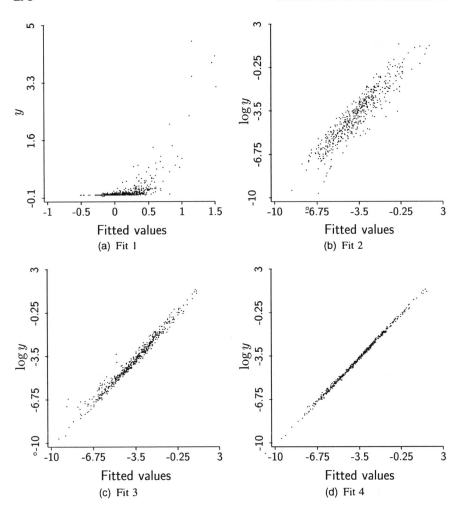

FIGURE 14.11 Response versus fitted values from four stages in the analysis of the environmental contamination data. (a) Original data. (b) log(y) and original predictors. (c) log(y) and transformed predictors. (d) Final fit.

14.6.2. Data analysis

As a baseline, Figure 14.11a gives a plot of the response versus the fitted values from the OLS linear regression on all 84 inputs. Letting η denote a basis for $\mathcal{S}_{y|x}$, this starting point is based on the rationale that if $E(x \mid \eta^T x)$ is a linear function of the value of $\eta^T x$ then from the Li–Duan Proposition the 84×1 vector of coefficient estimates $\hat{b}$ converges to a vector in $\mathcal{S}_{y|x}$. OLS estimation was used for convenience only.

Figure 14.11a indicates that the response variable is highly skewed, ranging over several orders of magnitude from 2×10^{-5} to 4.5. Figure 14.11b was

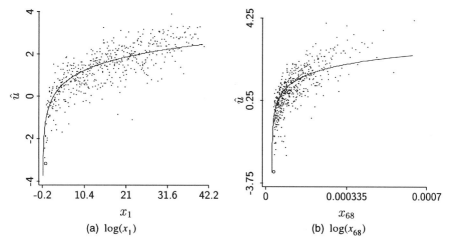

FIGURE 14.12 CERES plots for the first and sixty-eighth input variables with response $\log(y)$ for the data on environmental contamination. The superimposed curves, (a) $\log(x_1)$ and (b) $\log(x_{68})$, are the estimated input transformations.

constructed as Figure 14.11a, except that y was replaced by $\log(y)$. There is fairly strong linearity in at least one direction in the log scale with $R^2 = 0.86$. Figure 14.11b seems to be a reasonable starting point for further analysis. The transformation from y to $\log(y)$ does not change the objective of the analysis because $\mathcal{S}_{\log(y)|x} = \mathcal{S}_{y|x}$ (see Problem 6.1).

Application of SIR with 100 slices of five observations each to the regression with $\log(y)$ did not provide anything beyond the projection in Figure 14.11b. SIR strongly indicated that $\dim[\mathcal{S}_{\log(y)|x}] = 1$. The correlation between the first SIR predictor and the OLS fitted values is 0.99, so both procedures indicate essentially the same solution.

Because the inputs are essentially independent by design, conditional expectations of the form $E(x_j \mid x_k)$ should be constant, implying that CERES plots of Section 14.2.1 may be used to suggest appropriate input transformations. The two panels of Figure 14.12 show illustrative CERES plots for inputs 1 and 68 along with superimposed estimated power curves. The implications from Figure 14.12 are that $\dim[\mathcal{S}_{\log(y)|x}] > 1$ and that improvements may result by replacing x_1 and x_{68} with their logarithms. CERES plots were used iteratively, as described in Section 14.3.3, restricting the class of transformations to power transformations. This procedure fits a generalized additive model with a restricted class of smooths. Because powers seemed to work well in each case, there was little reason to allow more general smoothing of the plots. In total, 13 inputs were transformed in this way and, as it turned out, the only transformations used were the logarithm and the square root.

Following the transformations there seemed to be many inputs that had little if any effect on the log response. Using a backward elimination procedure, 54 input variables were removed in the hope of increasing the power of

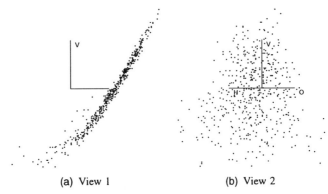

(a) View 1   (b) View 2

FIGURE 14.13   Two views of the 3D CERES plot for g_{68} and g_{69} in the regression with the transformed and reduced inputs. The horizontal screen axis for View 1 is $g_{68} + g_{69}$.

subsequent analysis. The variables removed were judged to be unimportant: Neither a CERES plot nor 2D or 3D added-variable plots showed dependence for any of the predictors that were removed. Recall from Chapter 7 that an added-variable plot gives a visual representation of the coordinate subspace for selected predictors. Let X_{30} denote the vector of the 30 remaining inputs with the transformations applied. A plot of the log response versus the OLS fitted values for the regression on X_{30} is shown in Figure 14.11c. A reduction in variance relative to Figure 14.11b is evident and $R^2 = 0.976$. Figure 14.11d, which is the result of further analysis, is discussed a bit later.

pHd was next applied to the residuals from the OLS linear regression of $\log(y)$ on X_{30}. The results indicated one clear direction, with the possibility of one or two more. Evidently, the associated central dimension-reduction subspace is at least two and possibly three dimensional. An inspection of the coefficients associated with the first pHd predictor indicates that the nonlinearity rests almost exclusively with the sum of the two transformed inputs $g_{68} = \log(x_{68})$ and $g_{69} = \log(x_{69})$. Two views of a 3D CERES plot for g_{68} and g_{69} are shown in Figure 14.13.

View 2 is provided for contrast. View 1 gives the 2D projection of best visual fit and strongly suggests that the structural dimension of the 3D plot is 1. The 3D CERES plot then confirms the indication of pHd that the sum $g_{68} + g_{69}$ is relevant. Because the coefficients of g_{68} and g_{69} are nearly identical in the visual fit of Figure 14.13a and because the projection in Figure 14.13a is well fit with a quadratic, these two predictors were replaced with linear and quadratic terms in their sum, $g_{68} + g_{69} = \log(x_{68}x_{69})$. It turned out that the decision to replace the pair of predictors (g_{68}, g_{69}) by their sum was exactly right: Following the analysis, an inspection of the LANL code revealed that the response depends on x_{68} and x_{69} only through their product $x_{68}x_{69}$.

Let $X_{30}^{(1)}$ denote the resulting set of 30 transformed inputs and let $e_{30}^{(1)}$ denote the residuals from the OLS linear regression of $\log(y)$ on $X_{30}^{(1)}$. Finally, let $X_{29}^{(1)}$

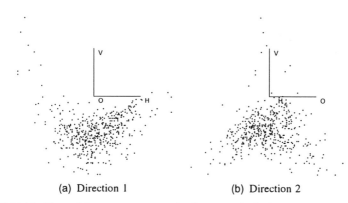

(a) Direction 1   (b) Direction 2

FIGURE 14.14 Plots of the response versus the first two principal Hessian directions from the regression on $X_{29}^{(1)}$ of the residuals from log(y) on $X_{30}^{(1)}$.

denote the vector of 29 predictors constructed from X_{30} by replacing (g_{68}, g_{69}) with the sum $g_{68} + g_{69}$. The input vectors $X_{30}^{(1)}$ and $X_{29}^{(1)}$ differ only by the quadratic term $(g_{68} + g_{69})^2$.

pHd was next applied to the regression of $e_{30}^{(1)}$ on $X_{29}^{(1)}$. Two significant pHd predictors pHd_1 and pHd_2 were indicated, implying that

$$\dim[\mathcal{S}_{e_{30}^{(1)}|X_{29}^{(1)}}] = 2.$$

The 3D plot of $e_{30}^{(1)}$ versus pHd_1 and pHd_2 seems fairly complicated, resembling a saddle with a high back. Plots of $e_{30}^{(1)}$ versus pHd_1 and pHd_2 are shown in Figure 14.14. These plots were smoothed and the smooths added to $X_{30}^{(1)}$, giving a new vector of 32 predictors $X_{32}^{(2)}$. Using a two-dimensional smooth may have produced somewhat better results.

Beyond this point little improvement seemed possible without considerable additional effort. A plot of log(y) versus the fitted values from the OLS regression of log(y) on $X_{32}^{(2)}$ is shown in Figure 14.11d. The corresponding $R^2 = 0.999$.

The final regression function is of the form

$$\mathrm{E}(\log(y) \mid X) \approx \beta_0 + \beta^T X_{29}^{(1)} + \alpha_1 \log^2(x_{68} x_{69})$$

$$+ g_1(\alpha_2^T X_{29}^{(1)}) + g_2(\alpha_3^T X_{29}^{(1)}) \qquad (14.21)$$

where g_1 and g_2 represent the smooths resulting from the second application of pHd. Further refinement may be desirable for a parsimonious predictive regression function but the present analysis seems sufficient for a reasonable idea about the important inputs. Figure 14.11d suggests that there is a dominant linear trend across the operability region in the transformed scale. This combined with the relative simplicity of the nonlinear terms in (14.21) and the approximate independence of the inputs can be used to assess the relative importance of the transformed inputs by partitioning $Var(y)$ in terms of the various components of the final model.

Based on a partition of $Var(y)$, the sum $g_{68} + g_{69} = \log(x_{68}x_{69})$ was judged to be the most important input, accounting for about 35 percent of the variation. The second most important predictor $\log(x_1)$ accounts for about 25 percent of $Var(y)$. Other predictors in X_{30} were partially ordered in three sets as follows: The three inputs

$$[\log(x_{24}), \log(x_{63}), \log(x_{84})]$$

account for about 17 percent of the variation and are more important than the four inputs

$$[\log(x_{35}), x_{48}, x_{67}, x_{83}]$$

which are in turn more important than the five inputs

$$[x_{19}, x_{54}, \log(x_{55}), \log(x_{61}), x_{65}].$$

The remaining inputs were judged to be of relatively minor importance.

PROBLEMS

14.1. What additional constraints could be placed on model (14.1) to allow the estimation of α_0 and g? How would the additional constraints change the construction of CERES plots?

14.2. Investigate the population behavior of a partial residual plot for x_2 when (14.1) is the "true" model, and $E(x_1 \mid x_2)$ is a nonlinear function of the value of x_2. In particular, give an expression for the expected population-adjusted response $E(u \mid x_2)$ and describe how nonlinearities in $E(x_1 \mid x_2)$ could influence the visualization of g.

14.3. Based on OLS estimation, obtain expressions for the regression function $E(u \mid w_j)$ and the variance function $Var(u \mid w_j)$ of the population CERES plot for each of the three predictors in model (14.11). Using these expressions, explain the appearance of the three CERES plots in Figure 14.6.

14.4. Recall that net-effect plots are designed to allow visualization of the regression of y on x_2 with x_1 fixed. Investigate the behavior of a net-effect plot for x_2 when (14.1) is the "true" model. Should a net-effect plot be expected to provide a visualization of $g(x_2)$ as a CERES plot is designed to do? Generally, contrast the behavior of net-effect and CERES plots for x_2 under (14.1).

CHAPTER 15

Graphics for Model Assessment

In previous chapters we studied mostly graphics for application in settings where a parsimoniously parameterized model is not available, focusing on the central dimension-reduction subspace as a characterization of the regression. In this chapter we turn to settings in which a model is available, and investigate graphics for assisting in an assessment of lack of fit.

Consider a generic regression model for $y \mid x$ represented by the CDF M that is specified up to a $q \times 1$ vector of unknown parameters θ. For example, M may represent the normal linear regression model

$$y \mid x = \beta_0 + \beta^T x + \sigma \varepsilon \qquad (15.1)$$

where $\theta = (\beta_0, \beta^T, \sigma)^T$, and ε is a standard normal random variable that is independent of x. Or M may represent a logistic regression model with

$$\log \frac{P(x)}{1 - P(x)} = \beta_0 + \beta^T x$$

where $P(x)$ is the probability of a "success." In any event, we would like a graphical assessment of the estimated model $\hat{M}$ obtained by replacing θ with a consistent estimate $\hat{\theta}$ under M.

When considering model assessment, it may be useful to distinguish between two different but related ideas. On the one hand, we could concentrate on model weaknesses, looking for information in the data to contradict the model. If no such contradictory information is found then we have some evidence in support of the model. Remedial action may be necessary otherwise. Model weaknesses are often investigated by using residual plots, which are discussed in Section 15.1. On the other hand, we could focus on model strengths, looking for ways in which the model may be adequate for the goals at hand. Graphics for this type of inquiry are developed in Section 15.2.

15.1. RESIDUAL PLOTS

Many standard graphical methods for detecting model weaknesses rely on plots of sample residuals $\hat{r}$ versus linear combinations of the predictors $a^T x$. When the model is correct or nearly so, we can often expect such plots to appear as if $\hat{r} \perp\!\!\!\perp a^T x$, except perhaps for dependences in small samples caused by substituting estimates for unknown parameters. Historically, the choice of the linear combinations has been mostly a matter of taste and prior information. Plots of residuals versus individual predictors or fitted values are common preferences.

15.1.1. Rationale

To explore the rationale for residual plots, consider a regression in which it is conjectured that

$$y \perp\!\!\!\perp x \mid E_M(y \mid x) \tag{15.2}$$

where E_M denotes expectation under the postulated model M. Here, we imagine that the value of the parameter θ is equal to the limit of $\hat{\theta}$ as $n \to \infty$. For use in practice M would be replaced by the estimated model $\hat{M}$. The methods of Chapter 4 could be used as a basis for investigating (15.2) when there are $p \le 2$ predictors. Different methods are required when $p > 2$, however, and this is where residuals may be useful.

Condition (15.2) is equivalent to

$$(y, E_M(y \mid x)) \perp\!\!\!\perp x \mid E_M(y \mid x)$$

which in turn is equivalent to

$$(r, E_M(y \mid x)) \perp\!\!\!\perp x \mid E_M(y \mid x)$$

where $(r, E_M(y \mid x))$ is any one-to-one function of $(y, E_M(y \mid x))$. As implied by the notation, r is intended as a residual computed under M. However, there are not yet any constraints on r, except for the one-to-one requirement just stated. Residuals are distinguished by the additional requirement that they be independent of the regression function when the model is true; that is, when $y \perp\!\!\!\perp x \mid E_M(y \mid x)$. In sum, under the location regression (15.2), a *population residual* r is defined to be any function of $(y, E_M(y \mid x))$ so that

- $(r, E_M(y \mid x))$ is a one-to-one function of $(y, E_M(y \mid x))$, and
- $r \perp\!\!\!\perp E_M(y \mid x)$ when the model is true.

The following proposition gives various implications of this setup.

Proposition 15.1. Let r be a population residual for the location regression (15.2). Then the following four conditions are equivalent:

a. $y \perp\!\!\!\perp x \mid E_M(y \mid x)$ and $r \perp\!\!\!\perp E_M(y \mid x)$
b. $r \perp\!\!\!\perp x \mid E_M(y \mid x)$ and $r \perp\!\!\!\perp E_M(y \mid x)$
c. $r \perp\!\!\!\perp x$
d. $r \perp\!\!\!\perp a^T x$ for all $a \in \mathbb{R}^p$.

Justification. The equivalence of conditions (c) and (d) follows immediately from Proposition 4.3, and the equivalence of (b) and (c) follows from Propositions 4.5 and 4.6. Finally, the equivalence of (a) and (b) was demonstrated earlier in this section. □

Proposition 15.1 forms a basis for using residual plots of the form $\{r, a^T x\}$ to check on model weaknesses. Because r was constructed so that $r \perp\!\!\!\perp E_M(y \mid x)$ when the model is correct, $r \perp\!\!\!\perp a^T x$ for all $a \in \mathbb{R}^p$ if and only if $y \perp\!\!\!\perp x \mid E_M(y \mid x)$. Thus, failure to find notable dependence in plots $\{\hat{r}, a^T x\}$ using sample residuals provides support for the model. Conversely, the finding that $\hat{r}$ is dependent on $a^T x$ for some a indicates a discrepancy between the model and the data.

In the case of the linear regression model (15.1), $E_M(y \mid x) = \beta_0 + \beta^T x$. Taking $r = y - \beta^T x$, condition (a) in Proposition 15.1 reduces to

$$y \perp\!\!\!\perp x \mid \beta^T x \qquad \text{and} \qquad (y - \beta^T x) \perp\!\!\!\perp \beta^T x.$$

The relationships between the equivalences of Proposition 15.1 simplify somewhat because of the special nature of the regression function.

15.1.2. Isomerization data

Bates and Watts (1988, pp. 55, 271) used the Hougen–Watson reaction rate model

$$y \mid x = \frac{\theta_1 \theta_3 (x_2 - x_3/1.632)}{1 + \theta_2 x_1 + \theta_3 x_2 + \theta_4 x_3} + \sigma\varepsilon \tag{15.3}$$

to fit 24 observations from a data set on the reaction rate y of the catalytic isomerization of n-pentane to isopentane (Carr 1960). The three predictors $x = (x_k)$ are the partial pressures of hydrogen x_1, n-pentane x_2 and isopentane x_3. The model M is described by (15.3) along with the additional conditions $E(\varepsilon) = 0$, $Var(\varepsilon) = 1$, and $\varepsilon \perp\!\!\!\perp x$. Population residuals can be defined as simply

$$r = y - E_M(y \mid x). \tag{15.4}$$

Clearly, $(r, E_M(y \mid x))$ is a one-to-one function of $(y, E_M(y \mid x))$ and, if M is true, $r \perp\!\!\!\perp E_M(y \mid x)$.

If the standard deviation σ in model (15.3) is a function of $E_M(y \mid x)$, then the residuals (15.4) are no longer appropriate because the condition $r \perp\!\!\!\perp E_M(y \mid x)$ fails. Instead, in such cases it becomes necessary to standardize the residuals,

$$r_s = \frac{y - E_M(y \mid x)}{\sigma(E_M(y \mid x))} \tag{15.5}$$

so that $r_s \perp\!\!\!\perp E_M(y \mid x)$ if the heteroscedastic model is true.

Figure 15.1 shows four plots that give information on the ordinary least squares fit of (15.3). The plot of the response versus the fitted values $\hat{y}$ in Figure 15.1a indicates a fairly strong relationship. We begin the residual analysis by inspecting the plot of the sample residuals $\hat{r}$ versus $\hat{y}$ shown in Figure 15.1b. The line on the plot is the fitted line from the ordinary least squares regression of $\hat{r}$ on $\hat{y}$. Its slope is nonzero because the sample residuals are not generally orthogonal to the fitted values in nonlinear regression, although they are so when the model is linear. Overall, the plot shows little information to suggest that r is dependent on $E_M(y \mid x)$ so we proceed by assuming that $r \perp\!\!\!\perp E_M(y \mid x)$. With this assumption, model (15.3) is true if and only if $r \perp\!\!\!\perp a^T x$ for all $a \in \mathbb{R}^3$.

The idea now is to inspect plots of the form $\{\hat{r}, a^T x\}$ for various values of the vector a to see if information can be found to contradict the model. Plots of $\hat{r}$ versus the individual predictors are useful because experience indicates that discrepancies between the model and the data are often manifest in such plots. Additionally, a may be set to the eigenvalues of $\hat{\mathrm{Var}}(x)$, or to other values as suggested by the model. For example, plots of $\hat{r}$ versus the linear combinations of the predictors in the numerator and denominator of the regression function for model (15.3) might be useful. The direction a might also be chosen randomly from the uniform distribution on the unit sphere in $\mathbb{R}^p$. From among the plots of $\hat{r}$ versus the individual predictors, the most evidence of dependence is seen with hydrogen as shown in Figure 15.1c. The curvature of the LOWESS smooth shown on the plot is suggestive of a discrepancy, but is far from compelling. Figure 15.1d is discussed in Section 15.1.4.

15.1.3. Using residuals in graphical regression

The standard practice of plotting residuals versus linear combinations $a^T x$ is based on a narrow connection with the regression as described in Proposition 15.1. Intuitively, support for the model increases with the number of residual plots $\{\hat{r}, a^T x\}$ considered and found acceptable. This is a useful notion, but because particular plots are not stipulated and there is no clearly defined endpoint, the methodology can be a bit elusive. If any residual plot is judged to exhibit dependence then the model is called into question and remedial action may be necessary. However, the residual plot itself may not provide direct information on how to improve the model. For example, if the

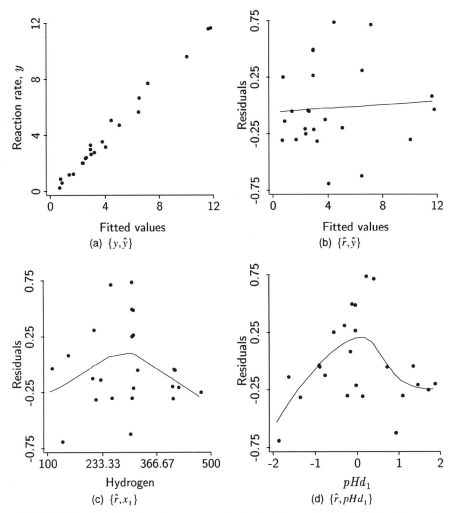

FIGURE 15.1 Four plots showing aspects of the fit of the isomerization data: Plot (d) contains the residuals $\hat{r}$ versus the first pHd predictor pHd_1 from the regression of $\hat{r}$ on the three predictors.

curvature in Figure 15.1c is judged sufficient to prompt remedial action, it does not necessarily follow that an adequate remedy can be found by confining attention to the way in which hydrogen enters the model. Introductory examples of this type of behavior were given in Chapter 1. Plotting procedures that are based on the central subspaces are more comprehensive than residual plots chosen routinely or spontaneously, and they provide a broader connection to the regression. Any of the methods discussed in previous chapters for inferring about a central subspace can be adapted for the residual regression of $\hat{r}$ on x, including graphical regression (see Chapter 7 and 9).

Graphical regression can be adapted for the study of the residual regression by first restricting the predictor vector for this regression so that no predictors are functionally related. This restriction involves no loss of generality and does not prohibit including functionally related terms in the model, but it does imply that two functionally related predictors would not normally be considered routinely in a graphical assessment of the residuals. Suppose that x can be written as a function of a vector w containing a minimal set of predictors in the following way,

$$x = \begin{pmatrix} x_1 \\ x_2 \end{pmatrix} = f(w) = \begin{pmatrix} f_1(w) \\ f_2(w) \end{pmatrix}$$

where x_j is $p_j \times 1$, $j = 1, 2$, $f_1(w) = x_1$, and f_1 is a one-to-one function. For example, suppose that $p = 3$, $x_1 = (x_{11}, x_{12})^T$ and $x_2 = x_{11} x_{12}$, where x_{11} and x_{12} are functionally independent. Then we may take w to be any one-to-one function of x_1.

Under this setup, $r \perp\!\!\!\perp x \mid E_M(y \mid x)$ if and only if $r \perp\!\!\!\perp w \mid E_M(y \mid w)$ because $E_M(y \mid x) = E_M(y \mid f_1(w)) = E_M(y \mid w)$. Thus there is no loss of generality in considering only the regression of $\hat{r}$ on w and neglecting all of the derived predictors in x_2. There are, of course, many choices for w. This flexibility is intended to allow w to be selected to satisfy the predictor conditions for the method under consideration. In particular, graphical regression works best when there are no strong nonlinear relations among the predictors, as discussed in Chapters 7 and 9.

For example, consider the naphthalene data introduced in Section 9.1, and suppose that we wish to check the fit of the full, second-order response surface model in the original three process variables $z = (z_j)$,

$$Y_N \mid z = \beta_0 + \beta^T z + z^T B z + \sigma \varepsilon$$

where Y_N is the conversion of naphthalene to naphthoquinone, and B is a symmetric 3×3 matrix of quadratic coefficients. In this case x consists of nine terms that depend on z: three linear terms, three quadratic terms, and three cross-product terms. However, for the purposes of model checking, it would appropriate to use the three transformed predictors $w = (AN, Btemp, Ctime)^T$ for the reasons given in Section 9.1. The graphical regression of $\hat{r}$ on w could then be used to infer about $S_{r|w}$ as illustrated in Chapter 9. Accordingly, as few as two plots—one 3D plot and one 2D plot—could be sufficient to conclude that $S_{r|w} = S(0)$ and thus that there is no evidence to contradict the model. If contradictory evidence is found, then a graphical regression analysis should produce an estimate of $S_{r|w}$ to guide remedial action.

For additional discussion of these ideas from a methodological view, see Cook and Weisberg (1994a, pp. 172–182). The *ALP* residual plots discussed by Cook and Weisberg (1994a, Section 11.4) correspond to basing the graphical regression of $\hat{r}$ on w on added-variable plots, as introduced in Section 7.4 and illustrated in Chapter 9.

15.1.4. pHd

Another option for residual plots is to choose the a's to be the principal Hessian directions from a pHd analysis of the regression of $\hat{r}$ on w, as described in Chapter 12. In this context, pHd can be used simply as a device for selecting p uncorrelated linear combinations of the predictors for the horizontal axes of residual plots without the formality of inferring about the central subspace $\mathcal{S}_{y|w}$, although the results of such inference may certainly be of value.

Returning to the analysis of the isomerization data based on the Hougen–Watson model, Figure 15.1d shows a plot of $\hat{r}$ versus the first pHd predictor from the regression of $\hat{r}$ on w. The spline smooth on the plot exhibits clear curvature, enough to cause some concern about possible model deficiencies. The first two pHd p-values determined by using the procedure described in Section 12.3.2 are about 0.07 and 0.45. The conclusions from the isomerization data are a little ambiguous. While there is some evidence to indicate a discrepancy between the model and the data, the evidence is not very strong. And it is not clear if there is room for appropriate remedial action to have substantial impact on the analysis. Although an inspection of various residual plots can be of value, the end result may be similar to that encountered in this example. There are several ways to proceed. For example, we could try to sharpen the methodology and thereby gain more information on the possibility of a significant model deficiency. Or we might take a more pragmatic approach and ask if the nature of the discrepancy requires remedial action. This latter possibility will be explored later in this chapter. For now we turn to another example of a somewhat different nature.

15.1.5. Residual plots: Kyphosis data

We now return to the kyphosis data introduced in Section 5.3.2. Recall that the response kyp is the presence $(kyp = 1)$ or absence $(kyp = 0)$ of kyphosis, and that the three predictors are age in months at time of surgery Age, the starting vertebrae $Start$, and number Num of vertebrae involved. Let $x = (x_k) = (Age, Start, Num)^T$. Since the response is binary, $kyp \mid x$ has a Bernoulli distribution with regression function $\mathrm{E}(kyp \mid x) = \mathrm{Pr}(kyp = 1 \mid x)$. The model M under consideration is the one developed by Hastie and Tibshirani (1990, p. 291) that was previously introduced in Section 5.4.2,

$$\gamma(x) = \log \frac{\mathrm{E}_\mathrm{M}(kyp \mid x)}{1 - \mathrm{E}_\mathrm{M}(kyp \mid x)}$$

$$= b_1 Age + b_2 Age^2$$

$$+ b_3 (Start - 12) J(Start > 12) \tag{15.6}$$

where J represents the indicator function.

According to the discussion of Section 15.1.1, residuals r must be constructed so that two requirements are satisfied: $(r, E_M(kyp \mid x))$ is a one-to-one function of $(kyp, E_M(kyp \mid x))$, and $r \perp\!\!\!\perp E_M(kyp \mid x)$ when the model is true. The first requirement can be satisfied easily. However, it seems impossible to satisfy the second requirement in this example. Because the response is binary, any nontrivial function of $(kyp, E_M(kyp \mid x))$ must necessarily be dependent on $E_M(kyp \mid x)$, except in uninteresting special cases like $kyp \perp\!\!\!\perp x$. For example, the population Pearson residual

$$r_P = \frac{kyp - E_M(kyp \mid x)}{\sqrt{E_M(kyp \mid x)(1 - E_M(kyp \mid x))}}$$

is often used for logistic regression problems. This residual is based on shifting the support for each $kyp \mid x$ so that $E_M(r_P \mid x) = 0$ and $\text{Var}(r_P \mid x) = 1$ if the model is correct. But, even if the model is correct, r_P is still dependent on $E_M(kyp \mid x)$ because the support of $r_P \mid x$ depends on $E_M(kyp \mid x)$. Similar comments apply to other types of residuals like those based on the deviance. In short, the previous requirements for a residual do not permit progress in this example.

However, progress is possible if we stipulate that

$$kyp \perp\!\!\!\perp x \mid E_F(kyp \mid x) \tag{15.7}$$

where E_F denotes expectation under the true CDF $F_{y|x}$. In the general development leading to Proposition 15.1, the class of possible models was constrained, but no constraints were placed on the distribution of $y \mid x$. Thus there were many ways in which the model could fail. Equation (15.7) constrains $F_{y|x}$ so that the true regression is characterized by its regression function.

Under (15.7), a necessary and sufficient condition for the model is

$$E_M(kyp \mid x) = E_F(kyp \mid x). \tag{15.8}$$

If we find evidence to contradict this condition, then we have evidence of a model discrepancy. Clearly, $E_F(kyp - E_M(kyp \mid x) \mid x) = 0$ if and only if (15.8) holds and thus we can check the model by plotting ordinary residuals $r = kyp - E_M(kyp \mid x)$ against linear combinations of the predictors as before. The Pearson or deviance residuals could also be used.

Figure 15.2 shows four residual plots for the kyphosis data. The first is a plot of the sample residuals versus $\hat{\gamma}$, where γ is as given in (15.6). The LOWESS smooth shown on the plot is an estimate of

$$E_F(kyp - E_M(kyp \mid x) \mid \gamma).$$

Since the smooth shows little variation, there is no clear evidence to contradict the model. The systematic pattern in the plotted points, which is due to the

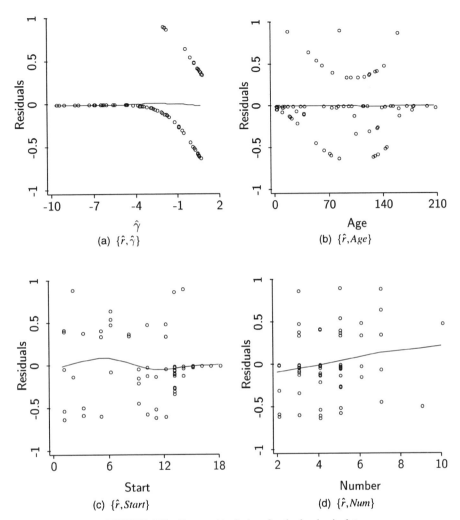

FIGURE 15.2 Four residual plots for the kyphosis data.

changing support of the residuals, is not directly relevant to the interpretation of the plot. Indeed, the changing support of the residuals can produce rather curious patterns of points in residual plots while still supporting the model. The residual plot for *Age* shown in Figure 15.2b is an example of this phenomenon. The plot for *Start*, Figure 15.2c, can be interpreted in a similar fashion. The plot for *Num*, which is not included in the model, shows an increasing trend that could indicate a model discrepancy, although the magnitude of the trend is notably influenced by the point with the largest value of *Num*. We will return to this example in Section 15.2.7 and develop additional evidence of lack of fit.

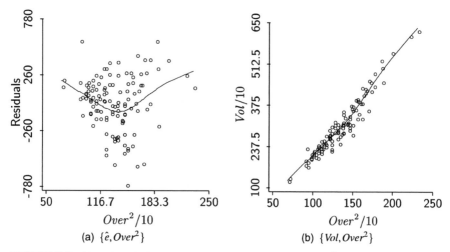

FIGURE 15.3 Two plots depicting the OLS regression of *Vol* on $Over^2$ from the haystack data. (a) Residuals versus $Over^2$. (b) Response *Vol* versus $Over^2$.

15.1.6. Interpreting residual plots

Failure to find contradictory information in residual plots does provide support for the model. But what have we really gained in such cases if we heed Box's (1979) admonition "All models are wrong, but some are useful?" Under the ideology implied by this statement, we should perhaps not be surprised when a model weakness is visually apparent in a residual plot. Further, the finding of a model weakness, or the belief that undetected weaknesses must be present, does not necessarily imply that the model is inadequate. For example, the reaction rate model (15.3) for the isomerization data could be adequate for prediction or other purposes, even if the curvature found in Figure 15.1d is truly indicative of a discrepancy between the model and the data.

Returning to the haystack data introduced in Section 4.1, consider predicting the volume *Vol* of a haystack from just *Over*. Shown in Figure 15.3a is a plot of the residuals versus $Over^2$ from the OLS fit of the model

$$Vol \mid Over = \beta_0 + \beta \times Over^2 + \sigma\varepsilon \qquad (15.9)$$

where $\mathrm{E}(\varepsilon) = 0$ and $\mathrm{Var}(\varepsilon) = 1$. The curvature in the smooth shown in Figure 15.3a suggests that the model does not capture all of the systematic features in the data and thus that remedial action may be in order. However, the plot of the response versus $Over^2$ shown in Figure 15.3b leaves quite a different impression. Because there is a strong linear relationship between *Vol* and $Over^2$, the OLS fit of (15.9) might be adequate for predicting *Vol* across the observed range of *Over*, particularly if a parsimonious predictive equation is important. The two plots shown in Figure 15.3 give quite different information: The discrepancies that seem clear in the residual plot are not nearly so in the response plot, while the strength of the relationship in the response plot is lost in the residual plot.

Residual plots can be helpful for detecting discrepancies between the model and the data, but they do not allow a straightforward assessment of model adequacy.

15.2. ASSESSING MODEL ADEQUACY

We now consider graphics for comparing selected characteristics of the true, unknown conditional CDF $F_{y|x}$ to the corresponding characteristics of $\hat{M}$, following the approach proposed by Cook and Weisberg (1997). In particular, we focus on comparing nonparametric estimates of the mean and variance of $y \mid x$ to the mean and variance computed from the estimated model $\hat{M}$, although the methods to be developed apply equally to other moments or to quantiles. The dimension of x is likely to be a complicating factor in any method for carrying out such comparisons. When $p = 1$, $E_F(y \mid x)$ can be estimated by smoothing the plot $\{y, x\}$. This smooth, say $\hat{E}_F(y \mid x)$, could then be compared to $E_{\hat{M}}(y \mid x)$ by superimposing both estimates of the regression function on a plot of y versus x. Here, $E_{\hat{M}}(y \mid x)$ is simply the regression function from the estimated model.

For example, consider the haystack data in combination with the OLS fit of model (15.9). The plot $\{Vol, Over^2\}$ is shown in Figure 15.4 with two smooths superimposed. The solid line is the LOWESS smooth of $\{Vol, Over^2\}$ with span 0.7. The dashed line is the estimated regression function $E_{\hat{M}}(Vol \mid Over)$ from the OLS fit of (15.9). The two curves seem quite close, although there is a systematic pattern to their difference. This difference $\hat{E}_F(y \mid x) - E_{\hat{M}}(y \mid x)$ is identical to the smooth of the residual plot shown in Figure 15.3a because the same value of the LOWESS smoothing parameter was used in each plot. Again we see that the discrepancy evident in Figure 15.3 may not imply an inadequate model because the variation between the fits shown in Figure 15.4 is small relative to the variation in the response. A direct extension of this idea is difficult if $p = 2$ and probably impossible with current technology if $p > 2$. Consequently, we have to find a method of reducing the plot dimensions when there are many predictors.

As a first step in dimension reduction we restrict x so that no predictors are functionally related, as described in Section 15.1.3. However, in the developments of this section, no restrictions are placed on the marginal distribution of the predictors, unless explicitly stated otherwise.

The following result, which can be demonstrated by using the idea in the justification of Proposition 4.3, suggests another way to restrict model assessment graphics. Let $F_{y|x}$ and $G_{y|x}$ be two conditional CDFs. Then $F_{y|x} = G_{y|x}$

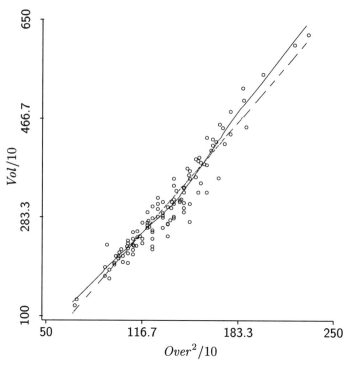

FIGURE 15.4   Scatterplot of $Vol/10$ versus $Over^2/10$ for the haystack data. The solid line is the LOWESS smooth of $Vol/10$ against $Over^2/10$ with span 0.7. The dashed line is the OLS fit from (15.9).

for all values of x in the sample space Ω_x if and only if $F_{y|a^T x} = G_{y|a^T x}$ for all values of $a^T x$ in $\{a^T x \mid a \in \mathbb{R}^p, \ x \in \Omega_x\}$. This result provides a starting point for developing graphics for model assessment because it suggests that we can focus on marginal models of the form $\hat{M}_{y|a^T x}$ to assess the adequacy of $\hat{M}_{y|x}$. The graphical problem then reduces to a comparison of available information on $F_{y|a^T x}$ with the corresponding information from $\hat{M}_{y|a^T x}$ for various a's. Ideas for selecting a's will be discussed shortly; for now we focus on methods for comparing $F_{y|a^T x}$ and $\hat{M}_{y|a^T x}$ with a fixed.

15.2.1. Marginal regression functions...

... *under* F

The marginal regression function $E_F(y \mid a^T x)$ is unknown, but can be estimated by smoothing y against $a^T x$, that is, smoothing a scatterplot of y versus $a^T x$. This was the method used to estimate the regression function for the haystack data.

...under $\hat{M}$

The marginal regression function implied by $\hat{M}$ can be obtained by averaging over the distribution of $x \mid a^T x$,

$$E_{\hat{M}}(y \mid a^T x) = E[E_{\hat{M}}(y \mid x) \mid a^T x] \tag{15.10}$$

where expectation operators without subscripts are used when computing moments with respect to a distribution involving only predictors. Exact calculation of $E_{\hat{M}}(y \mid a^T x)$ requires knowledge of the marginal distribution of x. While this information may be available occasionally, the marginal distribution of x will more generally be unknown. In such cases, $E_{\hat{M}}(y \mid a^T x)$ can be estimated by smoothing $E_{\hat{M}}(y \mid x)$ against $a^T x$.

Suppose, for example, that $p = 2$ and that the estimated model is of the form

$$y = b_0 + b_1 x_1 + b_2 x_2 + \hat{\sigma}\varepsilon \tag{15.11}$$

Then

$$E_{\hat{M}}(y \mid x) = b_0 + b_1 x_1 + b_2 x_2$$

and for $a = (1,0)^T$,

$$E_{\hat{M}}(y \mid a^T x) = E_{\hat{M}}(y \mid x_1) = b_0 + b_1 x_1 + b_2 E(x_2 \mid x_1). \tag{15.12}$$

The regression function $E_{\hat{M}}(y \mid x_1)$ depends on $E(x_2 \mid x_1)$. To estimate $E_{\hat{M}}(y \mid x_1)$, we could estimate $E(x_2 \mid x_1)$ by smoothing the scatterplot of x_2 against x_1 and substituting this smooth into (15.12). Or $E_{\hat{M}}(y \mid x_1)$ could be estimated directly by smoothing $E_{\hat{M}}(y \mid x)$ against $a^T x$. This latter choice amounts to smoothing a scatterplot of the fitted values under $\hat{M}$ against $a^T x$.

Marginal regression functions need not be confined to predictors that appear explicitly in the model. For illustration, suppose that $p = 3$ while the estimated model still has the form given in (15.11). The third predictor x_3 does not appear in (15.11), implying $y \perp\!\!\!\perp x_3 \mid E_M(y \mid x_1, x_2)$. Nevertheless, in some problems it may be desirable to check on excluded predictors by comparing a smooth of y against x_3 to the corresponding marginal regression function implied by the estimated model,

$$E_{\hat{M}}(y \mid x_3) = b_0 + b_1 E(x_1 \mid x_3) + b_2 E(x_2 \mid x_3).$$

As before, exact calculation of $E_{\hat{M}}(y \mid x_3)$ requires knowledge of the marginal distribution of the predictors. Alternatively, $E_{\hat{M}}(y \mid x_3)$ could be estimated by combining smooth estimates of the predictor regression functions, or by smoothing $E_{\hat{M}}(y \mid x_1, x_2)$ against x_3. In the case of model (15.3) for the isomerization data, the rather complicated nature of the regression function for this model makes the option of estimating $E_{\hat{M}}(y \mid a^T x)$ by smoothing the fitted values against $a^T x$ seem quite attractive.

15.2.2. Marginal variance functions...

...*under* F

Like $E_F(y \mid a^T x)$, the marginal variance function $\text{Var}_F(y \mid x)$ can be estimated by smoothing. For example, a first method might be based on smoothing twice: First, smooth y against $a^T x$ to obtain an estimate $\hat{E}_F(y \mid a^T x)$ of $E_F(y \mid a^T x)$, and then smooth $(y - \hat{E}(y \mid a^T x))^2$ against $a^T x$ to obtain an estimate $\widehat{\text{Var}}_F(y \mid a^T x)$ of $\text{Var}_F(y \mid x)$. Ruppert, Wand, Holst, and Hössjer (1997) studied variance function estimation by smoothing squared residuals, and proposed adjustments to account for the biasing effect of preliminary estimation of the regression function.

...*under* $\hat{M}$

A marginal variance function under the estimated model can be expressed as

$$\text{Var}_{\hat{M}}(y \mid a^T x) = E[\text{Var}_{\hat{M}}(y \mid x) \mid a^T x] + \text{Var}[E_{\hat{M}}(y \mid x) \mid a^T x]. \quad (15.13)$$

With sufficient information on the marginal distribution of x, this marginal variance function can be computed analytically. Otherwise, $\text{Var}_{\hat{M}}(y \mid a^T x)$ can be estimated by using smooths to estimate the two terms on the right of (15.13), given that both $E_{\hat{M}}(y \mid x)$ and $\text{Var}_{\hat{M}}(y \mid x)$ are available from the estimated model. For a generalized linear model, for example, these are just the estimated mean and variance functions. The average variance function $E[\text{Var}_{\hat{M}}(y \mid x) \mid a^T x]$ can be estimated by smoothing $\text{Var}_{\hat{M}}(y \mid x)$ against $a^T x$, while the variance of the mean function $\text{Var}[E_{\hat{M}}(y \mid x) \mid a^T x]$ can be estimated by using a variance smoother on a scatterplot of $E_{\hat{M}}(y \mid x)$ versus $a^T x$.

Under model (15.11), the variance function $\text{Var}_{\hat{M}}(y \mid x) = \hat{\sigma}^2$, and

$$\text{Var}_{\hat{M}}(y \mid a^T x) = \hat{\sigma}^2 + \text{Var}[b_1 x_1 + b_2 x_2 \mid a^T x].$$

If the predictors are normally distributed, then the second term on the right of this equation is constant, and $\text{Var}_{\hat{M}}(y \mid a^T x)$ is constant. If a falls in the subspace spanned by $(b_1, b_2)^T$, so that $a^T x$ gives essentially the fitted values, then $\text{Var}_{\hat{M}}(y \mid a^T x)$ is again constant. Otherwise, $\text{Var}_{\hat{M}}(y \mid a^T x)$ may depend on the value of $a^T x$, even if the homoscedastic model form (15.11) is correct.

As a second example, suppose that $\hat{M}$ specifies a logistic regression model where $y \mid x$ is a binomial $(n_x, P(x))$ random variable with

$$P(x) = [1 - \exp(-b_0 - b^T x)]^{-1}.$$

We treat y/n_x as the response, so the regression function is $P(x)$ and the variance function is $P(x)(1 - P(x))/n_x$. The marginal variance is then

$$\text{Var}_{\hat{M}}\left(\frac{y}{n_x}\,\middle|\, a^T x\right) = \text{E}\left(\frac{P(x)(1 - P(x))}{n_x}\,\middle|\, a^T x\right) + \text{Var}[P(x) \mid a^T x].$$

(15.14)

The first term on the right of this equation can be estimated by smoothing $P(x)[1 - P(x)]/n_x$ against $a^T x$, and the second term by applying a variance smooth to the scatterplot of $P(x)$ versus $a^T x$.

If a standard model for overdispersion holds (see, e.g., Collett 1991, Sec. 6.2), then the variance function will be of the form

$$\text{Var}_{\hat{M}}\left(\frac{y}{n_x}\,\middle|\, x\right) = P(x)(1 - P(x))[1 + (n_x - 1)\phi]/n_x$$

and if some of the n_x exceed 1 and the overdispersion parameter ϕ is positive, then the smooth of the binomial variance function will underestimate the variance in the marginal plot. If all the $n_x = 1$, then overdispersion is not observable, and examination of the variance functions contains no more information than examination of the regression functions since they are functionally related. This situation is the same as that encountered with the kyphosis data in Section 15.1.5.

15.2.3. Marginal model plots

We now have estimates of the regression and variance functions for the marginal data, and estimates of the marginal regression and variance functions implied by the estimated model $\hat{M}$. It is often useful to display this information in six curves,

- $\hat{\text{E}}_F(y \mid a^T x)$,
- $\text{E}_{\hat{M}}(y \mid a^T x)$,
- $\hat{\text{E}}_F(y \mid a^T x) \pm \text{sd}_F(y \mid a^T x)$, and
- $\text{E}_{\hat{M}}(y \mid a^T x) \pm \text{sd}_{\hat{M}}(y \mid a^T x)$,

where "$\pm$" in the last two items indicates two curves, and the standard deviation function $\text{sd}_{(\cdot)}(y \mid a^T x)$ is the square root of the corresponding estimated variance function, either $[\hat{\text{Var}}_F(y \mid a^T x)]^{1/2}$ or $[\text{Var}_{\hat{M}}(y \mid a^T x)]^{1/2}$. When superimposed on a scatterplot of y versus $a^T x$ to yield a *marginal model plot*, these six curves allow for an assessment of the adequacy of the marginal model for the marginal data. If the model is a close representation of $F_{y|x}$, we can expect $\text{E}_{\hat{M}}(y \mid a^T x) \approx \hat{\text{E}}_F(y \mid a^T x)$ and $\text{sd}_F(y \mid a^T x) \approx \text{sd}_{\hat{M}}(a^T x)$. Any bias in these estimates of the regression and standard deviation functions should be similar as long as the same value of the smoothing parameter is used for each curve.

Consequently, pointwise comparison of the curves makes sense because the bias should largely cancel. See Bowman and Young (1996) for further discussion of this point.

An assessment of the importance of a particular difference

$$E_{\hat{M}}(y \mid a^T x) - \hat{E}_F(y \mid a^T x)$$

could depend on the variation in the response. A difference of 1 unit may be unimportant if the standard deviation is 10 units, but important if the standard deviation is 0.1 units, for example. A comparison of the standard deviation curves $sd_F(y \mid a^T x)$ and $sd_{\hat{M}}(a^T x)$ allows one aspect of the marginal model to be contrasted with the same aspect of the marginal data. But a comparison of the standard deviation smooths with the difference in the regression smooths allows for an assessment of the importance of the difference. Superimposing the data on the plot permits a similar comparison.

Occasionally it may be difficult to distinguish between six curves, three from the marginal data and three from the marginal model, on a single plot. In such cases the regression and variance smooths could be displayed separately, with the data providing a background for assessing the importance of differences in the regression function.

The ability to discriminate between the model and the data can often be enhanced by using marginal model plots within subsets of the data determined by restrictions on the predictors. These *conditional marginal model plots* are constructed is the same way as (unconditional) marginal model plots, except the various smooths are restricted to a subset of the data. Note that the model in question is not to be refitted to the data subset. The marginal regression functions $E_{\hat{M}}(y \mid a^T x)$ and $\hat{E}_F(y \mid a^T x)$ are still constructed by smoothing $E_{\hat{M}}(y \mid x)$ against $a^T x$ and y against $a^T x$, but within the selected subset. Similar remarks apply to the marginal variance functions. To distinguish conditional marginal model plots from unconditional plots, the symbol χ will be used in various expressions to represent the predictor constraints. For example, $E_{\hat{M}}(y \mid a^T x; \chi)$ represents the marginal model regression function computed from the subset of the data that satisfies the constraint represented by χ.

Partition x into a vector of $p - 1$ predictors x_1 and a single predictor x_2. The constraint χ could represent $x_2 > c$, where c is a user-selected constant, or $x_2 = c$, or $x \neq c$, the latter two constraints being useful when x_2 is discrete. In any case, if $\hat{M}$ is a good approximation of F then it should be so in any subpopulation defined in this way.

The regression function for an unconditional marginal model plot has a straightforward relationship to regression functions from conditional marginal model plots,

$$E_{\hat{M}}(y \mid a^T x) = Pr(\chi \mid a^T x) E_{\hat{M}}(y \mid a^T x; \chi)$$

$$+ (1 - Pr(\chi \mid a^T x)) E_{\hat{M}}(y \mid a^T x; \text{ not } \chi).$$

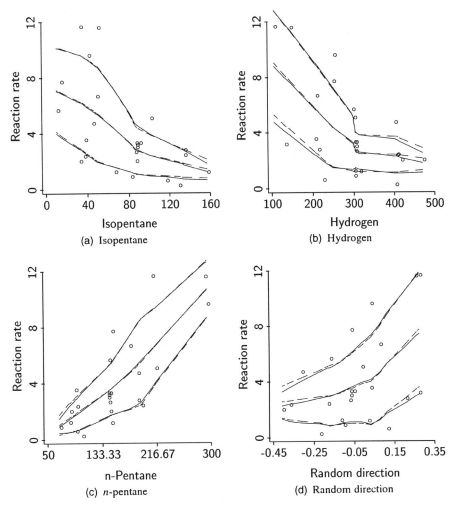

FIGURE 15.5 Four marginal model plots for the isomerization model. Marginal data ———; Marginal model – – –.

This expression also holds with $\hat{M}$ replaced by F. Thus, it is possible for $E_{\hat{M}}(y \mid a^T x)$ and $\hat{E}_F(y \mid a^T x)$ to be close, while $E_{\hat{M}}(y \mid a^T x; \chi)$ and $\hat{E}_F(y \mid a^T x; \chi)$ are considerably different.

15.2.4. Isomerization data again

Shown in Figure 15.5 are four marginal model plots for the Hougen–Watson model (15.3). The first three plots are for the individual predictors; the horizontal axis of the fourth plot is $a_u^T x$ where a_u was randomly chosen from the uniform distribution on the unit sphere. The marginal model plot for the fitted values would normally be of particular interest. However, in this example, the rather strong relationship between y and $\hat{y}$ shown in Figure 15.1 makes it difficult to distinguish between the various smooths when superimposed on a plot of the data. For this reason, the marginal model plot for the fitted values is not shown in this example.

The dashed lines in the plots of Figure 15.5 were constructed from the marginal regression and variance functions for the model, while the solid lines were constructed from smooths of the data showing on the plot. The middle curves are the regression smooths, while the upper and lower curves are the smooths for the standard deviation represented as deviations from the regression smooths. All estimated regression and variance functions in this section are based on LOWESS smooths with tuning parameter 0.6 or 0.7.

Some of the regression functions in Figure 15.5 may appear curved. Generally, curvature in the estimated regression functions of marginal model plots could be a consequence of the curvature in the regression function for the model, or curvature in the regression functions among the predictors as described in Section 15.2.1, and does not necessarily indicate a weakness in the model. Similarly, the heteroscedastic patterns in the marginal model plots for isopentane, hydrogen, and the random direction do not necessarily indicate heteroscedasticity of the errors in (15.3), even if the regression function is adequate.

A striking impression from the plots in Figure 15.5 is that, relative to the variation in the data, the marginal regression and variance function from the model are quite close to the corresponding functions estimated from the marginal data. Depending on the particular goals of the experiment, these plots support the idea that the Hougen–Watson model could be adequate for the isomerization data.

In the discussion of Section 15.1.4 on residual plots for the isomerization data, we reasoned that while pHd suggests the possibility of a significant discrepancy between the model and the data, the evidence is not very strong and its implications for model improvement are unclear. This issue can now be addressed directly by using the marginal model plot shown in Figure 15.6 for the first pHd direction. As in Figure 15.5, the marginal model regression and variance functions in Figure 15.6 are relatively close to the corresponding functions estimated from the marginal data, although a systematic pattern to their difference is evident. The difference between the regression smooths in Figure 15.6 is equal to the smooth on the residual plot for the first pHd direction shown in Figure 15.1d, because the same value of the LOWESS tuning parameter was used for all smooths. Thus, while Figure 15.1d and the associated pHd test suggest a discrepancy, the corresponding marginal model plot of Figure 15.6 shows that the apparent discrepancy is not substantial relative to the variation in the data, providing confirmation of the notion that the model could well be adequate.

If, as suggested by pHd, the regression of r on x has at most 1D structure then the estimated sufficient summary plot in Figure 15.1d is expected to

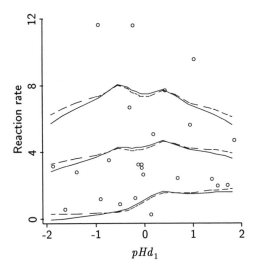

FIGURE 15.6 Marginal model plot for the first pHd direction pHd_1 from the isomerization data. Marginal data ———; Marginal model – – –.

capture all model weaknesses in the regression function and the subspace spanned by the first pHd direction provides an estimate of $S_{r|x}$. The marginal model plot of Figure 15.6 shows the discrepancy between the model and the data in the "worst" direction, the direction corresponding to the estimated basis for $S_{r|x}$ provided by pHd. Up to the limitations of pHd, the analysis associated with Figure 15.1d and Figure 15.6 gives a comprehensive check on the adequacy of the model.

For contrast with the previous results, we now consider the fit of the standard linear model. Let $\hat{e}$ denote the residuals from the OLS fit of the linear model

$$y \mid x = \beta_0 + \beta^T x + \varepsilon$$

to the isomerization data. Figure 15.7 shows a marginal model plot for the first pHd predictor from the regression of $\hat{e}$ on x. The curves on the plot were again determined by using LOWESS smooths, the dashed and solid curves corresponding to the linear model and the data, respectively. It seems clear that the summary of the data provided by the linear model is not nearly as good as that provided by model (15.3). The regression smooth, corresponding to the middle dashed line, noticeably underestimates the data at the ends of the horizontal axis and the variance smooth from the model is quite different from its data-based counterpart. The marginal model plots for the three predictors and several random directions also showed notable discrepancies, but none seemed as great as that displayed in Figure 15.7. In a sense, marginal model plots based on pHd can give an upper bound on the discrepancies to be expected in other plots.

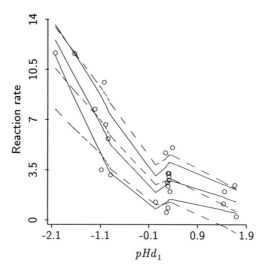

FIGURE 15.7 Marginal model plot for isomerization data based on the linear model. pHd_1 is the linear combination of the predictors for the first pHd direction from the regression of $\hat{e}$ on x. Marginal data ————; Marginal model – – –.

15.2.5. Reaction yield data

The data on reaction yield y, which were analyzed using graphical regression in Section 9.3, consist of 32 observations from a two-stage chemical process characterized by $p = 5$ predictors, linear transformations T_1 and T_2 of the reaction temperatures at the two stages, linear transformations Lt_1 and Lt_2 of the log times of reaction at the two stages, and a linear transformation C of the concentration of one of the reactants. Let $x = (T_1, T_2, Lt_1, Lt_2, C)^T$. To illustrate a relatively extreme case of model failure, consider summarizing the data with the OLS fit of the usual homoscedastic linear model. Let $\hat{e}$ denote the corresponding residuals. A pHd analysis of the regression of $\hat{e}$ on x indicates at least two significant directions; the first three pHd p-values determined by using the procedure described in Section 12.3.2 are 0.003, 0.054, and 0.057 (see Table 12.1). Figure 15.8 shows marginal model plots for the first two pHd predictors, pHd_1 and pHd_2. The smooths were constructed as described in the previous illustrations of this section. In each case, the marginal model seriously fails to describe the marginal data.

15.2.6. Tomato tops

The discussion of marginal model plots so far has been under the assumption that the data (y_i, x_i^T) are sampled independently from the distribution of (y, x^T). This assumption, which was used mainly to simplify the presentation

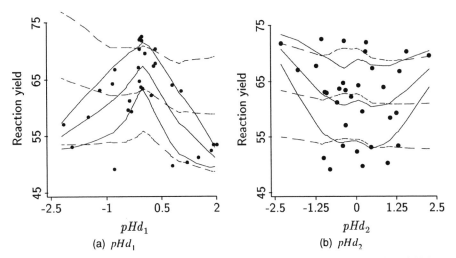

FIGURE 15.8 Marginal model plot for the first two pHd directions from the reaction yield data. Marginal data ————; Marginal model – – –.

a bit, is really unnecessary. Marginal model plots apply equally to designed experiments, where there may be no natural distribution for x, and to other situations where it may be desirable to condition on the observed values of the predictors. The only change is in interpretation: Conditional expectations involving only predictors are taken with respect to the observed data rather than with respect to some larger population.

For illustration, consider an experiment described by Box and Draper (1987, p. 385) on the dry matter yield of tomato tops (y in g dm/pot) in response to the application of various combinations of three nutrients, nitrogen N, phosphorus P, and potassium K. The design consists of a full 3^3 factorial plus 6 additional star points as in a central composite design, for a total of 33 design points. Each design point was replicated twice, resulting in a total of 66 observations. The analysis here uses coded levels, ± 1.5, ± 1, and 0, for the three factors as in Box and Draper (1987).

Consider the OLS fit of the usual linear regression model

$$y \mid x = \beta_0 + \beta^T x + \sigma\varepsilon$$

where $x = (N, P, K)^T$. Shown in Figure 15.9 are four marginal model plots for assessing this model. The first two are unconditional plots showing the marginal model fit (dashed line) against the marginal data (solid line) for the fitted values and potassium. In the plot for fitted values, the model noticeably underfits the data at the extremes and overfits a bit in the middle. In the plot for potassium, the model underestimates at the middle value of potassium, but overestimates the response at the star points. Although the marginal model deviates systematically from the marginal data, the deviations seem small rel-

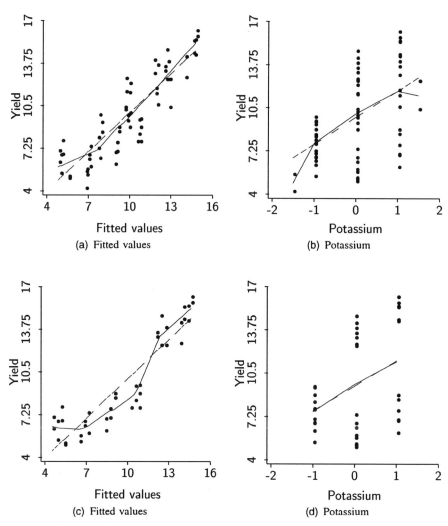

FIGURE 15.9 Unconditional (a and b) and conditional (c and d, $N \neq 0$) marginal model plots for tomato top yield. Marginal data ——— ; Marginal model – – –.

ative to the marginal variation, perhaps indicating that the model could be adequate.

However, conditional marginal model plots leave a different impression. Figures 15.9c and 15.9d show marginal model plots for the fitted values and potassium restricted to the subset of the data in which $N \neq 0$. The marginal model now seriously fails to capture the marginal data in the plot for the fitted values. There is nothing special about restricting attention to the subset of the data in which $N \neq 0$. Large deviations can also be seen when conditioning on other characteristics of N and on characteristics of P and K.

15.2.7. Marginal model plots: Kyphosis data

Residual plots for the kyphosis data were considered in Section 15.1.5. In this section we investigate what marginal model plots indicate about the adequacy of model (15.6). Since the response is binary, variance functions carry no more information about model adequacy than do regression functions, as previously mentioned near the end of Section 15.2.2. Consequently, only the regression functions will be displayed in marginal model plots.

As a baseline, we begin by considering the fit of the first-order model with

$$\gamma(x) = \log \frac{P(x)}{1 - P(x)} = \beta_0 + \beta^T x \qquad (15.15)$$

where $x = (Age, Start, Num)^T$. The fit of this model has a deviance of 61.4 on 77 degrees of freedom. Shown in Figures 15.10a and 15.10b are the marginal model plot for $\hat{\gamma}$ and for Age. The response in this example is binary and so the data fall on two horizontal lines, one for $kyp = 1$ and one for $kyp = 0$, in each marginal model plot. In contrast to marginal model plots for many-valued responses, the data in Figure 15.10 are not much help, except perhaps for showing when a smooth is responding to a particular feature of the empirical distribution of the variable on the horizontal axis. Smooths of the marginal data tend to be more sensitive to remote points than marginal model smooths, for example.

The curves in the marginal model plots of Figures 15.10a and 15.10b are the estimates of the marginal regression functions obtained by using LOWESS smooths with span 0.6. As in previous applications, the solid curves correspond to the marginal data, while the dashed curves correspond to the marginal model. The agreement between the marginal data and the marginal model does not seem very good in either figure. The plot for Age in Figure 15.10b suggests that the model overestimates $\Pr(kyp \,|\, Age)$ for the relatively young and old, and underestimates otherwise. The plot for $\hat{\gamma}$ is influenced by the point with the largest value of $\hat{\gamma}$, but with or without this point, the agreement between the marginal model and the marginal data in Figure 15.10a does not seem close.

Shown in Figures 15.10c and 15.10d are the marginal model plots for $\hat{\gamma}$ and Age from model (15.6). The fits are quite close, which suggests that the model could well be adequate. We could now explore marginal model plots for other directions looking for additional confirmation of the model. Instead, we turn to pHd.

Let $\hat{r}$ denote a typical deviance residual from model (15.6), and let h_1 denote the first pHd direction from the regression of $\hat{r}$ on the three original predictors, Age, $Start$, and Num. The first pHd predictor is thus $pHd_1 = h_1^T x$. The pHd analysis indicates 1D structure, since the p-value for h_1 is 0.05, while the other p-values are relatively large. The plot $\{\hat{r}, h_1^T x\}$ (not shown) exhibits nonlinear tendencies and a relatively remote point (number 52 in the

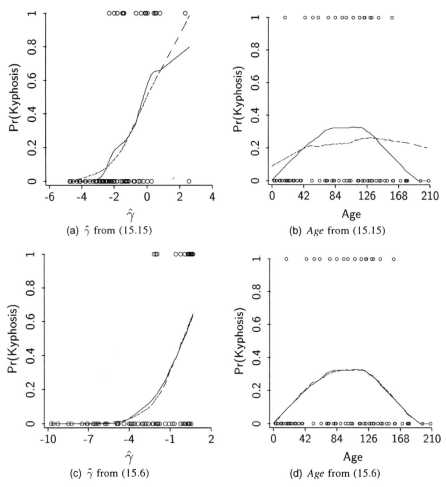

FIGURE 15.10 Marginal model plot for the kyphosis data. Marginal data ———; Marginal model – – –.

data set of 81 observations described in Section 5.3.2) that may be influencing h_1 and the corresponding test. Deleting this point decreases the p-value and sharpens the nonlinear trend in the plot of $\hat{r}$ versus $h_1^T x$, which is shown in Figure 15.11a along with a LOWESS smooth to aid visual impressions. The regression function for the plot exhibits a clear quadratic tendency, suggesting that some improvement in the Hastie–Tibshirani model (15.6) may be possible. The marginal model plot for $h_1^T x$ computed without case 52 is shown in Figure 15.11b. This plot suggests that there is a fairly large deviation between the marginal model and the marginal data, probably enough to justify revision of the model.

Recall that the model (15.6) in question contains three terms, Age, Age^2, and $(Start - 12)J(Start > 12)$. To check on the possible gains from revising the

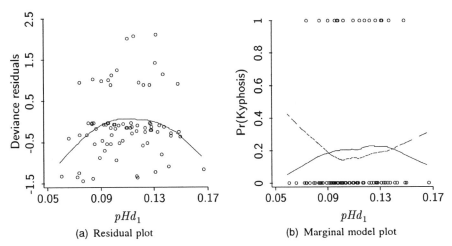

FIGURE 15.11 Two plots for assessing model (15.6). In each the horizontal axis is the first pHd predictor pHd_1 from the regression of the deviance residuals on *Age*, *Start*, and *Num* without case 52. Marginal data ———; Marginal model – – –.

model, three predictors were added, *Start*, *Num*, and $(h_1^T x)^2$, where h_1 was taken from the pHd analysis without case 52. However, case 52 was restored when fitting the logistic model with these six predictors. The estimated coefficients for *Age*, Age^2, *Num*, and $(h_1^T x)^2$ all exceed twice their respective standard errors. The estimated coefficient of *Start* is about 1.8 times its standard error, and the coefficient of $(Start - 12)I(Start > 12)$ is about 1.5 times its standard error. The deviance for model (15.6) is 52 on 77 degrees of freedom, while the deviance for the expanded model is 43.9 on 74 degrees of freedom. In short, there does seem to be reason to believe that the model may be improved.

As anticipated, case 52 is quite influential in the fit of the expanded model. However, the results with case 52 removed give even stronger support for the conclusion that improvement is possible. For example, the deviance for the expanded model drops to 33.9, while the deviance for (15.6) reduces to only 50.6.

Finally, the expanded model is used here only to support the conclusion that useful improvement in (15.6) may be possible. A different model might well result from further analysis.

For further discussion on the use of pHd in model checking, see Cook and Bura (1997).

PROBLEMS

15.1. In reference to the naphthalene data with response Y_N, investigate the fit of the full, second-order quadratic model in the original three process variables, as described in Section 15.1.3. Begin by inspecting a few residual plots, and then use marginal model plots. If revision of the model seems appropriate, use graphical regression to help understand the type of revision required. Does it seem easier to work in terms of the original process variables, or the three transformed predictors, *AN*, *Btemp*, and *Ctime*?

15.2. The graphical regression analysis of the naphthalene data in Section 9.1.1 ended with the linear combination of the predictors *ACB* given in (9.1). Let $E_{\hat{M}}(Y_N \mid ACB)$ represent a smooth of Y_N against *ACB* and, as implied by the notation, this is the regression function for the model. Construct a few marginal model plots comparing marginal regression functions implied by this model to the corresponding regression functions from the marginal data. How might pHd be adapted for this situation?

Bibliography

Aitchison, J. (1982). The statistical analysis of compositional data (with discussion). *Journal of the Royal Statistical Society Ser. B*, **44**, 139–177.

Aitchison, J. (1986). *The statistical analysis of compositional data*. New York: Chapman and Hall.

Altman, N. S. (1992). An introduction to kernel and nearest neighbor nonparametric regression. *The American Statistician*, **46**, 175–185.

Anderson, J. E., T. A. Louis, N. V. Holm, and B. Harvald (1992). Time-dependent association measures for bivariate survival distributions. *Journal of the American Statistical Association*, **87**, 641–650.

Anscombe, F. J. (1973). Graphs in statistical analysis. *The American Statistician*, **27**, 17–21.

Anscombe, F. J. and J. W. Tukey (1963). The examination and analysis of residuals. *Technometrics*, **5**, 141–160.

Asimov, D. (1985). The grand tour: A tool for viewing multidimensional data. *SIAM Journal on Scientific and Statistical Computing*, **6**, 123–143.

Atkinson, A. C. (1982). Regression diagnostics, transformations and constructed variables (with discussion). *Journal of the Royal Statistical Society Ser. B*, **44**, 1–36.

Atkinson, A. C. (1985). *Plots, Transformations and Regression*. Oxford: Oxford University Press.

Barnett, V. and T. Lewis (1994). *Outliers in Statistical Data* (3rd ed.). New York: Wiley.

Basu, D. and C. A. B. Pereira (1983). Conditional independence in statistics. *Sankhyā*, **45**, 324–337.

Bates, D. and D. G. Watts (1988). *Nonlinear Regression Analysis and its Applications*. New York: Wiley.

Bean, L. H. (1929). A simplified method of graphic curvilinear correlation. *Journal of the American Statistical Association*, **24**, 386–397.

Bean, L. H. (1930). Application of a simplified method of correlation to problems in acreage and yield variations. *Journal of the American Statistical Association*, **25**, 428–439.

Bean, L. H. (1940). The use of the short-cut graphic method of multiple correlation: Comment. *Quarterly Journal of Economics*, **55**, 318–331.

Becker, R. A. and W. S. Cleveland (1987). Brushing scatterplots. *Technometrics*, **29**, 127–142.

Becker, R. A., W. S. Cleveland, and M.-J. Shyu (1996). Visual design and control of trellis display. *Journal of Computational and Graphical Statistics*, **5**, 123–155.

Becker, R. A., W. S. Cleveland, and A. R. Wilks (1987). Dynamic graphics for data analysis (with discussion). *Statistical Science*, **2**, 355–395.

Beckman, R. J. and R. D. Cook (1983). Outlier.......s. *Technometrics*, **25**, 119–149.

Belsley, D. A., E. Kuh, and R. Welsch (1980). *Regression Diagnostics*. New York: Wiley.

Berk, K. N. (1998). Regression diagnostic plots in 3-D. *Technometrics*, **40**, 39–47.

Berk, K. N. and D. E. Booth (1995). Seeing a curve in multiple regression. *Technometrics*, **37**, 385–398.

Bowman, A. and S. Young (1996). Graphical comparison of nonparametric curves. *Applied Statistics*, **45**, 83–98.

Box, G. E. P. (1979). Robustness in the strategy of scientific model building. In R. Launer and G. Wilkinson (eds.), *Robustness in Statistics*, pp. 201–235. New York: Academic Press.

Box, G. E. P. (1980). Sampling and Bayes inference in scientific modeling and robustness (with discussion). *Journal of the Royal Statistical Society Series A*, **134**, 383–430.

Box, G. E. P. and D. R. Cox (1964). An analysis of transformations (with discussion). *Journal of the Royal Statistical Society Series B*, **26**, 211–246.

Box, G. E. P. and N. Draper (1987). *Empirical Model-Building and Response Surfaces*. New York: Wiley.

Brinkman, N. (1981). Ethanol fuel: A single-cylinder engine study of efficiency and exhaust emissions. *SAE Transactions*, **90**, 1410–1424.

Buja, A. and D. Asimov (1986). Grand tour methods: An outline. In *Computer Science and Statistics: Proceedings of the 17th Symposium on the Interface*, pp. 63–76. Amsterdam: Elsevier.

Bura, E. (1996). *Dimension Reduction via Inverse Regression*. Ph. D. Thesis, School of Statistics, University of Minnesota.

Bura, E. (1997). Dimension reduction via parametric inverse regression. In Y. Dodge (ed.), *L1-Statistical Procedures and Related Topics: IMS Lecture Notes-Monograph Series, Vol. 31*, 215–228. Hayward, CA: Institute of Mathematical Statistics.

Cambanis, S., S. Huang, and G. Simons (1981). On the theory of elliptically contoured distributions. *Journal of Multivariate Analysis*, **7**, 368–385.

Camden, M. (1989). *The Data Bundle*. Wellington, New Zealand: New Zealand Statistical Association.

Carr, D. B., R. J. Littlefield, W. L. Nicholson, and J. S. Littlefield (1987). Scatterplot matrix techniques for large N. *Journal of the American Statistical Association*, **82**, 424–436.

Carr, N. L. (1960). Kinetics of catalytic isomerization of n-pentane. *Industrial and Engineering Chemistry*, **52**, 391–396.

Carroll, R. J. and K. C. Li (1995). Binary regressors in dimension reduction models: A new look at treatment comparisons. *Statistica Sinica*, **5**, 667–688.

Carroll, R. J. and D. Ruppert (1988). *Transformations and Weighting in Regression*. London: Chapman and Hall.

Chambers, J. M., W. S. Cleveland, B. Kleiner, and P. Tukey (1983). *Graphical Methods for Data Analysis*. Boston: Duxbury Press.

Chambers, J. M. and T. J. Hastie (1992). *Statistical Models in S*. Pacific Grove: Wadsworth and Brooks/Cole.

Chatterjee, S. and A. S. Hadi (1988). *Sensitivity Analysis in Linear Regression*. New York: Wiley.

Cheng, C.-S. and K.-C. Li (1995). A study of the method of principal Hessian direction for analysis of data from designed experiments. *Statistica Sinica*, **5**, 617–639.

Chiaromonte, F. and R. D. Cook (1997). On foundations of regression graphics. *University of Minnesota, School of Statistics, Technical Report*, **616**, 1–33.

Cleveland, W. S. (1979). Locally weighted regression and smoothing scatterplots. *Journal of the American Statistical Association*, **74**, 829–836.

Cleveland, W. S. (1993). *Visualizing Data*. Summit, New Jersey: Hobart Press.

Cleveland, W. S., S. J. Devlin, and E. H. Grosse (1988). Regression by local fitting: Methods, properties, and computational algorithms. *Journal of Econometrics*, **37**, 87–114.

Cleveland, W. S. and M. E. McGill (1984). The many faces of a scatterplot. *Journal of the American Statistical Association*, **79**, 807–822.

Cleveland, W. S. and M. E. McGill (1988). *Dynamic Graphics for Statistics*. Belmont, CA: Wadsworth.

Cleveland, W. S., M. E. McGill, and R. McGill (1988). The shape parameter of a two-variable graph. *Journal of the American Statistical Association*, **83**, 289–300.

Collett, D. (1991). *Modelling Binary Data*. London: Chapman and Hall.

Cook, R. D. (1977). Detection of influential observations in linear regression. *Technometrics*, **19**, 15–18.

Cook, R. D. (1986). Assessment of local influence (with discussion). *Journal of the Royal Statistical Society Ser B*, **48**, 133–155.

Cook, R. D. (1987a). Influence assessment. *Journal of Applied Statistics*, **14**, 117–131.

Cook, R. D. (1987b). Parameter plots in nonlinear regression. *Biometrika*, **74**, 699–677.

Cook, R. D. (1987c). Software review of MacSpin. *The American Statistician*, **41**, 233–236.

Cook, R. D. (1992a). Graphical regression. In Y. Dodge and J. Whittaker (eds.), *Computational Statistics*, Volume 1, pp. 11–22. New York: Springer-Verlag.

Cook, R. D. (1992b). Regression plotting based on quadratic predictors. In Y. Dodge (ed.), *L1-Statistical Analysis and Related Methods*, pp. 115–128. Amsterdam: North Holland.

Cook, R. D. (1993). Exploring partial residual plots. *Technometrics*, **35**, 351–362.

Cook, R. D. (1994a). On the interpretation of regression plots. *Journal of the American Statistical Association*, **89**, 177–189.

Cook, R. D. (1994b). Using dimension-reduction subspaces to identify important inputs in models of physical systems. In *Proceedings of the Section on Physical and Engineering Sciences*, pp. 18–25. Alexandria, VA: American Statistical Association.

Cook, R. D. (1995). Graphics for studying net effects of regression predictors. *Statistica Sinica*, **5**, 689–708.

Cook, R. D. (1996a). Added-variable plots and curvature in linear regression. *Technometrics*, **38**, 275–278.

Cook, R. D. (1996b). Graphics for regressions with a binary response. *Journal of the American Statistical Association*, **91**, 983–992.

Cook, R. D. (1998). Principal Hessian directions revisited (with discussion). *Journal of the American Statistical Association*, **93**, 84–100.

Cook, R. D. and E. Bura (1997). Testing the adequacy of regression functions. *Biometrika*, **84**, 949–956.

Cook, R. D. and R. Croos–Dabrera (1998). Partial residual plots in generalized linear models. *Journal of the American Statistical Association*, **93**, 730–739.

Cook, R. D., D. Hawkins, and S. Weisberg (1992). Comparison of model misspecification diagnostics using residuals from least median of squares fits. *Journal of the American Statistical Association*, **87**, 419–424.

Cook, R. D. and C. J. Nachtsheim (1994). Re-weighting to achieve elliptically contoured covariates in regression. *Journal of the American Statistical Association*, **89**, 592–600.

Cook, R. D. and S. Weisberg (1982). *Residuals and Influence in Regression*. London: Chapman and Hall.

Cook, R. D. and S. Weisberg (1983). Diagnostics for heteroscedasticity in regression. *Biometrika*, **70**, 1–10.

Cook, R. D. and S. Weisberg (1989). Regression diagnostics with dynamic graphics (with discussion). *Technometrics*, **31**, 277–312.

Cook, R. D. and S. Weisberg (1990). Three-dimensional residual plots. In K. Berk and L. Malone (eds.), *Proceedings of the 21st Symposium on the Interface: Computing Science and Statistics*, pp. 162–166. Washington: American Statistical Association.

Cook, R. D. and S. Weisberg (1991a). Added variable plots in linear regression. In W. Stahel and S. Weisberg (eds.), *Directions in Robust Statistics and Diagnostics, Part I*, pp. 47–60. New York: Springer-Verlag.

Cook, R. D. and S. Weisberg (1991b). Discussion of "Sliced inverse regression" by K. C. Li. *Journal of the American Statistical Association*, **86**, 328–332.

Cook, R. D. and S. Weisberg (1994a). *An Introduction to Regression Graphics*. New York: Wiley.

Cook, R. D. and S. Weisberg (1994b). Transforming a response variable for linearity. *Biometrika*, **81**, 731–738.

Cook, R. D. and S. Weisberg (1997). Graphics for assessing the adequacy of regression models. *Journal of the American Statistical Association*, **92**, 490–499.

Cook, R. D. and N. Wetzel (1993). Exploring regression structure with graphics. *TEST*, **2**, 33–100.

Cox, D. R. (1958). *Planning of Experiments*. New York: Wiley.

Cox, D. R. (1972). Regression models and life tables (with discusssion). *Journal of the Royal Statistical Society Ser. B*, **34**, 187–202.

Cox, D. R. and D. Hinkley (1974). *Theoretical Statistics*. London: Chapman and Hall.

Cox, D. R. and E. J. Snell (1968). A general definition of residuals. *Journal of the Royal Statistical Society Ser. B*, **30**, 248–275.

Cox, D. R. and N. Wermuth (1996). *Multivariate Dependencies*. London: Chapman and Hall.

Cox, D. R. (1978). Some remarks on the role in statistics of graphical methods. *Journal of the Royal Statistical Society Ser. C*, **27**, 4–9.

Croxton, F. E. and D. J. Cowden (1935). *Practical Business Statistics*. New York: Prentice-Hall.

Davison, A. C. and E. J. Snell (1991). Residuals and diagnostics. In D. V. Hinkley and N. Reid (eds.), *Statistical Theory and Modeling: In Honor of Sir David Cox, FRS* pp. 83–106. London: Chapman and Hall.

Dawid, A. P. (1979a). Conditional independence in statistical theory (with discussion). *Journal of the Royal Statistical Society, Ser. B*, **41**, 1–31.

Dawid, A. P. (1979b). Some misleading arguments involving conditional independence. *Journal of the Royal Statistical Society, Ser. B*, **41**, 249–252.

Dawid, A. P. (1980). Conditional independence for statistical operations. *Annals of Statistics*, **8**, 598–617.

Dawid, A. P. (1984). Comment: Casual inference from messy data. *Journal of the American Statistical Association*, **79**, 22–24.

Draper, N. R. and H. Smith (1966). *Applied Regression Analysis*. New York: Wiley.

Draper, N. R. and H. Smith (1981). *Applied Regression Analysis* (2nd ed.). New York: Wiley.

Duan, N. and K. C. Li (1991). Slicing regression: A link free regression method. *Annals of Statistics*, **19**, 505–530.

Eaton, M. L. (1983). *Multivariate Statistics*. New York: Wiley.

Eaton, M. L. (1986). A characterization of spherical distributions. *Journal of Multivariate Analysis*, **20**, 272–276.

Eaton, M. L. and D. Tyler (1991). On Wielandt's inequality and its application to the asymptotic distribution of the eigenvalues of a random symmetric matrix. *Annals of Statistics*, **19**, 260–271.

Eaton, M. L. and D. Tyler (1994). The asymptotic distribution of singular values with application to canonical correlations and correspondence analysis. *Journal of Multivariate Analysis*, **50**, 238–264.

Enz, R. (1991). *Prices and Earnings Around the Globe*. Zurich: Union Bank of Switzerland.

Ernst, M. D., R. Guerra, and W. R. Schucany (1996). Scatterplots for unordered pairs. *The American Statistician*, **50**, 260–265.

Ezekiel, M. (1924). A method for handling curvilinear correlation for any number of variables. *The Journal of the American Statistical Association*, **19**, 431–453.

Ezekiel, M. (1930a). *Methods of Correlation Analysis* (1st ed.). New York: Wiley.

Ezekiel, M. (1930b). The sampling variability of linear and curvilinear regressions. *The Annals of Statistics*, **1**, 274–333.

Ezekiel, M. (1940). The use of the short-cut graphic method of multiple correlation: Further comment. *Quarterly Journal of Economics*, **55**, 331–346.

Ezekiel, M. (1941). *Methods of Correlation Analysis* (2nd ed.). New York: Wiley.

Fang, K. T., S. Kotz, and K. W. Ng (1990). *Symmetric Multivariate and Related Distributions*. London: Chapman and Hall.

Farebrother, R. (1990). The distribution of a quadratic form in normal variables. *Applied Statistics*, **39**, 294–309.

Fearn, T. (1983). A misuse of ridge regression in the calibration of near infrared reflectance instruments. *Applied Statistics*, **32**, 73–79.

Ferré, L. (1997). Dimension choice for sliced inverse regression based on ranks. *Student*, **2**, 95–108.

Ferré, L. (1998). Determining the dimension in sliced inverse regression and related methods. *Journal of the American Statistical Association*, **93**, 132–140.

Field, C. (1993). Tail areas of linear combinations of chi-squares and noncentral chi-squares. *Journal of Statistical Computation and Simulation*, **45**, 243–248.

Fienberg, S. E. (1979). Graphical methods in statistics. *The American Statistician*, **33**, 165–178.

Filliben, J. J. and K.-C. Li (1997). A systematic approach to the analysis of complex interaction patterns in two-level factorial designs. *Technometrics*, **39**, 286–297.

Fisherkeller, M. A., J. H. Friedman, and J. W. Tukey (1974). PRIM-9: An interactive multidimensional data display and analysis system. SLAC-PUB-1408. Stanford, CA: Stanford Linear Accelerator Center.

Flury, B. and H. Riedwyl (1988). *Multivariate Statistics: A Practical Approach*. London: Chapman and Hall.

Fox, J. (1991). *Regression Diagnostics.* Newbury Park: Sage Publications.

Franklin, N. L., P. H. Pinchbeck, and F. Popper (1956). A statistical approach to catalyst development, part 1: The effect of process variables in the vapor phase oxidation of naphthalene. *Transactions of the Institute of Chemical Engineers,* **34**, 280–293.

Green, P. J. and B. W. Silverman (1994). *Nonparametric Regression and Generalized Linear Models.* London: Chapman and Hall.

Gunst, R. F. and R. L. Mason (1980). *Regression Analysis and its Applications.* New York: Dekker.

Guttman, I. (1982). *Linear Models.* New York: Wiley.

Hall, P. and K.-C. Li (1993). On almost linearity of low dimensional projections from high dimensional data. *The Annals of Statistics,* **21**, 867–889.

Härdle, W. (1990). *Applied Nonparametric Regression.* Cambridge: Cambridge University Press.

Härdle, W. and A. B. Tsybakov (1991). Discussion of a paper by Li. *The Journal of the American Statistical Association,* **86**, 333–335.

Hastie, T. J. and R. J. Tibshirani (1990). *Generalized Additive Models.* London: Chapman and Hall.

Henderson, H. V. and P. Velleman (1981). Building multiple regression models interactively. *Biometrics,* **37**, 391–411.

Hougaard, P., B. Harvald, and N. V. Holm (1992). Measuring similarities between the lifetimes of adult danish twins born between 1881–1930. *Journal of the American Statistical Association,* **87**, 17–24.

Hsing, T. and R. Carroll (1992). An asymptotic theory of sliced inverse regression. *Annals of Statistics,* **20**, 1040–1061.

Huber, P. (1985). Projection pursuit (with discussion). *Annals of Statistics,* **13**, 435–525.

Huber, P. (1987). Experiences with three-dimensional scatterplots. *Journal of the American Statistical Association,* **82**, 448–453.

Ibrahimy, A. and R. D. Cook (1995). Regression design for one-dimensional subspaces. In C. P. Kitsos and W. G. Müller (eds.), *MODA4: Advances in Model-Oriented Data Analysis,* pp. 125–132. Heidelberg: Physica-Verlag.

Johnson, B. W. and R. E. McCulloch (1987). Added-variable plots in linear regression. *Technometrics,* **29**, 427–434.

Johnson, M. (1987). *Multivariate Statistical Simulation.* New York: Wiley.

Kafadar, K. (1985). Notched box-and-whisker plot. *Encyclopedia of Statistical Sciences,* **6**, 367–370.

Kay, R. and S. Little (1987). Transforming the explanatory variables in the logistic regression model for binary data. *Biometrika,* **74**, 495–501.

Kelker, D. (1970). Distribution theory of spherical distributions and a location-scale parameter generalization. *Sankhyā, Series A,* **32**, 419–430.

Kent, J. T. (1991). Discussion of a paper by Li. *The Journal of the American Statistical Association,* **86**, 336–337.

Koehn, U. and D. L. Thomas (1975). On statistics independent of a sufficient statistic: Basu's Lemma. *The American Statistician,* **29**, 40–42.

Kötter, T. (1996). An asymptotic result for sliced inverse regression. *Computational Statistics,* **11**, 113–136.

Larsen, W. A. and S. J. McClearly (1972). The use of partial residual plots in regression analysis. *Technometrics,* **14**, 781–790.

Lawrance, A. (1986). Discussion of "Assessment of local influence" by R. D. Cook. *Journal of the Royal Statistical Society Ser B,* **48**, 157–159.

Li, K. C. (1991). Sliced inverse regression for dimension reduction (with discussion). *Journal of the American Statistical Association*, **86**, 316–342.

Li, K. C. (1992). On principal Hessian directions for data visualization and dimension reduction: Another application of Stein's Lemma. *Journal of the American Statistical Association*, **87**, 1025–1040.

Li, K. C. (1997). Nonlinear confounding in high-dimensional regression. *Annals of Statistics*, **57**, 577–612.

Li, K. C. and N. Duan (1989). Regression analysis under link violation. *Annals of Statistics*, **17**, 1009–1052.

Malenbaum, W. and J. D. Black (1937). The use of the short-cut graphic method of multiple correlation. *Quarterly Journal of Economics*, **52**, 66–112.

Mallows, C. L. (1986). Augmented partial residual plots. *Technometrics*, **28**, 313–320.

Mallows, C. L. (1988). Letter to the editor: Response to O'Brien. *Technometrics*, **30**, 136.

Mansfield, R. and M. D. Conerly (1987). Diagnostic value of residual and partial residual plots. *The American Statistician*, **41**, 107–116.

Marron, J. S. and A. B. Tsybakov (1995). Visual error criteria for qualitative smoothing. *Journal of the American Statistical Association*, **90**, 499–507.

McCulloch, R. E. (1993). Discussion of "Exploring regression structure with graphics" by R. D. Cook and N. Wetzel. *TEST*, **2**, 84–86.

McGill, R., J. W. Tukey, and W. A. Larsen (1978). Variations of box plots. *The American Statistician*, **32**, 12–16.

McKay, M. D., R. J. Beckman, and W. J. Conover (1979). A comparison of three methods for selecting values of input variables in the analysis of output from computer code. *Technometrics*, **21**, 239–245.

McKean, J. W. and J. S. Sheather (1997). Exploring data sets using partial residual plots based on robust fits. In Y. Dodge (ed.), *L1-Statistical Procedures and Related Topics: IMS Lecture Notes–Monograph Series, Vol. 31*, pp. 241–255. Hayward, CA: Institute of Mathematical Statistics.

Meyers, M., J. Stollsteimer, and A. Wims (1975). Determination of gasoline octane numbers from chemical composition. *Analytical Chemistry*, **47**, 2301–2304.

Mosteller, F. and J. Tukey (1977). *Data Analysis and Regression*. Reading, MA: Addison-Wesley.

Mouchart, M. and J.-M. Rolin (1984). A note on conditional independence with statistical applications. *Statistica*, **54**, 557–584.

Nelder, J. A. (1989). Nearly parallel lines in residual plots. *The American Statistician*, **43**, 221–222.

O'Brien, C. M. (1988). Letter to the editor on Mallows (1986). *Technometrics*, **30**, 135–136.

O'Hara Hines, R. J. and E. M. Carter (1993). Improved added variable and partial residual plots for the detection of influential observations in generalized linear models. *Applied Statistics*, **42**, 3–16.

Okabe, A., B. Boots, and K. Sugihara (1992). *Spatial Tessellations*. New York: Wiley.

Pillai, K. (1955). Some new test criteria in multivariate analysis. *Annals of Mathematical Statistics*, **26**, 117–121.

Rice, J. A. (1988). *Mathematical Statistics and Data Analysis*. Pacific Grove, CA: Wadsworth & Brooks/Cole.

Ruppert, D., M. P. Wand, U. Holst, and O. Hössjer (1997). Local polynomial variance-function estimation. *Technometrics*, **39**, 262–273.

Schott, J. (1994). Determining the dimensionality in sliced inverse regression. *Journal of the American Statistical Association*, **89**, 141–148.

Scott, D. (1992). *Multivariate Density Estimation*. New York: Wiley.

Searle, S. (1988). Parallel lines in residual plots. *The American Statistician*, **42**, 211.

Seber, G. (1977). *Linear Regression Analysis*. New York: Wiley.

Seber, G. (1984). *Multivariate Observations*. New York: Wiley.

Serfling, R. J. (1980). *Approximation Theorems of Mathematical Statistics*. New York: Wiley.

Sheather, J. S. and J. W. McKean (1997). A comparison of procedures based on inverse regression. In Y. Dodge (ed.), *L1-Statistical Procedures and Related Topics: IMS Lecture Notes–Monograph Series, Vol. 31*, pp. 271–278. Hayward, CA: Institute of Mathematical Statistics.

Simonoff, J. S. (1996). *Smoothing Methods in Statistics*. New York: Springer.

St. Laurent, R. T. (1987). *Detecting Curvature in the Response in Regression*. Ph. D. thesis, School of Statistics, University of Minnesota.

St. Laurent, R. T. (1990). The equivalence of the Milliken-Graybill procedure and the score test. *The American Statistician*, **44**, 36–37.

Stein, C. (1981). Estimation of the mean of a multivariate normal distribution. *Annals of Statistics*, **9**, 1135–1151.

Stone, C. J. (1977). Consistent nonparametric regression (with discussion). *The Annals of Statistics*, **5**, 595–645.

Tierney, L. (1990). *LISP-STAT*. New York: Wiley.

Tufte, E. R. (1983). *The visual display of quantitative information*. Cheshire, CT: Graphics Press.

Tukey, J. W. (1949). One degree of freedom for nonadditivity. *Biometrics*, **5**, 232–242.

Tukey, J. W. (1990). Data-based graphics: Visual display in the decades to come. *Statistical Science*, **5**, 327–339.

Tukey, J. W. and P. A. Tukey (1983). Some graphics for studying four-dimensional data. In *Computer Science and Statistics: Proceedings of the 14th Symposium on the Interface*, **14**, 60–66.

Velilla, S. (1993). A note on multivariate Box–Cox transformations to normality. *Statistics and Probability Letters*, **17**, 315–322.

Velilla, S. (1995). Diagnostics and robust estimation in multivariate data transformations. *Journal of the American Statistical Association*, **90**, 945–951.

Waite, W. C. (1941). Place of, and limitations to the method. *Journal of Farm Economics*, **23**, 311–323.

Wand, M. P. and M. C. Jones (1995). *Kernel Smoothing*. London: Chapman and Hall.

Wang, P. C. (1985). Adding a variable in generalized linear models. *Technometrics*, **27**, 273–276.

Weisberg, S. (1985). *Applied Linear Regression* (2nd ed.). New York: Wiley.

Wellman, H. R. (1941). Application and uses of the graphic method of correlation. *Journal of Farm Economics*, **23**, 311–323.

Wetzel, N. (1996). Graphical modeling methods using CERES plots. *Journal of Statistical Computation and Simulation*, **54**, 37–44.

Wetzel, N. (1997). Interactive modeling methods for regression. In Y. Dodge (ed.), *L1-Statistical Procedures and Related Topics: IMS Lecture Notes–Monograph Series, Vol. 31*, pp. 279–286. Hayward, CA: Institute of Mathematical Statistics.

Whittaker, J. (1990). *Graphical Models in Applied Multivariate Statistics*. New York: Wiley.

Wood, A. (1989). An F-approximation to the distribution of a linear combination of chi-squared random variables. *Communication in Statistics, Part B—Simulation and Computation*, **18**, 1439–1456.

Wood, F. S. (1973). The use of individual effects and residuals in fitting equations to data. *Technometrics*, **15**, 677–695.

Zhu, L.-X. and K.-T. Fang (1996). Asymptotics for kernel estimate of sliced inverse regression. *Annals of Statistics*, **24**, 1053–1068.

Zhu, L.-X. and K. W. Ng (1995). Asymptotics of sliced inverse regression. *Statistica Sinica*, **5**, 727–736.

Author Index

Subject Index

WILEY SERIES IN PROBABILITY AND STATISTICS

ESTABLISHED BY WALTER A. SHEWHART AND SAMUEL S. WILKS

Editors
*Vic Barnett, Ralph A. Bradley, Noel A. C. Cressie, Nicholas I. Fisher,
Iain M. Johnstone, J. B. Kadane, David G. Kendall, David W. Scott,
Bernard W. Silverman, Adrian F. M. Smith, Jozef L. Teugels;
J. Stuart Hunter, Emeritus*

Probability and Statistics Section

*Now available in a lower priced paperback edition in the Wiley Classics Library.

*Now available in a lower priced paperback edition in the Wiley Classics Library.

*Now available in a lower priced paperback edition in the Wiley Classics Library.

*Now available in a lower priced paperback edition in the Wiley Classics Library.

*Now available in a lower priced paperback edition in the Wiley Classics Library.

Texts and References Section

AGRESTI · An Introduction to Categorical Data Analysis

ANDERSON · An Introduction to Multivariate Statistical Analysis, *Second Edition*

ANDERSON and LOYNES · The Teaching of Practical Statistics

ARMITAGE and COLTON · Encyclopedia of Biostatistics: Volumes 1 to 6 with Index

BARTOSZYNSKI and NIEWIADOMSKA-BUGAJ · Probability and Statistical Inference

BERRY, CHALONER, and GEWEKE · Bayesian Analysis in Statistics and Econometrics: Essays in Honor of Arnold Zellner

BHATTACHARYA and JOHNSON · Statistical Concepts and Methods

BILLINGSLEY · Probability and Measure, *Second Edition*

BOX · R. A. Fisher, the Life of a Scientist

BOX, HUNTER, and HUNTER · Statistics for Experimenters: An Introduction to Design, Data Analysis, and Model Building

BOX and LUCEÑO · Statistical Control by Monitoring and Feedback Adjustment

BROWN and HOLLANDER · Statistics: A Biomedical Introduction

CHATTERJEE and PRICE · Regression Analysis by Example, *Second Edition*

COOK and WEISBERG · An Introduction to Regression Graphics

COX · A Handbook of Introductory Statistical Methods

DILLON and GOLDSTEIN · Multivariate Analysis: Methods and Applications

DODGE and ROMIG · Sampling Inspection Tables, *Second Edition*

DRAPER and SMITH · Applied Regression Analysis, *Third Edition*

DUDEWICZ and MISHRA · Modern Mathematical Statistics

DUNN · Basic Statistics: A Primer for the Biomedical Sciences, *Second Edition*

FISHER and VAN BELLE · Biostatistics: A Methodology for the Health Sciences

FREEMAN and SMITH · Aspects of Uncertainty: A Tribute to D. V. Lindley

GROSS and HARRIS · Fundamentals of Queueing Theory, *Third Edition*

HALD · A History of Probability and Statistics and their Applications Before 1750

HALD · A History of Mathematical Statistics from 1750 to 1930

HELLER · MACSYMA for Statisticians

HOEL · Introduction to Mathematical Statistics, *Fifth Edition*

JOHNSON and BALAKRISHNAN · Advances in the Theory and Practice of Statistics: A Volume in Honor of Samuel Kotz

JOHNSON and KOTZ (editors) · Leading Personalities in Statistical Sciences: From the Seventeenth Century to the Present

JUDGE, GRIFFITHS, HILL, LÜTKEPOHL, and LEE · The Theory and Practice of Econometrics, *Second Edition*

KHURI · Advanced Calculus with Applications in Statistics

KOTZ and JOHNSON (editors) · Encyclopedia of Statistical Sciences: Volumes 1 to 9 wtih Index

KOTZ and JOHNSON (editors) · Encyclopedia of Statistical Sciences: Supplement Volume

KOTZ, REED, and BANKS (editors) · Encyclopedia of Statistical Sciences: Update Volume 1

KOTZ, REED, and BANKS (editors) · Encyclopedia of Statistical Sciences: Update Volume 2

LAMPERTI · Probability: A Survey of the Mathematical Theory, *Second Edition*

LARSON · Introduction to Probability Theory and Statistical Inference, *Third Edition*

LE · Applied Categorical Data Analysis

LE · Applied Survival Analysis

MALLOWS · Design, Data, and Analysis by Some Friends of Cuthbert Daniel

MARDIA · The Art of Statistical Science: A Tribute to G. S. Watson

MASON, GUNST, and HESS · Statistical Design and Analysis of Experiments with Applications to Engineering and Science

*Now available in a lower priced paperback edition in the Wiley Classics Library.

Texts and References (Continued)

MURRAY · X-STAT 2.0 Statistical Experimentation, Design Data Analysis, and Nonlinear Optimization

PURI, VILAPLANA, and WERTZ · New Perspectives in Theoretical and Applied Statistics

RENCHER · Methods of Multivariate Analysis

RENCHER · Multivariate Statistical Inference with Applications

ROSS · Introduction to Probability and Statistics for Engineers and Scientists

ROHATGI · An Introduction to Probability Theory and Mathematical Statistics

RYAN · Modern Regression Methods

SCHOTT · Matrix Analysis for Statistics

SEARLE · Matrix Algebra Useful for Statistics

STYAN · The Collected Papers of T. W. Anderson: 1943–1985

TIERNEY · LISP-STAT: An Object-Oriented Environment for Statistical Computing and Dynamic Graphics

WONNACOTT and WONNACOTT · Econometrics, *Second Edition*

WILEY SERIES IN PROBABILITY AND STATISTICS

ESTABLISHED BY WALTER A. SHEWHART AND SAMUEL S. WILKS

Editors
Robert M. Groves, Graham Kalton, J. N. K. Rao, Norbert Schwarz, Christopher Skinner

Survey Methodology Section

BIEMER, GROVES, LYBERG, MATHIOWETZ, and SUDMAN · Measurement Errors in Surveys

COCHRAN · Sampling Techniques, *Third Edition*

COX, BINDER, CHINNAPPA, CHRISTIANSON, COLLEDGE, and KOTT (editors) · Business Survey Methods

*DEMING · Sample Design in Business Research

DILLMAN · Mail and Telephone Surveys: The Total Design Method

GROVES and COUPER · Nonresponse in Household Interview Surveys

GROVES · Survey Errors and Survey Costs

GROVES, BIEMER, LYBERG, MASSEY, NICHOLLS, and WAKSBERG · Telephone Survey Methodology

*HANSEN, HURWITZ, and MADOW · Sample Survey Methods and Theory, Volume 1: Methods and Applications

*HANSEN, HURWITZ, and MADOW · Sample Survey Methods and Theory, Volume II: Theory

KASPRZYK, DUNCAN, KALTON, and SINGH · Panel Surveys

KISH · Statistical Design for Research

*KISH · Survey Sampling

LESSLER and KALSBEEK · Nonsampling Error in Surveys

LEVY and LEMESHOW · Sampling of Populations: Methods and Applications

LYBERG, BIEMER, COLLINS, de LEEUW, DIPPO, SCHWARZ, TREWIN (editors) · Survey Measurement and Process Quality

SKINNER, HOLT, and SMITH · Analysis of Complex Surveys

*Now available in a lower priced paperback edition in the Wiley Classics Library.

Cardiovascular Magnetic Resonance

Cardiovascular Magnetic Resonance

Warren J. Manning, MD

Section Chief, Non-invasive Cardiac Imaging
Beth Israel Deaconess Medical Center
Associate Professor of Medicine and Radiology
Harvard Medical School
Boston, Massachusetts

Dudley J. Pennell, MD

Professor of Cardiology
National Heart and Lung Institute
Imperial College
Clinical Director, CMR Unit
Royal Brompton Hospital
London, United Kingdom

CHURCHILL LIVINGSTONE

An Imprint of Elsevier Science
New York, Edinburgh, London, Philadelphia

CHURCHILL LIVINGSTONE
An Imprint of Elsevier Science

The Curtis Center
Independence Square West
Philadelphia, Pennsylvania 19106

Library of Congress Cataloging-in-Publication Data

Cardiovascular magnetic resonance/[edited by] Warren J. Manning,
Dudley J. Pennell.

p.   cm.

Includes bibliographical references and index.

ISBN 0–443–07519–0

1. Heart—Magnetic resonance imaging.   I. Manning, Warren J.
II. Pennell, Dudley J.
[DNLM: 1. Cardiovascular Diseases—diagnosis.   2. Coronary Angiography.
3. Heart Diseases—diagnosis.   4. Magnetic Resonance Imaging.
WG 141.R2 C267 2002]

RC683.5.M35 C37 2002   616.1'07548—dc21   2001017367

Acquisitions Editor: Stephanie Donley
Production Editor: Robin E. Davis
Production Manager: Mary Stermel
Illustration Specialist: Rita Martello
Book Designer: Steven Stave

CARDIOVASCULAR MAGNETIC RESONANCE   ISBN 0–443–07519–0

Churchill Livingstone and the Sail Boat Design are trademarks of Reed-Elsevier, Inc., registered in the United States of America and/or other jurisdictions.

Printed in the United States of America.

Last digit is the print number: 9   8   7   6   5   4   3

To the joys of my life—
Susan Gail, Anya, Sara, and Isaac
—WJM

To my parents Terence and Joan
for enduring support and pride in their autumn;
And my wife Elisabeth
for love and understanding in our spring
—DP

Contributors

Leon Axel, PhD, MD
Professor of Radiology and Medicine, Director of Cardiac Imaging, New York University Medical Center, New York, New York
Tagged Cardiovascular Magnetic Resonance of Systole

Frank M. Baer, MD
Assistant Professor of Medicine and Cardiology, University of Cologne School of Medicine and Dentistry; Attending Physician, Department of Internal Medicine III (Cardiology, Pneumonology, Angiology and Internal Critical Care Medicine), Cologne, Germany Medizinische Kliniken der Universitaet zu Koeln, Cologne, Germany
Myocardial Viability

Robert S. Balaban, PhD
Scientific Director, Laboratory Research Program, National Heart, Lung, and Blood Institute, National Institutes of Health, Bethesda, Maryland
The Physics of Image Generation by Magnetic Resonance

Nicholas G. Bellenger, MB, MRCP
Cardiology Specialist Registrar, CMR Unit, Royal Brompton Hospital, London, England
Assessment of Cardiac Function

René M. Botnar, PhD
Clinical Scientist, Beth Israel Deaconess Medical Center, Cardiology, Cardiac MR Center, Boston, Massachusetts
Coronary Magnetic Resonance Angiography— Methods

Paul A. Bottomley, PhD
Russell H. Morgan Professor of Radiology, Director of MR Research, Department of Radiology and Radiological Science, Johns Hopkins University School of Medicine, Baltimore, Maryland
Clinical Cardiac Magnetic Resonance Spectroscopy

Jens Bremerich, MD
Department of Radiology, Basel University Hospital, Basel, Switzerland
Cardiovascular Magnetic Resonance of Complex Congenital Heart Disease in the Adult

Michael Chwialkowski, PhD
Associate Professor of Radiology, University of Texas Southwestern Medical Center, Dallas, Texas
Normal Cardiac Anatomy, Orientation, and Function

Geoffrey D. Clarke, PhD
Associate Professor of Radiology, Radiological Sciences Division, University of Texas Health Science Center at San Antonio, San Antonio, Texas
Normal Cardiac Anatomy, Orientation, and Function

Peter G. Danias, MD, PhD
Assistant Professor of Medicine, Harvard Medical School; Director, Nuclear Cardiology, Beth Israel Deaconess Medical Center, Boston, Massachusetts
Coronary Magnetic Resonance Angiography— Methods; Coronary Magnetic Resonance Angiography—Clinical Results

Jörg F. Debatin, MD, MBA
Professor of Radiology, Chairman, University Hospital Essen, Institute of Diagnostic Radiology, Essen, Germany
Interventional Cardiovascular Magnetic Resonance

Albert de Roos, MD
Professor of Radiology, Leiden University Medical Center, Leiden, The Netherlands
Cardiovascular Magnetic Resonance of Simple Cardiovascular Defects

Robert R. Edelman, MD
Chairman of Radiology, Evanston Northwestern Healthcare; Professor of Radiology, Northwestern University School of Medicine, Chicago, Illinois
Magnetic Resonance Angiography: Aorta and Peripheral Vessels

William T. Evanochko, PhD
Associate Professor of Medicine, Center for NMR Research and Development, University of Alabama at Birmingham, Birmingham, Alabama
Cardiac Transplantation

David Firmin, MPhil, PhD
Reader in the Physics of Cardiovascular Magnetic Resonance, Imperial College of Science, Technology, and Medicine, London, United Kingdom
Blood Flow Velocity Assessment; The Use of Navigator Echoes in Cardiovascular Magnetic Resonance and Factors Affecting Their Implementation

Fatima Franco, MD
Cardiologist, Coimbra University Hospital, Coimbra, Portugal
Normal Cardiac Anatomy, Orientation, and Function

Herbert Frank, MD
Professor of Medicine, Director, Cardiac MRI, University of Vienna, Department of Cardiology, Vienna, Austria
Cardiac and Paracardiac Masses

Matthias G. Friedrich, MD
Head, Cardiovascular MR Working Group, Humboldt University Berlin, Berlin, Germany
Cardiovascular Magnetic Resonance in Cardiomyopathies

James W. Goldfarb, PhD
Assistant Professor of Medicine, University of Medicine and Dentistry of New Jersey; Director of Cardiovascular Magnetic Resonance Research, Robert Wood Johnson University Hospital, New Brunswick, New Jersey
Magnetic Resonance Angiography: Aorta and Peripheral Vessels

Maarten Groenink, MD
Department of Cardiology, Academic Medical Center, Amsterdam, The Netherlands
Cardiovascular Magnetic Resonance of Simple Congenital Cardiovascular Defects

Robert J. Gropler, MD
Associate Professor of Radiology, Medicine, and Biomedical Engineering, Washington University in St. Louis School of Medicine, St. Louis, Missouri
Magnetic Resonance Assessment of Myocardial Oxygenation

Willem A. Helbing, MD
Professor of Paediatric Cardiology, Erasmus University, Director, Department of Paediatric Cardiology, University Hospital Rotterdam, Sophia Children's Hospital, Rotterdam, The Netherlands; Leiden University Medical Center, Leiden, The Netherlands
Cardiovascular Magnetic Resonance of Simple Congenital Cardiovascular Defects

Charles B. Higgins M.D.
Professor of Radiology University of California, San Francisco Medical School; UCSF Medical Center, San Francisco, California
Cardiovascular Magnetic Resonance of Complex Congenital Heart Disease in the Adult

Paul R. Hilfiker, MD
Faculty, Institute of Diagnostic Radiology, University Hospital Zurich, Zurich, Switzerland
Interventional Cardiovascular Magnetic Resonance

Agnes E. Holland, MD
Fellow, Department of Radiology, New York University Medical Center, New York, New York
Magnetic Resonance Angiography: Aorta and Peripheral Vessels

Michael Jerosch-Herold, PhD
Assistant Professor of Radiology, University of Minnesota Medical School, Minneapolis, Minnesota
Myocardial Perfusion—Theory

Anastazia Jerzewski, MD, PhD
Fellow in Training in Cardiology, Leiden University Medical Center, Leiden, The Netherlands
Special Considerations for Cardiovascular Magnetic Resonance: Safety, Electrocardiographic Set-Up, Monitoring, Contraindications

Roberto Kalil-Filho, MD, FACC
Associate Professor of Cardiology, São Paulo Heart Institute, University of São Paulo School of Medicine, São Paulo, Brazil
Clinical Cardiac Magnetic Resonance Spectroscopy

Jennifer Keegan, PhD
Physicist, CMR Unit, Royal Brompton Hospital, London, England
The Use of Navigator Echoes in Cardiovascular Magnetic Resonance and Factors Affecting Their Implementation; Coronary Artery and Coronary Sinus Velocity and Flow

Jan T. Keijer, MD
Cardiologist-in-Training, Free University Hospital, Amsterdam, The Netherlands
Stress CMR—Clinical Myocardial Perfusion Imaging

Philip J. Kilner, MD, PhD
Consultant in Cardiovascular Magnetic Resonance, CMR Unit, Royal Brompton Hospital, London, United Kingdom
Valvular Heart Disease

Kraig V. Kissinger, BS, RT(R)(MR)
Senior MR Technologist, Cardiac MR Center, Beth Israel Deaconess Medical Center, Boston, Massachusetts
Coronary Magnetic Resonance Angiography—Methods

Malgorzata Knap, MD
Instructor of Cardiology, University of Rostock; University Hospital Rostock, Rostock, Germany
Thoracic Aortic Disease

Christopher M. Kramer, MD
Associate Professor of Medicine and Radiology, Director, Cardiac MRI, University of Virginia School of Medicine, Charlottesville, Virginia
Myocardial Infarction—Remodeling

Debiao Li, PhD
Associate Professor of Radiology and Biomedical Engineering, Northwestern University, Chicago, Illinois
Magnetic Resonance Assessment of Myocardial Oxygenation

João A. C. Lima, MD, FACC
Associate Professor of Medicine and Radiology, Johns Hopkins University School of Medicine, Director of Echocardiology and Cardiac MRI, Johns Hopkins Hospital, Baltimore, Maryland
Acute Myocardial Infarction

Christine H. Lorenz, PhD
MR Scientist, Siemens Medical Solutions, Erlangen Germany
Right Ventricular Anatomy and Function in Health and Disease

Warren J. Manning, MD
Section Chief, Non-invasive Cardiac Imaging, Beth Israel Deaconess Medical Center; Associate Professor of Medicine and Radiology, Harvard Medical School, Boston, Massachusetts
Coronary Magnetic Resonance Angiography—Methods; Coronary Magnetic Resonance Angiography—Clinical Results; Coronary Magnetic Resonance Angiography for Suspected Anomalous Coronary Artery Disease

Roderick W. McColl, PhD
Assistant Professor, University of Texas Southwestern Medical Center, Dallas, Texas
Normal Cardiac Anatomy, Orientation, and Function

Michael V. McConnell, MD, MSEE
Assistant Professor of Medicine, Clinical Director, Cardiovascular MRI, Division of Cardiovascular Medicine, Stanford University School of Medicine, Stanford, California
Coronary Magnetic Resonance Angiography for Suspected Anomalous Coronary Artery Disease

Raad H. Mohiaddin, MD, PhD, MRCP, FRCR, FESC
Consultant in CMR, Royal Brompton Hospital, CMR Unit, London, United Kingdom
Assessment of the Biophysical Mechanical Properties of the Arterial Wall; Valvular Heart Disease

Christoph A. Nienaber, MD
Professor of Medicine and Cardiology, Department of Internal Medicine, University of Rostock; Division Head, Cardiology, University Hospital Rostock, Rostock, Germany
Thoracic Aortic Disease

R. André Niezen, MD
Department of Radiology, Leiden University Medical Center, Leiden, The Netherlands
Cardiovascular Magnetic Resonance of Simple Congenital Cardiovascular Defects

William F. Oellerich, MD
Fellow in Training in Cardiology, Washington University School of Medicine, St. Louis, Missouri
Magnetic Resonance Assessment of Myocardial Oxygenation

Robert W. Parkey, MD
Professor of Radiology, University of Texas Southwestern Medical Center, Dallas, Texas
Normal Cardiac Anatomy, Orientation, and Function

Dudley J. Pennell, MD, FRCP, FACC, FESC
Professor of Cardiology, National Heart and Lung Institute, Imperial College of Science, Technology, and Medicine; Clinical Director, CMR Unit, Honorary Consultant, Royal Brompton Hospital, London, United Kingdom
Assessment of Cardiac Function; Stress CMR—Wall Motion; Coronary Artery and Coronary Sinus Velocity and Flow

Ronald M. Peshock, MD
Professor of Radiology and Internal Medicine, University of Texas Southwestern Medical Center, Dallas, Texas
Normal Cardiac Anatomy, Orientation, and Function

Gerald M. Pohost, MD
Professor of Cardiovascular Medicine, Director of the Center for NMR Research and Development, The University of Alabama at Birmingham, Birmingham, Alabama
Cardiac Transplantation

Carlos E. Rochitte, MD
Physician Investigator, University of São Paulo School of Medicine; Director of Cardiovascular MRI, São Paulo Heart Institute, São Paulo, Brazil
Acute Myocardial Infarction

Arno A. W. Roest, MSc
Research Fellow, Departments of Radiology, Cardiology and Pediatric Cardiology, Leiden University Medical Center, Leiden, The Netherlands; Research Fellow, Interuniversity Cardiology Institute of the Netherlands, Utrecht, The Netherlands
Cardiovascular Magnetic Resonance of Simple Congenital Cardiovascular Defects

Roxann Rokey, MD
Clinical Professor of Medicine, University of Wisconsin-Madison Medical School, Director, Adult Congenital Heart Disease, Department of Cardiology, Marshfield Clinic, Marshfield, Wisconsin
Cardiovascular Magnetic Resonance Evaluation of the Pericardium in Health and Disease

Udo P. Sechtem, MD
Associate Professor of Medicine and Cardiology, University of Tübingen, Germany; Head, Department of Cardiology and Pneumonology, Robert-Bosch-Krankenhaus, Stuttgart, Germany
Myocardial Viability

Daniel K. Sodickson, MD, PhD
Assistant Professor of Medicine, Harvard Medical School; Director, Laboratory for Biomedical Imaging Research, Beth Israel Deaconess Medical Center, Boston, Massachusetts
Clinical Cardiovascular Magnetic Resonance Imaging Techniques

Matthias Stuber, PhD
Senior Scientist, Cardiac MR Center, Beth Israel Deaconess Medical Center, Harvard Medical School, Boston, Massachusetts; Senior Clinical Scientist, Philips Medical Systems, Best, The Netherlands
CSPAMM Assessment of Left Ventricular Diastolic Function; Coronary Magnetic Resonance Angiography—Methods; Coronary Magnetic Resonance Angiography—Clinical Results

Jean-François Toussaint, MD, PhD
Associate Professor of Physiology, Faculté de Médecine Broussais Hotel-Dieu, Université Paris VI; Cardiologist, Service de Physiologie, Hôpital Européen Georges Pompidou, Paris, France
Atherosclerotic Plaque Imaging

Ernst E. van der Wall, MD, PhD
Professor of Cardiology, Leiden University Medical Center, Leiden, The Netherlands
Special Considerations for Cardiovascular Magnetic Resonance: Safety, Electrocardiographic Set-Up, Monitoring, Contraindications; Cardiovascular Magnetic Resonance of Simple Congenital Cardiovascular Defects

Albert C. van Rossum, MD, PhD
Professor of Cardiology, VU University Medical Centre (VUMC), Amsterdam, The Netherlands
Stress CMR—Clinical Myocardial Perfusion Imaging; Coronary Artery Bypass Graft Imaging and Assessment of Flow

G. Wesley Vick III, MD, PhD
Assistant Professor of Cardiology and Atherosclerosis, Departments of Pediatrics and Medicine, Baylor College of Medicine; Associate in Pediatric Cardiology, Texas Children's Hospital, Houston, Texas
Cardiovascular Magnetic Resonance Evaluation of the Pericardium in Health and Disease

Gustav K. von Schulthess, MD, PhD
Chair, Department of Nuclear Medicine, University Hospital Zurich, Zurich, Switzerland
Interventional Cardiovascular Magnetic Resonance

Eberhard Voth, MD
Lecturer in Nuclear Medicine, University of Cologne School of Medicine and Dentistry; Attending Physicians, Department of Nuclear Medicine, Medizinische Kliniken der Universitaet zu Koeln, Cologne, Germany
Myocardial Viability

Robert G. Weiss, MD
Associate Professor of Medicine and Radiology, Johns Hopkins University School of Medicine; Attending Physician, Cardiology Division, Johns Hopkins Hospital, Baltimore, Maryland
Clinical Cardiac Magnetic Resonance Spectroscopy

Simon Wildermuth, MD
Faculty, Institute of Diagnostic Radiology, University Hospital Zurich, Zurich Switzerland
Interventional Cardiovascular Magnetic Resonance

Norbert M. Wilke, MD
Assistant Professor of Radiology, University of Minnesota Medical School, Minneapolis, Minnesota
Myocardial Perfusion—Theory

Katherine C. Wu, MD
Assistant Professor of Medicine, Johns Hopkins University School of Medicine, Baltimore, Maryland
Acute Myocardial Infarction

Rolf Wyttenbach, MD
Chief, Magnetic Resonance Unit, Department of Radiology, Ospedale San Giovanni, Bellinzona, Switzerland
Cardiovascular Magnetic Resonance of Complex Congenital Heart Disease in the Adult

Foreword

Since William Harvey's discovery of the circulation, the assessment of cardiac structure and function has been a major goal of physicians responsible for the management of patients with heart disease. The first techniques were quite primitive—palpation of the arterial pulse, observation and direct auscultation of the precordium—followed by René Laënnec's development of the stethoscope, the first tool to amplify the senses. A paradigm shift occurred with the discovery of the x-ray by Wilhelm Roentgen at the end of the 19th century, allowing the heart to be imaged for the first time.

The 20th century can be considered the century of cardiac imaging, beginning with angiography and arteriography, both requiring the injection of contrast material through a catheter. The development and progressive refinement of noninvasive approaches—echocardiography, radionuclide imaging, and computed tomography—advanced cardiology enormously. Cardiovascular magnetic resonance imaging, however, may finally fulfill the dreams of cardiologists. Ths technique offers a veritable treasure trove of information about the heart and circulation. It not only provides superior displays of chamber structure but allows the identification of myocardial necrosis, infiltration, and fibrosis. Both valvular performance and ventricular function can be measured. Magnetic resonance angiography allows assessment of vascular function and tissue perfusion. Recently, both the walls and luminae of the major coronary arteries have been visualized. When combined with the ability to determine myocardial viability and to measure cardiac high-energy phosphate stores noninvasively by magnetic resonance spectroscopy, these techniques seem certain to improve enormously the care of patients with ischemic heart disease. It is no longer fanciful to think that coronary arteriography, now performed on more than 1.5 million patients in the United States each year, will become obsolete.

Just as it became necessary for physicians entrusted with the care of cardiac patients to become familiar with cardiac roentgenography at the beginning of the 20th century, it now has become vital for their successors to do the same with cardiac magnetic resonance imaging. Warren J. Manning and Dudley J. Pennell and their talented contributors to *Cardiovascular Magnetic Resonance* have taken a significant step toward enabling clinicians to accomplish this. They deserve the appreciation of cardiovascular specialists of all types for producing this detailed, authoritative, yet eminently readable book. A century ago, great strides were being made in cardiac roentgenography and the future promise was enormous. The situation is similar with cardiac magnetic resonance imaging today. This field is advancing at such a breathtaking speed that I can't wait for the second edition.

Eugene Braunwald, MD
Boston, Massachusetts

Preface

Cardiovascular Magnetic Resonance (CMR) is a new medical imaging field that has received tremendous interest because it combines unsurpassed image quality of the heart with new techniques that allow for probing of the cardiovascular system in novel ways. What surprises is the versatility of the high technology: blood flow, angiography, assessment of atherosclerosis, myocardial perfusion, focal necrosis, oxygen saturation, temperature, and chemical composition are among the measurements that are being refined for clinical use in addition to the well-known capabilities of CMR in defining anatomy and ventricular function. Such multiple potential, however, comes at a price. This technology is not learned and ably practiced in a trice. Professional didactic and clinical training is required for all newcomers to the field. The aim of this book is to provide instruction in the current clinical practice of CMR, while also highlighting areas of clinical potential that are presently in varying stages of development. If we succeed in drawing new investigators and clinicians to enter the field, or illuminating new areas for those already involved, then we will have achieved our objective: the healthy growth of competent and motivated practitioners in CMR for the benefit of clinical science, patient care, and the advancement of the field. The reader should be forewarned that CMR is a rapidly developing field. Thus, by the nature of the production process, no text can include the most recent developments, but the foundations provided will serve the reader for many years to come.

Contents

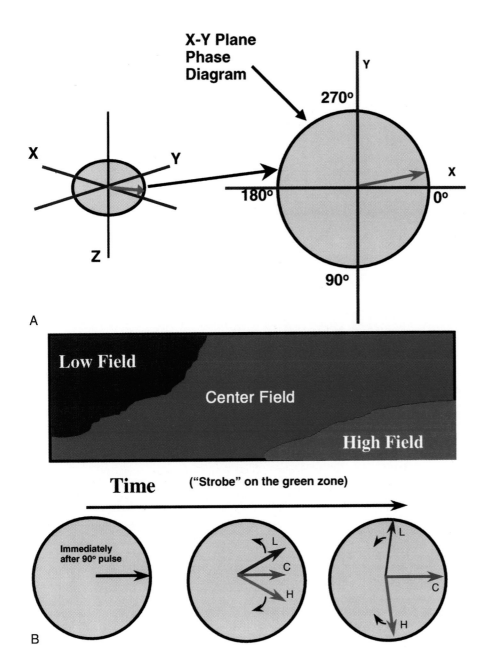

Figure 1–3. The phase diagram. *A*, A phase diagram is a view of the magnetization vectors in the X-Y plane. *B*, The magnetic field is variable in the sample shown at the top. Sample spins from the low field (L), center field (C), and high field (H) region are followed as a function of time after a 90° pulse using the phase diagram. The phase diagram is set up to present the phases of the spins relative to C, which is always normalized to the 0° position. The phase diagrams at the bottom are presented as a function of time after the B_1 field pulse. Note that the L vector lags behind the C vector whereas the H vector has a higher phase velocity than the other spins. This finding is due to the relative frequency of the spins caused by the differences in the magnetic field.

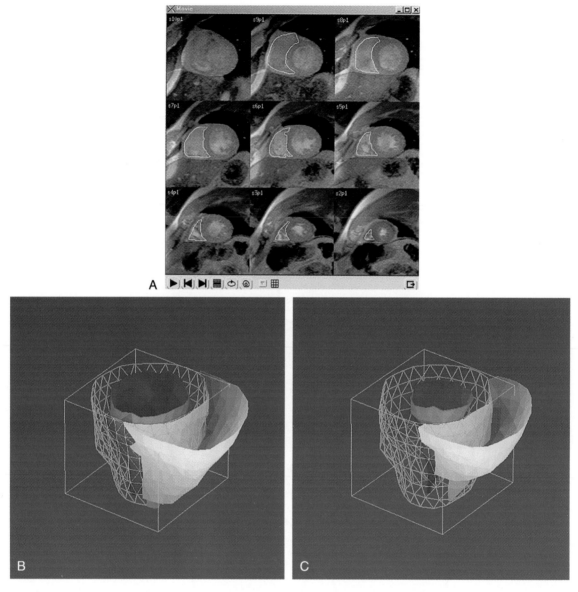

Figure 8–7. *A,* Right ventricular volume can be determined from the same set of short-axis views. The right ventricular chamber cross-sectional volume is determined for each slice and then summed over the length of the right ventricle. *B,* Three-dimensional reconstruction of the right ventricle from short-axis views at end diastole. The right ventricle is in yellow viewed from the base towards the apex. The left ventricular endocardium (red) and epicardium (green mesh) are shown for orientation. *C,* Three-dimensional reconstruction of the right ventricle from short-axis views at end systole.

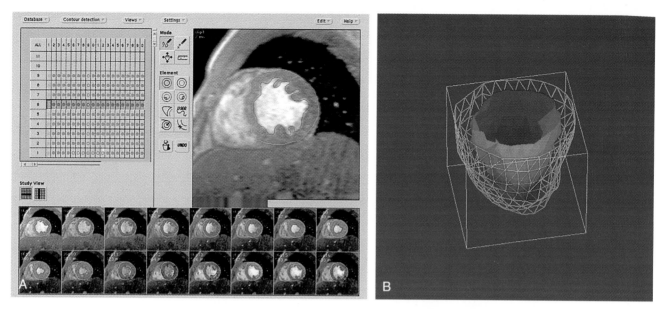

Figure 8–8. *A*, Left ventricular mass is estimated by determining the left ventricular wall volume in each slice, summing over all slices, and multiplying the total wall volume by the density of myocardium. *B*, Three-dimensional rendering of the endocardial outlines (red) and epicardial outlines (green mesh) used in the wall volume calculation to determine myocardial mass.

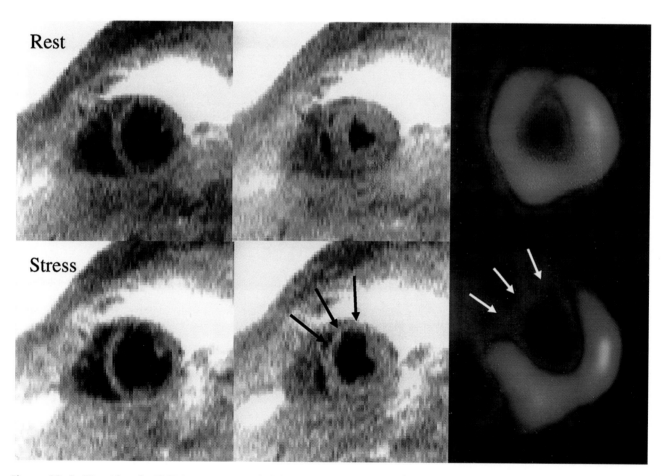

Figure 10–9. Dipyridamole CMR in a patient with left anterior descending coronary artery disease. In the top row are short-axis images before dipyridamole with postdipyridamole images below. End diastole is in the left column and end systole in the middle column. Left ventricular contraction is normal prior to vasodilation, but reduced in the anteroseptal region after dipyridamole (black arrows). The CMR abnormality is closely matched by the perfusion defect (white arrows on the color maps in the right column) seen during dipyridamole thallium myocardial perfusion tomography, which shows full reversibility. (From Pennell DJ, Underwood SR, Ell PJ, et al: Dipyridamole magnetic resonance imaging: A comparison with thallium-201 emission tomography. Br Heart J 64:362, 1990, with permission from the BMJ Publishing Group. Copyright © 1990 BMJ Publishing Group.)

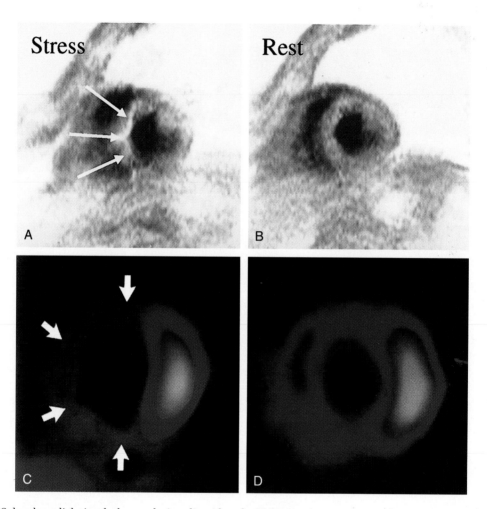

Figure 10–11. Subendocardial signal change during dipyridamole CMR seen in a patient with extensive septal ischemia (white arrows). The CMR images are at end diastole after dipyridamole *(A)* and before dipyridamole *(B)*. The gradient echo cine frames are displayed in reverse video format for easier appreciation of the myocardial signal, and the white subendocardial line therefore represents reduced signal on the original image. The reason for this finding remains unclear but might result from transmural shunting. The equivalent stress *(C)* and redistribution *(D)* thallium scans are shown on the bottom row, with the severe defect marked by white arrows on the stress image. The defect shows clear reversibility, which would have probably been greater with imaging at rest or with reinjection. The patient had no evidence of septal infarction.

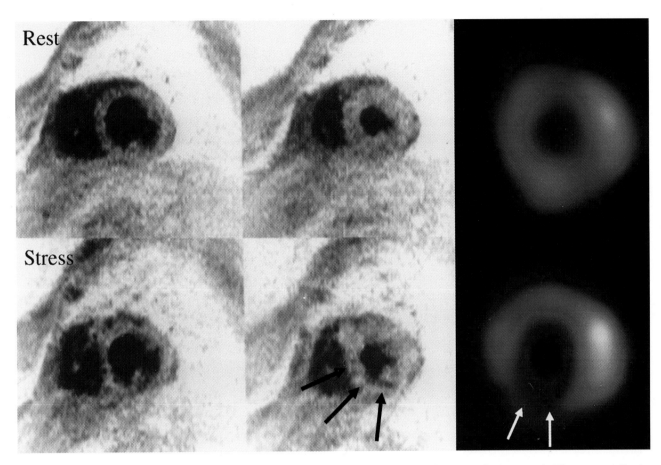

Figure 10–12. Dobutamine CMR of a patient with right coronary artery disease. The format is the same as in Figure 10–9. Resting contraction is normal but during dobutamine contraction is abnormal in the inferoseptal wall (black arrows), which matches the reversible perfusion defect (white arrows) seen during dobutamine thallium myocardial perfusion tomography. (From Pennell DJ, Underwood SR, Manzara CC, et al: Magnetic resonance imaging during dobutamine stress in coronary artery disease. Am J Cardiol 70:34, 1992, with permission from Excerpta Medica Inc.)

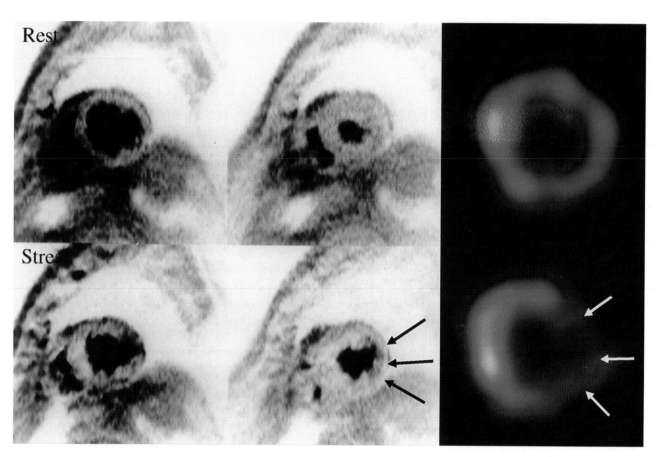

Figure 10–13. Dobutamine CMR of a patient with left circumflex artery disease. The format is the same as in Figure 10–9. Resting contraction is normal but during dobutamine contraction is abnormal in the lateral wall (black arrows), which matches the reversible perfusion defect (white arrows) seen during dobutamine thallium myocardial perfusion tomography. (From Pennell DJ, Underwood SR, Manzara CC, et al: Magnetic resonance imaging during dobutamine stress in coronary artery disease. Am J Cardiol 70:34, 1992, with permission from Excerpta Medica Inc.)

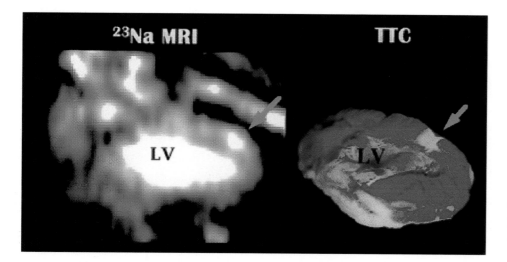

Figure 12–7. Sodium imaging in infarction. Long-axis ^{23}Na CMR image of a rabbit heart *(left)* and corresponding triphenyl tetrazolium chloride (TTC)-stained section *(right).* The nonviable, infarcted region has increased myocardial ^{23}Na image intensity (arrows). (From Kim RJ, Lima JAC, Chen EL, et al: Fast ^{23}Na magnetic resonance imaging of acute reperfused myocardial infarction: Potential to assess myocardial viability. Circulation 95:1877, 1997.)

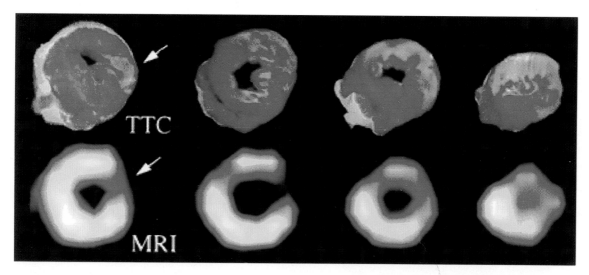

Figure 12–8. Potassium imaging in infarction. Sequential TTC-stained short-axis slices and corresponding ^{39}K CMR images (colorized) from a three-dimensional data set of a rabbit heart subjected to reperfused infarction. Histological sections and short-axis images are arranged as follows: left to right—base to apex, and lateral wall of the left ventricle is oriented to the right. Viable areas of myocardium stain brick red, whereas nonviable areas appear yellowish-white. Note the similarity of reduced ^{39}K image intensity and nonviable areas of myocardium. (From Fieno DS, Kim RJ, Rehwald WG, Judd RM: Physiological basis for potassium (^{39}K) magnetic resonance imaging of the heart. Circ Res 84:913, 1999.)

Figure 14–1. Wall thickness in acute and chronic infarction. *Left:* Left anterior descending (LAD) occlusion with 1-week-old anteroseptal infarct. Note that there is only minor wall thinning despite the fact that the infarct is transmural. The left ventricular (LV) cavity is filled with thrombus. *Right:* In contrast, there is extreme wall thinning in this chronic healed transmural infarct that occurred 6 months previously. Nitroblue tetrazolium stains viable cells blue. (From Braunwald E, Califf RM (eds): Atlas of Heart Diseases Vol. VIII: Acute Myocardial Infarction and Other Ischemic Syndromes. Philadelphia, Current Medicine, 1996, with permission.)

Figure 14–2. Preserved wall thickness in nontransmural infarction. Irrespective of age, nontransmural infarcts do not show severe wall thinning. This patient suffered an old nontransmural infarct in the inferolateral wall (short arrows) and a fresh anteroseptal nontransmural infarct (long arrows). Triphenyl tetrazolium chloride (TTC) fails to stain mature lateral scar and pale freshly infarcted muscle in anterior region. Viable myocardium is stained red. (From Braunwald E, Califf RM (eds): Atlas of Heart Diseases Vol. VIII: Acute Myocardial Infarction and Other Ischemic Syndromes. Philadelphia, Current Medicine, 1996, with permission.)

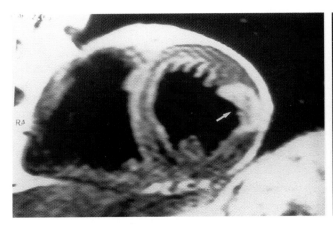

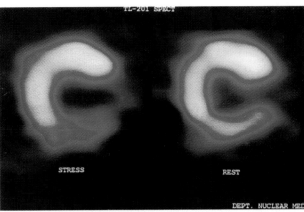

Figure 14–4. Small transmural myocardial infarct by CMR *(left)* and thallium single-photon emission computed tomography (SPECT) images *(right)*. The coronary angiogram showed an occluded left circumflex coronary artery, 75% stenosis of the LAD coronary artery, and a 90% stenosis of the right coronary artery. The ECG was normal. There is a well-defined focal area of high signal intensity at the lateral wall of the left ventricle (arrow) on this breath-hold fast spin-echo T2-weighted CMR. Note that there is some myocardium with normal signal intensity (viable myocardium) extending from the anterolateral region towards the infarct area near the epicardium. Such a viable border zone is often seen at necropsy. Thallium images reveal a corresponding fixed defect. However, structural details such as the epicardial rim of viable myocardium, which may be worth preserving by revascularizing the circumflex coronary artery, can only be seen on the high spatial resolution CMR. (From Lim TH, Hong M, Lee JS, et al: Novel application of breath-hold turbo spin-echo T2 MRI for detection of acute myocardial infarction. J Magn Reson Imag 7:996, 1997.)

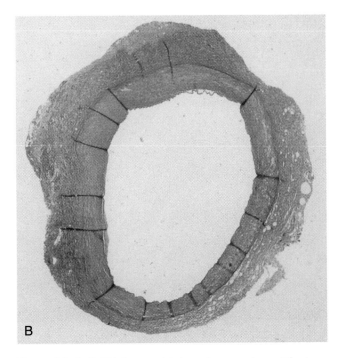

Figure 21–1. *B*, Trichrome staining of a normal carotid artery. See also Chapter 21, Figure 21–1*A*.

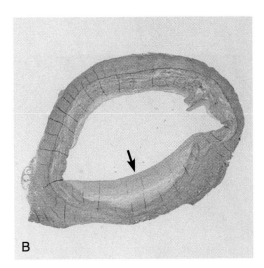

Figure 21–2. *B*, Trichrome staining showing the plaque (arrow) with the large fibrous cap and opposite intimal thickening. See also Chapter 21, Figure 21–2*A*.

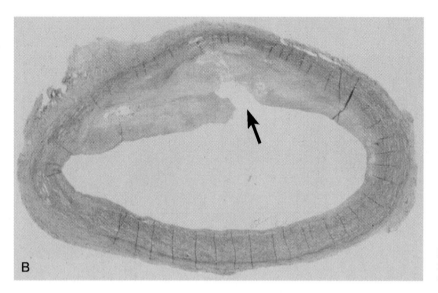

B

Figure 21-4. *B*, Trichrome staining showing the plaque with the rupture (arrow). See also Chapter 21, Figure 21-4*A*.

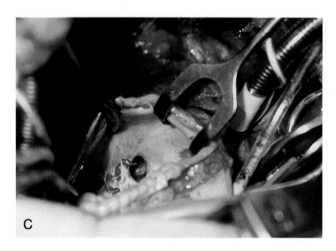

C

Figure 24-9. *C*, View from the right atrium into the right upper lung vein. See also Chapter 24, Figure 24-9*A* and *B*.

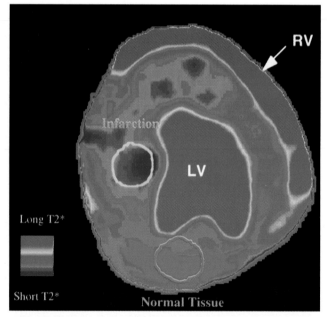

Figure 34-8. *A*, A calculated T2* image of a dog with an occluded left anterior descending artery (LAD) obtained after dipyridamole administration. Note the shorter T2* in the regions of the anterior wall and anterior papillary muscle that are supplied by the LAD than that in remote normal myocardial regions of the left ventricular wall. LV = left ventricle; RV = right ventricle. See also Chapter 34, Figure 34-8*B*.

BASIC PRINCIPLES OF CARDIOVASCULAR MAGNETIC RESONANCE

CHAPTER 1

The Physics of Image Generation by Magnetic Resonance*

Robert S. Balaban

Magnetic resonance imaging (MRI) is generally based on the detection of water and fat protons (¹H) in the body. Because these ¹H are in very high concentration in the body (~100 molar), even the weak nuclear magnetic resonance (NMR) signal can be used to create a distribution map, or image, of this nuclide. MRI depends on the detection of the intrinsic angular momentum or spin of protons, which is a basic property of matter. ¹H has one proton and no neutrons, giving it a net spin detectable in an MRI experiment. The frequency of the spin associated with a nuclide is related to the magnitude of the magnetic field (B_0) and its intrinsic gyromagnetic ratio, γ. ¹H has one of the highest γ values, making it a very effective NMR probe. When water containing the ¹H molecular magnets is placed in a large magnetic field (B_0), the ¹H magnets align with the applied magnetic field much like iron filings. In contrast to iron filings, the angular momentum of the ¹H nuclide results in a precession around the axis of the primary magnetic field at a frequency determined by γ. This rotation is why the nuclides in MRI are also referred to as "spins." The frequency of precession (v) is a fundamental property of the spins in a magnetic field and is defined by the *Larmor equation*:

---

*All material in this chapter is in the public domain, with the exception of any borrowed figures or tables.

$$v = \gamma B/2\pi \qquad (1)$$

where v is the precession frequency in cycles/s, and B is the applied magnetic field. The *Larmor equation* is the basis of MRI. In an MRI study, the magnetic field, B, varies slightly so as to be a linear function of position using specialized coils in the magnet, resulting in v being a linear function of position in the sample. Using this information, an image can be created,[1] as will be discussed later.

DETECTION OF THE NMR SIGNAL

How can the NMR signal from ¹H spins be detected? It is important to remember that all of water ¹H spins in a magnet are *not* rotating together around the axis of the magnetic field but are randomly distributed around this axis. This is illustrated in three dimensions in Figure 1–1. The net magnetic field is along the Z axis (parallel with B_0). The spins align along the Z axis, or longitudinal direction, like any magnet (see Figure 1–1A). However, the spins rotate around the Z axis in a completely random or incoherent fashion, resulting in no net signal to detect. A receiver coil placed to detect the precession in the X-Y plane (perpendicular to the main magnetic field) will not detect any signal because the spins are in different positions,

Figure 1–1. Orientation of molecular spins with main magnetic field (B_0). *A*, The spins orient with the main magnetic field, producing a net magnetic field aligned with B_0. *B*, The spins rotate around the B_0 field axis (Z axis) in a random form, resulting in no net oscillating field for detection with the coil oriented in the X-Y plane.

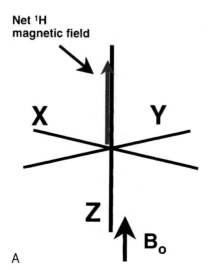
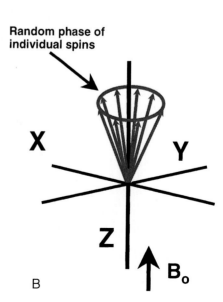

A   B

or phase, thereby canceling each other out, as seen in Figure 1–1B. The concept of phase will be discussed further later. Thus, if a person was placed in a magnet and a coil applied to his or her chest, no magnetic resonance (MR) signal would be detected from the ^{1}H in water or fat.

To detect an MR signal, we need to get all of water ^{1}H spins rotating together, in a coherent manner. To accomplish this task, a *weak* magnetic field (B_1) oscillating at the Larmor frequency is applied perpendicular to the main magnetic field, in this case along the Y axis (Figure 1–2). The frequency of the perpendicular field has to precisely match the resonance frequency of the water ^{1}H to efficiently rotate the spins independent of the stronger B_0 field. This process is analogous to exciting a piano string with a tuning fork. Only the string with the same

resonant frequency as the tuning fork will efficiently absorb the energy from the fork and resonate. The water (^{1}H) spins will rotate around the B_1 field following the right hand rule at a rate defined by the Larmor equation. Because the B_0 field in most cardiovascular scanners has a magnitude of 1.5 Tesla (15,000 gauss) whereas the excitation field is on the order of a few gauss, the rotational frequency around the B_1 field is much lower than around the B_0 field. In addition, the B_1 field is only applied transiently (milliseconds [ms]), long enough to rotate the spins into a plane perpendicular to B_0. This effect on the net magnetic field is illustrated in Figure 1–2A. The plane perpendicular to the B_0 field is called the transverse plane or X-Y plane. Magnetization in this plane is often referred to as transverse magnetization. A B_1 pulse that drives the magnetization com-

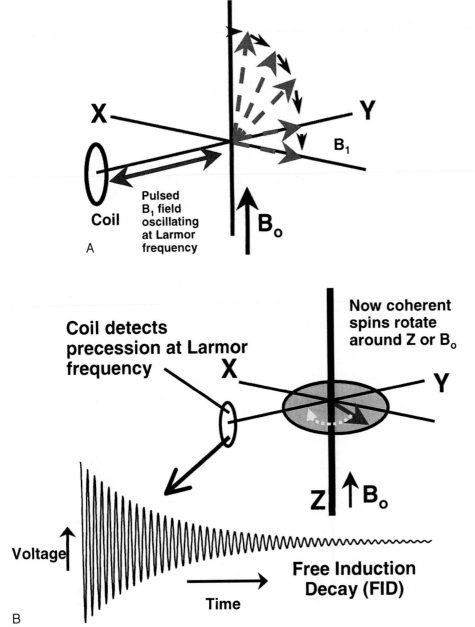

Figure 1–2. Effects of a perpendicular B_1 magnetic field. *A,* The oscillating B_1 field at the Larmor frequency of the spins is absorbed by the spins and causes a rotation around the B_1 field axis (Y axis). *B,* Once the B_1 field is removed, the spins continue to rotate around the Z axis at the Larmor frequency. The result is a coherent oscillating magnetic field that is detected with the coil as the free induction decay (FID) illustrated. The FID is a decaying sinusoidal voltage signal with a frequency equal to the Larmor frequency of the spins.

pletely into the transverse plane is called a 90°
pulse, referring to the angle the net ^{1}H magnetization
moved relative to the Z-axis. Once the B_1 field is
turned off, the only field causing a rotation is the
B_0. However, now the net magnetization of ^{1}H is in
the transverse plane and coherent for detection. As
the spins rotate around the B_0 field axis, an oscillating magnetic field can be detected with the coil
poised to detect the transverse magnetization as
shown in Figure 1–2B. Note that in this case the
same coil that excited the spins can be used to
subsequently detect the rotating transverse magnetization. Again, the frequency of this signal is defined
by the Larmor equation.

An example of the signal detected with a coil after
placing the spins in the transverse plane is also
shown in Figure 1–2B. The frequency of this signal
is usually determined using a Fourier transform
(FT), which converts time oscillating data into its
frequency components with only a few assumptions. The oscillating signal decays in amplitude
with time and is called a free induction decay (FID).
This decay in coherence and net magnitude in the
transverse plane is what is known as magnetic relaxation. The decay occurs because the spins absorb
the energy from the transient B_1 field, but when B_1
is turned "off," they must return to their original
state in equilibrium with the B_0 field, alone. That is,
the system must revert to completely random spins
aligned along the Z axis. This reversion occurs via
two processes. One is the release of the energy to
the environment, or lattice, which results in the
reestablishment of the magnetization along the Z
axis. This is basically a release of heat. This process
is known as spin-lattice relaxation and is called
the T1 or longitudinal relaxation rate. The T1 of
myocardium is ~880 ms at 1.5 Tesla (Table 1–1).

The second form of relaxation is that of the randomization of the phase of the spins. Phase is reviewed in Figure 1–3A and B and refers to the
relative position (not frequency) of the rotating net
magnetic field vector of the spins in the transverse
(X-Y) plane. A phase diagram is used to follow the
relative positions of the spin vectors. This type of
phase diagram looks directly down on the same

X-Y plane shown in Figure 1–1 showing the net
magnetization vector for each spin being examined
(Figure 1–3A). In Figure 1–3B a phase diagram is
used to show the changing phase relationships for
spins in a sample that has different magnetic fields
within it. The magnetic field strength increases from
blue to green to red (see color plates). This is actually a common problem in MRI that will be discussed further later. A phase diagram is a strobo-
scopic image because it is arbitrarily referenced to
one nuclide magnetic vector, in this case the spins
in the center region for Figure 1–3B. This strobo-
scopic effect is analogous to taking a flash picture
every time the rotating middle vector reaches the 0°
position. Using this approach the relative position
of the spin vectors, or phase, of the different nuclides can be easily seen. Immediately after a 90°
pulse, all of the spins have the same relative phase;
however, with time, the different spin frequencies
cause their relative phase to change. The higher B_0
field spins (red arrows) are advancing phase faster
than the green arrows, whereas the blue arrows at
low field are lagging behind both (see color plates).
By watching the time development of this process,
one can appreciate that the phase and the frequency
are related. The rate of change in phase is the frequency of rotation in the X-Y plane. High-frequency
spins change phase rapidly whereas slower frequency spins change phase more slowly. However,
the *initial starting points* of each of these signals are
independent of frequency. For example, after the
90° pulse, all of the spins have the same initial
phase (i.e., all start with the same phase at 0°). The
initial phase can be dependent on prior conditions
or how the spin is placed in the transverse plane.
Thus, despite the close relationship between phase
and frequency, there is one point in the FID where
the phase is independent of the frequency, the initiation point of the FID. Because all spins start at the
same point, this information is independent of the
frequency of the spin. We will find this property
useful in forming an MR image (discussed later).

Immediately after the 90° pulse, all of the spins
have the same phase. Because nature abhors order,
randomization is necessary to minimize the energy
in the system and return to equilibrium with the
main magnetic field, B_0. This randomization process
is a spin-spin or transverse relaxation and is termed
T2. It is called spin-spin relaxation because the
mechanism of the relaxation process is through the
interaction of spins in the sample with each other
below or at the Larmor frequency, making this
process dependent on the microscopic motions
within the sample. In a phase diagram the T2 process is shown as a decrease in the magnetic vector
amplitude as the spins lose coherence.

To understand many of the problems in cardiac
imaging, we need to further develop T2. In general,
three distinct processes contribute to the dephasing
of spins in the heart. The first mechanism is the true
spin-spin interaction (T2) that is unavoidable and
dependent on the molecular interactions within the

**Table 1–1. Approximate Values of Relaxation Times
at 1.5 T for Various Tissues***

Tissue	T1 (ms)	T2 (ms)
Myocardium	880	75
Blood	1200	360
Fat	260	110
Skeletal muscle	880	45
Lung	820	140

*T1 values are strongly field-strength dependent, and at lower fields, T1
values are shorter. T2 values vary minimally with field strength.
Data from Bottomley PA, Foster TH, Argersinger RE, Pfeifer LM: A
review of normal tissue hydrogen NMR relaxation times and mechanisms
from 1–100 MHz: Dependence on tissue type, NMR frequency, temperature, species, excision and age. Med Phys 11:425, 1984.

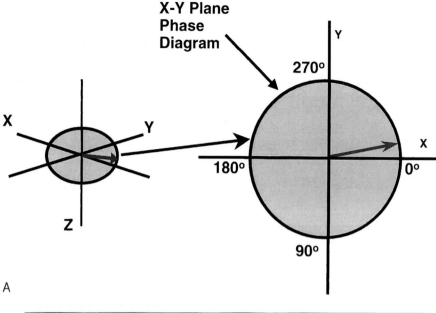

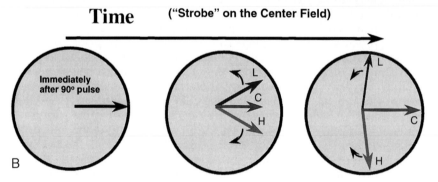

Figure 1–3. The phase diagram. *A*, A phase diagram is a view of the magnetization vectors in the X-Y plane. *B*, The magnetic field is variable in the sample shown at the top. Sample spins from the low field (L), center field (C), and high field (H) region are followed as a function of time after a 90° pulse using the phase diagram. The phase diagram is set up to present the phases of the spins relative to C, which is always normalized to the 0° position. The phase diagrams at the bottom are presented as a function of time after the B_1 field pulse. Note that the L vector lags behind the C vector whereas the H vector has a higher phase velocity than the other spins. This finding is due to the relative frequency of the spins caused by the differences in the magnetic field. (See also color plates.)

myocardium. Myocardial T2 is ~ 75 ms at 1.5 Tesla (T) (see Table 1–1). The second process is related to the homogeneity of the main magnetic field (B_0). As illustrated in Figure 1–3*B*, if the B_0 field is not homogeneous through the sample, the frequency of the spins in different regions will vary. This situation results in a randomization of the spin phases as they rotate at different frequencies, and a decrease in the net transverse magnetization like T2. However, this process can be reversed with B_1 field generated echo (see later discussion). This is an important process in the heart because the B_0 homogeneity is very poor as a result of the lung cavity and deoxygenated blood in the right heart. Finally, the heart also moves through this inhomogeneity as a result of contraction and respiratory motion. This process also contributes to different frequencies of

precession as a function of time, leading to a further dephasing of the heart water spins. The combined impact of all dephasing processes, including the molecular spin-spin interactions (T2), is called T2* (pronounced "T2 star"). The T2* of myocardium is ~40 ms at 1.5 Tesla in the human heart.[2] Later we will describe how to minimize or utilize these effects to make in vivo measurements.

In tissue, T2* relaxation processes frequently dominate magnetic relaxation of water protons. Thus, the decay rate of the FID is actually a measure of T2*. Because T2* is a more rapid process, it limits the time that we can detect the MR signal. Thus, it would be desirable to minimize the difference between T2 and T2*, which can be accomplished using a B_1 field generated echo. This type of echo is outlined in Figure 1–4, using the same

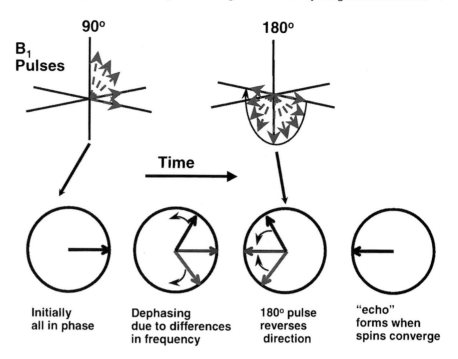

Figure 1–4. Effects of a 180° B_1 field pulse following a 90° pulse. Time is running from left to right. The spins are the same populations as in Figure 1–3B. The spins are first excited by a 90° pulse as in Figure 1–3. After some time to permit dephasing due to differences in frequency as in Figure 1–3B, a 180° pulse is applied. The 180° pulse results in the spins moving towards each other due to the difference in frequency, rather than away as was occurring before the 180° pulse. At a time equal to that which was required to dephase the spins, the spins refocus to form an echo of coherent magnetization.

convention as in Figure 1–3B. After the 90° pulse, we apply another B_1 field that rotates the magnetization 180° around the X axis. This 180° flip is achieved by applying the B_1 field for twice as long or with twice the power as the 90° pulse. This 180° flip causes the spins to rotate into the opposite sector of the transverse plane. On this side of the plane, the spins now drift together (rephase), taking the *same amount of time* that they took to dephase to rephase. This process is analogous to having two people back to back in a field and sending them walking apart (90° pulse) at some fixed rate (but different rates for each person). At some time later, we instruct them to reverse (180° pulse) their direction and to now walk towards each other at their same "individual" pace. The time it takes them to reach each other will be precisely the same amount of time they walked apart. This is the nature of the echo that effectively recaptures all of the coherent transverse magnetization that had been destroyed by the field inhomogeneity in the sample. An example of an echo detected with a coil is shown in Figure 1–5. Note that the echo has both a rephasing as well as a dephasing period. The spins continue to dephase after reaching the coherent echo due to ongoing field inhomogeneity, just as the people in the field would walk past each other and continue to "dephase." A subsequent 180° B_1 pulse could be used to refocus these spins again and again. This method of continually refocusing the magnetization with 180° pulses can be used in cardiac imaging (see later). Using repeated 180° pulses to refocus the spins is limited because 180° pulse will not recover the spin-spin relaxation (T2) processes occurring on the molecular level. Thus, when 180° pulses are used to refocus the magnetization, the magnetization still dephases, but at the much slower T2 rate.

These basic relaxation processes, T1, T2, and T2* are key in the generation of image contrast in MRI as well as guiding the optimal image sequence for gathering information on cardiac anatomy, function, and physiology. Myocardial relaxation properties change with edema associated with different disease states.[3] Generally, the T1 and T2 values are more prolonged with increasing water content. Myocardial remodeling or myocyte replacement with connective tissue also changes the water relaxation properties because the nature of the macromolecules in contact with water is critical for these relaxation processes. Finally, most exogenous MRI contrast agents work by shortening either T1 or T2*.[4] By appropriate modification of the imaging sequences, these changes in relaxation properties due to pathology or exogenous contrast agents can be highlighted in the MRI of the heart. How this is accomplished will be described later.

MAGNETIC RESONANCE IMAGING

To create an MR image you need three-dimensional information on the spatial coordinates (X, Y, and Z), and you need to determine the intensity of the MR signal for each of these coordinates. Thus, for a single image, we need to collect information on X, Y, Z position and signal amplitude. Examining the echo in Figure 1–5, we can detect a frequency, amplitude, and phase, which can be used to determine some spatial or amplitude information. But we have only three pieces of information and we need four. Clearly, not enough information to create a two- or three-dimensional image is available in a single FID. As we will see, this is a major limitation

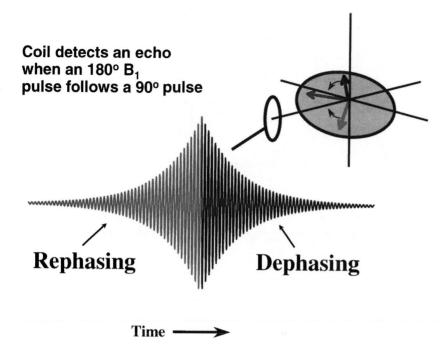

Coil detects an echo when an 180° B$_1$ pulse follows a 90° pulse

Rephasing   **Dephasing**

Time ⟶

Figure 1–5. The echo detected by the coil after a 90°–180° sequence.

of MRI, resulting in a relatively slow image acquisition rate compared with many other modalities.

The simplest imaging experiment can be divided into four stages, as shown in the block diagram in Figure 1–6: the slice select, the phase encode, refocusing echo, and frequency encode/readout. Each stage is used to encode the MR data with information on the position and the amplitude of the water proton (^{1}H) signal. For the purposes of this discussion, Z will be selected using the slice select period, Y information in the phase encode, and X information in the readout. In general, any of these dimensions could be evaluated using any of these steps; indeed, oblique imaging slicing through a tissue at 45°, or any other angle, is also possible. This latter property is almost a unique attribute of MRI. Each of these stages will be discussed separately to describe how a simple image is created.

In the slice select stage, one of the spatial coordinates is eliminated by only exciting magnetization in a selected slice of the sample. The slice selection process is illustrated in Figure 1–7. A linear magnetic field gradient is applied to the sample along the direction that a slice is going to be created (the Z axis in this example). The magnetic field gradient causes the spins along this gradient to have slightly

MRI Block Diagram

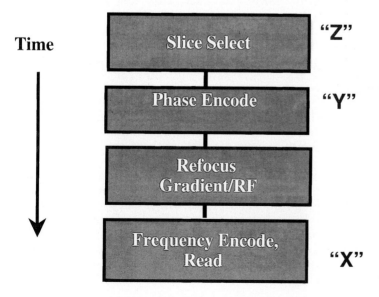

Time

Slice Select   **"Z"**

Phase Encode   **"Y"**

Refocus Gradient/RF

Frequency Encode, Read   **"X"**

Figure 1–6. General imaging scheme for collecting a simple MRI image.

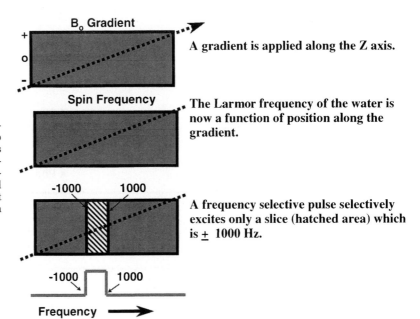

Figure 1–7. Slice select. The B_0 gradient is applied from a negative to positive value with zero being near the middle of the magnet. This process results in a linear distribution of frequencies as a function of position along the gradient. A frequency selective (± 1000 Hz) B_1 field pulse is applied to excite those spins in that frequency bandwidth, resulting in the selection of a slice to be placed in the transverse plane.

different frequencies as defined by the Larmor equation. Generally, these linear magnetic field gradients are applied $\pm$ a given field strength with zero being in the center of the magnet, though slice offsets from the center of the magnet are easy to apply by offsetting the center frequency of the pulse. If only a selected band of frequencies is excited (e.g., -1000 to $+1000$ Hz), a slice is put in the transverse plane corresponding to these frequencies, not affecting the rest of the sample. Thus, if a "Z" magnetic field is applied down a body from the head to the toe, the frequency of the water protons in the head will be slightly higher than the spins in the toe. By exciting an appropriate band of frequencies between the head and toe, a selective slice within the heart can be put into the transverse plane for detection.

To generate a controlled bandwidth for excitation, a sinx/x (sinc) function is played out in time, which results in a reasonably well defined slice in frequency, as shown in Figure 1–8. Here an FFT is used to show how the spins interpret the sinx/x pulse in frequency, as a reasonably well-defined band of frequencies providing a useful slice selective pulse. Many other modifications to the sinx/x pulse are possible to optimize the performance of this approach as well as slice definition.[5, 6] In summary, to selectively excite a slice, a linear magnetic field gradient is placed along the axis to be sliced. A sinx/x B_1 field pulse is applied, and only the predefined slice within the sample will be placed in the transverse plane for further modification to create an image.

Because we have limited information in the FID, the phase of the spins will be needed for more spatial information, but care must be used to ensure that the slice selection process does not influence the phase of the spins. Frequency contamination is usually not a problem because the frequency will almost immediately revert to the frequency driven by B_0 after the slice select gradient is turned off. However, after the 90° pulse the spins will dephase, depending on where they are in the slice (similar to that seen in Figure 1–3B) because a relatively wide bandwidth of frequencies is excited (± 1000 Hz). Thus, another magnetic field gradient of half the magnitude and of opposite direction to the slice selection gradient must be applied to eliminate the phase introduced by the slice selection process. The amplitude is half because the spins were in the transverse plane only halfway through the slice gradient duration. By applying the gradient in the opposite direction, the spins are forced back to the same phase as they had before application of the slice select gradient.

Figure 1–8. The frequency selective B_1 pulse. In order to generate the B_1 field pulse shown in Figure 1–7, a fast Fourier transform (FFT) is required. The FFT of a frequency square wave is approximated by a sinx/x function. The FFT is a fast Fourier transform of the desired frequency characteristics, which results in a time-varying signal that has the appropriate frequency characteristics.

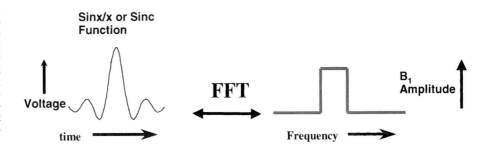

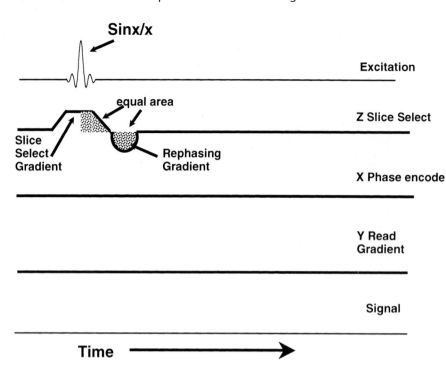

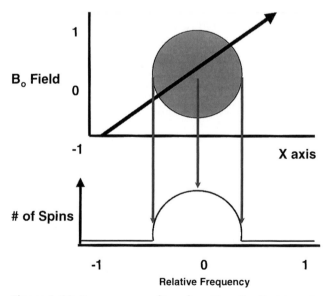

Figure 1–9. Pulse sequence diagram. Each action is given a separate time line. The excitation or B_1 field, as well as the slice select, phase encode and readout gradient are shown. At the bottom, the signal obtained by the coil is presented. This convention is also used in all subsequent figures. Only the slice select process is shown here; the other actions will be added in the later sections.

The slice selection process is summarized in the imaging pulse sequence diagram shown in Figure 1–9. This diagram presents all of the gradients and B_1 field excitation and data collection on a time line. This permits the visual display of the slice selection portion of the sequence. The slice selection gradient is first brought up to a stable level by a ramp. The gradient is maintained and the frequency (i.e., slice selective) selective B_1 field applied. Subsequently, the slice gradient is ramped back down and the rephasing gradient applied to remove the phase introduced by the slice select gradient. After the slice select excitation has been performed, only a slice of the sample has been placed in the transverse plane with no effect on the phase or frequency of the transverse magnetization within the slice.

We are left with encoding spatial information in the FID for the X and the Y coordinates. We will skip the phase encoding step at this point (see later in chapter) and move on to the refocusing and frequency encoding steps. To eliminate the effects of T2* on the signal, an echo is created by applying a 180° pulse with the excitation coil. This process results in an echo forming at a time equal to the time interval between the 90° slice select and the 180° pulse. If a gradient is applied before the echo has formed, and in the direction we wish to frequency encode (X axis in this example), then the frequency of the spins will reflect the position of the spins in X, as illustrated in Figure 1–10 for a simple sphere. The arrows from the edges and center of the sphere to the frequency axis point out the frequency of these spins within the sphere with the imaging magnetic field gradient applied. The position of these spins along the X axis is easily detected by determining the frequency of the spins while a gradient is applied along this axis. The

gradient played out during this time period is often referred to as the readout gradient.

The phase of the signal following the readout gradient is again a cause of concern. Because the readout gradient is on during the acquisition of the echo, it will cause phase shifts in the data along the X axis, and like the slice select gradient, could influence the phase information reserved for the Y dimension. The solution to this problem is the same. An X gradient that is half the area of the readout gradient and opposite in direction is applied before the readout gradient. This gradient is called a *de-*

Figure 1–10. Frequency encoding of position. Gradient is again applied from a negative to positive value across the sample, and the frequency of the spins is directly read out. The vertical arrows indicate the frequency of spins on either side of the sphere.

phasing gradient. The dephasing gradient causes the spins to dephase according to their position along the X axis. When the readout gradient is subsequently applied, spin frequency, relative to the X gradient, is reversed, and the spins refocus to form an echo at the center of the readout gradient at the peak of the 180° pulse generated echo. This echo is called a gradient recalled echo because the readout gradient is used to refocus the dephasing gradient applied before it. In summary, the position of the spins in the X axis is encoded in frequency by applying a static gradient along the X axis during the formation of a spin echo. The phase effects of this gradient are removed with the use of a gradient echo.

The frequency encoding component is added to the pulse diagram in Figure 1–11. First the dephasing gradient is applied with an area equal to half of the readout gradient. The read gradient is then ramped up in the opposite direction and held constant during the echo acquisition. After the data acquisition the gradient is ramped back down to zero.

Preserving the phase information is important for the determination of the Y spatial information. Spatial encoding in Y uses a process called phase encoding. While the spins are still in the transverse plane after the slice selection process, the spins can be phase encoded by transiently applying a magnetic field gradient along the axis chosen for phase encoding (Y axis in this case). Recall that to determine the position of a spin within the sample, a gradient is applied, and the resulting frequency reports its position; and frequency is simply a measure of how fast the phase is changing in time. The complication with using a phase measurement to determine the frequency of a spin is that only one phase point, the initial point, is useful within each

echo. Thus, to determine the frequency of the spin in the phase encoding axis, a series of echoes must be collected to determine how fast the phase of the spins is changing (i.e., frequency) with regard to the phase encoding gradient applied. To understand phase encoding, consider three points in different Y positions in a sample through the phase encoding process, as illustrated in Figure 1–12 where three points (A, B, and C) are identified in the Y axis of the sphere. To phase encode the spins for position in the Y, a magnetic field gradient is transiently applied along the Y axis. This gradient dephases the spins relative to their position in the Y axis because they will be moving at different frequencies in the transverse plane as a result of the change in the magnetic field. The spins at the center of the sphere, "A," experience no gradient field and continues to rotate with a phase induced by the slice selection process alone. The water spins in the higher field regions, "B" and "C," have an accelerated rate of phase change as a result of their high frequency. The "C" spins are experiencing twice the field of "B" and rotate twice as fast. When the gradient is turned off, the spins resume their precession rate relative to the B_0, but now they possess the "memory" of their Y position in phase. When the echo is finally collected during the read portion of the experiment, the phase of each of the spins is influenced by its position in Y by the brief time the frequency of the spins was modified by the pulsed Y gradient.

However, the frequency of the spins relative to the applied Y gradient, and therefore absolute position in Y, cannot be accurately determined from a single phase point because it is the *rate of **change*** in phase that determines the frequency. Thus, repeated measurements must be made to determine how fast the phase is changing in these different positions as

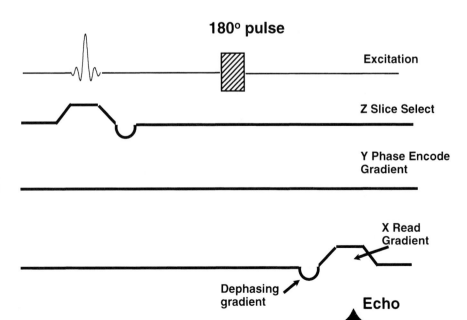

Figure 1–11. Pulse sequence diagram. The readout portion of the sequence has been added from Figure 1–10.

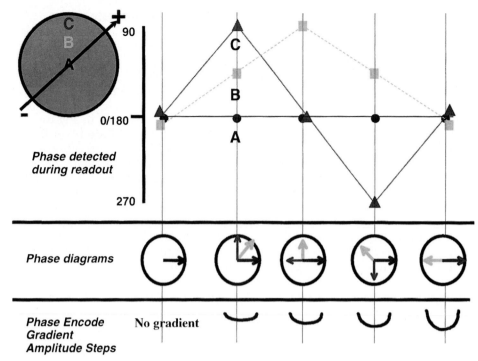

Figure 1–12. Phase encoding. Across a sphere a gradient is transiently applied as shown. The magnitude of the gradients in time is shown at the bottom of the figure. No gradient and four steps of gradient strength are being evaluated. The phase diagram of where the spins end up after the gradient has been applied is shown directly above the gradient magnitudes. In the uppermost graph the relative phase of the spins is plotted. C is the triangles, B is the squares, and A is the circles. A is at the center of the magnet and sees no field changes. Thus, the phase of A is constant. B is further and further dephased relative to A by the ever-increasing gradients. C sees the largest field difference and rapidly changes phase. The frequency of the spins is basically the rate at which the phase is changing relative to the gradient strength in this phase encoding scheme. As seen in the top figure, the rate of change in phase is highest for C, followed by B and then A as they are located along the gradient's direction.

a function of the Y gradient. The most obvious way of performing this task is to apply the Y gradient for different durations and to collect echoes for each of these times. By plotting the initial phase of the spins as a function of the duration the Y gradient is on, you could then determine the rate of phase change, or frequency. Knowing the slope of the gradient, one can then determine position. However, using the duration of the Y gradient is problematical because it would require the spins to spend a great deal of time in the transverse plane, resulting in T2 losses and other complications. Thus, a more efficient system was devised whereby instead of varying time, the *magnitude* of the Y gradient is varied. This system is based on the simple principle that a gradient that is twice as strong will dephase the spins to the same extent as leaving the gradient on for twice as long. This results in a method of phase encoding by maintaining constant the time in the transverse plane, but systematically varying the magnitude of the phase encoding gradient to generate the information required to determine frequency. Five such measurements are shown in Figure 1–12, along with the phase diagrams and gradient transient that resulted in the phase displacement of the spins. By collecting a series of echoes, all phase encoded with different gradient strengths in Y, the rate of phase change with gradient strength, or frequency, can be determined, as shown on the top graph of Figure

1–12, where the phase detected in the echo is shown as a function of the gradient strength. As seen in this figure, the "A" spin is unchanged because it always experiences the same magnetic field in the center of the magnet. The rate of change in phase of the "C" spin is twice that of the "B" spin, or the "C" frequency is twice that of "B," demonstrating that the "C" spin is further from the center position on the Y axis. The position of all of these spins in the Y dimension can be accurately determined if enough points are collected to accurately determine the frequency. Each one of these phase encoded steps requires a whole new echo acquisition. Thus, multiple data acquisitions are required for each MR image to determine the frequency of the spins along the phase encoded gradient. Therefore, this slow phase encoding process provides the last piece of information, Y axis position, needed to create the simplest MR image.

All of the steps are combined into a single acquisition scheme in Figure 1–13. A separate echo is collected for each phase encode step. The phase encode step can be applied any time after the spins are in the transverse plane. Immediately after the slice select process is the most common time. Frequently the phase encode step is applied at the same time that the rephasing gradient for the slice select is being played out. The precision of the determination of the phase encode direction frequency, and

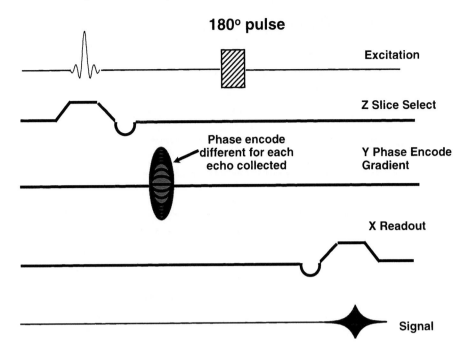

180° pulse

Excitation

Z Slice Select

Phase encode different for each echo collected

Y Phase Encode Gradient

X Readout

Signal

Figure 1–13. Pulse sequence diagram. The phase encoding steps are added immediately after the slice selection process. Note that no corrections for phase are added here because this signal is the one we want to persist to the readout section.

therefore position, is directly proportional to the number of phase encode steps collected. Usually this number is on the order of 64 to 512 phase encode steps.

To obtain the frequency in the read direction and phase direction, a two-dimensional (2-D) FFT is performed on the data that extracts the frequency in both dimensions. The raw data collected is referred to as "k-space" and is shown in Figure 1–14. The "ideal" k-space model is to take an infinite resolution image and perform a 2-D FFT on it to generate

the type of data collected in an MRI experiment. The resulting data can be viewed as the ideal k-space representation of the image that we try to attain with our pulse sequence to create the MR image. Naturally there are limits on the amount of data that can be collected. The resolution of MRI k-space data (and the resulting image) is usually limited by the number of phase encoded echoes that are collected because this collection takes the most amount of time. Many schemes to increase the efficiency of collecting the phase encoded data will be

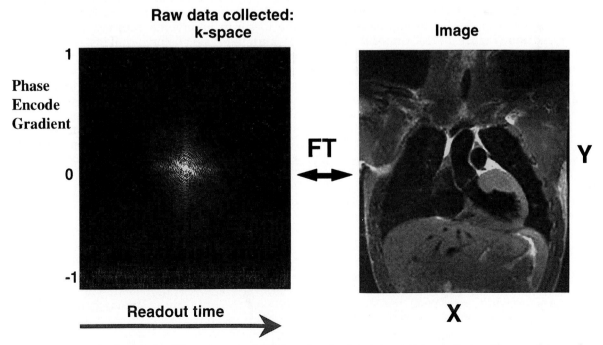

Raw data collected: k-space

Image

Phase Encode Gradient

1

0

-1

Readout time

FT

Y

X

Figure 1–14. Image of a heart. The FT converts the MRI time domain data into spatial coordinates. The raw data are known as k-space and ideally represent all of the spatial, spectral, and intensity data in the image.

TR and TE in a Spin Echo Sequence

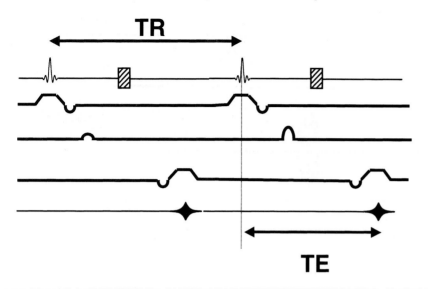

Figure 1–15. TR (repetition time) and TE (echo time) in a spin-echo sequence. See text for explanation.

discussed later. As mentioned earlier, in cardiac MRI, the number of phase encode steps varies from 64 to 512, depending on the speed that the data must be collected, the field of view that is being covered, and the resolution required to observe the structure of interest.

The sequence we constructed in Figure 1–13 is called a spin-echo sequence. There are several factors in this simple sequence that will change the contrast in the MR image based on the previously described relaxation processes. Because a 180° refocusing pulse is used in this sequence, the total time the spins stay in the transverse plane determines the amount of T2 relaxation that will occur. This time is called the echo time (TE) and is measured from the center of the slice select sinx/x pulse to the center of the refocused echo during the readout or data acquisition (Figure 1–15). Generally, the longer the TE, the more T2 contrast or T2 weighting is generated in the image. The effect of TE on the MR water signal from normal (T2 ~75 ms) and

infarcted (T2 ~100 ms) regions of the myocardium is shown in Figure 1–16A. Note that the signal in both regions decreases with increasing TE, but that the infarcted region relaxes more slowly. This situation results in a contrast, or increased difference in signal between the normal and the infarcted tissue as the TE is prolonged. By looking at the difference of the two tissue curves (dotted line), you can select a TE of about 50 ms to optimize the contrast between these two tissue types. Thus, by simply adjusting the imaging parameters, fundamental information on the heart structure can be obtained.

Another sequence timing parameter that influences the signal amplitude is based on T1 relaxation. For a 90° excitation pulse used in spin-echo imaging, one must wait ~5 times the T1 value of the tissue to permit the spins to completely relax back up the Z axis before applying another pulse to collect the entire NMR signal available. If a shorter time is used between slice selective pulses, the spins will add up the energy applied by each excita-

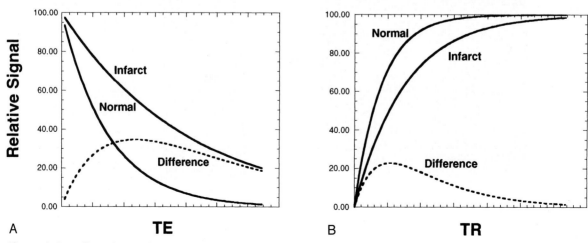

Figure 1–16. Effect of TE and TR on MRI signal amplitude. See text for explanation.

tion pulse, resulting in partial saturation of the spins and a reduction in the MR signal. This reduction in signal is dependent on the T1 of the sample. The longer the T1, the greater the reduction in signal because less of the magnetization can recover. The time between each slice excitation pulse is the critical factor in this sequence and is called the repetition time (TR) (see Figure 1–15). The impact of TR on signal amplitude is shown in Figure 1–16B for a spin-echo sequence. The effect of TR is illustrated for normal myocardium with a T1 of 880 ms and a chronic infarct with a T1 of 1.5 s. Note that the shorter the T1, the more *rapidly* the pulses can be applied and still maintain the MR signal. Also apparent in Figure 1–16B is that by varying the TR, the contrast or difference between tissues can be altered. Note that a TR of 1 s would be optimal in generating the largest difference (or contrast— "dotted" line) between the normal and the infarcted tissue. Thus, the TR can be used to vary the image contrast based on T1.

The relatively long times required for T1 relaxation is the major reason that spin-echo methods are slow to collect the phase-encoded information required to create an MR image. One approach to circumvent this problem is to use multiple 180° refocusing pulses and to collect many echoes during each slice selective excitation. This fast spin echo (FSE) approach is diagrammed in Figure 1–17 and is an important technique in cardiac imaging.[7] By applying a phase encoding gradient between each 180° pulse, each echo is phase encoded with a different magnitude of gradient, thereby providing the necessary phase data to create the image. For cardiac imaging, 16 to 64 echoes can be collected for each slice selective pulse, thereby reducing the time to collect a spin echo image by 16 to 64 times, respectively. Because large blocks of time are required to collect all of these echoes, this method is usually restricted to relatively motion free phases of the cardiac cycle, such as diastole. The inherent high signal-to-noise ratio of these FSE approaches provides very high resolution images of the myocardium. In addition, true T2 contrast can be generated by varying the time between the 180° pulses or the

number of the 180° pulses, the so-called echo train length. The longer the time between the 180° degree pulses, the more T2 weighting will occur.

To avoid the time delay needed for complete spin relaxation from a 90° slice selective pulse, lower flip angles can be used. In addition, eliminating the 180° pulse and associated echo time might be useful to image the beating heart with high temporal resolution. Ernst showed that as the flip angle is reduced, the TR can be shortened to provide the optimal signal-to-noise ratio per unit time. This results because a dynamic equilibrium is set up between the T1 relaxation processes and the excitation, which maintains a steady state amount of magnetization. This defines the parameter known as the Ernst angle:

$$\text{Flip angle} = \cos^{-1}(e^{-TR/T1}) \qquad (2)$$

If you wish to collect 64 phase encode lines in k-space and you want to collect them in 100 ms (TR) to freeze diastolic motion of the heart, assuming a myocardial T1 of 880 ms, the Ernst angle equation identifies an optimal angle of ~11°. Thus, data can be collected rather rapidly using an excitation pulse lower than the 90° pulse, which would require a TR of greater than 4 s for complete magnetization recovery.

To use low flip angles and short TR values, the time associated with the 180° pulse must be removed. This removal is accomplished by using the gradient recalled echo alone to refocus the magnetization at the right time for readout. This process is schematically illustrated in Figure 1–18. Here the overall TR of the sequence is greatly shortened, permitting the rapid acquisition of data as long as the flip angle is appropriately adjusted. This type of data acquisition is known as a gradient recalled echo (GRE).[8] Note that to keep the phase information intact, rewinding gradients are applied after each acquisition equal to and opposite to the phase encoding gradient and the last half of the readout gradient to prevent any phase information "bleeding" into subsequent acquisitions. One of the major limitations of the GRE approach is that the magneti-

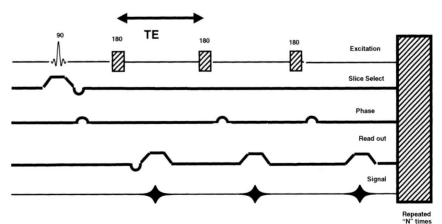

Figure 1–17. Pulse sequence diagram of a fast spin echo image acquisition scheme.

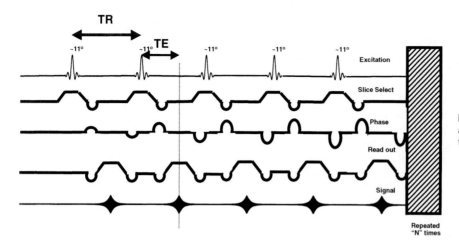

Figure 1–18. Pulse sequence diagram of a gradient recalled echo image acquisition scheme.

zation dephases according to the much more rapid T2*, which means that any field inhomogeneity within the chest may result in major losses in signal amplitude before the echo is collected. Thus, minimizing the TE with good gradient performance is critical. The role of gradient performance will be discussed further later.

The GRE could also be multiplied after the slice select pulse in a manner analogous to that described for the FSE sequence. In this mode, a train of GREs can be collected after the slice select pulse for as long as the T2* does not destroy all of the magnetization. This approach is called echo-planar imaging (EPI) and was among the first MR imaging methods described[9] along with projection reconstruction methods.[1] For tissues with long T2*, EPI can result in a complete image collection with a single slice select pulse and a train of gradient echoes (Figure 1–19). Because all the data are collected in a single shot and T1 recovery is not an issue, a 90° pulse is used. In addition, to minimize time, the dephasing of the read gradient is done in a rapid ramp before the actual read gradient. In general, this approach is not very useful in the heart because myocardial T2* is so short at ~30 ms. This process results in poor signal-to-noise ratio and image distortions. However, these limitations can be overcome by collecting only a few phase encode steps per slice select

excitation and repeating these processes at the same time in the cardiac cycle in subsequent heartbeats until a complete image can be created. This process is called a segmented k-space acquisition and can be expanded to cover the whole heart volume throughout the cardiac cycle. A schematic of an electrocardiogram (ECG)-gated segmented EPI sequence for cardiac imaging is shown in Figure 1–20. This system permits a rapid acquisition of data with a minimal time spent in the transverse plane. With modern gradient hardware, eight echoes can be collected per slice excitation with a TR on the order of 10 ms. Using this approach, imaging times on the order of 80 ms can be achieved in low resolution images (64 × 128 image resolution over a 15 × 32 cm field of view) without the magnetization ever spending more than 10 ms in the transverse plane.[10] To overcome the constant motion of the heart due to respiration and contraction, but still obtain accurate high-resolution images, some form of cardiac gating is required to fill in k-space over multiple heartbeats, because enough data cannot be collected in a single heartbeat. For most studies a breath-hold is sufficient to eliminate the respiratory motion,[11] and either prospective or retrospective gating is used to sort the data collected into the appropriate cardiac phases. Several phase encode steps are collected per heartbeat; thus, after several heartbeats, enough data

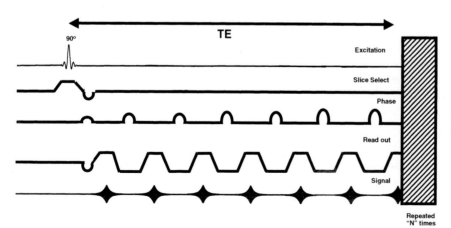

Figure 1–19. Pulse sequence diagram of an echo-planar image acquisition scheme.

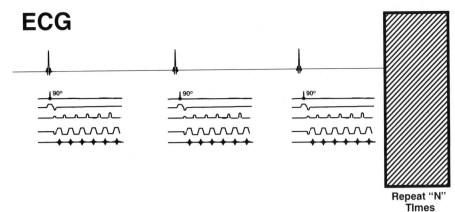

Figure 1–20. Pulse sequence diagram of a segmented echo-planar image acquisition scheme.

are collected to provide images of the heart throughout the entire cardiac cycle.[12] Accurate ECG gating and a regular rhythm are an absolute requirement.[13] Recent advances in these segmentation schemes have resulted in useful ungated image acquisition schemes that will be covered in other chapters.

In summary, MRI is the result of determining the spatial distribution of radiofrequency MR signals actively induced from water and fat protons in the body. Beyond the simple concentration differences between water and fat in different tissues, the molecular interactions of these protons with their environment result in fundamentally different magnetic relaxation times, T1 and T2. These differences are used to generate useful contrast concerning the morphology and physiology of the heart. In combination with specific contrast agents, direct measures of perfusion, blood flow, vascular permeability, and other parameters can be determined. The basics of imaging these spins are the use of known magnetic field gradients across the body that make the frequency of the spins a function of position within the body. This property is used to make selective body slices as well as to fill in the X and Y spin positions. The limited frequency information in a single MR acquisition limits the temporal resolution of MRI. The requirement for phase encoding one of the spatial dimensions requires multiple acquisitions, which slows the overall process. Numerous acquisition schemes are available to image the heart, with new approaches being developed at a remarkable rate as this technique becomes more and more clinically attractive.

References

1. Lauterbur P: Image formation by reduced local interactions: Examples employing nuclear magnetic resonance. Nature 242:190, 1973.
2. Reeder SB, Faranesh AZ, Boxerman JL, McVeigh ER: In vivo measurement of T2* and field inhomogeneity maps in the human heart at 1.5 T. Magn Reson Med 39:988, 1998.
3. Scholz TD, Martins JB, Skorton DJ: NMR relaxation times in acute myocardial infarction: Relative influence of changes in tissue water and fat content. Magn Reson Med 25:1120, 1992.
4. Mathur-De VR, Lemort M: Invited review: Biophysical properties and clinical applications of magnetic resonance imaging contrast agents. Br J Radiol 68:225, 1995.
5. Conolly S, Glover G, Nishimura D, Macovski A: A reduced power selective adiabatic spin-echo pulse sequence. Magn Reson Med 18:28, 1991.
6. Meyer CH, Pauly JM, Macovski A, Nishimura DG: Simultaneous spatial and spectral selective excitation. Magn Reson Med 15:287, 1990.
7. Hennig J, Nauerth A, Friedburg H: RARE imaging: A fast imaging method for clinical MR. Magn Reson Med 3:823, 1986.
8. Haase A, Frahm J, Matthaei D, et al: FLASH imaging: Rapid NMR imaging using low flip-angle pulses. Magn Reson Med 67:258, 1986.
9. Mansfield P, Pykett IL: Biological and medical imaging by NMR. J Magn Res 29:355, 1978.
10. Epstein FH, Wolff SD, Arai AE: Segmented k-space fast cardiac imaging using an echo-train readout. Magn Reson Med 41:609, 1999.
11. Edelman RR, Manning WJ, Burstein D, Paulin S: Coronary arteries: Breath-hold MR angiography. Radiology 181:641, 1991.
12. Feinstein JA, Epstein FH, Arai AE, et al: Using cardiac phase to order reconstruction (CAPTOR): A method to improve diastolic images. J Magn Reson Imaging 7:794, 1997.
13. Dimick RN, Hedlund LW, Herfkens RJ, et al: Optimizing electrocardiograph electrode placement for cardiac-gated magnetic resonance imaging. Invest Radiol 22:17, 1987.

CHAPTER 2

Clinical Cardiovascular Magnetic Resonance Imaging Techniques

Daniel K. Sodickson

The heart is an organ in continual motion. This simple fact leads to the principal challenge of cardiovascular magnetic resonance (CMR). The cyclical motion of the heart, and of the diaphragm on which it sits, adds new time scales and new constraints to the CMR procedure. If blurring of images is to be avoided, the acquisition of image data must occur during intervals that are relatively short when compared with the characteristic time scales of cardiac and respiratory motion. Thus, relatively high (<50 ms) temporal and spatial (<2 mm) resolutions are generally required for accurate definition of cardiac anatomy and ventricular function (higher spatial resolution is needed for coronary artery assessment). With progress in rapid CMR techniques, images of sufficient spatial resolution for many applications can now be acquired during a fraction of a single cardiac cycle. However, in many cases, more stringent requirements for spatial resolution and/or signal-to-noise ratio (SNR) necessitate longer acquisitions, and these constraints have led to the routine use of electrocardiograph (ECG) triggering in most CMR scans. For ECG-gated scans, time is measured by the number of cardiac cycles, and acquisition of data for high-resolution images is frequently divided up among multiple cardiac cycles. Respiratory motion then complicates the piecing together of segments acquired during different cycles, and various strategies of compensating for the effects of respiratory motion have been developed. Specific characteristics and technical requirements of pulse sequences and motion-compensation strategies vary depending upon the particular application. In this section, we will discuss certain basic principles and tools that form part of the common repertoire of CMR. We begin with a discussion of the general categories of imaging sequences and then discuss common methods for motion compensation, additional methods for characterization and quantification of motion, and finally some basic approaches to the generation and manipulation of contrast. Many of these subjects are explored in greater detail in subsequent sections and chapters. Here we emphasize general principles, some of which were introduced in Chapter 1, in order to provide some context for the choices of methods and sequence parameters in particular applications.

BASIC IMAGING SEQUENCES

The pulse sequence is the basic building block of the CMR procedure. It determines the timing of image acquisitions and is also the fundamental determinant of SNR, contrast-to-noise ratio (CNR), the nature and severity of image artifacts, and so forth. Two general categories of pulse sequences are used in CMR—the spin-echo and the gradient-echo sequences. These two general categories of sequences have particular features of interest.

A timing diagram for the prototypical spin-echo sequence is shown in Figure 2–1A. First, spins are excited by a 90° radio frequency (RF) pulse in the presence of a slice-selection gradient G_s. The spin magnetization is then refocused by a 180° pulse, which is also made slice-selective so as not to introduce unwanted signal components from inverted spins outside the initial slice. Following this, an MR signal is sampled in the presence of a readout gradient G_R, and the process is repeated with varying degrees of phase encoding (G_P) having been applied earlier in the sequence. One salient feature of spin-echo images is that blood flowing *rapidly* through the image plane will appear dark, whereas stationary tissue or stagnant flow within the slice will appear bright(er). This effect is a result of the slice-selective 90°–180° pulse pair, and is illustrated in Figure 2–1B. The initial 90° pulse sets up a "window" of excitation (depicted as a white slice between dark regions in Figure 2–1B), and only spins that *remain* in this window at the time of the 180° pulse are refocused and detected. Outflow from the chosen image plane results in "black blood" images, if flow is sufficiently fast for blood to traverse the image slice during the interpulse interval. Because arterial blood is pulsatile, acquisitions during diastole (low flow) may result in a "bright lumen" whereas systolic acquisitions may provide a "dark lumen." An example of a black-blood cardiac image is shown in Figure 2–2A. Other features of spin-echo sequences include a generally increased tissue contrast compared with gradient-echo sequences, and relatively reduced sensitivity to small magnetic field inhomogeneities caused, for example, by metallic implants (sternal wires, vascular clips, prosthetic valves, vascular stents) (see Figure 16–10).

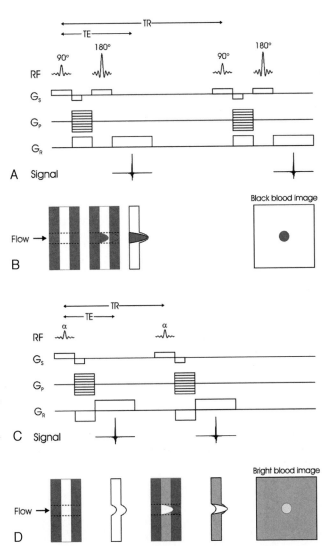

Figure 2–1. Timing diagrams and flow properties for prototypical spin-echo and gradient-echo CMR sequences. *A*, Spin-echo timing diagram. Radiofrequency (RF) pulse timings and amplitudes are shown schematically along with slice selection (G_s), phase encoding (G_p), and readout (G_R) gradients. Magnetic resonance (MR) signal acquisition occurs in the presence of the readout gradients. Two sequence cycles are shown, with sequence repetition time (TR) and signal echo time (TE) indicated above. *B*, "Black blood" flow properties of spin-echo sequences. The 90° pulse sets up a window of excitation (white column), which is subsequently refocused by the slice-selective 180° pulse. Excited spins leaving the slice and unexcited spins (dark-shaded areas) flowing into the slice, e.g., through a crossing blood vessel (dotted lines), are not refocused, resulting in a signal void in the image. *C*, Gradient-echo timing diagram. Once again, two sequence cycles are shown. RF pulses of flip angle α are used, where α is generally ≤90°. *D*, "Bright blood" flow properties of gradient-echo sequences. Because there is no slice-selective refocusing pulse following the initial excitation pulse, all of the excited spins are detected during the first sequence cycle. Slice-selective excitation in the next sequence cycle, however, results in partial saturation of all spins that have remained in the slice and have not had time to relax to equilibrium during the TR (light gray regions). Inflow of "fresh" unsaturated spins results in a comparatively bright signal contribution in the final image. In general, for flowing blood, the degree of signal attenuation in spin-echo sequences or signal enhancement in gradient-echo sequences depends upon the rate of flow, the direction of flow with respect to the image plane, the thickness of the slice, and the timing (TE, TR) of the sequence.

Refocusing of field inhomogeneities occurs naturally as a result of the 90°–180° pulse pair. Common variants on the basic sequence shown in Figure 2–1*A* include the acquisition of two distinct spin echoes after each 90° pulse in order to provide different degrees of relaxation contrast and flow compensation. Alternatively, multiple differently phase-encoded echoes may be acquired in a rapid train in order to increase imaging speed, in what are commonly referred to as fast spin echo (FSE) or turbo spin echo (TSE) methods. Spin-echo sequences are typically used for anatomical delineation of the cardiac mediastinum, pericardium, and great vessel structures. A summary of spin-echo features is included in Table 2–1.

Figure 2–1*C* shows a timing diagram for a prototypical gradient-echo imaging sequence. A single slice-selective RF pulse of flip-angle α (where α is generally ≤90°) is followed in rapid succession by a phase encoding gradient and a two-part readout gradient, in which reversal of the gradient polarity takes the place of the 180° spin-echo refocusing pulse in generating a signal echo. Acquisition of the entire k-space signal occurs by repetition of this α

pulse/gradient building block with different phase encoding gradient steps. As is indicated schematically in Figure 2–1*C*, the gradient-echo acquisition is generally faster than a spin-echo acquisition because less time is required between spin excitation and signal detection. Furthermore, the appearance of flowing blood is generally different for a gradient-echo than for a spin-echo sequence because there is no selective "window" of detection to match the excitation window. Rapid repetition of the α pulses, however, results in saturation effects that affect flowing blood and stationary tissue in much the opposite manner as the 90°–180° pulse pair of the spin-echo sequence.

If spins do not have sufficient time to relax back to their unexcited state (via T1 relaxation) during the interpulse interval, repeated α pulses lead to partial saturation of spins in the slice, as is indicated by the lighter grayed regions in Figure 2–1*D*. Blood flowing into the slice during the interpulse interval, however, has not been affected by earlier pulses, and therefore contributes a greater signal than the multiply excited stationary tissue (see Figure 2–1*D*). Thus, *inflow* of blood past the fixed window of *saturation* established by the slice-selective α pulses results in "bright blood" images, of which an example is shown in Figure 2–2B. Three-dimensional gradient-echo acquisitions, which are commonly used for rapid imaging of large volumes, can lead to reduced bright-blood contrast due to saturation of blood throughout the 3-D volume. Saturation effects, in combination with shorter acquisition times, are also responsible for the reduced tissue contrast in gradient-echo as compared with spin-echo sequences. Some approaches that have been taken to address this problem include flip-angle optimization to achieve the highest possible steady-state SNR and CNR, and magnetization preparation, in which contrast is added into the initial condition of the scan.

Gradient-echo imaging sequences tend to be more sensitive than spin-echo sequences to magnetic field inhomogeneities from sternal wires, vascular clips, prosthetic valves, and vascular stents. This is because spin refocusing in gradient-echo sequences is accomplished by the reversal of field gradients rather than by the application of 180° pulses. Because inhomogeneities in the main magnetic field

Table 2–1. General Features, Common Uses, and Selected Sequence Names of Spin-Echo and Gradient-Echo Sequences for Cardiac Imaging Applications

Spin Echo (SE)	Sequence Features: · Black-blood images (for rapidly flowing blood) · Good tissue contrast · Relative insensitivity to magnetic field inhomogeneities (clips, wires, stents, etc.) · Generally lower imaging speeds than gradient-echo sequences Common Uses: · Anatomic delineation of mediastinum and great vessels Variants: · Dual echo · Fast Spin Echo (FSE) or Turbo Spin Echo (TSE)
Gradient Echo (GRE)	Sequence Features: · Bright-blood images (for rapidly flowing blood) · Saturation effects, and potentially reduced inherent contrast when compared with spin-echo sequences (addressable with magnetization preparation techniques) · Relative sensitivity to magnetic field inhomogeneities · High imaging speeds possible Common Uses: · Coronary artery imaging · Ventricular function assessment · Myocardial perfusion assessment · Valvular motion · Valvular regurgitation (qualitative) Synonyms: · Field Echo · Gradient Field Echo · Gradient Recalled Echo Variants: · Fast Low-Angle Shot (FLASH) · Turbo-FLASH or Turbo Field Echo (TFE) · Echo Planar Imaging (EPI)

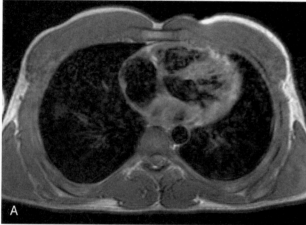

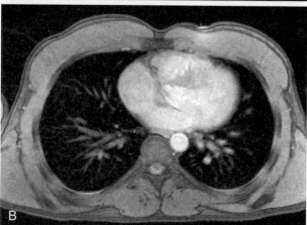

Figure 2–2. *A,* Black-blood CMR obtained in a transverse plane intersecting the heart using a spin-echo sequence. Blood in the cardiac chambers and the descending aorta appears substantially darker than myocardial tissue. *B,* Bright-blood image in the same plane using a gradient-echo sequence.

may be viewed as fixed field gradients, which are not affected by changes in the externally applied imaging gradients, gradient reversals will not in themselves remove field inhomogeneity effects.

Just as spin-echo sequences may be implemented in multiecho mode, multiple gradient echoes may also be acquired after each α pulse in a gradient-echo sequence. The limiting case of a single α pulse followed by a full set of rapid phase-encoded gradient echoes defines the ultrafast single-shot echo-planar imaging (EPI) sequence. Another gradient-echo variant, segmented EPI sequences, are intermediate between the single-shot and the single-echo cases, and they have been used, for example, to perform rapid volumetric imaging of the heart[1] or real-time imaging of cardiac contraction.[2, 3] The imaging speed advantages of EPI sequences are at least partially offset by a susceptibility to artifacts relating to differential relaxation between echoes and to mismatch between even and odd echoes.

Coronary artery imaging applications have tended to use gradient-echo sequences, as have studies of ventricular function and myocardial perfusion, and other applications in which imaging speed (temporal resolution) is of primary importance. Some of the basic features and uses of gradient-echo sequences are listed in Table 2–1, along with some of the more common sequence names used by the various CMR vendors. See Appendix II for a more comprehensive list of cross-vendor terminology.

MOTION COMPENSATION

Despite the considerable speed of some of the sequences we have just described, CMR acquisition times, especially for images with high spatial resolution, are often too long to avoid blurring due to cardiac and respiratory motion. To eliminate or minimize these artifacts, motion compensation is often required. Most of the approaches to motion compensation may be combined both with spin-echo and gradient-echo imaging sequences. Cardiac motion is the most rapid of the motions confronted by a practitioner of CMR and will be addressed first.

To borrow the language of photography, if a single "exposure" were too long to prevent blurring of a rapidly moving object, one would attempt to reduce the exposure time. Because CMR image acquisition speed, and hence effective "exposure time," generally cannot be reduced arbitrarily without adversely affecting image quality, the problem is addressed instead by dividing the single exposure into multiple shorter exposures. Taking advantage of the approximate regularity and repeatability of the normal cardiac cycle, most cardiac imaging applications now employ a technique akin to the principle of strobe photography—namely ECG triggering and k-

space segmentation (Figure 2–3). Initially described for liver imaging,[4] then subsequently demonstrated for cardiac applications in animals[5] and in humans,[6] k-space segmentation involves the division of MR data acquisition over multiple cardiac cycles, using the R wave of the ECG as a synchronizing signal. Different "pieces" of the MR data set are thereby acquired in successive cardiac cycles, with collection of each piece timed to occur at the same delay following an R wave, and hence at the same stage of cardiac contraction. This allows the full signal data set for an image of the heart at a given position to be pieced together from segments short enough to prevent blurring. Just as strobe photographs can track rapidly moving objects such as a bullet or freeze rapid periodic motions such as a hummingbird's wing using brief regularly spaced exposures, k-space segmentation allows the beating heart to be viewed in near still-frame. Figure 2–3 illustrates schematically the principle of k-space segmentation. Data acquisition (dark gray squares) occurs during a fixed acquisition window at a well-defined trigger delay following each R wave. Each cardiac cycle is devoted to the acquisition of a different set of data lines (vertical black lines below the ECG trace), and these lines are combined with previously acquired lines (vertical gray lines) to build up the full image data set. The ordering of lines may vary depending upon the imaging application, and two possible orderings are shown in Figure 2–3. Choice of the number of k-space lines to acquire per segment will also be influenced by the particular application: When high temporal resolution is required, for example, a smaller number of lines per segment is acquired, so as to allow a short acquisition window. In general, in a segmented scan, imaging time is defined by two parameters: the acquisition window within each cardiac cycle and the total number of cardiac cycles.

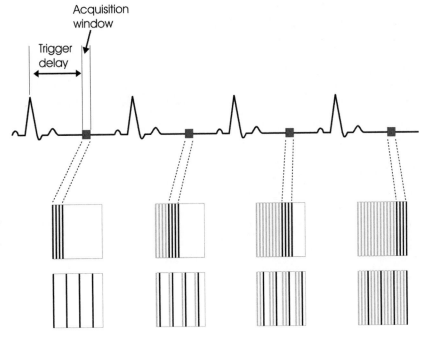

Figure 2–3. Electrocardiogram (ECG) triggering and k-space segmentation. Data collection occurs during a restricted acquisition window at a fixed trigger delay following each R wave on the ECG. Different segments of the data matrix required for image generation are acquired during different cardiac cycles, and the "pieces" are combined to yield a full data set. Two possible strategies for the order of data collection are shown in the two rows below the ECG tracing. In both cases, vertical black lines represent data acquired during the current cardiac cycle, whereas gray lines represent data from previous cycles.

The duration of the acquisition window determines the degree of blurring from cardiac motion: a window duration of less than 100 ms is generally required, and many would argue for 50 ms or less. Of course, the question of acceptably short acquisition windows also depends upon when during the cardiac cycle signal acquisition occurs—a quantity determined by the trigger delay following the R wave. Bulk cardiac motion is most rapid during ventricular systole, but also following atrial systole, with relatively little motion during isovolumic ventricular relaxation and middiastole.[7, 8]

Although prospective ECG gating is ideal, in some patients, R wave detection in the CMR environment can be relatively limited. For these situations, more sophisticated algorithms such as vector ECG[9] or peripheral pulse gating can be implemented. Preliminary reports suggest the former is quite robust, even in myopathic hearts, whereas the latter approach detects the peripheral pulsation in a digit, most commonly a finger. Because there is considerable delay between the R wave and the peripheral pulsation, peripheral pulse gating is more useful for diastolic imaging (or for systolic imaging during the subsequent heartbeat).

The ordering of k-space lines in segmented acquisitions may have an influence on relaxation effects and image artifacts. Artifacts may also occur when segments acquired in different cardiac cycles do not "line up," for example in the presence of arrhythmias or of uncompensated respiratory motion, both of which perturb the periodicity of cardiac motion. Arrhythmias constitute a difficult problem for cardiac imaging, and may necessitate ultrafast or real-time imaging as an ultimate solution. Fortunately, respiratory motion is easier to control, or at least to predict and detect.

The most straightforward approach to compensating for respiratory motion is to eliminate it temporarily by having the patient hold their breath during image acquisition. In cooperative patients, breath-holding has yielded high-quality images free of blurring in a number of cardiac applications.[6, 10, 11] An example of improved image quality in a breath-hold gradient-echo scan is shown in Figure 2–4. Breath-holding has a number of practical limitations, however. Patients—and especially those with underlying cardiac and/or respiratory disease—may not be able to hold their breath for the prolonged imaging times associated with high-resolution scans. Repeated breath-holding requires patient cooperation and leads to patient/operator fatigue. Coregistration of multiple breath-holds may be a nontrivial matter because repeatability has been shown to be poor in many subjects.[12–15] Furthermore, breath-holding itself does not necessarily eliminate all diaphragmatic motion, and the residual motion due to diaphragmatic drift may be sufficient to cause image artifacts.[16]

Other approaches to respiratory motion compensation involve monitoring of the respiratory motion itself. Respiratory bellows gating has been used in a number of applications.[14, 17, 18] This technique involves placement of an air-filled bellows between the chest wall and a rigid support structure or a circumferential belt. Chest wall expansion (inspiration) compresses the bellows and expands the circumferential belt, thereby producing a pressure tracing that can be used for timing of CMR acquisitions. Most commonly, end-expiratory acceptance is used. Respiratory bellows gating is combined with ECG triggering, and only cardiac cycles that occur in the same general region of the respiratory motion tracing are used for data acquisition. To the extent that chest wall expansion is a faithful measure of diaphragmatic and ultimately of cardiac position, this

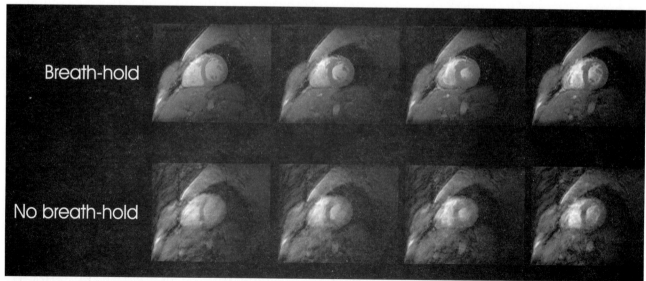

Figure 2–4. Short-axis images of the heart at various stages of the cardiac cycle, with and without breath-holding. An identical segmented k-space gradient-echo sequence was used to obtain both of these image sets. Respiratory motion artifacts are apparent in the images without breath-holding (*bottom row*).

gating ensures that the heart is at least approximately in the same absolute location for different k-space segments.

More accurate monitoring of respiratory motion is possible using navigator echo techniques (Figure 2–5; see Chapter 15). "Navigator echoes" refer to targeted CMR scans that serve to identify the position of some clearly visible anatomical structure whose relation to the heart is known and that can therefore be used to predict cardiac position for a subsequent full scan. Navigator echoes may be produced using selective two-dimensional excitations or crossed planar 90° and 180° excitations, both of which result in a column of excited magnetization extending across a high-contrast tissue interface such as the liver-lung or the heart-lung border. Before each potential acquisition (or following each actual acquisition), signal from such an excited column is rapidly compared with a reference echo in order to detect relative displacement of the interface, and the interface position is used for gating in a manner similar to the use of respiratory bellows

tracings described earlier. Two possible navigator positions, one at the right hemidiaphragm (RHD) and one at the left ventricular (LV) free wall, are shown as thick-bordered rectangles on a coronal scout image in Figure 2–5A, and are further localized on transverse scans in Figures 2–5B and 2–5C. Figure 2–5D shows a typical navigator echo tracing that might result in a free-breathing subject. The echo intensity profile runs vertically in this tracing, with time running horizontally from left to right. The position of the interface (topmost black-gray border) is seen to oscillate as a function of time as the subject breathes. Various computational algorithms may be used to detect the interface position from the oscillatory tracing, with a premium placed on processing speed as well as on accuracy. In addition, various acceptance-rejection criteria may be applied, with a gating window of 3 to 5 mm typically being chosen. In Figure 2–5D, the acceptance window is indicated by thick white lines, and gray bars at the bottom of the figure indicate accepted acquisition intervals. Scan efficiency (i.e., the num-

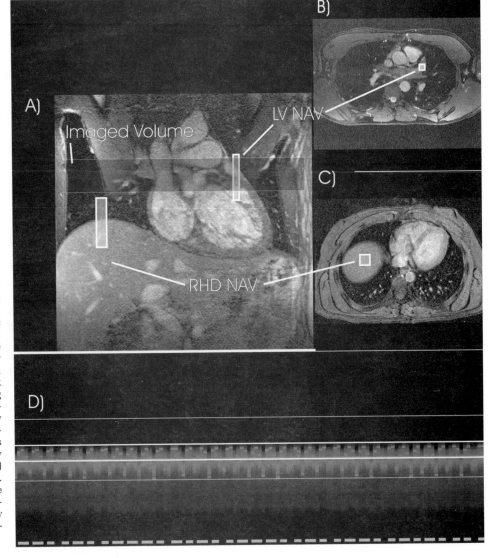

Figure 2–5. Navigator echo gating. One or more columns of magnetization traversing a tissue interface are excited using a tailored excitation sequence, and the displacements of the interface are calculated and used as measure of respiratory motion. *A*, Coronal scout image showing two possible placements of the navigator echo excitation, one through the right hemidiaphragm (RHD NAV), and one through the left ventricular free wall (LV NAV). The navigator excitation may or may not overlap the imaged volume. *B*, Transverse scout image showing the LV NAV in cross section. *C*, Transverse scout image in a more caudal plane showing the RHD NAV in cross section. *D*, Sample navigator echo tracing, with navigator signal on the vertical axis shown as a function of time on the horizontal axis. Respiration results in a cyclic motion of the tissue interface. Gray dots superimposed on the tracing represent automatically detected interface positions in successive cardiac cycles. Thick white lines indicate the gating acceptance window. In a prospective gating strategy, only cardiac cycles in which the interface position falls within this range (indicated also by gray bars at the bottom of the tracing) are used for data acquisition. Thin white lines on either side of the acceptance window indicate a range or "kernel" used by the automatic interface detection algorithm.

ber of cardiac cycles used for data acquisition divided by the total number of accepted *and rejected* cardiac cycles) depends upon the stringency of the gating criteria, the breathing pattern of the patient, and the relationship between the heart rate and the respiratory rate. Navigator gating strategies may be retrospective, in the sense that the decision to accept or reject each acquired segment is made after collection of a large and hopefully sufficient number of acquisitions; alternatively, prospective gating may be used, in which acceptance decisions are made during each cardiac cycle, and the scan is terminated as soon as all necessary k-space lines have been acquired. Single, dual, or multiple navigators may be used, with a resulting trade-off between image quality and time efficiency. (With multiple navigators acceptance efficiency is adversely impacted.) A great deal of research has been devoted in recent years to the characterization and optimization of patient breathing patterns,[16] to the ordering of k-space lines in navigator-gated scans,[19] and to the optimal anatomical positioning of navigator echoes for various applications.[14]

The use of navigator data is not limited to gating. Information on the relative position of the chosen interface, if it falls within an acceptable range, may be used to *correct* slice position prospectively. In other words, the position of the image plane can be adjusted during a scan to match the observed displacement of the navigator interface. This technique may allow for greater acceptance window (and resulting scan efficiency) without disturbing the spatial registration of data segments in a segmented scan. The concept of navigator correction was first proposed by Ehman and Felmlee in 1989,[20] and it has been used since in a number of cardiac applications.[15, 21] The precise nature of the required correction depends upon the location of the navigator. For example, if the navigator is placed on the RHD, the correlation between diaphragmatic displacement and displacement of the heart itself must be known.[22] A "fixed" relationship may not hold for all subjects. Alternatively, the navigator may be placed directly across the cardiac surface to be imaged (e.g., through the LV free wall), in which case a very close correlation between navigator interface position and cardiac position exists. Navigators may also be combined with repetitive breath-holds[15] (to improve slice registration) or with coached breathing (to improve scan efficiency).

With navigators, cardiac scans may be obtained during free-breathing, enhancing patient comfort and compliance, and facilitating scans in patients incapable of prolonged breath-holding. Because properly functioning navigators are not limited by breath-hold times or the need for interbreath-hold registration, one can easily increase spatial resolution through acquisition of additional k-space lines, or else increase SNR through signal averaging. Some potential disadvantages of navigator use include longer overall scans (depending upon the gating ef-

ficiency) and the need to designate some amount of time during each cardiac cycle for navigator excitation and processing (typically 20–65 ms).

There is one more principal motion of concern to the cardiac imager that we have not yet addressed: namely, the motion of flowing blood in the cardiac chambers and surrounding vessels. In some cases, flowing blood can produce serious ghosting artifacts due to motion of spins along magnetic field gradients and the accumulation of extra phase factors between the time of excitation and the time of signal detection. If bright-blood contrast is not required, artifacts from flowing blood may be reduced by presaturation of spins prior to their entrance into the slice of interest. Alternatively, the phase evolution of flowing spins may be compensated using a technique called gradient moment nulling. In this approach, the shape and sequence of imaging gradients are adjusted to cancel the extra phase that would normally accrue for spins moving in a constant field gradient. Different orders of motion (e.g., velocity vs. acceleration) may be compensated independently. Flow compensation generally adds somewhat to imaging time because additional gradient lobes must be included in the imaging sequence. Compensation will only be completely successful for comparatively well defined flow patterns such as laminar flow. Turbulent flow, on the other hand, involves a complicated mixture of many different orders of motion and tends to result in signal losses or "flow voids" in the turbulent regions.

The most straightforward way to minimize the effects of physiological motion, of course, is to image as rapidly as possible. A number of ultrafast imaging techniques have been investigated for cardiac applications over the years, including EPI and spiral imaging.[23] Progress in the design of gradient hardware (gradient strength and slew rates) promises to enhance imaging speed still further by allowing more rapid switching of the powerful gradients needed for spatial encoding and signal readout. Though improvements of this nature will doubtless be seen in years to come, some state-of-the-art scanners are already approaching basic physical and physiological limits on gradient switching rate and RF power deposition. It has recently been demonstrated that some of the task of spatial encoding may also be performed using RF coil arrays, allowing multiple lines of k-space to be acquired simultaneously in partially parallel imaging strategies such as the simultaneous acquisition of spatial harmonics (SMASH)[24] and sensitivity encoding (SENSE)[25] techniques. These partially parallel acquisitions allow many-fold improvements in imaging speed without violating constraints on gradient switching rate or RF power deposition, and therefore these techniques hold substantial promise for addressing the problem of motion in CMR.[26–28] Within the constraints of SNR, increased acquisition speed afforded by these methods can be used to enhance temporal and/or spatial resolution.

MOTION DETECTION AND ASSESSMENT

The investigation of cardiac function, valvular function, and hemodynamics all call for a detailed assessment of how the relevant structures move during the cardiac cycle. Both qualitative and quantitative techniques for motion assessment exist, taking advantage of some of the same principles as are used for the elimination of motion artifacts in anatomical scans.

For any given imaging speed, ungated scans may be sufficient to follow relatively slowly moving structures in real time. (This approach has been used, for example, with EPI[3] and spiral[29] imaging sequences.) If this is not possible at spatial resolutions of interest, the cine approach is generally used. Cine imaging employs ECG triggering and k-space segmentation, but multiple acquisition segments are performed one after the other in each cardiac cycle, with each segment occurring in a different phase of the cycle (Figure 2–6A). These multiple "exposures" are then pieced together across different cardiac cycles to yield a series of full images tracking the position of the heart during different phases of the cardiac cycle (Figure 2–6B). The cine approach is an effective tool for generation of moving pictures when the individual frames would otherwise take too long to acquire. It makes efficient use of each cardiac cycle—indeed, the available time in each cycle may be filled by as many distinct images as are compatible with sequence speed limits without adding to overall scan time.

By itself, cine imaging yields a qualitative picture of cardiac motion. It may, however, be combined with tagging techniques to allow more quantitative determinations. In cardiac tagging techniques (see Chapters 4 and 5), polar or rectangular grids are superimposed on the magnetization in the image plane just prior to the application of cine or real-time imaging sequences. Cardiac motion results in distortion of the grid lines, and detailed estimates of in-plane displacement and velocity may be obtained by tracking the movement and deformation of each grid element. Techniques for the creation of tagging grids generally involve combinations of RF pulses and gradients designed to produce periodic spatial modulations of magnetization. Such techniques include delays alternating with nutations for tailored excitation (DANTE),[30] spatial modulation of magnetization (SPAMM),[31] and complementary SPAMM (CSPAMM),[32] among others. Subsequent chapters are devoted to the SPAMM (Chapter 4) and CSPAMM (Chapter 5) approaches in particular, but the basic principle of cardiac tagging is illustrated in Figure 2–7. A tagging sequence is applied prior to data acquisition during each cardiac cycle of a cine imaging sequence (Figure 2–7A), and the tagged cine images (Figures 2–7B and C) are built up over multiple cycles in the usual way. Tagging studies have been used to characterize a variety of cardiac motions during systole in healthy volunteers and in patients, including long-axis shortening,[33] axial twist,[34] and more localized wall motion abnormalities associated with myocardial infarction.[35] More recently, alterations in diastolic untwisting

Figure 2–6. Cine imaging. *A*, Schematic cine imaging sequence. Multiple acquisition windows distributed through the cardiac cycle are used for segmented k-space data acquisition, resulting in a series of images tracking the heart through successive stages of contraction and relaxation. *B*, Sample short-axis cine images of the heart, obtained in a breath-hold using a gradient-echo cine CMR sequence.

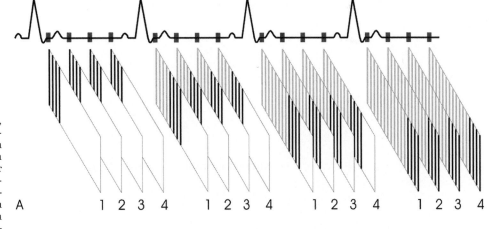

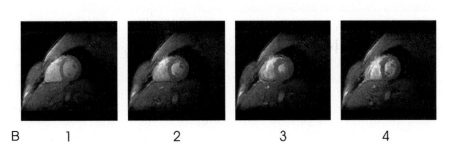

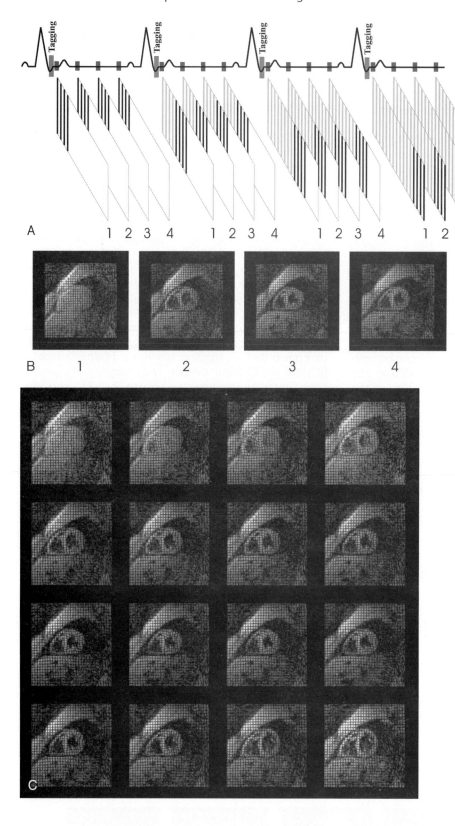

Figure 2–7. Cardiac tagging. *A*, Schematic tagging sequence. In each cardiac cycle, a tagging module is appended to the front of a cine sequence. *B*, Sample tagged short-axis cine images, obtained using the complementary spatial modulation of magnetization (CSPAMM) tagging technique.[32] *C*, Montage of 16 CSPAMM-tagged images spaced evenly throughout the cardiac cycle, and documenting in detail the sequence of cardiac contraction and relaxation. Time runs from left to right across each row, and from top to bottom between rows. The initially rectilinear tagging grid is distorted by cardiac contraction, finally returning to its original configuration at the end of the cardiac cycle.

have been found in patients with pathological hypertrophy and pressure overload due to aortic stenosis, whereas no alterations in either the systolic or the diastolic twisting pattern could be found in athletes with physiological hypertrophy.[36]

The phase of CMR images provides another quantitative handle on motion. The technique of phase velocity mapping exploits the fact that spins in orderly motion across a magnetic field gradient acquire a well-defined phase (this is the same principle used in the gradient moment nulling technique for flow velocity compensation). As is demonstrated in Figure 2–8, the basic pulse sequence building block for phase velocity mapping involves two gra-

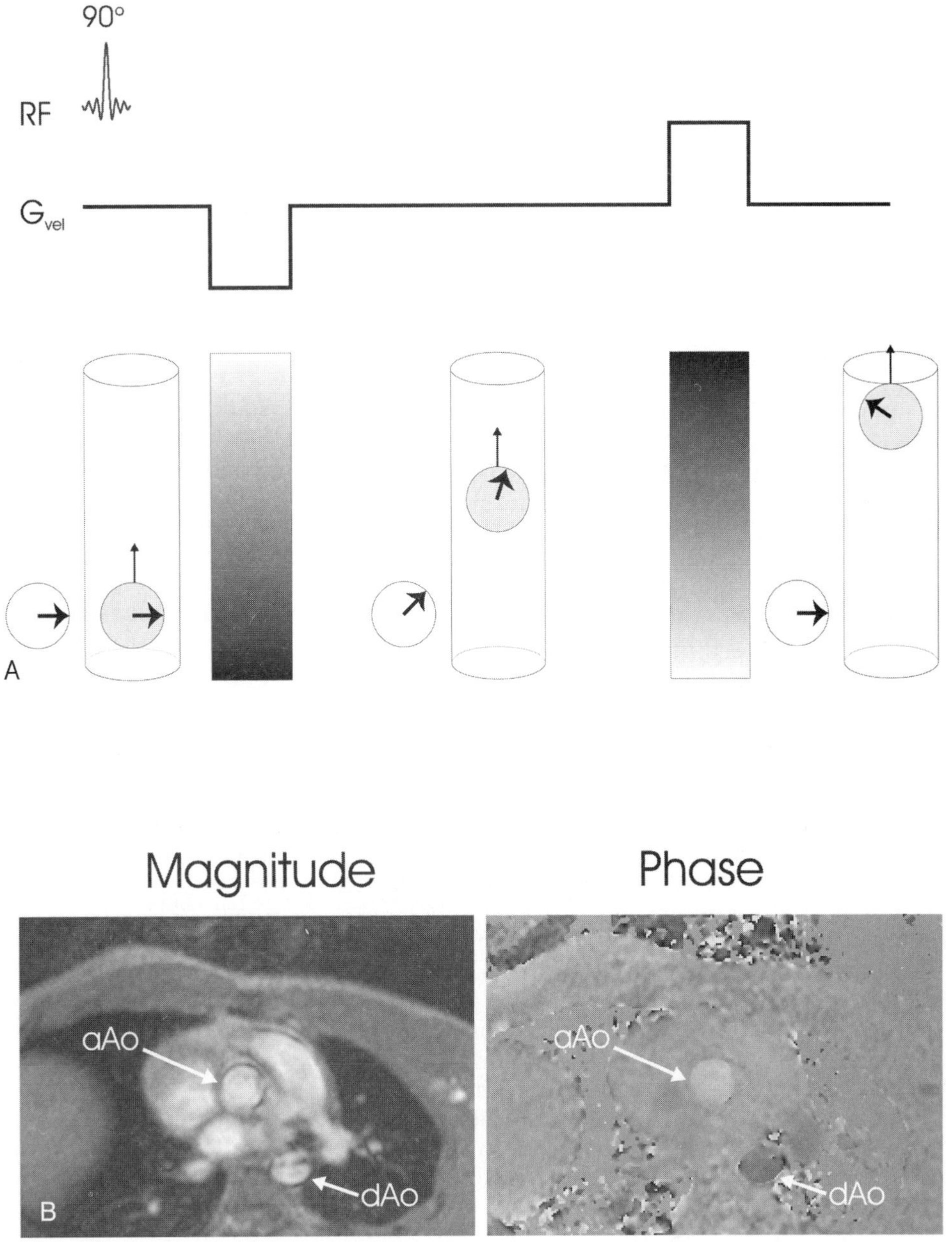

Figure 2–8. Phase velocity encoding. *A*, Basic pulse sequence building block for phase velocity encoding. Following a 90° radiofrequency (RF) excitation pulse, a bipolar velocity encoding gradient (G_{vel}) is applied. The spatial distribution of magnetic fields during gradient application is indicated by shaded bars below the sequence diagram, with the range from black to white indicating the range of field strengths from low to high. Stationary spins acquire equal and opposite phases during the two opposing portions of the gradient. Spins that are moving, for example in a blood vessel that runs in the direction of the velocity encoding gradient, are in different positions during application of the two gradient lobes, and are left with a net phase proportional to their velocity. *B*, Magnitude and phase images from a phase-contrast study of aortic flow. Blood in the ascending aorta (aAo) appears bright on the phase image, indicating flow in a cranial direction through this transverse image plane. Caudally directed flow in the descending aorta (dAo) appears dark in the phase image.

dients separated in time—one to establish a phase shift and the second of reversed polarity to refocus that phase shift. Stationary spins will acquire equal and opposite phases in the two gradients, and will be left with no net phase at the end of the sequence. Moving spins, on the other hand, will encounter different regions of the two gradients, and will be left with a net phase that is proportional to their velocity in the direction of the flow-encoding gradients (see Figure 2–8*A*). Phase-sensitive data collection and reconstruction will then yield an image whose magnitude describes the signal intensity and whose phase describes the velocity at each pixel (Figure 2–8*B*). The strength of the flow-encoding gradients and the interval between gradients sets the velocity scale which may reliably be detected with this method: For any given set of gradient parameters, there is a maximum measurable velocity

because motion that results in more than a 360° phase accumulation "wraps" back around to 0° and cannot be distinguished from slower motions. In practice, a reference image without flow-encoding gradients must generally also be acquired and its phases subtracted from those of the encoded image in order to account for baseline phase variations (due to slight field inhomogeneities across the image plane, for example) that might otherwise be interpreted as spurious velocities. In the cardiac imaging arena, phase velocity mapping is most commonly used for flow measurement, e.g., to quantitate blood flow in the great arteries[37, 38] or the coronary arteries.[39–41] In such applications, sensitivity of the velocity-encoding sequences to bulk cardiac motion as well as to flow may necessitate the use of velocity references in the heart tissues surrounding the arteries or valves of interest. In addition to its application for flow measurement, phase velocity mapping has been used to track the motion of cardiac muscle.[42] Quantitative flow measurements, on the other hand, may also be performed using time-of-flight techniques.[43]

TISSUE CONTRAST AND SIGNAL-TO-NOISE RATIO

A number of pulse sequence–specific effects on tissue contrast have been alluded to earlier. When additional contrast is needed in a particular scan, it is often added during a preparation phase prior to the pulse sequence building block of choice. We have already mentioned an example of presaturation to null the signal of and eliminate artifacts from flowing blood originating outside the target image slice. Spectrally selective saturation may also be performed, and this technique finds an important application in saturation of epicardial fat to facilitate coronary artery imaging. RF pulses tuned specifically to the fat resonance frequency are applied prior to the imaging sequence to reduce the signal from epicardial fat (along with all fat in the imaging plane) that might otherwise obscure signal from the coronary arteries. Spin relaxation contrast may also be manipulated by a number of mechanisms. T1 contrast may be enhanced using inversion recovery preparations, in which spins are inverted by RF pulses (spatially and spectrally selective or nonselective), and their relative signal contributions are modulated by their degree of return to equilibrium by the time the subsequent imaging sequence is initiated (Figure 2–9). Such approaches are commonly used for "first-pass" assessment of a CMR contrast agent to assess regional myocardial perfusion. T2 preparations using combinations of RF pulses also exist, and have been used to enhance CNR in cardiac scans by suppressing unwanted signal from myocardium and from deoxygenated blood

in cardiac veins.[44–46] Magnetization transfer between macromolecule-associated water and free water may be induced by off-resonance RF irradiation during a preparation phase, and may also be used to enhance tissue contrast.[47] All of these approaches operate by manipulating endogenous magnetization in blood and tissue to achieve the desired contrast effects. Exogenous contrast agents may also be used, but have been relatively limited in cardiac applications beyond that of assessment of regional myocardial perfusion (see Chapters 3 and 11) and myocardial viability (see Chapter 14).

An introduction to clinical CMR techniques would be incomplete without at least a brief mention of SNR. An SNR on the order of 5 or higher is generally required for visualization of small structures such as the coronary arteries, and the achievable spatial and temporal resolution for cardiac scans is in many cases limited by SNR considerations. The detailed balance of signal and noise in any particular application can be a complex subject. However, for our purposes, the subject of SNR may be divided into three main categories: general principles, sequence-specific effects, and hardware issues. One general principle is that SNR for a given imaging sequence is roughly proportional to the square root of the imaging time and inversely proportional to the voxel volume.[48] Thus, if imaging time is halved by doubling the readout gradient strength and the corresponding sampling bandwidth, SNR will fall by a factor of the square root of two (i.e., 1.41). If resolution is doubled in one dimension, the SNR will decrease by a factor of two. The comparison of SNR between different imaging sequences, on the other hand, can be dictated by sequence-specific effects, such as the magnitude of the flip angle (α) in gradient-echo sequences. Finally, the RF coil used for signal detection can have a profound effect on overall SNR. At the moderately high field strengths (e.g., 1.5 T) that have become commonplace for clinical CMR scanners, the

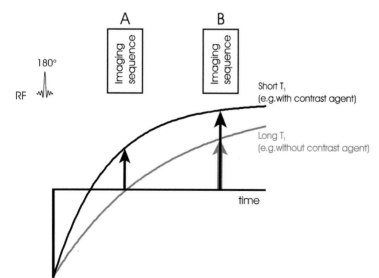

Figure 2–9. Inversion recovery contrast preparation sequence. A 180° pulse inverts spins, which subsequently relax back to equilibrium at a rate given by the local T1 relaxation time. Tissue with short T1, for example, tissue that is in contact with injected CMR contrast agents, relaxes more rapidly than tissue with longer T1, which provides a mechanism for contrast enhancement. For a given set of T1 values, contrast depends upon the time interval between the inversion pulse and application of the imaging sequence. In sequence A, for example, tissue with the longer relaxation time is at a zero crossing, and will be nearly invisible in the image. For a later sequence B, both schematic tissue components have relaxed further, and contrast will be reduced.

smaller area of surface coils, as compared with body coils, gives the surface coils an improved SNR over their sensitive volume and accounts for the extensive use of surface coils in CMR. For similar reasons, arrays of smaller coils, which can provide enhanced SNR over extended fields of view,[49] have begun to see increasing use in cardiac imaging.

CONCLUSION

We conclude this introduction with an example to demonstrate the reasoning that goes into choosing a particular sequence for a particular cardiac imaging application. The example we describe is taken from our laboratory's protocols for coronary artery imaging.[46, 50] Requirements of this application include a high spatial resolution sufficient to identify and trace the proximal and mid-portions of the coronary arteries, and a high temporal resolution to prevent motional blurring that might obscure these small structures. Coverage of a large volume is desirable to trace the potentially tortuous course of the arteries along the cardiac surface, and tissue contrast must be sufficient to identify the coronaries against a background of both muscle and epicardial fat. Figure 2–10A shows the elements of a sequence that meets many of these requirements.[46] A 3-D gradient-echo acquisition is used, in which both the coronary artery wall and its blood-filled lumen appear bright. The sequence is ECG triggered and k-space segmented, with data acquisition occurring during middiastole. Prior to the acquisition of each segment, a T2 preparation prepulse is applied to suppress myocardial and deoxygenated blood signal, thereby enhancing contrast between blood in the coronary artery lumen and surrounding tissues. This is followed by a 2-D selective navigator echo, which is processed in real time for use in prospective gating and correction, so that the scan can proceed during free-breathing. The navigator is intentionally placed in close proximity to the imaging portion of the sequence. Following the navigator (and concurrent with navigator analysis), a frequency-selective fat saturation pulse is applied to eliminate epicardial fat signal, and this is followed by application of the 3-D imaging sequence for acquisition of 8 lines of data per segment during an acquisition window of 60 ms per cardiac cycle. The combination of these elements yields images in which the course of the coronary arteries may clearly be discerned (Figure 2–10B).

In this section, we have described some of the fundamental elements that may be combined in a CMR scan in order to extract information useful for the assessment of cardiovascular anatomy and physiology. Subsequent sections and chapters address many of these elements in greater detail and explore some of the clinical applications for which CMR has particular power and promise.

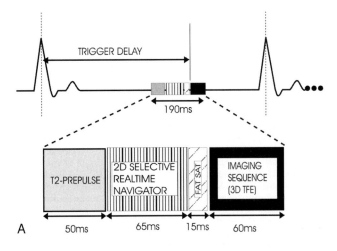

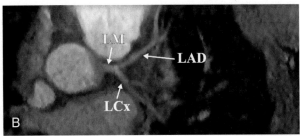

Figure 2–10. *A*, A coronary MRA sequence as reported by Botnar and colleagues.[46] Parameters of the sequence are as follows: TE = 2.5 ms; TR = 7.4 ms; flip angles are ramped to yield a constant weighting for each k-space line; total matrix size is 512 × 304 and field of view is 360 mm × 305 mm resulting in an in-plane spatial resolution of 0.7 mm × 1.0 mm. For comparison, the diameter of the proximal left and right coronary arteries is in the vicinity of 3 to 5 mm. Ten three-dimensional (3-D) partitions or slices are obtained with an effective slice thickness of 3 mm, for a total volume coverage of 30 mm. Typical total scan time ranges from 10 to 20 minutes depending upon subject heart rate and navigator efficiency. In the absence of segmentation and gating, the scan would require approximately 25 s (or 2.5 s per slice)—far longer than is tolerable in the presence of cardiac and respiratory motion. *B*, Multiplanar-reformatted image from a 3-D data set acquired using the sequence in *A*. The left main (LM), proximal-to mid-left anterior descending (LAD), and proximal- to mid-left circumflex (LCx) coronary arteries may all be discerned in the image.

References

1. Wielopolski PA, Manning WJ, Edelman RR: Single breath-hold volumetric imaging of the heart using magnetization-prepared 3-dimensional segmented echo planar imaging. J Magn Reson Imaging 4:403, 1995.
2. McKinnon GC: Ultrafast interleaved gradient-echo-planar imaging on a standard scanner. Magn Reson Med 30:609, 1993.
3. Stuber M, Scheidegger MB, Boesiger P: Realtime imaging of the heart. Fifth Scientific Meeting and Exhibition of the International Society for Magnetic Resonance in Medicine, Vancouver, BC, Canada, 1997, p 908.
4. Edelman RR, Wallner B, Singer A, et al: Segmented turboFLASH: Method for breathhold MR imaging of the liver with flexible contrast. Radiology 174:757, 1990.
5. Burstein D: MR imaging of coronary artery flow in isolated and in vivo hearts. J Magn Reson Imaging 1:337, 1991.

6. Edelman RR, Manning WJ, Burstein D, Paulin S: Coronary arteries: Breath-hold MR angiography. Radiology 181:641, 1991.

7. Wang Y, Vidan E, Bergman GW: Cardiac motion of coronary arteries: Variability in the rest period and implications for MR coronary angiography. Radiology 213:751, 1999.

8. Hofman MBM, Wickline SA, Lorenz CH: Quantification of in-plane motion of the coronary arteries during the cardiac cycle: Implications for acquisition window duration for MR flow quantification. J Magn Reson Imaging 8:568, 1998.

9. Fischer SE, Wickline SA, Lorenz CH: Novel real-time R-wave detection algorithm based on the vector cardiogram for accurate gated magnetic resonance acquisitions. Magn Reson Med 42:361, 1999.

10. Manning WJ, Li W, Boyle NG, Edelman RR: Fat-suppressed breath-hold magnetic resonance coronary angiography. Circulation 87:94, 1993.

11. Atkinson D, Burstein D, Edelman RR: Cine angiography of the heart in a single breath-hold with a segmented turboFLASH sequence. Radiology 178:357, 1991.

12. Liu YL, Riederer SJ, Rossman PJ, et al: A monitoring, feedback, and triggering system for reproducible breath-hold MR imaging. Magn Reson Med 30:507, 1993.

13. Wang Y, Christy PS, Korosec FR, et al: Coronary MRI with a respiratory feedback monitor: The 2D imaging case. Magn Reson Med 33:116, 1995.

14. McConnell MV, Khasgiwala VC, Savord BJ, et al: Comparison of respiratory suppression methods and navigator locations for MR coronary angiography. AJR Am J Roentgenol 168:1369, 1997.

15. McConnell MV, Khasgiwala VC, Savord BJ, et al: Prospective adaptive navigator correction for breath-hold MR coronary angiography. Magn Reson Med 37:148, 1997.

16. Taylor AM, Jhooti P, Wiesmann F, et al: MR navigator-echo monitoring of temporal changes in diaphragm position: Implications for MR coronary angiography. J Magn Reson Imaging 7:629, 1997.

17. Ehman RL, McNamara MT, Pallack M, et al: Magnetic resonance imaging with respiratory gating: Techniques and advantages. AJR Am J Roentgenol 143:1175, 1984.

18. Oshinski JN, Hofland L, Mukundan S Jr, et al: Two-dimensional coronary MR angiography without breath holding. Radiology 201:737, 1996.

19. Sachs TS, Meyer CH, Irarrazabal P, et al: The diminishing variance algorithm for real-time reduction of motion artifacts in MRI. Magn Reson Med 34:412, 1995.

20. Ehman RL, Felmlee JP: Adaptive technique for high-definition MR imaging of moving structures. Radiology 173:255, 1989.

21. Chuang ML, Chen MH, Khasgiwala VC, et al: Adaptive correction of imaging plane position in segmented k-space cine cardiac MRI. J Magn Reson Imaging 7:811, 1997.

22. Wang Y, Riederer SJ, Ehman RL: Respiratory motion of the heart: Kinematics and the implications for the spatial resolution in coronary imaging. Magn Reson Med 33:713, 1995.

23. Meyer CH, Hu BS, Nishimura DG, Macovski A: Fast spiral coronary artery imaging. Magn Reson Med 28:202, 1992.

24. Sodickson DK, Manning WJ: Simultaneous acquisition of spatial harmonics (SMASH): Fast imaging with radiofrequency coil arrays. Magn Reson Med 38:591, 1997.

25. Pruessmann KP, Weiger M, Scheidegger MB, Boesiger P: SENSE: Sensitivity encoding for fast MRI. Magn Reson Med 42:952, 1999.

26. Jakob PM, Griswold MA, Edelman RR, et al: Accelerated cardiac imaging using the SMASH technique. J Cardiovasc Magn Reson 1:153, 1999.

27. Weiger M, Pruessmann KP, Boesiger P: Cardiac real-time imaging using SENSE. SENSitivity Encoding scheme. Magn Reson Med 43:177, 2000.

28. Sodickson DK: Spatial encoding using multiple RF coils: SMASH imaging and parallel MRI. *In* Young IR (ed): Methods in Biomedical Magnetic Resonance Imaging and Spectroscopy. Chichester, John Wiley & Sons Ltd, 2000, p. 239.

29. Kerr AB, Pauly JM, Hu BS, et al: Real-time interactive MRI on a conventional scanner. Magn Reson Med 38:355, 1997.

30. Mosher TJ, Smith MB: A DANTE tagging sequence for the evaluation of translational sample motion. Magn Reson Med 15:334, 1990.

31. McVeigh ER, Atalar E: Cardiac tagging with breath-hold cine MRI. Magn Reson Med 28:318, 1992.

32. Fischer SE, McKinnon GC, Maier SE, Boesiger P: Improved myocardial tagging contrast. Magn Reson Med 30:191, 1993.

33. Rogers WJ Jr, Shapiro EP, Weiss JL, et al: Quantification of and correction for left ventricular systolic long-axis shortening by magnetic resonance tissue tagging and slice isolation. Circulation 84:721, 1991.

34. Rademakers FE, Buchalter MB, Rogers WJ, et al: Dissociation between left ventricular untwisting and filling. Accentuation by catecholamines. Circulation 85:1572, 1992.

35. Kramer CM, Rogers WJ, Theobald TM, et al: Remote noninfarcted region dysfunction soon after first anterior myocardial infarction. A magnetic resonance tagging study. Circulation 94:660, 1996.

36. Stuber M, Scheidegger MB, Fischer SE, et al: Alterations in the local myocardial motion pattern in patients suffering from pressure overload due to aortic stenosis. Circulation 100:361, 1999.

37. Klipstein RH, Firmin DN, Underwood SR, et al: Blood flow patterns in the human aorta studied by magnetic resonance. Br Heart J 58:316, 1987.

38. Kilner PJ, Yang GZ, Mohiaddin RH, et al: Helical and retrograde secondary flow patterns in the aortic arch studied by three-directional magnetic resonance velocity mapping. Circulation 88(5 Pt 1):2235, 1993.

39. Edelman RR, Manning WJ, Gervino E, Li W: Flow velocity quantification in human coronary arteries with fast, breath-hold MR angiography. J Magn Reson Imaging 3:699, 1993.

40. Keegan J, Firmin D, Gatehouse P, Longmore D: The application of breath hold phase velocity mapping techniques to the measurement of coronary artery blood flow velocity: Phantom data and initial in vivo results. Magn Reson Med 31:526, 1994.

41. Sakuma H, Saeed M, Takeda K, et al: Quantification of coronary artery volume flow rate using fast velocity-encoded cine MR imaging. AJR Am J Roentgenol 168:1363, 1997.

42. Pelc LR, Sayre J, Yun K, et al: Evaluation of myocardial motion tracking with cine-phase contrast magnetic resonance imaging. Invest Radiol 29:1038, 1994.

43. Poncelet BP, Weisskoff RM, Wedeen VJ, et al: Time of flight quantification of coronary flow with echo-planar MRI. Magn Reson Med 30:447, 1993.

44. Wright GA, Nishimura DG, Macovski A: Flow-independent magnetic resonance projection angiography. Magn Reson Med 17:126, 1991.

45. Brittain JH, Hu BS, Wright GA, et al: Coronary angiography with magnetization-prepared T2 contrast. Magn Reson Med 33:689, 1995.

46. Botnar RM, Stuber M, Danias PG, et al: Improved coronary artery definition with T2-weighted, free-breathing, three-dimensional coronary MRA. Circulation 99:3139, 1999.

47. Wolff SD, Balaban RS: Magnetization transfer contrast (MTC) and tissue water proton relaxation in vivo. Magn Reson Med 10:135, 1989.

48. Macovski A: Noise in MRI. Magn Reson Med 36:494, 1996.

49. Roemer PB, Edelstein WA, Hayes CE, et al: The NMR phased array. Magn Reson Med 16:192, 1990.

50. Stuber M, Botnar RM, Danias PG, et al: Double-oblique free-breathing high resolution three-dimensional coronary magnetic resonance angiography. J Am Coll Cardiol 34:524, 1999.

CHAPTER 3

Myocardial Perfusion—Theory

Michael Jerosch-Herold* and Norbert M. Wilke*

The concept of using a tracer and detecting its transit and distribution through the heart for the determination of myocardial perfusion is well established in nuclear cardiology,[1] contrast echocardiography,[2] and x-ray densitometry.[3] Cardiovascular magnetic resonance (CMR) has been noted for the anatomical detail, the good tissue contrast, and the excellent spatial and temporal resolution it provides. The great versatility of CMR, and the many subtleties related to the signal enhancement with contrast agents, has led to what at times appeared a confounding array of CMR choices for myocardial perfusion imaging. Both exogenous injected contrast agents and endogenous contrast mechanisms have been used to assess perfusion. To date the use of an exogenous contrast agent has been extensively validated[4, 5] and successfully applied in patient studies.[6–11] With later developments, it has become possible to combine the requirements for spatial and temporal resolution for myocardial perfusion imaging during the first pass, with multislice coverage, necessary to determine blood flow,[11] lesion severity, and infarct extent.[12] A consensus is emerging as to which CMR techniques represent the best choices for myocardial perfusion imaging. The need for quantitative analysis is also receiving increasing acceptance,[13] and we will discuss possible approaches for quantification of myocardial blood flow.[11, 14–17] It is hoped that this review of the theoretical foundations of myocardial perfusion imaging with CMR will convince the reader that the techniques have matured to a point where they are applicable in large-scale clinical studies.

THE PHYSIOLOGICAL BASIS FOR MEASURING MYOCARDIAL PERFUSION

The blood flow resistance of the coronary circulation is determined under normal conditions primarily by the myocardial microcirculation, meaning vessels that are smaller than approximately 300 μm in diameter. The adequate supply of oxygen and metabolites to the myocytes is tightly coupled to myocardial blood flow. Adequate and nearly constant blood flow is maintained through autoregulation and can compensate under resting conditions for up to 80 percent coronary artery diameter narrowing.[18] With more severe narrowing in an epicardial vessel, and in the absence of significant collateral flow, the distal perfusion bed is fully vasodilated even under resting conditions, and no augmentation of blood flow is feasible if the patient exercises or undergoes pharmacologically induced stress. In healthy subjects, myocardial blood flow can increase at least three- to fourfold with maximal vasodilation.[19] This increase means that differences in myocardial blood flow between a region subtended by a stenosed coronary artery and the territory of a normal coronary artery are significantly amplified with maximal vasodilation. Noninvasive myocardial perfusion imaging during pharmacological vasodilation (e.g., with adenosine or dipyridamole) rests on this physiological observation that the hemodynamic significance of a lesion is most apparent during maximal vasodilation.[18] A related measure could be obtained in the catherization laboratory with an intravascular Doppler flow probe by measuring the coronary flow reserve to assess lesion severity.[20] These functional tests of coronary and myocardial blood flow were prompted by the recognition that judging the lesion severity by luminal narrowing as seen on x-ray angiograms has serious shortcomings. Coronary arteriography as performed today under x-ray fluoroscopy is invasive and carries risks for the patient. For this reason noninvasive techniques such as CMR are being used to determine both the myocardial[11] and coronary blood flow reserve.[21, 22]

With the development of CMR as a new imaging modality for myocardial perfusion imaging, it has become possible to probe with sufficient spatial resolution for more subtle indicators of myocardial ischemia (Figure 3–1). Blood flow across the myocardial wall is not uniform, but instead favors the subendocardium to accommodate the higher workload and higher rate of oxygen consumption of the subendocardial layer. Under normal conditions the ratio of endocardial to epicardial blood flow is on the order of 1.15:1. With a coronary artery stenosis, blood flow is first diverted away from the subendocardial layer, and the endocardial to epicardial

*Support through RO1 HL 58876-01 from the National Institutes of Health, a Grant-in-Aid from the American Heart Association (NW), and a biomedical engineering grant from The Whitaker Foundation (MJ-H) is gratefully acknowledged.

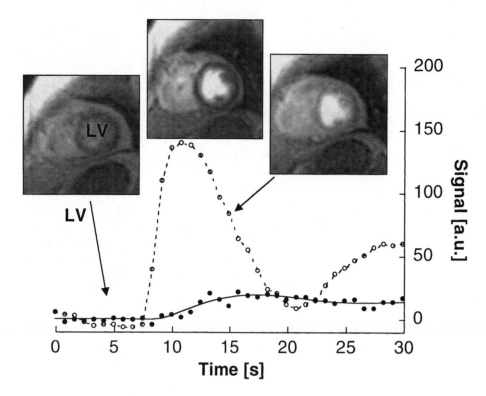

Figure 3–1. Example of CMR images acquired during the first pass of an extracellular contrast agent (Gd-DTPA) in a patient. Images were acquired in the short axis view at the level of the papillary muscle. The arrows show the correspondence between images and the time course of changes in signal intensity for a region of interest in the left ventricular cavity (open circles) and a myocardial segment (solid circles). Characteristics of the signal curve for a tissue region such as its increased dispersion, lower amplitude, and reduced rate of signal enhancement compared with the ventricle result from transit through the tissue microcirculation, where the volume of distribution is limited to either the extracellular or the intravascular space.

blood flow ratio is often less than 1, in particular under stress.[23] With myocardial ischemia the subendocardial layer is accordingly most susceptible to necrosis. This potential advantage of myocardial perfusion imaging can only be sufficiently appreciated if the imaging modality provides spatial resolution on the order of 2 mm or better. The spatial resolution of conventional imaging modalities such as positron-emission tomography (PET) and single photon emission computed tomography (SPECT) is not sufficient to detect blood flow deficits limited to the subendocardial layer (Figure 3–2). More specifically, the sensitivity of a myocardial perfusion imaging technique is directly related to its spatial resolution and the ability to discern transmural variations of flow.[24]

CMR offers the unique possibility of quantitatively assessing both perfusion and function with high accuracy. For the clinical management of patients this unique capability of CMR perfusion and function studies should have deep repercussions, once such protocols for the simultaneous measurement of myocardial blood flow and function have been widely disseminated. Bolli and colleagues[25] showed that even small differences in blood flow during ischemia result in large differences in postischemic function, suggesting that the ability to quantify flow in the low flow range is of importance to predict the probability of postischemic recovery. Herein lies the key to assessing hibernating or stunned myocardium, and distinguishing between these two conditions.

FIRST-PASS IMAGING WITH EXOGENOUS TRACERS

The most important choices for imaging during the first pass of an injected contrast agent relate to the pulse sequence protocol, the type of contrast agent that is being used, the mode of injection, and the application of postprocessing algorithms, including perfusion modeling. Nearly all CMR pulse sequences that have been used for myocardial perfusion imaging have in common that images are acquired in a fraction of a single heartbeat to freeze cardiac motion and capture the transit of the injected tracer with sufficient temporal resolution. The techniques can be further subdivided into T1 or T2* weighted, meaning that the largest changes in regional image intensity occur with T1 or T2* changes, respectively. Changes in T1 and T2* of protons are induced by the presence of contrast agent in tissue or blood. In the microcirculation, T2* changes produced by magnetic susceptibility changes in the blood pool may extend beyond the blood pool because the magnetic field gradients created on a microscopic scale by susceptibility differences extend beyond the region containing the contrast agent. Therefore the changes of a T2*-weighted signal observed with an intravascular contrast agent may suggest a larger volume of distribution than the true vascular volume.[26] By comparison, T1 changes are mediated by magnetic interactions between nuclear dipoles that are short ranged and are directly proportional to contrast agent concentration in

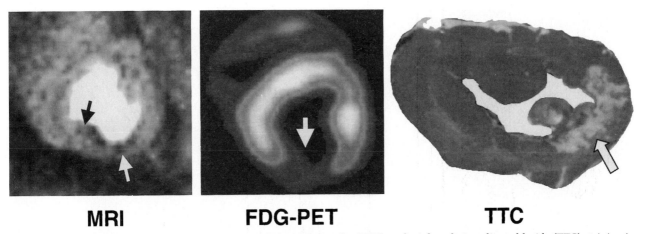

Figure 3–2. Short axis images from CMR, positron-emission tomography (PET), and triphenyltetrazolium chloride (TTC) staining in a porcine model in which obtuse marginal branches of the left circumflex coronary artery had been ligated 4 weeks before combined CMR and PET studies. The CMR image shown on the left was acquired during the first pass of an intravascular iron-oxide contrast agent (NC100150 injection, Nycomed) and corresponds to the highest peak signal enhancement in tissue. A subendocardial perfusion defect is present in the inferior wall (arrows), in agreement with TTC staining, shown at right. Fluorodeoxyglucose (FDG)-PET (middle image) was carried out 3 hours before the CMR study and shows a fixed defect in the posterior segment suggesting a transmural infarct (arrow). $^{13}NH_3$ PET images (not shown) also indicated irreversible damage in the posterior segment, in disagreement with the findings from CMR and TTC staining.

blood or tissue.[27, 28] Gradient-echo sequences with short echo times (TE) are most suitable for T1-weighted perfusion imaging of the heart because they are relatively immune to T2* effects and motion. Echo-planar imaging (EPI) has, with a few exceptions, been used for T2*-weighted perfusion imaging.

EPI has the allure of being the fastest imaging method for reducing motion artifacts (50–100 ms per image acquisition) and intrinsically provides strong T2* weighting. With EPI pulse sequences, numerous gradient echoes are generated by applying a rapidly alternating magnetic field gradient after a single radiofrequency pulse. This process speeds up image acquisition by a factor of at least two compared with ultrafast gradient echo sequences. An effective TE can be defined that characterizes the T2* weighting of the central k-space lines and T2* contrast in the image. This effective TE is about an order of magnitude longer for EPI sequences compared with fast gradient-echo sequences (30–40 ms vs. 1.2–2.0 ms for fast gradient-echo imaging). EPI is therefore suitable for ultrafast T2*-weighted imaging but has also been adapted for T1-weighted imaging because of its speed advantage.[29] Unfortunately, the longer effective TE of EPI sequences gives rise to magnetic susceptibility artifacts in the ventricular cavity, flow artifacts make it very difficult to obtain an accurate input function, and the artifacts can also overshadow the signal changes in the myocardium. As will be discussed later, an arterial input function is necessary for tracer kinetic modeling and quantification of tissue perfusion. The sensitivity to magnetic susceptibility effects can be reduced by generating a spin echo (SE) with a combination of 90° and 180° pulses, and by applying the EPI readout gradient during the formation and decay of the SE. The signal intensity in the ventricle is strongly attenuated with SE EPI, as can be seen in Figure 3–3, because rapid flow prevents complete refocusing of the transverse magnetization with a slice-selective 180° pulse. The measurement of an accurate arterial input function in the ventricle or the aorta continues to pose a problem for EPI. Because of this problem, the role of EPI in myocardial perfusion imaging will probably be confined to a qualitative evaluation of perfusion and infarct extent.

By contrast, fast gradient-echo imaging is well suited for T1-weighted, quantitative myocardial perfusion imaging. Gradient coils and amplifiers on a standard clinical 1.5 T CMR scanner produce maximum gradient amplitudes in the order of 20 to 30 mT/m with a slew-rate of 40 to 50 mT/m/ms. With these specifications one can run ultrafast gradient-echo sequences with a repetition time per phase-encoding step (repetition time, TR) of 2.0 to 2.5 ms, TE of 1.2 to 1.5 ms, a receiver bandwidth of 800 to 1000 Hz/pixel, and an inplane spatial resolution of 2 to 3 mm. Due to the short TE, the signal is quite insensitive to flow and magnetic susceptibility variations. The relatively low signal-to-noise ratio requires the use of a dedicated phase-array cardiac receiver coil for acceptable image quality. T1 weighting is achieved by preceding the magnetization with a nonselective 90° pulse followed by a gradient crusher pulse (saturation recovery preparation), or a nonselective 180° pulse (inversion recovery preparation). The saturation recovery preparation can be repeated in a sequential multislice sequence for each slice, which will result in the same degree of T1 weighting for each slice (Figure 3–4). With a linear k-space order for the phase-encoding steps, the fast gradient-echo readout only needs to be delayed by 10 to 20 ms after the saturation recovery preparation. Figure 3–5 shows

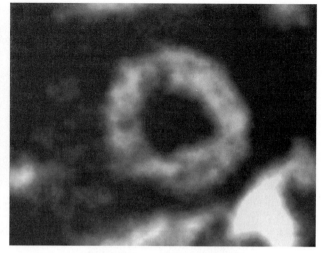

Pre Contrast   Peak of wash-in

Figure 3–3. T1-weighted spin echo echo-planar CMR images at precontrast and during peak signal enhancement with a bolus injection of Gd-DTPA (0.05 mmol/kg Magnevist, Schering AG). The images were acquired with an inversion-recovery prepared spin-echo echo-planar imaging (EPI) sequence at 0.5 T on a mobile CMR scanner using a 20-cm surface coil (echo time [TE] 40 ms; repetition time [TR] = one R-R interval; inversion time adjusted to 300 ms to null myocardial signal before contrast injection; slice thickness 20 mm; field-of-view 50 × 50 cm; read 128; phase 64). With this technique, no significant signal enhancement can be observed in the blood pool. (Images courtesy of J. R. Panting, P. D. Gatehouse, G. Z. Yang, D. N. Firmin, and D. J. Pennell, Royal Brompton Hospital, London, UK.)

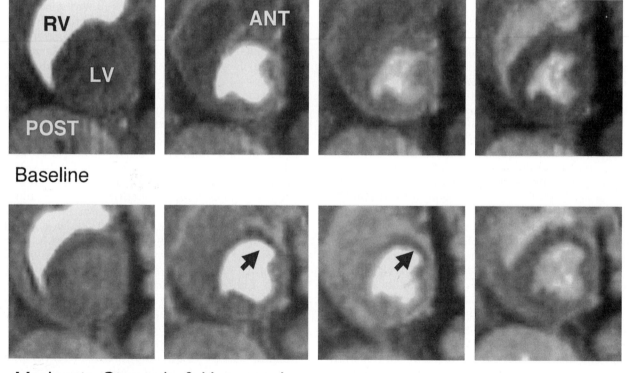

Figure 3–4. First-pass CMR images in the short axis view acquired with a fast gradient-echo sequence in a porcine model with graded stenosis in the left anterior descending (LAD) coronary artery. Images in the upper row were acquired during baseline, and those in the lower row with a moderate LAD stenosis (transstenotic pressure gradient of 30 mm Hg; aortic mean pressure 112 mm Hg; heart rate 112 bpm) and with adenosine-induced hyperemia (0.14 mg/kg/min IV adenosine). A saturation-recovery prepared fast gradient-echo sequence was used at 1.5 T (Siemens Magnetom Vision; TR/TE = 2.4/1.2 ms; flip angle = 18°; field-of-view = 300 mm; read 128, phase 64; slice thickness of 10 mm) to image during the first pass of bolus-injected intravascular contrast agent (Gadomer-17, Schering AG, 0.007 mmol/kg). With this intravascular agent, reduced signal enhancement in the anterior segment (arrows) was observed during moderate stenosis for approximately 2 minutes after injection.

Signal [a.u.]

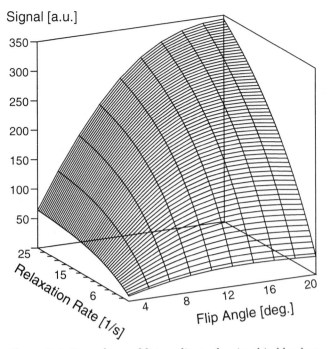

Figure 3–5. Dependence of fast gradient-echo signal in blood on the flip angle and the relaxation rate, calculated for TR/TE = 2.4/1.2 ms; 80 phase-encode steps, and assuming an intrinsic relaxation rate of 1.2 s^{-1} for blood. The graph shows that the dependence of the local signal on the relaxation rate is approximately linear over a wider range of relaxation rates when the flip angle is increased. Optimization of the sequence parameters is essential to obtain an accurate arterial input function.

how the signal intensity measured with the saturation recovery (SR) prepared fast gradient-echo sequence varies with R1 relaxation rate and flip angle. Sequence parameters such as the flip angle should be adjusted such that the signal changes are linearly proportional to the relaxation rate (i.e., contrast agent concentration) over the widest possible range.

Ultrafast T1 measurements may provide the most accurate measurements of contrast agent concentration, but current hardware on clinical scanners may not be up to the task of measuring T1 during contrast agent transit with single-heartbeat temporal resolution. Chen and colleagues developed such a T1 Fast Acquisition Relaxation Mapping (T1-FARM) method to obtain single-slice T1 maps of the heart with 1-s resolution.[30, 31] With a direct measurement of T1, the problems associated with the saturation of the signal with increasing contrast agent dosage can be avoided. This method should allow for a more accurate quantification of blood flow with usage of higher contrast agent dosages. Figure 3–6 shows two signal curves for a region of interest in the left ventricle measured during bolus injection of 0.075 mmol/kg of Gd-DTPA, with the T1-FARM and the saturation recovery prepared fast gradient-echo techniques, respectively. For better temporal resolution, to achieve multislice coverage, and to allow simpler image reconstruction, one will resort to pulse sequences that produce a T1-weighted signal

versus measuring multiple points of the magnetization recovery to quantify T1.

With ^{1}H CMR, the presence of a contrast agent is detected indirectly through the reduction in relaxation rates of ^{1}H nuclear spins. Because the contrast agent molecules are detected indirectly through their effect on the ^{1}H signal, a confounding effect is introduced by the effects of water exchange. Thus, even if a contrast agent is confined only to the vascular space where it reduces T1, the magnetization outside the vascular space can also relax at a faster rate due to the exchange of water protons between vascular and extravascular compartments. The T1 relaxation rate of a voxel comprising several compartmental spaces is therefore not only a function of the intrinsic T1 relaxation rates in each compartment, but also the rate of water exchange between compartments. Although the rate of water exchange might provide useful information about capillary permeability, one would in general rather deal with the situation of no exchange, where a simple relationship relates the contrast agent concentration and signal intensity (the opposite limit of fast exchange is probably less relevant for myocardial first-pass signal curves).[32] The primary effect of water exchange on the quantification of perfusion is an overestimate of the blood volume if the no-exchange assumption is used, and an underestimate of blood volume if the fast-exchange assumption is used.[33] With a saturation recovery prepared, ultrafast gradient-echo sequence, one can minimize the effects of water exchange by choosing a short TR, and high flip angle (e.g. TR < 4 ms and flip angle > 20°).[33] Higher flip angle pulses drive the magnetization faster during the relaxation recovery, thereby increasing the effective relaxation rate. Fewer water exchange steps take place during relax-

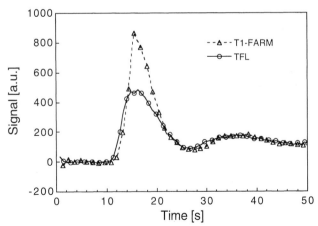

Figure 3–6. Comparison of signal time curves for region of interest in left ventricle obtained in a dog with quantitative T1 imaging (T1 Fast Acquisition Relaxation Mapping [T1-FARM]) and T1-weighted, saturation-recovery prepared gradient echo (TFL; TR/TE = 2.4/1.2 ms; flip angle = 18°) and a Gd-DTPA dosage of 0.075 mmol/kg at 1.5 T. The curves were normalized so the fast gradient-echo and T1-FARM recirculation peaks were equal. (Data courtesy of Z. Chen, C. A. McKenzie, and F. S. Prato, St. Joseph's Hospital, London, Ontario.)

ation recovery and the effects of exchange are reduced.[33] With a sufficiently high flip angle (20°) the no-exchange assumption is reasonable for the interpretation of the signal curves measured with an ultra-fast (TR = 2.4 ms) saturation recovery prepared fast gradient-echo sequence.[34]

ENDOGENOUS CONTRAST FOR THE ASSESSMENT OF MYOCARDIAL PERFUSION

Approaches have been developed for an assessment of myocardial perfusion that are not based on the use of an exogenous CMR contrast agent, but instead exploit endogenous contrast mechanisms related to blood flow and blood oxygenation. These approaches fall under the categories of spin-labeling,[35] blood-oxygen dependent (BOLD)[36] contrast (see Chapter 34), and magnetization transfer contrast (MTC).[37, 38] These techniques have to be considered technically more challenging than measurements with exogenous contrast agents, and generally offer only an indirect measure of blood flow.

With spin labeling, the spins are either inverted or saturated in a slab that is located upstream from the imaged slice.[35, 39, 40] The flow-dependent change of signal intensity due to the inflow of saturated or inverted spins into the image slice provides a measure of tissue perfusion. The spin-labeling method generally relies on the assumption that the net arterial blood flow in the myocardium follows the direction from base to apex.[40] Cardiac motion complicates the interpretation of signal changes in a spin-labeling experiment because the labeled spins can be transported into the imaged slice either through blood flow or by through-slice-plane motion of the heart. Water can be considered a freely diffusible tracer; that is, blood flow is flow limited and does not depend on compartment sizes and barrier permeabilities. For a highly diffusible tracer such as water, studies by Bassingthwaighte (personal communication) indicate that flow shunting from the arterial to the venous side has to be taken into account.

BOLD offers a measure of hemoglobin saturation reflecting regional oxygen supply and demand. Deoxyhemoglobin is paramagnetic, whereas oxyhemoglobin is only diamagnetic, which means that deoxyhemoglobin causes a considerably larger reduction of T2* than oxyhemoglobin and therefore a larger signal intensity attenuation in gradient-echo images.[41] Deoxyhemoglobin can be used as an endogenous intravascular tracer due to the tight coupling between oxygen demand and blood flow. BOLD contrast changes have been observed in the heart by Li and associates after administration of dipyridamole and dobutamine.[36] Li and associates concluded that the link between BOLD contrast and blood flow depends on the balance between blood flow and oxygen metabolism, that is, between oxygen supply and demand (see Chapter 34).

MTC is created by saturating the pool of restricted protons bound in macromolecules and observing the subsequent effects of magnetization exchange between these restricted protons and freely diffusible protons. The apparent first-order exchange rate between the restricted proton pool and the freely diffusible protons is related to tissue perfusion. Other factors causing changes in MTC are blood volume changes and tissue denaturation after infarction. It has been found that MTC is significantly decreased in infarcted myocardium, and that MTC is significantly greater in chronic infarcts compared with acute infarcts.[42] The exact relationship between MTC and blood flow remains a subject of investigation.

Spin-labeling, BOLD, and MTC CMR can be carried out in the steady state, meaning there is no need to capture with ultrafast imaging the transient signal changes observable in the myocardium after injection of an exogenous tracer. The physiological basis for the observed signal changes observed with these noninvasive techniques is still being investigated.

QUANTITATIVE EVALUATION OF MYOCARDIAL PERFUSION

The images acquired in a myocardial perfusion study with a bolus injection of contrast agent can be qualitatively evaluated based on the level of differential signal enhancement in the myocardium. With a T1-weighted imaging technique, myocardial segments not showing signal intensity enhancement during the first pass are interpreted as hypoperfused. This type of qualitative judgment of signal enhancement differences suffers from substantial observer bias, and small reductions in perfusion are missed. With an appropriate imaging technique such as a fast T1-weighted gradient echo sequence and low contrast agent dosages (e.g., 0.03 mmol/kg of Gd-DTPA for an IV bolus injection into an antecubital vein), the signal changes are proportional to the local contrast agent concentration. Under these conditions the signal curves can be interpreted as, or transformed into, contrast agent residue curves. Signal curves can be generated for user-defined regions of interest (ROI) but this is a somewhat cumbersome process. It is preferable to use computer programs that calculate parametric or myocardial sector maps depicting the variation of blood flow, for example in a short axis section of the heart.[43] A three-dimensional volumetric rendering of myocardial blood flow is being developed for this type of evaluation. CMR perfusion studies require a technically more sophisticated approach than PET for image segmentation and registration. By contrast to PET, CMR perfusion studies are not motion-averaged, the endocardial borders are better defined due to the superior spatial resolution, and the large signal enhancement in the ventricular cavity requires accurate image segmentation to avoid spillover effects.

QUANTIFICATION OF REGIONAL MYOCARDIAL BLOOD FLOW

The central volume theorem relates mean transit time (MTT) for a tracer, its distribution volume, and flow:

$$MTT = \frac{Volume}{Flow} \qquad (1)$$

The mean transit time is further defined by the first moment of a tracer concentration curve, $c(t)$:

$$MTT = \frac{\int_0^\infty t \cdot c(t) dt}{\int_0^\infty c(t) dt} \qquad (2)$$

A fundamental problem is that the MTT should be determined from an outflow concentration curve, not from the measured residue curve.[44] Because a tissue region of interest does not have a single, well-defined arterial input and venous output, and may actually comprise conduit vessels and venous output from adjacent regions, this distinction between residue curves and outflow curves tends to be fuzzy in practice. Reasonable agreement between the MTT determined from the measured signal curve and *relative* changes in myocardial blood flow has been observed when the arterial input is narrow in comparison to the tissue residue curve.[4, 15] Recirculation of the contrast agent gives rise to an additional delayed input that has to be excluded if the MTT is calculated directly from the measured residue curves. The input function is generally obtained from a region upstream from the region of interest, and in practice this means either a region in the ascending aorta[45] or the left ventricular blood pool.[16] This approximation is only appropriate if delay and dispersion in the vascular elements upstream from the region of interest are small, or if transport in the upstream vascular elements has been well characterized to account for it.

A measure of blood flow can be recovered when the input is dispersed by calculating the tissue residue curve for a hypothetical impulse input.[46] If we assume that the transport of tracer through the tissue is a linear and stationary process, we can express the measured residue curve $m(t)$ as a convolution of the measured input function, $c_{in}(t)$, with the residue impulse response, $R(t)$:

$$m(t) = F \int_0^t c_{in}(\tau) \cdot R(t - \tau) \cdot d\tau = R(t) \cdot c_{in}(t) \qquad (3)$$

$R(t)$ can be interpreted as the probability that a tracer molecule remains in the ROI up to time t. $R(t)$ decays with time from its initial amplitude as the probability that a tracer molecule remains in the ROI decreases. The initial amplitude of the residue impulse response is a measure of blood flow, whereas the MTT is given by the ratio of area over height of the residue impulse response[46]:

$$MTT = \frac{\int_0^\infty R(t) \cdot dt}{max[R(t)]} = \frac{Volume}{Flow} \qquad (4)$$

The task is now to recover $R(t)$ from the measured residue curve through *deconvolution* to estimate flow from the amplitude of $R(t)$.

Model-free deconvolution is sensitive to the noise in the measured data. This sensitivity to noise is substantially reduced if the deconvolution is constrained with a model for the flow of the tracer through tissue. Numerous tracer kinetic models have been developed to model and determine tissue blood flow. They can be subdivided into compartmental and spatially distributed models. For a compartmental model it is assumed that the tracer is well mixed within each compartment, meaning that the contrast agent concentration equilibrates instantaneously within each compartment. A two-compartment model originally developed by Sangren and Sheppard[47] to account for the exchange of tracer between the intravascular and interstitial spaces has been used for the absolute quantification of myocardial blood flow with an extracellular contrast agent.[48] The Kety model represents the most rudimentary compartmental model because the myocardium is treated as a single compartment. The impulse response for the Kety model is a single exponential function with parameters for blood flow, the fraction of contrast agent extracted into the myocardium, and the partition coefficient. The extraction fraction is not well defined during the first pass of Gd-DTPA because it increases over the initial 30 to 40 s and then decreases.[49] The Kety equation allows determination of only the product of flow and extraction fraction. Due to the variation in the extraction fraction during the first pass, and because blood flow and flux through the capillary barrier are not distinguished, it has not been proven that the Kety model provides a true measure of blood flow.

The rapid injection of a tracer will create a tracer concentration gradient from the arterial to the venous side. Spatially distributed models are described by *partial* differential equations that account for the variation of tracer concentrations within each tissue region as a function of time *and* at least one spatial variable. Spatially distributed models such as the one developed by Bassingthwaighte and associates[50, 51] include parallel pathways to account for the heterogeneity of blood flows. The number of parameters in these models tends to be rather large. Sensitivity analysis is called for to determine which (hopefully few) model parameters need to be adjusted for a best fit of the measured residue curves.[52] The number of degrees of freedom can be reduced

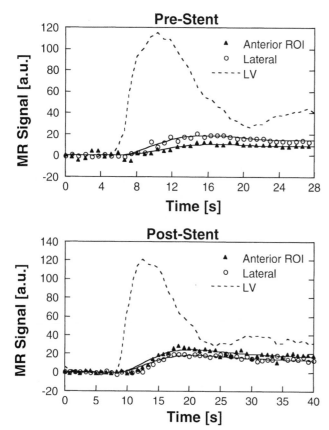

Figure 3–7. First-pass curves for the anterior and lateral segments obtained in a patient with fast gradient-echo CMR before and after stent placement. Catheterization showed that the patient had a 5-mm lesion causing a 50% LAD stenosis with a mild 13 mm Hg translesional pressure gradient. Fitting of the first-pass signal curves with a multipath, distributed model (MMID4, National Simulation Resource, University of Washington, Seattle) indicated that blood flow in the anterior segment improved from 0.6 ml/min/g (heart rate [HR] = 76 bpm) before angioplasty to 1.4 ml/min/g (HR = 73 bpm) after stent placement. This example demonstrates the ability to quantify incremental changes of myocardial blood flow using CMR in a patient to determine if tissue blood flow has been completely restored.

by using an intravascular contrast agent.[16, 53] The realism of spatially distributed models of blood tissue exchange is of advantage for the identification and quantification of physiological changes observed indirectly with tomographic imaging techniques such as CMR.[54] The parameters of compartmental models often do not have a clear physiological analog. Figure 3–7 shows an example of the application of the MMID4 model[50] for fitting tissue residue curves and determining absolute myocardial blood flow. The application of this model has been validated against blood flow measurements with radioisotope labeled microspheres.[53]

MYOCARDIAL PERFUSION RESERVE

The measurement of the myocardial perfusion reserve allows assessment of the functional signifi-

cance of a coronary artery lesion, or assessment of impaired microvascular reactivity.[55] It is determined from the ratio of perfusion during maximal hyperemia and rest. Wilke and coworkers have demonstrated that this ratio can be achieved using CMR first-pass measurements.[11] In the presence of an epicardial narrowing, the blood flow reserve in the coronary vessel is reduced, but may be closer to normal in the territory of the coronary vessel if the artery is fully collateralized. The maximum recruitable collateral flow corresponds to the difference between the coronary flow reserve and the myocardial perfusion reserve. Measurement of the perfusion reserve is also useful in the absence of coronary artery lesions to investigate coronary endothelial function. The endothelium plays a major role in the regulation of the vasomotor tone of large arteries through release of vasoactive substances that also have a marked effect on coronary resistance vessels. CMR first-pass imaging and PET[55] are the only noninvasive modalities for measuring the myocardial perfusion reserve.[11]

For determination of the myocardial perfusion reserve with CMR, (absolute) myocardial blood flow is measured during rest and stress with two consecutive first-pass studies. The measurements should be separated by 5 to 10 minutes to allow the contrast agent injected for the rest study to reach a semiequilibrium distribution. A previous low-dosage injection of contrast agent causes only a slight elevation of the background signal, which can be assumed constant during the 1- to 2-minute duration of a first-pass measurement.

REQUIREMENTS FOR QUANTITATIVE CMR PERFUSION IMAGING OF THE HEART

The requirements for optimizing the quantitative assessment of myocardial perfusion with an exogenous tracer, which can be either an extracellular or intravascular contrast agent, are summarized here:

1. Contrast agent injection: Requires a very rapid bolus injection for sensitive discrimination of blood flow changes (Figure 3–8). This injection can be achieved by using power injector and an injection rate of approximately 5 to 10 ml/s.
2. CMR: Spatial resolution should be less than 2.5 mm to resolve transmural variations in blood flow. The temporal resolution should be 1 to 2 images per heartbeat during rest studies, and 1 image/heartbeat for stress studies.
3. Residue detection: Recommend T1-weighted imaging with minimized sensitivity to water exchange to simplify conversion of signal time curves into residue curves. Use low-contrast agent dosages to maintain approximately linear relationship between signal and contrast agent concentration. T1-weighted imaging provides accurate arterial input

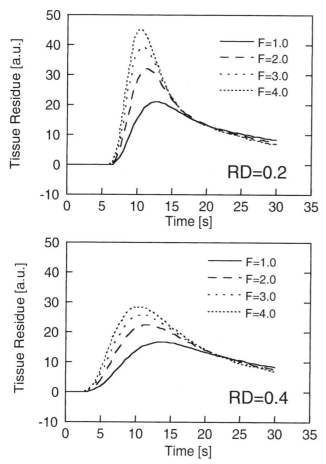

Figure 3–8. Tissue residue model curves calculated with a multipath, distributed model for gamma-variate input functions with a relative dispersion (RD) of 0.2 and 0.4, respectively. This example illustrates that a more compact contrast agent bolus will result in a higher myocardial signal enhancement and better discrimination of incremental changes in blood flow (F; 1.0 ml/min/g increments in this figure).

function—difficult or impossible to achieve with T2*-weighted imaging.

4. Modeling: Parametric deconvolution to minimize sensitivity to noise. Recommend use of physiologically realistic model that establishes a clear correspondence between changes in model parameters and physiological parameters (blood flow, blood volume, capillary permeability, etc.). Recirculation component should be treated as delayed input, instead of extrapolating first-pass component (e.g., with a gamma-variate function fit).

CONCLUSION

In the context of assessing myocardial, microcirculatory blood flow, the CMR first-pass technique can address several clinical issues, if the appropriate conditions for spatial and temporal resolution can be met. Valuable, quantitative, and largely observer-independent information related to the pathophysiology of ischemic heart disease becomes available with myocardial perfusion imaging. CMR

perfusion reserve measurements can be applied to assess the degree of atherosclerosis, the functional significance of coronary artery lesions, and to evaluate myocardial viability. In milder forms of ischemic heart disease and ischemic cardiomyopathies, CMR at rest and during stress is a good test to probe for more subtle perfusion limitations that may be confined to the subendocardium or have resulted in only moderate blood flow reductions. For these applications, quantitative CMR methods that provide a true measure of myocardial blood flow have been successfully developed and validated.

References

1. Nitzsche EU, Choi Y, Czernin J, et al: Non-invasive quantification of myocardial blood flow in humans. Circulation 93:2000, 1996.
2. Kaul S: Myocardial contrast echocardiography—15 years of research and development. Circulation 96:3745, 1997.
3. Haude M, Caspari G, Baumgart D, et al: Comparison of myocardial perfusion reserve before and after coronary balloon predilatation and after stent implantation in patients with post-angioplasty restenosis. Circulation 94:286, 1996.
4. Wilke N, Simm C, Zhang J, et al: Contrast-enhanced first pass myocardial perfusion imaging: Correlation between myocardial blood flow in dogs at rest and during hyperemia. Magn Reson Med 29:485, 1993.
5. Burstein D, Taratuta E, Manning WJ: Factors in myocardial "perfusion" imaging with ultrafast MRI and Gd-DTPA administration. Magn Reson Med 20:299, 1991.
6. Manning WJ, Atkinson DJ, Grossman W, et al: First-pass nuclear magnetic resonance imaging studies using gadolinium-DTPA in patients with coronary artery disease. J Am Coll Cardiol 18:959, 1991.
7. Atkinson DJ, Burstein D, Edelman RR: First-pass cardiac perfusion: Evaluation with ultrafast MR imaging. Radiology 174:757, 1990.
8. Wilke N, Maching T, Engels G: Dynamic perfusion studies by ultrafast MR imaging: Initial clinical results from cardiology. Electromedica 58:102, 1990.
9. Schaefer S, Tyen Rv, Saloner O: Evaluation of myocardial perfusion abnormalities with gadolinium-enhanced snapshot MR imaging in humans. Radiology 185:795, 1992.
10. Walsh EG, Doyle M, Lawson MA, et al: Multislice first-pass myocardial perfusion imaging on conventional clinical scanner. Magn Reson Med 34:39, 1995.
11. Wilke N, Jerosch-Herold M, Wang Y, et al: Myocardial perfusion reserve: Assessment with multisection, quantitative, first-pass MR imaging. Radiology 204:373, 1997.
12. Wendland MF, Saeed M, Masui T, et al: Echo-planar MR imaging of normal and ischemic myocardium with gadodiamide injection. Radiology 186:535, 1993.
13. Bianco JA, Alpert JS: Physiologic and clinical significance of myocardial blood flow quantitation: What is expected from these measurements in the clinical ward and in the physiology laboratory? Cardiology 88:116, 1997.
14. Diesbourg LD, Prato FS, Wisenberg G, et al: Quantification of myocardial blood flow and extracellular volumes using a bolus injection of Gd-DTPA: kinetic modelling in canine ischemic disease. Magn Reson Med 23:239, 1992.
15. Keijer JT, Rossum ACv, Eenige MJv, et al: Semiquantitation of regional myocardial blood flow in normal human subjects by first-pass magnetic resonance imaging. Am Heart J 130:893, 1995.
16. Kroll K, Wilke N, Jerosch-Herold M, et al: Accuracy of modeling of regional myocardial flows from residue functions of an intravascular indicator. Am J Physiol (Heart Circ Physiol) 40:H1643, 1996.
17. Jerosch-Herold M, Wilke N, Wang Y, Stillman AE: Can absolute myocardial blood flow be quantified with the MR first

pass technique and an extracellular contrast agent (abstract). 5th scientific meeting and exhibition, International Society for Magnetic Resonance in Medicine, Vancouver, 1997, p 841.

18. Gould KL, Kirkeeide RL, Buchi M: Coronary flow reserve as a physiologic measure of stenosis severity. J Am Coll Cardiol 15:459, 1990.

19. Wilson RF, Wyche K, Christensen BV, et al: Effects of adenosine on human coronary arterial circulation. Circulation 82:1595, 1990.

20. Wilson RF: Assessment of the human coronary circulation using a Doppler catheter. Am J Cardiol 67:44D, 1991.

21. Sakuma H, Blake LM, Amidon TM, et al: Coronary flow reserve: Noninvasive measurement in humans with breath-hold velocity encoded cine MR imaging. Radiology 198:745, 1996.

22. Clarke GD, Eckels R, Chaney C, et al: Measurement of absolute epicardial coronary artery flow and flow reserve with breath-hold cine phase-contrast magnetic resonance imaging. Circulation 91:2627, 1995.

23. Bache RJ, Schwartz JS: Effect of perfusion pressure distal to a coronary stenosis on transmural myocardial blood flow. Circulation 65:928, 1982.

24. Klocke FJ: Quantitative evaluation of coronary perfusion in man. Cathet Cardiovasc Diagn 1:349, 1975.

25. Bolli R, Zhu W-X, Thornby JI, et al: Time course and determinants of recovery of function after reversible ischemia in conscious dogs. Am J Physiol 254:H102, 1988.

26. Weisskoff RM, Zuo CS, Boxerman JL, Rosen BR: Microscopic susceptibility variation and transverse relaxation: Theory and experiment. Magn Reson Med 31:601, 1994.

27. Koenig SH, Spiller M, Brown R III, Wolf GL: Relaxation of water protons in the intra- and extracellular regions of blood containing Gd(DTPA). Magn Reson Med 3:791, 1986.

28. Strich G, Hagan PL, Gerber KH, Slutsky RA: Tissue distribution and magnetic resonance spin lattice relaxation effects of gadolinium-DTPA. Radiology 154:723, 1985.

29. Edelman RR, Li W: Contrast-enhanced echo-planar MR imaging of myocardial perfusion: Preliminary study in humans. Radiology 190:771, 1994.

30. Chen Z, Prato FS, McKenzie CA: T1 Fast Acquisition relaxation mapping (T1-FARM): An optimised reconstruction. IEEE Trans Med Imag 17:155, 1998.

31. Tong CY, Prato FS, Wisenberg F, et al: Techniques for the measurement of the local myocardial extraction efficiency for inert diffusible contrast agents such as gadopentate dimeglumine. Magn Reson Med 30:332, 1993.

32. Judd RM, Atalay MK, Rottman GA, Zerhouni EA: Effects of myocardial water exchange on T1 enhancement during bolus administration of MR contrast agents. Magn Reson Med 33:215, 1995.

33. Donahue KM, Weisskoff RM, Chesler DA, et al: Improving MR quantification of regional blood flow volume with intravascular T1 contrast agents: Accuracy, precision, and water exchange. Magn Reson Med 36:858, 1996.

34. Jerosch-Herold M, Wilke N, Stillman AE, Wilson RF: MR quantification of the myocardial perfusion reserve with a Fermi function model for constrained deconvolution. Med Phys 25:73, 1998.

35. Williams DS, Detre JA, Leigh JS, Koretsky AP: Magnetic resonance imaging of perfusion using spin inversion of arterial water. Proc Natl Acad Sci U S A 89:212, 1992.

36. Li D, Dhawale P, Rubin PJ, et al: Myocardial signal response to dipyridamole and dobutamine: Demonstration of the BOLD effect using a double-echo gradient-echo sequence. Magn Reson Med 36:16, 1996.

37. Balaban RS, Chesnick S, Hedges K, et al: Magnetization transfer contrast in MR imaging of the heart. Radiology 180:671, 1991.

38. Prasad PV, Burstein D, Edelman RR: MRI evaluation of myocardial perfusion without a contrast agent using magnetization transfer. Magn Reson Med 30:267, 1993.

39. Atalay MK, Reeder SB, Zerhouni EA, Forder JR: Blood oxygenation dependence of T1 and T2 in the isolated, perfused rabbit heart at 4.7T. Magn Reson Med 34:623, 1995.

40. Reeder SB, Atalay MK, McVeigh ER, et al: Quantitative cardiac perfusion: A noninvasive spin-labeling method that exploits coronary vessel geometry. Radiology 200:177, 1996.

41. Ogawa S, Lee TM, Kay AR, Tank DW: Brain magnetic resonance imaging with contrast dependent on blood oxygenation. Proc Natl Acad Sci U S A 87:9868, 1990.

42. Scholz TD, Hoyt RF, DeLeonardis JR, et al: Water-macromolecular proton magnetization transfer in infarcted myocardium: A method to enhance magnetic resonance image contrast. Magn Reson Med 33:178, 1995.

43. Jerosch-Herold M, Wilke N, Stillman AE, et al: Multi-slice, functional perfusion maps derived from magnetic resonance first pass images—validation studies in dogs (abstract). Circulation 92:I-316, 1995.

44. Weisskoff RM, Chesler D, Boxerman JL, Rosen BR: Pitfalls in MR measurement of tissue blood flow with intravascular tracers: Which mean transit time? Magn Reson Med 29:553, 1993.

45. Fritz-Hansen T, Rostrup E, Larsson HBW, et al: Measurement of the arterial concentration of Gd-DTPA using MRI: A step toward quantitative perfusion imaging. Magn Reson Med 36:225, 1996.

46. Clough AV, Al-Tinawi A, Linehan JH, Dawson C: Regional transit time estimation from image residue curves. Ann Biomed Eng 22:128, 1994.

47. Sangren WC, Sheppard CW: Mathematical derivation of the exchange of a labeled substance between a liquid flowing in a vessel and an external compartment. Bull Math Biophys 15:387, 1953.

48. Jerosch-Herold M, Wilke N: MR first pass imaging: Quantitative assessment of transmural perfusion and collateral flow. Int J Card Imaging 13:205, 1997.

49. Tong CY, Prato FS, Wisenberg F, et al: Measurement of the extraction efficiency and distribution volume for Gd-DTPA in normal and diseased canine myocardium. Magn Reson Med 30:337, 1993.

50. Bassingthwaighte JB, Chan IS, Wang CY, King RB: XSIM/MMID4—multi-path, multi-indicator dilution, 4-region model software, National Simulation Resource, Seattle, 1997.

51. Bassingthwaighte JB, Goresky CA: Modeling in the analysis of solute and water exchange in the microvasculature. *In* Renkin EM, Michel CC (eds): Handbook of Physiology—The Cardiovascular System. Bethesda, MD, American Physiology Society, 1984, p 549.

52. Bassingthwaighte JB, Chaloupka M: Sensitivity functions in the estimation of parameters of cellular exchange. Fed Proc 43:180, 1984.

53. Wilke N, Kroll K, Merkle H, et al: Regional myocardial blood volume and flow: First pass MR imaging with polylysine-gadolinium-DTPA. J Magn Reson Imaging 5:227, 1995.

54. Bassingthwaighte JB, Raymond GR, Chan JIS: Principles of tracer kinetics. *In* Zaret BL, Beller GA (eds): Nuclear Cardiology: State of the Art and Future Directions. St. Louis, Mosby-Year Book, 1993, p 3.

55. Beanlands RS, Muzik O, Melon P, et al: Noninvasive quantification of regional myocardial flow reserve in patients with coronary atherosclerosis using nitrogen-13 ammonia positron emission tomography. Determination of extent of altered vascular reactivity. J Am Coll Cardiol 26:1465, 1995.

CHAPTER **4**

Tagged Cardiovascular Magnetic Resonance of Systole

Leon Axel

The primary function of the heart is to contract and pump blood, thereby maintaining blood pressure and providing nutrients to the other organs of the body. However, many cardiac diseases can affect the systolic function of the heart. Assessment of cardiac function is thus important for diagnosis and treatment of heart disease. Conventional means of assessing cardiac function, including two-dimensional echocardiography, radionuclide ventriculography, invasive left ventriculography, and cine magnetic resonance imaging (MRI) have many limitations, particularly for the measurement of regional systolic function. Tagged MRI has the potential to significantly improve the evaluation of cardiac function, including the ability to provide novel information about regional function.

Quantitative assessment of cardiac function in clinical practice is currently largely confined to determination of the ejection fraction (EF), the fraction of the end-diastolic intraventricular volume that is ejected during ventricular systole. Although very useful as a measure of global function, the EF is only indirectly related to regional cardiac function. A patient may have significant regional dysfunction yet maintain an EF that is within normal limits.

Qualitative evaluation of regional ventricular function can be performed with tomographic imaging methods such as transthoracic or transesophageal echocardiography and cine MRI. With tomographic imaging, the motion of the endocardial surface or the changing thickness of the wall in the image plane can be followed through systole and visually assessed. However, this process does not account for the base-to-apex shortening that occurs with systole. The three-dimensional (3-D) motion of the curved heart wall through the fixed imaging plane can affect the apparent local motion in the two-dimensional (2-D) image. In addition, the lack of identifiable landmarks within the heart wall makes assessment of the transmural variation of motion, or any components of the motion other than radial, very limited.

For animal/research studies, small metal markers or pairs of ultrasonic transducers can be implanted in the ventricular wall to study intramural motion. However, the invasive nature of the implantation makes this method impractical for clinical use. Even

for research purposes, only a limited number of such implanted markers can be tracked within the wall of a given heart. In addition, the implantation process may significantly alter regional contractility.

The intrinsic motion sensitivity of MRI, coupled with the excellent tomographic images of the heart that MRI can provide, offers the possibility of a powerful new quantitative noninvasive tool to characterize regional systolic function. Two different approaches have been taken to studying regional cardiac motion with MRI: tagging methods and phase shift methods. With tagging methods, the magnetization of the heart (and surrounding structures) is locally perturbed in order to create MRI-visible markers (tags) within the heart wall that will move (shift) with the underlying myocardium. In phase shift methods, the magnetic resonance (MR) pulse sequence is modified so that the phase of the received MR signal is altered, depending on the local velocity of the wall. Although both approaches have their relative strengths and weaknesses, in this chapter, I will focus on tissue-tagging MRI methods, reviewing basic tagging techniques, general methods to extract motion data from tagged images, the analyses of tagged motion data, and reviewing clinical data obtained in health and disease. All of these technical methods are undergoing continuing development and it is to be expected that tagged MRI will become an increasingly useful tool for the study of cardiac function as these methods are further improved, analyses become more automated, and more experience is acquired with them.

TAGGED CARDIOVASCULAR MAGNETIC RESONANCE

The basic idea of using magnetization tagging to study motion with MR precedes MRI. For example, Singer[1] used local magnetization tagging of blood (with a surface coil) and measured the effect on the signal detected with another coil located downstream over the vessel. Similarly, various time-of-flight effects have been used to study blood flow with cardiovascular magnetic resonance (CMR). The initial tagging studies of cardiac motion used selective excitation, similar to the slice selective excita-

41

tion used to select the region to be imaged in CMR, to saturate the magnetization in a thin planar region (or a few such regions) perpendicular to the imaging plane prior to acquiring the CMR data.[2] The altered magnetization in the tagged region showed up as a dark linear region in the subsequent image where it intersected the imaging plane. If the underlying tissue moved between the times of tagging and imaging, the altered magnetization of the tagged region moved with it, revealing the tissue motion by the motion of the tag line in the resulting image. Such a noninvasive MR tag persists for times on the order of the T1 relaxation time, which is comparable to the cardiac cycle duration. Conventional CMR requires data acquisition over multiple heart beats; if the tagging is applied at a consistent time in the cardiac cycle (e.g., end diastole) the tag positions in the images acquired at different cardiac phases will reflect the corresponding interval motions of the heart wall. The accuracy of the magnetization tag motion as reflecting the motion of the underlying material has been validated, both with deformable phantoms[3, 4] and in vivo.[5, 6]

Although only a limited number of tag planes can be readily created with selective excitation techniques, a whole family of parallel tagging planes can be efficiently created simultaneously with spatial modulation of magnetization (SPAMM). In the simplest implementation of SPAMM, nonselective radiofrequency (RF) excitation is used first to create transverse magnetization throughout the imaging region, initially all with the same phase. Application of a magnetic field gradient (in the direction perpendicular to the desired tag planes) then causes evolu-

tion of a phase variation along the direction of the gradient. Finally, application of a second RF excitation pulse then turns the resulting periodic phase variation into a corresponding periodic variation in the magnetization.[7] This periodic variation of magnetization shows up as parallel dark bands in a subsequent image. This simple sequence of excitation-gradient-excitation produces a sinusoidal variation of magnetization, corresponding to fuzzy tag stripes. Longer sequences of excitation RF pulses interspersed with intervals of magnetic field gradient-induced phase evolution can be used to create sharper tags.[8] Creation of two perpendicular families of tag lines results in a tagging grid (Figure 4–1).[8] The size and spacing of the grid is limited by the resolution of the image. A resolution-limited set of tag lines can typically be created in a few milliseconds. The magnetization tagging is a "preconditioning" pulse sequence, analogous to an inversion or fat saturation pulse, and can be used in combination with essentially any desired imaging method.

Tag persistence is affected by the type of imaging used. When the slice being imaged experiences only one imaging excitation per heartbeat (as in conventional electrocardiogram [ECG]-gated spin echo or echo-planar imaging), the fading of the tagged grid will directly reflect the T1 relaxation time of the tissue. However, when the slice experiences multiple excitations per heartbeat, as with rapid cine CMR, the saturating effect of the imaging excitations will diminish the relative contrast between the tags and the background tissue, appearing as accelerated fading of the tags.[9] This impact can be minimized by reducing the flip angle of the gradient echo RF

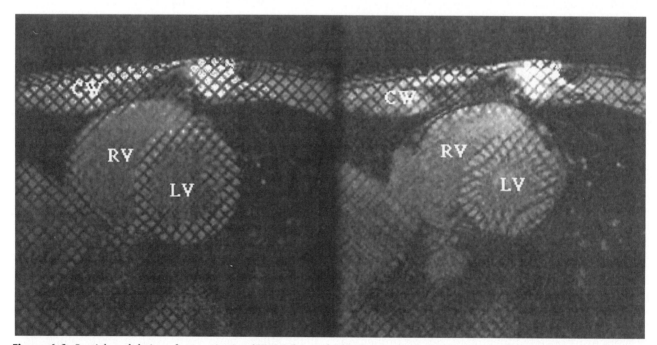

Figure 4–1. Spatial modulation of magnetization (SPAMM) tagged CMR images in short axis plane at different phases of the cardiac cycle (end-diastole, *left*, and end-systole, *right*), showing deformation of the tagging grid (dark lines) produced by corresponding motion of the underlying heart wall. The images were acquired with free breathing, using respiratory-gated, segmented k-space acquisition gradient echo CMR. CW = chest wall; LV = left ventricle; RV = right ventricle.

excitation pulses. Any blurring of the tagged images, such as due to respiratory motion during image acquisition or rapid cardiac motion during segmented k-space data acquisition, will also result in premature tag fading. This problem can be minimized by reducing the corresponding motion effects (e.g., breath-holding).

There are other variations on tagging. SPAMM can be used in combination with selective tagging to create spatially modulated tagging stripes.[10] Two images acquired with tagging patterns offset by half a wavelength can be subtracted from each other to create a difference image with apparently sharper and longer lasting tags, but at the cost of longer image acquisition time.[11] Imaging of tag planes perpendicular to the image plane gives information about components of motion in the image plane; tag planes oriented obliquely to the image plane will have tag image motion that reflect both in-plane and through-plane motions.[12, 13] Phase-contrast imaging techniques with sensitization to through-plane velocity can be used in conjunction with tagging for in-plane displacement to provide more complete information on wall motion.[14, 15] Alternatively, tagged images can be acquired in orthogonal planes (e.g., long-axis and short-axis images) to provide more complete 3-D wall motion information.[16]

ANALYSIS OF TAGGED IMAGES

Although qualitative visual assessment of the motion of the tags in the heart wall in successive phases of the cardiac cycle can be useful, particularly with a dynamic display of the images, the full potential of tagged CMR studies of cardiac function can be achieved only through quantitative analysis of the wall and tag motion. To achieve this, sequential positions of the wall contours and tags from the images must first be extracted. Next, one must effectively interpolate the motion of the heart wall between the tagged positions. Analysis of the regional motion can then be calculated to determine the regional distribution of such measures of the motion as displacement and deformation. Finally, one can create functional images, displaying the results of the regional motion analysis.

Extraction of tag and wall contour positions from the images is the most technically demanding and critical part of the analysis. The ultimate limiting factor is image quality. The sharper the image, and the higher the contrast between the wall and its adjacent structures and between the tags and the adjacent tissue, the easier and more reliable will be the extraction of the contour and tag positions. The most straightforward (although time consuming) approach to contour extraction is to manually select the positions of selected points along the desired contour and use a computer to fit a smooth curve through them. When image quality is suboptimal or when tags are near the endocardial surface, dynamic display of the images can help the user identify the position of the edge of the wall. This process can be partially automated by starting a generic initial contour curve near the desired contour and using image intensity–derived forces to guide the fitting of the curve to the contour; the use of the active contour (or snakes) approach permits interactively guiding the fitting process with user-controlled forces (Figure 4–2).[17] Thus, rather than independently define the contours in each phase, the position determined for the contour of the prior (or subsequent) cardiac phase can be used as the start-

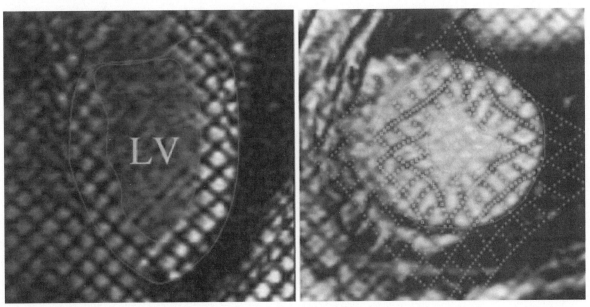

Figure 4–2. Representative steps in analysis of tagged images using SPAMMVU analysis program: Semiautomated extraction of left ventricle (LV) wall contours (long axis image) using "snakes" *(left)*, and semiautomated tracking of tagging grid (short axis image) using modified snake grid *(right)*.

ing location for the contour at the next (or prior) phase. The final series of extracted contours can then be stored in a file corresponding to that structure, for example, left ventricular endocardium.

The extraction of the tag locations in the images can be carried out in a manner similar to that described for the edge contours. Manual selection can be used to pick locations of points at the intersections of tags or of points along tag lines. When manual approaches are used to pick points along a tag line, a smooth curve can be fitted through them. Semiautomatic approaches can be used to speed up tag tracking. For example, the active contour approach described above for contour tracking can be adapted to track the tags (see Figure 4–2). The endocardial and epicardial contours of the heart wall can be used to delineate the region within which the tag line will be tracked. As with contour tracking, the final positions determined at one cardiac phase can be used as the initial iteration for the next (or prior) phase. A tag grid intersection location at one phase must be linked to the corresponding locations at other phases; each detected tag line must be similarly linked to its corresponding line positions in the image at other phases. Inasmuch as the initial SPAMM tag lines are created at a known orientation and spacing, this information can be used to create the initial starting locations for tag tracking. This information is particularly valuable if there is significant motion between the time the tags are created and the first image in the cine acquisition.

The tags faithfully follow the motion of the underlying tissue. However, only the motions of the tagged regions are directly revealed in the images. Thus, an interpolation method must be used to estimate the wall motion between the tags. If we use the positions of the intersections of a tag grid to divide the space between them into a set of triangles, and if we assume the strain (deformation) is homogeneous within each triangle, the motion of the vertices of each triangle can be used to calculate the rigid body motion (translation and rotation) and deformation (strain) within the triangle (Figures 4–3 through 4–5).[18]

If the tagged planes are initially created perpendicular to the image plane, the motion of each tag line reveals the component of motion perpendicular to the initial tag plane at the tagged locations in the image. Thus, each one-dimensional (1-D) family of tag lines provides a sampling of one component of the wall motion. Combining two image orientations (e.g., long- and short-axis images), with two families of 1-D tags in at least one image orientation set, can provide a full 3-D sampling of the wall motion. The sequential tag positions can be fitted to a finite element model of heart wall deformation. In finite element approaches, the motion of a mesh of nodal points in the wall (not necessarily corresponding to tagged points) is fitted to be consistent with the tag data. The motion of the wall in the elements between the modes is then described by smoothly varying shape functions that interpolate between the motions of the nodes.[19, 20] The deformation between tag points need not be assumed to be homogeneous. The corresponding distribution of rigid body

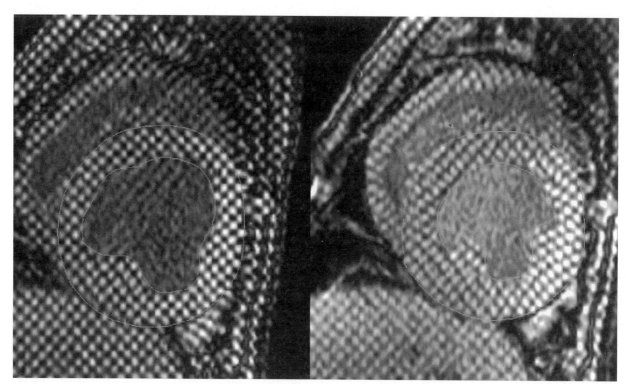

Figure 4–3. Representative high resolution breath-hold tagged images of a canine model of heart failure in the short axis plane at midventricular level showing end-diastole *(left)* and end-systole *(right)*.

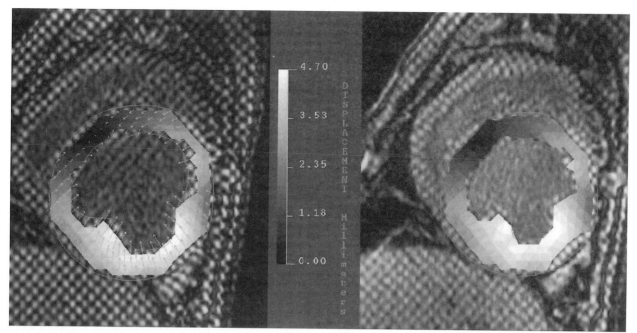

Figure 4–4. The corresponding pseudocolor/gray scale display of local average net systolic displacement between the two images in Figure 4–3 is shown displayed as an overlay on the images at those two times; greater brightness indicates greater displacement. Arrows show the displacements of the grid intersection points.

and strain components of the motion can then be determined from the finite element model. A relatively small number of finite elements can be used with higher order shape functions (e.g., cubic) and a suitable coordinate system (e.g., prolate spheroidal).[20] Alternatively, the motion can be modeled as a combination of a suitable basis set of smoothly varying parameter functions such as radial contrac-

tion and torsional motion,[21, 22] with a large number of simple finite elements.

The process described earlier can be applied to both ventricles (or any moving structure). The motion of the right ventricle (RV) is more difficult to model than that of the left ventricle (LV). The wall of the RV is typically thinner than that of the LV, and the geometry of the RV more complex. One

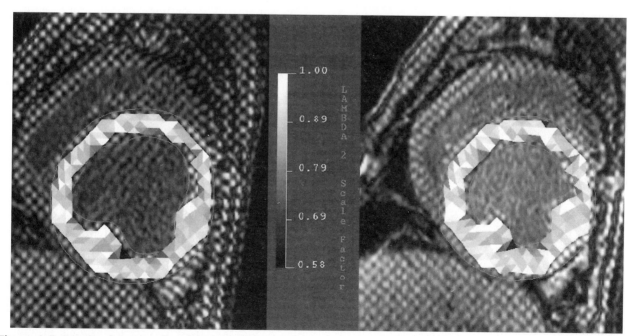

Figure 4–5. Functional images of results of tagged image motion analysis: Pseudocolor/gray scale display of local principal strain eigenvalues approximately corresponding to circumferential wall shortening for images in Figure 4–3; darker indicates more shortening. Note that regional strain is more uniform than displacement.

approach to RV motion analysis is to track the tag positions at the midportion of the RV freewall. The tag data from orthogonal image sets can be combined to reconstruct the contraction pattern considering the midwall as a deforming sheet, and the corresponding in-plane strain components can be calculated.[23, 24] With higher resolution imaging (or RV hypertrophy), the tag lines can be used to extract endocardial and epicardial motion information.[25] This process can be extended to the development of an integrated LV-RV wall motion model constructed from tagged CMR data.

DISPLAY OF MOTION ANALYSIS

The analysis of tagged CMR images of motion can result in very large data sets, with multiple motion variables calculated at numerous positions in the heart wall and evolving over time. Although the numerical data can be presented in tabular form and pooled by myocardial region, it is valuable to create a visual display of the spatial distribution of the motion variables as functional images. Display of the large and high-dimensional data sets that can result from analysis of tagged CMR can be difficult. Scalar quantities, such as the magnitude of local displacement or rigid body rotation, can be conveniently displayed as a corresponding pseudocolor map superimposed on a displayed surface (see Figures 4–4 and 4–5), although only one such quantity can be readily displayed at a time. Vector quantities, such as the magnitude and direction of displacement, can be displayed using an overlaid set of arrows at a suitable set of sampled locations. Deformation is a higher order quantity, a tensor, that cannot be displayed with a simple arrow. The change in length of a unit piece of deforming material will depend on the direction being considered. For example, there may be lengthening in the radial direction coincident with shortening in the circumferential and longitudinal directions. At a given location, there will be a direction of greatest lengthening and (perpendicular to it) a direction of greatest shortening. These directions are the "eigenvectors" and the corresponding relative changes in length are the "eigenvalues" or "principal values" of the deformation. In 3-D deformations, there is a third eigenvector, perpendicular to the greatest and least eigenvectors. One way to display a tensor quantity such as deformation is to display the eigenvectors, for example, as a crossed dihedral or trihedral icon; alternatively, lower dimensional variables derived from them, such as the eigenvalues, can be displayed. When considering components of the deformation in other than the principal directions, there will, in general, be additional shearing components of the motion to consider.

INITIAL RESULTS

Tagged CMR has revealed a fairly consistent pattern of normal regional variation in heart wall mo-

tion, which could previously only be demonstrated with implanted markers/invasive methods.[26] For example, normal ventricular contraction is characterized primarily by base-to-apex shortening, with little motion of the true apex. The magnitude of free wall contraction exceeds that of the septum, and endocardial thickening exceeds that of the epicardium.[27, 28] Although there is a fairly large regional variation in normal displacement, the regional deformation is more uniform, with the greatest lengthening in systole being approximately radially directed. There is also a normal torsional motion of the ventricle about its long axis, with the base and apex rotating in opposite directions (wringing motion). Some variation in the amount of this torsion within the wall at a given level along the axis is also present.

Left ventricular hypertrophy (e.g., due to chronic pressure overload) can impair contraction, with reduced regional deformation, even while the EF remains within normal limits.[29] Similarly, in hypertrophic cardiomyopathy, the normal regional deformation and torsion may be reduced.[30, 31]

Pharmacological stress, such as dobutamine loading, can be used to alter regional contraction[27] and may be useful to evaluate the effects of regional myocardial ischemia. With myocardial infarction, the region of altered contraction may extend beyond the region of the infarct itself[28, 32–36]; this finding may bear on the development of effects of infarction such as remodeling.

FUTURE GOALS

As general CMR methods develop and improve, their tagged versions will also improve. Higher spatial and temporal resolution imaging will permit better definition of regional motion patterns. There is a major need for more efficient, automated, and robust analysis methods, particularly for tag tracking. The ongoing development of higher resolution imaging will result in even bigger and denser tag data sets, making the need for better data extraction methods more acute. Inasmuch as the motion data provided by tagged CMR has not been previously available noninvasively, nor with such wide spatial sampling, the normal patterns of regional motion need to be more accurately defined in larger series. This information will provide a basis of comparison for the evaluation of patients with known or suspected heart disease.

Finally, these new methods for cardiac motion assessment must be clinically evaluated, perhaps in combination with other CMR methods, such as regional perfusion, to evaluate their potential role in clinical management of patients. Tagged CMR is already a useful tool for cardiac research; as the aforementioned goals are approached, it may become a useful clinical tool as well.

References

1. Singer JR: Blood flow rates by nuclear magnetic resonance measurements. Science 130:1652, 1959.
2. Zerhouni CA, Parish DM, Rogers WJ, et al: Human heart: Tagging with MR imaging, a method for noninvasive assessment of myocardial motion. Radiology 169:59, 1988.
3. Young AA, Axel L, Dougherty L, et al: Validation of MRI tagging to estimate material deformation. Radiology 188:101, 1993.
4. Moore CC, Reeder SB, McVeigh ER: Tagged MR imaging in a deforming phantom: Photographic validation. Radiology 190:765, 1994.
5. Yeon S, Reichek N, Palmon L, et al: Myocardial segment shortening in canine myocardial infarction using spatial modulation of magnetization and sonomicrometry (abstract). In Book of Abstracts. Berkeley, Society of Magnetic Resonance in Medicine, 1990, p. 270.
6. Lima JAC, Jeremy R, Grier WH, et al: Accurate systolic thickening by MRI with tissue tagging. Correlation with sonomicrometers in normal and ischemic myocardium. J Am Coll Cardiol 21:1741, 1993.
7. Axel L, Dougherty L: MR imaging of motion with spatial modulation of magnetization. Radiology 171:841, 1989.
8. Axel L, Dougherty L: Improved method of spatial modulation of magnetization (SPAMM) for MRI of heart wall motion. Radiology 172:349, 1989.
9. Reeder SG, McVeigh ER: Tag contrast in breath-hold cine cardiac MRI. Magn Reson Med 31:521, 1994.
10. Bolster BD, McVeigh ER, Zerhouni AE: Myocardial tagging in polar coordinates with the use of stripped tags. Radiology 177:769, 1990.
11. Fischer SE, McKinnon GC, Maier SE, Boesiger P: Improved myocardial tagging contrast. Magn Reson Med 30:191, 1993.
12. Axel L, Dougherty L: Three-dimensional MR imaging of heart wall motion (abstract). Radiology 173(P):223, 1989.
13. Pipes JG, Boes JL, Chenevert TC: Method for measuring three-dimensional motion with tagged MR imaging. Radiology 181:591, 1991.
14. Axel L, Dougherty L: System and method for magnetic resonance imaging of 3-dimensional heart wall motion with spatial modulation of magnetization 1990. U.S. Patent No. 5111820.
15. Perman WH, Creswell LL, Wyers SG, et al: Hybrid DANTE and phase-contrast imaging technique for measurement of three-dimensional myocardial wall motion. J Magn Reson Imaging 5:101, 1995.
16. Young AA, Axel L: Three-dimensional motion and deformation of the heart wall: Estimation with spatial modulation of magnetization—a model based approach. Radiology 185:241, 1992.
17. Kass M, Witkin A, Terzopoulos D: Snakes: Active contour models. International Journal of Computer Vision 1:321, 1988.
18. Axel L, Gonçalves R, Bloomgarden D: Regional heart wall motion: Two-dimensional analysis and functional imaging of regional heart wall motion with magnetic resonance imaging. Radiology 183:745, 1992.
19. Young AA, Axel L: Non-rigid heart wall motion using MR tagging. Proceedings of IEEE, Conference on Computer Vision and Pattern Recognition. Champaign, IL, June 15, 1992.
20. Young AA, Axel L: Three-dimensional motion and deformation of the heart wall: Estimation with spatial modulation of magnetization—a model based approach. Radiology 185:241, 1992.
21. Park J, Metaxas D, Young AA, Axel L: Deformable models with parameter functions for cardiac motion analysis from tagged MRI data. IEEE Trans Med Imag 15:278, 1996.
22. Park J, Metaxas D, Axel L: Analysis of left ventricular wall motion based on volumetric deformable models and MRI-SPAMM. Medical Image Analysis 1:53, 1996.
23. Young AA, Fayad ZA, Axel L: Right ventricular midwall surface motion and deformation using magnetic resonance tagging. Am J Physiol 271:H2677, 1996.
24. Fayad ZA, Ferrari VA, Kraitchman DL, et al: Right ventricular regional function using MR tagging: Normals vs. chronic pulmonary hypertension. Magn Reson Med 39:116, 1998.
25. Haber E, Metaxas DN, Axel L: Motion analysis of the right ventricle from MRI images. Proceedings of Medical Image Computing and Computer-Assisted Intervention—MICCAI, 1998, p 177.
26. Young AA, Imai H, Chang C-N, Axel L: Two-dimensional left ventricle motion during systole using MRI with SPAMM. Circulation 89:740, 1994.
27. Scott CH, St. John Sutton MG, Gusani N, et al: Effect of dobutamine on regional left ventricular function measured by tagged magnetic resonance imaging in normal subjects. Am J Cardiol 83:412, 1999.
28. Kramer CM, Lima JAC, Reichek N, et al: Regional differences in function within noninfarcted myocardium during left ventricular remodeling. Circulation 88:1279, 1993.
29. Palmon LC, Reichek N, Yeon SB, et al: Intramural myocardial shortening in hypertensive left ventricular hypertrophy with normal pump function. Circulation 89:122, 1994.
30. Young A, Kramer CM, Ferrari VA, et al: Three-dimensional left ventricular deformation in hypertrophic cardiomyopathy. Circulation 90:854, 1994.
31. Kramer CM, Reichek N, Ferrari VA, et al: Regional heterogeneity of function in hypertrophic cardiomyopathy. Circulation 90:184, 1994.
32. Lima JAC, Ferrari VA, Axel L, et al: Segmental motion and deformation of infarcted myocardium by magnetic resonance tissue tagging. Am J Physiol 268:H1304, 1995.
33. Kramer CM, Ferrari VA, Rogers WJ, et al: Angiotensin-converting enzyme inhibition limits dysfunction in adjacent noninfarcted regions during left ventricular remodeling. J Am Coll Cardiol 27:211, 1996.
34. Kraitchman DL, Wilke N, Hexeberg E, et al: Myocardial perfusion and function in dogs with moderate coronary stenosis. Magn Reson Med 35:771, 1996.
35. Marcus JT, Götte MJW, Van Rossum AC, et al: Myocardial function in infarcted and remote regions early after infarction in man: Assessment by magnetic resonance tagging and strain analysis. Magn Reson Med 38:803, 1997.
36. Kraitchman DL, Young AA, Bloomgarden DC, et al: Integrated MRI assessment of regional function and perfusion in canine myocardial infarction. Magn Reson Med 40:311, 1998.

CHAPTER 5

CSPAMM Assessment of Left Ventricular Diastolic Function

Matthias Stuber

Left ventricular diastolic function has been recognized as an important factor in the pathophysiology of many common cardiovascular diseases. Dilated and hypertrophic cardiomyopathies, coronary artery disease, and hypertension are all associated with abnormal left ventricular filling. Diastolic dysfunction has also been increasingly appreciated as a major cause of heart failure. Although invasive hemodynamic measures/assessment of diastole are considered the gold standard, echocardiographic methods have gained greater use in the clinical assessment of the time-dependent changes in diastolic left ventricular volume and dimensions due to their noninvasive acquisition and ease for serial assessments.

Cardiac Motion

During the cardiac cycle, the heart performs a complex wringing motion. During systole, the base and apex rotate in opposite directions with counterclockwise rotation at the apex and clockwise rotation at the base.[1] In parallel, the valvular plane (basal left and right ventricle) moves towards the apex. The lateral free wall of the right ventricle performs a more pronounced long-axis contraction than the lateral wall of the left ventricle.[2] During isovolumic relaxation, myofibrils return to their resting state from the contracted state. This process is accompanied by a rapid untwisting at the apex while volume and cavity shape of the heart almost remain unchanged. This rapid untwisting typically lasts less than 75 ms and precedes the filling phase of the ventricles. During this filling phase, practically no rotational components can be seen at the apex of the healthy heart.[3]

Assessment of Cardiac Rotation/ Motion—Non–Magnetic Resonance Methods

With echocardiographic imaging, the myocardium itself displays relatively poor internal structure due to the absence of structural landmarks, making quantification of parameters such as rotation, stress or strain quite limited. Several invasive methods have been published to examine cardiac motion during diastole. One approach is the surgical implantation of tantalum markers into the mid-wall of the myocardium.[4] In combination with x-ray angiography, the motion of these markers can then be recorded with high temporal and spatial resolution. Using such an approach, alterations in diastolic untwisting have been observed in heart transplant patients shortly before rejection.[5] Although this method is very powerful, it is invasive, requires ionizing radiation, and is inappropriate for clinical use. Alternative angiographic markers such as tracking of the bifurcations of the coronaries[6] suffer from the limited number of landmarks and their irregular geometrical distribution. Furthermore, they only provide motion information regarding epicardial layers of the myocardium.

Assessment of Cardiac Rotation/ Motion—Magnetic Resonance Methods

Similar to echocardiography, conventional cine cardiovascular magnetic resonance (CMR) imaging does not provide information regarding the internal structure of the myocardium in the images. However, CMR myocardial tagging techniques as proposed by Zerhouni[7] and Axel,[8] and further developed and refined by others (see Chapter 4), have been reported. In these methods, the magnetization of the muscle tissue is spatially modulated, or "tagged," by the application of a specific time series of radiofrequency (RF) pulses and magnetic field gradients. This tagging procedure is typically applied immediately after the detection of the R wave of the electrocardiogram (ECG). Subsequently, multiple heart phase images are acquired at different phases of the cardiac cycle. In these images the tags may be identified as dark lines or grids (Figure 5–1). Because these tags are spatially fixed with respect to the muscle tissue, local myocardial motion can be derived from the translation, rotation, and distortion of the tags on the myocardium. However, due to the relaxation effects, the tags rapidly fade and cannot be reliably detected after end systole (ap-

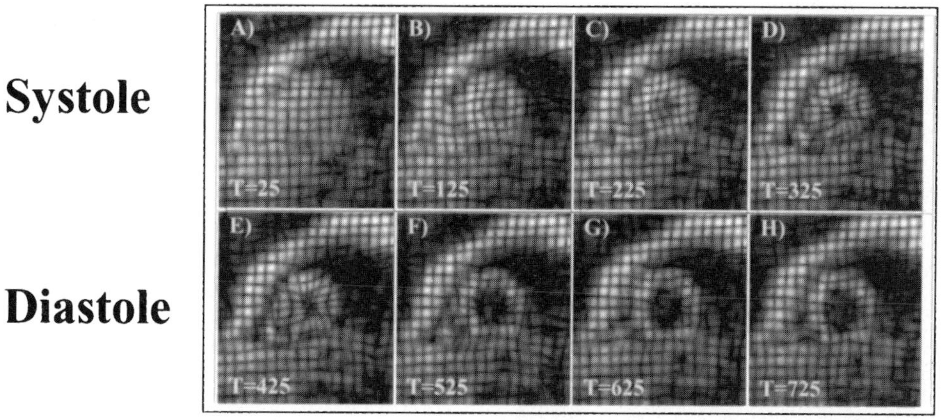

Systole

Diastole

Figure 5–1. Eight apical short-axis images (healthy subjects) from a time series of 20 acquired with complementary spatial modulation of magnetization (CSPAMM) myocardial tagging. The systolic images are shown in the upper row and the diastolic images in the lower row. The individual heart phase interval was 37 ms and line-tagged acquisitions were combined offline. During systole *(A–D)*, a counterclockwise rotation is seen. During diastole, the untwisting phase *(E and F)* precedes the filling phase *(G and H)* of the ventricles. The time (T; ms) following the R wave for each image is shown.

proximately 300 ms). This is a serious drawback for the quantification of systolic *and* diastolic dynamics of the heart wall. Another limitation is that this approach does not compensate for through-plane motion.

The problems associated with tag fading as well as through-plane motion[2] can be overcome by the application of complementary spatial modulation of magnetization (CSPAMM) CMR tagging approaches.[3, 9] This technique is based on a subtraction of two acquisitions. For both acquisitions, the magnetization of the tissue is modulated or tagged locally dependent in a thin slice and a thick slice is imaged during the subsequent imaging procedure. If the modulation function preceding the second acquisition is inverted with respect to the first tagging procedure, subtraction of the two acquisitions leads to signal that is derived only from the initially tagged thin slice. This approach serves to both prolong the persistence of the tags until late diastole and assure that the same tissue plane is imaged in the multiple heart phase images. In the following sections, the technique is described in more detail and initial results associated with diastolic motion components of the human heart are discussed.

METHODS

CSPAMM and Slice Following

CSPAMM myocardial tagging involves the periodic modulation of the magnetization in a thin slice of the myocardium (Figure 5–2, dz). A sinusoidal modulation of the magnetization is typically performed immediately after the detection of the R wave of the ECG. Subsequently, a thick slice (Figure 5–2, ds) encompassing the expected full extent of motion of the selected thin slice is imaged periodically (n heart phase images) during the cardiac cycle. The procedure consisting of labeling of a thin

slice and subsequent imaging of a thicker volume is performed twice with an inverted modulation of the magnetization for the second experiment. Subtraction of the two acquisitions leads to an image derived from the signal coming from the labeled part of the magnetization in the thin slice. Due to the subtraction procedure, the signal coming from the thick volume outside the tagged slice (Figure 5–2, ds) is suppressed. Therefore, the problem of through-plane motion is avoided and a projection of the same tissue elements is visualized in the images.

The signal from the tagged thin slice can be decomposed into two parts: the first part holding the tagging information and a second part (responsible for the fading of the tags) that is built up as a function of time. For CSPAMM, this second component is suppressed as well by the subtraction procedure. In consequence, only signal derived from the tagged component of the magnetization remains after subtraction and fading of the tags may be avoided. For a constant tag contrast and a maximized signal-to-noise ratio (SNR) in systolic and diastolic images, a series of ramped RF excitation angles have to be used.[9]

Typically, double oblique short-axis sections of the myocardium are tagged with a slice thickness of

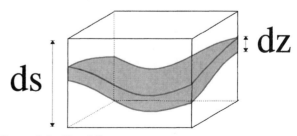

Figure 5–2. Slice following principle. An initially tagged planar slice of the thickness, dz, translates and distorts during the cardiac cycle. A volume of the thickness, ds, is imaged multiple times during the cardiac cycle. This volume has to encompass the potential extent of the motion of the tagged, thin slice.

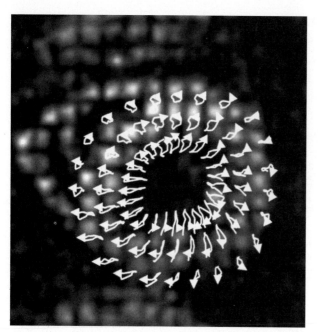

Figure 5–3. End-diastolic apical image acquired in a healthy adult subject. The grid-tagged image is overlaid with the corresponding local trajectories. The arrows start at the beginning of systole and end at end diastole.

6 to 8 mm. Subsequently, 16 to 20 heart phase images are acquired with a temporal resolution (Δt) of 35 ms. With this high temporal resolution, rapid motion components such as diastolic untwisting at the apex can be readily identified. Because the ratio of wanted to unwanted signal components has to be optimized, the thickness of the imaged volume (see Figure 5–2, ds) has to be reduced to a minimum. Therefore, it depends on the level of the tagged slice with respect to its level on the long axis. For basal left ventricular images, where a long-axis contraction of more than 20 mm may be expected for the lateral free wall of the right ventricle,[2] a slice thickness of 30 mm is typically chosen. For equatorial slices, 25 mm is appropriate, and at the apex, a slice thickness of 20 mm is used. For the suppression of breathing-induced motion artifacts in the images, a repetitive breath-hold scheme[3, 9] or single breath-hold techniques[10] can be applied.

Considering the location of the relevant tagging information in k-space, a reduced k-space acquisition scheme can be applied.[9, 11] Hereby, two sets of orthogonal line tagged images are acquired. Subsequent combination of these acquisitions results in grid tagged images (Figure 5–3). With this method, acquisition time is significantly reduced and image resolution perpendicular to the line tags is not affected.

Evaluation

For the extraction of motion data from the tagged time series of images, a specific image analysis procedure is used.[12, 13] This involves the identification of the tags in all heart phase images. With sophisticated automatic or semiautomatic algorithms, the tags may be identified with an accuracy that exceeds the image resolution.[14] If the grid intersection points are identified for all heart phase images, local trajectories on the myocardium (see Figure 5–3) are defined and motion-specific parameters (e.g., rotation, rotation velocity, radial or circumferential shortening or shear between epi- and endocardial muscle layers of the myocardium) may be derived. Using this approach, several studies of healthy volunteers and patients with myocardial infarction,[15] aortic stenosis with pathologically hypertrophied hearts, as well as athletes with physiological hypertrophy were investigated for apical untwisting during diastole.[16] Untwisting velocity and time to peak untwisting velocity can be determined as an index of diastolic function. Time to peak untwisting velocity (= Untwisting Time, Table 5–1) is defined as the time delay between (a) the point in time of minimum inner cavity area and (b) the maximum untwisting velocity (Figure 5–4A).

RESULTS

Images

In Figure 5–1, 8 heart phase images (out of a series of 20 with a temporal resolution of 35 ms) acquired at the apex of a healthy volunteer are presented.[10] The grid structure remains visible with a high contrast up to the last acquisition in late diastole (>700 ms). No fading of the tags is seen in the images. Therefore, the method is well suited for the quantification of diastolic heart wall motion.

Apical Rotation

At the apex of the healthy heart, a counterclockwise rotation during systole can be seen (see Figure 5–1 and Figure 5–4B, phase 1, 2). This systolic rotation is followed by a rapid untwisting during isovolumic relaxation (Figure 5–4B, phase 3). This untwisting phase is typically followed by the filling

Table 5–1. Peak Rotation at the Apex, Maximum Rotation Velocity During Diastolic Untwisting, and Untwisting Time*

Patients	Peak Rotation (Deg)	Untwisting Velocity (Deg/s)	Untwisting Time (% ES)
Aortic stenosis[16]	12 ± 5	80 ± 29	32 ± 6
Athletes[16]	6 ± 2	56 ± 8	17 ± 8
Volunteers[16]	7 ± 2	55 ± 17	16 ± 8

*For 12 aortic stenosis patients, 12 athletes, and 11 healthy volunteers. The untwisting time is related to the duration of systole. Data are expressed as mean ± 1 SD.
ES = end systole; Deg = degrees.

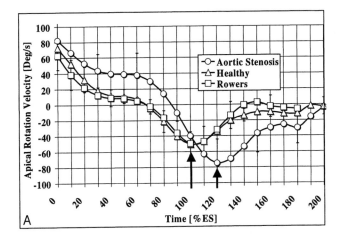

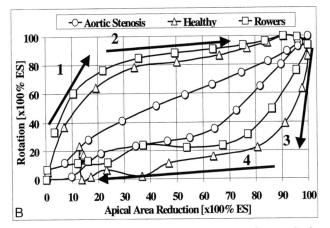

Figure 5–4. *A,* Cross-sectional apical rotation velocity of athletes, healthy athletic subjects, and patients with aortic stenosis. The values are mean values ± 1 SD. The time axis is normalized to the end-systolic (ES) point of the cardiac cycle. The time point of maximum diastolic untwisting velocity (arrows) is delayed in the patients when compared to the athletes or to the healthy controls. Apical rotation velocity is identical in athletes and controls. *B,* Left ventricular rotation-area loop (apical plane) in healthy controls, rowers, and patients with aortic stenosis. The loop is separated in isovolumic contraction (1), ejection (2), isovolumic relaxation (3), and filling of the left ventricle (4). Both rotation and area change of the inner lumen at the apex are related to their maximum values (=100%).

phase of the ventricles, where practically no rotational component is seen (Figure 5–4B, phase 4).[3, 17] An identical separation of early diastolic apical untwisting and the filling phase of the ventricles can be reported in the athletes with physiologically hypertrophied hearts. Neither the apical peak rotation angle nor diastolic untwisting time is changed in the athlete's heart when compared with the healthy controls (see Table 5–1). However, patients with aortic stenosis and pathologically hypertrophied hearts due to pressure overload show a completely different apical rotation pattern (see Figure 5–4). The end-systolic peak rotation is significantly increased ($p < 0.01$) in comparison to the athletes or the controls and during diastole, no separation of untwisting and filling can be seen. Untwisting and filling occur simultaneously and the point in time of maximum untwisting velocity is significantly de-

layed in these patients ($p < 0.01$) (see Figure 5–4A). Pressure overload hypertrophy results in addition of new sarcomeres in parallel with the existing ones.[18] Furthermore, an increase in the amount of collagen with a consequently increased elastic stiffness has been reported.[19] This rearranged fiber architecture together with the increased stiffness of the muscle tissue may explain the alterations of the diastolic rotation pattern with a prolonged and delayed apical untwisting in these patients. The prolongation of untwisting results in an overlap with early diastolic filling and presumably impedes normal filling. Thus, a prolongation of early diastolic untwisting may be responsible for the occurrence of diastolic dysfunction in these patients. In patients with myocardial infarction, a prolonged untwisting phase with an overlap of diastolic untwisting and filling has also been observed[20] and end-systolic apical peak rotation is usually severely reduced.

Limitations

The additional value of CMR tagging applied for the quantification of diastolic function remains to be investigated in comparison to gold standard techniques. At present, the CMR technique is not widely accessible to clinical cardiologists and the CSPAMM sequence is currently limited to one CMR vendor (Philips Medical Systems, Best, NL). The investigation of heart wall motion is still the subject of basic research, and appropriate parameters for the quantification of diastolic dysfunction and threshold values for normal subjects remain to be fully defined.

CONCLUSION

CSPAMM myocardial tagging is a noninvasive method for the quantification for local heart wall motion. Due to the suppressed fading of the tags and the accessibility to the diastolic phase of the cardiac cycle, CSPAMM is well suited for the characterization of the diastolic portion of the cardiac cycle. Moreover, by the application of a slice following procedure, the effects of through-plane motion can be suppressed and the same tissue elements can be tracked. Due to the relatively high temporal resolution of the data, rapid cardiac motion components such as apical diastolic untwisting can be recorded.

Preliminary findings suggest that pathological hypertrophy in patients with aortic stenosis can be differentiated from hypertrophy in athletic hearts.[16] Alterations in diastolic untwisting are seen in patients with pressure overload due to aortic stenosis, whereas none are observed in athletes' hearts. Even though the sizes of the athletes' hearts were significantly increased in comparison to the controls, apical diastolic untwisting remains unchanged. The present data derived from CSPAMM myocardial tagging suggest that alterations in the diastolic phase of the cardiac cycle may be recorded. Although this method requires off-line computer analysis, the technique may offer a new tool for the evaluation of diastolic wall relaxation in healthy and diseased states.

References

1. Maier SE, Fischer SE, McKinnon GC, et al: Evaluation of left ventricular segmental wall motion in hypertrophic cardiomyopathy with myocardial tagging. Circulation 86.1919, 1992.
2. Rogers WJ Jr, Shapiro EP, Weiss JL, et al: Quantification of and correction for left ventricular systolic long-axis shortening by magnetic resonance tissue tagging and slice isolation. Circulation 84:721, 1991.
3. Fischer SE, McKinnon GC, Scheidegger MB, et al: True myocardial motion tracking. Magn Reson Med 31:401, 1994.
4. Ingels NB Jr, Daughters GTd, Stinson EB, Alderman EL: Measurement of midwall myocardial dynamics in intact man by radiography of surgically implanted markers. Circulation 52:859, 1975.
5. Yun KL, Niczyporuk MA, Daughters GTD, et al: Alterations in left ventricular diastolic twist mechanics during acute human cardiac allograft rejection. Circulation 83:962, 1991.
6. Potel MJ, Rubin JM, MacKay SA, et al: Methods for evaluating cardiac wall motion in three dimensions using bifurcation points of the coronary arterial tree. Invest Radiol 18:47, 1983.
7. Zerhouni EA, Parish DM, Rogers WJ, et al: Human heart. tagging with MR imaging—a method for noninvasive assessment of myocardial motion. Radiology 169:59, 1988.
8. Axel L, Dougherty L: MR imaging of motion with spatial modulation of magnetization. Radiology 171:841, 1989.
9. Fischer SE, McKinnon GC, Maier SE, Boesiger P: Improved myocardial tagging contrast. Magn Reson Med 30:191, 1993.
10. Stuber M, Spiegel MA, Fischer SE, et al: Single breath-hold slice-following CSPAMM myocardial tagging. MAGMA 9:85, 1999.
11. McVeigh ER, Atalar E: Cardiac tagging with breath-hold cine MRI. Magn Reson Med 28:318, 1992.
12. Kumar S, Goldgof D: Automatic tracking of SPAMM grid and the estimation of deformation parameters from cardiac MR images. IEEE Trans Biomed Eng 13:122, 1994.
13. Bundy JM, Lorenz CH: TAGASIST: A post-processing and analysis tools package for tagged magnetic resonance imaging. Comput Med Imaging Graph 21:225, 1997.
14. Atalar E, McVeigh ER: Optimization of tag thickness for measuring position with magnetic resonance imaging. IEEE Trans Biomed Eng 13:152, 1994.
15. Nagel E, Stuber M, Lakatos M, et al: Cardiac rotation and relaxation after anterolateral myocardial infarction. Coron Artery Dis 11:261, 2000.
16. Stuber M, Scheidegger MB, Fischer SE, et al: Alterations in the local myocardial motion pattern in patients suffering from pressure overload due to aortic stenosis. Circulation 100:361, 1999.
17. Rademakers FE, Buchalter MB, Rogers WJ, et al: Dissociation between left ventricular untwisting and filling. Accentuation by catecholamines. Circulation 85:1572, 1992.
18. Grossman W, Jones D, McLaurin LP: Wall stress and patterns of hypertrophy in the human left ventricle. J Clin Invest 56:56, 1975.
19. Hess OM, Lavelle JF, Sasayama S, et al: Diastolic myocardial wall stiffness of the left ventricle in chronic pressure overload. Eur Heart J 3:315, 1982.
20. Matter C, Mandinov L, Kaufmann P, et al: Function of the residual myocardium after infarct and prognostic significance. Z Kardiol 86:684, 1997.

Blood Flow Velocity Assessment

David Firmin

The idea of mapping measurements of blood flow onto a magnetic resonance (MR) image was first discussed in an article by Singer in 1978.[1] The methods that followed could generally be categorized into time-of-flight or phase shift types and were based on the techniques that had previously been described for nonimaging nuclear magnetic resonance (NMR) flow studies.[2] A number of review articles have covered the subject and described the variety of different methods[3–5] that have now been used and validated both in vitro and in vivo. The interest in flow in MR has not been solely directed towards the goal of quantitative flow measurement. A large amount of effort has also been devoted to understanding the appearance of a flowing fluid on an image because this appearance can often be indicative of the type of flow present and therefore give important information in the diagnosis of a particular disorder. Also, the development of MR angiography techniques has required a full understanding of these effects.

In 1984, soon after the development of the first clinical MR scanners, there was an increase of interest in the quest for an MR method of imaging flow. Review articles were published and a number of techniques described.[6–10] In the following text there is a description of and a brief historical overview of the methods that have been used to measure blood flow in the heart and great vessels.

TECHNIQUES TO MEASURE BLOOD FLOW

Time-of-Flight Methods

There are two categories of time-of-flight techniques. The first, often known as wash-in/washout or flow enhancement methods, relies on the saturation or partial saturation of material in a selected slice or volume being replaced by fully magnetized "high signal" spins due to flow (Figure 6–1A). The second involves some form of tagging and then imaging to follow the motion of the tagged material (Figure 6–1B).

Singer and Crooks[11] adopted the first approach in an attempt to measure flow in the internal jugular veins although quantification was questionable because of other factors affecting the flow signal. The

first to describe a tagged time-of-flight approach were Feinberg and associates.[10] Their method involved a variation on a dual echo spin-echo sequence; the first 180° selected slice was displaced by 3 mm from the initial excitation slice and the second was displaced by 9 mm. The first 180° selection overlapped sufficiently with the 90° selection to produce a good anatomical image. The second 180° pulse selection did not overlap with that of the 90° or the first 180° pulse selection, and therefore produced no anatomical image but gave high signal from the blood that had experienced all the preceding radiofrequency pulses (i.e., that had passed between the different selected planes). Flow in the carotid and vertebral arteries of a volunteer's neck was identified with the technique, although flow velocities had to be within a specific range and could not be defined accurately.

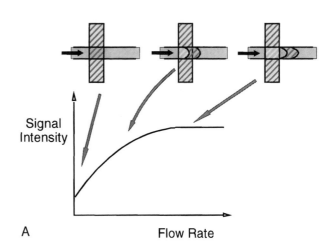

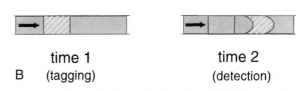

Figure 6–1. Schematic illustrating the time-of-flight approaches to CMR flow measurement. In the saturation wash-in method *(A)*, the signal is enhanced by an amount related to the inflow of fully magnetized blood. In the true time-of-flight method *(B)*, there is a time of spatial labeling or tagging followed after a defined period by a detection of the movement of the labeled blood.

Methods have also been described where the time-of-flight flow movement can be visualized directly on an image.[12, 13] The methods involved slice selection and frequency encoding being applied in the same axis. In this way, material that had moved in this axis between selection and readout would be displaced relative to the stationary material. These techniques were therefore making use of signal misregistration, an effect that is often seen as a problem in other methods of flow imaging. Another time-of-flight approach is to saturate a band of tissue, for example in a transverse plane, then to follow the progress of this dark band in the coronal or sagittal planes.[14] The major limitation of these saturation methods is that they are limited by the T1 of the various tissues being saturated. The contrast of the saturated blood will reduce with time, eventually making it difficult to measure accurately the distances traveled. Also, motion during the sampling gradients results in signal and thus image distortion.[15] In addition, for arterial flow measurements where cardiac gating is required, only two-dimensional (2-D) images can be acquired in a reasonable time so that only limited details of the flow profile can be studied.

Phase Flow Imaging Methods

Considerable knowledge had been gained on the measurement of flow from the phase of the NMR signal from the nonimaging studies following an original suggestion by Hahn in 1960 of a method of measuring the slow flow of currents in the sea.[16] In 1984 the first attempts of measuring blood flow were described by van Dijk[8] and Bryant and colleagues,[9] using methods based on the theory suggested by Moran 2 years earlier.[17] The imaging methods that followed fell broadly into two categories:

1. Phase contrast velocity mapping methods that mapped the phase of the signal directly in order to measure the flow.
2. Fourier flow imaging methods that phase encoded flow velocity to produce an image following Fourier transformation with spatial information resolved on one axis and velocity on another.

Both methods rely on the same principles that cause flowing material to attain a phase shift that is related to its motion. Figure 6–2 gives a simple illustration of these principles for a fluid flowing down a tube surrounded by stationary material. A bipolar gradient pulse is applied, consisting of first a positive magnetic field gradient, followed a certain time later by an equal but opposite negative magnetic field gradient in the direction of the flow. During the period of the positive gradient, flowing and stationary material in a particular location will take up a frequency shift that will depend on their position in the direction of the field gradient. When the gradient is turned off, the phase of the flowing and stationary materials can be considered to be equal.

In the period between the positive and negative gradients, the flowing material moves away from its stationary neighbor. During the period of the negative gradient, the stationary material will take up an equal but opposite frequency shift and return to the phase it had prior to the first gradient. However, the flowing fluid will take up a different frequency shift dependent on the distance it has moved; its final phase, therefore, also depends on this distance and hence its velocity.

The relationship between the phase of the signal and flow velocity is:

$$\phi = \gamma v \Delta A_g \tag{1}$$

where A_g is the area of one gradient pulse, Δ is the time between the centers of the two gradient pulses, v is the velocity and γ is the gyromagnetic ratio. A_g, Δ, and γ are all constants for a particular imaging sequence so that a quantitative measure of velocity can be determined if the phase shift can be measured.

The two principal approaches of utilizing the phase shift to produce a quantitative flow image, phase contrast velocity mapping and Fourier flow imaging, are discussed in the following paragraphs.

Phase Contrast Velocity Mapping

The early phase contrast velocity mapping methods[8, 9] used a spin-echo sequence, which was not ideal because of problems in repeating the sequence rapidly, and signal loss due to shear and other more complex flows. These problems were reduced and the methods were made clinically more useful, partly by the use of a gradient-echo sequence[18, 19] and, more importantly, by the introduction of velocity-compensated gradient waveforms.[20, 21]

Normally two images are acquired with different gradient waveforms in the direction of desired flow measurement. The difference in the waveforms is calculated to produce a well-defined velocity-related phase difference between the two images. A phase reconstruction is produced for each of the images that are then subtracted pixel by pixel to produce the final velocity map. This process of subtraction removes any phase variations that are not related to flow. The velocity phase sensitivity of the final image is normally set such that the expected velocity related phase shifts are within the range of $\pm\pi$. If a larger range of velocities is present, then aliasing or wrap-around will occur, resulting in ambiguous velocities being measured. This problem can be avoided by reducing the velocity sensitivity or can potentially be corrected by a process known as phase unwrapping (described later).

Figure 6–3A shows the phase velocity images of a series of time frames in the transverse plane at the level of the right pulmonary artery. Flow can be seen in the ascending (AA) and descending aortae (DA), the main pulmonary artery (MPA), and the superior vena cava (SVC). Flow-versus-time curves

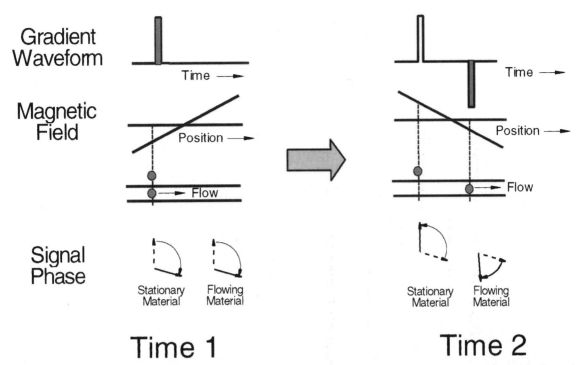

Figure 6–2. The principles of phase velocity encoding. At Time 1, positive magnetic field gradient is applied, which results in an equal frequency and associated phase shift for neighboring stationary and flowing spins. At Time 2, an equal but opposite magnetic field is applied. By this time, the spins in the flowing blood have separated from their original neighbors to be in a different strength of magnetic field during the gradient application. The result is that whereas the phase of the stationary spins will be returned to zero, the flowing spins will accumulate a phase shift proportional to the distance moved and hence the velocity.

throughout the cardiac cycle are presented in Figure 6-3*B*, and the stroke volume can be measured by integrating under the aortic or pulmonary flow curve. The technique has been validated both in vitro and in vivo[22] and is now routinely providing useful measurements in clinical and physiological flow studies.[23]

Fourier Flow Imaging

This method normally involves replacing the spatial phase encoding gradient with a bipolar velocity phase encoding gradient that is stepped through a range of defined amplitudes (Figure 6–4). The result of this process is the formation of an image with velocity information in one dimension. Stationary materials are positioned in the center of the image with faster velocities towards the edge.

The method was first described by Redpath and colleagues[24] in 1984; in this case, eight velocity phase encoding steps were added to a 2-D imaging sequence, to image velocities in a circle of fluid-filled tubing that rotated in the image plane. Different segments of the circle, each corresponding to different velocity ranges, were seen on the eight resultant images. A year later Feinberg and coworkers[25] applied the method both in vitro and in vivo and increased the velocity range and resolution by increasing the number of flow phase encoding steps. However, in this case to maintain a tolerable scan time, only one spatial phase encoding direction was used so that there was only spatial resolution in one

direction. The accuracy of the method was demonstrated using a phantom whereas the in vivo study, which showed the flow in the descending aorta, highlighted the problem of very high zero velocity signal from the large amount of stationary tissue imaged. In 1988 Hennig and coworkers[26] described a development of this method where the signal from stationary tissue was saturated and the sequence repeated much more rapidly. The issue of time precluding the use of spatial phase encoding remains a problem although 2-D radiofrequency pulses have been used successfully to locate signals within a column.[27, 28] More recently Luk Pat and colleagues[29] have demonstrated a method of real-time Fourier velocity imaging, which again used an excited column to localize the signals. In vivo aortic flow waveforms were presented with a temporal resolution of 33 ms.

IMPROVING THE ACCURACY OF PHASE CONTRAST VELOCITY MEASUREMENTS

The vast majority of MR flow imaging applications have used the method of phase contrast velocity mapping. The accuracy of this method is highly dependent on such factors as the flow pulsatility, the velocity, and the size and tortuosity of the vessel. One simple approach to improving the overall accuracy of the method is to adjust the velocity sensitivity of the sequence, so that the velocity re-

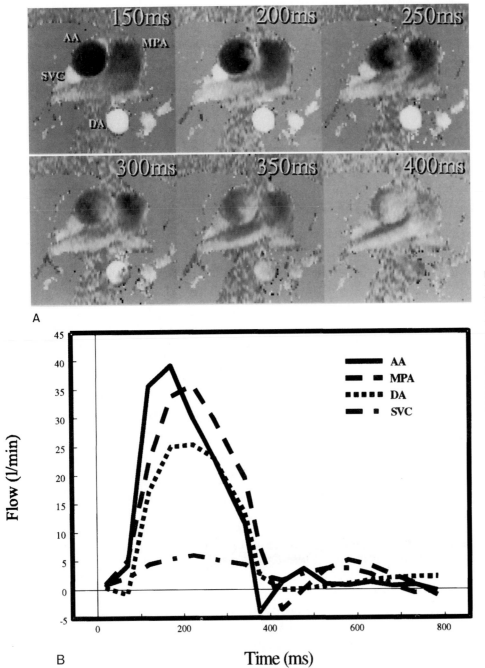

Figure 6–3. Flow velocity MR images of a transverse slice at the level of the right pulmonary artery showing head/foot flow velocities at six times in the cardiac cycle *(A)*, and *(B)*, a plot of measured volume flow versus time for the ascending aorta (AA), descending aorta (DA), main pulmonary artery (MPA), and the superior vena cava (SVC).

Spatial phase encoding is replaced by velocity phase encoding

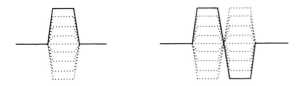

| Spatial phase encoding gradient waveform | Velocity phase encoding gradient waveform |

Figure 6–4. Schematic comparing the gradient waveforms required for spatial and velocity phase encoding.

lated phase shift is close to 2π for the maximum expected velocity. Buonocore[30] extended this approach by varying the velocity sensitivity during the cardiac cycle in the knowledge that the arterial flow velocity is high in systole but low in diastole. The accuracy can be improved even further by allowing a velocity-related phase shift greater than 2π that will result in aliasing, which can be corrected by use of a phase unwrapping algorithm (Figure 6–5).[31]

Another approach to improving accuracy has been suggested by Bittoun and colleagues.[32] The method is a combination of phase contrast velocity mapping and Fourier velocity imaging with a small number of velocity phase encoding steps. The final phase contrast velocity map is calculated from the best fit through the Fourier velocity-encoded result. One particular potential problem with this method, which can also affect conventional phase contrast velocity imaging, occurs if there is any beat-to-beat variation in flow velocity. As a result of the high velocity phase sensitivity used, significant phase variations can occur with resulting ghosting artifacts and loss of flow information.

Another method that was originally described to give a measure of velocity and flow quantification was phase contrast angiography. This technique again involves acquiring two image data sets with sequences having opposite phase velocity sensitivities; however, in this case the raw data are subtracted prior to reconstruction. This method has the advantage of subtracting out signal from stationary tissue, which removes errors caused by partial volumes where voxels contained a mixture of flowing and stationary tissue. The method is, however, generally less accurate because the signal and hence the velocity measurement is affected by factors such as inflow enhancement and intra-voxel dephasing (signal loss). In the mid 1990s Polzin and associates[33] suggested a method of combining this method with phase contrast velocity mapping, which they showed to be more accurate in a number of phantom studies. The methods are yet to be fully validated in vivo, however, and are likely to still be affected by problems of signal loss and motion, particularly when imaging small mobile vessels such as the coronary arteries.[34]

One of the most significant factors that can affect the accuracy of the flow measurement methods is that of flow-related signal loss. This problem is nor-

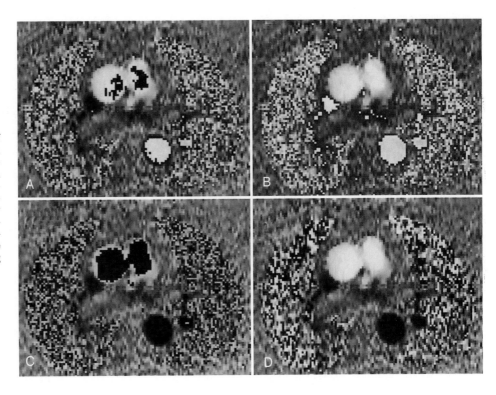

Figure 6–5. The method of phase unwrapping. *A* shows a systolic image where high velocities result in aliasing in both the positive and negative directions. *B* and *C* illustrate the adjustment of the velocity window to remove aliasing in the positive and negative directions, respectively. *D* shows the same image data following processing by the anti-aliasing algorithm.

mally a result of loss of phase coherence within a voxel and, eventually, will result in an inability to detect the encoded phase of the flow signal above the random phase of the background noise. Even if a velocity-compensated imaging sequence is used, the acceleration and even higher orders of motion present in complex flows will result in loss of phase coherence. Figure 6–6 shows an example of a long-axis image of a patient with mitral valve stenosis and mitral regurgitation. In this case, a region of blood signal is lost from the ventricle during diastole, as a result of the stenosis generating complex flows, and in the atrium during systole due to a regurgitant jet of flow through the valve. The size of the signal void is related to the echo time (TE). Partial signal loss, however, does not greatly affect the accuracy of the phase contrast velocity mapping measurement unless it is accompanied by partial volume errors. In the case where signal loss is the result of a spread of phase within a voxel, the mean phase will be detected, although this measurement will be affected by differential saturation effects.[35] The phase contrast velocity mapping techniques are most susceptible to signal loss of one form or another, although this problem can normally be minimized by appropriate gradient profile design. A good way to reduce signal loss is to use a symmetrical gradient waveform that nullifies phase shifts due to all the odd-order derivatives of position and then to shorten the sequence as much as possible to reduce the effects of the even-order derivatives.[36] Signal loss of the type described is much less of a problem for the Fourier flow imaging method. In this case the Fourier transform is used to separate out constituent velocities.

RAPID PHASE FLOW IMAGING METHODS

With the very rapid scanning hardware available today, it is possible to repeat a phase contrast velocity sequence so fast that low-resolution images can be acquired in a few hundred milliseconds or a high-resolution image in a breath-hold. The major problem is for pulsatile flow where the accuracy of the measurements and the temporal resolution can be limited if the acquisition period per cardiac cycle is too long. Also, if high spatial resolution is required, the cardiac motion of structures such as coronary arteries can cause blurring with subsequent error to flow measurement.[37] It is likely that more efficient k-space coverage methods such as interleaved spiral will be important for this reason.

Ultrafast flow imaging techniques have also been developed, either by combining a phase mapping type approach with imaging methods such as single-shot echo-planar and spiral[38, 39] or by imaging only one spatial dimension.[40] A compromise will generally have to be made in temporal or spatial resolution and probably also in signal-to-noise ratio. However, taking into account these constraints, the methods have generally been shown to be accurate. One complication with the echo-planar sequence is flow signal loss because of its inherent phase sensitivity even when additional flow compensation is applied. However, this method has been used for more qualitative flow imaging, showing flow disturbances, for example.[41]

The one-dimensional rapid acquisition mode, RACE (Real time ACquisition and velocity Evalua-

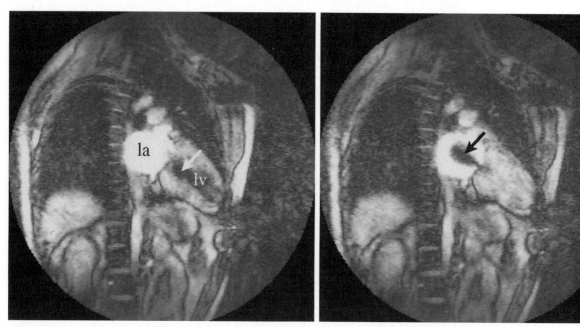

Figure 6–6. Diastolic and systolic vertical long axis images of a patient with a stenotic and regurgitant mitral valve. Signal loss (arrows) can be seen in the left ventricle (lv) during diastole (*left*) and the left atrium (la) during systole (*right*).

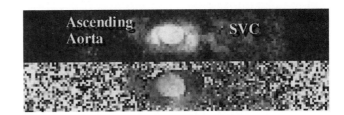

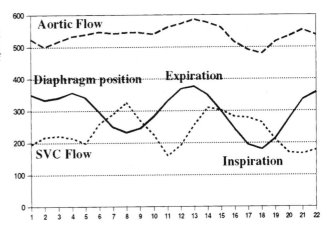

Figure 6–7. Real-time magnitude and flow images from the excited region containing the aorta and the superior vena cava (SVC). Below is a plot showing the variation in flow during a period of respiratory maneuver.

tion),[40] can be used to measure flow perpendicular to the slice. The technique can be repeated rapidly throughout the cardiac cycle in order to give near real time flow information. One problem with this type of approach is that data are acquired from a projection through the patient. This approach means that any signal overlapping with the flow signal will combine and introduce errors to the flow measurement. Several strategies have been suggested for localizing the signal in order to avoid this problem. They include spatial presaturation, projection dephasing (applying a gradient to suppress stationary tissue), and collecting a cylinder of data and multiple oblique measurements.

Yang and associates[42] used a 2-D radiofrequency excitation scheme to excite a narrow rectangular x-section column and used only 16 echoes to spatially resolve the other dimension in high resolution. This approach allowed real-time flow measurements to be acquired. The authors used the method to show the effect of controlled breathing on the flow in the ascending aorta and superior vena cava (Figure 6–7).

VISUALIZING FLOW AND FLOW PARAMETERS

The method used to visualize the MR flow data has depended on the method used for acquisition. For Fourier velocity measurement, each voxel may contain a range of measured velocities and the Fourier velocity image normally takes the form of a plot of velocity versus time or velocity versus position in one direction. Figure 6–8 shows an illustration of this method where velocity images were acquired

from a column of excited tissue including the descending aorta.[43] The front edge of the aortic pulse wave can be seen on successive frames as it travels down the vessel.

Phase contrast velocity images contain only one velocity measure per image voxel. These have historically been displayed using a gray scale such that flow in one direction tends towards white, flow in the other direction tends towards black, and stationary is mid gray as shown previously in Figure 6–3A. When flow is measured in more than one direction, more sophisticated methods of display can be used. Figure 6–9 shows a vector map of flow in the root of the aorta of a patient with an atherosclerotic aneurysm. This systolic image, shown alongside a pressure map (to be described below), nicely shows high-velocity flow impinging on the wall of the aneurysm. An alternative approach of representation would be to use the cine velocity images to calculate the path of a seed over time.[44]

The ability to study flow in such detail and at any site in the body is unique to MR. Because of this ability, a large amount of interest is being generated from those who wish to understand the physiology of blood flow and its interaction with blood vessels and the cardiac chambers. Despite the relatively poor spatial resolution, a number of groups have investigated methods of extracting a measure of the wall shear stress from the MR images. Both Oshinski and Oyre and their colleagues[45, 46] developed fitting methods to derive the velocity profile at sub-pixel distances from the vessel wall. Both groups presented expected values of stress, although it is difficult to suggest a method of validating the true accuracy of these measurements. Frayne and Rutt[47]

ECG Gate Delay

100ms 105ms 115ms 125ms 135ms 145ms 155ms 165ms 175ms

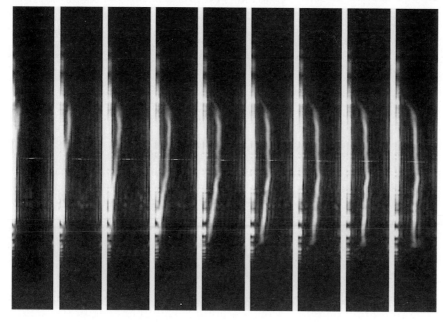

250mm FOV

0   168cm/s

Figure 6–8. Series of nine Fourier velocity images at 10-ms intervals showing the velocity pulse wave propagating down the descending aorta. ECG = electrocardiogram; FOV = field of view.

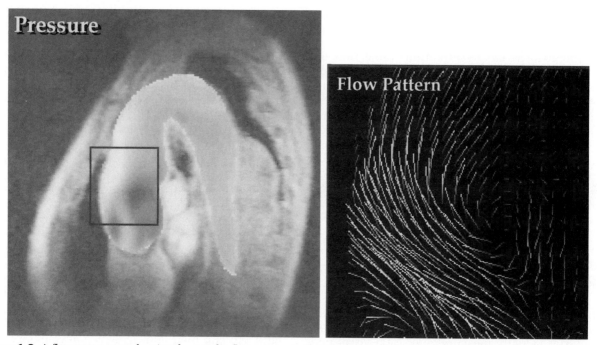

Figure 6–9. A flow vector map showing the systolic flow pattern in the aortic root of a patient with an atherosclerotic aneurysm. The associated image shows the corresponding pressure distribution.

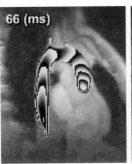
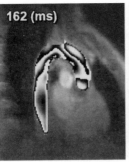
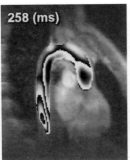
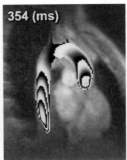
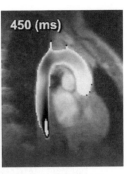

Figure 6–10. Selected frames from a cine series of calculated flow pressure maps showing the variation in pressure at different times in the cardiac cycle. Each gray scale band from white to black represents a pressure gradient of 1 mm Hg. A positive pressure gradient during systole reverses during the deceleration phase of diastole.

suggested an alternative approach, which potentially gave more information about the flow within a voxel that straddled the vessel wall. Their method used Fourier velocity encoding to distinguish the distribution of flow velocities, so that only the spatial location had to be considered.

There has also been considerable interest in the possibility of deriving pressure measurements from the MR images. Urchuk and colleagues[48] considered the vessel compliance and the flow pulse wave to calculate the pressure waveform and showed a good correlation with catheter pressure measurements made in a pig model. In contrast, Yang and coworkers[49] derived flow pressure maps from the cine phase contrast velocity maps using the Navier-Stokes equations. Figure 6–10 shows an example of the changing flow pressure around the aortic arch during the first half of the cardiac cycle. Figure 6–11 shows an interesting example of a flow pressure

map, showing the descending aorta in a patient who has had a Dacron graft repair of an aortic coarctation. In contrast to the rest of the aorta, no pressure gradient can be seen in the repaired region, possibly because of the reduced compliance.

To fully understand, visualize, and measure flow parameters in blood vessels, a method of acquisition is required that measures velocity in three dimensions and three directions over time, with a high spatial and temporal resolution. Even with the fastest gradient systems available today, this method presents a major problem in terms of acquisition time and data handling. A method has been suggested, however, using a combination of echo-planar and k-space view sharing that suggests that the acquisition times can be reduced to acceptable levels.[50]

References

1. Singer JR: NMR diffusion and flow measurement and an introduction to spin phase graphing. J Phys E: Sci Instrum 11:281, 1978.
2. Jones DW, Child TF: NMR in flowing systems. Adv Magn Reson 8:123, 1976.
3. Bradley, WG: Flow phenomenon in MR imaging. AJR Am J Roentgenol 150:983, 1988.
4. Alfidi RJ, Masaryk TJ, Haacke EM, et al: MR angiography of peripheral, carotid, and coronary arteries. AJR Am J Roentgenol 149:1097, 1987.
5. Firmin DN, Dumoulin C, Mohiaddin RH: Quantitative MR flow measurement. *In* Haacke EM, Potchen EJ, Gottschalk A, Siebert JE (eds): Magnetic Resonance Angiography: Concepts and Applications. St Louis, Mosby, 1993, p187.
6. Crooks LE, Kaufman L: NMR imaging of blood flow. Br Med Bull 40:167, 1984.
7. Axel L: Blood flow effects in magnetic resonance imaging. AJR Am J Roentgenol 143:1157, 1984.
8. van Dijk P: Direct cardiac NMR imaging of heart wall and blood flow velocity. J Comput Assist Tomogr 8:429, 1984.
9. Bryant DJ, Paynel JA, Firmin DN, Longmore DB: Measurement of flow with NMR imaging using a gradient pulse and phase difference technique. J Comput Assist Tomogr 8:588, 1984.
10. Feinberg DA, Crooks LE, Hoenninger J, et al: Pulsatile blood velocity in human arteries displayed by magnetic resonance imaging. Radiology 153:177, 1984.
11. Singer JR, Crooks LE: Nuclear magnetic resonance blood flow measurements in the human brain. Science 221:654, 1983.
12. Shimizu K, Matsuda T, Sakurai T, et al: Visualisation of

Figure 6–11. A systolic pressure map of the aortic arch in a patient with a Dacron repair. No pressure gradient is seen in the region of the repair.

Dacron Graft

Pressure

moving fluid: Quantitative analysis of blood flow velocity using MR imaging. Radiology 159:195, 1986.

13. Axel L, Shimakawa A, MacFall J: A time-of-flight method of measuring flow velocity by magnetic resonance imaging. Magn Reson Imaging 4:199, 1986.

14. Edelman RR, Mattle HP, Kleefield J, Silver MS: Quantification of blood flow with dynamic MR imaging and presaturation bolus tracking. Radiology 171:551, 1989.

15. Izen SH, Haacke EM: Measuring non-constant flow in magnetic resonance imaging. IEEE Trans Med Imaging 9:450, 1990.

16. Hahn EL: Detection of sea-water motion by nuclear precession. Geophys Res 65:776, 1960.

17. Moran PR: A flow zeugmatographic interlace for NMR imaging in humans. Magn Reson Imaging 1:197, 1982.

18. Young IR, Bydder GM, Payne JA: Flow measurement by the development of phase differences during slice formation in MR imaging. Magn Reson Med 3:175, 1986.

19. Ridgway JP, Smith MA: A technique for velocity imaging using magnetic resonance imaging. Br J Radiol 59:603, 1986.

20. Nayler GL, Firmin DN, Longmore DB: Blood flow imaging by cine magnetic resonance. J Comput Assist Tomogr 10:715, 1986.

21. Haacke EM, Lenz GW: Improving MR image quality in the presence of motion by using rephasing gradients. AJR Am J Roentgenol 148:1251, 1987.

22. Firmin DN, Nayler GL, Klipstein RH, et al: In vivo validation of MR velocity imaging. J Comput Assist Tomogr 11:751, 1987.

23. Mohiaddin RH, Pennell DJ: MR blood flow measurement: Clinical application in the heart and circulation. Cardiol Clin 16:161, 1998.

24. Redpath TW, Norris DG, Jones RA, Hutchinson MS: A new method of NMR flow imaging. Phys Med Biol 29:891, 1984.

25. Feinberg DA, Crooks LE, Sheldon P, et al: Magnetic resonance imaging and velocity vector components of fluid flow. Magn Reson Med 2:555–566, 1985.

26. Hennig J, Mueri M, Brunner P, Friedburg H: Quantitative flow measurement with the fast Fourier flow technique. Radiology 166:237, 1988.

27. Gatehouse PD, Link K, Bebbington MWP, et al: Pulse-wave and stenosis studies by cylinder excitation with Fourier velocity encoding (abstract). Proceedings of the Third Annual Meeting of the International Society of Magnetic Resonance, 1995, p 318.

28. Hardy CJ, Bolster BD Jr, McVeigh ER, et al: Pencil excitation with interleaved Fourier velocity encoding: NMR measurement of aortic distensibility. Magn Reson Med 35:814, 1996.

29. Luk Pat GT, Pauly JM, Hu BS, Nishimura DG: One-shot spatially resolved velocity imaging. Magn Reson Med 40:603,1998.

30. Buonocore MH: Blood flow measurement using variable velocity encoding in the RR interval. Magn Reson Med 29:790, 1993.

31. Yang GZ, Burger P, Kilner PJ, et al: Dynamic range extension of cine velocity measurements using motion registered spatio-temporal phase unwrapping. J Magn Reson Imaging 6:495, 1996.

32. Bittoun J, Bourroul E, Jolivet O, et al: High-precision MR velocity mapping by 3D-Fourier phase encoding with a small number of encoding steps. Magn Reson Med 29:674, 1993.

33. Polzin JA, Alley MT, Korosec FR, et al: A complex-difference phase-contrast technique for measurement of volume flow rates. J Magn Reson Imaging 5:129, 1995.

34. Frayne R, Polzin JA, Mazaheri Y, et al: Effect of and correction for in-plane myocardial motion on estimates of coronary-volume flow rates. J Magn Reson Imaging 7:815, 1997.

35. Polzin JA, Korosec FR, Wedding KL, et al: Effects of through-plane myocardial motion on phase-difference and complex-difference measurements of absolute coronary artery flow. J Magn Reson Imaging 6:113, 1996.

36. Firmin DN, Nayler GL, Kilner PJ, Longmore DB: The application of phase shifts in NMR for flow measurement. Magn Reson Med 14:230, 1990.

37. Hofman MB, van Rossum AC, Sprenger M, Westerhof N: Assessment of flow in the right human coronary artery by magnetic resonance phase contrast velocity measurements: Effects of cardiac and respiratory motion. Magn Reson Med 35:521, 1996.

38. Firmin DN, Klipstein RH, Hounsfield GL, et al: Echo-planar high-resolution flow velocity mapping. Mag Reson Med 12:316, 1989.

39. Gatehouse PD, Firmin DN, Collins S, Longmore DB: Real time blood flow imaging by spiral scan phase velocity mapping. Magn Reson Med 31:504, 1994.

40. Mueller E, Laub G, Grauman R, Loeffler W: RACE—Real time ACquisition and Evaluation of pulsatile blood flow on a whole body MRI unit (abstract). Proceedings of the Seventh Annual Meeting of the International Society of Magnetic Resonance in Medicine, 1988, p 729.

41. Kose K: One shot velocity mapping using multiple spin-echo EPI and its application to turbulent flow. J Magn Reson 92:631, 1991.

42. Yang GZ, Gatehouse PD, Mohiaddin RH, et al: Zonal echo-planar flow imaging with respiratory monitoring (abstract). Proceedings of the Fifth Annual Meeting of the International Society of Magnetic Resonance in Medicine, 1997, p 1885.

43. Gatehouse PD, Link K, Bebbington MWP, et al: Pulse-wave and stenosis studies by cylinder excitation with Fourier velocity encoding (abstract). Proceedings of the Third Annual Meeting of the International Society of Magnetic Resonance and the Twelfth Annual Meeting of the European Society for Magnetic Resonance in Medicine and Biology, 1995, p 318.

44. Napel S, Lee DH, Frayne R, Rutt BK: Visualizing three-dimensional flow with simulated streamlines and three-dimensional phase-contrast MR imaging. J Magn Reson Imaging 2:143, 1992.

45. Oshinski JN, Ku DN, Mukundan S Jr, et al: Determination of wall shear stress in the aorta with the use of MR phase velocity mapping. J Magn Reson Imaging 5:640, 1995.

46. Oyre S, Ringgaard S, Kozerke S, et al: Accurate noninvasive quantitation of blood flow, cross-sectional lumen vessel area and wall shear stress by three-dimensional paraboloid modeling of magnetic resonance imaging velocity data. J Am Coll Cardiol 32:128, 1998.

47. Frayne R, Rutt BK: Measurement of fluid-shear rate by Fourier-encoded velocity imaging. Magn Reson Med 34:378, 1995.

48. Urchuk SN, Fremes SE, Plewes DB: In vivo validation of MR pulse pressure measurement in an aortic flow model: Preliminary results. Magn Reson Med 38:215, 1997.

49. Yang GZ, Kilner PJ, Wood NB, et al: Computation of flow pressure fields from magnetic resonance velocity mapping. Magn Reson Med 36:520, 1996.

50. Firmin DN, Gatehouse PD, Yang GZ, et al: A 7-dimensional echo-planar flow imaging technique using a novel k-space sampling scheme with velocity compensation (abstract). Proceedings of the Fifth Annual Meeting of the International Society of Magnetic Resonance in Medicine, 1997, p 118.

Special Considerations for Cardiovascular Magnetic Resonance: Safety, Electrocardiographic Set-Up, Monitoring, Contraindications

Anastazia Jerzewski and Ernst E. van der Wall

During the last decade magnetic resonance (MR) has developed into an important diagnostic clinical tool in cardiology. Not only the anatomy of the heart but also its function, metabolism, perfusion, and, more recently, the proximal coronary artery status can be assessed with cardiovascular MR (CMR). MR offers some special advantages over other diagnostic imaging methods. First, MR does not use ionizing radiation. Second, the radiofrequency (RF) radiation penetrates bony structures and air without attenuation. Third, MR gives additional diagnostic information of tissue characteristics. Finally, MR provides three-dimensional (3-D) images or images of arbitrarily oriented slices.

However, when performing CMR, particular precautions have to be made. Because MR operates with high static and gradient magnetic fields, special safety regulations must be taken into account and certain contraindications have to be considered. In this chapter we review CMR with regard to safety, electrocardiographic set-up, patient monitoring, and contraindications.

SAFETY OF CARDIOVASCULAR MAGNETIC RESONANCE

General Issues

The CMR examination often takes longer than other diagnostic modalities (although this situation may well change with the advent of real-time imaging[1]), and the confined space in which the patient is placed is rather narrow, which some patients find uncomfortable. During CMR, communication with the patient may be difficult because of interfering noise from the gradient coils. On the other hand, CMR is entirely noninvasive. Overall the safety issues during CMR that might pose potential safety concerns[2] can be summarized into 9 areas:

1. biological effects of the static magnetic field
2. ferromagnetic attractive effects of the static magnetic field on certain devices
3. potential effects on the relatively slowly time-varying magnetic field gradients
4. effects of the rapidly varying RF magnetic fields, including RF power deposition concerns
5. auditory considerations due to noise from the gradients
6. safety considerations concerning superconducting magnet systems
7. psychological effects
8. side effects from intravenous MR contrast agents
9. patient safety during stress conditions

The safety concerns inherent to these issues are sequentially discussed.

Biological Effects

Many structures in animals and humans are affected by magnetic fields. Many potential biological effects and different magnitudes of magnetic fields have been examined, among which is the study of the effect of the field on cardiac contractility and function. Gulch and Lutz concluded that static magnetic fields used in MR do not constitute any hazard in terms of cardiac contractility.[3] These magnetic fields do not increase ventricular vulnerability as assessed by the repetitive response threshold and the ventricular fibrillation threshold.[4] In one of the investigations, however, the cardiac cycle length was found to be altered.[5] Numerous biological effects on other systems have been investigated extensively, and it may be concluded that no deleterious biological effects from static magnetic field strengths currently used in clinical MR (≤ 1.5 T) have yet been established. However, as in all aspects of safety monitoring for patients, further research needs to be continued in this area.

Ferromagnetism

The physical effect of the static magnetic field consists of a potential health hazard from the attrac-

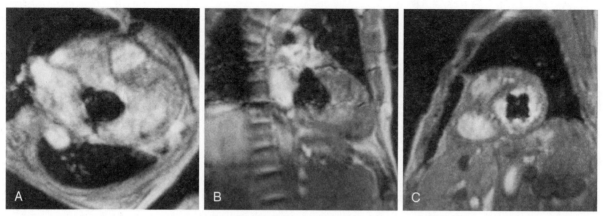

Figure 7–1. CMR of a patient with a Starr-Edwards mitral prosthetic valve. *A,* Horizontal long axis; *B,* Vertical long axis; *C,* Basal short axis. The dark artifact is obvious on the images, but does not interfere with assessment of ventricular function. These images were acquired at 0.5 T with a gradient echo cine sequence and an echo time (TE) of 14 ms. Shorter TE gradient echo sequences and spin echo sequences typically show less artifact. The Starr-Edwards valve causes the largest artifact because of the large amount of metal present in its construction. Other valves cause considerably less disturbance, especially the tissue valves. (Images courtesy of Dr. D. J. Pennell.)

tive effect on ferromagnetic objects. Ferromagnetic objects can be defined as those in which a strong intrinsic magnetic field can be induced when they are exposed to an external magnetic field. The existence of different kinds of scanners with different shielding makes the discussion of this topic even more crucial. When dealing with a static magnetic field two types of physical concerns exist.

First, there are concerns from forces exerted on ferromagnetic objects within, or close or distant proximity to the patient. These forces result in rotational (torque) and/or translational (attractive) motion of the object. Within the human body, a ferromagnetic metallic structure might be sufficiently attracted, or have a sufficient amount of torque exerted, to create a hazardous situation. These factors should be carefully considered before subjecting a patient with a ferromagnetic implant or material to CMR, particularly if the device is located in a potentially dangerous area of the body where movement or dislodgment of the device could injure the patient. Another potentially injurious effect is known as the projectile or missile effect. This term refers to the fact that ferromagnetic objects have the potential to gain sufficient speed during attraction to the magnet that the accumulated kinetic energy could be injurious or even lethal if the object were to strike a patient. Numerous studies have been performed to assess the ferromagnetic qualities of various metallic implants and materials.[6–10] The results indicate that patients with certain metallic implants or prostheses that are nonferromagnetic or are minimally deflected by static magnetic fields can safely undergo CMR. The literature on this topic has been extensively reviewed and compiled.[8] It should be noted, however, that there are common misconceptions about what types of objects are ferromagnetic. The most important misconception is that stainless steel is ferromagnetic, which it is not. Patients with most stainless steel implants can therefore be imaged safely (see special considerations

of pacemakers later in this chapter). However, the implant will result in local field distortion and resultant image artifact; for example, signal loss occurs around metallic prosthetic valves (Figure 7–1) and sternal wires (Figure 7–2) after bypass surgery, but this situation does not make the imaging hazardous. Non-stainless steel, which may be ferromagnetic, is not used for human implants, but is commonly used, for example, for scissors, stethoscopes, pen clips, and oxygen cylinders. Finally, batteries are typically attracted to the magnet, which is one of the problems associated with MR exams in patients with pacemakers.

The second type of physical concern deals with magnetically sensitive equipment, the functioning of which may be adversely affected by the magnetic field. The most common of these pieces of equipment is the cardiac pacemaker. Most pacemakers include a reed relay switch whereby the sensing mechanism can be bypassed and excitation in the asynchronous mode can occur. This switch is commonly activated when a magnet of sufficient strength is held over the pacemaker.[11] In addition to this problem, however, the function of cardiac pacemakers may be influenced by field strengths as low as 17 Gauss.[11] In practice, reed switch closure can be expected in all pacemakers placed in the bore of the scanner. Pacemaker function is considered again later in this chapter.

Rapidly Switched Magnetic Fields

CMR exposes the patient to rapid variations of magnetic fields by the transient application of magnetic gradients during imaging. The effect may be the induction of currents within the body, or any other electrical conductor, according to Faraday's Law. The current is dependent on the time rate of change of the magnetic field (dB/dt), the cross-

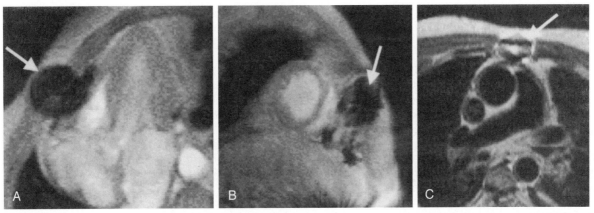

Figure 7–2. CMR of a patient with sternal wires after thoracotomy. The artifact is clearly seen (arrows) on the horizontal long axis *(A)* and short axis *(B)* gradient echo cine images, and to a much lesser extent on the transaxial spin echo image *(C)*. (Images courtesy of Dr. D. J. Pennell.)

sectional area of the conducting tissue loop, and the conductivity of the tissue. Biological effects of induced currents can be caused either by power deposition by the induced currents (thermal effects) or by direct effects of the current (nonthermal effects). Thermal effects owing to switching gradients are not believed to be clinically significant.[12–14] Possible nonthermal effects are stimulation of nerve or muscle cells. The threshold currents for nerve stimulation and ventricular fibrillation are known to be much higher than the estimated current densities induced under clinical CMR conditions. The echoplanar imaging method, however, involves more rapidly changing magnetic field gradients, and peripheral muscle stimulation in humans has been reported.[15] Such considerations have become more important as new technology has allowed the introduction of commercially available ultrafast gradient switching systems, and guidelines for maximum magnetic field variation are under development.

Radiofrequency (RF) Time Varying Field

The transmitted RF time varying field induces electrical currents within the tissue of the patient. The majority of this power is transformed into heat within the patient's tissue as a result of ohmic heating. The time varying magnetic gradients have the potential to cause either thermal or nonthermal biological effects. Distinction between these two is a matter of frequency, waveform shape, and magnitude. The discussion about nonthermal effects from RF magnetic fields is controversial because of questions of the relationship between chronic exposure over many years to low-frequency electromagnetic fields and the causation of cancer or developmental abnormalities. Recent evidence suggests that proximity to power lines is not injurious.[16] It should be remembered, of course, that acute exposure of a patient to short-term RF fields for a diagnostic CMR examination may be quite different from chronic exposure. The induced currents from RF magnetic fields are unable to cause nerve excitations. One of the difficulties faced by the field of medicine is proving that a procedure is noninjurious as a result of anecdotal adverse event case reports and publication bias towards nonneutral reports.[17] This issue is also faced by such well-established technology as ultrasound, of which safety concerns have been raised over acoustic exposure.[18]

By contrast to the insignificant thermal effects due to switched gradients, thermal effects as a result of the RF pulses are of significant concern. A general point of discussion is the appropriate safety regulations of levels of magnetic field strength in MR imaging. Application of the fundamental law of electrostimulation is well established, both on theoretical and experimental grounds. Application of this law, in combination with Maxwell's law, yields an equation called the fundamental law of magnetostimulation, which has the hyperbolic form of a strength-duration curve and allows an estimation of the lowest possible value of the magnetic flux density capable of stimulating nerves and muscles. Calculations have shown that the threshold for heart excitation is more than 200 times higher than for nerve and muscle stimulations, depending on pulse duration.[19] However, in clinical practice, some precautions are necessary. First and most importantly, the specific absorption rate (SAR) of the imaging sequence being operated is monitored by the scanner software and has to be kept below limits set by such bodies as the United States Food and Drug Administration (FDA). Second, circumstances that could enhance the possibility of heating injury should be avoided. This avoidance includes ensuring the prevention of loops, which could act as aerials within the scanner and enhance the heating effect locally. Patients should therefore not be allowed to cross their legs (loop via the pelvis), clasp their hands together (loop via the shoulder and upper chest), or even cross their fingers. The simple use of pillows prevents such problems. Other possible loops include the electrocardiogram

(ECG) leads, which should always be run out of the scanner parallel to the main field, and not looped across the chest. Finally, pacemaker leads make excellent antennae independent of the pacemaker itself. In most cases, MR in patients with pacemakers or permanent pacemaker leads is contraindicated, although some exceptions exist (see Pacemakers section later in the chapter). The pacemaker lead can heat significantly during MR and be a potential hazard. Another consideration in patients following cardiac surgery is the effect of retained epicardial pacemaker leads. These leads can be left in place after the surgery, and they might therefore act as an antenna during CMR. Studies have suggested that such short retained epicardial wires do not pose a significant clinical problem.[20, 21] Finally, the use of ECG electrodes, which are essential for cardiac gating, needs to be considered. Metallic ECG electrodes may cause burns during CMR.[22, 23] This risk can be reduced by using carbon fiber electrodes, and these electrodes have now become standard.

Auditory Considerations

During CMR examinations, the gradient coils and adjacent conductors produce a repetitive sound because they act essentially as loudspeakers, with current being driven through them, while they are in a magnetic field. Auditory considerations should therefore be taken into account when imaging a patient. The amplitude of this noise depends on factors such as the physical configuration of the magnet, pulse sequence type, timing specifications of the pulse sequence, and the amount of current passing through these coils.[24] In general, the amplitude of the generated noise from the clinical MR scanners remains between 65 and 95 dB. However, there have been reported instances of temporary hearing impairment as a result of MR. Magnet-safe headphones or wax earplugs are readily available, which have been shown to prevent hearing loss,[25] and these devices are in common use. Systems combining sound attenuation with the facility to play music of the patient's choice are also available. Research into the reduction of noise in MR scanners is ongoing, and the use of anti-noise is one area of interest.[26]

Superconducting System Issues

Most current superconducting MR scanners systems utilize liquid helium. The helium maintains the magnet coils in their superconducting state. Helium achieves the gaseous state at approximately $-269°C$ (4K). If for any reason the temperature within the cryostat rises, or in a system quench, the helium will enter the gaseous state. This situation means a marked increase in volume and thereby pressure within the cryostat. A pressure-sensitive valve is designed to give way to the gaseous helium,

which should be vented outside the MR scanner room. However, it is possible that some helium gas might be released into the imaging room. Asphyxia and frostbite are potential hazards if a patient is exposed to the helium vapor for a prolonged time, although there are no reports of such occurrence during clinical use in the medical community. For older scanners that still use a buffer of liquid nitrogen within the system (boils at 77°K), an oxygen monitor is recommended in the scanner room. Cryogen Dewars should be stored away from the scanner and in well-ventilated areas.

Psychological Effects

Claustrophobia or other psychological problems may be encountered in up to 10 percent of patients undergoing MR.[27] In our experience, the number is nearer to 2 to 4 percent, which can be reduced further to a small number of intractably anxious patients by the use of explanation, reassurance, and, when necessary, light sedation with, for example, 1 to 5 mg of IV diazepam.[28] In addition, the development of shorter magnets as well as open designs is proving to be helpful. Such feelings originate from a variety of factors including the restrictive dimensions of the scanner, duration of the examination, noise, and ambient conditions within the magnet bore.[29] Fortunately, adverse psychological effects to MR are usually transient. In a study by Weinreb and colleagues,[30] based on the experience of 450 patients undergoing MR and computed tomography (CT) examinations, it was clearly shown that patients often prefer the MR study, although MR took longer. Furthermore, the patient is placed into a confined space and there are difficulties in communicating with the patient during MR scanning because of the noise from the gradient coils and the necessity of eliminating all extraneous RF sources from the examination room. To a certain extent this problem can be avoided when the patient takes a prone position in the scanner,[31] facilitating the communication with the outside surroundings. Simple maneuvers such as mirrors also help in allowing the patient a clear view of the scanning room; in addition, allowing the anxious patient the opportunity to visit the scanner prior to the appointment to become familiar with the facility and staff can be helpful.

Gadolinium-Based Contrast Agents

The safety of contrast agents containing gadolinium currently on the market is extremely good. Gadopentetate dimeglumine (Gd-DTPA) is the best established, and its safety profile is well documented,[32, 33] but it should be noted that similar safety results have been shown with the other commercially available agents. The median lethal dose of Gd-DTPA is roughly 10 mmol/kg, which is 50 to

100 times the diagnostic dose, and shows the wide safety margin that the contrast agent enjoys. Patient tolerance of this drug is also high, and the prevalence of adverse reactions is approximately 2 percent, most of which are mild. Among the reactions related to the IV administration of this drug are transient headache, nausea, vomiting, local burning or cool sensation, and hives. There are only a few reported cases of fatalities that have been temporally associated with the administration of Gd-DTPA. The actual relationship between the deaths and the drug is uncertain. There have been incidents of anaphylactoid reactions associated with the IV injection,[34] although the frequency of such incidents appears to be approximately 1 per 100,000 doses. The safety margins with these agents appear to be considerably better than those of iodinated x-ray contrast agents and have the major advantage of being safe in patients with renal impairment. For CMR, these agents are used in order to increase contrast between blood and soft tissue, for cine imaging during functional studies, for angiography, to enhance cardiac tumors and cysts, to assess myocardial perfusion, and to examine for myocardial infiltration. In summary, FDA-registered gadolinium complexes such as Gd-DTPA can be safely used in cardiac patients.

Multiple new MR contrast agents are being developed and investigated. These agents are mainly gadolinium complexes, sometimes with novel binding molecules for special actions, but iron-based compounds are also being developed. Some of these agents are retained in the vascular system and do not leak into the extravascular space. This property suggests they may have clinical use for angiography, possibly in the coronaries, and for functional imaging.

Patient Safety During Stress Conditions

A concern with CMR stress studies has been the ability to handle emergency situations. Patient monitoring during stress conditions is a critical issue because myocardial ischemia can be provoked in patients with coronary artery disease. Commercial equipment exists to monitor noninvasive blood pressure, heart rate, oxygen saturation, and other vital parameters in CMR scanners. The most crucial difference when compared with conventional exercise testing outside a magnetic field is the lack of a diagnostic electrocardiogram, precluding the assessment of stress-induced ST segment changes. This situation holds both for conventional exercise using a specially adapted bicycle ergometer and pharmacologically induced stress. Under these circumstances, only heart rate can be monitored reliably. When performing pharmacological stress CMR (e.g., with dipyridamole, adenosine, or dobutamine) an experienced physician should be present during the examination, and appropriate treatments for complications should be in direct proximity of physician and patient. Dipyridamole (half-life 30 minutes) and adenosine (half-life 10 seconds) are both vasodilators. Both agents have similar side effects such as bronchospasm, hypotension, dysrhythmias, and bradycardia. In particular, adenosine infusion may promote atrioventricular heart block in a small percentage of patients (0.7 to 2.8 percent). This problem is usually asymptomatic and self-limiting. When patients are symptomatic, the short physical half-life of adenosine means that heart rhythm and symptoms can be restored very quickly by halting the infusion. As a suitable antagonist to both dipyridamole and adenosine, aminophylline may be given slowly in an initial dose of 50 mg IV up to a maximum of 250 mg if necessary. In case of persisting advanced heart block, 0.5 mg atropine IV should be administered up to a total dose of 3 mg. Dipyridamole and adenosine should not be given to patients with asthma. Dobutamine (half-life 2 minutes) is a beta-agonist leading to an increase in cardiac inotropy (contractility) and chronotropy (heart rate). Common side effects are cardiac pounding and palpitations, and less commonly dysrhythmias such as supraventricular tachycardia and nonsustained ventricular tachycardia. Dobutamine can be safely given to patients with asthma, but not those with ventricular irritability.[35] The actions of dobutamine can be counteracted by IV administration of a short-acting beta-blocking agent such as esmolol. In case of cardiac arrest or ventricular fibrillation, the recommendations should be followed according to published guidelines, such as proposed by the European Resuscitation Council.[36] In every CMR facility, an alarm system and a written flowchart should be visually available with the necessary instructions in case of emergency. It is necessary to be able to safely and quickly remove the patient out of the examination room (preferably within 20 seconds) to an area where emergency treatment can be performed safely away from the hazards of the magnetic field. A nonferromagnetic stretcher stored in the scanner room or a detachable scanner table is ideal for this. A cardiac arrest trolley must be maintained in close proximity to the scanner room, and all personnel should undergo regular training in cardiopulmonary resuscitation techniques. Regular checks should be made of both the resuscitation equipment and the alarm system.

PATIENT MONITORING AND ELECTROCARDIOGRAPHIC SET-UP

Patient monitoring during CMR poses problems that will not be familiar to users of other technologies, such as echocardiography. Ferrous metal, which is present in most monitoring equipment, can distort the magnetic field, and such an item has the potential to become a projectile. In addition, monitoring wires attached to the patient and that are leaving the scanner and passing to another room may act as antennae for stray RF signals. In addition,

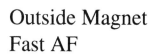

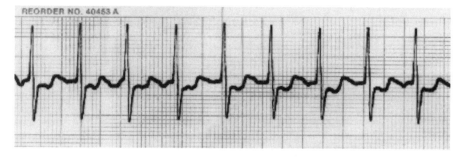

Outside Magnet
Fast AF

Inside Magnet

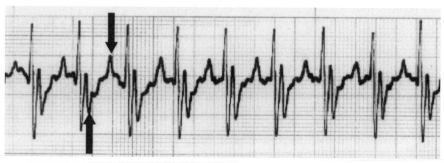

Figure 7–3. Example of the magnetohydrodynamic effect. The top trace was recorded in a patient with atrial fibrillation (AF) outside of the magnet, and the bottom trace while inside the magnet. Note the distortion of the ST segment and T waves (black arrows) caused by added potentials arising from systolic flow in the aorta. (Courtesy of Dr. D. J. Pennell.)

electrical equipment in the scanner room can act as a source of RF noise. All these disturbances may result in image degradation. Therefore, specific answers to these problems have been designed.

Commercially available MR-compatible monitoring such as ECGs, blood pressure, chest wall movements and general anesthesia equipment have been tested in several studies.[37, 38] Satisfactory monitoring can be obtained, and images obtained during its use can be adequately evaluated.[39] For some monitoring, simple solutions work, such as that from Roth and coworkers,[39] who measured arterial blood pressure outside the MR scanner by lengthening the rubber tubing connected to a blood pressure cuff. The most modern monitoring equipment eliminates the need for wires and tubes to leave the scanning room by using a microwave transmitter communicating with a slave display unit in the operating room.

CMR depends on a high-quality ECG signal for routine imaging, and each of the manufacturers has developed its own solution to the problems posed. Fiberoptic transmission of ECG signals for gating is now commonplace, which significantly reduces RF pulse artifacts in the ECG. Felblinger and associates showed that this type of system could yield signals almost free from interference,[40] during both conventional[41] and high gradient activity sequences such as during echo-planar imaging.[42] From this signal the authors also developed a method for respiration monitoring during CMR sequences. Third-party ECG solutions are also now being incorporated into the latest generation of scanners, and these scanners come with specific recommendations for ECG lead placement. Carbon fiber electrodes minimize the risk of burning that has been reported with standard metallic ECG electrodes.[22] Typical lead placement is

the result of compromise. A better signal results from widely spaced electrodes, but this placement results in more artifact from the gradients. In general, therefore, the leads are kept relatively close together, and on the left side, which reduces the magnetohydrodynamic effect (the effect of systolic aortic flow causing surface potentials on the ECG that distort the ST-T segment—Figure 7–3). A typical lead placement that is commonly adopted is shown in Figure 7–4. Some centers have found ECG gating using electrodes on the back to be successful, but this technique is not widely used. The ECG leads should not be allowed to form loops, which could present a burning hazard, and they should be braided together and brought out of the magnet aligned parallel to the bore to reduce electrical interference. Keeping the electrical cables short is valuable, and fiberoptic conversion modules are therefore often very close to the patient's chest. Switching between the ECG traces sometimes allows flexibility to reduce gating errors from tall T waves or electrical interference. One thing is certain, however, and that is that time spent ensuring that the ECG is stable and working correctly at the start of the scan is time very well spent.

An alternative technique to routine surface ECG recording has recently been described by Fischer and colleagues using vectorcardiography, and this advance has proven to be quite useful.[43] The system examines the 3-D orientation of the ECG signal and uses the calculated vector of the QRS complex as a filter mechanism to ignore electrical signals that are of a similar timing in the cardiac cycle or a similar magnitude, but of a different vector. The system as reported identified the QRS complex correctly in 100 percent of cases with 0.2 percent false positives.

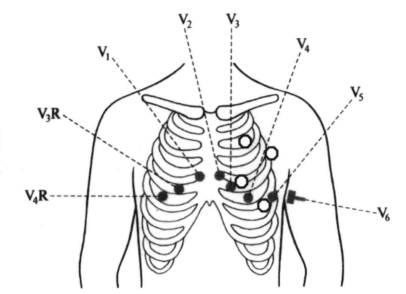

Figure 7–4. Typical electrode placement for CMR (open circles). The conventional chest leads are shown for comparison and the position of the four carbon ECG electrodes over the left chest is indicated with the black and white markers. (Courtesy of Dr. D. J. Pennell.)

If this system works to this level of accuracy in patients as well as normal individuals, it would represent a significant improvement for CMR in stabilizing this important gating signal.

Should interpretation prove to be impossible for technical reasons, it has been shown that a standard vascular Doppler can be used to monitor heart rate during CMR. The Doppler and telemetric ECG do not contain enough ferromagnetic material to cause visible image degradation. Jorgensen and coworkers evaluated whether patients could be monitored during CMR with 1.5 Tesla (T) machines in a manner that complies with monitoring standards.[44]

It should also be remembered that the high magnetic field can interfere with normal functioning of equipment, and this hazard applies not only to the monitoring equipment but also to smaller items such as infusion pumps used for stress testing. In general, the influence of the CMR scanner on nearby equipment depends upon several factors such as the strength of the MR magnetic field, the equipment's proximity to the scanner, the amount of ferromagnetic material of the equipment, and the design of its electrical circuitry.

Finally, simple devices such as closed-circuit television and a two-way intercommunication system also aid with monitoring by allowing a constant view of the patient and easy communication if the patient is in discomfort. However, the latter may be impaired during imaging because of the noise of the gradients. Inasmuch as sequences are now commonly being reduced in duration to a breath-hold, however, this limitation is becoming much less important.

CONTRAINDICATIONS TO CARDIOVASCULAR MAGNETIC RESONANCE

In general there are potential hazards and artifacts of ferromagnetic and nonferromagnetic materials in

CMR, for example with neurosurgical clips and ocular implants.[45–47] The (relative) contraindications of these materials for CMR are dependent on such factors as the degree of ferromagnetism, geometry of the material, gradient (for force), and field strength (for torque) of the imaging magnet and many other factors. Whenever there is a concern about safety in a patient with an implant, the CMR examination should be deferred until the device and/or issues are clarified.

General Issues

There are a number of circumstances in which CMR is better avoided because of reports of death or harm that has occurred. Prominent among these reports are patients with cerebral aneurysm clips, which have become dislodged during MR examinations, causing fatal cerebral hemorrhage. Modern clips are not ferromagnetic and are safe, but the problem is establishing the type of clip with certainty prior to performing CMR. In general, therefore, CMR in these patients should be avoided unless written information is available and appropriate advice from a neurosurgical center is obtained. Even if a patient has undergone a prior MR without adverse effects, the specific type and safety of the clip should be verified. There have been cases of bleeding in the eye in patients with previous injury with metallic shards (usually metal workers), and again a history of this should be sought. A skull x-ray can be helpful in cases of doubt. Electronic implants are the other major problem. These may malfunction or be damaged by CMR, which is therefore best avoided. This warning applies to cochlear implants, automatic cardioverter defibrillators, nerve stimulation units, and a number of other modern implants. In general, the main rule to observe is to determine the risk-benefit ratio of the proposed procedure. If the question at hand can only be answered by CMR, and if it is very important, then the risk-benefit may be positive for some patients (such has been demonstrated in selected patients with pacemakers[75]). In many clinical circumstances, however, the information required can be obtained by other means. Reference texts on the safety of specific devices are available.

For CMR there are a number of other specific device issues that require mention. Swan-Ganz catheters and temporary pacing wires in general preclude CMR, but sternal wires and vascular clips on bypass grafts do not present problems for safety, although localized artifacts occur on the images. Three specific areas are dealt with in more detail: stents, valve prostheses, and pacemakers.

Stents

The first intracoronary stents were implanted in human coronary arteries in 1986 by Sigwart and associates.[48] Over the past years the indications for intracoronary stents have expanded and the use of stents has grown dramatically. One of the reasons for this development is the reduced restenosis rate compared with conventional balloon angioplasty.[49–52] Because of the increased indications for intracoronary stents, the population with stents in situ who need to undergo CMR for various diagnostic reasons is rapidly expanding.

Coronary stents are metallic structures that remain in situ for life. As a result, there have been concerns regarding stent dislodgment during CMR, and possible local heating effects. There are several factors that determine the risk of these materials placed in a magnetic field. These factors include the ferromagnetism of the material, the strength of the static and gradient magnetic fields, the metal mass, and the geometry of the material.[10, 53] The most commonly used stents are made of stainless steel or tantalum, which are not ferromagnetic.[54] In vitro experiments performed by Scott and Pettigrew[55] evaluated and quantified the influence of magnetic fields used in clinical MR scanners on widely used coronary stents. They used a 1.5 T magnet and the results did not show any significant deflection of the stents used. In vivo investigations have been carried out in dogs with tantalum stents in the aorta using a 1.5 T scanner.[54] The purpose of the study was to examine the MR compatibility of tantalum stents and to evaluate the feasibility of using vascular MR imaging to evaluate the patency of stented vessel in vivo. In this study, the animals were repeatedly subjected to MR imaging in the first 8 weeks following implantation. No evidence was found of any stent migration even with the MR immediately following implantation in all animals. Angiographic evaluation demonstrated no interval development of luminal narrowing or thrombus in the region of the vascular stent. CMR artifacts produced by the stent were increased with longer echo times. Further experiments by Strohm and col-

leagues have shown that the common stents do not heat up during MR.[56] This finding accords with widespread clinical experience where stent imaging has been performed in the day after implantation or soon afterwards.[56a, 57, 58] A summary of the various coronary stents that have been examined is shown in Table 7–1.

Another issue regarding CMR imaging procedures in patients with coronary stents is the image distortion caused by the ferromagnetic properties of the stent (see Figure 16–22). In general, the greater the ferromagnetism of a metallic implant, the greater its magnetic susceptibility artifact. However, several studies have shown that MR angiography could be useful for noninvasive imaging and evaluation of coronary artery flow after stent placement.[59, 60]

It should be noted that stent manufacturers have taken a conservative line in some product information suggesting that CMR be delayed for 6 weeks after implantation, when endothelialization is complete.[61] However, such an approach ignores the fact that displacement of the stent by the magnetic field is negligible and far smaller than the deflection forces exerted by the beating heart. We conclude on the evidence available that CMR is a safe procedure in all patients with coronary stents immediately after implantation. This stenting issue brings again to our attention the outstanding editorial by Roger Ordidge and Gary Fullerton calling us to resist the "dangerous contemporary trend to extrapolate the lack of scientific evidence proving that a new medical device is safe to the conclusion that it is dangerous. This malformed logic has the unfortunate potential to harm patients by depriving them of the diagnostic and therapeutic procedures necessary to live productively."[62] This editorial accords with recent recommendations on wall stents.[63]

Table 7–1. Coronary Stents Reported To Be Safe for MR Examination

Stent Type	Reference
ACS Multi-link RX Duet	56a
ACS RX Multi-link	56a
AVE	56, 56b
Micro Stent	56a
BeStent	56, 56a
Crown	56b
Giantourco-Roubin	55, 56
Giantourco-Roubin II	56a
InFlow	56, 56a
InFlow Gold	56a
InStent	56
MAC-Stent	56a
Multilink	56, 56b
Palmaz-Schatz	55, 56, 56a
R-Stent	56a
Seaquence	56a
Strecker	55
Tenax-Stent	56
Wallstent	56, 56a
Wiktor	55
Wiktor GX	56a

Valvular Prostheses

Heart valve prostheses are all safe for CMR.[64] These most recent recommendations supersede those that suggested that pre-6000 series Starr-Edwards valves might cause problems during MR. This conclusion is backed up by numerous data. Several studies in 1.5 T to 2.35 T static magnetic fields have shown for a number of prosthetic valves that there is no hazardous deflection during exposure of the magnetic field.[53, 65, 66] Heating of small metallic implants was tested in a study reported by Davis and coworkers.[45] They found no significant increase in temperature in steel and copper clips, which were exposed to changing magnetic fields 6.4 times as strong as those expected to be used in the MR scanner. As a result, the currently used prosthetic valves do not constitute any contraindication to CMR.[67–69] It should, however, be realized that prosthetic material may lead to artifacts on CMR images (see Figure 7–1). To evaluate the influence of prosthetic valves on interpretation of the CMR images and on the capability of functional valve analysis, Bachmann and associates showed convincingly, in a group of 89 patients and 100 heart valve prostheses, that all patients could be imaged with CMR without any risk and that prosthesis-induced artifacts did not interfere with image interpretation.[68] In particular, physiologic valvular regurgitation was easy to differentiate from pathological or transvalvular regurgitation. DiCesare and colleagues studied 14 patients who were surgically treated with 9 biological and 7 mechanical aortic and mitral valves.[69] Three classes of artifacts were distinguished and graded as minimal, moderate, or significant. The biological valves gave minimal artifacts, and the mechanical valves showed only moderate artifacts. In all 16 prosthetic valves, CMR allowed adequate semiquantitative analysis of flow behind the valve.

Pacemakers

Patients with cardiac pacemakers are currently considered by many centers to be absolutely contraindicated for MR. Several deaths have occurred in patients who unknowingly had a pacemaker, were not monitored, and died during MR examination. However, as is often the case, such a dogmatic approach is not entirely correct, because there are many patients with pacemakers recorded in the literature who have undergone MR safely. Therefore, by all the normal rules of semantics, the presence of a pacemaker is not an *absolute* contraindication. However, it remains a fact that the presence of a pacemaker is a strong relative contraindication to scanning, and such procedures still require a great deal more research, and should only be undertaken after careful evaluation of the risk-benefit ratio to the patient, and only at experienced CMR centers.

The issues surrounding CMR of pacemakers are complex. Various generations and types of pacemakers and pacemaker leads are implanted in patients, and studies on the effects of MR on cardiac pacemakers, both in phantoms and in patients, have been performed.[70] In an in vitro study from Lauck and coworkers, it was concluded that no disturbances arise when the systems are tested in the asynchronous mode in 0.5 T MR imaging under standard examination conditions with ECG-triggered imaging.[70] In patients without permanent pacemaker dependency, complete inactivation in the "000" mode minimizes the possibility of RF interference,[71, 72] although induction voltage cannot be avoided in this mode.

Gimbel and associates investigated the effect of MR in five patients with permanent cardiac pacemakers, one of whom was pacemaker dependent.[73] A variety of pacing configurations were studied, but none of the patients experienced any torque or heat sensation. Four non–pacemaker-dependent patients remained in sinus rhythm throughout the MR imaging procedure. During and after MR imaging, all pacemakers continued to function normally except for one transient pause of 2 seconds towards the end of the imaging procedure. This occurred in the pacemaker-dependent patient, who had a unipolar dual-chamber device programmed in the "DOO" mode. The authors concluded that, when appropriate strategies are used, MR might be performed with an acceptable risk-benefit ratio for the patient.[73] Erlebacher and coworkers investigated the effect of MR on four different DDD pacemakers.[74] All units paced normally in the static magnetic field, but during MR imaging, all units malfunctioned. All malfunctions were a result of RF interference, whereas gradient and static magnetic fields had no effects. Thus, despite magnetic field strengths adequate to close pacemaker reed switches, RF interference during MR imaging may cause total inhibition of atrial and ventricular output in DDD pacemakers, and may also lead to dangerous atrial pacing at high rates. Pennell reported the study of four patients with pacemakers and urgent clinical problems who underwent CMR without significant problems in three patients (Figure 7–5), but was not attempted in one patient because the pacemaker switched into full output mode near to the magnet.[75] Recommendations were made that only non–pacemaker-dependent patients should be scanned, with the pacemaker in 000 mode if possible and a bipolar pacing lead.

In a study by Achenbach and associates, the effect of MR imaging on pacemakers and electrodes was investigated using phantoms.[76] Twenty-five electrodes were exposed in a 1.5 T scanner, with continuous registration of temperature at the tip of the electrode. Eleven pacemakers were exposed to MR imaging and the pacemaker output was monitored. Temperature increases of up to 63.1°C were observed. No pacemaker malfunctions were observed in the asynchronous mode (VOO/DOO). Inhibition or rapid pacing was observed during spin-echo MR imaging if the pacemakers were set to VVI or DDD mode. During scanning with gradient echo MR imaging, pacemaker function was not impaired. Pre-

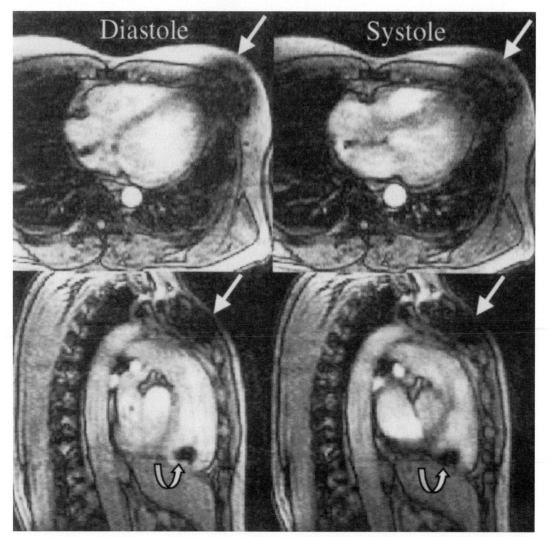

Figure 7–5. Images from a young patient with a pacemaker. The patient had suffered numerous ventricular fibrillation arrests. CMR suggested sarcoidosis (diagnostic images not shown). The pacemaker artifact can be seen in the transaxial gradient echo cine images in the top row, and in the right ventricular outflow tract cine images in the bottom row (straight arrows). The curved arrows show the artifact from the pacing lead in the apex of the right ventricle. Imaging was performed at 0.5 T with TE of 14 ms.[75] (Images courtesy of Dr. D. J. Pennell.)

liminary data suggest the heat generated in pacemaker leads may be impacted by the type of transmit coil and distance from the isocenter of the magnet.[76b]

Finally, pacemaker leads might serve as antennae, which could result in pacing the heart during scanning at the frequency of the applied imaging pulses. This action could potentially lead to hypotension and dysrhythmias. This effect was demonstrated in experiments and in several patients while positioned in an MR imaging system,[77–79] but this effect must be separated from excitation due to the pulse generator.

There are new approaches to the issue of pacing being developed in animals that may help us to understand the problems in humans. The feasibility and safety of transesophageal cardiac pacing during clinical MR imaging at 1.5 T has been tested both in vitro and in dogs by Hofman and colleagues, who developed an MR-compatible pacing catheter.[80] The authors concluded that transesophageal pacing dur-

ing MR imaging at low RF exposure allows the performance of cardiac stress studies to achieve stable heart rates. Jerzewski and coworkers reported on the development of an MR-compatible catheter for pacing the heart, tested both in vitro and in sheep.[81] Effective pacing with minor image distortion was observed during echo-planar MR imaging.

The major limitations of these studies are that there is far too little experience and too many types of pacemakers to make general statements about suitability for CMR. The conclusion should be that patients with pacemakers should not be scanned unless special circumstances arise, and then only in centers with special CMR expertise, monitoring, and cardiology backup.

CONCLUSION

In general, CMR is safe and no long-term ill effects have been reported. Very rapidly changing gradients may induce nerve excitations, which may result in muscle twitching, but clinical scanners operate below the threshold for such effects. The threshold of excitation of the myocardium lies about 200 times higher than that for other muscles, so the heart will not be stimulated by the rapid changing gradients. Most metallic implants, such as hip and knee prostheses, intracoronary stents, prosthetic valves and sternal sutures, present no health hazard because most materials used are nonferromagnetic. Intracranial clips, intraocular shards, and cochlear implants remain as contraindications. In general, patients with pacemakers and implanted cardiodefibrillators should not undergo CMR because of the unquantifiable risks. These recommendations are in agree-ment with the 1998 report of the Task Force of the European Society of Cardiology in collaboration with the Association of European Paediatric Cardiologists.[82]

References

1. Yang PC, Kerr AB, Liu AC et al: New real time interactive magnetic resonance imaging system complements echocardiography. J Am Coll Cardiol 32:2049, 1998.
2. Kanal E, Shellock FG, Talagala L: Safety considerations in MR imaging. Radiology 176:593, 1990.
3. Gulch RW, Lutz O: Influence of strong static magnetic fields on heart muscle contraction. Phys Med Biol 31:763, 1986.
4. Doherty JU, Whitman GJR, Robinson MD, et al: Changes in cardiac excitability and vulnerability in NMR fields. Invest Radiol 20:129, 1985.
5. Jehensen P, Duboc D, Lavergne T, et al: Change in human cardiac rhythm induced by a 2-T static magnetic field. Radiology 166:227, 1988.
6. New PFJ, Rosen BR, Brady TJ, et al: Potential hazards and artifacts of ferromagnetic and nonferromagnetic surgical and dental materials and devices in nuclear magnetic resonance imaging. Radiology 147:139, 1983.
7. Shellock FG, Crues JV: High-field strength MR imaging and metallic biomedical implants: An ex vivo evaluation of deflection forces. AJR Am J Roentgenol 151:389, 1988.
8. Shellock FG: MR imaging of metallic implants and materials: A compilation of the literature. AJR Am J Roentgenol 151:811, 1988.
9. Randall PA, Kohman LJ, Scalzetti EM, et al: Magnetic resonance imaging of prosthetic cardiac valves in vitro and in vivo. Am J Cardiol 62:973, 1988.
10. Teitelbaum GP, Bradley WG, Klein BD: MR imaging artefacts, ferromagnetism, and magnetic torque of intravascular filters, coils and stents. Radiology 166:657, 1988.
11. Pavlicek W, Geisinger M, Castle L, et al: The effects of nuclear magnetic resonance on patients with cardiac pacemakers. Radiology 147:149, 1983.
12. Bottomley PA, Edelstein WA: Power deposition in whole body NMR imaging. Med Phys 8:510, 1981.
13. Safety aspects of magnetic resonance imaging. In Schaefer DJ, Wehrli FW, Shaw D, Kneeland BJ (eds): Biomedical Magnetic Resonance Imaging: Principles, Methodology, and Applications. New York, VCH, 1988, p 553.
14. Extremely low frequency (ELF) magnetic fields. In Persson BRR, Stahlberg F (eds): Health and Safety of Clinical NMR Examinations. Boca Raton, CRC, 1989, p 49.
15. Cohen M, Weisskoff R, Rzedzian RR, Cantor HL: Sensory stimulation by time-varying magnetic fields. Magn Reson Med 14:409, 1990.
16. UK Childhood Cancer Study Investigators: Exposure to power-frequency magnetic fields and the risk of childhood cancer. Lancet 354:1925, 1999.
17. Easterbrook PJ, Berlin JA, Gopalan R, Matthews DR: Publication bias in clinical research. Lancet 337:867, 1991.
18. Newnham JP, Evans SF, Michael CA, et al: Effects of frequent ultrasound during pregnancy: A randomised controlled trial. Lancet 342:887, 1993.
19. Irnich W, Schmitt F: Magnetostimulation in MRI. Magn Reson Med 33:619, 1995.
20. Hartnell GG, Spence L, Hughes LA, et al: Safety of MR imaging in patients who have retained metallic materials after cardiac surgery. AJR Am J Roentgenol 168:1157, 1997.
21. Murphy KJ, Cohan RH, Ellis JH: MR imaging in patients with epicardial pacemaker wires. AJR Am J Roentgenol 172:727, 1999.
22. Boutin RD, Briggs JE, Williamson MR: Injuries associated with MR imaging: Survey of safety records and methods used to screen patients for metallic foreign bodies before imaging. AJR Am J Roentgenol 162:189, 1994.
23. Jones S, Jaffe W, Alvi R: Burns associated with electrocardiographic monitoring during magnetic resonance imaging. Burns 22:420, 1996.
24. Hurwitz R, Lane SR, Bell RA, Brant-Zawadzki MN: Acoustic analysis of gradient-coil noise in MR imaging. Radiology 173:545, 1989.
25. Brummett RE, Talbot JM, Charuhas P: Potential hearing loss resulting from MR imaging. Radiology 169:539, 1988.
26. McJury M, Stewart RW, Crawford D, Toma E: The use of active noise control (ANC) to reduce acoustic noise generated during MRI scanning: Some initial results. Magn Reson Imaging; 15:319, 1997.
27. Flaherty JA, Hoskinson K: Emotional distress during magnetic resonance imaging. N Engl J Med 320:467, 1989.
28. Francis JM, Pennell DJ: The treatment of claustrophobia during cardiovascular magnetic resonance; use and effectiveness of mild sedation. J Cardiovasc Magn Reson 2:139, 2000.
29. Quirk ME, Letendre AJ, Ciottone RA, Lingley JF: Anxiety in patients undergoing MR imaging. Radiology 170:463, 1989.
30. Weinreb JC, Maravilla KR, Peshock R, Payne J: Magnetic resonance imaging: Improving patient tolerance. AJR Am J Roentgenol 143:1285, 1984.
31. Hricak H, Amparo EG: Body MRI: Alleviation of claustrophobia by prone positioning. Radiology 152:819, 1984.
32. Goldstein HA, Kashanian FK, Blumetti RF, et al: Safety assessment of gadopentetate dimeglumine in U.S. clinical trials. Radiology 174:17, 1990.
33. Sullivan ME, Goldstein HA, Sansone KJ, et al: Hemodynamic effects of Gd-DTPA administered via rapid bolus or slow infusion: A study in dogs. AJNR 11:537, 1990.
34. Weiss KL. Severe anaphylactoid reaction after IV Gd-DTPA. Magn Reson Imaging 8:817, 1990.
35. Pennell DJ, Underwood SR, Ell PJ: Safety of dobutamine stress for thallium myocardial perfusion tomography in patients with asthma. Am J Cardiol 71: 1346, 1993.
36. Guidelines for advanced life support. A statement by the advanced life support working party of the European Resuscitation Council. Resuscitation 24:111, 1992.
37. Sellden H, de Chateau P, Ekman G, et al: Circulatory monitoring of children during anaesthesia in low-field magnetic resonance imaging. Acta Anaesthesiol Scand 34:41, 1990.
38. Lindberg LG, Ugnell H, Oberg PA: Monitoring of respiratory and heart rates using a fibre-optic sensor. Med Biol Eng Comput 30:533, 1992.
39. Roth JL, Nugent M, Gray JE, et al: Patient monitoring during magnetic resonance imaging. Anesthesiology 62:80, 1985.
40. Felblinger J, Boesch C: Amplitude demodulation of the electrocardiogram signal (ECG) for respiration monitoring and compensation during MR examinations. Magn Reson Med 38:129, 1997.
41. Felblinger J, Lehmann C, Boesch C: Electrocardiogram acquisition during MR examinations for patient monitoring and sequence triggering. Magn Reson Med 32:523, 1994.
42. Felblinger J, Debatin JF, Boesch C, et al: Synchronization device for electrocardiography-gated echo-planar imaging. Radiology 197:311, 1995.
43. Fischer SE, Wickline SA, Lorenz CH. Novel real-time R-wave detection algorithm based on the vectorcardiogram for

accurate gate magnetic resonance acquisitions. Magn Reson Med 42: 361, 1999.

44. Jorgensen NH, Messick JM, Gray J, et al: ASA monitoring standards and magnetic resonance imaging. Anesth Analg 79:1141, 1994.

45. Davis PL, Crooks L, Arakawa M, et al: Potential hazards in NMR imaging: Heating effects of changing magnetic fields and RF fields on small metallic implants. AJR Am J Roentgenol 137:857, 1981.

46. Laakman RW, Kaufman B, Han JS, et al: MR imaging in patients with metallic implants. Radiology 157:711, 1985.

47. Mechlin M, Thickman D, Kressel HY, et al: Magnetic resonance imaging of postoperative patients with metallic implants. AJR Am J Roentgenol 143:1281, 1984.

48. Sigwart U, Puel J, Mirkovitch V, et al: Intravascular stents to prevent occlusion and restenosis after transluminal angioplasty. N Engl J Med 316:70, 1987.

49. Fischman DL, Leon MB, Baim DS, et al: A randomized comparison of coronary stent placement and balloon angioplasty in the treatment of coronary artery disease. N Engl J Med 331:496, 1994.

50. Serruys PW, de Jaegere P, Kiemeneij F, et al: A comparison of balloon-expandable stent implantation with balloon angioplasty in patients with coronary artery disease. N Engl J Med 331:489, 1994.

51. Kimura T, Yokoi H, Nakagawa Y, et al: Three-year follow up after implantation of metallic coronary artery stents. N Engl J Med 334:561, 1996.

52. Carrozza JP, Kuntz RE, Levine MJ, et al: Angiographic and clinical outcome of intra-coronary stenting: Immediate and longterm results from a large single-center experience. J Am Coll Cardiol 20:328, 1992.

53. Shellock FG, Morisoli S, Kanal E: MR procedures and biomedical implants, materials and devices; an update. Radiology 189:587, 1993.

54. Matsumoto AH, Teitelbaum GP, Barth KH, et al: Tantalum vascular stents: In vivo evaluation with MR imaging. Radiology 170:753, 1989.

55. Scott NA, Pettigrew RI: Absence or movement of coronary stents after placement in a magnetic resonance imaging field. Am J Cardiol 73:900, 1994.

56. Strohm O, Kivelitz D, Gross, et al: Safety of implantable coronary stents during H-1 magnetic resonance imaging at 1.0 and 1.5T. J Cardiovasc Magn Reson 1:239, 2000.

56a. Kramer CM, Rogers WJ, Pakstis DL: Absence of adverse outcomes after magnetic resonance imaging early after stent placement for acute myocardial infarction. A preliminary study. J Cardiovasc Magn Reson 2:257, 2000.

56b. Hug J, Nagel E, Schnackenburg B, et al: Coronary arterial stents: Safety and artifacts during MR imaging. Radiology 216:781, 2000.

57. Nagel E, Hug J, Bunger S, et al: Coronary flow measurements for evaluation of patients after stent implantation. MAGMA 6:184, 1998.

58. Kramer CM, Rogers WJ, Reichek N, et al: Magnetic resonance contrast enhancement versus dobutamine tagging response for assessment of myocardial viability after infarction (abstract). J Am Coll Cardiol 33(suppl a):485A, 1999.

59. Duerinckx AJ, Atkinson D, Hurwitz R, et al: Coronary MR angiography after coronary stent placement. AJR Am J Roentgenol 165:662, 1995.

60. Kotsakis A, Tan KH, Jackson G: Is MRI a safe procedure in patients with coronary stents in situ? Int J Clin Practice 51:349, 1997.

61. Roubin GS, Robinson KA, King SBI, et al: Early and late results of intracoronary arterial stenting after coronary angioplasty in dogs. Circulation 76:891, 1987.

62. Ordidge R, Fullerton GD: Global call to action on MR safety. J Magn Reson Imaging 9:629, 1999.

63. Shellock FG, Shellock VJ: Metallic stents: Evaluation of MR imaging safety. AJR Am J Roentgenol 173:543, 1999.

64. Shellock F: Pocket Guide to MR Procedures and Metallic Objects: Update 1998. Philadelphia, Lippincott-Raven, 1998.

65. Hassler M, Le Bas JF, Wolf JE, et al: Effects of magnetic fields used in MRI on 15 prosthetic heart valves. J Radiol 67:661, 1986.

66. Soulen RL, Budinger TF, Higgins CB: Magnetic resonance imaging of prosthetic heart valves. Radiology 154:705, 1985.

67. Globits S, Higgins CB: Assessment of valvular heart disease by magnetic resonance imaging. Am Heart J 129:369, 1995.

68. Bachmann R, Deutsch HJ, Jungehulsing M, et al: Magnetic resonance tomography in patients with a heart valve prosthesis. Rofo 155:499, 1991.

69. DiCesare E, Enrici RM, Paparoni S, et al: Low field magnetic resonance imaging in the evaluation of mechanical and biological heart valve function. Eur J Radiol 20:224, 1995.

70. Lauck G, Smekal AV, Wolke S, et al: Effects of nuclear magnetic resonance imaging on cardiac pacemakers. PACE 18:1549, 1995.

71. Alagona P, Toole JC, Maniscalco BS, et al: Letter to the editor: Nuclear magnetic resonance imaging in a patient with a DDD pacemaker. PACE 12:619, 1989.

72. Inbar S, Larson J, Burt T, et al: Case report: Nuclear magnetic resonance imaging in a patient with a pacemaker. Am J Med Sci 305:174, 1993.

73. Gimbel JR, Johnson D, Levine PA, Wilkoff BL: Safe performance of magnetic resonance imaging on five patients with permanent cardiac pacemakers. PACE 19:913, 1996.

74. Erlebacher JA, Cahill PT, Pannizzo F, Knowles RJ: Effect of magnetic resonance imaging on DDD pacemakers. Am J Cardiol 57:437, 1986.

75. Pennell DJ: Cardiac magnetic resonance with a pacemaker in-situ: Can it be done (abstract). J Cardiovasc Magn Reson 1:72, 1999.

76. Achenbach S, Moshage W, Diem B, et al: Effects of magnetic resonance imaging on cardiac pacemakers and electrodes. Am Heart J 134:467, 1997.

76b. Luechinger R, Duru F, Zeijlemaker VA, et al: Heating effects of magnetic resonance imaging of the brain on pacemaker leads: Send/receive coils vs. receive-only coils (abstract). J Am Coll Cardiol 37:436A, 2001.

77. Hayes DL, Holmes DR, Gray JE: Effect of 1.5 tesla nuclear magnetic resonance imaging scanner on implanted permanent pacemakers. J Am Coll Cardiol 10:782, 1987.

78. Holmes DR, Hayes DL, Gray JE, Merideth J: The effects of magnetic resonance imaging on implantable pulse generators. PACE 9:360, 1986.

79. Fetter J, Aram G, Holmes DR, et al: The effects of nuclear magnetic resonance imagers on external and implantable pulse generators. PACE 7:720, 1984.

80. Hofman MBM, De Cock CC, Van der Linden JC, et al: Transesophageal cardiac pacing during magnetic resonance imaging: Feasibility and safety considerations. Magn Reson Med 35:413, 1996.

81. Jerzewski A, Pattynama PMT, Steendijk P, et al: Development of an MRI-compatible catheter for pacing the heart: Initial in vitro and in vivo results. J Magn Reson Imaging 6:948, 1996.

82. Task Force of the European Society of Cardiology, in collaboration with the Association of European Paediatric Cardiologists: The clinical role of magnetic resonance in cardiovascular disease. Eur Heart J 19:19, 1998.

CHAPTER **8**

Normal Cardiac Anatomy, Orientation, and Function

Ronald M. Peshock, Fatima Franco, Michael Chwialkowski, Roderick W. McColl, Geoffrey D. Clarke, and Robert W. Parkey

NORMAL CARDIAC ANATOMY

Cardiovascular magnetic resonance (CMR) can be used to obtain images of the heart in any plane. Thus, to define normal anatomy and function, it is important to define standard imaging planes in order to develop knowledge of normal anatomy, anatomic variants, and potential artifacts. Standard CMR planes have evolved from other techniques including body computed tomography (CT) imaging, echocardiography, and x-ray contrast angiography. The problem is often one of determining the appropriate plane as rapidly as possible to make the diagnosis. As with most other cardiac imaging techniques, it is important to know as much as possible regarding the clinical question prior to determining the protocol. All examinations therefore should be planned to answer a specific clinical question.

The basic imaging planes can be grouped into planes oriented with respect to the heart, such as horizontal and vertical long axis and short axis, and planes oriented with respect to the major axes of the body, such as the transaxial, sagittal, and coronal planes. Cardiac-oriented planes are essential for evaluation of cardiac chamber size and function and are familiar from other cardiac imaging techniques such as echocardiography. With CMR the position of these planes can be positioned very accurately. As shown in Figure 8–1*A*, a breath-hold scout image in the coronal or sagittal plane is the usual starting point. An axial scout (Figure 8–1*B*) is used to define the vertical long axis (also known as the two-chamber view, Figure 8–1*C*). The horizontal long axis (also known as the four-chamber view, Figure 8–1*D*) is then planned, and followed by the short axis (Figure 8–1*E–G*), which can be used to generate the left ventricular outflow tract view (Figure 8–1*H*), which is similar to the parasternal long axis view of echocardiography.

The main structures of normal cardiac anatomy in the coronal, axial, and sagittal planes are shown for spin-echo sequences in Figure 8–2*A–J*. There are many atlases of cross-sectional anatomy by CMR that can be helpful[1] and web sites (www.scmr.org) with interactive learning of the cross-sectional anat-

omy that are very useful teaching aids, and the reader is recommended to refer to these for further detailed analysis. From the standpoint of tissue characterization, the spin-echo images typically permit the differentiation of fat (white) from muscle (intermediate gray). Black regions in spin-echo CMR studies represent several tissues: air, bone, fibrous tissue, metal, or rapidly moving blood. It is important to note that if fluid moves relatively slowly (for example, in an aneurysm or a large pericardial effusion), its signal intensity will increase, which can mimic more solid tissue such as thrombus.

The placement of imaging planes, slice thickness, and in-plane resolution are determined by the size of the structure of interest. As has been indicated in previous chapters, presaturation bands can be added to remove specific artifacts. For example, in the evaluation of arrhythmogenic right ventricular dysplasia, it is important to obtain high-resolution spin-echo images of the anterior right ventricular wall that are free from respiratory artifact. This goal can be achieved by using a surface coil to improve signal-to-noise ratio compared with the standard body coil, and through the use of spatial and/or fat saturation to reduce artifacts from blood and chest wall motion. Breath-hold, double inversion recovery spin-echo techniques can also be very effective in removing respiratory artifacts.

Imaging planes oriented with respect to the principal axes of the body are particularly useful in the evaluation of the aorta, pericardium, anterior right ventricular wall, and paracardiac masses. Coronal images can also be quite useful because they present tomographic information in an orientation similar to the chest x-ray, which is familiar to most clinicians (see Figure 8–2*A* and *B*). In general, axial planes are also useful because they are familiar from CT (see Figure 8–2*C–H*). Specific vascular structures of interest evaluated well with axial imaging include the thoracic aorta and its branches, the pulmonary artery and veins, and the superior vena cava (see Figure 8–2*C* and *D*). Axial images through the heart can be particularly useful in the evaluation of the pericardium and right ventricular free wall (see Figure 8–2*E–H*). They are of limited value in the assessment of myocardial wall thickness and cham-

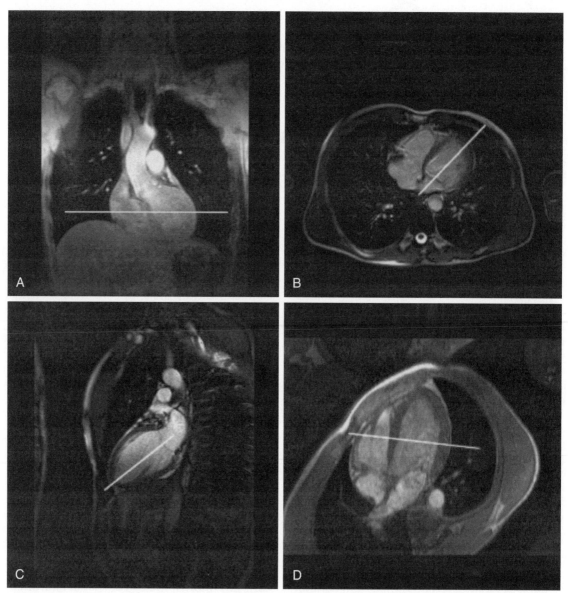

Figure 8–1. *A*, Scout image 1, coronal: Typical breath-hold image used to begin study (alternatively a sagittal image could be used). The white line indicates the location of an axial image used to locate the mitral valve plane and interventricular septum. *B*, Scout image 2, axial: Typical breath-hold image obtained to set up a vertical long axis image. The white line indicates the position of the vertical long axis (VLA) and is drawn to pass through the middle of the mitral valve and the ventricular apex. *C*, Vertical long axis: End-diastolic image from breath-hold cine CMR done in the position indicated on *B*. The white line indicates the position of the horizontal long axis (HLA) image and is drawn to pass through the apex and the leaflets of the mitral valve. *D*, Horizontal long axis: End-diastolic image from breath-hold cine CMR done in the position indicated by the line in *C*. The white line indicates the position of one short axis (SA) image and is drawn perpendicular to the ventricular septum and posterior left ventricular wall.

Illustration continued on opposite page

ber size because of the variable orientation of the heart relative to the principal axes of the body. Sagittal images are in general the least familiar to clinicians and are often more difficult to interpret (see Figure 8–2*I* and *J*). However, sagittal images are useful in depicting the right ventricular outflow tract and are therefore helpful in the evaluation of patients with congenital heart disease and right ventricular dysplasia. Oblique sagittal planes are useful in the evaluation of the thoracic aorta, and these planes can be easily defined from the transaxial images especially if three-point plane definition is

available using the arch and lower ascending and descending aorta as the reference points (Figure 8–2*K* and *L*). Black-blood images (Figure 8–2*M–P*) oriented along the functional axes introduced in Figure 8–1 can be particularly useful in the definition and tissue characterization of intracardiac and paracardiac masses. In addition, there is recent interest in using these planes with double-inversion recovery black-blood imaging in the evaluation of the valvular disease and the coronary artery wall.[2, 3, 3a]

The main findings using gradient echo (white-blood) cine CMR of the heart are shown in Figures

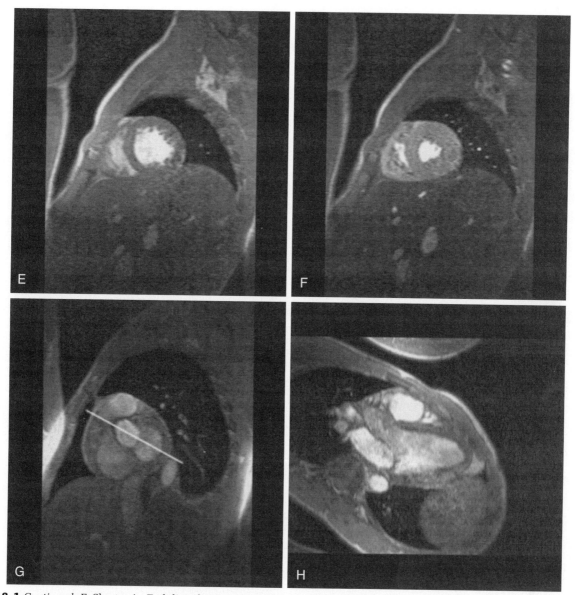

Figure 8–1 *Continued. E,* Short axis: End-diastolic image from breath-hold cine CMR done at the position indicated by the line in *D. F,* Short axis: End-systolic image from breath-hold cine CMR done at the position indicated by the line in *D. G,* Short axis at the level of the aortic valve: The white line indicates the position of the long axis (comparable to the parasternal long axis view in echocardiography). An additional short axis slice should be examined to ensure that the line passes through the center of the left ventricular cavity and apex. *H,* Long axis (comparable to the parasternal long axis view in echocardiography).

8–1 and 8–2*Q–U*. These cines are typically used to assess myocardial and valve function. In gradient echo images, blood and pericardial fluid appear white, muscle is again an intermediate gray, and air, bone, fibrous tissue, and metal are dark. The left ventricular outflow tract view, for example, is used to show the mitral and aortic valves (see Figure 8–2*R* and *T*). One advantage of CMR is the ability to precisely position the long axis plane through the aortic valve and apex (see Figure 8–1*G*) to avoid the foreshortening that can occur in contrast ventriculography or echocardiography. Short axis views are planned from the long axis views to span the entire left ventricle. At the base (Figure 8–2*R*), the aortic valve leaflets are clearly depicted, as is the interatrial septum. Thus, this view can be helpful in the

evaluation of the aortic valve disease, atrial septal defects, and atrial masses. The short axis views in Figure 8–2*S–U* are useful in the evaluation of ventricular size and regional function.

Coronary CMR requires yet another set of imaging planes to place the coronary arteries in tomographic slices. This subject is discussed in more detail in Chapter 16.

Anatomic Variants

Given the ability to obtain images in many planes, it is important to be aware of normal structures that may complicate interpretation of studies and

Text continued on page 83

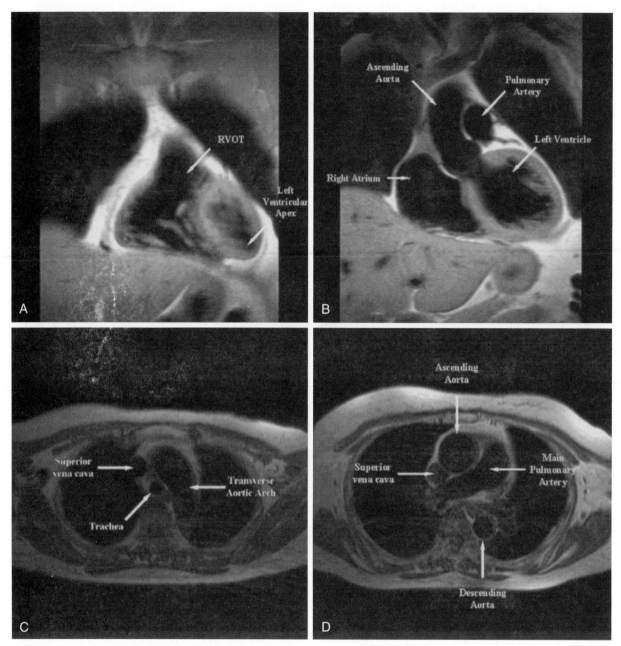

Figure 8–2. *A*, Coronal breath-hold, double inversion recovery, spin-echo image. Fat is white, myocardium is of intermediate gray intensity, and blood is dark. The slice is positioned anteriorly and cuts through the right ventricle, right ventricular outflow tract (RVOT), interventricular septum, and left ventricular apex. *B*, Coronal breath-hold, double inversion recovery, spin-echo image. The slice is positioned more posteriorly and cuts through the right atrium, ascending aorta, pulmonary artery, and left ventricle. The aortic valve leaflets are also demonstrated. *C*, Transverse conventional gated spin-echo image at the level of the transverse aortic arch. The trachea and superior vena cava are also demonstrated. This view is useful in the evaluation of possible aortic dissection. *D*, Transverse conventional gated spin-echo image at the level of the main pulmonary artery. This view is useful in evaluating the ascending aorta in patients with possible aortic dissection.

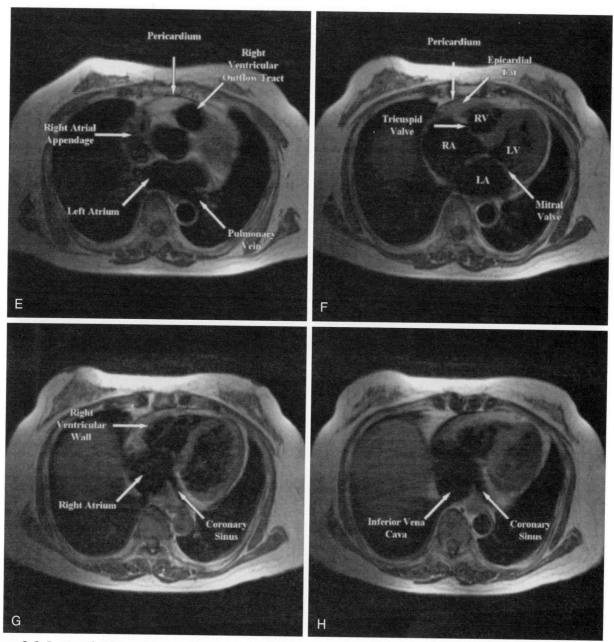

Figure 8–2 *Continued. E*, Transverse conventional spin-echo image at the level of the aortic valve. *F*, Transverse conventional spin-echo image at the level of the interatrial septum. The pericardium and epicardial fat are clearly demonstrated. This view can be useful in evaluating atrial masses and pericardial disease. RA = right atrium; RV = right ventricle; LA = left atrium; LV = left ventricle. *G*, Transverse conventional spin-echo image at the level of the coronary sinus. The right ventricular wall, epicardial fat, and pericardium are also demonstrated. This view can be helpful in evaluating patients for constrictive pericarditis and right ventricular dysplasia. *H*, Transverse conventional spin-echo image at the level of the entrance of the inferior vena cava into the right atrium.

Illustration continued on following page

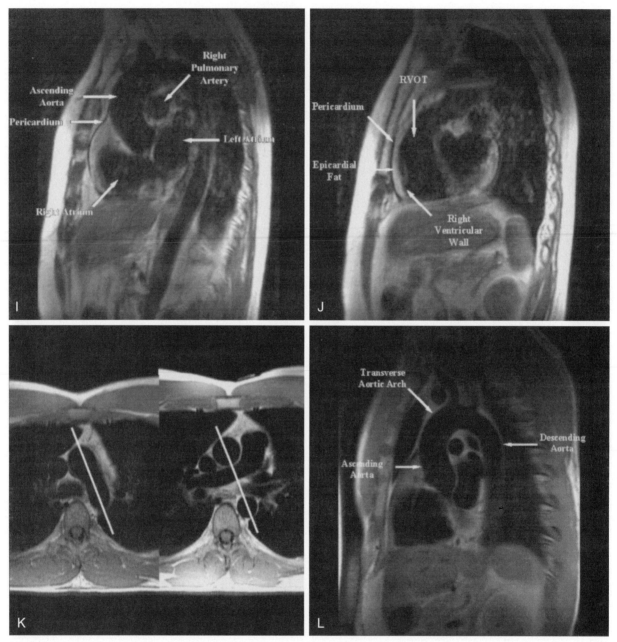

Figure 8–2 *Continued. I,* Sagittal conventional spin-echo image obtained through the ascending aorta. The pericardium is clearly demonstrated. This view can be helpful in the evaluation of the ascending aorta and pericardium. *J,* Sagittal conventional spin-echo image obtained through the right ventricular outflow tract. The view can be helpful in evaluating the pericardium, right ventricular outflow tract, and right ventricular wall. *K,* Transverse breath-hold, double inversion recovery, spin-echo images obtained at the level of the transverse portion of the aortic arch *(left)* and main pulmonary artery *(right).* The white line indicates the position of a parasagittal oblique plane used to obtain a "candy cane" view of the aorta *(L). L,* Parasagittal "candy cane" view of the aorta. The ascending, transverse, and descending aorta are seen in a single slice. The vessels to the head and neck are also well seen. This view can be helpful in the evaluation of aortic disease.

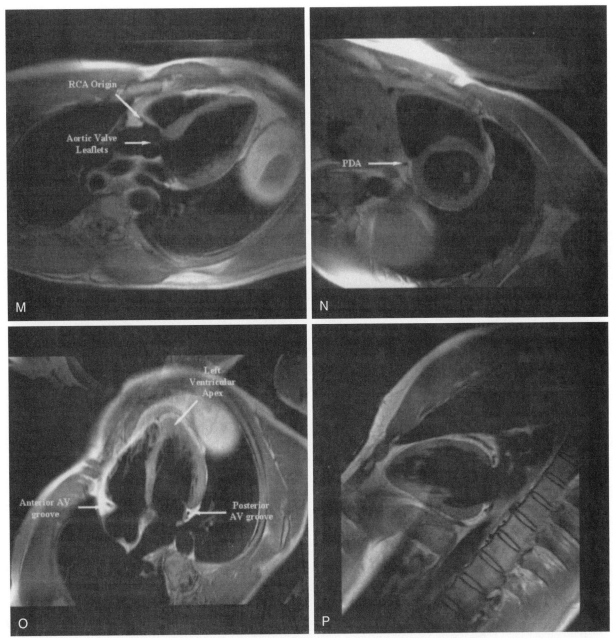

Figure 8–2 *Continued. M*, Long axis view using breath-hold double inversion recovery technique. This image is comparable to the long axis view in echocardiography. The left ventricle is well demonstrated, as is the right ventricle. The origin of the right coronary artery (RCA) is seen in the fat in the anterior atrioventricular (AV) groove. The aortic valve leaflets are also well seen. This view can be useful in the evaluation of hypertrophic cardiomyopathy with septal asymmetry. *N*, Short axis view using breath-hold double inversion recovery technique. The left and right ventricular walls are well demonstrated. In this image the posterior descending artery (PDA) is also seen in cross section in the posterior interventricular groove. *O*, Horizontal long axis or four-chamber view using breath-hold double inversion recovery technique. *P*, Vertical long axis or two-chamber view using breath-hold double inversion recovery technique.

Illustration continued on following page

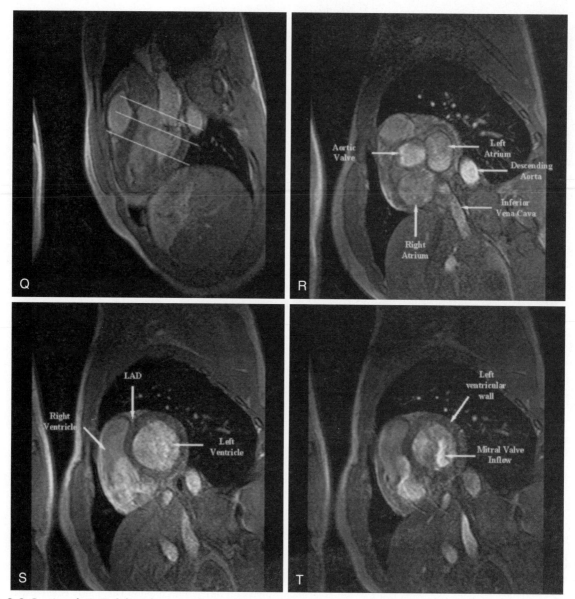

Figure 8–2 *Continued. Q,* End-diastolic image from a breath-hold gradient echo cine CMR sequence. The white lines indicate the locations of short axis imaging planes in subsequent panels. *R,* Systolic short axis from a breath-hold gradient echo cine CMR sequence obtained at the level of the aortic valve. The tri-leaflet aortic valve is open and additional structures are as labeled. *S,* End-diastolic short axis image obtained at the level of the right ventricle. The left anterior descending artery (LAD) is seen in the anterior interventricular groove. *T,* Mid-diastolic short axis image obtained at the level of the right ventricle. Mitral valve inflow is evident in the mid ventricle.

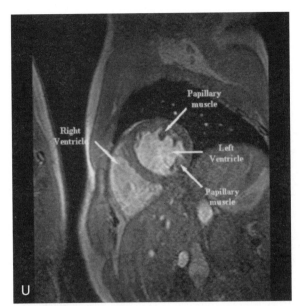

Figure 8–2 *Continued. U,* Diastolic short axis image obtained at the level of the mid right ventricle. The left ventricular papillary muscles and trabeculations are demonstrated.

anatomic variants. Several potential confusing features have been described:

- Prominence of the lateral border of the right atrial wall (Figure 8–3*A*): This structure is a prominence of the trabeculae carne (christa terminalis) and does not represent an atrial mass.[4]
- Lipomatous hypertrophy of the atrial septum (Figure 8–3*B*): In elderly women, there is often fat deposition in the atrial septum. This process spares the region of fossa ovalis and thus leads to the characteristic dumbbell described on echocardiography.[5, 6] This process is in general considered benign, but it is associated with atrial arrhythmias in older patients. More severe and extensive lipomatous hypertrophy does occur and may extend well outside the heart.[7]
- Superior pericardial recess (Figure 8–3*C*): The pericardium normally extends up the ascending aorta, and this space may contain fluid. This recess can be mistaken for aortic dissection or potentially an anomalous coronary vessel in coronary imaging.

Common Artifacts

A number of artifacts related to CMR can complicate interpretation of the images. These artifacts relate primarily to a number of features of CMR. The acquisition time is often relatively long compared to physiological processes, which leads to cardiac and respiratory motion artifacts. This problem must be recognized if present, and minimized at the acquisition stage if possible. Also, because the strength of the local magnetic field determines the position of an object in an MR image, if the local

magnetic field is altered, the position of the structure in the image is also altered. Therefore metal on or in the body can alter the local magnetic field, leading to distortion and local signal loss. Finally, hydrogen nuclei in fat see a slightly different magnetic field compared with hydrogen nuclei in water molecules because of the local chemical environment. This chemical shift is used in MR spectroscopy to differentiate one compound from another. However, in CMR, this results in what is known as a chemical shift artifact at the interface of water and fatty tissues. This artifact results from sharing within a pixel of fat and water components, leading to signal cancellation. Specific examples are given for each type of artifact.

Cardiac Motion Artifacts (Figure 8–4*A* and *B*). Except for single-shot echo-planar imaging (EPI) or other real-time imaging approaches, CMR requires gating to the electrocardiogram (ECG) or peripheral pulse. Problems with gating can result in ghosting and other noise that degrades the quality of the images. In general, focused efforts to obtain the best ECG possible before beginning scanning will minimize cardiac motion artifacts and save time. Surprisingly good quality images can be obtained in patients with atrial fibrillation, which may be related to the relatively consistent length of systole relative to changes in heart rate.[8] Ventricular bigeminy often results in poor images in that every other beat is activated differently, resulting in combining data from two different activation patterns. Many CMR systems provide arrhythmia rejection in an attempt to reduce these effects; however, use of these tools generally results in increased scan time because of rejection of cardiac cycles. Recently, vectorcardiographic techniques have been implemented to take advantage of the difference in the normal QRS vector and the vector of the artifact from the magnetohydrodynamic effect to improve ECG gating.[9]

Respiratory Motion Artifacts (see Figure 8–4*A* and *B*). Respiration is associated with significant motion of the heart. Motion in the craniocaudal direction is on the order of a centimeter in normal individuals.[10] This motion can result in significant image degradation with ghosting and blurring, particularly in those with inconsistent respiratory patterns. Strategies to reduce respiratory artifact include the use of breath-hold imaging, presaturation of the high-intensity signal from fat in the chest wall, and the use of respiratory gating. Respiratory gating using a bellows or by tracking the diaphragm position using a navigator echo[11] relies on accepting cardiac cycles only during some portion of the respiratory cycle. It can substantially improve image quality and can be useful in coronary imaging without breath-hold[12] and patients with heart failure.[13] However, it can increase the total scan time.

Metal Artifact (Figure 8–4*C* and *D*). Pieces of metal outside or inside the body alter the local magnetic field and can result in artifacts. Patients are screened carefully for the presence of metal, as de-

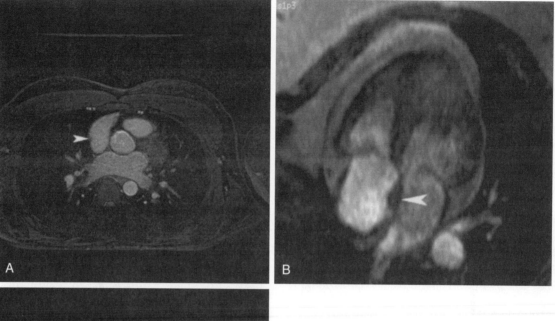

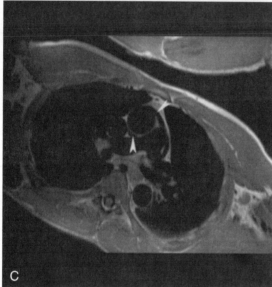

Figure 8–3. *A*, Transverse gradient echo image at the level of the aortic valve obtained using respiratory gating with a navigator echo. A right atrial ridge is noted on the lateral wall of the right atrium (arrow). This finding is normal and should not be mistaken for a right atrial mass. *B*, Single frame from horizontal long axis, cine CMR. Lipomatous hypertrophy of the atrial septum is demonstrated (arrow). There is fatty infiltration of the septum that does not involve the region of the fossal ovalis, resulting in the typical "dumbbell" appearance. *C*, Oblique double inversion recovery breath-hold image obtained at the level of the right pulmonary artery. The extension of the pericardial space both anterior and posterior to the ascending aorta is demonstrated (arrows). The pericardial recess should not be mistaken for evidence of aortic dissection.

scribed earlier. Despite vigilance, objects common in the hospital may still go with the patient into the scanner. Figure 8–4*C* shows an artifact related to a safety pin on the patient's gown. Note that signal loss and distortion are present in both the turbo-spin echo and field echo images. However, the severity of artifact is worse in the gradient echo images, severely compromising interpretation of the right ventricle and septum. Figure 8–4*D* shows the artifact related to sternal wires and bioprosthetic valve. Here the artifacts are less severe, permitting the evaluation of left ventricular function.

Chemical Shift Artifact (Figure 8–4*E* and *F*). This artifact occurs because the hydrogen nuclei in fat see a slightly different magnetic field than hydrogen nuclei in water because of the different chemical environment.[14] This process results in displacement of the fat signal in the frequency encoding direction relative to water and is accentuated with narrow bandwidth sequences, which can present a diagnostic problem in spin-echo imaging of patients with

suspected aortic dissection.[15] It can be addressed by using a wider bandwidth sequence or repeating the sequence with frequency encoding in the alternate direction, which will result in changing the orientation of the artifact and thus help exclude the presence of an aortic dissection. In echo-planar images, chemical shift effects lead to artifacts displaced in the phase encoding direction. As shown in Figure 8–4*F*, this effect can be minimized using multi-shot EPI techniques. In single-shot EPI with long acquisition times, the chemical shift effects can be quite large. For this reason, single-shot echo-planar images often employ fat saturation to suppress this artifact.

NORMAL CARDIAC SYSTOLIC AND DIASTOLIC FUNCTION

The management of cardiovascular disease is critically dependent on the assessment of cardiac function. Thus, every cardiac imaging technique has been used to assess systolic and diastolic function. There is now an extensive body of evidence to indicate that CMR provides highly accurate and reproducible assessments of global and regional cardiac function and is being increasingly recognized as the gold standard for the noninvasive evaluation of cardiac function.[16] The assessment of ventricular function is covered in Chapters 8, 9 and 23, and therefore only some abbreviated points are described here.

An important consideration in determining func-

tion is the temporal resolution or frame rate of the cine CMR sequence. It is generally accepted that a frame rate of at least 25 frames/s or a temporal resolution of 40 ms/frame is required to determine end-systole. Historically, contrast ventriculograms have been typically obtained at a frame rate of 30 frames/s or a temporal resolution of 33 ms. The frame rate in echocardiography is dependent on speed of ultrasound in the body and the distance of the heart from the transducer but is in the range of 20 to 30 frames/s (temporal resolution of 33–50 ms). With modern CMR scanners with high-performance

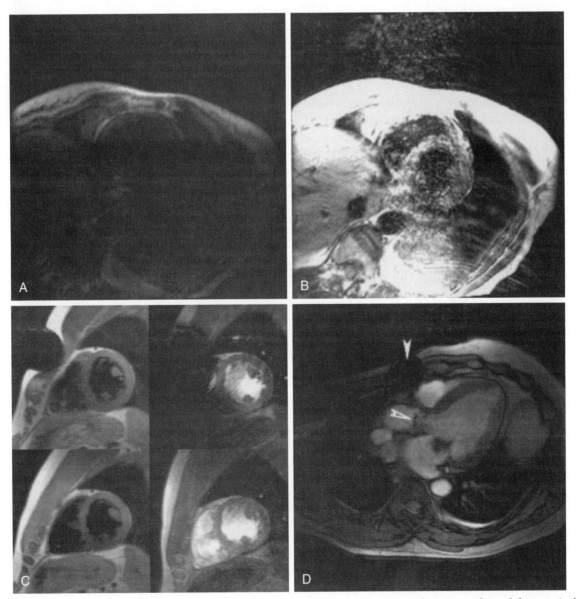

Figure 8–4. *A,* Artifacts due to respiration and poor gating: In this gated spin-echo image, there is mottling of the ventricular wall and loss of edge sharpness. *B,* The same image as in *A,* but with the window and level adjusted to accentuate the artifact. There are ghosts of the chest wall related to respiratory motion and additional artifact over the heart as a result of poor ECG gating. *C,* Metal artifact. Images on the upper left and right were obtained with a safety pin present on the subject's gown and after removal in the images on the lower right and left panels. The distortion is greatest in the gradient echo image in the upper right and would make interpretation very limited. The turbo spin echo image on the upper left is less distorted and could be interpreted. *D,* Metal artifact in a patient with prior aortic valve homograft. There is artifact related to the sternal wires (solid arrow) and modest artifact related to the homograft (open arrow). Mild mitral regurgitation is also present.

Illustration continued on following page

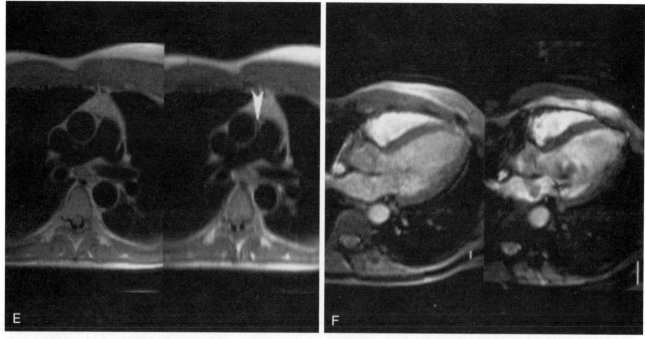

Figure 8–4 *Continued. E*, Chemical shift artifact: The image on the left is done with a relatively short signal acquisition time (wide bandwidth). The image on the right is done with a longer signal acquisition time (narrow bandwidth). This disparity accentuates the effect of the difference in chemical shift of water and fat, creating the artifactual space between the aortic wall and fat (arrow). The shift occurs in the frequency-encoding direction (right-left in these images). *F*, Chemical shift in echo-planar imaging (EPI). In EPI, the chemical shift occurs in the phase-encoding direction. The image on the left is obtained using a multishot EPI sequence with a relatively short EPI acquisition with each shot. The chemical shift artifact is indicated by the white line in the posterior chest wall. The image on the right is obtained using fewer shots with a longer EPI acquisition. The chemical shift is larger, as indicated by the longer white line posteriorly. The image is degraded by superimposition of anterior subcutaneous fat onto the heart. This problem can be addressed by adding fat saturation to the sequence.

gradients, frame rates of 20 to 25 frames/s for breath-hold sequences are possible, and can be extended up to very high frame rates (100 frames/s) for non–breath-hold sequences; however, there is usually a trade-off between temporal and spatial resolution so that at small fields of view, the time per frame will increase.

Left Ventricle

Assessment of ventricular function includes global and regional function. Assessment of global ventricular function is based on measuring changes in chamber volumes. These changes can be estimated from linear dimension measurements in echocardiography, but with CMR, more accurate measures of chamber volume can be made using two- and three-dimensional methods. The two-dimensional methods (area length technique, Figure 8–5A and B) have no advantages over echocardiography, and are not widely used.[17, 18, 18a] More accurate measures, particularly in deformed ventricles, which do not fit the prolate ellipsoid mode, can be obtained using the Simpson's rule technique. In this method typically short-axis images are obtained spanning the entire ventricle and the volume in each slice measured and summed over the entire ventricle.[18b] This approach has been shown to be highly accurate and reproducible in a large number

of studies and is widely used in research studies.[18a] Typical images used for the Simpson's rule method are shown in Figure 8–5C and D. If images are analyzed over the entire cardiac cycle, measures of ventricular volume over time can be obtained for ventricular ejection and filling rate analysis (Figure 8–5E and F).

Regional left ventricular function can be assessed both qualitatively, similar to echocardiography, or quantitatively. As shown in Figure 8–6A and B, the standard long axis, four-chamber, two-chamber, and short-axis views can be mapped onto the standard wall segment definitions used in echocardiography for qualitative assessment of wall thickening. Wall thickening can also be determined more quantitatively using centerline methods and other methods as are used with other techniques (Figure 8–6C and D). Importantly, myocardial tagging techniques (see Chapters 4 and 5) can be used to define myocardial contractility using strain without the need to identify endocardial or epicardial borders, which is a great advantage over these techniques (Figure 8–6E).

Right Ventricle

Historically, measurement of right ventricular volumes has been largely qualitative. This situation is due to the lack of a standard geometric model for

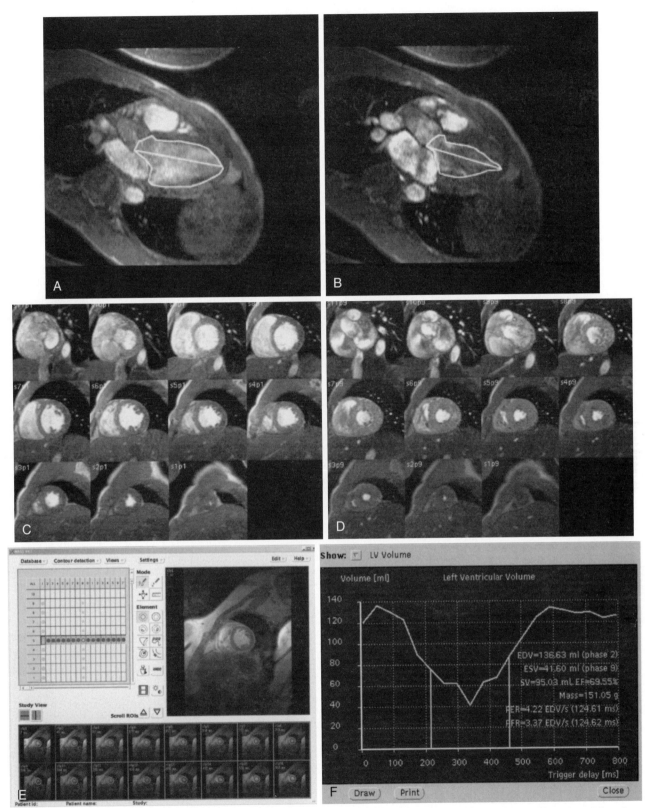

Figure 8–5. *A*, End-diastolic long axis image obtained from a breath-hold cine CMR sequence. The end-diastolic area and length are shown in white for use in a standard area length estimation of left ventricular volume. *B*, End-systolic long axis image from the same cine CMR sequence. The end-systolic area and length are shown in white. *C*, End-diastolic short axis images from a series of breath-hold cine CMR studies. Ventricular volume is determined by summing the volume in each slice over the entire length of the ventricle using Simpson's rule. *D*, End-systolic short axis images from the same series of breath-hold cine CMR studies. End-systolic volume is determined in a similar manner. Note that the motion of the base towards the apex with systole means that the aortic valve plane moves from s11 to s10 with systole. *E*, Typical software analysis tool used to facilitate the calculation of ventricular volumes and left ventricular mass. Semiautomated techniques can be used to speed complete analysis of a short axis data set. *F*, Typical graph of ventricular volume over the entire cardiac cycle obtained using semiautomated analysis.

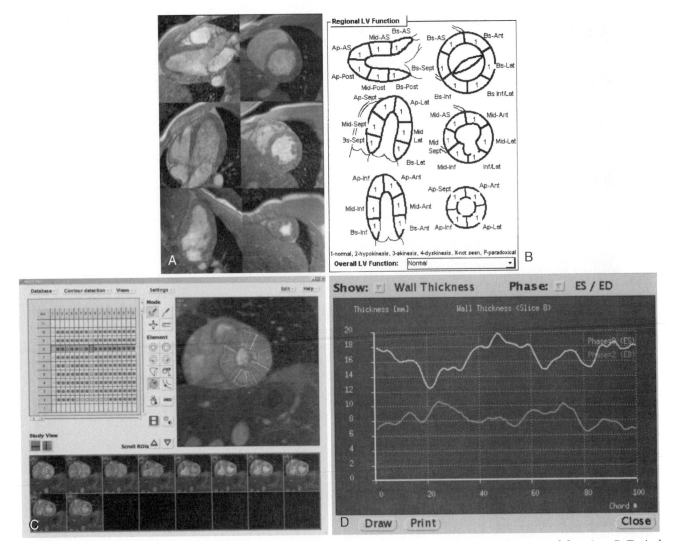

Figure 8–6. *A*, Images in each of the standard planes typically used for scoring wall thickening and segmental function. *B*, Typical form showing segment definitions used to score regional left ventricular function. The definitions are similar to those used in echocardiography. *C*, Assessment of wall thickening in the short axis view using a semiautomated method. Wall thickness is determined along each radial line over the cardiac cycle. *D*, Graph of the wall thickening for each chord for the short axis slice seen in *C*.

the right ventricle. An advantage of CMR is that the same Simpson's rule approach may be readily applied to the right ventricle (Figure 8–7; see color plates). Studies in ventricular casts have shown an excellent correlation between CMR and displacement measurements.[19] (See Chapters 8, 9, and 23.)

Stroke Volume

Stroke volume is the amount of blood ejected from the heart with each cardiac cycle. It can be readily calculated by subtracting the end-systolic volume from the end-diastolic volume (see Figure 8–5). Multiplying the stroke volume by the heart rate yields the cardiac output, typically reported in liters/minute. Studies comparing cardiac output determined by CMR with invasive thermodilution methods have shown good correlation using gradient echo and echo-planar techniques.[20–22] The car-

diac output determined in this way from the stroke volume is the same as the cardiac output determined in the catheterization laboratory. In the setting of aortic or mitral regurgitation, however, part of this volume does not result in the net delivery of blood to the periphery. In this setting, the Simpson's rule cardiac output based on the left ventricular stroke volume is greater than the forward flow (which can be measured using flow techniques (see Chapters 6 and 29). In this setting, the regurgitant volume can be determined by subtracting the forward flow from the apparent stroke volume.[23, 24]

Ventricular Mass

The mass of the left ventricular wall can be estimated by measuring the volume of the myocardium (Figure 8–8; see color plates) and multiplying it by the specific gravity of myocardium, 1.05 g/cm³. CMR

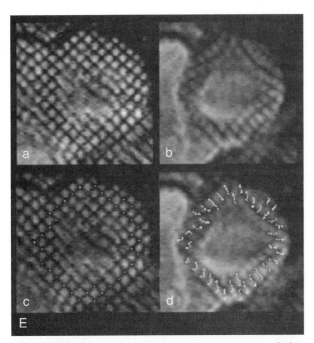

Figure 8–6 *Continued. E,* Example of short axis myocardial tagging (described in detail in subsequent chapters). The upper left panel **(A)** shows tagging at end-diastole; panel **B** shows the deformation that occurs with myocardial contraction; panel **C** illustrates a method for marking the intersection point of the tags at end-diastole, whereas panel **D** tracks the motion of each point over cardiac contraction.

provides accurate estimates of myocardial mass over a broad range of heart sizes both in animals and in man.[25–27] In cadaver heart studies, linear correlation analysis demonstrated a correlation coefficient of 0.99 with a standard error of 6.8 grams. Intraobserver, interobserver, and interstudy variability are excellent. This same approach has been used in the right ventricular mass with good results.[28, 29] Normal adult values for left and right ventricular volumes and mass by CMR have been published by Lorenz (Table 8–1).[30]

Aortic Flow

An important feature of MR is its sensitivity to motion, and this feature can be harnessed to allow the measurements of velocity and vessel flow without the constraints of Doppler methods. Although there is potential for artifacts, there is growing evidence that CMR is the gold standard for the measurement of vessel flow in vivo.[31–35] The approach for the measurement of aortic flow is straightforward. The coronal scout image is used, and the imaging plane is placed perpendicular to the direction of flow several centimeters above the aortic valve typically at approximately the level of the bifurcation of the main pulmonary artery (Figure 8–9A). The velocity encoding value should be chosen to be just above the anticipated maximum velocity (approximately 1.5 m/s in normals, and higher in patients with aortic valve disease). The recon-

structed images typically consist of a set of magnitude images that are used to determine the cross-sectional area of the aorta in each frame (Figure 8–9B). There is also a set of velocity map images in which the gray scale indicates the velocity of motion in each voxel (Figure 8–9C). Velocity is measured at each voxel across the vessel, integrated over the cross-sectional area of the vessel and then integrated over the cardiac cycle (Figure 8–9D and E). The integrated flow across the slice at each point in time can be graphed. Note that some retrograde flow is normal in the ascending aorta during early diastole due to closure of the aortic valve, diastolic ascent of the base of the heart, and diastolic coronary flow (Figure 8–9E).

Pulmonary Artery Flow

Pulmonary artery flow can be measured using the techniques similar to those for aortic flow. This is particularly valuable in evaluation of patients with left-to-right intracardiac shunting to determine the ratio of pulmonary to systemic flow (Qp/Qs), which has good correlation with invasive techniques.[36, 37] Depending on the pulmonary artery orientation, the location of the flow images can be planned from the axial and/or sagittal scout with a perpendicular plane positioned several centimeters distal to the pulmonary valve. The velocity profile is then integrated over the cross-sectional area of the artery over time to determine the volume flow per cardiac cycle in a manner similar to that used in the ascending aorta.

NORMAL VALVULAR FUNCTION

Assessment of valve function involves evaluation of morphology, motion, competence, and effects on ventricular function. Until recently, the use of CMR in the evaluation of valvular heart disease has been limited.[38, 39] Imaging cardiac valve morphology poses significant problems for CMR. First, the normal valve is a thin, fibrous structure often less than a millimeter thick, leading to potential for partial volume effects. Second, it is constantly in motion. In conventional CMR, the image is acquired over a number of cardiac cycles, requiring that the valve return to the same position with each cycle. With breath-hold imaging, one can obtain very high resolution images, indicating that normal valve motion appears to be quite reproducible over a limited number of cycles when the effects of respiratory motion are removed. However, vegetations and other valve pathology are characterized by valve motion, which is less reproducible, leading to loss of signal and motion artifacts. Third, valve pathology frequently involves fibrosis and calcification, both of which are characterized by loss of signal on CMR, making it difficult to detect, particularly in spin-echo imaging. Lastly, regions of turbulence are associated with loss

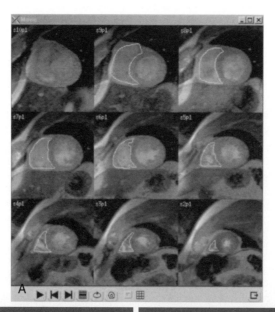

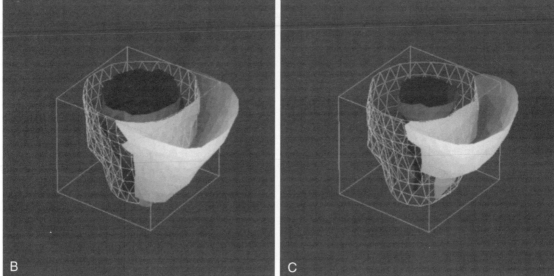

Figure 8–7. See color plates. *A,* Right ventricular volume can be determined from the same set of short axis views. The right ventricular chamber cross-sectional volume is determined for each slice and then summed over the length of the right ventricle. *B,* Three-dimensional reconstruction of the right ventricle from short axis views at end-diastole. The right ventricle is in yellow viewed from the base towards the apex. The left ventricular endocardium (red) and epicardium (green mesh) are shown for orientation. *C,* Three-dimensional reconstruction of the right ventricle from short axis views at end-systole. The right ventricle is in yellow viewed from the base towards the apex. The left ventricular endocardium (red) and epicardium (green mesh) are shown for orientation.

of signal on CMR, which may lead to overestimation of the extent of abnormality on gradient echo imaging. In spite of these concerns, recent reports indicate that CMR can be used to obtain high-resolution images of valves using a breath-hold, double-inversion recovery technique.[40] An example of a normal valve image obtained using this technique is shown in Figure 8–10*A*.

However, imaging the valvular abnormality is only one part of the evaluation of the patient with valvular disease. It is essential to quantify the functional severity of the lesion and to determine its impact on ventricular size and function. Full details of assessing valvular abnormalities are detailed in Chapter 30, but some general comments are useful. As noted earlier, CMR is highly accurate in de-

termining ventricular dimensions and volumes. In addition, phase contrast quantitative flow techniques provide effective means for determining velocity and blood flow. Thus, in addition to demonstrating that valvular disease is present, CMR can be used to quantify the degree of dysfunction and to determine its effect on ventricular size and function. Valve pressure gradients estimated using CMR correlate well with ultrasound measurements.[41, 42] In addition, measurements of aortic valve area and cardiac output by CMR agree with measurements of valve area at catheterization.[43] Regurgitant jets are generally well visualized due to turbulence creating signal loss in gradient echo cines (Figure 8–10*C*). The qualitative severity of regurgitation has been estimated from the size of the region of signal loss

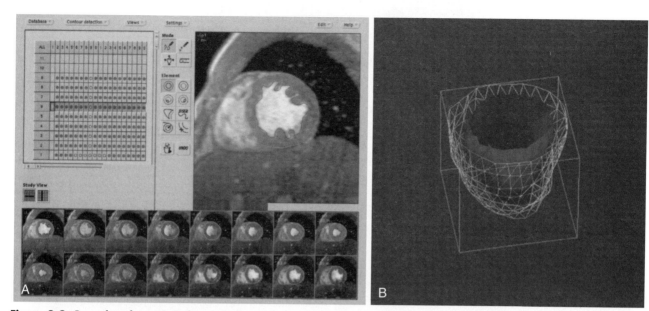

Figure 8–8. See color plates. *A*, Left ventricular mass is estimated by determining the left ventricular wall volume in each slice, summing over all slices, and multiplying the total wall volume by the density of myocardium. *B*, Three-dimensional rendering of the endocardial outlines (red) and epicardial outlines (green mesh) used in the wall volume calculation to determine myocardial mass.

Table 8–1. Normal Values for Cardiac Volume and Mass in Adults

Gender-Related LV Parameters*		
Parameter	**Males**	**Females**
LVEDV (ml)	136 ± 30 (77–195)	96 ± 23 (52–141)
RVEDV (ml)	157 ± 35 (88–227)	106 ± 24(58–154)
LVES volume (ml)	45 ± 14 (19–72)	32 ± 9 (13–51)
RVES volume (ml)	63 ± 20 (23–103)	40 ± 14 (12–68)
LV mass (g)	178 ± 31 (118–238)	125 ± 26 (75–175)
RVFWM (g)	50 ± 10 (30–70)	40 ± 8 (24–55)
LVEF (%)	67 ± 5 (56–78)	67 ± 5 (56–78)
RVEF (%)	60 ± 7 (47–74)	63 ± 8 (47–80)
LVSV (ml)	92 ± 21 (51–133)	65 ± 16 (33–97)
RVSV (ml)	95 ± 22 (52–138)	66 ± 16 (35–98)
CO (L/min)	5.8 ± 3.0 (2.82–8.82)	4.3 ± 0.9 (2.65–5.98)
Normalized to BSA*		
Parameter	**Males**	**Females**
LVEDV/BSA (ml/m²)	69 ± 11 (47–92)	61 ± 10 (41–81)
RVEDV/BSA (ml/m²)	80 ± 13 (55–105)	67 ± 10 (48–87)
LV mass/BSA (g/m²)	91 ± 11 (70–113)	79 ± 8 (63–95)
RVFWM/BSA (g/m²)	26 ± 5 (16–36)	25 ± 4 (18–33)
LVSV/BSA (ml/m²)	47 ± 8 (32–62)	41 ± 8 (26–56)
RVSV/BSA (ml/m²)	48 ± 8 (32–64)	42 ± 8 (27–57)
CO/BSA (1/min/m²)	3.0 ± 0.6 (1.74–4.20)	2.8 ± 0.5 (1.75–3.80)

*(mean ± 1 SD) with 95% confidence intervals (1.96 SD) in parentheses.

LVEDV = left ventricular end-diastolic volume; RVEDV = right ventricular end-diastolic volume; LVES = left ventricular end-systolic; RVES = right ventricular end-systolic; LV mass = left ventricular total mass, including septum (including papillary muscles); RVFWM = right ventricle free wall mass (no septum); LVEF = left ventricular ejection fraction; RVEF = right ventricular ejection fraction; LVSV = left ventricular stroke volume; RVSV = right ventricular stroke volume; CO = cardiac output; BSA = body surface area.

Modified from Lorenz CH, et al: Normal human right and left ventricular mass, systolic function, and gender differences by cine magnetic resonance imaging. J Cardiovasc Magn Reson 1:7, 1999.

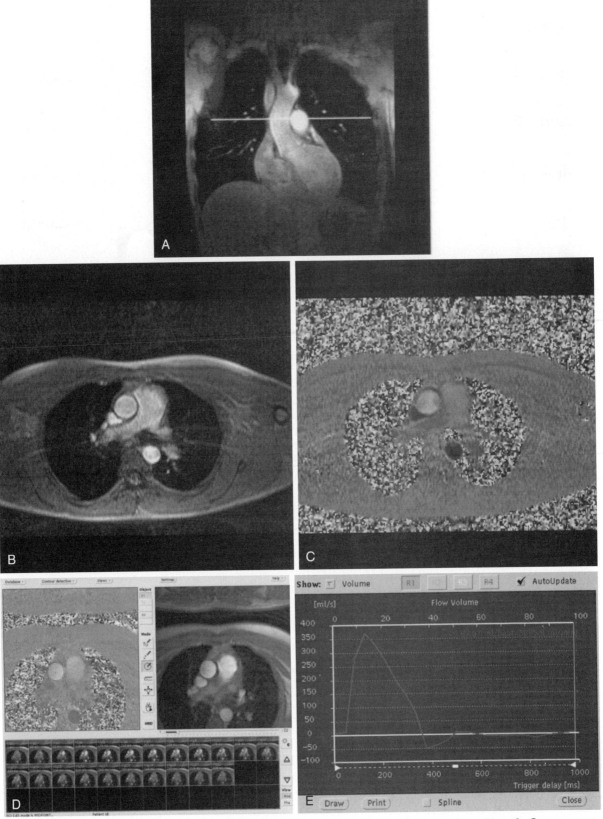

Figure 8–9. *A*, Coronal scout image for measuring aortic flow. The white line indicates the anatomic position of a flow sequence. The plane is positioned at the level of the pulmonary artery well above the aortic valve and perpendicular to the walls of the aorta. *B*, Magnitude reconstruction from the flow sequence positioned in *A*. The plane is transverse and positioned at the level of the pulmonary artery. The ascending aorta is seen anteriorly. *C*, Velocity map reconstruction from the same flow sequence shown in *B*. The gray scale in this image indicates the velocity of motion toward the head as bright and away as dark. *D*, Typical data set for semiautomated analysis. The area of the ascending aorta is determined in each frame and the velocity over the area is integrated to calculate flow volume per frame. *E*, Graph of flow volume over the cardiac cycle in the ascending aorta. The forward stroke volume is calculated by integrating the flow over the cardiac cycle as indicated in the upper right.

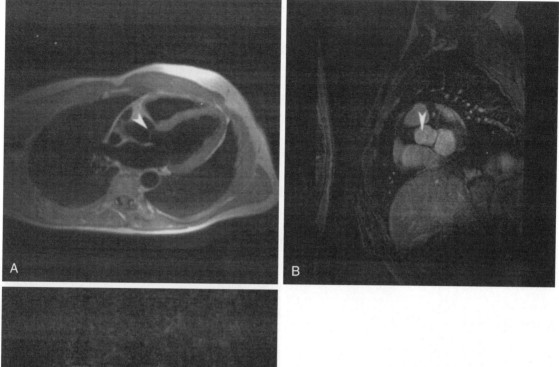

Figure 8–10. *A,* Double inversion recovery spin echo long axis image. The aortic valve leaflets are demonstrated (arrow). *B,* Image from a gradient echo coronary imaging sequence with navigator echo. The aortic valve leaflets are indicated by the arrow. *C,* Frame from a cine CMR in a patient with aortic insufficiency. There is loss of signal in the region of the regurgitant jet. The size of the signal loss may be related to the echo time and other sequence parameters.

on cine CMR, but the assessment is subject to many confounding factors including the echo time. Quantitative flow techniques are more useful in quantifying the regurgitant volume, for example, by measuring retrograde flow in diastole in the aorta.[44–46] Interestingly, measures of chamber volume and cardiac output by CMR in patients with atrial fibrillation agree well with invasive measures.[47] Mitral regurgitation has also been studied using quantitative measures.[48, 49] The presence of a prosthetic valve is not a contraindication for CMR except in the case of probable valve dehiscence.[50] Deutsch and associates found that CMR and transesophageal echo were comparable in the evaluation of regurgitant flow in patients with prosthetic valves.[51]

NORMAL CORONARY ARTERY FLOW

With the development of methods for coronary CMR, it has become possible to measure absolute coronary artery flow using phase contrast methods similar to those used in the aorta. It is important to realize that CMR provides the first method for the noninvasive measurement of absolute coronary artery flow. Although resting coronary flow is of interest, it is typically not depressed until a severe coronary stenosis is present. Thus, it is important to measure coronary flow under conditions of stress so that one can determine if flow increases normally. Typically, coronary flow can increase by a factor of 3 to 5 times resting flow. This increase is termed the coronary flow reserve and reflects normal coronary vasodilatation. Clarke and coworkers compared CMR measurement of absolute coronary artery flow with measures obtained using an ultrasonic transit time probe placed around the artery in an animal model.[52] There was good agreement between CMR and the invasive measure, and the differentiation of critical from noncritical stenoses on the basis of alterations in coronary flow reserve was possible. Later work in humans demonstrated good agreement

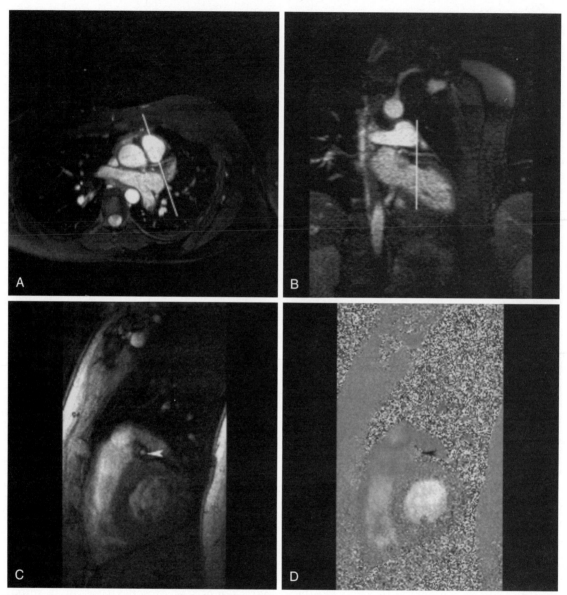

Figure 8–11. *A,* Transverse navigator echo image of the left anterior descending coronary artery (LAD). The white line indicates the location of the perpendicular plane used for the coronary flow measurement and is perpendicular to the artery both in this plane and in the paracoronal plane shown in *B.* Paracoronal navigator echo image of the LAD obtained by placing the imaging plane along the course of the artery as defined in *A.* The white line indicates the location of the perpendicular plane used for the coronary flow measurement. *C,* Magnitude reconstruction from a breath-hold, cine coronary flow sequence obtained at the position indicated in *A* and *B.* The LAD is seen in cross section in the center of the image (white arrow). *D,* Velocity map reconstruction from the same breath-hold, cine coronary flow sequence shown in *C.* The gray scale in this image indicates the velocity of motion with motion toward the viewer as white. This image in early diastole shows flow in the LAD as seen in cross section in the center of the image (black arrow).

between measurements of coronary flow reserve obtained by CMR and those obtained in the catheterization laboratory using an invasive intracoronary flow wire.[53] CMR techniques have also been shown to be useful in the assessment of coronary artery bypass graft patency and flow.[54] Saphenous vein grafts are typically of larger diameter and more fixed in position than native coronary arteries, facilitating imaging.

The general approach to the measurement of left anterior descending coronary artery (LAD) flow is shown in Figure 8–11. A navigator echo guided axial image is obtained to locate the LAD. An oblique coronal image is then obtained along the course of the coronary artery. Once the coronary artery is located, the imaging plane is typically placed perpendicular to the vessel for the measurement of flow. Typically, several baseline measurements are made, followed by measurements during adenosine infusion. The accuracy of coronary flow measurements has been shown to improve as the number of frames obtained per cardiac cycle increase. In addition, correction for motion of the coronary through the imaging plane has also been shown to improve the accuracy of the measurement.

CONCLUSION

CMR can be used to clearly delineate cardiac structure and assess function. Like any imaging technique, it is important to have a strategy for imaging and be familiar with the normal anatomy and potential artifacts. With this knowledge in hand, CMR can be a very effective tool in the evaluation of patients with cardiovascular disease.

References

1. El-Khoury GY, Bergman RA, Montgomery WJ: Sectional Anatomy by MRI. New York, Churchill Livingstone, 1995.
2. Arai AE, Epstein FH, Bove KE, Wolff SD: Visualization of aortic valve leaflets using black blood MRI. J Magn Reson Imaging 10:771, 1999.
3. Fayad ZA, Fuster V, Fallon JT, et al: Noninvasive in vivo human coronary artery lumen and wall imaging using black-blood magnetic resonance imaging. Circulation 102:506, 2000.
3a. Botnar RM, Stuber M, Kissinger KV, et al: Non-invasive coronary vessel wall and plaque imaging with magnetic resonance imaging. Circulation 102:2582, 2000.
4. Jarvinen VM, Kupari MM, Hekali PF, Poutanen VP: Right atrial MR imaging studies of cadaveric atrial casts and comparison with right and left atrial volumes and function in healthy subjects. Radiology 191:137, 1994.
5. Mortele KJ, Mergo PJ, Williams WF: Lipomatous hypertrophy of the atrial septum: Diagnosis with fat suppressed MR imaging. J Magn Reson Imaging 8:1172, 1998.
6. Shirani J, Roberts WC: Clinical, electrocardiographic and morphologic features of massive fatty deposits ("lipomatous hypertrophy") in the atrial septum. J Am Coll Cardiol 22:226, 1993.
7. Noma M, Kikuchi Y: Images in cardiovascular medicine. Lipomatous hypertrophy of the atrial septum. Circulation 100:684, 1999.
8. Hundley WG, Meshack BM, Willett DL, et al: Quantitation of left ventricular volumes, ejection fraction and cardiac output in patients with atrial fibrillation by cine magnetic resonance imaging: A comparison with invasive measurements. Am J Cardiol 78:1119, 1996.
9. Fischer SE, Wickline SA, Lorenz CH: Novel real-time R-wave detection algorithm based on the vectorcardiogram for accurate gated magnetic resonance acquisitions. Magn Reson Med 42:361–370, 1999.
10. Danias PG, Stuber M, Botnar RM, et al: Relationship between motion of coronary arteries and diaphragm during free breathing: Lessons from real-time MR imaging. AJR Am J Roentgenol 172:1061, 1999.
11. Taylor AM, Jhooti P, Wiesmann F, et al: Magnetic resonance navigator-echo monitoring of temporal changes in diaphragm position: Implications for magnetic resonance coronary angiography. J Magn Reson Imaging 7:629, 1997.
12. Botnar RM, Stuber M, Danias PG, et al: Improved coronary artery definition with T2-weighted, free-breathing, three-dimensional coronary MRA. Circulation 99:3139, 1999.
13. Bellenger NG, Gatehouse PD, Rajappan K, et al: Left ventricular quantification in heart failure by cardiovascular MR using prospective respiratory navigator gating: Comparison with breath-hold acquisition. J Magn Reson Imaging 11:411, 2000.
14. Babcock EE, Brateman L, Weinreb JC, et al: Edge artifacts in MR images: Chemical shift effect. J Comput Assist Tomogr 9:252, 1985.
15. Lotan CS, Cranney GB, Doyle M, Pohost GM: Fat-shift artifact simulating aortic dissection on MR images. AJR Am J Roentgenol 152:385, 1989.
16. Peshock RM, Willet D, Sayad D, et al: Quantitation of cardiac function by MRI. In Boxt LM (ed): MRI Clinics of North America, May 1996.
17. Dodge HT, Sandler H, Ballew DW, et al: The use of biplane angiocardiography for the measurement of left ventricular volume in man. Am Heart J 60:762, 1960.
18. Sandler H, Dodge HT: The use of single plane angiocardiograms for the calculation of left ventricular volume in man. Am Heart J 75:325, 1968.
18a. Chuang ML, Hibberd MG, Beaudin RA, et al: Importance of imaging method over imaging modality in noninvasive determination of left ventricular volumes and ejection fraction: Assessment by two and three dimensional echocardiography and magnetic resonance imaging. J Am Coll Cardiol 35:477, 2000.
18b. Strohm O, Schulz-Menger J, Pilz B, et al: Measurement of left ventricular dimensions and function in patients with dilated cardiomyopathy. J Magn Reson Imaging 13:367, 2001.
19. Boxt LM, Katz J, Kolb T, et al: Direct quantitation of right and left ventricular volumes with nuclear magnetic resonance imaging in patients with primary pulmonary hypertension. J Am Coll Cardiol 19:1508, 1992.
20. Utz JA, Herfkens RJ, Heinsimer JA, et al: Cine MR determination of left ventricular ejection fraction.
21. Culham JAG, Vince DJ: Cardiac output by MR imaging: An experimental study comparing right ventricle and left ventricle with thermodilution. J Can Assoc Radiol 39:247, 1988.
22. Hunter GJ, Hamberg LM, Weisskoff RM, et al: Measurement of stroke volume and cardiac output within a single breath hold with echo-planar MR imaging. J Magn Reson Imaging 4:51, 1994.
23. Hundley WG, Li HF, Hillis LD, et al: Quantitation of cardiac output with velocity-encoded, phase-difference magnetic resonance imaging: Validation with invasive measurements. Am J Cardiol 75: 1250, 1995.
24. Hundley WG, Li HF, Willard JE, et al: Magnetic resonance imaging assessment of the severity of mitral regurgitation: A comparison with invasive techniques. Circulation 92:1151, 1995.
25. Franco F, Dubois SK, Peshock RM, Shohet RV: Magnetic resonance imaging accurately estimates cardiac mass in a transgenic mouse model of cardiac hypertrophy. Am J Physiol 43:H679, 1998.
26. Maddahi J, Crues J, Berman DS, et al: Noninvasive quantification of left ventricular myocardial mass by gated proton

nuclear magnetic resonance imaging. J Am Coll Cardiol 10:682, 1987.

27. Katz J, Milliken MC, Stray-Gundersen J, et al: Estimation of human myocardial mass with MR imaging. Radiology 169:495, 1988.

28. Katz J, Whang J, Boxt LM, and Barst RJ: Estimation of right ventricular mass in normal subjects and in patients with primary pulmonary hypertension by nuclear magnetic resonance imaging. J Am Coll Cardiol 21:1475, 1993.

29. McDonald KM, Parrish T, Wennberg P, et al: Rapid, accurate and simultaneous noninvasive assessment of right and left ventricular mass with nuclear magnetic resonance imaging using the snapshot gradient method. J Am Coll Cardiol 19:1601, 1992.

30. Lorenz CH, Walker ES, Morgan VL, et al: Normal human right and left ventricular mass, systolic function, and gender differences by cine magnetic resonance imaging. J Cardiovasc Magn Reson 1:7, 1999.

31. Stahlberg F, Sondergaard L, Thomsen C, Henriksen O: Quantification of complex flow using MR phase imaging—a study of parameters influencing the phase/velocity relation. Magn Reson Imaging 10:13, 1992.

32. Tang C, Blatter DD, Parker DL: Accuracy of phase-contrast flow measurements in the presence of partial volume effects. J Magn Reson Imaging 3:377, 1993.

33. Tang C, Blatter DD, Parker DL: Accuracy of phase-contrast flow measurements in the presence of partial volume effects. Magn Reson Med 3:377, 1993.

34. Buonocore MH, Bogren H: Factors influencing the accuracy and precision of velocity-encoded phase imaging. Magn Reson Med 26:141, 1992.

35. Hundley WG, Li HF, Hillis LD, et al: Quantitation of cardiac output with velocity-encoded, phase-difference magnetic resonance imaging: Validation with invasive measurements. Am J Cardiol 75:1250, 1995.

36. Brenner LD, Caputo GR, Mostbeck G, et al: Quantification of left to right atrial shunts with velocity-encoded cine nuclear magnetic resonance imaging. J Am Coll Cardiol 20:1246, 1992.

37. Hundley WG, Li HF, Lange RA, et al: Assessment of left-to-right intracardiac shunting by velocity-encoded, phase-difference magnetic resonance imaging: A comparison with oximetric and indicator dilution techniques. Circulation 91:2955, 1995.

38. Duerinckx AJ, Higgins CB: Valvular heart disease. Radiol Clin North Am 32:613–630, 1994.

39. Globits S, Higgins CB: Asessment of valvular heart disease by magnetic resonance imaging. Am Heart J 129:369, 1995.

40. Arai AE, Epstein FH, Bove KE, Wolff SD: Visualization of aortic valve leaflets using black blood MRI. J Magn Reson Imaging 10:771, 1999.

41. Kilner PJ, Firmin DN, Rees RSO, et al: Valve and great vessel stenosis: Assessment with magnetic resonance jet velocity mapping. Radiology 178:229, 1991.

42. Eichenberger AC, Jenni R, von Shulthess GK: Aortic valve pressure gradients in patients with aortic valve stenosis: Quantification with velocity-encoded cine MR imaging. AJR Am J Roentgenol 160:971, 1993.

43. Sondergaard L, Hildebrandt P, Lindvig K, et al: Valve area and cardiac output in aortic stenosis: Quantification by magnetic resonance velocity mapping. Am Heart J 127:1156, 1993.

44. Dulce MC, Mostbeck GH, O'Sullivan M, et al: Severity of aortic regurgitation: Interstudy reproducibility of measurements with velocity-encoded cine MR imaging. Radiology 185:235, 1992.

45. Honda N, Machida K, Hashimoto M, et al: Aortic regurgitation: Quantitation with MR imaging velocity mapping. Radiology 186:189, 1993.

46. Walker PG, Oyre S, Pedersen EM, et al: A new control volume method for calculating valvular regurgitation. Circulation 92:579, 1995.

47. Hundley WG, Meshack BM, Willett DL, et al: Quantitation of left ventricular volumes, ejection fraction and cardiac output in patients with atrial fibrillation by cine magnetic resonance imaging: A comparison with invasive measurements. Am J Cardiol 78:1119, 1996.

48. Fujita N, Chazouilleres AF, Hartiala JJ, et al: Quantification of mitral regurgitation by velocity-encoded cine nuclear magnetic resonance imaging. J Am Coll Cardiol 23:95, 1994.

49. Hundley WG, Li HF, Willard JE, et al: Magnetic resonance imaging assessment of the severity of mitral regurgitation: A comparison with invasive techniques. Circulation 92:1151, 1995.

50. Shellock FG, Kanal E: Magnetic Resonance: Bioeffects, Safety, and Patient Management. 2nd ed. Philadelphia, Lippincott-Raven, 1996.

51. Deutsch HJ, Bachmann R, Sechtem U, et al: Regurgitant flow in cardiac valve prostheses: Diagnostic value of gradient echo nuclear magnetic resonance imaging in reference to transesophageal two-dimensional color doppler echocardiography. J Am Coll Cardiol 19:1500, 1992.

52. Clarke GD, Eckels R, Chaney C, et al: Measurement of absolute epicardial coronary artery flow and flow reserve using breath-hold cine phase contrast magnetic resonance imaging. Circulation 91:2627, 1995.

53. Hundley WG, Clarke GD, Lange RA, et al: Magnetic resonance imaging measurement of absolute coronary artery flow and flow reserve in humans. Circulation 93:1502, 1996.

54. Hoogendoorn LI, Pattynama PMT, Buis B, et al: Noninvasive evaluation of aortocoronary bypass grafts with magnetic resonance flow mapping. Am J Cardiol 75:845, 1995.

ISCHEMIC HEART DISEASE

Assessment of Cardiac Function

Nicholas G. Bellenger and Dudley J. Pennell

The accurate and reproducible assessment of cardiac function is a fundamental aim of noninvasive cardiac imaging, and forms the foundation upon which much of the assessment and management of myocardial dysfunction, ischemia, viability, remodeling, valvular, and other cardiac disorders are based. In this chapter we discuss the importance of the measurement of global cardiac function, compare techniques, and provide a practical step-by-step guide to its assessment by cardiovascular magnetic resonance (CMR).

THE POPULATION IMPACT OF CARDIAC DYSFUNCTION

Cardiac dysfunction can result from a broad spectrum of organ-specific and multisystem disorders. The defining property in common with all these disorders is the impairment of the ventricle to eject blood, which in its broadest sense, and without discussing semantics, is known as heart failure. Heart failure is common, with approximately 1.5 to 2.0 percent of the population younger than 65 years of age being affected, rising to 6 to 10 percent of those who are older than 65 years.[1] It afflicts 4.8 million people in the United States, with 400,000 to 700,000 new cases developing each year. It is the leading cause of hospital admission in individuals older than 65 years of age, and, despite aggressive treatment, 40,000 patients die of heart failure each year.[2] It is also an enormous consumer of health care budgets in the Western world, and treatments to prevent its occurrence, to slow its progression, or to prevent repeated hospitalizations can have an important economic impact.

THE IMPORTANCE OF MEASURING CARDIAC FUNCTION

A single assessment of cardiac function can provide important diagnostic and prognostic information, whether in the setting of postinfarction recovery,[3] left ventricular hypertrophy,[4] or chronic heart failure.[5, 6] As well as improving symptoms and quality of life, treatment of heart failure aims to decrease the likelihood of disease progression, resulting in costly hospital admissions and ultimately premature death. To allow early evaluation and alteration of an individual patient's management, serial studies need to be performed using a technique that not only is accurate but also has good interstudy reproducibility. This principle also applies when considering study populations of trial therapies.

TECHNIQUES FOR ASSESSING CARDIAC FUNCTION

Bedside clinical assessment of cardiac function is generally poor[7] and electrocardiogram (ECG) findings are nonspecific, although the presence of an entirely normal ECG has a 95 percent likelihood of normal systolic function.[8] In the search for a better assessment, it is worth considering the ideal imaging technique. This technique would provide a noninvasive, accurate, and reproducible assessment of cardiac function without exposure to ionizing radiation or exogenous contrast agent. It would be widely available and time and cost effective. Although no technique fully meets all of these criteria, there are clear differences between modalities that merit discussion.

Echocardiography

Echocardiography is a widely available but less than an ideal imaging technique for quantifying cardiac function because the image acquisition is operator and acoustic window dependent.[9] The quantification of ventricular function is limited by a priori geometric assumptions that may provide a reasonable assessment in the normal ventricle but are less reliable in remodeled hearts as a result of complex irregular shape changes.[10, 11] In M-mode echocardiography, for example, ventricular diameters are measured in the minor axis (Figure 9–1), and volumes are obtained by cubing these values (thereby magnifying the errors).[12–14] Functional estimates, such as ejection fraction and fractional shortening, are then derived from these values. This method assumes that a single view is representative of all the myocardial segments and that contraction is uniform throughout the ventricle. Two-dimensional (2-D) echocardiography provides the opportunity to derive the cardiac function from cardiac volumes by

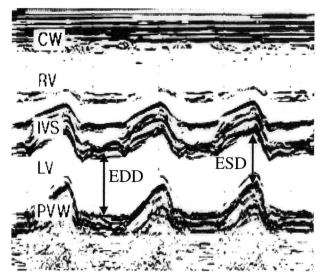

Figure 9–1. M-mode echo recording at the mid-ventricular level in a patient with tricuspid regurgitation and pulmonary hypertension showing paradoxical systolic motion of the interventricular septum and a dilated right ventricle. CW = chest wall; RV = right ventricle; IVS = interventricular septum; LV = left ventricle; PVW = posterior ventricular wall; EDD = end-diastolic dimension; ESD = end-systolic dimension.

the area length and Simpson's method of discs (Figure 9–2).[12] This procedure, however, relies on good visualization of the entire endocardial border, which is frequently not possible. For example, in a recent multicenter study that required good-quality echocardiograms as an entry criterion, the core laboratory was unable to perform a confident 2-D ejection fraction analysis in 31 percent of patients.[14] Overall, the practical difficulties of quantifying global function by echocardiography are underlined by the fact that in the "real world" it is often visually estimated by the clinician performing the imaging. This method is highly subjective but can be clinically valid with experience.[15] The advent of three-dimensional (3-D) echocardiography is answering some of the questions that limit this technology, by removing the need for geometric assumptions,[16] but issues of acoustic windows and practicality in clinical practice remain to be answered in due course as more experience is gained. Echocardiography is particularly limited with regard to quantitative assessment of right ventricular volume and systolic function.

Nuclear Cardiology

Nuclear cardiology with radionuclide ventriculography is commonly used to measure left and right ventricular function by measuring the ejection fraction, but it has relatively low spatial and temporal resolution, and preparation and scanning times are relatively prolonged.[17] In addition, ventricular volumes are difficult to measure and are rarely performed clinically (though they are used occasionally for research[18]) and ventricular mass cannot be ob-

tained. The use of gated perfusion single photon emission computed tomography (SPECT) has allowed the development of 3-D solutions to ventricular function, and this method is now achieving widespread use, especially in the United States. This procedure is useful when perfusion needs to be assessed and adds prognostic value to the perfusion assessment,[19] but it is not being performed solely to assess ventricular function, and to date there are no reports of its use for ventricular remodeling, despite its good reproducibility.[20] The ventricular volumes are reported as reliable,[21] but there is concern over their accuracy in both small and large ventricles because of the limited spatial resolution and the problems of assigning a ventricular border in areas of transmural infarction and thinning where counts are very low.[22] For both nuclear cardiology techniques, the need for repeated radionuclide doses in follow-up studies is problematic, especially for research, for which radiation exposure must be justified in a milieu of competing technologies and public pressure in general to limit radiation burdens.

Cardiovascular Magnetic Resonance (CMR)

CMR has some fundamental advantages over other imaging techniques, which have fueled the growing

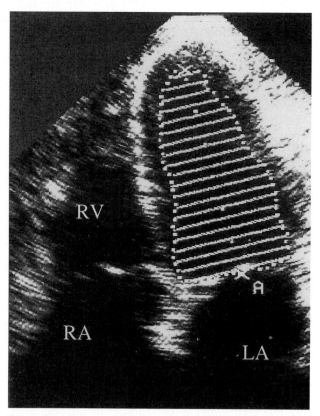

Figure 9–2. Echo measurement by Simpson's method of discs. This method produces a more accurate measurement than M-mode or area and length assessment, but relies on good endocardial border definition. RV = right ventricle; RA = right atrium; LA = left atrium; A = mitral annulus.

enthusiasm for its use in clinical practice and research. CMR offers accurate and reproducible tomographic, static, or cine images of high spatial and temporal resolution in any desired plane without exposure to exogenous contrast agents or ionizing radiation. As such, horizontal and vertical long-axis and short-axis views with high spatial and temporal resolution can be easily and rapidly acquired to allow a visual, qualitative assessment of function, similar to echocardiography. The main advantage of CMR, however, lies in its quantitative accuracy and reproducibility.

There are two main methods for measuring the end-diastolic (ED) and end-systolic (ES) volumes. The earliest semiquantitative method was an adaptation of the echo area-length where the volume of the left ventricle is assumed to constitute an ellipsoid. With a single apical four-chamber view, the area (A) of the left ventricular endocardial border can be traced, as well as the length (L) from the apex to the mitral annulus. The volume for systole and diastole is then easily calculated: Volume = $0.85A^2/L$. A more accurate adaptation uses both long axis planes (vertical and horizontal long-axis—VLA, HLA) to measure two perpendicular areas and lengths: Volume = $(0.85 \times$ area A $\times$ area B$)/$ smaller length. Although this method offers a simple and time-efficient volume analysis, it suffers from the same limitations of echocardiography, namely, the need for geometric assumptions and the inability to take account of regional differences in wall motion.[16] An example of a distorted heart following myocardial infarction is shown in Figure 9–3.

A better method of measuring volumes, and thereby function, is by the use of Simpson's rule. A stack of contiguous tomographic slices is acquired that encompasses the entire left ventricle (LV). At

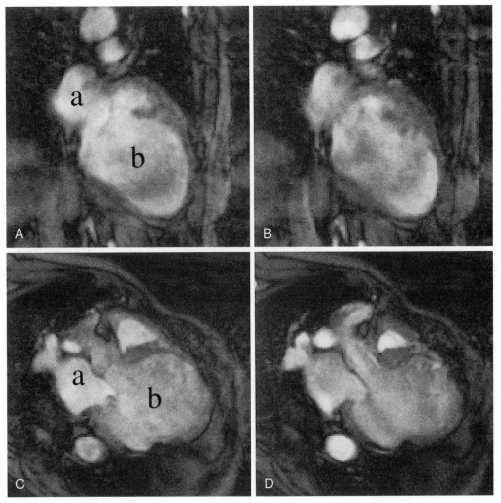

Figure 9–3. A patient with ischemic heart disease in whom the left ventricle no longer conforms to geometric assumptions in either diastole (*A* and *C*) or systole (*B* and *D*) frames. Both the vertical long axis (VLA; *A* and *B*) and the horizontal long axis (HLA; *C* and *D*) views are illustrated. a = left atrium; b = left ventricle.

Long Axis of Left Ventricle   Short Axis

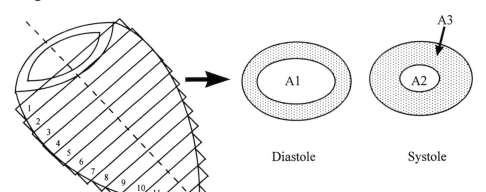

Figure 9–4. Schematic of the left ventricular short axis slices that encompass the entire left ventricle. Each slice is acquired as a cine. A1 = end-diastolic endocardial area, A2 = end-systolic endocardial area, A3 = myocardium.

one time, it was common to use a stack of transverse images for this measurement, but although this method makes it easy to define the mitral valve plane and thereby the true base of the LV,[23] it is also subject to considerable partial volume effects, especially in the inferior wall. For this reason, short axis slices are now commonly employed, and nearly all sites specializing in CMR have now adopted

this practice (Figure 9–4). The ventricular volume is equal to the sum of the endocardial areas multiplied by the distance between the centers of each slice (Figure 9–5). The volumes obtained by this method are independent of geometric assumptions and dimensionally accurate.[16, 24, 25]

In order to achieve full 3-D coverage of the ventricle using conventional free-breathing gradient-echo

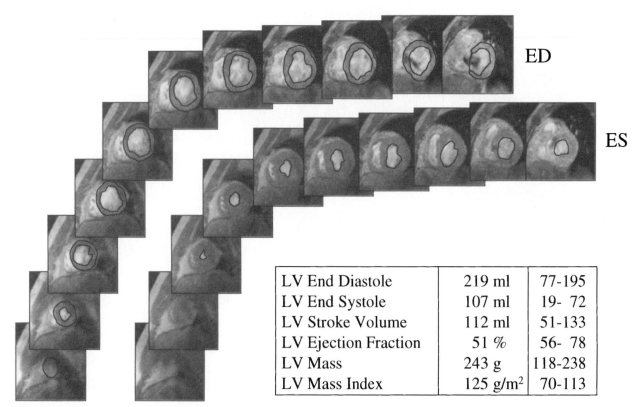

LV End Diastole	219 ml	77-195
LV End Systole	107 ml	19- 72
LV Stroke Volume	112 ml	51-133
LV Ejection Fraction	51 %	56- 78
LV Mass	243 g	118-238
LV Mass Index	125 g/m^2	70-113

Figure 9–5. Series of end-diastolic (ED) and end-systolic (ES) slices from contiguous short axis cines that encompass the left ventricle, from base to apex, in a patient with ventricular dilatation and hypertrophy from chronic aortic regurgitation. The epicardial and endocardial borders are traced and the summation values shown in the left column, with normal values in the right column. Note that there is one more image at end-diastole (11) than at end-systole (10), reflecting the need to allow for the systolic descent of the atrioventricular ring, as described in the text. Note also the ingress of the left ventricular (LV) outflow tract in the most basal end-diastolic image, where there is no ventricular mass, and the open ends of the LV myocardial horseshoe are joined together in order to form the appropriate volume.

cine sequences, a total scanning time of 30 minutes or more is required, but on modern scanners with fast imaging, a single cine can be acquired in just one breath-hold of 8 to 12 seconds, allowing the whole stack of images to be acquired in 5 to 10 minutes.[14] This method has the considerable additional advantage of reducing breathing and movement artifact. In patients unable to hold their breath consistently, or who are orthopneic, solutions using the same sequence combined with navigator echo imaging have been shown to be successful, during free-breathing.[26] The most modern scanners with ultrafast capability can acquire the complete 3-D ventricular data set in a single breath-hold,[27] whereas a 3-D solution using an intravascular contrast agent has also been reported.[28] Advances in acquisition sequences have also started to play an important role with the elimination of blood saturation artifact with the use of steady-state free procession (SSFP) cine imaging, with the magnetization driven to steady state, which makes the cines independent of inflow enhancement, thereby greatly improving the blood to myocardium contrast, especially in the long axis planes.[29] This sequence runs at its best with ultrafast gradients because a very short repetition time (TR) is required to reduce the sensitivity of the sequence to movement artifact.

In addition to the LV, it is important to remember the right ventricle (RV), inasmuch as its function is also known to be an important determinant of prognosis, both in coronary artery disease[30] and in congenital heart and pulmonary disease. Global RV function is difficult to assess adequately by echocardiography, whereas radionuclide ventriculography suffers from assumptions concerning projection of overlapping structures, unless research techniques such as first pass techniques with ultrashort half-life isotopes are used. Such problems are not experienced by CMR, and RV function and mass are well characterized.[31, 32] The RV is discussed in greater detail in Chapter 23.

ACCURACY AND REPRODUCIBILITY OF CARDIOVASCULAR MAGNETIC RESONANCE

It is now widely accepted that CMR offers the reference standard for the noninvasive assessment of cardiac function, being both accurate[24, 25, 33, 34] (Figures 9–6 through 9–9) and reproducible in normal and abnormal ventricles.[35–39] Much validation work was done using conventional non-breath-hold cine CMR on older scanners, however, and currently breath-hold sequences are now routinely employed in nearly all centers. More recent studies have been performed to examine whether this change has impacted the quality of the results, and these studies show that the reproducibility of both techniques is similar.[40–43] This finding is illustrated in Figure 9–10,

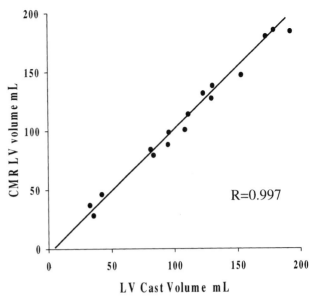

Figure 9–6. Validation of left ventricular volume measurements by CMR. CMR-derived volume (vertical scale) is compared with the displacement volume of the left ventricular casts. There is excellent accuracy and a close relation of the regression line to identity. (Data from Rehr RB, Malloy CR, Filipchuk NG, et al: Left ventricular volumes measured by MR imaging. Radiology 156:717, 1985.)

which shows the intraobserver, interobserver, and interstudy variability for volume and functional assessment by breath-hold CMR in dilated and normal ventricles, compared with conventional cine CMR.[43]

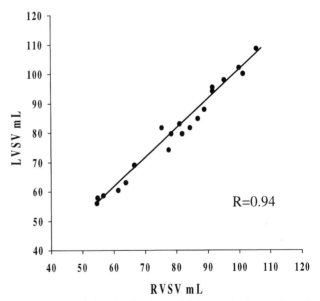

Figure 9–7. Validation of right and left ventricular stroke volumes in vivo using CMR. Using Simpson's rule, there is an excellent agreement between the stroke volumes (SV) of the right and left ventricles, which is strong in-vivo evidence that both measurements are accurate, because both are equivalent in vivo in the absence of valve regurgitation or shunting. (Data from Longmore DB, Klipstein RH, Underwood SR, et al: Dimensional accuracy of magnetic resonance in studies of the heart. Lancet 1:1360, 1985.)

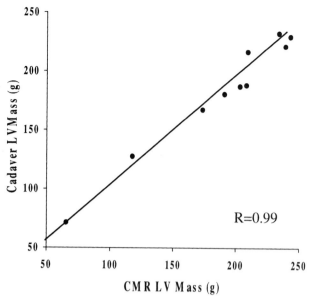

Figure 9–8. Validation of left ventricular mass with comparison of CMR-derived mass and cadaver human hearts. (Data from Katz J, Milliken MC, Stray-Gunderson J, et al: Estimation of human myocardial mass with MR imaging. Radiology 169:495, 1988.)

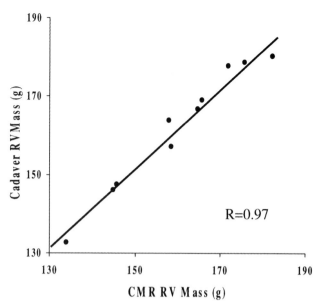

Figure 9–9. Validation of right ventricular mass with comparison of CMR-derived mass and postmortem bovine hearts. (Data from Katz J, Whang J, Boxt LM, Barst RJ: Estimation of right ventricular mass in normal subjects and in patients with primary pulmonary hypertension by nuclear magnetic resonance imaging. J Am Coll Cardiol 21:1475, 1993.)

Normal values for volumes and mass of left and right ventricles are shown in Tables 9–1 and 9–2.

The excellent reproducibility of CMR versus echocardiography can be illustrated by considering the sample size required for a drug trial designed to show a 10 g decrease in LV mass with antihypertensive treatment. In a direct comparison of CMR with echocardiography for reproducibility, it has been shown that for an 80 percent power and a p value of 0.05, the sample size required would be 505 patients with 2-D echo but only 14 patients using CMR.[44] Similarly, we have found that to show a 5 percent difference in ejection fraction with a 90 percent power and a p value of 0.05 would require only 7 normal subjects or 5 patients with dilated ventricles.[43] This reproducibility has significant implications for research and in particular for pharmaceutical companies, where CMR offers a more cost- and time-effective research tool (Figure 9–11).

A PRACTICAL GUIDE TO FUNCTIONAL CARDIOVASCULAR MAGNETIC RESONANCE

In modern medicine a balance must be struck between the information gained from an investigation and the resources it demands. The following protocol is designed to be as efficient as possible in gaining the volumetric data from the ventricles of the heart.[17] Figure 9–12 illustrates the sequence of pilot images used to achieve imaging in the long axis of the left ventricle, and thereby the true short axis. A coronal pilot is first taken and used to acquire transverse pilots, which show both the mitral valve and the apex of the left ventricle. By taking a plane though the center of the mitral valve (halfway between the back end of the septum and the back

Table 9–1. Normal Values in Adults for Left Ventricular (LV) Volumes and Mass*

| Parameter | Men | | Women | |
	Absolute	*Normalized to BSA*	*Absolute*	*Normalized to BSA*
LVEDV	136 ± 30 [77–195] ml	69 ± 11 [47–92] ml/m²	96 ± 23 [52–141] ml	61 ± 10 [41–81] ml/m²
LVESV	45 ± 14 [19–72] ml	23 ± 5 [13–33] ml/m²	32 ± 9 [13–51] ml	21 ± 5 [11–31] ml/m²
LVSV	92 ± 21 [51–133] ml	47 ± 8 [32–62] ml/m²	65 ± 16 [33–97] ml	41 ± 8 [26–56] ml/m²
LVEF	67 ± 5 [56–78] %		67 ± 5 [56–78] %	
LVM	178 ± 31 [118–238] g	91 ± 11 [70–113] g/m²	125 ± 26 [75–175] g	79 ± 8 [63–95] g/m²

*Values are quoted as mean ± 1 standard deviation, with the 95% confidence interval for the normal range in brackets.
BSA = body surface area; EDV = end-diastolic volume; ESV = end-systolic volume; SV = stroke volume; EF = ejection fraction; M = mass, including papillary muscles.
Data from Lorenz CH, Walker ES, Morgan VL, et al: Normal human right and left ventricular mass, systolic function and gender differences by cine magnetic resonance imaging. J Cardiovasc Magn Reson 1:7, 1999.

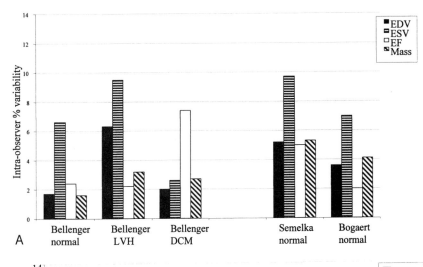

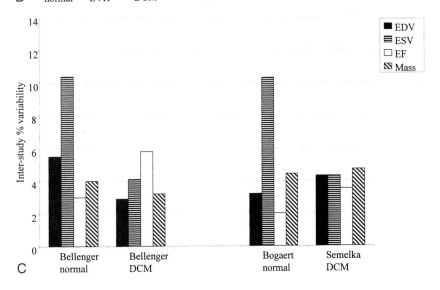

Figure 9–10. The intraobserver *(A)*, interobserver *(B)*, and interstudy *(C)* percentage variability for end-diastolic volume (EDV), end-systolic volume (ESV), ejection fraction (EF), and LV mass. Results using breath-hold segmented k-space gradient-echo cine CMR from our center in patients with heart failure and dilated ventricles (DCM) and left ventricular hypertrophy (LVH) are compared with normals (Bogaert) and traditional, slower gradient-echo cine imaging (Semelka).[36] Overall, the results are very similar (probably favoring the breath-hold imaging), both between techniques and between different population groups.

end of the lateral wall) and the tip of the apex, the vertical long axis (VLA) is acquired. This VLA is used to plan the horizontal long axis (HLA), by again using a plane through the center of the mitral valve (halfway between the back of the anterior and inferior walls) and the tip of the apex. It should be noted that it is common to find centers describing planes that are parallel to the septum for the VLA, and parallel to the inferior wall for the HLA, but these dictates are not correct because they are likely

Table 9–2. Normal Values in Adults for Right Ventricular (RV) Volumes and Mass*

Parameter	Men		Women	
	Absolute	*Normalized to BSA*	*Absolute*	*Normalized to BSA*
RVEDV	157 ± 35 [88–227] ml	80 ± 13 [55–105] ml/m²	106 ± 24 [58–154] ml	67 ± 10 [48–87] ml/m²
RVESV	63 ± 20 [23–103] ml	32 ± 8 [16–48] ml/m²	40 ± 14 [12–68] ml	26 ± 6 [20–32] ml/m²
RVSV	95 ± 22 [52–138] ml	48 ± 8 [32–64] ml/m²	66 ± 16 [35–98] ml	42 ± 8 [27–57] ml/m²
RVEF	60 ± 7 [47–74] %		63 ± 8 [47–80] %	
RVM	50 ± 10 [30–70] g	26 ± 5 [16–36] g/m²	40 ± 8 [24–55] g	25 ± 4 [18–33] g/m²

*Values are quoted as mean ± 1 standard deviation, with the 95% confidence interval for the normal range in brackets.

BSA = body surface area; EDV = end-diastolic volume; ESV = end-systolic volume; SV = stroke volume; EF = ejection fraction; M = mass of the RV free wall.

Data from Lorenz CH, Walker ES, Morgan VL, et al: Normal human right and left ventricular mass, systolic function and gender differences by cine magnetic resonance imaging. J Cardiovasc Magn Reson 1:7, 1999.

to lead to the long axis plane not passing through the center of the basal ring of the LV, and they also may lead to an offset from the tip of the apex, which are both undesirable. This difficulty can lead to problems planning the short axis cuts to adequately cover the full extent of left and right ventricles, and in addition it reduces the reproducibility of the short axis plane positioning for repeated studies. Finally, the short axis slices are placed on the HLA to encompass the heart. In order to achieve the most reliable results, which are the most reproducible, attention to detail is required. First, if the short axis cuts are to be acquired using a breath-hold cine sequence, the VLA and HLA must also have been acquired using a breath-hold, and at end-expiration. Second, when using breath-hold techniques, it is generally more reproducible to ask patients to hold their breath at end-expiration, rather than elsewhere in the respiratory cycle, and this instruction applies to the pilot images as well as the short- and long-axis cines.[45] Third, the first short axis plane should be placed at the base of the heart covering the most basal portion of the LV and RV just forward of the

atrioventricular ring (see Figure 9–12*D*), and it should be placed on the end-diastolic HLA image. Finally, further short axis planes should then be planned to move apically from this plane until the apex is encompassed (see Figure 9–12*E*). Although it is possible to acquire the VLA and HLA as single pilot images instead of cines, there is little practical merit in this method inasmuch as the time for two breath-hold cines is small, and the contraction pattern in these two planes is very useful during qualitative assessment of ventricular function. In the future with the development of 3-D postprocessing software solutions for analyzing the cines, these long axis cines will become mandatory. It is also worth noting that in the future, full 3-D analysis of atria as well as ventricles may be simple and practical with automated analysis, in which case cines encompassing the entire heart would be acquired.

Technical Tips

- The average segmented k-space gradient echo breath-hold required to acquire 16 phases with a

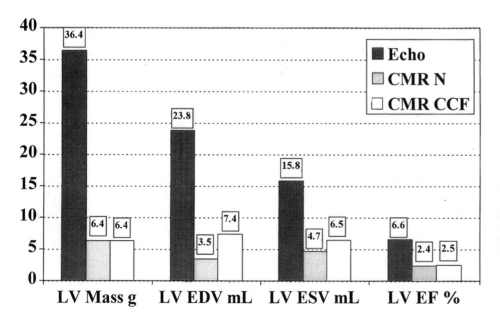

Figure 9–11. Comparison of the interstudy reproducibility of CMR and recently published values for two-dimensional echocardiography for parameters of ventricular remodeling.[69] Although these results are from different patient populations, CMR appears to offer a marked improvement in reproducibility, which translates into reduced sample sizes for remodeling and hemodynamic studies, which require repeated volume and mass parameter assessments.[43] N = normals; CCF = congestive cardiac failure.

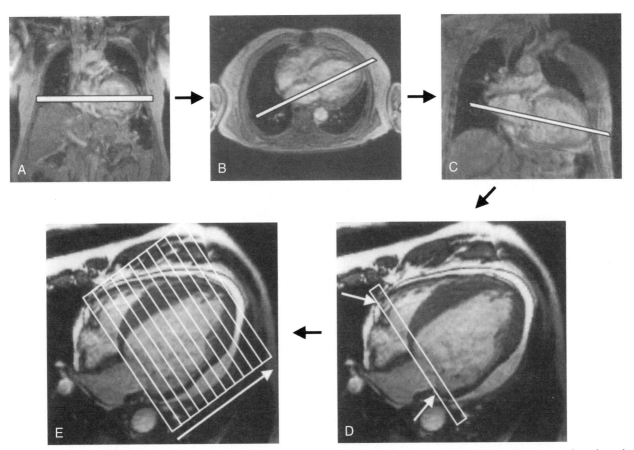

Figure 9–12. Pilot images used to achieve the true short axis of the left ventricle. A coronal pilot *(A)* is first acquired, and used to pilot the transverse image *(B)*. The vertical long axis *(C)* is obtained from this image and subsequently the horizontal long axis *(D)*, upon which the short-axis cines are placed. Note the placement of the first short-axis plane to encompass the rear of right and left ventricular myocardium (arrows). The full short-axis cine series covering both ventricles *(E)* is shown.

phase encoded grouping (PEG) of 6 is approximately 15 to 20 seconds. Although this length does not constitute a problem for most patients, those with orthopnea may find it difficult. By increasing the PEG, the breath-hold time can be reduced. A compromise is reached, however, because fewer phases will be captured with a higher PEG. Fewer than 11 phases will give inadequate information on wall motion and may not cover end systole precisely. Decreasing the field of view will also reduce the breath-hold time but may result in some wraparound occurring at the edges of the image. If this wraparound remains remote from the heart, it may be considered an acceptable compromise. The PEG size and the number of phases acquired are related through the TR of the sequence. Modern CMR scanners with faster gradients allow a shorter TR and echo time (TE), which improves this compromise, and as scanners improve, such compromises should become a problem of the past. Typically, best results are seen with the time between cine phases being 50 ms or less, which yields a cine with approximately 16 frames in clinical practice. This result is not always achievable with older scanners with breath-holding.

- Analysis of the short axis slices is relatively

straightforward provided that the quality of the images is reasonable (mainly dependent on accurate ECG gating and good breath-holding). The main source of error is in separating the ventricles from the atria. Identifying this basal slice is made more difficult by the through-plane descent of the atrioventricular (AV) ring in systole, which is usually about 1 cm. This difficulty makes the placement of the first, most basal short axis slice very important. By ensuring that this basal slice is carefully positioned on the end-diastolic, end-expiratory breath-hold HLA image just forward of the AV ring, the first short axis cine will by definition contain end-diastolic volume and mass within both ventricles. However, at end systole, the basal slice will include only atrium as a result of descent of the AV ring, and in general the systolic area in the basal cine is not included in the analysis of the systolic volume. In general, the next slice down contains both end-diastolic and end-systolic data. An alternative approach to this rigorous method is to oversample with short axis slices into the atrium and attempt to retrospectively differentiate ventricle from atrium based on the degree of descent of the AV ring in the long axis images, and whether the chamber dilates or contracts in systole. In general, we prefer not to oversample but to ensure the first basal slice is acquired correctly because this method leads to a reproducible approach and is more time efficient for both acquisition and analysis, and because oversampling relies more heavily on good image quality to differentiate atrium from ventricle.

- Papillary muscles and endocardial trabeculae should be *excluded* from the LV volume and included in the LV mass. Although there is no clear consensus at present, LV mass is usually taken from the end-diastolic images. Unpublished data show that LV mass by CMR varies by a small amount from end diastole to end systole (B. Cowan, personal communication), and this may be due to expulsion of intramyocardial blood into the venous system. The reproducibility of LV diastolic volumes is in general better than at end systole because the volumes are larger, which is probably as good a reason as any for working from the end-diastolic images to determine mass. However, in addition, following the aforementioned routine, there may be doubt as to whether LV mass is present in the most basal LV slice at end systole, if its quality is less than ideal, but by definition, LV mass is always present at end diastole in the most basal slice.

- A slice thickness of 8 mm at 1.5 Tesla (T) provides adequate spatial resolution without overly increasing the number of slices, and thereby the analysis time. It also limits partial volume effects. A 2-mm slice gap is commonly used to allow easy calculation of volumes because the center of each slice is then 1 cm apart. There are no formal studies to aid the choice of slice thickness, but 8 mm is a reasonable consensus. Some centers prefer thinner slices, but it is important to maintain signal to noise and image quality. 3-D imaging may eventually allow more partitions with thinner slices.

- For the very best interstudy reproducibility for follow-up studies and drug trials, it is necessary to go the extra mile in the analysis, and always have the first set of cines with the regions of interest (ROIs) on screen alongside the follow-up cines during the analysis. At least a printout on paper of the first study and the ROIs used for analysis is required for comparison. For this reason, we routinely print out the diastolic and systolic ROIs in a systematic way for every patient on whom volumes are analyzed.

- It should be noted that this field is still being defined, and as new approaches are being generated regularly, these recommendations may change in the future. In particular, as mentioned above, it is highly likely that automated postprocessing of the cines will become the norm in the near future, but with a 3-D analytical approach to fully take into account the motion in systole of the AV ring by incorporating VLA and HLA cines to define accurately the mitral valve position at each phase in the cardiac cycle. In our opinion, none of the 2-D automatic approaches on the market in 2001 fully addressed this issue with a complete solution, although they clearly can save time and effort in drawing the ROIs, and are a useful aid. A fully robust solution for the postprocessing is the Holy Grail of this field, and there is every reason to be optimistic that it will soon be solved. Enhanced contrast between the blood pool and endocardium afforded by SSFP acquisition may facilitate automated methods.

OTHER CARDIOVASCULAR MAGNETIC RESONANCE MEASURES OF GLOBAL FUNCTION—BRIEF SYNOPSIS

Systolic

There are other ways to derive important functional information from the heart than from volumetry alone. For example, flow in the major vessels is easily measured with great accuracy using velocity mapping.[46] The aortic flow measured over a complete cardiac cycle represents the LV stroke volume. The RV stroke volume can likewise be found from flow in the pulmonary artery. Cardiac output can be easily derived from this formula: Cardiac output = stroke volume × heart rate. The measurement of stroke volume is useful in valve disease,[47] and for determining the need for surgery in cardiovascular shunting,[48] and is noninvasive and quantitative.[49] The peak flow velocity and acceleration, which are derived from the flow curves, have been used to determine the burden of ventricular ischemia during dobutamine stress (see Chapter 10).[50]

Diastolic

For diastolic function, velocity mapping of the transmitral E wave agrees with that of echocardiography but tends to underestimate the A wave, due to beat-to-beat variability that occurs in the diastolic period.[51] A decrease in peak E/A ratio, however, can be demonstrated by CMR in cases of reduced ventricular compliance. Cine CMR also allows direct visualization of the abnormal diastolic filling that has been shown in dilated ventricles with inflow directed not toward the apex but toward the free wall, giving rise to a well-developed circular flow pattern turning back toward the septum and outflow tract, persisting through diastole.[52] Studies of myocardial velocity in diastole both at rest and during dobutamine have also been performed using CMR, but experience with this technique is very limited.[53] Tagging in diastole has also been examined,[53a] but again more experience is needed.[54] More details of the assessment of diastolic function can be found in Chapters 5 and 10.

REGIONAL FUNCTION

Cine CMR sequences allow a qualitative assessment of regional cardiac function in similar manner to echocardiography, but with improved image quality and fewer nonvisualized segments. In addition, it is possible to image routinely in the true long and short axis of the heart without compromises resulting from restricted angulation resulting from awkward acoustic windows. This wall motion analysis can be performed at rest, with low-dose dobutamine for the detection of viable myocardium,[55, 56] and high-dose dobutamine for detection of ischemia.[57, 58] Studies using real-time CMR are now being published, and these too show that CMR is superior to echocardiography in patients with limited acoustic access.[59, 60]

Several methods based on cine CMR have been suggested to provide quantitative assessment of wall motion and wall thickening.[61–66] In reality, however, myocardial dynamics are more complicated than simple thickening and 2-D motion because of a complex interaction of contraction, expansion, twisting, and through-plane motion. This is best measured by the process of tagging,[67] which can provide a 3-D solution and is discussed in detail in Chapters 4 and 5.

THE FUTURE

CMR offers a rapid, accurate, and reproducible assessment of cardiac function that is free of geometric assumptions and is truly noninvasive. At present, however, CMR analysis of cardiac function is time intensive, and experience is required. Automated border detection systems that incorporate both the long and short axis views, and thereby overcome the difficulties in basal slice differentiation, are under development. This process may be aided by the use of improved cine sequences such as SSFP or intravascular contrast agents.[68] Experience in CMR is growing, and advanced specialized cardiac hardware is becoming more available. As CMR continues to become faster, it also becomes more cost effective for this function, and there is a general belief that this simple technique will become the standard tool in clinical and research practice.

References

1. Schocken DD, Arrieta MI, Leaverton PE, Ross EA: Prevalence and mortality rate of congestive heart failure in the United States. J Am Coll Cardiol 20:301, 1992.
2. Anonymous: From the Centers for Disease Control and Prevention. Mortality from congestive heart failure—United States, 1980–1990. JAMA 271: 813, 1994.
3. White HD, Norris RM, Brown MA, et al: Left ventricular end-systolic volume as the major determinant of survival after recovery from myocardial infarction. Circulation 76:44, 1987.
4. Dunn FG, Pringle SD: Sudden cardiac death, ventricular arrhythmias and hypertensive left ventricular hypertrophy. J Hypertens 11: 1003, 1993.
5. Levy D, Garrison RJ, Savage DD, et al: Prognostic implications of echocardiographically determined left ventricular mass in the Framingham heart study. N Engl J Med 332: 1561, 1990.
6. Ghali JK, Liao Y, Cooper RS: Influence of left ventricular geometric patterns on prognosis in patients with or without coronary artery disease. J Am Coll Cardiol 31:1635, 1998.
7. Marantz PR, Tobin JN, Wassertheil-Smoller S, et al: The relationship between left ventricular systolic function and congestive heart failure diagnosed by clinical criteria. Circulation 77:607, 1988.
8. O'Keefe JH, Zinsmeister AR, Gibbons RJ: Value of normal electrocardiographic findings in predicting resting left ventricular function in patients with chest pain and suspected coronary artery disease. Am J Med 86:658, 1989.
9. Allison JD, Flickinger FW, Wright CJ, et al: Measurement of left ventricular mass in hypertrophic cardiomyopathy using MRI: Comparison with echo. Magn Reson Imaging 11:329, 1993.
10. Kronik G, Slany J, Mosslacher H: Comparative value of eight M-mode echocardiographic formulas for determining left ventricular stroke volume. Circulation 60:1308, 1979.
11. Teichholz LE, Kreulen T, Herman MV, Gorlin R: Problems in echocardiographic volume determinations: Echocardiographic-angiographic correlations in the presence or absence of asynergy. Am J Cardiol 37:7, 1976.
12. Bellenger NG, Pennell DJ: Comparison of nuclear and non-nuclear techniques for the combined assessment of perfusion and function. In Germano G, Berman DS (eds): Clinical Gated Cardiac SPECT. New York, Futura, 349, pp 3 1999.
13. Bellenger NG, Marcus N, Davies LC, et al: Left ventricular function and mass after orthotopic heart transplantation: A comparison of cardiovascular magnetic resonance with echocardiography. J Heart Lung Transplant 19:444, 2000.
14. Bellenger NG, Burgess M, Ray SG, et al: on behalf of the CHRISTMAS Steering Committee and Investigators: Comparison of left ventricular ejection fraction and volumes in heart failure by two-dimensional echocardiography, radionuclide ventriculography and cardiovascular magnetic resonance: Are they interchangeable? Eur Heart J 21:1387, 2000.
15. Amico AF, Lichtenberg GS, Reisner SA, et al: Superiority of visual versus computerized echocardiography estimation of radionuclide left ventricular ejection fraction. Am Heart J 118:1259, 1989.
16. Chuang ML, Hibberd MG, Salton CJ, et al: Importance of imaging method over imaging modality in noninvasive determination of left ventricular volumes and ejection fraction: Assessment by two- and three-dimensional echocardiography and magnetic resonance imaging. J Am Coll Cardiol 35:477, 2000.
17. Bellenger NG, Francis JM, Davies CL, et al: Establishment and performance of a magnetic resonance cardiac function clinic. J Cardiovasc Magn Reson 2:15, 2000.
18. Gaudron P, Eilles C, Kugler I, Ertl G: Progressive left ventricular dysfunction and remodeling after myocardial infarction. Potential mechanisms and early predictors. Circulation 87:755, 1993.
19. Sharir T, Germano G, Kavanagh PB, et al: Incremental prognostic value of post-stress left ventricular ejection fraction and volume by gated myocardial perfusion single photon emission computed tomography. Circulation 100:1035, 1999.
20. Johnson LL, Verdesca SA, Aude WY, et al: Postischemic stunning can affect left ventricular ejection fraction and regional wall motion on post-stress gated sestamibi. J Am Coll Cardiol 30:1641, 1997.
21. Iskandrian AE, Germano G, van Decker W, et al: Validation of left ventricular volume measurements by gated SPECT 99mTc-labeled sestamibi imaging. J Nucl Cardiol 5:574, 1998.
22. Anagnostopoulos C, Gunning MG, Pennell DJ, et al: Resting regional myocardial motion and thickening assessed by ECG-gated Tc99m-MIBI emission tomography and by magnetic resonance imaging. Eur J Nucl Med 23:909, 1996.
23. Buser PT, Auffermann W, Holt WW, et al: Noninvasive evalu-

ation of global left ventricular function with use of cine nuclear magnetic resonance. J Am Coll Cardiol 13:1294, 1989.

24. Longmore DB, Klipstein RH, Underwood SR, et al: Dimensional accuracy of magnetic resonance in studies of the heart. Lancet 1:1360, 1985.

25. Rehr RB, Malloy CR, Filipchuk NG, et al: Left ventricular volumes measured by MR imaging. Radiology 156:717, 1985.

26. Bellenger, NG, Gatehouse PD, Rajappan K, et al: Left ventricular quantification in heart failure by CMR using prospective respiratory navigator gating: Comparison with breath-hold acquisition. J Magn Reson Imaging 11:411, 2000.

27. Motooka M, Matsuda T, Kida M, et al: Single breath-hold left ventricular volume measurement by 0.3-sec turbo fast low-angle shot MR imaging. AJR Am J Roentgenol 172:1645, 1999.

28. Alley MT, Napel S, Amano Y, et al: Fast 3D cardiac cine MR imaging. J Magn Reson Imaging 9:751, 1999.

29. Fang W, Pereles FS, Bundy J, et al: Evaluating left ventricular function using real-time TrueFISP: A comparison with conventional MR techniques (abstract). J Cardiovasc Magn Reson 1:310, 1999.

30. Zehender M, Kasper W, Kauder E, et al: Right ventricular infarction as an independent predictor of prognosis after acute inferior myocardial infarction. N Engl J Med 328:981, 1993.

31. Rominger MB, Bachmann GF, Pabst W, Rau WS: Right ventricular volumes and ejection fraction with fast cine MR imaging in breath-hold technique: Applicability, normal values from 52 volunteers, and evaluation of 325 adult cardiac patients. J Magn Reson Imaging 10:908, 1999.

32. Lorenz CH, Walker ES, Graham TP, Powers TA: Right ventricular performance and mass by use of cine MRI late after atrial repair of transposition of the great arteries. Circulation 92(suppl II): 233, 1995.

33. Katz J, Milliken MC, Stray-Gunderson J, et al: Estimation of human myocardial mass with MR imaging. Radiology 169:495, 1988.

34. Katz J, Whang J, Boxt LM, Barst RJ: Estimation of right ventricular mass in normal subjects and in patients with primary pulmonary hypertension by nuclear magnetic resonance imaging. J Am Coll Cardiol 21:1475, 1993.

35. Semelka RC, Tomei E, Wagner S, et al: Normal left ventricular dimensions and function: Interstudy reproducibility of measurements with cine MR imaging. Radiology 174:763, 1990.

36. Semelka RC, Tomei E, Wagner S, et al: Interstudy reproducibility of dimensional and functional measurements between cine magnetic resonance imaging studies in the morphologically abnormal left ventricle. Am Heart J 119:1367, 1990.

37. Pattynama PM, Lamb HJ, van der Velde EA, et al: Left ventricular measurements with cine and spin-echo MR imaging: A study of reproducibility with variance component analysis. Radiology 187:261, 1993.

38. Shapiro EP, Rogers WJ, Beyar R, et al: Determination of left ventricular mass by MRI in hearts deformed by acute infarction. Circulation 79:706, 1989.

39. Lorenz CH, Walker ES, Morgan VL, et al: Normal human right and left ventricular mass, systolic function and gender differences by cine magnetic resonance imaging. J Cardiovasc Magn Reson 1:7, 1999.

40. Bogaert JG, Bosmans HT, Rademakers FE, et al: Left ventricular quantification with breath hold MR imaging: Comparison with echocardiography. MAGMA 3:5, 1995.

41. Sakuma H, Fujita N, Foo TKF: Evaluation of left ventricular volume and mass with breath hold cine MR imaging. Radiology 188:377, 1993.

42. Bloomgarden DC, Fayad ZA, Ferrari VA, et al: Global cardiac function using fast breath-hold MRI: Validation of new acquisition and analysis techniques. Magn Reson Med 37:683, 1997.

43. Bellenger NG, Davies LC, Francis JM, et al: Reduction in sample size for studies of remodelling in heart failure by the use of cardiovascular magnetic resonance. J Cardiovasc Magn Reson 2:271, 2000.

44. Bottini PB, Carr AA, Prisant M, et al: Magnetic resonance imaging compared to echocardiography to assess left ventricular mass in the hypertensive patient. Am J Hypertens 8: 221, 1995.

45. Taylor AM, Jhooti P, Wiesmann F, et al: MR navigator-echo monitoring of temporal changes in diaphragm position: Implications for MR coronary angiography. J Magn Reson Imaging 7:629, 1997.

46. Mohiaddin RH, Longmore DB: Functional aspects of cardiovascular magnetic resonance imaging: Techniques and application. Circulation 88:264, 1993.

47. Underwood SR, Klipstein RH, Firmin DN, et al: Magnetic resonance assessment of aortic and mitral regurgitation. Br Heart J 56:455, 1986.

48. Taylor AM, Stables RH, Poole-Wilson PA, Pennell DJ: Definitive clinical assessment of atrial septal defect by magnetic resonance imaging. J Cardiovasc Magn Reson 1:43, 1992.

49. Hundley WG, Li HF, Lange RA, et al: Assessment of left-to-right intracardiac shunting by velocity-encoded, phase-difference magnetic resonance imaging. A comparison with oximetric and indicator dilution techniques. Circulation 91: 2955, 1995.

50. Pennell DJ, Firmin DN, Burger P, et al: Assessment of magnetic resonance velocity mapping of global ventricular function during dobutamine infusion in coronary artery disease. Br Heart J 74:163, 1995.

51. Karwatowski SP, Brecker SJ, Yang GZ, et al: Mitral valve flow measured by cine MR velocity mapping in patients with ischemic heart disease: Comparison with Doppler echocardiography. J Magn Reson Imaging 5:89, 1995.

52. Mohiaddin R, Hasegawa M: Flow pattern in the dilated ischaemic left ventricle studied by MRI in healthy volunteers and in patients with myocardial infarction. J Magn Reson Imaging 5:493, 1995.

53. Karwatowski SP, Mohiaddin RH, Yang GZ, et al: Regional myocardial velocity imaged by magnetic resonance in patients with ischaemic heart disease. Br Heart J 72:332, 1994.

53a. Stuber M, Scheidegger HB, Fischer SE, et al: Alterations in the local myocardial motion pattern in patients suffering from pressure overload due to aortic stenosis. Circulation 100:361, 1999.

54. Rademakers FE, Buchalter MB, Rogers WJ, et al: Dissociation between left ventricular untwisting and filling. Accentuation by catecholamines. Circulation 85:1572, 1992.

55. Baer FM, Voth E, Schneider CA, et al: Comparison of low-dose dobutamine-gradient-echo magnetic resonance imaging and positron emission tomography with [18F]fluorodeoxyglucose in patients with chronic coronary artery disease. A functional and morphological approach to the detection of residual myocardial viability. Circulation 91:1006, 1995.

56. Dendale PA, Franken PR, Waldman GJ, et al: Low-dosage dobutamine magnetic resonance imaging as an alternative to echocardiography in the detection of viable myocardium after acute infarction. Am Heart J 130:134, 1995.

57. Pennell DJ, Underwood SR, Manzara CC, et al: Magnetic resonance imaging during dobutamine stress in coronary artery disease. Am J Cardiol 70:34, 1992.

58. Nagel E, Lehmkuhl HB, Bocksch W, et al: Noninvasive diagnosis of ischemia induced wall motion abnormalities with the use of high dose dobutamine stress MRI. Comparison with dobutamine stress echocardiography. Circulation 99:763, 1999.

59. Yang PC, Kerr AB, Liu AC, et al: New real time interactive magnetic resonance imaging system complements echocardiography. J Am Coll Cardiol 32:2049, 1998.

60. Hundley WG, Hamilton CA, Thaomas MS, et al: Utility of fast cine magnetic resonance imaging and display for the detection of myocardial ischemia in patients not well suited for second harmonic stress echocardiograpy. Circulation 100:1697, 1999.

61. van Rugge FP, van der Wall EE, Spanjersberg SJ, et al: Magnetic resonance imaging during dobutamine stress for detection and localization of coronary artery disease. Quantitative wall motion analysis using a modification of the centerline method. Circulation 1994 90:127, 1994.

62. Sechtem U, Sommerhoff BA, Markiewicz W, et al: Regional

left ventricular wall thickening by magnetic resonance imaging: Evaluation in normal persons and patients with global and regional dysfunction. Am J Cardiol 59:145, 1987.

63. Pflugfelder PW, Sechtem U, White RD, et al: Quantification of regional myocardial function by rapid (cine) MRI. AJR Am J Roentgenol 150:525, 1988.

64. van Rugge FP, van der Wall EE, Spanjersberg SJ, et al: Magnetic resonance imaging during dobutamine stress for detection and localization of coronary artery disease. Quantitative wall motion analysis using a modification of the centerline method. Circulation 90:127, 1994.

65. van Rugge FP, Holman ER, van der Wall EE, et al: Quantitation of global and regional left ventricular function by cine magnetic resonance imaging during dobutamine stress in normal human subjects. Eur Heart J 14:456, 1993.

66. Buser PT, Auffermann W, Holt WW, et al: Non-invasive evaluation of the global left ventricular function using cine MRI. J Am Coll Cardiol 13:1294, 1989.

67. Kramer C, Rogers WJ, Theobald TM, et al: Remote noninfarcted region dysfunction soon after first anterior myocardial infarction. A magnetic resonance tagging study. Circulation 94:660, 1996.

68. Taylor AM, Panting JR, Keegan J, et al: Use of the intravascular contrast agent NC100150 injection in spin echo and gradient echo imaging of the heart. J Magn Reson Imaging 1:23, 1999.

69. Otterstad JE, Froeland G, St John Sutton M, Holme I: Accuracy and reproducibility of biplane two-dimensional echocardiographic measurements of left ventricular dimensions and function. Eur Heart J 18:507, 1997.

CHAPTER 10

Stress CMR—Wall Motion

Dudley J. Pennell

Stress testing has been the noninvasive technique of choice for the diagnosis of coronary artery disease since the introduction of the exercise test by Master and Oppenheimer in 1929.[1] Modern evaluation techniques, however, have advanced considerably since the development of the electrocardiogram (ECG); and there are now several ways of assessing myocardial ischemia including myocardial scintigraphy (single photon and positron emission), echocardiography, electron beam computed tomography (CT), and cardiovascular magnetic resonance (CMR). Although the expense of these tests has risen, there is no doubt that diagnostic accuracy has also improved considerably with the obviation of the need to interpret surface electrical signals whose genesis and variation remain incompletely understood and in which changes mimicking ischemia are commonly caused by diagnoses unrelated to coronary disease. It is not only the diagnosis that has been improved; the assessment of prognosis is vastly better with these techniques than by the ECG. The largest literature on this topic comes from nuclear cardiology,[2, 3] but increasing data are becoming available on stress echocardiography.[4] In this chapter we review the use of CMR for the analysis of stress-induced ischemia as represented by a focal wall motion abnormality. Perfusion stress CMR is addressed in Chapter 11.

EXERCISE STRESS

Reports of the use of dynamic exercise with CMR are limited, and there are none for the diagnosis of coronary artery disease, although nonferromagnetic exercise devices are now available commercially for fitting to the rear of the magnet for more conventional stress imaging (Figure 10–1).[5, 6] In general, supine exercise becomes awkward as the workload increases, which leads to movement artifact. This problem is often exacerbated by hyperventilation and the relatively unpleasant sensation of exercising in a confined environment. Fortunately, some of these issues are now being addressed by the use of real-time imaging techniques,[7] and there may be a resurgence of interest in the imaging of exercise-induced ischemia in the near future. Ultrafast techniques have already been used in anticipation of these developments, but without real-time visualization of the results. Using spiral flow velocity mapping[8] in normal individuals,[6] Mohiaddin imaged aortic flow from a single heartbeat and showed significant increases in mean and peak aortic flow, with a reduction in the time to peak flow (Figure 10–2).[6] Scheidegger used an echo-planar readout during a 10-heartbeat breath-hold in a study of 10 normal subjects to examine wall motion with tagging, and flow measurements in the aorta.[9] Movement artifact became excessive above an exercise level of 65W for multi-heartbeat acquisitions, but it was possible to stress up to 130W with single-heartbeat acquisitions. Oshinski examined five normal individuals and five patients with coronary artery disease using exercise in the magnet with restraints across the hips and shoulder supports to restrict motion.[10] It was possible to achieve 65 percent of

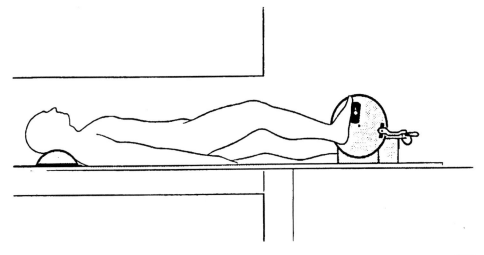

Figure 10–1. Diagrammatic representation of an exercise apparatus used for CMR stress studies made from nonferromagnetic materials. The potential for inducing patient movement during the scan is obvious and can be limited using shoulder and other restraints. (From Mohiaddin RH, Gatehouse PD, Firmin DN: Exercise related changes in aortic flow measured by magnetic resonance spiral echo-planar phase-shift velocity mapping. J Magn Reson Imaging 5:159, 1995. Reprinted by permission of Wiley-Liss, Inc., a subsidiary of John Wiley & Sons, Inc.)

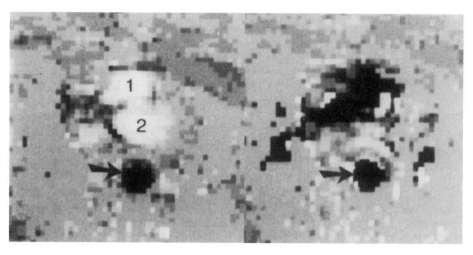

Figure 10–2. Spiral echo-planar flow imaging performed at baseline *(top, left)* and during exercise *(top, right)*. The black arrows indicate the descending thoracic aorta, from where the flow measurements were taken. Flow towards the head is in white, zero velocity gray, and flow towards the feet is in black. The flow graph below shows that the flow and the rate of upstroke of flow velocity both increase during exercise. The left and right ventricular outflow tracts are numbered 1 and 2, respectively and appear black during exercise because of velocity aliasing. (From Mohiaddin RH, Gatehouse PD, Firmin DN: Exercise related changes in aortic flow measured by magnetic resonance spiral echo-planar phase-shift velocity mapping. J Magn Reson Imaging 5:159, 1995. Reprinted by permission of Wiley-Liss, Inc., a subsidiary of John Wiley & Sons, Inc.)

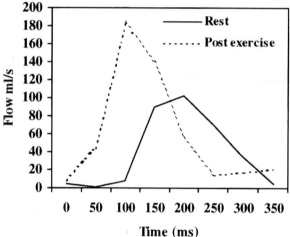

maximum predicted heart rate for age in all subjects, and fast gradient-echo acquisitions were used to obtain six short axis slices at rest and three during stress to estimate the ejection fraction. The stress images were acquired during a 6- to 8-second breath-hold during temporary suspension of exercise. The ejection fraction rose from 55 percent to 62 percent in normal individuals but remained at 44 percent in patients with coronary artery disease; however, scan quality was insufficient for regional wall motion analysis. These ejection fraction responses are similar to those long established by radionuclide ventriculography.[11] For exercise in the magnet to become a clinical reality, however, there is a need for improvement in image resolution, and direct comparison will need to be made with the pharmacological stress techniques that have been highly successful with other imaging modalities.

There are a few reports of the use of magnetic resonance spectroscopy to investigate ischemia. Prone exercise during magnetic resonance spectroscopy has been reported for the investigation of cardiac metabolism during exercise in normal volunteers.[12] Handgrip exercise has been used to diagnose ischemia by examining the ratio of adenosine triphosphate (ATP) to phosphocreatine (PCr). Weiss and coworkers showed that during ischemia there

was a fall in the PCr/ATP ratio that is not present in normal subjects (Figure 10–3) and that this change is abolished by successful revascularization (Figure 10–4).[13] Yabe and colleagues took this concept further and compared the changes seen with the results of thallium imaging.[14] The ratio fell during handgrip exercise in patients with reversible thallium defects but not in fixed defects (Figure 10–5). This finding confirms clinically the association with ischemia and suggests that with further development, the assessment of stress-induced metabolic changes might play a useful role in the research and clinical assessment of coronary disease.

PHARMACOLOGICAL STRESS

The most suitable alternative to exercise for stress in the magnet is the use of pharmacological stress. The use of these agents is well documented in the nuclear cardiology literature,[15] and their use is widespread.[16] They are particularly valuable for the one third of patients who fail to exercise maximally,[17] in whom the sensitivity of detecting coronary disease may be markedly diminished.[18] There are a number of agents that are licensed for use including the vasodilators (dipyridamole and adenosine), and beta-agonists (dobutamine and arbutamine), although the regulatory situation varies around the world. For the first three agents there is considerable clinical validation and experience from other imaging modalities that can be drawn upon for CMR. Arbutamine has not been used for CMR and will not be considered further here inasmuch as its mode of action is similar to dobutamine but with a longer half-life (8 minutes), and it is supplied with a closed-loop computer-controlled delivery system that is not ideally suited for operation in the magnetic field.[19]

Vasodilators

Dipyridamole and adenosine exert their action through a common final pathway. Dipyridamole raises the interstitial levels of adenosine by inhibiting adenosine breakdown via the adenosine deaminase pathway, and it inhibits the facilitated uptake

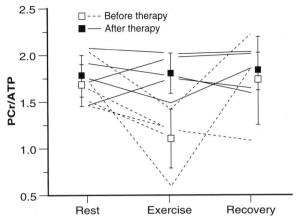

Figure 10–4. Five of the patients shown in Figure 10–3 underwent revascularization for their coronary artery disease, and this graph shows the preoperative and postoperative differences in response to handgrip stress. The PCr/ATP ratio during stress after revascularization is normalized compared with before, confirming that the observed metabolic abnormality results from ischemia, which is reversible. (From Weiss RG, Bottomley PA, Hardy CJ, Gerstenblith G: Regional myocardial metabolism of high energy phosphates during isometric exercise in patients with coronary artery disease. N Engl J Med 323:1593, 1990. Copyright © 1990 Massachusetts Medical Society. All rights reserved.)

of adenosine into cells for incorporation into DNA (Figure 10–6). The vasodilator dipyridamole has been used extensively and in large series has been shown to be very safe.[20] Dipyridamole and adenosine, however, have very different molecular structures (Figure 10–7). Dipyridamole is simple to administer in a 4-minute infusion of 0.56 mg/kg and causes an increase in human coronary flow velocity of up to six times baseline.[21] In the presence of significant coronary stenoses, myocardial flow heterogeneity is induced, which can be readily detected by perfusion techniques. However, reduction in perfusion pressure distal to a stenosis, reduction in collateral flow, and redistribution of flow from subendocardium to subepicardium can cause ischemia and wall motion abnormalities (Figure 10–8),[22] which can be detected by CMR. The main clinical problem with dipyridamole is its long $t_{1/2}$ of effect, which is 30 minutes,[23] which results in prolonged side effects and the relatively frequent need to give aminophylline as an adenosine receptor antagonist for their relief. Adenosine may be given directly at a dose at 140 μg/kg/min, and it causes similar changes in coronary flow.[24] Its side effects are shorter lived because of its $t_{1/2}$ of only 4 seconds, but it is less effective at provoking wall motion abnormalities.[25, 26] It is therefore reserved for perfusion studies for both nuclear cardiology[27] and CMR.[28]

Beta-Agonists

Of the beta-agonists that are available (see Figure 10–7), most have been used at some time for cardiac imaging, but only dobutamine has a combination of

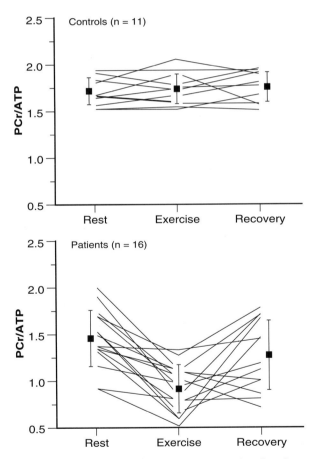

Figure 10–3. Changes during handgrip stress in the phosphocreatine/adenosine triphosphate (PCr/ATP)-ratio in controls (upper panel) and patients with coronary artery disease (lower panel). There is no change in the ratio with the controls, but a significant fall in the ratio with the patients, which recovers after cessation of exercise. (From Weiss RG, Bottomley PA, Hardy CJ, Gerstenblith G: Regional myocardial metabolism of high energy phosphates during isometric exercise in patients with coronary artery disease. N Engl J Med 323:1593, 1990. Copyright © 1990 Massachusetts Medical Society. All rights reserved.)

A

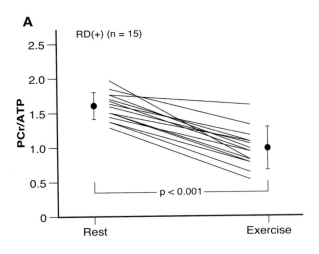

B

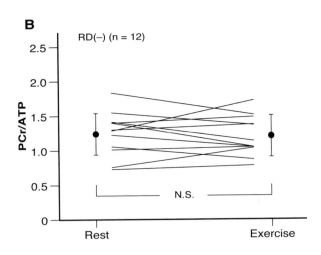

C

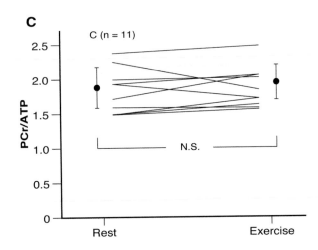

Figure 10–5. The PCr/ATP ratio falls during exercise in patients with reversible ischemia (*A*, RD+), but not in patients with fixed thallium defects (*B*, RD−, which represents infarction without ischemia), or normal controls (*C*, C). Note that the resting level of PCr/ATP is significantly lower in the patients with reversible and fixed thallium defects than controls. N.S. = not significant. (From Yabe T, Mitsunami K, Okada M, et al: Detection of myocardial ischemia by ^{31}P magnetic resonance spectroscopy during handgrip exercise. Circulation 89:1709, 1994, with permission.)

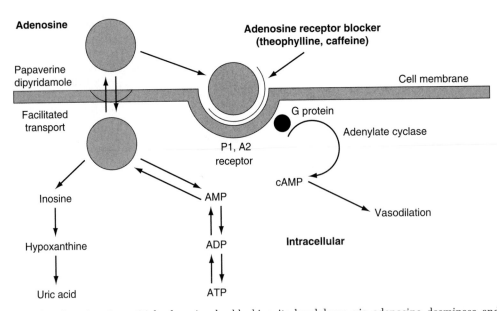

Figure 10–6. Dipyridamole raises interstitial adenosine by blocking its breakdown via adenosine deaminase and inhibiting the facilitated transport mechanism into cells. The increased level of interstitial adenosine acts on the A2 receptor to cause vasodilation. This binding is competitively antagonized by caffeine and drugs such as aminophylline. Adenosine can be given directly, and under these circumstances the effect is greatly potentiated by concomitant dipyridamole treatment. The electrophysiological effects of adenosine are mediated by the A1 receptor and include slowed atrioventricular conduction and hyperpolarization of atrial cells. AMP = adenosine monophosphate; ADP = adenosine diphosphate; ATP = adenosine triphosphate; cAMP = cyclic adenosine monophosphate.

properties suitable for widespread use. These properties are that it produces hemodynamic effects that are similar to exercise, it has a low arrhythmogenicity, and it has a good tolerance to peripheral vein infusion. Other agents such as isoproterenol have been used with CMR for stress testing in animals.[29] Like dipyridamole, dobutamine may be used for both perfusion[30, 31] and wall motion imaging.[32] It is infused in doses up to 40 μg/kg/min, usually in 3-minute increments of 10 μg/kg/min. The endpoints of the stress test include symptoms, significant hypotension, and arrhythmias. In patients with a poor heart rate response to dobutamine due to high parasympathetic drive or the use of beta blockers, additional atropine has been shown to be helpful,[33–35] and is given in divided doses of 0.25 mg up to 2 mg. The risk of atropine intoxication, however, must be borne in mind when it is used.[36] The safety of dobutamine stress has been demonstrated to be good,[37–39] although vasodilator stress probably has a slightly lower level of complications.[20, 40] Dobutamine exerts its effects by increasing myocardial oxygen demand above availability in the setting of acute ischemia.[41] It also increases coronary flow[42, 43] and lowers perfusion pressure distal to coronary stenoses causing heterogeneous myocardial perfusion,[44]

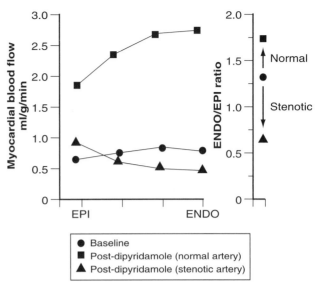

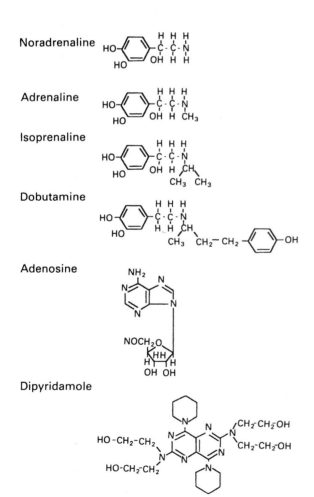

Figure 10–7. The structural formulas of the commonly available beta-agonists, adenosine, and dipyridamole. Dobutamine is synthetic and has the most complex molecule of the beta-agonists.

Figure 10–8. Changes in transmural blood flow with dipyridamole measured in dogs with microspheres. At baseline (circles) the transmural flow is even. After dipyridamole, the myocardium supplied by a normal coronary artery (squares) shows a significant flow increase up to 5 times with preferential flow to the subendocardium (ENDO). In the territory supplied by a stenosed coronary artery (triangles), the flow response is muted with preferential flow to the subepicardium (EPI) with an absolute decrease in flow to the subendocardium. Therefore the ENDO/EPI flow ratio falls. This is termed transmural steal and may result in ischemia. (Data from Mays AE, Cobb FR: Relationship between regional myocardial blood flow and thallium-201 distribution in the presence of coronary artery stenosis and dipyridamole induced vasodilation. J Clin Invest 73:1359, 1984.)

which may also be redirected to the subepicardium.[45] Dobutamine may also increase flow resistance at the site of a stenosis.[45] Both dipyridamole and dobutamine have been used for stress wall motion imaging by CMR. Dobutamine, however, has a number of advantages in the magnet, including operator-controlled level of stress, a short $t_{1/2}$ of 120 s, physiological effects mimicking exercise more closely than dipyridamole, and stress-induced tachycardia, which considerably shortens the stress imaging period when conventional CMR techniques are used.

Choice of Pharmacological Agent

It is likely that the role of pharmacological vasodilatation and stress in the magnet will grow for the functional assessment of heart disease. This will require familiarity of the strengths of each technique and the important contraindications.[15, 46] The vasodilators should not be administered to asthmatics because of the risk of provocation of severe bronchospasm.[47, 48] Adenosine infusion in sinoatrial disease may lead to sinus arrest, and should also be avoided.[49] Caffeine is a competitive antagonist of adenosine and should be avoided for at least 12 hours prior to scanning or attenuated vasodilatation will occur. The contraindications to dobutamine are the same as for dynamic exercise. Dobutamine is competitively antagonized by beta blockers, and in some centers these drugs are routinely stopped prior to diagnostic imaging. However, this practice may occasionally lead to withdrawal angina, and the patients need to be warned of this possibility. Other centers take the approach that atropine can be used to overcome the problem at the time of the stress test. Both approaches have their advocates, and there is no clear consensus, which probably indicates that both approaches are reasonable. However, it is in practice much easier to leave patients on their treatment. The vasodilators remain first-choice agents for studies of myocardial perfusion and coronary flow because dobutamine causes less coronary hyperemia, and the induced tachycardia causes problems with temporal resolution. Dobutamine is better suited to wall motion and global ventricular studies, where myocardial ischemia is more reliably provoked by increased myocardial oxygen demand.

In this chapter, we consider the role of pharmacological stress for regional and global wall motion studies, and myocardial perfusion. Coronary flow and myocardial perfusion studies during stress are described in Chapters 11 and 19.

PHARMACOLOGICAL STRESS WALL MOTION STUDIES

Dipyridamole Cardiovascular Magnetic Resonance

Pennell first reported the induction of stress regional wall motion abnormality in coronary artery disease using CMR with dipyridamole.[50] Dipyridamole was infused at a dose of 0.56 mg/kg with a 10-mg bolus after 10 minutes to ensure continuing effect during the conventional gradient echo cine imaging at 0.5 T in the vertical and horizontal long axes and two short axis planes. The technique was subsequently performed in a series of 40 patients. The sensitivity for a new wall motion abnormality was found to be only 67 percent compared with areas of reversible ischemia assessed by thallium tomography, despite 23 of the patients having had previous infarction (Figure 10–9).[51] The sensitivity was 62 percent for detection of significant coronary artery disease as defined by coronary angiography. Further analysis showed that there was a particular inability to detect smaller areas of ischemia (Figure

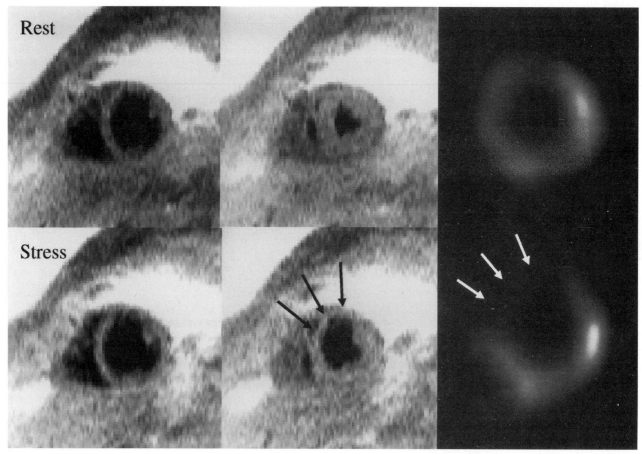

Figure 10–9. Dipyridamole CMR in a patient with left anterior descending artery disease. The images are shown in reverse video format so that the blood is dark. In the top row are short axis images before dipyridamole with postdipyridamole images below. End-diastole is in the left column and end-systole in the middle column. Left ventricular contraction is normal prior to vasodilation, but reduced in the anteroseptal region after dipyridamole (black arrows). The CMR abnormality is closely matched by the perfusion defect (white arrows on the color maps in the right column; see color plates) seen during dipyridamole thallium myocardial perfusion tomography, which shows full reversibility. (From Pennell DJ, Underwood SR, Ell PJ, et al: Dipyridamole magnetic resonance imaging: A comparison with thallium-201 emission tomography. Br Heart J 64:362, 1990, with permission from the BMJ Publishing Group. Copyright © 1990 BMJ Publishing Group.)

10–10). The procedure was well tolerated, but side effects from the dipyridamole, both cardiac and noncardiac, were common, and imaging time was prolonged with this conventional cine technique, taking up to 20 minutes after the dipyridamole infusion. A small but significant 4 percent signal reduction was seen in ischemic myocardium, which was visually appreciable in 38 percent of ischemic segments. When seen visually it occurred predominantly in the subendocardium (Figure 10–11). The signal changes were not explicable by changes in relaxation times with ischemia because these changes occur over a longer time frame, nor was it related to hypokinesis, which would be expected to increase the myocardial signal, and the likeliest explanation was thought to be a relatively lower myocardial blood content due to shunting.[52] Later studies have shown that dipyridamole infusion causes an increase in T2*, which would be expected to increase signal intensity.[53, 54] This effect is thought to be due to reduced myocardial venous deoxyhemoglobin concentration, which occurs as myocardial oxygen supply exceeds demand, which reduces effects from susceptibility. Interestingly, dobutamine failed to change the T2*, suggesting that increased oxygen supply was matched to increased demand.[54] Further work is required to determine whether more than one effect is operating, but it would appear that imaging with T2* weighting can be used to assess the balance of myocardial oxygen supply and demand. (See also Chapter 34.)

Other studies have shown wall motion abnormalities induced by dipyridamole, although with varying levels of sensitivity. Casolo studied 10 patients at 0.5 T, of whom 7 had had previous infarction, and infused 0.7 mg/kg dipyridamole over 5 minutes with comparison of wall motion changes with Tc[99m]-MIBI perfusion tomography and angiography.[55] Cine gradient echo imaging with a single midventricular short axis slice was used, and the sensitivity of detection of disease compared with both MIBI scanning and angiography was 100 percent. Baer studied 23 patients with no resting wall motion abnormality at 1.5 T,[56] allowing a more confident estimation of sensitivity of detection of ischemia in individual arterial territories. Two midventricular short axis slices were imaged during a dose of 0.75 mg/kg dipyridamole over 10 minutes. The overall detection rate of coronary artery disease was 78 percent compared with angiography, and the sensitivity for one- and two-vessel disease was 69 percent and 90 percent. Although the higher sensitivity of these results might be explained by the higher dose of dipyridamole used, the side effects proved to be problematic, which may have been exacerbated by the mild sedation with diazepam that was used, which potentiates the action of dipyridamole.[57] In a further study of 33 patients, a sensitivity of 84 percent for detection of coronary artery disease was shown, with agreement between MIBI single photon emission computed tomography (SPECT) and CMR of 90 percent between segments for abnormality.[58] Interestingly, the specificity for dipyridamole CMR was marginally better than MIBI SPECT in the inferior wall abnormalities associated with right coronary artery stenosis (89% vs. 80%), which might reflect the known problems with inferior attenuation in nuclear techniques.[59] Bremerich compared 12 patients using both dipyridamole (0.56 mg/kg) wall motion and perfusion CMR with MIBI SPECT, and showed a 93 percent concordance with scintigraphy but only a 67 percent sensitivity for detecting coronary artery stenosis.[60] His results improved when the results from the perfusion CMR were combined with the wall motion assessment.

Dipyridamole has also been used to investigate coronary disease in other ways other than with conventional cine imaging, and in comparison with other techniques. A comparison between CMR and transesophageal echo (TEE) has been reported in 35 patients, of whom 29 had coronary artery disease.[61] All patients had been shown to have poor echo windows for transthoracic echocardiography and a nondiagnostic exercise ECG. The sensitivity for detection of disease was 90 percent for TEE and 83 percent for CMR (p = ns), with specificity of 100 percent for both techniques. There was significant correlation between the techniques in a quantitative analysis for wall thickening, but there were a significantly higher number of abnormal chords by CMR than TEE. In addition, 75 percent of patients indicated that they preferred the CMR to the TEE. These results suggest a benefit from the higher spatial resolution of CMR combined with its greater patient acceptability. Zhao attempted to quantify di-

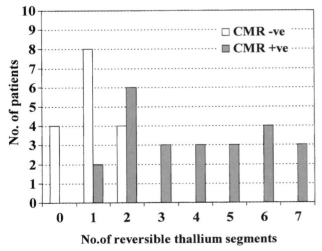

Figure 10–10. The likelihood of development of abnormal wall motion during dipyridamole CMR was strongly influenced by the extent of ischemia shown on the dipyridamole thallium scan. Smaller defects often did not have accompanying wall motion abnormalities that were detectable by CMR (maximum number of thallium segments was 9). +ve = abnormal; −ve = normal. (From Pennell DJ, Underwood SR, Ell PJ, et al: Dipyridamole magnetic resonance imaging: A comparison with thallium-201 emission tomography. Br Heart J 64:362, 1990, with permission from the BMJ Publishing Group. Copyright © 1990 BMJ Publishing Group.)

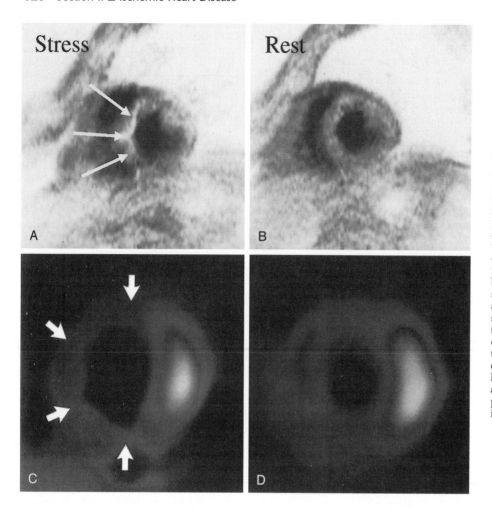

Figure 10–11. Subendocardial signal change during dipyridamole CMR seen in a patient with extensive septal ischemia (white arrows). The CMR images are at end-diastole after dipyridamole *(A)* and before dipyridamole *(B)*. The gradient echo cine frames are displayed in reverse video format for easier appreciation of the myocardial signal, and the white subendocardial line therefore represents reduced signal on the original image. The reason for this finding remains unclear but might result from transmural shunting. The equivalent stress *(C)* and redistribution *(D)* thallium scans are shown on the bottom row (see color plates), with the severe defect marked by white arrows on the stress image. The defect shows clear reversibility, which would have probably been greater with imaging at rest or with reinjection. The patient had no evidence of septal infarction.

pyridamole-induced wall motion changes imaged using a breath-hold cine technique in 16 patients without myocardial infarction.[62] Using receiver operating characteristic (ROC) curve analysis, superior results were shown for a quantitative analysis using percent wall thickening as the test parameter. Comparison with thallium myocardial perfusion imaging showed equivalent results that were insensitive but specific (thallium 69%/80% and CMR 80%/75%). The studies are summarized in Table 10–1.

Dobutamine Cardiovascular Magnetic Resonance

Dobutamine CMR clinical studies are summarized in Table 10–2. Pennell reported the first clinical use of dobutamine CMR,[63] which showed a considerable improvement in sensitivity, when compared with the results with dipyridamole in similar patients.[51] Conventional gradient echo cine imaging was performed in 25 patients in the vertical and horizontal long axes and two short axis planes during dobutamine infusion up to 20 μg/kg/min. The increase in heart rate had the advantage of shortening the imaging time during stress to 10 to 15 minutes. Of the 22 patients with significant coronary artery disease, 21 had reversible ischemia identified by dobutamine thallium tomography, and of these 20 (95%) had reversible myocardial wall motion abnormalities (Figures 10–12 and 10–13). This represented a sensitivity for detection of significant coronary artery disease of 91 percent. There was a close

Table 10–1. Sensitivity Data for Dipyridamole Stress CMR in Detection of Coronary Artery Disease

Author	Year	Dose (mg/kg)	No. of Patients	Sensitivity (%)
Pennell[51]	1990	0.56	40	67
Casolo[55]	1991	0.7	10	100
Baer[56]	1992	0.75	23	78
Baer[58]	1993	0.75	33	84
Zhao[62]	1997	—	16	79
Bremerich[60]	1997	0.56	19	80
Summary			132	79

Table 10–2. Sensitivity and Specificity Analysis for Dobutamine Stress CMR in Detection of Coronary Artery Disease (CAD)

Author	Year	Dose (μg/kg/min)	No. of Patients (CAD/normal)	Sensitivity (%)	Specificity (%)
Pennell[63]	1992	20	25 (22/3)	91	—
van Rugge[65]	1993	20	45 (37/8)	81	100
van Rugge[66]	1994	20	39 (33/6)	91	80
Baer[67]	1994	20	28 (28/0)	85	—
Baer[68]	1994	20	35 (35/0)	84	—
Nagel[69]	1999	40 + atropine	172 (109/63)	86	86
Hundley[69a]	1999	40 + atropine	41 (35/6)	83	83
Summary			**385**	**86**	**84**

concordance in site and extent of the perfusion and wall motion abnormalities, with 96 percent agreement at rest, 90 percent during stress, and a 91 percent agreement for the assessment of reversible ischemia (Figure 10–14). There were no significant differences between CMR and thallium in the detection or location of coronary stenoses, but determination of specificity was not reliable with the small patient numbers. A small (9.2%) reduction in signal was found in the ischemic segments. Areas of signal reduction were seen in half of the patients with a new wall motion abnormality, but occasional areas of reduced signal were seen in nonischemic segments, and therefore the specificity of signal reduction for ischemia was reduced. Overall, the dobutamine was well tolerated in the magnet, but cardiac and noncardiac side effects were common. Images contained more artifacts as the heart rate increased. Intraventricular turbulence during ejection occurred, causing intense signal loss in the myocar-

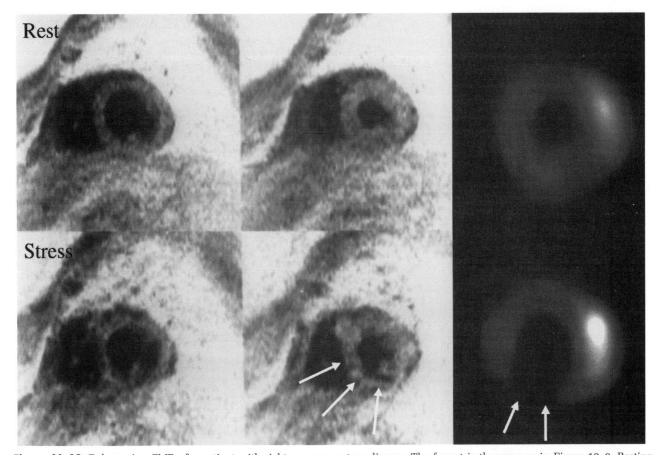

Figure 10–12. Dobutamine CMR of a patient with right coronary artery disease. The format is the same as in Figure 10–9. Resting contraction is normal but during dobutamine contraction is abnormal in the inferoseptal wall (long arrows), which matches the reversible perfusion defect (short arrows) seen during dobutamine thallium myocardial perfusion tomography (see color plates). (From Pennell DJ, Underwood SR, Manzara CC, et al: Magnetic resonance imaging during dobutamine stress in coronary artery disease. Am J Cardiol 70:34, 1992, with permission from Excerpta Medica Inc.)

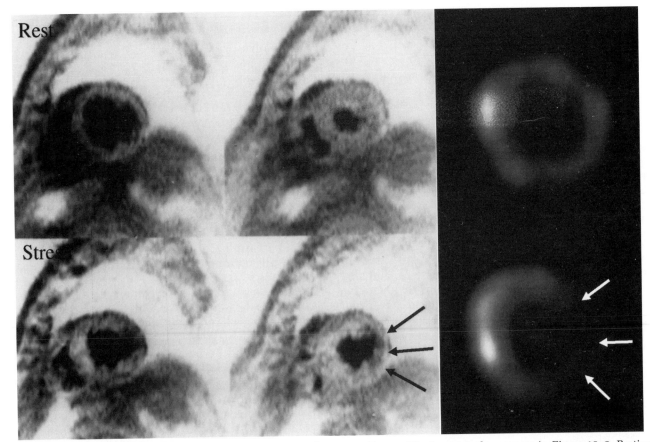

Figure 10–13. Dobutamine CMR of a patient with left circumflex artery disease. The format is the same as in Figure 10–9. Resting contraction is normal but during dobutamine contraction is abnormal in the lateral wall (black arrows), which matches the reversible perfusion defect (white arrows) seen during dobutamine thallium myocardial perfusion tomography (see color plates). (From Pennell DJ, Underwood SR, Manzara CC, et al: Magnetic resonance imaging during dobutamine stress in coronary artery disease. Am J Cardiol 70:34, 1992, with permission from Excerpta Medica Inc.)

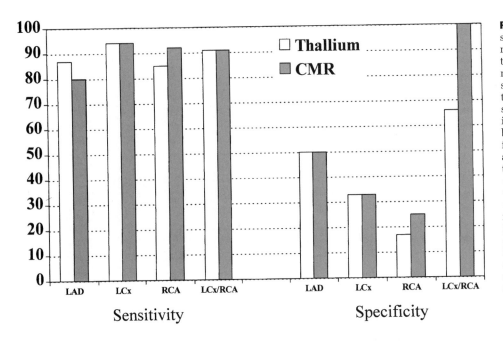

Figure 10–14. Comparison of sensitivity of detection of coronary artery disease by thallium tomography with CMR of wall motion during dobutamine stress. The specificity values in this study were low as a result of small patient numbers and the inclusion of patients with left bundle branch block. CMR performed as well as thallium imaging in the study. LAD = left anterior descending artery; LCx = left circumflex artery; RCA = right coronary artery LCx/RCA = combined left circumflex and right coronary artery territories. (Data from Pennell DJ, Underwood SR, Manzara CC, et al: Magnetic resonance imaging during dobutamine stress in coronary artery disease. Am J Cardiol 70:34, 1992.)

dium, and obliteration of the cavity in the apical short axis view occurred. A lower echo time was helpful, but some of the stress cines were of low quality.

van Rugge reported findings in normal individuals and subsequently in patients with coronary artery disease. In 23 normal subjects, dobutamine was given to a maximum dose of 15 μg/kg/min.[64] Imaging was performed at 1.5 T using conventional gradient echo cines with eight short axis slices covering the left ventricle from base to apex. Normal ranges were established using a quantitative analysis for global ventricular function and regional wall thickening during rest and stress in the true short axis plane. By recording wall thickening in 20 segments around each short axis slice, the investigators could display the thickening graphically to show regional variation, with comparison of rest and dobutamine stress. van Rugge then reported on patients with coronary artery disease, using both qualitative[65] and quantitative[66] analyses. In the qualitative study of 45 patients, 37 patients had coronary artery disease and 30 (81%) showed a wall motion abnormality with dobutamine stress. The specificity in this series was 100 percent. The results were better than exercise electrocardiography (70/63%) or dobutamine electrocardiography (51/63%). Single-, double-, and triple-vessel disease were detected with 75 percent, 80 percent, and 100 percent sensitivity, respectively. In the quantitative study, 39 patients without rest wall motion abnormality and 10 normal volunteers were stressed with dobutamine to 20 μg/kg/min. A short axis stress cine that was judged to show wall motion abnormality was analyzed with a modified centerline method. In these preselected cuts, stress wall motion was considered abnormal if 4 or more adjacent chords (100 encompassed the ventricle) showed systolic wall thickening below 2 SD of that obtained from the normal subjects. This method resulted in a 91 percent sensitivity and an 80 percent specificity for detection of disease. Single-, double-, and triple-vessel disease was detected with 88 percent, 91 percent, and 100 percent sensitivity, respectively, whereas the sensitivity of detection of individual coronary artery stenosis was 75 percent, 87 percent, and 63 percent for the left anterior descending, right coronary, and left circumflex arteries, respectively. No direct comparison of the quantitative and qualitative methods was given to determine in the same patients if the time-consuming task of endocardial and epicardial tracing yielded a significant diagnostic improvement. However, the advantages of a quantitative analysis in the reduction of observer variability are clear, but it is not clear whether the quantification is significantly disturbed by the presence of resting wall motion abnormalities. In addition, quantification approaches with tagging as described later may prove easier to implement.

Baer studied 28 patients with coronary artery disease but without previous infarction,[67] at 1.5 T, with a peak dobutamine dose of 20 μg/kg/min. Although the peak rate pressure product was lower during dobutamine CMR than exercise electrocardiography, the relative sensitivities were 85 percent and 77 percent. Single-vessel and multivessel disease was detected with 73 percent and 100 percent sensitivity respectively. The sensitivity and specificity for detection of disease in the individual arteries was 87 percent and 100 percent in the left anterior descending, 78 percent and 88 percent for the right coronary, and 62 percent and 93 percent for the left circumflex arteries.

In another study by Baer of 35 consecutive patients with coronary artery disease, comparison was made between dobutamine CMR and MIBI SPECT.[68] The sensitivity of CMR and SPECT was 84 percent and 87 percent, respectively for the detection of disease. Comparison of the detection of individual artery ischemia was very similar between the techniques, indicating that CMR could be successfully used in the assessment of disease in comparison with a well-established clinical standard.

Finally, a seminal paper comparing stress CMR and stress echocardiography has been published by Nagel, which contains the largest comparative group examined to date of 208 consecutive patients.[69] All patients were examined by echocardiography with harmonic imaging using the latest technology, and a state of the art CMR scanner using breath-hold gradient echo cines. The sensitivity for CMR was 86 percent compared with 74 percent by echo, and the specificity for CMR was 86 percent compared with 70 percent by echo (both comparisons $p < 0.05$, Figure 10–15). For both tests 18 patients could not be examined. For echo, the main reason was poor image quality, and for CMR, claustrophobia and obesity were the main problems. A subgroup analysis of patients with reduced echo image quality showed that CMR was particularly advantageous at defining the presence of coronary disease (Figure 10–16) in this population. The image quality of CMR was therefore demonstrated to be a major issue in the confidence of interpretation of stress testing in general, which is where CMR is strong. This issue was further validated by Hundley,[69a] who performed dobutamine CMR among 153 patients with poor echo image quality, demonstrating the high sensitivity and specificity of dobutamine stress CMR in this group. The issue of problems of interpretation variability is well known to echocardiographers and is strongly influenced by image quality.[70]

The issue of quantification of in vivo wall motion has plagued cardiology for many years, and CMR offers one possible approach that remains under investigation at present—myocardial tagging during stress (see Chapter 4). Although little work has been reported using dobutamine with CMR tagging for the diagnosis of coronary artery disease, the pattern of normal left ventricular response during breath-hold cines has been described. Mean circumferential shortening has been shown to increase from 21 percent at baseline to 26 percent at 10 μg/kg/min before significant increases in blood pressure or

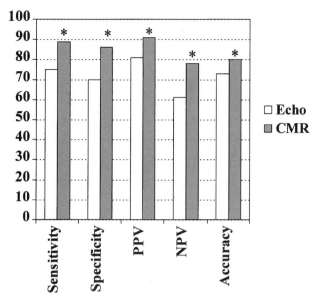

Figure 10–15. Results of stress CMR versus stress echocardiography (echo) in a large comparative study. The diagnostic values are on a patient basis, not by artery or territory. The data suggest significant diagnostic superiority of CMR. PPV = positive predictive value; NPV = negative predictive value; *p < 0.05. (Data from Nagel E, Lehmkuhl HB, Bocksch W, et al: Noninvasive diagnosis of ischemia induced wall motion abnormalities with the use of high dose dobutamine stress MRI. Comparison with dobutamine stress echocardiography. Circulation 99:763, 1999, with permission.)

heart rate, and not to increase with higher doses.[71] Circumferential shortening velocity, however, increased linearly with dobutamine dose from 4.4 mm/s to 9.8 mm/s at 20 µg/kg/min. The latter measurement may therefore be the more appropriate parameter to use for examining for regional contractile dysfunction (Figure 10–17). Direct measurements of the principal strains (E1 and E2) in the left ventricle have also shown significant changes with dobuta-

mine (see Figure 10–17), and analysis of variations in contraction around the heart known to exist at rest have been shown with this technique not to be exacerbated with dobutamine.[72] Another parameter that has been examined is torsion of the left ventricle. Buchalter studied the effect of ischemia on the canine heart and showed that dobutamine increased torsion, but that regional ischemia not only reduced torsion locally but could also affect torsion in remote myocardium due to the complex fiber arrangement of the ventricle.[73] It has also been suggested that untwisting could be used as a marker for dysfunction.[74]

Thus, groups who have studied wall motion during dobutamine stress now report good results with excellent patient tolerance, and the main problem of the duration of the study has been solved with the use of breath-hold CMR. Comparison with thallium imaging and stress echocardiography shows excellent correlation with the former, and significant improvement over the latter. In addition, for the future, the combination of perfusion and wall motion in the same dobutamine stress CMR study has also been evaluated with good results, which may prove valuable especially for hibernation studies.[75, 76] We also await with great interest the possible role of quantification of myocardial contraction as a further adjunct to dobutamine stress CMR.

OTHER STRESS CARDIOVASCULAR MAGNETIC RESONANCE TECHNIQUES TO ASSESS ISCHEMIA

Myocardial Velocity Imaging

The steady-state hemodynamics generated by infusion of dobutamine allows imaging of other as-

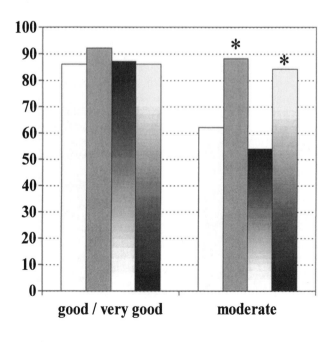

□ sensitivity (Echo)

▨ sensitivity (CMR)

▤ specificity (Echo)

▨ specificity (CMR)

* p<0.01

Figure 10–16. Comparison between stress echocardiography and stress CMR grouped according to the quality of the stress echocardiography. The CMR showed particular diagnostic superiority when stress echocardiography imaging was suboptimal. (Data from Nagel E, Lehmkuhl HB, Klein C, et al: Influence of image quality on the diagnostic accuracy of dobutamine stress magnetic resonance imaging in comparison with dobutamine stress echocardiography for the noninvasive detection of myocardial ischemia. Z Kardiol 88:622, 1999, with permission.)

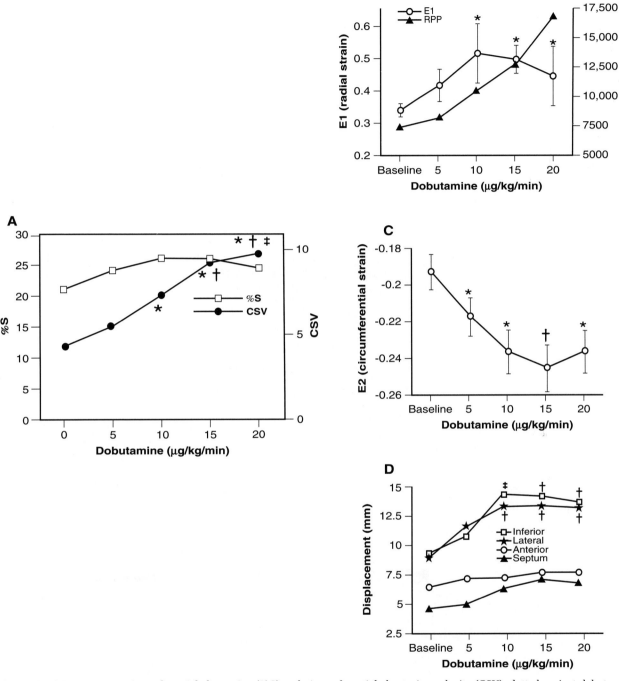

Figure 10–17. *A*, Percent circumferential shortening (%S) and circumferential shortening velocity (CSV) plotted against dobutamine dose. %S plateaus at 10 μg/kg/min, but CSV shows a relatively linear change with dose. This finding suggests that it might be a useful parameter for quantifying contractile function. (*$p < 0.002$; †$p < 0.001$; ‡$p < 0.001$) *B–D*, Data showing the changes of E1 (radial strain or an approximation to wall thickening) and E2 (circumferential strain or an approximation to circumferential shortening) with increasing doses of dobutamine. The displacement values can be seen to be less useful for measuring myocardial contractility. (*$p < 0.01$; †$p < 0.03$; ‡$p < 0.02$) RPP = rate pressure product. (*A*, from Power TP, Kramer CM, Shaffer AL, et al: Breath-hold dobutamine magnetic resonance myocardial tagging: Normal left ventricular response. Am J Cardiol 80:1203, 1997, with permission from Excerpta Medica Inc.; *B–D*, from Scott CH, St. John Sutton MG, Gusani N, et al: Effect of dobutamine on regional left ventricular function measured by tagged magnetic resonance imaging in normal subjects. Am J Cardiol 83:412, 1999, with permission from Excerpta Medica Inc.)

pects of cardiac function during stress using conventional imaging techniques with longer image acquisition times. Karwatowski has studied ventricular long axis motion before and after dobutamine stress in normal subjects[77] and in patients with coronary artery disease.[78] Long axis motion of the left ventricle is thought to be a particularly sensitive indicator of contractile dysfunction because the myocardial fibers in the subendocardium are aligned longitudinally and the subendocardium is the first portion of the myocardium to be affected by reduced perfusion.[79] The technique uses velocity mapping of the myocardium in the short axis plane just below the mitral annulus, with through-plane velocity sensitization (0.3–0.5m/s) to measure the long axis velocities (Figure 10–18). Karwatowski and colleagues defined normal long axis dynamics in 31 normal subjects.[77] The peak velocity of long axis motion always occurred in early diastole, and

significant heterogeneity occurred around the ventricular wall with greatest velocities in the lateral wall. A mean figure for long axis velocity was generated by considering the myocardial slice as a whole, or regional velocities were calculated by dividing the slice into 16 segments. These figures can be graphically displayed (Figure 10–19).

Following this study of normal subjects, Karwatowski and colleagues studied 9 normal subjects and 25 patients with coronary artery disease before and during dobutamine stress.[78] The study concentrated on diastolic function, because highest long axis velocities occur at this time, and abnormalities of left ventricular function during ischemia may occur first in diastole.[80] Diastolic function was assessed by measuring the time to peak early diastolic velocity, and at this time point the mean velocity of the myocardium, and the maximum and minimum regional segmental velocity. The time to peak diastolic

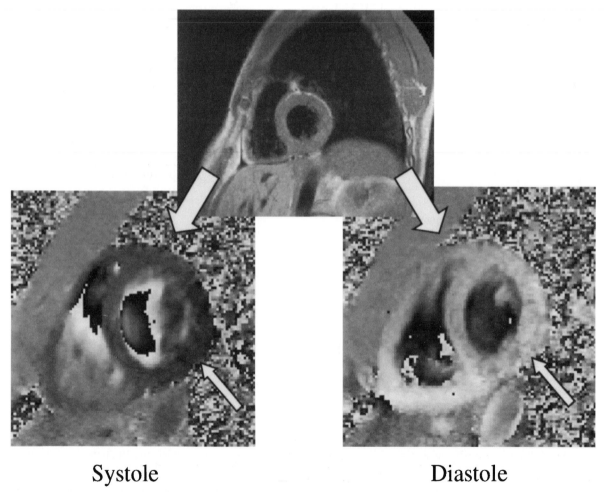

Systole   Diastole

Figure 10–18. Figure showing a short axis spin-echo image and myocardial velocity maps in the same plane in early systole *(left)* and early diastole *(right)*. Mid gray represents stationary velocity. The myocardium is relatively thin in early systole and in a dark shade (small arrow), representing increased velocities towards the apex. In early diastole, the myocardium is thicker with white shading of the myocardium (small arrow), showing recoil towards the base. Aliasing of the ventricular blood velocities occurs because the velocity window is set very low for the myocardium. Regional variations in velocity may be determined from segments around the myocardium. (From Karwatowski SP, Mohiaddin RH, Yang GZ, et al: Noninvasive assessment of regional left ventricular long axis motion using magnetic resonance velocity mapping in normal subjects. J Magn Reson Imaging 4:151, 1994. Reprinted by permission of Wiley-Liss, Inc., a subsidiary of John Wiley & Sons, Inc.)

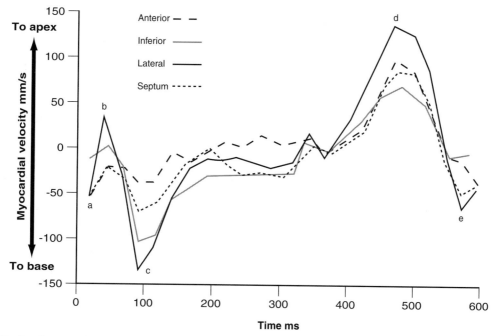

Figure 10–19. Regional myocardial velocity can be represented in graphical form to illustrate quantitative and regional abnormalities. Point a represents displacement of the heart with shape change; point b, isovolumic contraction prior to the descent of the base towards the apex in systole, point c, peak systolic long axis velocity followed by rapid motion of the base away from the apex; point d, peak early diastolic velocity; and point e, brief period of movement back towards the base. Measurements of peak diastolic timing of velocity are made at point d. (From Karwatowski SP, Mohiaddin RH, Yang GZ, et al: Noninvasive assessment of regional left ventricular long axis motion using magnetic resonance velocity mapping in normal subjects. J Magn Reson Imaging 4:151, 1994. Reprinted by permission of Wiley, Liss, Inc., a subsidiary of John Wiley & Sons, Inc.)

velocity decreased in both normals and patients from baseline to low-dose dobutamine (5–7.5 μg/kg/min), and from low- to high-dose dobutamine (10–15 μg/kg/min). The mean long axis velocity increased in the normal subjects with low-dose dobutamine and remained elevated with the high dose. In the patients with reversible ischemia, however, 62 percent developed a reduced mean long axis velocity. Those with previous infarction but no reversible ischemia behaved similarly to normal subjects. Regional changes in long axis velocity were also examined. In the normal subjects, regional long axis velocity increased with low-dose dobutamine, but with high dose some reduction was seen, particularly in the inferoseptal wall. In patients with reversible ischemia, 62 percent developed abnormal regional velocities with dobutamine. Overall, 67 percent of patients with reversible ischemia had an abnormal global or regional response to stress. The patients with anterior ischemia were more likely to develop abnormal velocity values during stress, because of the greater contribution to mean myocardial long axis velocity from these segments. Therefore, there was a lower sensitivity for uncovering inferior ischemia.

Technical advances have now introduced new methods for the acquisition of myocardial velocity data that might prove useful. Segmented phase contrast gradient echo imaging permits acquisition of myocardial velocities within a period of a breathhold with high resolution. This method improves image quality and significantly reduces velocity arti-

facts from respiratory motion.[81] In addition, some interesting work from Arai and associates has been presented where components of velocity accounting for rigid body movement of the heart (translation and rotation) are calculated from short axis myocardial velocity maps and then subtracted to yield velocity maps showing pure radial velocity in systole and diastole.[82] These uncontaminated myocardial velocity maps might be useful for stress imaging in the quantification of myocardial mechanics. Using myocardial velocities has been suggested as an alternative to myocardial tagging,[83, 84] and, in addition, strain rate can be derived from the myocardial velocity data.[85] To date there are no reports of stress studies.

Global Ventricular Function

Dobutamine infusion also allows the measurement of aortic flow during stress by CMR velocity mapping. Absolute flow is measured from simultaneous area and velocity measurements of the ascending aorta yield. By also measuring the heart rate and blood pressure during stress, the stroke volume, cardiac output, aortic acceleration, cardiac power output, and flow wave velocity may be calculated. Pennell studied normal subjects and patients with coronary artery disease to determine which parameters were predictive of the extent of reversible myocardial ischemia.[86]

In normal individuals, an increase in all parameters was seen with dobutamine stress, except for the stroke volume, which rose and then fell, and the diastolic blood pressure, which remained unchanged. In patients with coronary artery disease, the qualitative pattern of change in parameters was similar, except that a fall in peak flow acceleration at peak stress was found to occur significantly more frequently in patients with moderate and severe ischemia than in patients with mild or no ischemia (Figure 10–20). The quantitative change from baseline to peak stress in five parameters was significantly related to the extent of myocardial ischemia (peak flow acceleration, peak flow, cardiac power output, maximum dobutamine dose tolerated, and

systolic blood pressure). Using multivariate analysis, the peak flow acceleration was found to be the most predictive variable ($p < 0.001$), and alone explained 58.4 percent of the variation in the observed myocardial ischemia (Figure 10–21). Only the cardiac power output retained predictive significance after allowing for the peak flow acceleration, but its contribution to the predictive accuracy of the model was small (4.2%). Attempts to reproduce this type of study using exercise have now been reported.[6, 87]

This study showed that an assessment of global ventricular function by CMR could be performed during dobutamine stress. There have been very few other attempts to study stress global function, which

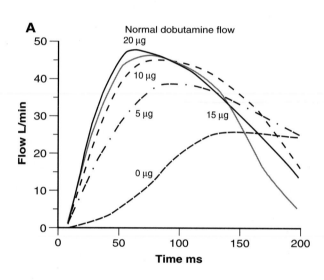

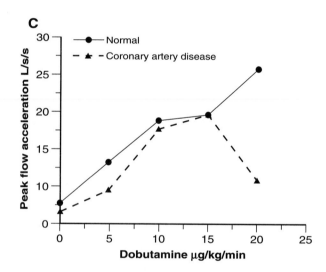

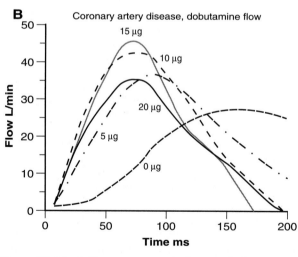

Figure 10–20. *A,* Flow curves acquired in systole from a patient with normal coronary arteries and no ischemia. Each dose increment shows an increase in the peak flow value, but also in the rate of upstroke of flow, known as flow acceleration. Note that systole reduces in length as the heart rate increases with higher dobutamine doses. *B,* Flow curves from a patient with two-vessel disease and ischemia on thallium myocardial perfusion scintigraphy. The flow responses are similar up to 10 µg/kg/min but flow acceleration falls at the 15-µg dose, and both flow and flow acceleration are markedly blunted at the 20-µg dose. This finding indicates the effect of significant ischemia on the ejection process. In *C,* the peak flow acceleration at each dose of dobutamine has been plotted against dobutamine dose for a normal patient and a patient with significant coronary artery disease. Note that the peak flow acceleration drops markedly with the onset of ischemia at 20 µg/kg/min in this case. The increase in peak flow acceleration from baseline to peak stress is a useful marker of ischemic burden, as shown in Figure 10–21. (*A* and *B,* from Pennell DJ, Underwood SR: The cardiovascular effects of dobutamine assessed by magnetic resonance imaging. Postgrad Med 67(suppl 1): S1, 1991, with permission from the BMJ Publishing Group. Copyright © 1991 BMJ Publishing; *C,* from Pennell DJ, Underwood SR: Stress cardiac magnetic resonance imaging. Am J Cardiac Imaging 5:139, 1991.)

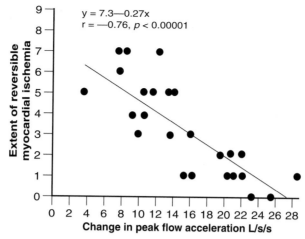

Figure 10–21. The change in peak flow acceleration from baseline to peak stress is inversely related to the extent of reversible myocardial ischemia (maximum number of segments was 9) and can therefore be used as an assessment of global burden of ischemia on ventricular function during stress. (From Pennell DJ, Firmin DN, Burger P, et al: Assessment of magnetic resonance velocity mapping of global ventricular function during dobutamine infusion in coronary artery disease. Br Heart J 74:163, 1995, with permission from the BMJ Publishing Group. Copyright © 1990 BMJ Publishing Group.)

is surprising bearing in mind that it is standard practice with radionuclide ventriculography. However, the aortic flow CMR technique is unable to measure the ejection fraction, although the ejection fraction has been reported in stress wall motion studies. With dipyridamole, Pennell found a nonsignificant difference between ischemic and nonischemic patients (+1% vs. +3%).[51] Pennell also showed a significant reduction in ejection fraction during ischemia using dobutamine in patients only (−6.1%), accompanied by reductions in ventricular volumes and stroke volumes.[63] Dobutamine has also been shown to increase ejection fraction in normals.[64] There is a need for further investigation into the potential value of assessing global ventricular function using CMR.

Dobutamine Magnetic Resonance Spectroscopy

There are a few reports of the use of dobutamine to study the myocardium during magnetic resonance spectroscopy. The first reports examined phosphorus spectra in normal subjects and patients with dilated cardiomyopathy.[88] No change was identified in the PCr/ATP ratio during moderate dobutamine stress. However, a later study of 20 normal subjects at rest and during dobutamine-atropine stress sufficient to triple the rate-pressure product[89] showed that at peak stress, PCr fell by 21 percent and ATP fell by 9 percent. The ratio of PCr to ATP fell from 1.42 to 1.22, indicating that myocardial high-energy phosphate metabolism is altered mainly at high workloads. The same group then compared controls with elite cyclists with left ventricular hypertrophy during dobutamine stress.[90] Both groups showed a reduction in PCr/ATP ratio matching that of the previous study, and indicating that the left

ventricular hypertrophy in these cases was not associated with demonstrable metabolic abnormality with stress suggesting that the hypertrophy in these cases was a truly physiological phenomenon. This very interesting new approach to the investigation of stress-induced abnormality has some way to go before it could be used widely, but it remains a valuable and interesting research tool.

PRACTICAL CONSIDERATIONS

There is no substitute for training in a center that is experienced in stress CMR in order to learn the techniques in the same way that stress echocardiography requires a period of training to climb the learning curve.[91, 92] There are, however, certain basic issues that are common to all centers in the performance of stress CMR with regard to set-up, performance, and monitoring. The patient clearly needs to be comfortable lying in the magnet and can anticipate a scan of between 30 and 60 minutes depending on the dobutamine dose required. An intravenous cannula will be in place with a long line attached to an infusion pump. The pump must be in a position where it works correctly despite the ambient magnetic field. Ideally, if located in the magnet room, it should be secured in place to prevent any possibility of entering the magnet. The infusion pump should be tested regularly in situ for correct function. It is attached to the long line at the start of the scan, and a very low infusion rate is begun to keep the vein open. Full cardiovascular monitoring must be in place. A continuous ECG rhythm display must be available in the scanning room to identify significant arrhythmias, although it is not useful for identification of ST segment changes because of the magnetohydrodynamic effect. Noninvasive blood pressure monitoring is also required and should be on the opposite arm to that which has the infusion cannula. Some centers routinely use pulse oximetry as well. The most important other features for safety are constant verbal communication with the patient for symptoms and rapid display of the breath-hold cines to determine whether new wall motion abnormalities are occurring. This display is valuable because these changes occur prior to ECG changes and symptoms in the physiological changes occurring with ischemia known as the angina cascade.[93-96] Full resuscitation equipment should be available with personnel in attendance with up-to-date training in its use. Dobutamine (and the vasodilators) not uncommonly cause nausea, and patients should preferably have abstained from eating for a few hours prior to the test. This practice helps to minimize the feeling and reduces the dangers of vomiting should it occur. Criteria for ending the stress test include intolerable angina, dyspnea, nausea, or other symptoms, significant arrhythmias, hypertension above 240/120 mm Hg, decrease in systolic blood pressure of more than 40 mm Hg, or other severe adverse event. Reversal of the beta-agonist effects of dobutamine can be achieved with intravenous beta blockade, and a commonly stocked short-acting drug for this purpose is esmolol. At the end of the test, angina can also be treated with nitrate spray if necessary.

Imaging is often performed with breath-hold segmented k-space gradient echo cines with the technology currently available, although hybrid gradient echo sequences with echo-planar readouts are also being used, as the contrast is good, and the increased speed allows for an increased number of phases to be acquired during tachycardia. This may change, however, and real-time approaches[7] or navigator cine imaging[97] may allow the patient to breathe freely during the acquisition. Ideally temporal resolution at rest should be 40 to 50 ms, but as the heart rate increases a temporal resolution of 30 ms is more suitable. Depending on system performance, the approximate imaging parameters include a TE of 3 ms, TR of 6 ms, flip angle 25°, slice thickness 8 mm, and spatial resolution 1.5 × 3 mm. Typically to achieve this goal requires a modern 1.5 T system with a phased array coil. Conventional cine imaging also works but with a time penalty, but this investigation can be performed on lower field systems.

Imaging should be performed in the long and short axes of the heart. This imaging includes the vertical long axis (defined from the plane joining the apex and the mid-mitral valve on a transaxial scout), the horizontal long axis (defined from the plane joining the apex and the mid-mitral valve on the vertical long axis), and three short axis planes (basal, mid, and apical defined from the horizontal long axis). These cines should be displayed according the American Heart Association guidelines,[98] so that the images are readily interpreted by clinicians familiar with conventional tomographic imaging techniques. It is particularly helpful if the scanner software allows simple reselection of existing imaging planes for the purpose of repeating the planes at each stage of dobutamine stress. The cines should be displayed as soon as possible (ideally in real time) after acquisition and examined for wall motion abnormality. This goal is greatly facilitated by viewing each cine with sequential doses on the same screen, as has been known in echocardiography for some time. However, because of the number of planes imaged by CMR, a multi-cine matrix of running cines is required. Custom solutions to this problem are available and the manufacturers will undoubtedly follow suit. This solution augments the safety of the procedure because wall motion abnormalities can be diagnosed early with the onset of ischemia. Real-time imaging would allow this diagnosis to be made even faster, but the quality of the breath-hold cine still probably makes the diagnosis easier to obtain.

CONCLUSION

A number of stress CMR techniques are being developed that will compete with scintigraphic and echocardiographic techniques for the evaluation of reversible myocardial ischemia. Wall motion assessment using dobutamine is the most developed and has been validated now in a number of centers. CMR has a number of inherent advantages over stress echocardiography, however, with better resolution, reproducible image positioning, true long and short axis imaging, and contiguous parallel slices. A simple and rapid technique for quantification of wall motion would greatly help clinical use, and tagging may provide this, but it is at an early stage of clinical development. Other CMR techniques remain experimental at present. Long axis velocity imaging and aortic flow imaging during dobutamine stress have been performed in only one center and their overall sensitivity needs further evaluation, whereas spectroscopy is still complex and also at an early stage of evaluation. Exercise stress in the magnet remains difficult despite some modest success with ultrafast imaging, and therefore there is every likelihood that pharmacological stress will continue to be the technique of choice for stress CMR in the long term. Once the rapid display of the stress cines has been improved, with or without the addition of real-time imaging, stress CMR needs to be subjected to comparative multicenter trials to validate its clinical role in comparison with stress echocardiography and myocardial perfusion imaging.

References

1. Master AM, Oppenheimer ET: A simple exercise tolerance test for circulatory efficiency with standard tables for normal individuals. Am J Med Sci 177:223, 1929.
2. Brown KA: Prognostic value of myocardial perfusion imaging: State of the art and new developments. J Nucl Cardiol 3:516, 1996.
3. Ladenheim ML, Pollock BH, Rozanski A, et al: Extent and severity of myocardial hypoperfusion as predictors of prog-

nosis in patients with suspected coronary artery disease. J Am Coll Cardiol 7:464, 1986.

4. Olmos LI, Dakik H, Gordon R, et al: Long term prognostic value of exercise echocardiography compared with exercise T1-201, ECG and clinical variables in patients evaluated for coronary artery disease. Circulation 98:2679, 1998.

5. Schaefer S, Peshock RM, Parkey RW, Willerson JT: A new device for exercise MR imaging. AJR 147:1289, 1986.

6. Mohiaddin RH, Gatehouse PD, Firmin DN: Exercise related changes in aortic flow measured by magnetic resonance spiral echo-planar phase-shift velocity mapping. J Magn Reson Imaging 5:159, 1995.

7. Nayak KS, Pauly JM, Nishimura DG, Hu BS: Rapid ventricular assessment using real-time interactive multislice MRI. Magn Reson Med 95:371, 2001.

8. Gatehouse PD, Firmin DN, Collins S, Longmore DB: Real time blood flow imaging by spiral scan phase velocity mapping. Magn Reson Med 31:504, 1994.

9. Scheidegger MB, Stuber M, Pederson EM, et al: Methodological and technical aspects of physiological (ergometer) stress for cardiovascular examinations in MR scanners (abstract). Proceedings of the Int Soc Magn Reson Med, 1997, p 899.

10. Oshinski JN, Ferichs F, Doyle JA, et al: Exercise stress measurements of cardiac performance using an MR compatible cycle ergometer (abstract). Proceedings of the Int Soc Magn Reson Med, 1997, p 900.

11. Udelson JE, Leppo JA: Single photon myocardial perfusion imaging and exercise radionuclide angiography in the detection of coronary artery disease. *In* Murray IPC, Ell PJ (eds): Nuclear Medicine in Clinical Diagnosis and Management. London, Churchill Livingstone, 1994, p 1129.

12. Conway MA, Bristow JD, Blackledge MJ, et al: Cardiac metabolism during exercise in healthy volunteers measured by ^{31}P magnetic resonance spectroscopy. Br Heart J 65:25, 1991.

13. Weiss RG, Bottomley PA, Hardy CJ, Gerstenblith G: Regional myocardial metabolism of high energy phosphates during isometric exercise in patients with coronary artery disease. N Engl J Med 323:1593, 1990.

14. Yabe T, Mitsunami K, Okada M, et al: Detection of myocardial ischemia by ^{31}P magnetic resonance spectroscopy during handgrip exercise. Circulation 89:1709, 1994.

15. Pennell DJ: Cardiac stress in nuclear medicine. *In* Murray IPC, Ell PJ (eds): Nuclear Medicine in Clinical Diagnosis and Management. 2nd ed. London, Churchill Livingstone, 1998.

16. Pennell DJ, Prvulovich E, Tweddel A, Caplin J: Nuclear cardiology in the UK: British Nuclear Cardiology Society survey 1994. Nucl Med Commun 19:305, 1998.

17. Zoghbi WA: Use of adenosine echocardiography for diagnosis of coronary artery disease. Am Heart J 122:285, 1991.

18. Iskandrian AS, Heo J, Kong B, Lyons E: Effect of exercise level on the ability of thallium-201 tomographic imaging in detecting coronary artery disease: Analysis of 461 patients. J Am Coll Cardiol 14:1477, 1989.

19. Kiat H, Iskandrian AS, Villegas BJ, et al: Arbutamine stress thallium-201 single-photon emission computed tomography using a computerised closed-loop delivery system—multicenter trial for evaluation of safety and diagnostic accuracy. J Am Coll Cardiol 26:1159, 1995.

20. Lette J and the Multicenter Dipyridamole Safety Study Investigators: Safety of dipyridamole testing in 73,806 patients: The multicenter dipyridamole safety study. J Nucl Cardiol 2:3, 1995.

21. Wilson RF, Laughlin DE, Ackell PH, et al: Transluminal, subselective measurement of coronary artery blood flow velocity and vasodilator reserve in man. Circulation 72:82, 1985.

22. Fung AY, Gallagher KP, Buda AJ: The physiological basis of dobutamine as compared with dipyridamole stress interventions in the assessment of critical coronary stenosis. Circulation 76:943, 1987.

23. Brown G, Josephson MA, Petersen RD, et al: Intravenous dipyridamole combined with isometric handgrip for near maximal coronary flow in patients with coronary artery disease. Am J Cardiol 48:1077, 1981.

24. Wilson RF, Wyche K, Christensen BV, et al: Effects of adenosine on human coronary arterial circulation. Circulation 82:1595, 1990.

25. Nguyen T, Heo J, Ogilby D, Iskandrian AS: Single photon emission computed tomography with thallium-201 during adenosine induced coronary hyperaemia: Correlation with coronary arteriography, exercise thallium imaging and two-dimensional echocardiography. J Am Coll Cardiol 16:1375, 1990.

26. Marwick T, Willemart B, Dhondt AM, et al: Selection of the optimal nonexercise stress for the evaluation of ischemic regional myocardial dysfunction and malperfusion—Comparison of dobutamine and adenosine using echocardiography and Tc-99m-MIBI single photon emission computed tomography. Circulation 2:345, 1993.

27. Pennell DJ, Mavrogeni S, Forbat SM, et al: Adenosine combined with dynamic exercise for myocardial perfusion imaging. J Am Coll Cardiol 25:1300, 1995.

28. Panting JR, Gatehouse PD, Yang GZ, et al: Echo planar magnetic resonance myocardial perfusion imaging: Parametric map analysis and comparison with thallium SPECT. J Magn Reson Imaging 13:192, 2001.

29. Pettigrew RI, Martin S, Eisner R, et al: Detection of partial coronary artery stenosis with isoproterenol stress cine MRI in dogs: Validation by on-line ultrasonic crystals and flow probes (abstract). Proceedings of the Soc Magn Reson Med, 1991, p 243.

30. Pennell DJ, Underwood SR, Swanton RH, et al: Dobutamine thallium myocardial perfusion tomography. J Am Coll Cardiol 18:1471, 1991.

31. Pennell DJ, Underwood SR, Ell PJ: Safety of dobutamine stress for thallium myocardial perfusion tomography in patients with asthma. Am J Cardiol 71:1346, 1993.

32. Sawada SG, Segar DS, Ryan T, et al: Echocardiographic detection of coronary artery disease during dobutamine infusion. Circulation 83:1605, 1991.

33. Fioretti P, Poldermans D, Salustri A, et al: Atropine increases the accuracy of dobutamine stress echocardiography in patients taking beta blockers. Eur Heart J 15:355, 1994.

34. Weissman NJ, Levangie MW, Guerrero JL, et al: Effect of beta-blockade on dobutamine stress echocardiography. Am Heart J 131:698, 1996.

35. McNeill AJ, Fioretti PM, El-Said ME-S, et al: Dobutamine stress enhanced sensitivity for detection of coronary artery disease by addition of atropine to dobutamine stress echocardiography. Am J Cardiol 70:41, 1992.

36. Picano E, Mathias W, Pingitore A, et al: Safety and tolerability of dobutamine-atropine stress echocardiography: A prospective multicentre study. Lancet 344:1190, 1994.

37. Mertes H, Sawada SG, Ryan T, et al: Symptoms, adverse effects and complications associated with dobutamine stress echocardiography. Experience in 1118 patients. Circulation 88:15, 1993.

38. Secknus MA, Marwick TH: Evolution of dobutamine echocardiography protocols and indications: Safety and side effects in 3,011 studies over 5 years. J Am Coll Cardiol 29:1234, 1997.

39. Dakik HA, Vempathy H, Verani MS: Tolerance, hemodynamic changes, and safety of dobutamine stress perfusion imaging. J Nucl Cardiol 3:410, 1996.

40. Cerqueira MD, Verani MS, Schwaiger M, et al and the Investigators of the Multicenter Adenoscan Trial: Safety profile of adenosine stress perfusion imaging: Results from the adenoscan multicenter trial registry. J Am Coll Cardiol 23:384, 1994.

41. Willerson JT, Hutton I, Watson JT, et al, Influence of dobutamine on regional myocardial blood flow and ventricular performance during acute and chronic myocardial ischemia in dogs. Circulation 53:828, 1976.

42. Vasu MA, O'Keefe DD, Kapellakis GZ, et al: Myocardial oxygen consumption: Effects of epinephrine, isoproterenol, dopamine, norepinephrine and dobutamine. Am J Physiol 235: 237, 1978.

43. Fowler MB, Alderman EL, Oesterle SN, et al: Dobutamine and dopamine after cardiac surgery: Greater augmentation

of myocardial blood flow with dobutamine. Circulation 70 (suppl I):103, 1984.

44. Meyer SL, Curry GC, Donsky MS, et al: Influence of dobutamine on hemodynamics and coronary blood flow in patients with and without coronary artery disease. Am J Cardiol 38:103, 1976.

45. Warltier DC, Zyvlowski M, Gross GJ, et al: Redistribution of myocardial blood flow distal to a dynamic coronary arterial stenosis by sympathomimetic amines. Comparison of dopamine, dobutamine and isoproterenol. Am J Cardiol 48:269, 1981.

46. Pennell DJ: Pharmacological cardiac stress: When and how? Nucl Med Commun 15:578, 1994.

47. Homma S, Gilliland Y, Guiney TE, et al: Safety of intravenous dipyridamole for stress testing with thallium imaging. Am J Cardiol 59:152, 1987.

48. Taviot B, Pavheco Y, Coppere B, et al: Bronchospasm induced in an asthmatic by the injection of adenosine. Presse Med 15:1103, 1986.

49. Pennell DJ, Mahmood S, Ell PJ, Underwood SR: Bradycardia progressing to cardiac arrest during adenosine thallium myocardial perfusion imaging in covert sino-atrial disease. Eur J Nucl Med 21:170, 1994.

50. Pennell DJ, Underwood SR, Longmore DB: The detection of coronary artery disease by magnetic resonance imaging using intravenous dipyridamole. J Comput Assist Tomogr 14:167, 1990.

51. Pennell DJ, Underwood SR, Ell PJ, et al: Dipyridamole magnetic resonance imaging: A comparison with thallium-201 emission tomography. Br Heart J 64:362, 1990.

52. Mays AE, Cobb FR. Relationship between regional myocardial blood flow and thallium-201 distribution in the presence of coronary artery stenosis and dipyridamole induced vasodilation. J Clin Invest 73:1359, 1984.

53. Niemi P, Poncelet BP, Kwong KK, et al: Myocardial signal changes associated with flow stimulation in blood oxygenation sensitive magnetic resonance imaging. Magn Reson Med 36:78, 1996.

54. Li D, Dhawale P, Rubin PJ, et al: Myocardial signal response to dipyridamole and dobutamine: Demonstration of the BOLD effect using a double echo gradient echo sequence. Magn Reson Med 36:16, 1996.

55. Casolo GC, Bonechi F, Taddei T, et al: Alterations in dipyridamole induced LV wall motion during myocardial ischaemia studied by NMR imaging. Comparison with Tc-99m-MIBI myocardial scintigraphy. G Ital Cardiol 21:609, 1991.

56. Baer FM, Smolarz K, Jungehulsing M, et al: Feasibility of high dose dipyridamole magnetic resonance imaging for detection of coronary artery disease and comparison with coronary angiography. Am J Cardiol 69:51, 1992.

57. Kenakin TP: The potentiation of cardiac responses to adenosine by benzodiazepines. J Pharmacol Exp Ther 222:752, 1982.

58. Baer FM, Smolarz K, Theissen P, et al: Identification of haemodynamically significant coronary stenoses by dipyridamole magnetic resonance imaging and 99mTc methoxyisobutyl-isonitrile SPECT. Int J Card Imaging 9:133, 1993.

59. Perault C, Loboguerrero A, Liehn JC, et al: Quantitative comparison of prone and supine myocardial SPECT MIBI images. Clin Nucl Med 20:678, 1995.

60. Bremerich J, Buser P, Bongartz G, et al: Noninvasive stress testing of myocardial ischemia: Comparison of GRE-MRI perfusion and wall motion analysis to 99mTc-MIBI SPECT, relation to coronary angiography. Eur Radiol 7:990, 1997.

61. Fedele F, Rosaanio S, Tocchi M, et al: Comparison of cine magnetic resonance imaging and multiplane transesophageal echocardiography during dipyridamole stress for detecting coronary artery disease: Qualitative and quantitative analysis (abstract). Circulation 94:I–180, 1996.

62. Zhao S, Croisille P, Janier M, at al: Comparison between qualitative and quantitative wall motion analyses using dipyridamole stress breath-hold cine MRI in patients with severe coronary artery stenosis. Magn Reson Imaging 15:891, 1997.

63. Pennell DJ, Underwood SR, Manzara CC, et al: Magnetic resonance imaging during dobutamine stress in coronary artery disease. Am J Cardiol 70:34, 1992.

64. van Rugge FP, Holman ER, van der Wall EE, et al: Quantitation of global and regional left ventricular function by cine magnetic resonance imaging during dobutamine stress in normal human subjects. Eur Heart J 14:456, 1993.

65. van Rugge P, van der Wall EE, de Roos A, Bruschke AVG: Dobutamine stress magnetic resonance imaging for detection of coronary artery disease. J Am Coll Cardiol 22:431, 1993.

66. van Rugge FP, van der wall EE, Spanjersberg SJ, et al: Magnetic resonance imaging during dobutamine stress for detection and localisation of coronary artery disease. Quantitative wall motion analysis using a modification of the centerline method. Circulation 90:127, 1994.

67. Baer FM, Voth E, Theissen P, et al: Gradient echo magnetic resonance imaging during incremental dobutamine infusion for the localisation of coronary artery stenosis. Eur Heart J 15:218, 1994.

68. Baer FM, Voth E, Theissen P, et al: Coronary artery disease: Findings with GRE MR imaging and Tc-99mm-methoxyisobutyl-isonitrile SPECT during simultaneous dobutamine stress. Radiology 193:203, 1994.

69. Nagel E, Lehmkuhl HB, Bocksch W, et al: Noninvasive diagnosis of ischemia induced wall motion abnormalities with the use of high dose dobutamine stress MRI. Comparison with dobutamine stress echocardiography. Circulation 99:763, 1999.

69a. Hundley WG, Hamilton CA, Thomas MS, et al: Utility of fast cine magnetic resonance imaging and display for the detection of myocardial ischemia in patients not well suited for second harmonic stress echocardiography. Circulation 100:1697, 1999.

70. Hoffmann R, Lethen H, Marwick T, et al: Analysis of interinstitutional observer agreement in interpretation of dobutamine stress echocardiograms. J Am Coll Cardiol 27:330, 1996.

71. Power TP, Kramer CM, Shaffer AL, et al: Breath-hold dobutamine magnetic resonance myocardial tagging: Normal left ventricular response. Am J Cardiol 80:1203, 1997.

72. Scott CH, St John Sutton MG, Gusani N, et al: Effect of dobutamine on regional left ventricular function measured by tagged magnetic resonance imaging in normal subjects. Am J Cardiol 83:412, 1999.

73. Buchalter MB, Rademakers FE, Weiss JL, et al: Rotational deformation of the canine left ventricle measured by magnetic resonance tagging: Effects of catecholamines, ischemia and pacing. Cardiovasc Res 28:629, 1994.

74. Rademakers FE, Buchalter MB, Rogers WJ, et al: Dissociation between left ventricular untwisting and filling. Accentuation by catecholamines. Circulation 85:1572, 1992.

75. Kraitchman DL, Wilke N, Hexeberg E, et al: Myocardial perfusion and function in dogs with moderate coronary stenosis. Magn Reson Med 35:771, 1996.

76. Hartnell G, Cerel A, Kamalesh M, et al: Detection of myocardial ischemia: Value of combined myocardial perfusion and cineangiographic MR imaging. AJR Am J Roentgenol 163:1061, 1994.

77. Karwatowski SP, Mohiaddin RH, Yang GZ, et al: Noninvasive assessment of regional left ventricular long axis motion using magnetic resonance velocity mapping in normal subjects. J Magn Reson Imaging 4:151, 1994.

78. Karwatowski SP, Forbat SM, Mohiaddin RH, et al: Regional left ventricular long axis function in controls and patients with ischaemic heart disease pre and post angioplasty (abstract). Circulation 88(suppl):I-83, 1993.

79. Jones CJH, Raposo L, Gibson DG: Functional importance of the long axis dynamics of the left ventricle. Br Heart J 63:215, 1990.

80. Reduto LA, Wickermeyer WJ, Young JB, et al: Left ventricular diastolic performance at rest and during exercise in patients with coronary artery disease. Assessment with first pass radionuclide ventriculography. Circulation 63:1228, 1981.

81. Hennig J, Markl M, Peschl S, et al: Measurement of myocardial wall motion with segmented breath-hold phase contrast gradient echo imaging (abstract). Procedure of the Int Soc Magn Reson Med 1997, p 390.

82. Arai AE, Gaither CC, Epstein FH, et al: Velocity gradient of phase contrast MRI to quantify regional contractile abnormalities with myocardial infarction (abstract). Proceedings of the Int Soc Magn Reson Med 1997, p 384.

83. Ligamenei A, Hardy PA, Powell KA, et al: Validation of cine phase contrast MR imaging for motion analysis. J Magn Reson Imaging 5: 331, 1995.

84. Pelc NJ, Drangova M, Pelc LR, et al: Tracking of cyclic motion with phase contrast cine MR velocity data. J Magn Reson Imaging 5:339, 1995.

85. Wedeed VJ: Magnetic resonance imaging of myocardial kinematics: Techniques to detect, localise and quantify the strain rates of active human myocardium. Magn Reson Med 27:52, 1992.

86. Pennell DJ, Firmin DN, Burger P, et al: Assessment of magnetic resonance velocity mapping of global ventricular function during dobutamine infusion in coronary artery disease. Br Heart J 74:163, 1995.

87. Niezen RA, Doornbos J, de Boer RW, et al: Great vessel flow studies with MRI at rest and during physical exercise (abstract). Proceedings of the Soc Magn Reson Med, 1997, p 867.

88. Schaefer S, Schwartz GG, Steinmann SK, et al: Metabolic response of the human heart to inotropic stimulation: In vivo phosphorus-31 studies of normal and cardiomyopathic myocardium Magn Reson Med 25:260, 1992.

89. Lamb HJ, Beyerbacht HP, Ouwerkerk R, et al: Metabolic response of normal human myocardium to high dose atropine-dobutamine stress studied by ^{31}P-MRS. Circulation 96:2969, 1997.

90. Pluim BM, Lamb HJ, Kayser HW, et al: Functional and metabolic evaluation of the athlete's heart by magnetic resonances, imaging and dobutamine stress magnetic resonance spectroscopy. Circulation 97:666, 1998.

91. Varga A, Picano E, Dodi C, et al: Madness and method in stress echo reading. Eur Heart J 20:1271, 1999.

92. Picano E, Lattanzi F, Orlandini A, et al: Stress echocardiography and the human factor: The importance of being expert. J Am Coll Cardiol 17:666, 1991.

93. Nesto RW, Kowalchuk GJ: The ischemic cascade: Temporal sequence of hemodynamic, electrocardiographic and symptomatic expressions of ischemia. Am J Cardiol 57: 23C, 1987.

94. Battler A, Froehlicher VF, Gallagher KP, et al: Dissociation between regional myocardial dysfunction and ECG changes during ischaemia in the conscious dog. Circulation 62:735, 1980.

95. Sugishita Y, Koseki S, Matsuda M, et al: Dissociation between regional myocardial dysfunction and ECG changes during myocardial ischaemia induced by exercise in patients with angina pectoris. Am Heart J 106:1, 1983.

96. Beller GA: Myocardial perfusion imaging for detection of silent myocardial ischaemia. Am J Cardiol 61(suppl): 22F, 1988.

97. Hoffman MBM, van Rossum AC, Sprenger M, Westerhof N: Assessment of flow in the right coronary artery by magnetic resonance phase contrast velocity measurement: effects of cardiac and respiratory motion. Magn Reson Med 35:521, 1996.

98. American College of Cardiology, American Heart Association, Society of Nuclear Medicine: Policy statement. Standardisation of cardiac tomographic imaging. Circulation 86: 338, 1992.

CHAPTER 11

Stress CMR—Clinical Myocardial Perfusion Imaging

Jan T. Keijer and Albert C. van Rossum

MYOCARDIAL PERFUSION IMAGING

Rationale

The challenges for noninvasive perfusion imaging are the sensitive detection of an abnormality, localization so that abnormalities can be related to a responsible coronary stenosis (the culprit lesion), and the assessment of the functional severity of a coronary stenosis. On the basis of this knowledge and in combination with the information from coronary arteriography, prognosis can be estimated and adequate therapy can be instituted. This information allows important clinical management to be planned, which includes risk stratification, evaluation of therapeutic interventions, and follow-up of patients with coronary artery disease (CAD). With 5 million myocardial perfusion scans performed annually in the United States alone, it is clear that perfusion imaging is a major contributor to cardiological practice, and the development of cardiovascular magnetic resonance (CMR) perfusion techniques could play a substantial role in improving the quality of the results and reducing the population radiation burden.

The Gold Standard: Positron-Emission Tomography

The present gold standard for assessment of myocardial perfusion in humans is positron-emission tomography (PET).[1–3] Its principal merits include reasonably good spatial resolution (in-plane $\approx 6 \times 6$ mm), the ability to quantify regional myocardial perfusion, and the capacity to perform tomography. Furthermore, other aspects of myocardial metabolism can also be studied with related techniques. A disadvantage is that the extraction of some tracers (^{13}N ammonia, rubidium 82) is inversely related to flow, and myocardial uptake is dependent not only on flow but also on metabolism.[2, 3] ^{15}O-labeled water is a better flow tracer, but it yields poor quality images for visual inspection. The necessity to have an on-site cyclotron for production of some radiolabeled tracers (^{13}N ammonia, [15]O water) currently contributes to the high cost and limits routine clinical use.[3] Radiation exposure is an inherent disadvantage of both PET and more conventional radionuclide scintigraphy.

Radionuclide Scintigraphy

In current clinical practice, tomographic scintigraphy (single-photon emission computed tomography, SPECT) uses tracers such as thallium 201 (^{201}Tl), and the technetium-99m–labeled compounds (sestamibi and tetrofosmin). Results are good, with high sensitivities (80–90%) for detection of CAD with somewhat lower specificity (75–85%).[4–14] In general, regional tracer uptake is assessed and significant reductions in regional uptake relative to maximum uptake are considered to reflect perfusion defects. Drawbacks are attenuation of tracer signal (all, especially ^{201}Tl, due to low photon energy), long half-life limiting the maximal dose (^{201}Tl), and high hepatobiliary uptake hampering correct interpretation of the inferior wall (sestamibi and tetrofosmin). Although other imaging techniques have also addressed the possibility of assessing myocardial perfusion, including electron beam computed tomography,[15–17] contrast echocardiography,[18–24] and videodensitometry,[25–29] none has yet been successful in clinical practice.

Potential

It has been known for some time that myocardial infarction and ischemia increase tissue longitudinal relaxation time (T1) and transverse relaxation time (T2),[30] which can lead to signal changes in the myocardium. Spin-echo CMR has been applied in order to demonstrate myocardial infarction,[31–33] but increased signal from the myocardium is only a nonspecific marker for perfusion abnormalities, and in general the techniques required to assess perfusion have concentrated on using CMR contrast agents such as the gadolinium diethylenetriamine pentaacetic acid (Gd-DTPA) compounds[34, 35] that shorten T1 relaxation time. Gd-DTPA diffuses from the intravascular space to the interstitial edema that is associated with myocardial infarction. Washout of Gd-

DTPA from normal tissue enhances the contrast between normal and infarcted tissue.[36, 37] Its maximum contrast enhancing activity can be observed by imaging 20 to 30 minutes after intravenous administration. In this type of imaging the *steady state* of the distribution of a contrast agent requires little temporal resolution, but is generally not capable of assessing myocardial ischemia. Therefore, *dynamic* or first-pass imaging has been developed to provide information on myocardial perfusion from analysis of myocardial signal changes during the wash-in of the contrast agent after bolus injection.

Subsecond CMR techniques[38, 39] provide the required temporal resolution to image the first pass of the contrast agent through the myocardium (acquisition time $\approx$ 350 ms/image, $\approx$ 1 image/s) with good spatial resolution ($\approx 2 \times 3$ mm)[40, 41] that is sufficient to allow transmural resolution within the myocardium.[40–42] When the possibility of perfusion is combined with myocardial function, blood flow in the coronary arteries, and myocardial metabolism,[43–48] the added value of CMR perfusion imaging as part of an integrated package of coronary disease assessments is clear.

THE CMR APPROACH TO PERFUSION

Perfusion abnormalities require visualization during stress as well as during a state of rest, but exercise is often not practical in the magnet; therefore, CMR perfusion imaging is usually performed during pharmacological stress. The rest study is commonly performed first and imaging takes place after the bolus of Gd-DTPA is given. After a brief period for equilibration of the first Gd-DTPA dose, the infusion of stress agent is started and the second bolus of Gd-DTPA given. This procedure makes a very rapid protocol.

The technique therefore is simply that of an impulse response technique that is similarly used in first-pass investigations in nuclear medicine and cardiac catheterization. After a bolus injection of the contrast agent into the blood, the enhancement pattern of the myocardium is evaluated. In practice, an imaging plane is selected in the short or long axis of the heart, and images are acquired with high frequency, usually one image per slice per heartbeat. For multislice imaging, this procedure entails multiple complete images per second, and higher quality equipment is therefore needed. The contrast agent is injected as a bolus usually into a peripheral vein, and for reproducible results, a power injector is often used. Some centers use breath-holding to preserve a consistent heart position for the first portion of the first pass, although other solutions to suppress diaphragmatic motion include navigator-guided slice following and postprocessing techniques to eliminate any significant movement. Imaging is usually done during systole, despite the fact that heart motion affects the images, because the myocardium is thicker than in diastole and it is easier to see regional and transmural signal differences. Normally perfused myocardium enhances more than hypoperfused myocardium and reaches peak intensity faster (Figure 11–1*A*). However, interpretation of single perfusion images is sometimes hampered by the occurrence of artifacts due to large differences in magnetic susceptibility between (not yet enhanced) myocardium and (maximally) enhanced blood in the ventricular cavity. This difference can result in dark rings especially at the septal endocardial border, or along the phase encoding direction.[49] These dark areas are known as Gibb's rings, and result from truncation of data acquisition in the raw data space, which leads to a loss of high-frequency spatial information in the images. Such artifacts are relatively short lived (unlike true perfusion defects). Gibb's ringing can be improved by sequence optimization, such as reducing the echo time and other maneuvers, and with the latest software available in modern CMR scanners, it is much less of a problem than it has been in the past. However, it is better to include the whole sequential series of perfusion images in the analysis, and to bear this possibility in mind when reviewing the data.

The signal intensity changes in the myocardium caused by the contrast agent reflect changes in the concentration of the contrast agent. For a consistent analysis, therefore, it is important that there should be a consistent relationship between the signal intensity and the contrast agent concentration. Provided this relationship is so, indicator-dilution theory[50–52] allows the calculation of myocardial blood flow from the dynamic information of the concentration of a blood-borne indicator substance during its first pass through the heart. Thus, quantitative information on myocardial perfusion can be gained from the dynamics of contrast wash-in and washout. In order to extract these data, regions of interest (ROI) are drawn in the myocardial wall, and signal intensity versus time (SI-time) curves can then be constructed (Figure 11–1*B*).[53] Curve fitting is applied to derive various perfusion-related parameters (Figure 11–1*C*),[54] which is important because of noise in the data. For quantitative analysis, model fitting is also required to allow for the various compartments of distribution of the contrast agent in the heart and their relative kinetics.

The parameters that can be derived include the amplitude of the curve (peak signal), the time to peak enhancement, the slope of the wash-in curve, the area under the SI-time curve, the mean transit time of the contrast agent (MTT, a parameter that is inversely related to flow), and many others. Many of these parameters can be used for a qualitative analysis, but for true quantitative analysis further prerequisites concerning the imaging system, the indicator used, and the method of injection of the indicator have to be fulfilled (Table 11–1):[50, 51, 55, 56]

1. Sufficient temporal resolution of the imaging technique to image changes in contrast agent concentration.

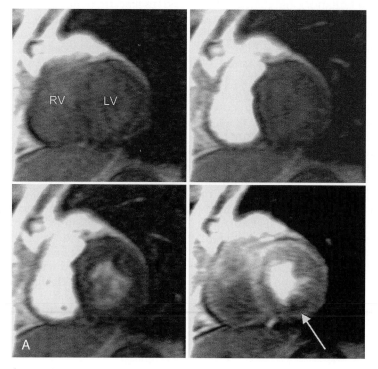

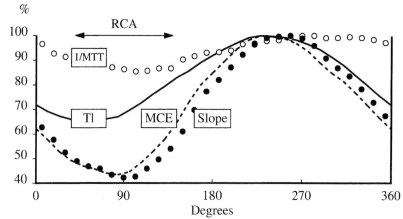

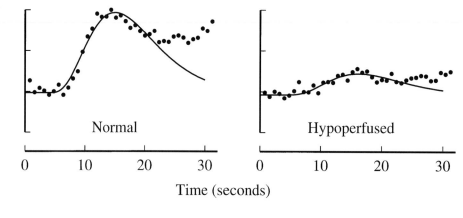

Figure 11–1. *See legend on opposite page*

Table 11–1. General Requirements of Indicator-Dilution Theory for Application to Densitometric Perfusion Imaging

Imaging Technique	Contrast Agent	Method of Injection
Detect concentration changes	Homogeneous distribution in blood	Rapid bolus
High temporal resolution	Confined to intravascular space	Small volume
High spatial resolution	Linear relation concentration–SI	Near myocardium
	No effect on hemodynamics	

SI = signal intensity.

2. A homogeneous distribution of the contrast agent in the blood with a linear relation between its concentration and measured signal intensity.

3. No effect of contrast agent on hemodynamics.

4. A negligible volume of the contrast agent compared with the vascular volume.

5. Recirculation is negligible.

6. Minimal dispersion of the contrast bolus (achieved by a rapid injection close to the myocardium, e.g., a right atrial injection).

In clinical practice, distortion of the input bolus (the impulse) is an important limiting factor for quantitative assessment of myocardial perfusion (the response). Recirculation, dispersion (due to slow and/or peripheral injection, low ejection fraction, valvular regurgitation), and diffusion of the contrast agent have deleterious effects on the myocardial signal intensity versus time curve (Figure 11–2). Differences between high temporal resolution imaging with central bolus injection and lower temporal resolution with peripheral bolus injection can be observed by comparing Figures 11–1B and 11–7. To circumvent the effects of the myocardial input, different models[57–59] have been developed to deconvolve the myocardial response curves with the input curves (obtained from ROI in the left ventricular cavity). One of the limitations of these models is the uncertainty regarding the concentration of the contrast agent in the blood and the variable relation between concentration and signal intensity.[40, 60] Another is the amount of contrast agent that diffuses

into the interstitium.[61] Many of these issues are addressed in more detail in Chapter 3.

CARDIOVASCULAR MAGNETIC RESONANCE PERFUSION TECHNIQUES

Many CMR sequences have been proposed for myocardial perfusion imaging. The main issue in the design of CMR sequences for imaging the dynamics of contrast passage is the trade-off between temporal resolution, spatial resolution, and acquisition time. The combination of CMR sequence and type and dosage of the contrast agent used determines the effect on CMR signal intensity of the myocardium. To obtain images that are acquired in the same phase of the cardiac cycle, imaging is performed using electrocardiogram (ECG) triggering.

The *fast gradient-echo sequences* are T1-weighted sequences[38, 39]: These sequences produce signal enhancement in the myocardium when used in combination with a predominantly T1-relaxing contrast agent such as Gd-DTPA.[34] They typically have a preparatory inversion pulse followed by a delay (inversion time, TI) (Figure 11–3). This delay is generally chosen such that the myocardial signal in baseline images before injection of contrast material is nulled. After the delay, radiofrequency (RF) pulses with a low ($\approx 15°$) angle (α, excitation angle) enable acquisition of the image. The time between the first RF pulse and the center of the readout gradient is the echo time (TE). Image intensity is determined by the sum of TI and the time point at which the central k-space lines are acquired. Changes in heart rate cause variations in signal intensity in the images,[56, 62] and therefore arrhythmias may have a deleterious effect on the images. This problem can be circumvented by adding a magnetization preparation pulse to drive the proton spins into a well-defined magnetization state.[63, 64]

T2* sequences can be used in combination with susceptibility contrast agents. However, these sequences typically have a long TE, thereby making the technique rather susceptible to cardiac motion.[65] Although successful imaging has been achieved with this technique, experience in clinical practice is very limited.

Echo-planar imaging (EPI) is an approach in

Figure 11–1. *A*, Example of sequential midventricular short axis CMR images in a patient with a 99% right coronary artery (RCA) stenosis during dipyridamole stress: Baseline *(top left)*, arrival of Gd-DTPA in the right ventricle *(top right)*, left ventricle *(bottom left)* and maximum contrast enhancement in the myocardium *(bottom right)*. A perfusion defect is visible in the inferior wall. RV = right ventricle; LV = left ventricle. *B*, The signal intensity (SI)–time curve in hypoperfused myocardium shows a delay and decrease in peak SI compared with normal myocardium. *C*, Circumferential profiles of CMR perfusion parameters can be constructed and compared with ^{201}Tl activity. The CMR profiles were derived from individual curve fits in 30 radial regions. The perfusion abnormality is illustrated by a decrease in CMR parameters and ^{201}Tl activity in the perfusion bed of the right coronary artery (RCA). 1/MTT appears to have less distinguishing capacity between normal and hypoperfused tissue than MCE and slope, which may be due to its dependence on the myocardial input function (bolus width and recirculation effects) and on vascular volume. 1/MTT = inverse mean transit time; Tl = [201]Tl activity; MCE = maximum myocardial contrast enhancement; Slope = slope of the wash-in curve. (Data from Keijer JT, van Rossum AC, van Eenige MJ, et al: Magnetic resonance imaging of regional myocardial perfusion in patients with single vessel coronary artery disease: Quantitative comparison with [201]Thallium-SPECT and coronary angiography. Proceedings 4th annual meeting ISMRM 1996, p 180 [abstract]).

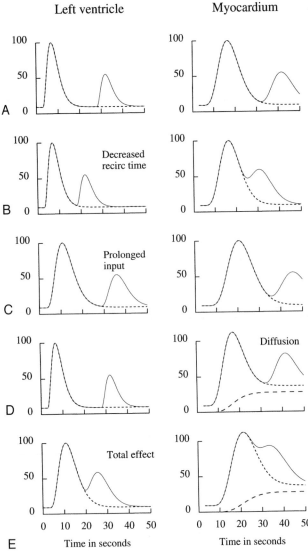

Figure 11–2. The effects of recirculation, dispersion, and diffusion on the myocardial signal intensity–time curve. In each graph, the solid line represents the measured outcome and the dashed line represents the true first-pass information. From top to bottom: *A*, Ideal left ventricular SI-time curve *(left)* and resulting myocardial SI-time curve *(right)*. *B*, Decrease in recirculation time. The descending part of the myocardial SI-time curve is clearly affected by recirculation. *C*, Prolongation of the myocardial input. *D*, Diffusion of Gd-DTPA to the extravascular space. The dashed line at the bottom of the graph represents the Gd-DTPA remaining in the extravascular space. *E*, Cumulative effects of recirculation, prolonged input, and diffusion.

which all data required for a single image are acquired with a single RF pulse.[66–67a] Acquisition times vary from 50 to 100 ms. This technique is therefore capable of imaging the transit of a contrast agent in multislice mode. Changing the imaging parameters yields T1-weighted or T2-weighted EPI sequences. Both gradient-echo and spin-echo type sequences have been used. Spin-echo imaging yields black blood images, which makes interpretation of the myocardial signal change easier and reduces blood-related artifacts, but also removes the input function, making the technique amenable to qualitative

analysis only. Currently, these techniques are not in wide use, but early clinical results are promising.

Multislice Imaging

For clinical decision-making, information on myocardial perfusion is best obtained from multiple (three or more) myocardial short-axis levels and/or long-axis levels. Using repetitive contrast boluses (one per level) is rather time consuming because approximately 10 to 15 minutes must be allowed for most extravascular contrast agents to wash out of the myocardium. It is much more time efficient to image multiple slices during a single first-pass bolus. An additional advantage of imaging multiple slices simultaneously is that every slice of myocardium has the same input of contrast agent, which makes comparison between slices easier. In order to image several myocardial levels simultaneously, several approaches have been taken to improve temporal resolution[68–72] of fast CMR imaging techniques. The common characteristic of these approaches is that they preferentially acquire the information that determines image contrast (the central, low order k-space lines).

The so-called "keyhole" technique acquires the low k-space lines, which determine image contrast and gross appearance,[68–70] and the more detailed high spatial frequency information is either filled in from a reference image or filled in at a slower rate than the low k-space lines. This approach can be applied to several CMR techniques. The major limitation of this technique is that temporal resolution improves at the cost of spatial resolution. Another approach is to acquire a reference image with a full field of view and acquire the dynamic images with a reduced field of view.[71] In this way, a decrease in acquisition time can be achieved. This technique is sensitive to changes in heart position (caused by respiration, or through-plane motion) because the baseline image will be different from the dynamic images. Finally, the k-space information can also be acquired in a random fashion, with the lower order lines visited more frequently.[72]

CARDIOVASCULAR MAGNETIC RESONANCE CONTRAST AGENTS

CMR contrast agents are generally used to enhance the signal of objects of interest or to visualize blood. The effect of these agents on CMR signal intensity mainly depends on the magnetic field strength, the CMR sequence used, water proton exchange rates between tissue compartments, and on the local concentration of the contrast agent, which is influenced by blood concentration, flow rate, diffusion into the interstitium, and relative intra- and extracellular volume fractions.[56, 61, 62] Although the predominance of their effects on T1 and T2 relax-

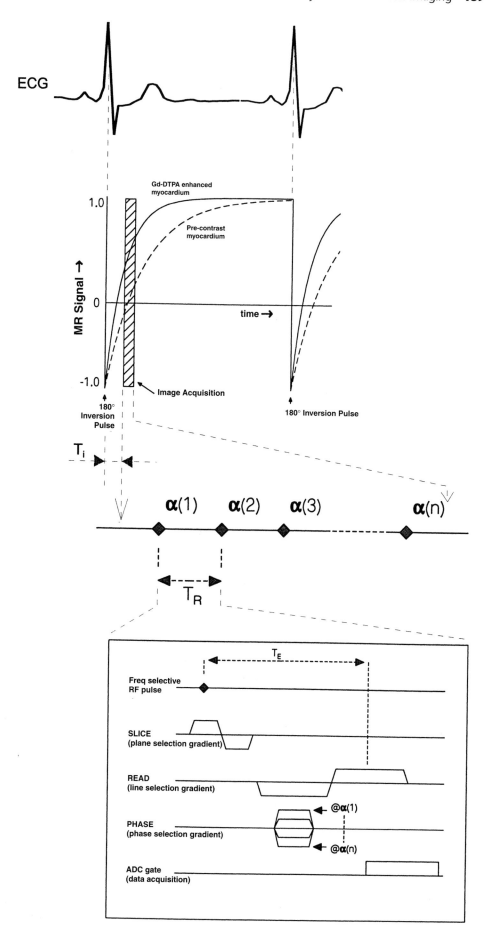

Figure 11–3. Segmented k-space gradient-echo CMR sequence. An inversion pulse is applied to invert magnetization; the time between inversion and the time of image acquisition (TI) determines signal intensity and is generally set at such a level that baseline intensity of myocardium is nulled. Gd-DTPA causes shortening of T1-relaxation, leading to increased magnetization and enhanced signal intensity. α = flip angle; TE = echo time, the time between radiofrequency (RF) pulse and center of image acquisition; TR = repetition time, the time between the consecutive RF pulses; ECG = electrocardiogram.

ation is dose dependent, CMR contrast agents can be classified in two general types[73, 74]:

1. The paramagnetic agents, like the compounds that contain the metal ion gadolinium, are currently in routine clinical use. These compounds increase the relaxation rate of surrounding protons.[62] For subsecond CMR techniques, this process results in shortening of T1 relaxation times and hence increased signal intensity on CMR images. In the low-dosage range (below 0.05 mmol/kg), this agent displays a relatively linear relation between concentration and signal intensity.[40, 55, 74] Most Gd-DTPA compounds available for application in humans are extracellular agents that readily diffuse into the interstitium[42] and may thus complicate differentiation of normally from abnormally perfused myocardium. On the other hand, this diffusion results in better contrast enhancement (higher signal-to-noise ratio) of the myocardium. This problem is to some extent ameliorated by the use of intravascular contrast agents,[57, 75, 76] although it should be noted that relaxed water will still diffuse out of the intravascular space into the myocardium, even if the contrast agent is held within the blood pool. Several approaches to making intravascular agents have been performed. One approach labels the contrast agent to a substance that remains in the blood pool such as albumin (e.g., MS-325 [EPIX Medical, Inc, Cambridge, MA]). Another uses an iron-based compound that is sufficiently large so that it does not pass through endothelial gaps very easily (NC100150 [Clariscan] from Nycomed Amersham, used at low dose for a predominant T1 effect). In animals, other solutions have also been possible, such as the use of polylysine, but this is not available for human use. The best type of contrast agent (intravascular or diffusible) for perfusion imaging remains to be demonstrated, and in particular, the role of intravascular agents in humans needs clarification.[77]

2. The susceptibility agents, e.g., compounds that contain dysprosium or iron oxide in higher dose, introduce local inhomogeneity in the magnetic field on a macroscopic scale by inducing large fluctuations of magnetic moment between the blood and the intracellular compartment.[78–80] The fluctuations are due to the large magnetic moment of the susceptibility agent itself. These fluctuations cause a reduction in T2* on neighboring hydrogen nuclei, which results in loss of signal intensity. If the susceptibility has a permanent magnetization, it is called a ferromagnetic contrast agent. If the magnetization is not permanent, the agent is called superparamagnetic. These agents are still limited to experimental trials and are not in routine clinical use.

Whether contrast agents will cause an increase or a decrease in signal intensity is related to the dose of the contrast agent and to the CMR imaging technique. For example, a low dose of a paramagnetic agent will increase signal intensity when using a T1-weighted sequence, whereas a high dose of this same agent will decrease signal intensity when us-ing a T2*-weighted sequence. Some CMR techniques are very sensitive to susceptibility effects, such as the gradient-echo and EPI techniques. Consequently, the same dose of a susceptibility agent causes more signal attenuation on gradient-echo or EPI images than on spin-echo images.

PHARMACOLOGICAL STRESS

Adenosine and dipyridamole are the most common agents used in myocardial perfusion imaging.[81] They decrease vascular resistance and consequently increase coronary flow to four to five times resting level.[82] Due to the maximum vasodilation, coronary autoregulation is lost. Myocardium that is perfused by a stenotic coronary artery will receive insufficient flow compared with the fully vasodilated, normal myocardium, thus leading to flow heterogeneity. Furthermore, vasodilation in the normal myocardium may reduce perfusion pressure to the ischemic area (coronary steal).[83–85] In as much as flow heterogeneity occurs before ischemia in the cascade of physiological events leading to angina, these agents are well suited for perfusion imaging because of the inherent sensitivity of the technique. The generally accepted infusion protocol for adenosine is 140 µg/kg/min given for 6 minutes. Side effects are associated with the generalized vasodilation and include headache, dizziness, flushing, abdominal discomfort and also bronchospasm in patients with asthma.[86] These effects are short due to its short half-life (<10 s) and are reversed by simply stopping the infusion or if necessary by the intravenous administration of aminophylline. Higher doses have an inhibiting effect on the sinoatrial node and on the atrioventricular (AV) node, potentially leading to sinus exit block, sinus arrest, and AV block, which may be considered its major adverse effects when used in combination with perfusion imaging.[82] Furthermore, it has a negative inotropic effect on the ventricular myocardium.[87]

Dipyridamole decreases the smooth muscle tone of the small-resistance arterioles by increasing plasma levels of adenosine.[88] It blocks the transmembrane transport and reuptake of adenosine into myocardial, endothelial, and blood cells and it prevents the inactivation of adenosine by inhibiting the enzyme adenosine deaminase. It also inhibits phosphodiesterase, which results in accumulation of cyclic adenosine monophosphate (AMP), causing vasodilation. Its action is slower and it lasts much longer than adenosine. The maximum effect starts 3 minutes after a 4-minute infusion of 0.56 mg/kg and continues for at least 10 minutes. The increase in myocardial blood flow is comparable to that of adenosine but may show more variation.[82] Side effects are similar, but may be tolerated better than those of adenosine because of slower onset. Side effects can be reversed by intravenous aminophylline. The choice between adenosine and dipyridamole for myocardial perfusion imaging may therefore be made on practical grounds (slow hand injection or continuous infusion system) in combination with the imaging protocol (intended duration of the hyperemia).[89] In general the controllability of adenosine and shorter duration of action have made it popular with physicians. Patients also prefer adenosine.

Another substance frequently used in cardiac stress imaging is dobutamine. Like most catecholamines, dobutamine increases heart rate and the force of contraction, thus causing oxygen demand to outstrip coronary supply.[85] This process leads to myocardial ischemia and consequently to abnormal myocardial wall motion. Logically, this agent is preferred when assessment of wall motion abnormalities is the objective (see Chapter 10) or when dipyridamole and adenosine are contraindicated.

EXPERIMENTAL AND CLINICAL STUDIES

Wilke and coworkers[40] demonstrated a good correlation between inverse mean transit time (MTT, Figure 11–4) and myocardial blood flow using pharmacological vasodilation to improve the detection of myocardial ischemia. In an experimental study in dogs with left atrial contrast injection, inverse MTT was shown to have a positive, linear relation with microsphere-assessed myocardial blood flow in dogs (r = 0.89), after correction for changes in vascular volume. Additional to 1/MTT, slope of the SI-time curve was also related to blood flow in this study.

Atkinson and associates[90] applied subsecond contrast-enhanced CMR in isolated perfused rat hearts. In an occlusive infarct model, perfused and nonperfused myocardium were demonstrated to have different contrast enhancement. They also applied the

first-pass technique in five healthy subjects, and other groups confirmed the technical feasibility of contrast-enhanced CMR in humans.[55, 91]

Manning and colleagues[92] measured resting peak signal intensity of the first pass of Gd-DTPA in 20 patients with greater than 90 percent single-vessel CAD. They found a decrease in peak signal intensity and a lower slope of the SI-time curve in myocardium perfused by a stenotic vessel. After revascularization, peak signal intensity recovered to normal values in 9/10 patients. The significance of this study is that it indicates the possibility to detect perfusion abnormalities at rest in patients with a critical coronary artery stenosis that normalized after revascularization. Kraitchman and coworkers[93] combined CMR perfusion imaging with CMR tissue tagging and reported reduced regional deformation and motion in myocardial areas with altered perfusion (lower upslope of the SI-time curve), indicating the possibility of CMR to perform an integrated quantitative examination of myocardial perfusion and function.

Comparison of Stress Imaging: Radionuclide Scintigraphy Versus Cardiovascular Magnetic Resonance

Several investigators have compared stress CMR perfusion with radionuclide methods. Schaefer and colleagues[94] measured first-pass signal intensity changes at four time points after bolus injection and related these to planar ^{201}Tl scintigraphy. Signal intensities were measured in the centerline of the myocardium, yielding a circumferential profile. The limits of normal perfusion were set using a control group of four normal subjects. In 6 patients with CAD, perfusion defects were assessed by both methods in 6/6 patients. A subsequent study by Hartnell and associates[95] reported a 73 percent agreement in the visual assessment of perfusion defects when comparing subsecond CMR after peripheral bolus injection and planar ^{201}Tl scintigraphy in 15 patients, with coronary angiography as the gold standard.

Klein and coworkers[96] compared first-pass CMR with sestamibi SPECT. In 5 patients they found agreement in 19/25 (76%) regions after visual analysis. Quantitative analysis did not improve these results. Only 10 consecutive images were obtained after peripheral bolus injection, thereby possibly missing peak signal intensity.

Eichenberger and associates[97] imaged multiple tomographic planes in one acquisition after one peripheral bolus injection. This study was limited by a suboptimal time resolution (every third heartbeat) and furthermore, input correction was done from nonlinear signal intensity in the left ventricular cavity. Using the slope of the wash-in curve, they detected ischemic regions in six out of eight patients as compared to ^{201}Tl-SPECT and bypass surgery.

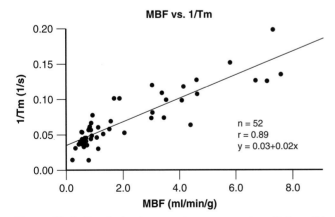

MBF vs. 1/Tm

n = 52
r = 0.89
y = 0.03+0.02x

Figure 11–4. Correlation between inverse mean transit time (1/Tm) in dogs, calculated from the SI-time curve, and microsphere blood flow (MBF) measured with radioactive microspheres as the gold standard. Each point represents an individual measurement. The transit times are corrected for the volume change with hyperemia in nonischemic regions. (Adapted from Wilke N, Simm C, Zhang J, et al: Contrast-enhanced first-pass myocardial perfusion imaging: Correlation between myocardial blood flow in dogs at rest and during hyperemia. Magn Reson Med 29:485, 1993.)

Walsh and colleagues[69] accomplished a reduction in acquisition time to 130 ms using an interpolated keyhole technique that dynamically samples only central k-space. This technique allowed simultaneous imaging of three to four short-axis levels with one venous injection. Qualitative analysis was performed after subtraction of blood pool intensity; 28/45 patients had 100 percent concordance (24/28 with perfusion defects, 4 normal) and 17 were discordant. In this study, peripheral bolus injection was claimed to permit visualization of not only major but also minor perfusion defects.

In a comparative study between first-pass CMR and sestamibi SPECT in patients with single-vessel disease, Matheijssen and coworkers[98] found an agreement of 90 percent in localization of perfusion defects between SPECT and two-level CMR in 10 patients, albeit through visual analysis only. In this study, peripheral bolus injection was performed. Maximum signal intensity and slope of the wash-in curve were found to be indicative of perfusion abnormalities when using a linear fit. Gamma variate fitting of all first-pass data points, however, did not result in any significant findings. A comparative study between ^{201}Tl-SPECT and first-pass CMR showed moderate correlations between contrast-enhanced CMR and ^{201}Tl-SPECT, with regard to extent ($r = 0.59$, $p < 0.04$) and severity ($r = 0.73$, $p < 0.005$) of dipyridamole-induced perfusion defects[53] and a sensitivity and specificity of 71 percent and 71 percent, respectively, in the assessment of perfusion defects. Wilke and associates[99] found a correlation ($r = 0.8$) between (Doppler) coronary flow reserve and myocardial perfusion reserve in patients with nonsignificant CAD using multislice imaging. The myocardial perfusion reserve was defined as the impulse response amplitude for hyperemic flow divided by the impulse response amplitude for basal flow. The myocardial impulse response was deconvoluted using the left ventricular SI-time curve as input function. Also, the relative perfusion index, defined as the ratio of maximum impulse amplitudes in normally perfused and hypoperfused myocardium, correlated well with microsphere blood flow ($r = 0.88$).

In summary, CMR is capable of imaging myocardial perfusion. Changes in the dynamic wash-in of contrast agent may sometimes be difficult to assess visually and artifacts may disturb visual analysis of single images. Therefore, some form of quantitative or semiquantitative analysis is preferred. Quantitation of myocardial blood flow using CMR is feasible in the animal setting, employing central bolus injection, which allows straightforward derivation of MTT as the perfusion parameter. In clinical practice, however, a peripheral bolus injection is preferred, but this practice allows only semiquantitation, using such parameters as the slope and the peak of the wash-in curve, which represent the rate and maximum of contrast enhancement, respectively. Deconvolution of the myocardial impulse response is a method to circumvent a variable myocardial input as seen with peripheral bolus injection. The agreement between CMR and scintigraphic techniques is reasonable. Due to the differences in study design and interpretation, there are no consistent data available on sensitivity and specificity of CMR perfusion imaging in larger patient groups, and multicenter studies are needed. The trend in recent clinical studies[39, 100] is to use peripheral bolus injection and sacrifice some temporal resolution in order to image multiple short-axis levels of myocardium.

CARDIOVASCULAR MAGNETIC RESONANCE EVALUATION OF TRANSMURAL DIFFERENCES

The subendocardium is more vulnerable to ischemia than the subepicardium. This vulnerability can be explained by higher metabolic demands, as well as by higher vascular resistance in the subendocardium. Systolic extravascular compressive forces are higher in the subendocardium than in the subepicardium. This finding emphasizes the dependence of the subendocardium on diastolic flow. More importantly, in systole, transmural pressure decreases to a greater extent in the subendocardium than in the subepicardium, thereby decreasing vascular diameter and increasing resistance,[101] affecting the following diastolic inflow. The increased resistance and the higher metabolic demand in the myocardium result in preferential vasodilation in the subendocardium. Consequently, the recruitable reserve in flow[102] by further decreasing vascular tone is lower in the subendocardium.[103]

The high spatial resolution of CMR makes it well suited for the study of transmural variations in myocardial perfusion (Figure 11–5). The possibility to assess transmural variations in myocardial blood flow with first-pass CMR has previously been indicated by Wilke and colleagues in dogs[40] and in patients.[41] Lima and coworkers[42] described different transmural patterns in myocardial SI in patients with reperfused and nonreperfused myocardial infarction, illustrating the potential of CMR to visualize transmural differences in myocardial perfusion. In a clinical study of 22 patients with single-vessel CAD by Keijer and associates,[104] it was shown that CMR is capable of at least approximation of the transmural redistribution of myocardial perfusion during pharmacological stress (Figure 11–6): Subendocardial myocardial contrast enhancement in areas perfused by a stenotic coronary artery decreased during hyperemia (0.89 ± 0.18 vs. 0.74 ± 0.15, $p < 0.003$) and was lower in subendocardium than in subepicardium (0.74 ± 0.15 vs. 0.84 ± 0.21, $p < 0.02$). Contrast enhancement in these areas was normalized to contrast enhancement in normal myocardium. The parameters of slope and 1/MTT paralleled maximum contrast enhancement (MCE). Other techniques potentially capable of imaging subendocardial perfusion are ultrafast computed tomography[16] and contrast echocardiography,[23, 24] but neither

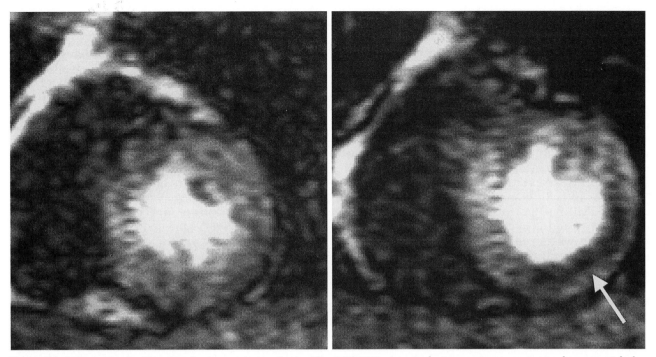

Figure 11–5. Example of midventricular images in a patient with a 99% RCA stenosis during maximum contrast enhancement, before *(left)* and after *(right)* dipyridamole. The perfusion abnormality (indicated by the arrow) is predominant in the subendocardium.

is in clinical use in a significant fashion, and both have significant limitations (limited imaging planes, image quality, radiation burden, or acoustic access).

EVALUATION OF CORONARY INTERVENTIONS

The functional result of a coronary revascularization procedure can be evaluated by assessing the coronary flow reserve or the myocardial perfusion reserve, using pharmacological stress. Coronary flow can be measured using the intracoronary Doppler guidewire, whereas myocardial perfusion can be studied with SPECT, PET, videodensitometry, or contrast echocardiography. More recently, results with CMR have been reported. The first to report on the use of contrast-enhanced CMR in patients after revascularization were Manning and colleagues.[92] They used a single-slice inversion-recovery segmented k-space gradient-echo technique with a temporal resolution of 1 image per 3 to 4 s, and an intravenous bolus injection of Gd-DTPA of 0.04 mmol/kg. CMR at rest was performed in 12 patients with coronary artery stenoses with more than 80 percent luminal diameter narrowing, of whom 4 patients underwent repeat CMR 0.5 to 2 months after revascularization. After revascularization, peak

Figure 11–6. The ratios of signal in abnormal and normal myocardium of the CMR perfusion parameter MCE (maximum contrast enhancement) at three transmural levels in a group of patients. MCE in subendocardial myocardium (ENDO) decreased during dipyridamole (DIPY), and was lower than in subepicardial myocardium (EPI), indicating dipyridamole-induced subendocardial hypoperfusion. A/N = abnormal to normal myocardium.

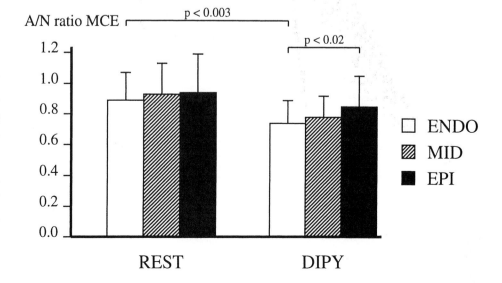

signal intensity relative to peak ventricular signal intensity of normally perfused myocardium did not change, whereas in myocardium perfused by stenosed arteries, this relative value was significantly lower before and normalized after revascularization.

Lauerma and associates[100] studied 11 patients with single-vessel LAD stenoses with more than 80 percent luminal diameter narrowing, of whom 4 had a total occlusion without infarction (presumably with collateral supply). The same CMR technique was used covering the left ventricle with 3 short axis slices, with repeat image acquisition every 3 to 6 seconds after 0.05 mmol/kg Gd-DTPA intravenously. CMR was performed after dipyridamole stress before and 3 months after revascularization. Comparison was made with ²⁰¹Tl-SPECT. Peak contrast-to-noise ratios (peak of normal–abnormal myocardium/background noise) and enhancement rates (slope of the first-pass curve fit) were measured. The contrast ratio increased during stress prerevascularization and normalized after revascularization (Figure 11–7). Slope decreased during stress before revascularization and normalized after revascularization. Furthermore, revascularization decreased the defect sizes both in scintigraphy and CMR.

Keijer and coworkers[105] applied contrast-enhanced CMR during dipyridamole-induced hyperemia before and 3 hours after coronary angioplasty in 15 patients with single-vessel CAD with stenoses ranging from 75 percent to 99 percent. A single-slice inversion-recovery segmented k-space gradient-echo technique was used with a temporal resolution of 1 image per heartbeat. A bolus of 0.03 mmol/kg Gd-

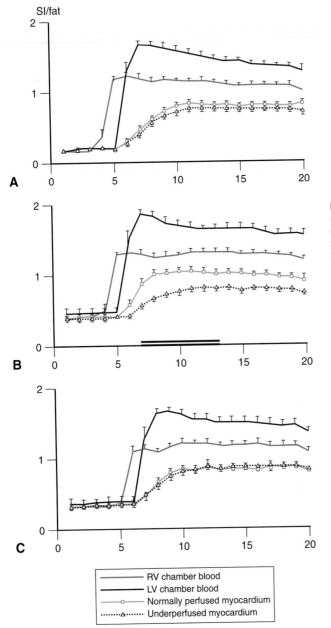

Figure 11–7. Plots of temporal signal changes in right ventricular (RV) and left ventricular (LV) chamber blood and in perfusion beds of normal and diseased coronary arteries during transit of 0.05 mmol/kg Gd-DTPA. Signal intensities were normalized to subcutaneous fat. Gated images were acquired at same level every 4 to 6 heartbeats, with mean temporal resolution of 4.4 ± 0.6 s (range, 3.4 to 6.0 s). Plots represent mean ± SEM curves of 11 patients *(A)* before treatment at rest, *(B)* before treatment during pharmacologically induced stress, and *(C)* 3 months after revascularization at stress. Curves of normal and underperfused myocardial regions differ significantly only in plot b, where black bar on x axis (images 7 through 13) indicates $p < 0.01$. Note the difference in curve shape (no downslope due to peripheral contrast injection and low temporal resolution) after peak signal intensity compared with Figure 11–1*B* (central injection and high resolution). (Adapted from Lauerma K, Virtanen K, Sipila LM, et al: Multislice MRI in assessment of myocardial perfusion in patients with single vessel proximal left anterior descending coronary artery disease before and after revascularization. Circulation 96:2859–2867, 1997.)

DTPA was injected into the right atrium. From the first-pass curves, MCE was derived. In 14 out of 15 patients, there was an increase in the subendocardial ratio of abnormally to normally perfused myocardium (A/N of MCE). The mean subendocardial A/N increased from 0.75 ± 0.14 to 0.86 ± 0.13 ($p < 0.02$). The subendocardial/subepicardial ratio (endo/epi) of MCE improved in 9 out of 15 patients. The mean endo/epi ratio of MCE in abnormal myocardium was 0.93 ± 0.13 before and 1.01 ± 0.12 ($p = 0.066$) after angioplasty. The mean endo/epi ratio in normal myocardium was 1.01 ± 0.14 before and 1.01 ± 0.12 after angioplasty (NS; $p < 0.03$ abnormal vs. normal myocardium before percutaneous transluminal coronary angioplasty [PTCA]). The failure of the A/N MCE to return completely to normal may reflect the phenomenon known from PET studies that coronary flow reserve may not necessarily normalize completely immediately after PTCA.[106, 107] Several mechanisms have been proposed to explain this finding. These include increase of resting myocardial flow, embolization of material during PTCA toward the distal microcirculation, and temporary dysfunction of the autoregulatory system due to prolonged decrease in perfusion pressure.

DELAYED CONTRAST ENHANCEMENT FOR ASSESSMENT OF PROGNOSIS AFTER INFARCTION

Finally, Wu and coworkers[108] investigated the prognostic value of contrast-enhanced CMR using a delayed imaging approach. They investigated 44 patients 10 ± 6 days after myocardial infarction. Microvascular obstruction was defined as hypoenhancement 1 to 2 minutes after contrast injection. Patients with microvascular obstruction had more cardiovascular events during a follow-up of 16 ± 5 months than patients without microvascular obstruction (45% vs. 9%, $p = 0.016$). The risk of adverse events increased with infarct size (as assessed by hyperenhancement of myocardium 10 minutes after contrast injection) 30, 43, and 71 percent for small, midsized, and large infarctions, respectively ($p < 0.05$). The use of CMR contrast agents to assess viability is addressed in more detail in Chapter 14.

Thus, clinical evidence is available indicating that during pharmacological stress, CMR first-pass perfusion imaging using Gd-DTPA differentiates normal from underperfused myocardium before revascularization and identifies recovery of perfusion after revascularization in selected patient groups (with normal systolic left ventricular function). Furthermore, delayed contrast-enhanced imaging may be of prognostic value. The feasibility of CMR to measure perfusion in the subendocardium may help to increase the sensitivity for detection of perfusion abnormalities, although the parameters used clinically are at best semiquantitative.

CONCLUSION

The potential of CMR to image myocardial perfusion with high temporal and spatial resolution has evolved into a few practical applications. In patients, CMR perfusion imaging has the capacity to detect, localize, and semiquantitate myocardial perfusion. Further study is warranted to establish the diagnostic accuracy of the method in larger patient groups and to evaluate the most practical (multislice) patient protocol. The feasibility of CMR perfusion imaging to assess subendocardial perfusion and to evaluate coronary interventions may prove particularly useful in future clinical practice. Absolute quantification of myocardial perfusion seems the prerequisite of animal studies (this process requires central bolus injection of contrast agent). Multicenter studies using a uniform approach are anxiously awaited, but studies from single centers are showing very promising results.[109, 110]

References

1. Bol A, Melin A, Vanoverschelde L, et al: Direct comparison of [13N]ammonia and [15O]water estimates of perfusion with quantification of regional myocardial flow by microspheres. Circulation 87:512, 1993.
2. Grover-McKay M, Ratib O, Schwaiger M, et al: Detection of coronary artery disease with positron-emission tomography and rubidium-82. Am Heart J 123:646, 1992.
3. Schwaiger M, Muzik O: Assessment of myocardial perfusion by positron emission tomography. Am J Cardiol 67:35D, 1991.
4. Maddahi J, Van Train K, Prigent F, et al: Quantitative single photon emission computed thallium-201 tomography for detection and localization of coronary artery disease: Optimization and prospective validation of a new technique. J Am Coll Cardiol 14:1689, 1989.
5. Van Train KF, Maddahi J, Berman DS, et al: Quantitative analysis of tomographic stress thallium-201 myocardial scintigrams: A multicenter trial. J Nucl Med 31:1186, 1990.
6. Okada R, Glover D, Gaffney T, et al: Myocardial kinetics of technetium-99m hexakis-2-methoxy-2-methylpropylisonitrile. Circulation 77:491, 1988.
7. Melon PG, Beanlands RS, DeGrado TR, et al: Comparison of technetium-99m sestamibi and thallium-201 retention characteristics in canine myocardium. J Am Coll Cardiol 20:1277, 1992.
8. Gray WA, Gewirtz H: Comparison of 99mTc-teboroxime with thallium for myocardial imaging in the presence of a coronary artery stenosis. Circulation 84:1796, 1991.
9. Phillips DJ, Henneman RA, Merhige ME: Rapid diagnosis of coronary disease using dipyridamole and teboroxime washout imaging (abstract). J Am Coll Cardiol 23:255A, 1994.
10. Sinusas AJ, Shi QX, Saltzberg MT, et al: Technetium-99m tetrofosmin to assess myocardial blood flow; experimental validation in an intact canine model of ischemia. J Nucl Med 35:664, 1994.
11. Higley B, Smith FW, Smith T, et al: Technetium-99m-1,2{bis(2-ethoxyethyl)phosphine}ethane: Human biodistribution, dosimetry and safety of a new myocardial perfusion imaging agent. J Nucl Med 34:30, 1993.
12. Jain D, Wackers FJ, Mattera J, et al: Biokinetics of 99m Tc-tetrofosmin, myocardial perfusion imaging agent: Implications for a one day imaging protocol. J Nucl Med 34:1254, 1993.

13. Tamaki N, Takahashi N, Kawamoto M, et al: Myocardial tomography using technetium-99m-tetrofosmin to evaluate coronary artery disease. J Nucl Med 35:594, 1994.

14. Zaret BL, Rigo P, Wackers FJ, et al: Myocardial perfusion imaging with ^{99m}Tc tetrofosmin. Comparison to ^{201}Tl imaging and coronary angiography in a phase III multicenter trial. Circulation 91:313, 1995.

15. Rumberger JA, Feiring AJ, Lipton MJ, et al: Use of ultrafast computed tomography to quantitate regional myocardial perfusion: A preliminary report. J Am Coll Cardiol 9:59, 1987.

16. Canty JM, Brody A, Klocke FJ: Quantitative assessment of reductions in subendocardial and full-thickness regional flow reserve during progressive coronary stenosis using ultrafast computed tomography. J Am Coll Cardiol 15:161A, 1990.

17. Weiss RM, Otoadese EA, Noel MP, et al: Quantitation of absolute regional myocardial perfusion using cine computed tomography. J Am Coll Cardiol 23:1186, 1994.

18. Lang RM, Feinstein SB, Feldman T, et al: Contrast echocardiography for evaluation of myocardial perfusion: Effects of coronary angioplasty. J Am Coll Cardiol 8:232, 1986.

19. Perchet H, Dupouy P, Duval-Moulin AM, et al: Improvement of subendocardial myocardial perfusion after percutaneous transluminal coronary angioplasty. A myocardial contrast echocardiographic study with correlation between myocardial contrast reserve and doppler coronary reserve. Circulation 91:1419, 1995.

20. Kaul S, Kelly P, Oliner JD, et al: Assessment of regional myocardial blood flow with myocardial contrast two-dimensional echocardiography. J Am Coll Cardiol 13:468, 1989.

21. Porter TA, D'Sa A, Turner C, et al: Myocardial contrast echocardiography for the assessment of coronary blood flow reserve: Validation in humans. J Am Coll Cardiol 21:349, 1993.

22. Skyba DM, Jayaweera AR, Goodman NC, et al: Quantification of myocardial perfusion with myocardial contrast echocardiography during left atrial injection of contrast. Implications for venous injection. Circulation 90:1513, 1994.

23. Cheirif J, Zoghbi WA, Bolli R, et al: Assessment of regional perfusion by contrast echocardiography, II: Detection of changes in transmural and subendocardial perfusion during dipyridamole-induced hyperemia in a model of critical stenosis. J Am Coll Cardiol 14:1555, 1989.

24. Lim YJ, Nanto S, Masuyama T, et al: Visualization of subendocardial myocardial ischemia with myocardial contrast echocardiography in humans. Circulation 79:233, 1989.

25. Pijls NHJ, Uijen GJH, Hoevelaken A, et al: Mean transit time for the assessment of myocardial perfusion by videodensitometry. Circulation 81:1331, 1990.

26. Eigler NL, Schuelen H, Whiting JS, et al: Digital angiographic impulse response analysis of regional myocardial perfusion. Estimation of coronary flow, flow reserve, and distribution volume by compartment transit time measurement in a canine model. Circ Res 68:870, 1991.

27. Schuelen H, Eigler NL, Whiting JS: Digital angiographic impulse response analysis of regional myocardial perfusion. Detection of autoregulatory changes in nonstenotic coronary arteries induced by collateral flow to adjacent coronary arteries. Circulation 89:1004, 1994.

28. Pijls NJ, Aengevaren WRM, Uijen GJH, et al: Concept of maximal flow ratio for immediate evaluation of percutaneous transluminal coronary angioplasty result by videodensitometry. Circulation 83:854, 1991.

29. Schuelen H, Eigler NL, Zeiger AM, et al: Digital angiographic assessment of the physiological changes to the regional microcirculation induced by successful coronary angioplasty. Circulation 90:163, 1994.

30. Williams ES, Kaplan JI, Thatcher F, et al: Prolongation of proton spin lattice times in regionally ischemic tissue from dog hearts. J Nucl Med 21:449, 1980.

31. Higgins CB, Herkens R, Lipton MJ, et al: Nuclear magnetic resonance imaging of acute myocardial infarction in dogs: Alterations in magnetic relaxation times. Am J Cardiol 52:184, 1983.

32. Aisen AM, Buda AJ, Zotz RJ, et al: Visualization of myocardial infarction and subsequent coronary reperfusion with MR using a dog model. Magn Reson Imaging 5:399, 1987.

33. Pflugfelder PW, Wisenberg G, Prato FS, et al: Serial imaging of canine myocardial infarction by in vivo nuclear magnetic resonance. J Am Coll Cardiol 7:843, 1986.

34. Weinmann HJ, Brasch RC, Press WR, et al: Characteristics of gadolinium-DTPA complex: A potential NMR contrast agent. AJR Am J Roentgenol 142:619, 1984.

35. Schmiedl U, Moseley ME, Ogan MD, et al: Comparison of initial biodistribution patterns of Gd-DTPA and Albumin-(Gd-DTPA) using rapid spin-echo MR imaging. J Comput Assist Tomogr 11(2):306, 1987.

36. Eichstaedt HW, Felix R, Dougherty FC, et al: Magnetic resonance imaging in different stages of myocardial infarction using the contrast agent Gadolinium-DTPA. Clin Cardiol 9:527, 1986.

37. Nishimura T, Kobayashi H, Ohara Y, et al: Serial assessment of myocardial infarction by using gated MR-imaging and Gd-DTPA. AJR Am J Roentgenol 150:531, 1988.

38. Frahm J, Merboldt KD, Bruhn H, et al: 0.3-Second FLASH MRI of the human heart. Magn Reson Med 13(1):150, 1990.

39. Cohen MS, Weiskoff RM: Ultra-fast imaging. Magn Reson Med 9:1, 1991.

40. Wilke N, Simm C, Zhang J, et al: Contrast-enhanced first-pass myocardial perfusion imaging: Correlation between myocardial blood flow in dogs at rest and during hyperemia. Magn Reson Med 29:485, 1993.

41. Wilke N, Jerosch-Herold M, Stillman AE, et al: Myocardial perfusion reserve and transmural perfusion in patients with coronary artery disease (abstract). Proceedings of the 3rd annual meeting ISMRM, 1995, p 1400.

42. Lima JAC, Judd RM, Bazille A, et al: Regional heterogeneity of human myocardial infarcts demonstrated by contrast-enhanced MRI. Potential mechanisms. Circulation 92:1117, 1995.

43. Higgins CB, Saeed M, Wendland M, et al: Evaluation of myocardial function and perfusion in ischemic heart disease. Magn Reson Mater Phys Biol Med 2/3:177–178, 1994.

44. Van Rugge FP, Van der Wall EE, Spanjersberg S, et al: Magnetic resonance imaging during dobutamine stress for detection and localization of coronary artery disease. Quantitative wall motion analysis using a modification of the centerline method. Circulation 90:127, 1994.

45. Hofman MBM, Van Rossum AC, Sprenger M, Westerhof N: Assessment of flow in the right human coronary artery by magnetic resonance phase contrast velocity measurements: Impact of cardiac and respiratory motion. Magn Reson Med 35:521, 1996.

46. Hundley WG, Lange RA, Clarke GD, et al: Assessment of coronary flow and flow reserve in humans with magnetic resonance imaging. Circulation 93:1502, 1996.

47. Sakuma H, Blake LM, Amidon TM, et al: Coronary flow reserve: Noninvasive measurement in humans with breath-hold velocity-encoded cine MR imaging. Radiology 198:745, 1996.

48. Weiss RG, Bottomley PA, Hardy CJ, Gerstenblith G: Regional myocardial metabolism of high energy phosphates during isometric exercise in patients with coronary artery disease. N Engl J Med 323:1593, 1990.

49. Albert MS, Huang W, Lee JH, Patlak CS: Susceptibility changes following bolus injections. Magn Reson Med 29(5):700, 1993.

50. Meier P, Zierler K: On the theory of the indicator-dilution method for measurement of blood flow and volume. J Appl Physiol 6:731, 1954.

51. Zierler KL: Circulation times and the theory of indicator-dilution methods for determining blood flow and volume. In Handbook of Physiology. Vol 1. Washington, DC, American Physiological Society, 1962, 585.

52. Bloomfield DA: Dye curves: The theory and practice of indicator-dilution. Baltimore, University Park Press, 1974.

53. Keijer JT, van Rossum AC, van Eenige MJ, et al: Magnetic resonance imaging of regional myocardial perfusion in patients with single vessel coronary artery disease: Quantita-

tive comparison with [201]Thallium-SPECT and coronary angiography (abstract). Proceedings of the 4th annual meeting ISMRM, 1996, p 180.

54. Thompson HK, Starmer CF, Whalen RE, et al: Indicator transit time considered as a gamma variate. Circ Res 14:502, 1964.

55. Keijer JT, Van Rossum AC, Van Eenige MJ, et al: Semiquantitation of regional myocardial blood flow in normal human subjects using first pass magnetic resonance imaging. Am Heart J 130:893, 1995.

56. Burstein D, Taratuta E, Manning WJ: Factors in myocardial perfusion imaging with ultrafast MRI and Gd-DTPA administration. Magn Reson Med 20:299, 1991.

57. Diesbourg LD, Prato FS, Wisenberg G, et al: Quantification of myocardial blood flow and extracellular volumes using a bolus injection of Gd-DTPA: Kinetic modeling in canine ischemic disease. Magn Reson Med 23:239, 1992.

58. Wilke N, Kroll K, Merkle H, et al: Regional myocardial blood volume and flow via MR first pass imaging in concert with polylysine-gadolinium-DTPA. J Magn Reson Imaging 5:227, 1995.

59. Bauer WR, Hiller KH, Roder F, et al: Magnetization exchange in capillaries by microcirculation affects diffusion-controlled spin-relaxation: A model which describes the effect of perfusion on relaxation enhancement by intravascular contrast agents. Magn Reson Med 35:43, 1996.

60. Keijer JT, Bax JJ, Van Rossum AC, et al: Myocardial perfusion imaging: Clinical experience and recent progress in radionuclide scintigraphy and magnetic resonance imaging. Int J Card Imaging 13:415, 1997.

61. Tong CY, Prato F, Wisenberg G, et al: Measurement of the extraction efficiency and distribution volume for Gd-DTPA in normal and diseased canine myocardium. Magn Reson Med 30:337, 1993.

62. Higgins CB, Saeed M, Wendland M: Contrast enhancement for the myocardium. Magn Reson Med 22:347, 1991.

63. Tsekos NV, Zhang Y, Merkle H, et al: Fast anatomical imaging of the heart and assessment of myocardial perfusion with arrhythmia insensitive magnetization preparation. Magn Reson Med 34:530, 1995.

64. Laub G, Simonetti O: Assessment of myocardial perfusion with saturation-recovery Turbo-FLASH sequences. Proceedings of the ISMRM, New York, 1996, p 179.

65. Hu X, Kim S-G: A new T2* weighting technique for magnetic resonance imaging. Magn Reson Med 30:512, 1993.

66. Edelman RR: Contrast-enhanced echo-planar MR imaging of myocardial perfusion: Preliminary study in humans. Radiology 190:771, 1994.

67. Wendland MF, Saeed M, Masui T, et al: Echo-planar imaging of normal and ischemic myocardium with gadodiamide injection. Radiology 186:535, 1993.

67a. Panting JR, Gatehouse PD, Yang GZ, et al: Echo-planar magnetic resonance myocardial perfusion imaging: Parametric map analysis and comparison with thallium SPECT. J Magn Reson Imaging 13:192, 2001.

68. Hu X: On the "keyhole" technique. J Magn Reson Imaging 31:691, 1994.

69. Walsh EG, Doyle M, Lawson M, et al: Multislice first-pass myocardial perfusion imaging on a conventional clinical scanner. Magn Reson Med 34:39, 1995.

70. Walsh EG, Doyle M, Foster RE, Pohost GM: Rapid cardiac imaging with turbo BRISK. Magn Reson Med 37:410, 1997.

71. Hu X, Parrish T: Reduction of field of view for dynamic imaging. Magn Reson Med 31:691, 1994.

72. Parrish T, Hu X: Continuous update with random encoding (CURE). A new strategy for dynamic imaging. Magn Reson Med 33(3):326, 1995.

73. Muehler A: Assessment of myocardial perfusion using contrast enhanced MRI: Current status and future developments. Magn Reson Mater Phys Biol Med 3(1):21, 1995.

74. Wilke N, Jerosch-Herold M, Stillman AE, et al: Concepts of perfusion imaging in magnetic resonance imaging. Magn Reson Quart 10:249, 1994.

75. Schuhmann-Giampieri G, Schmit-Willich, Frenzel T, et al: In vivo and in vitro evaluation of Gd-DTPA-polylysine as a macromolecular contrast agent for magnetic resonance imaging. Invest Radiol 26:969, 1991.

76. Dolan RP, Pottumarthi VP, Wielopolski PA, et al: First-pass myocardial imaging with MS-325, an intravascular MRI contrast agent. Proceedings of the 4th annual meeting ISMRM, 1996, p 686.

77. Crnac J, Schmidt MC, Theissen P, Sechtem U: Assessment of myocardial perfusion by magnetic resonance imaging. Herz 22:16, 1997.

78. Weissleder R, Elizondo G, Wittenberg J, et al: Ultrasmall superparamagnetic iron oxide: Characterization of a new class of contrast agents for MR imaging. Radiology 175:489, 1990.

79. Chambon C, Clement O, Leblanche A, et al: Superparamagnetic ion oxides as positive MR contrast agents: In vitro and in vivo evidence. Magn Reson Imaging 11:509, 1993.

80. Wiener EC, Brechbiel MW, Brothers H, et al: Dendrimer-based metal chelates: A new class of magnetic resonance imaging contrast agents. Magn Reson Imaging 31:1, 1994.

81. Van Rugge FP, Van der Wall EE, Bruschke AVG: New developments in pharmacologic stress imaging. Am Heart J 24:468, 1992.

82. Chan SY, Brunken RC, Czernin J, et al: Comparison of maximal myocardial blood flow during adenosine infusion with that of intravenous dipyridamole in normal men. J Am Coll Cardiol 20:979, 1992.

83. Rossen JD, Quillen JE, Lopez AG, et al: Comparison of coronary vasodilation with intravenous dipyridamole and adenosine. J Am Coll Cardiol 18:485, 1991.

84. Meerdink DJ, Okada RD, Leppo JA: The effect of dipyridamole on transmural blood flow gradients. Chest 96:400, 1989.

85. Vasu MA, O'Keefe DD, Kapellakis GZ, et al: Myocardial oxygen consumption: Effects of epinephrine, isoproterenol, dopamine, norepinephrine and dobutamine. Am J Physiol 235:H237, 1978.

86. Belardinelli L, Linden J, Berne RM: The cardiac effects of adenosine. Prog Cardiovasc Dis 32:73, 1989.

87. Wilson RF, Wyche K, Christensen BV, et al: Effects of adenosine on human coronary arterial circulation. Circulation 82:1596, 1990.

88. Fitzgerald GA: Dipyridamole. N Engl J Med 316:1247, 1987.

89. Wackers FTJ: Which pharmacological stress is optimal? A technique-dependent choice. Circulation 87:646, 1993.

90. Atkinson DJ, Burstein D, Edelman RR: First-pass cardiac perfusion: Evaluation with ultrafast MR-imaging. Radiology 174:757, 1990.

91. Van Rugge FP, Boreel JJ, Van der Wall EE, et al: Cardiac first-pass and myocardial perfusion in normal subjects assessed by subsecond Gd-DTPA enhanced MR imaging. J Comput Assist Tomogr 15:959, 1991.

92. Manning WJ, Atkinson DJ, Grossman W, et al: First-pass nuclear magnetic resonance imaging studies using gadolinium-DTPA in patients with coronary artery disease. J Am Coll Cardiol 18:959, 1991.

93. Kraitchman DL, Wilke N, Hexeberg E, et al: Myocardial perfusion and function in dogs with moderate coronary stenosis. Magn Reson Med 35:771, 1996.

94. Schaefer S, Van Tyen R, Saloner D: Evaluation of myocardial perfusion abnormalities with gadolinium-enhanced snapshot MR-imaging in humans. Radiology 185:795, 1992.

95. Hartnell G, Cerel A, Kamalesh M, et al: Detection of myocardial ischemia: Value of combined myocardial perfusion and cineangiographic MR imaging. AJR Am J Roentgenol 163:1061, 1994.

96. Klein MA, Collier BD, Hellman RS, et al: Detection of chronic coronary artery disease: Value of pharmacologically stressed, dynamically enhanced Turbo-Fast Low-Angle Shot MR images. AJR Am J Roentgenol 161:257, 1993.

97. Eichenberger AC, Schuiki E, Koechli VD, et al: Ischemic heart disease: Assessment with Gadolinium-enhanced ultrafast MR imaging and dipyridamole stress. J Magn Reson Imaging 4:425, 1994.

98. Matheijssen NAA, Louwerenburg HW, Van Rugge FP, et al: Comparison of ultrafast dipyridamole magnetic resonance

imaging with dipyridamole SestaMIBI SPECT for detection of perfusion abnormalities in patients with single vessel coronary artery disease: Assessment by quantitative model fitting. Magn Reson Med 35:221, 1996.

99. Wilke N, Jerosch-Herold M, Wang Y, et al: Myocardial perfusion reserve: Assessment with multisection quantitative, first pass MR imaging. Radiology 204:373, 1997.

100. Lauerma K, Virtanen K, Sipila LM, et al: Multislice MRI in assessment of myocardial perfusion in patients with single vessel proximal left anterior descending coronary artery disease before and after revascularization. Circulation 96:2859, 1997.

101. Hoffman JIE: Transmural myocardial perfusion. Prog Cardiovasc Dis 29:429, 1987.

102. Hoffman JIE: Maximal coronary flow and the concept of coronary vascular reserve. Circulation 70:153, 1984.

103. Canty JM, Klocke FJ: Reduced regional myocardial perfusion in the presence of pharmacological vasodilator reserve. Circulation 71:370, 1985.

104. Keijer JT, Van Rossum AC, Wilke N, et al: Magnetic resonance imaging of myocardial perfusion in single vessel coronary artery disease: Implications for transmural assessment of myocardial perfusion (abstract). Proceedings of the 4th annual meeting ISMRM, 1996, p 680.

105. Keijer JT, Van Rossum AC, Van Eenige MJ, et al: Improvement of subendocardial perfusion can be assessed by contrast-enhanced first-pass MRI. Circulation 96(suppl):I–132, 1997.

106. Wilson RF, Johnson MR, Marcus ML, et al: The effect of coronary angioplasty on coronary flow reserve. Circulation 77:873, 1988.

107. Uren NG, Crake T, Lefroy DC, et al: Delayed recovery of coronary resistive vessel function after coronary angioplasty. J Am Coll Cardiol 21:612, 1993.

108. Wu KC, Zerhouni EA, Judd RM, et al: Prognostic significance of microvascular obstruction by magnetic resonance imaging in patients with acute myocardial obstruction. Circulation 97:765, 1998.

109. Al-Saadi N, Nagel E, Gross M, et al: Noninvasive detection of myocardial ischemia from perfusion reserve based on cardiovascular magnetic resonance. Circulation 101:1379, 2000.

110. Schwitter J, Nanz D, Kneifel S, et al: Assessment of myocardial perfusion in coronary artery disease by magnetic resonance: A comparison with positron emission tomography and coronary angiography. Circulation 103:2230, 2001.

CHAPTER 12

Acute Myocardial Infarction

Katherine C. Wu, Carlos E. Rochitte, and João A. C. Lima

In the treatment of patients with acute myocardial infarction, reperfusion therapy has significantly improved survival.[1-3] Improved patient prognosis in part results from myocardial salvage, which leads to smaller infarcts and preservation of left ventricular function.[4-6] However, a number of important issues remain to be fully resolved in the clinical setting. The benefits of late reperfusion, beyond the point of myocardial salvage, have been suggested and numerous studies support the open artery hypothesis.[7-9] When and if optimal reperfusion has occurred is difficult to determine clinically,[10] and even when a patent infarct-related epicardial coronary artery is demonstrated angiographically, this evidence can be an inadequate marker of reperfusion at the tissue level.[10-12] This phenomenon, known as no-reflow, arises from the inhomogeneity of myocardial injury during ischemia, which may preclude optimal reperfusion because of microvessel occlusion. In addition, the assessment of the extent and transmurality of myocardial injury, and the detection of unresolved reversible components such as stunning or hibernation, are also important to determine the risk of remodeling and the possibilities for partial restoration of ventricular function with revascularization.

Following acute reperfused myocardial infarction, the injured myocardial territory may be divided into three distinct regions according to the extent of myocardial ischemia experienced during coronary occlusion.[12-14] At the extreme periphery of the territory at risk, myocardial cells may be protected from the ischemic insult because of prompt opening of collateral vessels. However, as the core of the territory at risk is approached, areas of myocardial tissue are present that have been subjected to varying degrees of ischemia but are salvaged by reperfusion.[12-14] This myocardial salvage results from coronary reflow at the level of the infarct-related epicardial coronary artery and/or, to a much lesser extent, by delayed opening of collateral channels. Depending upon the magnitude of the ischemic insult during coronary occlusion, this portion of the entire underperfused territory may exhibit decreased regional function (myocardial stunning) and/or varying degrees of cellular injury.[15-17] Cells may undergo necrosis with loss of membrane permeability and integrity, mitochondrial membrane rupture, and depletion of high-energy intracellular stores.[12] Finally, at the very core of the infarcted region, profound ischemia at the time of coronary occlusion and reperfusion may result in collapse and occlusion of the microvasculature.[12, 13] This region of microvascular obstruction with impaired blood flow, despite recanalization of the epicardial coronary artery, has been termed the no-reflow or low-reflow territory.[12, 13] Until recently, the clinical study of the heterogeneous effects of ischemic injury could only be characterized through the use of histopathologic methods and was thus limited to experimental laboratory conditions. Magnetic resonance (MR) techniques, and others, are now starting to play a role in improving our understanding, which in the human situation has been limited.

THE ROLE OF MAGNETIC RESONANCE IN DEMONSTRATING THE HETEROGENEITY OF MYOCARDIAL INFARCTION

Hyperenhanced Regions and Infarct Size

Recent developments in magnetic resonance imaging (MRI) and spectroscopy (MRS) have resulted in hardware and software techniques[18-20] that are now being developed in a dedicated way for the study of the cardiovascular system (cardiovascular MR—CMR), and this development has enabled improved assessment and differentiation of the different types of myocardial injury caused by coronary occlusion. Combining these techniques with contrast enhancement by paramagnetic agents, which shorten T1 relaxation times, facilitates the reliable measurement of infarct size in experimental models[20-24] and in patients.[25, 26] Earlier contrast-enhanced studies employed spin-echo CMR techniques to measure infarct size,[20, 22, 23] but these studies required prolonged imaging times, and image quality and infarct delineation were less good. More recent studies, however, have utilized gradient-echo techniques, which have yielded significant imaging improvements and are achievable during a few breath-holds.[21, 24, 25]

When CMR is used, the reperfused infarcted myocardium is characteristically seen as a hyperenhanced region on late imaging (>10 minutes following contrast bolus; Figure 12–1A–C). Because reperfused infarcted myocardium takes up a higher concentration of gadolinium compared with noninfarcted tissue,[22] this hyperenhancement is at least partly due to higher gadolinium concentrations. Two mechanisms potentially explain the increased contrast concentrations in the hyperenhanced regions. First, the volume of distribution for the contrast molecules within the imaging voxels may increase. Theoretically, this increase may result from interstitial edema and/or disruption of the myocyte membrane.[27] This role of this mechanism is supported by the results of several experimental studies,[28–30] as well as analyses of studies performed in patients with acute myocardial infarction.[26] Second, contrast molecule kinetics may be abnormal in infarcted/reperfused regions compared with normal regions. Support for this latter mechanism comes from work by Kim and coworkers, who used an isolated rabbit heart model of ischemia/reperfusion in which perfusate contrast agent concentration was directly manipulated and contrast recirculation

eliminated.[24] These investigators found that regions of hyperenhancement were characterized by delayed contrast washout.[24] When examined under electron microscopy, these same bright areas had extensive regions of sarcomere membrane rupture.[24] It was postulated that the main reason for contrast hyperenhancement in infarcted territories is delayed contrast washout, possibly because sarcolemmal rupture could affect myocardial contrast kinetics by hindering diffusion in and out of cells; the contribution of increased volume of distribution of contrast was found to be only mild.[24] These results are further supported by several in vivo experimental,[23, 31] and clinical studies.[26]

Prior studies performed in the canine model of acute reperfused myocardial infarction to validate infarct size measurements by contrast-enhanced CMR[21–23] have consistently shown very good correlation coefficients (r = 0.88–0.93) but an overestimation of 8 to 15 percent when compared with the histopathological gold standard of triphenyl tetrazolium chloride (TTC) tissue enzyme staining technique.[32] This finding led to conjecture as to whether

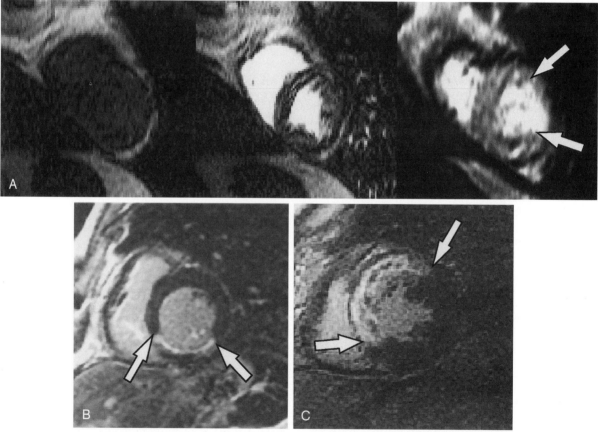

Figure 12–1. CMR hyperenhancement: infarcted region. *A,* This series of three CMR images is from a patient with an anterolateral infarct. The first image, precontrast, shows homogeneous signal from myocardium and intracavitary blood. The second image, taken immediately after contrast, shows bright ventricular cavities before myocardial contrast penetration. The third image was taken 520 seconds after contrast, at which time the intensity of the infarcted area (between arrows) resembles that of blood and is most discernible from noninfarcted myocardium. (From Wu KC, Zerhouni EA, Judd RM, et al: The prognostic significance of microvascular obstruction by magnetic resonance imaging in patients with acute myocardial infarction. Circulation 97:765, 1998, with permission.) *B,* This image was taken 600 seconds following contrast bolus administration and demonstrates subendocardial hyperenhancement in an individual who sustained an inferior infarction (between arrows). *C,* This image was obtained 610 seconds following contrast bolus administration and delineates an extensive hyperenhanced region of anteroseptal infarction (between arrows) in a patient.

tissue edema at the periphery of an infarct could account for such an overestimation (i.e., that hyperenhancement occurs in reversibly injured regions surrounding acute infarcts). Several early reports supported this hypothesis.[22, 33–35] Others postulated that the differences were due to intrinsic limitations of the TTC method, which relies on the loss of intramitochondrial co-factors as the marker for cellular death.[36, 37] However, the few preliminary clinical studies available in the literature that compared CMR with nuclear methods[26] or serum enzymes[25] have confirmed the accuracy of infarct size determination by CMR.

Two recent experimental studies directly addressed this issue.[24, 38] In their isolated rabbit heart model of ischemia and reperfusion, Kim and coworkers found that the spatial extent of MR hyperenhancement varied, depending upon the signal intensity threshold used to define injured regions.[24] An intensity threshold of 2 SD above that of remote regions led to overestimation of TTC-negative regions whereas a cutoff of 4 SD resulted in underestimation. This suggested that there may be a significant contribution of partial volume effects with thick slice widths leading to infarct edge blurring and overestimation of infarct size. In a subsequent paper, the same group used a canine model to compare the contrast enhancement patterns seen in acute reperfused infarction with those seen in severe but reversible ischemic injury and in chronic infarction.[38] They found that reversibly injured regions did not exhibit myocardial contrast hyperenhancement. In the acutely reperfused animals, when high-resolution thin slices (0.5 × 0.5 × 0.5 mm) are used, the location, spatial extent, and three-dimensional (3-D) shape of hyperenhanced regions

are essentially identical to those of the irreversibly injured regions defined by TTC. Furthermore, when a composite image is obtained from summing 16 high-resolution slices, thereby yielding an effective slice thickness of 8 mm (typical of prior studies), the partial volume effect could be demonstrated.

In the setting of chronic infarction (8 weeks following infarction), there continues to be hyperenhancement of the infarcted area.[38] The investigators found that from 3 days to 8 weeks after acute reperfused infarction, the absolute volume of myocardial hyperenhancement decreases by a factor of 3.4; however, others have reported a similar degree of infarct shrinkage during the same time period due to the healing process and transition from myocyte necrosis and collagenous scar formation.[39] Hence, it can be concluded that CMR hyperenhanced regions represent nonviable tissue and can be used to accurately measure infarct size.

Recently, the extent of CMR hyperenhanced regions was shown to be predictive of the occurrence of untoward clinical cardiovascular events in patients with acute myocardial infarction.[40] In a prospective observational study involving 44 patients, patients with infarct sizes greater than 30 percent of the ventricle by CMR had a significantly worse prognosis than patients with mid-sized infarcts (between 18 and 30%) or small infarcts (<18%; Figure 12–2). Direct infarct size measurement was found to be a better predictor of long-term outcome than left ventricular ejection fraction. These results echo a similar prior study utilizing nuclear techniques to directly measure infarct size, which also demonstrated the superiority of this approach when compared with ejection fraction as a predictor of postinfarct prognosis.[41]

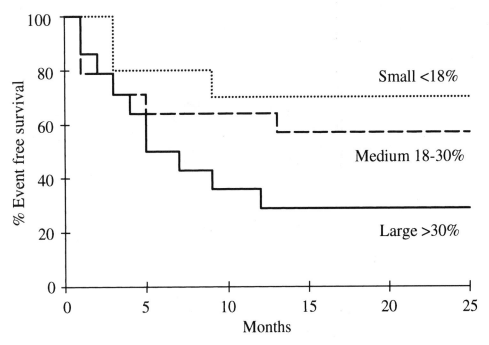

Figure 12–2. Event-free survival curves following infarction. Patients are grouped by CMR infarct size and events were defined as a clinical course without cardiovascular death, reinfarction, congestive heart failure, stroke, or unstable angina requiring hospitalization. (From Wu KC, Zerhouni EA, Judd RM, et al: The prognostic significance of microvascular obstruction by magnetic resonance imaging in patients with acute myocardial infarction. Circulation 97:765, 1998, with permission.)

Hypoenhanced Regions and Microvascular Obstruction

In contrast to the peripheral regions of the infarcted territory, the infarct core may experience the no-reflow phenomenon when flow down the epicardial infarct-related artery is restored. Upon reperfusion, sequestration of neutrophils in the microvasculature can lead to microvessel occlusion by erythrocytes, leukocytes, and cellular debris, thereby thwarting the return of flow at the microvessel level.[12, 13] Until recently, this process could be assessed only by methods such as radioactive microspheres and other histopathological techniques, which could only be performed as the terminal phase of an experimental study, and not clinically. However, recent advances in the field of noninvasive cardiac imaging have enabled the serial assessment of this phenomenon by CMR,[21, 23, 24, 26, 40, 42] echocardiography,[11, 42–44] and nuclear cardiology,[45, 46] thereby facilitating a much greater understanding of its pathophysiological and prognostic significance.

Microvascular obstruction in patients with acute myocardial infarction has been well documented by contrast-enhanced CMR and is visualized as a dark area within the infarcted territory early (1–4 minutes) after intravenous contrast administration (Figure 12–3A and B).[26, 40] This hypoenhanced region within the infarct core results from delayed contrast penetration relative to the surrounding reperfused infarcted tissue, which contains patent microvasculature.[24] Kim and coworkers demonstrated by electron microscopy that within the core of the infarcted region, there is a 37-fold higher number of capillaries with red cell stasis and intravascular neutrophil accumulation.[24] Thus, in these regions, there is a functional reduction in capillary density that could lead to slower contrast penetration by two mechanisms. First, there would be a reduced capillary surface area available for solute transport, and second, once in the interstitium, solute molecules would have to traverse greater diffusion distances to fill the extravascular space. Consequently, these areas of microvascular obstruction would be expected to be characterized by delayed wash-in of contrast and appear as hypoenhanced regions early after contrast bolus administration.

The location of these no-reflow regions at the infarct core arises because of a greater extent of local ischemia during coronary occlusion that leads to simultaneous endothelial and myocardial cell death. Endothelial injury precipitates the local activation and adhesion of neutrophils and the release of proinflammatory mediators,[47, 48] which leads to microvessel thrombosis. Biopsy specimens obtained from CMR hypoenhanced regions in experimental animals[24] have revealed the typical histomorphological findings documented in the basic studies that originally described and characterized the no-reflow phenomenon.[12, 13] In addition, other studies performed in the canine model have validated the presence and extent of CMR-determined regions of microvascular obstruction as compared with radioactive microsphere blood flow measurements and thioflavin-S,[21, 23] both of which are standard histopathological methods of quantifying no-reflow.[12, 13, 49]

More recently, detailed studies have been performed comparing contrast-enhanced CMR with contrast echocardiography against microspheres and thioflavin-S in a canine acute infarction model.[50] It was found that regions of microvascular obstruction as depicted by CMR, contrast echocardiography, and

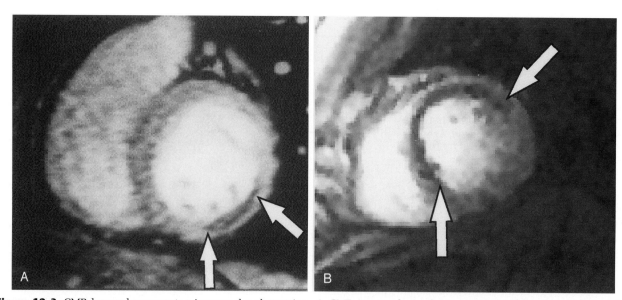

Figure 12–3. CMR hypoenhancement: microvascular obstruction. *A,* CMR image, obtained 1 to 2 minutes following contrast bolus, from a patient with an inferolateral infarct and subendocardial microvascular obstruction (note the hypoenhanced area between the arrows). (From Lima JAC, Judd RM, Bazille A, et al: Regional heterogeneity of human myocardial infarcts demonstrated by contrast-enhanced MRI. Circulation 92:1117, 1995, with permission.) *B,* CMR image obtained 1 to 2 minutes following contrast bolus in a patient with an anteroseptal infarct and extensive subendocardial microvascular obstruction showing as hypoenhancement (between arrows).

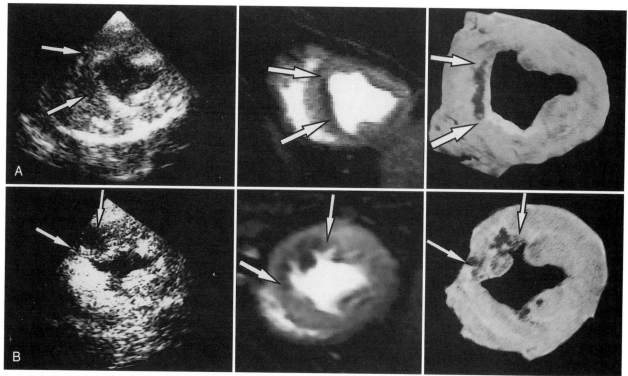

Figure 12–4. Microvascular obstruction: different techniques. Regions of microvascular obstruction (between arrows) as imaged by contrast echocardiography *(left)*, CMR *(center)*, and thioflavin-S *(right)* from corresponding myocardial cross-sectional slices (images in *A* and *B* are from two separate animals). (From Wu KC, Kim RJ, Bluemke DA, et al: Quantification and time course of microvascular obstruction by contrast-enhanced echocardiography and magnetic resonance imaging following acute myocardial infarction and reperfusion. J Am Coll Cardiol 32:1756, 1998. Reprinted with permission from the American College of Cardiology.)

thioflavin-S corresponded in spatial location (Figure 12–4). However, the extent of the regions quantified by the three techniques differed: The area of microvascular obstruction measured by CMR was smaller than that by thioflavin-S, which in turn was smaller than that by contrast echocardiography (Figure 12–5). A potential explanation was found in corresponding microsphere maps that allowed the quantification of regions receiving different thresholds of reduced blood flow. Regions of microvascular obstruction defined by CMR had a blood flow rate that was less than 40 percent of the flow to the remote, noninfarcted region. Corresponding regions defined by contrast echocardiography had flow rates less than 60 percent of remote flow, whereas those areas defined by thioflavin-S had flow less than 50 percent of normal.

These experiments highlighted several of the advantages of CMR.[50] Because CMR allows precise multislice imaging, microvascular obstruction can be measured relative to the entire left ventricle. Furthermore, the greater spatial resolution afforded by CMR results in visually more distinct regions of microvascular obstruction with better border definition. Moreover, the infarcted region is seen by CMR as a late-appearing hyperenhanced region, and thus infarct size is readily measurable. With contrast echocardiography, administration of coronary vasodilators is required for assessment of infarct size.[51] Despite these differences, both CMR and contrast

echocardiography can effectively quantify microvascular obstruction. However, it is clear that the techniques have different thresholds for detecting flow reduction inasmuch as CMR can detect more severe flow reduction than contrast echocardiography. As

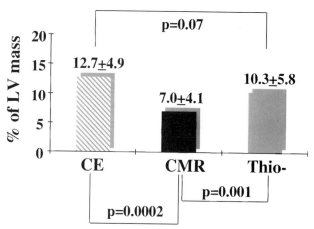

Figure 12–5. Microvascular obstruction: comparison of extent. Extent of microvascular obstruction as quantified by contrast echocardiography (CE), CMR, and thioflavin-S. LV = left ventricular. (From Wu KC, Kim RJ, Bluemke DA, et al: Quantification and time course of microvascular obstruction by contrast-enhanced echocardiography and magnetic resonance imaging following acute myocardial infarction and reperfusion. J Am Coll Cardiol 32:1756, 1998. Reprinted with permission from the American College of Cardiology.)

markers for degree of microvascular damage, CMR is perhaps more specific and contrast echocardiography more sensitive. Ambrosio and associates characterized the infarct core as having severe endothelial cell injury and intracapillary erythrocyte stasis.[13] Adjacent regions have progressively fewer damaged endothelial cells with interspersion of areas with normal microvessels. Thus, within an infarcted territory, the infarct core is likely characterized by complete microvessel occlusion. Surrounding it are rims of tissue with microvessels with progressively less obstruction with gradual interspersion of larger and larger areas with normal microvasculature. CMR may thus detect more severe degrees of microvascular damage than contrast echocardiography. Whether this difference is significant clinically remains unknown and requires further studies.

Microvascular obstruction occurs because the inflammatory response following profound ischemia and/or reperfusion leads to the collapse of the myocardial microvasculature, thereby leading to incomplete reperfusion.[12, 13, 36, 52] Incomplete reperfusion may be the consequence of lack of recanalization of the infarct-related artery, arterial reocclusion, and/or underperfusion of a large territory with profound ischemia at the infarct core.[10, 26, 43] What role reperfusion plays as a contributing factor to the genesis of microvascular obstruction remains controversial.[36, 52] This question is particularly significant given the time course of this phenomenon. After reperfusion, the region of microvascular obstruction nearly triples in size from 2 to 48 hours.[21] This increase has been documented by both CMR and radioactive microspheres in a canine model of reperfused infarction and extends the findings of previous investigators who had demonstrated expansion of the no-reflow region up to 3.5 hours after reperfusion.[13] It remains unclear whether the stimulus for the development of microvascular obstruction originates during coronary occlusion exclusively or if reperfusion plays an active role in progression of the phenomenon. In addition to showing increasing no-reflow, the same study documented an increase in infarct size over the same time frame.[21] Thus, similar factors could lead to myocyte injury well beyond the time of reperfusion.

The presence of microvascular obstruction predicts unfavorable postinfarction prognosis in patients who have sustained an acute myocardial infarction (Figure 12–6). This finding has been corroborated by studies using contrast-enhanced CMR,[26, 40, 42] and contrast echocardiography.[42, 43, 53] The reasons for the deleterious influence of microvascular obstruction on short-[26, 42, 43, 53] and long-term[40] prognosis are incompletely understood but may relate to a propensity for greater left ventricular postinfarction remodeling. Left ventricular remodeling is associated with more frequent cardiovascular complications (especially heart failure) following infarction and is directly related to the magnitude of microvascular obstruction early after experimental[54]

and clinical acute infarction,[53] as well as 6 months after the acute event.[40] The mechanisms by which microvascular obstruction induces greater ventricular remodeling remain unknown. Possibilities include the potentiation of wall thinning and infarct expansion early after infarction,[55–57] as well as potential impairment of infarct healing,[9, 58] given the association between presence of microvascular obstruction and greater transmural scar formation 6 months after infarction.[40] A recent study by Gerber and colleagues used a canine model of acute reperfused myocardial infarction to examine the effects of increasing extents of microvascular obstruction on myocardial strain as determined by 3-D CMR tagging.[59] Between 6 and 48 hours after ischemia-reperfusion, it was found that (1) both infarct size and extent of microvascular obstruction predicted left ventricular enlargement; and (2) the extent of microvascular obstruction correlated with altered myocardial strains in both the infarcted and adjacent noninfarcted territories. Specifically, increasing amounts of microvascular obstruction were associated with greater reductions in myocardial strain (i.e., reduced passive stretching or increased myocardial stiffness) within the infarcted territory. In adjacent, noninfarcted regions, myocardial strains were also reduced in proportion to the extent of microvascular obstruction. This alteration of strains in the adjacent, noninfarcted area corroborates the work of earlier investigators.[57] These findings suggest that increasing amounts of microvascular obstruction led to greater tissue stiffness, which may adversely affect left ventricular remodeling and segmental function.[59] In the adjacent, noninfarcted region, the reduced expansibility of the myocardium leads to increased local wall stress distribution[60, 61] that may result in lengthening of noninfarcted tissue, as seen early in the remodeling process.[57] Consequently, increasing amounts of microvascular obstruction may negatively influence contraction of the adjacent, noninfarcted myocardium and thereby cause increased remodeling of the left ventricle.[59]

CARDIOVASCULAR MAGNETIC RESONANCE AND MYOCARDIAL VIABILITY FOLLOWING ACUTE INFARCTION

The ability to assess myocardial viability following acute reperfused infarction is clinically important. CMR approaches to viability assessment in acute and chronic settings have included (1) employing resting CMR studies and focusing on end-diastolic wall thickness (only useful in chronic infarction after local remodeling of the infarct area), systolic wall thickening, and signal intensity without contrast enhancement[62, 63] and (2) assessing contractile reserve during dobutamine stimulation using changes in wall thickness from untagged images.[63–65] Recently, 3-D tagged CMR has been

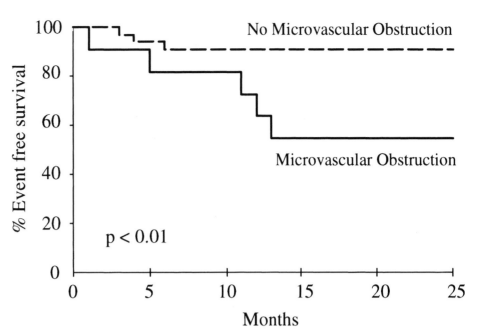

Figure 12–6. Microvascular obstruction: prognosis. Event-free survival (clinical course without cardiovascular death, reinfarction, CHF, or stroke) for patients with and without CMR microvascular obstruction. (From Wu KC, Zerhouni EA, Judd RM, et al: The prognostic significance of microvascular obstruction by magnetic resonance imaging in patients with acute myocardial infarction. Circulation 97:765, 1998, with permission.)

coupled with dobutamine challenge as a means of precisely quantifying myocardial 3-D deformation and strain.[66] This approach was found to be accurate and precise and may be a more sensitive method for characterizing viability.[66] The extensive work relating CMR contrast enhancement patterns to tissue perfusion following acute reperfused infarction suggests another technique. Following myocardial infarction, the size and shape of regions that exhibit delayed contrast hyperenhancement are identical to regions of irreversible injury.[21, 23, 24, 26, 38] Regions that do not hyperenhance with contrast are viable.[38] Furthermore, in experiments performed by Kim and colleagues, it was found that a dissociation may occur between the area of impaired wall thickening by cine CMR and the area of hyperenhancement.[38] Thus, the combination may be used to define viability without the need to measure stress function; for example: (1) acute myocardial infarction (hyperenhanced region with contractile dysfunction); (2) injured but viable myocardium (no region of hyperenhancement but contractile dysfunction present); and (3) normal myocardium (region of normal function with no hyperenhancement).[38]

Another approach to assessing myocardial viability is based on the fact that normal myocyte function depends upon active cellular maintenance of electrochemical gradients. Hence, cellular injury or dysfunction manifests itself by disruption of these ionic concentration gradients. Specifically, in viable myocytes, intracellular sodium concentration is much lower than that of the extracellular space because of active outward transport pumps. Irreversible ischemia causes a permanent increase in intracellular sodium concentration, whereas reversible ischemia results in a two-to-fivefold linear increase in sodium concentration, which returns to baseline with reperfusion.[67] Theoretically, one could therefore measure tissue sodium content to distinguish viable from

nonviable myocardium following ischemia.[68] It has been shown that CMR can be used to obtain in vivo ^{23}Na images of the heart in animals using a high-field (4.7 Tesla) magnet,[68] and in human volunteers using a 1.5 Tesla scanner.[69] In their rabbit model of acute reperfused infarction, Kim and associates found that infarcted regions defined by TTC had a significant elevation in ^{23}Na image intensity by CMR (Figure 12–7) compared with viable regions, supporting the presence of elevated tissue sodium content.[68] They also performed ^{23}Na MRS to further show that sodium concentration was indeed higher in nonviable than in viable myocardium.[68] Moreover, they adapted rapid gradient-echo MR techniques, used typically for proton imaging, to the sodium nucleus and were thus able to reduce imaging times to a few minutes.[68]

Kim and associates went on to further investigate the correlation between elevated ^{23}Na CMR image intensity and histochemical measurements of infarct size as well as possible mechanisms for increased tissue sodium concentration.[70] In a large animal (canine) and small animal (rabbit) model of ischemia/reperfusion, they found a close correlation between infarct size by TTC and extent of regions with elevated ^{23}Na image intensity. When high-resolution ex vivo imaging was used the extent of regions with elevated ^{23}Na image intensity was identical to pathological determination of the infarcted territory. In contrast, severe but reversible ischemic injury did not result in elevated ^{23}Na image intensity. They also used MRS, electron probe x-ray microanalysis, and morphometric analysis to examine possible pathophysiological mechanisms for the observed increase in tissue sodium concentration. In infarcted regions, it was found that total tissue sodium concentration was elevated and was associated with increased intracellular sodium and an elevated intracellular sodium/potassium ratio. At the same

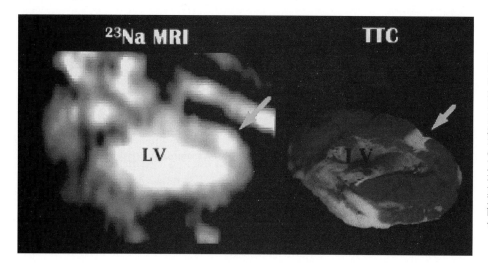

Figure 12–7. Sodium imaging in infarction. Long-axis ^{23}Na CMR image of a rabbit heart *(left)* and corresponding triphenyl tetrazolium chloride (TTC)-stained section *(right)*. The nonviable, infarcted region has increased myocardial ^{23}Na image intensity (arrows). (See also color plates.) (From Kim RJ, Lima JAC, Chen EL, et al: Fast ^{23}Na magnetic resonance imaging of acute reperfused myocardial infarction: Potential to assess myocardial viability. Circulation 95:1877, 1997, with permission.)

time, there was only a minor increase in extracellular volume in the infarcted region. This finding suggests that the mechanism of increased ^{23}Na image intensity following reperfused infarction is the loss of myocyte ionic homeostasis that leads to intracellular sodium accumulation.

Similar studies have now been performed that demonstrate the feasibility of ^{39}K CMR imaging of the heart (Figure 12–8).[71] During ischemia, there is efflux of potassium from the myocytes that occurs with a time course similar to that of the loss of contractile function.[72, 73] It is the increase in extracellular potassium concentration that may lead to ischemic ventricular arrhythmias.[74] Irreversible ischemia leads to a profound reduction in intracellular potassium concentration that is closely related to the rise in serum creatine kinase levels.[75, 76] The relative distributions of potassium across the intracellular and extracellular spaces therefore are

closely related to myocardial metabolism and provide the basis for the utility of ^{39}K CMR imaging. Fieno and colleagues found that (1) in irreversibly injured regions, there is a reduction in ^{39}K CMR image intensities; (2) the location and spatial extent of regions of ^{39}K image intensity correlated well with histologically determined infarct size (see Figure 12–8); and (3) regions of reduced ^{39}K image intensity are associated with the loss of intracellular potassium.[71] Hence, ^{39}K CMR also has the potential to be used clinically to distinguish between viable and nonviable myocardium following acute reperfused infarction.

CONCLUSION

Recent developments in noninvasive imaging techniques underscore the complexity and heteroge-

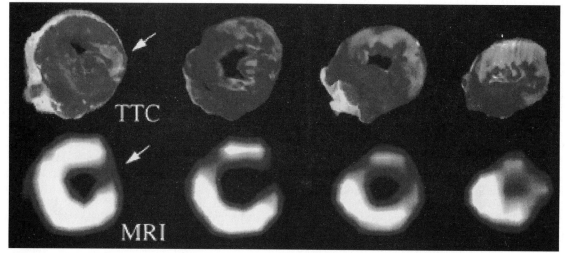

Figure 12–8. (See color plates.) Potassium imaging in infarction. Sequential TTC-stained short-axis slices and corresponding ^{39}K CMR images (colorized) from three-dimensional data set of a rabbit heart subjected to reperfused infarction. Histological sections and short-axis images are arranged as follows: left to right—base to apex, and lateral wall of the left ventricle is oriented to the right. Viable areas of myocardium stain brick red, whereas nonviable areas appear yellowish-white. Note the similarity of reduced ^{39}K image intensity and nonviable areas of myocardium. (From Fieno DS, Kim RJ, Rehwald WG, Judd RM: Physiological basis for potassium (^{39}K) magnetic resonance imaging of the heart. Circ Res 84:913, 1999, with permission.)

neity of myocardial infarcts. Although extensive insight into the pathophysiological mechanisms of acute coronary thrombosis and reperfusion has been elucidated through the use of such technology, much remains unknown. CMR is particularly well suited to the study of coronary artery disease and myocardial infarction, and its use should lead to further advances in the future. The possibility of combining detailed studies of myocardial function, viability, perfusion, and sodium and potassium metabolism with the noninvasive assessment of coronary anatomy[77] and epicardial coronary artery blood flow[78] highlights the diagnostic potential of CMR.

References

1. Gruppo Italiano Per Lo Studio Della Streptochinasi Nell'Infarto Miocardico (GISSI): Effectiveness of intravenous thrombolytic treatment in acute myocardial infarction. Lancet 1:397, 1986.
2. ISIS-2 (Second International Study of Infarct Survival) Collaborative Group: Randomized trial of intravenous streptokinase, oral aspirin, both or neither among 17,187 cases of suspected acute myocardial infarction:ISIS-2. Lancet 2:349, 1988.
3. Grines CL, Browne KF, Marco J, et al: A comparison of immediate angioplasty with thrombolytic therapy for acute myocardial infarction. N Engl J Med 328:673, 1993.
4. Gersh BJ, Anderson JL: Thrombolysis and myocardial salvage: Results of clinical trials and the animal paradigm-paradox or predictable? Circulation 88:296, 1993.
5. Guerci AD, Gerstenblith G, Brinker JA, et al: A randomized trial of intravenous tissue plasminogen activator for acute myocardial infarction with subsequent randomization to elective coronary angioplasty. N Engl J Med 317:1613, 1987.
6. Sheehan FH, Doerr R, Schmidt WG, et al: Early recovery of left ventricular function after thrombolytic therapy for acute myocardial infarction: An important determinant of survival. J Am Coll Cardiol 12:289, 1988.
7. Hochman JS, Choo H: Limitation of myocardial infarct expansion by reperfusion independent of myocardial salvage. Circulation 75:299, 1987.
8. Lavie CJ, O'Keefe JH, Chesebro JH, et al: Prevention of late ventricular dilatation after acute myocardial infarction by successful thrombolytic reperfusion. Am J Cardiol 66:31, 1990.
9. Nidorf SM, Siu SC, Galambos G, et al: Benefit of late coronary reperfusion on ventricular morphology and function after myocardial infarction. J Am Coll Cardiol 21:683, 1993.
10. Lincoff AM, Topol EJ: Illusion of reperfusion: Does anyone achieve optimal reperfusion during acute myocardial infarction? Circulation 87:1792, 1993.
11. Ito H, Tomooka T, Sakai N, et al: Lack of myocardial perfusion immediately after successful thrombolysis. Circulation 85:1699, 1992.
12. Kloner RA, Ganote CE, Jennings RB: The "no-reflow" phenomenon after temporary occlusion in the dog. J Clin Invest 54:1496, 1974.
13. Ambrosio G, Weisman HF, Mannisi JA, Becker LC: Progressive impairment of regional myocardial perfusion after restoration of postischemic blood flow. Circulation 80:1846, 1989.
14. Braunwald E, Kloner RA: Myocardial reperfusion: A double-edged sword. J Clin Invest 76:1713, 1985.
15. Braunwald E, Kloner RA: The stunned myocardium: Prolonged, postischemic ventricular dysfunction. Circulation 66:1146, 1982.
16. Bolli R: Mechanism of myocardial "stunning." Circulation 82:723, 1990.
17. Marban E: Myocardial stunning and hibernation: The physiology behind the colloquialisms. Circulation 83:681, 1991.
18. Fram J, Merboldt KD, Bruhm H, et al: A 0.3-second FLASH MRI of the human heart. Magn Reson Med 13:150, 1990.
19. Manning WJ, Atkinson DJ, Grossman W, et al: First-pass nuclear magnetic resonance imaging studies of patients with coronary artery disease. J Am Coll Cardiol 18:959, 1991.
20. Goldman MR, Brady TJ, Pykett IL, et al: Quantification of experimental myocardial infarction using nuclear magnetic resonance imaging and paramagnetic ion contrast enhancement in excised canine hearts. Circulation 66:1012, 1982.
21. Rochitte CE, Lima JAC, Bluemke DA, et al: The magnitude and time course of microvascular obstruction and tissue injury after acute myocardial infarction. Circulation 98:1006, 1998.
22. Schaeffer S, Malloy CR, Katz J, et al: Gadolinium-DTPA-enhanced nuclear magnetic resonance imaging of reperfused myocardium: Identification of the myocardial bed at risk. J Am Coll Cardiol 12:1064, 1988.
23. Judd RM, Lugo-Olivieri CH, Arai M, et al: Physiological basis of myocardial contrast enhancement in fast magnetic resonance images of 2-day-old reperfused canine infarcts. Circulation 92:1902, 1995.
24. Kim RJ, Chen EL, Lima JAC, Judd RM: Myocardial Gd-DTPA kinetics determine MRI contrast enhancement and reflect the extent and severity of myocardial injury after acute reperfused infarction. Circulation 94:3318, 1996.
25. Holman ER, van Jongergen HPW, van Dijkman PRM, et al: Comparison of magnetic resonance imaging studies with enzymatic indexes of myocardial necrosis for quantification of myocardial infarct size. Am J Cardiol 71:1036, 1993.
26. Lima JAC, Judd RM, Bazille A, et al: Regional heterogeneity of human myocardial infarcts demonstrated by contrast-enhanced MRI. Circulation 92:1117, 1995.
27. Jennings RB, Murray CE, Steenbergen CJ, Reimer KA: Development of cell injury in sustained acute ischemia. Circulation 82(suppl II):II2, 1990.
28. Saeed M, Wendland MF, Masui T, Higgins CB: Reperfused myocardial infarctions on T1- and susceptibility-enhanced MRI: Evidence for loss of compartmentalization of contrast media. Magn Reson Med 31:31, 1994.
29. Diesbourg LD, Prato FS, Wisenberg G, et al: Quantification of myocardial blood flow and extracellular volumes using a bolus injection of Gd-DTPA: Kinetic modeling in canine ischemic disease. Magn Reson Med 23:239, 1992.
30. Schwitter J, Saeed M, Wendland MF, et al: Influence of severity of myocardial injury on distribution of macromolecules: Extravascular versus intravascular gadolinium-based magnetic resonance contrast agents. J Am Coll Cardiol 30:1086, 1997.
31. Peshock RM, Malloy CR, Buja LM, et al: Magnetic resonance imaging of acute myocardial infarction: Gadolinium diethylenetriamine pentaacetic acid as a marker of reperfusion. Circulation 74:1434, 1986.
32. Fishbein MC, Meerbaum S, Rit J, et al: Early phase acute myocardial infarct size quantification: Validation of the triphenyl tetrazolium chloride tissue enzyme staining technique. Am Heart J 101:593, 1981.
33. Nishimura T, Yamada Y, Hayashi M, et al: Determination of infarct size of acute myocardial infarction in dogs by magnetic resonance imaging and gadolinium-DTPA: Comparison with indium-111 antimyosin imaging. Am J Physiol Imaging 4:83, 1989.
34. Masui T, Saeed M, Wendland MF, Higgins CB: Occlusive and reperfused myocardial infarcts: MR imaging differentiation with nonionic Gd-DTPA-BMA. Radiology 181:77, 1991.
35. Baer FM, Theissen P, Schneider CA, et al: Magnetic resonance tomography imaging techniques for diagnosing myocardial vitality. Herz 19:51, 1994.
36. Miura T: Does reperfusion induce myocardial necrosis? Circulation 82:1070, 1990.
37. Horneffer PJ, Healy B, Gott VL, Gardner TJ: The rapid evolution of a myocardial infarction in an end-artery preparation. Circulation 39:V39, 1987.
38. Kim RJ, Fieno DS, Parrish TB, et al: Relationship of MRI delayed contrast enhancement to irreversible injury, infarct age, and contractile function. Circulation 100:1992, 1999.

39. Reimer KA, Jennings RB: The changing anatomic reference base of evolving myocardial infarction. Underestimation of myocardial collateral blood flow and overestimation of experimental anatomic infarct size due to tissue edema, hemorrhage and acute inflammation. Circulation 60:866, 1979.

40. Wu KC, Zerhouni EA, Judd RM, et al: The prognostic significance of microvascular obstruction by magnetic resonance imaging in patients with acute myocardial infarction. Circulation 97:765, 1998.

41. Mahmarian JJ, Mahmarian AC, Marks GF, et al: Role of adenosine thallium-201 tomography for defining long-term risk in patients after acute myocardial infarction. J Am Coll Cardiol 25:1333, 1995.

42. Asanuma T, Tanabe K, Ochiai K, et al: Relationship between progressive microvascular damage and intramyocardial hemorrhage in patients with reperfused anterior myocardial infarction. Circulation 96:448, 1997.

43. Ragosta M, Camarano G, Kaul S, et al: Microvascular integrity indicated myocellular viability in patients with recent myocardial infarction: New insights using myocardial contrast echocardiography. Circulation 89:2562, 1994.

44. Ito H, Okamura A, Iwakura K, et al: Myocardial perfusion patterns related to thrombolysis in myocardial infarction perfusion grades after coronary angioplasty in patients with acute anterior wall myocardial infarction. Circulation 93:1993, 1996.

45. Schofer J, Montz R, Mathey DG: Scintigraphic evidence of the "no-reflow" phenomenon in human beings after coronary thrombolysis. J Am Coll Cardiol 5:593, 1985.

46. Jeremy RW, Links JM, Becker LC: Progressive failure of coronary flow during reperfusion of myocardial infarction: Documentation of the no-reflow phenomenon with position emission tomography. J Am Coll Cardiol 16:695, 1990.

47. Albelda SM, Smith CW, Ward PA: Adhesion molecules and inflammatory injury. FASEB J 8:504, 1994.

48. Nathan C, Srimal S, Farber CEA: Cytokine-induced respiratory burst of human neutrophils: Dependence on extracellular matrix proteins and CD11/CD18 integrins. J Cell Biol 109:1341, 1989.

49. Krug A, De Rochemont WM, Korb G: Blood supply of the myocardium after temporary coronary occlusion. Circ Res 19:57, 1966.

50. Wu KC, Kim RJ, Bluemke DA, et al: Quantification and time course of microvascular obstruction by contrast-enhanced echocardiography and magnetic resonance imaging following acute myocardial infarction and reperfusion. J Am Coll Cardiol 32:1756, 1998.

51. Firschke C, Lindner JR, Goodman NC, et al: Myocardial contrast echocardiography in acute myocardial infarction using aortic root injections of microbubbles in conjunction with harmonic imaging: Potential application in the cardiac catheterization laboratory. J Am Coll Cardiol 29:207, 1997.

52. Becker LC, Ambrosio G: Myocardial consequences of reperfusion. Prog Cardiovasc Dis 30:23, 1987.

53. Ito H, Maruyama A, Iwakura K, et al: Clinical implications of the "no reflow" phenomenon: A predictor of complications and left ventricular remodeling in reperfused anterior wall myocardial infarction. Circulation 93:223, 1996.

54. Rochitte CE, Melin JA, Bluemke DA, et al: The extent of microvascular obstruction best predicts left ventricular remodeling after acute myocardial infarction. Circulation 96:1–305, 1997.

55. Erlebacher JA, Weiss JL, Eaton LW, et al: Late effects of acute infarct dilation on heart size: A two-dimensional echocardiographic study. Am J Cardiol 49:1120, 1982.

56. Pfeffer MA, Braunwald E: Ventricular remodeling after myocardial infarction: Experimental observations and clinical implications. Circulation 81:1161, 1990.

57. Kramer CM, Lima JAC, Reichek N, et al: Regional differences in function within noninfarcted myocardium during left ventricular remodeling. Circulation 88:1279, 1993.

58. Richard V, Murray CE, Reimer KA: Healing of myocardial infarcts in dogs: Effects of late reperfusion. Circulation 92:1891, 1995.

59. Gerber BL, Rochitte CE, Melin JA, et al: Microvascular obstruction and left ventricular remodeling 48 hours after acute myocardial infarction. Circulation 101:2734, 2000.

60. Bogen DK, Rabinowitz SA, Needleman A, et al: An analysis of the mechanical disadvantage of myocardial infarction in the canine left ventricle. Circ Res 47:728, 1980.

61. Lima JAC, Becker LC, Melin JA, et al: Impaired thickening of nonischemic myocardium during acute regional ischemia in the dog. Circulation 71:1048, 1985.

62. Klow NE, Smith HJ, Gullestad L, et al: Outcome of bypass surgery in patients with chronic ischemic left ventricular dysfunction. Predictive value of MR imaging. Acta Radiol 38:76, 1997.

63. Bax JJ, de Roos A, van Der Wall EE: Assessment of myocardial viability by MRI. J Magn Reson Imaging 10:418, 1999.

64. Sechtem U, Voth E, Baer F, et al: Assessment of residual viability in patients with myocardial infarction using magnetic resonance techniques. Int J Card Imaging 9:31, 1993.

65. Higgins CB: Prediction of myocardial viability by MRI. Circulation 99:727, 1999.

66. Croisille P, Moore CC, Judd RM, et al: Differentiation of viable and nonviable myocardium by the use of three-dimensional tagged MRI in 2-day-old reperfused canine infarcts. Circulation 99:284, 1999.

67. Jennings RB, Sommers HM, Kaltenbach JP, West JJ: Electrolyte alterations in acute myocardial ischemic injury. Circ Res 14:260, 1964.

68. Kim RJ, Lima JAC, Chen EL, et al: Fast ^{23}Na magnetic resonance imaging of acute reperfused myocardial infarction: Potential to assess myocardial viability. Circulation 95:1877, 1997.

69. Parrish TB, Fieno DS, Fitzgerald SW, Judd RM: Theoretical basis for sodium and potassium MRI of the human heart at 1.5 T. Magn Reson Med 38:653, 1997.

70. Kim RJ, Judd RM, Chen EL, et al: Relationship of 23-Na magnetic resonance image intensity to infarct size after acute reperfused myocardial infarction. Circulation 100:185, 1999.

71. Fieno DS, Kim RJ, Rehwald WG, Judd RM: Physiological basis for potassium (39K) magnetic resonance imaging of the heart. Circ Res 84:913, 1999.

72. Hill JL, Gettes LS: Effect of acute coronary artery occlusion on local myocardial extracellular K$^+$ activity in swine. Circulation 61:768, 1980.

73. Kleber AG: Resting membrane potential, extracellular potassium activity, and intracellular sodium activity during acute global ischemia in isolated perfused guinea pig hearts. Circ Res 52:442, 1983.

74. Janse MJ, Wit AL: Electrophysiological mechanisms of ventricular arrhythmias resulting from myocardial ischemia and infarction. Physiol Rev 69:1049, 1989.

75. Conrad GL, Rau EE, Shine KI: Creatine kinase release, potassium-42 content, and mechanical performance in anoxic rabbit myocardium. J Clin Invest 64:155, 1979.

76. Johnson RN, Sammel NL, Norris RM: Depletion of myocardial creatine kinase, lactate dehydrogenase, myoglobin, and K$^+$ after coronary artery ligation in dogs. Cardiovasc Res 15:529, 1981.

77. Manning WJ, Li W, Edelman RR: A preliminary report comparing magnetic resonance coronary angiography with conventional angiography. N Engl J Med 328:828, 1993.

78. Clarke GD, Eckels R, Chaney C, et al: Measurement of absolute epicardial coronary artery flow and flow reserve with breath-hold cine phase-contrast magnetic resonance imaging. Circulation 91:2627, 1995.

CHAPTER 13

Myocardial Infarction—Remodeling

Christopher M. Kramer

Cardiovascular magnetic resonance (CMR) has greatly expanded our understanding of postinfarction remodeling. Left ventricular (LV) remodeling after myocardial infarction occurs in response to the initial injury and myocyte loss. During the subsequent healing phase after the initial injury, infarct expansion can occur.[1] This infarct expansion arises as a result of reduced tensile strength in the infarcted segment and involves *thinning* of the affected myocardial wall segment along with an *increase* in its endocardial surface area. Infarct size, infarct location, preload, afterload, and contractile pull from noninfarcted segments all impact on the degree to which infarct expansion occurs.[2] Once infarct expansion has occurred, the left ventricle experiences a considerable mechanical and energetic disadvantage because the resultant increase in ventricular size is at the expense of increased wall tension.[3] Within *noninfarcted* segments of the LV wall there is cell slippage and myocyte hypertrophy,[4] causing lengthening of these regions.

McKay and colleagues coined the term *LV remodeling* in a serial invasive left ventriculographic study[5] in which they observed that the degree of dilation observed in noninfarcted regions was correlated to the size of the initial infarct and was not due to higher left ventricular filling pressures. Subsequent left ventriculographic studies have documented that the late increase in heart size after anterior myocardial infarction (MI) is due to elongation of the noninfarcted segment, and not due to further infarct expansion.[6] This increase in cardiac size late after MI has been shown to correlate with increased mortality.[7, 8]

Imaging modalities used to evaluate changes in LV size and function after myocardial infarction include invasive left ventriculography,[5, 6] noninvasive two-dimensional (2-D) echocardiography,[9–12] and cine computed tomography.[13] CMR is ideally suited to assess LV remodeling because, as a tomographic technique, CMR allows comprehensive data regarding the entire deformed ventricle to be obtained. Most importantly, the operator can image in any desired plane, facilitating the acquisition of a complete data set. CMR spectroscopy has also added to the understanding of global and regional cardiac energetics during LV remodeling. In addition, CMR has played an important role in the investigation of the effects of pharmacological and nonpharmacological therapies of the remodeling process.

CARDIOVASCULAR MAGNETIC RESONANCE EVALUATION OF SIZE DURING LEFT VENTRICULAR REMODELING

Previous studies have documented the importance of LV volumes after infarction in assessing prognosis.[7] White and associates[8] established end-systolic volume as the most powerful predictor of long-term outcome after myocardial infarction. The tomographic capability of CMR allows coverage of the left ventricle from apex to base (Figure 13–1), and thus CMR is being increasingly used to monitor the natural history of LV remodeling after large MI, both in animal models[14] and in man.

Various investigators have validated the CMR measurement of LV mass after MI in comparison with LV mass at autopsy and found excellent correlations (r 0.93–0.99).[15–17] In an ovine model of anteroapical infarction induced by coronary ligation[17] (Figure 13–2), a tomographic set of short-axis CMR images (Figure 13–3) analyzed by modified Simpson's rule was used to follow LV mass, end-diastolic, and end-systolic volumes in the first 6 months after the initial event. A stepwise increase in LV mass and end-diastolic and end-systolic volumes was found. The increase in LV end-diastolic volume was disproportionate such that the LV volume/mass ratio increased over the 6-month time period. Infarct wall thickness fell over this time period, whereas no change was noted in wall thickness in noninfarcted regions. These noninfarcted regions demonstrated an increase in segment length over the 6-month time period, suggesting that the hypertrophy that occurs during LV remodeling is of the eccentric type.[18]

In man, cine CMR has been used to measure LV volumes and mass after infarction. Its reproducibility has been validated by Matheijssen and coworkers[19] in a study of 7 patients after anterior infarction with 2 to 4 percent intraobserver and interobserver variability. In a study of 61 patients 1, 4, and 26 weeks after nonreperfused infarction,[20] Konermann and colleagues used cine CMR to evaluate changes in LV morphology. Between week 1 and 26 in the group as a whole, LV end-diastolic volume index (LVEDVI) increased from 74 ± 23 ml/m^2 to 85 ± 28 ml/m^2 and the LV end-systolic volume index (LVESVI) increased from 40 ± 19 to 51 ± 29 ml/

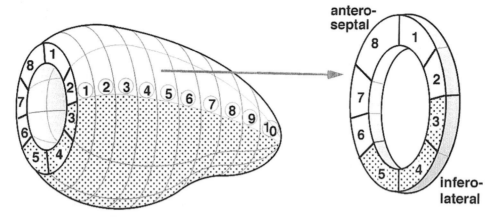

Figure 13–1. Diagram demonstrating the tomographic approach to cardiovascular magnetic resonance (CMR) of the left ventricle (LV). Short axis slices are imaged, spanning the LV from base (slice 1) to apex (slice 10), and a segmental approach to analysis can be performed. In this example, each short axis slice is divided into 8 segments. The inferolateral segments are shaded. (From Dubach P, Myers J, Dziekan G, et al: Effect of exercise training on myocardial remodeling in patients with reduced left ventricular function after myocardial infarction: application of magnetic resonance imaging. Circulation 95:2060, 1997.)

m^2. Most of the increase was found in the subgroup of 32 patients who had an anterior infarction. No change in LV stroke volume index was found over the course of the study. LV mass increased in the group as a whole from 246 ± 66 g to 276 ± 80 g. The change in volume-to-mass ratio was directly related to enzymatic infarct size: decreasing in smaller infarcts, remaining unchanged in moderate-sized infarcts, and increasing in larger infarcts.

CMR was used in a study of 26 patients imaged on day 5 ± 2 and week 8 ± 1 after first anterior infarction.[21] All had single-vessel left anterior descending (LAD) disease and had received reperfusion therapy, but had regional LV dysfunction and an initial ejection fraction of 50 percent or less. The LV mass index trended downwards during the 8-week period, falling from 109 ± 19 g/m² to 102 ± 18 g/m². LV end-diastolic volume indices increased from 83 ± 24 ml/m² to 96 ± 27 ml/m² (Figure 13–4), whereas the end-systolic volume did not change (52 ± 20 ml/m² at day 5 to 54 ± 24 ml/m² at week 8). LV ejection fraction increased during this time period from 39 ± 12 percent to 45 ± 14 percent. Therefore, despite a significant increase in LV end-diastolic volume, global LV function im-

Figure 13–2. Long-axis cine image at end systole of the left ventricle (LV) of a sheep 8 weeks after anteroapical infarction induced by coronary ligation. The LV apex (arrow) is markedly thinned and is dyskinetic. Noninfarcted tissue in the mid and basal septum (S) and lateral wall (L) can be seen thickening in this systolic image. The apex of the right ventricle (RV) is likewise thinned in this image.

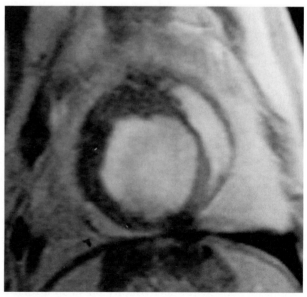

Figure 13–3. Short-axis end-diastolic cine image at the apex of the LV in a sheep at 8 weeks after anteroapical infarction induced by coronary ligation. Note the thinned transmurally infarcted tissue from 1 to 6 o'clock in the image. The epicardial and endocardial LV borders can be easily contoured for measurement of LV mass and end-diastolic volume.

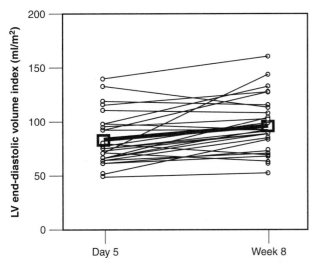

Figure 13–4. Graph demonstrating the change in left ventricular end-diastolic volume index (LVEDVI) in 26 patients after first reperfused anterior infarction, from day 5 to week 8 following myocardial infarction. The mean LVEDVI increased from 82 ± 24 ml/m² to 96 ± 27 ml/m² (see open squares). (From Kramer CM, Rogers WJ, Theobald TM, et al: Dissociation between changes in intramyocardial function and left ventricular volumes in the 8 weeks after first anterior myocardial infarction. J Am Coll Cardiol 30:1625, 1997. Reprinted with permission from the American College of Cardiology.)

proved in the 8-week postinfarction period. By multivariate analysis, the only significant predictor of an increase in LVEDVI over the study period was peak creatine kinase as a measure of infarct size.

CARDIOVASCULAR MAGNETIC RESONANCE EVALUATION OF FUNCTION DURING LEFT VENTRICULAR REMODELING

CMR methods used to study the natural history of regional LV function during remodeling after infarction include cine CMR,[20] three-dimensional (3-D) wall thickening analysis,[22] and tagging.[17, 21, 23] In the study of Konermann and colleagues,[20] endocardial motion towards the centroid of the LV cavity was measured, which the authors termed *motility*. Within the zone of infarction (defined as a region of systolic thickening < 2 mm), mean motility fell during the 6-month follow-up from 2.0 ± 1.6 to 0.5 ± 2.9 mm with the extent dependent on infarct size, but independent of anterior or nonanterior location. Within noninfarcted regions, motility decreased in the group as a whole, falling from 7.1 ± 2.4 to 6.3 ± 2.7 mm from week 1 to 26 and the decline was most marked in anterior infarcts. Wall thickening overall did not change within noninfarcted regions.

Holman and associates performed cine CMR in 25 patients 3 weeks after anterior infarction and analyzed functional data with a 3-D wall thickening approach using a centerline[22] (Figure 13–5). Wall thickening was reduced compared with a normal data base in myocardial territories perfused by the LAD and left circumflex (LCx) arteries and in the LAD compared with the other territories within the patient group. The quantity of dysfunctional myocardium, at this single time point after infarction, when any myocardial stunning should have resolved, correlated well with an enzymatic estimate of infarct size.

In the ovine model of LV remodeling, we[17] used CMR tagging[24, 25] to evaluate regional function in circumferential and longitudinal planes at baseline and 1 and 8 weeks and 6 months after MI. Shortening within infarcted regions was reduced throughout the study period. A persistent difference in intramyocardial shortening was found between noninfarcted regions adjacent to and remote from the infarct border. Function in adjacent noninfarcted regions fell markedly at 1 week after infarction and improved partially by 8 weeks after infarction, but remained depressed relative to baseline and to remote regions. Moulton and coworkers used the same model and CMR tagging to demonstrate that adjacent or border zone fibers were stretched during isovolumic systole, which contributed to reduced fiber shortening during systolic ejection.[26]

A study using CMR tagging has demonstrated mild dysfunction within remote noninfarcted regions in 28 patients on day 5 ± 2 after first reperfused anterior infarction.[23] Function within the apex, anterior wall, and septum was depressed relative to a normal data base, as expected. Basal lateral percent intramyocardial circumferential shortening (%S) measured 17 ± 5 percent, significantly less than 22 ± 7 percent in normals. In addition mid inferior %S (12 ± 10%) tended to be lower than that of normals (19 ± 5%).

When this patient group was followed for 8 weeks after anterior infarction,[21] improvement in regional function was found in both infarcted and noninfarcted regions (Figure 13–6). Apical %S improved from 9 ± 6 percent to 13 ± 5 percent, as it did in mid anterior (6 ± 6% to 10 ± 7%) and mid septal regions (8 ± 7% to 12 ± 6%). The dysfunction demonstrated on day 5 in remote mid inferior and basal lateral regions resolved by 8 weeks. Ejection fraction increased overall from 39 ± 12 percent to 45 ± 14 percent. However, LVEDVI increased during this time period, as noted above (see Figure 13–4). Therefore, as in the ovine model, the improvement in regional function from week 1 to week 8 after infarction was uncoupled from changes in global LV volume.

CARDIOVASCULAR MAGNETIC RESONANCE EVALUATION OF ENERGETICS DURING LEFT VENTRICULAR REMODELING

CMR spectroscopy has been used in animal models to define the metabolic alterations that occur within remodeled myocardium after myocardial injury. McDonald and associates[27] studied a canine model of LV remodeling after transmyocardial direct

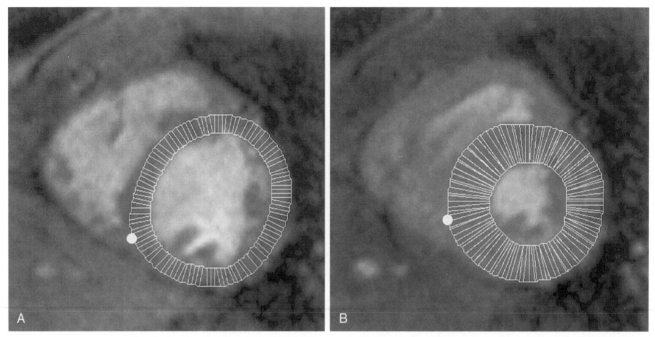

Figure 13–5. Gradient-echo short-axis cine images demonstrating the quantitation of wall thickening from images at the midpapillary level at end diastole *(A)* and end systole *(B)* using the centerline method. The endocardial and epicardial contours are shown as well as 100 equidistant chords, each representing the wall thickness at that point, that are placed perpendicular to the endocardial contour. The increase in length of each chord is calculated in order to calculate wall thickening. The starting points (white dots) on the end-diastolic and end-systolic images are used to correct for rotational motion. (From Holman ER, Buller VG, de Roos A, et al: Detection and quantification of dysfunctional myocardium by magnetic resonance imaging: A new three-dimensional method for quantitative wall-thickening analysis. Circulation 95:924, 1997.)

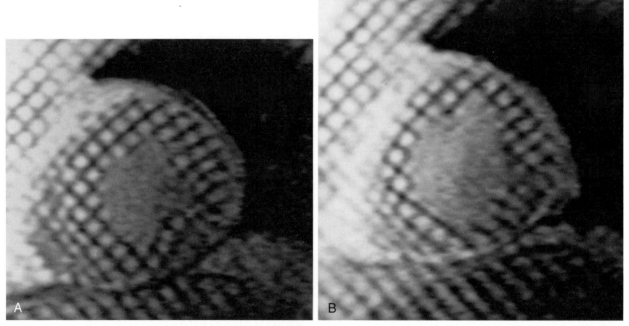

Figure 13–6. A pair of end-systolic apical MR tagged short-axis images from a single patient on day 4 *(A)* and week 8 *(B)* after anterior infarction. The right ventricular apex and septum are in the 6 to 9 o'clock position on the images, the anterior wall from 9 to 12 o'clock, the lateral wall from 12 to 3 o'clock; and the inferior wall from 3 to 6 o'clock. Reduced deformation of the tags is seen in the anterior wall and septum in both images. In the week 8 image *(B)*, the anterior wall and septum have thinned and the end-systolic LV cavity is larger than that of the day 4 image. (From Kramer CM, Rogers WJ, Theobald TM, et al: Dissociation between changes in intramyocardial function and left ventricular volumes in the 8 weeks after first anterior myocardial infarction. J Am Coll Cardiol 30:1625, 1997. Reprinted with permission from the American College of Cardiology.)

current shock that causes localized myocardial necrosis to approximately 20 percent of the anteroapical LV and leads to LV remodeling over the ensuing 12 months. These investigators used a surface coil over the lateral wall, remote from the direct region of myocardial damage and performed spatially localized ^{31}P spectroscopy. The ratio of creatine phosphate (PCr) to adenosine triphosphate (PCr/ATP), a measure of high-energy phosphate stores, was reduced in the subendocardium and subepicardium of the remodeled left ventricle. In the same model,[28] pacing reduced the subendocardial/subepicardial blood flow ratio, which was associated with a fall in the subendocardial PCr/ATP ratio and an increase in ratio of inorganic phosphate to PCr (Figure 13–7). This finding suggests that redistribution of blood flow may play a role in the alterations in myocardial high-energy phosphate levels in remodeled myocardium. The extent of remodeling, as indexed by the increase in end-diastolic volume and mass over time, correlated with the subendocardial PCr/ATP ratio.

In a study using ^{31}P spectroscopy in isolated, perfused remodeled rat myocardium after infarction induced by coronary ligation, Friedrich and colleagues demonstrated similar changes in high-energy phosphates.[29] The creatine phosphate was reduced in noninfarcted myocardium compared with control hearts and the extent of reduction correlated with the size of infarction. However, ATP content was unchanged within noninfarcted myocardium. The PCr/ATP ratios were reduced compared with controls. These investigators postulated that assessment of high-energy phosphates may be an excellent marker of the magnitude of dysfunction within noninfarcted myocardium. Finally, Zhang and coworkers performed a study in a porcine model of infarction and LV remodeling that supports this concept.[30] Their group studied 18 animals after infarction, 6 of whom developed clinical congestive heart failure while the other 12 developed asymptomatic LV dysfunction. The PCr/ATP ratio in remodeled noninfarcted myocardium was decreased, especially within the subendocardium. The decrease in PCr/ATP in the animals with heart failure was transmural and greater than animals with asymptomatic LV dysfunction and remodeling. Therefore, the extent of bioenergetic abnormalities reflects the severity of LV dysfunction in the remodeled left ventricle.

CONTRAST-ENHANCED CARDIOVASCULAR MAGNETIC RESONANCE IMAGING AND PREDICTORS OF LEFT VENTRICULAR REMODELING

Contrast-enhanced CMR has been used to demonstrate abnormal contrast uptake patterns in infarcted

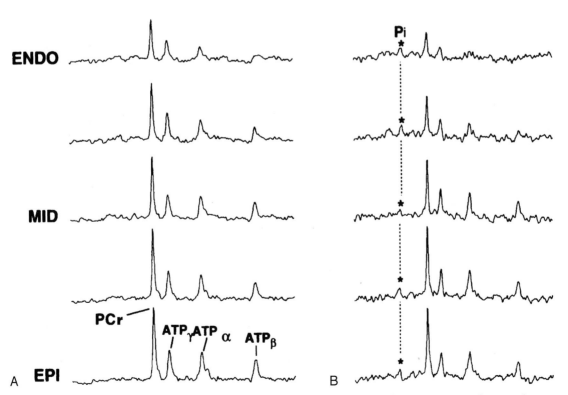

Figure 13–7. Transmurally localized ^{31}P-NMR spectra that show high energy-phosphate levels in a canine heart with LV remodeling after localized myocardial damage after DC shock. Panel *A* shows levels during sinus rhythm and panel *B* is during pacing at 240 beats per minute. Note that pacing caused a significant decrease of the PCr/ATP ratio and an increase of ΔPi/PCr in these hearts. At baseline, the PCr/ATP ratio is lower in remodeled hearts than in controls. ATP = adenosine triphosphate; PCr = creatine phosphate; Pi = inorganic phosphate; Endo = subendocardial voxel; Mid = midwall; Epi = subepicardial voxel. (From Zhang J, McDonald KM: Bioenergetic consequences of left ventricular remodeling. Circulation 92:1011, 1995.)

myocardium. Investigators have demonstrated that hypoenhancement on first-pass imaging is a marker of microvascular damage due to the no-reflow phenomenon.[31] In dogs with reperfused infarction, Gerber and associates demonstrated that the extent of the hypoenhancement or microvascular obstruction correlated with the increase in LV end-diastolic volume in the first 10 days after infarction.[32] In humans, Wu and colleagues demonstrated that the size of the area of microvascular obstruction by contrast-enhanced CMR predicted the risk of cardiovascular morbidity and mortality.[33]

CARDIOVASCULAR MAGNETIC RESONANCE EVALUATION OF THERAPY OF LEFT VENTRICULAR REMODELING

Animal Studies

Due to the precision with which CMR can be used to measure LV mass and volumes, it has been increasingly used to evaluate pharmacological therapy of LV remodeling after infarction. One of the first studies to use CMR in this regard was that of Saeed and associates,[34] in which a rat model was used to study the effect of cilazapril, an angiotensin-converting enzyme (ACE) inhibitor, on wall thickness and chamber diameter of the postinfarct left ventricle. Cilazapril preserved wall thickness in the infarct-containing segment of the anterior and septal walls and limited the increase in chamber surface area, a surrogate of LV volume. CMR techniques and the ovine model of LV remodeling were employed to demonstrate that ACE inhibition during the first 8 weeks after anteroapical infarction limited the increase in LV end-diastolic volume noted in untreated controls.[35] The limitation in LV dilatation was associated with maintenance of regional function in noninfarcted myocardium adjacent to the infarction, suggesting that dysfunction in this region may be an important determinant of the remodeling process.

A group of investigators at the University of Minnesota have used CMR to assess LV remodeling in the canine model of DC current–induced myocardial necrosis in several studies. In one,[16] they randomized dogs to nitrate therapy (n = 10) or control (n = 17) and studied them at baseline and 1 and 16 weeks after injury. Cine CMR demonstrated that nitrate therapy prevented the increase in LV mass seen in control animals at 1 week after damage and the increase in LV mass and end-diastolic volume seen at 16 weeks. In another study using CMR in the same model at 16 weeks after DC shock,[36] these investigators demonstrated that high-dose ACE inhibition limited the increase in LV mass and end-diastolic volume seen in controls, whereas neither low-dose ACE inhibition, α_1 receptor blockade, nor angiotensin II type 1 receptor blockade were effective.

To further characterize the mechanisms underlying the reduction in remodeling with ACE inhibition, McDonald and coworkers studied the canine model by CMR before and 4 weeks after DC shock in the setting of a bradykinin antagonist added to ACE inhibition.[37] They found bradykinin antagonism counteracted the reduction in LV mass seen with ACE inhibition alone, suggesting that bradykinin may be largely responsible for the antigrowth effect of ACE inhibition. No change in LV volumes was noted at the 4-week time point in any of the groups.

These investigators used similar methods to evaluate therapy with the ACE inhibitor captopril and the beta blocker metoprolol begun in the chronic stage (11 months after DC shock) and continued for 3 months.[38] They found that both therapies were associated with reduction in LV mass and LV end-diastolic volume compared with untreated controls during the treatment period. Using the ovine model of anteroapical infarction, we demonstrated that when beta blockade was added to ACE inhibition during LV remodeling, that overall LV end-diastolic volume did not change, but LV ejection fraction did improve over the first 8 weeks compared with ACE inhibition alone.[39] In the same ovine model, the most powerful pharmacological regimen for attenuating LV remodeling and preserving regional and global function is the combination of ACE inhibition and angiotensin receptor antagonism.[40]

Human Studies

CMR has begun to make inroads in clinical trials of LV remodeling. Schulman and colleagues[41] randomized 43 patients with a Q wave acute MI within 24 hours of symptom onset to intravenous enalaprilat or placebo and then oral therapy for 1 month after infarction. Twenty-three of the patients underwent CMR at 1 month after infarction for evaluation of infarct expansion. Short-axis imaging was performed and an end-diastolic slice that encompassed the papillary muscles was selected and contoured, and anterior and posterior endocardial segment lengths were measured as defined by the papillary muscles. Infarct expansion index was defined as the ratio of the infarct to noninfarct endocardial segment length. Other parameters measured included infarct segment length and wall thickness. ACE inhibitor therapy was associated with a reduced infarct segment length (7.9 ± 1.0 cm versus 10.6 ± 0.9 cm in controls) and a lower infarct expansion index (1.1 ± 0.3 vs. 1.8 ± 0.3 in controls). The greatest impact of ACE inhibitor therapy on infarct expansion was found in the subgroup with anterior infarcts.

CMR was utilized in the study of Johnson and associates[42] of 35 patients after first acute Q wave MI with a left ventricular ejection fraction (LVEF) greater than 40 percent. Studies were performed at 1 week and 3 months after MI. Stacked short-axis

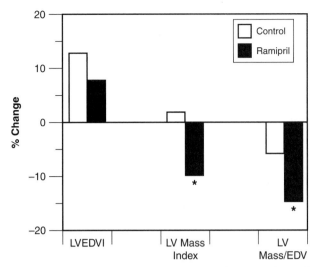

Figure 13–8. Graph depicting changes in left ventricular end-diastolic volume index (LVEDVI), LV mass index (LVMI), and LV mass/end-diastolic volume (LVM/EDV) ratio in controls (white) and ramipril-treated patients (black). *$p < 0.001$ for change versus baseline. Note the significant fall in LVMI and LVM/EDV ratio in the treated group. (From Johnson DB, Foster RE, Barilla F, et al: Angiotensin-converting enzyme inhibitor therapy affects left ventricular mass in patients with ejection fraction > 40% after acute myocardial infarction. J Am Coll Cardiol 29:49, 1997. Reprinted with permission from the American College of Cardiology.)

cine CMR slices from apex to base were summed to measure LV end-diastolic and end-systolic volume index and mass. Therapy with the ACE inhibitor ramipril contributed to a fall in LV mass index (from 82 ± 18 to 79 ± 23 g/m^2) whereas there was no significant change in control patients (77 ± 15 to 79 ± 23 g/m^2) (Figure 13–8). No significant change in LVEDVI was noted in either the treatment group or controls in these patients with mild LV dysfunction at baseline. The same group[43] used CMR to study 29 patients with an LVEF less than 40 percent treated with ACE inhibition and demonstrated that between day 5 and 3 months after MI, LV mass increased, LV end-diastolic volume increased, and ejection fraction improved despite an increase in calculated end-systolic wall stress.

The latter studies demonstrate that differences in LV structure due to pharmacological therapy of acute infarction can be demonstrated with CMR in relatively small-sized patient cohorts. Studies using 2-D echocardiography in larger infarcts have required a much larger group of patients (on the order of several hundred) to demonstrate quantifiable differences between treatment with ACE inhibitors and controls.[44, 45]

A recent study documenting the beneficial effects of early infarct artery patency on LV remodeling was done using CMR to document changes in LV mass and volumes over time.[46] Sixteen patients with an occluded LAD 10 days after anterior infarction were randomized to angioplasty of the LAD, either within the following 2 weeks or on a delayed basis (3 months later). Immediate angioplasty resulted in improved LV ejection fraction, regional systolic func-

tion, and reduced end-systolic volume. Improvements were not seen in the delayed therapy group.

CMR has also been used in studies of the effects of nonpharmacological therapies of LV remodeling after infarction. Dubach and associates[47] studied 25 patients with LV dysfunction after infarction (EF 32 $\pm$ 6%) and randomized them to 2 months of exercise in a rehabilitation program or control. Cine CMR was performed and LV volumes, mass, and ejection fraction measured. At the end of the training period, no differences were noted between groups in any of the aforementioned parameters. In addition, no differences were noted within groups over time. Exercise capacity increased in the treated group, but no deleterious effects on global parameters of LV remodeling were found, contrary to previously published data.[48]

CONCLUSION

CMR is particularly well suited to following changes in LV size, shape, and regional and global function after infarction due to its three-dimensional nature and ability to accurately define the deformed left ventricle. Investigators are increasingly using CMR in animal and clinical studies to study the effects of pharmacological and other interventions on the remodeling process. With its inherent precision and accuracy, CMR should allow studies of these interventions to occur with more power, in fewer subjects, increasing their cost effectiveness.

References

1. Schuster EH, Bulkley BH: Expansion of transmural myocardial infarction: A pathophysiologic factor in cardiac rupture. Circulation 60:1532, 1979.
2. Jugdutt BI, Warnica JW: Intravenous nitroglycerin therapy to limit myocardial infarct size, expansion, and complications. Circulation 78:906, 1988.
3. Pfeffer JM, Pfeffer MA, Fletcher PJ, Braunwald E: Progressive ventricle remodeling in the rat with myocardial infarction. Am J Physiol 260:H1401, 1991.
4. Olivetti G, Capasso JM, Meggs LG, et al: Cellular basis of chronic ventricular remodeling in rats. Circ Res 68:856, 1991.
5. McKay RG, Pfeffer MA, Pasternak RC, et al: Left ventricular remodeling following myocardial infarction: A corollary to infarct expansion. Circulation 74:693, 1986.
6. Mitchell GF, Lamas GA, Vaughan DE, Pfeffer MA: Left ventricular remodeling in the year after first anterior myocardial infarction. J Am Coll Cardiol 19:1136, 1992.
7. Hammermeister KE, DeRouen TA, Dodge HT: Variables predictive of survival in patients with coronary disease: Selection by univariate and multivariate analyses from the clinical, electrocardiographic, exercise, arteriographic, and quantitative angiographic evaluations. Circulation 59:421, 1979.
8. White HD, Norris RM, Brown MA, et al: Left ventricular end-systolic volume as the major determinant of survival after recovery from myocardial infarction. Circulation 76:44, 1987.
9. Eaton LW, Weiss JL, Bulkley BH, et al: Regional cardiac dilatation after acute myocardial infarction. N Engl J Med 300:57, 1979.
10. Erlebacher JA, Weiss JL, Eaton LW, et al: Late effects of acute infarct dilation on heart size: A two dimensional echocardiographic study. Am J Cardiol 49:1120, 1982.

11. Gaudron P, Eilles C, Kugler I, Ertl G: Progressive left ventricular dysfunction and remodeling after myocardial infarction: Potential mechanisms and early predictors. Circulation 87:755, 1993.

12. Picard MH, Wilkins GT, Ray PA, Weyman AE: Natural history of left ventricular size and function after acute myocardial infarction. Circulation 82:484, 1990.

13. Rumberger JA, Behrenbeck T, Breen JR, et al: Nonparallel changes in global left ventricular chamber volume and muscle mass during the first year after transmural myocardial infarction in humans. J Am Coll Cardiol 21:673, 1993.

14. Franco F, Thomas GD, Giroir B, et al: Magnetic resonance imaging and invasive evaluation of development of heart failure in transgenic mice with myocardial expression of tumor necrosis factor-alpha. Circulation 99:448, 1999.

15. Shapiro EP, Rogers WJ, Beyar R, et al: Determination of left ventricular mass by magnetic resonance imaging in hearts deformed by acute infarction. Circulation 79:706, 1989.

16. McDonald KM, Francis GS, Matthews J, et al: Long-term oral nitrate therapy prevents chronic ventricular remodeling in the dog. J Am Coll Cardiol 21:514, 1993.

17. Kramer CM, Lima JAC, Reichek N, et al: Regional function within noninfarcted myocardium during left ventricular remodeling. Circulation 88:1279, 1993.

18. Grossman W, Jones D, McLaurin LP: Wall stress and patterns of hypertrophy in the human left ventricle. J Clin Invest 56:56, 1975.

19. Matheijssen NA, Baur LH, Reiber JH, et al: Assessment of left ventricular volume and mass by cine magnetic resonance imaging in patients with anterior myocardial infarction: Intra-observer and inter-observer variability on contour detection. Int J Card Imaging 12:11, 1996.

20. Konermann M, Sanner BM, Horstmann E, et al: Changes of the left ventricle after myocardial infarction—estimation with cine magnetic resonance imaging during the first six months. Clin Cardiol 20:201, 1997.

21. Kramer CM, Rogers WJ, Theobald TM, et al: Dissociation between changes in intramyocardial function and left ventricular volumes in the 8 weeks after first anterior myocardial infarction. J Am Coll Cardiol 30:1625, 1997.

22. Holman ER, Buller VG, de Roos A, et al: Detection and quantification of dysfunctional myocardium by magnetic resonance imaging: A new three-dimensional method for quantitative wall-thickening analysis. Circulation 95:924, 1997.

23. Kramer CM, Rogers WJ, Theobald T, et al: Remote noninfarcted region dysfunction soon after first anterior myocardial infarction: A magnetic resonance tagging study. Circulation 94:660, 1996.

24. Zerhouni E, Parrish D, Rogers WJ, et al: Human heart: tagging with MR imaging—a method for noninvasive measurement of myocardial motion. Radiology 169:59, 1988.

25. Axel L, Dougherty L: MR imaging of motion with spatial modulation of magnetization. Radiology 171:841, 1989.

26. Moulton MJ, Downing SW, Creswell LL, et al: Mechanical dysfunction in the border zone of an ovine model of left ventricular aneurysm. Ann Thorac Surg 60:986, 1995.

27. McDonald KM, Yoshiyama M, Francis GS, et al: Myocardial bioenergetic abnormalities in a canine model of left ventricular dysfunction. J Am Coll Cardiol 23:786, 1994.

28. Zhang J, McDonald KM: Bioenergetic consequences of left ventricular remodeling. Circulation 92:1011, 1995.

29. Freidrich J, Apstein CS, Ingwall JS: ³¹P nuclear magnetic resonance spectroscopic imaging of regions of remodeled myocardium in the infarcted rat heart. Circulation 92:3527, 1995.

30. Zhang J, Wilke N, Wang Y, et al: Functional and bioenergetic consequences of postinfarction left ventricular remodeling in a new porcine model. Circulation 94:1089, 1996.

31. Judd RM, Lugo-Olivieri CH, Arai M, et al: Physiological basis of myocardial contrast enhancement in fast magnetic resonance images of 2-day-old reperfused canine infarcts. Circulation 92:1902, 1995.

32. Gerber BL, Rochitte CE, Melin JA, et al: Microvascular obstruction and left ventricular remodeling early after acute myocardial infarction. Circulation 101:2734, 2000.

33. Wu KC, Zerhouni EA, Judd RM, et al: Prognostic significance of microvascular obstruction by magnetic resonance imaging in patients with acute myocardial infarction. Circulation 97:765, 1998.

34. Saeed M, Wendland MF, Seelos K, et al: Effect of cilazapril on regional left ventricular wall thickness and chamber dimension following acute myocardial infarction: In vivo assessment using MRI. Am Heart J 123:1472, 1992.

35. Kramer CM, Ferrari VA, Rogers WJ, et al: Angiotensin converting enzyme inhibition limits dysfunction in adjacent noninfarcted regions during left ventricular remodeling. J Am Coll Cardiol 27:211, 1996.

36. McDonald KM, Garr M, Carlyle PF, et al: Relative effects of alpha 1-adrenoceptor blockade, converting enzyme inhibitor therapy, and angiotensin II subtype 1 receptor blockade on ventricular remodeling in the dog. Circulation 90:3034, 1994.

37. McDonald KM, Mock J, D'Aloia A, et al: Bradykinin antagonism inhibits the antigrowth effect of converting enzyme inhibition in the dog myocardium after discrete transmural myocardial necrosis. Circulation 92:2043, 1995.

38. McDonald KM, Rector T, Carlyle PF, et al: Angiotensin-converting enzyme inhibition and beta-adrenoceptor blockade regress established ventricular remodeling in a canine model of discrete myocardial damage. J Am Coll Cardiol 24:1762, 1994.

39. Kramer CM, Nicol PD, Rogers WJ, et al: β-blockade improves adjacent regional sympathetic innervation during postinfarction remodeling. Am J Physiol 277:H1429, 1999.

40. Mankad S, d'Amato T, Reichek N, et al: Combining angiotensin II receptor antagonism and angiotensin converting enzyme inhibition further attenuates post infarction remodeling. Circulation 2001 (in press).

41. Schulman SP, Weiss JL, Becker LC, et al: Effect of early enalapril therapy on left ventricular function and structure in acute myocardial infarction. Am J Cardiol 76:764, 1995.

42. Johnson DB, Foster RE, Barilla F, et al: Angiotensin-converting enzyme inhibitor therapy affects left ventricular mass in patients with ejection fraction > 40% after acute myocardial infarction. J Am Coll Cardiol 29:49, 1997.

43. Foster RE, Johnson DB, Barilla F, et al: Changes in left ventricular mass and volumes in patients receiving angiotensin-converting enzyme inhibitor therapy for left ventricular dysfunction after Q-wave myocardial infarction. Am Heart J 136:269, 1998.

44. St. John Sutton M, Pfeffer MA, Plappert T, et al: Quantitative two-dimensional echocardiographic measurements are major predictors of adverse cardiovascular events after acute myocardial infarction. Circulation 89:68, 1994.

45. Bellenger NG, Davies LC, Francis JM, et al: Reduction in sample size for studies of remodelling in heart failure by the use of cardiovascular magnetic resonance. J Cardiovasc Magn Reson 2:271, 2000.

46. Pfisterer ME, Buser PT, Osswald S, et al: Time dependence of left ventricular recovery after delayed recanalization of an occluded infarct-related coronary artery: Findings of a pilot study. J Am Coll Cardiol 32;97, 1998.

47. Dubach P, Myers J, Dziekan G, et al: Effect of exercise training on myocardial remodeling in patients with reduced left ventricular function after myocardial infarction: Application of magnetic resonance imaging. Circulation 95:2060, 1997.

48. Jugdutt BI, Michororski BL, Kappagoda CT: Exercise training after anterior Q wave myocardial infarction: Importance of regional left ventricular function and topography. J Am Coll Cardiol 12:362, 1988.

CHAPTER 14

Myocardial Viability

Udo P. Sechtem, Frank M. Baer, and Eberhard Voth

The identification of residual myocardial viability in a patient with regional or global severe left ventricular systolic dysfunction is of clinical importance to plan the therapeutic strategy, because revascularization of dysfunctional but viable myocardium may improve left ventricular function.[1] Several imaging techniques have been shown to be successful in detecting myocardial viability, and these techniques include left ventricular angiography using appropriate interventions,[2] perfusion scintigraphy, positron emission tomography (PET),[3] and echocardiography.[4] More recently, cardiovascular magnetic resonance (CMR) techniques have been applied to identify viable myocardium and distinguish it from myocardial necrosis and scar.[5] This chapter reviews the current knowledge of how these techniques can be used in humans to guide clinical decision-making and predict recovery of function after revascularization of dysfunctional myocardium.

FEATURES OF VIABLE MYOCARDIUM DETECTABLE BY CARDIOVASCULAR MAGNETIC RESONANCE

Scar Formation and Left Ventricular Wall Thickness

The most common clinical definition of viable dysfunctional myocardium and also the most practical one is the following: Myocardium is viable if it shows severe dysfunction at baseline but recovers function with time either spontaneously (myocardial stunning) or following mechanical revascularization (hibernating myocardium). Clinically, stunned myocardium may be found in patients with early reperfusion of an infarct-related artery. If there is no residual high-grade stenosis, blood flow at rest will be normal and the myocardium will recover spontaneously after a few days. Patients with hibernating myocardium often present with severe multivessel coronary disease, globally depressed left ventricular function, and prominent dyspnea but surprisingly little angina. This type of dysfunction is often more chronic, and previous myocardial infarction may or may not be reported in the history.

Pathology may reveal subendocardial infarction in the former case, whereas in the latter case regions of transmural scar, regions with predominantly subendocardial scar, and those with mixtures of scar and viable myocardium can be found. Severe regional wall thinning is the hallmark of transmural chronic myocardial infarction (Figure 14–1). However, wall thinning may require local infarct healing and remodeling, which takes up to 4 months.[6]

In contrast to the severe thinning of chronic transmural scar, acute and subacute transmural infarcts may not yet have reached the stage of thinning because local infarct remodeling is incomplete (see Figure 14–1). In contrast to transmural myocardial infarction, which may or may not appear thinned depending on infarct age, nontransmural infarcts do not develop severe thinning (Figure 14–2). Some thinning may, however, be observed, depending on the degree to which the endocardially located infarct extends through the wall. Even in chronic subendocardial infarcts of more than 4 months of age, extreme wall thinning such as seen in transmural infarcts is not observed. Therefore, the finding of preserved diastolic myocardial wall thickness in a patient with a known chronic infarct of more than 4 months of age will likely represent nontransmural infarction with a substantial epimyocardial rim of viable tissue surrounding the endocardial scar. If the infarct has been more recent than 4 months, the feature of end-diastolic wall thickness cannot be used to distinguish between viable and nonviable myocardium.

Contractile Reserve of Viable Myocardium

A well-known feature of viable myocardium is augmented contractility in response to a suitable stimulus.[7] Such stimuli include sympathomimetic agents[7] or postextrasystolic potentiation.[2] In contrast, necrotic or scarred tissue will not respond to such stimulation. Today, the most widely used mode of stimulation is to infuse low doses (2.5–10 μg/kg/min) of dobutamine. If a contractile reserve can be elicited, the responsive myocardium will usually recover function after appropriate revascularization.[8]

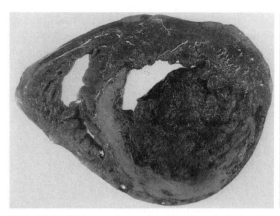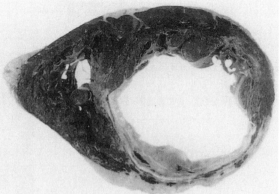

Figure 14–1. Wall thickness in acute and chronic infarction. *Left:* Left anterior descending (LAD) occlusion with 1-week-old anteroseptal infarct. Note that there is only minor wall thinning despite the fact that the infarct is transmural. The left ventricular (LV) cavity is filled with thrombus. *Right:* In contrast, there is extreme wall thinning in this chronic healed transmural infarct that occurred 6 months previously. Nitroblue tetrazolium stains viable cells blue. (See also color plates.) (From Braunwald E, Califf RM (eds): Atlas of Heart Diseases Vol. VIII: Acute Myocardial Infarction and Other Ischemic Syndromes. Philadelphia, Current Medicine, 1996.)

Noninvasive Observation of Tissue Edema

Irreversible myocardial damage occurs after 30 to 120 minutes of ischemia. Very early changes can be observed by electron microscopy, and these changes include intracellular edema and swelling of the entire cell, including the mitochondria. The sarcolemma ruptures, and there is free exchange between the extra- and intracellular compartments. In some infarcts, light microscopy reveals changes just a few hours after the onset of ischemia, and these changes are most pronounced at the periphery of the infarct. After 8 hours, there is edema of the interstitium and

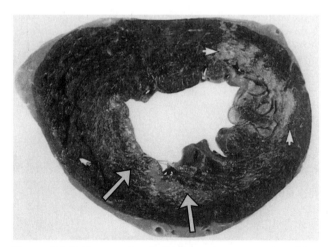

Figure 14–2. Preserved wall thickness in nontransmural infarction. Irrespective of age, nontransmural infarcts do not show severe wall thinning. This patient suffered an old nontransmural infarct in the inferolateral wall (short arrows) and a fresh anteroseptal nontransmural infarct (long arrows). Triphenyl tetrazolium chloride (TTC) fails to stain mature lateral scar and pale freshly infarcted muscle in anterior region. Viable myocardium is stained red. (See also color plates.) (From Braunwald E, Califf RM (eds): Atlas of Heart Diseases Vol. VIII: Acute Myocardial Infarction and Other Ischemic Syndromes. Philadelphia, Current Medicine, 1996.)

infiltration of the infarct zone by neutrophils, and red blood cells become evident. Small blood vessels undergo necrosis, and karyolysis of muscle cell nuclei can be observed.[9] Plugging of capillaries by erythrocytes is most pronounced in the center of the infarct. Interestingly, some intact muscle fibers can be observed even in firm connective tissue scar. If reperfusion can be achieved at an early stage, the resulting infarcts contain a mixture of necrosis and hemorrhage within zones of irreversibly injured myocytes. Islands of viable myocardium may be interspersed with zones of necrosis.

Myocardial edema is associated with prolonged relaxation times that lead to characteristically increased signal intensity changes on cardiovascular magnetic resonance (CMR) images.[10] However, because the resolution of CMR images is far inferior to light microscopy, mixtures of zones of necrosis and viable tissue cannot be displayed in sufficient detail. In contrast, it is possible with modern CMR techniques to distinguish between endocardially located zones of necrosis and epicardial rims of viable tissue.

The No-Reflow Phenomenon and Early Hypoenhancement with CMR Contrast Agents

A feature of the central necrotic region within a myocardial infarct is intracapillary red blood cell stasis.[11] Plugging of the capillaries leads to tissue hypoperfusion. This hypoperfusion is primarily related to functional capillary density rather than microvascular flow rates.[12] A decrease in functional capillary density results in a prolonged wash-in time constant. This lack of reperfusion after restoration of flow in epicardial vessel is known as the *no-reflow phenomenon*. When the myocardium is imaged by CMR early after injection of gadolinium diethylenetriamine pentaacetic acid (Gd-DTPA), no-reflow zones appear dark, as compared with the subepicardial rim regions of the infarct.[13]

Late Hyperenhancement in Infarcted Tissue Related to Gd-DTPA

Rupture of myocyte membranes leads to an increased volume of distribution of CMR contrast agents with a corresponding increase in the effective voxel concentration of such agents.[14, 15] This volume of distribution is influenced by the effects of infarct healing and increases for up to 6 days following reperfusion.[16] In addition, infarct tissue time constants are prolonged as compared with normal tissue, and clearance of CMR contrast agents from infarct tissue does not follow blood or normal tissue clearance. Thus, although early hypoenhancement of infarcted regions after injection of contrast material is due to delayed contrast penetration,[13] late hyperenhancement in infarction is due to both increased volume of distribution and slow contrast washout.[12, 17] The enhancement pattern seen will depend on regional differences in tissue wash-in/washout kinetics, as well as the time after injection of contrast when the image is acquired. Late Gd-DTPA enhancement has now been extensively validated in animal and human studies, and with improved imaging sequences (notably the use of inversion recovery to null signal from normal myocardium), the signal-to-noise ratio of enhanced to unenhanced tissue is dramatically higher than with previous sequences, at approximately 500 percent.[18] This capacity has led to greatly improved image quality and a substantial increase in the clinical use of the technique. In animal experiments, the area of late Gd-DTPA contrast enhancement has been shown to correlate closely with areas of infarction,[18] and for the first time in vivo, high-quality imaging of regional scar is possible.

High-Energy Phosphates and Viability

The primary energy reserve in living myocardial cells is stored in the form of creatine phosphate and adenosine triphosphate (ATP). Depletion of total myocardial creatine, creatine phosphate, and ATP follows severe ischemic injury as shown in biopsy samples obtained from patients during cardiac surgery[19] or necropsy.[20] Using ^{31}P magnetic resonance spectroscopy (MRS), one can measure the myocardial content of phosphocreatine and ATP.[21] This technique is, however, hampered by its slow intrinsic sensitivity and low metabolite concentrations, which have restricted studies to large myocardial voxels (approximately 30 ml) near the anterior chest wall.[22] ^{1}H MRS has a higher sensitivity than ^{31}P MRS and has the ability to detect the total pool of

phosphorylated plus unphosphorylated creatine in skeletal and cardiac muscle. Therefore ^{1}H MRS has a 20-fold sensitivity improvement compared with ^{31}P MRS of phosphorylated creatine. Consequently, ^{1}H MRS allows one to metabolically interrogate small voxels of less than 10 ml in all regions of the left ventricle including the posterior wall. Moreover, this task can be accomplished on clinical CMR systems with field strengths of 1.5 T.[23]

Sodium and Potassium Cardiovascular Magnetic Resonance

Although much of this work has been performed in animals, there are now some early human results, and therefore this brief section is included for completeness. When cell membranes break down at the time of infarction, there are two biochemical changes with inorganic elements that can be detected using CMR. The first is the increase in sodium concentration, which occurs because the volume of muscle previously occupied by the intact cell has low sodium concentration. Sodium imaging using CMR therefore shows a bright signal at the site of acute infarction.[24–26] By contrast, the potassium concentration, which is high in the intracellular space but low extracellularly, falls markedly with the loss of cell membrane integrity, and potassium imaging shows a dark area therefore with infarction.[27, 28] Such imaging requires a multifrequency CMR system, which is very limited in availability, but the technique has the potential to distinguish acute from chronic infarction on the basis of these acute cation fluxes, and it is conceivable that clinical application might be found.

CARDIOVASCULAR MAGNETIC RESONANCE IN ACUTE MYOCARDIAL INFARCTION

Signal Intensity Changes on Spin-Echo Images

Myocardial edema characterizes acute myocardial necrosis. On T2-weighted spin-echo images, the increased water content leads to an increase in signal intensity (Figure 14–3). In animal models, a good correlation between water content and T2 relaxation time or T2-weighted signal intensity, respectively, has been described.[29] Moreover, the area at risk measured by CMR correlates well with that determined at pathology.[29] T2-weighted spin-echo images acquired early after myocardial infarction (within 10 days) demonstrate the infarct site as a region of high signal intensity as compared with normal myocardium.[30] However, there are several pitfalls to this technique, including the necessity to differentiate signal from slowly flowing blood in the ventricle

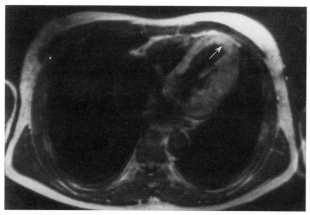

Figure 14–3. T2-weighted transverse spin-echo image (echo time [TE] = 30 ms) in a patient with a 10-day-old anteroseptal myocardial infarction. Note the increased signal intensity in the distal, septal, and apical regions. The arrow points to a region of slow blood flow near the thinned anterior wall.

from increased signal intensity from a region of infarction, and to recognize artifactual variation of signal intensity in the myocardium due to respiratory motion or residual cardiac motion. To enhance the usefulness of T2-weighted spin-echo CMR for detailed characterization of infarction, Johnston and associates[31] developed a velocity, compensated spin-echo pulse sequence. Using this sequence, they correctly identified the location of myocardial infarction by its characteristic high signal intensity in all 10 patients examined.[31] Using this technique, it was also possible to visualize remnants of viable tissue, because most patients had a mixture of transmural and nontransmural injury. Moreover, heterogeneous distribution of signal intensity within the infarction suggested the presence of hemorrhage.

This group extended their observations by comparing CMR findings with those obtained from adenosine ^{201}Tl single-photon emission computed tomography (SPECT) imaging at approximately 6 days after the infarction. Viable myocardium was defined with CMR as a segment with increased signal intensity but preserved wall thickening. Ten of 11 patients with redistribution on the thallium images in the infarct region had preserved wall thickening by CMR. Of 13 patients with fixed defects (the depth of the defect was not reported), 6 had preserved wall thickening by CMR.[32] A good correlation between thallium planar scintigraphy and T2-weighted spin-echo CMR was also noted by Krauss and colleagues.[33] Of 20 patients with a documented myocardial infarction (mean 4 days old), thallium imaging detected infarcts in 18 patients, whereas CMR was successful in 17 patients. T2-weighted spin-echo CMR can be performed very easily and thus permits serial follow-up of patients. There is a gradual reduction of signal intensity of the infarct area over time accompanied by a concentration of the bright signal to the subendocardium of the infarct region over 3 months. This finding corresponds to

the well-known sequence of events described by pathologists, with infarct healing from the periphery of the infarct towards the center. Patients who are readmitted with acute coronary syndromes related to the same infarct region show an increase in signal intensity on follow-up studies.[34]

Another improvement in image quality was described by Lim and coworkers.[35] T2-weighted spin-echo CMR was obtained within 10 heartbeats during breath-hold, and signal from inflowing blood flow was suppressed by using appropriate prepulses. Areas of high signal intensity with CMR corresponded to fixed perfusion defects on thallium SPECT images in 85 percent of segments. The size of the infarct correlated well to that measured by thallium SPECT. A primary advantage of CMR is the improved spatial resolution (Figure 14–4), which permits visualization of rims of viable tissue in the epicardial portion of the left ventricular wall. This improved image quality has been helpful, but in clinical practice it has largely been superseded by inversion recovery late enhancement imaging with Gd-DTPA.

Contrast-Enhanced Studies Using Spin-Echo Cardiovascular Magnetic Resonance

Detection of areas of acute myocardial infarction can be improved by using Gd-DTPA.[36–39] Findings from animal studies indicating improved detection of acute myocardial infarcts were confirmed in humans with a study by deRoos and co-workers of 5 patients 2 to 17 days after myocardial infarction before and after administration of 0.1 mmol/kg Gd-DTPA.[38] Contrast between normal and infarcted myocardium was greatest 20 to 30 minutes after Gd-DTPA injection. The precontrast intensity ratio between infarcted and normal myocardium was 1.1 at echo time (TE) = 30 ms and was 1.4 at TE = 60 ms ($p < 0.05$). The postcontrast intensity ratio at TE = 30 ms was 1.6, which was not statistically different from the ratio at TE = 60 ms precontrast but which was significantly higher than the ratio at TE = 30 ms ($p < 0.01$). These researchers confirmed in larger patient populations of up to 45 patients that the detectability of acute myocardial infarction was similar on precontrast images at TE = 60 ms and with Gd-DTPA–enhanced CMR at the shorter TE of 30 ms.[17, 37] Image quality, however, was superior using the Gd-DTPA–enhanced short-TE technique.

In order to use Gd-DTPA–enhanced spin-echo CMR for distinguishing between infarcted and viable myocardium, it is important to know how well areas of signal enhancement correlate with the extent of myocardial necrosis. Myocardium that is reperfused very early after a brief occlusion and is hence viable cannot be distinguished from normal myocardium on the basis of Gd-DTPA enhancement.[40] In contrast, irreversibly injured myocardium shows more signal enhancement than either normal

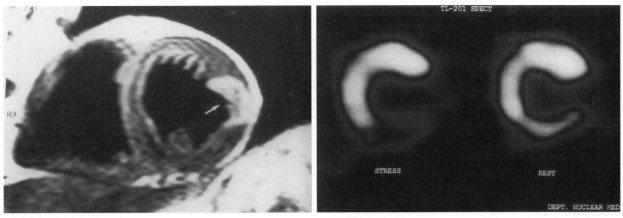

Figure 14–4. Small transmural myocardial infarct by CMR *(left)* and thallium single photon emission tomography (SPECT) images *(right)*. The coronary angiogram showed an occluded left circumflex coronary artery, 75% stenosis of the LAD coronary artery, and a 90% stenosis of the right coronary artery. The ECG was normal. There is a well-defined focal area of high signal intensity at the lateral wall of the left ventricle (arrow) on this breath-hold fast spin-echo T2-weighted CMR. Note that there is some myocardium with normal signal intensity (viable myocardium) extending from the anterolateral region towards the infarct area near the epicardium. Such a viable border zone is often seen at necropsy. Thallium images reveal a corresponding fixed defect. However, structural details such as the epicardial rim of viable myocardium, which may be worth preserving by revascularizing the circumflex coronary artery, can only be seen on the high spatial resolution CMR. (See also color plates.) (From Lim TH, Hong MK, Lee JS, et al: Novel application of breath-hold turbo spin-echo T2 MRI for detection of acute myocardial infarction. J Magn Reson Imaging 7:996, 1997. Copyright 1997. Reprinted with permission of Wiley-Liss, Inc., a subsidiary of John Wiley & Sons, Inc.)

or reversibly injured myocardium. Data from animal studies suggest that the Gd-DTPA–enhanced region of reperfused myocardium on spin-echo CMR closely correlates with the myocardial bed at risk.[41]

Different patterns of signal enhancement in the infarct region after intravenous injection of 0.1 mmol/kg Gd-DTPA were described in patients studied more than 1 month after onset of acute myocardial infarction.[42] A conventional spin-echo technique with TE of 30 and 70 ms was employed before and 5 to 10 minutes after Gd-DTPA. Enhanced regions were classified into four types (Figure 14–5):

Viability Pattern 1   *Viability Pattern 2*   *Viability Pattern 3 =Scar Pattern*

Figure 14–5. Viability patterns on conventional electrocardiogram (ECG) gated T2-weighted spin-echo CMR (TE = 70 ms). Imaging was performed 10 minutes after injection of 0.1 mmol/kg Gd-DTPA. Viability pattern 1 shows nontransmural enhancement of infarct area. Viability pattern 2 is characterized by an almost homogeneous transmural enhancement. Viability pattern 3, which is usually associated with scar, shows transmural and marginal enhancement and less enhancement of the endocardial portion of the myocardium. Because there is no contraction in this region, slow blood flow (high signal intensity) can be seen adjacent to the infarct zone. RV = right ventricle; LV = left ventricle. (From Yokota C, Nonogi H, Miyazaki S, et al: Gadolinium-enhanced magnetic resonance imaging in acute myocardial infarction. Am J Cardiol 75:577, 1995. Reprinted with permission from Excerpta Medica Inc.)

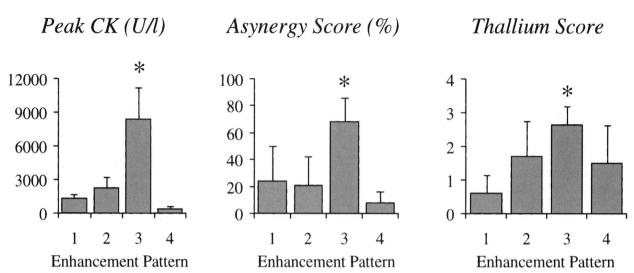

Figure 14–6. Clinical findings and enhancement patterns shown in Figure 14–5. Peak creatine kinase (CK) values were significantly higher in type 3 patients as were the asynergy score and the thallium score. (Modified from Yokota C, Nonogi H, Miyazaki S, et al: Gadolinium-enhanced magnetic resonance imaging in acute myocardial infarction. Am J Cardiol 75:577, 1995. Reprinted with permission from Excerpta Medica Inc.)

nontransmural (type 1); transmural and homogeneous (type 2); transmural and marginal (type 3); and no enhancement (type 4). These patterns were related to clinical data obtained from serial creatine kinase measurements, thallium imaging, and left ventricular angiography. In type 3 patients, peak creatine kinase levels, thallium score, and percent asynergy perimeter were significantly higher than in patients with the other three patterns (Figure 14–6). Therefore the type 3 enhancement pattern correlated best with nonviable myocardium. The mechanism of this type of enhancement of the endocardial layer was not related to the patency of the infarct-related artery or the development of collaterals as shown by x-ray coronary angiography in all patients. A more likely explanation for this pattern is a decrease in viable muscle and an increase in interstitial fibrous tissue in the inner layer. A pattern of enhancement limited to the endocardial layer was associated with the lowest thallium score, indicating the presence of the largest amounts of viable myocardium.

Similar findings were reported by Dendale and coworkers, who related perfusion patterns after Gd-DTPA administration to recovery of wall motion under dobutamine stress, which was used as the gold standard for the presence of viable myocardium.[43] They described enhancement patterns as subendocardial, transmural, or doughnut types. The subendocardial or absent infarct enhancement patterns were related to functional recovery under stress in 31 of 37 infarct segments. In contrast, transmural infarct enhancement was correlated with nonviable myocardium in 10 of 17 infarct segments. Again, marginal or doughnut enhancement was associated with the largest extent of damage and absence of viability. In both studies, image quality was

not optimal because conventional spin-echo CMR was employed. Image quality may be improved by using fast spin-echo sequences but this finding has not yet been reported. Another limitation of determining viability from contrast-enhanced spin-echo CMR is that enhancement patterns may be influenced by several factors: residual blood supply to the infarcted region, size and depth of the infarct, and the interval between the onset of myocardial infarction and the CMR examination. Other factors include different echo times and different doses of contrast material.

Gd-DTPA Studies Using Perfusion Cardiovascular Magnetic Resonance

Ultrafast CMR allows the imaging of the passage of a bolus of contrast material through the heart and especially the left ventricular myocardium (see also Chapter 11).[41] Perfusion CMR was performed by van Rugge and colleagues,[44] who showed that patients with healed myocardial infarction showed less signal intensity enhancement (50% vs. 134% in normal myocardium, $p < 0.01$). Moreover, the rate of signal increase in infarcted myocardium was significantly lower than in normal myocardium (5.2 ± 2.2 vs. 19.0 ± 10 s^{-1}). This study, however, did not specifically address the question of myocardial viability. Lima and co-workers studied 22 patients with recent myocardial infarction using perfusion CMR.[13] Time-intensity curves obtained from infarcted and noninfarcted regions were correlated with coronary anatomy and left ventricular function. Two perfusion patterns were observed in infarcted regions by comparison with the normal myocardial pattern. The first pattern was seen in all but one patient and showed persistent myocardial hyperenhancement within the infarcted region up to 10 minutes after contrast injection. The second pattern was seen in 10 patients in whom this hyperenhanced region surrounded a subendocardial area of the decreased signal at the center of the infarcted region (Figure 14–7). These patients had coronary occlusion at angiography, Q waves in the electrocardiogram (ECG), and greater regional dysfunction by echocardiography. The extent and location of perfusion abnormalities detected by CMR correlated well with the extent and location of fixed thallium SPECT defects. It is conceivable that hypoenhancement in the central infarct region as observed on CMR perfusion studies reflects the same mechanism of microvascular obstruction as the marginal enhancement or doughnut pattern observed after Gd-DTPA injection by spin-echo CMR. Animal studies demonstrated that hypoenhanced regions correlate closely with no-reflow regions as demonstrated by thioflavin-S injection,[12, 45, 46] and this correlation is best at levels of perfusion of less than 0.1 ml/g/min.[47] As expected in dogs, which have ample collaterals, the spatial extent to the zone of hyperenhancement is smaller than the risk region. In early experiments, the zone of hyperenhancement was highly correlated with 2,3,5-triphenyltetrazolium chloride (TTC)–negative regions, but was 12 percent larger.[47] This overestimation has been shown in subsequent studies to result from partial volume effects, and when this effect is corrected for, there is a very close relationship between infarct size and the area of gadolinum enhancement.[18]

In the center of the infarct region, myocytes and capillaries may undergo necrosis simultaneously because of profound and sustained ischemia. In that situation, capillaries become occluded by dying blood cells and debris to the extent that even with restoration of epicardial blood flow, the infarct core will not promptly reperfuse. This area of microvascular obstruction is called the *no-reflow* region.[11] Microvascular obstruction following acute infarction correlates with greater myocardial damage by echocardiography and poorer global left ventricular function in the early postinfarction phase.[48] Whereas delineation of microvascular damage by contrast echocardiography usually requires cardiac catheterization,[48] this phenomenon can be easily observed noninvasively by perfusion CMR following the injection of Gd-DTPA.[46] Regions with profound microvascular obstruction appear as dark subendocardial zones surrounded by hyperenhanced infarcted or injured myocardium and correspond to experimentally produced no-reflow regions.[46] Wu and coworkers studied 44 patients by using perfusion CMR.[46] Almost all of these patients had thrombolysis or direct angioplasty. To study left ventricular remodeling, 17 patients underwent repeated CMR 6 months after the initial study. Microvascular obstruction was defined as hypoenhancement seen 1 to 2 minutes after contrast injection. Infarct size was assessed as percent left ventricular mass hyperenhanced 5 to 10 minutes after contrast. Patients with microvascular obstruction (n = 11) had more cardiovascular events than those without (45% vs. 9%, p = 0.02). The risk of adverse cardiovascular events increased with infarct extent (30%, 43%, and 71% for small [n = 10], mid-sized [n = 14], and large [n = 14] infarcts, $p < 0.05$; Figure 14–8). Even after infarct size was controlled for, the presence of microvascular obstruction remained a prognostic marker for postinfarction complications. Among those returning for follow-up imaging, microvascular obstruction was predictive of left ventricular remodeling (Figure 14–9). How is the presence of microvascular obstruction related to postinfarction prognosis? The structural alterations associated with microvascular obstruction may render the myocardium more vulnerable to the forces that produce remodeling of the left ventricle.[48] Larger left ventricular volumes after myocardial infarction carry an independent risk for cardiac events.[49] However, these data indicate that the presence of microvascular obstruction may carry a risk independent of increased left ventricular volumes. Further studies using CMR will be necessary to define whether this adverse prognosis can be improved by recanalization of the infarct-related artery.[50]

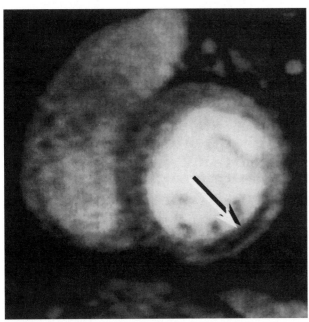

Figure 14–7. Patient with occlusion of the left circumflex coronary artery and failed thrombolysis who was treated with rescue angioplasty within 6 hours after the onset of chest pain. The CMR perfusion image shows a central zone of reduced signal enhancement (arrow) in the subendocardial half of the left ventricular wall surrounded by a region of hyperenhancement that corresponds in location and extent to a fixed defect seen on the thallium study obtained the day before CMR (not shown here). (From Lima JAC, Judd RM, Bazille A, et al: Regional heterogeneity of human myocardial infarcts demonstrated by contrast-enhanced MRI. Potential mechanisms. Circulation 92:1117–1125, 1995.)

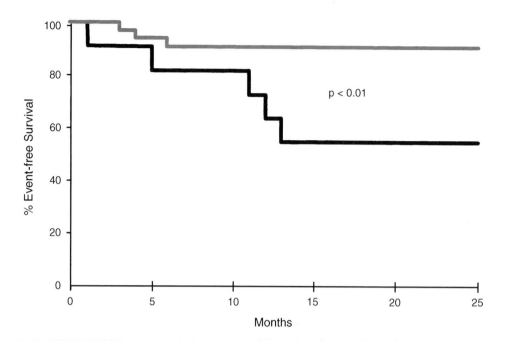

Figure 14–8. Event-free survival (clinical course without cardiovascular death, reinfarction, congestive heart failure, or stroke) for patients with (top line) and without (bottom line) CMR microvascular obstruction. (From Wu KC, Zerhouni EA, Judd RM, et al: Prognostic significance of microvascular obstruction by magnetic resonance imaging in patients with acute myocardial infarction. Circulation 97:765, 1998.)

Late Enhancement with Gd-DTPA in Acute Infarction

The latest work in the area of viability in acute infarction has concentrated on the use of late myocardial enhancement after injection of Gd-DTPA in animals. This technique appears to be attractive because of the high image quality, simplicity, and high resolution allowing clear demarcation of the transmural extent of necrosis and scar. In acute infarction, after the first-pass and the early enhancement phases, which show perfusion and microvas-

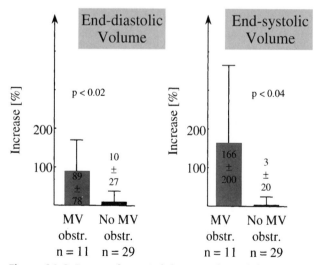

Figure 14–9. Percent change in left ventricular end-diastolic and end-systolic volumes early and 6 months after infarction. There is a significantly larger increase of both parameters in the group with microvascular (MV) obstruction by CMR as compared to the group without this finding. (Modified from Wu KC, Zerhouri EA, Judd RM, et al: Prognostic significance of microvascular obstruction by magnetic resonance imaging in patients with acute myocardial infarction. Circulation 97:765, 1998.)

cular obstruction, respectively, comes the late enhancement period from 5 to 15 minutes after injection. The less the *transmural* extent of hyperenhancement, the greater the amount of surviving subepicardial rim of viable myocardium, which if dysfunctional can be expected to recover either spontaneously (stunning) or after revascularization (hibernation).[51] The results suggest that hyperenhancement of greater than 50 percent of the myocardial transmural extent is associated with a low likelihood of recovery of function. There are other reports in humans with acute infarction, suggesting that spontaneous recovery occurs after acute infarction in hyperenhanced regions, which may appear to be contrary to the aforementioned findings.[52, 53] However, these latter studies examined enhancement in a different way, within the first few minutes after Gd-DTPA injection, and did not differentiate the transmural extent of hyperenhancement in their analysis. These confounding factors make direct comparison of the results between the studies of various investigators difficult. The hypoenhanced regions in this study did not recover function,[52, 53] as would be expected inasmuch as they represent areas of necrosis associated with microvascular obstruction, and this finding was concordant with previous results. Overall, opinion remains that with high resolution imaging, significant transmural hyperenhancement at 5 to 15 minutes reflects nonviable myocardium, and results from other centers in the near future should clarify the situation further. A recent paper has shown that the increased partition coefficient in acute infarctions can be used to demonstrate hyperenhancement during a constant infusion of Gd-DTPA, and thus predict viability.[54]

Although there has been some discussion as to whether late Gd-DTPA enhancement in acute infarction might overestimate infarct size, recent data suggest that the enhanced area closely matches the TTC–negative staining infarcted area in dogs at all stages after infarction (4 hours, 1, 3, and 10 days, and 8 weeks).[55] The change in size of the Gd-DTPA enhancement with time reflects infarct resorption and scar formation, and not a reduction in enhancement of at risk area. This finding accords with recent results showing that reversible injury does not cause late enhancement.[18, 55] Late enhancement can also be used to define myocardial salvage after infarction, by comparing the at-risk zone with the actual area of infarction, and showing that the shorter the occlusion period the lesser the Gd-DTPA enhancement and the greater the functional recovery.[51]

Wall Thickness and Wall Thickening Measurements

After an acute ischemic event, structural changes occur within the infarct zone, and infarct healing with scar formation is completed by approximately 3 to 4 months.[6] Thinning of the infarct region may occur early, especially in large anterior myocardial infarcts. The consequence is an increase in the size of the infarcted segment, known as infarct expansion.[56] However, infarct expansion usually does not occur in patients with open infarct-related arteries,[57] which are encountered more often today with the widespread use of thrombolysis and acute angioplasty of the infarct artery. Therefore transmural necrosis and nontransmural necrosis may have the same wall thickness early after myocardial infarction. Both conditions may also be associated with complete absence of resting contractility early after the acute event. Consequently, observing anatomy and function of the left ventricle at rest by CMR may not be helpful to detect residual viability. However, even a small amount of wall thickening in a region of interest indicates the presence of residual contracting cells and hence of viable myocardium.

Measurements of left ventricular wall thickening by CMR are probably more accurate than echocardiographic measurements.[58] However, as with all cross-sectional imaging techniques, the complex motion of the heart in relation to the body axes makes it impossible to observe exactly the same portion of myocardium during systole and diastole in the same image. CMR tagging techniques (see Chapter 4) permit tracing of identical portions of the myocardium, and wall thickening measurements by CMR using this technique have been shown to be as accurate as the current gold standard, ultrasonic crystals sewn to the heart.[59]

Inotropic Reserve and CMR Tissue Tagging

If no wall thickening is present or the amount of wall thickening is so small as to leave serious doubt about the potential for recovery of regional ventricular function, inotropic stimulation with low-dose (2.5–10 μg/kg/min) dobutamine can be employed with CMR to assess residual viability in patients with recent infarction.[60] Contractile reserve demonstrated by CMR was strongly associated with abnormalities of fatty acid metabolism as demonstrated by beta-methyl-iodophenyl-pentadecanoic acid single-photon emission tomography.[60] The combination of tagging with dobutamine stress has been evaluated in normal subjects[61, 62] and has now been shown to be reliable in predicting recovery from acute infarction,[63] which allows a more quantitative approach than visual assessment of improved contraction. Animal studies using tagging are concordant.[64] In addition, recent work looking at subepicardial function using tagging after infarction suggests that recovery in this area is crucial to improved regional function,[65] and this finding is in agreement with the findings using late Gd-DTPA enhancement—that for functional recovery, greater than 50 percent of the transmural extent of the myocardial wall must be viable (nonenhancing).

CARDIOVASCULAR MAGNETIC RESONANCE IN CHRONIC MYOCARDIAL INFARCTION

As mentioned earlier, chronic myocardial infarcts are structurally different from acute myocardial infarcts. The most obvious anatomic difference is that chronic transmural infarcts may be very thin due to infarct expansion and remodeling.[66] Consequently, this feature can be detected by CMR and can be used to distinguish between chronic transmural scar and residual viable myocardium in the infarct area. However, caution must be used when observing a small area of pronounced wall thinning in order to not assume that the entire region perfused by an occluded coronary artery is completely scarred. Frequently, myocardial cells in the border zone survive, and ischemia of this border zone alone may cause substantial symptoms in a patient. Therefore, in a patient with single-vessel disease, previous myocardial infarction, and anginal symptoms, restoration of blood flow by reestablishing patency of the occluded artery may be justified despite evidence of complete necrosis in the center of the infarct zone.[67]

Myocardial Wall Thickness as a Feature of Viable Myocardium

The hypothesis that thinned and akinetic myocardium represents chronic scar has been tested by comparing CMR findings with those obtained by PET and SPECT in identical myocardial regions.[68, 69] Comparison of CMR with scintigraphy is easily accomplished because identical regions can be matched due to the three-dimensional nature of both techniques.

In order to define transmural scar by left ventricular end-diastolic wall thickness, a cutoff value of 5.5 mm was selected. This value corresponded to the mean end-diastolic wall thickness in normal individuals -2.5 SD. It also corresponded well to the wall thickness of less than 6 mm found in a histopathologic study of transmural chronic scar.[66] Regions with a mean end-diastolic wall thickness of less than 5.5 mm had a significantly reduced fluorodeoxyglucose (FDG) uptake as compared to regions with an end-diastolic wall thickness of 5.5 mm or greater (Figure 14–10).[69] In 29 of 35 patients studied, the diagnosis of viability based on FDG uptake was identical to the one based on myocardial morphology as assessed by CMR. Importantly, relative FDG uptake did not differ between segments with systolic wall thickening at rest or akinesia at rest, as long as wall thickness was preserved. These findings were extended in another patient population who underwent revascularization and control CMR at 3 months after revascularization.[70] Of 125 segments (in 43 patients with chronic infarcts) with an end-diastolic wall thickness less than 5.5 mm, only 12 segments recovered (corresponding to a negative predictive accuracy of 90% for the finding of end-diastolic wall thinning to predict transmural scar). In contrast, the positive predictive accuracy was 62 percent for preserved end-diastolic wall thickness of 5.5 mm or greater for predicting the presence of viable myocardium with the potential for recovery. The most likely explanation for this finding is that the amount of viable myocardium cannot be directly visualized on gradient-echo CMR. However, it is the amount of viable myocardium that is present in a

particular region of the left ventricle that determines whether the segment will recover function or not. Regions with preserved wall thickness may contain very small rims of epicardially located viable myocardium and yet not exhibit substantial wall thinning. Nevertheless, such a small rim of viable myocardium may not be sufficient to result in improved wall thickening after revascularization (Table 14–1). Reduced end-diastolic wall thickness was also found to be a strong predictor of irreversibly damaged tissue in a study employing resting transthoracic echocardiography in patients with healed Q wave anterior wall infarcts.[71] This study, which used recovery of function after revascularization for defining myocardial viability, found a predictive value of 87 percent for a pattern of increased acoustic reflectance combined with reduced end-diastolic wall thickness.[71]

The relationship between end-diastolic wall thickness and viability has been disputed by Perrone-Filardi and colleagues,[72] who found FDG uptake on PET images largely independent of regional end-diastolic wall thickness. However, this study included recent and chronic infarcts and used a suboptimal conventional spin-echo technique with a short echo time of 20 ms to measure wall thickness. More recently, thallium uptake was correlated with end-diastolic and end-systolic left ventricular wall thickness as measured from cine CMR in patients with acute and healed myocardial infarcts.[73] These authors found that end-systolic wall thickness correlates better with normalized thallium activity than end-diastolic wall thickness. Normalized thallium activity for the 17 patients with chronic infarcts is shown in Figure 14–11. Regions with normalized thallium activity of less than 50 percent showed wall thickening values of 12.3 ± 30.6 percent and wall thickening was only slightly larger in

Table 14–1. Comparison of Magnetic Resonance Findings Based on Dobutamine-Induced Systolic Wall Thickening and End-Diastolic Wall Thickness with Postrevascularization Recovery of Systolic Wall Thickening

| MRI Post | Dobutamine CMR (10 μg/kg per min) (407 akinetic segments) | | | Rest CMR (407 akinetic segments) | | |
| | Dobutamine-Induced SWT | | | Preserved-DWT | | |
	Viable	Scar	Total	Viable	Scar	Total
Recovery	155	33	188	176	12	188
Scar	42	177	219	106	113	219
	197	210	407	282	125	407

CMR post = cardiovascular magnetic resonance after successful revascularization; SWT = systolic wall thickening; DWT = end-diastolic wall thickness.

Adapted from Baer FM, Theissen P, Schneider CA, et al: Dobutamine magnetic resonance imaging predicts contractile recovery of chronically dysfunctional myocardium after successful revascularization. J Am Coll Cardiol 76:1002, 1995.

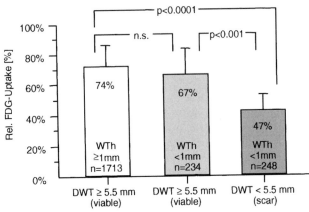

Figure 14–10. Normalized uptake of ^{18}F fluorodeoxyglucose (FDG) uptake stratified by left ventricular end-diastolic wall thickness as measured from gradient-echo CMR. Regions with preserved wall thickening (WTh) of ≥ 1 mm and preserved end-diastolic wall thickness (DWT) of ≥ 5.5 mm had a similar relative FDG uptake as regions without wall thickening (WTh < 1 mm) but preserved end-diastolic wall thickness. In contrast, akinetic regions (WTh < 1 mm) with reduced end-diastolic wall thickness had significantly reduced FDG uptake (mean 47%), indicating scar formation. (Data from Baer FM, Voth E, Schneider CA, et al: Comparison of low-dose dobutamine-gradient-echo magnetic resonance imaging and positron emission tomography with [18F] fluorodeoxyglucose in patients with chronic coronary artery disease. Circulation 91:1006, 1995.)

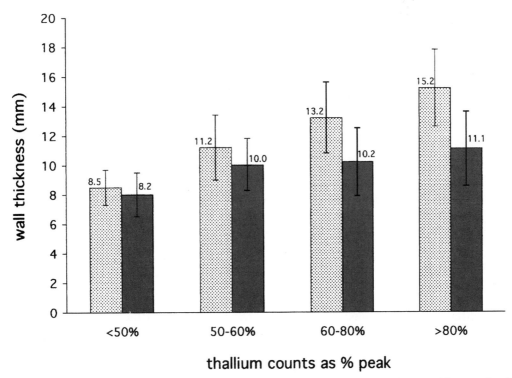

Figure 14–11. Normalized thallium activity for all segments of 17 patients studied late after infarction (chronic phase) classified into four groups based on percentiles with the corresponding CMR end-systolic (ES, stippled bars) and end-diastolic (ED, solid bars) wall thicknesses. There is a smaller difference among the thallium levels for ED wall compared with ES wall thickness, indicating that some regions showed wall thickening by CMR. (Adapted from Lawson MA, Johnson LL, Coghlan L, et al: Correlation of thallium uptake with left ventricular wall thickness by magnetic resonance imaging in patients with acute and healed myocardial infarcts. Am J Cardiol 80:434, 1997.)

regions with thallium activities between 50 and 60 percent (13.8 ± 27.0%). From a receiver-operating curve, an end-systolic wall thickness value of 9.8 mm gave a sensitivity of 90 percent and a specificity of 94 percent for identifying regions with a normalized thallium uptake of less than 50 percent. In our opinion, the study does not really contradict our finding that end-diastolic wall thickness is able to distinguish between scarred and viable myocardium. The patient population of Lawson[73] differed in one important aspect from our[70] patients: It included also patients with hypokinesia, whereas only patients with akinesia were included in our study. Obviously, any degree of wall thickening relates to the presence of contracting and hence viable cells in the region of interest. Nevertheless, it is undoubtedly true that end-systolic wall thickness gives even larger differences between normal and viable myocardium if one includes zones that are still contracting as viable. However, viable zones in our studies did not contract, and systolic wall thickness was therefore not better than end-diastolic wall thickness to distinguish between viable myocardium and scar.

Contractile Reserve During Low-Dose Dobutamine Infusion

Although severely reduced end-diastolic wall thickness is very helpful in identifying myocardium that is unlikely to recover after mechanical revascularization because it is completely scarred, the predictive value of a preserved end-diastolic wall thickness for predicting recovery of function following revascularization is disappointingly low. However, CMR offers the possibility to measure wall thickening not only at rest but also during low- and high-dose[74] dobutamine infusion. Until recently, a protocol with acquisition of cine CMR images in multiple short axes and two long axes sections at rest and at 5 and 10 μg/kg/min dobutamine required an imaging time of more than 60 minutes. The advent of fast CMR sequences now permits completion of the same protocol within 30 to 45 minutes, and image quality is often better with breath-hold cine CMR images than with non–breath-hold techniques. The sensitivity of dobutamine CMR for detection of viable myocardium as defined by a normalized FDG uptake on PET images is 81 percent with a specificity of 95 percent.[69] When recovery of wall thickening was considered to be the gold standard, the sensitivity of dobutamine CMR in predicting recovery of function after revascularization was 89 percent at a specificity of 94 percent. The latter analysis was patient related, which is clinically more meaningful than a segment-by-segment analysis.[70]

We also presented data on the relative value of dobutamine CMR and dobutamine transesophageal echocardiography (TEE).[75] Normalized FDG uptake on PET images was used as the standard against which both techniques were compared. The sensitivity and the specificity of dobutamine TEE and dobutamine CMR for FDG PET–defined myocardial viability were 77 percent versus 81 percent and 94 percent versus 100 percent, respectively. Thus, both imaging techniques provide similar accuracy. When choosing the appropriate technique, patient acceptance becomes an important consideration. Although claustrophobia may be a problem with CMR, only a small fraction of patients is affected. In contrast, many patients do not like the experience of a TEE examination. On the other hand, there is a clear cost advantage for TEE because the echocardiography probe costs only a fraction of a CMR scanner, and additional investment is not necessary.

Late Gd-DTPA Enhancement in Chronic Infarction

Similar principles apply to the assessment of myocardial infarction in the chronic phase as shortly after the event. Late enhancement of infarction also occurs in chronic infarcts,[76, 77] because of the continuing increase in partition coefficient, and delayed contrast agent kinetics. In a study of 71 subjects, 40 patients with healed myocardial infarction were prospectively enrolled after enzymatically proven necrosis and were imaged 3 ± 1 months (n = 32) and/or 14 ± 7 months (n = 19) later.[77] They were compared with 20 patients with nonischemic cardiomyopathy, and 11 normal volunteers. Twenty-nine of 32 healed infarction patients (91%) with 3-month-old infarcts (13 non–Q wave) and all 19 with 14-month-old infarcts (8 non–Q wave) exhibited regional hyperenhancement. In patients in whom the infarct-related artery was determined at angiography, 24 of 25 patients with 3-month-old infarcts (96%) and all 14 with 14-month-old infarcts had hyperenhancement in the corresponding territory. None of the 20 patients with nonischemic cardiomyopathy or the 11 volunteers exhibited regional hyperenhancement. Regardless of the presence or absence of Q waves, the majority of patients with hyperenhancement had only nontransmural involvement. Normal left ventricular contraction was visualized in 7 (22%) patients with 3-month-old infarcts and 3 (16%) with 14-month-old infarcts, but in these cases, hyperenhancement was limited to the subendocardium. Therefore the presence, location, and transmural extent of healed Q wave and non–Q wave myocardial infarction is accurately determined by contrast-enhanced CMR, and distinguishing ischemic from nonischemic causes for ventricular dysfunction proved reliable.

In remodeled ventricles, the wall may be very thin and show complete scar formation. With modern therapy, however, such as angiotensin converting enzyme inhibitors and beta blockers, remodeling as a process is becoming attenuated in the extent to which thinning occurs and in its frequency even after large infarctions.[78] Therefore, in modern cohorts, the wall may not be thinned with transmural enhancement as may have been in the past. In any event, however, there is evidence that once transmural replacement with enhancement is greater than

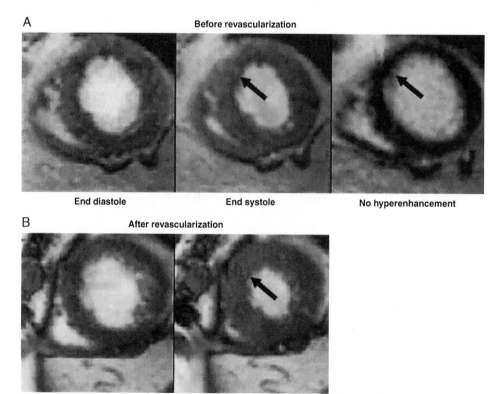

A **Before revascularization**

End diastole   End systole   No hyperenhancement

B **After revascularization**

End diastole   End systole

Figure 14–12. Reversible myocardial dysfunction. *A,* End-diastolic and end-systolic frames from a gradient echo cine acquisition, and the same slice imaged using the delayed Gd-DTPA enhancement technique. Although there is significant anteroseptal (arrow) dysfunction, no enhancement is seen, and the area is therefore both viable and hibernating. *B,* Improved contraction is seen after revascularization. (From Kim RJ, Wu E, Rafael A, et al: The use of contrast-enhanced magnetic resonance imaging to identify reversible myocardial dysfunction. N Engl J Med 343:1445, 2000, with permission. Copyright © 2000 Massachusetts Medical Society. All rights reserved.)

50 percent, there is little likelihood for functional recovery, and this has now been shown in both animals[51] and humans (Figures 14–12 to 14–15).[79]

Magnetic Resonance Spectroscopy

The hallmark of viable myocardium is the presence of high-energy phosphates within the cell. Inasmuch as phosphorus-31 (^{31}P) MRS is the only available technique to observe high-energy phosphates noninvasively in vivo, it can be employed to detect and quantify this sign of life within a myocardial region. By quantifying the amount of high-energy phosphate compounds, it is possible to determine the amount of viable myocardium present in the region of interest. Yabe and colleagues evaluated patients with reversible and irreversible thallium defects on exercise-redistribution studies.[21] All patients had a severe stenosis of the left anterior descending coronary artery. MRS spectra were localized by one-dimensional chemical shift imaging with slice selection in the sagittal direction. The volume of interest in this study was in the order of 30 cm³. Quantification of spectra was done by using a vial of hexamethylphosphoric triamide for comparison. Representative spectra from the three groups are shown in Figure 14–16. Phosphocreatine (PCR) content was significantly lower in the group without thallium redistribution (which may indicate absence of viability) and in the group with reversible defects (indicating residual viability) as compared with a group of 11 healthy subjects. The

ATP concentration, however, was significantly lower than in normals only in the group without thallium redistribution (see Figure 14–16). Although much overlap was found between groups, this study demonstrated that quantitative MRS measurements are possible in patients after myocardial infarction and that MRS can be used in the clinical setting to gain information about the presence of myocardial viability. Nevertheless, the technique described by Yabe is not clinically helpful because volumes of interest usually incorporate mixtures of scar, normal, and ischemically injured viable myocardium. Only surface coils in close contact with a heart beating outside the chest in animal experiments provide sufficient resolution with ^{31}P MRS to permit high-resolution spectroscopic imaging.[80]

Proton (^{1}H) MRS has higher sensitivity than ^{31}P MRS and is able to detect the total pool of phosphorylated plus unphosphorylated creatine in both skeletal and cardiac muscle. ^{1}H MRS offers about a 20-fold net theoretical sensitivity improvement compared with ^{31}P MRS of phosphorylated creatine. This finding is due to the higher sensitivity of ^{1}H MRS, the higher concentration of total creatine, and the higher content of ^{1}H in the creatine N-methyl resonance at 3.0 ppm. Consequently, ^{1}H MRS allows for the first time at magnetic field strengths of many clinical CMR systems the metabolic interrogation of small voxels (<10 ml) in all regions of the left ventricle. In contrast, ^{31}P MRS, in addition to its large voxel size, is also restricted to interrogating the anterior wall only. Inasmuch as the entire ventricle can be examined by ^{1}H MRS, comparison of viable and nonviable tissue is possible within the same patient.

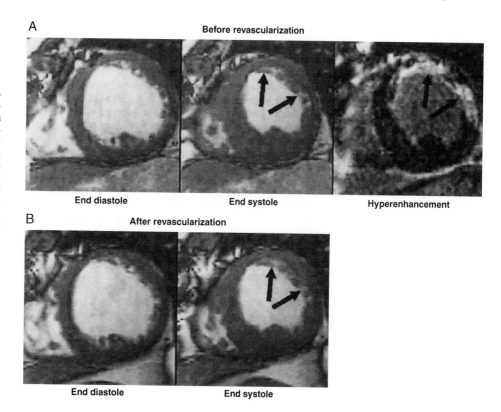

Figure 14–13. Irreversible myocardial dysfunction. *A,* End-diastolic and end-systolic frames from a gradient echo cine acquisition, and the same slice imaged using delayed Gd-DTPA enhancement technique. There is significant anterior and lateral (arrows) dysfunction, and near transmural enhancement in the same areas. *B,* No change in contraction is seen after revascularization. (From Kim RJ, Wu E, Rafael A, et al: The use of contrast-enhanced magnetic resonance imaging to identify reversible myocardial dysfunction. N Engl J Med 343:1445, 2000, with permission. Copyright © 2000 Massachusetts Medical Society. All rights reserved.)

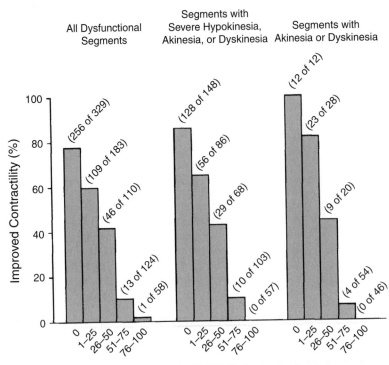

Figure 14–14. Likelihood of improvement in myocardial segments according to the extent of transmural enhancement on delayed Gd-DTPA CMR and the severity of the wall-motion abnormality. Segments with transmural enhancement greater than 50% have a very low likelihood of recovery. Conversely, segments with enhancement of 0–25% have a high likelihood of recovery. Those segments with enhancement from 25–50% have a variable response, but they formed only approximately 12% of the total number of segments. (From Kim RJ, Wu E, Rafael A, et al: The use of contrast-enhanced magnetic resonance imaging to identify reversible myocardial dysfunction. N Engl J Med 343:1445, 2000.)

Thus, the patient can serve as his or her own control. Bottomley and Weiss were the first to employ proton spectroscopy in patients with remote myocardial infarction (longer than 1 month).[23] In a dog model, they established that enzymatic degradation of creatine in heart extracts resulted in the complete disappearance of the ^{1}H N-methyl resonance peak at 3.0 ppm. Figure 14–17 shows representative myocardial ^{1}H spectra from anterior and posterior left ventricular regions in a patient with an anteroapical infarction. Figure 14–18 shows the raw creatine-to-water signal ratios from individual patients and controls. Myocardial creatine is significantly reduced in infarction. However, some overlap between noninfarcted myocardium and infarcted myocardium can be seen. This finding may be due to the fact that some of the patients had some viable myocardium in the infarct region because residual viability was not excluded on the basis of other imaging studies. Nevertheless, this study showed for the first time that it was possible to measure creatine within myocardial regions small enough to make this technique clinically useful. Further work needs to establish the usefulness of creatine measurements by ^{1}H MRS as compared with established CMR techniques.

CONCLUSION

CMR techniques provide a variety of novel methods of obtaining information on residual viability after acute and chronic myocardial infarction. Indirect signs of viability that can be observed by CMR are the absence of increased signal intensity on spin-echo CMR in a myocardial region involved in a recent infarct, any sign of wall thickening at rest (which is detectable with high accuracy by cine CMR), wall thickening after stimulation by low-dose dobutamine, and preserved wall thickness. In contrast, myocardial necrosis is characterized by high signal intensity on spin-echo images, enhancement (possibly with a low-intensity core region due to no-reflow) of the infarct area after injection of Gd-

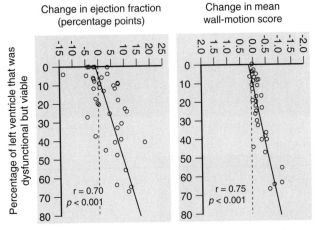

Figure 14–15. The change in wall-motion score *(right)* and ejection fraction *(left)* versus the percentage of viable but dysfunctional myocardium. There is a significant relationship, demonstrating the importance of recovery of function on the amount of residual viability present. (From Kim RJ, Wu E, Rafael A, et al: The use of contrast-enhanced magnetic resonance imaging to identify reversible myocardial dysfunction. N Engl J Med 343:1445, 2000.)

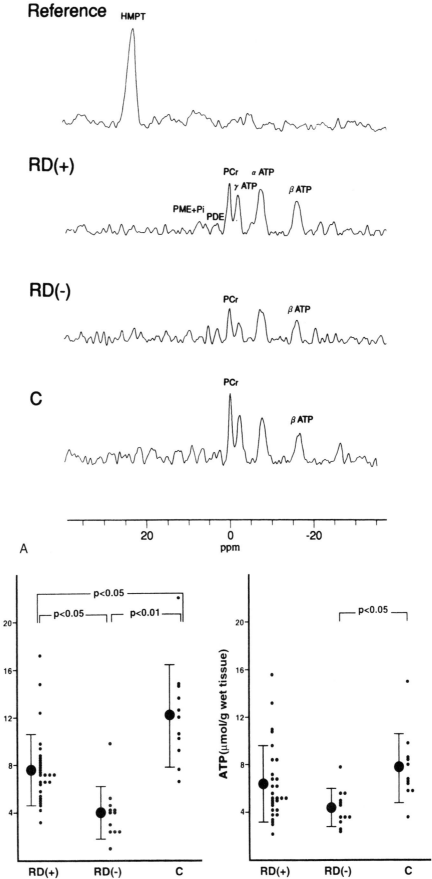

Figure 14–16. *A*, ³¹P MRS of control (C) and typical patients with redistribution (RD+) and without redistribution (RD−) on 3-hour postexercise thallium SPECT images. The RD− patient had a lower myocardial PCr content compared with the RD+ patient and the normal control person. Moreover, the RD+ patient had a lower PCr content than the normal control person. HMPT = hexamethylphosphoric triamide (used as the standard for quantification); PCr = phosphocreatine; PME = phosphomonoesters; Pi = inorganic phosphate; PDE = phosphodiesters; ATP = adenosine triphosphate. *B, Left*, Phosphocreatine content in the three groups RD−, RD+, and control. By analysis of variance (ANOVA) there were significant differences between the three groups. *Right*, ATP content in the three groups. Subjects in the RD− group had significantly lower myocardial ATP content than those in group C ($p < 0.05$). However, no significant differences existed between RD+ and C groups. (From Yabe T, Mitsunami K, Inubushi T, Kinoshita M: Quantitative measurements of cardiac phosphorus metabolites in coronary artery disease by 31P magnetic resonance spectroscopy. Circulation, 92:15, 1995.)

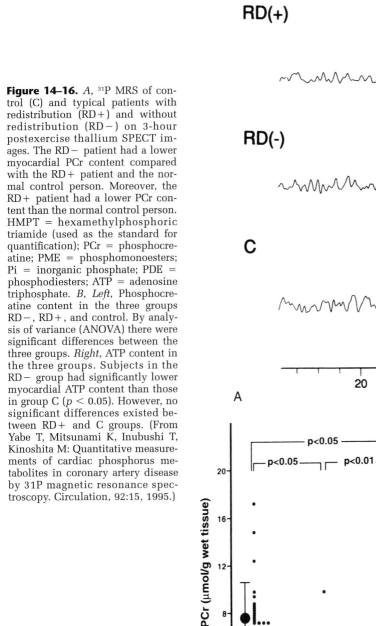

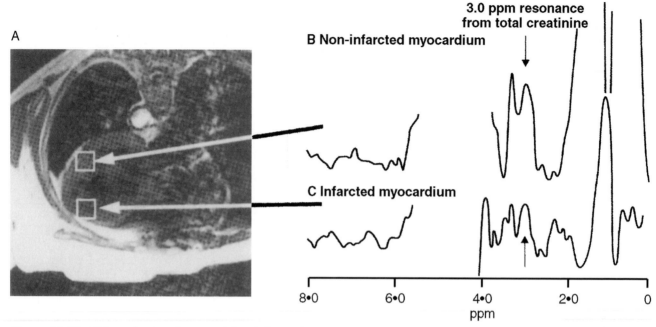

Figure 14–17. ECG gated spin-echo CMR *(A)* and short echo-time stimulated-echo acquisition mode (STEAM) localized magnetic resonance spectra from noninfarcted *(B)* and infarcted *(C)* myocardium in anterior left ventricle of a 56-year-old man with anterior myocardial infarction, septal and inforelateral akinesis, and dyskinesis. (From Bottomley PA, Weiss RG: Non-invasive magnetic-resonance detection of creatine depletion in non-viable infarcted myocardium. Lancet 351:714, 1998, with permission. Copyright © 1998 The Lancet Ltd.)

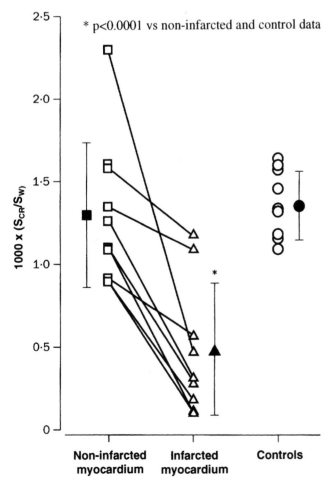

Figure 14–18. Myocardial creatine-to-water ratio (S_{CR}/S_W) measured by MRS in voxels from noninfarcted and infarcted myocardium of patients and controls. (From Bottomley PA, Weiss RG: Non-invasive magnetic-resonance detection of creatine depletion in non-viable infarcted myocardium. Lancet 351:714, 1998.)

DTPA, reduced wall thickness (in chronic infarcts), and absence of a contractile reserve during dobutamine stimulation. Direct observation of the presence of high-energy phosphates and measurement of total creatine is possible using MRS. Evaluation of viable myocardium is most important in patients with left ventricular dysfunction because these patients can gain most from revascularization if substantial amounts of myocardium are present. Revascularization in these patients will improve systolic function and hence prognosis. In contrast, transplantation should be reserved for those patients for whom all other therapeutic options have been excluded.[81] Unfortunately, the available information about the value of CMR in identifying patients with global left ventricular dysfunction who have a high likelihood of profiting from a coronary revascularization is scarce. Preliminary evidence suggests that low-dose dobutamine CMR is at least as accurate as low-dose dobutamine echocardiography but less sensitive in identifying viable regions than thallium SPECT.[82, 83] Further studies including the results of revascularization of these patients are needed before dobutamine CMR can be employed clinically in this important patient subgroup. In patients with regional left ventricular dysfunction, however, in whom the need for revascularization needs to be established, dobutamine CMR is the best validated technique of all the techniques presented in this chapter. Therefore, the high image quality and the ease with which viable and nonviable myocardium can be differentiated make CMR an attractive alternative to dobutamine echocardiography in centers with experience in performing pharmacological stress testing within the magnet. Depiction of zones of acute necrosis either by T2-weighted spin-echo CMR or by observing delayed contrast enhancement after application of Gd-DTPA carries substantial promise for detailed studies of the effects of different treatment strategies for acute and chronic myocardial infarction, as well as for treatment in the early and late postinfarction phase.

References

1. Rahimtoola SH: The hibernating myocardium. Am Heart J 117:211, 1989.
2. Popio KA, Gorlin R, Bechtel D, Levine JA: Postextrasystolic potentiation as a predictor of potential myocardial viability: Preoperative analysis compared with studies after coronary bypass surgery. Am J Cardiol 39:944, 1977.
3. Dilsizian V, Rocco TP, Freedman NMT, et al: Enhanced detection of ischemic but viable myocardium by the reinjection of thallium after stress-redistribution imaging. N Engl J Med 323:141, 1990.
4. Pierard LA, De Landsheere CM, Berthe C, et al: Identification of viable myocardium by echocardiography during dobutamine infusion in patients with myocardial infarction after thrombolytic therapy: Comparison with positron emission tomography. J Am Coll Cardiol 15:1021, 1990.
5. Sechtem U, Voth E, Baer F, et al: Assessment of residual viability in patients with myocardial infarction using magnetic resonance techniques. Int J Card Imaging 1:31, 1993.
6. Mallory GK, White PD, Salcedo-Galger J: The speed of healing of myocardial infarction: A study of the pathologic anatomy in 72 cases. Am Heart J 18:647, 1939.
7. Heusch G, Schulz R: Hibernating myocardium: A review. J Mol Cell Cardiol 28:2359, 1996.
8. Cigarroa CG, de Filippi CR, Brickner ME, et al: Dobutamine stress echocardiography identifies hibernating myocardium and predicts recovery of left ventricular function after coronary revascularization. Circulation 88:430, 1993.
9. Schoen FJ: The heart. In Cotran RS, Kumar V, Robbins SL (eds): Pathologic Basis of Disease. Philadelphia, WB Saunders, 1994, p 517.
10. Pflugfelder PW, Wisenberg G, Prato FS, et al: Early detection of canine myocardial infarction by magnetic resonance imaging in vivo. Circulation 71:587, 1985.
11. Ambrosio G, Weisman HF, Mannisi JA, Becker LC: Progressive impairment of regional myocardial perfusion after initial restoration of postischemic blood flow. Circulation 80:1846, 1989.
12. Kim RJ, Chen EL, Lima JAC, Judd RM: Myocardial Gd-DTPA kinetics determine MRI contrast enhancement and reflect the extent and severity of myocardial injury after acute reperfused infarction. Circulation 94:3318, 1996.
13. Lima JA, Judd RM, Bazille A, et al: Regional heterogeneity of human myocardial infarcts demonstrated by contrast-enhanced MRI. Potential mechanisms. Circulation 92:1117, 1995.
14. Pereira RS, Prato FS, Sykes J, Wisenberg G: Assessment of myocardial viability using MRI during a constant infusion of Gd-DTPA: Further studies at early and late periods of reperfusion. Magn Reson Med 42:60, 1999.
15. Flacke SJ, Fischer SE, Lorenz CH: Measurement of the gadopentetate dimeglumine partition coefficient in human myocardium: Normal distribution and elevation in acute and chronic infarction. Radiology 218:703, 2001.
16. Tong CY, Prato FS, Wisenberg G, et al: Measurement of the extraction efficiency and distribution volume for Gd-DTPA in normal and diseased canine myocardium. Techniques for the measurement of the local myocardial extraction efficiency for inert diffusible contrast agents such as gadopentetate dimeglumine. Magn Reson Med 30:337, 1993.
17. Matheijssen NAA, de Roos A, van der Wall EE, et al: Acute myocardial infarction: Comparison of T2-weighted and T1-weighted gadolinium-DTPA enhanced MR imaging. Magn Reson Med 17:460, 1991.
18. Kim RJ, Fieno DS, Parrish TB, et al: Relationship of MRI delayed contrast enhancement to irreversible injury, infarct age, and contractile function. Circulation 100:1992, 1999.
19. Ingwall JS, Kramer MF, Fifer MA, et al: The creatine kinase system in normal and diseased human myocardium. N Engl J Med 313:1050, 1985.
20. Delanghe J, De Buyzere M, De Scheerder I, et al: Creatine determinations as an early marker for the diagnosis of acute myocardial infarction. Ann Clin Biochem 25:383, 1988.
21. Yabe T, Mitsunami K, Inubushi T, Kinoshita M: Quantitative measurements of cardiac phosphorus metabolites in coronary artery disease by ^{31}P magnetic resonance spectroscopy. Circulation 92:15, 1995.
22. Bottomley PA: MR spectroscopy of the human heart: The status and the challenges. Radiology 191:593, 1994.
23. Bottomley PA, Weiss RG: Non-invasive magnetic-resonance detection of creatine depletion in non-viable infarcted myocardium. Lancet 351:714, 1998.
24. Parrish TB, Fieno DS, Fitzgerald SW, Judd RM: Theoretical basis for sodium and potassium magnetic resonance MRI at 1.5 T. Magn Reson Med 38:653, 1997.
25. Kim RJ, Lima JA, Chen EL, et al: Fast ^{23}Na magnetic resonance imaging of acute reperfused myocardial infarction. Potential to assess myocardial viability. Circulation 95:1877, 1997.
26. Kim RJ, Judd RM, Chen EL, et al: Relationship of elevated ^{23}Na magnetic resonance image intensity to infarct size after acute reperfused myocardial infarction. Circulation 100:185, 1999.
27. Parrish TB, Fieno DS, Fitzgerald SW, Judd RM: Theoretical basis for sodium and potassium magnetic resonance imaging of the human heart at 1.5T. Magn Reson Med 38:653, 1997.

28. Fieno DS, Kim RJ, Rehwald WG, Judd RM: Physiological basis for potassium (^{39}K) magnetic resonance imaging of the heart. Circ Res 84:913, 1999.

29. Garcia DD, Oliveras J, Gili J, et al: Analysis of myocardial oedema by magnetic resonance imaging early after coronary artery occlusion with or without reperfusion. Cardiovasc Res 27:1462, 1993.

30. Fisher MR, McNamara MT, Higgins CB: Acute myocardial infarction: MR evaluation in 29 patients. AJR Am J Roentgenol 148:247, 1987.

31. Johnston DL, Wendt RE, Mulvagh SL, Rubin H: Characterization of acute myocardial infarction by magnetic resonance imaging. Am J Cardiol 69:1291, 1992.

32. Johnston DL, Gupta VK, Wendt RE, et al: Detection of viable myocardium in segments with fixed defects on thallium-201 scintigraphy: Usefulness of magnetic resonance imaging early after acute myocardial infarction. Magn Reson Imaging 11:949, 1993.

33. Krauss XH, Van der Wall EE, Doornbos J, et al: Value of magnetic resonance imaging in patients with a recent myocardial infarction: Comparison with planar thallium-201 scintigraphy. Cardiovasc Intervent Radiol 12:119, 1989.

34. Thompson RC, Liu P, Brady TJ, et al: Serial magnetic resonance imaging in patients following acute myocardial infarction. Magn Reson Imaging 9:155, 1991.

35. Lim TH, Hong MK, Lee JS, et al: Novel application of breath-hold turbo spin-echo T2 MRI for detection of acute myocardial infarction. J Magn Reson Imaging 7:996, 1997.

36. van Rossum AC, Visser FC, van Eenige MJ, et al: Value of gadolinium-diethylene-triamine pentaacetic acid dynamics in magnetic resonance imaging of acute myocardial infarction with occluded and reperfused coronary arteries after thrombolysis. Am J Cardiol 65:845, 1990.

37. van Dijkman PR, van der Wall EE, de Roos A, et al: Gadolinium-enhanced magnetic resonance imaging in acute myocardial infarction. Eur J Radiol 11:1, 1990.

38. de Roos A, Doornbos J, van der Wall EE, van Voorthuisen AE: MR imaging of acute myocardial infarction: Value of Gd-DTPA. AJR 150:531, 1988.

39. van Dijkman PR, Doornbos J, de Roos A, et al: Improved detection of acute myocardial infarction by magnetic resonance imaging using gadolinium-DTPA. Int J Card Imaging 5:1, 1989.

40. McNamara MT, Tscholakoff D, Revel D, et al: Differentiation of reversible and irreversible myocardial injury by MR imaging with and without gadolinium-DTPA. Radiology 158:765, 1986.

41. Schaefer S, Malloy CR, Katz J, et al: Gadolinium-DTPA-enhanced nuclear magnetic resonance imaging of reperfused myocardium: identification of the myocardial bed at risk. J Am Coll Cardiol 12:1064, 1988.

42. Yokota C, Nonogi H, Miyazaki S, et al: Gadolinium-enhanced magnetic resonance imaging in acute myocardial infarction. Am J Cardiol 75:577, 1995.

43. Dendale P, Franken PR, Block P, et al: Contrast enhanced and functional magnetic resonance imaging for the detection of viable myocardium after infarction. Am Heart J 135:875, 1998.

44. van Rugge FP, van der Wall EE, van Dijkman PR, et al: Usefulness of ultrafast magnetic resonance imaging in healed myocardial infarction. Am J Cardiol 70:1233, 1992.

45. Rochitte CE, Lima JAC, Bluemke DA, et al: Magnitude and time course of microvascular obstruction and tissue injury after acute myocardial infarction. Circulation 98:1006, 1998.

46. Wu KC, Zerhouni EA, Judd RM, et al: Prognostic significance of microvascular obstruction by magnetic resonance imaging in patients with acute myocardial infarction. Circulation 97:765, 1998.

47. Judd RM, Lugo OC, Arai M, et al: Physiological basis of myocardial contrast enhancement in fast magnetic resonance images of 2-day-old reperfused canine infarcts. Circulation 92:1902, 1995.

48. Ito H, Maruyama A, Iwakura K, et al: Clinical implications of the "no reflow" phenomenon. A predictor of complications and left ventricular remodeling in reperfused anterior wall myocardial infarction. Circulation 93:223, 1996.

49. White HD, Norris RM, Brown MA, et al: Left ventricular end-systolic volume as the major determinant of survival after recovery from myocardial infarction. Circulation 76:44, 1987.

50. Pfisterer ME, Buser P, Osswald S, et al: Time dependence of left ventricular recovery after delayed recanalization of an occluded infarct-related coronary artery: Findings of a pilot study. J Am Coll Cardiol 32:97, 1998.

51. Hillenbrand HB, Kim RJ, Parker MA, et al: Early assessment of myocardial salvage by contrast enhanced magnetic resonance imaging. Circulation 102:1678, 2000.

52. Rogers WJ, Kramer CM, Geskin G, et al: Early contrast-enhanced MRI predicts late functional recovery after reperfused myocardial infarction. Circulation 99:744, 1999.

53. Kramer CM, Rogers WJ, Mankad S, et al: Contractile reserve and contrast uptake pattern by magnetic resonance imaging and functional recovery after reperfused myocardial infarction. J Am Coll Cardiol 36:1835, 2000.

54. Pereira RS, Wisenberg G, Prato FS, Yvorchuk K: Clinical assessment of myocardial viability using MRI during a constant infusion of Gd-DTPA. MAGMA 11:104, 2000.

55. Fieno DS, Kim RJ, Chen EL, et al: Contrast-enhanced magnetic resonance imaging of myocardium at risk: Distinction between reversible and irreversible injury throughout infarct healing. J Am Coll Cardiol 36:1985, 2000.

56. Pirolo JS, Hutchins GM, Moore GW: Infarct expansion: Pathologic analysis of 204 patients with a single myocardial infarct. J Am Coll Cardiol 7:349, 1986.

57. Ito H, Yu H, Tomooka T, et al: Incidence and time course of left ventricular dilation in the early convalescent stage of reperfused anterior wall acute myocardial infarction. Am J Cardiol 73:539, 1994.

58. Sayad DE, Willett DL, Bridges WH, et al: Noninvasive quantitation of left ventricular wall thickening using cine magnetic resonance imaging with myocardial tagging. Am J Cardiol 76:985, 1995.

59. Lima JAC, Jeremy R, Guier W, et al: Accurate systolic wall thickening by nuclear magnetic resonance imaging with tissue tagging: Correlation with sonomicrometers in normal and ischemic myocardium. J Am Coll Cardiol 21:1741, 1993.

60. Dendale P, Franken PR, van der Wall EE, de Roos A: Wall thickening at rest and contractile reserve early after myocardial infarction: Correlation with myocardial perfusion and metabolism. Coron Artery Dis 8:259, 1997.

61. Power TP, Kramer CM, Shaffer AL, et al: Breath-hold dobutamine magnetic resonance myocardial tagging: Normal left ventricular response. Am J Cardiol 80:1203, 1997.

62. Scott CH, St John Sutton MG, Gusani N, et al: Effect of dobutamine on regional left ventricular function measured by tagged magnetic resonance imaging in normal subjects. Am J Cardiol 83:412, 1999.

63. Geskin G, Kramer CM, Rogers WJ, et al: Quantitative assessment of myocardial viability after infarction by dobutamine magnetic resonance tagging. Circulation 98:217, 1998.

64. Croisille P, Moore CC, Judd RM, et al: Differentiation of viable and nonviable myocardium by the use of three-dimensional tagged MRI in 2 day old reperfused canine infarcts. Circulation 99:284, 1999.

65. Bogaert J, Maes A, van de Werf F, et al: Functional recovery of subepicardial myocardial tissue in transmural myocardial infarction after successful reperfusion. An important contribution to the improvement of regional and global left ventricular function. Circulation 99:36, 1999.

66. Dubnow MH, Burchell HB, Titus JL: Postinfarction left ventricular aneurysm. A clinicomorphologic and electrocardiographic study of 80 cases. Am Heart J 70:753, 1965.

67. Braunwald E, Kloner RA: Myocardial reperfusion: A double-edged sword? J Clin Invest 76:1713, 1985.

68. Baer FM, Smolarz K, Jungehulsing M, et al: Chronic myocardial infarction: Assessment of morphology, function, and perfusion by gradient echo magnetic resonance imaging and 99mTc-methoxyisobutyl-isonitrile SPECT. Am Heart J 123:636, 1992.

69. Baer FM, Voth E, Schneider CA, et al: Comparison of low-dose dobutamine-gradient-echo magnetic resonance imaging and positron emission tomography with [18F]fluorodeoxy-

glucose in patients with chronic coronary artery disease. A functional and morphological approach to the detection of residual myocardial viability. Circulation 91:1006, 1995.

70. Baer FM, Theissen P, Schneider CA, et al: Dobutamine magnetic resonance imaging predicts contractile recovery of chronically dysfunctional myocardium after successful revascularization. J Am Coll Cardiol 31:1040, 1998.

71. Faletra F, Crivellaro W, Pirelli S, et al: Value of transthoracic two-dimensional echocardiography in predicting viability in patients with healed Q-wave anterior wall myocardial infarction. Am J Cardiol 76:1002, 1995.

72. Perrone-Filardi P, Bacharach SL, Dilsizian V, et al: Metabolic evidence of viable myocardium in regions with reduced wall thickness and absent wall thickening in patients with chronic ischemic left ventricular dysfunction. J Am Coll Cardiol 20:161, 1992.

73. Lawson MA, Johnson LL, Coghlan L, et al: Correlation of thallium uptake with left ventricular wall thickness by cine magnetic resonance imaging in patients with acute and healed myocardial infarcts. Am J Cardiol 80:434, 1997.

74. Nagel E, Lehmkuhl HB, Bocksch W, et al: Noninvasive diagnosis of ischemia induced wall motion abnormalities with the use of high dose dobutamine stress MRI. Comparison with dobutamine stress echocardiography. Circulation 99:763, 1999.

75. Baer FM, Voth E, LaRosee K, et al: Comparison of dobutamine transesophageal echocardiography and dobutamine magnetic resonance imaging for detection of residual myocardial viability. Am J Cardiol 78:415, 1996.

76. Ramani K, Judd RM, Holly TA, et al: Contrast magnetic resonance imaging in the assessment of myocardial viability

77. Wu E, Judd RM, Vargas JD, et al: Visualisation of presence, location, and transmural extent of healed Q-wave and non-Q-wave myocardial infarction. Lancet 357:21, 2001.

78. Johnson DB, Foster RE, Barilla F, et al: Angiotensin converting enzyme inhibitor therapy affects left ventricular mass in patients with ejection fraction >40% after acute myocardial infarction. J Am Coll Cardiol 29:49, 1997.

79. Kim RJ, Wu E, Rafael A, Chen EL, et al: The use of contrast-enhanced magnetic resonance imaging to identify reversible myocardial dysfunction. N Engl J Med 343:1445, 2000.

80. von Kienlin M, Rosch C, Le Fur Y, et al: Three-dimensional ^{31}P magnetic resonance spectroscopic imaging of regional high-energy phosphate metabolism in injured rat heart. Magn Reson Med 39:731, 1998.

81. Costanzo MR, Augustine S, Bourge R, et al: Selection and treatment of candidates for heart transplantation: A statement for health professionals from the Committee on Heart Failure and Cardiac Transplantation of the Council on Clinical Cardiology, American Heart Association. Circulation 92:3593, 1995.

82. Zamorano J, Delgado J, Almeria C, et al: Comparison of low-dose dobutamine echocardiography vs. dobutamine magnetic resonance imaging for assessment of myocardial viability: A functional and morphologic approach (abstract). Eur Heart J 19(suppl):207, 1998.

83. Zamorano J, Delgado J, Almeria C, et al: Comparison of low-dose dobutamine magnetic resonance imaging vs. thallium-201 resting redistribution for assessment of myocardial viability: A functional and morphologic approach (abstract). Eur Heart J 19(suppl):206, 1998.

15

The Use of Navigator Echoes in Cardiovascular Magnetic Resonance and Factors Affecting Their Implementation

David Firmin and Jennifer Keegan

Respiration has been shown to be an important factor influencing the quality of cardiovascular magnetic resonance (CMR) images. In addition to the cardiac motion, which can be addressed reasonably well by electrocardiogram (ECG) triggering, respiratory motion moves the position and distorts the shape of the heart by several millimeters between inspiration and expiration. In 1991, Atkinson and Edelman[1] illustrated the detrimental affects of breathing on the quality of cardiac studies by showing improved detail in breath-hold segmented k-space acquisitions compared with conventional non–breath-hold images. Although breath-holding produces images that are relatively free of respiratory motion artifact, it is not without problems. The breath-hold position may vary from one breath-hold scan to the next, giving rise to misregistration effects and it may also vary during the breath-hold period itself,[2] resulting in image blurring and artifacts. In addition, the scan parameters are limited by the need to perform imaging within the duration of a comfortable breath-hold period. For a number of patients, this period may be very short.

An alternative to breath-holding is to monitor the respiratory motion throughout the data acquisition period and to correct the data for that motion, either in real time or through postprocessing, with the efficacy of both techniques being strongly dependent on the accuracy of the method of motion assessment. During normal tidal respiration, the superior-inferior motion of the diaphragm is approximately 4 to 5 times the anterior-posterior motion of the chest wall,[3] and diaphragm motion is consequently the most sensitive measure of respiratory motion. In 1989, Ehman and Felmlee[4] were the first to introduce navigator echoes for measuring the displacement of a moving structure and to demonstrate their use in determining diaphragm motion during respiration. The navigator echo is the signal from a column of material oriented perpendicular to the direction of the motion to be monitored, which on Fourier transformation results in a well-defined edge of the moving structure. The navigator echoes may be interleaved with the imaging sequence and consequently enable the motion to be determined throughout the data acquisition period.

In CMR, there have been a number of developments in ways of using the navigator measurement to reduce the problems of respiratory motion. In this chapter we initially discuss these developments, then consider the various choices that have been studied in the implementation of navigators and discuss the importance of these choices. We should note at this stage that there are many variables in the application and use of navigator echoes and although there have been some attempts to study these variables, it is unlikely that we are near to optimizing their application.

USE OF NAVIGATOR INFORMATION

There are two distinct ways of using navigator echoes to reduce the problems of respiratory motion in CMR, these being multiple breath-holding with feedback and free-breathing methods. The first of these methods uses the navigator information to provide visual feedback of the diaphragm position to subjects to allow them to repeatedly hold their breath at the same point.[5] The second uses the navigator echo measurement as an input to some form of respiratory gating algorithm while the patient breathes normally. Figure 15–1 demonstrates actual respiratory trace data in a subject when performing multiple breath-holds and when free-breathing. In both cases, a navigator acceptance window, typically 5 mm wide, is defined and all data acquired when the navigator is outside of this window are ignored. The resulting image is therefore comprised of data acquired over a narrow range of respiratory positions. The respiratory or scan efficiency is defined as the percentage of ECG triggers that fall within the navigator acceptance window, and it is a measure of the data rejection rate, which in turn determines the overall scan duration. As the navigator acceptance window is reduced, the rejection rate increases and the scan efficiency falls. Figure 15–2 demonstrates the residual diaphragm displacements that occurred during data acquisitions performed with conventional breath-holding, breath-holding

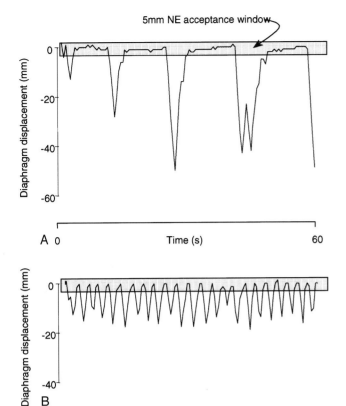

Figure 15–1. Navigator echo (NE) respiratory trace data during *(A)* multiple breath-holding with navigator feedback and *(B)* free breathing. In each case, the shaded region shows the position of a 5-mm navigator acceptance window outside of which data is rejected. (Adapted from Taylor AM, Keegan J, Jhooti P, et al: Difference between normal subjects and patients with coronary artery disease for three different MR coronary angiography respiratory suppression techniques. J Magn Reson Imag 9:786, 1999. Reprinted by permission of Wiley-Liss, Inc., a subsidiary of John Wiley & Sons, Inc.)

with navigator feedback, and navigator free-breathing in normal subjects.[6] Both navigator techniques result in images acquired over a reduced range of diaphragm displacements compared with those acquired using repeated conventional breath-holding and, in addition, allow a longer overall scan time. This method allows for averaging of data to improve the signal-to-noise ratio, increasing the k-space coverage for improved spatial resolution and increasing the temporal resolution by reducing the number of views acquired per cardiac cycle.

Multiple Breath-Hold Methods

Wang and colleagues[7] were the first to demonstrate the use of a respiratory feedback monitor to reduce misregistration artifacts in consecutive breath-hold segmented k-space gradient echo coronary artery images and to demonstrate improved image quality from averaging scans acquired over multiple breath-holds. On informed healthy volunteers, this technique has been shown to produce good results with reasonable scan efficiencies.[8]

However, a period of training is required, and the process can be problematic, particularly with patients, many of whom have difficulty holding their breath due to a combination of illness and anxiety.[6] Although it may be expected that breath-holding with respiratory feedback may enable the completion of a cardiac study much more quickly than when using the free-breathing methods described later, the time required for training and the required rest periods between breath-holds result in the overall examination times being longer than anticipated. In fact, in a group of patients with coronary artery disease, there was no significant difference between the overall examination times with the two techniques,[6] although the same study showed that in a group of normal healthy subjects, multiple breath-holding resulted in a reduction of 20 percent.

Free-Breathing Methods

Free-breathing methods have the advantage that they require relatively little cooperation from the subject being scanned. The main disadvantage is the potential for respiratory drift, which can cause considerably reduced scan efficiency.[9] Recently, therefore, most effort has gone towards improving the scan efficiency in this approach.

Much of the early work used retrospective respiratory gating,[10] which, although inherently inefficient, remains the method of choice in those scanners not capable of prospective control of the data acquisition. In this method, the data acquisition is oversampled, typically by a factor of 5, and then retrospectively sorted so that the final image is constructed from data acquired over the narrowest possible range of respiratory positions. Hofman and associates[11] have shown that using this approach, the image quality of three-dimensional (3-D) coronary acquisitions was improved over those acquired with multiple averages. However, the scan efficiency is poor (20% for an oversampling ratio of 5), and although the final image is constructed from the narrowest respiratory window possible, the range is highly dependent on the subject's breathing pattern during the acquisition and may still be unacceptably high.

Following the introduction of prospective control techniques, navigators have been used with a simple accept/reject algorithm in which data may be acquired depending on whether the navigator measurement is within a predefined acceptance window. Oshinski and coworkers[12] were the first investigators to demonstrate high-quality coronary artery images using such an approach. The problem with this method, however, is that for reasonably high scan quality, a narrow acceptance window of 5 mm or less is required, which generally results in relatively poor scan efficiency. This problem is compounded by the fact that, as has been noted above, many subjects and patients undergo a drift in diaphragm position over time,[9] such that the pre-

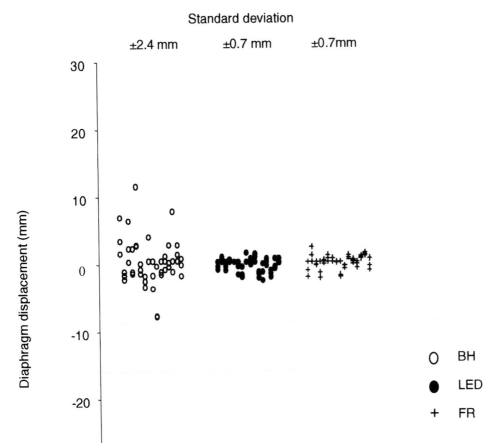

Figure 15–2. Mean diaphragm displacements in 17 normal subjects for conventional breath-holding (open circles), for breath-holding with navigator feedback (closed circles) and for free breathing (plus marks). The navigator controlled studies used a 5-mm navigator acceptance window. (Adapted from from Taylor AM, Keegan J, Jhooti P, et al: Differences between normal subjects and patients with coronary artery disease for three different MR coronary angiography respiratory suppression techniques. J Magn Reson Imag 9:786, 1999. Reprinted by permission of Wiley-Liss, Inc., a subsidiary of John Wiley & Sons, Inc.)

defined acceptance window becomes less and less suitable as the study progresses. The Diminishing Variance Algorithm (DVA) overcomes this problem because it does not use a predefined acceptance window.[13] With this method, one complete scan is acquired and the navigator positions saved for each line of data. At the end of the initial scan, the most frequent diaphragm position during that scan is determined, and a process of reacquiring lines of data that were acquired with diaphragm positions furthest offset from this position begins so that, as time progresses, the range of diaphragm positions for data comprising the final set is considerably reduced. As well as the lack of requirement of an acceptance window, this method has the advantage that an image can be reconstructed at any time after the initial data set is complete.

Another alternative to the simple accept/reject algorithm, which can both improve image quality and scan efficiency, is to use a k-space ordering that depends upon diaphragm position. Two similar approaches have been suggested, both of which use the fact that the center of k-space appears to be more sensitive to motion than the edges.[14] Jhooti and colleagues developed a phase encode ordering method that used a dual acceptance window of 5 mm for the center of k-space and 10 mm for the outer regions.[15] This approach enabled a much greater scan efficiency than other methods, while retaining scan quality (Table 15–1 and Figure 15–3). An alternative method developed by Sinkus and Börnert initially to address general respiratory motion,[16] and more recently applied to imaging of the coronary arteries, used a tailored acceptance window through k-space as opposed to phase encode ordering to obtain a very similar result.[17] Both of these phase ordering techniques use a predefined navigator acceptance window and, as such, suffer from problems of reduced scan efficiency when the

Table 15–1. Image Quality Scores and Scan Efficiencies† for Three-Dimensional Magnetic Resonance Angiography*

	Phase Ordered	ARA	DVA†	RRG
Image Quality Mean Score	4.4	4.7	6.6	6.8
Scan Efficiency	72 (± 11.6)	48 (± 11.5)	72 (± 11.6)	20

*Mean image quality scores (1 = excellent, 10 = very poor) and scan efficiencies† (± SD) for data acquired using phase ordering, an accept/reject algorithm (ARA), the diminishing variance algorithm (DVA), and retrospective respiratory gating (RRG) in 15 subjects.

†Scan time for the DVA technique set to that of the phase ordered technique.

Adapted from Jhooti P, Keegan J, Gatehouse PD, et al: 3D coronary artery imaging with phase reordering for improved scan efficiency. Magn Reson Med 41:555, 1999.

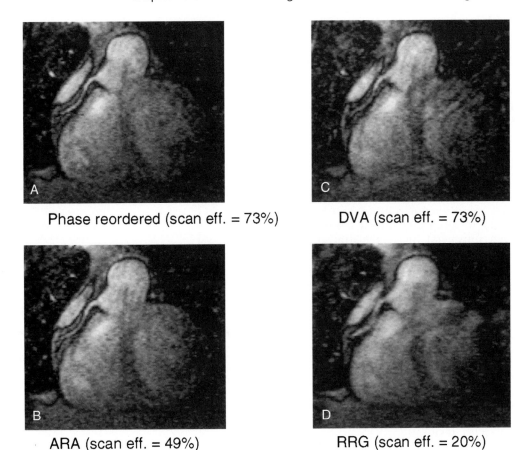

Phase reordered (scan eff. = 73%)   DVA (scan eff. = 73%)

ARA (scan eff. = 49%)   RRG (scan eff. = 20%)

Figure 15–3. A single slice from a 3D CMR data set showing a long section of the right coronary artery. The phase ordered images *(A)* are of comparable image quality to those acquired with the accept/reject algorithm (ARA) *(B)* and better than those acquired with both the diminishing variance algorithm (DVA) *(C)* and retrospective respiratory gating (RRG) *(D)*. The scan efficiency is also significantly higher for phase ordering than for both ARA and RRG techniques. (Adapted from Jhooti P, Keegan J, Gatehouse PD, et al: 3D coronary artery imaging with phase reordering for improved scan efficiency. Magn Reson Med 41:556, 1999. Reprinted by permission of Wiley-Liss, Inc., a subsidiary of John Wiley & Sons, Inc.)

respiratory pattern changes during the study acquisition. This problem has been recently addressed by Jhooti and colleagues, who have developed a technique that combines the benefits of phase ordering with an automatic window selection that enables the acquisition of high-quality coronary artery images in the shortest time possible for the breathing pattern pertaining at that time.[18] Finally, a 3-D k-space reordering acquisition scheme, reported by Huber and associates, may offer superior results, but it remains to be tested in a large patient population.[18a]

NAVIGATOR ECHO IMPLEMENTATION

Method of Column Selection

Two methods have been used for the generation of a navigator echo:

1. In the spin-echo technique, a spin-echo signal is generated from the column of material formed by the intersection of two planes, one excited by a 90° radiofrequency (RF) pulse, and the other by a 180° RF pulse. The column cross section may be either rectangular or rhomboidal, depending upon the orientation of the two planes. The advantages of this approach are that it is very robust and produces an extremely well defined column. It cannot, however, be repeated rapidly, and care has to be taken that the column selection planes do not impinge on the region of interest.

2. The alternative approach is to use a selective two-dimensional (2-D) RF pulse to excite a column of approximately circular cross section.[19] Although this approach is much more sensitive to factors such as shimming errors, which can potentially cause blurring and distortion of the column, it has the advantages that, with a reduced flip angle, it can be repeated more rapidly and the in-plane navigator artifact is less extensive.

Both methods have been used routinely for studies on coronary magnetic resonance angiography (MRA) without any reported problems.

Correction Factors

In CMR, navigator echoes are most frequently used to measure the position of the diaphragm.

However, the motion of the heart is not straightforward, and only the inferior border, which sits on the diaphragm, will move to the same extent, whereas superiorly, the relative motion will be reduced. This phenomenon was first studied by Wang and colleagues,[3] who measured the displacements of the right coronary artery root, the origin of the left anterior descending artery, and the superior and inferior margins of the heart relative to that of the diaphragm in 10 healthy subjects. For the right coronary artery origin, the mean ($\pm$ SD) relative displacement (or correction factor) was 0.57 ± 0.26. McConnell and coworkers[20] first used this correction factor to track the position of the imaging slice during breath-holding and showed improved image registration relative to untracked acquisitions. In free-breathing studies, the correction factor was first applied by Danias and associates,[21] who showed that tracked image quality was maintained as the navigator acceptance window increased from 3 mm to 7 mm, whereas in untracked images, it significantly decreased. This technique, called real-time prospective slice-following, is now used routinely for both 2-D and 3-D methods of acquisition. Of note, however, is the relatively high standard deviation of the correction factor noted earlier, which reflects considerable intersubject variation in the degree of cardiac motion with respiration. The accuracy of slice-following techniques will obviously depend upon the accuracy of the correction factor implemented. In 1997, Taylor and colleagues[22] showed how a subject-specific factor could be rapidly measured using end-inspiratory and end-expiratory breath-hold scans prior to the coronary imaging protocol. Figure 15–4 shows the relationship between the motion of the right hemidiaphragm and the coronary ostia measured in one subject with the slope of the graph giving the correction factor. Figure 15–5 shows two examples of subjects with very different correction factors, illustrating how a wider acceptance window can be used, thus improving scan efficiency. The need for a subject-specific correction factor has further been confirmed in 3-D coronary MRA where its use was found to yield optimal image quality.[23]

An additional or alternative approach to the real-time slice following described earlier is to use a postprocessing adaptive motion correction technique[4] to retrospectively correct an image for movement occurring during the data acquisition. This technique, which can be used to correct a 2-D acquisition for in-plane displacement or a 3-D acquisition for both in-plane and through-plane displacements, may not initially appear to be an attractive option, but it has the advantages of allowing the correction factor to be optimized for each individual patient and provides an alternative approach to those centers with scanners that do not have a real-time decision-making capability. This approach has been implemented with both segmented k-space gradient

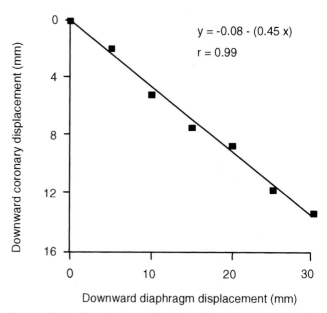

Figure 15–4. Plot of superior/inferior right coronary artery displacement against superior/inferior diaphragm displacement for a single subject. The gradient of the linear regression line is the subject-specific correction factor. (Modified from Taylor AM, Keegan J, Jhooti P, et al: Calculation of a subject-specific adaptive motion-conversion factor for improved real-time navigator echo-gated magnetic resonance coronary angiography. J Cardiovasc Magn Reson 1:131, 1999, courtesy of Marcel Dekker, Inc.)

echo[24] and interleaved spiral[25] coronary MRA acquisitions with promising results.

Column Positioning

The degree of diaphragm motion detected by the navigator echo is dependent on the positioning of the navigator column. The dome of the right hemidiaphragm is higher than that of the left, and the two move coherently with respiration but to differing degrees.[26] Motion of the diaphragm is also greater posteriorly than anteriorly (anterior and dome excursions being 56% and 79% respectively of posterior excursions) and, at the level of the dome, is greater laterally than medially.[27] The correction factor implemented in real-time slice following or postprocessing adaptive motion correction as described earlier will therefore be strongly dependent on the positioning of the navigator column and further supports the use of a subject-specific factor as described in the previous section.

McConnell and associates[28] investigated the effects of varying the navigator location on the image quality of coronary artery studies. Navigators were positioned through the dome of the right hemidiaphragm, through the posterior portion of the left hemidiaphragm, through the anterior and posterior left ventricular walls, and through the anterior left ventricular wall, as shown in Figure 15–6. The advantage of the latter navigator position in close proximity to the coronary artery is that it would

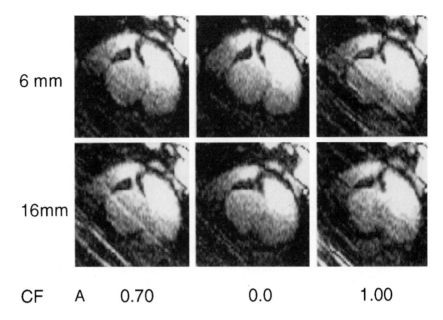

6 mm

16mm

CF   A   0.70   0.0   1.00

Figure 15–5. Right coronary artery origin images acquired with navigator acceptance windows of 6 mm and 16 mm in subjects with subject-specific correction factors (CFs) of (A) 0.7 and (B) 0.25. For both subjects, images were also acquired with CFs of 0 and 1. In the absence of slice-following (CF = 0), image quality is reduced as the navigator acceptance window increases from 6 mm to 16 mm. When slice-following with a subject-specific CF is used however, image quality is maintained. (Modified from Taylor AM, Keegan J, Jhooti P, et al: Calculation of a subject-specific adaptive motion-conversation factor for improved real-time navigator echo-gated magnetic resonance coronary angiography. J Cardiovasc Magn Reson 1:131, 1999, courtesy of Marcel Dekker, Inc.)

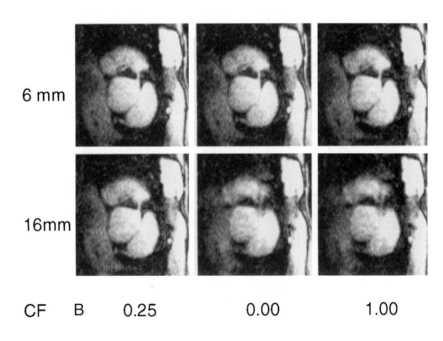

6 mm

16mm

CF   B   0.25   0.00   1.00

remove the need for a correction factor, as described in the previous section, relating the navigator echo measured displacement to the coronary artery motion. The results are summarized in Table 15–2 and show no significant differences in the image quality scores obtained with varying navigator location, although there was a tendency, which did not reach statistical significance, for the anterior left ventricular wall navigator scans to be longer in duration. One of the problems of monitoring the heart itself is the complex anatomy, making it more difficult to find a suitable position for the navigator column. It is hoped that more sophisticated methods of positioning the column will further improve this method of monitoring cardiac motion.

Multiple Column Orientations

It has been shown that there is a linear relationship between the superior-inferior and anterior-posterior motions of the heart, with the superior-inferior motion being approximately five times that of the anterior-posterior motion.[3] For this reason, the real-time slice-following methods first used by McConnell and coworkers[20] and by Danias and associates[21] included a correction for anterior-posterior motion of the heart, assuming it to be equal to 20 percent of the superior-inferior motion. Unfortunately, there is not always such a strong relationship between the directions of motion of the heart with respiration. Sachs and colleagues showed this prob-

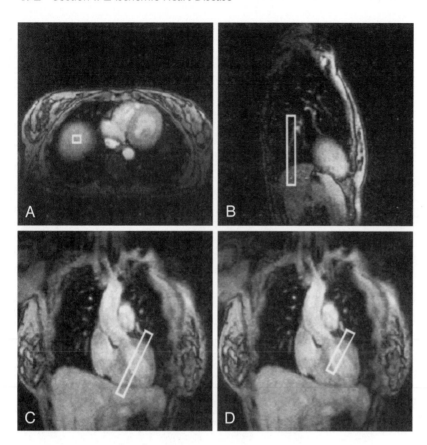

Figure 15–6. Navigator column locations positioned on transverse *(A)*, sagittal *(B)*, and coronal *(C and D)* pilot images: *(A)* through the dome of the right hemi-diaphragm, *(B)* through the posterior left hemi-diaphragm, *(C)* through both anterior and posterior left ventricular walls, and *(D)* through the anterior left ventricular wall. (Modified from McConnell MV, Khasgiwala VC, Savord BJ, et al: Comparison of respiratory suppression methods and navigator locations for MR coronary angiography. AJR Am J Roentgenol 168:1369, 1997.)

lem by using three navigators to measure the inferior-superior, anterior-posterior, and right-left motions of the heart.[29] Figure 15–7 shows an example from this study illustrating the scatter of inferior-superior, right-left, and anterior-posterior measurements, made over a period of approximately 10 minutes. The group went on to compare the use of one, two, and three navigators for imaging the right coronary artery and showed an improvement when multiple directions of motion were considered. This improvement in image quality, however, has to be offset against the main disadvantage, which is that the scan efficiency is reduced, potentially introducing more problems associated with long-term drift in the breathing pattern.

Navigator Timing

Navigator timing is one of the more important parameters; however, flexibility to alter timing is often limited by the computing architecture of the scanner being used (see following discussion). Figure 15–8 illustrates the three main alternatives, these being (1) pre-, (2) pre- and post-, and (3) navigators repeated regularly throughout the cardiac cycle. A simple pre-navigator provides the highest scan efficiency when a navigator acceptance window is used, but may not be reliable if there was a sudden change in breathing between the navigator measurement and the image data acquisition. Pre- and post-navigators overcome this problem, but of

TABLE 15–2. Image Quality Scores, Registration Errors, and Total Scan Times*

Parameter	Right Diaphragm Navigator	Left Diaphragm Navigator	Left Ventricle Navigator	Anterior LV Wall Navigator
Image Quality Score (0–4)	2.3 ± 0.1	2.3 ± 0.1	2.4 ± 0.1	2.2 ± 0.2
Registration Error (mm)				
Craniocaudal	0.5 ± 0.1	0.4 ± 0.1	0.6 ± 0.1	0.4 ± 0.1
Anteroposterior	0.3 ± 0.1	0.3 ± 0.1	0.3 ± 0.1	0.4 ± 0.1
Total Scan Time† (sec)	294 ± 28	314 ± 30	342 ± 62	427 ± 111

*Image quality scores (0 = very poor, 4 = excellent), registration errors, and total scan times for different navigator column positions during free-breathing MR coronary angiography. There were no statistically significant differences between the navigator column locations. Data are presented as mean ± standard error of the mean (SEM); LV = left ventricle.

†Total scan time is the time from start to finish for 6 scans.

Adapted from McConnell MV, Khasgiwala VC, Savord BJ, et al: Comparison of respiratory suppression methods and navigator locations for MR coronary angiography. AJR Am J Roentgenol 168:1369, 1997.

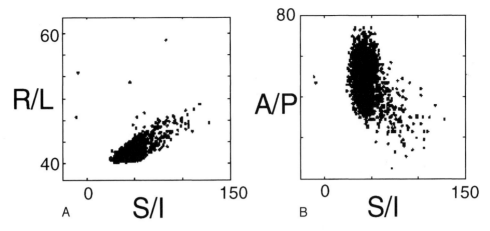

Figure 15–7. Graphs showing *(A)* right/left (R/L) and *(B)* anterior/posterior (A/P) navigator echo measurements as a function of superior/inferior (S/I) navigator echo measurements in a healthy subject. (Image courtesy of Dr. Todd Sachs, Stanford University.)

course, also reduce the scan efficiency. In our experience, the use of pre-navigators only produces acceptable results for free-breathing studies, whereas multiple breath-hold acquisitions certainly require both pre- and post-navigators. An important factor, which depends on the computer hardware and architecture, is that the time required after the navigator acquisition before the start of the imaging sequence needs to be quite brief. Particularly if pre-navigators only are being used, the longer this time interval, the greater the potential for errors caused by interim respiratory motion. Also, for ECG R wave–triggered scans, this finding may have implications on the minimum gating delay that can be obtained, and for short gating delay or cine scans. Post-only navigators can be used as an alternative. This approach has recently been implemented in left ventricular function studies, where it was found that image quality in a group of heart failure patients was significantly improved over conventional breath-hold scans.[30]

Repeated navigators allow for improved cine or multi-slice imaging and also provide some potential for estimating the inter-navigator respiratory mo-

tion. The potential problem with this is that the navigator signal-to-noise ratio could be reduced and this may affect the accuracy of the navigator edge detection. In addition, as the time for navigator output increases, the time for imaging decreases and the number of phases or slices that can be acquired will be reduced.

Precision of Navigator Measurement

Commonly, a spatial resolution of 1 mm is used along the navigator echo column, for example, having a field-of-view of 512 mm and sampling 512 points on the navigator readout. However, the precision of the measurement is dependent to a large extent on the signal-to-noise ratio of the navigator measurement. The most important factor affecting signal-to-noise is the coil arrangement. If, for example, a single coil is used for imaging and navigator detection, it has to be big enough to cover both the imaging area of interest and the region of navigator detection. On the other hand, if phased-array coils

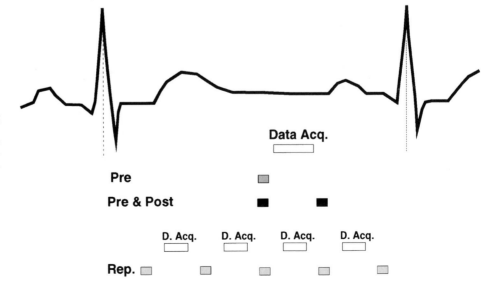

Figure 15–8. Diagram showing the timings of navigators for pre, pre and post, and repeated navigator echo controlled acquisitions relative to the data acquisition (Data Acq.) period.

are used, it can be possible to position one coil specifically for navigator detection, possibly over the region of the right diaphragm. Another important factor in the precision of the measurement is the quality of the edge on the navigator trace. To obtain a well-defined edge of the diaphragm, for example, it is important to have a reasonably small column cross section and to position it through the dome of the diaphragm, so that the column is perpendicular to the diaphragm edge rather than more posteriorly where motion is greatest. Finally, the diaphragm edge may be detected by edge detection, correlation, or least-squares fit algorithms. For rapid tracking (repetition time < 100 ms), the signal-to-noise ratio in the diaphragm trace would be too poor for simple edge detection methods to succeed. Of the remaining two techniques, the least-squares fit method has been shown to be more resistant to the effects of noise and to the diaphragm profile deformation that occurs during respiration than the correlation method and would be the technique of choice.[31] However, most navigator techniques acquire only one or two navigators per cardiac cycle, and in such cases, signal-to-noise ratios are relatively high and edge detection algorithms generally adequate.

Computer Architecture

The computer architecture of modern CMR scanners can be very complex, generally incorporating three main computers. The host computer runs the user interface and allows connection to the image data base and so forth, the reconstruction computer is a dedicated rapid processor for reconstruction of the CMR image data, and the scan computer allows control and adjustment of parameters associated with the scanning sequence. The architecture of these computers can significantly affect the potential and usefulness of navigator echoes. On many systems, for example, the navigator signal must be reconstructed and processed on the reconstruction computer but the measurement made must be passed through the host to control the parameters on the scan computer. This arrangement inevitably adds a variable and unknown delay dependent on other tasks being performed by the host operating system. To overcome this difficulty, either a direct and rapid link is required allowing transfer of data from the reconstruction to the scan computer or the scan computer itself must be capable of acquiring and reconstructing the navigator data, so that there is no data transfer required.

CONCLUSION

Navigator echoes have been shown to be an important method of monitoring respiration, which has been used for defining the position of the heart, enabling improved coronary and other cardiac imaging. The fact that there have been a limited number of studies and there are so many parameters and variables involved in their use suggests that an optimal method of application may not yet have been developed. Certainly with future system development there will be minimal cost involved, in imaging time or other factors, in obtaining this positional information, and it would therefore seem worthwhile to collect and use it where appropriate. At present, the methods are not totally robust probably because of their relative lack of sophistication.

One of the major advantages is that navigators allow images to be acquired during free respiration, which takes away the requirement of patient cooperation. They also allow longer acquisition times, enabling higher spatial and temporal resolution, increasing the potential of more sophisticated techniques such as detailed flow imaging.[32] A balance has to be met, however, and the imaging time should not be increased so much that increased respiratory drift cancels any potential benefit.

With development enabling more general measures of cardiac position as well as rotation of the heart, the navigator echo should have an important role in the future of CMR.

References

1. Atkinson DJ, Edelman RR: Cineangiography of the heart in a single breath hold with a segmented turboFLASH sequence. Radiology 178:357, 1991.
2. Holland AE, Goldfarb JW, Edelman RR: Diaphragmatic and cardiac motion during suspended breathing: Preliminary experience and implications for breath-holding. Radiology 209:483, 1998.
3. Wang Y, Riederer SJ, Ehman RL: Respiratory motion of the heart: Kinematics and the implications for the spatial resolution in coronary imaging. Magn Reson Med 33: 713, 1995.
4. Ehman RL, Felmlee JP: Adaptive technique for high-definition MR imaging of moving structures. Radiology 173: 255, 1989.
5. Liu YL, Riederer SJ, Rossman PJ, et al: A monitoring, feedback, and triggering system for reproducible breath-hold MR imaging. Magn Reson Med 30:507, 1993.
6. Taylor AM, Keegan J, Jhooti P, Gatehouse PD, et al: Differences between normal subjects and patients with coronary artery disease for three different MR coronary angiography respiratory suppression techniques. J Magn Reson Imaging 9:786, 1999.
7. Wang Y, Grimm RC, Rossman PJ, et al: 3D coronary MR angiography in multiple breath-holds using a respiratory feedback monitor. Magn Reson Med 34:11, 1995.
8. Keegan J, Gatehouse PD, Taylor AM, et al: Coronary artery imaging on a mobile 0.5 Tesla scanner: Implementation of real-time navigator-echo controlled segmented k-space FLASH and interleaved spiral sequences. Magn Reson Med 41:392, 1999.
9. Taylor AM, Jhooti P, Wiesmann FW, et al: MR navigator-echo monitoring of temporal changes in diaphragm position: Implications for MR coronary angiography. J Magn Reson Imaging 7:629, 1997.
10. Lenz GW, Haacke EM, White RD: Retrospective cardiac gating: A review of technical aspects and future directions. Magn Reson Imaging 7:445, 1989.
11. Hofman MB, Paschal CB, Li D, et al: MRI of coronary arteries: 2D breath-hold vs 3D respiratory-gated acquisition. J Comput Assist Tomogr 19:56, 1995.
12. Oshinski JN, Hofland L, Mukundan S, et al: Two-dimensional

coronary MR angiography without breath-holding. Radiology 201:737, 1996.

13. Sachs TS, Meyer CH, Irarrazabal P, et al: The diminishing variance algorithm for real-time reduction of motion artifacts in MRI. Magn Reson Med 34:412, 1995.

14. Maki JH, Prince MR, Londy FJ, Chenevert TL: The effects of time varying intravascular signal intensity and k-space acquisition order on three-dimensional MR angiography image quality. J Magn Reson Imaging 6:642, 1996.

15. Jhooti P, Keegan J, Gatehouse PD, et al: 3D coronary artery imaging with phase reordering for improved scan efficiency. Magn Reson Med 41:555, 1999.

16. Sinkus R, Börnert P: Motion pattern adapted real-time respiratory gating. Magn Reson Med 41:148, 1999.

17. Sinkus R, Börnert P: Extension of real-time MR gating to cope with changes in motion pattern: Making MR gating autarkic. Proceedings of the 6th Scientific Meeting of ISMRM, Sydney, 1998, p 2127.

18. Jhooti P, Gatehouse PD, Keegan J, et al: Phase ordering with automatic window selection (PAWS): A novel motion-resistant technique for 3D coronary imaging. Magn Reson Med 43:470, 2000.

18a. Huber ME, Hengesbach D, Botnar RM, et al: Motion artifact reduction and vessel enhancement for free-breathing navigator-gated coronary MRA using 3D k-space reordering. Magn Reson Med 45:645, 2001.

19. Pauly J, Nishimura D, Macovski A: A k-space analysis of small tip-angle excitation. J Magn Reson 81:43, 1989.

20. McConnell MV, Khasigawala VC, Savord BJ, et al: Prospective adaptive navigator correction for breath-hold MR coronary angiography. Magn Reson Med 37:148, 1997.

21. Danias PG, McConnell MV, Khasigawala VC, et al: Prospective navigator correction of image position for coronary MR angiography. Radiology 203:733, 1997.

22. Taylor AM, Keegan J, Jhooti P, et al: Calculation of a subject-specific adaptive motion-correction factor for improved real-time navigator echo-gated magnetic resonance coronary angiography. J Cardiovasc Magn Reson 1:131, 1999.

23. Nagel E, Bornstedt A, Schnackenburg B, et al: Optimisation of real-time adaptive navigator correction for 3D magnetic resonance coronary angiography. Magn Reson Med 42;408, 1999.

24. Wang Y, Ehman RL: Retrospective adaptive motion correction for navigator-gated 3D coronary MR angiography. J Magn Reson Imaging 11:208, 2000.

25. Keegan J, Gatehouse PD, Yang GZ, Firmin DN: Adaptive motion correction of interleaved spiral images and velocity maps: Implications for coronary imaging. Proceedings of the 15th Annual Meeting of the ESMRMB. MAGMA 6(suppl 1):67, 1998.

26. Korin HW, Ehman RL, Riederer SJ, et al: Respiratory kinematics of the upper abdominal organs: A quantitative study. Magn Reson Med 23:172, 1992.

27. Gierada DS, Curtin JJ, Erickson SJ, et al: Diaphragmatic motion: Fast gradient recalled echo MR imaging in healthy subjects. Radiology 194:879, 1995.

28. McConnell MV, Khasgiwala VC, Savord BJ, et al: Comparison of respiratory suppression methods and navigator locations for MR coronary angiography. AJR Am J Roentgenol 168:1369, 1997.

29. Sachs TS, Meyer CH, Pauly JM, et al: The real-time interactive 3D-DVA for robust coronary MRA. IEEE Trans Med Imaging 19:73, 2000.

30. Bellenger NG, Gatehouse PD, Rajappan K, et al: Left ventricular quantification in heart failure by cardiovascular MR using prospective respiratory navigator gating: Comparison with breath-hold acquisition. J Magn Reson Imaging 11:411, 2000.

31. Wang Y, Grimm RC, Felmlee JP, et al: Algorithms for extracting motion information from navigator echoes. Magn Reson Med 36:117, 1996.

32. Nagel E, Bornstedt A, Hug J, et al: Noninvasive determination of coronary blood flow velocity with magnetic resonance imaging: Comparison of breath-hold and navigator techniques with intravascular ultrasound. Magn Reson Med 41:544, 1999.

Coronary Magnetic Resonance Angiography—Methods

René M. Botnar, Matthias Stuber, Peter G. Danias, Kraig V. Kissinger, and Warren J. Manning

Despite improvements in early detection and prevention, coronary artery disease remains one of the major causes of morbidity and mortality in the Western world. The current gold standard for the diagnosis of coronary disease is the contrast x-ray coronary angiogram. Each year, over 1,000,000 diagnostic procedures are performed both in the United States[1] and Europe. Despite numerous noninvasive tests to determine the presence or absence of disease prior to referral for angiography, up to 40 percent of patients referred for x-ray coronary angiography are found to have no significant disease.[2] Because mechanical revascularization of proximal coronary disease has the greatest impact on survival, it would be desirable to have a noninvasive method that allowed for direct visualization of the proximal/mid native coronary vessels for the accurate identification and/or exclusion of focal disease.

In this chapter we review the obstacles to and general strategies for acquiring coronary magnetic resonance angiography (MRA). (Chapter 17 reviews the clinical results obtained using the methods described in this chapter.) Unlike x-ray angiography, MRA neither exposes the patient to potentially harmful ionizing radiation nor requires iodinated contrast. In addition to providing high spatial resolution, MRA allows for coronary image acquisition in any imaging plane and is not associated with any short- or long-term side effects. As with more general MRA of the aorta and renal arteries (see Chapter 29), coronary MRA is primarily based on blood flow and can be divided into "bright blood" and "black blood" methods. Bright-blood techniques have received the greatest attention, with coronary blood contrast provided by inflow of unsaturated blood and differences in tissue relaxation rather than an exogenous magnetic resonance (MR) contrast agent.

The general approaches described in this chapter are available on current MR units across all vendor platforms, but specific nuances may be vendor specific (e.g., navigator implementation). Though beyond the scope of this chapter, cardiovascular MR (CMR) also has the potential to noninvasively image the coronary vessel wall,[3] including detection of subclinical atherosclerotic plaque,[4, 5] thereby giving new insights into the development and progress of subclinical atherosclerosis.

OBSTACLES TO CORONARY MAGNETIC RESONANCE ANGIOGRAPHY

Over the past decade, MRA has become widely recognized as the premier clinical noninvasive imaging technique for the assessment of larger arteries (and veins) in the body, including the aorta (dissection, aneurysm, coarctation) and carotid arteries (stenoses). The coronary arteries, however, present a particular challenge due to their small caliber, tortuosity (particularly among patients with left ventricular hypertrophy), surrounding signal from epicardial fat and myocardium, and near-constant motion related to both the cardiac cycle and normal respiration.

Cardiac Motion

Bulk cardiac motion is a major impediment to coronary MRA and can be separated into motion related to cardiac contraction/relaxation and that due to superimposed diaphragmatic motion/respiration. The magnitude of both components may greatly exceed the coronary artery dimensions, thereby leading to blurring of the coronary vessel if adequate motion-suppressive methods are not used. To account for bulk cardiac motion, accurate electrocardiographic (ECG) synchronization is an absolute necessity. Less robust peripheral pulse detection methods lead to inferior results.

Several investigators have characterized coronary motion both with CMR[6–8] and with x-ray coronary angiography.[9, 10] There is a triphasic displacement pattern of both coronary arteries during the cardiac cycle (Figure 16–1). The magnitude of motion is approximately twice as large for the right as for the left coronary artery. In-plane coronary motion is maximal at early to midsystole (when the ventricles are contracting) with motion again becoming promi-

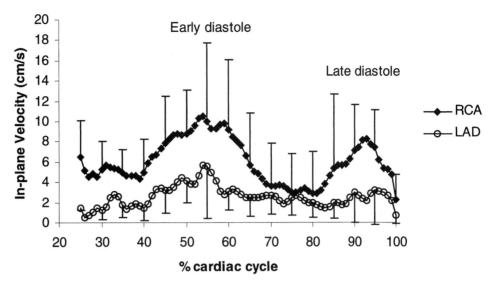

Figure 16–1. Graph of in-plane right (RCA) and left anterior descending (LAD) coronary artery motion during the cardiac cycle. The x-axis displays time as a percentage of the RR interval (cardiac cycle). (Courtesy of Yong Kim, MD, PhD.)

nent during early diastole (when the ventricles rapidly fill with blood). During isovolumic relaxation, approximately 350 to 400 ms after the R wave, and again at middiastole (of which the timing and duration depend on the RR cycle length), the motion of the coronary arteries is relatively quiescent.[6] The third component of rapid motion follows atrial systole.

Because bright-blood coronary MRA depends on inflow of unsaturated spins, middiastole is generally preferred for image acquisition because it also represents a period of rapid coronary blood flow (~30 cm/s). The duration of middiastolic diastasis is inversely related to heart rate and dictates the data acquisition interval each RR interval.[6–8] Although a heart rate–dependent formula may be sufficient for most subjects, there is considerable patient-to-patient variation. The acquisition of a cine scout image allows for determination of a patient-specific quiescent period and appears preferred for the more rigorous demands of coronary vessel wall imaging.[7] A coronary MRA acquisition duration of less than 60 to 100 ms during each cardiac cycle results in improved image quality.[11] With tachycardia, the duration needs to be further minimized (<50 ms).

Respiratory Motion

Among the major obstacles to coronary MRA are the artifacts resulting from bulk cardiac motion associated with respiration. During normal respiration, the predominant dwell time is end-expiration, with diaphragmatic excursion during the respiratory cycle approaching 30 mm.[12] Acquisition of coronary MRA during the entire respiratory cycle results in ghosting, blurring, and image degradation. Several different approaches have been used (Table 16–1), including the use of breath-holding, multiple averages during free-breathing, respiratory bellows, coached breathing with (visual or auditory) feedback, and free-breathing with MR navigators. Al-

though sustained breath-holds appear to have a role in healthy volunteers and motivated patients, free (uncoached) breathing with diaphragmatic MR navigators is preferred for most patients, particularly for patients with coexistent pulmonary disease[12]

Breath-Hold Methods

Though heavily dependent on patient cooperation, initial approaches for visualizing the proximal native coronary arteries used prolonged (15–20 s) end-expiratory breath-holds to suppress respiratory motion.[13] Breath-holding offers the distinct advantages of rapid imaging and ease of implementation in compliant subjects. However, the total MRA acquisition time is limited by the length of the breath-hold, and both slice registration errors and end-breath-hold cranial drift of the diaphragm are common.[12–18] Diaphragmatic drift may result in image

---

Table 16–1. Respiratory Suppression Methods

Breath-Holding

Sustained end-expiratory breath-hold (inspiratory breath-hold)
Multiple brief breath-holds/coached breathing
Breath holds with visual or auditory feedback
Hyperventilation
Supplemental oxygen

Free Breathing

Multiple averages
Chest wall bellows
CMR Navigators
    Navigator location: Right hemidiaphragm
                             Left hemidiaphragm
                             Basal left ventricle
                             Coronary of interest
                             Anterior thorax
Prone vs. supine imaging
Single vs. multiple navigators
Prospective vs. retrospective navigator triggering
Navigator triggering vs. navigator gating with real-time motion correction

blurring whereas slice registration errors result in apparent gaps between the segments of the coronary arteries during two-dimensional (2-D) coronary MRA. These gaps may be misinterpreted as signal voids due to coronary stenoses. In addition, the use of signal enhancement techniques such as signal averaging or fold-over suppression are restricted by the time constraint of a single breath-hold. Although supplemental oxygen and hyperventilation (alone or in combination) can effectively prolong the breath-hold duration, these methods may not be appropriate in cardiac patients, and both diaphragmatic drift and slice registration errors persist.[17] Finally, patient and operator fatigue ensue as the study progresses and/or the number of breath-holds increases. As a slight variation, multiple brief breath-holds[19] and coached breath-holding with visual or audible feedback[20-22] have been used effectively in motivated patients.

Free-Breathing Methods

Averaging and Respiratory Bellows

Initial free-breathing approaches used multiple signal averaging to minimize motion artifacts.[23] This averaging approach is reasonable for relatively low (>1–2 mm) spatial resolutions, but is inadequate for accurate stenosis detection. Another early free-breathing approach utilized a thoracic respiratory bellows to gate to the end-expiratory position,[18, 24] with promising results when compared with breath-holding.[24] Enhancements included respiratory feedback monitoring.[21] Subsequently, more accurate (and flexible) MR navigators displaced bellows gating.

Magnetic Resonance Navigators— Gating Alone

The use of free-breathing with respiratory navigators (see Chapter 15), first proposed by Ehman and Felmlee,[25] serves to overcome the time constraint and cooperation imposed by breath-hold approaches. The use of navigator gating has received considerable attention and varies from simple to complex (see Table 16–1) implementations. In principle, the MR navigator monitors the motion of an interface. It can be positioned at any interface that accurately reflects respiratory motion, including the dome of the right hemidiaphragm (Figure 16–2),[21, 24, 26] the left hemidiaphragm, the anterior chest wall, or directly through the anterior free wall of the left ventricle.[18, 27] Data are accepted only when the diaphragm-lung (or myocardium-lung) interface falls within a user-defined window (usually 3–5 mm), positioned around the end-expiratory level of the interface (see Figure 16–2). Studies suggest that the bulk of cardiac motion related to breathing is in the superior-inferior axis[28] and that most single-navigator locations lead to similar image quality.[18, 27]

Thus, the right hemidiaphragm has become preferred due to the ease in identifying the interface from a series of coronal, sagittal, and transverse scout images. The gating process can either be prospective[18, 24] (i.e., determined before coronary data acquisition and offering the opportunity to correct for slice position) or retrospective[29-31] (i.e., following data acquisition, but before image reconstruction). With navigator gating (without tracking), a 3-mm end-expiratory diaphragmatic window is commonly used with data collected on average from one third of RR intervals (33% navigator efficiency).[18]

Magnetic Resonance Navigators— Gating and Slice Tracking

From CMR studies of cardiac borders, Wang and associates noted that the dominant impact of respiration on cardiac position is in the superior-inferior direction. At the end-expiratory position, the relationship between diaphragmatic and cardiac motion is ~0.6 for the right coronary artery and ~0.7 for the left coronary artery.[28] Advances in computer processing and knowledge of this relatively fixed relationship offer the opportunity for prospective navigator gating with real-time tracking[26, 32] and the use of wider gating windows and shorter scan time (increased navigator efficiency). With real-time tracking implementations, a 5-mm diaphragmatic gating window is often used with a navigator efficiency close to 50 percent[32] (vs. ~33% with 3-mm window and navigator gating alone). Coronary MRA with real-time navigator tracking has been shown to minimize registration errors (as compared with breath-holding) with maintained or improved image quality both for 2-D and three-dimensional (3-D) approaches.[27, 32] Free-breathing navigator methods are particularly well suited for longer 3-D coronary MRA approaches, combining the postprocessing benefits of thin adjacent slices with submillimeter spatial resolution afforded by improved signal-to-noise ratio (SNR). Although patient cooperation is reduced with navigator/free-breathing methods, diaphragmatic drift and patient motion remain relevant issues.[12, 16, 17] Though less comfortable, prone imaging appears to reduce these problems and also serves to minimize the chest wall motion (not accounted for by diaphragmatic navigator) and improve navigator efficiency.[33] Since the navigator data are intended to reflect the position of the heart during the subsequent acquisition period, *positioning of the navigator immediately prior to the data acquisition block and rapid navigator analysis* are very important.[34] Finally, sophisticated navigator algorithms have also been implemented to more efficiently collect important k-space profiles[35-37] (see Chapter 15).

Magnetic Resonance Navigators— Adaptive Averaging

Recently, preliminary data for a novel respiratory suppressive approach that does not utilize ECG gat-

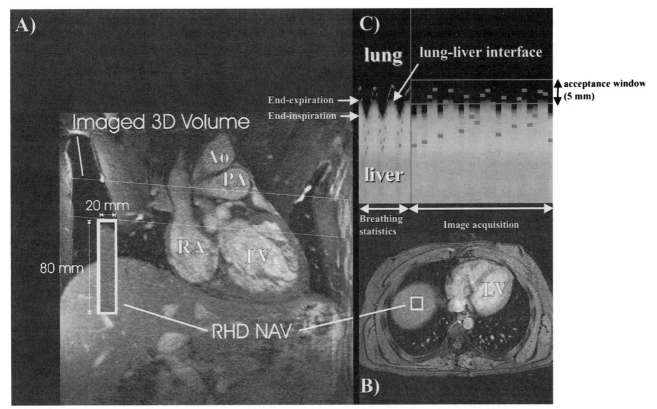

Figure 16–2. Coronal *(A)* and transverse *(B)* thoracic image with identification of navigator location at the dome of the right hemidiaphragm (RHD NAV). *(C)* Respiratory motion of the lung-diaphragm interface recorded using a two-dimensional (2-D) selective navigator. The maximum excursion between end-inspiration and end-expiration in this example is ~11 mm, but can vary greatly among individuals. The "dots" on the left side of the figure (Breathing statistics) indicate the position of the lung-liver interface with the scale magnified by a factor of 4. Data are only accepted (image acquisition) if the lung-liver interface is within the "acceptance window" of 5 mm. Data acquisition with the navigator outside of the gating window is rejected. Ao = ascending aorta; PA = pulmonary artery; RA = right atrium; LV = left ventricle.

ing, breath-holds, or navigator gating suggest that it may prove useful, particularly for those with an irregular heart rhythm or irregular breathing pattern. With real-time imaging and adaptive averaging,[38] cross-correlation is used to automatically identify those real-time imaging frames in which a coronary vessel is present, and to determine the location of the vessel within each frame. This information is then used for selective averaging of frames to increase the SNR and to improve vessel visualization.

Spatial Resolution

With free-breathing approaches, spatial resolution is primarily limited by the SNR. There are several methods that provide increasing SNR including signal averaging, 3-D imaging, MR contrast agents, novel k-space acquisition schema, and dedicated receiver coils. The normal adult proximal coronary arteries are 3 to 5 mm in diameter, as compared with 30 to 35 mm for the ascending aorta, 5 to 8mm for the carotid artery, and 4 to 7 mm for the renal artery. Were SNR and acquisition duration not limiting, one might argue that coronary MRA requires x-ray angiographic spatial resolution. Presently, submillimeter (700–1000 μm in-plane) coronary MRA

spatial resolution is readily achievable, but these do not approach angiographic resolutions of ~300 μm. Even with submillimeter in-plane spatial resolution, the CMR image has a thickness of 2 to 3 mm as compared with the x-ray angiogram, which is a projection.

Spatial resolution requirements for coronary MRA depend on whether the goal is to simply identify the ostial take-off and proximal course of the coronary artery (as in cases of suspected congenital anomalous coronary disease), or whether the goal is to identify focal stenoses. Figure 16–3 displays an x-ray coronary angiogram displayed at (1) 300 μm, (2) 500 μm, (3) 1000 μm, and (4) 2000 μm spatial resolution. At 500 μm and 1000 μm resolutions, the focal coronary stenoses are readily visible, whereas at in-plane resolutions greater than 1000 μm, only the gross/overall coronary artery is visible, potentially allowing for identification of anomalous coronary vasculature, but not focal stenoses. Thus, it appears that submillimeter spatial resolution is necessary to visually identify focal stenoses.

While submillimeter in-plane spatial resolution appears important, relatively thick (1.5–3 mm) slices are used for coronary MRA to allow for adequate SNR, resulting in anisotropic voxels. This anisotropic voxel size may lead to vessel blurring with

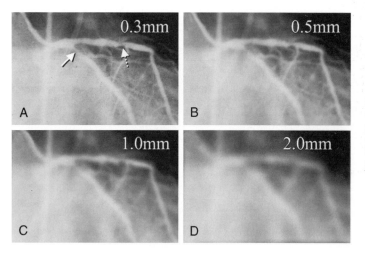

Figure 16–3. X-ray coronary angiogram displayed at *(A)* 300 μ, *(B)* 500 μ, *(C)* 1000 μ, and *(D)* 2000 μ in-plane spatial resolution. The focal coronary stenoses of the proximal left circumflex coronary artery (solid white arrow) and the proximal left anterior descending coronary artery (dotted arrow) are only appreciated with image resolution of <1000 μ.

oblique and multiplanar reconstructions. The use of isotropic voxel size has been shown to be advantageous for reconstructions from intracranial and aortic MRA data sets[39, 40] as well as for coronary MRA[41] (Figure 16–4), with resultant improved vessel sharpness. Quantitative coronary MRA methods have been reported for diameter assessment of normal vessels and for comparison of vessel sharpness[11] but not for quantification of focal stenoses.

Finally, in-plane spatial resolution requirements for coronary MRA mandate maximizing SNR by the use of appropriate cardiac receiver coils. As SNR is reduced sharply with distance of the organ of interest from the receiver coil, these cardiac-specific coils are optimized for the size of the heart and the distance of the heart from the chest wall. Fortunately, the right coronary artery, left main, and left anterior descending coronary arteries are relatively anterior structures. Nearly all vendors have cardiac specific phased array coils that allow flexibility (use of single coil or multiple anterior coils or in combination with posterior elements) and enhanced SNR as compared with use of the body coil as the receiver. Use of these cardiac-specific coils should be

the standard for all coronary MRA and nearly all CMR studies.

Suppression of Signal from Surrounding Tissue

Suppression of Signal from Epicardial Fat

The intrinsic contrast between the coronary blood pool and the surrounding tissue (myocardium, epicardial fat) can be manipulated using the inflow effect (unsaturated protons entering the imaging field between successive radiofrequency [RF] pulses) and/or by the application of CMR prepulses. In most subjects, the coronary arteries are surrounded by epicardial fat. Fat has a relatively short T1 and resultant MR signal intensity similar to that of flowing blood. Fat saturation prepulses are used to selectively suppress signal from surrounding fat so as to allow visualization of the underlying coronary arteries. This process is often accomplished with a frequency-selective prepulse that minimizes

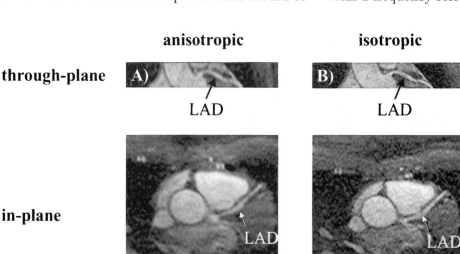

Figure 16–4. Through-plane (*A* and *B*) and in-plane (*C* and *D*) views of the left anterior descending coronary artery (LAD) with anisotropic and isotropic voxel size. In both the in-plane and through-plane views, improved vessel definition is observed in the isotropic images. (From Botnar RM, Stuber M, Kissinger KV, Manning WJ: Free-breathing 3D coronary MRA: The impact of "isotropic" image resolution. J Magn Reson Imaging 11:389, 2000.)

fat signal and thereby allows for visualization of the coronary vessels.[13, 23] (In the case of spiral imaging, a spectral spatial RF excitation pulse may be used, which selectively excites water.[42])

Suppression of Signal from Myocardium (and Deoxygenated Blood Pool)

In addition to epicardial fat, the coronary arteries run in close proximity to the epimyocardium. Myocardium and coronary blood have relatively similar T1 relaxation values (850 ms and 1200 ms, respectively) making delineation of the coronary arteries difficult for 3-D techniques, where blood exchange (inflow effect) is reduced between successive RF excitations. There are different methods that can be used to enhance the contrast between the coronary arteries and myocardium. Most promising are prepulses such as T2 preparation[11, 43] or magnetization transfer contrast (MTC).[23, 44] Because the T2 relaxation time of coronary arterial blood (T2 = 250 ms) and myocardium (T2 = 50 ms) are substantially different, the application of a T2 preparation prepulse serves to suppress myocardial signal, with relative preservation of signal from coronary arterial blood. As an added benefit, deoxygenated blood has T2 relaxation time (T2 = 35ms) close to myocardium. Thus, the T2 prepulse also suppresses signal from deoxygenated blood in the cardiac veins. This suppression is particularly helpful if there is minimal epicardial fat and/or the great cardiac vein runs in close proximity to the left anterior descending (LAD) and circumflex (LCX) coronary arteries (Figure 16–5). The application of MTC prepulses also serves to suppress myocardial signal.[23, 44] Thus, in combination with fat saturation and T2 (or MTC) prepulses, the coronary lumen appears bright and the surrounding tissue (including fat, myocardium) has reduced signal intensity.

CORONARY MAGNETIC RESONANCE ANGIOGRAPHY ACQUISITION SEQUENCES

Although much progress has been made over the past decade, consensus regarding the ideal coronary MRA sequence has not been established. Coronary MRA sequences can be conceptualized as a building block of components that include (1) cardiac (ECG) gating for cardiac motion suppression, (2) respira-

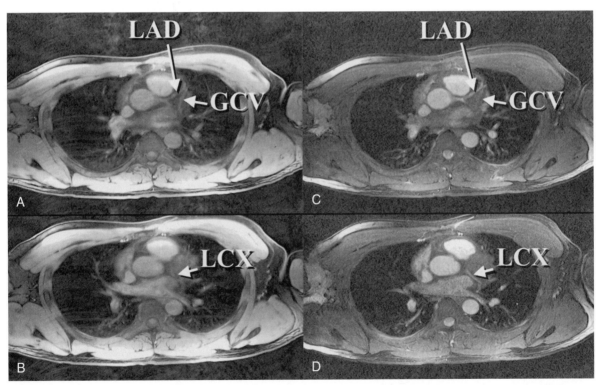

Figure 16–5. Baseline images (*A* and *B*) and T2 prepulse-enhanced images (*C* and *D*) from a three-dimensional (3-D) coronary data set. Application of the T2 prepulse suppresses signal from cardiac muscle as well as skeletal muscle in the anterior chest and back. Suppression of venous blood in the great cardiac vein (GCV) and a reduction of respiratory motion and flow artifacts is also apparent (*C* and *D*). Both data sets have fat saturation with minimal signal from subcutaneous and epicardial fat. LAD = left anterior descending coronary artery; LCX = left circumflex coronary artery. (From Botnar RM, Stuber M, Danias PG, et al: Improved coronary artery definition with T2-weighted free-breathing 3D-coronary MRA. Circulation 99:3139, 1999.)

tory motion suppression (breath-holding, respiratory bellows, navigators), (3) prepulses to enhance contrast-to-noise ratio (CNR) of the coronary arterial blood from surrounding tissue (fat saturation, T2 preparation, MTC, selective tagging of blood in the aortic root, exogenous contrast agents), and (4) image acquisition that optimizes coronary arterial SNR. The image acquisition schema (Table 16–2) include time-of-flight (TOF) bright-blood (segmented k-space), black-blood (fast spin-echo and dual inversion), which may be implemented as 2-D (typically breath-hold) and 3-D (generally non–breath-hold) acquisitions. Although these more conventional acquisitions have received the most attention, novel rapid acquisition methods (e.g., echo planar, spirals, steady state free precession, aortic root tagging, spatial harmonics [SMASH and SENSE]) have and will continue to receive increasing attention (see Chapters 1 and 2), but their clinical role remains to be defined and will therefore only briefly be mentioned.

Conventional Spin-Echo Coronary Magnetic Resonance Angiography

Early attempts to image the coronary arteries using conventional ECG-gated spin-echo (black-blood) coronary CMR were met with limited success. Though occasionally successful for identifying coronary ostia (Figure 16–6), this approach was not reliable for assessment of anomalous vessels or disease. In one of the first studies, Lieberman and colleagues[45] used ECG-gated spin-echo CMR and were able to visualize portions of the native coronary arteries in only 30 percent of 23 subjects. Subsequently, Paulin and coworkers[46] used similar methodology in six patients who had undergone x-ray coronary angiography. Despite acquiring data during ventricular systole, the absence of respiratory motion suppression, and data acquisition over several

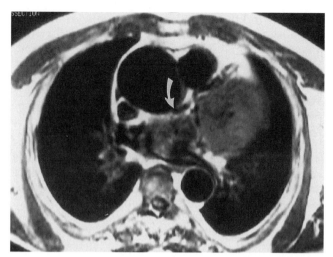

Figure 16–6. Conventional electrocardiographic-gated, multiphase spin-echo transverse CMR at the level of the aortic root/ostium of the left main coronary artery (curved white arrow).

minutes, the origin of the left main coronary artery was seen in all six (100%) subjects and the ostium of the right coronary artery in four (67%) of six subjects. No stenoses were visualized in either report.

2-D and 3-D Segmented k-Space Gradient-Echo Coronary MRA

Advances in hardware (gradient strength/slew rate, receiver coils) and software (motion suppression, prepulses, acquisition schemes) over the past decade have allowed for successful coronary MRA of the proximal regions of the major epicardial coronary arteries in all subjects in sinus rhythm. Currently, the vast majority of reported coronary MRA methodologies have utilized bright blood time-of-flight (TOF) CMR approaches using 2-D or 3-D segmented k-space gradient-echo sequences. Though exogenous contrast agents have been studied (and have become the standard for many noncoronary MRA applications), their role in coronary MRA is presently undefined.

2-D Segmented k-Space Gradient Echo Coronary MRA

The first robust approach to coronary MRA was the 2-D segmented k-space gradient-echo acquisition scheme first described in an isolated heart and in vivo animal model by Burstein[47] and subsequently in humans by Edelman and associates.[13, 14] With this approach, multiple phase encoding steps are acquired during each cardiac cycle. As initially implemented, eight phase encoding steps (repetition time of 14 ms; temporal resolution of 8 × 14 = 112 ms) were acquired during each of 16 successive heartbeats using an incremental flip angle series and fat saturation prepulse. Middiastolic data were ac-

---

Table 16–2. Coronary MRA Methods

Black Blood
Spin echo
Dual inversion fast spin echo

Bright Blood
Segmented gradient echo
  2-D breath-hold
  3-D free breathing (or breath-hold)
Segmented echo planar
Spiral acquisitions
Steady state free precession
Contrast-enhanced coronary MRA
  (Extracellular and intravascular agents)
Aortic root tagging methods

Parallel Imaging Techniques
SMASH
SENSE

quired during a 12- to 16-second (16-heartbeat) breath-hold to complete the 128 × 256 matrix (field of view 240 mm, resulting in 1.9 × 0.9 mm spatial resolution).

After a coronal or sagittal scout to define the position of the basal heart, a series of 10 to 15 overlapping transverse images (each requiring a single breath-hold) are acquired at the take-off of the right and left coronary arteries (Figure 16–7). Due to variability in the diaphragmatic position among breath-holds, repetitive images acquired at the same position may display different regions of the coronary artery. The left main and proximal/mid left anterior descending coronary artery are generally well seen in the transverse plane. For imaging of the right and left circumflex coronary arteries, a single oblique image is then acquired along the major axis of the right and/or left circumflex as defined in the transverse plane (Figure 16–8). Breath-hold variability and coronary vessel tortuosity both contribute to the need for 30 to 40 breath-holds for a complete examination. This number of breath-holds may be reduced by the incorporation of navigator with gating and real-time tracking.[18, 32, 48]

Subsequent improvements in gradient strength/slew rate have allowed implementation of this 2-D approach using repetition times of 6 or 8 ms. This method facilitates other options including (1) data acquisition during an abbreviated 6- to 8-second breath-hold (with maintained temporal and spatial resolution); (2) maintained breath-hold duration with fewer phase encoding steps during each RR interval (helpful if the heart rate is rapid); (3) maintained breath-hold duration and acquisition dura-

tion with enhanced in-plane spatial resolution (at the expense of SNR); (4) combinations of the above.

Similar breath-hold 2-D segmented k-space acquisitions may also be used to image the larger diameter coronary artery bypass grafts (Figure 16–9) (see Chapter 20). These bypass grafts are less mobile during the cardiac cycle (as compared with the native coronary arteries) with predominant flow during ventricular systole. This finding facilitates data acquisition during a longer period (150–200 ms) within each RR interval and with less rigorous requirements for respiratory motion suppression (wider navigator gating window or respiratory bellows gating). In addition, because saphenous vein bypass flow is predominantly systolic, and TOF methods are dependent on inflow of unsaturated spins, imaging during late ventricular systole is preferred. Susceptibility artifacts from bypass graft markers and vascular clips can hinder CMR acquisitions (Figure 16–10).

3-D Segmented k-Space Gradient-Echo Coronary MRA

The superior SNR and postprocessing capabilities of 3-D coronary MRA make it particularly attractive, but the prolonged data acquisition period (far exceeding the duration that most patients can sustain a breath-hold) and reduced contrast due to attenuation of the blood inflow effect have hampered the development of 3-D coronary MRA. These hurdles were removed with the development of free-breathing/navigator methods and the application of magnetization transfer[23, 44] and T2 preparatory pre-

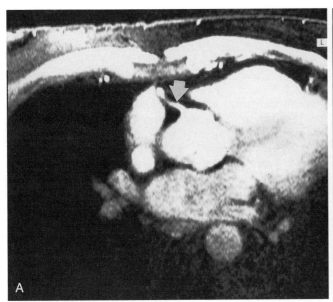
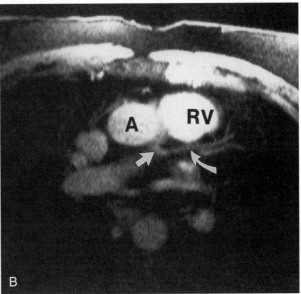

Figure 16–7. *A,* Breath-hold transverse coronary MRA in a healthy subject at the level of the take-off of the proximal right coronary artery (white arrow). *B,* Breath-hold transverse coronary MRA of the left main (straight arrow) and left anterior descending (LAD; curved arrow) coronary arteries in a 47-year-old patient with angiographically normal vessels. Signal from a cardiac vein is seen parallel with the LAD. A = aorta; RV = right ventricle. (*A,* from Manning WJ, Li W, Boyle N, Edelman RR: Fat-suppressed breath-hold magnetic resonance coronary angiography. Circulation 87:94, 1993. *B,* from Manning WJ, Li W, Edelman RR: A preliminary report comparing magnetic resonance coronary angiography with conventional contrast angiography. N Engl J Med 328:828, 1993, with permission. Copyright © 1993 Massachusetts Medical Society. All rights reserved.)

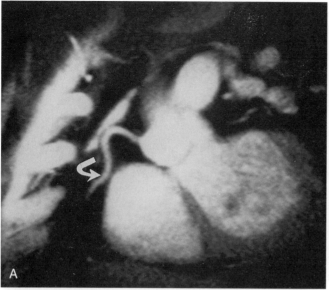

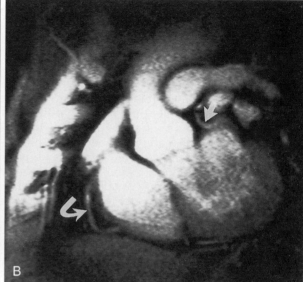

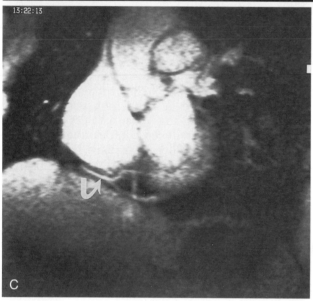

Figure 16–8. A–C, Series of breath-hold single-oblique coronary MRA images taken along the major axis at the level of the proximal right coronary artery as defined in Figure 16–7A. Each image was acquired during a single breath-hold. The proximal (A), mid (B), and distal (C) right coronary artery (curved arrows) is seen. The left circumflex (B; straight arrow) is often seen in these acquisitions. (From Manning WJ, Li W, Boyle N, Edelman RR: Fat-suppressed breath-hold magnetic resonance coronary angiography. Circulation 87:94, 1993.)

pulses,[11, 43] resulting in the widespread acceptance of 3-D coronary MRA as the current standard at most CMR centers.

Because data from a volume of tissue surrounding the coronary arteries are acquired, the set-up of free-breathing navigator 3-D coronary MRA requires less operator intervention than repetitive breath-hold methods.[49] We have found it to be imperative that the timing and respiratory suppression method for gating of all scout images be coherent with the coronary imaging sequences (Figure 16–11).[49] For our first scout we use an ECG-triggered, free-breathing, multislice 2-D segmented gradient echo thoracic acquisition with nine transverse, nine coronal, and nine sagittal interleaved acquisitions. From this data set, the navigator position at the dome of the right hemidiaphragm as well as location of the base of the heart can be readily identified (see Figure 16–2). A second ECG-triggered 3-D fast gradient-echo–echo-planar imaging (EPI) scout is then acquired

with navigator gating about a volume that includes the coronary arteries. From this second scout, the location of the ostial take-off of the left main and right coronaries can be delineated. Subsequently, a 3-D volume is prescribed in the transverse plane centered about the left main coronary artery using the same ECG delay and navigator parameters as the prior (second) scout. Typically, a 30-mm slab with 20 overlapping slices is acquired using a segmented k-space gradient echo acquisition (repetition time 7 ms, 8 to 12 phase encoding lines/RR interval) with submillimeter in-plane spatial resolution and a temporal resolution of less than 80 ms/heartbeat (Figure 16–12).[11, 49] For imaging of the right coronary artery, transverse images from the second scout depicting the proximal, mid, and distal right coronary artery are identified. Using a three-point planscan software tool (Figure 16–13),[49] the imaging plane passing through all three coordinates of the right coronary artery is thereby identified, and the 3-D coronary

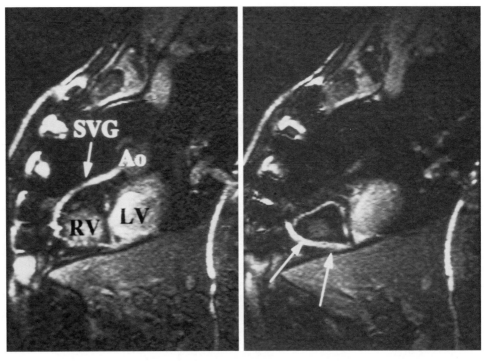

Figure 16–9. Breath-hold coronary MRA of a non-diseased/patent saphenous vein bypass graft (SVG). Two adjacent images show the graft (arrows) extending from its aortic origin to the distal touchdown on the posterior descending coronary artery. RV = right ventricle; LV = left ventricle; Ao = ascending aorta.

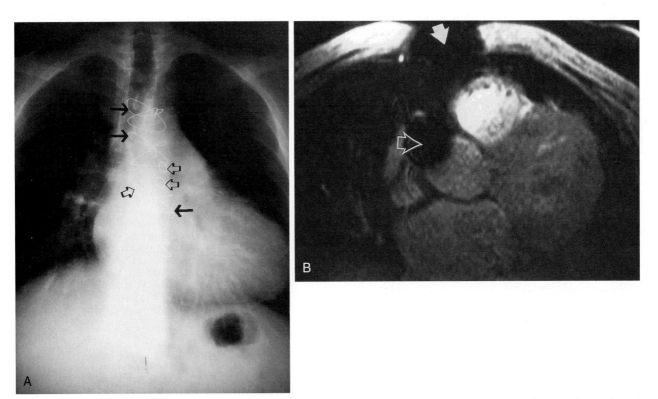

Figure 16–10. *A*, Posterior-anterior (PA) chest x-ray in a patient with prior coronary artery bypass graft. Note the sternal wires (straight black arrows) as well as the coronary artery bypass graft markers (open black arrows). *B*, Transverse coronary MRA in the same patient. Note the large artifacts (signal voids) related to the sternal wires (solid white arrow) and bypass graft markers (open white arrow). The size of the artifacts is related to the type of graft marker and coronary MRA sequence. Barium and tantalum markers result in the smallest artifacts.

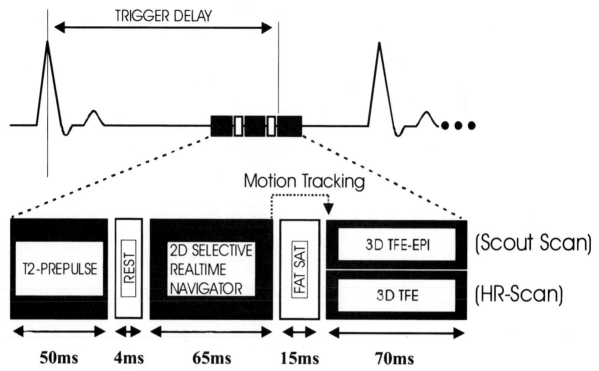

Figure 16–11. Schematic of the CMR pulse sequence for scout scanning (Scout Scan) and subsequent high-resolution coronary MRA (HR-Scan). The elements of the sequence (T2-prepulse, anterior chest saturation prepulse [REST], fat saturation prepulse [FAT SAT], and the 3-D imaging sequence) are shown in temporal relationship to the electrocardiogram (ECG) and trigger delay. TFE = turbo field echo; EPI = echo-planar imaging. (From Stuber M, Botnar RM, Danias PG, et al: Double-oblique free-breathing high resolution 3D coronary MRA. J Am Coll Cardiol 34:524, 1999. Reprinted with permission from the American College of Cardiology.)

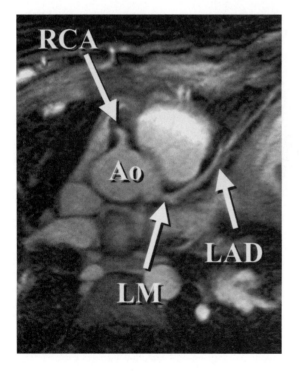

Figure 16–12. Left coronary system and proximal right coronary artery (RCA) acquired in a healthy adult subject during free-breathing using real-time navigator technology. The double-oblique acquisition displays the ascending aorta (Ao), the left main (LM), and the left anterior descending (LAD). (From Stuber M, Botnar RM, Danias PG, et al: Double-oblique free-breathing high resolution 3D coronary MRA. J Am Coll Cardiol 34:524, 1999. Reprinted with permission from the American College of Cardiology.)

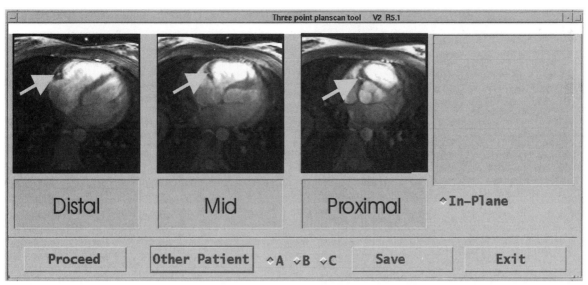

Figure 16–13. Definition of the double-oblique imaging plane for the right coronary artery (RCA). Transverse images at three different anatomical levels are acquired during a low resolution scout scan and displayed. On all three levels, the user manually identifies the RCA (arrows) and the software automatically prescribes the imaging plane passing through all three data points. (From Stuber M, Botnar RM, Danias PG, et al: Double-oblique free-breathing high resolution 3D coronary MRA. J Am Coll Cardiol 34:524, 1999. Reprinted with permission from the American College of Cardiology.)

sequence is repeated in this orientation (Figure 16–14). The left circumflex coronary artery is often seen in the transverse (left) data set or lies in a plane parallel with right coronary artery and is therefore seen on the double-oblique right coronary dataset. Each submillimeter 3-D segmented gradient-echo acquisition is typically 10 to 14 minutes in duration (assuming a navigator efficiency of 40–55%).

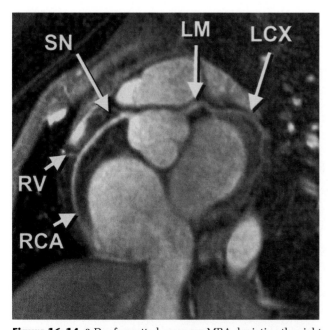

Figure 16–14. 3-D reformatted coronary MRA depicting the right coronary artery (RCA), left main (LM) and left circumflex coronary artery (LCX). The in-plane spatial resolution is 0.7 × 1.0 mm. The sinus node branch (SN) and an acute marginal (RV) branch are also seen. (From Stuber M, Botnar RM, Danias PG, et al: Double-oblique free-breathing high resolution 3D coronary MRA. J Am Coll Cardiol 34:524, 1999. Reprinted with permission from the American College of Cardiology.)

Breath-Hold and Free-Breathing 3-D Segmented Echo-Planar Coronary MRA

The beauty of EPI CMR is the time efficient coverage of k-space (see Chapter 1). However, flow- and susceptibility-related phase errors can severely degrade image quality. With the introduction of segmented EPI techniques,[50] these phase errors are minimized. For fast breath-hold[51–55] or free-breathing[56] (Figure 16–15) 3-D coronary MRA, 2 to 4 excitation pulses are followed by a short EPI readout train (e.g., 5–9 echoes), thereby taking advantage of EPI speed while keeping the echo and acquisition time short to minimize artifacts related to blood flow and motion.

For patients with an irregular breathing pattern, or if navigator gating is not robust, a long (20–30 second) breath-hold 3-D segmented-EPI approach has been implemented by Wielopolski and colleagues for use with and without an exogenous MR contrast agent.[54] This approach (volume coronary angiography with targeted scans—VCATS) has lower spatial resolution than that described above for free-breathing segmented gradient-echo approaches and acquires data during a longer period (100 ms) from each RR interval, but does allow for acquisition of images of the major coronary vessels in fewer than 13 breath-holds. A breath-hold 3-D scout data set is loaded into a multiplanar platform to plan subsequent double-oblique VCATS images of individual coronary arteries (Figure 16–16). As with the free-breathing navigator approach, it is important to acquire all the images (scout and 3-D coronary) using the same ECG trigger delay and respiratory suppression methodology. The addition of navigator tracking to prolonged breath-holding

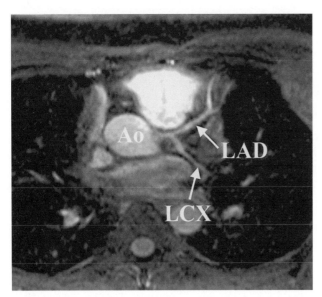

Figure 16–15. Reformatted coronary MRA of the left anterior descending (LAD) and left circumflex (LCX) coronary artery in a healthy subject using a free-breathing segmented echo-planar approach. Ao = ascending aorta. (From Botnar, RM, Stuber M, Danias PG, et al: A fast 3D approach for coronary MRA. J Magn Reson Imaging 10:821, 1999.)

approaches[55] may be beneficial. Free-breathing segmented-EPI approaches have also been described.[56]

CORONARY MAGNETIC RESONANCE ANGIOGRAPHY— ADVANCED METHODS

Despite the many advances over the past decade, SNR and the speed of data acquisition remain some-

what limiting for coronary MRA. To overcome these hurdles, several groups are working on novel approaches, including spiral acquisition, MR contrast agents, and tagging of blood in the aortic root, black-blood coronary MRA, and rapid acquisition schemes such as SMASH and SENSE.

Spiral Coronary Magnetic Resonance Angiography

Meyer[57] first reported on the use of spiral coronary MRA. Advantages of spiral acquisitions include a more efficient filling of k-space, enhanced SNR, and favorable flow properties (see Chapter 1). Spirals have more complex reconstruction algorithms which has led to limited implementations. Similar to echo planar, a single-shot k-space trajectory can be employed, but interleaved spiral imaging appears favorable and is well suited for coronary artery imaging.[57–59] Such an approach may be implemented in a breath-hold (2-D) or with free-breathing/navigator gating.[59, 60] Data suggest that as compared with conventional cartesian approaches, single spiral acquisitions (per RR interval) afford an approximately threefold improvement in SNR[60, 60a] (Figure 16–17). Acquiring two spirals for each RR interval will halve the acquisition time while maintaining superior SNR (vs. cartesian acquisition). Due to their complexity, large clinical studies have not been implemented, but because the increase in SNR exceeds that reported with other approaches, spiral coronary MRA appears particularly promising.

Balanced FFE (TrueFISP) Coronary MRA

As discussed, bright-blood, TOF coronary MRA methods are heavily dependent on the inflow of

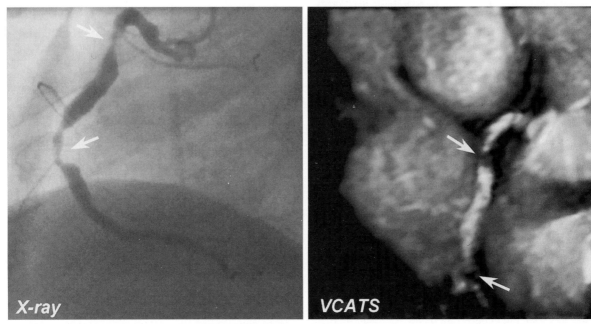

Figure 16–16. Patient with right coronary artery (RCA) disease. *B,* The double oblique coronary MRA using VCATS approach depicting the proximal and mid RCA. Two areas with decreased signal intensities (arrows), corresponding to two focal stenoses as displayed on the CMR and the x-ray angiogram (arrows)*(A).* (Courtesy of Piotr Wielopolski, PhD.)

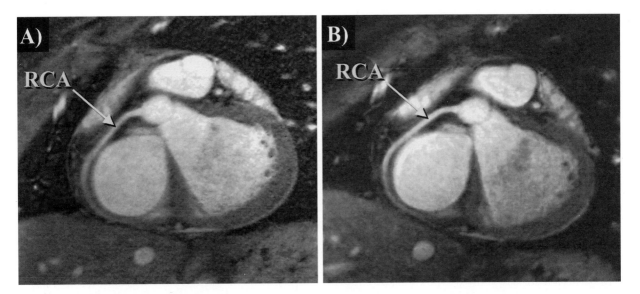

Figure 16–17. Double-oblique coronary MRA of the right coronary artery (RCA) acquired with a 3-D fat and muscle suppressed (T2 preparation) cartesian *(A)* and spiral *(B)* k-space sampling technique. Due to more efficient k-space sampling, the spiral technique results in a threefold signal-to-noise improvement compared with the conventional cartesian segmented k-space technique.[60]

unsaturated protons/blood into the imaging plane to provide contrast within the coronary blood pool. If, however, there is slow flow, saturation effects will cause loss of signal. Advances in CMR hardware and software now allow for the acquisition of high-quality functional images of the heart using steady state free precession (SSFP) or FISP,[61–63] an approach that is *not* dependent on inflow of unsaturated spins for contrast. Using this sequence, enhanced contrast is found between the ventricular blood-pool and the myocardium, together with high SNR. The technique has been introduced by various CMR vendors as "TrueFISP," "FIESTA," or "Balanced FFE." Very preliminary studies suggest a potential role for Balanced FFE in coronary MRA applications.[64]

Contrast-Enhanced Coronary Magnetic Resonance Angiography

Bright blood, TOF coronary MRA methods are heavily dependent on the inflow of unsaturated protons/blood into the imaging plane. If, however, there is slow flow, saturation effects will cause loss of signal. Furthermore, vessel wall, plaque and thrombus can have signal intensities similar to that of coronary blood. With contrast-enhanced MRA, blood signal enhancement is primarily based on the intravascular T1 relaxation rate and therefore might allow for true lumen imaging. MR contrast agents have already been found to be clinically useful for abdominal, aortic, renal, and peripheral MRA (see Chapter 29), but the previously described unique

constraints for coronary MRA have limited the application of these agents.

Exogenous MR contrast agents can be subcategorized into extracellular and intravascular agents. Extracellular paramagnetic contrast agents (gadolinium chelates) have been used for first-pass studies through the coronary bed.[54, 65, 66] The effective T1 relaxation depends on the relaxivity of gadolinium and its local concentration (see Chapter 29), with prominent, but transient shortening of blood T1 relaxation (T1 = 1200 ms at 1.5 T) to less than 100 ms during first passage of the bolus through the vascular bed. Rapid vascular equilibration and extravasation into the extravascular space (and lowering of myocardial T1) subsequently ensues. To identify the timing of the peak gadolinium concentration (minimal T1), a test dose is often used with imaging of the aortic root.[65] Under development are several intravascular CMR contrast agents, thereby affording longer scan times with free-breathing navigator gating. Using a nonselective 180° inversion pulse with imaging when the longitudinal magnetization of myocardium crosses the null point, we have examined the intravascular agent MS-325 (Epix Medical, Inc) and compared the results with our non–contrast-enhanced T2 prepulse coronary MRA sequence.[67] Free-breathing, 3-D coronary MRA with MS-325 demonstrated a 60 percent improvement in CNR with similar SNR (Figure 16–18). Taylor[68] performed 2-D segmented k-space gradient-echo imaging using three (2, 3, and 4 mg/kg) doses of NC100150 (Nycomed Amersham Imaging, Oslo, Norway), a novel, very small particle superparamagnetic iron oxide contrast agent. The agent effectively lowered blood T1 to less than 100 ms for more than

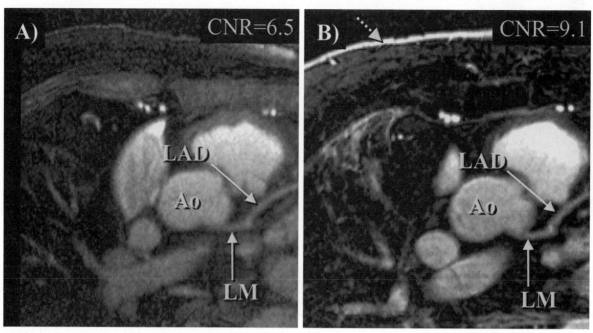

Non Contrast Agent
(T2Prep)

Contrast Agent
(MS-325)

Figure 16–18. Transverse coronary MRA acquired without *(A)* and with *(B)* an intravascular CMR contrast agent (MS-325, EPIX Medical Inc., Cambridge, MA) demonstrating improved vessel delineation after *(B)* injection of the blood pool agent. The contrast between coronary blood and myocardium after administration of the blood pool agent is approximately 50% higher than in a similar scan using T2 contrast enhancement *(A)*. A side effect of the intravascular MR contrast agent is skin line enhancement (dotted arrow). LAD = left anterior descending; LM = left main; Ao = ascending aorta; CNR = contrast-to-noise ratio. (From Stuber M, Botnar RM, Danias PG, et al: Contrast agent-enhanced, free-breathing, three-dimensional coronary magnetic resonance angiography. J Magn Reson Imaging 10:790, 1999.)

2 hours, and provided superior SNR and CNR as compared with noncontrast methods. Both these (and other intravascular) agents remain under investigation at this time.

Coronary Magnetic Resonance Angiography Tagging Methods

Though segmented k-space gradient echo approaches have received the most attention, tagging methods have some useful attributes. Wang[69] was among the first to report on selective tagging of blood in the aortic root at end systole using a 2-D inversion pulse. After a wash-in time of 300–600 ms, the tagged blood entered the proximal coronary vessels and was imaged with a 2- to 3-cm thick slab in either the transverse or oblique projection. The entire acquisition occurred during a 24-s breath-hold. This area has recently been reexamined using a 3-D interleaved segmented spiral approach with free-breathing, yielding impressive results (Figure 16–19).[70] Though dependent on blood flow velocity, such an approach may be an alternative for assessing proximal coronary arteries and the patency of intracoronary stents.

Dual Inversion Fast Spin-Echo Coronary Magnetic Resonance Angiography

Early black-blood, spin-echo coronary MRA met with limited success,[9, 45] and subsequent investigators focused on segmented gradient-echo bright-blood methods. These bright blood approaches portray the coronary lumen as bright (high signal intensity) and the surrounding tissue (including fat and myocardium) as dark or with reduced signal intensity with great dependence on inflow of unsaturated protons for adequate contrast. Though successful, bright-blood TOF coronary MRA suffers from difficulties in the accurate delineation of luminal stenosis because focal turbulent flow may result in artifactual darkening.[71] Vessel lumen diameter may therefore be underestimated and appear biased with respect to conventional x-ray angiography. In addition, thrombus, vessel wall, and various plaque components may appear with high signal intensity on bright-blood coronary MRA,[72] thereby obscuring identification of the focal stenosis. Thus, a black-blood spin-echo–based coronary MRA technique that exclusively displays the coronary blood pool may offer advantages for coronary MRA.

Using an ECG-triggered, navigator free-breathing

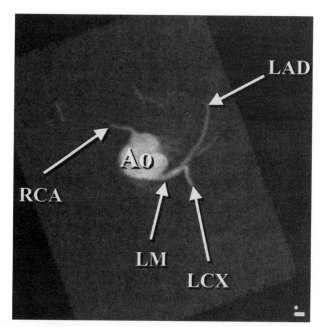

Figure 16–19. Transverse projection coronary MRA (maximum intensity projection). The image was acquired using a 2-D selective aortic spin-labeling pulse (= "spin tagging," wash-in time = 200 ms) in conjunction with real-time navigator technology for free-breathing data acquisition and an interleaved segmented 3-D spiral imaging sequence. The image exclusively displays the labeled blood pool in the ascending aorta (Ao), the left main (LM), the left coronary circumflex (LCX), the left anterior descending (LAD), and the proximal right coronary artery (RCA). Signal of the surrounding tissue including myocardium, epicardial fat, the great vessels, and the chest wall is entirely suppressed.

maps of multicoil arrays, thereby allowing for image reconstruction following acquisition of a fraction of k-space. The remaining k-lines are then reconstructed by using the sensitivity information of the individual coil elements. The cost is expected loss of SNR, with a large benefit of imaging speed (similar SNR considerations hold true for faster and stronger gradient systems). These two novel parallel imaging techniques can be combined with any imaging sequence, and promising results have already been reported with 3-D coronary MRA.[77]

SPECIAL CONSIDERATIONS: INTRACORONARY STENTS

The recognition of superior long-term patency following percutaneous coronary stent implantation has resulted in the widespread use of intracoronary stents for elective and emergent needs. Typically made from high-grade stainless steel, tantalum, or alloy, these stents pose a particular problem for CMR. Although the attractive force and local heating are negligible at 1.5 T,[78–82] the local susceptibility artifact that leads to signal voids/artifacts in close proximity to the stent can be substantial (Figure 16–22). The signal void is sequence dependent, being relatively larger with gradient-echo methods. This signal void precludes direct evaluation of intrastent and peri-stent coronary integrity although

dual inversion fast 2-D spin-echo MRA, investigators have successfully acquired submillimeter black-blood coronary MRA (Figure 16–20).[73] The timing of the inversion pulse is specifically chosen so as to null signal from the coronary blood pool. The superior CNR between the coronary blood pool and surrounding tissue has allowed for 3-D implementation with in-plane spatial resolution of below 500 μm.[74] Black-blood methods appear to be particularly advantageous for patients with metallic implants such as vascular clips. These metallic objects are a source of local artifacts due to local magnetic field distortion. The size of the artifacts is accentuated with gradient-echo/bright-blood coronary MRA while minimized on black-blood approaches (Figure 16–21).

Parallel Imaging Techniques

The advent of parallel imaging techniques such as SMASH[75] and SENSE[76] offer a novel method for markedly abbreviating the time required for image acquisition (see Chapter 2). Both SMASH and SENSE require minimal hardware modifications. Traditionally, imaging speed was dependent on gradient performance. In contrast, these parallel imaging methods take advantage of the sensitivity

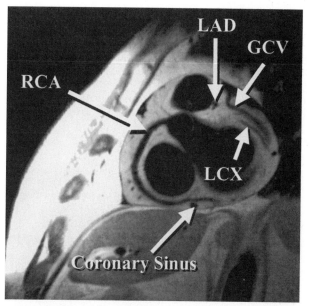

Figure 16–20. Free-breathing black-blood coronary MRA acquired with a free-breathing navigator gated and corrected 2-D fast spin echo imaging sequence in conjunction with a dual-inversion prepulse for signal nulling of the coronary and ventricular blood pool. The right coronary artery (RCA), the left circumflex (LCX), the great cardiac vein (GCV), and perpendicular views of the left anterior descending coronary artery (LAD) and the coronary sinus are displayed in a double-oblique view. (From Stuber M, Botnar RM, Kissinger KV, Manning WJ: Free breathing black-blood coronary magnetic resonance angiography. Radiology 219:278, 2001, with permission.)

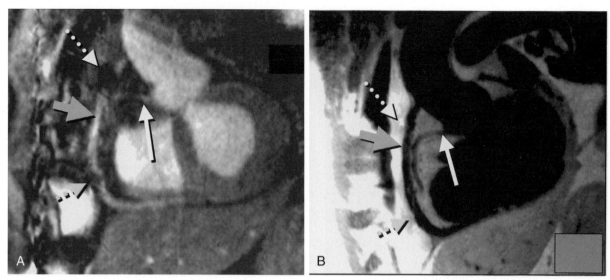

Figure 16–21. 65-year-old male subject with a 100% proximal right coronary artery (RCA) occlusion (solid arrow) and RCA bypass graft (bold gray arrow). The conventional bright-blood coronary MRA *(A)* is compared with the black-blood coronary MRA *(B).* Both images were acquired in the same double-oblique view in parallel to the RCA bypass graft. Local artifacts induced by a vascular clip (dotted arrow) and a sternal wire (dashed arrow) obscure the bright-blood coronary MRA. These artifacts are minimized on the black-blood coronary MRA in which a long continuous segment of the RCA bypass graft is visualized. (From Stuber M, Botnar RM, Kissinger KV, Manning WJ: Free breathing black-blood coronary magnetic resonance angiography. Radiology 219:278, 2001, with permission.)

assessment of blood flow/direction proximal and distal to the stent may provide indirect evidence of patency.

FUTURE DEVELOPMENTS

Current coronary MRA research is focusing on the advanced methods described in this chapter. The goal is to provide a novel noninvasive test that allows for screening for major proximal and mid coronary artery disease. Though once thought near impossible, several groups have now successfully imaged the coronary vessel wall and plaque.[3–5] These novel approaches will continue to bring in-

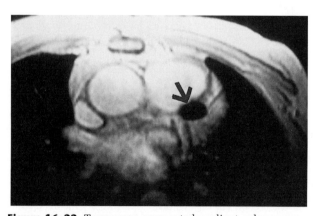

Figure 16–22. Transverse, segmented gradient-echo coronary MRA at the level of the left anterior descending coronary artery. No fat saturation was used. Note the large signal void (black arrow) corresponding to the site of a proximal left anterior descending coronary artery stent. (Courtesy of Andre Duerinkx, MD, PhD.)

tense interest and promise to the coronary MRA arena for many years to come.

References

1. Heart and Stroke Facts: 1999 Statistical Supplement. Dallas, American Heart Association, 1999.
2. Budoff MJ, Georgiou D, Brody A, et al: Ultrafast computed tomography as a diagnostic modality in the detection of coronary artery disease: A multicenter study. Circulation 93:898, 1996.
3. Meyer CH, Hu B, Macovski A, Nishimura DG: Coronary vessel wall imaging. In: ISMRM 6th Annual Meeting, Sydney, Australia. Berkeley, CA, International Society for Magnetic Resonance in Medicine, 1998, p 15.
4. Fayad ZA, Fuster V, Fallon J, et al: Noninvasive in vivo human coronary artery lumen and wall imaging using black-blood magnetic resonance imaging. Circulation 102:506, 2000.
5. Botnar RM, Stuber M, Kissinger KV, et al: Non-invasive coronary vessel wall and plaque imaging using MRI. Circulation 102:2582, 2000.
6. Hoffman MB, Wickline SA, Lorenz CH: Quantification of in-plane motion of the coronary arteries during the cardiac cycle: Implications for acquisition window duration for MR flow quantification. J Magn Reson Imaging 8:568, 1998.
7. Kim WY, Botnar RM, Stuber M, et al: Patient-specific diastolic acquisitions are required for free breathing right coronary MR vessel wall imaging (abstract). J Cardiovasc Magn Reson (in press), 2001.
8. Sodickson DK, Chuang ML, Khasgiwala VC, Manning WJ: In-plane motion of the left and right coronary arteries during the cardiac cycle. In: ISMRM, 5th Annual Meeting, Vancouver. Berkeley, CA, International Society for Magnetic Resonance in Medicine, 1997, p 910.
9. Paulin S: Coronary angiography: A technical, anatomical and clinical study. Acta Radiol Suppl (Stockholm) S233:1, 1964.
10. Wang Y, Vidan E, Bergman GW: Cardiac motion of coronary arteries: Variability in the rest period and implications for coronary MR angiography. Radiology 213:751, 1999.
11. Botnar, RM, Stuber M, Danias PG, et al: Improved coronary

artery definition with T2-weighted free-breathing 3D-coronary MRA. Circulation 99:3139, 1999.

12. Taylor AM, Jhooti P, Wiesmann F, et al: MR navigator-echo monitoring of temporal changes in diaphragm position: Implications for MR coronary angiography. J Magn Reson Imaging 7:629, 1997.

13. Edelman RR, Manning WJ, Burstein D, Paulin S: Coronary arteries: Breath-hold MR angiography. Radiology 181:641, 1991.

14. Manning WJ, Li W, Boyle N, Edelman RR: Fat-suppressed breath-hold magnetic resonance coronary angiography. Circulation 87:94, 1993.

15. Manning WJ, Li W, Edelman RR: A preliminary report comparing magnetic resonance coronary angiography with conventional contrast angiography. N Engl J Med 328:828, 1993.

16. Holland AE, Goldfarb JW, Edelman RR: Diaphragmatic and cardiac motion during suspended breathing: Preliminary experience and implications for breath-hold MR imaging. Radiology 209:483, 1998.

17. Danias PG, Stuber M, Botnar RM, et al: Navigator assessment of breath-hold duration: Impact of supplemental oxygen and hyperventilation. AJR Am J Roentgenol 171:395, 1998.

18. McConnell MV, Khasgiwala VC, Savord BJ, et al: Comparison of respiratory suppression methods and navigator locations for MR coronary angiography. AJR Am J Roentgenol 168:1369, 1997.

19. Doyle M, Scheidegger MB, De Graff RG, et al: Coronary artery imaging in multiple 1-sec breath holds. Magn Reson Imaging 11:3, 1993.

20. Liu YL, Riederer SJ, Rossman PJ, et al: A monitoring, feedback, and triggering system for reproducible breath-hold MR imaging. Magn Reson Med 30:507, 1993.

21. Wang Y, Christy PS, Koresec FR, et al: Coronary MRI with a respiratory feedback monitor: the 2D imaging case. Magn Reson Med 33:116, 1995.

22. Wang Y, Grimm RC, Rossman PJ, et al: 3D coronary MR angiography in multiple breath-holds using a respiratory feedback monitor. Magn Reson Med 34:11, 1995.

23. Li D, Paschal CB, Haacke EM, Adler LP: Coronary arteries: Three-dimensional MR imaging with fat saturation and magnetization transfer contrast. Radiology 187:401, 1993.

24. Oshinski JN, Hofland L, Mukundan S Jr, et al: Two-dimensional coronary MR angiography without breath holding. Radiology 201:737, 1996.

25. Ehman RL, Felmlee JP: Adaptive technique for high-definition MR imaging of moving structures. Radiology 173:255, 1989.

26. Sachs TS, Meyer CH, Hu BS, et al: Real-time motion detection in spiral MRI using navigators. Magn Reson Med. 32:639, 1994.

27. Stuber M, Botnar RM, Danias PG, et al: Submillimeter three-dimensional coronary MR angiography with real-time navigator correction: Comparison of navigator locations. Radiology. 212:579, 1999.

28. Wang Y, Riederer SJ, Ehman RL: Respiratory motion of the heart: Kinematics and the implications for the spatial resolution of coronary MR imaging. Magn Reson Med 34:412, 1995.

29. Li D, Kaushikkar S, Haacke EM, et al: Coronary arteries: three-dimensional MR imaging with retrospective respiratory gating. Radiology 201:857, 1996.

30. Post JC, van Rossum AC, Hofman MB, et al: Three-dimensional respiratory-gated MR angiography of coronary arteries: Comparison with conventional coronary angiography. AJR Am J Roentgenol 166:1399, 1996.

31. Hofman MBM, Paschal CB, Li D, et al: MRI of coronary arteries: 2D breath-hold vs. 3D respiratory-gated acquisition. J Comput Assist Tomogr 19:56, 1995.

32. Danias PG, McConnell MV, Khasgiwala VC, et al: Prospective navigator correction of slice position for coronary magnetic resonance angiography. Radiology 203:733, 1997.

33. Stuber M, Danias PG, Botnar RM, et al: Superiority of prone position in free-breathing coronary MRA in patients with coronary artery disease. J Magn Reson Imaging 13:185, 2001.

34. Spuentrup E, Botnar RM, Kissinger KV, et al: Impact of navigator timing parameters and navigator spatial resolution on

free breathing 3D coronary magnetic resonance imaging. J Magn Reson Imaging (in press), 2001.

35. Sachs TS, Meyer CH, Irarrazabal P, et al: The diminishing variance algorithm for real-time reduction of motion artifacts in MRI. Magn Reson Med 34:412, 1995.

36. Jhooti P, Wiesmann F, Taylor AM, et al: Hybrid ordered phase encoding (HOPE): An improved approach for respiratory artifact reduction. J Magn Reson Imaging 8:968, 1998.

37. Huber ME, Hengesbach D, Botnar RM, et al: Motion artifact reduction and vessel enhancement for free-breathing navigator-gated coronary MRA using three dimensional k-space reordering. Magn Reson Med 45:645, 2001.

38. Hardy CJ, Saranthan M, Zhu Y, Darrow RD: Coronary angiography by real-time MRI with adaptive averaging. Magn Reson Med 44:940, 2000.

39. Anzalone N, Triulzi F, Scotti G: Acute subarachnoid haemorrhage: 3D time-of-flight MR angiography versus intra-arterial digital angiography. Neuroradiology 37:257, 1995.

40. Krinsky G., Weinreb J: Gadolinium-enhanced three-dimensional MR angiography of the thoracoabdominal aorta. Semin Ultrasound CT MR 17:280, 1996.

41. Botnar RM, Stuber M, Kissinger KV, Manning WJ: Free-breathing 3D coronary MRA: The impact of "isotropic" image resolution. J Magn Reson Imaging 11:389, 2000.

42. Meyer CH, Hu BS, Nishimura DG, Macovski A: Fast spiral coronary artery imaging. Magn Reson Med 28:202, 1992.

43. Brittain JH, Hu BS, Wright GA, et al: Coronary angiography with magnetization-prepared T2 contrast. Magn Reson Med 33:689, 1995.

44. Balaban RS, Ceckler TL: Magnetization transfer contrast in magnetic resonance imaging. Magn Reson Q 8:116, 1992.

45. Lieberman LM, Botti RI, Nelson AD: Magnetic resonance of the heart. Radiol Clin North Am 22:847, 1984.

46. Paulin S, von Schulthess GK, Fossel E, et al: MR imaging of the aortic root and proximal coronary arteries. AJR Am J Roentgenol 148:665, 1987.

47. Burstein D: MR imaging of coronary artery flow in isolated and in vivo hearts. J Magn Reson Imaging 1:337, 1991.

48. McConnell MV, Khasgiwala VC, Savord BJ, et al: Prospective adaptive navigator correction for breathhold MR coronary angiography. Magn Reson Med 37:148, 1997.

49. Stuber M, Botnar RM, Danias P, et al: Double-oblique free-breathing high resolution 3D coronary MRA. J Am Coll Cardiol 34:524, 1999.

50. McKinnon GC: Ultrafast interleaved gradient-echo-planar imaging on a standard scanner. Magn Reson Med. 30:609, 1993.

51. Wielopolski PA, Manning WJ, Edelman RR: Single breath-hold volumetric imaging of the heart using magnetization-prepared 3-dimensional segmented echo planar imaging. J Magn Reson Imaging 4:403, 1995.

52. Slavin GS, Riederer SJ, Ehman RL: Two-dimensional multishot echo-planar coronary MR angiography. Magn Reson Med. 40:883, 1998.

53. Börnert P, Jensen D: Coronary artery imaging at 0.5 T using segmented 3D echo planar imaging. Magn Reson Med 34:779, 1995.

54. Wielopolski PA, van Geuns RJ, de Feyter PJ, Oudkerk M: Breath-hold coronary MR angiography with volume-targeted imaging. Radiology. 209:209, 1998.

55. Stuber M, Botnar RM, Danias PG, et al: Breath-hold three dimensional coronary MRA using real-time navigator technology. J Cardiovasc Magn Reson 1:233, 1999.

56. Botnar, RM, Stuber M, Danias PG, et al: A fast 3D approach for coronary MRA. J Magn Reson Imaging 10:821, 1999.

57. Meyer CH, Hu BS, Nishimura DG, Macovski A: Fast spiral coronary artery imaging. Magn Reson Med 28:202, 1992.

58. Thedens DR, Irarrazaval P, Sachs TS, et al: Fast magnetic resonance coronary angiography with a three-dimensional stack of spirals trajectory. Magn Reson Med 41:1170, 1999.

59. Börnert P, Aldefeld B, Nehrke K. Improved 3D spiral imaging for coronary MR angiography. Magn Reson Med 45:172, 2001.

60. Börnert P, Stuber M, Botnar R, et al: 3D spiral vs. cartesian coronary magnetic resonance angiography. Magn Reson Med (in press), 2001.

60a. Taylor AM, Keegan J, Jhooti P, et al: A comparison between segmented k-space flash and interleaved spiral MR coronary angiography sequences. J Magn Reson Imaging 11:394, 2000.

61. Oppelt A, Graumann R, Barfuss H, et al: FISP—a new fast MRI sequence. Electromedica 54:15, 1986.

62. Heid O: True FISP cardiac fluoroscopy (abstract). *In:* ISMRM, 4th Annual Meeting. Berkeley, CA, International Society for Magnetic Resonance in Medicine, 1:320, 1997.

63. Deimling M, Heid O: True FISP imaging with inherent fat cancellation (abstract). *In:* ISMRM, 7th Annual Meeting. Berkeley, CA, International Society for Magnetic Resonance in Medicine, 3:1500, 2000.

64. Stuber M, Boernert P, Botnar RM, et al: Free-breathing balanced FFE coronary magnetic resonance angiography (abstract). *In:* ISMRM, 8th Annual Meeting. Berkeley, CA, International Society for Magnetic Resonance in Medicine, 2001, p. 515.

65. Zheng J, Li D, Bae KT, et al: Three-dimensional gadolinium-enhanced coronary magnetic resonance angiography: Initial experience. J Cardiovasc Magn Reson 1:33, 1999.

66. Goldfarb JW, Edelman RR: Coronary arteries: Breath-hold gadolinium-enhanced, three-dimensional MR angiography. Radiology 206:830, 1998.

67. Stuber M, Botnar RM, Danias PG, et al: Contrast agent-enhanced, free-breathing, three-dimensional coronary magnetic resonance angiography. J Magn Reson Imaging 10:790, 1999.

68. Taylor AM, Panting JR, Keegan J, et al: Safety and preliminary findings with the intravascular contrast agent NC100150 injection for MR coronary angiography. J Magn Reson Imaging 9:220, 1999.

69. Wang WJ, Hu BS, Macovski A, Nishimura DG: Coronary angiography using fast selective inversion recovery. Magn Reson Med 18:417, 1991.

70. Stuber M, Boernert P, Spuentrup E, et al: Three-dimensional projection coronary magnetic resonance angiography (abstract). *In:* ISMRM, 8th Annual Meeting. Berkeley, CA, International Society for Magnetic Resonance in Medicine, 2001, p. 177.

71. Evans AJ, Blinder RA, Herfkens RJ, et al: Effects of turbulence on signal intensity in gradient echo images. Invest Radiol 23:512, 1988.

72. Jara H, Yu BC, Caruthers SD, et al: Voxel sensitivity function description of flow-induced signal loss in MR imaging: Implications for black-blood MR angiography with turbo spin-echo sequences. Magn Reson Med 41:575, 1999.

73. Stuber M, Botnar RM, Kissinger KV, Manning WJ: Free breathing black-blood coronary magnetic resonance angiography. Radiology 219:278, 2001.

74. Stuber M, Botnar RM, Spuentrup E, et al: Three-dimensional high resolution fast spin-echo coronary magnetic resonance angiography. Magn Reson Med 45:206, 2001.

75. Sodickson DK, Manning WJ: Simultaneous acquisition of spatial harmonics (SMASH): Fast imaging with radiofrequency coil arrays. Magn Reson Med 38:591, 1997.

76. Pruessmann KP, Weiger M, Scheidegger MB, Boesiger P: SENSE: Sensitivity encoding for fast MRI. Magn Reson Med 42:952, 1999.

77. Sodickson D, Stuber M, Botnar R, et al: Accelerated coronary MR angiography in volunteers and patients using double-oblique 3D acquisitions combined with SMASH. J Cardiovasc Magn Reson 1:260, 1999.

78. Shellock FG, Shellock VJ: Metallic stents: Evaluation of MR imaging safety. AJR Am J Roentgenol 173:543, 1999.

79. Scott NA, Pettigrew RI: Absence of movement of coronary stents after placement in a magnetic resonance imaging field. Am J Cardiol 73:900, 1994.

80. Hug J, Nagel E, Bornstedt A, et al: Coronary arterial stents: Safety and artifacts during MR imaging. Radiology 216:781, 2000.

81. Strohm O, Kivelitz D, Gross W, et al: Safety of implantable coronary stents during ^{1}H-magnetic resonance imaging at 1.0 and 1.5T. J Cardiovasc Magn Reson 1:239, 1999.

82. Kramer CM, Rogers WJ, Palestis DL: Absence of adverse outcomes after magnetic resonance imaging early after stent placement for acute myocardial infarction. J Cardiovasc Magn Reson 2:257, 2001.

CHAPTER 17

Coronary Magnetic Resonance Angiography—Clinical Results

Peter G. Danias, Matthias Stuber, and Warren J. Manning

Since the late 1980s, tremendous progress has occurred in the field of coronary magnetic resonance angiography (MRA). In this chapter, we present clinical results using the methodologies described in the prior chapter (see Chapter 16). Currently, the largest body of reported clinical experience has been using the electrocardiographic (ECG)-triggered segmented k-space gradient-echo approach. Both two-dimensional (2-D) and three-dimensional (3-D) techniques have been applied with either breath-holding or free-breathing navigator methods for suppression of respiratory motion artifacts. These studies serve as the primary focus of this chapter.

CORONARY MAGNETIC RESONANCE ANGIOGRAPHY OF NORMAL CORONARY ARTERIES

Although conventional ECG-gated spin-echo MR was occasionally successful at imaging the native coronary arteries,[1, 2] it was the breath-hold 2-D segmented k-space gradient-echo approach described by Burstein[3] and Edelman[4] that first offered a robust coronary MRA method for imaging the native coronary arteries. As implemented across vendor platforms, the left main, left anterior descending, and right coronary arteries are visualized in nearly all compliant subjects (Table 17–1). These early reports[5–9] had reduced (67–77%) success for imaging of the left circumflex coronary artery, a finding likely related to the use of an anterior surface coil (with reduced signal from the posteriorly directed left circumflex coronary artery) in these studies. The extent of contiguous coronary artery visualization in healthy subjects included up to 89 mm of the right coronary artery, the entire left main coronary artery, up to 62 mm of the left anterior descending coronary artery, and up to 38 mm for the left circumflex coronary artery (Figure 17–1). Despite relatively limited spatial resolution, normal proximal coronary artery diameter compared favorably with values derived from x-ray angiography and pathology (Figure 17–2).[5, 14, 15] Subsequently, volumetric, 3-D segmented k-space gradient-echo studies reported successful visualization of all four major vessels in nearly all subjects (see Table 17–1).[11–14] In addition to improved visualization of the coronary arteries, another distinct advantage of the 3-D approach was improved visualization length (Figure 17–3) as compared with 2-D approaches. With 3-D coronary MRA, similar coronary length measurements are found among healthy adults and patients with angiographic coronary artery disease.[18] Preliminary ex-

Table 17–1. Successful Visualization of the Native Coronary Arteries Using 2-D and 3-D Segmented k-Space Gradient-Echo Coronary MRA

Investigator	Technique	Resp Comp	# Sub	RCA	LM	LAD	LCX
Manning et al[5]	2-D GRE	BH	25	100%	96%	100%	76%
Pennell et al[6]	2-D GRE	BH	26	95%	95%	91%	76%
Duerinckx and Urman[7]	2-D GRE	BH	20	100%	95%	86%	77%
Sakuma et al[8]	2-D GRE cine	BH	18	100%	100%	100%	67%
Masui et al[9]	2-D GRE	BH	13	85%	92%	100%	92%
Davis et al[10]	2-D GRE	BH	33*	100%	100%	100%	100%
Li et al[11]	3-D GRE	Mult averages	14	100%	100%	86%	93%
Post et al[12]	3-D GRE	Retro Nav G	20	100%	100%	100%	100%
Wielopolski et al[13]	3-D Seg EPI	BH	32	100%	100%	100%	100%
Botnar et al[14]	3-D GRE	Pro Nav G/C	13	97%	100%	100%	97%

*Including 18 heart transplant recipients.
Resp Comp = respiratory compensation; Sub = subjects; RCA = right coronary artery; LM = left main coronary artery; LAD = left anterior descending coronary artery; LCX = left circumflex coronary artery; GRE = gradient echo; Seg EPI = segmented echo–planar imaging; BH = breath-hold; Retro Nav G = retrospective navigator-gated; Pro Nav G/C = prospective navigator gating with real time motion correction.

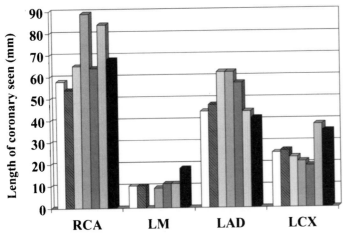

Figure 17–1. Contiguous length of coronary artery visualized using two-dimensional (2-D) breath-hold segmented k-space coronary MRA.

perience with intravascular contrast agents has demonstrated no significant improvement in coronary length visualization.[19]

ANOMALOUS CORONARY ARTERY IDENTIFICATION

The ability to visualize the origin and proximal native coronary arteries led to early investigations into the use of coronary MRA for the identification and characterization of anomalous coronary arteries (see Chapter 18). Coronary MRA has several advantages in the diagnosis of coronary anomalies. In addition to being noninvasive and not requiring ionizing radiation or iodinated contrast agents, coronary MRA provides a 3-D roadmap of the entire mediastinum (compared with the 2-D projection provided by conventional invasive x-ray angiography) in which the user can subsequently acquire and/or reconstruct an image at any orientation. The vast majority of early studies used a 2-D breath-hold

segmented k-space gradient-echo approach,[20–26a] though most centers now utilize 3-D non–breath-hold navigator coronary MRA because of superior reconstruction capabilities afforded by 3-D data sets.

Early reports of coronary MRA to visualize anomalous coronary arteries included case report confirmation of x-ray angiographic data.[24–26] Subsequently, there have been at least four published series[20–23] of patients who underwent blinded comparison of 2-D breath-hold gradient-echo coronary MRA data with x-ray angiography. These studies have uniformly reported excellent accuracy, including several instances in which coronary MRA was determined to be superior to x-ray angiography (Table 17–2).[20–23, 26a] At experienced CMR centers, clinical coronary MRA is now the preferred test for patients in whom anomalous disease is suspected, known anomalous disease needs to be further clarified, or if the patient has another cardiac anomaly associated with coronary anomalies (e.g., tetralogy of Fallot).

In a somewhat analogous fashion, 2-D breath-hold coronary MRA has also been utilized to define the altered coronary artery orientation in the cardiac transplant population.[10] Among cardiac transplant recipients, coronary MRA has documented a $+25^{\circ}$ anterior (clockwise) ostial rotation, likely explaining the more complex coronary engagement during x-ray angiography. As compared with a healthy control group, a similar contiguous length of native coronary artery is seen (see Figure 17–1).

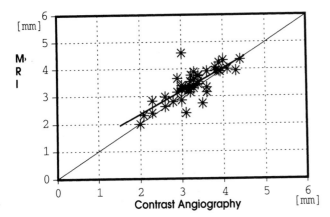

Figure 17–2. Scatterplot from x-ray angiographically normal coronary arteries comparing coronary MRA and x-ray angiographic diameters. (From Scheidegger MB, Vassalli G, Hess OM, et al: Validation of coronary artery MR angiography: Comparison of measured vessel diameters with quantitative contrast angiography (abstract). *In:* Book of Abstracts. Berkeley, CA, Society of Magnetic Resonance, 1994, p 497.)

Table 17–2. Anomalous Coronary MRA

Investigator	# Patients	Correctly Classified Anomalous Vessels
McConnell et al[20]	15	14 (93%)
Post et al[21]	19	19 (100%)*
Vliegen et al[22]	12	11 (92%)†
Taylor et al[23]	25	24 (96%)
Razmi et al[26a]	12	12 (100%)

*Including 3 originally misclassified by x-ray angiography.
†Including 5 patients unable to be classified by x-ray angiography.

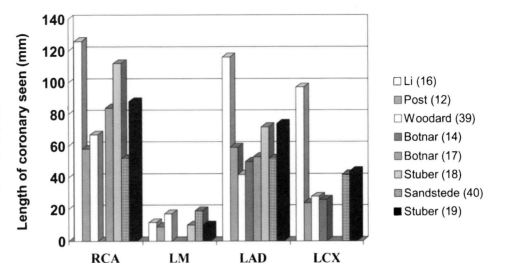

Figure 17–3. Contiguous length of coronary artery visualized using three-dimensional (3-D) breath-hold segmented k-space coronary MRA. RCA = right coronary artery; LM = left main coronary artery; LAD = left anterior descending coronary artery; LCX = left circumflex coronary artery.

CORONARY ARTERY ANEURYSMS/KAWASAKI DISEASE

Though relatively uncommon, coronary artery aneurysms are receiving increasing attention for assessment by coronary MRA. The vast majority of acquired coronary aneurysms are felt to be due to mucocutaneous lymph node syndrome (Kawasaki disease), a generalized vasculitis of unknown etiology usually occurring in children younger than 5 years of age. Infants and children with this syndrome may show evidence of myocarditis and/or pericarditis with coronary artery aneurysms developing in approximately 20 percent that cause both short- and long-term morbidity and mortality.[27] Approximately half of the children with coronary aneurysms during the acute phase of the disease have normal-appearing vessels by angiography 1 or 2 years later.[28–30] Among children, cardiac ultrasound is usually adequate for diagnosing and following aneurysms, but echocardiography is deficient after adolescence. These young adults are therefore often referred for monitoring by serial x-ray angiography. In case reports of patients with known disease, coronary MRA has been reported to adequately characterize the presence and site of coronary aneurysms in patients with Kawasaki disease (Figure 17–4).[31–33] Although data from larger series are needed before

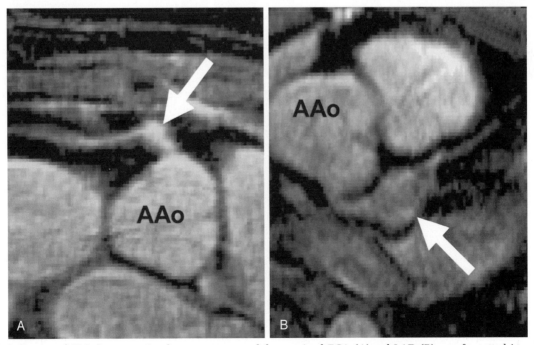

Figure 17–4. Coronary MRA demonstrating large aneurysms of the proximal RCA *(A)* and LAD *(B)* on reformatted images (arrows). Flow-related signal within the aneurysms is homogeneous, with no evidence of thrombosis. Aao = ascending aorta. (From Flacke S, Setser RM, Barger P, et al: Coronary aneurysms in Kawasaki's disease detected by magnetic resonance coronary angiography. Circulation 101:E156, 2000.)

this process can be advocated clinically, coronary MRA may be reasonable at experienced CMR centers.

CORONARY MAGNETIC RESONANCE ANGIOGRAPHY FOR IDENTIFICATION OF FOCAL CORONARY STENOSES

Although clinical coronary MRA for anomalous coronary artery disease (and coronary artery bypass graft patency) are widely accepted, data are currently not sufficient to support widespread clinical coronary MRA for identification of focal coronary stenoses. Thus, until final data from large series and multicenter trials (using common hardware and software) are available, coronary MRA for identification of focal stenosis should be considered an investigative procedure, restricted to selected CMR centers with focused expertise.

Using gradient-echo bright-blood coronary MRA methods, rapidly moving laminar blood flow appears bright, whereas areas of stagnant flow and/or focal turbulence appear dark due to local saturation (stagnant flow) or dephasing (turbulence) (Figure 17–5). Areas of focal stenoses appear as varying severity of signal voids in the coronary MRA, with the severity of the signal loss related to the angiographic stenosis (Figure 17–6).[34] However, bright-blood coronary MRA methods sometimes may be misleading. If there is slow blood flow distal to a stenosis, there may be complete loss of signal in the segment distal to the lesion, despite the absence of a total occlusion (Figure 17–7). Similarly, because these gradient-echo sequences are not sensitive to the *direction* of blood flow, a total occlusion with adequate retrograde (or antegrade collateral) blood flow to the distal segment may result in signal in the postocclusion lumen.

Due to time constraints of the breath-hold, the 2-D breath-hold coronary MRA technique has relatively

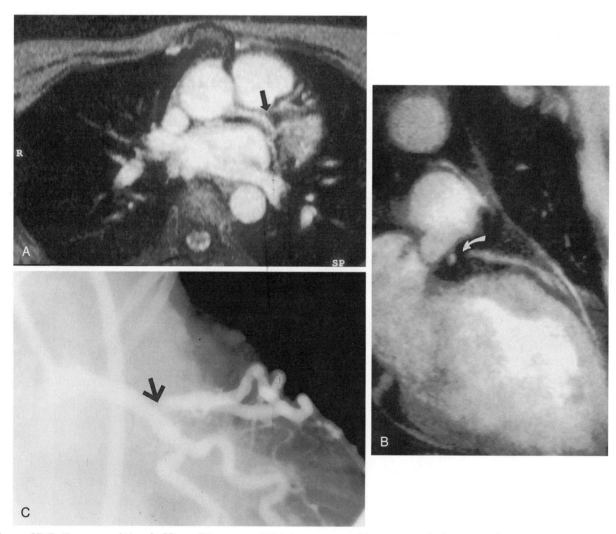

Figure 17–5. Transverse *(A)* and oblique *(B)* coronary MRA in a 45-year-old woman with chest pain demonstrating a signal void (arrows) in the proximal left anterior descending (LAD) coronary artery. Visualization of the more distal LAD and diagonal vessel are seen in the image. *C,* Corresponding right anterior oblique caudal x-ray angiogram confirms the tight proximal LAD stenosis (black arrow).

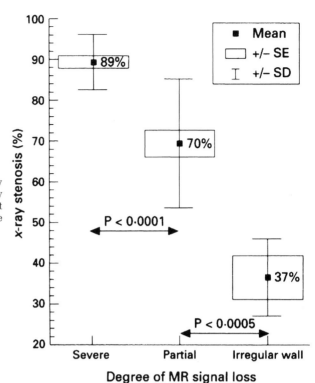

Figure 17–6. Breath-hold segmented k-space gradient-echo coronary MRA comparison of degree of focal CMR signal loss vs. x-ray coronary artery diameter stenosis. (From Pennell DJ, Bogren HG, Keegan J, et al: Assessment of coronary artery stenosis by magnetic resonance imaging. Heart 75:127, 1996.)

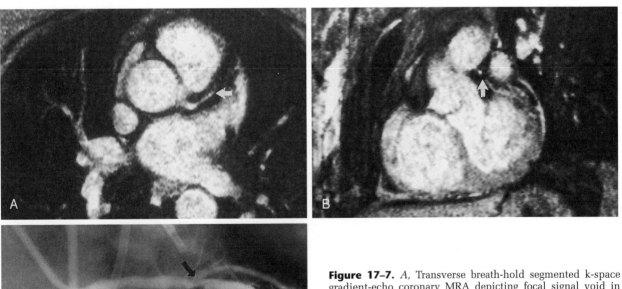

Figure 17–7. *A*, Transverse breath-hold segmented k-space gradient-echo coronary MRA depicting focal signal void in the proximal left anterior descending (LAD) coronary artery (white arrow). The more distal LAD was not visualized on inferior and superior images. *B*, Single-oblique coronary MRA demonstrating a signal void in the proximal left circumflex (LCX) coronary artery (white arrow). *C*, Corresponding x-ray angiogram demonstrating the severe proximal LCX and LAD stenoses (black arrows). Neither vessel is occluded.

Table 17–3. Coronary MRA for Identification of Focal Coronary Stenoses: 2-D Breath-Hold Studies

Investigator	#Subjects	# (%) Vessels w/Stenoses	For ≥50% Diam Stenosis	
			Sensitivity	*Specificity*
Manning[35]	39	52 (35%)	90% (71–100)	92% (78–100)
Duerinckx[7]	20	27 (34%)	63% (0–73)	(37–82)
Pennell et al[34]	39	55 (35%)	85% (75–100)	
Post et al[36]	35	35 (28%)	63% (0–100)	89% (73–96)
Nitatori et al[37]	57*		87%	94%
	13†		43%	90%

*With ≥90% diameter stenosis.
†With 50–75% diameter stenosis.
w/ = with; Diam = diameter.

limited in-plane spatial resolution, but nevertheless has successfully demonstrated proximal coronary stenoses in several clinical studies (Table 17–3). The distance from the vessel origin to the focal stenosis on coronary MRA compares favorably with x-ray angiography (Figure 17–8).[34] Although each of these investigators used a breath-hold 2-D segmented k-space gradient-echo approach, other clinical (presence of arrhythmias, prevalence of disease) and technical (MR vendor, echo time, receiver coils, timing of the acquisition, acquisition duration, breath-hold maneuvers) parameters varied greatly. No multicenter 2-D coronary MRA study using a uniform hardware and software approach has yet been reported.

With the increasing availability of CMR navigators (see Chapter 15), most CMR centers have migrated to 3-D gradient-echo coronary MRA for ease in patient acceptance (free-breathing), improved signal-

to-noise ratio, and facilitated multiplanar reconstructions afforded by 3-D coronary MRA. As with 2-D gradient-echo methods, a focal stenosis/turbulence appears as a signal void along the course of the vessel (Figure 17–9). Data from several single center sites have now been published, primarily using retrospective navigators (Table 17–4). Early studies used very prolonged acquisition times (260 ms),[12] with later investigators utilizing acquisition intervals of less than 120 ms.[38–43] These single-center reports are very encouraging, with sensitivity and specificity of up to 90 percent for proximal coronary disease.[43, 43a] Overall, stenosis sensitivity has been found to be similar for source as compared with projection images,[39] and both methods are currently used. Our preference is to make diagnoses with the source images and to use the projection images to visually convey our findings to the referring physician. A multicenter 3-D coronary MRA study of 117 patients using common hardware and software was recently completed.[44] Preliminary reports suggest a sensitivity of greater than 90 percent for identifying a patient with coronary artery disease and a very high sensitivity and specificity for identifying patients with left main and three vessel disease.[44]

For very motivated patients who can cooperate with prolonged (>30 s) breath-holds, breath-hold 3-D coronary MRA has been reported to have potential value. Regenfus and coworkers[45] studied 50 patients referred for diagnostic coronary angiography, including 72 percent with x-ray coronary artery disease. Three-dimensional segmented k-space coronary MRA was performed with an extracellular CMR contrast agent. Eighty-two percent of patients were able to sustain the 32-heartbeat breath-hold with an overall sensitivity and specificity of 86 percent and 91 percent, respectively. Van Geuns and colleagues studied 38 patients using the volume coronary angiography with targeted volumes (VCATS) approach requiring an average of 10 21-heartbeat breath-holds.[46] Studies were completed in 34 (90%) of subjects, including 23 (68%) with x-ray angiographic coronary artery disease. Sixty-nine percent of coronary MRA segments were deemed interpretable. Of these, the overall diagnostic accuracy of coronary MRA was 92 percent, sensitivity 68 percent (50–77), and specificity 97 percent (94–100).

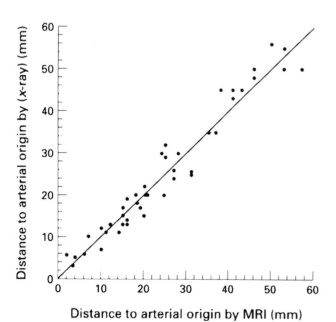

Figure 17–8. Scatterplot comparing the distance from the coronary origin to the stenosis as measured by x-ray and magnetic resonance coronary angiography. (From Pennell DJ, Bogren HG, Keegan J, et al: Assessment of coronary artery stenosis by magnetic resonance imaging. Heart 75:127, 1996.)

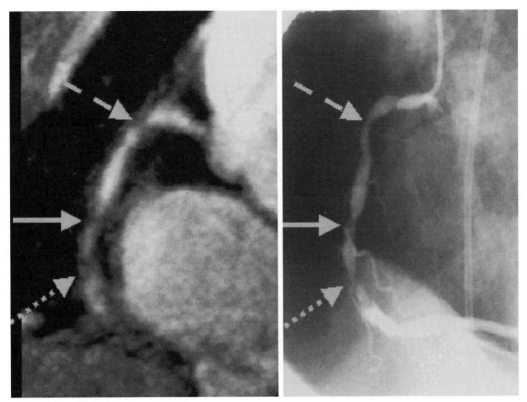

Figure 17–9. Coronary MRA *(A)* and corresponding x-ray angiography *(B)* in a patient with coronary artery disease. Coronary MRA was obtained during free-breathing and shows good agreement with the x-ray angiographic disease (arrows). (From Stuber M, Botnar RM, Danias PG, et al: Double-oblique free-breathing high resolution three-dimensional coronary magnetic resonance angiography. J Am Coll Cardiol 34:524, 1999. Reprinted with permission from the American College of Cardiology.)

CORONARY MAGNETIC RESONANCE ANGIOGRAPHY FOR CORONARY ARTERY BYPASS GRAFT ASSESSMENT

In comparison with the native coronary arteries, reverse saphenous vein and internal mammary grafts are relatively easy to image due to their relatively stationary position and larger lumen diameter. Furthermore, their predictable and less convoluted course has allowed imaging of bypass grafts even with conventional CMR techniques. With schematic knowledge of the origin and touchdown site, conventional free-breathing ECG-gated 2-D spin-echo[47–50] and gradient-echo[50–52] CMR in the transverse plane have both been used to reliably assess bypass graft *patency* (see Chapter 20). Patency is determined by visualizing a patent graft lumen in at least two contiguous transverse levels along its

Table 17–4. Coronary MRA for Identification of Focal Coronary Stenoses: 3-D Gradient-Echo with Free-Breathing and Retrospective Navigators

Investigator	# Subjects	# (%) Vessels w/Stenoses	For ≥50% Diam Stenosis	
			Sensitivity	*Specificity*
Post et al[12]	20	21 (27%)	38% (0–57)	95% (85–100)
Müller et al[38]	35		83%*	94%*
Woodard et al[39]	10	10 (100%)	70%	
Sandstede et al[40]	30	30 (100%)	81%†	89%†
Van Geuns et al[41]	32		50% (50–55)‡	91%‡ (73–95)
Huber et al[42]	40	20 (50%)	73% (25–100)	50% (25–82)
Sardanelli et al[43]	42	40% of segments	82% (57–100)	89% (72–100)
			90% proximal	90% proximal
Moustapha[43a]	25	10 (43%)	90% proximal	92% proximal

*Excluding 5 patients for "lack of cooperation" and 15 segments for being uninterpretable.
†Based on 23 (77%) with high-quality scans.
‡Based on 74% of coronary artery segments analyzable by coronary MRA.
w/ = with; Diam = diameter.

expected course (presenting as signal void for spin-echo techniques and bright signal for gradient-echo approaches [Figure 17–10]), one can conclude that there is flow present through the bypass graft, and therefore that it is patent. If a patent lumen is only seen at one level (e.g., for spin-echo techniques, a signal void is seen at only one level), a graft is considered indeterminate. If a patent graft is not seen at any level, the graft is considered occluded. Combining spin-echo and gradient-echo imaging in the same patient does not appear to improve accuracy.[50] Contrast-enhanced coronary MRA has

also been described for the assessment of graft patency.[54, 55] An overview of the clinical studies assessing bypass graft patency is presented in Table 17–5. These graft imaging techniques have proved useful in at least two trials using CMR patency rate as a surrogate endpoint for bypass surgical techniques.[56, 57]

Limitations of coronary MRA bypass graft assessment include the inability to differentiate severely diseased yet patent grafts, as well as difficulties related to local signal loss/artifact due to implanted metallic objects (hemostatic clips, ostial graft mark-

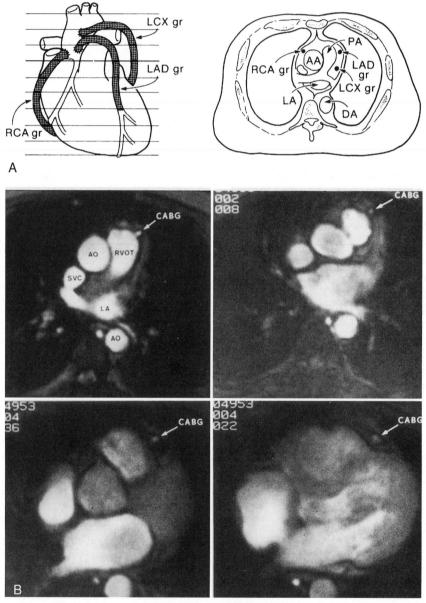

Figure 17–10. *A,* Typical location of coronary artery bypass grafts (CABG; RCA gr, LAD gr, LCX gr) originating from the aortic root (AA) and anastomosing with the distal native coronary arteries with location of contiguous transverse slices. *B,* Transverse electrocardiogram (ECG)-gated conventional gradient-echo coronary MRA demonstrating patent flow (white arrows) on multiple contiguous levels indicating the graft (CABG) is patent. PA = pulmonary artery; DA = descending aorta; LA = left atrium; RVOT = right ventricular outflow tract; SVC = superior vena cava; AO = aorta. (*A,* from Rubinstein RI, Askenase AD, Thickman D, et al: Magnetic resonance imaging to evaluate patency of aortocoronary bypass grafts. Circulation 76:786, 1987. *B,* from Aurigemma GP, Reichek N, Axel L, et al: Noninvasive determination of coronary artery bypass graft patency by cine magnetic resonance imaging. Circulation 80:1595, 1989.)

Table 17–5. Sensitivity, Specificity, and Accuracy of Coronary MRA for Assessment of Coronary Artery Bypass Graft Patency

Investigator	Technique	# Grafts	Patency	Sens	Spec	Accuracy
White et al[47]	2-D Spin-echo	72	69%	86%	59%	78%
Rubenstein et al[48]	2-D Spin-echo	47	62%	90%	72%	83%
Jenkins et al[49]	2-D Spin-echo	41	63%	89%	73%	83%
Galjee et al[50]	2-D Spin-echo	98	74%	98%	85%	89%
White et al[51]	2-D GRE	28	50%	93%	86%	89%
Aurigemma et al[52]	2-D GRE	45	73%	88%	100%	91%
Galjee et al[50]	2-D GRE	98	74%	98%	88%	96%
Engelmann et al[53]	2-D GRE	55	100% (IMA)	100%		100%
			66% (SVG)	92%	85%	89%
Wintersperger et al[54]	CE-3D GRE	76	79%	95%	81%	95%
Vrachliotis et al[55]	CE-3D GRE	45	67%	93%	97%	95%

IMA = internal mammary artery graft; SVG = saphenous vein graft; Sens = sensitivity; Spec = specificity; GRE = gradient-echo; CE = contrast-enhanced.

ers, sternal wires, coexistent prosthetic valves and supporting struts or rings, and graft stents) (see Figure 16–10).

CONCLUSION

Coronary MRA is currently clinically useful for the identification/characterization of anomalous coronary arteries and for assessment of coronary artery bypass graft patency. Technical and methodological advances (see Chapter 16) with more experience in defined disease will no doubt facilitate improved visualization of the native coronary anatomy. The clinical utility of coronary MRA for native vessel disease is already encouraging for exclusion of proximal coronary disease and for identification of patients with left main and three-vessel disease. Final data from multicenter trials will better define the clinical role of coronary MRA for evaluation of proximal native atherosclerosis in patients with known or suspected coronary artery disease.

References

1. Lieberman LM, Botti RI, Nelson AD: Magnetic resonance of the heart. Radiol Clin North Am 22:847, 1984.
2. Paulin S, von Schulthess GK, Fossel E, et al: MR imaging of the aortic root and proximal coronary arteries. AJR Am J Roentgenol 148:665, 1987.
3. Burstein D: MR imaging of coronary artery flow in isolated and in vivo hearts. J Magn Reson Imaging 1:337, 1991.
4. Edelman RR, Manning WJ, Burstein D, Paulin S: Coronary arteries: Breath-hold MR angiography. Radiology 181:641, 1991.
5. Manning WJ, Li W, Boyle NG, Edelman RR: Fat-suppressed breath-hold magnetic resonance coronary angiography. Circulation 87:94, 1993.
6. Pennell DJ, Keegan J, Firmin DN, et al: Magnetic resonance imaging of coronary arteries: Technique and preliminary results. Br Heart J 70:315, 1993.
7. Duerinckx A, Urman MK: Two-dimensional coronary MR angioghraphy: Analysis of initial clinical results. Radiology 193:731, 1994.
8. Sakuma H, Caputo GR, Steffens JC, et al: Breath-hold MR cine angiography of coronary arteries in healthy volunteers: Value of multiangle oblique imaging planes. AJR Am J Roentgenol 163:533, 1994.
9. Masui T, Isoda H, Mochizuki T, et al: MR angiography of the coronary arteries. Radiat Med 13:47, 1995.
10. Davis SF, Kannam JP, Wielopolski P, et al: Magnetic resonance coronary angiography in heart transplant recipients. J Heart Lung Transplant 15:580, 1996.
11. Li D, Paschal CB, Haacke EM, Adler LP: Coronary arteries: Three-dimensional MR imaging with fat saturation and magnetization transfer contrast. Radiology 1897:401, 1993.
12. Post JC, van Rossum AC, Hofman MBM, et al: Three-dimensional respiratory-gated MR angiography of coronary arteries: Comparison with conventional angiography. AJR Am J Roentgenol 166:1399, 1996.
13. Wielopolski PA, van Geuns RJM, de Feyter PJ, Oudkerk M: Breath-hold coronary MR angiography with volume targeted imaging. Radiology 209:209, 1998.
14. Botnar RM, Stuber M, Danias PG, et al: Improved coronary artery definition with T2-weighted, free-breathing, three-dimensional coronary MRA. Circulation 99:3139, 1999.
15. Scheidegger MB, Vassalli G, Hess OM, et al: Validation of coronary artery MR angiography: Comparison of measured vessel diameters with quantitative contrast angiography (abstract). In: Book of Abstracts. Berkeley, CA, Society of Magnetic Resonance, 1994, p 497.
16. Li D, Kaushikkar S, Haacke EM, et al: Coronary arteries: Three dimensional MR imaging with retrospective respiratory gating. Radiology 201:857, 1996.
17. Botnar RM, Stuber M, Danias PG, et al: A fast 3D approach for coronary MRA. J Magn Reson Imaging 10:821, 1999.
18. Stuber M, Botnar RM, Danias PG, et al: Double-oblique free-breathing high resolution three-dimensional coronary magnetic resonance angiography. J Am Coll Cardiol 34:524, 1999.
19. Stuber M, Botnar RM, Danias PG, et al: Contrast agent-enhanced, free-breathing three-dimensional coronary magnetic resonance angiography. J Magn Reson Imaging 10:790, 1999.
20. McConnell MV, Ganz P, Selwyn AP, et al: Identification of anomalous coronary arteries and their anatomic course by magnetic resonance coronary angiography. Circulation 92:3158, 1995.
21. Post JC, van Rossum AC, Bronzwaer JG, et al: Magnetic resonance angiography of anomalous coronary arteries. A new gold standard for delineating the proximal course? Circulation 92:3163, 1995.
22. Vliegen HW, Doornbos J, de Roos A, et al: Value of fast gradient echo magnetic resonance angiography as an adjunct to coronary angiography in detecting and confirming the course of clinically significant coronary artery anomalies. Am J Cardiol 79:773, 1997.
23. Taylor AM, Thorne SA, Rubens P, et al: Coronary artery imaging in grown up congenital heart disease: Complementary role of magnetic resonance and x-ray coronary angiography. Circulation 101:1670, 2000.

24. Manning WJ, Li W, Cohen SI, et al: Improved definition of anomalous left coronary artery by magnetic resonance coronary angiography. Am Heart J 130:615, 1995.

25. Machado C, Bhasin S, Soulen RL: Confirmation of anomalous origin of the right coronary artery from the left sinus of Valsalva with magnetic resonance imaging. Chest 104:1284, 1993.

26. Doorey AJ, Willis JS, Blasetto J, Goldenberg EM: Usefulness of magnetic resonance imaging for diagnosing an anomalous coronary artery coursing between aorta and pulmonary trunk. Am J Cardiol 74:198, 1994.

26a. Razmi RM, Meduri A, Chun W, et al: Coronary magnetic resonance angiography (CMRA): The gold standard for determining the proximal course of anomalous coronary arteries (abstract). J Am Coll Cardiol 37:380, 2001.

27. Dajani AS, Taubert KA, Takahashi M, et al: Guidelines for long-term management of patients with Kawasaki disease. Circulation 89:916, 1994.

28. Akagi T, Rose V, Benson LN, et al: Outcome of coronary artery aneurysms after Kawasaki disease. J Pediatr 121:689, 1992.

29. Kato H, Ichinose E, Yoshioka F, et al: Fate of coronary aneurysms in Kawasaki diseases: Serial coronary angiography and long-term follow-up study. Am J Cardiol 49:1758, 1982.

30. Suzuki A, Miyagawa-Tomita S, Nakazawa M, Yutani C: Remodeling of coronary artery lesions due to Kawasaki disease: Comparison of arteriographic and immunohistochemical findings. Jpn Heart J 41:245, 2000.

31. Flacke S, Setser RM, Barger P, et al: Coronary aneurysms in Kawasaki's disease detected by magnetic resonance coronary angiography. Circulation 101:e156, 2000.

32. Duerinckx AJ, Troutman B, Allada V, Kim D: Coronary MR angiography in Kawasaki disease. AJR Am J Roentgenol 168:114, 1997.

33. Molinari G, Sardanelli F, Zandrino F, et al: Coronary aneurysms and stenosis detected with magnetic resonance coronary angiography in a patient with Kawasaki disease. Ital Heart J 1:368, 2000.

34. Pennell DJ, Bogren HG, Keegan J, et al: Assessment of coronary artery stenosis by magnetic resonance imaging. Heart 75:127, 1996.

35. Manning WJ, Li W, Edelman RR: A preliminary report comparing magnetic resonance coronary angiography with conventional angiography. N Engl J Med 328:828, 1993.

36. Post JC, van Rossum AC, Hofman MBM, et al: Clinical utility of two-dimensional magnetic resonance angiography in detecting coronary artery disease. Eur Heart J 18:426, 1997.

37. Nitatori T, Yokoyama K, Hachiya J, et al: Comparison of 2D coronary MR angiography with conventional angiography-difference in imaging accuracy according to the severity of stenosis. Asian Oceanian J Radiol 3:15, 1998.

38. Müller MF, Fleisch M, Kroeker R, et al: Proximal coronary artery stenosis: Three-dimensional MRI with fat saturation and navigator echo. J Magn Reson Imaging 7:644, 1997.

39. Woodard PK, Li D, Haacke EM, et al: Detection of coronary stenoses on source and projection images using three-dimensional MR angiography with retrospective respiratory gating: Preliminary experience. AJR Am J Roentgenol 170:883, 1998.

40. Sandstede JJW, Pabst T, Beer M, et al: Three-dimensional MR coronary angiography using the navigator technique compared with conventional coronary angiography. AJR Am J Roentgenol 172:135, 1999.

41. van Geuns RJ, de Bruin HG, Rensing BJ, et al: Magnetic resonance imaging of the coronary arteries clinical results from three dimensional evaluation of a respiratory gated technique. Heart 82:515, 1999.

42. Huber A, Nikolaou K, Gonshior P, et al: Navigator echo-based respiratory gating for three-dimensional MR coronary angiography: Results from healthy volunteers and patients with proximal coronary artery stenoses. AJR Am J Roentgenol 173:95, 1999.

43. Sardanelli F, Molinari G, Zandrino F, Balbi M: Three-dimensional, navigator-echo MR coronary angiography in detecting stenoses of the major epicardial vessels, with conventional coronary angiography as the standard of reference. Radiology 214:808, 2000.

43a. Moustapha AI, Pereyra M, Muthupillai R, et al: Coronary magnetic resonance angiography using a free breathing T2 weighted, three dimensional gradient echo sequence with navigator respiratory and ECG gating can be used to detect coronary artery disease (abstract). J Am Coll Cardiol 37:380, 2001.

44. Danias PG, Kim WY, Stuber M, et al: Coronary magnetic resonance angiography: A prospective international multicenter study (abstract). Eur Heart Soc (in press), 2001.

45. Regenfus M, Ropers D, Achenbach S, et al: Noninvasive detection of coronary artery stenosis using contrast enhanced three-dimensional breath-hold magnetic resonance coronary angiography. J Am Coll Cardiol 36:44, 2000.

46. van Geuns RJM, Wielopolski PA, de Bruin, et al: MR coronary angiography with breath-hold targeted volumes: Preliminary clinical results. Radiology 217:270, 2000.

47. White RD, Caputo GR, Mark AS, et al: Coronary artery bypass graft patency: Noninvasive evaluation with MR imaging. Radiology 164:681, 1987.

48. Rubinstein RI, Askenase AD, Thickman D, et al: Magnetic resonance imaging to evaluate patency of aortocoronary bypass grafts. Circulation 76:786, 1987.

49. Jenkins, JPR, Love HG, Foster CJ, et al: Detection of coronary artery bypass graft patency as assessed by magnetic resonance imaging. Br J Radiol 61:2, 1988.

50. Galjee MA, van Rossum AC, Doesburg T, et al: Value of magnetic resonance imaging in assessing patency and function of coronary artery bypass grafts. An angiographically controlled study. Circulation 93:660, 1996.

51. White, RD, Pflugfelder PW, Lipton MJ, et al: Coronary artery bypass grafts: Evaluation of patency with cine MR imaging. AJR Am J Roentgenol 150:1271, 1988.

52. Aurigemma GP, Reichek N, Axel L, et al: Noninvasive determination of coronary artery bypass graft patency by cine magnetic resonance imaging. Circulation 80:1595, 1989.

53. Engelmann MG, Knez A, von Smekal A, et al: Non-invasive coronary bypass graft imaging after multivessel revascularisation. Int J Cardiol 76:65, 2000.

54. Wintersperger, BJ, Engelmann MG, von Smekal A, et al: Patency of coronary bypass grafts: Assessment with breath-hold contrast-enhanced MR angiography—Value of a non-electrocardiographically triggered technique. Radiology 208:345, 1998.

55. Vrachliotis TG, Bis KG, Aliabadi D, et al: Contrast-enhanced breath-hold MR angiography for evaluating patency of coronary artery bypass grafts. AJR Am J Roentgenol 168:1073, 1997.

56. O'Regan DJ, Borland JAA, Chester AH, et al: Assessment of human long saphenous vein function with minimally invasive harvesting with the Mayo Stripper. Eur J Cardiothor Surg 12:428, 1997.

57. Bidstrup BP, Underwood SR, Sapsford RN: Effect of aprotinin (Trasylol) on aortocoronary bypass graft patency. J Thorac Cardiovasc Surg 105:147, 1993.

CHAPTER 18

Coronary Magnetic Resonance Angiography for Suspected Anomalous Coronary Artery Disease

Michael V. McConnell and Warren J. Manning

CLINICAL BACKGROUND

Though rare and usually benign, specific congenital coronary anomalies are a recognized cause of myocardial ischemia and sudden cardiac death, especially among young adults.[1] Data regarding the prevalence of anomalous coronary arteries in the normal population are primarily derived from coronary angiography data among adults referred for diagnostic coronary angiography because of chest pain, and the prevalence has been estimated at approximately 0.85 percent.[2–4] Engel and associates[2] reviewed the data from 4250 adults without known congenital heart disease referred for x-ray coronary angiography. Fifty-one (1.2%) patients had at least one anomalous origin of a coronary artery. In a larger series of 7000 patients, Kimbiris and coworkers[3] reported 45 (0.6%) cases of coronary artery anomalies. Similarly, Chaitman and colleagues[4] reported 31 (0.8%) cases from 3750 patients referred for x-ray angiography. As might be anticipated, anomalous coronary arteries are more common among patients with other forms of congenital heart disease,[5] especially those with tetralogy of Fallot, complete transposition of the great arteries, congenitally corrected transposition, and bicuspid aortic valves. Among patients with a congenitally bicuspid aortic valve, the right and left coronary arteries may be oriented 180 degrees apart, as compared with the more normal 120- to 150-degree orientation. Should these patients require surgery, information regarding coronary orientation may be relevant for guiding the choice of aortic valve prosthesis.

Coronary Artery Embryology. During the 7th week of development and shortly after the aorta is formed by division of the truncus arteriosus, coronary artery buds are seen. Transient, abortive "coronary buds" sometimes arise from the pulmonary trunk as well.[6] Once established, the coronary arteries rapidly mature, with large branches extending over the surface of the heart. In the normal situation, the right coronary artery (RCA) travels in the right atrioventricular groove, whereas the left main coronary artery bifurcates within 1 cm of its origin to form the left anterior descending (LAD) coronary artery, extending into the interventricular groove, and the left circumflex (LCX) coronary artery, which travels in the left atrioventricular groove.

Specific Coronary Anomalies. With normal coronary anatomy (Figure 18–1), the coronary arteries originate from the aortic sinus closest to their eventual path. Therefore, the RCA originates from the right coronary sinus and the left main coronary artery originates from the left coronary sinus. In contrast, anomalous coronaries typically originate on the contralateral side to their destination (or less commonly from the noncoronary sinus or pulmonary artery) and must cross to the other side of the aorta. As a result, the anomalous vessels are longer, often without branching vessels within the anomalous segment.

The most common anomalous coronary artery condition, accounting for over 50 percent of anomalous coronary artery conditions in the adult, is anomalous origin of the LCX from the *right* sinus of Valsalva or RCA.[2, 3, 7, 8] The anomalous LCX may arise from a common ostium with the RCA, from a proximal branch of the RCA, or from a separate ostium.[8] In most situations, the anomalous LCX follows a benign, *retro*aortic course (Figure 18–2) and enters the proximal left atrioventricular groove as if it had originated as a proximal branch of the left main coronary artery. The second most common anomaly, occurring in 20 to 25 percent of ectopic origins, is one in which the RCA originates from the *left* sinus of Valsalva (Figure 18–3). In this situation, the anomalous RCA often follows an anterior course, running *between* the aorta and pulmonary artery. Anomalous origin of the left main or LAD from the *right* sinus of Valsalva is among the rarest of anomalies. The anomalous vessel may follow one of four proximal pathways: (1) interarterial (i.e., between the aorta and pulmonary trunk, as in Figure 18–4); (2) posterior or retroaortic; (3) anterior to the pulmonary artery; or (4) intraseptal, in which the proximal vessel is intramyocardial. Sometimes noted, though often not reported as an anomaly in all series, is the benign absence of the left main with direct origin of the LAD and LCX from the left sinus of Valsalva.[9]

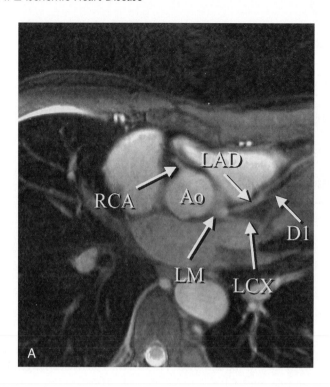

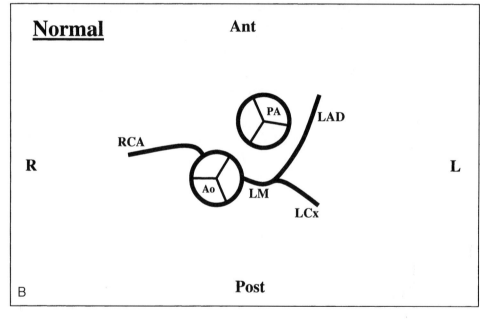

Figure 18–1. Normal coronary artery orientation. *A,* Transverse reconstruction from a three-dimensional T2-prepared gradient echo data set depicting the normal orientation of the ascending aorta (Ao), proximal right coronary artery (RCA) and left main (LM) coronary artery, left anterior descending artery (LAD) with first diagonal (D1) and left circumflex artery (LCX). *B,* Schematic of normal anatomic relationships. Ao = aortic valve; PA = pulmonic valve; Ant = anterior; Post = posterior.

Finally, there are other very rare coronary anomalies associated with hemodynamic manifestations. As previously mentioned, the entire left coronary system can arise from the pulmonary artery.[10, 11] Because of the significant hemodynamic and oxygen delivery impact of this lesion, this anomaly typically presents in infancy or early childhood with myocardial infarction and/or congestive heart failure and has substantial morbidity and mortality. Coronary fistulas (Figure 18–5), with connections to the coronary sinus,[12] pulmonary artery, or the right ventricle, may also be hemodynamically important, especially if there is substantial left-to-right shunting.[12–14]

Clinical and Hemodynamic Significance. Although the incidence of coronary anomalies is very low, the impact on premature cardiac morbidity and mortality among young adults in the population is not trivial. A recent prospective study[15] following young athletes (as part of a screening program for hypertrophic cardiomyopathy) has implicated an anomalous coronary artery in over 12 percent of sudden death cases among athletes.

The clinical significance of the common coronary artery anomalies primarily relates to their association with myocardial ischemia or infarction, as well as syncope and sudden cardiac death.[4, 16–20] This premature cardiac morbidity and mortality in young

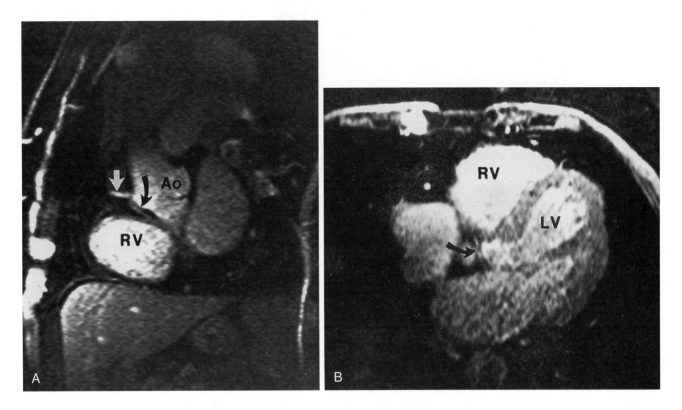

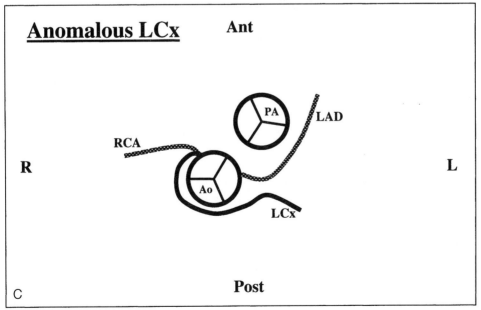

Figure 18–2. Single-oblique *(A)* and transverse coronary *(B)* MRA. Note the normal RCA (white arrow) with the anomalous LCX (curved black arrows) originally adjacent to the proximal RCA and traversing in a "benign," retroaortic manner. *C,* Schematic of this anomaly. (*A* and *B* from McConnell MV, Ganz P, Selwyn AP, et al: Identification of anomalous coronary arteries and their anatomic course by magnetic resonance coronary angiography. Circulation 92:3163, 1995, reprinted with permission of the American Heart Association.)

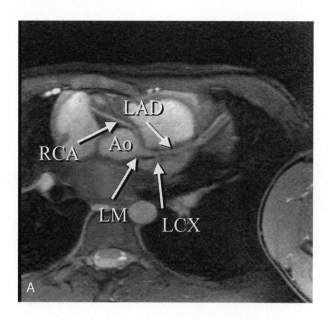

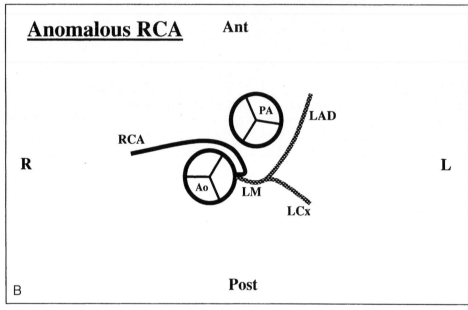

Figure 18–3. *A,* Coronary MRA from a 14-year-old boy who presented with atypical exertional chest pain. An anomalous RCA is seen originating from the left coronary sinus immediately adjacent to the origin of the left main (LM). The anomalous vessel then travels in a "malignant fashion" between the aorta (Ao) and right ventricular outflow tract. *B,* Schematic of this anomaly.

adults is most common in the setting of vigorous physical exercise. Clinical and pathological data suggest that the proximal course of the anomalous vessel is the primary determinant of hemodynamic significance. In particular, anomalous coronary arteries that course between the aorta and pulmonary trunk (see Figures 18–3 and 18–4) have been shown to be associated with exercise-induced ischemia, myocardial infarction, and sudden cardiac death in young patients without evidence of coronary atherosclerosis.[18] There are several postulated mechanisms. The most widely accepted hypothesis is that the aorta and pulmonary artery impinge on the proximal anomalous vessel, especially during exercise when there is aortic and pulmonary artery dilation in response to increased cardiac output, coincident with increased coronary flow demand.[21] Another contributing mechanism to flow limitation may be a

sharp turn or bend at the origin of the anomalous vessel from the contralateral sinus of Valsalva. Again, with increased cardiac output, a "kink" may transiently develop in the anomalous segment, thereby limiting blood flow.[4, 22] Finally, the anomalous origin can be associated with a slit-like ostial stenosis that impairs coronary blood flow. The interarterial course (between the aorta and pulmonary artery) of an anomalous left main coronary artery is considered the highest risk for ischemia and sudden death. Although the three noninterarterial pathways of the left main are considered lower risk,[19] they have been associated with documented cases of myocardial infarction and sudden death in the absence of atherosclerosis.[20] An anomalous right coronary artery, which is typically interarterial, has also been documented to be at risk for myocardial ischemia and infarction.[18] In the absence of atherosclero-

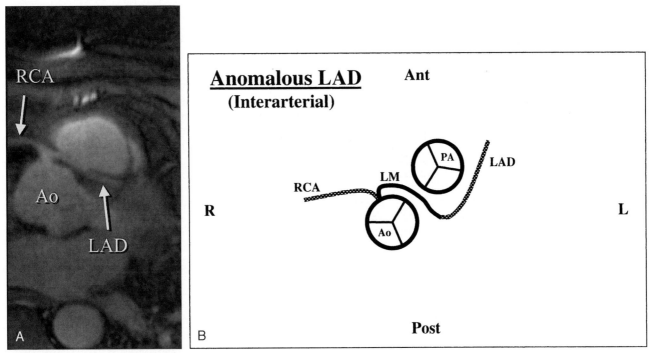

Figure 18–4. *A,* Coronary MRA obtained in a 57-year-old man with exertional angina. Note the anomalous LAD originating from the proximal RCA and coursing in a "malignant fashion" between the aorta (Ao) and pulmonary artery. *B,* Schematic of this anomaly.

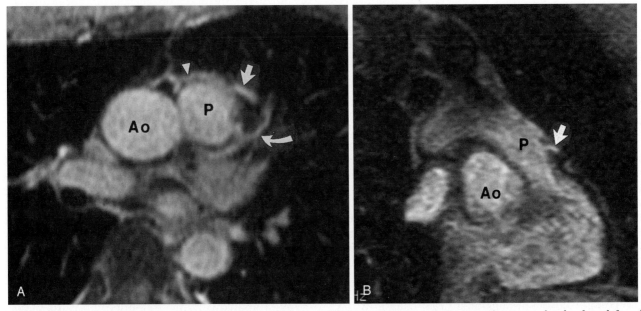

Figure 18–5. Coronary MRA from a 41-year-old man who had previously undergone embolization of coronary fistulas from left and right coronary arteries and draining into the pulmonary artery. *A,* Transverse coronary MRA demonstrating a prominent vessel (straight white arrow) originating from the LAD (curved white arrow) and terminating in the anterior aspect of the pulmonary artery (P). A smaller vessel is seen originating from the right (white arrowhead). *B,* Single oblique coronary MRA demonstrating termination of the fistula (white arrow) in the pulmonary artery. (Courtesy of Drs. Michael D. Black and Robert J. Herfkens.)

sis, there has been no risk ascribed to the retroaortic anomalous circumflex.

DIAGNOSIS OF ANOMALOUS CORONARY ARTERY DISEASE

Echocardiography. In the infant and young child, acoustic windows are generally excellent, allowing transthoracic echocardiography with a high-frequency transducer to successfully define the origin and proximal course of the native coronary arteries.[23] In the adolescent and adult, however, progressive rib ossification and thoracic/lung growth often lead to inadequate transthoracic echocardiographic visualization.[24] In small series and case reports, transesophageal echocardiography (TEE), a moderately invasive test in which there is close proximity of the high-frequency transducer to the coronary ostia, has been used to image coronary anomalies. The largest TEE series included nine patients with coronary anomaly.[25] Although TEE did confirm the origin and initial course of the anomalous coronary vessel, TEE image acquisition and analysis were performed with prior knowledge of the angiographic findings. Thus, the ability of TEE to identify coronary anomalies in a blinded analysis remains unknown.

X-ray Angiography. For several decades, the gold standard for the premorbid diagnosis of coronary artery anomalies was invasive x-ray coronary angiography. As previously mentioned, defining both the presence and the proximal course with respect to the aorta and pulmonary artery are essential, inasmuch as these findings are used to determine treatment. Accurate x-ray coronary angiography usually requires selectively engaging each vessel. Commonly, the anomalous origin may not be engaged during a standard procedure, sometimes leading to the erroneous assumption that the vessel is occluded. After sheath removal and later review of the left ventriculogram, an anomalous vessel may then be suspected. Misclassification by x-ray angiography of the proximal course of anomalous coronary arteries is a well-described phenomenon[26] although visual aids, including placement of a right heart catheter in the pulmonary artery, have been developed to try to minimize this problem.[20] Owing to its invasive nature and expense, however, invasive x-ray coronary angiography poses some risk to the patient and is impractical for screening purposes in young adults presenting with typical or atypical exertional chest pain. More importantly, x-ray angiography provides only a two-dimensional (2-D) projection of the complex three-dimensional (3-D) course of the anomalous vessel.

Magnetic Resonance Angiography. The use of magnetic resonance angiography (MRA) of the coronary arteries[27–30] (see Chapter 16) has several advantages in the diagnosis of coronary anomalies. MRA is noninvasive and does not require ionizing radiation or iodinated contrast agents. Most importantly,

it provides a 3-D road map (i.e., not the 2-D projection of invasive x-ray angiography) in which the user can acquire an image at any orientation. Coronary MRA is also particularly suited to imaging the larger proximal coronary vessels.[27–30]

Early reports of MRA to visualize anomalous coronary arteries included case report confirmation of x-ray angiographic data.[31–33] There are at least two published series of patients who underwent blinded comparison of coronary MRA results with x-ray angiography. We reported on 16 patients with anomalous coronaries who underwent 2-D, electrocardiogram (ECG)-gated, segmented k-space gradient-echo coronary MRA with fat suppression and breath-holding.[34] Investigators interpreting the coronary MRA were blinded to the coronary anomaly and all clinical data. Coronary MRA correctly identified the anomalous vessel in 93% of cases (using conventional x-ray coronary angiography as the gold standard). The proximal course of the anomalous vessel was not well seen in one case with a small, nondominant anomalous RCA and in another case with a small anomalous retroaortic LCX. Post and associates[35] reported on 38 subjects, including 19 patients with anomalous coronaries and 19 control patients without coronary anomalies. A similar 2-D breath-hold MRA sequence was used with overlapping slices obtained at the level of the aortic root and along the atrioventricular groove. Coronary MRA identified all subjects with anomalous vessels, as well as the course of the proximal segment of the anomalous vessel. Importantly, there were three cases in which there was disagreement between x-ray angiography and coronary MRA regarding the proximal course of the vessel. Side-by-side review of these cases led to the consensus that coronary MRA was correct in all three instances. These investigators concluded that coronary MRA should be considered the new gold standard for this condition. Subsequently, Vliegen and colleagues[36] reported on 12 patients with coronary anomalies, including 5 patients referred for CMR because the x-ray angiogram was inconclusive with regards to the anatomic course of the anomalous vessel. A breath-hold 2-D approach was also used. Coronary MRA provided the diagnosis in all 5 subjects for whom the x-ray angiogram had been inconclusive. Of 7 patients in which the diagnosis had been established by x-ray, the x-ray diagnosis was changed in one patient as a result of the MRA data. The x-ray in this one subject had suggested the anomalous RCA coursed anterior to the pulmonary artery, whereas the MRA (and subsequent surgery) documented its retroaortic course.

Taylor and coworkers[37] extended the value of coronary MRA for identification of anomalous vessels by performing a study in 25 patients with congenital cardiac anomalies often associated with anomalous coronary arteries—including tetralogy of Fallot and congenitally corrected transposition of the great arteries. Respiratory gated coronary MRA was used and compared in a blinded manner to conventional x-ray coronary angiography, with consensus diagnosis. Coronary MRA had a 92 percent sensitivity and 100 percent specificity for detecting coronary anomalies. These investigators also found that coronary MRA was superior to x-ray angiography for defining the proximal course of the vessels.

Limitations and Pitfalls of Magnetic Resonance Angiography. Although MRA is noninvasive and without significant short- or long-term risk, some studies are still suboptimal. Patients may have contraindications, such as pacemakers, intracranial clips or certain other metal implants, or they may be claustrophobic. Most coronary MRA protocols require a regular heart rhythm and consistent breath-holding or regular respiratory rhythm. New techniques for respiratory gating or rapid acquisition of the entire cardiac volume remove the need for repetitive breath-holding.[37–41] Real-time cardiac CMR can be performed regardless of cardiac rhythm or breathing.[42, 43]

There are several pitfalls that we and others have encountered that must be considered when interpreting coronary MRA for anomalous vessels. The anomalous vessel may be quite small and thereby challenge the spatial resolution of coronary MRA, as in the two cases in which we could not identify the proximal vessel course. Higher resolution and contrast-enhanced coronary MRA techniques may overcome this limitation.[43–45] The retroaortic pericardium may be mistaken for an anomalous circumflex. The one case from our study[34] in which the anomalous vessel was incorrectly identified was initially interpreted as a retroaortic anomalous circumflex. On later review, it was noted that this linear retroaortic structure was present over multiple transverse slices (i.e., too large for a vessel) and was pericardial tissue that on a single slice mimicked an anomalous circumflex.

CLINICAL IMPLICATIONS

The ability of coronary MRA to noninvasively identify coronary artery anomalies has several potential clinical applications. First, coronary MRA may be used to evaluate the 3-D path of anomalous coronary vessels identified by x-ray angiography, particularly when there is uncertainty as to whether an anomalous vessel follows a hemodynamically significant course anterior to the aorta and posterior to the pulmonary artery. Second, coronary MRA could distinguish an occluded vessel from an anomalous vessel in which a vessel could not be engaged by conventional angiography. Third, it should be considered among patients with congenital cardiac anomalies associated with coronary anomalies (especially prior to cardiac interventions). Finally, though yet unproved, coronary MRA may be helpful in selected cases of young patients with unexplained syncope or chest discomfort for whom a coronary anomaly is included in the differential diagnosis. Although cost would be a factor in considering this technique as a screening test for competitive athletes,[15] it would have the advantage over echocardiography in more comprehensively detecting coronary as well as myocardial abnormalities.

CONCLUSION

Coronary artery anomalies are a rare but important cause of cardiac morbidity and mortality, especially in adolescents and young adults. Coronary MRA is well suited to noninvasively detect anomalous coronaries and to define their proximal course and thereby guide therapy. With increasing clinical experience and continued technological improvements, coronary MRA will likely emerge as the gold standard for the diagnosis of this condition.

References

1. Cheitlin MD, De Castro MD, McAllister HA: Sudden death as a complication of anomalous left coronary origin from the anterior sinus of Valsalva: A not-so-minor congenital anomaly. Circulation 50:780, 1974.
2. Engel HJ, Torres C, Page HL: Major variations in anatomical origin of the coronary arteries: Angiographic observations in 4,250 patients without associated congenital heart disease. Cathet Cardiovasc Diagn 1:157, 1975.
3. Kimbiris D, Iskandrian AS, Segal BL, Bermis CE: Anomalous aortic origin of coronary arteries. Circulation 58:606, 1978.
4. Chaitman BR, Lesperance J, Saltiel J, Bourassa MG: Clinical angiographic, and hemodynamic findings in patients with anomalous origin of the coronary arteries. Circulation 53:122, 1976.
5. Dabizzi RP, Teodori G, Barletta GA, et al: Associated coronary and cardiac anomalies in the tetralogy of Fallot. An angiographic study. Eur Heart J 11:692, 1990.
6. Perloff J: The Clinical Recognition of Congenital Heart Disease. 3rd ed. Philadelphia, WB Saunders, 1987, p 663.
7. Leberthson RR, Dinsmore RE, Bharati S, et al: Aberrant coronary artery origin from the aorta: Diagnosis and clinical significance. Circulation 50:774, 1974.
8. Page HL Jr, Engel HJ, Campbell WB, Thomas CS Jr: Anomalous origin of the left circumflex coronary artery. Circulation 50:768, 1974.
9. Dicicco BS, McManus BM, Waller BF, Roberts WC: Separate aortic ostium of the left anterior descending and left circumflex coronary arteries from the left aortic sinus of Valsalva (absent left main coronary artery). Am Heart J 104:153, 1982.
10. Wilson CL, Dlabal PW, Holeyfield RW, et al: Anomalous origin of left coronary artery from pulmonary artery. Case report and review of literature concerning teen-agers and adults. J Thorac Cardiovasc Surg 73:887, 1977.
11. Douard H, Barat JL, Laurent F, et al: Magnetic resonance imaging of an anomalous origin of the left coronary artery from the pulmonary artery. Eur Heart J 9:1356, 1988.
12. Kugelmass AD, Manning WJ, Piana RN, et al: Coronary arteriovenous fistula presenting as congestive heart failure. Cathet Cardiovasc Diagn 26:19, 1992.
13. Ueno T, Nakayama Y, Yoshikai M, et al: Unique manifestations of congenital coronary artery fistulas. Am Heart J 124:1388, 1992.
14. Vandenbossche JL, Felice H, Grivegnee A, Englert M: Noninvasive imaging of left coronary arteriovenous fistula. Chest 93:885, 1988.
15. Corrado D, Basso C, Schiavon M, Thiene G: Screening for hypertrophic cardiomyopathy in young athletes. N Engl J Med 339:364, 1998.

16. Levin DC, Fellows KE, Abrams HL: Hemodynamically significant primary anomalies of the coronary arteries: Angiographic aspects. Circulation 58:25, 1978.

17. Liberthson RR, Dinsmore RE, Fallon JT: Aberrant coronary artery origin from the aorta. Report of 18 patients, review of literature and delineation of natural history and management. Circulation 59:748, 1979.

18. Kragel AH, Roberts WC: Anomalous origin of either the right or left main coronary artery from the aorta with subsequent coursing between aorta and pulmonary trunk: Analysis of 32 necropsy cases. Am J Cardiol 62:771, 1988.

19. Roberts WC, Dicicco BS, Waller BF, et al: Origin of the left main from the right coronary artery or from the right aortic sinus with intramyocardial tunneling to the left side of the heart via the ventricular septum. The case against clinical significance of myocardial bridge or coronary tunnel. Am Heart J 104:303, 1982.

20. Serota H, Barth CW III, Seuc Cam Vandormael M, et al: Rapid identification of the course of anomalous coronary arteries in adults: The "dot and eye" method. Am J Cardiol 65:891, 1990.

21. Barth CW III, Roberts WC: Left main coronary artery originating from the right sinus of Valsalva and coursing between the aorta and pulmonary trunk. J Am Coll Cardiol 7:366, 1986.

22. Roberts WC, Siegel RJ, Zipes DM: Origin of the right coronary artery from the left sinus of Valsalva and its functional consequences. Am J Cardiol 49:863, 1982.

23. Schmidt KG, Cooper MJ, Silverman NH, Stanger P: Pulmonary artery origin of the left coronary artery: Diagnosis by two-dimensional echocardiography, pulsed Doppler ultrasound and color flow mapping. J Am Coll Cardiol 11:396, 1988.

24. Douglas PS, Fiolkoski J, Berko B, Reichek N: Echocardiographic visualization of coronary artery anatomy in the adult. J Am Coll Cardiol 11:565, 1988.

25. Fernandes F, Alam M, Smith S, Khaja F: The role of transesophageal echocardiography in identifying anomalous coronary arteries. Circulation 88:2532, 1993.

26. Ishikawa T, Brandt PW: Anomalous origin of the left main coronary artery from the right anterior aortic sinus: Angiographic definition of anomalous course. Am J Cardiol 55:770, 1985.

27. Wang SJ, Hu BS, Macovski A, Nishimura DG: Coronary angiography using fast selective inversion recovery. Magn Reson Med 18:417, 1991.

28. Manning WJ, Li W, Edelman RR: A preliminary report comparing magnetic resonance coronary angiography with conventional contrast angiography. N Engl J Med 328:828, 1993.

29. Pennell DJ, Keegan J, Firmin DN, et al: Magnetic resonance imaging of coronary arteries: Technique and preliminary results. Br Heart J 70:315, 1993.

30. Duerininckx AJ, Bogaert J, Jiang H, Lewis BS: Anomalous origin of the left coronary artery: Diagnosis by coronary MR angiography. AJR 164:1095, 1995.

31. Manning WJ, Li W, Cohen SI, et al: Improved definition of anomalous left coronary artery by magnetic resonance coronary angiography. Am Heart J 130:615, 1995.

32. Machado C, Bhasin S, Soulen RL: Confirmation of anomalous origin of the right coronary artery from the left sinus of Valsalva with magnetic resonance imaging. Chest 104:1284, 1993.

33. Doorey AJ, Willis JS, Blasetto J, Goldenberg EM: Usefulness of magnetic resonance imaging for diagnosing an anomalous coronary artery coursing between aorta and pulmonary trunk. Am J Cardiol 74:198, 1994.

34. McConnell MV, Ganz P, Selwyn AP, et al: Identification of anomalous coronary arteries and their anatomic course by magnetic resonance coronary angiography. Circulation 92:3158, 1995.

35. Post JC, van Rossum AC, Bronzwaer JG, et al: Magnetic resonance angiography of anomalous coronary arteries. A new gold standard for delineating the proximal course? Circulation 92:3163, 1995.

36. Vliegen HW, Doornbos J, de Roos A, et al: Value of fast gradient echo magnetic resonance angiography as an adjunct to coronary angiography in detecting and confirming the course of clinically significant coronary artery anomalies. Am J Cardiol 79:773, 1997.

37. Taylor AM, Thorne SA, Rubens P, et al: Coronary artery imaging in grown up congenital heart disease: Complementary role of magnetic resonance and x-ray coronary angiography. Circulation 101:1670, 2000.

38. Sachs TS, Meyer CH, Hu BS, et al: Real-time motion detection in spiral MRI using navigators. Magn Reson Med 32:639, 1994.

39. McConnell MV, Khasgiwala VC, Savord BJ, et al: Comparison of respiratory suppression methods and navigator locations for MR coronary angiography. AJR 168:1369, 1997.

40. Wielopolski PA, van Geuns RJ, de Feyter PJ, Oudkerk M: Breath-hold coronary MR angiography with volume-targeted imaging. Radiology 209:209, 1998.

41. Goldfarb JW, Edelman RR: Coronary arteries: Breath-hold, gadolinium-enhanced, three-dimensional MR angiography. Radiology 206:830, 1998.

42. Yang PC, Kerr AB, Liu AC, et al: New real-time interactive cardiac magnetic resonance imaging system complements echocardiography. J Am Coll Cardiol 32:2049, 1998.

43. Meyer CH, Hu BS, Kerr AB, et al: High-resolution multislice spiral coronary angiography with real-time interactive localization. Proceedings of the Fifth Meeting of the International Society for Magnetic Resonance in Medicine, 1997, p 439.

44. Botnar, RM, Stuber M, Danias PG, et al: Improved coronary artery definition with T2-weighted free-breathing 3D-coronary MRA. Circulation 99:3139, 1999.

45. Stuber M, Botnar RM, Danias PG, et al: Double-oblique free-breathing high resolution 3D coronary MRA. J Am Coll Cardiol 34:524, 1999.

CHAPTER 19

Coronary Artery and Coronary Sinus Velocity and Flow

Jennifer Keegan and Dudley J. Pennell

Coronary stenosis may be observed during coronary magnetic resonance angiography (MRA) as an area of reduced signal caused by the reduced luminal area or turbulent flow, and whereas both the degree and the extent of signal loss are indicative of the severity of the stenosis,[1] accurate quantification has not proved possible. However, both phasic coronary artery blood flow and flow velocity may be affected by the presence of stenoses, and the ratio of coronary artery flow under maximal vasodilation to that at rest (the coronary flow reserve) is a good indicator of the physiological significance of the coronary obstruction. This chapter examines the current and potential future status of cardiovascular magnetic resonance (CMR) measurements of coronary velocity and flow, which may develop clinical value as part of a comprehensive CMR examination.

MR can be used to quantify blood flow noninvasively, and since the 1980s, it has been used for the assessment of phasic blood velocity and flow in a wide range of cardiovascular applications,[2] but application to the coronary arteries is more challenging. One of the main problems is the small size of the coronary arteries (typically < 4 mm diameter) which, for current levels of in-plane resolution, results in only a few pixels across the vessel. This dilemma has implications for the accurate measurement of both vessel cross-sectional area and blood flow velocity. Vessel tortuosity is a further problem, giving rise to difficulties in accurately aligning the vessel so that flow is truly through-plane or in-plane, as required, and this problem is exacerbated by the movement of the arteries with both the cardiac and respiratory cycles. In addition, the temporal resolution of the velocity encoding sequence must be sufficiently good to minimize the blurring of the vessel due to motion within the period of acquisition and to resolve the phasic velocity profile. The low peak flow velocities in normal coronary arteries at rest (typically < 25 cm/s) present a further challenge and require sensitive velocity windows, whereas in the presence of stenoses, high velocities are present together with complex flow, which may lead to signal loss. The combination of these problems is formidable and it is only in the last 5 years that CMR techniques have started to generate significant results. There are direct and in-direct methods for assessing coronary blood flow, and both will be examined, although it is clear that the future lies in developing accurate direct techniques.

INDIRECT ASSESSMENT OF TOTAL CORONARY FLOW AND FLOW RESERVE

Two indirect approaches for assessing total coronary blood flow and flow reserve by CMR techniques have been reported, the first from velocity mapping of cardiac venous outflow and the second from velocity mapping in the aortic root.

Coronary Sinus Flow

Velocity mapping of coronary venous outflow is less problematic than velocity mapping in the coronary arteries, primarily because the coronary sinus has a much larger diameter (typically 7–10 mm), and also because signal loss effects are less prevalent because flow is less prone to turbulence. In the human heart, coronary sinus flow approximates to total coronary flow because approximately 96 percent of left ventricular venous blood flow drains into the right atrium via the coronary sinus.[3] In a study of 24 healthy volunteers, van Rossum and associates[4] first showed that blood flow in the distal coronary sinus could be measured using CMR. The study was performed at 0.6 T using a sequence with a 17-ms echo time (TE) and a velocity window of ±60 cm/s, a temporal resolution of less than 50 ms, and an in-plane pixel size of 1.6 × 1.6 mm. The 7.5-mm thick oblique coronal imaging plane was piloted from a transverse image obtained during iso-volumetric contraction so that coronary sinus movement during the cardiac cycle was predominantly an in-plane translation with the direction of flow through-plane (Figure 19–1). For each time point in the cardiac cycle, the cross-sectional area of the coronary sinus was defined on the magnitude image and the mean velocity within the contours found on the corresponding velocity map. Instantaneous blood flow was calculated by multiplying the mean

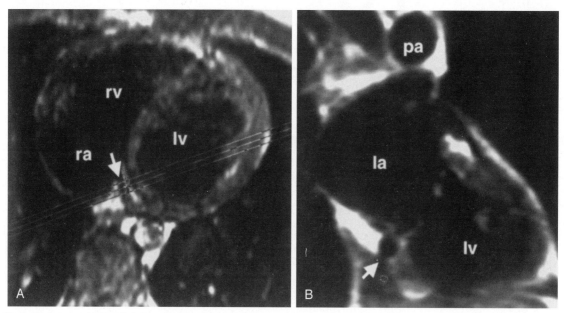

Figure 19–1. Transverse image *(A)* demonstrating positioning of oblique coronal slice *(B)* for through-plane sinus flow assessment. Arrows indicate the coronary sinus ostium. rv = right ventricle; lv = left ventricle; la = left atrium; ra = right atrium; pa = pulmonary artery. (Modified from van Rossum AC, Visser FR, Hofman MBM, et al: Global left ventricular perfusion: Noninvasive measurement with cine MR imaging and phase velocity mapping of coronary venous outflow. Radiology 182:685, 1992, with permission.)

velocity by the cross-sectional area, and mean blood flow was calculated by integration over the entire cardiac cycle. The velocity mapping sequence was validated against a float flowmeter for laminar flow in a 10-mm diameter tube, and it was shown that CMR velocity mapping had an overall accuracy of 5 percent. Figure 19–2 shows a typical flow profile in a healthy subject. The profiles were generally found to be biphasic, with a mean peak of 260 ml/min at 200 ms after the R wave and a second peak of 1100 ml/min at 500 ms. In 37 percent of the subjects, reverse flow was seen immediately after the R wave. The mean volumetric flow over the cardiac cycle

was 145 ml/min with a mean flow velocity of 2.1 cm/s. The coronary sinus cross-sectional area varied from 1.9 cm² in systole to 0.5 cm² in diastole. The phasic blood flow in the sinus may have been expected to be predominantly systolic, as in other venous structures, but the authors suggested that the thin, compliant walls of the sinus together with its drainage into the right atrium, where the pressure varies considerably, may be responsible for the predominantly diastolic flow profile. These findings have been supported by other independent studies, both directly, using an ultrasonic transit-time technique to measure phasic volumetric coronary sinus

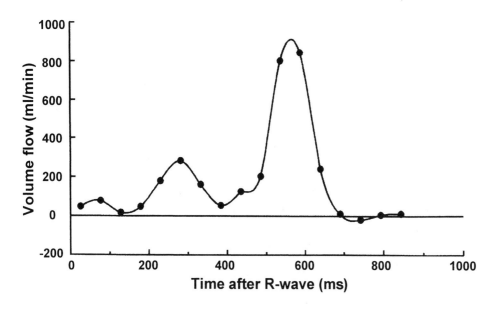

Figure 19–2. Biphasic blood flow profile in the coronary sinus of a healthy volunteer. In this example, blood flow per cardiac cycle was 2.8 ml and mean blood flow was 160 ml/min. (Modified from van Rossum AC, Visser FR, Hofman MBM, et al: Global left ventricular perfusion: Noninvasive measurement with cine MR imaging and phase velocity mapping of coronary venous outflow. Radiology 182: 685, 1992, with permission.)

flow in conscious dogs,[5] and indirectly, by observing areas of signal void caused by accelerating and turbulent flow near the entrance of the coronary sinus in the right atrium in early diastole in conventional CMR cine imaging in normal subjects.[6] Promising though this early work was, it was realized that there were limitations to the study: First, the errors were introduced by cardiac and respiratory motion, the former being minimized by careful selection of the imaging plane; second, the small number of pixels covering the sinus (typically 5 across the diameter in diastole) results in considerable partial volume averaging in edge pixels, which is problematical for the accurate assessment of sinus cross-sectional area and for the determination of the mean sinus flow velocity. Despite these limitations, the measurements of mean sinus blood flow were similar to those found by continuous thermodilution in normal subjects (120 ml/min),[7] and the technique appears worth further evaluation.

Although this technique has not yet been widely applied by other groups, there are some limited reports of its use. Kawada and coworkers reported results in a study of 9 healthy volunteers and 29 patients with hypertrophic cardiomyopathy (HCM).[8] They also assessed the possibility of using a segmented k-space[9] approach to the acquisition, so that all data could be acquired in less than 25 seconds, rather than the 4 minutes using conventional cine imaging. This method had the advantage that the acquisition could be performed in a single breath-hold, thereby eliminating the effects of respiratory

motion in cooperative patients. Data were acquired from a 5-mm thick oblique coronal plane using a sequence with a TE of 5 ms and a repeat time (TR) of 15 ms. The velocity window implemented was ± 100 cm/s. Four reference and four velocity-encoded views were acquired per data segment, giving a segment duration of 120 ms, but the temporal resolution was effectively improved by view sharing, a technique whereby data are generated at intermediate time points from the preceding and following data segments.[10] Hence, depending on the RR interval, velocity maps were obtained at up to 14 phases in the cardiac cycle. The authors observed the same biphasic velocity and flow profiles in both healthy volunteers and HCM patients with no significant difference between the baseline coronary blood flow per unit mass of myocardium in the two groups (0.74 vs. 0.62 ml/min/g, respectively). However, after the intravenous administration of 0.56 mg/kg dipyridamole, the increase in coronary blood flow was less in the HCM patients than in the healthy volunteers (1.03 vs. 2.14 ml/min/g), resulting in significantly different coronary flow reserves for the two groups (1.7 vs. 3.0, respectively, $p < 0.01$; Figure 19–3). The authors concluded that the technique might prove useful in the clinical evaluation of HCM. Finally, Schwitter reported results of comparing CMR coronary sinus flow measurements with positron emission tomography (PET) in 16 normal persons and 9 patients after heart transplantation.[11] Coronary sinus flow reserve measured using dipyridamole correlated well between CMR and PET,

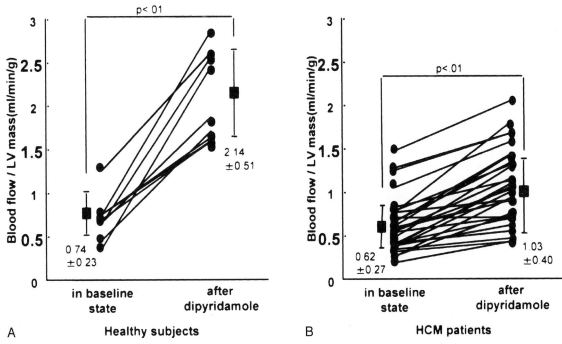

Figure 19–3. Graphs of the coronary blood flow per gram of myocardial mass before and after dipyridamole induced vasodilation. The increase in coronary blood flow was more substantial in (A) healthy volunteers than in (B) patients with hypertrophic cardiomyopathy (HCM). LV = left ventricle. (From Kawada N, Sakume H, Yamakado T, et al: Hypertrophic cardiomyopathy: MR measurement of blood flow and vasodilator flow reserve in patients and healthy subjects. Radiology 211:129, 1999, with permission.)

though with some variability. Estimates of myocardial perfusion in ml/min/g were also good comparing CMR and PET, but some assumptions regarding the mass of myocardium draining into the coronary sinus were made.

Aortic Root Flow Measurements

It has been proposed that coronary flow reserve might also be derived from flow measurements made in the ascending aorta, which is less susceptible than the coronary arteries to cardiac and respiratory motion and partial volume effects.[12] Coronary diastolic flow, which represents the bulk of coronary artery flow, can be estimated as the retrograde flow in the ascending aorta during systole and diastole minus the antegrade flow during diastole. In this study, a variable velocity encoding window was implemented to maintain the accuracy of velocity measurements during periods of both high flow in systole (window = 200 cm/s) and low flow in diastole (window = 30 cm/s).[13] Although it was suggested that the assessment of absolute diastolic coronary flow by this technique is inaccurate because of known errors in the velocity mapping technique, it was argued that these errors should be the same both before and after vasodilation and should therefore be eliminated from the assessment of diastolic coronary flow reserve, defined as the difference (rather than the ratio) between the two measurements. In 7 patients with abnormal myocardial perfusion scintigraphy, the diastolic coronary flow reserve was −50 ml/min, compared with 260 ml/min in 8 healthy volunteers. The principle of this technique was later refined by taking into account the motion of the coronary arteries during the cardiac cycle.[14] A model was developed describing the flow through five transverse parallel aortic slices covering from the base of the aortic valve to above the level of the coronary ostia. This problem was then solved mathematically and in 5 healthy subjects, it was shown that the standard error in the measurement of total coronary artery flow was approximately 90 ml/min or 30 percent of total coronary artery flow. The error in coronary flow reserve is higher still because the errors in the baseline and maximal vasodilation flows are additive. In addition, the assumption that flow is predominantly diastolic, although true in normal subjects, may not hold in the presence of disease, thereby introducing further unknown errors into the technique. A more direct approach to coronary flow assessment is therefore desirable.

DIRECT ASSESSMENT OF CORONARY ARTERY VELOCITY

The problems associated with the assessment of flow velocity and flow in the coronary arteries are outlined in the introduction to this chapter and preclude the application of standard CMR techniques. The major advance for CMR coronary blood flow techniques was the shortening of the sequence duration such that the acquisition could be achieved in a single breath-hold, effectively eliminating respiratory artifact. Since this time, further advances with navigator-echo monitoring of the diaphragm position have allowed data to be acquired over multiple reproducible breath-holds or during free-breathing, enabling increased spatial and temporal resolution as well as data averaging, resulting in higher signal-to-noise ratios. In this section, we describe the approaches that have been used to assess coronary flow velocity and flow.

Bolus Tagging Techniques

The first report detailing the imaging of coronary artery flow was made in rat and mice hearts using bolus tracking.[15] In this technique, based on previous in vivo studies in the aorta and the carotid arteries,[16] a section of blood in the aortic root above the coronary ostia is tagged by the application of a slice-selective presaturation pulse, and imaging of the coronaries is performed after a delay. At this time, the tagged volume of blood has washed into the coronaries, where it is seen as a signal void. The mean velocity of the tagged blood can be calculated from the distance of the tag movement and the wash-in delay. The technique has been developed for multibolus tracking using stimulated echoes[17] and its application demonstrated in a 3-mm diameter tube with laminar flow and in a perfused rat heart. Using this approach, an image of multiple tag pulses (typically three), can be obtained simultaneously to show the coronary artery tree, each having a different wash-in time. Again, the extent of the arterial pathway seen depends upon the flow velocities in the tagged volumes and on the wash-in delays,[17a] with short delays required for visualization of the proximal portions and longer delays for the mid- and distal portions.

Echo-Planar Time-of-Flight Technique

Echo-planar techniques are an attractive option for coronary artery investigations because of their fast imaging times, which reduce the effects of both cardiac and respiratory motion. In 1993, Poncelet first detailed the use of an echo-planar single-shot time-of-flight technique to assess coronary artery flow velocity in 11 normal subjects.[18] In this study, performed at 1.5 T, fat-suppressed 5- to 10-mm short-axis cardiac slices were acquired with an in-plane pixel size of 1.5 × 3 mm, each taking approximately 95 ms. Prior to the 90° slice selection radiofrequency pulse, another 90° pulse was applied to saturate a thick band centered on the imaging slice. Increasing time delays were programmed between the saturation and slice select pulses, which causes the signal intensity in the vessel changes to be sensitive to blood wash-in through the slice. The rate at which it changed was used to calculate the blood velocity. An example of images obtained at a single time point after the R wave is shown in Figure 19–4. For this timing in the cardiac cycle, images (b) through (d) show that blood flow is too slow to wash in to the image slice when the saturation delays were less than 80 ms. However, as the delay is increased from 80 ms (e) to 170 ms (g), increasing amounts of blood wash-in occurred, and by 200 ms (h), flow is fast enough for wash-in to be complete. The images for each time point in the cardiac cycle were acquired within a single breath-hold. The average velocity profile for all 11 subjects is shown in Figure 19–5, which shows the expected peak flow in early diastole. As can be seen, it was not possible to measure the velocities at less than 200 ms from the R wave with this technique, inasmuch as this time period is required for the application of saturation delays. This situation was not regarded to be a problem for normal subjects because coronary flow is predominantly diastolic. A further restriction of the technique is that the long TE of the sequence (28 ms) is such that signal loss is likely to be a problem at sites of turbulent flow. At such sites, there would also be a breakdown in the assumption of laminar flow required for the velocity calcula-

tions from the wash-in data. The authors reported that in 9 subjects imaged during continuous hand and lower extremity exercise, 8 showed a 52 percent increase in diastolic velocity (increase over exercise period). Isometric exercise did not produce a maximal physiological stress response, however, and more recently, coronary flow velocity reserve has been measured in 10 healthy subjects after the infusion of 0.56 mg/kg dipyridamole.[19] In these subjects, the peak diastolic velocity was observed to increase from 22 cm/s to 90 cm/s, resulting in a coronary flow velocity reserve of 3.9, with the velocities returning to baseline after the administration of aminophylline, an established adenosine receptor antagonist.

In this study,[18] the authors chose to acquire low-resolution single-shot coronary images rather than using a segmented approach to build up higher resolution images over a number of cardiac cycles. This technique was prompted by their observation of considerable variability in the beat-to-beat position of the left anterior descending artery during both long and short breath-hold periods. This variability usually corresponded to a downward drift in the vessel position as the breath-hold continued and could be by as much as 6 mm, which is greater than the vessel diameter. It was unclear, however, how much of this movement was due to poor breath-holding and how much was due to beat-to-beat vari-

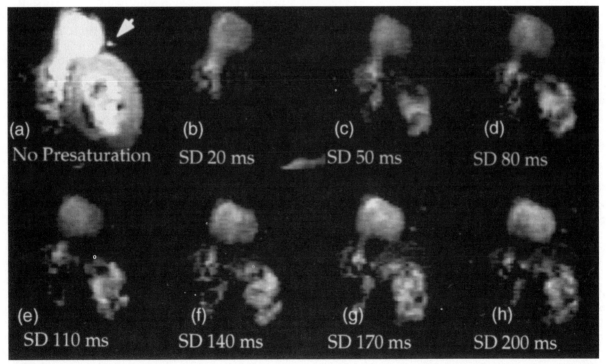

Figure 19–4. Echo-planar time-of-flight images of a short axis slice acquired 350 ms after the R wave with the left anterior descending artery perpendicular to the imaging plane (arrow), acquired in a single breath-hold. The initial image *(a)* is obtained with no saturation pulse and acts as a reference. Images *(b)* to *(h)* are acquired at the same time after the R wave but with a thick slab saturation prepulse at increasingly large saturation delays (SDs) before the image slice selection pulse. The intensity change in the artery as a function of SD is used to calculate the blood flow velocity, on the assumption that the velocity profile is laminar. Images *(a)* to *(h)* were all acquired within the same single breath-hold. (Modified from Poncelet BP, Weisskoff RM, Weeden VJ, et al: Time of flight quantification of coronary flow with echo-planar MRI. Magn Reson Med 30:447, 1993. Reprinted by permission of Wiley-Liss, Inc., a subsidiary of John Wiley & Sons, Inc.)

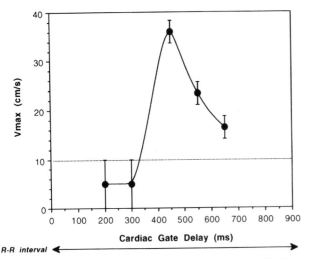

Figure 19–5. Time-of-flight coronary flow velocity profile from 11 normal subjects. For each subject and for each gating delay, a series of images with different saturation delays was acquired as shown in Figure 19–4. (Modified from Poncelet BP, Weisskoff RM, Weeden VJ, et al: Time of flight quantification of coronary flow with echo-planar MRI. Magn Reson Med 30:447, 1993. Reprinted by permission of Wiley-Liss, Inc., a subsidiary of John Wiley & Sons, Inc.)

ations in cardiac contraction. Regardless of the cause, it would introduce blurring, which is largely avoided in a single-shot image. Beat-to-beat variations in flow, as opposed to spatial position, still have an effect on the measured velocity because individual images need to be acquired with different saturation delays in consecutive cardiac cycles in order to generate the wash-in curve. The measured velocity will therefore be affected by changes over the breath-hold period. The importance of this study lies in its pioneering and largely successful approach to a previously unsolved problem. It has largely been superseded by phase velocity mapping approaches, as discussed in the next sections, which are more robust and have higher availability.

Gradient-Echo Phase Velocity Mapping—Breath-Hold Techniques

Using a velocity-encoded segmented k-space gradient-echo technique, velocity maps may be acquired in a single breath-hold, thereby reducing respiratory motion artifact. To accomplish this goal, the segment duration is typically of the order of 100 ms. In addition to limiting the temporal resolution of the sequence, the need to perform data acquisition within a single breath-hold limits the number of phase encoding steps that can be acquired, which in turn limits the spatial resolution and the signal-to-noise ratio in the resulting images. These issues introduce problems in applying the technique in the coronaries.

The first through-plane coronary artery phase velocity maps were reported by Edelman and cowork-

ers.[20] An example of such an acquisition is demonstrated in Figure 19–6. The velocity maps were acquired at 1.5 T using a fat-suppressed sequence consisting of a reference velocity compensated gradient waveform followed by a sensitized (velocity window = 150 cm/s) gradient waveform, each repeated four times in a cardiac cycle. The TE was 8 ms and the repetition time (TR) 13 ms, resulting in a segment duration of 104 ms. The sequence was validated against a Doppler flowmeter in vitro using both constant and mildly pulsatile flow in an 8-mm diameter tube and in vivo, against a standard nonsegmented velocity mapping sequence in the descending aorta of three volunteers. Velocity maps were acquired over 24 cardiac cycles with in-plane pixel dimensions of 1.4 × 0.8 mm. In healthy subjects, the mean flow velocities at rest in the mid-portion of the right and left anterior descending coronary arteries were 10 cm/s and 21 cm/s, respectively. These values are lower than those found in Doppler flow-wire studies, which is to be expected, inasmuch as the small number of pixels across the vessel results in partial volume averaging of the velocity profile. In four subjects who received an intravenous administration of adenosine, velocities increased by at least a factor of 4, which suggested that CMR had the potential to assess the flow response to vasodilation. Following this finding, there have been further validation studies. Grist and associates[21] assessed coronary flow velocity in 10 normal subjects and showed significant increases during hyperemia similar to published values. Kessler and colleagues[22] compared CMR to intracoronary Doppler in 15 patients with coronary disease and found a good correlation, although CMR measurements showed significantly lower velocities than were seen by Doppler (9 vs. 12 cm/s; $p < 0.001$). Shibata and coworkers[23] found lower values of velocity by CMR compared with Doppler guidewire (13 vs. 32 cm/s; $p < 0.01$), with good correlation. Similar results have been found by Furber and colleagues[24] in patients after myocardial infarction (maximum velocity 27 vs. 36 cm/s). The overall picture emerges that, in general, CMR underestimates coronary flow velocity using current techniques but can track changes well and measure flow reserve reasonably.

Another approach to quantifying stenosis severity is to perform in-plane coronary artery velocity mapping with a view to measuring an increased velocity at the site of a lumen narrowing. In-plane coronary artery velocity mapping was first performed in normal subjects by Keegan and colleagues[25] using a sequence with a TE of 10 ms, a TR of 20 ms, and a segment duration of 160 ms, and acquisition in early diastole. The in-plane resolution was 1.6 × 0.8 mm and the data acquired over breath-holds of 24 to 32 cardiac cycles. The velocity sensitivity used was 50 cm/s and was achieved by the phase map subtraction of two images, one sensitized to flow velocities of +100 cm/s and the other sensitized to flow velocities of −100 cm/s. This method was shown to result in fewer blood flow artifacts than subtracting

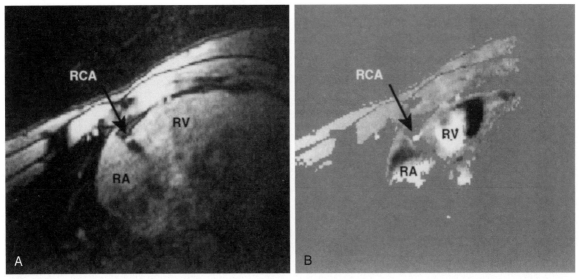

Figure 19–6. Mid-diastolic (gating delay = 550 ms) oblique transverse breath-hold acquisition showing through-plane right coronary artery *(A)* magnitude image and *(B)* corresponding velocity map. The velocity map shows a peak flow velocity of 13 cm/s. RV = right ventricle; RA = right atrium; RCA = right coronary artery.

an image sensitized to 50 cm/s from a reference nonsensitized image. Through-plane and in-plane velocities measured with this technique were validated in vitro against a standard nonsegmented velocity encoding sequence in a 5.6-mm diameter tube with pulsatile flow having a maximum velocity of 30 cm/s and a maximum rate of change of velocity of 126 cm/s^2, comparable to those values expected in normal human coronary arteries at rest. Phantom work was also carried out to show the ability of the technique to measure a velocity increase at the sites of mild, moderate, and severe area reducing stenoses and hence to quantify severity. Examples of in-plane left anterior descending and right coronary artery velocity maps in healthy subjects are shown in Figure 19–7. Although this approach appears to be promising, the tortuous pathways and small caliber of the coronary vessels result in partial-volume type effects being more problematic for in-plane than for through-plane velocity mapping. In addition, in-plane studies require a high degree of reproducibility in the breath-holding position, which is difficult to achieve without techniques such as navigator-echo monitoring. Despite these problems, velocity increases have been observed at the sites of area-reducing stenoses, an example being shown in Figure 19–8, where a CMR image and velocity map of the right coronary artery of a patient with moderate and severe area reducing stenoses are presented, together with the corresponding x-ray contrast angiogram.

Gradient-Echo Phase Velocity Mapping—Navigator Techniques

All of the aforementioned studies have used breath-holding as a means of respiratory motion

control, limiting the sequence parameters to allow the entire data set to be acquired within the duration of a single breath-hold and requiring a high degree of patient cooperation, which is not always forthcoming. In addition, inter and intra variations in the breath-holding position may be problematical and hemodynamic changes that occur secondary to breath-holding, such as increased intrathoracic pressure and raised heart rate, may themselves alter the blood flow being measured. A navigator-echo approach under either prospective or retrospective control would enable data to be acquired during free-breathing and avoid these problems. Furthermore, the temporal resolution of the sequence could be improved by reducing the segment duration albeit at the expense of prolonged scan time. The influence of improved temporal resolution was investigated by Hofman and associates,[26] who compared the use of a segmented breath-hold technique (segment duration 126 ms) with a retrospective respiratory gating technique[27] (reference/velocity sensitized view pair duration 32 ms) for the assessment of flow velocity, vessel cross-sectional area, and volume flow in the right coronary artery of six healthy volunteers. Eight data averages were acquired and the data reconstructed without retrospective respiratory gating for four and eight averages (resulting in decreasing degrees of ghosting and improving signal-to-noise ratios) and with retrospective gating for four averages. The residual displacement of diaphragm positions in the latter image set was 3.9 mm. The in-plane spatial resolution was 0.8 × 1.6 mm and the velocity sensitivity ±25 cm/s. The slice orientation was varied throughout the cardiac cycle according to localizer images acquired at different cardiac phases, thereby ensuring that the imaging slice was always perpendicular to the direction of flow. Vessel regions of interest were obtained semi-

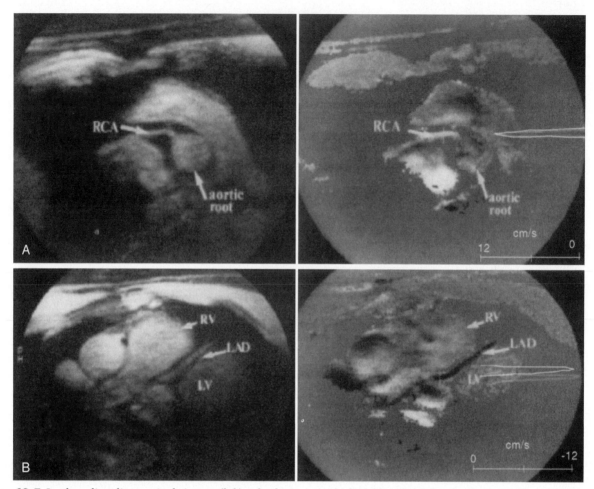

Figure 19–7. In-plane diastolic magnitude images *(left)* and velocity maps *(right)* with superimposed velocity profiles of *(A)* a proximal right coronary artery (peak velocity = 12 cm/s) and *(B)* a left anterior descending artery (peak velocity = 12 cm/s). LV = left ventricle; RV = right ventricle; RCA = right coronary artery; LAD = left anterior descending artery. (Modified from Keegan J, Firmin D, Gatehouse PD, Longmore D: The application of breath-hold phase velocity mapping techniques to the measurement of coronary artery blood flow velocity: Phantom data and initial in vivo results. Magn Reson Med 31:526, 1994, with permission.)

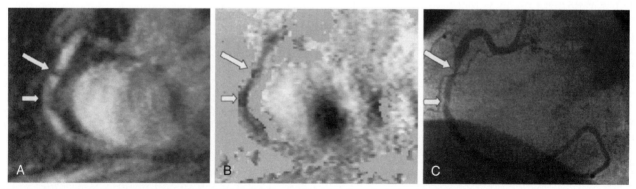

Figure 19–8. *(A)* CMR magnitude image and *(B)* velocity map of an in-plane right coronary artery showing signal loss and increased flow velocity at the sites of moderate and severe area reducing stenoses (arrows) and *(C)* corresponding x-ray contrast angiogram. (Modified from Mohiaddin RH, Pennell DJ: Magnetic resonance imaging of flow in the cardiovascular system. *In* Reichek N (ed): Cardiology Clinics. Philadelphia, WB Saunders, 1998, p 161, with permission.)

automatically from the magnitude images of the velocity-compensated data set by means of by a seed-growing algorithm and a magnitude threshold set at 50 percent of the difference between the magnitude signal of the surrounding tissue and the vessel center. Velocity profiles generated from the retrospectively gated data of all six subjects (uncorrected for through-plane velocity of the vessel itself) are shown in Figure 19–9 and demonstrate considerable similarity with a peak in systole, a minimum at end systole, and a second peak in early diastole. The in-plane displacement of the vessel, measured on the same images, was also found to vary considerably (Figure 19–10), with a peak displacement in systole, a second peak in early diastole, and minimal displacement (of the order of the spatial resolution of the image) in mid to end diastole. The authors have since reported a similar pattern of movement for the left anterior descending artery,[28] data that have been confirmed by others.[28a, b] During times of peak vessel displacement, motion blurring of the artery occurs and the breath-hold segmented images are consequently of poorer quality than those acquired with the respiratory gated short segment duration sequence (Figure 19–11A). In diastole, however, when the in-plane displacement of the vessel is low, the two techniques generate images of comparable quality (Figure 19–11B). This same effect results in the breath-hold segmented k-space gradient-echo sequence overestimating the instantaneous vessel

cross-sectional area by as much as a factor of 4, with the average increase in the time-averaged cross-sectional area being 90 percent. The nongated technique with four and eight averages also overestimates the time-averaged cross-sectional areas and, as a result, the time-averaged mean velocities within the vessel are significantly lower.

A comparison between breath-hold segmented k-space gradient-echo phase velocity mapping and Doppler flow-wire techniques has recently been performed in 26 angiographically normal coronary artery segments.[29] Both breath-holding and real-time slice-followed[30] navigator-echo–controlled free-breathing CMR techniques (temporal resolution 140 ms and 46 ms, respectively) were performed within 24 hours of the invasive procedure and the maximal coronary flow velocities determined in a 2×2 pixel area. Both CMR techniques were found to significantly underestimate flow velocity, although the correlations with the invasive measurements were strong (r = 0.7 and 0.86, respectively). The navigator-echo–controlled technique was significantly more accurate than the breath-hold technique ($p <$ 0.02), although this improvement was at the expense of prolonged total acquisition time. The underestimation of blood flow velocity by both techniques is largely due to the partial volume averaging of the spatial flow profile in the CMR studies together with insufficient temporal sampling of the flow profile. The decreased accuracy of the breath-

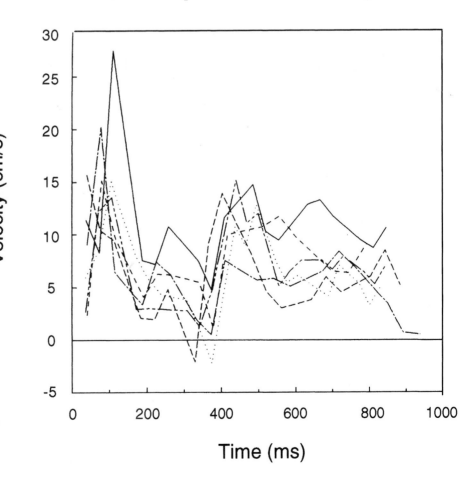

Figure 19–9. Through-plane cross-sectional averaged velocity in the right coronary artery as a function of time in the cardiac cycle as measured on a respiratory gated acquisition in six subjects. (Modified from Hofman MBM, van Rossum AC, Sprenger M, Westerhof N: Assessment of flow in the right human coronary artery by magnetic resonance phase contrast velocity measurement: Effects of cardiac and respiratory motion. Magn Reson Med 35:521, 1996. Reprinted by permission of Wiley-Liss, Inc., a subsidiary of John Wiley & Sons, Inc.)

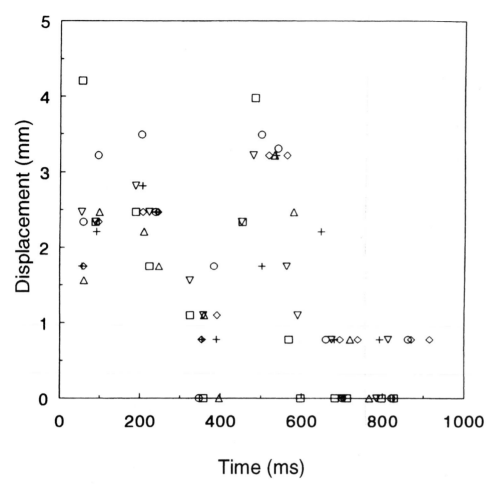

Figure 19–10. In-plane displacement of right coronary artery through the cardiac cycle as measured on a respiratory gated acquisition (oversampling factor = 8). Different markers represent data from different subjects (n = 6). (Modified from Hofman MBM, van Rossum AC, Sprenger M, Westerhof N: Assessment of flow in the right human coronary artery by magnetic resonance phase contrast velocity measurement: Effects of cardiac and respiratory motion. Magn Reson Med 35:521, 1996, with permission.)

hold technique compared with the navigator-echo free-breathing technique is likely to be due to a combination of the longer acquisition window of the former together with hemodynamic changes resulting from the breath-holding procedure itself. Figure 19–12 shows an example of the results obtained. Although still underestimating flow velocities, the significantly improved accuracy of the navigator-echo free-breathing technique compared with the breath-hold technique provides a further step towards the CMR assessment of coronary flow parameters albeit at the expense of prolonged acquisition times.

Interleaved Spiral Phase Velocity Mapping

As described earlier, interleaved spiral imaging is an alternative technique for generating high-resolution images of the coronary arteries,[31] which, when compared with those acquired using a segmented k-spae gradient-echo approach, have higher signal-to-noise ratios and better temporal resolution.[32, 33] Interleaved spiral phase velocity maps of the coronary arteries have been acquired in diastole[34] and throughout the cardiac cycle.[35] The latter study showed that the improved temporal resolution of

the spiral sequence enabled better detection of the peak phasic coronary blood flow and the shorter readout period resulted in less vessel blur, this being particularly apparent at times of high vessel mobility, in early systole and early diastole. More recently, navigator-echo–controlled interleaved spiral cine phase velocity mapping has been used to generate coronary flow velocity curves in normal subjects,[34] an example being shown in Figure 19–13. This study was acquired in 112 s under free-breathing conditions on a 0.5 T scanner and demonstrates the phasic nature of coronary blood flow velocity. The relatively rapid scan time should enable the acquisition of a similar study under pharmacological stress and hence an estimate of coronary flow reserve. This technique shows considerable promise but more research is needed to investigate how the blurring of off-resonance material[36] and flow direction–sensitive "implosion/explosion" artifacts in regions of poor homogeneity[37] affect the measurements.

CORONARY FLOW AND CORONARY FLOW RESERVE

The assessment of coronary flow, rather than flow velocity, is more difficult due to partial volume ef-

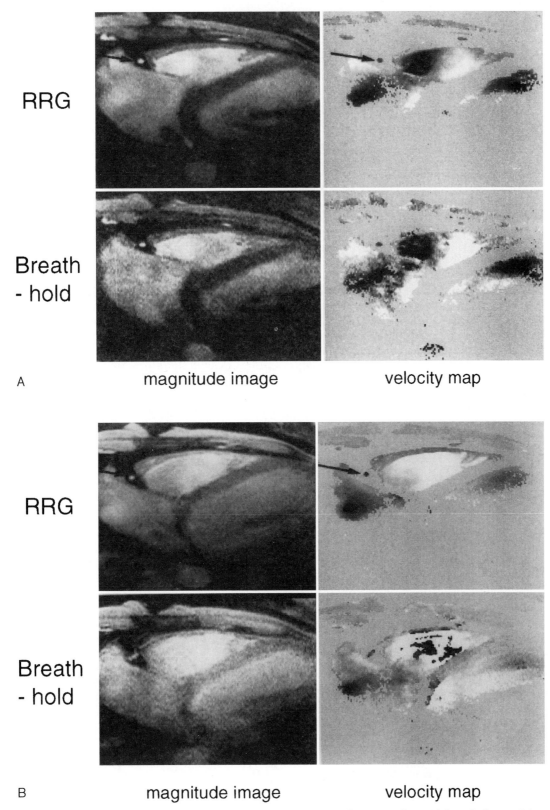

Figure 19–11. *A,* Through-plane magnitude images *(left)* and velocity maps *(right)* acquired in mid-diastole (gating delay 760 ms) for retrospective respiratory gated acquisitions *(top,* RRG) and breath-hold acquisitions *(bottom). B,* Through-plane magnitude images and velocity maps in the same subject as *(A)* acquired in early systole gating delay 90 ms. Note the blurring in the breath-hold image in systole due to coronary motion. (Modified from Hofman MBM, van Rossum AC, Sprenger M, Westerhof N: Assessment of flow in the right human coronary artery by magnetic resonance phase contrast velocity measurement: Effects of cardiac and respiratory motion. Magn Reson Med 35:521, 1996, with permission.)

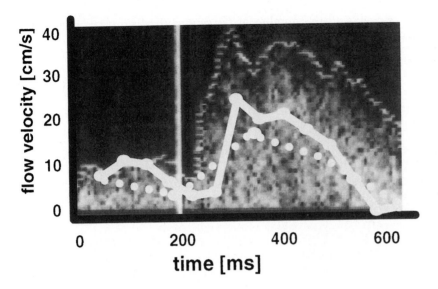

Figure 19–12. Original tracing of an invasively determined flow curve of the left coronary artery in comparison with noninvasively determined flow curves of the same patient. Full line = real time adaptive navigator correction technique; dotted line = breath-hold technique. (Modified from Nagel E, Bornstedt A, Hug J, et al: Noninvasive determination of coronary blood flow velocity with magnetic resonance imaging: Comparison of breath-hold and navigator techniques with intravascular ultrasound. Magn Reson Med 41:544, 1999. Reprinted by permission of Wiley-Liss, Inc., a subsidiary of John Wiley & Sons, Inc.)

fects at the vessel boundary that result in overestimation of the vessel cross-sectional area. The first directly validated measurements of coronary artery flow and flow reserve using segmented k-space gradient-echo phase velocity mapping were performed by Clarke and colleagues in dogs in 1995.[38] Nonmagnetic perivascular ultrasound probes were placed around the isolated left anterior descending and circumflex arteries in eight ventilated dogs. A subcritical stenosis was generated in the left anterior descending artery by placement of a Lexan constrictor.

Breath-holding was achieved by temporarily turning the ventilator off. Cine phase-velocity mapping was performed using a sequence with a TE of 11 ms and a TR of 19 ms with 2 to 3 reference/velocity-sensitized view pairs per data segment. The segment duration was therefore 76 to 115 ms, allowing the acquisition of 4 to 6 images over the cardiac cycle. Images were acquired with an in-plane pixel size of 0.7 to 0.9 mm over breath-holds of up to 40 seconds duration. The velocity sensitivity was ±138 cm/s. Data were acquired both before and after the admin-

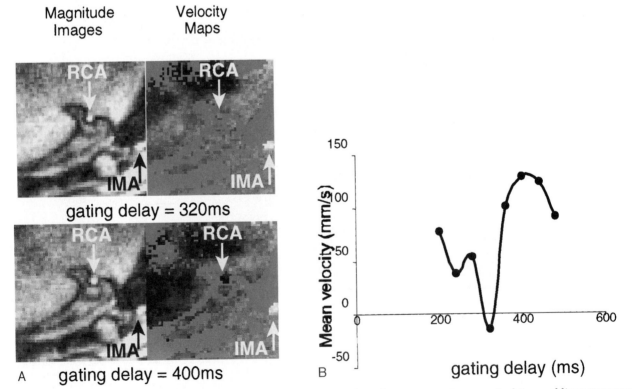

Figure 19–13. Interleaved spiral magnitude images and velocity maps of a right coronary artery acquired in an oblique transverse plane at times of minimum hand maximum flow velocity (gating delays of 320 ms and 400 ms, respectively) *(A)*, and flow velocity throughout the cardiac cycle *(B)*. IMA = internal mammary artery.

istration of adenosine. A region of interest was drawn around the artery of interest in the magnitude image and the mean velocity in that region on the corresponding velocity map calculated. In this study, the size of the region of interest was kept constant from frame to frame and only its position changed, as the authors felt that the spatial resolution in the CMR image was insufficient to track phasic changes in the coronary arterial diameter. The results of a typical experiment are shown in Figure 19–14A, and a plot of CMR flow reserve versus that measured by ultrasound is shown in Figure 19–14B. For the circumflex and left anterior descending arteries, the mean CMR measured flow reserves were 2.6 and 1.4, respectively ($p = 0.011$) compared with ultrasound measured values of 2.6 and 1.4 respectively ($p = 0.002$). The authors concluded that although the breath-hold period in this study is not feasible for the majority of patients and the temporal and spatial resolution limited, the results of the phase velocity mapping technique agreed well with Doppler ultrasound and suggested that it could be developed further for the measurement of flow in the major epicardial coronary arteries.

This same technique was used to measure flow in the left anterior descending artery of 12 subjects both before and after adenosine prior to assessment with an intracoronary Doppler flow-wire during cardiac catheterization.[39] In this study with an in-plane pixel size of 0.8 to 1 mm, an in-plane presaturation pulse was applied prior to each cine acquisition, suppressing the signal from tissue in the slice and enabling inflowing blood to be visualized with improved contrast. Depending on the heart rate, 4 to 5 cine phases were acquired per cardiac cycle with

the number of phase encoding steps reduced to ensure complete data acquisition in a breath-hold of up to 25 seconds. A respiratory gating belt was used to ensure the consistency of the breath-hold positions. Excellent agreement was reported between CMR and ultrasound coronary flow and flow reserve measurements, with the limits of agreement being 23 and − 25 ml/min and 0.6 and −0.6, respectively.

Sakuma and associates[40] used a similar technique to measure coronary flow in dogs and velocity profiles in the left anterior descending artery of eight healthy volunteers both before and after dipyridamole.[41] However, the use of view sharing enabled the effective temporal resolution of the imaging sequence to be improved to 64 ms and 7 to 13 cine images were acquired in an end-expiratory breath-hold of 24 heartbeats. Correction of the coronary blood flow velocity for the through-plane velocity of the vessel as a whole was also performed by assuming the velocity of the vessel was the same as that in an area of adjacent myocardium. This figure averages to zero over the cardiac cycle as a whole and is therefore unimportant for the assessment of mean coronary blood flow parameters, but may vary from as much as + 20 cm/s in systole to 10 cm/s in diastole[42] and consequently has important implications for the assessment of instantaneous flow and flow velocity.[43] Volume flow data were not calculated because the authors felt that the small number of pixels across the vessel prevented the accurate assessment of cross-sectional area. An example of velocity profiles both before and after dipyridamole are shown in Figure 19–15. The average baseline peak diastolic flow velocity was 15 cm/s rising to 46 cm/s after dipyridamole. The average coronary

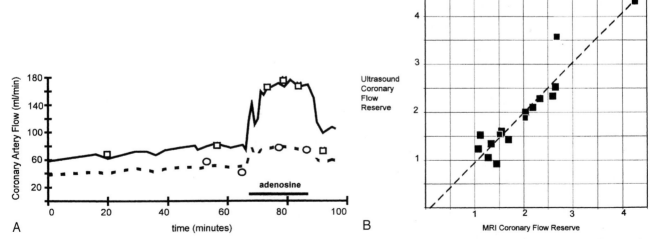

Figure 19–14. Validation of coronary flow reserve in dogs. On the left (A) is a graph of the ultrasound measurements of flow in the left circumflex artery (solid line) and left anterior descending artery (dashed line). Two baseline cine velocity magnetic resonance image sets were also acquired in the left anterior descending artery (open circle) and left circumflex artery (open square). Magnetic resonance velocity cine image sets were also recorded in each vessel during adenosine infusion. On the right (B) is a scatterplot of linear regression of the 16 magnetic resonance–estimated and ultrasound-measured coronary flow reserve measurements, which produced a regression line with a slope of 1.04, intercept of 0.1, and correlation coefficient of 0.94. (Modified from Clarke GD, Eckels R, Chaney C, et al: Measurement of absolute epicardial coronary artery blood flow and flow reserve with breath-hold cine phase-contrast magnetic resonance imaging. Circulation 91:2627, 1995, with permission.)

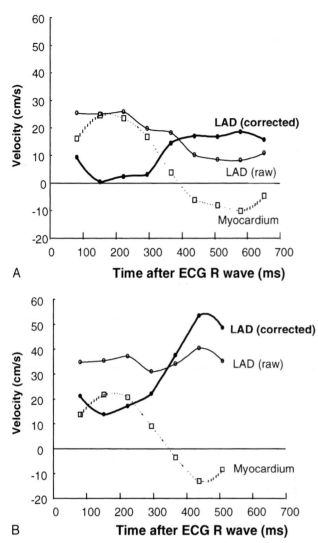

Figure 19–15. Phasic blood flow velocity profiles before and after correction for vessel movement both before (upper graph, *A*) and after (lower graph, *B*) administration of dipyridamole in a healthy subject. Note the important effect that the subtraction of the myocardial velocity has on the apparent coronary flow velocity. LAD = left anterior descending artery; ECG = electrocardiogram. (Modified from Sakuma H, Blake LM, Amidon TM, et al: Coronary flow reserve: Non-invasive measurement in humans with breath-hold velocity encoded cine MR imaging. Radiology 198:745, 1996, with permission.)

flow reserve was 3.1. The interstudy reproducibility before and after pharmacological stress (the absolute difference in the two measurements divided by their mean) was 9.5 percent and 6.8 percent. Interobserver reproducibility was comparable. A similar study performed at the same center in 12 healthy volunteers using sustained handgrip exercise to induce vasodilation likewise showed an increase in the peak diastolic flow velocity from 21 cm/s to 31 cm/s.[44]

In a further feasibility study, coronary volume flow profiles were measured by a similar technique in normal volunteers.[45] Again, by using view-sharing techniques, 7 to 13 images could be obtained per cardiac cycle over breath-holds of up to 20 seconds, depending on the R-R interval. In this study, the in-

plane resolution was 0.9 × 1.4 mm. Regions of interest were drawn around the vessel using the magnitude image as a guide. Within this region, automated edge detection was employed to determine the exact vessel area, using a magnitude threshold of 35 percent. The region of interest was adjusted (in both position and size) for each frame in the cardiac cycle. Three acquisitions were performed before and after dipyridamole. The mean area covered by the left anterior descending artery was 16 mm[2] before dipyridamole, rising to 18 mm[2] after dipyridamole. Peak velocity, mean coronary flow velocity, and mean flow all increased significantly after dipyridamole, with the coronary flow reserve being 5.0 compared with a coronary velocity reserve of 3.5. There are a number of potential reasons for the discrepancy between the coronary reserves measured with volume flow and peak velocity, the most likely being that the calculations made from peak velocity measurements are based on single pixel values with higher statistical noise that are uncorrected for the through-plane movement of the vessel itself.

Wedding and coworkers[46] also compared the CMR measurements of coronary flow and flow reserve with those obtained from an ultrasonic transit-time probe in dogs. In this study, the CMR results were obtained from phase-difference processing with 30 percent and 50 percent magnitude thresholds and from complex difference processing,[47, 48] both with[49] and without modification for in-plane vessel motion. The complex difference technique takes into account partial volume averaging of edge pixels and, given the small diameter of the coronary arteries and the relatively large pixel size, was expected to have performed better than the phase difference processing. In fact, mean CMR flow measurements correlated well with the ultrasound data (r ≈ 0.92 for all techniques), but both the complex difference technique and the 50 percent magnitude threshold phase difference technique systematically underestimated flow. The phase difference technique with the 30 percent magnitude threshold provided the best agreement, similar to that reported in other studies,[38, 39] but was very sensitive to vessel boundary identification.

Finally, two recent studies have been reported by Hundley and associates.[50, 51] In the first, 30 patients were studied to determine if coronary flow reserve could identify significant left anterior descending or left main coronary disease.[50] A coronary flow reserve of less than 1.7 identified stenosis greater than 70 percent, with a sensitivity of 100 percent and a specificity of 83 percent. In the second similar study, a flow reserve of less than 2 was shown to predict restenosis after angioplasty with a sensitivity of 100 percent and a specificity of 82 percent for luminal narrowing of greater than 70 percent.[51]

CONCLUSION AND FUTURE DEVELOPMENTS

A number of feasibility studies have now shown that CMR techniques have the potential to assess coronary artery flow velocity, flow, and flow reserve, with the most commonly used technique being phase velocity mapping. Although several feasibility studies have shown good correlation of CMR data with Doppler flow-wire and ultrasound transit-time techniques, the robustness of the technique needs to be improved and larger scale validation studies need to be performed. To summarize the requirements, any technique must take into account the following factors:

1. *Spatial resolution.* For the assessment of volume flow, the spatial resolution must be sufficient to measure the cross-sectional area throughout the cardiac cycle and to minimize partial volume averaging of flow velocities in edge pixels. These partial volume effects become larger as the vessel becomes smaller and the relative number of pixels in the boundary increases. The degree to which measured velocities are affected depends strongly on the relative magnitude of the stationary material included in the boundary pixels. Several studies have investigated these effects in small vessels. Hofman and coworkers[52] compared phase contrast measurements of time-averaged volume flow in the femoral arteries of dogs with measurements made by an ultrasonic transit time meter and found that the proportional difference between the techniques was 0.8 percent when the number of pixels across the vessel diameter was just 3. Tarnawski and associates[53] similarly found good agreement between measured and actual laminar flows in phantoms when the number of pixels across the phantom diameter varied from 2.5 to 10, provided that the region of interest defined was larger than the actual phantom cross-sectional area. Tang and coworkers[54] found that for both in vitro and in vivo studies, the volume flow accuracy increases with resolution, as expected, and that errors are less than 10 percent when the ratio of pixel size to vessel radius is less than 0.5. This was also seen by Wolf and colleagues,[55] who showed a percentage error of 9 percent for 5 pixels across the vessel diameter. Sondergaard and coworkers[56] showed a less than 18 percent error in measured flow rate for 4 pixels across the vessel diameter in vitro.

2. *Temporal resolution.* The temporal resolution of the sequence employed should be adequate to resolve the biphasic velocity profile of coronary blood flow, this ability being determined by the number of views acquired per data segment in a segmented k-space gradient-echo acquisition. For the accurate assessment of time-averaged parameters, measurements should be made throughout the entire cardiac cycle, the accuracy increasing as the number of cine frames increases.[57] For segmented k-space gradient-echo acquisitions, both these points favor the acquisition of small numbers of views per data segment although this method leads to long breath-hold periods or to the need for navigator free-breathing techniques. Alternatively, more rapid and efficient methods of k-space coverage may be employed such as interleaved rectilinear or spiral echo-planar. A further reason to reduce the data acquisition duration is to minimize blurring of the vessel due to movement in the acquisition window, which leads to overestimation of the vessel cross-sectional area through partial volume averaging of edge pixels and to underestimations in the mean flow velocity.

3. *Velocity sensitivity.* The maximum velocity in normal coronary arteries at rest is typically less than 25 cm/s. The window used for resting studies should therefore be narrow, ≤50 cm/s, so as to be able to measure these low velocities with maximum accuracy but without aliasing. For maximal vasodilation studies, the window should be increased accordingly.

4. *Through-plane movement of vessel.* As has been noted, through-plane movement of the vessel through the cardiac cycle can considerably affect instantaneous measurements of blood flow velocity and volume flow in that vessel, although time-averaged measurements throughout the cardiac cycle are unaffected because the through-plane movement of the vessel averages to zero. These errors are potentially reduced by implementing a complex difference rather than a phase difference approach to velocity assessment.[47, 48, 58]

In addition to these factors, it should also be borne in mind that beat-to-beat variations in the position in and flow through the artery will inevitably influence the results and that these variations may be exacerbated by breath-holding, particularly if prolonged. A free-breathing technique making use of navigator echoes may therefore be the best approach and would also allow higher spatial and temporal resolution images albeit with reduced scan efficiency and consequently longer scan duration. As discussed earlier, such a technique has recently been shown to significantly improve the accuracy of CMR velocity assessment[29] compared with long segment duration breath-holding studies. The long scan times could potentially be reduced by employing a real-time phase encode reordering technique whereby the most significant central lines of k-space are acquired with the diaphragm position, as measured by a navigator echo, within a narrow range and the outermost lines acquired with the diaphragm position within a larger range, a technique that has already been successfully applied to coronary imaging.[58] Furthermore, real-time slice-following may also be implemented and could potentially result in improved scan efficiency by allowing the use of larger navigator windows without reduction in image quality.

References

1. Pennell DJ, Bogren H, Keegan J, et al: Assessment of coronary artery stenosis by magnetic resonance imaging. Heart 75:127, 1996.
2. Mohiaddin RH, Pennell DJ: Magnetic resonance imaging of flow in the cardiovascular system. *In* Reichek N (ed): Cardiology Clinics. Philadelphia, WB Saunders, 1998, p 161.

3. Hood WB: Regional drainage of the human heart. Br Heart J 30:105, 1968.

4. van Rossum AC, Visser FR, Hofman MBM, et al: Global left ventricular perfusion: Noninvasive measurement with cine MR imaging and phase velocity mapping of coronary venous outflow. Radiology 182:685, 1992.

5. Canty JM, Brooks A: Phasic volumetric coronary venous outflow patterns in conscious dogs. Am J Physiol 258:H1457, 1990.

6. Mirowitz SA, Lee JKT, Guterrez FR, et al: Normal signal-void patterns in cardiac cine MR images. Radiology 176:49, 1990.

7. Ganz W, Tamura K, Marcus HS, et al: Measurement of coronary sinus blood flow by continuous thermodilution in man. Circulation 54:181, 1971.

8. Kawada N, Sakuma H, Yamakado T, et al: Hypertrophic cardiomyopathy: MR measurement of blood flow and vasodilator flow reserve in patients and healthy subjects. Radiology 211:129, 1999.

9. Edelman RR, Wallner B, Singer A, et al: Segmented TurboFLASH: Method for breath-holding MR imaging of the liver with flexible contrast. Radiology 177:515, 1990.

10. Foo TK, Bernstein MA, Aisen AM: Improved ejection fraction and flow velocity estimates with use of view sharing and uniform repetition time excitation with fast cardiac techniques. Radiology 195:471, 1995.

11. Schwitter J, DeMarco T, Kneifel S, et al: Magnetic resonance-based assessment of global coronary flow and flow reserve and its relation to left ventricular functional parameters: a comparison with positron emission tomography. Circulation. 101:2696, 2000.

12. Bogren HG, Buonocore MH: Measurement of coronary artery flow reserve by magnetic resonance velocity mapping in the aorta. Lancet 341:899, 1993.

13. Buonocore MH: Blood flow measurement using variable velocity encoding in the RR interval. Magn Reson Med 29:790, 1993.

14. Buonocore M: Estimation of total coronary artery flow using measurements of flow in the ascending aorta. Magn Reson Med 32:602, 1994.

15. Burstein D: MR imaging of coronary artery flow in isolated and in vivo hearts. J Magn Reson Imaging 1:337, 1991.

16. Edelman R, Mattle HP, Keefield J, Silver MS: Quantification of blood flow with dynamic MR imaging and presaturation bolus tracking. Radiology 171:551, 1989.

17. Chao H, Burstein D: Multibolus stimulated echo imaging of coronary artery flow. J Magn Reson Imaging 7:603, 1997.

17a. Stuber M, Boernert P, Spuentrop E, et al: Three-dimensional projection coronary magnetic resonance angiography (abstract). Int Soc Magn Reson Med 2001 (in press).

18. Poncelet BP, Weisskoff RM, Weeden VJ, et al: Time of flight quantification of coronary flow with echo-planar MRI. Magn Reson Med 30:447, 1993.

19. Firmin DN, Poncelet BP: Echo-planar imaging of the heart. *In* Schmitt F, Stehling MK, Turner R (eds): Echo-Planar Imaging—Theory, Technique and Application. Berlin, Springer-Verlag, 1998, p 389.

20. Edelman RR, Manning WJ, Gervino E, Li W: Flow velocity quantification in human coronary arteries with fast breath-hold MR angiography. J Magn Reson Imaging 3:699, 1993.

21. Grist TM, Polzin JA, Bianco JA, et al: Measurement of coronary blood flow and flow reserve using magnetic resonance imaging. Cardiology 88:80, 1997.

22. Kessler W, Moshage W, Galland A, et al: Assessment of coronary blood flow in humans using phase difference MR imaging. Comparison with intracoronary Doppler flow measurement. Int J Card Imaging 14:179, 1998.

23. Shibata M, Sakuma H, Isaka N, et al: Assessment of coronary flow reserve with fast cine phase contrast magnetic resonance imaging: comparison with measurement by Doppler guide wire. J Magn Reson Imaging 10:563, 1999.

24. Furber AP, Lethimonnier F, Le Jeune JJ, et al: Noninvasive assessment of the infarct-related coronary artery blood flow velocity using phase-contrast magnetic resonance imaging after coronary angioplasty. Am J Cardiol 84:24, 1999.

25. Keegan J, Firmin D, Gatehouse PD, Longmore D: The application of breath hold phase velocity mapping techniques to the measurement of coronary artery blood flow velocity: Phantom data and initial in vivo results. Magn Reson Med 31:526, 1994.

26. Hofman MBM, van Rossum AC, Sprenger M, Westerhof N: Assessment of flow in the right human coronary artery by magnetic resonance phase contrast velocity measurement: Effects of cardiac and respiratory motion. Magn Reson Med 35:521, 1996.

27. Lenz GW, Haacke EM, White RD, et al: Retrospective respiratory gating: A review of technical aspects and future directions. Magn Reson Imaging 7:445, 1989.

28. Hofman MBM, Wickline SA, Lorenz CH: Quantification of in-plane motion of the coronary arteries during the cardiac cycle: Implications for acquisition window duration for MR flow quantification. J Magn Reson Imaging 8:568, 1998.

28a. Kim WY, Botnar RM, Stuber M, et al: Patient-specific diastolic acquisitions are required for free breathing right coronary MR wall imaging (abstract). J Cardiovasc Magn Reson 2001 (in press).

28b. Wang Y, Vidan E, Bergman GW: Cardiac motion of coronary artery variability in the rest period and implications for coronary MR angiography. Radiology 213:751, 1999.

29. Nagel E, Bornstedt A, Hug J, et al: Noninvasive determination of coronary blood flow velocity with magnetic resonance imaging: Comparison of breath-hold and navigator techniques with intravascular ultrasound. Magn Reson Med 41:544, 1999.

30. Danias PG, McConnell MV, Khasigawala VC, et al: Prospective navigator correction of image position for coronary MR angiography. Radiology 203:733, 1997.

31. Meyer CH, Hu BS, Nishimura DG, Macovski A: Fast spiral coronary artery imaging. Magn Reson Med 28:202, 1992.

32. Keegan J, Gatehouse PD, Taylor AM, et al: Coronary artery imaging in a 0.5Tesla scanner: implementation of real-time navigator echo controlled segmented k-space FLASH and interleaved spiral sequences. Magn Reson Med 41:392, 1999.

33. Taylor AM, Keegan J, Jhooti P, et al: A comparison between segmented k-space FLASH and interleaved spiral MR coronary angiography sequences. J Magn Reson Imaging 11:394, 2000.

34. Keegan J, Gatehouse P, Yang GZ, Firmin D: Interleaved spiral cine coronary artery velocity mapping. Magn Reson Med 43:787, 2000.

35. Hofman MBM, Groen J, van Muiswinkel A, et al: Spiral acquisition increases temporal resolution and reduces motion artefacts in breath-hold coronary flow measurements. Proceedings of the 6th annual meeting of the International Society of Magnetic Resonance in Medicine, 1998, p 2142.

36. Yudilevich E, Stark H: Spiral sampling in magnetic resonance imaging—the effect of inhomogeneities. IEEE Trans Med Imaging MI-6:337, 1987.

37. Gatehouse PD, Firmin DN: Flow distortion and signal loss in spiral imaging. Magn Reson Med 41:1023, 1999.

38. Clarke GD, Eckels R, Chaney C, et al: Measurement of absolute epicardial coronary artery blood flow and flow reserve with breath-hold cine phase-contrast magnetic resonance imaging. Circulation 91:2627, 1995.

39. Hundley GW, Lange RA, Clarke GD, et al: Assessment of coronary arterial flow and flow reserve in humans with magnetic resonance imaging. Circulation 93:1502, 1996.

40. Sakuma H, Saeed M, Takeda K, et al: Quantification of coronary artery volume flow rate using fast velocity encoded cine MR imaging. AJR Am J Roentgenol 168:1363, 1997.

41. Sakuma H, Blake LM, Amidon TM, et al: Coronary flow reserve: Non-invasive measurement in humans with breath-hold velocity encoded cine MR imaging. Radiology 198:745, 1996.

42. Karwatowaki SP, Mohiaddin RH, Yang GZ, et al: Noninvasive assessment of regional left ventricular long axis motion using magnetic resonance velocity mapping in normal subjects. J Magn Reson Imaging 4:151, 1994.

43. Scheidegger M, Hess O, Boesiger P: Assessment of coronary flow over the cardiac cycle and diastolic-to-systolic flow ratio with correction for vessel motion (abstract). Proceedings of the Society of Magnetic Resonance, 1994, p 498.

44. Globits S, Sakuma H, Shimakawa A, et al: Measurement of coronary blood velocity during handgrip exercise using breath-hold velocity encoded cine magnetic resonance imaging. Am J Cardiol 79:234, 1997.

45. Davis CP, Liu P, Hauser M, et al: Coronary flow and coronary flow reserve measurements in humans with breath-hold magnetic resonance phase contrast velocity mapping. Magn Reson Med 37:537, 1997.

46. Wedding KL, Grist TM, Folts JD, et al: Coronary flow and flow reserve in canines using MR phase difference and complex difference processing. Magn Reson Med 40:656, 1998.

47. Polzin JA, Alley MT, Korosec FR, et al: A complex-difference phase-contrast technique for measurement of volume flow rates. J Magn Reson Imaging 5:129, 1995.

48. Polzin JA, Korosec FR, Wedding KL, et al: Effects of through-plane myocardial motion on phase difference and complex difference measurements of absolute coronary artery flow. J Magn Reson Imaging 1:113, 1996.

49. Frayne R, Polzin JA, Mazaheri Y, et al: Effect of and correction for in-plane myocardial motion on estimates of coronary-volume flow rates. J Magn Reson Imaging 7:815, 1997.

50. Hundley WG, Hamilton CA, Clarke GD, et al: Visualization and functional assessment of proximal and middle left anterior descending coronary stenoses in humans with magnetic resonance imaging. Circulation 99:3248, 1999.

51. Hundley WG, Hillis LD, Hamilton CA, et al: Assessment of coronary arterial restenosis with phase contrast magnetic resonance imaging measurements of coronary flow reserve. Circulation 101:2375, 2000.

52. Hofman MBM, Visser FC, van Rossum AC, et al: In vivo validation of magnetic resonance blood volume flow measurements with limited spatial resolution in small vessels. Magn Reson Med 33:778, 1995.

53. Tarnawski M, Porter DA, Graves MJ, et al: Flow determination in small diameter vessels by magnetic resonance imaging. Proceedings of the 8th meeting of the International Society of Magnetic Resonance in Medicine 1989, p 896.

54. Tang C, Blatter DD, Parker DL: Accuracy of phase contrast flow measurements in the presence of partial-volume effects. J Magn Reson Imaging 3:377, 1993.

55. Wolf RL, Ehman RL, Riederer SJ, Rossman PJ: Analysis of systematic and random error in MR volumetric flow measurements. Magn Reson Med 30:82, 1993.

56. Sondergaard L, Stahlberg F, Thomsen C, et al: Accuracy and precision of MR velocity mapping in measurement of stenotic cross-sectional area, flow rate and pressure gradient. J Magn Reson Imaging 3:433, 1993.

57. Clarke GD, Hundley WG, McColl RW, et al: Velocity-encoded, phase-difference cine MRI measurements of coronary artery flow: Dependence of flow accuracy on the number of cine frames. J Magn Reson Imaging 6:733, 1996.

58. Jhooti P, Keegan J, Gatehouse PD, et al: 3D coronary artery imaging with phase reordering for improved scan efficiency. Magn Reson Med 41:555, 1999.

CHAPTER 20

Coronary Artery Bypass Graft Imaging and Assessment of Flow*

Albert C. van Rossum

Since the introduction of coronary artery bypass grafting (CABG) by Favoloro in 1968,[1] an increasing number of these surgical procedures have been performed during the last three decades. In the United States approximately 1 in every 1000 persons undergoes CABG, with an estimated 300,000 operations annually.[2] Generally, the saphenous vein is used for sequential grafting to distal branches of the right and circumflex coronary artery, and to diagonal branches of the left anterior descending coronary artery. The left internal mammary artery (IMA) is frequently used as an arterial conduit to the left anterior descending artery (LAD) and its diagonal branches. Other arterial conduits include the right internal mammary artery and the right gastro-epiploic artery, which may be placed to the right coronary artery.

The long-term results of aortocoronary bypass surgery depend largely on the maintenance of graft patency.[3] About 25 percent of venous grafts occlude within 1 year, half of these within 2 weeks of surgery. In the following 5 years, the annual occlusion rate is approximately 2 percent, increasing to 5 percent yearly thereafter. Thus, 50 to 60 percent of venous grafts are occluded after 10 years.[4] The responsible mechanisms for occlusion are considered to be thrombosis in the early weeks after surgery, followed by intimal proliferation during the first year, and progressive atherosclerosis in the later stages. Atherosclerotic changes develop in only a small percentage of patients with IMA grafts. Arterial grafts occlude less frequently, up to 5 percent in the first year and 20 to 30 percent after 10 years, leading to an improved long-term survival.[5, 6]

From these figures it follows that there is a strong need for diagnostic procedures that can evaluate CABG patency and function during postoperative follow-up. In many patients these evaluations have to be made several times in a lifetime.

IMAGING MODALITIES CAPABLE OF EVALUATING GRAFTS

Selective x-ray angiography is the routine procedure and gold standard for assessment of CABG patency and stenosis, but it is invasive and bears a limited risk associated with the use of ionizing radiation and administration of iodinated contrast material. An important advantage of selective angiography is the simultaneous assessment of the status of the native coronary artery system. Also, using the Doppler-tipped guidewire, physiological information with respect to graft function can be obtained by measuring diastolic-to-systolic flow velocity ratios at rest, and flow velocity reserve after pharmacologically induced hyperemia.[7]

Noninvasive or semi-invasive techniques capable of directly evaluating CABG patency include computed tomography (CT), electron beam tomography, and cardiovascular magnetic resonance (CMR).[8–13] Two-dimensional (2-D) Doppler echocardiography is restricted to evaluation of grafts placed on the left anterior descending coronary artery.[14–16] A unique feature of CMR is that, in addition to standard imaging of morphology, blood flow can be quantified within the grafts. Thus, the true function of a graft can be determined noninvasively.[17]

CARDIOVASCULAR MAGNETIC RESONANCE OF BYPASS GRAFTS

In recent years, several CMR techniques have been introduced to evaluate aortocoronary bypass grafts (Table 20–1).[18] Generally speaking, pulse sequences developed for imaging of coronary arteries can also be applied to imaging of bypass grafts, but the reverse is not necessarily true. CMR of proximal graft segments is easier than of distal segments or native coronary arteries, because the former are less subject to cardiac motion and are not embedded in fat or in direct contact with the myocardium, thereby yielding a higher contrast. However, this advantage disappears when imaging distal graft segments that have insertion sites on the native coronary arteries. The majority of clinical studies available report on imaging of proximal vein grafts. Only a minority address CMR of distal graft segments and of arterial grafts, where there are additional problems in obtaining good image quality due to the artifacts associated with the abundant use of metallic hemostatic clips in IMA grafts. Also, vein grafts are larger and therefore easier to image than arterial grafts.

---

*Published in part in van Rossum AC, Bedaux WLF, Hofman MBM: Morphologic evaluation of coronary artery bypass conduits. J Magn Reson Imaging 10:734, 1999.

Table 20–1. Detection of Bypass Graft Patency According to Different Cardiovascular Magnetic Resonance (CMR) Techniques

Reference	Technique	No. of Grafts	Sensitivity %	Specificity %	Accuracy %
White et al[19]	SE-CMR	65	91	72	86
Rubinstein et al[20]	SE-CMR	44	92	85	89
Jenkins et al[21]	SE-CMR	60	90	90	90
Frija et al[22]	SE-CMR	52	98	78	94
White et al[23]	cine-CMR	28	93	86	89
Aurigemma et al[24]	cine-CMR	45	88	100	91
Galjee et al[25]	SE-CMR	98	98	85	96
	cine-CMR		98	88	96
	combined		98	76	94
Kessler et al[32]	3-D navigator	19	87	100	89
Vrachliotis et al[35]	3-D CE MRA, ECG-trig.	44	93	97	95
Wintersperger et al[34]	3-D CE MRA non-ECG trig.	76	95	81	92
Kalden et al[29]	HASTE	59	95	93	95
	3-D CE MRA, ECG-trig.		93	93	93

SE = spin-echo; CE = contrast-enhanced; MRA = magnetic resonance angiography; ECG-trig. = electrocardiogram-triggered; HASTE = Half-Fourier Acquisition Single-shot Turbo spin-Echo sequence.

TECHNIQUES AND RESULTS

Conventional Spin-Echo and Gradient-Echo Imaging

The assessment of saphenous vein aortocoronary bypass graft patency has been a relatively early indication for CMR studies. Several groups have reported the feasibility of visualizing graft patency using conventional electrocardiogram (ECG)-triggered multislice spin-echo techniques.[19–22] On spin-echo (SE) images, patent grafts appear in consecutive imaging planes as conduits with a signal void, whereas stenotic grafts with slow flow or occluded grafts appear with intermediate signal intensity (Figure 20–1). With x-ray angiography as the method of reference, the sensitivity in predicting graft patency ranged from 90 to 98 percent with a specificity from 72 to 90 percent (see Table 20–1).

Using conventional gradient-echo CMR, that is, non-breath-hold with a relatively long echo time (TE) and repetition time (TR), the sensitivity was in the same order of magnitude (88–98%), with a specificity somewhat higher (86–100%).[23–25] On gradient-echo images, blood within patent grafts appears bright (Figure 20–2). On SE images, signal voids from metal clips, stents, calcifications, and thickened pericardium can falsely mimic graft patency. These artifacts are more easily recognized on gradient-echo images because they are larger, thus decreasing the number of false-positive patent grafts (specificity in predicting graft patency goes up). On the other hand, one might expect the number of false-positive occlusions to increase (sensitivity goes down).

Two-Dimensional Breath-Hold Coronary Magnetic Resonance Angiography

This technique was first described for imaging of the native coronary arteries,[26] but it can also be applied for visualization of bypass grafts. Within a breath-hold of 16 to 20 heartbeats, a segmented gradient-echo image is acquired with 4- or 5-mm slice thickness and an in-plane resolution of approximately 1.0 × 1.4 mm using a surface coil. To cover the three-dimensional (3-D) course of a bypass graft, repetitive breath-holding is necessary, which makes the technique highly patient dependent. Especially when multiple grafts or sequential grafts are to be evaluated, the procedure is time consuming. Hartnell and coworkers[27] reported a high susceptibility to metallic materials introduced during CABG, which importantly affected the diagnostic accuracy. Post and associates[28] used this technique to evaluate sequential grafts (Figure 20–3). A high sensitivity and specificity was found for predicting patency of proximal, mid, and distal graft segments to the branches of the right coronary artery, but the accuracy decreased in distal segments to branches of the LAD and was poor in distal segments to the left circumflex system.

Another 2-D breath-hold approach uses a multislice Half-Fourier Acquisition Single-shot Turbo spin-Echo sequence (HASTE). Seven T2-weighted images are generated within a breath-hold of approximately 14 seconds, with a 1.3 × 1.4 mm in-plane resolution and 5-mm slice thickness. Using this technique, Kalden and colleagues[29] reported a sensitivity and specificity of 95 percent and 93 percent, respectively, in predicting graft patency. Also

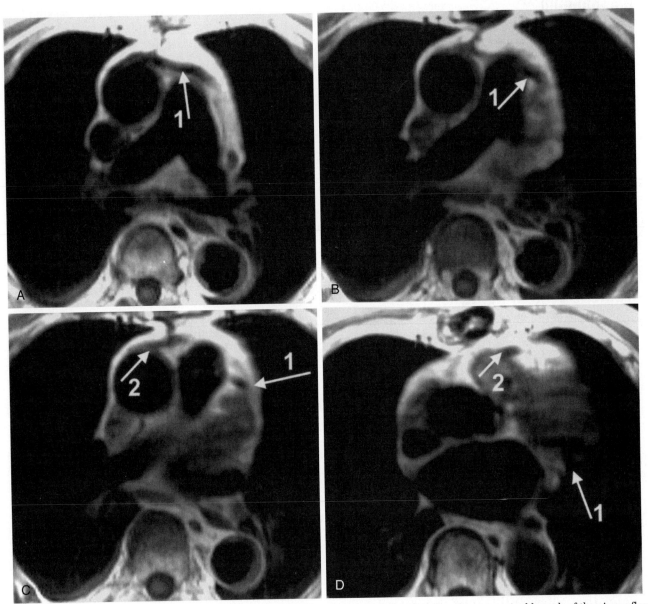

Figure 20–1. *A–D*, Series of four transverse spin-echo images of a sequential vein graft to the obtuse marginal branch of the circumflex artery (1), and a second vein graft to the left anterior descending artery (2). Patent grafts show low signal intensity.

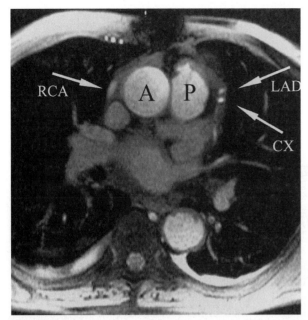

Figure 20–2. Transverse image obtained with gradient-echo technique. Cross-sections of patent grafts to right coronary (RCA), left anterior descending (LAD), and circumflex (CX) arteries are indicated and demonstrate a high signal intensity. (From van Rossum AC, Galjee MA, Post JC, Visser CA: A practical approach to CMR of coronary artery bypass graft patency and flow. Int J Card Imaging 13:199, 1997, with permission from Kluwer Academic Publishers.)

a high percentage (83%) of distal graft anastomoses were revealed. Because the sequence is less susceptible to artifacts induced by metallic implants than gradient-echo sequences, a remarkably high accuracy was found in detecting arterial graft patency, i.e., a sensitivity of 90 percent and a specificity of 100 percent. However, these figures must be interpreted with care, inasmuch as the likelihood of occluded arterial grafts is low. In that study it is unclear how many of the 14 reported IMA grafts were occluded. Personal experience obtained with the HASTE sequence indicates an excellent visualization of coronary arteries and bypass grafts, but rather poor detection of disease (Figure 20–4). Accordingly, Kalden and colleagues[29] noted that only 2 out of 8 hemodynamically significant graft stenoses were detected using HASTE.

Three-Dimensional Respiratory Gated Coronary Magnetic Resonance Angiography

A 3-D data set of truly contiguous slices may be obtained with respiratory gated gradient-echo techniques. Gating to the respiratory cycle is achieved by using navigators that monitor the diaphragm position.[30] CMR data acquired within a preset acceptance window of the respiration-induced diaphragm excursion are used for image reconstruction. The patient is allowed to breathe freely without the need

for repetitive breath-holding, at the expense of an increase in imaging time. Several refinements of this gating procedure have been developed for imaging of the coronary arteries.[31]

Kessler and coworkers[32] used respiratory gated 3-D magnetic resonance angiography (MRA) with navigator guiding and retrospective data processing for imaging of bypass grafts. In a relatively small number (19 grafts), 13 out of 15 patent grafts and 4 out of 4 occluded grafts were correctly classified.

Three-Dimensional Contrast-Enhanced Breath-Hold Coronary Magnetic Resonance Angiography

Breath-hold contrast-enhanced MRA is a relatively new technique, first applied in imaging of the aorta.[33] The T1-shortening effect of the contrast agent on the blood allows obtaining high vascular contrast, using short TR/TE gradient-echo sequences with high field-strength gradient systems. After intravenous injection of an extracellular contrast agent, preferably with the aid of a CMR-compatible power injector, a 3-D spoiled gradient-echo sequence with short TR/TE (4.4/1.4 ms or even shorter) is applied. Within a breath-hold of approximately 30 seconds, a 3-D volume slab of 6- to 9-cm thickness is imaged, consisting of 24 to 32 contiguous slice partitions. Prior to acquiring the 3-D volume-slab, a single-slice 2-D single-shot gradient-echo sequence is used to time the arrival in the aorta of a contrast agent test-bolus. To maximize the contrast-enhancing effect, acquisition of the central k-space lines of the 3-D imaging data is set to coincide with peak contrast arrival. This timing is achieved by introducing a time delay between contrast injection and start of the imaging sequence (delay = arrival time − ½ or ⅓ of acquisition time). The spatial resolution would typically be 1 × 1.5 mm in-plane, and 2- to 3-mm section thickness, depending on field-of view, matrix size, number of partitions, and slab thickness. Each partition of the 3-D acquisition yields a source image. Studies may be evaluated by reading the source images and by postprocessing techniques such as maximum intensity projection, and planar or curved reformatting (Figure 20–5).[18]

Evaluation of aortocoronary bypass grafts using 3-D contrast-enhanced MRA has been reported without and with ECG triggering.[29, 34, 35] Sensitivity, specificity, and accuracy for predicting graft patency varied between 93 and 95 percent, 81 and 97 percent, and 92 and 95 percent, respectively (see Table 20–1). The lower specificity was found in the non-ECG-triggered study.[34] Theoretically, one would expect ECG-triggered acquisitions to be superior for visualizing the sites of graft insertion on native coronary arteries. Although Kalden and colleagues[29] specifi-

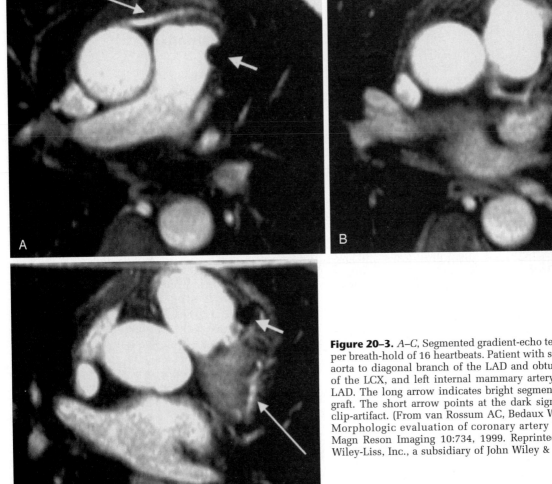

Figure 20–3. *A–C,* Segmented gradient-echo technique, one image per breath-hold of 16 heartbeats. Patient with sequential graft from aorta to diagonal branch of the LAD and obtuse-marginal branch of the LCX, and left internal mammary artery (IMA) graft to the LAD. The long arrow indicates bright segments of the sequential graft. The short arrow points at the dark signal loss of the IMA clip-artifact. (From van Rossum AC, Bedaux WLF, Hofman MBM: Morphologic evaluation of coronary artery bypass conduits. J Magn Reson Imaging 10:734, 1999. Reprinted by permission of Wiley-Liss, Inc., a subsidiary of John Wiley & Sons, Inc.)

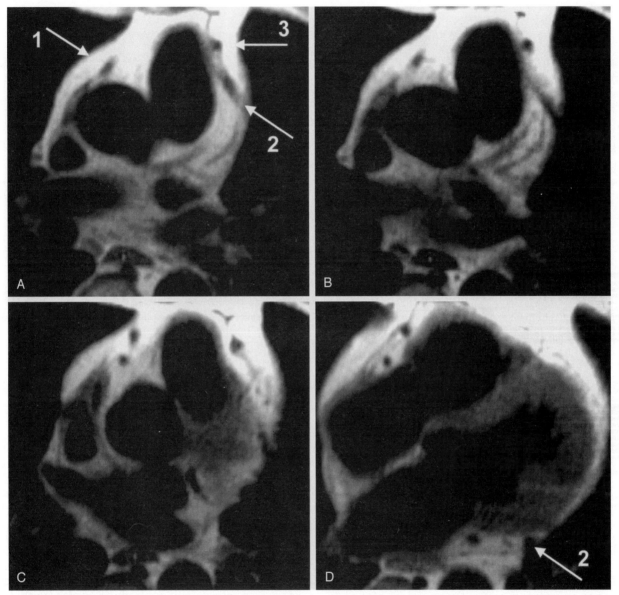

Figure 20–4. Half-Fourier Acquisition Single-shot Turbo spin-Echo sequence (HASTE) sequence, four out of five images obtained in a single breath-hold *(A–D)*. Patient with single vein graft to RCA (1), sequential vein graft to first diagonal branch of the LAD and obtuse marginal branch of the CX (2), and left internal mammary artery (LIMA) to the LAD (3). On *A* and *B* the native left main coronary artery, the LAD and the great cardiac vein are visualized. On *C* and *D* the right coronary artery is seen inferior to the graft in the right atrioventricular groove.

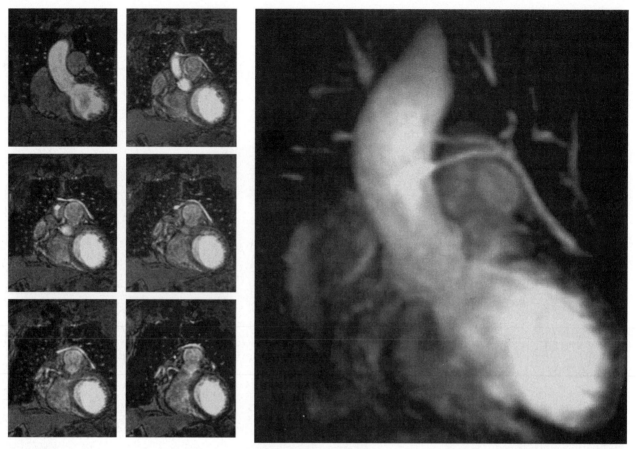

Figure 20–5. Gadolinium-enhanced 3-D breath-hold MRA, ECG-triggered. The six images on the left are several of the source images, demonstrating the proximal course of two sequential vein grafts to LAD and CX perfusion territory, respectively. The right image is obtained through post-processing using a maximum intensity projection (MIP).

cally addressed this issue, they succeeded in pursuing distal anastomoses in only 64 percent. This problem might have been caused partly by the relatively long data collection window of 560 milliseconds within each cardiac cycle, leading to residual blurring by cardiac contraction. The use of a shorter acquisition window of 120 milliseconds has been implemented, but so far was reported in coronary arteries of healthy volunteers only.[36]

IMAGING STRATEGY

The CMR imaging strategy and image interpretation are facilitated when the surgical report is known prior to the CMR procedure with respect to the number of grafts and insertion sites. Grafts descending to the perfusion area of the left circumflex artery (LCX), including anterolateral and obtuse marginal branches, generally originate most superior from the ascending aorta (Figure 20–6).[25] The graft to the perfusion area of the LAD, including diagonal branches, originates inferiorly compared with most LCX grafts. Both types of grafts cross the pulmonary artery trunk in a left lateral and inferior course. Then, the LCX graft proceeds to posterior and inferior, and the LAD graft to left and anterior.

The graft to the perfusion area of the right coronary artery (RCA) generally has the lowest origin from the ascending aorta and runs anteriorly or to the lateral side of the right atrium. For assessment of patency and measurement of flow within a graft, CMR can best be performed at the proximal part of the graft. At this level, most grafts have a straight course, are the least susceptible to motion, and are easily distinguished from the native coronary vessels.

Initially, we performed the examination with the patients positioned prone on a surface coil. We believe that this position was an important factor in the success of our early results, because it decreased respiratory chest wall motion. However, with the advent of segmented k-space sequences and phased-array coils, images can now be obtained within a breath-hold. Thus, respiratory motion is minimized and image quality improved, even when the patient is in a supine position, thereby discarding the need for the uncomfortable prone positioning. The surface coil must be centered at the middle of the sternum, which is somewhat more superior than in a standard cardiac study.

Because the course of a graft is more or less predictable, the choice of imaging planes can be fairly well standardized. An effective approach is first to

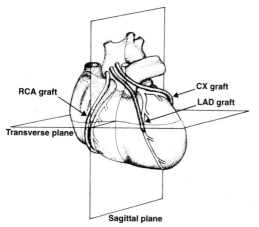

Figure 20–6. Diagram showing typical course of vein grafts from ascending aorta to coronary artery insertion sites, relative to sagittal and transverse imaging planes. RCA = right coronary artery; CX = circumflex artery; LAD = left anterior descending artery. (From Galjee MA, van Rossum AC, Doesburg T, et al: Value of magnetic resonance imaging in assessing patency and function of coronary artery bypass grafts: An angiographically controlled study. Circulation 93:660, 1996, with permission.)

acquire a set of multislice transverse images covering the ascending aorta and superior part of the heart. According to our experience, one or two breath-hold multislice series using the HASTE sequence will be most informative (see Figure 20–4). The images just superior to the pulmonary artery trunk will demonstrate in-plane views of proximal parts of LCX and LAD grafts, whereas the images at a lower level show cross sections of grafts to the

left and right coronary arteries. Once the proximal course of the grafts has been localized, one may proceed with high-resolution single or multislice 2-D imaging, in orientations following the more distal course of the grafts. The advantage of this approach is the short image reconstruction time and immediate availability of the images. The disadvantage is that it requires patient cooperation and may be subject to misregistration due to inconsistent breath-holding. Alternatively 3-D MRA techniques may be used to acquire larger imaging slabs covering the course of the grafts. The acquisition time is shorter and the resolution higher. However, the reconstruction time is longer than in 2-D techniques and interpretation requires some form of postprocessing. Notwithstanding these limitations, the 3-D approaches are likely to become first choice with improving technology.

MAGNETIC RESONANCE QUANTIFICATION OF BYPASS GRAFT FLOW AND FLOW RESERVE

Flow velocity and volume flow in bypass grafts can be measured by applying velocity-encoded phase-contrast cine CMR sequences, thus allowing assessment of graft function in addition to a morphologic evaluation (Figure 20–7).[37] Galjee and associates[25] demonstrated that adequate velocity profiles throughout the cardiac cycle could be obtained in 85 percent of angiographically patent vein grafts,

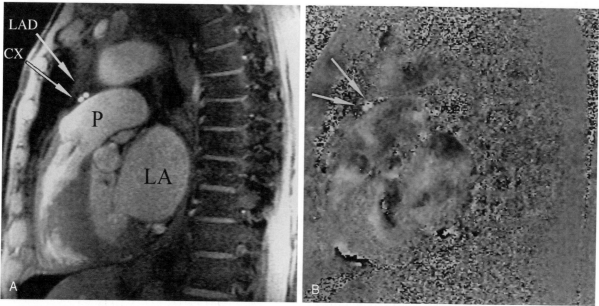

Figure 20–7. Oblique-sagittal image obtained with velocity-encoded phase-contrast technique to measure flow. *A,* Magnitude reconstruction of CMR signal depicting cross sections of two grafts anterior and superior of the main pulmonary artery. *B,* Corresponding reconstruction of the phase of the CMR signal, where brightness of each pixel is proportional to flow-velocity and mid-gray equals zero flow-velocity. Within the grafts the bright signal (white) indicates high cross-sectional velocities. LAD = left anterior descending artery; CX = circumflex artery; P = pulmonary artery; LA = left atrium. (From van Rossum AC, Galjee MA, Post JC, Visser CA: A practical approach to CMR of coronary artery bypass graft patency and flow. Int J Card Imaging 13:199, 1997, with permission from Kluwer Academic Publishers.)

Volume flows
Corrected for RR interval

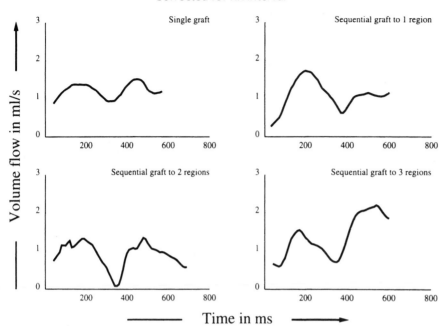

Figure 20–8. Plots of volume-flow patterns in single and sequential saphenous vein grafts, normalized for heart rate. Typical patterns consist of a first flow peak in systole and a second peak in diastole. (From Galjee MA, van Rossum AC, Doesburg T, et al: Value of magnetic resonance imaging in assessing patency and function of coronary artery bypass grafts: An angiographically controlled study. Circulation 93:660, 1996, with permission.)

using non–breath-hold MR velocity mapping. Graft flow velocity was characterized by a biphasic pattern, with one peak in systole and a second peak in diastole (Figure 20–8).[25] Similar findings have been obtained using invasive Doppler guidewire approaches and transthoracic Doppler echocardiography.[7, 14] CMR-assessed volume flow of grafts with three sequential anastomoses significantly differed from single grafts.

More recently, breath-hold segmented k-space sequences have been applied to measure flow velocity and volume flow at rest and after pharmacological stress. A preliminary report by Voigtländer and coworkers[38] demonstrated the feasibility of measuring the flow reserve in bypass grafts and its potential to differentiate between patent and stenotic grafts. In 21 grafts without stenoses compared with 6 grafts with greater than 75 percent stenoses, a flow velocity reserve was found of 2.6 ± 1.5 versus 0.8 ± 0.4 ($p < 0.005$) and a flow reserve of 2.9 ± 1.9 versus 1.2 ± 0.5 ($p < 0.05$).

Using non–breath-hold and breath-hold techniques, functional results were also obtained in IMA grafts, despite the imaging artifacts due to metallic

clips.[39, 40] The diastolic/systolic peak velocity ratio was found to be higher in IMA grafts than in native IMA. Preliminary findings in IMA grafts by Kawada and colleagues[41] in 18 patients without and 5 patients with 75 percent or greater stenosis indicate a higher sensitivity and specificity for detection of IMA graft stenosis using measurements of the mean flow rate and diastolic/systolic peak flow ratio at rest than using the flow reserve after administration of dipyridamole.

Thus, measuring mean flow, diastolic/systolic flow profiles, and flow reserve of coronary artery bypass grafts may become helpful in noninvasively differentiating between a nonsignificantly or significantly obstructed graft (Figure 20–9).

LIMITATIONS

So far, CMR of coronary artery bypass grafts has been limited to demonstrating patency versus total occlusion only. Evidence in documenting nonocclusive graft disease has not been provided with the techniques used to date. CMR measurements of

Figure 20–9. Patient with single vein graft inserting into the LAD, and sequential vein graft inserting on posterior descending and posterolateral branch of the RCA. *A,* Curved planar reformat of gadolinium-enhanced 3-D MRA of graft inserting on LAD. The dashed line indicates the orientation of the plane perpendicular to the proximal graft segment, used for velocity-encoded cine CMR. *B,* X-ray angiography of distal graft segment demonstrating irregular aspect of luminal borders and tight insertion on LAD. *C,* Cross-sectional averaged velocities measured at multiple phases throughout the cardiac cycle using velocity-encoded cine CMR. Calculated volume-flow was 38 ml/min at rest and 68 ml/min during adenosine stress, yielding a flow reserve of 1.8. *D,* Curved planar reformat of proximal part of the sequential graft. Dashed line indicating plane of MR flow acquisition. *E,* X-ray angiography reveals a tight stenosis (arrow) in graft segment between the two distal insertion sites. *F,* Volume-flow calculated from CMR flow velocity measurements was 50 ml/min at rest and 51 ml/min during adenosine stress, yielding a flow reserve of 1.0. (From van Rossum AC, Bedaux WLF, Hofman MBM: Morphologic evaluation of coronary artery bypass conduits. J Magn Reson Imaging 10:734, 1999. Reprinted by permission of Wiley-Liss, Inc., a subsidiary of John Wiley & Sons, Inc.)

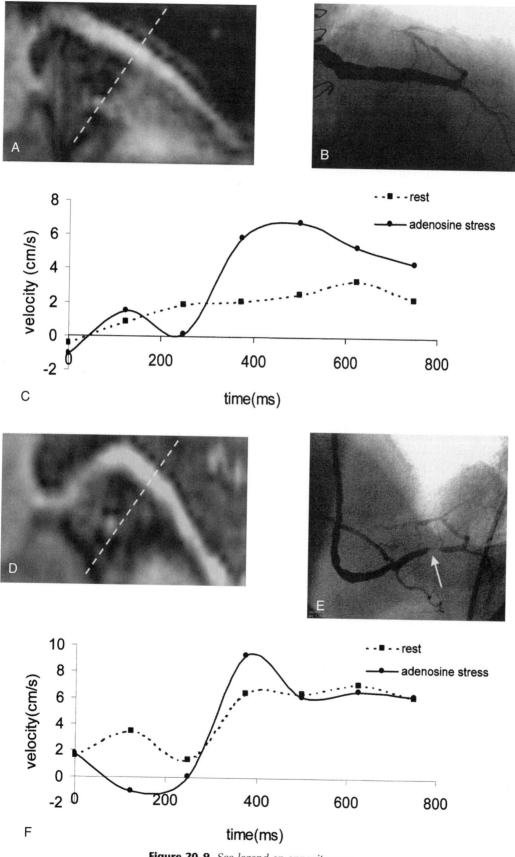

Figure 20–9. *See legend on opposite page*

blood flow at rest and under stress, with calculation of the diastolic/systolic flow ratio and the coronary flow reserve, may become helpful in addition to anatomic imaging to determine the functional status of a diseased bypass graft and its recipient coronary artery. Furthermore, most of the reported studies have been confined to imaging of proximal graft segments and only few data are available with respect to assessing patency of segments beyond the first coronary anastomosis in sequential bypass grafts. Imaging of the distal graft segments requires a higher spatial resolution and signal-to-noise ratio than can be obtained with currently reported techniques. Another point of consideration is the problem associated with CMR of arterial grafts. IMA grafts are increasingly used because of the improved long-term patency and patient survival, but they have been excluded from most CMR studies due to local metallic clip artifacts. Flow reserve measurements proximal to the clip artifacts may be of diagnostic help or, alternatively, methods to obtain hemostasis without use of ferromagnetic clips would be helpful.

However, even a clear demonstration of graft segment patency or narrowing will often not suffice for clinical decision-making. In most patients, there is also a need to know the status of the native coronary arteries. Narrowing of the recipient coronary artery may have developed beyond the anastomosis of a patent graft segment, and progression of disease in other native coronary arteries must be excluded. Thus, CMR does not eliminate the need for conventional x-ray coronary angiography when coronary reinterventions are under consideration.

INDICATIONS

A clinical indication for CMR of bypass grafts may therefore exist only in patients in whom there is no immediate need to know the status of the native coronary arteries. Such a category consists, for example, of patients with chest pain shortly after CABG surgery. Also late after CABG surgery, the information whether the grafts are patent and function well might be helpful in deciding to postpone coronary angiography in patients with ambiguous thoracic pain or mild anginal complaints. Noninvasive monitoring of flow parameters may then be useful to detect a gradual increase of graft stenosis and decide to proceed to x-ray angiography and stent placement, before the onset of a total occlusion.

Furthermore, CMR can be used as a screening procedure before angiography, indicating the number of grafts to be visualized. This process might considerably shorten the angiographic procedure. A useful indication also appears to be the assessment of patency of grafts that are not visualized at conventional angiography. Although often this technique will indicate proximal occlusion of the graft, failure of the catheter to fit with the aortic graft anastomosis may result in a false diagnosis of graft occlusion. In cases of doubt, MRA will rapidly confirm or discard this diagnosis. Also, when angiography has demonstrated a graft stenosis, it can be helpful for further management to assess the flow reserve of a graft. There are also occasions when CMR is useful for the definition of the complications of vein grafting such as aneurysm of the graft and helping in surgical management.[42]

Finally, vein graft imaging has a role in research of bypass techniques and has been used to demonstrate the efficacy of aprotinin[43] during operation, and in comparison of techniques for vein graft stripping,[44] by demonstrating patency rates without the need for invasive angiography, which is an important ethical issue.

Notwithstanding these indications, the majority of patients after CABG require evaluation with respect to a renewed coronary intervention, either by angioplasty including stents or by a CABG-redo. Unless CMR will also provide more detailed information regarding the status of the native coronary arteries, a wide application is unlikely to occur. Continuing improvement in CMR of coronary arteries and bypass grafts may be expected from new developments in hardware, including more powerful gradient systems, more refined pulse sequence design, and the use of intravascular contrast agents.

CONCLUSION

There is clear evidence that conventional SE and gradient-echo CMR is capable of assessing patency of coronary artery bypass grafts. With more recently introduced breath-hold 2-D and contrast-enhanced 3-D techniques, the predictive accuracy has further improved with sensitivities and specificities in the 90 percent range. Limitations arise with regard to assessing obstructive disease and evaluating distal segments of sequential grafts, due to insufficient spatial resolution, low signal-to-noise ratio, and cardiac motion. Imaging of arterial grafts is often complicated by the metallic clip artifacts. Adding information on graft flow patterns and flow reserve using velocity-encoded cine CMR may help to reduce some of the problems. Clinically, these functional measurements may become of use in noninvasive monitoring of gradually increasing graft narrowing. However, apart from a few exceptions, the majority of the patients undergo evaluation of their grafts because they are considered for a reintervention by angioplasty or CABG surgery. In these cases, information on the status of the native coronary arteries is required. A broader clinical use of CMR in the evaluation of patients with coronary artery bypass grafts may therefore be expected only with further improvement in CMR techniques for coronary angiography.

References

1. Favaloro RG: Saphenous vein autograft replacement of severe segmental coronary artery occlusion. Operative technique. Ann Thorac Surg 5:334, 1968.

2. Gersh BJ, Braunwald E, Rutherford JD: Chronic coronary artery disease: Coronary artery bypass graft surgery. *In* Braunwald E (ed): Heart Disease: A Textbook of Cardiovascular Medicine. 5th ed. Philadelphia, WB Saunders, 1997, p 1316.

3. Chesebro JH, Clements IP, Fuster V, et al: A platelet-inhibitor-drug trial in coronary-artery bypass operations. Benefit of perioperative dipyridamole and aspirin therapy on early post-operative vein-graft patency. N Engl J Med 307:73, 1982.

4. Henderson WG, Goldman S, Copeland JG, et al: Antiplatelet or anticoagulant therapy after coronary artery bypass surgery. A meta-analysis of clinical trials. Ann Intern Med 111:743, 1989.

5. van der Meer J, Hillege HL, van Gilst WH, et al: A comparison of internal mammary artery and saphenous vein grafts after coronary artery bypass surgery: No difference in 1-year occlusion rates and clinical outcome. Circulation 90:2367, 1994.

6. Cameron A, Davis KB, Green G, Schaff HV: Coronary bypass surgery with internal-thoracic-artery grafts: Effects on survival over a 15-year period. N Engl J Med 334:216, 1996.

7. Bach R, Kern M, Donohue T, et al: Comparison of phasic flow velocity characteristics of arterial and venous coronary artery bypass conduits. Circulation 88:133, 1993.

8. Stanford W, Galvin JR, Skorton DJ, Marcus ML: The evaluation of coronary bypass graft patency: Direct and indirect techniques other than coronary arteriography. AJR Am J Roentgenol 156:15, 1991.

9. Stanford W, Brundage BH, MacMillan R, et al: Sensitivity and specificity of assessing coronary bypass graft patency with ultrafast computed tomography: Results of a multicenter study. J Am Coll Cardiol 12:1, 1988.

10. Tello R, Costello P, Ecker C, Hartnell G: Spiral CT evaluation of coronary artery bypass graft patency. J Comput Assist Tomogr 17:253, 1993.

11. Engelmann MG, Von Smekal A, Knez A, et al: Accuracy of spiral computed tomography for identifying arterial and venous coronary graft patency. Am J Cardiol 80:569, 1997.

12. Achenbach S, Moshage W, Ropers D, et al: Noninvasive, three-dimensional visualization of coronary artery bypass grafts by electron beam tomography. Am J Cardiol 79:856, 1997.

13. van Rossum AC, Galjee MA, Doesburg T, et al: The role of magnetic resonance in the evaluation of functional results after CABG/PTCA. Int J Card Imaging 9:59, 1993.

14. Fusejima K, Takahara Y, Sudo Y, et al: Comparison of coronary hemodynamics in patients with internal mammary artery and saphenous vein coronary artery bypass grafts: A noninvasive approach using combined two-dimensional and Doppler echocardiography. J Am Coll Cardiol 15:131, 1990.

15. Takagi T, Yoshikawa J, Yoshida K, Akasaka T: Noninvasive assessment of left internal mammary artery graft patency using duplex Doppler echocardiography from supraclavicular fossa. J Am Coll Cardiol 22:1647, 1993.

16. Voudris V, Athanassopoulos G, Vassilikos V, et al: Usefulness of flow reserve in the left internal mammary artery to determine graft patency to the left anterior descending coronary artery. Am J Cardiol 83:1157, 1999.

17. Galjee MA, van Rossum AC, Doesburg T, et al: Quantification of coronary artery bypass graft flow by magnetic resonance phase velocity mapping. Magn Reson Imaging 14:485, 1996.

18. Van Rossum AC, Bedaux WLF, Hofman MBM: Morphologic evaluation of coronary artery bypass conduits. J Magn Reson Imaging 10:734, 1999.

19. White RD, Caputo GR, Mark AS, et al: Coronary artery bypass graft patency: Noninvasive evaluation with MR imaging. Radiology 164:681, 1987.

20. Rubinstein RI, Askenase AD, Thickman D, et al: Magnetic resonance imaging to evaluate patency of aortocoronary bypass grafts. Circulation 76:786, 1987.

21. Jenkins JPR, Love HG, Foster CJ, et al: Detection of coronary artery bypass graft patency as assessed by magnetic resonance imaging. Br J Radiol 61:2, 1988.

22. Frija G, Schouman-Claeys E, Lacombe P, et al: A study of coronary artery bypass graft patency using MR imaging. J Comput Assist Tomogr 13:226, 1989.

23. White RD, Pflugfelder PW, Lipton MJ, Higgins CB: Coronary artery bypass grafts: Evaluation of patency with cine MR imaging. AJR Am J Roentgenol 150:1271, 1988.

24. Aurigemma GP, Reichek N, Axel L, et al: Noninvasive determination of coronary artery bypass graft patency by cine magnetic resonance imaging. Circulation 80:1595, 1989.

25. Galjee MA, van Rossum AC, Doesburg T, et al: Value of magnetic resonance imaging in assessing patency and function of coronary artery bypass grafts: An angiographically controlled study. Circulation 93:660, 1996.

26. Manning WJ, Li W, Boyle NG, Edelman RR: Fat-suppressed breath-hold magnetic resonance coronary angiography. Circulation 87:94, 1993.

27. Hartnell GG, Cohen MC, Charlamb M, et al: Segmented k-space magnetic resonance angiography for the detection of coronary artery bypass graft patency. Book of Abstracts of the 4th Meeting of the ISMRM, New York, 1996, Vol 1, p 178.

28. Post JC, van Rossum AC, Bronzwaer JGF, et al: Magnetic resonance angiography of sequential aortocoronary bypass grafts. Circulation 96(suppl):I-133, 1997.

29. Kalden P, Kreitner KF, Wittlinger T, et al: Assessment of coronary artery bypass grafts: Value of different breath-hold MR imaging techniques. AJR Am J Roentgenol 172:1359, 1999.

30. Hofman MBM, Paschal CB, Li D, et al: CMR of coronary arteries, 2D breath-hold versus 3D respiratory-gated acquisition. J Comput Assist Tomogr 19:56, 1995.

31. Danias PG, McConnell MV, Khasgiwala VC, et al: Prospective navigator correction of image position for coronary MR angiography. Radiology 203:733, 1997.

32. Kessler W, Achenbach S, Moshage W, et al: Usefulness of respiratory gated magnetic resonance coronary angiography in assessing narrowings ≥ 50% in diameter in native coronary arteries and in aortocoronary bypass conduits. Am J Cardiol 80:989, 1997.

33. Prince MR, Narasimham DL, Stanley JC, et al: Breath-hold gadolinium-enhanced MR angiography of the abdominal aorta and its major branches. Radiology 197:785, 1995.

34. Wintersperger BJ, Engelmann MG, von Smekal A, et al: Patency of coronary artery bypass grafts: Assessment with breath-hold contrast-enhanced MR angiography—Value of a non-electrocardiographically triggered technique. Radiology 208:345, 1998.

35. Vrachliotis TG, Bis KG, Aliabadi D, et al: Contrast-enhanced breath-hold MR angiography for evaluating patency of coronary artery bypass grafts. AJR Am J Roentgenol 168:1073, 1997.

36. Goldfarb JW, Edelman RR: Coronary arteries: Breath-hold, gadolinium-enhanced, three-dimensional MR angiography. Radiology 206:830, 1998.

37. Van Rossum AC, Galjee MA, Post JC, Visser CA: A practical approach to CMR of coronary artery bypass graft patency and flow. Int J Cardiac Imaging 13:199, 1997.

38. Voigtländer T, Kreitner KF, Wittlinger T, et al: MR measurement of flow reserve in coronary grafts (abstract). J Cardiovasc Magn Reson 1:275, 1999.

39. Debatin JF, Strong JA, Sostman HD, et al: MR characterization of blood flow in native and grafted internal mammary arteries. J Magn Reson Imaging 3:443, 1993.

40. Sakuma H, Globits S, O'Sullivan M, et al: Breath-hold MR measurements of blood flow velocity in internal mammary arteries and coronary artery bypass grafts. J Magn Reson Imaging 6:219, 1996.

41. Kawada N, Sakuma H, Cruz BC, et al: Noninvasive detection of significant stenosis in the coronary artery bypass grafts using fast velocity-encoded cine CMR (abstract). J Cardiovasc Magn Reson 1:261, 1999.

42. Warner OJ, Ohri SK, Pennell DJ, Smith PLC: Magnetic resonance coronary artery imaging for redo cardiac surgery. Ann Thor Surg 62:1513, 1996.

43. Bidstrup BP, Underwood SR, Sapsford RN: Effect of aprotinin (Trasylol) on aorto-coronary bypass graft patency. J Thorac Cardiovasc Surg 105:147, 1993.

44. O'Regan DJ, Borland JAA, Chester AH, et al: Assessment of human long saphenous vein function with minimally invasive harvesting with the Mayo stripper. Eur J Cardiothor Surg 12:428, 1997.

21

Atherosclerotic Plaque Imaging

Jean-François Toussaint

Many lesion varieties affect the arterial wall, from nonprotruding fatty streaks to more complex lesions consisting of lipid, smooth muscle, fibroblasts, and calcification. The morphology and composition of arterial segments containing atheroma is of considerable importance. Plaques of different morphology (e.g., concentric or eccentric) have different effects on the arterial wall, such as the potential for thrombosis and the effect of arterial spasm.[1] The lipid content may also affect the propensity for fissuring, ulceration, and thrombosis.[2] Other properties of atheroma that may be affected by the lipid content are the short- and long-term outcome of angioplasty, and the potential for regression.[3] More recently the importance of the thickness of the fibrous cap in preventing fissuring and thrombosis has been recognized.[4] The importance of assessing plaque constituents, as well as the widespread prevalence of atherosclerotic vascular disease, has given rise to the need for a noninvasive imaging examination capable of plaque characterization.

Numerous techniques are now available for imaging atherosclerotic vascular disease, but clinical assessment of human atherosclerosis and its progression currently depends predominantly on evaluation of plaques by conventional x-ray angiography or surface B-mode ultrasound. However, angiographic studies have clearly demonstrated that luminal morphometry is not able to predict the occurrence of infarction[5–7] or unstable angina.[8] High-grade stenoses occlude more frequently[9–11] but are less likely to cause acute myocardial infarction,[10, 12] whereas moderate stenoses occlude less often[13] but are more likely to cause acute infarction.[5, 7, 9, 10] Intravascular ultrasound and angioscopy have recently improved the interrogation of arterial walls, but these techniques are invasive and are limited in their ability to determine biochemical composition. Magnetic resonance (MR), a new noninvasive imaging tool for this application, is capable of discriminating plaque components on the basis of chemical composition, molecular motion, diffusion, physical state, or water content.

Several means of assessing plaque composition have been developed using MR with ^{13}C-MR spectroscopy[14]; proton (^{1}H) imaging with T1,[15] or T2 contrast[16]; water diffusion techniques[17]; and magnetization transfer.[18] T2 contrast has been used to discriminate in vivo wall components in normal and atheromatous arteries,[19] and test plaque resistance with in vitro models of angioplasty and atherectomy.[20] The arterial regions that can be identified with such techniques include the media, adventitia, perivascular fat, lipid-rich core, fibrous cap, and calcification.

ATHEROMATOUS LIPID COMPOSITION AND MAGNETIC RESONANCE SPECTROSCOPY

The predominant fibrous tissue components of atheromatous plaque are connective tissue matrix proteins such as collagen, elastin, and proteoglycans. Most of the fibrous tissue components are in the cap that covers the lipid-rich center. Lipid components of plaque are a complex mixture of cholesterol crystals, cholesterol esters, free cholesterol, and phospholipids in oily phase.[21] These lipids may appear either extracellularly or intracellularly in foam cells, and the concentration of free cholesterol increases with advancing development of the lesions.

^{13}C-MR spectroscopy (MRS) allows characterization of lipid composition. It provides more information regarding chemical constituents than ^{1}H-MRS, and has been used for structural and dynamic studies of cholesterol esters (CE), triglycerides (TG), and phospholipids (PL). This information is useful for understanding the biochemistry of the pathological processes involved, as well as the signal found on T1- and T2-weighted MR imaging. Hamilton and Cordes[22] examined plasma lipoproteins and intact atherosclerotic plaques in humans, and showed that atheroma had spectral characteristics very similar to thermally denatured low density lipoproteins (LDL), suggesting a comparable chemical composition. The importance of CE phase transitions (liquid, smectic, or solid) was important in determining the spectral characteristics of the lipids and indicated that most of the atheromatous CE was derived from nonmetabolized lipoproteins.

Using this technique, investigators have shown that the mean saturation ratio of unsaturated to polyunsaturated fatty acids (UFA/PUFA) of lesions composed of uncomplicated, nonulcerated fibrous plaques is low, whereas complex and more stenotic lesions have a higher ratio (a 42% increase).[14] A decrease of the cholesterol ester resonance corresponding to the carbon atoms in positions 19 and 21 can also be shown. The ^{13}C peaks are predominantly derived from the mobile atheromatous lipids, which generate relatively narrow resonances. This suggests that the relation between UFA/PUFA, the carbon-19 and carbon-21 peaks, and stenosis does not apply to the total lipid content (including solid-state lipids such as free cholesterol crystals) but only to the mobile component. The importance of characterizing this component (the "soft" lipids) is explained by its probable contribution to plaque vulnerability and its role in the processes leading to plaque rupture through abnormal distribution of circumferential stress.

Alterations in fatty acid saturation and cholesterol esters in atheroma have been described and attributed to lipid and lipoprotein oxidation. This process has been previously studied with ^{1}H and ^{13}C-MRS by using a model of LDL peroxidation.[23] These studies showed an increase in the UFA/PUFA ratio as a result of oxidation, similar to the findings in lesions of increasing severity, resulting from a PUFA decrease without changes in UFA.[14]

Further support for the reduction in fatty acid saturation resulting from lipoprotein oxidation comes from other studies, which showed the effects of oxidation on fatty acyl chain double bonds, demonstrating a 55 percent loss of polyunsaturated fatty acid chains. The loss of PUFA has been ascribed to their low resistance to oxidation and may have clinical implications with respect to atherosclerosis prevention. Their low content also influences the cytotoxic effects of oxidized LDL and alters the lipid phase transition and fluidity.

PLAQUE ANALYSIS BY PROTON MAGNETIC RESONANCE IMAGING

For clinical applicability, MR imaging is more useful at present than MRS. Proton MR can be used to discriminate arterial wall (Figure 21–1A and B) and plaque components (Figure 21–2A and B) on the basis of chemical composition, which may help in the future to determine the factors of plaque rupture, such as circumferential stress,[24] vulnerability,[4] or thrombogenicity.[25]

Atheroma has been characterized with MR by the acquisition of the lipid signal, acquired with a T1-weighted sequence using either a nonselective, a lipid-selective, a modified Dixon pulse, or a combination of these.[15, 26–29] These studies were conceived to image plaque lipids with long T2 and short T1 relaxation times, similar to adipocyte triglycerides, such as occur in subcutaneous tissue. However, cholesterol and cholesterol esters are the predominant lipids in atherosclerotic plaques, in solid (crystal) or smectic (liquid-crystalline) states, with MR relaxation constants differing from triglycerides. Unlike the triglycerides, these cholesterol compounds are associated with a short T2 when compared with the collagenous cap and the normal media. Consequently, bright areas on non–frequency-selective T2-weighted images in arterial walls do not correspond to lipid-rich regions, but to regions predominantly composed of fibers in media or fibrous caps, whereas the cholesterol species are dark. Therefore, lipid-rich and collagen layers can be localized through changes in water T2 in the same sequence that is useful in discriminating the lipid pool from the fibrous cap.

Atherosclerotic plaque components can also be discriminated in vitro using high-resolution [1]H-MR imaging without frequency-selective sequences. Spectral measurements show a lipid/water peak ratio of 0.1 inside the atheromatous core. It is therefore important to consider that any chemical shift technique based on lipid frequency selection and aimed at imaging the lipid core of atherosclerotic plaque will face the inherent problem of a 90 percent lower signal-to-noise ratio than techniques based on water proton imaging.

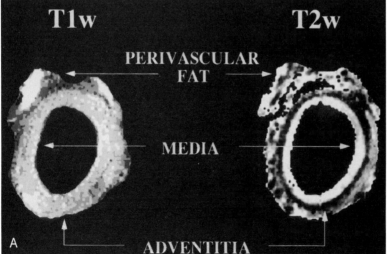
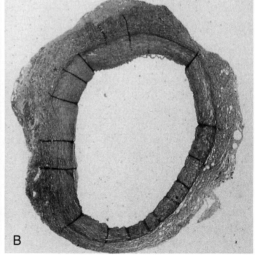

Figure 21–1. *A*, In vitro imaging of a normal carotid artery at 9.4 T (T1w: TR = 700 ms, TE = 3 ms; T2w: TR = 2 s, TE = 50 ms; resolution: 156 × 156 μm per pixel; slice thickness: 600 μm). The T2w image shows enhanced contrast between media and adventitia. *B*, Trichrome staining of this normal vessel for comparison (see color plates). T1w = T1-weighted, T2w = T2-weighted; TE = echo time; TR = repetition time.

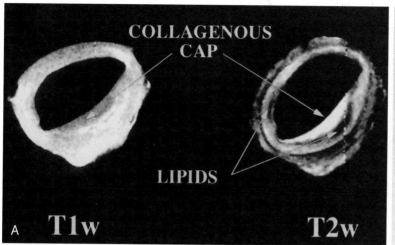
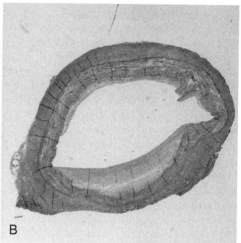

Figure 21–2. *A*, In vitro imaging of a fibrofatty carotid plaque with a complete collagenous cap, as clearly shown in the T2w image. T2 contrast also reveals a small lipid infiltration opposite the plaque (arrows). *B*, Trichrome staining showing the plaque with the large fibrous cap and opposite intimal thickening for comparison (see color plates).

However, early results using chemical shift imaging (CSI) have been validated against histopathological findings in vitro and applied in vivo. Mohiaddin and associates[26] published the first report of the use of the Dixon chemical shift selective sequence for the imaging of atherosclerotic lipids. The lipid content in atheromatous plaques in necropsic human arteries was assessed by MR and compared with histology. The distribution of lipid within the plaque and between intima and media was also noted. The findings of chemical shift imaging agreed well with histological examination both for total lipid content and for distribution within each plaque. The technique was also applied for evaluation of the aortoiliac region of patients with peripheral vascular disease in comparison with the findings in healthy volunteers.[30] The majority of aortic atheromatous plaques studied were fibrous by MR, but in this study there was no histological validation. This methodology showed encouraging results for atherosclerosis imaging by MR. However, there are important issues for its use. Several factors need careful attention if this technique is to be used routinely. Motion artifact can be a problem in long scans, and subtraction of two images makes this problem more severe unless interleaved acquisition of the images is used. If signal is obtained from slow-moving blood, it may mimic or obscure areas of atheroma. This difficulty can be avoided by using a presaturation band to reduce the blood signal or acquiring images during diastole, where blood is almost stagnant and gives an unmistakably high signal. Another factor influencing the ease of CSI is magnetic field strength and homogeneity. The absolute chemical shift (in Hertz) is linearly related to field strength, with greater separation obtained between the water and fat peaks at higher fields. High field would therefore make the technique more reliable for static specimens. At all field strengths, however, careful adjustment of shimming for good homogeneity is vital. Knowledge of the quality of the

main field is necessary before embarking on these studies.[31, 32] Another factor to be considered is that eddy currents created by the gradient pulses can affect the field homogeneity during the sequence.

T2 Imaging

Several investigators have examined the plaque lipid component using methods that suppress water, incorporate water and lipid in a multiparameter data set, or use chemical shift imaging with long T2 suppression.[33, 34] In the plaque, there is a short T2 component (T2 = 2 ms) corresponding to 14 percent of the MR-visible lipid signal, which represents an ordered state, but the larger part of the visible lipids (86%) has a T2 of 22 ms, which corresponds to a less ordered state such as liquid. The ordered component likely results from core lipids with a low content of triglycerides, and a high concentration of cholesterol esters and free cholesterol (the latter being either in a monohydrate crystal form or complexed with phospholipids).

Several factors may contribute to shorten water T2 in the atheromatous core. These include (1) the susceptibility differences from the micellar structure of parietal lipoproteins; (2) the more numerous or more exposed hydrophilic sites resulting from lipoprotein oxidation[35]; (3) a longer contact time of the hydrophilic sites of cholesteryl esters and the water molecules, either from -C=O on the fatty acid chain (which may explain the contrast between adventitia and media) or from -OH of C10 and C18 on the cholesterol ring, with a further interchange between bound water layers and free water.

Two groups have addressed the issue of short T2 lipids and developed methods to improve in vitro imaging of the lipid component in atheromatous plaque. Altbach and colleagues[36, 37] examined aortic plaque using a stimulated echo diffusion weighted sequence and showed improved water suppression and lipid visualization. This technique, however, also faces the problem of low signal-to-noise ratio and long acquisition times, which is not a limiting factor in the water imaging sequence used with T2 contrast. Gold and coworkers[33] implemented a back-projection technique with long T2 lipid suppression to image the short component, using fresh atheromatous human aortic tissue suspended in solutions of agarose and manganese chloride and heated to body temperature. The sequence detected species with a T2 between 150 μs and 9 ms. The vessel wall and plaque components could be identified by means of their MR characteristics, and there was good correlation with histology. Calcification and fibrous tissue appeared with signal loss or attenuated signal on the Dixon water image, and perivascular and intimal lipid appear with a high signal on the lipid image. However, the development of this technique for in vivo studies may be limited by low signal-to-noise ratio, and a great increase in acquisition time necessitated by electrocardiographic (ECG) gating. These limitations do not apply for water imaging because at 1.5 T, a contrast-to-noise ratio of 40 can be produced for atheromatous core versus normal media with a total acquisition time of less than 4 minutes.

T1-Weighted Imaging

Pearlman and associates[34] presented ¹H-spectra of atherosclerotic plaques at high field (6.3, 8.5, and 11.7 Tesla). T1-weighted images identify calcifications because these regions appear as low-intensity zones in all MR sequences (Figure 21–3) due to low water content.[15] Contrast-to-noise ratio for calcified regions versus other components is higher in T1-weighted than T2-weighted images.

DIFFUSION IN PLAQUE COMPONENTS

Because water diffusion results from random motion, measuring displacements of water by calculating the apparent diffusion coefficient (D) can probe the microstructure of the environment in which these displacements take place. With a pulsed field gradient sequence, one can measure water diffusion in atherosclerotic components and diffusion isotropy in the lipid core at the plaque shoulder (where destruction of the fibrous components can be accentuated under the action of macrophage metalloproteinases[38]). Using a technique in which 6 serial spin-echo images are created at 9.4 T with two diffusion gradients applied 1 ms apart from the nonselective refocusing pulse, investigators have shown that water diffusion is limited and behaves isotropically in the lipid core of atheromatous plaques[17]: D values are much lower in that region than in any of the other components of normal or diseased arterial walls. The mechanisms for altered diffusion in such atheromatous structures are not known, but they may result from the presence of oxidized lipoproteins, which restrict diffusion by allowing water penetration into their micellar structure, or from the presence of lipids in smectic phase, which may structure water to a higher degree than matrix proteoglycans or proteins do. The alteration of water diffusion in lipid cores may help to further discriminate this major component of susceptible plaques by MR.

MAGNETIZATION TRANSFER

As described above, atheroma contains short T2 compounds,[16] and these species and the stiffness of some components may allow a magnetization transfer (MT) effect inasmuch as numerous potential sites of exchange are present at the surface of collagen and elastin fibers, and in oxidized lipoproteins of the lipid core. Using a method developed by Pachot-Clouard and colleagues,[18] a transfer rate of 30 to 35 percent has been demonstrated in the fibrous cap and media. A 20 to 25 percent rate has been shown in the atheromatous core. MT images of plaques seem to provide a contrast similar to the T2-

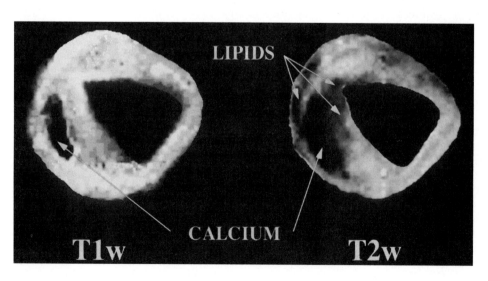

LIPIDS

CALCIUM

T1w   T2w

Figure 21–3. In vitro T1w and T2w images of a left coronary artery. Calcification is black on both sequences. The lipid infiltration is well discriminated as a low-intensity region on the T2w image that is not seen on the T1w image. The lower part of the stenotic lesion is composed of collagen fibers with the same intensity as the normal media.

weighted images.[39] The aim is now to optimize this contrast for MT imaging in vivo, which could be useful when T2 contrast is not easily implemented, as in fast imaging.

APPLICATION OF MAGNETIC RESONANCE ATHEROMA STUDIES

Animal Models

The value of MR is in describing progression of experimental atherosclerosis (fibrous caps, necrotic cores, and intraplaque hemorrhage resulting from balloon injury to the abdominal aorta).[40] Skinner and associates[40] used high-resolution cardiovascular magnetic resonance (CMR) to serially image advanced lesions of atherosclerosis in the rabbit abdominal aorta. They demonstrated that progression of disease resulted in increased lesion mass and intralesion complications with decreased arterial lumen. Images acquired in vivo correlated with fine structure of the lesions of atherosclerosis, including the fibrous cap, necrotic core, and lesion fissures, as verified by gross examination, dissection microscopy, and histology.[40] The regression effect of a hypocholesterolemic diet in rabbits has also been assessed.[41] Transgenic apolipoprotein-E knockout mice, which develop aortic and coronary lesions very similar to human ones, can be studied at high field (9.4 T), with the demonstration of the rapid progression of atheromatous plaques.[42] Pharmacological companies have found such approaches valuable because the follow-up of animal studies is noninvasive, and costs are reduced from a reduction in the number of sacrificed animals.

In Vitro Study of the Effects of Angioplasty

It has proved possible to apply CMR techniques in the study of angioplasty in vitro. Using T2 contrast, which combines morphometry and tissue characterization, one can image the effects of cardiac interventional techniques to test the resistance of the fibrous cap to radial compression, investigate the consequences of balloon inflation on calcified and noncalcified plaques, and compare the effects of in vitro angioplasty and atherectomy on the fibrous components.[20] In a study of this type, Toussaint and associates[20] studied three types of plaque: Group A consisted of plaques with a complete collagenous cap (type Va lesions[43]); group B plaques had no cap (type IV lesions); and Group C consisted of calcified plaques. In vitro angioplasty was performed by radial compression on unfixed samples in a saline bath at 37°C (Figure 21–4) to specifically study the compressive effects of angioplasty on the plaque components. A single balloon inflation at 6 atmospheres was performed for the noncalcified plaques. Images were acquired at 9.4 T before and after the

procedures, using conventional spin-echo sequences: T1-weighted (T1w): TR = 600 ms, TE = 3 ms; T2-weighted (T2w): TR = 2 s, TE = 50 ms; field of view = 2 cm, resolution = 156 × 156 × 600 μm. Using a computerized planimetry program, the luminal area, the plaque area, the areas of collagenous cap and lipid core, the area surrounded by the external elastic lamina (EEL area), and the maximal cap thickness were measured. The obstruction ratio was defined as the ratio of the plaque area divided by the EEL area.

After balloon compression, in group A, no significant reduction was found in the plaque, cap, or lipid core. These plaques with a complete collagenous cap were not modified, but the disease-free segment was stretched. Significant plaque reduction was observed after radial compression of group B plaques. This finding was mainly due to the reduction in lipid core, which was partly extruded into the lumen, and partly redistributed into the wall. Calcified lesions from group C show large dissections at the shoulder of the plaque where the maximal stiffness gradient is usually found. Angioplasty did not affect large collagenous caps and did not change the lesion size. Atherectomy produced lacerations and flaps on the luminal side of the cap.

Most of the plaque fractures during balloon angioplasty occur in a region of highest circumferential stress, usually at the shoulder of the plaque.[44, 45] The shoulder is a junction site where the collagenous cap, which can be as much as 5 times stiffer than normal intima, produces a large stiffness gradient with the lipid core, and thereby plays an important role in plaque susceptibility to rupture.[44, 46] The resistance of a plaque to radial compression depends on the distribution of stress over the different plaque components. Lee and coworkers[47, 48] studied the stress-strain relation in dissected fibrous caps under uniaxial compression. They showed that plaques did not rupture under uniform compression if a cap occupied 50 percent of the lesion volume. Using a finite element analysis, they also demonstrated that variation of collagenous cap and lipid core thicknesses alters the distribution of circumferential stress in the vessel wall, which determines plaque stability.[24]

A thick complete collagenous cap is not altered in situ under a 6 atmosphere radial compression. This cap prevents any alteration of the underlying atheroma and redistributes stress over the disease-free segments. In a clinical situation, one would expect that the effect of balloon inflation on such vessels would create a dissection of the intima media, which may be considered a mechanism of successful dilation.[49] Stretching of the disease-free segments rather than stretching of the plaque may be one mechanism of balloon angioplasty in lesions with a complete collagenous cap.[50]

These results give broad support to the contention that an intact collagenous cap largely reduces atheroma compressibility. This study, emphasizing the critical role of the collagenous cap, suggests that it may be important to characterize in vivo the chemical composition of atherosclerotic lesions before

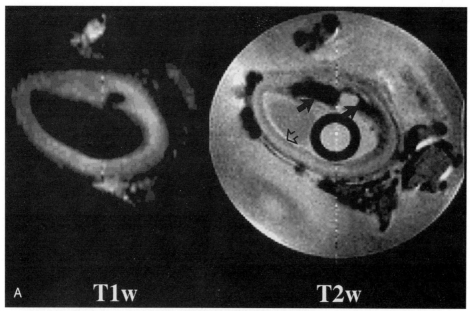

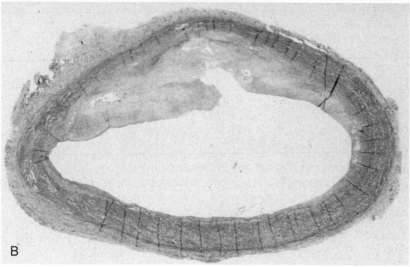

Figure 21–4. *A,* T1w and T2w images of a ruptured fatty plaque without a collagenous cap. The large lipid core is shown as a short T2 region (dark areas with black arrows), and the disease-free wall as a long T2 layer (open arrowhead). The contours of the ruptured core are well seen on the T1w. *B,* Trichrome staining showing the plaque with the rupture for comparison (see color plates).

proceeding with an intravascular intervention. In order to predict coronary plaque rupture for a particular patient, we need to develop a biochemical imaging technique, sensitive to the heterogeneous chemistry of a plaque, and particularly capable of discriminating in vivo fibrous cap from lipid core (see Figure 21–3). Recent technological developments in CMR may help to distinguish these components with high resolution in vivo[51] and therefore improve our ability to select the most appropriate interventional method. Improving our comprehension of the three-dimensional response of atherosclerotic plaques to these procedures should limit their adverse effects on the disease-free wall segments.

In Vivo Cardiovascular Magnetic Resonance

Human plaque structure can be studied in situ with images of advanced lesions in carotid arteries from patients referred for endarterectomy (Figure 21–5).[19] The T2 of various plaque components can be calculated in vivo before surgery and compared with values obtained in vitro after surgery. Yuan and associates[52] have also reported the measurement of atherosclerotic plaque burden in patients undergoing endarterectomy, showing that in vivo CMR provides an accurate assessment of histological plaque burden. More recently, the first attempts of using CMR to image coronary wall plaques have been reported,[51, 53, 54, 55] and further work in this area is awaited with great interest. Thus, atherosclerotic plaque components can be discriminated in vivo using CMR in a clinical scanner with routinely available pulse sequences, allowing the fine description of normal and pathological walls and the quantification of the plaque size. This advance may allow studies of plaque progression and regression to be planned in vivo in humans. Early data regarding in vivo quantification of subclinical aortic atherosclerosis in population-based cohorts and the relationship of disease to clinical risk factors have also been reported.[56, 57]

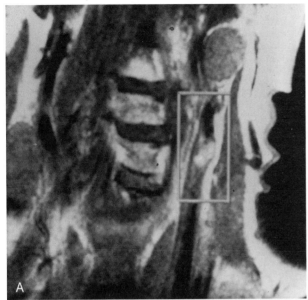

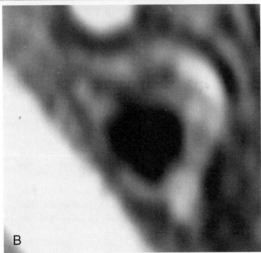

Figure 21–5. *A,* In vivo T1w coronal section of a suboccluded left carotid artery. The upper black area in the white box is calcification. *B,* Transaxial in vivo T2w image of the left common carotid stenosis downstream of the site of maximal stenosis. From the lumen (black center) to adventitia, three successive layers are defined: two long T2 high-signal layers (56 ms and 54 ms) surround a short T2 low-signal region (32 ms). Adventitia is the short T2 dark ring around the vessel. This finding was confirmed by both in vitro images and histology.

MAGNETIC RESONANCE AND HIGH-FREQUENCY INTRAVASCULAR ULTRASOUND

Intravascular ultrasound allows the study of coronary lesions. It provides some prognostic criteria to predict plaque evolution and restenosis after angioplasty. However, discrimination of the fatty and collagenous components may be difficult. A protocol was recently designed to compare the results of CMR and high-frequency ultrasound.[58] The spectral analysis of the detected signal allows the calculation of several parameters such as attenuation and retrodiffusion. However, high-frequency ultrasound

has a low penetration (around 0.5 mm), which limits it to intravascular studies. Comparison of parametric images of atheromatous lesions with CMR T2 maps at 3T and ultrasound attenuation maps at 30 to 50 MHz[59] revealed a significant correlation between plaque values of T2 and attenuation values. The study also showed differences between the T2 of media and collagen, a result that might favor high-field imaging for a better discrimination of plaque components in the future.

WHAT MAY CARDIOVASCULAR MAGNETIC RESONANCE CHANGE IN ATHEROSCLEROSIS DIAGNOSIS?

From all the imaging sequences discussed, water T2 contrast, which permits an excellent tissue discrimination and is easy to implement, is the most promising technique for future clinical investigations. Magnetization transfer may also help this discrimination with ultrafast imaging sequences.

Several technical problems, however, may still limit in vivo applications. First, motion artifacts from both luminal (slow-moving blood and turbulence) and parietal surfaces (systolic wave propagation) must be minimized by methods such as inflow saturation or ECG gating. Second, images will require high resolution and short acquisition times without a significant reduction in signal-to-noise ratio, which may require higher field strength. Finally, the design of an optimized receiver coil for carotid or iliac imaging, or intravascular utilization,[60] may further improve signal-to-noise ratio and allow for a limited field of view. With these developments, MR microscopy of atherosclerotic vessels could be added to MR angiography for a better estimation of plaque vulnerability. The long-range goal is to (1) guide therapeutic decisions and avoid acute ischemic syndromes by determining the plaque susceptibility to rupture; (2) focus treatment on high-risk plaques; and (3) avoid invasive treatment of stable plaques that are not discernible by angiographic techniques alone.

A new capability of discriminating atheromatous components would allow large-scale clinical studies of plaque progression, regression, and stabilization, and their effect on clinical events.[56] This advance could improve our understanding of the slow evolution of atherosclerosis in the early decades of life and may provide a basis for screening young patients. Longitudinal studies of vascular interventional therapies may benefit from these results. This tool could also allow the study of deep injury, plaque growth (edge vs. shoulder), arterial remodeling, and compensatory enlargement concepts in animal models of rupture and thrombosis. With dietary or pharmacological studies, these new techniques could avoid unnecessary and costly histological sampling and may finally provide new grounds for intravascular imaging.

The combination of MR angiography for luminal morphometry, three-dimensional velocity mapping for the analysis of peristenotic blood flow stream lines, determination of flow reserve in terminal arteries, perfusion measurement based on T1 changes, determination of circumferential stress, and tissue characterization may provide a complete description of the atherosclerotic process and its dynamic consequences. Finally, identification of intraplaque hemorrhage and acute thrombosis[61] may allow the study of the cascade of events leading from plaque rupture to arterial thrombosis, acute ischemia, and infarction, which may provide newer methods for prediction and prevention of these life-threatening events.

ACKNOWLEDGMENTS

This work has been supported in part by the Harold M. English Fund from the Harvard Medical School, la Société Française de Cardiologie, l'Institut Electricité Santé, and le Fonds d'Etudes et de Recherche du Corps Médical des Hôpitaux de Paris.

References

1. Wissler RW, Vesselinovitch D, Davis HR, et al: A new way to look at atherosclerotic involvement of artery wall and the functional effects. Ann N Y Acad Sci 454:9, 1985.
2. Smith E: Development of atherosclerotic plaque. *In* Shillingford J, Birdwood G (eds): Impact of Research on the Practice of Cardiology. London, British Heart Foundation 1986, p 1.
3. Potkin BN, Roberts WC: Effects of percutaneous transluminal coronary angioplasty on atherosclerotic plaques and relation of plaque composition and arterial size to outcome. Am J Cardiol 62:41, 1988.
4. Falk E: Why do plaques rupture? Circulation 86:III.30, 1992.
5. Little WC, Constantinescu M, Applegate RJ, et al: Can coronary angiography predict the site of subsequent myocardial infarction in patients with mild to moderate coronary artery disease? Circulation 78:1157, 1988.
6. Fishbein MC, Siegel RJ: How big are coronary atherosclerotic plaques that rupture? Circulation 84:2662, 1996.
7. Ambrose JA, Tannenbaum MA, Alexopoulos D, et al: Angiographic progression of coronary artery disease and the development of myocardial infarction. J Am Coll Cardiol 12:56, 1988.
8. Ambrose JA, Winters SL, Arora RR, et al: Angiographic evolution of coronary artery morphology in unstable angina. J Am Coll Cardiol 7:472, 1986.
9. Falk E, Shah PK, Fuster V: Coronary plaque disruption. Circulation 92:657, 1995.
10. Lesperance J, Theroux P, Hudon G, Waters D: A new look at coronary angiograms: Plaque morphology as a help to diagnosis and to evaluate angiograms. Int J Card Imaging 10:75, 1994.
11. Alderman EL, Corley SD, Fisher LD, et al: Five year angiographic follow-up of factors in a model of coronary artery stenosis and thrombosis. Am J Physiol 265:H1787, 1993.
12. Lichtlen PR, Nikutta P, Jost S, et al: Anatomical progression of coronary artery disease in humans as seen by prospective, repeated, quantitated coronary angiography. relation to clinical events and risk factors. The INTACT study group. Circulation 86:828, 1992.
13. Dacanay S, Kennedy HL, Uretz E, et al: Morphological and quantitative angiographic analysis of progression of coronary stenoses: A comparison of Q-wave and non-Q-wave myocardial infarction. Circulation 90:1739, 1994.
14. Toussaint JF, Southern JF, Fuster V, Kantor HL: ^{13}C-NMR spectroscopy of human atherosclerotic lesions: Relation between fatty acid saturation, cholesterol ester content, and luminal obstruction. Arterioscler Thromb Vasc Biol 14:1951, 1994.
15. Kaufman L, Crooks LE, Sheldon PE, et al: Evaluation of NMR imaging for detection and quantification of obstructions in vessels. Invest Radiol 17:554, 1982.
16. Toussaint JF, Southern JF, Fuster V, Kantor HL: T2 contrast for NMR characterization of human atherosclerosis. Arterioscl Thromb Vasc Biol 15:1533, 1995.
17. Toussaint JF, Southern JF, Fuster V, Kantor HL: Water diffusion properties of human atherosclerosis and thrombosis measured by pulse field gradient NMR. Arterioscler Thromb Vasc Biol 17:542, 1997.
18. Pachot-Clouard M, Vaufrey F, Darasse L, Toussaint JF: Magnetization transfer characteristics in atherosclerotic plaque components assessed by adapted binomial preparation pulses. MAGMA 7:9, 1998.
19. Toussaint JF, LaMuraglia GM, Southern JF, et al: Magnetic resonance images lipid, fibrous, calcified, hemorrhagic, and thrombotic components of human atherosclerosis in vivo. Circulation 94:932, 1996.
20. Toussaint JF, Jang IK, Southern JF, et al: Behavior of atherosclerotic plaque components after in vitro angioplasty and atherectomy studied by high-field MR imaging. Magn Reson Imaging 16:175, 1998.
21. Lundberg B: Chemical composition and physical state of lipid deposits in atherosclerosis. Atherosclerosis 56:93, 1985.
22. Hamilton JA, Cordes EH: Lipid dynamics in human low density lipoproteins and human aortic tissue with fibrous plaques. J Biol Chem 254:5435, 1979.
23. Bradamante S, Barenghi L, Guidici GA, Vergani C: Free radicals promote modifications in plasma high-density lipoprotein: Nuclear magnetic resonance analysis. Free Radic Biol Med 12:193, 1992.
24. Cheng GC, Loree HM, Kamm RD, et al: Distribution of circumferential stress in ruptured and stable atherosclerotic lesions. A structural analysis with histopathological correlation. Circulation 87:1179, 1993.
25. Davies MJ, Richardson PD, Woolf N, et al: Risk of thrombosis in human atherosclerotic plaque: Role of extracellular lipid, macrophage, and smooth muscle cell content. Br Heart J 69:377, 1993.
26. Mohiaddin RH, Firmin DN, Underwood SR, et al: Chemical shift magnetic resonance imaging of human atheroma. Br Heart J 62:81, 1989.
27. Vinitski S, Consigny PM, Shapiro MJ, et al: Magnetic resonance chemical shift imaging and spectroscopy of atherosclerotic plaque. Invest Radiol 26:703, 1991.
28. Herfkens RJ, Higgins CB, Hricak H, et al: Nuclear magnetic resonance imaging of atherosclerotic disease. Radiology 148:161, 1983.
29. Merickel MB, Carman CS, Brookeman JR, et al: Identification and 3-D quantification of atherosclerosis using magnetic resonance imaging. Comput Biol Med 18:89, 1988.
30. Mohiaddin RH, Sampson C, Firmin DN, Longmore DB: Magnetic resonance morphological, chemical shift and flow imaging in peripheral vascular disease. Eur J Vasc Surg 5:383, 1991.
31. Yeung HN, Kormos DW: Separation of true fat and water images by correcting magnetic field inhomogeneity in situ. Radiology 159:783, 1986.
32. Yang GZ, Firmin DN, Mohiaddin RH, et al: Inhomogeneity correction for two point Dixon chemical shift imaging (abstract). Proceedings of the Society for Magnetic Resonance Medicine, 1992, p 3819.
33. Gold GE, Pauly JM, Glover GH, et al: Characterization of atherosclerosis with a 1.5T imaging system. J Magn Reson Imaging 3:399, 1993.
34. Pearlman JD, Zajicek J, Merickel MB, et al: High-resolution 1H NMR spectral signature from human atheroma. Magn Reson Med 7:262, 1988.
35. Witztum JL, Steinberg D: Role of oxidized low density lipoprotein in atherogenesis. J Clin Invest 88:1785, 1991.

36. Altbach MI, Mattingly MA, Brown MF, Gmitro AF: Magnetic resonance imaging of lipid deposits in human atheroma via a stimulated-echo diffusion-weighted technique. Magn Reson Med 20:319, 1991.
37. Trouard TP, Altbach MI, Hunter GC, et al: MRI and NMR spectroscopy of the lipids of atherosclerotic plaque in rabbits and humans. Magn Reson Med 38:19, 1997.
38. Libby P: Molecular basis of the acute coronary syndromes. Circulation 91:2844, 1995.
39. Pachot-Clouard M, Darrasse L, Vaufrey F, Toussaint JF: Magnetization transfer imaging for assessment of unstable atherosclerotic plaque components (abstract). Proceedings of the International Society for Magnetic Resonance Medicine, 1998, p 590.
40. Skinner MP, Yuan C, Mitsumori L, et al: Serial MRI of experimental atherosclerosis detects lesion fine structure, progression, and complications in vivo. Nat Med 1:69, 1995.
41. McConnell MV, Aikawa M, Maier SE, et al: MRI of rabbit atherosclerosis in response to dietary cholesterol lowering. Arterioscler Thromb Vasc Biol 19:1956, 1999.
42. Fayad ZA, Fallon JT, Shinnar M, et al: Noninvasive in vivo high-resolution magnetic resonance imaging of atherosclerotic lesions in genetically engineered mice. Circulation 98:1541, 1998.
43. Fuster V: Lewis A Conner Memorial Lecture. Mechanisms leading to myocardial infarction: insights from studies of vascular biology. Circulation 90:2126, 1994.
44. Lee RT, Loree HM, Cheng GC, et al: Computational structure analysis based on intravascular ultrasound imaging before in vitro angioplasty: Prediction of plaque fracture location. J Am Coll Cardiol 21:777, 1993.
45. Laerum F, Castaneda-Zuniga WR, Rysavy J, et al: The site of arterial rupture in transluminal angioplasty: An experimental study. Radiology 144:769, 1982.
46. Richardson PD, Davies MJ, Born GVR: Influence of plaque configuration and stress distribution on fissuring of coronary atherosclerotic plaques. Lancet 2:941, 1989.
47. Lee RT, Grodsinsky AJ, Frank EH, et al: Structure-dependent dynamic mechanical behavior of fibrous caps from human atherosclerosis plaques. Circulation 83:1764, 1991.
48. Lee RT, Richardson GS, Loree HM, et al: Prediction of mechanical properties of human atherosclerotic tissue by high-frequency intravascular ultrasound imaging. An in vitro study. Arterioscl Thromb 12:1, 1992.
49. Waller BF, Orr CM, Pinkerton CA, et al: Coronary balloon angioplasty dissections: "The good, the bad, and the ugly." J Am Coll Cardiol 20:701, 1992.
50. Hjemdal-Monsen CE, Ambrose JA, Borrico S, et al: Angiographic patterns of balloon inflation during percutaneous transluminal coronary angioplasty: Role of pressure-diameter curves in studying distensibility and elasticity of the stenotic lesion and the mechanism of dilation. J Am Coll Cardiol 16:569, 1990.
51. Meyer CH, Hu BS, Macovski A, Nishimura DG: Coronary vessel wall imaging (abstract). Proceedings of the International Society for Magnetic Resonance Medicine, 1998, p 15.
52. Yuan C, Beach KW, Smith LH, Hatsukami TS: Measurement of atherosclerotic carotid plaque size in-vivo using high resolution magnetic resonance imaging. Circulation 98:2666, 1998.
53. Fayad ZA, Fuster V, Fallon JT, et al: Noninvasive in vivo human coronary artery lumen and wall imaging using black blood magnetic resonance imaging. Circulation 102:506, 2000.
54. Worthley SG, Helft G, Fuster V, et al: Noninvasive in vivo MRI of experimental coronary artery lesions in a porcine model. Circulation 101:2956, 2000.
55. Botnar RM, Stuber M, Kissinger KV, et al: Noninvasive coronary vessel wall and plaque imaging with magnetic resonance imaging. Circulation 102:2582, 2000.
56. Jaffer FA, O'Donnell CJ, Larson M, et al: MRI assessment of aortic atherosclerosis in an asymptomatic population: The Framingham Heart Offspring MRI pilot study. Circulation 102:II-458, 2000.
57. Chan SK, Jaffer FA, Botnar RM, et al: MRI aortic atherosclerosis reproducibility project [abstract]. Circulation 102:II-460, 2000.
58. Toussaint JF, Bridal SL, Raynaud JS, et al: Atherosclerotic plaque vulnerability studied by magnetic resonance and ultrasound parametric images. Circulation 96:I-235, 1997.
59. Raynaud JS, Bridal SL, Toussaint JF, et al: Characterization of atherosclerotic plaque components by high-resolution quantitative MR and US imaging. J Magn Reson Imaging 8:622, 1998.
60. Zimmermann-Paul GG, Quick HH, Vogt P, et al: High-resolution intravascular MRI: Monitoring of plaque formation in heritable hyperlipidemic rabbits. Circulation 99:1059, 1999.
61. Johnstone MT, Perez A, Stewart R, et al: In-vivo magnetic resonance imaging of thrombosis [abstract]. J Am Coll Cardiol 37:400A, 2001.

CHAPTER 22

Assessment of the Biophysical Mechanical Properties of the Arterial Wall

Raad H. Mohiaddin

Arteries are thin-walled elastic tubes, the diameter of which varies with the pulsating pressure. In addition, they propagate pressure and flow waves, created by the ejection of blood by the heart, at a velocity that is largely determined by the elastic properties of the arterial wall. The vascular wall can be deformed by pressure and shear stress forces exerted by the blood as well as the tethering imposed by the surrounding tissues. The biophysical mechanical properties of the arterial wall play an important role in the pathogenesis of cardiovascular diseases. Sclerosis (or stiffness), for example, is an important aspect of the atherosclerotic vascular disease that has long been appreciated both in experimental disease in animals[1] and in man,[2] and regression of the disease leads to reduced stiffness.[3, 4] Rupture of atherosclerotic plaque (a common initiating mechanism of acute myocardial infarction) and aortic dissection can be viewed as mechanical failures in the diseased vessels. In addition, interventional procedures such as angioplasty are often effective because of mechanical injury to the vessel wall, although injury itself may lead to restenosis. A number of common conditions are associated with changes in arterial mechanical properties, although the importance of these changes is not always recognized.

Systemic hypertension is almost always associated with altered mechanical properties of the peripheral vasculature. A decrease in aortic compliance results in an increase in aortic systolic pressure, a decrease in aortic diastolic pressure, and therefore an increase in pulse pressure. Left ventricular-vascular coupling is an important determinant of left ventricular performance, and its measurement takes into account the pulsatile load imposed on the left ventricle as well as the systemic vascular resistance.[5] Metabolic disorders such as Ehlers-Danlos and Marfan syndromes,[6] diabetes mellitus,[7] familial hypercholesterolemia,[8] and growth hormone deficiency[9] are also known to alter arterial compliance.

Similarly, the distensibility of the pulmonary artery is reduced in pulmonary hypertension. Postmortem studies of pulmonary arterial strips have shown that the extensibility of the pulmonary trunk is decreased in pulmonary arterial hypertension.[10]

The wall of an artery becomes less extensible the more it is stretched from its natural length. An increased stretching of the circumference of the vessel will diminish the distensibility. When the pulmonary artery resistance increases, the vessels become more distended and less distensible. Indirect measurements of pulmonary artery compliance have suggested that pulmonary arterial distensibility decreases with rising pulmonary artery pressure.[11, 12] Cardiovascular magnetic resonance (CMR) provides the means to study pulmonary arterial distensibility and flow noninvasively and directly in patients with pulmonary arterial hypertension.[13, 14]

Finally, it is known that changes in aortic function can affect other important areas of cardiovascular function. For example, aortic distensibility is an important factor in the determination of myocardial perfusion. A decrease in aortic distensibility has been shown to significantly alter the distribution of transmural myocardial blood flow and decrease the subendocardial/subepicardial flow ratio.[15] Thus, decreased aortic distensibility may increase the risk of subendocardial ischemia in the presence of coronary artery stenosis, left ventricular hypertrophy, or both.

In this chapter, the clinical importance of arterial biophysical function and its assessment by CMR is examined in detail, as a complement to Chapter 21.

ARTERIAL STRUCTURE

A normal artery consists of three morphologically distinct layers. The intima consists of a single continuous layer of endothelial cells bounded peripherally by a fenestrated sheet of elastic fibers. The media consists entirely of diagonally oriented smooth muscle cells, surrounded by variable amounts of collagen, elastin, and proteoglycans. The adventitia consists predominantly of fibroblasts intermixed with smooth muscle cells loosely arranged between bundles of collagen and proteoglycans. Each structural component has its own characteristic properties. Smooth muscle is the physiologically active element, and by contracting or developing force, it can alter the diameter of the vessel or the tension in the wall. The other components are essentially passive in their mechanical behavior. Elastin, which can be stretched to as much as 300 percent of its length at rest without rupturing,[16] behaves mechanically more similar to a linear elastic material such as rubber than other connective tissue components. When elastin fibers are stretched and release, they return promptly to their original state. Elastin fibers are important for maintaining normal pulsatile behavior, but they fracture at very low stresses and are probably much less important in determining the overall strength of the vessel wall. Collagen fibers, on the other hand, are much stiffer and much stronger. The proportion of these components varies from artery to artery. In the thoracic aorta, the elastin forms 60 percent of total fibrous element, and collagen forms 40 percent. The collagen proportion increases with increasing distance from the heart, reaching 30 percent elastin and 70 percent collagen in the extrathoracic vessels.[17] The collagen/elastin ratio increases with age, which is one reason why vascular stiffness increases with age. The wall of the human thoracic aorta is supplied by vasa vasorum and grows by increasing the number of lamellar units. The abdominal aorta, in comparison, is avascular because it lacks vasa vasorum and grows by increasing the thickness of each lamellar unit. The avascular thickness and the elevated tension per lamellar unit of the abdominal aorta are thought to predispose it to atherosclerosis.

The distensibility of a blood vessel depends on the proportions and interconnections of these materials and on the contractile state of the vascular smooth muscle. Elasticity is the ability of a material to return to its original shape and dimensions after deformation, the deformation being proportional to the force applied. This proportionality was first described by Hooke in 1676 and is known as Hooke's law. The point at which Hooke's law ceases to apply is known as the elastic limit. When a solid has been deformed beyond this point, it cannot regain its original form and acquires a permanent distortion. With larger loads still, the yield point is reached when the deformation continues to increase without further load and usually rapidly leads to breakage. In purely elastic bodies, stress (the force per unit area that produces deformation) produces its characteristic strain (the deformation of a stressed object) instantaneously, and strain vanishes immediately on removal of the stress. Some materials, however, require a finite time to reach the state of deformation appropriate to the stress and a similar time to regain their unstressed shape. Blood vessels typically exhibit such behavior, which is called viscoelasticity.

DEFINITION OF VASCULAR WALL STIFFNESS

Vascular mechanics have been described using different elastic moduli and assumptions, and for different purposes. Several approaches have been described that use clinically available methods for in vivo characterization of the stiffness of the vessel wall. The ability to measure vascular stiffness has been greatly improved by the recent advances in imaging, such as high frequency ultrasound and CMR.

The relation between vascular wall deformation (strain) and the pressure exerted on the inner surface of the vascular wall (stress) is commonly used for the measurement of arterial wall biophysical properties (elastic modulus). A plethora of terminology for the description of different elastic moduli, which can be confusing, has been described.[18] The pressure-strain elastic modulus of the arterial wall (E_p), Young's modulus, named for the 19th century English physician and physicist, Thomas Young,[18] is commonly used. This elastic modulus, which applies to an open-ended vessel in the absence of reflection, is defined as: $E_p = 2\Delta P/(\Delta V/V)$. This is the fractional distensibility ($\Delta V/V$) of the arterial lumen per unit pulse pressure ΔP.

Arterial compliance, C, which is defined as the change in volume (ΔV) per unit change in pressure (ΔP), also has been used in the literature. It has been argued that this definition is appropriate to measurement of ventricular compliance and not to the compliance of an open-ended arterial segment. For the latter, the inverse of Young's modulus has been suggested ($1/E_p$).

The average arterial compliance of a particular vessel pathway can also be determined by measuring the speed of propagation of the pressure or flow waves in the vessel pathway. The velocity of such waves depends principally on the distensibility of the vessel wall. Flow waves are propagated in much the same way as the pressure wave. The propagation of flow waves has not been studied as extensively as that of pressure waves, partly because, unlike flow, accurate methods of pulsatile pressure measurements have been available for a long time, and partly because the distinction between flow wave velocity and blood velocity has not always been clearly recognized. Blood velocity means the speed of an average drop of blood, whereas flow wave velocity means the speed with which motion is transmitted.

MEASUREMENT OF ARTERIAL WALL STIFFNESS

Arterial stiffness, which describes the resistance of arterial wall to deformation, is difficult to measure because of the complex mechanical behavior of arterial wall. In vitro human arterial compliance has been measured from pressure-volume curves in postmortem arteries.[19–22] In vivo estimation of arterial wall compliance is more difficult, however, and has been performed using indirect and invasive techniques, including pulse wave velocity measurements in animals and in humans,[23, 24] the pressure-radius relationship using the Peterson transformer coil in animals,[25] x-ray contrast angiography in humans,[26, 27] and sonography.[28–30] The unique contribution of CMR to the assessment of arterial wall mechanics is discussed in the following paragraphs.

CARDIOVASCULAR MAGNETIC RESONANCE OF REGIONAL AORTIC COMPLIANCE

CMR provides a direct noninvasive method of studying regional aortic compliance.[31, 32] High-resolution cine imaging or electrocardiogram (ECG)-gated spin-echo imaging in a plane perpendicular to the ascending and/or descending aorta allows measurement of aortic cross-sectional area during systole and diastole. The lumen of the aorta is outlined manually on the computer screen to measure the change in aortic area (ΔA) between diastole and systole. Regional aortic compliance *(C)* (μl/mm Hg,

m^2/N) is calculated from the change in volume (ΔV = ΔA × slice thickness) of the aortic segment divided by the aortic pulse pressure (ΔP) measured by a sphygmomanometer (Figures 22–1 and 22–2).[33] Automatic measurement of aortic cross-sectional area is also possible.[34] Other indices of aortic stiffness that can be derived from these measurements include distensibility, Young's elastic modulus, and stiffness index β ([systolic blood pressure/diastolic blood pressure]/area strain).

Measurement of regional aortic compliance by CMR is calculated from the change in volume of an aortic segment and from aortic pulse pressure estimated by a sphygmomanometer at the level of the brachial artery. The accuracy of the indirect measurement of the pressure change needed to compute compliance is limited because it ignores the changes in the pressure wave as it propagates through the arterial tree. Despite the limitations of the pressure measurement, there is a good correla-

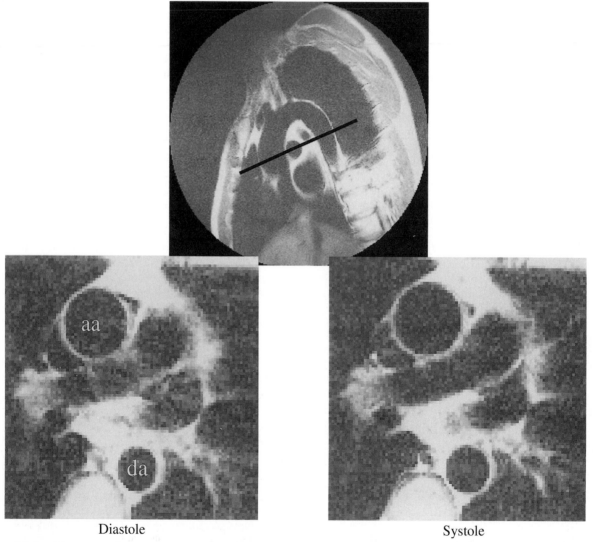

Diastole

Systole

Figure 22–1. Oblique sagittal image of the ascending aorta, arch, and descending thoracic aorta showing the sites where flow wave velocity and regional compliance are measured *(top)*. In the bottom row, the oblique transverse plane shown in the top image is represented in diastole *(left)* and systole *(right)*. This figure shows the change in area of a 38-year-old normal volunteer. aa = ascending aorta; da = descending aorta.

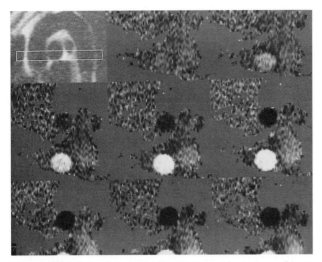

Figure 22–2. Cine velocity mapping in a plane equivalent to that shown in Figure 22–1. The first frame was acquired 50 ms after the R wave of the electrocardiogram (ECG) and represents the onset of left ventricular systole. The velocity maps indicate zero velocity as mid-gray, caudal velocities in the descending aorta as light gray to white, and cranial velocities as darker shades of gray to black, with gray level intensity proportional to velocity.

coworkers[36] used the comb-excited Fourier velocity-encoded method previously reported by Dumoulin and associates[37] to measure local arterial wave speed in the femoral artery in healthy men. In this method, simultaneous Fourier velocity-encoded data from multiple stations was acquired. The technique em-

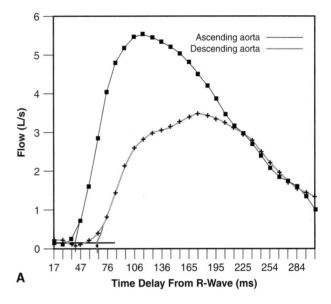

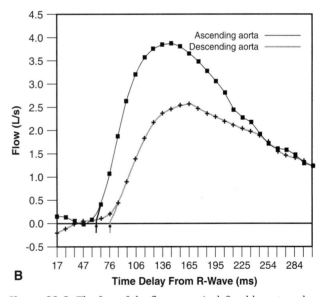

Figure 22–3. The foot of the flow wave is defined by extrapolation of the rapid upstroke of the flow wave to the baseline, and this process is performed for flow in both the ascending and descending aorta. The transit time needed for the flow wave to propagate from a point in the mid-ascending aorta to a point in the mid-descending aorta can then be measured. The distance around the arch from the plane in the ascending aorta to the plane in the descending aorta can be measured from the oblique sagittal image shown in Figure 22–1, and the flow wave velocity is calculated by division of this distance by the propagation time. The transit time in *A* comes from a normal subject with good compliance, and the example in *B* comes from an elderly patient with poor compliance. Note that the transit time is significantly shorter in the poor-compliance patient. (From Mohiaddin RH, Firmin DN, Longmore DB: Age-related changes of human aortic flow wave velocity measured non-invasively by magnetic resonance imaging. J Appl Physiol 74:492, 1993, with permission.)

tion between measurement of regional aortic compliance and measurement of global compliance from the speed of the propagation of the flow wave within the vessel.[35]

CARDIOVASCULAR MAGNETIC RESONANCE OF FLOW WAVE VELOCITY

Mohiaddin and colleagues[35] showed the feasibility of using magnetic resonance (MR) phase-shift velocity mapping to measure aortic flow wave velocity in man. Cine two-dimensional phase-shift velocity maps were acquired with high temporal resolution in a plane perpendicular to the ascending and descending aorta, and the time taken for the flow wave to travel between the two points was measured (see Figure 22–1). The instantaneous flow (liter/s) in the ascending and descending aorta can be calculated from aortic cross-sectional area and the mean velocity within that area. Pulse wave velocity (PWV) was calculated in m/s from the transit time (T) of the foot of the flow wave (see Figure 22–2) and from the distance (D) between the two points obtained from an oblique sagittal spin-echo image (Figure 22–3). The distance is determined manually on the computer screen by drawing a line in the center of the aorta joining the two points. The foot of the flow wave was defined by extrapolation of the rapid upstroke of the flow wave to the base line, and PWV is defined as

$$PWV = D/T$$

Others have used different MR flow imaging techniques to assess arterial compliance. Tarnawski and ploys a comb excitation radiofrequency pulse that excites an arbitrary number of slices. This method causes the signals from the spin in a particular slice to appear at a position in the phase encoding direction, which is the sum of the spin velocity and an offset arising from the phase increment given to that excitation slice. Acquisition of spin velocity information occurs simultaneously for all slices, permitting the calculation of wave velocities arising from the pulsatile flow.

Hardy and coworkers[38] studied aortic flow wave velocity using a two-dimensional MR selective-excitation pulse to repeatedly excite a cylinder of magnetization in the aorta, with magnetization readout along the cylinder axis each time. A toggled bipolar flow-encoding pulse was applied prior to readout, to produce a nondimensional phase-contrast flow image. Cardiac gating and data interleaving were employed to improve the effective time resolution to 2 ms. Wave velocities were determined from the slope of the leading edge of flow measured on the resulting M-mode velocity image. Aortic pulse wave velocity was also measured by the same group using a combination of cylinder of magnetization with Fourier-velocity encoding and readout gradients applied along the cylinder axis (aorta),[39] with the advantage of eliminating partial volume effects that hindered their previous approach, but the Fourier method has the drawback that it is no longer real-time and errors occur due to accumulation of flow data over several (typically 16–32) cardiac cycles.

In vitro experiments showed MR measurements of pulse wave velocity in a tube phantom to be very reproducible and in good agreement with pulse wave velocity measurements made with a pressure catheter.[40]

THE CLINICAL USE OF CARDIOVASCULAR MAGNETIC RESONANCE FOR ASSESSING ARTERIAL WALL STIFFNESS

Mohiaddin and colleagues[41] were the first to use CMR for measurement of aortic compliance. They demonstrated that aortic compliance in asymptomatic subjects falls with age, and that patients with coronary artery disease have abnormally low compliance (Figure 22–4). The results suggest a possible role for compliance in the assessment of cardiovascular fitness and the detection of coronary artery disease. Because there is overlap between normal compliance and compliance in patients with coronary artery disease above the age of 50, the test cannot have perfect sensitivity and specificity. Below the age of 50, however, there is much less overlap and the test is more specific. Abnormally low aortic compliance has also been demonstrated among patients with aortic coarctation,[42] and in patients with Marfan syndrome.[43] The same group also showed the feasibility of using MR velocity map-

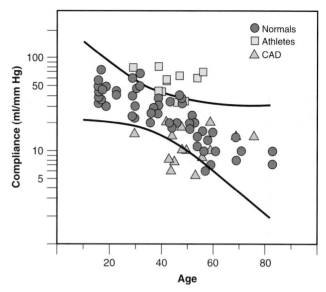

Figure 22–4. Ascending aortic compliance displayed using a logarithmic scale and plotted against age in three groups: Normals (circles), athletes (triangles), and patients with coronary artery disease (CAD) (squares). The 95% confidence limits are shown for the normals. The athletes' compliance is abnormally high, and in coronary disease patients it is abnormally low. (From Mohiaddin RH, Underwood SR, Bogren HG, et al: Regional aortic compliance studied by magnetic resonance imaging: The effects of age, training, and coronary artery disease. Br Heart J 62:90, 1989, with permission.)

ping for measurement of aortic flow wave velocity. Aortic flow wave velocity increased linearly with age, and there was a significant difference between the youngest decade and the oldest decade studied. Flow wave velocity was negatively correlated with regional ascending aortic compliance measured in the same subjects (Figure 22–5).

Regression of atheroma with reduction of cholesterol levels is recognized to occur, but less is known about reversal of sclerosis. Noninvasive indices of sclerosis have largely been based on carotid ultrasound measurements. Forbat and colleagues[44] measured aortic compliance, coronary calcification, and carotid intimal-medial thickness during reduction of cholesterol level in hypercholesterolemic patients with and without coronary artery disease. All received fluvastatin for 1 year. Aortic compliance was assessed using CMR, and coronary calcification score was determined by electron beam computed tomography. Carotid intimal-medial thickness was measured by carotid ultrasound. The authors showed an improvement in aortic compliance over 1 year, which indicates that the lipid changes induced by fluvastatin (an increase in high-density lipoprotein [HDL] level, decrease in low-density lipoprotein level, and improvement in the ratio of low- to high-density lipoprotein [LDL/HDL]) beneficially influenced vascular pathophysiology. In those patients studied with carotid ultrasound, carotid intimal-medial thickness decreased from 1.09 to 0.87 mm ($p = 0.004$), corroborating these results.

Kupari and coworkers[45] measured aortic elastic

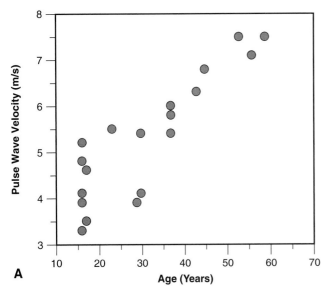

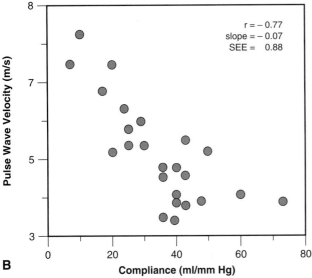

Figure 22–5. Pulse wave velocity is directly proportional to age *(A)*, and inversely proportional to regional aortic compliance *(B)*. (From Mohiaddin RH, Firmin DN, Longmore DB: Age-related changes of human aortic flow wave velocity measured non-invasively by magnetic resonance imaging. J Appl Physiol 74:492, 1993, with permission.)

modulus by CMR in asymptomatic subjects and correlated these measurements with physical activity, ethanol consumption, systolic blood pressure, fasting blood lipids, and serum insulin. They showed that the average value of the ascending and descending aortic elastic modulus was associated positively and statistically significantly with blood pressure, physical activity, serum insulin, and HDL. The elastic modulus was associated negatively with LDL/HDL cholesterol ratio. No association between aortic elastic modulus and either smoking or ethanol consumption was demonstrated. The same group demonstrated a higher aortic elastic modulus in patients with Marfan syndrome than normal healthy subjects, indicating a relative decrease in the distensi-

bility of the thoracic aorta.[46] Kupari and associates also demonstrated that aortic flow wave velocity was more reproducible (inter- and intraobserver) than measurement of the pulsatile aortic area change or the elastic modulus.

Adams and coworkers demonstrated abnormal aortic distensibility and stiffness index in patients with Marfan syndrome using CMR.[47] Beta-adrenergic blocking agents may reduce the rate of aortic root dilation and the development of aortic complications in patients with Marfan syndrome. This may be due to beta blocker–induced changes in aortic stiffness. To investigate this, Groenink and colleagues[48] used CMR to measure aortic distensibility and aortic pulse wave velocity to assess aortic stiffness in Marfan syndrome and healthy volunteers before and after beta blocker therapy. They showed that in both groups, mean blood pressure decreased significantly but only the Marfan syndrome patients had a significant increase in aortic distensibility at multiple levels and a significant decrease in pulse wave velocity after beta blocker therapy.

Savolaninen and associates [49] used CMR and indirect brachial artery blood pressure measurements to assess aortic elastic modulus (E_p) in patients with essential hypertension before and after 3 weeks and 6 months of therapy with cilazapril (an angiotensin-converting enzyme inhibitor) or atenolol (a beta$_1$-adrenergic blocker). The authors concluded that 6 months of treatment with either cilazapril or atenolol reduced the stiffness of the ascending aorta in essential hypertension. No statistically significant differences between the effects of the two drugs were observed. Honda and colleagues[50] used CMR to measure aortic distensibility in patients with systemic hypertension and demonstrated that the antihypertensive drugs nicardipine and alacepril have a beneficial effect on aortic distensibility. Resnick and associates[51] assessed aortic distensibility, left ventricular mass index, abdominal fat (subcutaneous and visceral), and free magnesium levels in the brain and skeletal muscle by CMR. In patients with essential hypertension, the following were concluded: (1) systolic hypertension and increased left ventricular mass index may result from arterial stiffness; (2) arterial stiffness may be one mechanism by which abdominal visceral fat contributes to cardiovascular risk; and (3) decreased magnesium contributes to arterial stiffness in hypertension.

Chelsky and coworkers[52] used CMR to measure aortic compliance in nine premenopausal women before and after menotropin therapy. They demonstrated a short-term rise in estrogen induced by menotropin treatment was associated with an increase in aortic compliance. Aortic size was not significantly increased within this time frame.

Bogren and colleagues[53] used CMR to study pulmonary artery distensibility in healthy volunteers and in patients with pulmonary arterial hypertension. The distensibility was found to be significantly lower in pulmonary arterial hypertension than in normal subjects, but there was no age-related differ-

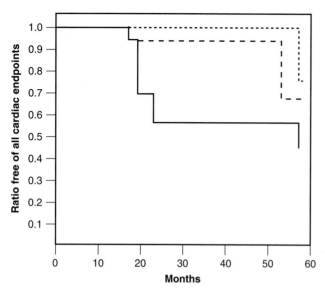

Figure 22–6. Relationship of aortic stiffness to occurrence of any cardiac endpoint, with the study population divided into terciles ($p = 0.001$). (From Stefanides C, Dernellis J, Tsiamis E, et al: Aortic stiffness as a risk factor for recurrent acute coronary events in patients with ischaemic heart disease. Eur Heart J 21:390, 2000, with permission.)

ence. They also demonstrated a small retrograde flow (2%) in the pulmonary trunk of normal subjects close to the pulmonic valve. Antegrade plug flow occurred in most normal subjects but varied among individuals. There were also other variations in the flow pattern among normal individuals. All patients with pulmonary arterial hypertension had a markedly irregular antegrade and retrograde flow and a large retrograde flow (average 26%).

Finally, it has recently been shown by Stefanides and colleagues[54] that aortic stiffness has prognostic power in determining the likelihood of future cardiac events (Figure 22–6). This interesting study merits further examination because it uses a simple marker of disseminated arterial disease that is easy to measure in large populations.

CONCLUSION

Atherosclerosis consists of two components, the most often discussed being atherosis, relating to plaque genesis and composition. Sclerosis is often the forgotten partner in clinical practice, but vessel wall stiffening has now been shown to be useful both diagnostically and prognostically. CMR is an ideal tool for its investigation, and further studies defining the clinical role of sclerosis parameters can be expected in the future.

References

1. Band W, Goedhard WJ, Knoop AA: Comparison of effects of high cholesterol intake on viscoelastic properties of the thoracic aorta in rats and rabbits. Atherosclerosis 18:163, 1973.
2. Banga I, Balo J: Elasticity of the vascular wall. 1. The elastic distensibility of the human carotid as a function of age and arteriosclerosis. Acta Physiol Acad Sci Hung 20–21:237, 1961.
3. Farrar DJ, Green HD, Wanger WD, Bond MG: Reduction in pulse wave velocity and improvement of aortic distensibility accompanying regression of atherosclerosis in the Rhesus monkey. Circ Res 47:425, 1980.
4. Farrar DJ, Bond GM, Riley WA, Sawyer JK: Anatomic correlates of aortic pulse wave velocity and carotid artery elasticity during atherosclerosis progression and regression in monkeys. Circulation 83:1754, 1991.
5. Isnard RN, Pannier BM, Laurent S, et al: Pulsatile diameter and elastic modulus of the aortic arch in essential hypertension: A non-invasive study. J Am Coll Cardiol 13:399, 1989.
6. Handler CE, Child A, Light NM: Mitral valve prolapse, aortic compliance, and skin collagen in joint hypermotility syndrome. Br Heart J 54:501, 1985.
7. Lehman ED, Gosling RG, Sonksen PH: Arterial compliance in diabetes. Diabetes Care 9:27, 1986.
8. Lehman ED, Watts GF, Gosling RG: Aortic distensibility and hypercholesterolaemia. Lancet 340:1171, 1992.
9. Lehman ED, Hopkins KD, Weissberger AJ, et al: Aortic distensibility in growth hormone deficient adults. Lancet 341:309, 1993.
10. Harris P, Heath D: The relation between structure and function in the blood vessels of the lung in pulmonary hypertension. In Harris P, Heath D (eds): The Human Pulmonary Circulation. 3rd ed. London, Churchill Livingstone, 1986, p 284.
11. Reuben SR: Compliance of the human pulmonary arterial system in disease. Circ Res 29:40, 1971.
12. Harris P, Heath D, Apostolopoulos A: Extensibility of the pulmonary trunk in heart disease. Br Heart J 27:660, 1965.
13. Paz R, Mohiaddin RH, Longmore DB: Magnetic resonance assessment of pulmonary trunk: Anatomy, flow, pulsatility and distensibility. Eur Heart J 14:1524, 1993.
14. Mohiaddin RH, Paz R, Theodoropolus S, et al: Magnetic resonance characterization of pulmonary arterial blood flow following single lung transplantation. J Thorac Cardiovasc Surg 101:1016, 1991.
15. Ohtsuka S, Kakihana M, Watanabe H, et al: Chronically decreased aortic distensibility causes deterioration of coronary perfusion during increased left ventricular contraction. J Am Coll Cardiol 24:1406, 1994.
16. Mukherjee DP, Kagan HM, Jordan RE, Franzblau C: Effect of hydrophobic elastin ligands on the stress-strain properties of elastin fibers. Connect Tissue Res 4:177, 1976.
17. Harkness ML, Harkness RD, McDonald DA: The collagen and elastin content of the arterial wall in the dog. Proc R Soc [Biol] 146:541, 1957.
18. Lee RT, Kamm RD: Vascular mechanics for the cardiologist. J Am Coll Cardiol 23:1289, 1994.
19. Bergel DH: The dynamic elastic properties of the arterial wall. J Physiol 156:458, 1961.
20. Hardung V: Vergleichende messungen der dynamischen elastizitat und Viskositat von blutegfassen, kautschuk und synthetischen elastomeren. Helv Physiol Acta 11:194, 1953.
21. Learoyd BM, Taylor MG: Alterations with age in the viscoelastic properties of human arterial walls. Circ Res 18:278, 1966.
22. Remington JW: Hysteresis loop, behaviour of the aorta and other extensible tissues. Am J Physiol 180:83, 1955.
23. Bramwell JC, Hill AV, McSwiney BA: The velocity of the pulse wave in man in relation to age as measured by hot-wire sphygmograph. Heart 10:233, 1923.
24. Hallock P: Arterial elasticity in man in relation to age as evaluated by the pulse wave velocity method. Arch Intern Med 54:770, 1934.
25. Remington JW: Pressure-diameter relations of the in vivo aorta. Am J Physiol 203:440, 1962.
26. Luchsinger PC, Sachs M, Patel D: Pressure-radius relationship in large blood vessels of man. Circ Res 11:885, 1962.

27. Stefanadis C, Stratos C, Boudoulas H, et al: Distensibility of the ascending aorta: Comparison of invasive and non-invasive techniques in healthy men and in men with coronary artery disease. Eur Heart J 11:990, 1990.

28. Gosling RG, King DH: Arterial assessment by Doppler-shift ultrasound. Proc R Soc Med 67:447, 1974.

29. Kok WEM, Peters RJG, Prins MH, et al: Contribution of age and intimal lesion morphology to coronary artery wall mechanics in coronary artery disease. Clin Sci 89:239, 1995.

30. Hanrath P, Heinzt B, vom Dahl J, et al: Evaluation of segmental elastic properties of the aorta in normotensive and medically treated patients by intravascular ultrasound. *In* Boudoulas P, Toutouzas PK, Wooley C (eds): Functional Abnormality of the Aorta. New York, Futura, 1996, p 221.

31. Mohiaddin RH, Underwood SR, Bogren HG, et al: Regional aortic compliance studied by magnetic resonance imaging: The effects of age, training, and coronary artery disease. Br Heart J 62:90, 1989.

32. Chien D, Saloner D, Laub G, Anderson CM: High resolution cine MRI of vessel distension. J Comput Assist Tomogr 18:576, 1994.

33. Mohiaddin RH, Underwood SR, Bogren HG, et al: Regional aortic compliance studied by magnetic resonance imaging: The effects of age, training, and coronary artery disease. Br Heart J 62:90, 1989.

34. Rueckert D, Burger P, Yang GZ, et al: Automatic tracking of the aorta in cardiovascular MR images using deformable models. IEEE Trans Med Imaging 16:581, 1997.

35. Mohiaddin RH, Firmin DN, Longmore DB: Age-related changes of human aortic flow wave velocity measured non-invasively by magnetic resonance imaging. J Applied Physiol 74:492, 1993.

36. Tarnawski M, Cybulski G, Doorly D, et al: Noninvasive determination of local wavespeed and distensibility of the femoral artery by comb-excited Fourier velocity-encoded magnetic resonance imaging: Measurements on athletic and nonathletic human subjects. Heart Vessels 9:194, 1994.

37. Dumoulin CL, Doorly DJ, Caro CG: Quantitative measurement of velocity at multiple positions using comb excitation and Fourier velocity encoding. Magn Reson Med 29:44, 1993.

38. Hardy CJ, Bolster BD, McVeigh ER, et al: A one-dimensional velocity technique for NMR measurement of aortic distensibility. Magn Reson Med 31:513, 1994.

39. Hardy CJ, Bolster BD Jr, McVeigh ER, et al: Pencil excitation with interleaved Fourier velocity encoding: NMR measurement of aortic distensibility. Magn Reson Med 35:814, 1996.

40. Bolster BD Jr, Atalar E, Hardy CJ, McVeigh ER: Accuracy of arterial pulse-wave velocity measurement using MR. J Magn Reson Imaging 8:878, 1998.

41. Mohiaddin RH, Underwood SR, Bogren HG, et al: Regional aortic compliance studied by magnetic resonance imaging: The effects of age, training, and coronary artery disease. Br Heart J 62:90, 1989.

42. Rees RSO, Somerville J, Ward C, et al: Magnetic resonance imaging in late post-operative assessment of coarctation of the aorta. Radiology 173:499, 1989.

43. Manzara CC, Mohiaddin RH, Pennell DJ, et al: Magnetic resonance assessment of thoracic aorta in Marfan's syndrome (abstract). Circulation 82(suppl III):497, 1990.

44. Forbat SM, Naoumova RP, Sidhu PS, et al: The effect of cholesterol reduction with fluvastatin on aortic compliance, coronary calcification and carotid intimal-medial thickness: A pilot study. J Cardiovasc Risk 5:1, 1998.

45. Kupari K, Hekali P, Keto P, et al: Relation of aortic stiffness to factors modifying the risk of atherosclerosis in healthy persons. Arterioscler Thromb 14:386, 1994.

46. Kupari K, Keto P, Hekali P, et al: Cine magnetic resonance imaging in the assessment of aortic distensibility. *In* Boudoulas P, Toutouzas PK, Wooley C (eds): Functional Abnormality of the Aorta. New York, Futura 1996, p 247.

47. Adams JN, Brooks M, Redpath TW, et al: Aortic distensibility and stiffness index measured by magnetic resonance imaging in patients with Marfan's syndrome. Br Heart J 73:265, 1995.

48. Groenink M, de Roos A, Mulder BJ, et al: Changes in aortic distensibility and pulse wave velocity assessed with magnetic resonance imaging following beta-blocker therapy in the Marfan syndrome. Am J Cardiol 82:203, 1998.

49. Savolainen A, Keto P, Poutanen VP, et al: Effects of angiotensin-converting enzyme inhibition versus beta-adrenergic blockade on aortic stiffness in essential hypertension. J Cardiovasc Pharmacol 27:99, 1996.

50. Honda T, Hamada M, Shigematsu Y, et al: Effect of antihypertensive therapy on aortic distensibility in patients with essential hypertension: Comparison with trichlormethiazide, nicardipine and alacepril. Cardiovasc Drugs Ther 13:339, 1999.

51. Resnick LM, Militianu D, Cunnings AJ, et al: Direct magnetic resonance determination of aortic distensibility in essential hypertension: Relation to age, abdominal visceral fat, and in situ intracellular free magnesium. Hypertension 30 (3 Pt 2):654, 1997.

52. Chelsky R, Wilson RA, Morton MJ, et al: Alteration of ascending thoracic aorta compliance after treatment with menotropin. Am J Obstet Gynecol 176:1255, 1997.

53. Bogren HG, Klipstein RH, Mohiaddin RH, et al: Pulmonary artery distensibility and blood flow patterns: A magnetic resonance study of normal subjects and of patients with pulmonary arterial hypertension. Am Heart J 118:990, 1989.

54. Stefanides C, Dernellis J, Tsiamis E, et al: Aortic stiffness as a risk factor for recurrent acute coronary events in patients with ischaemic heart disease. Eur Heart J 21:390, 2000.

RIGHT VENTRICULAR ANATOMY AND FUNCTION

CHAPTER 23

Right Ventricular Anatomy and Function in Health and Disease

Christine H. Lorenz

Accurate noninvasive assessment of right ventricular (RV) mass and systolic function is important for many pediatric and adult patients. Unfortunately, *quantitative* RV assessments using two-dimensional (2-D) echocardiography or radionuclide ventriculography are limited due to the required geometric assumptions regarding RV anatomy or overlap of other cardiac chambers.[1–4] By contrast, cardiovascular magnetic resonance (CMR) allows the acquisition of true RV short-axis images encompassing the entire RV with high spatial and temporal resolution, thereby providing highly accurate quantitative RV mass and functional data.[5–8] Recently, three-dimensional (3-D) echocardiography has been compared with CMR for the evaluation of RV function, and improved results compared with 2-D echocardiography have been obtained.[9, 10] Three-dimensional echocardiography techniques, however, remain limited to research centers, whereas CMR is now widely available and provides for easier border recognition.

The clinical implications of accurate noninvasive assessment of RV function are diverse. Patients with chronic RV pressure or volume overload due to pulmonary hypertension or intracardiac shunts often develop RV dilation and/or systolic dysfunction. This situation is common among children and adults with corrected or uncorrected congenital heart disease. For example, 10 to 25 percent of patients who undergo surgical correction of transposition of the great arteries develop right heart failure over the succeeding two decades. Among patients with left heart failure due to coronary or valvular heart disease, RV systolic dysfunction often accompanies or precedes left ventricular (LV) failure, and the prognostic value of RV dysfunction exceeds that of LV systolic function.[11] Finally, improved understanding of the RV response to pressure and volume overload may lead to more optimal surgical and medical treatment algorithms. This chapter aims to summarize the features of the normal right ventricle and magnetic resonance techniques for assessing RV function, as well as to give examples of the importance of RV data in disease.

NORMAL RIGHT VENTRICULAR ANATOMY

The right ventricle is a thin, highly trabeculated structure with a continuum of muscle bands that rotate by ~160° from the epicardium to the endocardium.[12] The principal axis of these fibers is oblique to the long axis of the right ventricle. This RV fiber arrangement appears to be maintained across mammalian species from the marmot to the elephant.[12] On gross inspection, the right ventricle appears to be wrapped around the left ventricle. In the normal adult, the total RV free wall mass is 26 ± 5 gm/m^2 and the free wall thickness (<6 mm) is approximately half of LV wall thickness.[13] Due to the much greater LV systolic pressure, the interventricular septum bulges into the RV cavity.[14] The right ventricle has prominent septomarginal trabeculae passing from the apical portion of the ventricular septum to the anterior wall of the morphologic right ventricle, and the most developed of these near midventricle is called the moderator band. The insertion point of the septal leaflet of the tricuspid valve is more apically placed than the septal leaflet of the mitral valve. The depiction of the moderator band and the different levels of insertion of the septal leaflets are important diagnostic features for identification of the right ventricle. Another distinctive feature of the right ventricle is its infundibulum or conus that separates the tricuspid and pulmonary valves.

IMAGING STRATEGIES FOR CARDIOVASCULAR MAGNETIC RESONANCE OF THE RIGHT VENTRICLE

CMR of the right ventricle can be performed at rest or during physiological/pharmacological stress.[15] Accurate electrocardiographic (ECG) gating with minimal ectopy is essential (or real-time CMR should be considered). For anatomical assessment of the right ventricle, spin-echo sequences (including turbo spin-echo, half-Fourier acquisition single-shot turbo spin-echo [HASTE], or spin-echo/echo-planar imaging [EPI]) have been used (Figure 23–1) to evaluate RV size, pulmonary artery dimensions, and possible fatty replacement/infiltration of the RV free wall as seen in some cases of RV dysplasia (see Chapter 31).[16] These techniques are routinely available on commercial scanners and are relatively robust.

Functional evaluation of the right ventricle has

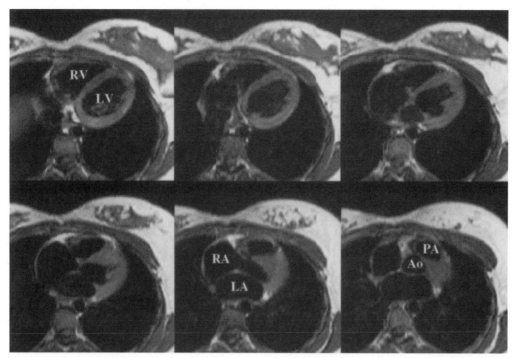

Figure 23–1. Transverse spin-echo CMR in a normal subject. The excellent depiction of the RV free wall against the black-blood pool and the black pericardium is clear. Note epicardial fat distribution around the right coronary artery and left anterior descending artery in the anterior atrioventricular groove and interventricular sulcus, respectively. Two features of the normal RV are seen: the muscular outflow tract, and the more apical insertion of the tricuspid valve in the septum (compared with the septal leaflet of the mitral valve). RV = right ventricle; LV = left ventricle; RA = right atrium; LA = left atrium; Ao = aorta; PA = pulmonary artery.

usually been performed using gradient-echo cine imaging, either in breath-hold (with k-space segmentation) or free-breathing (with conventional k-space acquisition) modes. The right ventricle is well depicted in the transverse plane, which also depicts the tricuspid valve. Cine acquisitions in the transverse plane have been used for analysis of RV systolic function, with good agreement with pulmonary flow and LV stroke volume,[17] and are more accurate than RV inflow or outflow tract views (Figure 23–2).[18] However for *quantitative* assessment of the right ventricle, true short-axis images are preferred[5]

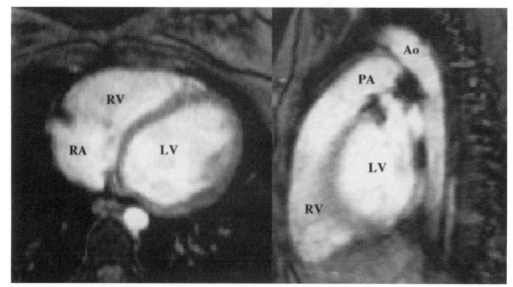

Figure 23–2. Gradient-echo cine CMR of the RV. The image on the left shows the diastolic frame from a cine in the transaxial plane. This plane is invaluable for examining regional RV wall motion, and when this is important, as in arrhythmogenic RV cardiomyopathy, multiple transaxial cines in contiguous planes are acquired. The right image is a diastolic frame from a cine through the RV outflow tract aligned in an oblique sagittal plane. This is useful for examining the function of the RV free wall from the apex to the pulmonary valve, and for visualizing pulmonary regurgitation. RV = right ventricle; LV = left ventricle; RA = right atrium; Ao = aorta; PA = pulmonary artery.

because of partial volume effects that particularly affect the anterior and inferior walls. It is important to ensure that the most basal part of the free wall of the right ventricle is included in the most basal cine slice, inasmuch as it is easily truncated. Most studies reporting quantitative RV functional parameters have performed measurements in the true RV short axis, an orientation that is slightly different from that of the LV short axis.[19]

The following steps describe an efficient protocol for the quantitative assessment of the right ventricle (Figure 23–3). An alternative standardized method is described in Chapter 9.

1. Transverse gradient-echo or EPI scout to determine the long axis of the left ventricle (see Figure 23–3A).

2. Gradient-echo scout image (scan time < 2 sec-

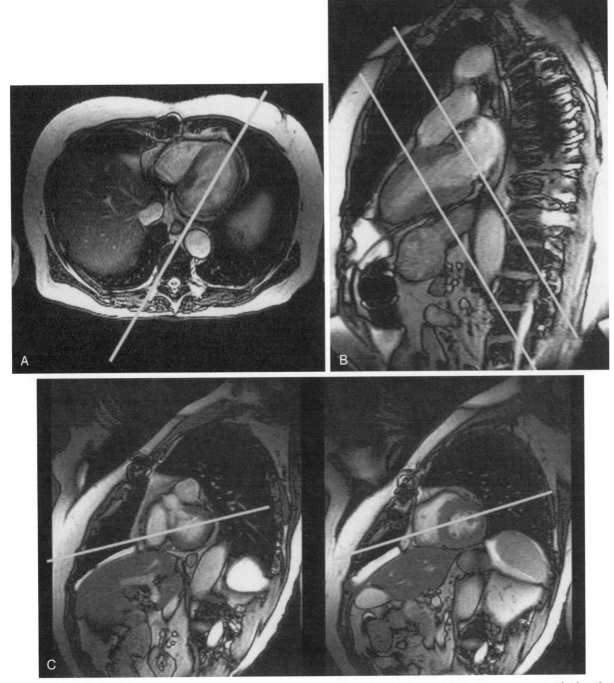

Figure 23–3. Demonstration of scan planning technique used to assess RV function and mass: *(A)* Transverse survey with identification of vertical long axis through the LV apex and mitral valve plane; *(B)* vertical long-axis scout image resulting in identification of "pseudo" short-axis plane parallel to mitral valve; *(C)* resulting pseudo short-axis view and identification of four-chamber view *(D)*.

Illustration continued on following page

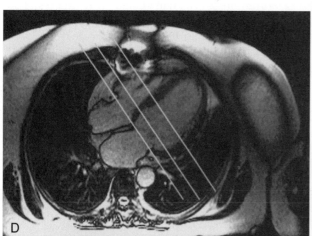

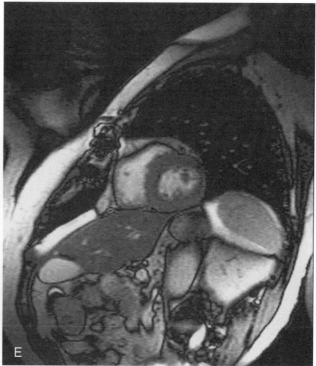

Figure 23–3 *Continued.* Placement of true short-axis slices across atrioventricular valve plane and extending to apex for measurement of LV and RV parameters *(D)*; *(E)* example of midventricular short-axis cine image at end diastole.

onds) along the long axis of the left ventricle defined by a line from the LV apex and through the mid point of the mitral valve (see Figure 23–3B).

3. Gradient-echo scout images (scan time < 2 seconds each) at the mitral valve plane and a midventricular level, perpendicular to a line bisecting the center of the mitral valve plane and the apex on the vertical long-axis view (see Figure 23–3B).

4. Gradient-echo scout image (scan time < 2 seconds) bisecting the right ventricle and the center of the LV cavity and passing through the tip of the RV apex to define a rotated four-chamber view (uses the short-axis scout images from the previous step in combination with the vertical long-axis scout image) (see Figure 23–3C).

5. Cine gradient-echo image acquisition in the four-chamber view with temporal resolution approximately 25 to 40 milliseconds, and a slice thickness of 8 mm for identification of the location of the atrioventricular valve plane at end diastole and end systole (scan time 15–20 seconds for a breath-hold and 1–2 minutes for non–breath-hold acquisition; see Figure 23–3D).

6. Cine gradient-echo acquisitions using contiguous 7- to 10-mm thick double-oblique slices in the short-axis plane, with the most basal slice placed across the atrioventricular valve plane at end diastole, covering both ventricles through the apex (see Figure 23–3E). The most basal slice bisects the atrioventricular groove on both the right ventricle and left ventricle. One approach is to acquire a fixed number of contiguous slices (10–12) over the ventricle and adjust the slice thickness as needed. Contig-

uous slices with no gap are used to minimize errors due to partial volume effects, especially problematic at the base of the heart. Temporal resolution should be 25 to 40 milliseconds.

Analysis of the images consists of the application of Simpson's rule by manual or semiautomatic planimetry of the endocardial borders of each ventricle at both end diastole and end systole, and of the epicardial borders at end diastole. Care must be taken to exclude the right and left atria as they come into the basal imaging planes during systole. The RV stroke volume is calculated as the difference between the RV end-diastolic and end-systolic volumes. RV ejection fraction is calculated as the RV stroke volume divided by the RV end-diastolic volume. Mass is calculated as the volume of tissue occupied by the free wall multiplied by an assumed density of 1.05 gm/cc. RV mass measurements agree well with autopsy studies (Figure 23–4). Measurements of both RV parameters are reproducible with this technique to within 5 percent.[13] If all images throughout the cardiac cycle are planimetered, filling and ejection rates can also be calculated.

NORMAL RIGHT VENTRICULAR VOLUMES AND SYSTOLIC FUNCTION

Although published reports of normal RV volumes, systolic function, and mass are few in comparison to reports regarding the left ventricle, there

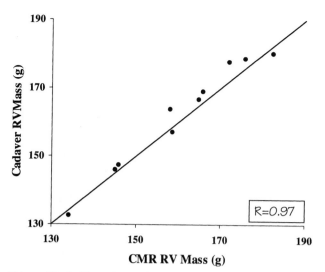

Figure 23-4. The relation between right ventricle (RV) mass measured at autopsy, and by CMR. There is good agreement with the line of identity. (Adapted from Katz J, Whang J, Boxt LM, Barst RJ: Estimation of right ventricular mass in normal subjects and in patients with pulmonary hypertension by nuclear magnetic resonance imaging. J Am Coll Cardiol 21:1475, 1993.)

have been several human series describing the normal characteristics of RV size and function using autopsy,[20, 21] echocardiography,[22] x-ray angiography,[3] CMR,[6, 7, 23] and ultrafast computed tomography (CT).[24, 25] The results of these studies are summarized in Table 23-1. Results have also been reported in animals.[26] We have reported on RV parameters utilizing a series of contiguous, short-axis stack of breath-hold gradient-echo cine images from 75 healthy young adult subjects.[13] These gender-specific data are summarized in Table 23-2 (with and without normalization to body surface area).

Another CMR approach that may be particularly valuable for quantifying *regional* RV free wall sys-

tolic function is myocardial tagging.[27–33] Klein and associates applied myocardial MR tagging to the normal RV free wall in humans.[29] The RV free wall was divided into three (inferior, mid, and superior) segments in each of three slices (apical, mid, and basal) to allow for a detailed analysis of the motion and contraction. Percent segmental shortening (PSS) was used to measure the amount of contraction, and a vector analysis was used to quantify the trajectory of the RV free wall throughout systole. Segmental shortening increased monotonically through time to an average of 12 percent across all segments at the base, 14 percent at the midventricle, and 16 percent at the apex of the heart. The trajectory of the RV free wall was characterized by a wave of motion towards the septum and outflow tract. Naito and colleagues used MR tagging to determine PSS in nine volunteers at the midventricular slice.[30] They found a PSS of 6.7 percent in the superior wall segment and 20 percent for the mid-wall segment. In slight contrast, in a series of 5 healthy adults, Fayad reported PSS values of 24.7 percent in the mid-wall segment of the midventricular slice and 28.7 percent in the mid-wall segment of apical slice.[31] Using CMR tagging, a 3-D reconstruction of RV contraction was described by Young and coworkers.[33] This technique demonstrated circumferential contraction as the right ventricle moved apically, while the primary contraction of the right ventricle was tangential to its own surface plane. A rocking or twisting motion was also noted, similar to that described by Klein and associates[29] and by Stuber and coworkers.[32] Fayad and colleagues also studied RV tagging in patients with chronic pulmonary hypertension.[34] Regional short-axis shortening was reduced in patients as compared with healthy controls. The greatest reductions in shortening were found in the outflow tract and basal septal region.

Among comparative animal studies, there has

Table 23-1. Right Ventricular Volume, Systolic Function, and Free Wall Mass by Autopsy and Imaging

Reference	Modality	Parameter	Literature Value
Katz et al[6]	CMR	RV mass/BSA (n = 10)	23.3 ± 1.4 g/m²
Doherty et al[7]	CMR	RV mass (male, n = 10)	45 ± 8 g
Rominger et al[23]	CMR	RV end-diastolic volume/BSA	78 ± 15 ml/m²
		RV end-systolic volume/BSA	30 ± 9 ml/m²
		RV stroke volume/BSA	48 ± 9 ml/m²
		RV ejection fraction	$62 \pm 6\%$
Fulton et al[20]	Autopsy	RV mass (male and female)	46 g (23–68 g)
Hangartner et al[21]	Autopsy	RV mass (females)	40 ± 8 g (26–57 g)
		RV mass (males)	56 ± 14 g (34–87 g)
Pietras et al[3]	X-ray angio	RV end-diastolic volume/BSA	76 ± 14 ml/m²
		RV end-systolic volume/BSA	33 ± 8 ml/m²
Hajduczok et al[24]	Ultrafast CT	RV mass (male, n = 7)	55 ± 3 g
		LV mass/RV mass (male, n = 7)	3.2 ± 0.2
Wachspress et al[25]	Ultrafast CT	RV end-diastolic volume/BSA	76 ± 19 ml/m²
		RV end-systolic volume/BSA	35 ± 13 ml/m²
		RV ejection fraction	$55 \pm 6\%$

CMR = cardiovascular magnetic resonance; angio = angiography; CT = computed tomography; RV = right ventricle; BSA = body surface area; LV = left ventricle.

Adapted from Lorenz CH, Walker ES, Morgan VL, et al: Normal human right and left ventricular mass, systolic function and gender differences by cine magnetic resonance imaging. J Cardiovasc Magn Reson 1:7, 1999.

Table 23–2. Right Ventricular Anatomical and Functional Indexes

	All (n = 75)	Males (n = 47)	Females (n = 28)
RV end-diastolic volume	138 ± 40 (59–217) ml	157 ± 35 (88–227) ml	106 ± 24 (58–154) ml
	75 ± 13 (49–101) ml/m²	80 ± 13 (55–105) ml/m²	67 ± 10 (48–87) ml/m²
RV end-systolic volume	54 ± 21 (12–96) ml	63 ± 20 (23–103) ml	40 ± 14 (12–68) ml
Interventricular septal mass	54 ± 13 (28–80) g	61 ± 11 (40–82) g	42 ± 8 (26–58) g
	30 ± 4 (21–38) g/m²	31 ± 4 (23–39) g/m²	27 ± 4 (20–34) g/m²
RV free wall mass	46 ± 11 (25–67) g	50 ± 10 (30–70) g	40 ± 8 (24–55) g
	26 ± 5 (17–34) g/m²	26 ± 5 (16–36) g/m²	25 ± 4 (18–33) g/m²
RV ejection fraction (%)	61 ± 7 (47–76)	60 ± 7 (47–74)	63 ± 8 (47–80)
RV stroke volume	84 ± 24 (37–131) ml	95 ± 22 (52–138) ml	66 ± 16 (35–98) ml
	46 ± 8 (30–62) ml/m²	48 ± 8 (32–64) ml/m²	42 ± 8 (27–57) ml/m²
RV FWM/RV EDV (g/ml)	0.35 ± 0.06 (0.23–0.47)	0.33 ± 0.06 (0.21–0.44)	0.38 ± 0.05 (0.28–0.48)
RV cardiac output (liters/min)	5.2 ± 1.4 (2.42–8.05)	5.8 ± 3.0 (2.82–8.82)	4.3 ± 0.9 (2.65–5.98)

*Mean ± 1 SD (95% confidence intervals) values without and with normalization for body surface area are provided.
RV = right ventricle; FWM = free wall mass; EDV = end-diastolic volume.
Adapted from Lorenz CH, Walker ES, Morgan VL, et al: Normal human right and left ventricular mass, systolic function and gender differences by cine magnetic resonance imaging. J Cardiovasc Magn Reson 1:7, 1999.

been a very good correlation with x-ray and implanted RV free wall markers. In a canine study utilizing implanted markers, Meier and associates found 16 percent shortening in the apex, 16 percent shortening in a midventricular slice, and 12 percent shortening in the pulmonary conus wall segment in transverse markers.[27] In another canine study with ultrasound crystals, Raines and colleagues found 13 percent shortening in the inflow tract and 21 percent shortening in the outflow tract.[28]

Though presumed by some, it is not yet known whether CMR tagging provides clinically relevant data beyond that provided by more standard cine CMR measures. Therefore, further studies are needed to better define the clinical role of RV tagging.

CARDIOVASCULAR MAGNETIC RESONANCE ASSESSMENT OF RIGHT VENTRICULAR ANATOMY AND FUNCTION IN DISEASE

Pulmonary Hypertension and Lung Transplantation

The high accuracy and interstudy reproducibility of CMR for *quantitative* assessment of RV mass and volume allow for serial studies on the same patient. Cine gradient-echo CMR has been used to study pulmonary hypertension[35, 36] and the progression of RV failure,[34] and CMR has been used to confirm the diagnosis of cor pulmonale when RV mass measurements exceed 60 g.[37] Pulmonary flow patterns are known to be abnormal in pulmonary hypertension,[38] which may affect RV afterload, whereas diastolic function has also been found to be abnormal by CMR using tricuspid flow patterns in pulmonary fibrosis.[39] We and others have used CMR to determine the time course of changes in ventricular mass and function after lung transplantation.[40–42] Our

study included 43 patients with pulmonary hypertension being evaluated for lung transplantation, 11 patients *early* (<6 months), *intermediate* (6–18 months) or *late* (>18 months) following lung transplantation and 65 healthy subjects. RV ejection fraction normalized in the early post–lung transplantation period. Other early changes included a decrease in RV end-diastolic volume to below normal levels, with persistence at this level even in the late studies. RV mass also regressed early, but remained increased as compared with healthy control subjects (in contrast to LV mass, which was depressed prior to transplantation and increased in the posttransplant period). Therefore RV anatomical normalization was found later than functional normalization, and RV mass remained increased even after late (>18 months) posttransplantation studies. The etiology of the disparate responses remains speculative, and may reflect rapid LV adjustments to preload changes and more protracted RV adjustment to afterload changes.

Congenital Heart Disease

A comprehensive summary of the role of CMR in congenital heart disease is detailed in other chapters (see Chapters 24 and 25). Assessment of RV size, location, and connections is important (Figure 23–5) as well as function and pulmonary flow.[43] In this chapter, discussion is limited to an example of atrial repair of transposition of the great vessels. In atrial repair of transposition, a baffle is constructed in the atria to collect the systemic venous return and pass it through the anatomical left ventricle (pulmonic ventricle) to the pulmonary artery. Blood returning from the lungs through the pulmonary veins crosses around the baffle in the atria and enters the anatomical right ventricle (systemic ventricle) to the aorta.

Over the past two decades, the arterial switch has become the preferred corrective procedure, but there are a large number of patients alive today who un-

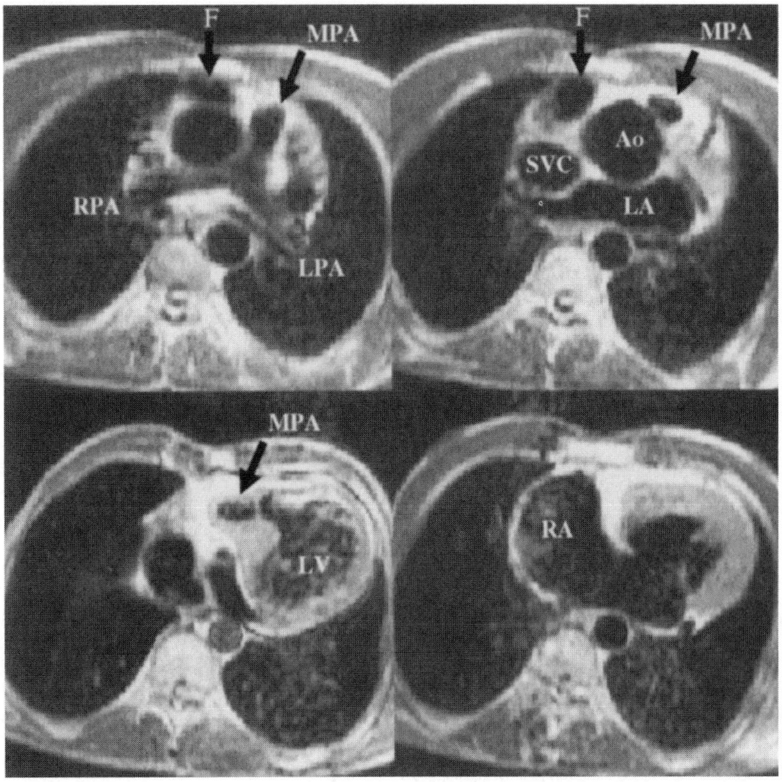

Figure 23–5. Transverse spin-echo CMR of a patient with tricuspid atresia. The RV is vestigial and connected to the LV through a ventricular septal defect. The main pulmonary artery (MPA) arises from the small RV and is small compared with the aorta. However, it does divide into two small pulmonary arteries on right and left (RPA and LPA). The right atrium (RA) is connected to a Fontan conduit (F), which connects to the RPA (connection not shown in these images). This example shows how well the contiguous tomographic approach to abnormalities of the vasculature and chambers can be used to elucidate complex problems. RV = right ventricle; LV = left ventricle; LA = left atrium; Ao = aorta; SVC = superior vena cava.

derwent successful atrial repair. For this group, the long-term adaptation of the anatomical right ventricle (systemic ventricle) after atrial repair remains a subject of major concern. To examine this issue, we examined the typical late adaptation of the right ventricle and left ventricle following atrial repair using CMR and an age-matched healthy control population.[8] The goal was to establish a baseline "normal" range of function for these patients that could then be used to compare patients with RV dysfunction and failure. Longitudinal studies should then prove useful in determining the mechanisms of late RV failure.

Cine CMR was performed in 22 patients who were 8 to 23 years out from atrial repair of transposition of the great arteries.[8] Results were compared to that of 24 age- and gender-matched healthy adults. We found elevated RV mass, decreased LV mass and interventricular septal mass, normal RV end-diastolic volume, and minimally depressed RV ejection fraction. Only 2 (9%) of patients had clinical RV dysfunction, and both of these had increased RV mass. Based on these early data, we hypothesized that inadequate hypertrophy did not appear to be the cause of late dysfunction in this patient group. The results of this study summarize the ability of CMR to establish quantitative criteria by which one can detect early dysfunction compared with a well-compensated status for a particular group of patients with complex congenital heart disease.

Right Ventricular Assessment in Heart Failure

During exercise, cardiac output can increase by a factor of 4 to 6 before there is a significant rise in pulmonary artery pressures. Therefore, the compliance of the pulmonary arterial system is important to minimize the energy requirements of the right ventricle. In left heart failure, atrial pressure rises, forcing open more pulmonary capillaries. After all reserve capillaries are open, the increase in pulmonary pressure leads to an increased load on the right ventricle. Therefore the function of the right ventricle during exercise in coronary disease and heart failure is important.

Several investigators have examined the prognostic value of RV function in patients with advanced heart failure of various causes (Figures 23–6 and 23–7). Di Salvo and colleagues studied 67 patients with heart failure with ischemic (46%) or dilated (54%) cardiomyopathy referred for cardiac transplantation.[11] Subjects underwent bicycle stress testing with radionuclide ventriculography and were followed for up to 180 weeks. Augmentation of RV ejection fraction or LV ejection fraction during exercise was not predictive of survival. However, a RV ejection fraction of greater than 35 percent at rest and with exercise predicted overall survival. Maximal oxygen consumption was also predictive of survival, with a modest correlation between RV ejection fraction and maximal oxygen consumption. Similarly, Polak and coworkers studied 36 patients with biventricular dysfunction and LV ejection fraction less than 40 percent.[44] Depressed RV ejection fraction (<35%) was associated with increased mortality. Positive correlates were the severity of LV dysfunction and pulmonary hypertension. These data suggest that quantitative RV volumetric assessment has prognostic value and should be performed in all CMR studies performed on patients with heart failure.

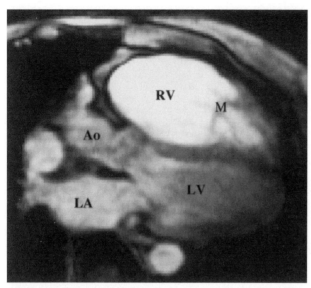

Figure 23–6. Transverse gradient-echo image of severe RV dilatation and dysfunction in a patient with advanced arrhythmogenic RV cardiomyopathy. Note the large RV and posterior displacement of the LV. RV = right ventricle; LV = left ventricle; LA = left atrium; Ao = aorta; M = moderator band.

Right Ventricular Assessment in Ischemic Heart Disease

Isolated RV infarction is relatively rare, but concurrent RV infarction in the setting of an inferior infarction due to proximal right coronary occlusion[45, 46] occurs in up to half of inferior LV infarctions.[47] It is known that RV infarction is less likely in patients with preinfarction angina.[48] RV in-

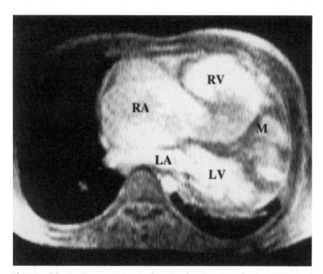

Figure 23–7. Transverse gradient-echo image of a patient with Uhl syndrome (parchment RV syndrome). In this rare condition, the RV cavity is markedly dilated and the RV free wall is paper thin. In this example, the RV is massively dilated and there was free tricuspid regurgitation from annular dilatation. There is a very prominent moderator band (M), and the LV is displaced posteriorly, with marked compression of the left lung. RV = right ventricle; LV = left ventricle; RA = right atrium; LA = left atrium.

farction can be detected and evaluated in extent using contrast-enhanced gadolinium CMR, which is described in the chapter on viability (Figure 23–8; see Chapter 14). With RV infarction, there is an overall loss of contractile mass, and flow patterns become abnormal.[49] If the inferior interventricular septum is also involved, there is also a loss of septal augmentation of RV function. Bueno studied 198 patients with acute inferior infarction.[46] Forty-one percent also had a coexistent RV infarction. Mortality in the RV infarction group was 47 percent as compared with only 10 percent for those without RV infarction. Cardiogenic shock occurred in 32 percent of the patients with combined inferior and RV infarction and in only 5 percent of those with isolated inferior infarction. The worse prognosis in RV infarction was also found by Zehender, with a 31 percent in-hospital mortality rate vs. 6 percent ($p < 0.001$) in those with or without RV infarction.[45] Reperfusion of acute RV infarcts by primary angioplasty has been shown to greatly improve RV function, and the mortality in patients with unsuccessful compared with successful reperfusion was poor (58 vs. 2%; $p = 0.001$).[50] These studies highlight the importance of characterizing the right ventricle in the setting of acute infarction, and suggest a useful role for CMR.

CONCLUSION

A greater appreciation of the important role of the right ventricle in acquired as well as congenital heart disease is now emerging, and CMR is the ideal noninvasive tool for the comprehensive evaluation of RV mass as well as global and regional RV function. Real-time assessment of RV function in response to physiological and pharmacological stress is also readily available on commercial scanners.[51]

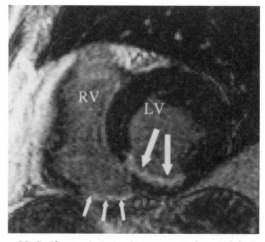

Figure 23–8. Short-axis inversion recovery late gadolinium enhancement image of a patient with known inferior myocardial infarction (large arrows). Note however, there is also enhancement of the inferior surface of the RV (small arrows) indicated concomitant RV infarction. RV = right ventricle; LV = left ventricle.

However, because few centers have experience with RV evaluation, further clinical studies are needed to establish standards for the best use of CMR for predicting patient outcome and the role of serial evaluation in the case of known RV dysfunction.

ACKNOWLEDGMENTS

I would like to thank Dr. S.A. Wickline for discussion of the role of the right ventricle in ischemic heart disease and Dr. D.J. Pennell for assistance with the figures.

References

1. Ohsuzu F, Handa S, Kondo M, et al: Thallium-201 myocardial imaging to evaluate right ventricular overloading. Circulation 61:620, 1980.
2. Baker BJ, Scovil JA, Kane JJ, Murphy ML: Echocardiographic detection of right ventricular hypertrophy. Am Heart J 505:611, 1983.
3. Pietras RJ, Kondos GT, Kaplan D, Lam W: Comparative angiographic right and left ventricular volumes. Am Heart J 109:321, 1985.
4. Dehmer GJ, Firth DG, Hillis LD, et al: Non geometric determination of right ventricular volumes from equilibrium blood pool scans. Am J Cardiol 49:79, 1982.
5. Pattynama PMT, Lamb HJ, van der Velde EA, et al: Reproducibility of magnetic resonance imaging-derived measurements of right ventricular volumes and myocardial mass. Magn Reson Imaging 13:53, 1995.
6. Katz J, Whang J, Boxt LM, Barst RJ: Estimation of right ventricular mass in normal subjects and in patients with pulmonary hypertension by nuclear magnetic resonance imaging. J Am Coll Cardiol 21:1475, 1993.
7. Doherty NE, Fujita N, Caputo GR, Higgins CB: Measurement of right ventricular mass in normal and dilated cardiomyopathic ventricles using cine magnetic resonance imaging. Am J Cardiol 69:1223, 1992.
8. Lorenz CH, Walker ES, Graham TP, Powers TA: Right ventricular performance and mass in adults late after atrial repair of transposition of the great arteries using cine magnetic resonance imaging. Circulation 92(suppl II):233, 1995.
9. Fujimoto S, Mizuno R, Nagakawa Y, et al: Estimation of the right ventricular volume and ejection fraction by transthoracic three-dimensional echocardiography. A validation study using magnetic resonance imaging. Int J Cardiac Imaging 14:385, 1998.
10. Vogel M, Gutberlet M, Dittrich S, et al: Comparison of transthoracic three dimensional echocardiography with magnetic resonance imaging in the assessment of right ventricular volume and mass. Heart 78:127, 1997.
11. Di Salvo TG, Mathier M, Semigran MJ, Dec GW: Preserved right ventricular ejection fraction predicts exercise capacity and survival in advanced heart failure. J Am Coll Cardiol 25:1143, 1995.
12. Armour JA, Randall WC: Structural basis for cardiac function. Am J Physiol 218:1517, 1970.
13. Lorenz CH, Walker ES, Morgan VL, et al: Normal human right and left ventricular mass, systolic function and gender differences by cine magnetic resonance imaging. J Cardiovasc Magn Reson 1:7, 1999.
14. Guyton AC: The pulmonary circulation. In: Textbook of Medical Physiology, 7th ed. Philadelphia, WB Saunders, 1986, p 287.
15. Roest AAW, Kunz P, Lamb HJ, et al: Biventricular response to supine physical exercise in young adults assessed with ultrafast magnetic resonance imaging. Am J Cardiol 87:601, 2001.
16. Blake LM, Scheinmann MM, Higgins CB: MR features of arrhythmogenic right ventricular dysplasia. AJR Am J Roentgenol 162:809, 1994.
17. Helbing WA, Rebergen SA, Maliepaard C, et al: Quantification of right ventricular function with magnetic resonance imaging in children with normal hearts and with congenital heart disease. Am Heart J 130:828, 1995.
18. Jauhiainen T, Jarvinen VM, Hekali PE, et al: MR gradient echo volumetric analysis of human cardiac casts: Focus on the right ventricle. J Comput Assist Tomogr 22:899, 1998.
19. American College of Cardiology, American Heart Association, Society of Nuclear Medicine policy statement. Standardisation of cardiac tomographic imaging. Circulation 86:338, 1992.
20. Fulton RM, Hutchinson EC, Morgan-Jones A: Ventricular weight in cardiac hypertrophy. Br Heart J 4:413, 1952.
21. Hangartner JRW, Marley NJ, Whitehead A, et al: The assessment of cardiac hypertrophy at autopsy. Histopathology 9:1295, 1985.
22. Daniels SR, Meyer RA, Liang Y, Bove KE: Echocardiographically determined left ventricular mass index in normal children, adolescents and young adults. J Am Coll Cardiol 12:703, 1988.
23. Rominger MB, Bachmann GF, Pabst W, Rau WS: Right ventricular volumes and ejection fraction with fast cine MR imaging in breath-hold technique: Applicability, normal values from 52 volunteers and evaluation of 352 adult cardiac patients. J Magn Reson Imaging 10:908, 1999.
24. Hajduczok ZD, Weiss RM, Stanford W, Marcus ML: Determination of right ventricular mass in humans and dogs with ultrafast cardiac computed tomography. Circulation 82:202, 1990.
25. Wachspress JD, Clark NR, Untereker WJ, et al: Systolic and diastolic performance in normal human subjects as measured by ultrafast computed tomography. Cathet Cardiovasc Diagn 15:277, 1988.
26. Stauffer NR, Greenberg SB, Marks LA, et al: Validation of right ventricular volume measurements by magnetic resonance imaging in small hearts using a fetal lamb model. Invest Radiol 30:87, 1995.
27. Meier GD, Bove AA, Santamore WP, Lynch PR: Contractile function in canine right ventricle. Am J Physiol 239:H794, 1980.
28. Raines RA, LeWinter MM, Covell JW: Regional shortening patterns in canine right ventricle. Am J Physiol 231:1395, 1976.
29. Klein SS, Graham TP, Lorenz CH: Noninvasive delineation of normal right ventricular contractile motion with MRI myocardial tagging. Ann Biomed Eng 26:756, 1998.
30. Naito H, Arisawa J, Yamagami H, et al: Assessment of right ventricular regional contraction and comparison with left ventricle in normal humans: A cine magnetic resonance study with presaturation myocardial tagging. Br Heart J 74:186, 1995.
31. Fayad ZA, Kraitchman DL, Ferrari VA, Axel L: Right ventricular regional function in normal subjects using magnetic resonance tissue tagging. In: Book of Abstracts. Berkeley, CA, Society of Magnetic Resonance, 1994, p 1504.
32. Stuber M., Fischer SE, Nagel E, et al: Assessment of right ventricular motion. In: Book of Abstracts. Berkeley, CA, Society of Magnetic Resonance, 1994, p 1502.
33. Young AA, Fayad ZA, Axel L: Right ventricular midwall surface motion and deformation using magnetic resonance tagging. Am J Physiol 271:H2677, 1996.
34. Fayad ZA, Ferrari VA, Kraitchman DL, et al: Right ventricular regional function using MR tagging: Normals versus chronic pulmonary hypertension. Magn Reson Med 39:116, 1998.
35. Boxt LM, Katz J, Kolb T, et al: Direct quantitation of right and left ventricular volumes with nuclear magnetic resonance imaging in patients with primary pulmonary hypertension. J Am Coll Cardiol 19:1508, 1992.
36. Saito H, Dambara T, Aiba M, et al: Evaluation of cor pulmonale on a modified short-axis section of the heart by magnetic resonance imaging. Am Rev Respir Dis 146:1576, 1992.
37. Pattynama PMT, Willems LNA, Smit AH, et al: Early diagnosis of cor pulmonale with MR imaging of the right ventricle. Radiology 182:375, 1992.

38. Bogren HG, Klipstein RH, Mohiaddin RH, et al: Pulmonary artery distensibility and blood flow patterns: A magnetic resonance study of normal subjects and of patients with pulmonary arterial hypertension. Am Heart J 118:990, 1989.

39. Kroft LJ, Simons P, van Laar JM, de Roos A: Patients with pulmonary fibrosis: Cardiac function assessed with MR imaging. Radiology 216:464, 2000.

40. Frist WH, Lorenz CH, Walker ES, et al: MRI complements standard assessment of right ventricular function after lung transplantation. Ann Thorac Surg 60:268, 1995.

41. Moulton JM, Creswell LL, Ungacta FF, et al: Magnetic resonance imaging provides evidence for remodeling of the right ventricle after single-lung transplantation for pulmonary hypertension. Circulation 94(suppl II):312, 1996.

42. Lorenz CH, Loyd JE, Klein SS, et al: Characterization of different time courses of left and right ventricular recovery after lung transplantation. J Am Coll Cardiol 29(suppl A):23A, 1997.

43. Rebergen SA, Ottenkamp J, Doornbos J, et al: Postoperative pulmonary flow dynamics after Fontan surgery: Assessment with nuclear magnetic resonance velocity mapping. J Am Coll Cardiol 21:123, 1993.

44. Polak JF, Holman BL, Wynne J, Colucci WS: Right ventricular ejection fraction: An indicator of increased mortality in patients with congestive heart failure associated with coronary artery disease. J Am Coll Cardiol 2:217, 1983.

45. Zehender M, Kasper W, Kauder E, et al: Right ventricular infarction as an independent predictor of prognosis after acute inferior myocardial infarction. N Engl J Med 328:981, 1993.

46. Bueno H, Lopez-Palop R, Bermejo J, et al: In-hospital outcome of elderly patients with acute inferior myocardial infarction and right ventricular involvement. Circulation 96:436, 1997.

47. Kinch JW, Ryan TJ: Right ventricular infarction. N Engl J Med 330:1211, 1994.

48. Shiraki H, Yoshikawa T, Anzai T, et al: Association between preinfarction angina and a lower risk of right ventricular infarction. N Engl J Med 338:941, 1998.

49. Kayser HW, van der Geest RJ, van der Wall EE, et al: Right ventricular function in patients after acute myocardial infarction assessed with phase contrast MR velocity mapping encoded in three directions. J Magn Reson Imaging 11:471, 2000.

50. Bowers TR, O'Neill WW, Grines C, et al: Effect of reperfusion on biventricular function and survival after right ventricular infarction. N Engl J Med 338:933, 1998.

51. Setser RM, Fischer SE, Lorenz CH: Quantification of left ventricular function with magnetic resonance images acquired in real-time. J Magn Reson Imaging 12:430, 2000.

STRUCTURAL CARDIOVASCULAR DISEASE

Cardiovascular Magnetic Resonance of Simple Congenital Cardiovascular Defects

Arno A. W. Roest, R. André Niezen, Maarten Groenink, Willem A. Helbing, Ernst E. van der Wall, and Albert de Roos

The evaluation of congenital heart and large vessel disease is one of the well-established clinical applications of cardiovascular magnetic resonance (CMR).[1] Accurate determination of cardiac anatomy and function is crucial for patient management at initial diagnosis as well as during follow-up of patients after repair of cardiovascular malformations. CMR has proven effectiveness in the assessment of biventricular function and large vessel flow after surgical correction of complex cardiovascular disease, where other imaging techniques have limited possibilities to provide comprehensive evaluation of function and flow.[2] Echocardiography may be hampered in this category of patients by the presence of scar tissue, rib and chest deformations, and interposed lung tissue. Cardiac catheterization is not suited for the routine follow-up of (corrected) cardiac malformations because of its invasiveness and limitations due to radiation; however, it may still be required when information on pulmonary pressures and vascular resistance is of prime importance for patient management. Eventually, CMR may help to reduce the number of invasive procedures during the follow-up of patients with congenital cardiovascular abnormalities.[3]

Spin-echo CMR is a black-blood technique and is used to assess the cardiac and vascular anatomy under investigation, whereas gradient-echo CMR is a white-blood technique and is reserved for assessment of ventricular function or to study flow phenomena across stenoses and to depict valvular disease (Figure 24–1). Flow mapping is useful to quantify flow in large vessels as well as across valves.[4] It can also be used to quantify the flow velocity in case of a stenosed vascular segment or valve. Flow measurements can be readily performed in vascular areas, which may not routinely be accessible by echo-Doppler (Figure 24–2). CMR can measure flow directly and routinely in the aorta and pulmonary circulation, thereby allowing quantification of shunt lesions that manifest themselves by a discrepancy between aortic and pulmonary flow. For example, atrial level left-to-right shunts can be assessed by quantifying the stroke flow in the aorta and pulmonary artery, thereby allowing direct shunt

quantification with high precision and accuracy.[5] CMR techniques currently accomplish complete evaluation of many clinical relevant parameters in patients with congenital heart and large vessel disease. A number of these parameters were not available on a routine basis with more traditional imaging modalities. In this chapter, we review the value of CMR for the evaluation of a number of simple congenital cardiovascular defects. For complex congenital abnormalities, the reader is referred to Chapter 25.

SPECIFIC CARDIOVASCULAR MALFORMATIONS

Aortic Coarctation

Coarctation of the aorta most commonly occurs as a discrete stenosis of the proximal descending aorta,

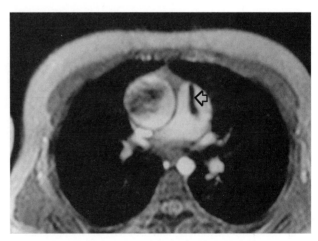

Figure 24–1. Transverse gradient-echo CMR image obtained during systole in a patient with pulmonic valve stenosis. Note the dark vertical slit-like jet (arrow) in the dilated pulmonary artery, revealing the presence of a significant stenosis of the pulmonary valve. (From Niezen RA, Helbing WA, van der Wall EE, de Roos A: Congenital heart disease assessed with magnetic resonance imaging. *In* Taveras JM, Ferruci JI (eds): Radiology, 2nd ed. Philadelphia, Lippincott-Raven, 1997, with permission.)

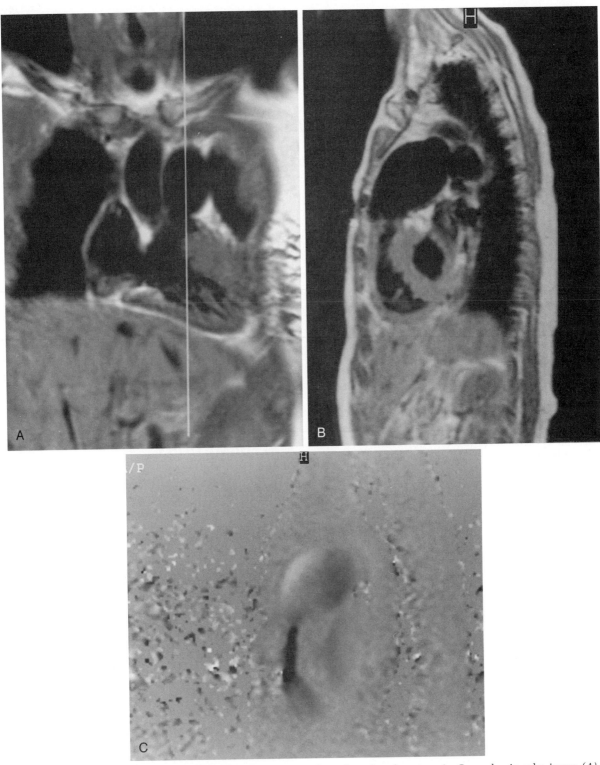

Figure 24–2. Patient with pulmonary regurgitation after treatment for pulmonic valve stenosis. Coronal spin-echo image *(A)* shows the dilated pulmonary artery. The plan-scan for sagittal image acquisition through the pulmonary artery is indicated by the line in *A*. The resulting sagittal CMR image is demonstrated in *B*. Note again the marked poststenotic dilation of the pulmonary trunk. In *C*, which is the same image orientation as in *B*, a sagittal in-plane CMR flow map is shown. Note the dark jet stream originating from the level of the pulmonary valve and passing into the right ventricle, indicating marked pulmonary valve insufficiency. (From Roest AAW, Helbing WA, van der Wall EE, de Roos A: Postoperative evaluation of congenital heart disease by magnetic resonance imaging. J Magn Reson Imaging 10:656, 1999, with permission.)

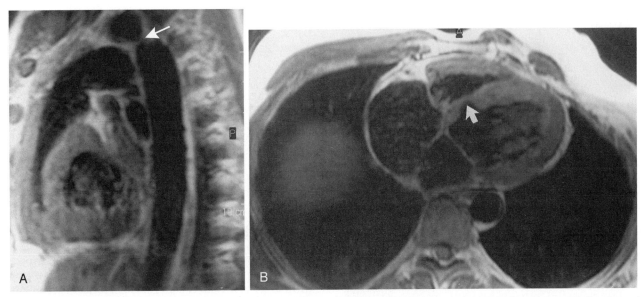

Figure 24–3. Sagittal spin-echo CMR image through the aorta *(A)* demonstrating the site and extent of aortic coarctation (arrow). Transverse spin-echo CMR image *(B)* at midventricular level illustrates the site of repair of a ventricular septal defect (arrow). Note also marked hypertrophy of the right ventricle owing to a residual left-to-right shunt through the ventricular septal defect.

just opposite to the (former) insertion of the ductus arteriosus (juxtaductal location). The gross morphology of the coarctation may vary from a discrete narrowing to a long-segment stenosis. CMR is now considered to be the standard noninvasive technique for evaluation of aortic coarctation, particularly in older children and adults (Figure 24–3A). Associated abnormalities such as arch hypoplasia, bicuspid aortic valve, and ventricular septal defect can also be assessed with CMR. Significant narrowing will impair blood flow into the descending aorta and therefore collateral vessels are required to reestablish aortic flow distal of the coarctation. The intercostal arteries, lateral thoracic, internal mammary, anterior spinal, and epigastric arteries can all serve as collateral pathways.

CMR readily identifies the coarctation site and extent, involvement of arch vessels, poststenotic dilatation, and dilated collateral vessels.[6] CMR is also well suited to assess restenosis after surgical repair and complications such as postoperative pseudoaneurysms. More recently, magnetic resonance angiography (MRA) has emerged as a valuable technique to assess the thoracic aorta. Contrast-enhanced MRA of the aorta is now routinely performed in older children and adults, resulting in high-quality aortograms with the aid of an intravenous infusion of gadolinium to shorten the arterial blood T1 (see also Chapters 26 and 29).[7] The image contrast is based on T1 relaxation, thereby minimizing flow artifacts and problems due to slow flow. We currently use a breath-hold, gadolinium-enhanced, three-dimensional turbo-field-echo MRA sequence with maximum arterial contrast enhancement during acquisition of the central portion of k-space and can image the thoracic aorta in 20 seconds (Figure 24–4). This acquisition is performed without electrocardiographic (ECG) gating.

Gradient-echo cine CMR can identify the jet flow across the coarctation directly and may be used to estimate the severity of stenosis based on the length of the flow void. Furthermore, peak jet velocity across the coarctation as measured by CMR provides comparable estimates of the pressure gradient to those obtained from continuous wave Doppler echocardiography by application of the modified Bernoulli formula.[8, 9] Moreover, flow mapping has proved to be a valuable adjunct to assess the severity of coarctation by measuring the flow in the proximal and distal descending thoracic aorta.[8, 10] It is recognized that CMR is the only technique allowing reliable flow measurements in the descending aorta. This approach for assessing coarctation severity is based on measurement of flow in the distal aorta above the diaphragm and comparing that to flow measured near the coarctation site. Retrograde flow in collateral channels will increase distal aortic flow depending on coarctation severity, thereby providing a direct estimate of the hemodynamic severity of the coarctation. Information on collateral flow in coarctation may be helpful for planning treatment options and monitoring patient outcome after surgical repair.

Pseudocoarctation

Narrowing at the level of the isthmus without an associated hemodynamically significant obstruction has been termed nonobstructing coarctation or pseudocoarctation. In the fetal period, it is a normal finding because the main flow is through the arterial duct. After closure of the duct, this segment should reach its normal size. In this malformation, the aortic arch is also markedly elongated, which results in kinking of the aorta. CMR is helpful to distin-

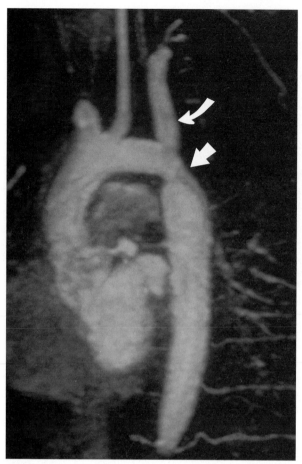

Figure 24–4. CMR angiogram of the aorta in a patient with mild coarctation. Note discrete narrowing (arrow) just distal to a dilated left subclavian artery (curved arrow). Note also the excellent demonstration of collateral vessels.

guish true coarctation and pseudocoarctation by demonstrating aortic arch anatomy as well as the functional consequences of the anatomic abnormality. Flow mapping can help to exclude impairment to flow at the level of the aortic arch anomaly by demonstrating the absence of a pressure gradient and the absence of collateral flow.

Vascular Rings

CMR displays the anatomy of vascular rings in great detail. In particular, depiction of the relationship between aortic arch anomalies and surrounding structures such as the bronchial tree and esophagus has diagnostic implications (Figure 24–5). A double aortic arch is the most common and serious cause for a vascular ring. This type of vascular ring encircles both the trachea and esophagus and may cause problems related to compression of both structures. Noting the spatial relationship between the vascular structures and trachea and esophagus may also be helpful for diagnosing a vascular sling, where an anomalous left pulmonary artery arises from the right pulmonary artery. The anomalous pulmonary

vessel courses between the trachea and esophagus back to the left lung (pulmonary sling anatomy). Vascular airway compression has been reported in a number of situations, including in patients with agenesis of the right lung.[11] Of particular interest is the pulmonary sling anatomy in association with right lung agenesis, which is characterized by coursing of the left pulmonary artery between the trachea and esophagus posterior to the left main bronchus (Figure 24–6). Recognition of this anomaly may be critical for planning surgery due to the association of vascular airway compression and intrinsic airway stenosis.[11]

Marfan Syndrome

Development of aortic aneurysms is of major prognostic importance in patients with Marfan syndrome.[12] Prophylactic surgery, performed when aortic aneurysm size reaches certain values (depending on localization and family history),[13, 14] has contributed to increased survival.[15] Therefore, annual investigations are required in most patients to follow changes in aortic diameter, in particular of the aortic root. Investigation is routinely performed with echocardiography in daily clinical practice, but both the nature of the aortic root and specific conditions in Marfan syndrome may require CMR for reliable imaging. Because of the trilobar aspect of the aortic root at the level of the sinus of Valsalva and the occurrence of isolated ectasia of one sinus, it does not seem appropriate to measure diameter, but rather area, which can only be imaged from a short-axis view through the aortic root. Chest deformities, so often present in Marfan syndrome, may hamper such a view echocardiographically, whereas position of the heart in the left thoracic cavity and extreme tortuosity of the aorta may distort standard echocardiographic imaging planes. Moreover, only part of the aortic arch and thoracic descending aorta can be visualized with echocardiography. Computed tomography may be useful to examine the aorta but seems less suitable for imaging of the aortic root because of its position (bent forward and to the left) and the need to acquire diastolic (triggered) images. CMR can replace more invasive studies to follow changes in aortic diameter in patients with Marfan syndrome. Aortic size can be measured with great accuracy noninvasively by means of standard spin-echo CMR (Figure 24–7) or with gadolinium-enhanced MRA. In addition, aortic regurgitation, aortic flow wave velocity, and regional aortic compliance can be assessed by CMR. Flow wave velocity is related to aortic compliance and may have clinical utility for follow-up of aortic disease in patients with Marfan syndrome (see Chapter 22).[16]

Atrial Septal Defect

Atrial septal defects are a common cause of a left-to-right shunt, which are sometimes not detected

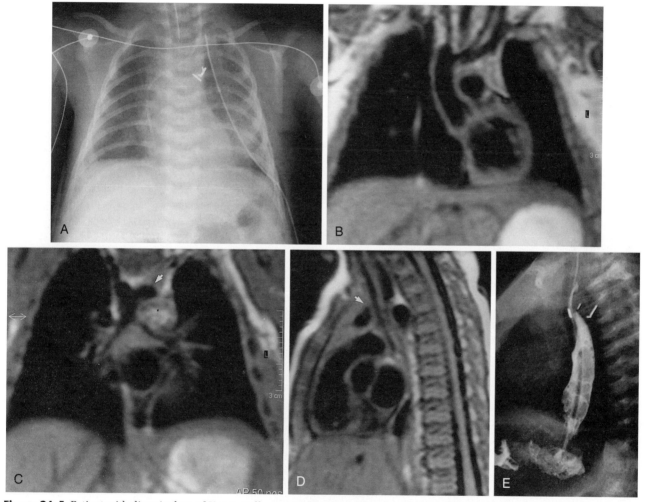

Figure 24–5. Patient with diverticulum of Kommerell compressing the esophagus after attempted surgical repair of double aortic arch. Chest film after surgery reveals some vascular clips *(A)*. The coronal spin-echo CMR image *(B)* displays the large right aortic arch and arch vessels. Coronal CMR image *(C)* more posteriorly identifies the diverticulum originating from the aorta (arrow). Sagittal CMR image *(D)* shows the posterior compression of the esophagus by the diverticulum (arrow). The corresponding esophagram is shown for comparison *(E)*.

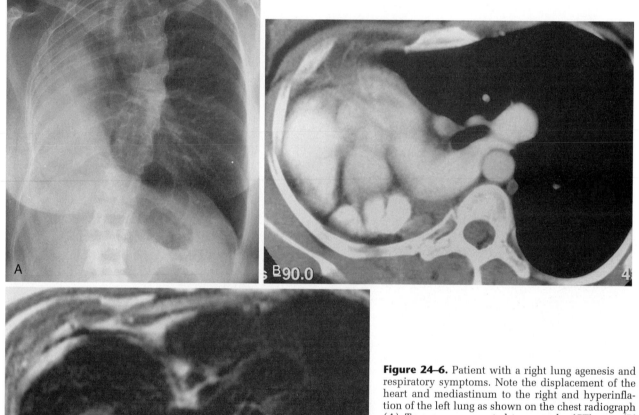

Figure 24–6. Patient with a right lung agenesis and respiratory symptoms. Note the displacement of the heart and mediastinum to the right and hyperinflation of the left lung as shown on the chest radiograph *(A).* Transverse computed tomography (CT) scan *(B)* and CMR scan *(C)* at the level of the pulmonary artery reveal pulmonary sling anatomy. Note the left pulmonary artery crossing from the right side between airway and esophagus, posterior to the left main bronchus.

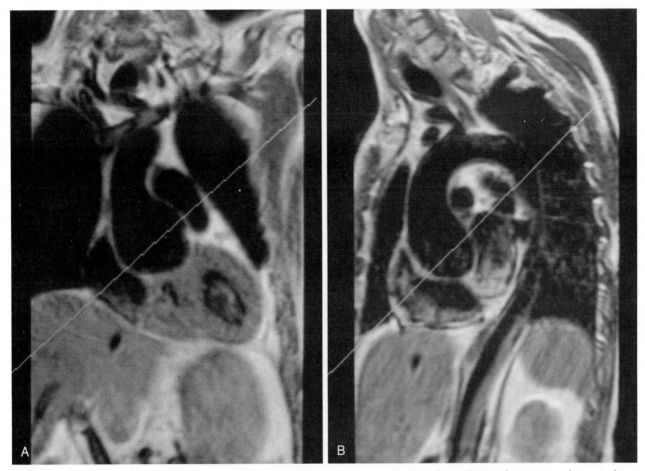

Figure 24–7. Coronal *(A)* and oblique sagittal *(B)* spin-echo CMR images in patient with Marfan syndrome. Note the typical pear-shaped appearance of the aortic root. The aortic aneurysm is accurately measurable from the CMR images. Lines indicate the plan-scan for image orientation perpendicular to the ascending aorta.

until adulthood. Other causes of left-to-right shunting include ventricular septal defects, patent ductus arteriosus, aortopulmonary window, and partial or total anomalous pulmonary venous return. Echocardiography is the first-line tool for detecting the presence of an atrial septal defect. Several types of atrial septal defect can be recognized (Figure 24–8). The most common type is the ostium secundum or fossa ovalis defect, which is located in the central part of the atrial septum. The sinus venosus defect is situated high on the septum, just below the entrance of the superior vena cava into the right atrium. This type may be associated with anomalous drainage of the right upper lobe pulmonary vein into the right atrium (Figure 24–9). The ostium primum defect is situated low on the atrial septum in close relationship with the atrioventricular valves. This defect belongs to the spectrum of atrioventricular septal defects and is accompanied by an abnormal position and structure of the atrioventricular valves.

The presence of an atrial septal defect can be established definitively by using cine gradient-echo CMR techniques to identify the atrial septal defect jet across the septum,[16a] with visualization of low signal on the right atrial side of the septum. The definition of such a flow-related signal void can be enhanced by applying a spatial saturation slab at the inflow region of the atrial septal defect.[16b] A defect diameter is derived from the maximum width of the transseptal flow from multiple parallel and intersecting cine CMR acquisitions.[16c]

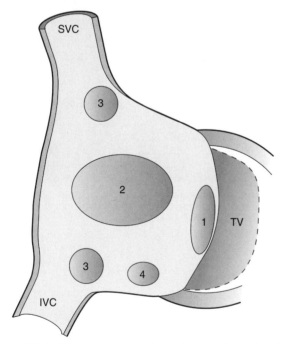

Figure 24–8. Most common locations of the atrial septal defect, viewed from the right atrium. 1 = ostium primum defect with cleft tricuspid (± mitral) valve (also known as atrioventricular septal defect); 2 = ostium secundum defect or fossa ovalis defect; 3 = sinus venosus defect; 4 = sinus coronarius defect; TV = tricuspid valve; SVC = superior vena cava; IVC = inferior vena cava.

The contribution of CMR in evaluating patients with left-to-right shunt lies in assessing its functional significance by determining the flow through the shunt. In atrial septal defects, the shunt flow can be extracted from the discrepancy between right and left ventricular stroke volumes. Shunt flow can also be determined by directly measuring aortic and pulmonary flow with flow mapping, resulting in the quantification of the shunt volume (Figure 24–10).[5, 17–18a] Direct measurement of flow across the atrial septal defect is also possible.[16c, 18b] Furthermore, the effects in right ventricle size can be adequately quantified.

Ventricular Septal Defect

The ventricular septal defect (VSD) is the most common congenital heart malformation and the major cause of left-to-right shunts. It can occur as an isolated anomaly or in combination with other cardiac malformations, like a coarctation of the aorta, tetralogy of Fallot, double outlet right ventricle, or truncus arteriosus. A VSD is classified according to the part of the ventricular septum in which it is located.[19] The four parts of the ventricular septum are the inlet septum, the trabecular septum, the outlet septum, and the membranous septum. A defect can occur in each part of the ventricular septum.

A defect in the membranous part of the ventricular septum is the most common form of VSD (see Figure 24–3B). The defect usually extends into other (muscular) parts of the septum (perimembranous VSD). A defect in the outlet septum is usually caused by malalignment of this part to the ventricular septum. This defect is associated with other malformations as in the tetralogy of Fallot (Figure 24–11) or the double-outlet right ventricle. Defects of the trabecular part of the septum are often an extension of the defects of the inlet or membranous part of the septum. Sometimes a number of defects are situated in the trabecular part towards the apex of the heart. The use of CMR to determine these small defects in the trabecular part of the septum is limited.[19]

For the surgical management in complex cardiac malformations, it is of the utmost importance to know the spatial relationship of the VSD to the orifices of the great arteries. Depending on the location of the VSD, the surgical treatment can vary from a simple patch, inserted from the right atrium (Figure 24–12), to a complicated biventricular correction.

As for the atrial septal defect, the first-line diagnostic modality is Doppler echocardiography. Particularly in VSDs in complex malformations, CMR is superior to Doppler echocardiography in the visualization of the spatial relationship with surrounding structures, atrioventricular (AV)-valves, and the great arteries. Furthermore CMR may provide exact quantification of the left-to-right shunt. To determine the location of the VSD, the spin-echo MR is used and images are made in the three

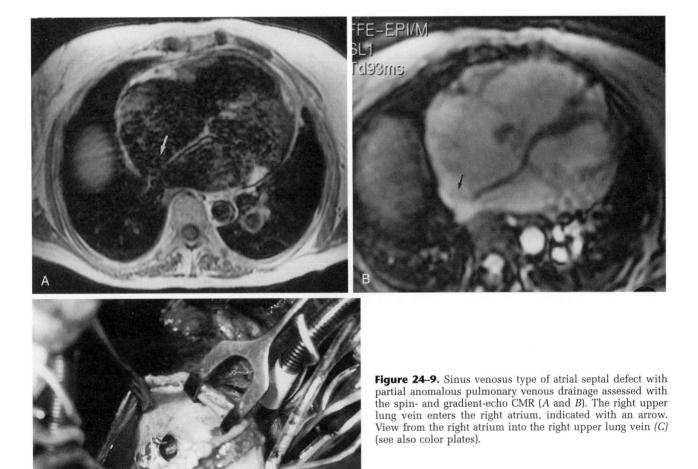

Figure 24–9. Sinus venosus type of atrial septal defect with partial anomalous pulmonary venous drainage assessed with the spin- and gradient-echo CMR (*A* and *B*). The right upper lung vein enters the right atrium, indicated with an arrow. View from the right atrium into the right upper lung vein *(C)* (see also color plates).

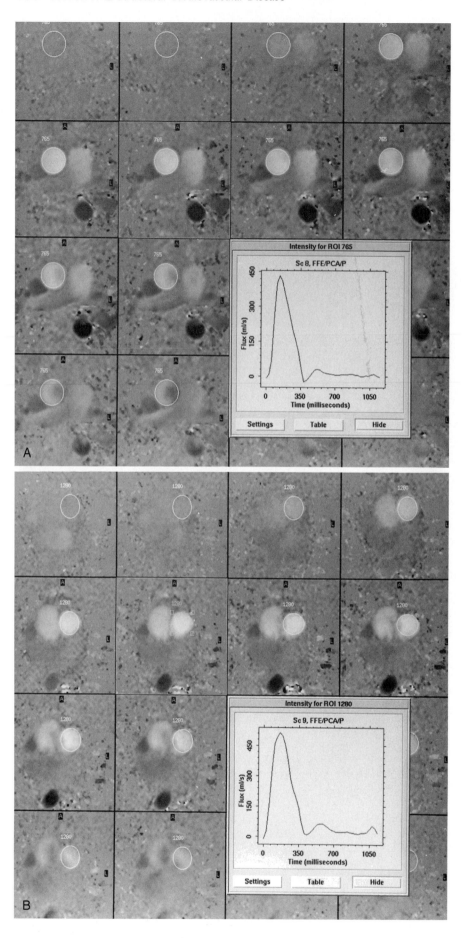

Figure 24–10. Chart of CMR flow mapping of both the ascending aorta *(A)* and the pulmonary trunk *(B)* just above the semilunar valves.

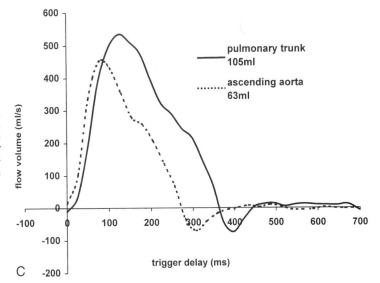

Figure 24–10 *Continued.* By subtracting both curves, one can quantify the amount of shunting per heartbeat, and the Qp/Qs in this case was 1.7 *(C)*. ROI = regions of interest. (*C,* From Roest AAW, Helbing WA, van der Wall EE, de Roos A: Postoperative evaluation of congenital heart disease by magnetic resonance imaging. J Magn Reson Imaging 10:656, 1999, with permission.)

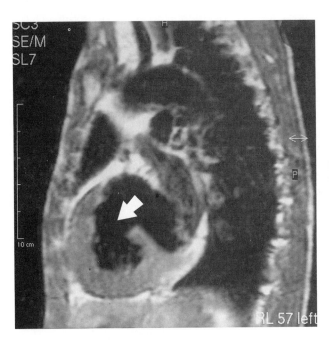

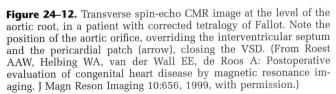

Figure 24–11. A sagittal spin-echo CMR image of a patient with an extreme tetralogy of Fallot. The outlet ventricular septal defect (VSD) is clearly visualized (arrow).

Figure 24–12. Transverse spin-echo CMR image at the level of the aortic root, in a patient with corrected tetralogy of Fallot. Note the position of the aortic orifice, overriding the interventricular septum and the pericardial patch (arrow), closing the VSD. (From Roest AAW, Helbing WA, van der Wall EE, de Roos A: Postoperative evaluation of congenital heart disease by magnetic resonance imaging. J Magn Reson Imaging 10:656, 1999, with permission.)

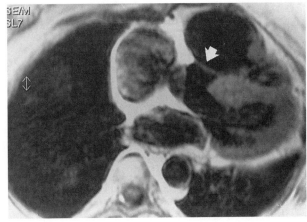

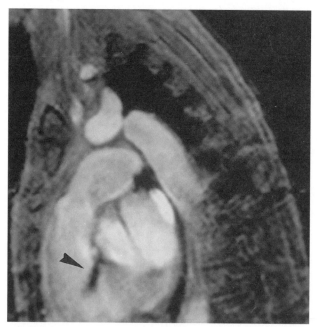

Figure 24–13. Sagittal gradient-echo CMR image obtained during systole in a patient after correction of a coarctation of the aorta. The area of signal loss (arrow) in the right ventricle from the interventricular septum is caused by turbulent flow across a perimembranous VSD.

orthogonal planes, of which the transverse plane is the most useful for the identification of all the aforementioned types of VSDs.[19] When precise information is needed on the shape and dimensions of the VSD and on the spatial relation of the VSD and the great arteries, these images should be completed with en face images of the VSD, made by planning a series of slices parallel to the VSD.[20] Gradient-echo MR is used to visualize a jet caused by turbulent blood flow through the VSD (Figure 24–13). As mentioned above, CMR is perfectly suited for the measurement of the flow volumes through the aorta and pulmonary trunk, from which the shunt ratio can be extracted.[17, 18a] This shunt ratio is used for determining the hemodynamic significance of the left-to-right shunt and is important for the management of the VSD.

Anomalous Pulmonary Venous Drainage

In this malformation, the pulmonary veins do not enter the left atrium, but some or all of the pulmonary veins are connected to the right atrium or to a systemic vein such as the vena cava superior or the vena cava inferior, directly or by way of a pulmonary confluence. In the normal heart, the entrance of the pulmonary veins into the left atrium can be clearly visualized with CMR in the transverse plane.[20a] In patients with total or partial anomalous pulmonary drainage, the pulmonary veins and their entrance to the right atrium or systemic vein can be distinguished and a collecting pulmonary vein can

be seen, connecting to the right atrium or systemic vein. Cine CMR in combination with contrast-enhanced MRA is particularly helpful for identifying anomalous pulmonary drainage.[20b]

Valvular Heart Disease

Abnormalities of the valves are frequently seen in patients with a congenital cardiac defect, either as a congenital malformation or as a result of the surgical treatment. The advantage of CMR is that it can be used for visualization of anatomical valve abnormalities, as well as for the assessment of functional valve abnormalities. Anatomical valve abnormalities frequently seen are the atresia of one of the valves (Figure 24–14), Ebstein's anomaly, bicuspid semilunar valve, single AV valve, and congenital mitral valve malformations. As a result of these anatomical valve abnormalities or because of surgical intervention, functional valve abnormalities occur, which can be divided into a valve being (1) too narrow, causing a stenosis, and/or (2) insufficient, causing regurgitation of blood (Figure 24–15). Both stenosis and regurgitation can be quantified with flow mapping. Measurement of the maximum velocity of the blood flow distal to the stenosis provides an estimation of the pressure gradient over the stenosis, indicating the severity of the stenosis.[21] In case of insufficiency of a valve, the volume of forward and backward flow can be measured, allowing quantification of regurgitation volume (Figure 24–16).[22] As a result of stenosis or insufficiency, the cardiac chamber situated upstream of the abnormal valve will react to the increased workload. Pressure overload resulting from stenosis will cause hypertrophy of the myocardium, whereas chronic regurgitation through an insufficient valve will cause an enlargement of the cardiac chamber because of the volume overload. With gradient-echo CMR, the se-

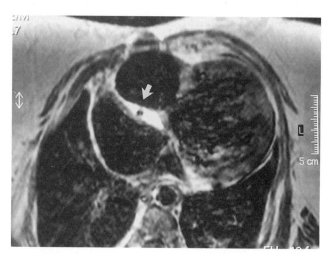

Figure 24–14. Transverse spin-echo CMR image at the level of the atrioventricular valves in a patient with atresia of the tricuspid valve. The arrow indicates the right coronary artery surrounded by fat, located at the site of the atretic tricuspid valve.

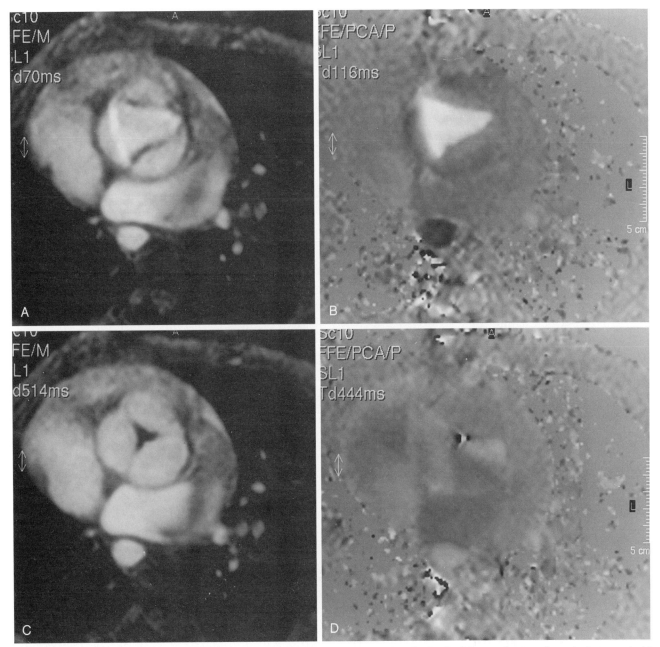

Figure 24–15. Gradient-echo images and flow mapping in the ascending aorta at the level of the semilunar valves, during systole (*A* and *B*) and diastole (*C* and *D*). Note the three cusps of the aortic valve in the gradient-echo images. With both magnitude and flow techniques it can be visualized that the cusps fail to close properly in diastole, causing aortic regurgitation.

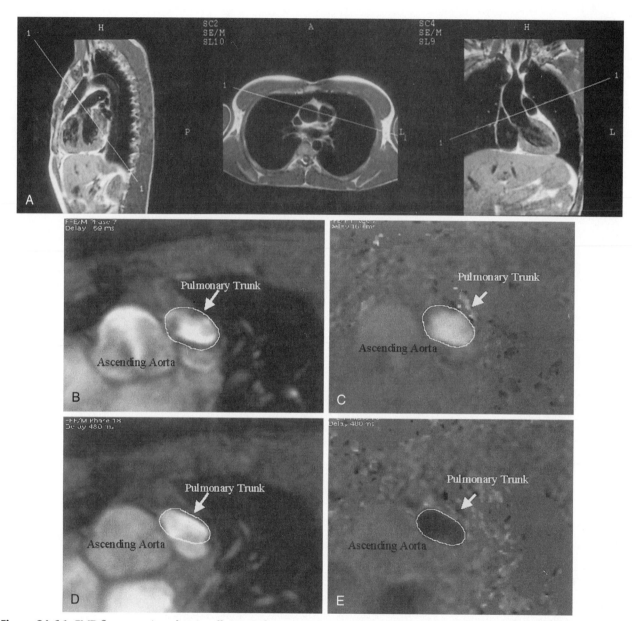

Figure 24–16. CMR flow mapping of an insufficient pulmonary valve. *A,* The scanning plane. Magnitude *(B)* and phase *(C)* images of the insufficient valve, obtained during systole. Forward flow appears white and backward flow is black on the phase images. Magnitude *(D)* and phase *(E)* images of the regurgitant valve in diastole. Note the dark area in the phase image, indicating backward flow, during diastole.

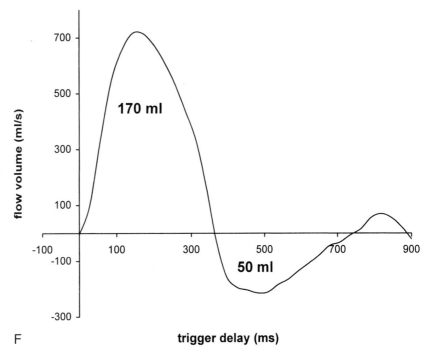

Figure 24–16 *Continued. F,* Graphical reconstruction of one heart phase, triggered from the R-wave of the ECG. During systole there is a forward flow of 170 ml and during diastole a backward flow of 50 ml, resulting in a net flow of 120 ml and a regurgitant fraction of 30% (=50/170). (From Roest AAW, Helbing WA, van der Wall EE, de Roos A: Postoperative evaluation of congenital heart disease by magnetic resonance imaging. J Magn Reson Imaging 10:656, 1999, with permission.)

verity and progression of the aforementioned secondary abnormalities can be analyzed by measuring the volumes of the left and right ventricles and myocardial mass.[23]

References

1. De Roos A, Rebergen SA, van der Wall EE: Congenital heart disease assessed with magnetic resonance techniques. *In* Skorton DJ, Schelbert HR, Wolf GL, Brundage BH (eds): Marcus Cardiac Imaging. A Companion to Braunwald's Heart Disease. Philadelphia, WB Saunders, 1996, p 672.
2. Helbing WA, Bosch HG, Maliepaard C, et al: Comparison of echocardiographic methods with magnetic resonance imaging for assessment of right ventricular function in children. Am J Cardiol 76:589, 1995.
3. Geva T, Vick W, Wendt RE, et al: Role of spin-echo and cine magnetic resonance imaging in presurgical planning of heterotaxy syndrome. Circulation 90:348, 1994.
4. Rebergen SA, Van der Wall EE, Doornbos J, De Roos A: Magnetic resonance measurement of velocity and flow: Technique, validation and cardiovascular applications. Am Heart J 126:1439, 1993.
5. Brenner LD, Caputo GR, Mostbeck GH, et al: Quantification of left-to-right atrial shunts with velocity-encoded cine nuclear magnetic resonance imaging. J Am Coll Cardiol 20:1246, 1992.
6. Von Schulthess GK, Higashino SM, Higgins SS, et al: Coarctation of the aorta: MR imaging. Radiology 158:469, 1986.
7. Prince MR, Narasimham DL, Jacoby WT, et al: Three-dimensional gadolinium-enhanced MR angiography of the thoracic aorta. AJR Am J Roentgenol 166:1387, 1996.
8. Mohiaddin RH, Kilner PJ, Rees S, Longmore DB: Magnetic resonance volume flow and jet velocity mapping in aortic coarctation. J Am Coll Cardiol 22:1515, 1993.
9. Oshinski JN, Parks WJ, Markou CP, et al: Improved measurement of pressure gradients in aortic coarctation by magnetic resonance imaging. J Am Coll Cardiol 28:1818, 1996.
10. Steffens JC, Bourne MW, Saluma H, et al: Quantification of collateral blood flow in coarctation of the aorta by velocity encoded cine magnetic resonance imaging. Circulation 90:937, 1994.
11. Newman B, Gondor M: MR evaluation of right pulmonary agenesis and vascular airway compression in pediatric patients. AJR Am J Roentgenol 168:55, 1997.
12. Roman MJ, Rosen SE, Kramer-Fox R, Devereux RB: Prognostic significance of the pattern of aortic root dilation in the Marfan syndrome. J Am Coll Cardiol 22:1470, 1993.
13. Finkbohner R, Johnston D, Crawford ES, et al: Marfan syndrome. Long-term survival and complications after aortic aneurysm repair. Circulation 91:728, 1995.
14. Silverman DI, Gray J, Roman MJ, et al: Family history of severe cardiovascular disease in Marfan syndrome is associated with increased aortic diameter and decreased survival. J Am Coll Cardiol 26:1062, 1995.
15. Silverman DI, Burton KJ, Gray J, et al: Life expectancy in the Marfan syndrome. Am J Cardiol 75:157, 1995.
16. Groenink M, de Roos A, Mulder BJM, et al: Changes in aortic distensibility and pulse wave velocity assessed with magnetic resonance imaging following beta-blocker therapy in the Marfan syndrome. Am J Cardiol 82:203, 1998.
16a. Thiessen P, Sechtem U, Mennicken U, et al: Noninvasive diagnosis of atrial septal defects and anomalous pulmonary venous return by magnetic resonance imaging. Nuklearmedizin 28:172, 1989.
16b. Hartnell GG, Sassower M, Finn JP: Selective presaturation magnetic resonance angiography: New method for detecting intracardiac shunts. Am Heart J 126:1032, 1993.
16c. Holmvang G: A magnetic resonance imaging method for evaluating atrial septal defects. J Cardiovasc Magn Reson 1:59, 1999.
17. Mohiaddin RH, Underwood R, Romeira L, et al: Comparison between cine magnetic velocity mapping and first-pass radionuclide angiocardiography for quantitating intracardiac shunts. Am J Cardiol 75:529, 1995.
17a. Arheden H, Holmqvist C, Thilen U, et al: Left-to-right cardiac shunts: Comparison of measurements obtained with MR velocity mapping and with radionuclide angiography. Radiology 211:453, 1999.
18. Sieverding L, Jung WI, Klose U, Apitz J: Noninvasive blood flow measurement and quantification of shunt volume by cine magnetic resonance in congenital heart disease. Pediatr Radiol 22:48, 1992.
18a. Beerbaum P, Körperich H, Barth P, et al: Non-invasive left-to-right shunt in pediatric patients: Phase contrast cine magnetic resonance imaging compared with invasive oximetry. Circulation 103:2476, 2001.

18b. Taylor AM, Stables RH, Pool-Wilson PA, Pennell DJ: Definitive clinical assessment of atrial septal defect by magnetic resonance imaging. J Cardiovasc Magn Reson 1:43, 1999.

19. Didier D, Higgins CB: Identification and localization of ventricular septal defect by gated magnetic resonance imaging. Am J Cardiol 57:1363, 1986.

20. Yoo SJ, Sea JW, Lim TH, et al: MR anatomy of ventricular septal defect in double-outlet right ventricle with situs solitus and atrioventricular concordance. Radiology 181:501, 1991.

20a. White C, Baffa J, Naney P, et al: Anomalies of pulmonary veins: Usefulness of spin-echo and gradient-echo MR images. AJR Am J Roentgenol 170:1365, 1998.

20b. Ferrari VA Scott CH, Holland GA, et al: Ultrafast three-dimensional contrast enhanced magnetic resonance angiography and imaging in the diagnosis of partial anomalous pulmonary venous drainage. J Am Coll Cardiol 27:1120, 2001.

21. Kilner PhJ, Manzara CC, Mohiaddin RH, et al: Magnetic resonance jet velocity mapping in mitral and aortic valve stenosis. Circulation 87:1239, 1993.

22. Rebergen SA, Chin JGJ, Ottenkamp J, et al: Pulmonary regurgitation in the late postoperative follow-up of tetralogy of Fallot: Volumetric quantification by MR velocity mapping. Circulation 88(part 1):2257, 1993.

23. Niezen RA, Helbing WA, van der Wall EE, et al: Biventricular systolic function and mass studied with MR imaging in children with pulmonary regurgitation after repair for tetralogy of Fallot. Radiology 201:135, 1996.

CHAPTER 25

Cardiovascular Magnetic Resonance of Complex Congenital Heart Disease in the Adult

Rolf Wyttenbach, Jens Bremerich, and Charles B. Higgins

Evaluation and characterization of congenital heart disease is a common clinical indication for cardiovascular magnetic resonance (CMR) in many institutions.[1] The number of adults with congenital heart disease continues to increase, a result of prolongation of survival produced by palliative and corrective surgical procedures.

The role of CMR in assessment of congenital heart disease is primarily defined by the use and the limitations of two-dimensional (2-D) echocardiography, which has become the accepted standard for evaluation of most forms of congenital heart disease. Advantages of CMR compared with echocardiography include (1) the entire thorax can be imaged in sequential high-resolution tomograms; (2) the information represents a continuous three-dimensional (3-D) data set, which can be obtained or reconstructed in any imaging plane; and (3) the superior depiction of the central pulmonary arteries,[2] the systemic and pulmonary veins,[3] as well as the aorta.[4]

The major requirement for evaluation of congenital heart disease is the precise depiction of cardiovascular anatomy. Electrocardiogram (ECG)-gated spin-echo CMR can be utilized with high diagnostic accuracy for assessment of morphologic features in simple and complex congenital heart disease.[5]

Cine gradient-echo CMR permits accurate measurements of functional parameters such as right and left ventricular stroke volume, ejection fraction, regional wall motion, and wall thickening. Moreover, cine CMR can depict the jet flow associated with valvular lesions and intracardiac shunts. With velocity-encoded CMR, measurement of flow velocity and flow volume is possible, allowing quantification of pulmonary artery flow, shunt lesions, valvular regurgitation, and ventricular filling. The breadth of noninvasive capabilities of these CMR techniques makes CMR ideal for sequential studies of patients with simple and complex congenital lesions.

In this chapter we discuss the CMR appearance of cardiovascular morphology by using a segmental approach, which represents the most rational method to evaluate complex congenital heart disease. The segmental approach is based on the mor-phological identification of the great arteries, the atria, the ventricles, and the visceroatrial relationship as well as the connection among these structures. Moreover, we briefly review morphological and functional evaluation of complex congenital heart disease by the use of CMR in the pre- and postoperative condition.

ATRIAL MORPHOLOGY AND DETERMINATION OF SITUS

Axial spin-echo CMR extending from the cardiac base to the dome of the liver depicts segmental cardiovascular anatomy. A segmental approach is based on the localization of the three cardiac segments (atria, ventricles, and great arteries), the type of atrioventricular (AV) and ventriculoarterial (VA) connections, and the detection of associated anomalies (shunts, valve atresia, etc.).[6, 7] This segmental approach to congenital heart disease provides a precise description of cardiac morphology, allowing accurate diagnosis of congenital heart disease obtained by CMR examination.[8]

Atrial situs solitus indicates the normal circumstance in which the morphological right (systemic venous) atrium is positioned on the right of the spine and the morphological left (pulmonary venous) atrium is positioned to the left of the spine. In atrial situs inversus, the mirror image of the normal situation occurs with the anatomical right atrium on the left side and anatomical left atrium on the right side. The morphological right atrium has distinctive anatomical characteristics, including a broad-based, triangular appendage, whereas the morphological left atrium contains an appendage with a narrow ostium and a more tubular configuration.[9] The right atrium can also be identified by its connection to the inferior vena cava. In virtually all individuals, the side of the inferior vena cava defines the side of the right atrium. Furthermore, the situs of the atria can be defined by the visceral situs because both are nearly always concordant.[7] Thus, the morphological right atrium is determined by the side of the short main bronchus and the liver, whereas the long main bronchus, the spleen, the stomach, and the aorta define the morphological left atrium.

In situs solitus, the left pulmonary artery courses cranially over the left main bronchus, and the right pulmonary artery runs anterior and slightly inferior to the right bronchus.

VENTRICULAR MORPHOLOGY AND ISOMERISM

Axial images at the midventricular level permit definition of the ventricular loop. The morphological right ventricle (RV) and morphological left ventricle (LV) have different anatomical characteristics allowing their differentiation. The morphologic RV can be recognized by its anterior and right-sided location (D-ventricular loop). The RV has prominent septomarginal trabeculae passing from the apical portion of the ventricular septum to the anterior wall of the morphological right ventricle. The insertion point of the septal leaflet of the tricuspid valve is more anterior (towards the cardiac apex) than is the septal leaflet of the mitral valve. The depiction of different levels of insertion of the septal leaflets is an important diagnostic feature for distinction of the ventricles.[10] Probably the most reliable sign for defining the RV is the presence of an infundibulum or conus that separates the tricuspid and pulmonary valves, best appreciated on axial images. The morphological LV is located posteriorly and to the left. Its septal leaflet of the mitral valve inserts more distant from the cardiac apex than does the septal leaflet of the tricuspid valve. The anatomical LV has a smoothly contoured apical portion of the ventricular septum and is characterized by a lack of complete muscular infundibulum. Therefore, the morphological LV is characterized by a direct (fibrous) continuity between mitral and aortic valves.

The relationship of the great vessels can be determined by axial sections through the base of the heart. The ascending aorta can be identified by its continuity with the aortic arch and the brachiocephalic arteries. The main pulmonary artery is characterized by its bifurcation into the right and left main pulmonary arteries. Normally the aorta lies to the right and posterior of the pulmonary trunk. The ascending aortic diameter is usually slightly larger than the main pulmonary artery.

The term isomerism refers to the situation in which both atria have features of the right atrium or the left atrium. In general, both atria develop with the same side as the thoracic and abdominal viscera (visceral-atrial rule). Therefore, bilateral left pulmonary artery anatomy with the artery passing over the left bronchus indicates left-sided isomerism (bilateral leftsidedness). This condition is associated with polysplenia. These findings are best depicted on coronal sections. Conversely, bilateral right pulmonary anatomy indicates right-sided isomerism (bilateral rightsidedness), associated with asplenia.

A previous report showed that ECG-gated spin-

echo CMR is highly accurate for determination of relationships among the great arteries, visceroatrial situs, and type of ventricular loop in patients with congenital heart disease.[5] Surgical planning may be altered in some patients with heterotaxy syndrome based on novel data supplied by CMR (not available by echocardiography or catheterization).[11]

ABNORMALITIES OF ATRIOVENTRICULAR CONNECTION

Once the atrial and ventricular morphology is determined, the next step is to determine whether the AV connection is concordant or discordant. Concordance is defined by morphological right atrium connection to the morphological RV and morphologic left atrium connection to the morphologic LV, *regardless of the positions of the atria and the ventricles.* In contrast, discordance means the morphological right atrium drains to the morphological LV and the morphological left atrium to the morphological RV, again irrespective of the chamber position within the chest. An example of AV (and VA) discordance is corrected transposition (L-transposition), which is described in further detail later. AV discordance in conjunction with VA concordance is a rare malformation, called *isolated ventricular inversion.* Other abnormalities of AV connections in which the terms *concordant* and *discordant* are not appropriate include the following: (1) double-inlet atrioventricular connection in which both atria are connected to a single ventricular chamber; (2) straddling AV valves in which one of the valves overlies the septum and drains blood into both ventricles; and (3) AV valve atresia in which one of the valves does not form and the AV ring may be replaced by fat (Figure 25–1).[12]

ABNORMALITIES OF VENTRICULOARTERIAL CONNECTIONS

Transposition

The transposition of the great arteries (TGA) is one of the most common cyanotic congenital cardiac malformations. CMR has been shown to define pathoanatomy of transposition and other abnormalities of VA connections accurately in several studies.[5, 8, 10, 13, 14] In complete or D-transposition of the great arteries, VA discordance exists in the presence of AV concordance. The ventricles receive the blood from the correct atrium, whereas the pulmonary artery is connected to the LV and the aorta with the RV. CMR of the normal great artery anatomy at the base of the heart has the aorta posterior and to the right of the pulmonary artery. Contrary to normal anatomy, the axial images at the base of the heart in

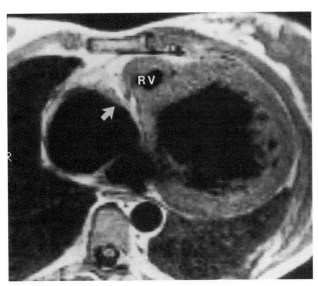

Figure 25–1. Axial spin-echo CMR of a patient with tricuspid atresia and hypoplastic right ventricle (RV). A solid bar of muscle and fat (arrow) separates the right atrium and the hypoplastic outflow chamber of the RV. (From Higgins CB: Congenital heart disease. *In* Higgins CB, Hricak H, Helms CA [eds]: Magnetic Resonance Imaging of the Body. 3rd ed. Philadelphia, Lippincott-Raven, 1996, with permission.)

TGA demonstrate the anterior position of the aorta relative to the main pulmonary artery (Figure 25–2). The aorta is located to the right of the pulmonary artery in D-transposition and to the left of the pulmonary artery in L-transposition. In patients with transposition of the great vessels, sagittal images can demonstrate the anterior aorta arising from the morphological RV and the posterior pulmonary artery arising from the morphological LV.[8]

L-transposition of the great arteries, also known as congenitally corrected transposition, results when the developing embryonic cardiac tube bends initially towards the left, rather than to the right—a so-called "L-loop." In L-transposition, both AV and VA connections are discordant. This results in an aorta positioned anterior and leftward to the pulmonary artery (Figure 25–3). In addition, the aorta arises from the left-sided morphological RV and the pulmonary artery from the right-sided morphological LV because the position of the ventricles is inverted (L-loop) in L-transposition. Therefore, systemic venous blood is pumped to the lungs by the morphological LV via the pulmonary arteries, and oxygenated blood is pumped to the systemic circulation by the morphological RV via the aorta. These anatomical features can be readily assessed by ECG-gated axial spin-echo CMR. In the coronal plane, the left-sided ascending aorta typically forms the upper heart border with L-transposition. Furthermore, CMR has the capabilities to demonstrate anomalies associated with TGA. Axial cine CMR can assess the presence and severity of atrial and ventricular septal defects. Additionally, the severity of valvular and subvalvular stenosis as well as pulmonary artery stenosis or atresia can be determined by CMR.[15]

Reversal in muscle thickness and shape of the RV compared with the LV, which is characteristic for transposition, can also be demonstrated by CMR. In this respect CMR is now regarded as the most accurate method to quantify ventricular mass. Consequently, cine CMR can be used to determine the LV mass in children or adults in whom an arterial switch procedure is under consideration (Figure 25–4).

Double-Outlet Right Ventricle

Double-outlet right ventricle (DORV) is defined as an abnormal VA connection where more than half of both aorta and pulmonary artery arise from the morphologic RV.[16, 17] On axial CMR, DORV is typically characterized by side-by-side positioning of the great arteries at the semilunar valve level, with the aorta to the right of the pulmonary artery, although this relationship may be variable.[18] An additional important feature of this condition is that neither semilunar valve is in direct fibrous continuity with the mitral valve. There is a complete rim of muscle separating both semilunar valves from the anterior mitral valve leaflet. On transverse CMR, this side-by-side positioning of two muscular circles in the outflow region of the RV is diagnostic for DORV (Figure 25–5). Coronal images define the side-by-side relationship of the aorta and the pulmonary artery at the level of the semilunar valves, and their origin from the RV. The LV is shown to be separated from the semilunar valves.

In this condition, the only outlet for the LV becomes the requisite ventricular septal defect (VSD). In order to adequately plan a surgical repair, it is important to determine the relative relationship of the VSD to the great vessels.[19, 20] This relationship can usually be assessed on axial images. The VSD may be localized near the aortic valve (subaortic VSD), the pulmonic valve (subpulmonic VSD; Bing-Taussig anomaly), or both semilunar valves (doubly committed VSD). It may also be found remote from both semilunar valves (uncommitted VSD). DORV may be associated with valvular or subvalvular pulmonic stenosis. In particular, if DORV arises in combination with a subaortic VSD, differentiation from tetralogy of Fallot may be difficult clinically and angiographically. Axial CMR provides direct visualization of this type of DORV by showing a complete circle of muscle separating the aortic from the mitral valve, which is better than having to indirectly infer this diagnosis from the distance between aortic and mitral valve as shown on left ventriculography.[21]

CMR can be used to determine the size and location of the VSD relative to the great arteries, and to define subpulmonary and/or subaortic stenosis, spatial relationships of the great vessels, and status of the pulmonary arteries and the aortic arch (Figure 25–6). These features are important for clinical and surgical management. Additionally, DORV may be

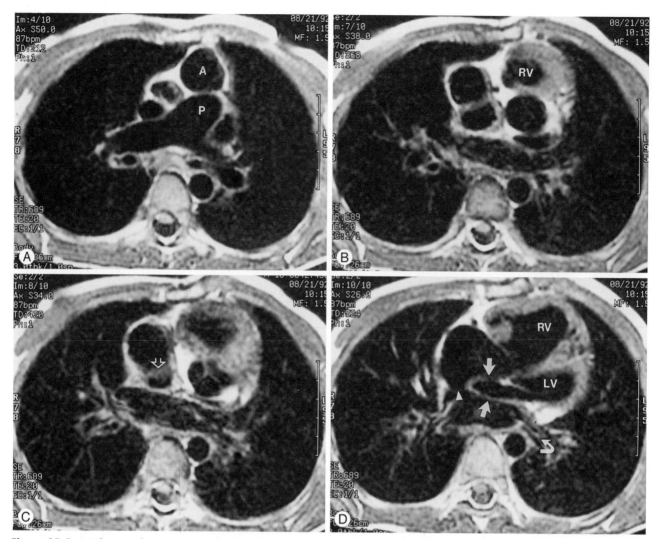

Figure 25–2. Axial spin-echo CMR in a patient with D-transposition of the great arteries after Senning procedure. *A* demonstrates the anterior and slightly rightward position of the ascending aorta (A) compared with the pulmonary artery (P). The aorta is connected to the RV, and the pulmonary artery originates from the left ventricle (LV) *(B–D)*. The intraatrial baffle (*D*, arrows), which isolates the mitral valve from the pulmonary venous drainage (*D*, curved arrow), is also seen. Pulmonary venous blood enters the posterior part of the pulmonary venous atrium and flows anteriorly (*D*, arrowhead) across the tricuspid valve. Note the narrowing of the upper limb of the baffle (open arrow) in image *C* compared with image *B*. (*D* from Higgins CB: Congenital heart disease. *In* Higgins CB, Hricak H, Helms CA [eds]: Magnetic Resonance Imaging of the Body. 3rd ed. Philadelphia, Lippincott-Raven, 1996, with permission.)

associated with atresia of the right AV valve. Because this condition will affect surgical repair, it is crucial to identify both AV valves in the axial plane.[22]

Truncus Arteriosus

Persistent truncus arteriosus results from failure of division of the embryonic truncus into a separate aorta and pulmonary artery. This abnormality is rare, representing less than 1 percent of all congenital heart disease.[23] On axial, sagittal, and coronal images, a single large vessel can be demonstrated arising from the expected position of the semilunar valves, just above the VSD (Figure 25–7). The truncus arteriosus gives rise to aorta, pulmonary, and coronary arteries. Based on configuration of the pul-

monary arteries, three different types of truncus arteriosus are recognized. Type I is the most common. It is characterized by a short main pulmonary artery arising from the truncus. Bidirectional shunting occurs at the truncus, resulting in early cyanosis and eventual congestive heart failure. Axial T1-weighted spin-echo CMR can define the anatomy of the truncus and the pulmonary arteries.[4] Sagittal and coronal planes are useful for demonstrating the origin of the pulmonary arteries from the truncus. For precise assessment of pulmonary artery caliber a reduction of slice thickness to 3 mm is needed. Furthermore, CMR can demonstrate ventricular size and wall thickness as well as associated abnormalities such as VSD, right aortic arch, or interrupted aortic arch. Axial and sagittal CMR are most effective for postoperative evaluation of caliber of the conduit between RV and the pulmonary artery (Rastelli procedure).

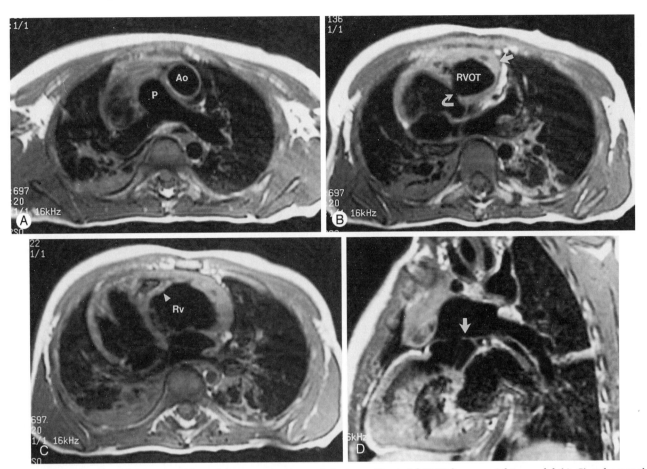

Figure 25–3. Patient with L-transposition of the great arteries and dextrocardia. Axial CMR from cranial to caudal *(A–C)* and coronal oblique image *(D)*. At the level of the pulmonary bifurcation *(A)*, the aorta (Ao) is situated to the left and anterior of the main pulmonary artery (P). Image *B* demonstrates the origin of the transposed aorta from the RV outflow tract (RVOT), which lies to the left. Note the muscular infundibulum of the RVOT (straight arrow) and the small ventricular septum defect (curved arrow). The pulmonary artery originates from the LV. Ventricular inversion is demonstrated in image *C* with the morphologic RV to the left of the morphologic LV (L-loop). Note the moderator band of the morphologic RV *(C*, arrowhead). The coronal oblique image *D* shows a supravalvular stenosis of the main pulmonary artery (arrow).

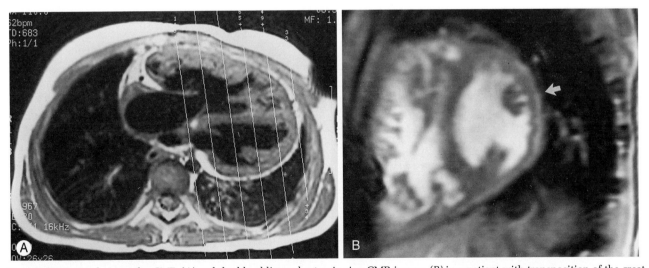

Figure 25–4. Axial spin-echo CMR *(A)* and double-oblique short-axis cine CMR images *(B)* in a patient with transposition of the great arteries in whom CMR was done prior to arterial switch operation to quantify LV mass. Image *A* demonstrates the orientation of the imaging planes in order to define the cardiac short-axis planes necessary for measurement of LV mass. Cine CMR image *B* shows an end-diastolic short-axis plane at midventricular level. Note the thin left ventricular free wall (arrow) resulting from support of the low-pressure pulmonary circuit.

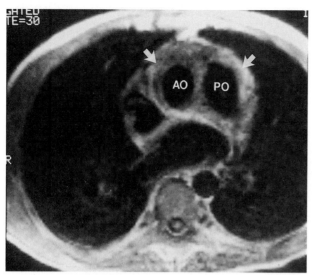

Figure 25–5. Axial image of an adult patient with double-outlet RV (DORV). At the conus region, the aortic (AO) and the pulmonary outflow tract (PO) lie side by side. Muscular tissue surrounds the aortic and pulmonary outflow tract, indicating DORV (arrows). (From Higgins CB: Congenital heart disease. *In* Higgins CB, Hricak H, Helms CA [eds]: Magnetic Resonance Imaging of the Body. 3rd ed. Philadelphia, Lippincott-Raven, 1996, with permission.)

Narrowing of the conduit, stenosis at origins of the right or left pulmonary arteries, and pseudoaneurysm are complications that can be readily demonstrated by CMR.[24] Truncal insufficiency, which is common, may be detected by the use of cine CMR.

TETRALOGY OF FALLOT

Consecutive transverse tomograms (ECG-gated spin-echo; 5- to 7-mm slice thickness) through the entire heart and pulmonary hili demonstrate (1) RV hypertrophy; (2) unequal division of the outflow tracts with enlarged and anteriorly displaced aorta; (3) membranous VSD; and (4) multilevel narrowing of infundibulum, pulmonary annulus, and main and central pulmonary arteries. Infundibulum and pulmonary annulus are best depicted on sagittal tomograms. The degree of pulmonary stenosis varies. In extreme cases, the pulmonary trunk may not be identifiable. Severe pulmonary stenosis and atresia are usually associated with numerous and large collateral channels arising from the aorta, principally the descending aorta, and proceeding to the pulmonary hili.[25] These vessels may be demonstrated on flow-sensitive transversal cine CMR at the level of the carina or on coronal images.

Stenoses of the central pulmonary arteries are frequent in tetralogy and not uncommonly remain after initial surgical correction (Figure 25–8). These stenoses are best depicted on a set of very thin tomograms with 3-mm slice thickness acquired in a plane parallel to the long axis of the right and left pulmonary arteries. The image plane should be parallel to the long axis of the right or left pulmonary artery.

Most adult patients with tetralogy of Fallot have already undergone one or more corrective surgeries. CMR is an ideal technique for monitoring these patients after surgery. Cine CMR is used to monitor RV volumes, mass, and ejection fraction. Velocity-encoded cine CMR has been used to monitor pulmonary regurgitation, which occurs in most patients after total correction of the anomaly (see Figure 25–8). One study has shown that velocity-encoded CMR quantitation of pulmonic valve regurgitation correlates with RV volumes, ejection fraction, and RV mass.[26]

RV failure can occur in tetralogy. RV mass or functional parameters (e.g., ejection fraction and

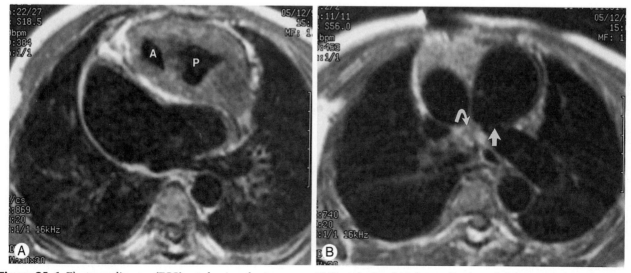

Figure 25–6. Electrocardiogram (ECG)-gated spin-echo transverse CMR at the level of the cardiac base *(A)* demonstrates side-by-side relationship of the aorta (A) and the main pulmonary artery (P) in a patient with DORV. Note the complete muscular ring surrounding both great arteries. Image *B* shows a stenosis of the left pulmonary artery (straight arrow) and an occluded right pulmonary artery (curved arrow).

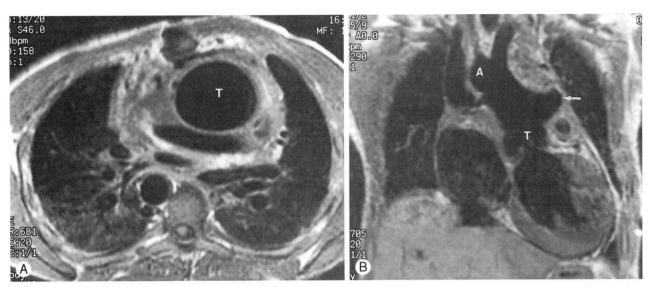

Figure 25–7. Axial *(A)* and coronal *(B)* spin-echo CMR in a patient with truncus arteriosus show a single great artery (T) at the base of the heart and the origin of the main pulmonary artery *(B,* arrow) from the left side of truncus. Note the right-sided aortic arch (A), which is frequently associated with this anomaly. *(B* from Higgins CB: Congenital heart disease. *In* Higgins CB, Hricak H, Helms CA [eds]: Magnetic Resonance Imaging of the Body. 3rd ed. Philadelphia, Lippincott-Raven, 1996, with permission.)

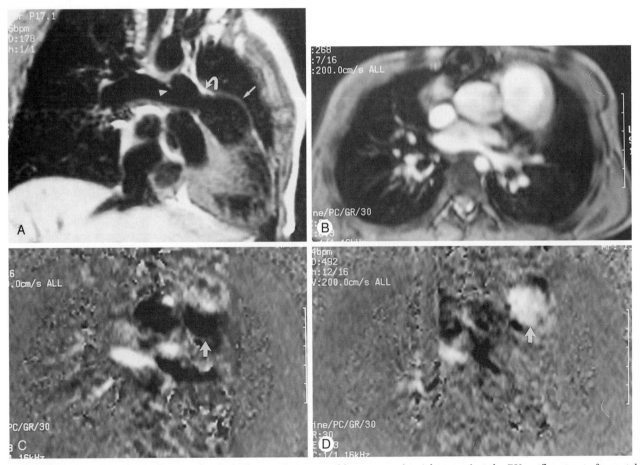

Figure 25–8. *A,* Oblique coronal spin-echo CMR demonstrating a mild aneurysm (straight arrow) at the RV outflow tract after patch repair of tetralogy of Fallot. Note the narrowing of the supravalvular pulmonary artery (curved arrow) and the right pulmonary artery (arrowhead). *B–D,* axial magnitude *(B)* and phase images *(C* and *D)* of velocity-encoded cine CMR acquisition near the origin of the great vessels. The phase images indicate the direction and velocity of blood flow in the pulmonary artery by a gray scale level. Blood flow directed cranially (antegrade) is represented by black coloration, whereas caudally (retrograde) blood flow is represented by white coloration. Note the change in direction of blood flow from antegrade in systole *(C)* to retrograde in diastole *(D)* in the main pulmonary artery *(C* and *D,* arrows), representing pulmonary regurgitation after repair of tetralogy of Fallot.

stroke volume) can be readily calculated from a stack of consecutive cine MR images acquired in the short axis of the heart covering the entire RV. RV mass calculated from such CMR images has been shown to correlate with the width of the QRS complex. A wide QRS complex is a harbinger of RV arrhythmias.

EBSTEIN ANOMALY OF THE TRICUSPID VALVE

Ebstein anomaly is an uncommon congenital developmental abnormality of the tricuspid valve that has a wide spectrum of pathological anatomy. Although the diagnosis of Ebstein anomaly is usually diagnosed from 2-D surface echocardiography, CMR may be helpful in defining the pathological anatomy and tailoring the surgical approach for the patient. Axial CMR images are most informative. In addition to abnormal septal leaflet insertion, the tricuspid valve is almost always dysplastic. In fact, the dominant feature is dysplasia of the valve rather than displacement.[27] The anterior leaflet usually attaches normally to the AV junction and is enlarged,[28] which can be depicted on CMR images in the majority of cases.[29] Distally, the anterior leaflet may be attached to an abnormal anterolateral papillary muscle and therefore be mobile; it may have a continuous muscular connection with restricted mobility; or it may be broadly plastered to the anterior wall of the RV and therefore not readily distinguishable on CMR.[28, 29] Classically, the septal and posterior tricuspid leaflets are displaced downward in Ebstein anomaly.[30] This displacement is best appreciated on axial and coronal images, respectively.[31] Both leaflets, however, may be deficient or absent and therefore not detectable on CMR.[29, 32]

Planning of surgical repair must include assessment of morphology and function of the atrialized and the hemodynamically effective portion of the RV, best achieved on coronal T1-weighted spin-echo and cine CMR. Reconstruction of the tricuspid valve by using a prosthetic ring and vertical plication of the right atrium and the AV annulus may be complemented by an additional plication of the atrialized ventricle, depending on size and function of the atrialized ventricle.[33] Moreover, cine CMR enables assessment of tricuspid regurgitation, tricuspid stenosis, and shunts through an associated atrial septal defect.

COMPLEX VENTRICULAR ABNORMALITIES (SINGLE VENTRICLE)

An early paper stressed the advantage of CMR as compared with invasive angiography for the definition of segmental anatomy and other defects in patients with a single ventricle.[34] The specific goals of

CMR in complex ventricular abnormalities are (1) determination of visceral situs; (2) assessment of type of ventricular loop, morphology of the predominant ventricle (right, left, or primitive), and position of the rudimentary ventricle; (3) definition of the AV and VA connections; (4) determination of the size of the interventricular communication; and (5) definition of connections of systemic and pulmonary veins and arteries.

Axial ECG-gated spin-echo CMR with 7-mm slice thickness are most useful for evaluation of complex cardiac anomalies (Figure 25–9).[35, 36] AV connections can be determined as double-inlet, absent left AV, or absent right AV connection. Stenoses or regurgitation can be detected on cine CMR as flow void caused by spin dephasing in turbulent blood flow. After identification of the AV connection, the ventricular morphology needs to be determined. Distinction between dominant left or dominant right ventricle is usually possible by axial and coronal CMR. When there is no detectable muscle separating either AV valve from the adjacent semilunar valve, the chamber is considered an LV. The position of a rudimentary RV is usually anterior and superior to the dominant ventricle, whereas the rudimentary LV usually is posterior and inferior to the dominant ventricle. A dominant LV is most common in adult patients. A primitive type of single ventricle has morphological features characteristic of neither the RV nor the LV. The communication between the dominant and the rudimentary ventricle can be assessed on both T1-weighted spin-echo and cine CMR.

POSTOPERATIVE EVALUATION

Today, there are thousands of patients with complex congenital heart disease who have survived to adult life after various palliative and corrective procedures. These patients require monitoring at regular intervals using imaging studies. Surface echocardiography is usually the initial technique used for this purpose, but it is not as effective in adults. Moreover, many surgical procedures involve supracardiac as well as extracardiac structures. Because the supracardiac structures are sometimes not well depicted by echocardiography, CMR is now becoming recognized as more comprehensive for postoperative monitoring of older children and adults after complex surgical procedures.[37–42]

Several reports confirm the effectiveness of CMR for postoperative evaluation of complex congenital heart disease, suggesting that CMR may obviate the serial use of postoperative invasive studies in many cases.[24, 37–39] CMR not only is capable of visualizing cardiac and extracardiac morphology but it also can quantify blood flow in the pulmonary arteries and conduits. Compared with echocardiography, CMR has the advantage of superior demonstration of conduits and anastomosis at the level of the great arteries and it is unaffected by postsurgical changes or

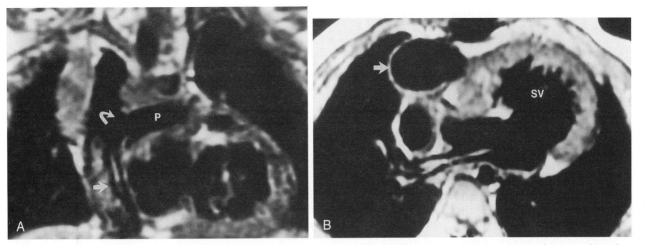

Figure 25–9. *A,* The coronal spin-echo CMR demonstrates the modified extracardiac Fontan connection (arrow) between the inferior vena cava and the right pulmonary artery (P) in an adult patient with single ventricle. The anastomosis between the superior vena cava and the right pulmonary artery is also shown (curved arrow). *B,* The transverse image shows the single left ventricle (SV) and the extracardiac Fontan connection (arrow).

graft material that can make echocardiography difficult.[37, 40] In addition, CMR has been found to be effective for monitoring pulmonary arterial status postoperatively and to be superior to echocardiography for the evaluation of the pulmonary arteries.[41, 42]

Many adult patients with TGA were treated with Mustard and Senning procedures (see Figure 25–2). This atrial switch procedure is accomplished by a complex atrial baffle to direct blood flow from pulmonary veins across the tricuspid valve into the RV and then to the aorta. Blood flow from the superior and inferior vena cava flows through the baffle and then across the mitral valve into the pulmonary artery. In postoperative follow-up of the Mustard procedure, CMR in the coronal plane is particularly useful for evaluation of the superior systemic venous channel, which is the most common site of systemic venous obstruction. The sagittal and transaxial planes can depict obstruction of pulmonary venous return or narrowing of the connection between the dorsal and ventral parts of the pulmonary venous atrium.

In both procedures, the RV continues to work against systemic load, which may eventually result in RV dysfunction and tricuspid insufficiency because the RV and the tricuspid valve are not structured for systemic pressure load.[43] Consequently, cine CMR can be used to monitor RV mass, volumes, stroke volume, and ejection fraction. Cine CMR is also useful to detect and qualitatively estimate the severity of tricuspid regurgitation.

Currently the favored procedure for TGA is the Jatene or arterial switch procedure, in which the aorta and the pulmonary artery are transected above the sinus portion and switched in order to redirect blood flow. The coronary arteries are then transplanted onto the neo-aorta. This surgery is usually done in the neonatal period. However, many older children or adults previously treated with the Senning or Mustard procedure are now candidates for

the arterial switch in order to avert or relieve RV pressure overload and or RV failure. CMR has a major role in the long-term follow-up of patients after surgical repair for TGA.[39, 44] CMR can assess various complications after atrial switch procedure, including baffle leaks, systemic or pulmonary venous obstructions, LV outflow tract obstruction, and tricuspid regurgitation.

Transverse and sagittal spin-echo images are used for assessment of great vessel anatomy after the Jatene procedure. Because this procedure is frequently performed in the neonatal period, thin 3-mm sections are preferred. Postoperatively, the position of the aorta posterior to the main pulmonary artery results occasionally in a proximal stenosis of the right and/or left pulmonary artery, which can be clearly demonstrated by axial CMR images. In addition, CMR can visualize other complications such as narrowing of the RV outflow region, dilatation of the aortic root, and supravalvular aortic stenosis.[24, 44] Compared with echocardiography, CMR has been shown to be similar for depicting stenoses of the RV outflow region. However, CMR was superior in detection of proximal pulmonary artery stenosis (41% vs. 94%).[44] Furthermore, CMR is unaffected by postsurgical changes and/or graft material or an inadequate acoustic window that may render technically deficient echocardiographic studies.

The Fontan procedure consists of an anastomosis between the right atrium or atrial appendage and the main pulmonary artery (Figure 25–10).[45, 46] Numerous variations of the Fontan procedure have now been devised. Many patients initially have a bidirectional Glenn shunt (superior vena cava to right pulmonary artery anastomosis) followed months or years later by placement of a conduit from the inferior vena cava to the right or left pulmonary artery (see Figure 25–9). Therefore, systemic venous blood is forwarded directly to the pulmonary circulation, thus bypassing the functional sin-

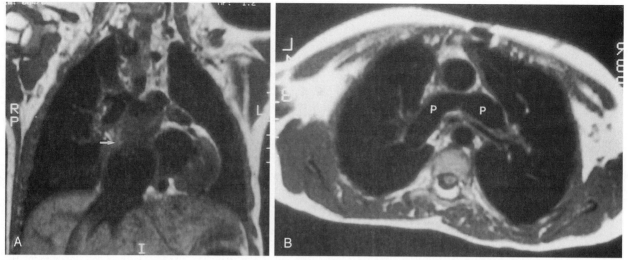

Figure 25–10. Coronal *(A)* and axial *(B)* spin-echo CMR of a patient following Fontan procedure for tricuspid atresia. The connection between the right atrial appendage and the right pulmonary artery is widely patent (arrow). The pulmonary arteries (P) are normal in size and without stenoses. (From Higgins CB: Congenital heart disease. *In* Higgins CB, Hricak H, Helms CA [eds]: Magnetic Resonance Imaging of the Body. 3rd ed. Philadelphia, Lippincott-Raven, 1996, with permission.)

gle ventricle, which is functioning as a systemic pumping chamber. The concomitant atrial septal defect is also closed. The major indications for the Fontan procedure are tricuspid atresia or severe stenosis, single ventricle, and hypoplastic left heart. CMR can be utilized to demonstrate the size of the atriopulmonary connection and to recognize the presence of obstruction.[40] Axial and coronal images are usually effective for this purpose. Besides obstruction of the conduit, complications of the Fontan operation include residual atrial septal defects and systemic venous hypertension. The former can be diagnosed by cine CMR; the latter may result in right atrial enlargement, venous stasis, pleural and peritoneal effusions, as well as edema. Severe right atrial enlargement may even compress the right pulmonary veins at the entrance to the left atrium.[24] Determination of the size of the pulmonary arteries is important in patients undergoing Fontan reconstruction because it is considered a major indicator of prognosis. CMR has been shown to be useful to determine pulmonary artery size in these patients and to be superior to echocardiography.[41]

Surgical repair of tetralogy of Fallot is achieved by closing the VSD and enlarging the pulmonary outflow tract by using patches. Axial CMR scans can demonstrate abnormalities in the RV outflow tract including residual stenosis or aneurysmal patch dilatation (see Figure 25–8). In the case of concomitant pulmonary artery atresia, surgical repair is more complex, necessitating systemic-to-pulmonary shunts in order to allow blood flow to the lungs and to promote growth of pulmonary vessels. The subclavian-to-pulmonary artery shunt (Blalock-Taussig) that was used in the past has been largely replaced by the modified Blalock shunt, representing a graft connecting the aorta or brachiocephalic artery with the pulmonary artery. An earlier report

showed the usefulness of ECG-gated CMR to assess size, course, patency of Blalock-Taussig, Glenn, and aortopulmonary shunts.[24] Coronal and axial images are particularly useful to demonstrate systemic-to-pulmonary shunts as well as potential complications including stenosis or occlusion of the shunt.

Repair of pulmonary atresia may also be accomplished by placing a valve conduit between the RV and the main pulmonary artery (Rastelli conduit). Sagittal CMR images are most effective to visualize the proximal anastomosis to the RV and the distal anastomosis to the pulmonary artery. Possible complications include false aneurysms at the anastomosis and stenosis or kinking of the conduit.[37]

Whereas spin-echo CMR is used for morphological evaluation of postoperative congenital heart disease, cine CMR and velocity-encoded cine CMR allow functional assessment of surgical baffles, conduits, and valvular function. In patients who had undergone the Mustard or Senning operation for TGA, cine CMR was capable of evaluating pulmonary and systemic venous connections, as well as RV function, tricuspid, and mitral regurgitation.[47, 48] Flow quantification with velocity-encoded cine CMR improved the evaluation of venoatrial connections after the Mustard or Senning procedure.[49] In addition, CMR velocity mapping has been used successfully to assess tricuspid volume flow in patients with the Mustard or Senning operation, which often showed abnormal tricuspid flow patterns.[50] Velocity-encoded cine CMR also provides accurate information of pulmonary flow volume and velocity after Fontan surgery. Velocity-encoded cine CMR can be used to assess the volume of retrograde flow that exists in patients after the Fontan procedure.[51] Velocity-encoded CMR has also been used to quantify the volume of pulmonary regurgitation after patch repair of tetralogy of Fallot,[52] and to estimate pressure gradients across ventriculopulmonary (Rastelli) conduits.[37]

EVALUATION OF FUNCTION IN CONGENITAL HEART DISEASE

In patients with untreated or repaired congenital heart disease, cine CMR and velocity-encoded cine CMR are attractive methods for quantifying left and right ventricular function and volumetric flow, respectively. Standard or breath-hold (segmented k-space) cine CMR provides sequential images through the cardiac cycle, which can be viewed as a cine loop. Typically, cine CMR uses short repetition times (20 to 35 ms), short echo time (4 to 20 ms), and low flip angles (35 to 60°) to acquire usually 16 or more phases (~50 ms temporal resolution) evenly spaced through the cardiac cycle.[53] Cine CMR images display high signal intensity in areas of normal blood flow. However, turbulent flow, which may occur in stenosis, regurgitation, or shunt lesions, causes a signal loss within the blood pool, therefore rendering these lesions readily visible on cine CMR. The size of the signal void, however, is strongly dependent on the echo time.[54] In addition, LV and RV mass and function can be measured precisely using a 3-D cine CMR (see Figure 25–4). Unlike echocardiography or cineangiography, CMR does not rely on geometric assumptions of calculation based on partial sampling of the cardiac volume. This 3-D data set from end-diastolic and end-systolic tomograms encompassing both ventricles allows measurement of RV and LV mass, volumes, stroke volume, and ejection fraction.

In contrast to typical adult cardiologic studies that focus on the LV, the RV is often of particular interest in adults with congenital heart disease. Either transverse or double-oblique short-axis tomograms are used to quantify ventricular volumes and global function. End-diastolic and end-systolic measurements acquired through the whole stack (Simpson's rule[55]) of images provide end-diastolic volume (EDV) and end-systolic volume (ESV), stroke volume (SV = EDV − ESV) and ejection fraction (EF = SV/EDV) for both the RV and LV. In normal individuals, LV and RV stroke volume are near equal (±10%).[56] Therefore, differences in ventricular stroke volume can be used to quantify valvular regurgitation and shunt lesions. For example, in isolated pulmonary or tricuspid regurgitation, the difference between RV stroke volume and LV stroke volume corresponds to the regurgitant volume, whereas LV stroke volume is greater than the RV stroke volume in patients with patent ductus arteriosus and patients with aortic or mitral regurgitation. Measurements of volume and function of the ventricles have been shown to be highly reproducible on sequential studies in patients with morphologically abnormal ventricles.[57] Thus, cine CMR seems to be highly attractive for serial assessment of ventricular volumes and function.

Velocity-encoded cine CMR provides direct measurement of aortic and pulmonary artery flow and therefore measures the effective stroke volume of both ventricles. In the absence of valvular regurgitation, this method can be used to determine the volumes of shunts. For example, with atrial septal defects and partial anomalous pulmonary venous connection, the difference between RV stroke volume and LV stroke volume is equal to the left-to-right shunt volume. Likewise, velocity-encoded CMR can be used to quantify the shunt volume in patent ductus arteriosus by calculating the difference between LV stroke volume as measured in the ascending aorta and RV stroke volume as measured in the main pulmonary artery. Velocity-encoded cine CMR has been successfully used in patients with various congenital shunt lesions.[58] The accuracy and reproducibility of this method for measuring pulmonary blood flow/systemic blood flow (Qp/Qs) in left-to-right shunts has been previously assessed.[59, 60]

Velocity-encoded CMR can measure blood flow in the main pulmonary artery as well as separately in the right and left pulmonary artery, being one of the only techniques with the capability to quantify right and left pulmonary flow separately.[61] This method can be used to determine the percentage of blood flow directed to each lung and thereby assess the hemodynamic significance of stenoses of either pulmonary artery.

Tricuspid regurgitation and shunts may be detected and quantified during the same examination.[62] One study has also demonstrated the feasibility and the accuracy of velocity-encoded cine CMR for measurement of tricuspid flow in healthy subjects and in patients after the Mustard or Senning procedure. In the latter group, abnormal diastolic tricuspid flow patterns were found, which may be of importance because RV diastolic filling abnormalities may precede systolic dysfunction.[63]

Velocity-encoded cine CMR can also be used to determine the severity of aortic coarctation by quantifying collateral flow in the descending aorta. Measurements of volume flow are performed in the proximal descending aorta just below the site of the coarctation and in the distal descending aorta near the diaphragm. In normal volunteers, the flow volume decreases steadily from the proximal to the more distal parts of the descending aorta due to antegrade flow into the intercostal arteries. In patients with hemodynamically significant coarctation of the aorta, the normal flow pattern in branches of the descending aorta is reversed, resulting in an increase in flow from the proximal to the distal descending aorta due to retrograde collateral flow mainly through intercostal arteries. Velocity-encoded CMR can also be used for identification of patients with a mismatch between the severity of anatomical obstruction and collateral flow, which may be of importance for planning surgery.[64] Velocity-encoded cine CMR has been successfully used to demonstrate abnormalities in the volume flow in the descending aorta following coarctation repair. These abnormalities are probably related to resistance to flow imposed by the coarctation segment and may represent an additional index for monitoring the hemodynamic significance of coarctation before and after intervention.[65] In addition, a 1996 study has demonstrated that CMR including velocity-encoded cine CMR allows a comprehensive evaluation of aortic coarctation by determining location and severity of stenosis as well as pressure gradients (with application of modified Bernoulli formula) across the coarctation segment.[66]

CONCLUSION

Due to the dramatic progress in palliative and corrective surgery over the past decades, an increasing number of patients with complex congenital heart disease survive to adulthood. Other patients with a milder form of congenital heart disease may become symptomatic only as adults. Patients with congenital heart disease often require serial follow-up studies for evaluation of morphology and function. CMR is a noninvasive method providing excellent anatomical detail of cardiac structures and the great vessels, especially when echocardiographic access to the area of interest is limited. Cine CMR allows for assessment of valvular and shunt lesions and for the measurement of ventricular volumes without having to rely on geometric assumptions. Velocity-encoded CMR can be used to measure flow velocity and volumes in the heart and the great arteries. CMR is particularly useful for postoperative evaluation in patients with repaired congenital heart disease, especially after placement of intraatrial baffles or extracardiac conduits. Furthermore, CMR can assess RV function, which is of particular interest to many patients with congenital heart disease and which may be difficult for other imaging modalities. Thus, CMR is a noninvasive imaging technique allowing the comprehensive assessment of cardiovascular anatomy and function in patients with complex congenital heart disease.

References

1. Higgins CB, Caputo GR: Role of MR imaging in acquired and congenital cardiovascular disease. AJR Am J Roentgenol 161:13, 1993.
2. Gomes AS, Lois JF, Williams RG: Pulmonary arteries: MR imaging in patients with congenital obstruction of the right ventricular outflow tract. Radiology 174:51, 1990.
3. Masui T, Seelos KC, Kersting-Sommerhoff BA, Higgins CB: Abnormalities of the pulmonary veins: Evaluation with MR imaging and comparison with cardiac angiography and echocardiography. Radiology 181:645, 1991.
4. Gomes AS, Lois JF, George B, et al: Congenital abnormalities of the aortic arch: MR imaging. Radiology 165:691, 1987.
5. Kersting-Sommerhoff BA, Diethelm L, Teitel DF, et al: Magnetic resonance imaging of congenital heart disease: Sensitivity and specificity using receiver operating characteristic curve analysis. Am Heart J 118:155, 1989.
6. Shinebourne EA, Macartney FJ, Anderson RH: Sequential chamber localization: Logical approach to diagnosis in congenital heart disease. Br Heart J 41:327, 1976.
7. Van Praagh R: The importance of segmental situs in the diagnosis of congenital heart disease. Semin Roentgenol 20:254, 1985.
8. Didier D, Higgins CB, Fisher M, et al: Congenital heart disease: Gated MR imaging in 72 patients. Radiology 158:227, 1986.
9. Anderson RH, Ho SY: The tomographic anatomy of the normal and malformed heart. In Higgins CB, Silvermann NH, Kersting-Sommerhoff BA, Schmidt K (eds): Congenital Heart Disease: Echocardiography and Magnetic Resonance Imaging. New York, Raven, 1990, p 1.
10. Guit GL, Bluemm R, Rohmer J, et al: Levotransposition of the aorta: Identification of segmental cardiac anatomy using MR imaging. Radiology 161:376, 1986.
11. Geva T, Wesley Vick G III, Wendt RE, Rockey R: Role of spin echo and cine magnetic resonance imaging in presurgical planning of heterotaxy syndrome. Circulation 90:348, 1994.
12. Fletcher BD, Jacobstein MD, Abramowsky CR, et al: Right atrioventricular valve atresia: Anatomic evaluation with MR imaging. AJR Am J Roentgenol 148:671, 1987.
13. Higgins CB, Byrd BF III, Farmer DW, et al: Magnetic resonance imaging in patients with congenital heart disease. Circulation 70:851, 1984.
14. Mayo JR, Roberson D, Sommerhoff B, et al: MRI of double outlet right ventricle. J Comput Assist Tomogr 14:336, 1990.
15. Kersting-Sommerhoff BA, Sechtem UP, Higgins CB: Evaluation of pulmonary artery supply by nuclear magnetic resonance imaging in patients with pulmonary atresia. J Am Coll Cardiol 11:166, 1989.
16. Lev M, Bharati S, Meng CCL, et al: A concept of double-outlet right ventricle. J Thorac Cardiovasc Surg 64:271, 1972.
17. Wilcox BR, Ho SY, Macartney FJ, et al: Surgical anatomy of double-outlet right ventricle with situs solitus and atrioventricular concordance. J Thorac Cardiovasc Surg 82:405, 1981.
18. Mayo JR, Roberson D, Sommerhoff B, Higgins CB: MR imaging of double outlet right ventricle. J Comput Assist Tomogr 14:336, 1990.
19. Patrick DL, McGoon DC: An operation for double outlet right ventricle with transposition of the great arteries. J Cardiovasc Surg 9:537, 1968.
20. Sridaromont S, Feldt RH, Ritter DG, et al: Double outlet right ventricle: Hemodynamic and anatomic correlations. Am J Cardiol 38:85, 1976.
21. Higgins CB: Congenital heart disease. In Higgins CB, Hricak H, Helms CA (eds): Magnetic Resonance Imaging of the Body. 3rd ed. Philadelphia, Lippincott-Raven, 1996, p 461.
22. Smith WL Jr, Stanford W, Skorton DJ, Wolf GL: Assessment of congenital heart disease by nuclear magnetic resonance imaging. In Skorton DJ (ed): Marcus Cardiac Imaging: A Companion to Braunwald's Heart Disease. 2nd ed. Vol 2. Philadelphia, WB Saunders, 1996, p 886.
23. Donnelly LF, Higgins CB: MR imaging of conotruncal abnormalities. AJR Am J Roentgenol 166:925, 1996.
24. Kersting-Sommerhoff BA, Seelos KC, Hardy C, et al: Evaluation of surgical procedures for cyanotic congenital heart disease by using MR imaging. AJR Am J Roentgenol 155:259, 1990.
25. Berry BE, McGoon DC, Ritter DG, Davis GD: Absence of anatomic origin from heart of pulmonary arterial supply. Clinical application of classification. J Thorac Cardiovasc Surg 68:119, 1974.
26. Niezen RA, Helbing WA, van der Wall EE, et al: Biventricular systolic function and mass studied with MR imaging in children with pulmonary regurgitation after repair for tetralogy of Fallot. Radiology 201:135, 1996.
27. Becker AE, Becker MJ, Edwards JE: Pathologic spectrum of dysplasia of the tricuspid valve. Features in common with Ebstein's malformation. Arch Pathol 91:167, 1971.
28. Anderson KR, Lie JT: Pathologic anatomy of Ebstein's anomaly of the heart revisited. Am J Cardiol 41:739, 1978.
29. Link KM, Herrera MA, D'Souza VJ, Formanek AG: MR imaging of Ebstein anomaly: Results in four cases. AJR Am J Roentgenol 150:363, 1988.

30. Markiewicz W, Sechtem U, Higgins CB: Evaluation of the right ventricle by magnetic resonance imaging. Am Heart J 113:8, 1987.
31. Choi YH, Park JH, Choe YH, Yoo SJ: MR imaging of Ebstein's anomaly of the tricuspid valve. AJR Am J Roentgenol 163:539, 1994.
32. Lev M, Liberthson RR, Joseph RH, et al: The pathologic anatomy of Ebstein's disease. Arch Pathol 90:334, 1970.
33. Carpentier A, Chauvaud S, Mace L, et al: A new reconstructive operation for Ebstein's anomaly of the tricuspid valve. J Thorac Cardiovasc Surg 96:92, 1988.
34. Kersting-Sommerhoff BA, Diethelm L, Stanger P, et al: Evaluation of complex ventricular anomalies with magnetic resonance imaging. Am Heart J 120:133, 1990.
35. Higgins CB, Byrd BFd, Farmer DW, et al: Magnetic resonance imaging in patients with congenital heart disease. Circulation 70:851, 1984.
36. Higgins CB, Byrd BFd, McNamara MT, et al: Magnetic resonance imaging of the heart: A review of the experience in 172 subjects. Radiology 155:671, 1985.
37. Martinez JE, Mohiaddin RH, Kilner PJ, et al: Obstruction of extracardiac ventriculopulmonary conduits: Value of nuclear magnetic resonance imaging with velocity mapping and Doppler echocardiography. J Am Coll Cardiol 20:338, 1992.
38. Hirsch R, Kilner PJ, Connelly MS, et al: Diagnosis in adolescents and adults with congenital heart disease: Prospective assessment of individual and combined roles of MRI and transesophageal echocardiography. Circulation 90:2937, 1994.
39. Soulen RL, Donner RM, Capitanio M: Postoperative evaluation of complex congenital heart disease by magnetic resonance imaging. Radiographics 7:975, 1987.
40. Sampson C, Martinez J, Rees S, et al: Evaluation of Fontan's operation by magnetic resonance imaging. Am J Cardiol 65:819, 1990.
41. Fogel MA, Donofrio MT, Ramaciotti C, et al: Magnetic resonance and echocardiographic imaging of pulmonary artery size throughout stages of Fontan reconstruction. Circulation 90:2927, 1994.
42. Duerinckx AJ, Wexler L, Banerjee A, et al: Postoperative evaluation of pulmonary arteries in congenital heart surgery by magnetic resonance imaging: Comparison with echocardiography. Am Heart J 128:1139, 1994.
43. Mee RB: Severe right ventricular failure after Mustard or Senning operation. Two stage repair: Pulmonary artery banding and switch. J Thorac Cardiovasc Surg 92:385, 1986.
44. Blankenberg F, Rhee J, Hardy C, et al: MRI vs echocardiography in the evaluation of the Jatene procedure. J Comput Assist Tomogr 18:749, 1994.
45. Fontan F, Baudet E: Surgical repair of tricuspid atresia. Thorax 26:240, 1971.
46. Fontan F, Deville C, Quaegebeur J, et al: Repair of tricuspid atresia in 100 patients. J Thorac Cardiovasc Surg 85:647, 1983.
47. Rees S, Sommerville J, Warnes C, et al: Comparison of magnetic resonance imaging with echocardiography and radionuclide angiography in assessing cardiac function and anatomy following Mustard's operation for transposition of the great arteries. Am J Cardiol 61:1316, 1988.
48. Chung KJ, Simpson IA, Glass RF, et al: Cine magnetic resonance imaging after surgical repair in patients with transposition of the great arteries. Circulation 77:104, 1988.
49. Sampson C, Kilner PJ, Hirsch R, et al: Venoatrial pathways after the Mustard operation for transposition of the great

arteries: Anatomic and functional MR imaging. Radiology 193:211, 1994.
50. Rebergen SA, Helbing WA, van der Wall EE, et al: MR velocity mapping of tricuspid flow in healthy children and in patients who have undergone Mustard or Senning repair. Radiology 194:505, 1995.
51. Rebergen SA, Ottenkamp J, Doornbos J, et al: Postoperative pulmonary flow dynamics after Fontan surgery assessment with nuclear magnetic resonance velocity mapping. J Am Coll Cardiol 21:123, 1993.
52. Rebergen SA, Chin JGJ, Ottenkamp J, et al: Pulmonary regurgitation in the late postoperative follow-up of tetralogy of Fallot; volumetric quantitation by nuclear magnetic resonance velocity mapping. Circulation 88:2257, 1993.
53. Sechtem U, Pflugfelder PW, White RD, et al: Cine MR imaging: Potential for the evaluation of cardiovascular function. AJR Am J Roentgenol 148:239, 1987.
54. Ross MJ, Botnar RM, Kissinger KV, et al: Evaluation of aortic insufficiency by breath-hold magnetic resonance imaging: Impact of echo time on visually apparent signal void areas. J Cardiovasc Magn Reson 1:343, 2000.
55. Chuang ML, Hibberd MG, Beaudin RA, et al: Importance of imaging method over imaging modality in noninvasive determination of left ventricular volumes and ejection fraction: Assessment by two and three-dimensional echocardiography and MRI. J Am Coll Cardiol 35:477, 2000.
56. Sechtem U, Pflugfelder PW, Gould RG, et al: Measurement of right and left ventricular volumes in healthy individuals with cine MR imaging. Radiology 163:697, 1987.
57. Semelka RC, Tomei E, Wagner S, et al: Interstudy reproducibility of dimensional and functional measurements between cine magnetic resonance studies in the morphologically abnormal left ventricle. Am Heart J 119:1376, 1990.
58. Rees S, Firmin D, Mohiaddin R, et al: Application of flow measurements by magnetic resonance velocity mapping to congenital heart disease. Am J Cardiol 64:953, 1989.
59. Brenner LD, Caputo GR, Mostbeck GH, et al: Quantification of left-to-right atrial shunts with velocity-encoded cine nuclear magnetic resonance imaging. J Am Coll Cardiol 20:1246, 1992.
60. Sieverding L, Jung WI, Klose U, Apirz J: Noninvasive blood flow measurement and quantification of shunt volume by cine magnetic resonance in congenital heart disease. Pediatr Radiol 22:48, 1992.
61. Caputo GR, Kondo C, Masui T, et al: Determination of right and left lung perfusion with oblique angle velocity-encoded cine MR imaging: In vitro and in vivo validation. Radiology 180:693, 1991.
62. Theissen P, Kaemmerer H, Sechtem U, et al: Magnetic resonance imaging of cardiac function and morphology in patients with transposition of the great arteries following Mustard procedure. Thorac Cardiovasc Surg 39(suppl):221, 1991.
63. Nishimura RA, Housmans PR, Hatle LK, Tajik AJ: Assessment of diastolic function of the heart: Background and current applications of Doppler echocardiography. I. Physiologic and pathophysiologic features. Mayo Clin Proc 64:71, 1989.
64. Steffens JC, Bourne MW, Sakuma H: Quantification of collateral blood flow in coarctation of the aorta by velocity encoded cine magnetic resonance imaging. Circulation 90:937, 1994.
65. Mohiaddin RH, Kilner PJ, Rees S, Longmore DB: Magnetic resonance volume flow and jet velocity mapping in aortic coarctation. J Am Coll Cardiol 22:1515, 1993.
66. Oshinski JN, Parks WJ, Markou CP, et al: Improved measurement of pressure gradients in aortic coarctation by magnetic resonance imaging. J Am Coll Cardiol 28:1818, 1996.

CHAPTER 26

Thoracic Aortic Disease

Christoph A. Nienaber and Malgorzata Knap

The anatomical and functional characteristics of the aorta, which may at first glance appear relatively straightforward, are now recognized to be complex. Recent insights from modern imaging technology and from a better understanding of the hydraulic principles associated with the variety of diseases affecting the aorta have helped the medical community to realize the multiple facets of in vivo aortic pathology as well as its varied clinical presentation.

Diagnostic modalities such as transesophageal echocardiography (TEE), cardiovascular magnetic resonance (CMR), and computed tomography (CT), including spiral CT and electron beam CT (EBT), have all been shown to be useful to interrogate the aorta, both in chronic disease and in acute aortic syndromes. Although x-ray contrast angiography is still considered a gold standard in acute and chronic aortic syndromes, in most centers it has been relegated to a secondary role, after the emergence of the noninvasive techniques, most importantly CMR, with their high sensitivity, specificity, and practical advantages.[1–7] However, none of the noninvasive diagnostic modalities listed above is ideal for all patients, and for a given individual, knowledge of both accuracy and limitations in the presenting clinical scenario is required.[6–11] Although the information content of CMR may greatly overlap with established methods such as echocardiography, CT, or angiography, CMR is more accurate and comprehensive. Although the cost effectiveness of CMR has not been proven in all areas,[12] CMR is the preferred modality in selected areas of disease of the aorta, such as aneurysm, dissection and its precursors, congenital and inherited heart diseases, and, in particular, for postoperative follow-up of aortic repair and cardiac malformations.[2–4] This chapter focuses on the possibilities and emerging advantages of CMR with respect to a spectrum of aortic pathologies.

TECHNICAL ASPECTS OF CARDIOVASCULAR MAGNETIC RESONANCE

CMR provides noninvasive information on anatomy, function, and blood flow with no exposure to ionizing radiation and often no need for contrast material. Simple spin-echo techniques can demonstrate the anatomy of both the proximal and distal anastomoses of aortic grafts and serve as a measure of quality control after corrective surgery, including complex operations such as aortic reconstruction and the Jatene or Norwood procedures.[13–16] The motion of blood within the lumen produces a high contrast between the blood and surrounding stationary tissue, on spin-echo images, although slowly moving blood will exhibit an increased intraluminal signal.[19, 21] A dissection is best seen when there is rapid blood flow in both the true and false lumens so that the intimal flap appears as a linear structure between the signal void of the two lumens (Figure 26–1). If the flow in the false lumen is slow, the flap will be outlined by a signal void on one side and

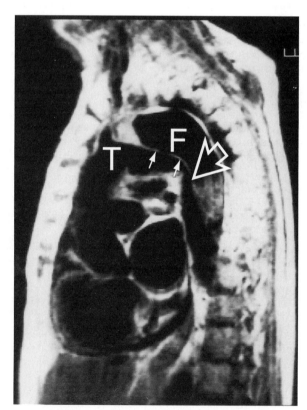

Figure 26–1. Spin-echo CMR image in sagittal orientation showing the typical feature of a type B dissection. The dissecting flap (solid arrows) separates the true (T) from false (F) lumen and an entry is identified (open arrow). Besides the clear anatomical picture of the aorta and the great arch vessels, the adjacent structures in the thorax are clearly defined.

an increased intraluminal signal on the other.[22, 23] Slow flow in the false lumen can sometimes resemble thrombus, but in such cases, velocity mapping or cine gradient-echo imaging may be used to clearly distinguish slow flow from thrombus formation.[24] When rates of blood flow in the true and false lumens differ only slightly and exhibit similar signal intensity, it may be difficult to distinguish the true from the false lumen. It should be remembered, however, that blood flow does not distinguish true from false lumen, but rather the presence of an endothelial lining of the true lumen and continuity with the aortic valve orifice.

CMR produces high-quality images in the transverse, coronal, sagittal, and oblique planes of the aorta that facilitate the diagnosis of a dissection, provide better definition of its location and extent, and may reveal involvement of the arch vessels. CMR is also well suited for the evaluation of patients with preexisting complex aortic disease, such as aortic aneurysm or prior aortic graft repair, because it can accurately distinguish dissection from other aortic processes.[25] Moreover, at the level of the aortic root, the ostia of the coronary arteries and involvement of the dissection can be visualized, which may be useful for surgical planning.

Magnetic resonance angiography (MRA) is also frequently used in the assessment of the aorta (see Chapter 29). The two principles used to visualize flowing blood are *time-of-flight* and *phase-contrast* MRA. Time-of-flight MRA applies a radiofrequency pulse to the tissue volume being sampled, then detects the unexcited, fresh protons of blood that enter the imaging field with flow. Phase-contrast MRA depends on phase differences in the immediate chemical environment of mobile (flowing) protons compared with stationary (nonflowing) protons. Both techniques have been used to generate clinically useful diagnostic images of the normal and diseased aorta without exogenous contrast. By acquiring the proton signals within a volume of tissue, and using a computer to project the multiple sections into one image, one can produce an angiogram-like image of all vessels containing flowing blood (maximum intensity projection).

In addition, contrast enhancement may be used to augment the MRA and is commonly used. During the arterial first pass of a gadolinium diethylenetriamine pentaacetic acid (Gd-DTPA) bolus through the anatomical structures under examination, rapid three-dimensional (3-D) MRA can be performed with high resolution.[26] Various image processing methods such as multiplanar reformatting, maximum intensity projections, and shaded surface displays can be used to visualize the vessels of interest (Figure 26–2). Rapid breath-hold data acquisition has become possible with the use of high-performance gradient systems, utilizing short repetition (TR) and echo (TE) times. Triggering techniques are available to synchronize the scan to the bolus transit time.[27] The advantage of cross-sectional imaging over conventional contrast aortography is obvious

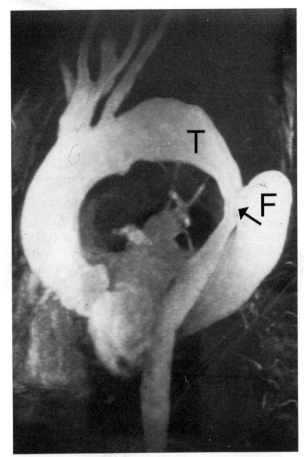

Figure 26–2. Contrast-enhanced magnetic resonance angiogram (ceMRA) after injection of Gd-DTPA in sagittal orientation. The image depicts a subacute type B dissection with the MRA clearly identifying both two lumina and the exact site of the communication (arrow) between the true (T) and the false (F) lumen by a typical jet pattern.

in the evaluation of thoracic and abdominal aortic aneurysms. The normal cross-sectional area of the thoracic aorta varies with subject age (Table 26–1). Aneurysms are easily displayed with contrast-en-

Table 26–1. Age-Related Changes in the Cross-Sectional Areas (Mean ± SD) of Ascending Aorta (AA), Aortic Arch (AR), and Descending Thoracic Aorta (DA)

Age (years)	AA (cm²/m²)	AR (cm²/m²)	DA (cm²/m²)
10–19	2.13 ± 0.35	1.65 ± 0.24	1.09 ± 0.21
20–29	2.33 ± 0.53	1.62 ± 0.19	1.23 ± 0.21
30–39	2.64 ± 0.48	2.17 ± 0.33	1.57 ± 0.20
40–49	3.27 ± 0.60	1.96 ± 0.37	1.89 ± 0.22
50–59	3.64 ± 0.40	2.26 ± 0.35	1.88 ± 0.19
>60	4.82 ± 1.56	2.91 ± 0.90	2.82 ± 0.90

The measurements were normalized for body surface area.
From Mohiaddin RH, Kilner PJ, Pennell DJ: Aortic diseases. *In* Pohost GM, O'Rourke RA, Shah PM, Berman DS (eds): Imaging in Cardiovascular Disease. Philadelphia, Lippincott Williams and Wilkins, 2000, p 821.

Table 26–2. Diagnostic Evidence of Aortic Dissection

Direct Signs
　Visualization of double lumen
　Visualization of intimal flap
Indirect Signs
　Compression of true lumen by false lumen
　Thickening of aortic wall
　Aortic insufficiency
　Branch vessel abnormality
　Ulcer-like projection from aortic wall
　Pericardial effusion
　Involvement of coronary artery ostia

hanced 3-D MRA (ceMRA), allowing not only the measurement of the dimension of the aneurysm, but also the presence of mural thrombus, intramural hematoma, and the involvement of adjacent side branches. Planning for stent interventions is easily performed with this information. Serial post–contrast-enhanced spin-echo images are especially helpful to interrogate the composition of the aortic wall.[28] Although intramural hematoma has no luminal component, the full picture of aortic dissection is easily recognized with MRA. Beyond identifying just the presence of aortic dissection, it is also prognostically important to identify its location and extent (Table 26–2). The site of an intimal perforation

and the communication between the true and the false lumen of a dissection may be identified by jet formation after ceMRA (see Figure 26–2). ceMRA is also useful for developmental abnormalities and postsurgical corrections. Arch deformities such as double aortic arch formations, aberrant left subclavian artery syndrome, right-sided or left-sided aortic arch, coarctation, and pseudocoarctation are easily visualized (Figure 26–3). In aortic coarctation, 3-D MRA may not just delineate the exact location and dimension of the stenosis but also provide functional information based on flow-sensitive sequences.

However, CMR has some disadvantages: CMR is not available in all hospitals; the need to move the patient to the scanner may cause logistical problems; patient access is limited; and the examination time may be longer than with TEE (especially in acute cases). These disadvantages are unlikely to result in increased individual risk, but may hinder utilization at some institutions (Table 26–3). With appropriate precautions, even unstable or ventilated patients may be safely examined with careful monitoring of oxygen saturation, blood pressure, and electrocardiogram (ECG) with close voice communication.[6] Technical improvement such as ultrafast new generation scanners with real-time imaging or open CMR scanners will improve patient access and

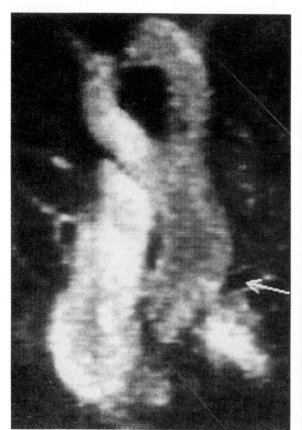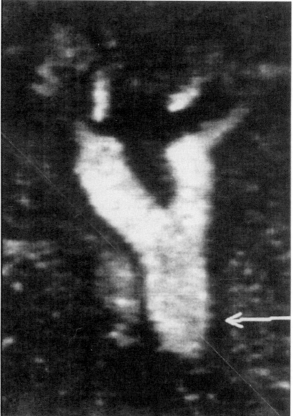

Figure 26–3. MRA after injection of Gd-DTPA bolus in breath-hold gradient-echo technique, revealing a double aortic arch after axial 3D reformatting of the images. The ascending aorta (left) ends in a double arch that eventually unifies toward the descending aorta (right).

Table 26–3. Cardiovascular Magnetic Resonance in Suspected Aortic Dissection

Advantages
Noninvasive
No iodinated contrast material
No radiation exposure
High-quality images in any projection
Useful in patients with preexisting complicated aortic disease
High reproducibility
Magnetic resonance angiography (MRA) with high resolution
Assessment of aortic insufficiency
Identification and characterization of pericardial effusion
Assessment of branch vessels and coronary ostia

Disadvantages
Patient transport to scanner
Limited monitoring during scan session
Additional time for flow mapping and ceMRA

reduce imaging time. An important safety advantage of MRA is that gadolinium agents are safer than iodinated x-ray contrast agents, because they are not nephrotoxic and have a much lower incidence of allergic reactions.[28] As a rule, MRA is also considered ideal for young patients in whom repeated radiation exposure is an issue, in patients with compromised renal function, in any patient with an aortic syndrome in a stable clinical situation, or in patients with significant risk factors for an acute aortic syndrome. Having recognized the aforementioned issues, however, in patients with suspected aortic rupture, hemodynamic instability, or continuous pain syndrome, where speed is a critical factor in patient management, clinicians have logistical reasons for using the fastest available technique, which is often TEE, performed at the bedside or in the emergency department, or sometimes contrast-enhanced CT. However, of all mentioned technologies considered to be useful to interrogate the aorta, CMR with the addition of ceMRA has undoubtedly the highest potential to emerge as a new gold standard.[29, 30]

DISSECTION OF THE THORACIC AORTA

With an estimated mortality rate of 1 to 2 percent per hour in the first 24 to 48 hours after onset, aortic dissection is a medical emergency.[31, 32] Dissections can occur throughout the length of the aorta, and two described classification systems exist. The De Bakey nomenclature is based on the anatomical site of the intimal tear and the extent of the resulting dissection.[33] In a type I dissection, the intimal tear originates in the ascending aorta and the dissecting hematoma extends past the origin of left subclavian artery. Type II dissections are confined to the ascending aorta. Type III dissections begin after the origin of the left subclavian artery and extend distally. The Stanford classification is conceptually founded on prognostic grounds. Type A dissections involve the ascending aorta, regardless of the intimal tear, and type B dissections spare the ascending aorta and often imply a better prognosis (Figure 26–4).[34] In general, acute dissections involving the ascending aorta require emergent surgical intervention, while descending aortic dissections may often be successfully managed with medical therapy alone.[32, 35] Thus, rapid detection and accurate diagnosis of relevant anatomical detail is critical for successful management.

The conventional gold standard until recently for diagnosis of aortic dissection had been x-ray aortography, which has a reported sensitivity of 80 to 90 percent, specificity of 90 to 100 percent, and positive predictive value of approximately 95 percent.[5–7, 35] Unfortunately, x-ray aortography is an invasive procedure that carries a small but real risk of complications. In addition, it requires a fully equipped catheterization laboratory and may be delayed by patient transportation and laboratory preparation. In the past several years, less invasive diagnostic techniques such as CT, echocardiography (including TEE), and CMR have been studied and

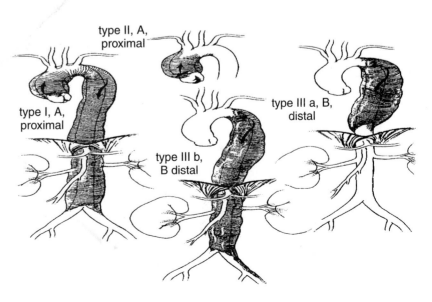

Figure 26–4. Schematic demonstration of the various types of classification of aortic dissection. Whereas the De Bakey classification (I, II, IIIa, b) focuses on the anatomical extent, the Stanford classification (A, B) highlights the involvement of the ascending aorta and the prognostic aspects of dissection.

type II, A, proximal

type I, A, proximal

type III b, B distal

type III a, B, distal

compared.[7, 8] CT requires intravascular iodinated contrast and provides little information on coronary anatomy, great vessels, and thrombus formation in the false lumen, and no information on aortic insufficiency but it is accurate for the diagnosis of dissection, with sensitivity ranging from 79 to 93 percent and specificity from 86 to 100 percent (Table 26–4).[6–10]

Cardiovascular Magnetic Resonance

Both CMR and ceMRA are emerging as the most accurate methods for detecting aortic dissection in hemodynamically stable patients with suspicion of an acute aortic syndrome. Unlike x-ray angiography

and CT, CMR yields images in multiple planes, a feature that greatly enhances its diagnostic capabilities; moreover, no iodinated contrast media are needed, and cine gradient-echo CMR permits detection and quantification of aortic regurgitation, identification of luminal tears by a characteristic flow signal, and evidence of thrombus formation.[6, 35, 36]

In our experience, ceMRA after bolus injection of Gd-DTPA has proven extremely useful. Based on the administration of Gd-DTPA and bolus tracking during data acquisition, the technique exploits the contrast-induced T1-shortening effects and avoids saturation problems with slow flow or turbulence-induced signal voids. With the use of ultrafast gradients, acquisition may be performed at breath-hold. A gradient-echo 3-D sequence is used with and without intravenous contrast to create

Table 26–4. Classification and Epiphenomena of Thoracic Aortic Dissection

Finding	Sensitivity	Specificity	Accuracy	Positive Predictive Value	Negative Predictive Value
Type A Dissection					
TE	78.1	86.7	84.1	71.4	90.3
TEE	96.4‡	85.7	90.0	81.8	97.3
CT	82.6	100*	94.9	100*	93.3
CMR	100‡	98.6§	99.0‡	96.8§	100
Type B Dissection					
TTE	10.0	100¶	80.4	100¶	80.0
TEE	100†	96.4	97.1	88.2	100‡
CT	96.0†	88.9	97.1	80.0	98.0‡
CMR	96.5†	100¶	99.9‡	100¶	98.7‡
Site of Entry					
TTE	26.23	100	71.0	100	67.7
TEE	72.7†	100	86.8	100	79.5
CT	—	—	—	—	—
CMR	88.0†	100	95.2‡	100	92.6‡
Thrombus Formation					
TTE	11.8	100	72.7	100	71.7
TEE	68.4†	100	91.3‡	100	89.3‡
CT	92.0†	95.6	94.4‡	92.0	95.6‡
CMR	98.2†	98.5	95.2‡	97.0	94.4‡
Aortic Regurgitation					
TTE	96.9	94.7	95.4	88.6	98.6
TEE	100	95.3	97.1	92.9	100
CT	—	—	—	—	—
CMR	83.2	100	96.6	100	96.8
Pericardial Effusion					
TTE	75.0	100	98.2	100	98.1
TEE	100	100	100	100	100
CT	100	100	100	100	100
CMR	100	100	100	100	100
Intramural Hematoma					
TTE	15	45	35	16	47
TEE	80	90	86	92	86
CT	84	90	87	92	88
CMR	95	90	93	81	96

*$p < 0.05$
†$p < 0.01$
‡$p < 0.05$ (vs TTE, TEE)
§$p < 0.01$ (vs TTE, TEE)
TTE = transthoracic echocardiography; TEE = transesophageal echocardiography; CT = computed tomography; CMR = cardiovascular magnetic resonance.

maximum intensity projections; echo and repetition time are typically 1.9 ms and 4.0 ms, respectively, but these parameters may be shorter on the most modern scanners. With an in-plane resolution of 1.1 × 1.6 mm and a slice thickness of 2 to 4 mm, imaging of 64 interpolated contiguous slices using 1/2 k-space data acquisition in phase encoding direction takes 20 to 28 seconds on a 1.5 T magnet with a fast gradient system and a body array coil. ECG gating is often not used, but may have advantages for patients who can sustain a longer breath-hold.[36a] The ultrafast scanners can complete the imaging in shorter time frames still. Both routine CMR and ceMRA are extremely useful for rapid detection of both type A and type B aortic dissection. The procedure requires no arterial access and can take as little as 15 minutes to perform irrespective of the location of dissection. Longer examination times, when including gradient echo or velocity mapping, may be necessary for proximal dissections and complicated cases. As part of a complete CMR examination, regional left ventricular function, pericardial effusion, and aortic regurgitation can also be assessed. CMR can image intramural hemorrhage, luminal tears, stagnant blood flow versus thrombosis, and periaortic blood (see Table 26–4). Moreover, CMR appears as an excellent method for long-term patient follow-up after medical or surgical treatment for ascending or descending dissections. The costs of CMR are in the range of CT, but more expensive than TEE, and an analysis of cost effectiveness is certainly required.

Although both the sensitivity and specificity of CMR approach 100 percent, it does have several important limitations, especially in the very acute clinical scenario. First, CMR requires patient transportation to the magnet and immobility for image acquisition. However, with continuous ECG and pressure monitoring, voice communication, and with a cardiologist present, no patient is in fact isolated and no side effects or substantial risks have been encountered.[6, 7] In general, CMR can be considered as safe as the other noninvasive methods and certainly associated with less stress, blood pressure fluctuations, and less risk of hypotension and nephrotoxicity from contrast material. Second, metallic objects (pacemakers, cardioverters) cannot be used near the CMR scanner. Third, CMR is considered more accurate and comprehensive, and less prone to artifact, but lacks the portability of TEE and the widespread availability of CT. Finally, coronary artery involvement, which occurs in 10 to 20 percent of proximal dissections, is not yet completely definable by routine CMR. Some surgeons still require coronary angiography before proximal repair of a dissected aorta; there is, however, mounting evidence that this time-consuming and potentially dangerous procedure is not justified in acute proximal dissection considering the urgency of repair and the possibility of visual inspection of the coronary ostia by the surgeon.[6, 7] The technology of MRA, however, is advancing rapidly, and coronary arterial imaging may soon become available on a routine basis.[37, 38]

Choice of Imaging Modality

Each of the current noninvasive imaging modalities has advantages and limitations in the evaluation of suspected aortic dissection. In selecting the study of choice, one must begin by considering what diagnostic information needs to be obtained. First and foremost the study must confirm or refute the diagnosis of dissection. Second, the study must determine whether or not the dissection involves the ascending aorta (type A) or is confined to the descending aorta (type B). Third, a number of anatomical features of the dissection should be identified, including its extent, the sites of entry and reentry, the presence of thrombus in the false lumen, the extent of branch vessel involvement, the presence and severity of aortic insufficiency, the presence of a pericardial effusion, and the presence of ostial coronary artery involvement (see Tables 26–2 and 26–4). It is also important to consider the accuracy of the diagnostic information obtained, inasmuch as a false-negative diagnosis may result in avoidable death while a false-positive diagnosis might lead to unnecessary surgery.

According to the present information, CMR and multiplane TEE are the most sensitive modalities, with both performing better than x-ray aortography. The sensitivities of aortography, CT, and CMR are all quite high, whereas the specificity of TEE may be comparable only when a strict definition of a positive study is applied (Table 26–5). Finally, availability, speed, safety, and cost should be taken into consideration when comparing various modalities. Aortography is rarely immediately available, requires transport of the patient, is a lengthy study, has the associated risks of both an invasive study and intravenous iodinated contrast, and is the most expensive. Yet it may be necessary in selected patients being considered for combined surgical treatment, especially those with a high likelihood of coronary artery disease or evidence of involvement of major arterial trunks arising from the aorta. CT scanning has the advantage that it is more easily obtained in less time and is noninvasive, but it is overall less accurate than the other techniques. CMR is usually less available in most hospitals, requires transportation of the patient, and is considered undesirable for unstable patients or those requiring very close monitoring. Meanwhile, multiplane TEE is readily obtained, quick to complete at the bedside and thus ideal for unstable patients, and the least costly of the four imaging techniques. Therefore, multiplane TEE with its accuracy, safety, speed, and convenience, is often considered the study of first choice in cases of suspected dissection. In some institutions, TEE has currently assumed this role, with many surgeons taking patients to the operating room based on the diagnostic findings of TEE alone,

Table 26–5. Diagnostic Potential of TTE, TEE, XCT, and CMR for the Detection of Thoracic Aortic Dissections

Dissection		Sensitivity (%)	Specificity (%)	Accuracy (%)	pos. PV (%)	neg. PV (%)
Ascending aorta	TTE	94.7	81.5	86.9	78.2	95.6
	TEE	100	82.1	89.4	79.2	100
	XCT	78.9	100*	91.5	100	85.5
	CMR	100	100*	100	100	100
Aortic arch	TTE	26.1	93.5	79.5	85.0	80.5
	TEE	92.3	93.9	93.5	85.7	96.9
	XCT	92.8	93.9	93.6	86.7	96.9
	CMR	92.8	100	97.9	100	97.0
Descending aorta	TTE	41.7†	100	66.7	100	56.2
	TEE	100	95.4	97.9	96.1	100
	XCT	88.0	86.4	87.2	88.0	86.4
	CMR	100	100	100	100	100

*$p < 0.05$ versus TTE and TEE.
†$p < 0.01$ versus TEE, XCT, and CMR.
Percentages are calculated on the basis of all included individuals with assessable findings.
PV = predictive value; TTE = transthoracic echocardiography; TEE = transesophageal echocardiography; XCT = x-ray computed tomography; CMR = cardiovascular magnetic resonance.
From Kram HB, Wohlmuth DA, Appel PL, Shoemaker WC: Clinical and radiographic indications for aortography in blunt chest trauma. J Vasc Surg 6:168, 1987.

especially when the likelihood of coronary disease and any compromise of major arterial trunks is low (Figure 26–5). These options may render coronary angiography obsolete in the routine work-up of aortic dissection. The optional approach to detecting dissection of the thoracic aorta should be a noninvasive strategy using CMR in all hemodynamically stable patients and TEE in patients too unstable for transportation.[7, 11] Comprehensive and detailed evaluation can thus be reduced to a single noninvasive imaging modality in the evaluation of suspected aortic dissection.[6]

Although CMR may be less practical than TEE for the evaluation of patients presenting with acute suspected aortic dissection, it is well suited for studying patients with stable or chronic dissections. The extraordinary accuracy of CMR with its high-quality images may make it the current gold standard for defining aortic anatomy in such patients. It appears an accepted policy to follow patients after successful evaluation and surgical treatment by CMR to identify any subsequent aneurysm formation or extension of dissection (Table 26–6).

Moreover, CMR and ceMRA have proven to be of particular importance for the individual morphometry of aortic lesions considered for endovascular covered stent-graft treatment. MRA appears the optimal diagnostic tool to custom-design an individual stent-graft for lesions such as aneurysms,[38a] aortic dissections, and localized aortic ulcers likely to penetrate the aortic wall. With a customized aortic endoprosthesis, such pathomorphological entities are becoming more frequently an ideal target for interventional treatment rather than surgical approaches with a considerably high mortality and morbidity (Figures 26–6 and 26–7).

Considering the need for rapid diagnosis, local hospital logistics, skills, and resources may also play an important role in the selection of the imaging modality; smaller community hospitals may only offer CT scanning, which is quite acceptable for the rapid screening of patients with suspected acute aortic dissection. Although aortography has had a time-honored role in the diagnosis of aortic dissection, the newer noninvasive techniques proved to be very accurate and useful in this regard. Further advances in noninvasive testing such as ce-MRA and 3D TEE will likely result in complete displacement of x-ray aortography from the evaluation of patients with suspected aortic dissection.

AORTIC INTRAMURAL HEMORRHAGE

An important differential diagnosis of aortic dissection is intramural hemorrhage (IMH), which usually presents with a similar clinical picture and risk

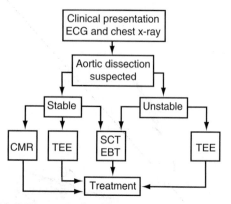

Figure 26–5. Diagnostic algorithm in suspected aortic dissection. ECG = electrocardiogram; CMR = cardiovascular magnetic resonance; TEE = transesophageal echocardiography; SCT = spiral computed tomography; EBT = electron beam computed tomography.

Table 26–6. Logistics of Diagnostic Interrogation of the Aorta

Advantage	Aortography	CT	CMR	TEE
Readily available	Sometimes	Yes	Sometimes	Yes
Rapid	Yes	Very	Fairly	Very
Performed at bedside	No	No	No	Yes
Noninvasive	No	Yes	Yes	Yes
Intravenous iodinated contrast agent	Yes	Yes	No	No
Costs	High	Reasonable	Moderate	Reasonable

CT = computed tomography; CMR = cardiovascular magnetic resonance; TEE = transesophageal echocardiography.
Adapted from Cigarroa JE, Isselbacher EM, De Sanctis RW, Eagle KA: Diagnostic imaging in the evaluation of suspected aortic dissection. N Engl J Med 328:35, 1993.

profile as overt aortic dissection. The noninvasive tomographic modalities such as CMR, CT, and TEE may be used to identify IMH with no luminal component, which is considered an imminent precursor of aortic dissection.[39, 40] IMH was first described in 1920 as "dissection without initial tear" and was originally considered a distinct entity at necropsy.[41] However, with high-resolution tomographic imaging, the in vivo diagnosis of IMH is now feasible and the data suggest that IMH is a precursor of dissection,[40–41] inasmuch as there is a high (>30%) rate of progression to overt dissection. Both CMR and CT can identify IMH in 13 percent of patients with acute aortic syndromes, this percentage being almost identical to autopsy results of 13 percent with no identifiable intimal tear.[40] Because of these findings in vivo before death, IMH appears more likely to be a variant of dissection than a separate entity.[40, 43, 44] Typical epiphenomena of dissection such as aortic insufficiency and pericardial or pleu-

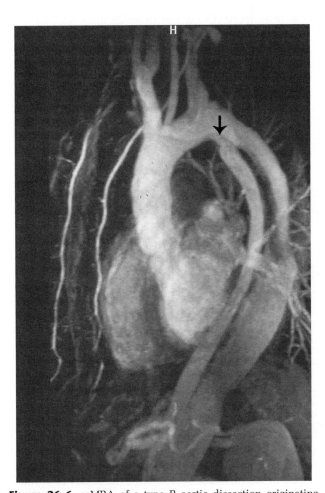

Figure 26–6. ceMRA of a type B aortic dissection originating from the aortic arch region. The communication (arrow) between true and false lumen is clearly identified and serves as orientation for subsequent stent-graft placement.

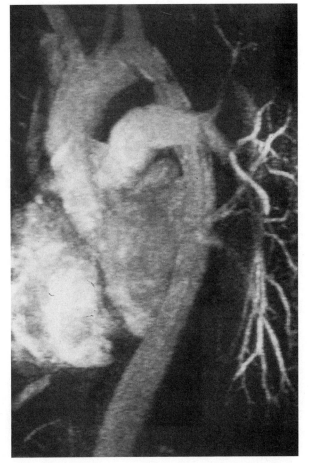

Figure 26–7. Follow-up ceMRA 7 days after stent-graft placement in the same patient as Figure 26–6 shows a completely sealed proximal entry to the thrombosed false lumen. The diameter of the true lumen is normalized and the descending aorta is reconstructed.

ral effusion may also occur in IMH.[42, 44–46] Moreover, arterial hypertension is the most frequent predisposing factor for IMH as well as for overt dissection, with an incidence of 67 to 84 percent.[45–48] Spontaneous rupture of aortic vasa vasorum, especially of nutrient vessels to the media layer, has been suggested to initiate the process of aortic disintegration without an intimal tear.[49, 50] With a pathogenesis that explains the high rate of progression to overt aortic dissection, and a prognosis and survival that are similar to aortic dissection, urgent diagnosis of IMH is very important (Figure 26–8). The extent of IMH can be considerable with abnormal aortic wall thickness of up to 30 mm both asymmetrical or symmetrical in circumference, with an extent of 3 to 30 cm. It should be noted that angiography is of no use for IMH because there is no intraluminal component.

Diagnostic Approach to IMH

The diagnosis of IMH relies on the visualization of intramural blood and/or evidence of localized increased wall thickness.[51] The high density of fresh hematoma on CT appears specific for IMH. CMR techniques, however, not only visualize the blood in the wall but also allow an assessment of the age of the hematoma based on signal changes caused by the formation of methemoglobin (see Figure 26–8). High signal intensity within the aortic wall on T1 spin-echo CMR suggests subacute IMH, whereas acute IMH (early stage of hemorrhage) may be determined on T1 images due to the isodense appearance of blood and aortic wall.[40, 43, 52] Acute IMH (early stage) is also well imaged on T2 weighted images due to high initial signal intensity of blood, whereas

blood of 1 to 5 days of age has lower signal intensity on T2 images. TEE has also been emphasized as a diagnostic tool; however, the differentiation of IMH from severe atherosclerosis with local wall thickening may be difficult, and IMH may only be diagnosed retrospectively with serial evaluation (resolution or progression to dissection). Moreover, false-positive findings of local thickening on tangential scans and around the hemiazygos vein may be more likely with TEE. Both TEE and CT may result in false-positive and false-negative findings (pathological wall thickening without hematoma), whereas the segmental extent of IMH is usually correctly defined with CMR. Although TEE has an excellent sensitivity to detect aortic dissection, the definite distinction between IMH and normal findings may require a second tomographic modality such as CT or CMR because a false-negative result (or false exclusion of IMH) is more likely to be avoided with independent morphological information (see Table 26–4).

In conclusion, as a precursor of dissection, IMH requires diagnostic attention by use of high-resolution tomographic imaging; due to its physical properties, CMR may play a prominent role not only to diagnose IMH but also to assess its age and differentiate IMH from mural thrombosis; angiography certainly is not diagnostic. Given the poor results and unfavorable outcome with medical treatment, early surgical repair should be considered for all patients with ascending aortic involvement (type A IMH) and for any patient with recurrent pain. Conversely, surgery may not be required in patients with IMH of the descending aorta. Yet, all patients may benefit from serial CMR follow-up to rule out progression, regardless of treatment strategy, due to new lesions

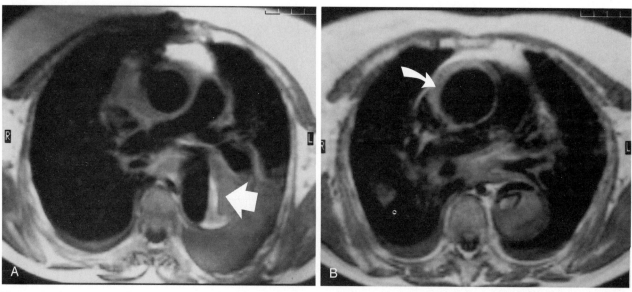

Figure 26–8. Transverse horizontal T1-weighted CMR image at the level of the pulmonic artery bifurcation 7 days after initial presentation with acute symptoms suggestive of aortic dissection. There is clear definition of a chronic type B dissection with almost complete thrombosis (straight arrow) of false lumen in the descending aorta. In addition, the ascending aorta reveals a crescent-shaped anterior wall thickening (curved arrow) suggestive of recent intramural hematoma by signal enhancement. There is no evidence of progression to a luminal component such as a flap or tear or dissecting membrane in the ascending aorta.

or spontaneous relapses even after surgical repair.[53, 54]

AORTIC ULCERS

Aortic ulcers occur in the presence of aortic atherosclerosis and may mimic subacute dissection, or may develop without major symptoms. In contrast to IMH, aortic ulcers are characterized on angiography by focal contrast enhancement beyond the confines of the aortic lumen but communicating with the lumen.[50, 55] At present, there is no consensus on the prognosis and outcome of aortic ulcers. Both IMH and ulcers are unrelated to intimal lacerations as in acute aortic dissection.[56] Lacerations and IMH usually occur at points of greatest hydraulic stress (right lateral ascending aorta or adjacent to the ligamentum arteriosum), whereas penetrating ulcers are typically found in the descending or abdominal aorta. Conversely, discrete penetrating atheromatous ulcers (giant ulcers) have been suspected as one cause of intramural bleeding.[52] In such a chronic setting, the hematoma is confined to the rim adjacent to the ulcer. Both CMR and TEE have helped to elucidate the pathogenic background and the complex anatomical peculiarities of these important features, but accurate diagnosis of penetrating ulcers is sometimes difficult and no test is ideal.[56] CT may demonstrate the surrounding hematoma and displaced calcifications in most cases; in addition to this finding, CMR scanning can differentiate subacute intramural hematoma from chronic intraluminal thrombus. No large-scale studies of CMR and aortic ulcers are currently available, but due to its versatility and imaging features, CMR appears best suited[56a] to characterize this form of aortic pathology as a constellation of diffuse aortic atherosclerosis, an ulceration with focal wall thickening, and missing evidence of aortic dissection (Figure 26–9).

TRAUMA TO THE AORTA

A traumatic aortic tear is a separate entity and results from shearing of the aorta, unlike the focal penetration of a pseudoaneurysm. This devastating event results in immediate death in approximately 90 percent of cases. In those fortunate enough to survive, the adventitia is usually intact and prevents exsanguination. If left untreated, 90 percent of these survivors will die.[57, 58] Therefore, it is imperative to establish the diagnosis quickly. Because aortic tears are usually caused by rapid deceleration forces following motor vehicle accidents, these patients typically have other life-threatening injuries that make CMR impractical. However, CMR is an appropriate technique in patients suspected of having a chronic or missed aortic tear. The tear is usually found in the area of the ligamentum arteriosus. It has the configuration of a localized saccular aneurysm, usually with an associated periaortic hematoma. The left anterior oblique and coronal planes graphically demonstrate the tear. T1-weighted spin-echo images should be used for the imaging for the sake of time and fewer artifacts (Figure 26–10).

Partial or complete aortic transection also results from blunt thoracic trauma or deceleration injury. In those who survive until hospital admission, the chest x-ray shows abnormalities in 90 percent of patients with aortic rupture. Both CT and CMR are not too difficult to perform in the acute setting. Aortography remains the accepted standard for the diagnosis and evaluation of aortic trauma, but requires transport of the injured patient to the catheterization laboratory for a procedure that often lasts longer than an hour and has a 10 percent complication rate in this setting.[59, 60] TEE can also visualize the aorta rapidly in an injured patient. The examination can proceed at the bedside or in the operating room, often during other diagnostic or therapeutic procedures. Studies comparing TEE with other diagnostic modalities in small patient groups, however,

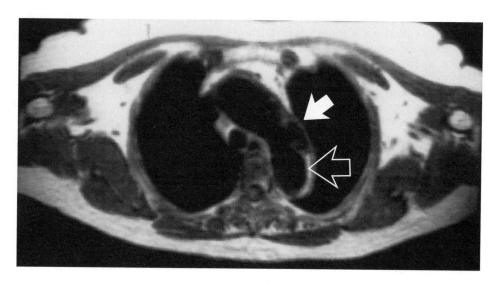

Figure 26–9. Transversal CMR using spin-echo technique revealing a subtle ulcer-like dissection (solid arrow) in the aortic arch that was not detected by any other method. Adjacent to the intimal laceration is evidence of methemoglobin formation (open arrow), indicating a subacute intramural hematoma by high-signal intensity.

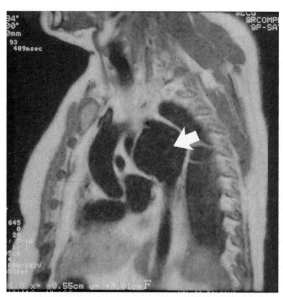

Figure 26–10. Spin-echo CMR in coronal orientation showing an aneurysm with subtotal rupture of the descending thoracic aorta and bleeding into the left pleural space. The patient was in stable hemodynamic condition and surgically treated within 10 days.

have produced mixed results; in the largest trial to date, 69 patients with suspected aortic trauma underwent both TEE and aortography soon after admission.[61] TEE revealed thoracic aortic injury in seven patients, and these injuries were all confirmed either at surgery or autopsy. Aortography in this group yielded two false-negative results and had an overall sensitivity of 67 percent. Both the sensitivity and specificity of TEE were 100 percent. TEE, however, has several limitations in the setting of aortic trauma. Protocols for airway management and cervical spine precautions have not been fully addressed, and the procedure could conceivably worsen a pre-existing or unstable neck injury. Disruption of the small portion of the ascending aorta obscured by the bronchus may escape detection, and small intimal tears may also be missed. Finally, pneumothorax (which is not uncommon in thoracic trauma) can limit evaluation of the descending aorta.[5, 62] TEE may help screen patients with thoracic aortic trauma to identify those who require angiography (and could perhaps replace aortography as the sole preoperative test in some cases), but long-term, large-population studies are necessary to assess the technique's best use in this setting.

THORACIC AORTIC ANEURYSM

Regional dilatation of a blood vessel can represent either a true aneurysm or a false aneurysm. The walls of true aneurysms involve all layers of the aortic wall and result from the degeneration of the elastin fibers within the media. False aneurysms, or pseudoaneurysms, do not contain a complete wall, but rather a contained perforation of the vessel wall

with penetration of the intima and media. The adventitia and perivascular connective tissue contain the process, thereby preventing exsanguination. Pseudoaneurysms commonly have a narrow "neck" leading to the aneurysm.[63]

Seventy percent of the aneurysms located in the thoracic aorta are associated with severe atherosclerosis. True aneurysms usually result from the atherosclerotic process of infiltration and damage to the aortic media. Pseudoaneurysms form after focal penetration of the intima and media by trauma (20–25% of thoracic aneurysms) or by infection (about 5% of thoracic aneurysms). Atherosclerosis may result in the formation of pseudoaneurysms if a true aneurysm ruptures but is contained by the adventitia and periaortic tissues. Thoracic aortic aneurysms are common, occurring in about 10% of autopsies. Atherosclerotic aneurysms are usually fusiform, involving long segments of the aorta. Saccular aneurysms are less common, but, like fusiform aneurysms, they are most often the result of atherosclerosis; infection, trauma, and degenerative disease may also result in saccular aneurysms. Aneurysms localized to the ascending aorta tend to be saccular in shape. Often they are caused by annuloaortic ectasia,[64] the Marfan syndrome, syphilis, or poststenotic dilatation resulting from aortic stenosis. The sinotubular junction is preserved in aortic stenosis but markedly dilated in the other entities. Pseudoaneurysms can be seen superimposed on poststenotic true aneurysms as a result of infective endocarditis of the aortic valve. If isolated to the descending aorta, saccular aneurysms may be traumatic or infective, but atherosclerosis remains the leading cause. Saccular and sinus Valsalva aneurysms in the aortic root very often have associated aortic insufficiency.[65] The clear delineation of the location, extent, and shape of an aneurysm, its relation to branch vessels and adjacent structures, and its associated complicating factors, such as rupture, periaortic hematoma or infection, hemopericardium, or aortic valve insufficiency, are all of importance.

CMR is effective in identification and characterization of thoracic aortic aneurysms as well as in evaluation of their pathophysiological consequences. The entire thoracic aorta is demonstrated in sagittal or near sagittal planes, allowing for assessment of location and extent of aneurysms or masses (Figure 26–11). In contrast to transaxial imaging, the oblique plane allows precise determination of lumen diameter. The relation of the branch vessels to the aneurysm is easily demonstrated with this imaging plane. The coronal plane graphically delineates ascending aortic aneurysms and associated aortic insufficiency or hemopericardium with the use of spin-echo and phase-mapping CMR techniques.[65–67]

Optimal treatment of thoracic aneurysms requires accurate definition of size, anatomy, rate of growth, presence of dissection or thrombus, and involvement of adjacent structures including the aortic valve. Only CMR is comprehensive and consistently

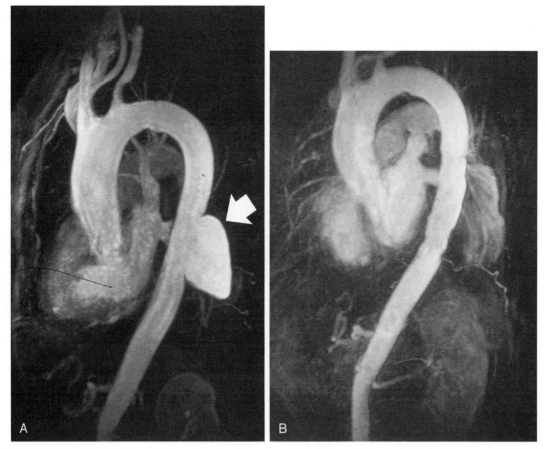

Figure 26–11. ceMRA after Gd-DTPA showing a fist-size thoracic aneurysm (arrow) that had developed over 9 months after a minor trauma. The MRA morphology allows exact quantitation and planning of a stent-graft procedure *(B)*.

provides all this information. Aortography may underestimate the size of aneurysms if significant intraaortic thrombus is present and may fail to fully define saccular structures because of stagnant intraaneurysmal blood flow. CT scanning is noninvasive and suitable for serial examinations but requires considerable iodinated contrast and provides only axial images. Transthoracic echocardiography (TTE) usually visualizes the ascending aorta adequately but may not fully image the arch and is less helpful for the descending aorta, as is TEE for the abdominal aorta.

Both multiplane TEE and CMR are well suited for serially evaluating any thoracic aortic aneurysm; in addition to measuring aneurysm dimensions, both multiplane TEE and CMR can also detect fistulas, false channels, intraluminal thrombosis, and aortic valvular insufficiency.[20, 36, 67] At present no data from a large trial comparing the value of CMR and TEE for thoracic aneurysms have been reported; there is, however, consensus that CMR is accurate and versatile but most likely not as cost effective in serial studies. Although CMR promises to emerge as the standard in the diagnosis and management of thoracic aortic aneurysms, cost considerations have to be evaluated before recommending widespread use of CMR scanning to follow stable patients with subcritical aneurysms.

INHERITED AORTIC DISEASE

In children, TTE does not always provide adequate imaging of cardiac and aortic pathology, including valvular abnormalities, subvalvular and supravalvular aortic stenosis, or right-sided aortic arches. Because surface ultrasonography often proved to be incomplete in congenital disease, recently developed pediatric TEE probes have been developed to improve the diagnostic spectrum. TEE in young patients, as in adults, is especially helpful for imaging posterior structures, including assessment of potential atrial septal defects, anomalous venous connections, and abnormalities of the great vessels. As the age and size of patients with congenital heart disease increase, the likelihood of complete and optimal transthoracic imaging gradually declines, and both CMR and TEE assume a more important role. Both techniques detect aortic coarctation, right-sided aortic arch, congenital bicuspid aortic valve, supravalvular and subvalvular aortic stenosis, patent ductus arteriosus, aorticopulmonary window, transposition of the great vessels, and truncus arteriosus.[16, 18, 68–72]

Of the inherited connective tissue diseases, the Marfan syndrome, which results from mutations in genes coding for the glycoprotein fibrillin, is an autosomal dominant trait with protein manifestations affecting the cardiovascular, skeletal, and ocular systems. Those patients with cardiovascular involvement have a decreased life expectancy, primarily from progressive aortic root dilation and subsequent aortic insufficiency, dissection, and rupture.[74–76] In early asymptomatic stages of aortic root dilatation, chronic beta-blockade has been established as an effective means of slowing the progressive enlargement and delaying the associated morbidity.[73] TTE provides adequate detail of the heart and aortic root in most instances, but occasionally, images are suboptimal and TEE or CMR becomes necessary. In suspected dissection, multiplane TEE (or CMR) should be performed urgently, because aortic rupture, acute valvular regurgitation, and hemopericardium occur in a high proportion of patients. Even after a successful surgical intervention, experienced centers suggest a baseline postsurgical CMR or CT scan, followed by serial imaging in 6 months and subsequent annual intervals to screen for new focal aneurysms or dissections. The use of Gd-DTPA–enhanced aortic MRA is particularly valuable for detecting leakage.[76a]

Ehlers-Danlos syndrome encompasses at least 10 distinct disorders, most of which affect the skin and little else. In one of the exceptions, type IV Ehlers-Danlos syndrome, the gene coding for type III procollagen is abnormal. The resulting deficiency of this collagen component causes decreased strength of arterial connective tissues. Spontaneous aortic or arterial rupture is the most common cause of death in these patients. Arteriography is quite hazardous because of vascular fragility, and less invasive vascular imaging is preferable. Therefore, CMR and multiplane TEE may be helpful in cases of suspected aortic dilatation or disruption.[77, 78]

CONGENITAL AORTIC ANOMALIES

There are a variety of congenital anomalies involving the thoracic aorta. Those that result from maldevelopment of the conotruncus (i.e., persistent truncus arteriosus, transposition of the great arteries, and tetralogy of Fallot) are almost always present in conjunction with intracardiac defects, most of which are suitable for presurgical and postsurgical follow-up imaging using CMR (Figure 26–12). However, these anomalies are best viewed as congenital heart defects rather than aortic anomalies. The most common anomalies are discussed here with the Edward's hypothetical double aortic arch model used as an aid in understanding.[79]

Left Aortic Arch

The normal left aortic arch forms as a result of a break in the Edward's hypothetical double aortic arch distal to the right subclavian artery. The distal aspect of the right dorsal arch regresses. On occasion, it does not regress completely and forms an

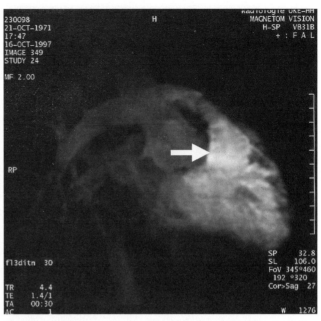

Figure 26–12. Noncontrast MRA in a patient 17 years after surgical correction of a D-type transposition of the great arteries. The MRA resolved the question of the anatomical nature of the Rastelli-type shunt (arrow) between the right ventricle and the pulmonary arteries. In contrast to TEE, MRA also excluded stenosis in this shunt structure.

aortic diverticulum along the right lateral aspect of the descending aorta at its junction with the transverse arch. The proximal half of the right aortic arch migrates inferiorly to join the ascending aorta. In the process, the ostia of the right subclavian and common carotid arteries fuse to form the innominate artery. The left arch remains intact, with separate ostia for the left common carotid and left subclavian arteries. The right ductus is almost always obliterated, but the left ductus remains patent, supplying blood to the descending aorta.

The most common thoracic aortic arch anomaly, occurring in approximately 0.5 percent of the population, is the left aortic arch with an aberrant right subclavian artery. It results from a break in the hypothetical double aortic arch between the right common carotid artery and the right subclavian artery. The right ductus is obliterated. The right subclavian artery is incorporated into the descending aorta at its junction with the transverse arch. The right subclavian can arise directly from the descending aorta or it can arise from an aortic diverticulum (the diverticulum of Kommerell). The aberrant right subclavian artery courses posterior to the esophagus, but, because a vascular ring is not intact, there is no associated dyspnea or dysphagia. Identification of an aberrant right subclavian artery is usually an incidental finding on CMR evaluation of the thoracic spine or the chest. If there is a clinical suspicion that needs to be confirmed, CMR is the imaging modality of choice. Thin-section (4-mm thick slices with a 1-mm gap) spin-echo transaxial images through the arch and proximal descending aorta demonstrate the origin and proximal aspect of the aberrant vessel as it passes posterior to the esophagus. Coronal spin-echo CMR using thin slices through the junction of the transverse and descending aortas demonstrates the origin and superior diagonal course of the aberrant vessel. Usually, the presence of a diverticulum of Kommerell can be determined with the use of a body coil; rarely a dedicated surface coil is required (Figure 26–13). In cases of coarctation with an associated aberrant subclavian artery, the location of the subclavian ostium relative to the coarctation site is best demonstrated with coronal or oblique left spin-echo CMR. In cases of juxtaductal coarctation, it may be difficult to determine whether the aberrant subclavian artery serves as a collateral vessel. In these cases, phase-mapping CMR can demonstrate the direction of blood flow within the aberrant artery.[80–82]

There is an increased incidence of an aberrant right subclavian artery both with tetralogy of Fallot and coarctation. With regard to the latter, if the aberrant artery arises distal to the coarctation, it serves as a major collateral vessel. In this situation, rib notching, if present, involves only the left hemithorax.

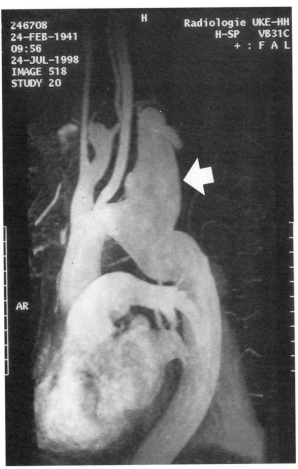

Figure 26–13. ceMRA with rare evidence of a Kommerell diverticulum (arrow) at the aortic-subclavian branch point in oblique sagittal orientation.

Right Aortic Arch

The right aortic arch passes to the right of the trachea and can descend either to the right or left of the thoracic spine. If the arch descends to the right of the spine, it usually recrosses the midline at the level of the diaphragm so as to be to the left of the lumbar spine. In nearly all cases, there is obliteration of the right ductus arteriosus. The mirror-image right aortic arch is formed by a break in the left dorsal arch distal to the ductus arteriosus. The first artery to arise from the transverse arch is the left innominate artery. It is, in turn, followed by the right common carotid artery and then the right subclavian artery. The arch passes to the right of the trachea and descends on the right side of the spine. The left ductus arteriosus usually arises from the left subclavian artery, and there is no vascular ring. In rare cases, the break occurs between the left subclavian artery and left ductus arteriosus. In these instances, the ductus arises from an aortic diverticulum and passes posterior to the trachea and esophagus, thereby forming a vascular ring. Almost all cases of mirror-image right aortic arch occur in conjunction with cyanotic congenital heart disease, most notably with tetralogy of Fallot. Because tetralogy of Fallot is a much more common congenital defect than persistent truncus arteriosus (10–12% vs. 1.5–2% of congenital heart disease), the mirror-image right aortic arch is more commonly seen in association with tetralogy of Fallot. The most common type of right aortic arch anomaly is the right arch with an aberrant left subclavian artery, which occurs in approximately 0.1 percent of the population.[79]

There are two types of right aortic arch with an aberrant left subclavian artery. In the more common type, the arch passes to the right of the trachea and descends on the right side of the thoracic spine. The left subclavian arises from the proximal descending aorta, and the left ductus arteriosus arises from the left subclavian artery, thereby creating a vascular ring (see Figure 26–3). With the other variety, the arch passes to the right of the trachea and turns to the left to pass behind the esophagus before descending to the left of the spine. The left subclavian artery and left ductus arteriosus usually arise from an aortic diverticulum, completing the vascular ring. This variety is also known as a *right circumflex retroesophageal arch*.

Right aortic arches are best evaluated with transaxial and coronal spin-echo CMR. In mirror-image right aortic arches, these views also serve to assess the heart for the associated intracardiac defects. Phase mapping is useful to assess the degree of infundibular stenosis and to detect collateral vessels to the pulmonary arteries. Aberrant left subclavian arteries are evaluated in essentially the same manner as aberrant right subclavian arteries.[81, 83, 84]

Double Aortic Arch

This thoracic aortic anomaly results from a hypothetical persistent double aortic arch model. The ascending aorta lies anterior to the trachea. It divides into two arches, which pass on either side of the trachea before fusing posterior to the esophagus to form the descending aorta. The descending aorta typically runs to the left of the spine. The right aortic arch is usually the more dominant of the two arches. It is larger in caliber and more cephalic in position. A separate common carotid artery, subclavian artery, and ductus arteriosus arise from each arch. The right ductus is almost always obliterated. This configuration usually results in a tight vascular ring that almost always requires surgical intervention to alleviate symptoms of dyspnea and dysphagia.

There are two varieties of double aortic arch. Type I, the most common type, has two patent aortic arches. In type II, the right arch is patent, but a portion of the left arch is atretic. It is uncommon for either type of double aortic arch to be found in conjunction with an intracardiac defect. Although the right aortic arch with an aberrant left subclavian artery is the most common course of a vascular ring, the anatomical configuration typically results in a loose ring and rarely requires surgical intervention. Although much less common, the double aortic arch results in a tight ring that almost always necessitates surgical transection.

Double aortic arches are best studied with coronal and transaxial spin-echo CMR and transaxial phase-mapping CMR. The coronal views demonstrate the larger caliber dominant right aortic arch and its more cephalic location relative to the nondominant left aortic arch. The amount of esophageal and tracheal constriction caused by the vascular ring is graphically demonstrated by transaxial cine CMR at the level of the inverted "U" configuration of the double aortic arch.[84, 85]

Aortic Coarctation

Aortic coarctation is a common congenital anomaly that results from an abnormality in the aortic media and refers to a discrete enfolding of the posterolateral wall of the aorta in the region of the ligamentum or ductus arteriosus. This usually discrete phenomenon occurs just distal to the ductus and is also labeled *postductal coarctation*. Because it is usually asymptomatic in the neonatal period, it is also referred to as *adult coarctation*. This anomaly is easily identified on *anatomical* spin-echo MR, *functional* gradient-echo scans,[85a] and MRA (Figure 26–14). Aortic coarctation is often seen in association with a bicuspid aortic valve.

Coarctation may occur proximal to the ductus and present itself shortly after birth; this variety has been termed *preductal* or *infantile coarctation*. It is less common than adult coarctation and is usually associated with hypoplasia of the arch between the left subclavian artery and the ductus of the aortic isthmus. In utero, during systole, blood from the ascending aorta passes preferentially into the branch

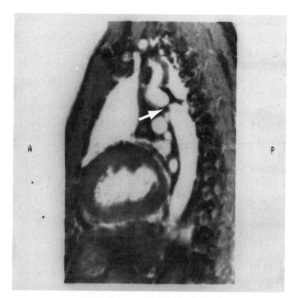

Figure 26–14. ceMRA after Gd-DTPA revealing an adult-type coarctation with critical luminal narrowing (arrow) of the aorta.

arteries. Blood also passes from the left pulmonary artery into the descending aorta through the ductus arteriosus, and there is also reflux of blood into the isthmus. Because the coarctation is located proximal to the ductus, there is no reflux of blood into the segment of the aorta between the ductus and the left subclavian arteries. Hence, there is not enough exposure to blood volume to ensure normal development of the aortic isthmus. Occasionally, the coarctation site is found proximal to the left subclavian artery. With both forms, there is usually dilatation of the descending aorta distal to the coarctation. As a result of the obstruction caused by the coarctation, collateral vessels develop to increase flow into the descending aorta. The intercostal arteries serve as a major source of collateral flow. The increased flow through these vessels results in their dilatation. This dilatation, in turn, can result in notching along the inferior aspect of the ribs, which usually takes 8 to 10 years to become significant enough to be observed on a chest radiograph. If there is an aberrant right subclavian artery that originates distal to the coarctation, it serves as a major collateral vessel, and rib notching only occurs on the left. If the left subclavian artery arises distal to the coarctation, it serves as a major collateral, and unilateral rib notching occurs on the right.

With its multiplanar image acquisition, large field of view, and dynamic quantitative flow imaging capacity, CMR appears the modality of choice for evaluation of coarctation. The left oblique sagittal view centered on the middle of the ascending and descending aortas is an ideal orientation that may also demonstrate associated bicuspid aortic valve disease or ventricular septal defects. This finding is important because there is a high association of bicuspid aortic valve and ventricular septal defects with coarctation.[81, 85, 86] The isthmus can be narrowed without any associated obstruction to blood flow, a situation termed *pseudocoarctation*. In addition, there is usually elongation of the arch, which results in redundancy and kinking of the aorta. These features, as well as associated ventricular septal defects and bicuspid aortic valve, are easily diagnosed with coronal and left anterior sagittal static and dynamic CMR. Finally, CMR is also valuable for serial evaluation of patients following percutaneous or surgical repair of aortic coarctation.

AORTITIS

Although there are many causes of aortitis, Takayasu arteritis is the type most often studied with CMR because of the diffuse stenotic nature of the disease, which often makes vascular access nearly impossible. CMR typically demonstrates diffuse wall thickening of the thoracic and abdominal aorta as well as stenosis and occlusion of the branch vessels. There is often asymmetrical thickening of the aortic wall caused by fibrosis.[88, 88a] This finding can lead to the erroneous diagnosis of pseudoaneurysm on aortography because the technique provides only a luminogram and is blind to the actual aortic wall. Patients with aortitis are best studied by spin-echo and dynamic cine CMR in oblique left orientation, transaxial, and coronal planes. ceMRA is often useful for the evaluation of branch vessels.

PARAAORTIC DISEASE

Thoracic CMR is used mainly to visualize cardiac and vascular structures, but it can also detect other intrathoracic pathology such as paracardiac and pericardial tumors, pleural effusions, collapsed or consolidated lungs, hiatal hernias, and paraaortic masses. Fluid-filled structures are easily distinguished from solid masses, and cardiac or vascular tumor invasion can sometimes be demonstrated. Although TEE may also provide useful images of such paraaortic masses, it is relatively untested and clearly has not displaced CT or CMR. Tissue characterization techniques, especially with CMR technology, may provide information about tumors and other structures, whereas it appears less likely that TEE will become the standard for this purpose.[87, 88]

CONCLUSION

In the scenario of acute aortic syndromes and suspected aortic dissection, 2D echocardiography should include the multiplane TEE approach to improve sensitivity and assess the anatomical extent of the lesion. Both TEE and CMR are feasible, reliable, and safe even in severely ill and potentially unstable patients with acute or chronic thoracic aortic dissection. CMR yields optimal reliability and may emerge as the future gold standard of sensitivity and specificity of thoracic aortic dissection irrespective of its location and possibly for many other diseases involving the aorta. CMR provides exact anatomical mapping of the entire aorta and adjacent tissues due to free choice of representative planes within reasonable time and at no risk. Disadvantages of current CMR technology such as patient transportation, limited patient access, and longer examination time than TEE (especially in acute cases) are unlikely to result in a significantly increased individual risk, but may currently be considered a limitation. Technical improvement such as open CMR scanners with better patient access and faster gradients or softwave approaches to facilitate more rapid acquisition may alleviate two of these drawbacks. The use of improved surface coils may improve spatial resolution and the signal-to-noise ratio in the future, and significantly shorten the examination time. As a clinical routine, we recommend the use of TEE in any emergency setting for immediate diagnostic evaluation of hemodynamically unstable and patients with a contraindication to CMR. CMR appears to be the method of choice in stable patients with suspected acute or subacute dissections and is certainly the best method for serial follow-up studies after surgical or percutaneous repair or in chronic cases. The comprehensive information from anatomical mapping using multiplane TEE or CMR may not only render invasive x-ray angiographic techniques obsolete in the clinical scenario of suspected aortic dissection but could eventually enable surgical interventions solely guided by noninvasive imaging modalities.

ACKNOWLEDGMENTS

The authors want to express their thanks to Mrs. Jeannette Hoffmann for preparation of the manuscript and Miss Dörte Oestreich and Miss Kathrin Niemetz for art and phototechnical support.

References

1. Stein HL, Steinberg I: Selective aortography: The definitive technique for diagnosis of dissecting aneurysm of the aorta. AJR Am J Roentgenol 102:333, 1968.
2. Hayashi K, Meaney TF, Zelch JV, Tarar R: Aortographic analysis of aortic dissection. AJR Am J Roentgenol 122:769, 1974.
3. Wilbers CR, Carrol CL, Hnilica MA: Optimal diagnostic imaging of aortic dissection. Tex Heart Inst J 17:271, 1990.
4. Singh H, Fitzgerald E, Ruttley MST: Computed tomography: The investigation of choice for aortic dissection? Br Heart J 56:171, 1986.
5. Enia F, Ledda G, Lo Mauro R, et al: Utility of echocardiography in the diagnosis of aortic dissection involving the ascending aorta. Chest 95:124, 1989.
6. Nienaber CA, von Kodolitsch Y, Nicolas V, et al: The diagnosis of thoracic aortic dissection by noninvasive imaging procedures. N Engl J Med 328:1, 1993.
7. Cigarroa JE, Isselbacher EM, De Sanctis RW, Eagle KA: Diagnostic imaging in the evaluation of suspected aortic dissection. N Engl J Med 328:35, 1993.
8. Cheitlin MD: Commentary on Anderson MW, Higgins CB: Should the patient with suspected acute dissection of the aorta have CMR, CAT scan, or aortography as the definitive

study? *In* Cheitlin MD, Brest AM (eds): Dilemmas in Cardiology. Philadelphia, FA Davis, 1990, p 293.

9. Ballal RS, Nanda NC, Gatewood R, et al: Usefulness of transesophageal echocardiography in assessment of aortic dissection. Circulation 84:1903, 1991.

10. Erbel R, Engberding R, Daniel W, et al and the European Cooperative Study Group for Echocardiography: Echocardiography in diagnosis of aortic dissection. Lancet 1:457, 1989.

11. Galvin IF, Black IW, Lee CL, Horton DA: Transesophageal echocardiography in acute aortic transection. Ann Thorac Surg 51:310, 1991.

12. Abernethy LJ, Szczepura AK, Fletcher J, et al: Cost effectiveness of cardiovascular magnetic resonance. Br Med J 304:183, 1992.

13. Nienaber CA, Rehders TC, Fratz S: Detection and assessment of congenital heart disease with MR techniques. J Cardiovasc Magn Reson 1:169, 1999.

14. Bornemeier RA, Weinberg PM, Fogel MA: Angiographic, echocardiographic and 3D cardiovascular magnetic resonance of extracardiac conduits in congenital heart disease. Am J Cardiol 78:713, 1996.

15. Sampson C, Martinez J, Rees S, et al: Evaluation of Fontan's operation by cardiovascular magnetic resonance. Am J Cardiol 65:819, 1990.

16. Kersting-Sommerhoff BA, Seelos KC, Hardy C, et al: Evaluation of surgical procedures for cyanotic congenital heart disease by using MR imaging. AJR Am J Roentgenol 155:259, 1990.

17. Mitchell MM: Evaluation of aortic disease. *In* Dittrich HC (ed): Clinical Transesophageal Echocardiography. St. Louis: Mosby-Year Book, 1992, p 97.

18. Hirsch R, Kilner PJ, Connelly MS, et al: Diagnosis in adolescents and adults with congenital heart disease. Prospective assessment of individual and combined roles of cardiovascular magnetic resonance and transesophageal echocardiography. Circulation 90:2937, 1994.

19. Solomon SL, Brown JJ, Glazer HS, et al: Thoracic aortic dissection: Pitfalls and artifacts in MR imaging. Radiology 177:223, 1990.

20. Sechtem U, Pflugfelder PW, Whithe RD, et al: Cine MR imaging: Potential for the evaluation of cardiovascular function. AJR Am J Roentgenol 148:239, 1987.

21. Nienaber CA, Spielmann RP, von Kodolitsch Y, et al: Diagnosis of thoracic aortic dissection: Cardiovascular magnetic resonance versus transoesophageal echocardiography. Circulation 85:434, 1992.

22. Amparo EG, Higgins CB, Hricak H, Sollitto R: Aortic dissection: Cardiovascular magnetic resonance. Radiology 155:399, 1985.

23. Goldman AP, Kotler MN, Scanlon MH, et al: The complementary role of cardiovascular magnetic resonance, Doppler echocardiography, and computed tomography in the diagnosis of dissecting thoracic aneurysms. Am Heart J 111:970, 1986.

24. White RD, Ullyot DJ, Higgins CB: MR imaging of the aorta after surgery for aortic dissection. AJR Am J Roentgenol 150:87, 1988.

25. Cambria RP, Brewster DC, Moncure AC, et al: Spontaneous aortic dissection in the presence of coexistent or previously repaired atherosclerotic aortic aneurysm. Ann Surg 208:619, 1988.

26. Prince MR: Gadolinium-enhanced MR aortography. Radiology 191:155, 1994.

27. Hany TF, McKinnon GC, Pfammatter T, Debatin JF: Optimization of contrast timing for breathhold 3D MR-angiography. J Magn Reson Imaging 16:901, 1998.

28. Holland GA, Dougherty L, Carpenter JP, et al: Breathhold ultrafast 3D gadolinium-enhanced MR angiography of the aorta and the renal and other visceral abdominal arteries. AJR Am J Roentgenol 166:971, 1996.

29. Bogaert J, Meyns B, Rademakers FE, et al: Follow-up of aortic dissection: Contribution of MR angiography for evaluation of the abdominal aorta and its branches. Eur Radiol 7:695, 1997.

30. Leung DA, Debatin JF: 3D contrast-enhanced MRA of the thoracic vasculature. Eur Radiol 7:981, 1997.

31. Hirst AE, Johns VJ, Kime SW Jr.: Dissecting aneurysm of the aorta: A review of 505 cases. Medicine 37:217, 1958.

32. Nienaber CA, von Kodolitsch Y: Meta-analysis of changing mortality pattern in thoracic aortic dissection. Herz 17:398, 1992.

33. DeBakey ME, McCollum CH, Crawford ES, et al: Dissection and dissecting aneurysms of the aorta: Twenty-year follow-up of five hundred twenty-seven patients treated surgically. Surgery 92:1118, 1982.

34. Daily PO, Trueblood W, Stinson EB, et al: Management of acute aortic dissections. Ann Thorac Surg 10:237, 1977.

35. Eagle KA, DeSanctis RW: Aortic dissection. Curr Probl Cardiol 14:231, 1989.

36. Feilcke G, Münster F, Siglow V, et al: Dynamic cardiovascular magnetic resonance of aortic regurgitation: Comparison with contrast aortography and color Doppler echocardiography. Z Kardiol 82:585, 1993.

36a. Arpasi PJ, Bis KG, Shetty AN, et al: MR angiography of the thoracic aorta with an electrocardiographically triggered breath-hold contrast enhanced sequence. Radiographics 20:107, 2000.

37. Stuber M, Botnar RM, Danias PG, et al: Double oblique free-breathing high resolution 3D coronary MRA. J Am Coll Cardiol 34:524, 1999.

38. Woodard PK, Li D, Haacke EM, et al: Detection of coronary stenoses on source and projection images using three-dimensional MR angiography with retrospective respiratory gating: Preliminary experience. AJR Am J Roentgenol 170:883, 1998.

38a. Stables RH, Mohiaddin R, Panting J, et al: Images in cardiovascular medicine. Exclusion of an aneurysmal segment of the thoracic aorta with covered stents. Circulation 101:1888, 2000.

39. Mohr-Kahaly S, Erbel R, Kearney P, et al: Aortic intramural hemorrhage visualized by transoesophageal echocardiography: Findings and prognostic implications. J Am Coll Cardiol 23:658, 1994.

40. Nienaber CA, von Kodolitsch Y, Petersen B, et al: Intramural hemorrhage of the thoracic aorta. Diagnostic and therapeutic implications. Circulation 92:1465, 1995.

41. Krukenberg E: Beiträge zur Frage des Aneurysma dissecans. Beitr Pathol Anat Allg Pathol 67:329, 1920.

42. Wolverson MK, Crepps LF, Sundaram M, et al: Hyperdensity of recent hemorrhage at body computed tomography: Incidence and morphologic variation. Radiology 148:779, 1983.

43. Wilson SK, Hutchins GM: Aortic dissecting aneurysms: Causative factors in 204 subjects. Arch Pathol Lab Med 106:175, 1982.

44. Robbins RC, McManus RP, Mitchel RS, et al: Management of patients with intramural hematoma of the thoracic aorta. Circulation 88(suppl 11):II-1, 1993.

45. Kodolitsch Y, Spielmann RP, Petersen B, et al: Intramural hemorrhage as a precursor of aortic dissection. Z Kardiol 84:939, 1995.

46. Murray JG, Maisali M, Flamm SD, et al: Intramural hematoma of the thoracic aorta: MR image findings and their prognostic implications. Radiology 204:349, 1997.

47. Larson EW, Edwards WD: Risk factors for aortic dissection: A necropsy study of 161 patients. Am J Cardiol 43:849, 1984.

48. Sütsch G, Jenni R, von Segesser L, Turina M: Predictability of aortic dissection as function of aortic diameter. Eur Heart J 12:1247, 1991.

49. Gore I: Pathogenesis of dissecting aneurysm of the aorta. Arch Pathol Lab Med 53:142, 1952.

50. Kazerooni EA, Bree RL, Williams DM: Penetrating atherosclerotic ulcers of the descending thoracic aorta: Evaluation with CT and distinction from aortic dissection. Radiology 183:759, 1992.

51. Wolff KA, Herold CJ, Tempany CM, et al: Aortic dissection: Atypical patterns seen at MR imaging. Radiology 181:489, 1991.

52. von Kodolitsch Y, Nienaber CA: Intramural hemorrhage of the thoracic aorta: Natural history, diagnostic and prognostic profiles of 209 cases with in vivo diagnosis. Z Kardiol 87:797, 1998.

53. Nienaber CA, Röttig K, Brockhoff CJ, et al: Noninvasive fol-

low-up in thoracic aortic dissection: Transesophageal echocardiography versus cardiovascular magnetic resonance. Circulation 90:I-585, 1994.

54. Rofsky NM, Weinreb JC, Grossi EA, et al: Aortic aneurysm and dissection: Normal MR imaging and CT findings after surgical repair with the continuous-suture graft-inclusion technique. Radiology 186:195, 1993.

55. Hussain S, Glover JL, Bree R, Bendick PJ: Penetrating atherosclerotic ulcers of the thoracic aorta. J Vasc Surg 9:710, 1989.

56. von Kodolitsch Y, Nienaber CA: Penetrating ulcer of the thoracic aorta: Natural history, diagnostic, and prognostic profiles. Z Kardiol 87:917, 1998.

56a. Hayashi H, Matsuoka Y, Sakamoto I, et al: Penetrating atherosclerotic ulcer of the aorta: Imaging features and disease concept. Radiographics 20:995, 2000.

57. Sturm JT, Biliar TR, Dorsey JS, et al: Risk factors for survival following surgical treatment of traumatic aortic rupture. Ann Thorac Surg 39:418, 1985.

58. Delrossi AJ, Cernaianu AC, Madden CD, et al: Traumatic disruptions of the thoracic aorta: Treatment and outcome. Surgery 108:864, 1990.

59. Eddy AC, Nance DR, Goldman MA, et al: Rapid diagnosis of thoracic aortic transection using intravenous digital subtraction angiography. Am J Surg 159:500, 1990.

60. Kram HB, Wohlmuth DA, Appel PL, Shoemaker WC: Clinical and radiographic indications for aortography in blunt chest trauma. J Vasc Surg 6:168, 1987.

61. Kearney PA, Smith W, Johnson SB, et al: Use of transesophageal echocardiography in the evaluation of traumatic aortic injury. J Trauma 34:696, 1993.

62. Blanchard DG, Dittrich HC, Mitchell M, McCann HA: Diagnostic pitfalls in transesophageal echocardiography. J Am Soc Echocardiogr 5:525, 1992.

63. Goarin JP, Le Bret F, Riou B, et al: Early diagnosis of traumatic thoracic aortic rupture by transeosphageal echocardiography. Chest 103:618, 1993.

64. Kersting-Sommerhoff BA, Sechtem UP, Schiller NB, et al: MR imaging of the thoracic aorta in Marfan patients. J Comput Assist Tomogr 11:633, 1987.

65. von Kodolitsch Y, Simic O, Nienaber CA: Aneurysms of the ascending aorta: Diagnostic features and prognosis in patients with Marfan's syndrome versus hypertension. Clin Cardiol 21:817, 1998.

66. von Schulthess GK, Augustiny N: Calculation of T2 values versus phase imaging for the distinction between flow and thrombus in MR imaging. Radiology 164:549, 1987.

67. Globits S, Frank H, Mayr H, et al: Quantitative assessment of aortic regurgitation by cardiovascular magnetic resonance. Eur Heart J 13:48, 1992.

68. Ryan K, Sanyal RS, Pinheiro L, Nanda NC: Assessment of aortic coarctation and collateral circulation by biplane transesophageal echocardiography. Echocardiography 9:277, 1992.

69. Simpson IA, Chung KJ, Glass RF, et al: Cine cardiovascular magnetic resonance for evaluation of anatomy and flow relations in infants and children with coarctation of the aorta. Circulation 78:142, 1988.

70. Bank ER, Aisen AM, Rocchini AP, Hernandez RJ: Coarctation of the aorta in children undergoing angioplasty; pretreatment and posttreatment MR imaging. Radiology 162:235, 1987.

71. Mohiaddin RH, Kilner PJ, Rees RSO, Longmore DB: Magnetic resonance volume flow and jet velocity mapping in aortic coarctation. J Am Coll Cardiol 22:1515, 1993.

72. Parsons JM, Baker EJ, Anderson RH, et al: Double-outlet right ventricle: Morphologic demonstration using nuclear cardiovascular magnetic resonance. J Am Coll Cardiol 18:168, 1991.

73. Pyeritz RE, McKusick VA: The Marfan syndrome: Diagnosis and management. N Engl J Med 300:772, 1979.

74. DeBakey ME, McCollum CH, Crawford ES, et al: Dissection and dissecting aneurysms of the aorta. Surgery 92:1118, 1982.

75. Schaefer S, Peshock RM, Mallot CR, et al: Nuclear cardiovascular magnetic resonance in Marfan's syndrome. J Am Coll Cardiol 9:70, 1987.

76. von Kodolitsch Y, Raghunath M, Dieckmann C, Nienaber CA: The Marfan syndrome: Diagnosis of the cardiovascular manifestations. Z Kardiol 87:161, 1998.

76a. Fattori R, Descovich B, Bertaccini P, et al: Composite graft replacement of the ascending aorta: Leakage detection with gadolinium-enhanced MR imaging. Radiology 212:573, 1999.

77. Steinmann B, Superti-Furga A, Joller-Jemelka HI, et al: Ehlers-Danlos syndrome type IV: A subset of patients distinguished by low serum levels of the amino-terminal propeptide of type III procollagen. Am J Med Genet 34:68, 1989.

78. Byers PH: Ehlers-Danlos syndrome. In Scriver CR, Beaudet AL, Sly W, Valle D (eds): The Metabolic Basis of Inherited Disease. 6th ed. New York, McGraw-Hill, 1989, p 2824.

79. Stewart JR, Kincaid OW, Edwards JE: An Atlas of Vascular Rings and Related Malformations of the Aortic Arch System. Springfield, IL, Charles C Thomas, 1964.

80. Didier D, Higgins CB, Fisher MR, et al: Congenital heart disease: Gated MR imaging in 72 patients. Radiology 158:227, 1986.

81. Fletcher BD, Jacobstein MD: CMR of congenital abnormalities of the great arteries. AJR Am J Roentgenol 146:941, 1986.

82. Barkovich AJ: Techniques and methods in pediatric cardiovascular magnetic resonance. Semin Ultrasound CT MR 9:186, 1988.

83. Bisset GS: Pediatric applications of cardiovascular magnetic resonance. J Thorac Imag 4:51, 1989.

84. Jaffe RB: Cardiovascular magnetic resonance of vascular rings. Semin Ultrasound CT MR 11:206, 1990.

85. Burrows PE: Cardiovascular magnetic resonance of the aorta in children. Semin Ultrasound CT MR 11:221, 1990.

85a. Riquelme C, Laissy JP, Menegazzo D, et al: MR imaging of coarctation of the aorta and its postoperative complications in adults: Assessment with spin-echo and cine-MR imaging. Magn Reson Imaging 17:37, 1999.

86. Boxer RA, LaCorte MA, Singh S, et al: Nuclear cardiovascular magnetic resonance in evaluation and follow-up of children treated for coarctation of the aorta. J Am Coll Cardiol 7: 1095, 1986.

87. Freedberg RS, Kronzon I. Rumanick WM, et al: The contribution of cardiovascular magnetic resonance to the evaluation of intracardiac tumors diagnosed by echocardiography. Circulation 77:96, 1988.

88. Lund JT, Ehman RL, Julsrud PR, et al: Cardiac masses: Assessment by MR imaging. AJR Am J Roentgenol 152:469, 1989.

88a. Yamada I, Nakagawa T, Himeno Y, et al: Takayasu arteritis: Diagnosis with breath-hold contrast-enhanced three-dimensional MR angiography. J Magn Reson Imaging 11:481, 2000.

Cardiac and Paracardiac Masses

Herbert Frank

CMR ASSESSMENT OF CARDIAC AND PARACARDIAC MASSES

Cardiovascular magnetic resonance (CMR) provides a noninvasive and three-dimensional (3-D) assessment of masses involving the cardiac chambers, the pericardium, and the extracardiac structures. CMR has become an established method to yield complementary diagnostic information and to guide cardiac surgeons in the design of an appropriate therapeutic strategy. Furthermore, CMR allows characterization of some tumor tissue. Although transthoracic echocardiography and/or transesophageal echocardiography are generally adequate for identification of intracavitary masses, CMR has a role in further characterizing these tumors and for identification of tumors of the myocardium or those invading the heart. The goal of CMR for assessing cardiac and paracardiac masses includes (1) to confirm or to exclude a mass suspected by x-ray or echocardiography; (2) to assess the location, mobility, and its relationship to surrounding tissues; (3) to image the degree of vascularity; (4) to distinguish solid from fluid lesions; and (5) to determine tissue characteristics and the specific nature of a mass.

Technical Considerations

For adequate image quality with reduced motion artifacts, CMR is generally performed using electrocardiogram (ECG) gating.[1] Alternatively, the more recently developed navigator technique allows combined ECG and respiratory triggering.[2] Spin-echo (SE) sequences provide detailed morphological information of the heart, the great vessels, and the adjacent structures.[3] For T1-weighted pulse sequences, the echo time (TE) is usually 20 to 30 ms, and the repetition time (TR) is dependent on the R-R interval. A longer TR is used for T2-weighted sequences and the TE is typically 50 to 90 ms.[4] T1-weighted images provide a better signal-to-noise ratio (SNR) and excellent soft tissue contrast between epicardial fat, myocardium, and rapidly flowing blood.[3] T2-weighted images have an increased image contrast, which may be helpful for tissue characterization; however, T2-weighted images in general have a lower SNR.

The fast or turbo spin-echo (TSE) technique com-

bines the acquisition of multiple profiles per excitation, with the multislice mode resulting in a marked reduction in imaging time. The contrast of those images is similar to that of SE images with the same TR and an equivalent TE. Fat is usually brighter with TSE than regular SE pulse sequences. TSE permits acquisition of a T2-weighted scan in a fraction of time compared with the conventional SE sequence. Furthermore, this technique has a reduced susceptibility to motion artifacts and is insensitive to field inhomogeneity. The combination with inversion recovery (IR) and breath-hold techniques will certainly lead to improved image quality and tissue characterization, which need to be assessed in the future.[5]

When combined with an IR technique, the signal from fat can be suppressed by using a short inversion time prior to the spin-echo pulse. Most tissues have a T1 relaxation time longer than that of fat, resulting in a signal increase on T1- and T2-weighted images. This technique leads to better contrast in these images. Applications of the IR technique are the distinct recognition of fatty tissues and the improved visualization of structures that are surrounded by fatty tissue, which often makes images analysis complicated.

For evaluation of cardiac function and tumor mobility, gradient-echo techniques (TE 4–12 ms) are recommended.[6, 7] Gradient-echo images are characterized by bright signal intensity from rapidly flowing blood, which is useful for the differentiation of thrombus or the assessment of turbulent flow in case of valvular regurgitation or intracardiac shunts. Due to the low soft tissue contrast, visualization of cardiac masses is not as adequate as with SE techniques, unless they are intraluminal.

Table 27–1 presents general technical parameters for evaluation of cardiac and paracardiac masses.

Contrast Agents

The most commonly used CMR contrast agent uses the paramagnetic element gadolinium in a complex (chelate) with other molecules to reduce toxicity. The commonest of these is diethylene-triamine-pentaacetic acid (Gd-DTPA),[8] although a number of other agents are now also available. The distribution of all currently available intravenous

Table 27–1. Typical Technical Parameters at 1.5 Tesla for Evaluation of Cardiac or Paracardiac Masses

Sequence	Technical Features	Indication	Advantages/Disadvantages
T1-weighted spin echo	8–12 slices; thickness = 6–10 mm; NSA = 2; TE = 20–25 ms; TR = shortest; time: 4–6 mm; axial, sagittal, and coronal orientation	Defining anatomical structures, delineation of the mass to the adjacent tissue, visualization of vascular walls	A: Excellent soft tissue contrast D: Respiratory and flow artifacts
T2-weighted spin echo, double echo	8–12 slices; thickness = 6–10 mm; NSA = 2; TE = 50–90 ms; TR = every 2nd or 3rd heartbeat; time = 5–8 min; axial orientation	Detecting the nature of cardiac masses by abnormal T2 values	A: Better tissue contrast, demonstration of fluid components D: Very long examination time, lower SNR and increased motion artifacts
T1- and T2-weighted turbo spin echo	Examination time: 2–4 min, also breath-hold technique		A: Shorter imaging times, reduced motion artifacts D: Lower SNR
Short inversion recovery	Inversion pulse, combination with spin echo or TSE	Suppression of fatty tissue	A: Eliminating artifacts from fat signal
T1-weighted spin echo with contrast agent administration	Axial orientation, antecubital Gd-DTPA administration (Dose: 0, 1 mmol/kg)	Signal intensity behavior of suspected masses, assessment of the degree of vascularity	To assess invasive and infiltrative components of a tumor
Cine gradient echo with flow compensation	1–3 Slices; thickness = 8–10 mm, TE = 9 ms, TR = shortest, 16 phases; NSA = 2; time = 1–3 min.	Imaging the hemodynamic effects of a mass, i.e., mobility, transvalvular flow, differentiation between blood flow and thrombus, identify turbulent flow regions	A: Functional information D: Low soft tissue contrast
Cine gradient echo EPI	Breath-hold technique; 1–3 slices; thickness = 8–10 mm; TE = shortest; TR = shortest; time = 10–40 s	Equal to conventional gradient echo	A: Shorter acquisition time, reduced motion artifacts D: Lower SNR

NSA = numbers of signal averaged; TE = echo time; TR = repetition time; TSE = turbo spin-echo; SNR = signal-to-noise ratio; EPI = echo-planar imaging.

Modified from Hoffmann U, Globits S, Frank H: Cardiac and paracardiac masses—Current opinion in the diagnostic evaluation by magnetic resonance imaging. Eur Heart J 19:166, 1998.

contrast agents is extracellular, and contrast is enhanced predominantly on T1-weighted images. The normal concentration is 0.1 mmol per kilogram body weight.[9] Gd-DTPA provides a better delineation of the mass by different enhancement of myocardium and tumor tissue due to variation of tissue vascularity.[10] Intravascular contrast agents are in development.

Tissue Characterization

Unlike two-dimensional (2-D) echocardiography, CMR has the potential for tissue characterization by comparing the T1 and T2 values of the mass to a reference tissue.[11] That is based on the observation that significant differences exist in the proton density and T1 and T2 relaxation time. Because fat shows very constant T1 and T2 values, it has been used as a reference tissue.[12] Previous studies have tried to determine specific T1 and T2 relaxation times of different tissues[13]; however, precise etiologic diagnoses are not possible.[14, 15] It should be mentioned that IR spin-echo sequences are now recommended to quantify tissue characteristics because they have the advantage of providing more accurate T1 data (~4% dispersion) compared with conventional spin-echo sequences (20% disper-

sion).[16] Inhomogeneity, tumor infiltration of adjacent structures, and hemorrhagic peritumorous pericardial effusion are suspicious for malignancy, whereas uniform tissue signal usually indicates benign masses.[17] Only lipomas, fibromas, pericardial cysts, and angiomas can be identified by their signal intensity and behavior.

BENIGN TUMORS OF THE HEART

Primary tumors of the heart are very rare, with an incidence between 0.0017 and 0.19 percent in unselected patients at autopsy. Three quarters of the tumors are benign and nearly half of them are myxomas, with another 10 percent benign lipomas.[18–21] Rhabdomyomas, fibromas, hemangiomas, teratomas, and mesotheliomas are found less frequently. Granular cell tumors, neurofibromas, and lymphangiomas are very rare.[22] Table 27–2 presents CMR features of common cardiac tumors.

Myxoma

Myxomas comprise 30 to 50 percent of all primary cardiac tumors and usually occur sporadically between the third and sixth decade of life.[23] In about

Table 27–2. CMR Features of Common Cardiac Tumors

	T1-Weighted Spin-Echo	T2-Weighted Spin-Echo	Gradient-Echo	Enhancement After Administration of Gd-DTPA
Myxoma	Intermediate varying SI; calcified areas; hypointense; hemorrhage: increased SI	Low SI, especially in iron-containing myxomas	Very low SI compared to the surrounding blood pool	Hyperintense
Lipoma/Lipomatous hypertrophy	Brightest SI similar to subcutaneous fat; using fat presaturation technique: reduced SI	Intermediate SI paralleling to subcutaneous fat	Nonspecific	Nonspecific
Fibroma	Intermediate to slightly hyperintense SI compared to myocardium; when calcification (hypointense) and hemorrhage (hyperintense) are present: heterogeneous	Decrease in SI compared to T1	Nonspecific	Slight and heterogeneous
Rhabdomyoma	Homogeneous, slightly lower SI than myocardium	Strong increased SI	Very low compared to myocardium	Nonspecific
Hemangioma	Intermediate SI	Increased SI (due to slow flowing blood in the tumor vessels), higher than myocardium	Nonspecific	Significant increasing SI, heterogeneous
Intravenous leiomyomatosis	Similar to myocardium	Similar to myocardium	Nonspecific	Nonspecific
Pericardial cysts (simple fluid)	Lowest SI, flow void	Highest SI	Nonspecific	Signal enhancement, visualization of intracystic septae
(proteinaceous fluid)	Low SI, but higher than in normal fluid, no flow void	High SI		
Angiosarcomata	Central hyperintense spot (blood vessels, hemorrhage or necrosis), surrounded by intermediate SI regions	No change	Nonspecific	Strong
Lymphoma	Isointense to hypointense to cardiac muscle	Isointense to myocardium	Nonspecific	Heterogeneous with less enhancing central regions
Liposarcoma	Bright SI equal to subcutaneous fat, but heterogeneous; decrease in SI when fat presaturation is used	Not published	Not published	Not published
Leiomyosarcoma	High SI, slightly higher than liver parenchyma, but not as high as subcutaneous fat, homogeneous, commonly connected with proteinaceous pericardial effusion	Not published	Not published	Not published
Thrombus	Intermediate, often slightly higher SI than myocardium, slightly higher than blood	Surrounding slowly flowing blood becomes higher SI than thrombus, contrast between thrombus and myocardium is further accentuated	Thrombus has the lowest SI	No signal enhancement, unless the thrombus is organized
Fresh	High SI (oxyhemoglobin)			
Chronic (older than 2 weeks)	Higher SI (deoxyhemoglobin)	Decreased SI		

SE = spin-echo; SI = signal intensity.
From Hoffman U, Globits S, Frank H: Cardiac and paracardiac masses—Current opinion in the diagnostic evaluation by magnetic resonance imaging. Eur Heart J 19:166, 1998.

75 percent, myxomas originate from the left atrium, and in 15 to 20 percent from the right atrium. They usually develop from the interatrial septum close to the fossa ovalis. Only a few myxomas are located in the ventricles. The histological structure shows typically a myxoid matrix, large blood vessels at the base, and often cysts and areas of hemorrhage and calcification.[24] They are generally polypoid, often pedunculated, round or oval with a smooth surface, often covered with thrombi, and they range in tumor size between 1 and 15 cm in diameter.[17] Clinical symptoms appear as a consequence of embolism or intracardiac obstruction and are determined by size, location, and mobility of the myxoma.[25, 26] Most

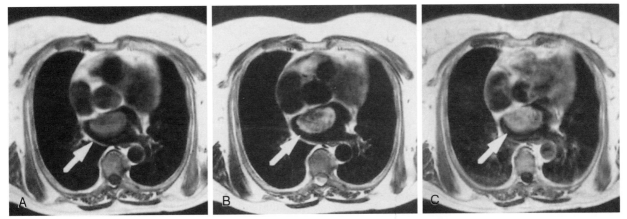

Figure 27–1. Axial spin-echo CMR of a large myxoma (arrow) in the left atrium using *(A)* T1-weighted and *(B)* T2-weighted spin-echo and *(C)* after administration of Gd-DTPA. Note the typical signal increase and inhomogeneous appearance of the myxoma at T2 and the moderate enhancement after Gd-DTPA administration.

commonly, they are first identified by transthoracic echocardiography in patients referred for suspected cardiac source of embolism.

On CMR, myxomas (Figures 27–1 and 27–2) are mainly diagnosed by the typically pedunculated, gelatinous, and prolapsing appearance and certain signal characteristics.[27] Therefore, cine display should be obtained to show the mobility of the tumor.[28] Due to the endocardial origin, myxomas are characterized by an intermediate but variable signal on spin-echo images, similar to that of myocardium.

On GRE images, myxomas often have a low signal intensity, which is caused by partial calcification. Therefore the tumor can usually be distinguished from the higher signal of the surrounding slowly flowing blood.[29, 30] Intratumorous areas of subacute or chronic hemorrhage are typically characterized by high signal intensity on both short and long echo times.[31] Myxomas show a moderately high contrast enhancement after intravenous administration of Gd-DTPA, which is due to their high vascularity.[9, 29]

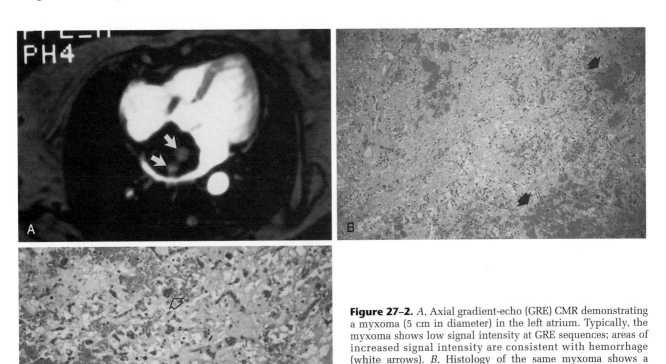

Figure 27–2. *A,* Axial gradient-echo (GRE) CMR demonstrating a myxoma (5 cm in diameter) in the left atrium. Typically, the myxoma shows low signal intensity at GRE sequences; areas of increased signal intensity are consistent with hemorrhage (white arrows). *B,* Histology of the same myxoma shows a typical myxoid matrix and hemorrhage (black arrows), as already diagnosed by CMR, and *(C)* siderophages (open arrows).

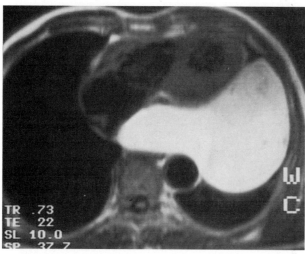

Figure 27–3. Oblique, axial T1-weighted spin-echo CMR of a large lipoma in a child, which arises close to the diaphragm with typical bright signal intensity.

Lipoma

Cardiac lipomas (Figures 27–3 and 27–4) are the second most frequent benign tumor of the heart.[22] True lipomas are encapsulated, contain neoplastic fat cells, and occur in young age.[32, 33] About 50 percent arise subendocardially, 25 percent subepicardially, and 25 percent from the myocardium.[34] Subepicardial lipomas may become quite large and may alter cardiac function, resulting in dyspnea or fatigue,[35] and involvement of the coronary arteries has been reported.[36] Endocardial lipomas commonly arise from the interatrial septum and are located in the right atrium. Arrhythmias due to myocardial infiltration have been reported.[37] Lipomatous hypertrophy of the atrial septum is histologically characterized by infiltration of lipomatous cells between atrial muscle fibers. Unlike true lipomas, they are unencapsulated and contain lipoblasts as well as mature fat cells.[38] This condition has been described in older, overweight patients who frequently have atrial fibrillation.[39]

On CMR, lipomas are characterized by bright signal intensity on T1-weighted images and a slight decrease in signal intensity on T2-weighted images similar to subcutaneous fat.[4] The administration of Gd-DTPA is not needed because signal intensity will remain unchanged. A decrease in signal intensity using fat presaturation technique verifies the diagnosis. In lipomatous hypertrophy a bilobed atrial septum thickening with a signal intensity comparable to subcutaneous fat on T1- and T2-weighted

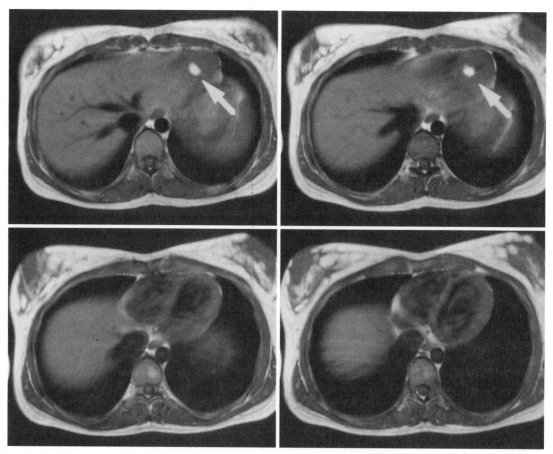

Figure 27–4. Consecutive axial T1-weighted spin-echo images in a 28-year-old woman with a lipomatous structure in the left ventricle (arrow). This structure was primarily diagnosed as a lipoma; however, a possible liposarcoma could not be absolutely excluded. A second CMR study after 8 months confirmed the unchanged lipomatous structure as evidence of a benign lipoma.

images can be visualized. In contrast to the benign lipomatous hypertrophy of the interatrial septum, the very rare liposarcoma shows usually infiltration, inhomogeneities, and fast tumor growth.[41]

Fibroma

Fibromas occur primarily in infants and children. They are congenital tumors frequently discovered in young adults.[42] Typically, fibromas are located intramyocardial within the ventricular septum. The left ventricle is more often involved than the right ventricle. Surgical excision is recommended even in asymptomatic patients,[43] due to the potential risk of sudden death caused by arrhythmias.

On T1-weighted spin-echo images, fibromas are isointense or slightly hyperintense compared to skeletal muscle. Due to the short T2 relaxation time of fibrous tissue, fibromas show a decrease in signal intensity relative to the myocardium from T1- to T2-weighted spin-echo images.[44, 45] A possible problem in diagnosing fibrous tissue might be the presence of fibromuscular elements within the right atrium. Small nodular soft tissue structures that are isointense to myocardium and nodules or linear strands in the right atrium are commonly visible and may simulate a tumor.[46] These structures represent variable degrees of remnants of the crista terminalis and the Chiari network in humans.[47]

Rhabdomyoma

Rhabdomyomas are congenital tumors mainly diagnosed in newborn infants. Usually they arise from the ventricular myocardium at multiple locations. In about 50 percent of patients, tuberous sclerosis can be diagnosed.[48] Rhabdomyomas typically have a solid and homogeneous appearance, which is hypointense to the myocardium on T1-weighted spin-echo images and slightly hyperintense on T2-weighted spin-echo CMR.[49]

Cardiac Hemangioma

There are only a few cases reported of arteriovenous hemangiomas of the interventricular septum of the heart. The location of the tumor is predominantly the right or left ventricle; in addition, septal involvement and multiple locations have been reported.[50] The distinction between a hemangioma and a vascular malformation can be difficult.[51] On CMR, hemangiomas are characterized as a region of increased signal intensity on T1-weighted spin-echo images compared with the myocardium due to slow flowing blood. After intravenous administration of Gd-DTPA, the vascular nature of the tumor can be easily visualized.[52] Figure 27–5 exemplifies the typical appearance of a hemangioma of the right ventricular wall after administration of Gd-DTPA.

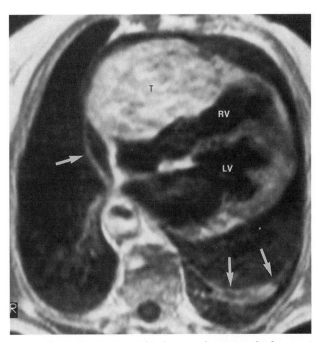

Figure 27–5. Axial T1-weighted spin-echo CMR of a hemangioma after administration of Gd-DTPA in a 85-year-old woman. Compared with the myocardium, the tumor (T) shows inhomogeneous, hyperintense signal intensity with infiltration of the right ventricular myocardium and the right atrial wall. Note the pericardial effusion (white arrows). RV = right ventricle; LV = left ventricle.

Leiomyomatosis with Intracardiac Extension

The intravenous leiomyomatosis is a rare pathological entity and all tumors have been observed in women, most of whom were white and premenopausal.[53] The tumor arises either from a uterine myoma, or from the wall of the vessel.[54] The tumor generally appears as a large mobile mass in the right atrium. Because the preoperative evaluation should include assessment of all cardiac chambers and the region of the inferior vena cava, CMR can be considered to be a primary diagnostic method.[55] Due to the myomatous tissue, signal intensity characteristics are similar to that of muscle.

Other Benign Tumors

There are a large number of very unusual conditions that affect the heart causing tumors. These may be infective, such as echinococcus cysts (Figure 27–6), or inflammatory (Behçet's syndrome mimicking myxoma) and other rare causes.

MALIGNANT TUMORS OF THE HEART

Nearly 25 percent of cardiac tumors are malignant tumors. Metastases are 20 to 40 times more common

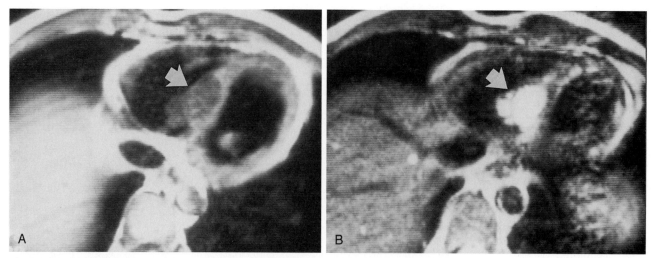

Figure 27–6. Axial *(A)* T1-weighted and *(B)* T2-weighted spin-echo of an echinococcal cyst located in the interventricular septum (arrow). Signal increase of the cyst is typically found on the T2-weighted spin-echo images.

than primary malignant tumors and appear in 6 percent of postmortem autopsies in malignant disease. The most common primary malignant cardiac tumors are various sarcomas and lymphomas.[56, 57]

Due to the small number of studied cardiac malignancies, and the differences in tumor age, vascularity, and a widespread variability in water content, a reliable tissue differentiation of cardiac malignancies is still not possible.[58, 59] Distinct features of malignant tumors are the presence of necrosis, calcification, a high degree of vascularity, infiltration of the adjacent tissues, inhomogeneous appearance, and peritumorous edema.[60, 61] Table 27–2 presents CMR features of common cardiac tumors.

Sarcoma

There exist various histological types of sarcoma, such as angiosarcomas, leiomyosarcomas, and liposarcomas.[22, 57] An angiosarcoma (Figures 27–7 and 27–8) is the most common primary malignant cardiac tumor.[22] It is usually located within the right atrium and arises from the interatrial septum. By contrast, other types of sarcoma also occur in the left side of the heart, where they are often clinically mistaken for myxoma. Typically an angiosarcoma has a polymorphic appearance, with a central region of hyperintensity, consistent with necrosis, and moderate signal intensity in peripheral regions in T1- and T2-weighted spin-echo images.[30] Due to the high degree of vascularity, signal enhancement is seen after intravenous administration of Gd-DTPA.

Seventy-five percent of primary leiomyosarcomas arise from the inferior vena cava, but they also have been reported with an origin of the superior vena cava.[62] This neoplasm demonstrates a signal intensity on T1-weighted spin-echo images slightly higher than liver parenchyma, but not as bright as the adjacent mediastinal fat. The advantage of CMR imaging is the ability to assess the tumor extension both in the superior vena cava and in the heart chambers.[63]

Liposarcomas often have a pericardial origin, and CMR is able to detect this pericardial mass with heterogeneous high signal intensity and epicardial infiltration. After administration of Gd-DTPA, liposarcomas may only show a slight signal enhancement.[64]

Lymphoma

On T1- and T2-weighted spin-echo images, lymphomas appear isointense or hypointense to cardiac muscle. After administration of Gd-DTPA, lymphomas appear heterogeneous with less enhancing central regions consisting of necrosis.[59] Postmortem studies have shown a cardiac involvement in up to 25 percent of lymphoma; however, in vivo diagnosis is still rare.[60]

Metastatic Tumors

There are three ways in which noncardiac tumors may invade the heart: (1) direct mediastinal infiltration of the myocardium, as in lung cancer (Figure 27–9), breast cancer, or mediastinal lymphomas; (2) metastasis by systemic tumors as occurs with melanoma, lymphoma, leukemia, and sarcoma; or (3) transvenous spreading from the inferior vena cava as in primary renal or hepatic tumors, and transvenous spreading from the superior vena cava in case of lung cancer.[22, 35]

INTRACARDIAC THROMBUS FORMATION

Intracardiac thrombus formation is often located in the left atrium (Figure 27–10) in case of chronic

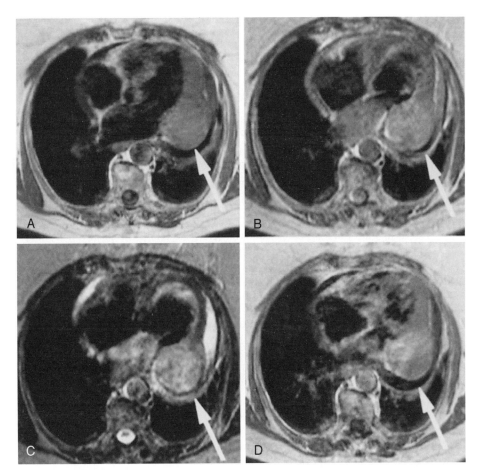

Figure 27–7. Axial spin-echo CMR of a sarcoma on *(A)* T1-weighted image, *(B)* spin-density image, *(C)* late T2-weighted image, and *(D)* after administration of Gd-DTPA. The tumor arises from the left ventricular lateral wall close to the posterior leaflet of the mitral valve. The sarcoma appears heterogeneous and slightly hyperintense to the myocardium on T1 with increasing signal intensity on the T2-weighted images. After the administration of Gd-DTPA, the sarcoma shows a central region of hyperintensity. Note the small pericardial effusion (arrow).

atrial fibrillation and dilatation or in the left ventricle in cases of severe left ventricular systolic dysfunction.[65] The diagnosis of cardiac thrombus is clinically important to identify, inasmuch as patients are at risk of systemic or pulmonary embolization.[66, 67] However, in most cases, the diagnosis of cardiac thrombi is coincidental and patients are

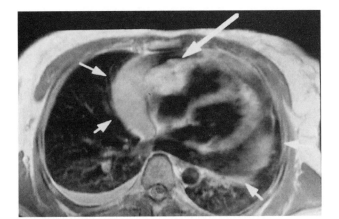

Figure 27–8. Axial spin-echo CMR of an angiosarcoma after the administration of Gd-DTPA, which is located within the right atrium close to the tricuspid valve. There is signal enhancement of the angiosarcoma (large arrow) due to the high degree of vascularity. The small arrows demarcate the pericardial effusion. A small right pleural effusion is also present.

asymptomatic. Despite the fact that 2-D transesophageal echocardiography is the method of choice for atrial thrombus diagnosis and 2-D transthoracic echocardiography for left ventricular thrombi, false-positive rates as high as 28 percent[68] in detection of left ventricular thrombi and 59 percent in left atrial thrombi[69] have been reported.

On CMR images, fresh thrombi on T1-weighted spin-echo images often have a higher signal intensity than myocardium, and the contrast is further accentuated on T2-weighted spin-echo images consistent with a high amount of hemoglobin.[65] However, depending on the age of the thrombus, other signal intensities are possible. After 1 or 2 weeks, paramagnetic compounds in the organizing thrombus such as deoxyhemoglobin and methemoglobin cause T1 and T2 shortening, which may result in increased signal intensity in T1-weighted and decreased signal intensity in T2-weighted images.[70] Chronic organized thrombi are of low signal intensity because of loss of water and protons. A problem concerning differentiation between thrombus and slow flowing blood occurs especially in laminated or immobile thrombi.[71] Compared with thrombus formation, slow flowing blood shows an increasing signal intensity on T2-weighted images.[44, 72] On GRE images, thrombus always has the lowest signal intensity compared with other cardiac structures, whereas flowing blood appears brightest.[70] If

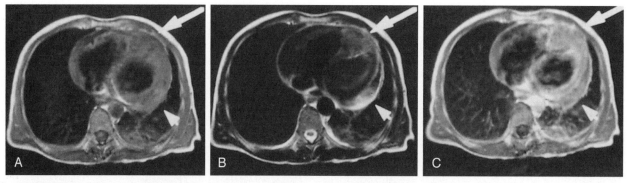

Figure 27–9. Axial T1-weighted spin-echo CMR of a bronchogenic metastasis located at the apex *(A)* on T1-weighted image, *(B)* on T2-weighted image, and *(C)* after the administration of Gd-DTPA. The metastasis shows only a slightly increased signal intensity after the administration of Gd-DTPA; however, it can easily be differentiated from the normal myocardium on T2-weighted images (large arrow). The small arrow indicates the hemorrhagic pericardial effusion, typically seen with cardiac metastasis.

thrombi contain calcification, they appear more heterogeneous.[31] To differentiate thrombus from tumor, intravenous administration of Gd-DTPA may be helpful. Thrombi usually do not show signal enhancement with Gd-DTPA (unless they are already organized).[9] Compared with other diagnostic procedures, CMR and computed tomography (CT) offer a similar sensitivity of about 90 percent with a slightly better specificity when compared with 2-D surface echocardiography for the diagnosis of left ventricular thrombi.[71]

PERICARDIAL LESIONS

CMR evaluation of pericardial neoplasm in most cases involves identification of abnormal anatomical

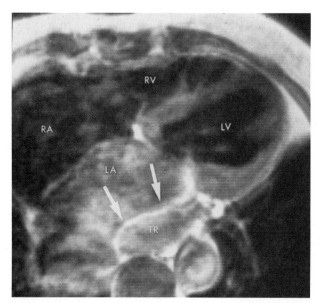

Figure 27–10. Axial T1-weighted spin-echo CMR of a large thrombus (TR) located in the left atrium after the administration of Gd-DTPA. The contrast agent is still attached to the thrombus surface (arrows), delineating the entire thrombus. Note also the increased signal in the left atrium of this patient with atrial fibrillation indicating slow-flowing blood. LV = left ventricle; RV = right ventricle; LA = left atrium; RA = right atrium.

structures and boundaries rather than characterization of relative tissue intensities. A few exceptions are included in the differential diagnosis of mediastinal masses, such as fibroma, lipoma, and pericardial cysts. The value of CMR for evaluation of potential neoplasm lies largely in treatment planning and particularly preoperative assessment. The loss of normal anatomical boundaries is an important sign of neoplasm. Neoplastic involvement of the pericardium results in focal and diffuse obliteration of the normal pericardial signal. In the case of malignancy adjacent to cardiac structures, visualization of the pericardial line is an indication that pericardial invasion has not occurred. The reader is also directed to Chapter 28.

Tumors

The diagnosis of pericardial tumors includes benign lipoma or pericardial cysts and some rare primary malignancies such as teratoma, mesothelioma, bronchial pheochromocytoma, lymphoma, fibrosarcoma, and angiosarcoma.[73, 74] Hemorrhagic effusions result from erosion into intrapericardiac vessels or myocardial wall, with possible acute or subacute tamponade. Lipoma and pericardial fat pads are easily distinguished by the high signal of fat tissue.[75]

Pericardial Metastasis

The involvement of the pericardium by metastatic disease at autopsy is much higher than clinically suspected, ranging from 1.5 to 22 percent in incidence.[76] Of malignant pericardial disease, 80 percent is associated with lung or breast cancer, leukemia, and lymphoma.[77] Metastatic involvement of the pericardium is characterized by large effusions out of proportion to the amount of tumor present and is the most frequent cause of tamponade. Focal or diffuse plaque-like thickening of the pericardium may occur with signal enhancement after administration of Gd-DTPA.

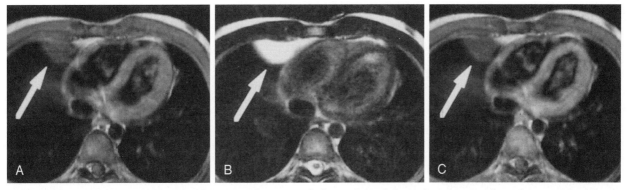

Figure 27–11. Axial T1-weighted spin-echo CMR of a pericardial cyst (arrow) located anterior to the right ventricle *(A)* on T1-weighted image *(B)* on T2-weighted image, and *(C)* after the administration of Gd-DTPA. Highest signal intensity can be typically detected on the T2-weighted images.

Cysts

Pericardial cysts (Figures 27–11 and 27–12) are rare lesions and are commonly located in the right pericardiophrenic angle.[78, 79] Pericardial cysts are usually filled with a clear fluid. The patients are generally asymptomatic, and the lesion is often discovered either on a routine chest film or as a loculated anterior effusion on surface echocardiography. The appearance is typically stable over a long time. In most cases no cardiac surgery is necessary.[80] On CMR, pericardial cysts appear as paracardiac masses with long T1 and T2 values and flow void, indicating fluid-filled structures. They have low signal intensity on T1-weighted and increased signal intensity on T2-weighted images.[81] After the administration of Gd-DTPA, intracystic septae may be observed. In addition, a line of low signal intensity, representing the pericardial layer, can be often visualized. The significant advantage of CMR is its ability to differentiate these lesions from other mediastinal masses and avoid explorative surgery to determine the diagnosis.

CONCLUSION

CMR techniques have contributed significantly to the ability to detect cardiac and paracardiac masses

and play an important role in the diagnostic evaluation that is complementary to echocardiography. CMR, due to its larger field of view, adds diagnostic information by assessing extracardiac components of a mass, such as mediastinal involvement and extension into large pulmonary vessels. CMR allows the exclusion of hiatus hernia, a tortuous descending aorta, or a bronchogenic cyst, which can mimic cardiac tumors. CMR findings are helpful in characterizing paracardiac masses and in guiding therapeutic strategies. Tissue characterization by CMR is limited to the diagnosis of myxomas, fibromas, thrombi, pericardial cysts, and fatty tissue. The signal features of malignant tumors are equivocal and do not permit a tissue diagnosis. However, the inhomogeneous appearance of signal enhancement after the administration of Gd-DTPA, the infiltrative components of a tumor, and a hemorrhagic pericardial effusion makes the diagnosis of tumor malignancy more likely.

References

1. Lanzer P, Barta C, Botvinick EH, et al: ECG-synchronized cardiac MR imaging. Method and evaluation. Radiology 155:681, 1985.
2. Wang Y, Rossman PJ, Grimm RC, et al: Navigator-echo-based real time respiratory gating and triggering for reduction of respiration effects in three-dimensional coronary MR angiography. Radiology 198:55, 1996.
3. Crooks LE, Barker B, Chang H, et al: Magnetic resonance imaging strategies for heart studies. Radiology 153:459, 1984.
4. Edelstein WA, Bottomley PA, Hart HR, Smith LS: Signal, noise, and contrast in nuclear magnetic resonance imaging. J Comput Assist Tomogr 7:391, 1983.
5. Seelos KC, Smekal von A, Vahlensieck M, et al: Cardiac abnormalities: Assessment with T2-weighted turbo spin-echo MR imaging with electrocardiogram gating at 0.5T. Radiology 189:517, 1993.
6. Pettigrew RI: Cardiovascular imaging techniques. *In* Stark DD, Bradley WB (eds): Magnetic Resonance Imaging. St. Louis, Mosby-Year Book, 1992, p 1605.
7. Sechtem U, Pflugfelder PW, White RD, et al: Cine MR imaging. Potential for the evaluation of cardiovascular function. AJR Am J Roentgenol 148:239, 1987.
8. Carr DH, Brown J, Bydder GM, et al: Gadolinium-DTPA as a contrast agent in MR. Initial experience in 20 patients. AJR Am J Roentgenol 143:215, 1984.
9. Weinmann HJ, Lanaido M, Mutzel W: Pharmakokinetics of

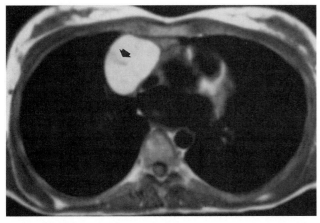

Figure 27–12. Pericardial cyst on a T2-weighted spin-echo CMR with an intracystic septum (arrow).

Gd-DTPA/dimeglumine after IV injection. Physiol Chem Phys Med NMR 16:167, 1984.

10. Funari M, Fujita N, Peck WW, Higgins CB: Cardiac tumors. Assessment with Gd-DTPA enhanced MR imaging. Comp Assist Tomogr 15:953, 1991.

11. Amparo EG, Higgins CB, Farmer D, et al: Gated MR of cardiac and paracardiac masses. Initial experience. Am Heart J 143:1151, 1984.

12. Dooms GC, Hricak H, Margulis AR, Geer G: MR imaging of fat. Radiology 158:51, 1986.

13. Schulthess GK, McMurdo K, Tscholakoff D, et al: Mediastinal masses. MR imaging. Radiology 158:289, 1986.

14. Tscholakoff D: MRT in the diagnosis of cardiovascular system and lung. Acta Med Austriaca 3:61, 1986.

15. Schmidt HC, Tscholakoff D, Hricak H, Higgins CB: MR image contrast and relaxation times of solid tumors in the chest, abdomen, and pelvis. J Comput Assist Tomogr 9:738, 1985.

16. Walker PM, Marie PY, Danchin N, Bertrand A: Comparison of T1 estimation techniques in cardiac MR. Magn Reson Imaging 12:43, 1994.

17. Hall RJ, Cooley DA, Mac Allister HR Jr, Frazier OH: Neoplastic heart disease. In Hurst JW (ed): The Heart, Arteries and Veins. 7th ed. New York, McGraw-Hill, 1990, p 1382.

18. Benjamin HG: Primary fibromyxoma of the heart. Arch Pathol 27:950, 1939.

19. Heath D: Pathology of cardiac tumors. Am J Cardiol 21:315, 1968.

20. Straus R, Merliss R: Primary tumor of the heart. Arch Pathol 39:74, 1945.

21. Wold LE, Lie JT: Cardiac myxomas. A clinicopathologic profile. Am J Pathol 101:219, 1980.

22. Mc Allister HA Jr, Fenoglio JJ Jr: Tumors of the cardiovascular system. In Atlas of Tumor Pathology. 2nd series. Fascicle 15. Washington, DC, Armed Forces Institute of Pathology, 1978, pp 1–20.

23. Reynen, K: Cardiac myxomas. N Engl J Med 333:1610, 1995.

24. Pritchard RW: Tumors of the heart. Review of the subject and report of one hundred and fifty cases. Arch Pathol 51:98, 1951.

25. St John Sutton MG, Mercier L-A, Giuliani ER, Lie JT: Atrial myxomas. A review of clinical experience in 40 cases. Mayo Clin Proc 55:371, 1980.

26. Peters MN, Hall RJ, Cooley DA, et al: The clinical syndrome of atrial myxoma. JAMA 230:695, 1974.

27. Lund JT, Ehman RL, Julsrud PR, et al: Cardiac masses. Assessment by MR. AJR Am J Roentgenol 152:469, 1989.

28. Go RT, O'Donnell JK, Underwood DA, et al: Comparison of gated cardiac MR and 2D echo of intracardiac neoplasms. AJR Am J Roentgenol 145:21, 1985.

29. Semelka RC, Shoenut JP, Wilson ME, et al: Cardiac masses. Signal intensity features on spin-echo, gradient-echo, gadolinium-enhanced-spin-echo, and TurboFLASH images. J Magn Reson Imaging 2:415, 1992.

30. Gomes AS, Lois JF, Child JS, et al: Cardiac tumors and thrombus. Evaluation with MR imaging. AJR Am J Roentgenol 149:895, 1987.

31. Roos A, Weijers E, Duinen S, Wall EE: Calcified right atrial myxoma demonstrated by MR. Chest 95:478, 1989.

32. Reyes CV, Jablokow VR: Lipomatous hypertrophy of the cardiac interatrial septum. A report of 38 cases and review of the literature. Am J Clin Pathol 5:785, 1979.

33. Crocker DW: Lipomatous infiltrates of the heart. Arch Pathol Lab Med 102:69, 1978.

34. Fine G: Neoplasms of the pericardium and heart. In Gould SE (ed): Pathology of the Heart and Blood Vessels. 3rd ed. Springfield, Charles C Thomas, 1968, p 865.

35. Moulton AL, Jaretzki A III, Bowman OF, et al: Massive lipoma of the heart. NY State J Med 76:1820, 1976.

36. Hananouchi GI, Goff WB: Cardiac lipoma. Six-year follow-up with MR characteristics, and a review of the literature. Magn Reson Imaging 8:825, 1990.

37. Conces DJ, Vix VA, Tarver RD: Diagnosis of a myocardial lipoma by using CT. AJR Am J Roentgenol 153:725, 1989.

38. Hutter Am Jr, Page DL: Atrial arrhythmias and lipomatous hypertrophy of the cardiac interatrial septum. Am Heart J 82:16, 1971.

39. Kluge WF: Lipomatous hypertrophy of the interatrial septum. Northwest Med 68:25, 1969.

40. Levine RA, Weyman AE, Dinsmore RE, et al: Noninvasive tissue characterisation. Diagnosis of lipomatous hypertrophy of the atrial septum by nuclear MR. J Am Coll Cardiol 7:688, 1986.

41. Applegate PM, Tajik AJ, Ehman RL, et al: Two-dimensional echocardiographic and MR observations in massive lipomatous hypertrophy of the atrial septum. Am J Cardiol 59:489, 1987.

42. Burke AP, Rosado CM, Templeton PA, Virmani R: Cardiac fibroma. Clinicopathologic correlates and surgical treatment. J Thorac Cardiovasc Surg 108:862, 1994.

43. Oliva PB, Breckinridge JC, Johnson ML, et al: Left ventricular outflow obstruction produced by a pedunculated fibroma in a newborn. Clinical, angiographic, echocardiographic and surgical observations. Chest 74:590, 1978.

44. Winkler M, Higgins CB: Suspected intracardiac masses. Evaluation with MR imaging. Radiology 165:117, 1987.

45. Gamsu G, Stark DD, Webb WR, et al: MR of benign mediastinal masses. Radiology 151:709, 1984.

46. Meier RA, Hartnell GG: MR of right atrial pseudomass. It is really a diagnostic problem? J Comput Assist Tomogr 18:398, 1994.

47. Edwards JE: Congenital malformations of the heart and great vessels. In Gould SE (ed): Pathology of the Heart. Springfield, IL, Bannerstone House, 1960, p 61.

48. Abushaban L, Denham B, Duff D: Ten year review of cardiac tumors in childhood. Br Heart J 70:166, 1993.

49. Hoffmann U, Globits S, Frank H: Cardiac and paracardiac masses—Current opinion in the diagnostic evaluation by magnetic resonance imaging. Eur Heart J 19:166, 1998.

50. Brizard C, Latremouille C, Jebara VA, et al: Cardiac hemangiomas. Ann Thorac Surg 56:390, 1993.

51. Newell JD II, Eckel C, Davis M, Tadros NB: MR appearance of an arteriovenous hemangioma of the interventricular septum. Cardiovasc Intervent Radiol 11:319, 1988.

52. Soberman MS, Plauth WH, Winn KJ, et al: Hemangioma of the right ventricle causing outflow obstruction. J Thorac Cardiovasc Surg 96:307, 1988.

53. Spellacy WN, Maire WJ, Buhi WC: Plasma growth hormone and estradiol. Levels in woman with uterine myomas. Obstet Gynecol 40:829, 1972.

54. Bassish MS: Mesenchymal tumors of the uterus. Clin Obstet Gynecol 17:51, 1974.

55. Rosenberg JM, Marvasti MA, Obeid A, et al: Intravenous leiomyomatosis. A rare cause of right sided cardiac obstruction. Eur J Cardiothorac Surg 2:58, 1988.

56. Roberts WC, Glancy DL, DeVita VT: Heart in malignant lymphoma. A study of 196 autopsy cases. Am J Cardiol 22:85, 1968.

57. Silverman J, Olwin JS, Graettinger JS: Cardiac myxomas with systemic embolization. Review of the literature and report of a case. Circulation 26:99, 1962.

58. Zeitler E, Kaiser W, Schuierer G, et al: MR of aneurysms and thrombi. Cardiovasc Intervent Radiol 8:321, 1986.

59. Dorsay TA, Ho VB, Rovira MJ, et al: Primary cardiac lymphoma. CT and MR findings. J Comp Assist Tomogr 17:978, 1993.

60. Hendrick RE, Raff U: Image contrast and noise. In Stark DD, Bradley WB (eds): Magnetic Resonance. St. Louis, Mosby-Year Book, 1992, pp 109–144.

61. Tazelaar HD, Locke TJ, McGregor CGA: Pathology of surgically excised primary cardiac tumors. Mayo Clin Proc 67:957, 1992.

62. Lupetin AR, Dash N, Beckman I: Leiomyosarcoma of the superior vena cava. Diagnosis by cardiac gated MR. Cardiovasc Intervent Radiol 9:103, 1986.

63. Somers K, Lotte F: Primary lymphosarcoma of the heart. Review of the literature and report of 3 cases. Cancer 3:449, 1960.

64. Garrigue S, Robert F, Roudaut R, Bonnet J: Assessment of non-invasive imaging techniques in the diagnosis of heart liposarcoma. Eur Heart J 16:139, 1995.

65. Dooms GC, Higgins CB: MR imaging of cardiac thrombi. J Comp Assist Tomogr 10:415, 1986.

66. Visser CA, Kan G, Meltzer RS, et al: Embolic potential of left ventricular thrombus after myocardial infarction. A two-dimensional echocardiographic study of 119 patients. J Am Coll Cardiol 5:1276, 1985.

67. Hamby RI, Wisoff BG, Davison ET, et al: Coronary artery disease and left ventricular mural thrombi. Clinical, hemodynamic, and angiocardiographic aspects. Chest 66:488, 1974.

68. Barakos JA, Brown JJ, Higgins CB: MR imaging of secondary cardiac and paracardiac lesions. AJR Am J Roentgenol 153:47, 1989.

69. Feigenbaum H: Coronary artery disease. *In* Feigenbaum H (ed): Echocardiography. 4th ed. Philadelphia, Lea and Febiger, 1986, pp 489–494.

70. Jungehülsing M, Sechtem U, Theissen P, et al: Left ventricular thrombi. Evaluation with spin-echo and gradient-echo MR imaging. Radiology 182:225, 1992.

71. Sechtem U, Theissen P, Heindel W, et al: Diagnosis of left ventricular thrombi by MR and comparison with angiocardiography, computed tomography and echocardiography. Am J Cardiol 64:1195, 1989.

72. Menegus MA, Greenberg MA, Spindola-Franco H, Fayemi A: MR of suspected atrial tumors. Am Heart J 123:1260, 1992.

73. Stark DD, Higgins CB, Lanzer P, et al: Magnetic resonance imaging of the pericardium. Normal and pathologic findings. Radiology 150:469, 1984.

74. Hort W, Braun H: Untersuchungen ueber Groesse, Wandstärke und mikroskopischen Aufbau des Herzbeutels unter normalen und pathologischen Bedingungen. Arch Kreislaufforsch 38:1, 1962.

75. Amparo EG, Higgins CB, Farmer D, et al: Gated MR of cardiac and paracardiac masses: Initial experience. AJR Am J Roentgenol 143:1151, 1984.

76. Mocanda R, Kotler MN, Churchill RJ, et al: Multimodality approach to pericardial imaging. Cardiovasc Clin 17:409, 1986.

77. Theoligides A: Neoplastic cardiac tamponade. Semin Oncol 5:181, 1978.

78. Fraser RG, Pare JAP: Diagnosis of Disease of the Chest. 2nd ed. Philadelphia, WB Saunders, 1977, p 656.

79. Le Roux BT: Pericardial coelomic cysts. Thorax 14:27, 1977.

80. Roberts WC, Spray TL: Pericardial heart disease. A study of its causes, consequences and morphologic features. Cardiovasc Clin 7:11, 1976.

81. Sechtem U, Tscholakoff D, Higgins CB: MR of the abnormal pericardium. AJR Am J Roentgenol 147:245, 1986.

Cardiovascular Magnetic Resonance Evaluation of the Pericardium in Health and Disease

G. Wesley Vick III and Roxann Rokey

ANATOMY OF THE NORMAL PERICARDIUM

The pericardium is composed of two distinct layers.[1-4] The inner layer, termed the *visceral* pericardium, is normally a one-cell-thick layer of mesothelial cells. The visceral pericardium covers the surface of the heart, overlying subepicardial fat. The outer pericardial layer, termed the *parietal* pericardium, is a fibrous structure that encloses the heart and extends to cover portions of the proximal great vessels. The parietal pericardium is composed of three sublayers—the serosa, the fibrosa, and the epipericardium. The serosa, like the visceral pericardium, is formed by mesothelial cells with numerous microvilli. The fibrosa is composed of collagen and elastic fibrils. It provides much of the pericardium's strength and elasticity. The outer epipericardial layer contains nerves, lymphatic vessels, blood vessels, and adipose tissue, in addition to some elastic and collagen fibrils. Pericardial attachments from the outer epipericardial layer extend to the diaphragm, to the manubrium, and to the vertebral column. These attachments, in conjunction with attachments to the great vessels, constrain bulk cardiac motion within the chest cavity. In health, intrapericardial pressure is slightly negative and the two layers of the pericardium are separated by a fluid layer produced by the visceral pericardium. This straw-colored pericardial fluid, usually 10 to 50 ml, acts as a lubricant to minimize frictional forces. Physiological pericardial fluid has a lower protein concentration than plasma.

DISEASES OF THE PERICARDIUM

Pericarditis

A variety of disorders can lead to pericardial inflammation and pericarditis.[5-7] A multitude of infectious agents, including viral, bacterial, mycobacterial, and fungal, can involve the pericardium and produce local inflammation. Patients with uremia frequently develop pericardial inflammation, which resolves with initiation of more intense dialysis. Pericardial inflammation may occur following large myocardial infarction (Dressler syndrome), after surgical pericardiotomy performed during cardiac surgery (postpericardiotomy syndrome), or after chest wall trauma. Malignancies may metastasize to the pericardium, and radiation therapy to the pericardium may induce focal inflammation. Pericarditis is frequently seen in autoimmune and connective tissue diseases. In many cases, the causation of acute pericarditis cannot be determined, and the etiology is termed *idiopathic/presumed viral.*

Constrictive Pericarditis

Though uncommon, all etiologies leading to acute pericardial inflammation can also lead to chronic injury/pericardial thickening.[8, 9] Sometimes, the original causative inflammation may be subclinical/asymptomatic. Progressive pericardial thickening and resultant constraint on ventricular filling/constriction can be the result of repeated acute inflammatory insults or develop long after the initial disorder responsible for the acute inflammation has resolved. Classically, fibrous scarring and adhesions of both pericardial layers leads to obliteration of the pericardial space. Early ventricular filling remains unimpeded, but mid and late diastolic filling is abruptly reduced due to the inability of the ventricles to fill because of physical constraints imposed by the rigid, thickened, and sometimes calcified pericardium. Severe pericardial constriction results in a constellation of symptoms and hemodynamic findings termed *constrictive pericarditis.*[5, 6] The symptoms of constrictive pericarditis include abdominal discomfort with distension and ascites, peripheral edema, weight loss, fatigue, lassitude, and muscle wasting. The hemodynamic findings of constrictive pericarditis include increased systemic venous pressure, rapid early diastolic ventricular filling with minimal filling during mid and late diastole, equilibration of diastolic pressures in all cardiac chambers, and increase of systemic venous pressure with inspiration (Kussmaul's sign). Although constrictive pericarditis is a progressive disorder in many cases, instances of transient constrictive pericarditis have been described. The symptoms and many of the hemodynamic findings of constrictive pericarditis are similar to those produced by restrictive cardiomyopathy. Differentiating between restrictive cardiomyopathy and constrictive pericarditis is a common clinical problem.[10]

Pericardial Effusions

Although 10 to 50 ml of bland pericardial fluid may be normal, an increased volume of pericardial fluid or fluid related to a pathological process (blood) may lead to hemodynamic compromise. As with pericarditis, pericardial effusions have a wide range of etiologies and varying degrees of clinical importance, dependent on their size, rate of development, and composition. Very large or rapidly progressing pericardial effusions can severely impair cardiac filling and result in cardiovascular collapse, a situation known as cardiac tamponade.[5–7] Prompt recognition and percutaneous pericardiocentesis can be lifesaving.

Pericardial Cysts

Pericardial cysts are rare congenital anomalies that often come to medical attention when a right-sided paracardiac mass is identified on chest roentgenogram (CXR) or a loculated pericardial effusion is seen around the right atrium on transthoracic echocardiography. Pericardial cysts usually contain transudative fluid and are benign.[11]

Pericardium in Malignancy

Primary malignancies of the pericardium are rare.[12] The most frequently encountered primary pericardial malignancy is a mesothelioma. Other primary pericardial tumors include lipomas, fibrosarcomas, and teratomas. More commonly, the pericardium is a site of metastatic involvement in patients with malignant neoplasms—including melanoma and malignancies of the breast, lung, and colon.

Congenital Absence of the Pericardium

Congenital complete or partial absence of the pericardium is another rare manifestation of pericardial disease.[13] Complete absence of the pericardium is usually asymptomatic and benign. In contrast, partial or localized absence of the pericardium may be associated with focal herniation and subsequent strangulation. The left side of the pericardium is more commonly affected, and has been associated with chest pain and herniation of the left atrial appendage or left ventricle through the defect. Sudden death from cardiac strangulation may result. Congenital absence of the right side of the pericardium is extremely rare. Herniation of the right atrial appendage or right ventricle can occur through a right-sided congenital pericardial defect, and the right lung can herniate into the pericardial cavity through such a defect. Because of reported complications, elective surgical repair is often recommended when asymptomatic partial defects are confirmed.

IMAGING TOOLS FOR PERICARDIAL ASSESSMENT

The CXR is often the first imaging abnormality among patients with a significant pericardial effusion. Typically, the CXR shows an enlarged cardiac silhouette with "water bottle" appearance and loss of anatomical contours. Sometimes, the presence of a pericardial effusion can be inferred from an CXR when pericardial fat lines are noted to be separated. When pericardial calcification is present, it is often apparent on CXR[5–7] but easier to identify by fluoroscopy.

The electrocardiogram (ECG) can often suggest the presence of pericardial disease. Diffuse ST segment elevation, with upward concavity, is characteristic in acute pericarditis. Large pericardial effusions and constrictive pericarditis are associated with tachycardia and low (≤5 mm) limb and low (≤10 mm) precordial QRS voltage. Electrical alternans can be an important sign of cardiac tamponade. Atrial fibrillation is common in chronic constrictive pericarditis, a rhythm that can impede R wave synchronization of cardiovascular magnetic resonance (CMR), x-ray computed tomographic, and radionuclide angiographic image acquisitions.[5–7]

Two-dimensional (2-D) surface echocardiography is the most widely employed imaging modality to evaluate patients with suspected pericardial disease. Circumferential pericardial effusion detection and qualitative assessment of effusion size are highly accurate. 2-D–directed transmitral and transtricuspid Doppler is also an excellent tool for confirming the presence of tamponade physiology. 2-D imaging and Doppler echocardiography can also be helpful in the evaluation of patients with suspected constrictive pericarditis.[5–7] However, echocardiography has many limitations. Acoustic windows are not optimal in many subjects, echocardiography cannot effectively characterize pericardial effusions, it is often limited among patients with loculated effusions, and it is generally unreliable for assessment of pericardial thickening. Purulent effusions, chylous effusions, serous effusions, and unclotted blood can all produce echo-free spaces of similar echocardiographic appearance. Clotted blood in the pericardial space is echogenic and therefore may not be identified as an effusion on 2-D echocardiography. Epicardial fat may simulate pericardial effusion on 2-D echocardiograms. Surface echocardiography can sometimes misclassify pleural effusions as pericardial effusions, and vice versa. Transthoracic echocardiography has generally been unreliable for determining pericardial thickness. The parietal pericardium is echogenic, and its apparent thickness on 2-D echocardiography is substantially affected by the settings of the echocardiographic instrument. Some investigators have found transesophageal echocardiography (TEE) to provide improved accuracy for pericardial thickness determination.[14]

X-ray computed tomography (CT) is a widely accepted imaging tool for evaluating pericardial thickness and pericardial calcification.[15–17] Most clinical CT scanners do not synchronize image acquisition with the cardiac cycle and have too long an acquisition time to effectively "freeze" cardiac motion. Hence, cardiac CT images have some degree of blurring. In many cases, however, the pericardium is visible as a thin line between the epicardium and the mediastinal fat. The pericardium is often not visible laterally, because of the absence of fat between the heart and adjacent lung. Thickened pericardium is difficult to differentiate from pericardial effusion by x-ray CT, particularly if the pericardial effusion has a high protein content. Effusions of high protein content have x-ray attenuation properties similar to fibrous pericardium. Thus, patients examined by CT and found to have a thickened pericardium would also benefit from assessment by echocardiography or CMR. More rapid, electron beam x-ray computed tomography (EBCT), which is becoming more widely available, enables rapid (100 ms) image acquisitions, thereby substantially reducing cardiac blurring. This technique improves pericardial resolution, but still may have limited ability to differentiate an exudative pericardial effusion from thickened pericardium.[17, 18] For suspected pericardial calcification, EBCT is likely the best noninvasive modality.

Although radionuclide angiography with technetium-99m sestamibi can sometimes demonstrate the presence of large pericardial effusions, spatial resolution is relatively low. First-pass radionuclide angiography has been successfully employed to differentiate between a constrictive pericarditis and a restrictive cardiomyopathy on the basis of ventricular filling dynamics.[19] Clinically, radionuclide methods are less commonly employed.

CARDIOVASCULAR MAGNETIC RESONANCE OF THE NORMAL PERICARDIUM

On T1-weighted spin-echo CMR, the normal pericardium appears as a low-intensity circumferential band between the high-intensity mediastinal fat and the medium-intensity epicardium (Figure 28–1).[20, 21]

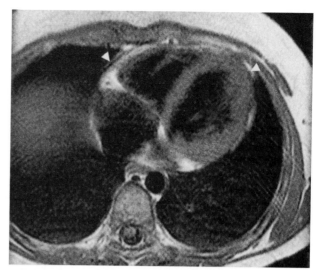

Figure 28–1. T1-weighted spin echo CMR of a normal heart and pericardium in the transverse plane. White arrowheads indicate position of the normal, thin black pericardium.

Because the visceral layer of normal human pericardium is only one cell thick (and thus cannot be resolved by CMR methods) it is likely that the low-intensity region originates from the parietal pericardium. This region of low-intensity signal, composed of fibrous tissue, may not be uniform in thickness. It is often most prominent over the right ventricle and around the inferior and apical surface of the left ventricle. It is often not distinguishable from adjacent low-intensity lung tissue over the right atrium and lateral wall of the left ventricle. Clinical studies have found that the thickness of this low-intensity region averages 1 to 2 mm for normal individuals. However, there is a relatively wide standard deviation of thickness of this low-intensity band in normals, so that a thickness of 3 mm or less is considered normal. The average thickness of the low-intensity region is relatively large in comparison to pathological studies of normal human pericardial thickness, which have found average parietal pericardial thickness to be between 0.4 and 1 mm.[22] The normal pericardium is also not uniform in thickness on pathological examination, with greater thickness at junctures with the pericardial attachments.

The discrepancy between CMR and histological pericardial thickness in normal individuals is likely related to partial volume averaging, inclusion of pericardial fluid in the CMR measurement, chemical shift artifact, and bulk cardiac/pericardial motion during image acquisition. Pericardial fluid is not uniformly distributed. Normal, transudative pericardial fluid has a low signal intensity on T1-weighted spin-echo images, thereby contributing to the pericardial signal seen on T1-weighted images. Chemical shift artifact is another potential contributor to the thickness of the low-intensity region. Fat and water have slightly different resonance frequencies because carbon and oxygen bind hydrogen with different affinity, resulting in differential electron shielding effects for their bound protons. At interfaces between tissues with a relatively large fat content (such as epicardial or mediastinal fat) and a relatively small fat content (such as myocardium), a low signal intensity interface/border is often present due to this chemical shift effect. Finally, motion artifacts may also increase the thickness of the low-intensity region. The parietal pericardium is relatively fixed in position by strong ligamentous attachments, whereas the normal heart muscle moves substantially within the pericardial sac during the cardiac cycle. Phase discontinuities will be created between myocardial and adjacent pericardial and mediastinal voxels by this differential motion of the heart and pericardium during the cardiac cycle. Such phase discontinuities will tend to reduce the intensity of the border zone between the myocardium and adjacent tissue.

Though thought to be less accurate, ECG-gated gradient-echo CMR may also be used to determine pericardial thickening. A low-intensity region is also often seen between the intermediate-intensity mediastinal fat and the intermediate-intensity epicardium.[23, 24] This low-intensity region tends to be thicker than on corresponding T1-weighted spin-echo CMR. As with the spin-echo images, there are a number of potential origins for the low-intensity band. Some of the signal from this band undoubtedly originates from the fibrous parietal pericardium. As with spin-echo CMR, chemical shift artifact and motion artifact may all contribute to the apparent excess thickness of this low-intensity band between the epicardium and mediastinal fat. Mobile pericardial fluid will have a high signal intensity on cine gradient-echo CMR (Figure 28–2). Thus, physiological pericardial fluid can be readily distinguished from thickened pericardium by the cine gradient-echo technique. In general, fat suppression should *not* be used because it limits the ability to distinguish between the parietal pericardium and pericardial fluid.

The transverse sinus of the pericardium is frequently visualized between the ascending aorta and the pulmonary artery on axial or transverse magnetic resonance images, and between the left atrium and right pulmonary artery in sagittal and coronal sections. The transverse pericardial sinus is a normal structure and should not be mistaken for an anomalous vessel (see Chapter 18) or aortic dissection.[25]

CARDIOVASCULAR MAGNETIC RESONANCE OF PERICARDIAL THICKENING AND CONSTRICTIVE PERICARDITIS

As mentioned earlier, pericardial thickening can occur subsequent to a variety of conditions. Furthermore, the pericardial thickening frequently occurs in a heterogeneous fashion. Many investigators have shown that CMR is a highly sensitive and specific imaging tool for identifying focal pericardial thickening.[26–28] Despite the multifactorial origin of the characteristic signal intensity changes seen in the anatomical region of the pericardium on CMR, substantial thickening of this region is strongly associated with surgically confirmed pericardial thickening (Figures 28–3 and 28–4). It is important to remember that pericardial thickening can be demonstrated in asymptomatic patients and therefore does *not* by itself define the need for treatment/intervention. However, demonstration of pericardial thickening in the presence of clinical symptoms and hemodynamic findings consistent with constrictive pericarditis serves as an important confirmation of the diagnosis. Thus, CMR can be of valuable assis-

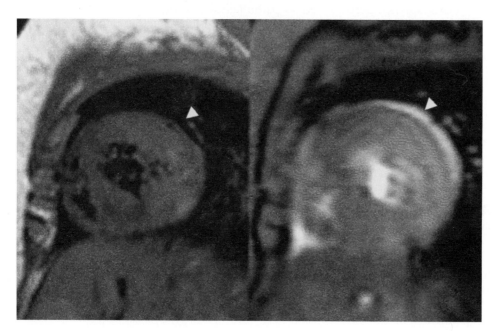

Figure 28–2. *Left*, Oblique short-axis T1-weighted spin-echo CMR in a patient with small pericardial effusion. The effusion is visible as a low-intensity region on the image (white arrowhead) and may be misinterpreted as pericardial thickening. *Right*, gradient-echo CMR in the same orientation. The effusion (bright signal; white arrowhead) is readily apparent on the gradient-echo image.

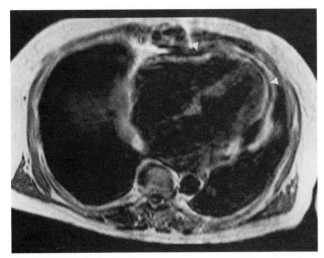

Figure 28–3. T1-weighted spin-echo image of heart in transverse plane in patient with dyspnea on exertion and constrictive physiology. White arrowheads indicate thickened pericardium.

tance in evaluating patients for suspected constrictive pericarditis. It should be noted that constrictive pericarditis can occasionally occur without abnormal thickening of the pericardium. Thus, although a normal CMR estimate of pericardial thickness makes constrictive pericarditis unlikely, it does not, by itself, completely exclude the diagnosis.

When spin-echo sequences suggest the presence of pericardial thickening, it is advisable to perform a cine gradient-echo sequence in the same region/ orientation. Due to low signal on spin-echo imaging associated with pericardial fluid, distinguishing between focal pericardial thickening and small to moderate size pericardial effusions can be difficult with spin-echo CMR alone. Although the cine gradient-echo method tends to overestimate the size of pericardial thickness (compared with the spin-echo method), the cine gradient-echo technique is exquisitely sensitive to pericardial effusions (Figure 28–5, see Figure 28–2). Using cine gradient-echo CMR, pericardial effusions are bright, with an intensity similar to blood, and are readily distinguishable from surrounding myocardium and mediastinal fat, which are relatively dark.

Because not all pericardial thickening is associated with hemodynamic compromise, a hemodynamic and functional assessment are necessary in addition to the imaging examination. This assessment is typically performed with cardiac catheterization. Functional CMR methods analogous to Doppler echocardiography and radionuclide characterization of ventricular filling as well as CMR tagging may be of assistance. Early diastolic ventricular filling is increased in constrictive pericarditis, whereas early diastolic filling is impaired in restrictive cardiomyopathy. Quantitative volumetric multiphasic studies of ventricular flow and volume can now be performed with CMR and should be of assistance in making the distinction between restrictive cardiomyopathy and constrictive pericarditis.[29, 30]

Preliminary data suggest that CMR myocardial tagging may distinguish between pericardial adhesions and thickened pericardium that is functioning normally.[31, 32] With pericardial thickening alone, there is normal slippage between the pericardium and ventricular epicardium. This slippage is not present with constrictive pericarditis (Figure 28–6). Rather, the pericardium in these patients is firmly adherent to the epicardium and does not slip during

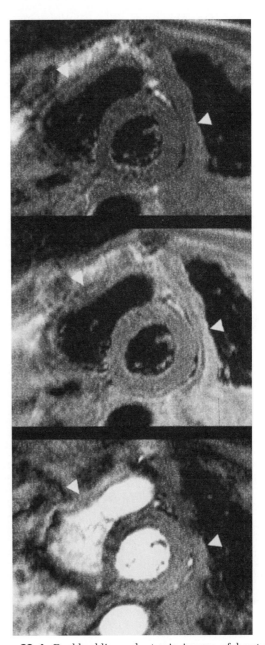

Figure 28–4. Double-oblique short-axis images of heart in a patient with constrictive pericarditis. *Top,* T1-weighted spin-echo image with standard repetition time (TR) (one R-R cycle) demonstrating thickened pericardium (white arrowhead). *Middle,* T1-weighted spin-echo image with long TR. Note that the long TR sequence makes the thickened pericardium more evident (white arrowhead). *Bottom,* Gradient-echo CMR. The gradient-echo image confirms the absence of substantial pericardial effusion. The white arrowhead again delineates the thickened pericardium.

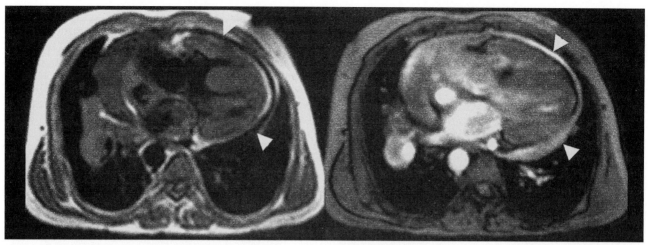

Figure 28–5. *Left,* Transverse T1-weighted spin-echo CMR in a patient with pulmonary valve atresia, ventricular septal defect, and small pericardial effusion. The effusion is visible as a low-intensity region on the image (white arrowheads), but could be confused with pericardial thickening. *Right,* gradient-echo CMR in the same patient. The enhanced signal corresponding to a pericardial effusion is readily apparent (white arrowheads).

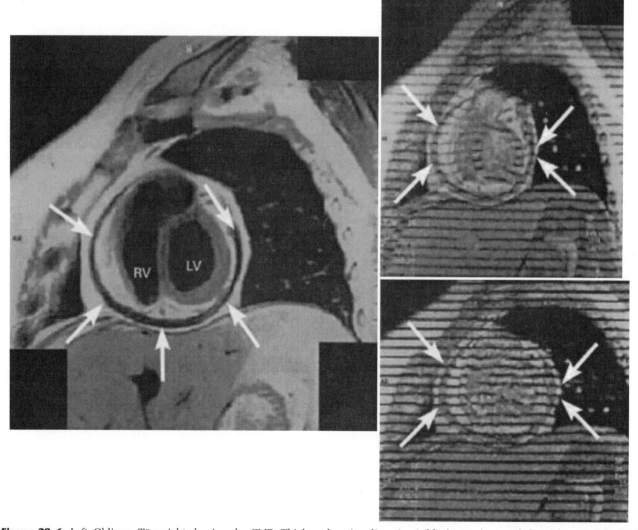

Figure 28–6. *Left,* Oblique, T2-weighted spin-echo CMR. Thickened pericardium is visible (arrows) around the right ventricle (RV) and left ventricle (LV). Tagged cine CMR at end systole (*top right*) and end diastole (*bottom right*). The tagged grids were generated at end diastole and remained unchanged at end systole, indicating tight adhesion between the pericardium and underlying myocardium. (From Kohjima S, Yamada N, Goto Y. Diagnosis of constrictive pericarditis by tagged cine magnetic resonance imaging. N Engl J Med 341:373, 1999, with permission. Copyright © 1999 Massachusetts Medical Society. All rights reserved.)

the cardiac cycle. As with pericardial thickening, extensive pericardial adhesions may be present in the *absence* of constrictive physiology.

Several findings often associated with constrictive physiology are also readily appreciated by CMR. These findings include inferior vena cava enlargement, ascites, and pleural effusions. Left atrial enlargement is present in some cases. Ventricular cavity size is often decreased in patients with constrictive pericarditis. A subset of patients with constrictive pericarditis has sizeable pericardial effusions. These patients are said to have *effusive-constrictive pericarditis*.[33] Pericardial calcification, seen in chronic pericarditis from many etiologies, is relatively difficult to detect by CMR methods, and may be better appreciated by fluoroscopy or EBCT.

Myocardial fibrosis can develop in patients with chronic constrictive pericarditis. CMR in such patients will demonstrate a thin-walled and poorly contractile left ventricle. Identification of myocardial fibrosis is important because patients with coexistent myocardial fibrosis often do not respond favorably to surgical stripping/pericardiectomy.[5, 34]

CARDIOVASCULAR MAGNETIC RESONANCE OF PERICARDIAL EFFUSIONS—IDENTIFICATION, QUANTITATION, CHARACTERIZATION

Although 2-D transthoracic echocardiography is the noninvasive method of choice for identification for most circumferential pericardial effusions, CMR can often provide useful adjunctive information. In addition to assessment of patients with technically limited echocardiographic studies, CMR is useful for (1) detecting localized/loculated pericardial effusions, especially in postthoracotomy patients, (2) differentiating between extensive epicardial fat and pericardial effusions, (3) differentiating between pleural and pericardial effusions, (4) differentiating between a fibrotic pericardium and a pericardial effusion, and (5) characterizing pericardial fluid composition.

Very small amounts of pericardial fluid can be detected by CMR. Animal studies suggest that a systolic effusion thickness of 5 mm or more represents an abnormal accumulation of fluid.[35] CMR-determined pericardial fluid volume correlates well with effusion size by echocardiography and with effusion volume determined by pericardiocentesis.[36] If precise noninvasive quantitation of pericardial fluid contents is desired, rapid CMR acquisition methods now allow for volumetric determination of pericardial contents by use of Simpson's rule. The wide field of view of CMR and lack of attenuation by lung tissue make CMR ideal for definitively evaluating regions that cannot be unambiguously delineated by echocardiography.

Because of motion effects, small pericardial effusions tend to have low signal intensities on spin-echo images regardless of effusion type. Thus, small effusions are relatively difficult to characterize. But, in many cases, CMR can effectively characterize the composition of moderate to large accumulations of pericardial fluid (Figure 28–7). With T1-weighted spin-echo CMR, chylous effusions have a characteristically very high signal intensity, greater than myocardium. This is because chylous effusions have T1 values similar to fat, which has a short T1 relative to myocardium. In contrast, transudative effusions will have low signal intensity on T1-weighted spin-echo sections because transudative effusions contain relatively little protein and few cells, thus a long T1. As compared with transudative effusions, exudative effusions contain a greater amount of protein and cells. Consequently, exudate effusions have a relatively shorter T1 and a higher signal intensity on T1-weighted spin-echo sections than transudative effusions. Determination of the effusion T1 can be estimated by performing an initial ECG-gated spin-echo CMR and then repeating the sequence at a multiple of the RR interval. This T1 estimate can then be compared to standard values for serum and combinations of serum with various cell concentrations.[35, 36]

In neurological and extracardiac MRI studies, T2-weighted sequences have been particularly helpful for defining the relative age of a hematoma and/or hemorrhage. A complex set of factors are responsible for the CMR properties of hemorrhagic regions, including formation of methemoglobin and red cell lysis. However, standard ECG-gated T2-weighted spin-echo sequences are highly sensitive to cardiac motion and are frequently disappointing in their anatomical definition of the heart, pericardium, and pericardial effusions. The signal intensity of pericardial fluid on such sequences is generally more influenced by motion than by the intrinsic magnetic properties of the pericardial fluid.[35] Fast spin-echo methods for obtaining T2-weighted images appear to be less subject to motion artifact and may be useful in characterization of hemorrhagic pericardial effusions.[37]

CARDIOVASCULAR MAGNETIC RESONANCE OF PERICARDIAL AND PARACARDIAC MASSES

Pericardial cysts, lipomas, lymphomas, and teratomas can present as pericardial and paracardiac masses. With the wide field of view afforded by CMR, these masses can be accurately localized, and accurate information about their 3-D structure and orientation in the thorax can be obtained. ECG-gated CMR gives improved definition of heart wall borders and improved soft tissue contrast in comparison with standard x-ray CT so that these masses can be differentiated from heart muscle. Pericardial cysts (see Figure 28–7) often contain transudative fluid, with a low protein concentration and no blood. The

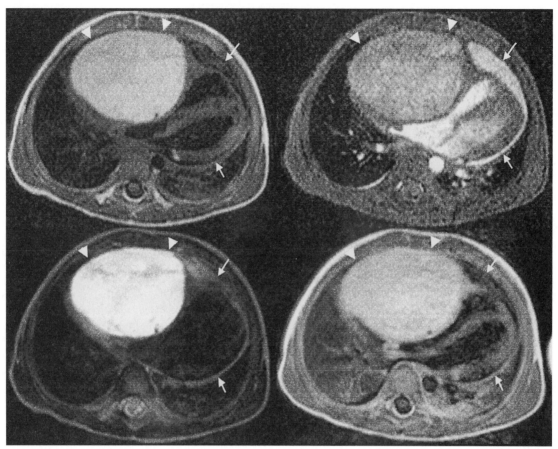

Figure 28–7. Transverse midventricular spin-echo CMR of a large pericardial cyst demonstrating effect of different MR contrast techniques. *Top left,* T1-weighted spin-echo image with TR of one R-R. Note the right ventricular compression by the large cyst (white arrowheads). There is also a moderate size transudative (low-intensity) pericardial effusion (white arrows). Note the substantially different signal intensities of the pericardial effusion and the pericardial cyst. *Top right,* gradient-echo CMR at the same level. Note the different signal intensities of the pericardial effusion and pericardial cyst. *Bottom left,* T2-weighted spin-echo CMR with marked enhancement within the pericardial cyst. The pericardial effusion is less well defined than on the two upper images. *Bottom right,* T1-weighted spin-echo CMR following gadolinium-DTPA infusion. The pericardial fluid still has a relatively low signal intensity, indicating that it has relatively low protein and cellular content (transudative). In contrast, the cyst has a high signal intensity, consistent with a high protein content. At surgery, the high protein content of the cyst was confirmed.

result is low signal intensity on T1-weighted spin-echo sequences and high signal intensity on cine gradient-echo images. However, pericardial cysts can occasionally contain fluid with a high protein content. Lipomas will have characteristic high signal intensity on T1-weighted sequences.[38] Mesotheliomas and other malignant tumors typically enhance with gadolinium-DTPA.[39]

CARDIOVASCULAR MAGNETIC RESONANCE OF CONGENITAL ABSENCE OF THE PERICARDIUM

Complete congenital absence of the left pericardium is usually readily apparent on CMR studies. In this benign condition, the heart is prominently displaced into the left hemithorax. In these patients, the lung is often interposed between the diaphragm and the heart and between the aorta and pulmonary artery, creating a distinctive hypointense region on the CMR sections. Cine CMR studies may demon-

strate prominent systolic anterior motion of heart due to the lack of pericardial constraint. Although there is also failure of the CMR to visualize the left pericardium, poor visualization of the left pericardium is not infrequent, particularly in patients without substantial pericardial fat. Therefore, simple lack of visualization of the left pericardium on CMR is *not* specific for absence of the left pericardium. Herniation of the left atrial appendage through a partial defect of the left pericardium has been demonstrated by CMR.[40] The left atrial appendage may be substantially enlarged in these cases. CMR in herniation of the left ventricle through a partial defect of the left pericardium has also been reported. A circumscribed crease in the left ventricular wall suggests strangulation.[41]

References

1. Holt JP: The normal pericardium. Am J Cardiol 26:455, 1970.
2. Ferrans VJ, Ishihara T, Roberts WC: Anatomy of the pericar-

dium. *In* Reddy PS, Leon DF, Shaver JA (eds): Pericardial Disease. New York, Raven Press, 1982, p 1.

3. Rokey R, Vick GW III, Wendt RE III: Assessment of pericardial disease using nuclear magnetic resonance imaging techniques. *In* Pohost GM: Cardiovascular Applications of Magnetic Resonance. Mount Kisco, NY, Futura, 1993, p 17.

4. Ishihara T, Ferrans VJ, Jones M: Histologic and ultrastructural features of normal human parietal pericardium. Am J Cardiol 46:744, 1980.

5. Lorell BH: Pericardial diseases. *In* Braunwald E (ed): Heart Disease. A Textbook of Cardiovascular Medicine. Philadelphia, WB Saunders, 1997, p 1478.

6. Shabetai R: The Pericardium. New York, Grune & Stratton, 1981.

7. Fowler NO: The Pericardium in Health and Disease. Mount Kisco, NY, Futura, 1985.

8. Blake S, Bonar S, O'Neill H, et al: Aetiology of chronic constrictive pericarditis. Br Heart J 50:273, 1983.

9. Cameron J, Oesterle SN, Baldwin JC, et al: The etiologic spectrum of constrictive pericarditis. Am Heart J 113:354, 1987.

10. Vaitkus PT, Kussmaul WG: Constrictive pericarditis versus restrictive cardiomyopathy: A reappraisal and up of diagnostic criteria. Am Heart J 122:1431, 1991.

11. Feigin DS, Fenoglio JJ, McAllister HA, et al: Pericardial cysts: A radiologic:pathologic correlation and review. Radiology 125:15, 1977.

12. Wilding G, Green HL, Longo DL, et al: Tumors of the heart and pericardium. Cancer Treat Rev 15:165, 1988.

13. Glover L, Barcia A, Reeves TJ: Congenital absence of the pericardium: A review of the literature with demonstration of previously unreported fluoroscopic findings. AJR Am J Roentgenol 106:542, 1969.

14. Ling LH, Oh JK, Tei C, et al: Pericardial thickness measured with transesophageal echocardiography: Feasibility and potential clinical usefulness. J Am Coll Cardiol 29:1317, 1997.

15. Lipton MJ, Higgins CB, Boyd DP: Computed tomography of the heart: Evaluation of anatomy and function. J Am Coll Cardiol 5(suppl 1):55S, 1985.

16. Hoit BD: Imaging the pericardium. Cardiol Clin 8:587, 1990.

17. Stanford W, Thompson BH: Cardiac masses and pericardial disease: Imaging by electron-beam computed tomography. *In* Skorton DJ, Schelbert HR, Wold GL, Brundage BH (eds): Marcus Cardiac Imaging: A Companion to Braunwald's Heart Disease. Philadelphia, WB Saunders, 1996, p 863.

18. Oren RM, Grover-McKay M, Stanford W, et al: Accurate preoperative diagnosis of pericardial constriction using cine computed tomography. J Am Coll Cardiol 22:832, 1993.

19. Furber A, Pezard P, Jeune JJ, et al: Radionuclide angiography and magnetic resonance imaging: Complementary non-invasive methods in the diagnosis of constrictive pericarditis. Eur J Nucl Med 22:1292, 1995.

20. Sechtem U, Tscholakoff D, Higgins CB: MRI of the normal pericardium. AJR Am J Roentgenol 147:239, 1986.

21. Stark DD, Higgins CB, Lanzer P, et al: Magnetic resonance imaging of the pericardium: Normal and pathologic findings. Radiology 150:469, 1984.

22. Ferrans VH, Ishihara T, Roberts WC: Anatomy of the pericardium. *In* Reddy PS, Leon DF, Schaver JA (eds): Pericardial Disease. New York, Raven Press, 1982, p 15.

23. Hartnell GG, Hughes LA, Ko JP, et al: Magnetic resonance imaging of pericardial constriction: Comparison of cine MR angiography and spin-echo techniques. Clinical Radiology 51:268, 1996.

24. Bogaert J, Duerinckx AJ: Appearance of the normal pericardium on coronary MR angiograms. J Magn Reson Imaging 5:579, 1995.

25. Im JG, Rosen A, Webb WR, et al: MR imaging of the transverse sinus of the pericardium. AJR Am J Roentgenol 150:79, 1988.

26. Sechtem U, Tscholakoff D, Higgins CB: MRI of the abnormal pericardium. AJR Am J Roentgenol 147:245, 1986.

27. Masui T, Finck S, Higgins CB: Constrictive pericarditis and restrictive cardiomyopathy: Evaluation with MR imaging. Radiology 182:369, 1992.

28. Hunter GJ, Paquet E: Cardiomyopathies, cardiac masses, and pericardial disease: Value of magnetic resonance in diagnosis and management. *In* Skorton DJ, Schelbert HR, Wold GL, Brundage BH (eds): Marcus Cardiac Imaging: A Companion to Braunwald's Heart Disease. Philadelphia, WB Saunders, 1996, p 744.

29. Mohaiddin RH, Amanuma M, Kilner PJ, et al: MR Phase-shift velocity mapping of mitral and pulmonary venous flow. J Comput Assist Tomogr 15:237, 1991.

30. Hartiala JJ, Mostbeck GH, Foster E, et al: Velocity-encoded cine MRI in the evaluation of left ventricular diastolic function. Measurement of mitral valve and pulmonary vein flow velocities and flow volume across the mitral valve. Am Heart J 125:1054, 1993.

31. Holmvang G, Dinsmore RE: Magnetic resonance imaging with tagging in constrictive pericarditis. J Am Coll Cardiol 29:24A, 1997.

32. Kohjima S, Yamada N, Goto Y: Diagnosis of constrictive pericarditis by tagged cine magnetic resonance imaging. N Engl J Med 341:373, 1999.

33. Hancock EW: Subacute effusive-constrictive pericarditis. 43:183, 1971.

34. Reinmuller R, Doppman JL, Lossner J: Constrictive pericardial disease: Prognostic significance of a nonvisualized left ventricular wall. Radiology 156:753, 1985.

35. Rokey R, Vick GW III, Boli R, et al: Assessment of experimental pericardial effusion using nuclear magnetic resonance imaging techniques. Am Heart J 121:1161, 1991.

36. Mulvaugh SL, Rokey R, Vick GW III, et al: Usefulness of nuclear magnetic resonance imaging for evaluation of pericardial effusions, and comparison with two-dimensional echocardiography. Am J Cardiol 64:1002, 1989.

37. Seelos KC, von Smekal A, Vahlensieck M, et al: Cardiac abnormalities: Assessment with T2-weighted turbo spin-echo MR imaging with electrocardiogram gating at 0.5 T. Radiology 189:517, 1993.

38. King SJ, Smallhorn JF, Burrows PE: Epicardial lipoma: Imaging findings. AJR Am J Roentgenol 160:261, 1993.

39. Ohnishi J, Shiotani H, Ueno H, et al: Primary pericardial mesothelioma demonstrated by magnetic resonance imaging. Jpn Circ J 60:898, 1996.

40. Altman CA, Ettedgui JA, Wozney P, et al: Noninvasive diagnostic features of partial absence of the pericardium. Am J Cardiol 63:1536, 1989.

41. Gassner I, Judmaier W, Fink C, et al: Diagnosis of congenital pericardial defects, including a pathognomic sign for dangerous apical ventricular herniation, on magnetic resonance imaging. Br Heart J 74:60, 1995.

CHAPTER 29

Magnetic Resonance Angiography—Aorta and Peripheral Vessels

Agnes E. Holland, James W. Goldfarb, and Robert R. Edelman

Magnetic resonance angiography (MRA) has become widely accepted for the evaluation of abnormalities throughout multiple vascular territories. Advances have allowed MRA to rapidly progress from a developmental technique to a noninvasive clinical imaging tool providing vital information in the care of patients every day.

Blood that flows through magnetic field gradients and radiofrequency (RF) fields produces signal changes that can be used to distinguish blood vessels from stationary surrounding tissue. Pulse sequences that exploit the effect of blood motion in order to directly visualize blood vessels, without the use of a contrast agent, include the time-of-flight (TOF)[1] and phase contrast (PC)[2] MRA techniques. These noncontrast MRA approaches are excellent for imaging of vessels in healthy subjects with normal, continuous, laminar flow. However, in patients with vascular disease that disturbs the laminar pattern and direction of flow, the quality of TOF and PC images is degraded. If a paramagnetic contrast agent is administered intravenously, the depiction of vessels is no longer dependent on blood inflow and motion. The use of contrast agents alleviates many of the problems encountered with flow-dependent MRA techniques.[3, 4]

The paramagnetic contrast agents are injected intravenously and the image data are collected as the contrast agent flows through the vascular territory of interest, depicting the blood vessels with a brief high signal intensity. This allows imaging of a large field of view (FOV) that encompasses an extensive region of vascular anatomy. The development of high-powered gradient technology has resulted in significantly shorter acquisition times, making it possible to acquire an entire three-dimensional (3-D) high-resolution volume data set in a single breath-hold using fast gradient-echo techniques.

Contrast-enhanced (CE) MRA also has many advantages over conventional x-ray angiography. It is noninvasive, providing high-resolution images without the need for arterial access, which may account for the popularity and widespread use of CE MRA throughout the world. Another advantage is that the paramagnetic contrast agents used for CE MRA have a much more favorable safety profile than the conventional iodinated contrast agents used in x-ray angiography.[5, 6]

This chapter provides a brief review of the underlying basic principles of MRA and provides an overview of the various MRA techniques and their application, with examples in clinical practice.

BASIC PRINCIPLES AND TECHNIQUES

The essence of MRA is creating high contrast between flowing blood and surrounding stationary tissue. To distinguish vessels from surrounding stationary tissue, either the intrinsic mechanisms of blood flow can be used, or a contrast agent can be added intravenously. Techniques depending on the intrinsic mechanisms of blood flow include the *TOF* and the *PC* angiography. The *gadolinium-enhanced 3-D techniques* involve the addition of contrast material to visualize blood vessels.

Time-of-Flight Magnetic Resonance Angiography

TOF methods depend on the inflow enhancement of flowing blood and the relative saturation of stationary tissue to produce high contrast. The signal from the background tissue is suppressed by multiple RF pulses, such that the magnetic spins associated with background tissue do not have enough time to regain their longitudinal magnetization. This saturation of spins from the stationary tissue makes stationary tissue appear dark in TOF images. The bright signal intensity of flowing blood is the consequence of the continuous inflow of fresh unsaturated blood into the imaging slice, which produces more signal than the surrounding stationary tissue, which was repeatedly exposed to the same RF pulse. This effect is known as the flow-related enhancement.[7, 8] The flow-related enhancement can be maximized by using sequences with a short repetition time (TR). The imaging slice should be thin and oriented perpendicular to the direction of flow. Cardiac gating to maximal flow can be used to optimize the flow and consequently the blood signal in arteries and even veins.

Phase Contrast Magnetic Resonance Angiography

In PC angiography, the contrast between flowing blood and the surrounding stationary tissue is generated by the flow-induced phase shift of the moving blood. When a magnetic spin in flowing blood is subjected to a gradient magnetic field, it acquires a phase shift that is proportional to its velocity.[9–11] In the presence of a pair of bipolar gradient lobes, the stationary spins do not develop a net phase shift, whereas spins moving along the direction of the magnetic field gradient develop a net phase shift. Because the flow-induced phase shifts are proportional to the signal intensity, flow direction and velocity can be derived directly from PC data.

Gadolinium-Enhanced Three-Dimensional Angiography

Instead of relying on blood flow, as with TOF and PC techniques, the gadolinium-enhanced MRA techniques utilize the addition of a gadolinium chelate to create intravascular signal. The gadolinium chelates are paramagnetic contrast agents that preferentially shorten the T1 relaxation time of blood in proportion to their local concentration. With this strategy, blood can be imaged irrespective of flow and image contrast is based on T1 relaxation rather than on flow effects. Therefore, flow artifacts and in-plane saturation effects, which occur with the TOF and PC techniques, are largely eliminated.[3] In addition, these gadolinium-enhanced techniques allow in-plane imaging of vessels, so that the number of image sections required to image an extensive region of vascular anatomy is strongly reduced.[12]

Gadolinium Chelates

Gadolinium is a paramagnetic ion that shortens the longitudinal or spin-lattice (T1) relaxation times of nearby protons.[13] When added to blood, it shortens the T1 relaxation time of blood according to equation 1:

$$1/T1 = 1/1200 \text{ ms} + R_1[Gd] \qquad [1]$$

where 1200 ms = T1 of blood without gadolinium at 1.5 T, R_1 = relaxivity of gadolinium, and [Gd] = gadolinium concentration in the blood.

Gadolinium itself is toxic and must be bound to a chelator, such as diethylenetriamine pentaacetic acid (DTPA), to be used safely in humans. Most of the currently available Food and Drug Administration (FDA)–approved gadolinium chelates are extra-

cellular agents. They readily diffuse through the capillary walls into the extravascular space. Consequently, imaging of arteries has to be done during the initial arterial phase of the contrast injection to minimize signal from gadolinium that crossed into surrounding stationary tissues and/or passed through the capillary system and into veins.

Gadolinium chelates can also be bound to larger molecules, such as albumin, that do not readily pass through the capillary membranes. These agents are the so-called blood pool contrast agents, which are currently not yet FDA approved but are receiving considerable attention.

The safety profile of the gadolinium chelates is much more favorable than that of the conventional iodinated contrast agents.[5, 6, 14] With gadolinium chelates, the incidence of adverse events is extremely low and idiosyncratic reactions are very rare.[4] In addition, gadolinium contrast agents have no nephrotoxicity, even when administered at high doses.[15–17] This finding is particularly relevant for the evaluation of patients with renal failure.

Three-Dimensional Imaging Pulse Sequence

A fast 3-D spoiled gradient-echo sequence is used, with TR = 3 to 8 ms, echo time (TE) = 1 to 3 ms, and flip angle = 20 to 40°. The imaging volume can be acquired in any desired orientation and the acquisition time ranges from 20 to 40 seconds. Scan times on this order allow imaging to be done during a prolonged breath-hold. At postprocessing, the 3-D nature of the data set allows viewing of the data from any desired angle, yielding multiple images in various anatomical orientations.

Vessel Brightness (T1) Versus Gadolinium Dose

An adequate amount of paramagnetic contrast agent should be administered to ensure that the T1 of blood is well below the T1 of the brightest surrounding tissue. This precaution ensures that blood will appear bright when compared with background tissue. As the background tissue with the shortest T1 relaxation time is usually fat (T1 = 280 ms at 1.5 T), a sufficient dose of gadolinium chelate should be injected at a sufficient rate to ensure that the T1 relaxation time of the arterial blood is well under 280 ms.[12]

The blood T1 relaxivity for various doses of gadolinium chelates can be calculated from equation [1] and plotted (Figure 29–1). (The currently available gadolinium chelates have a relaxivity of approximately 4.5 mMolar⁻¹s⁻¹ at 1.5 T.)

Currently available FDA-approved gadolinium contrast agents can readily cross the capillary membranes. A few minutes after the injection, there already is a redistribution of the contrast agent into the extracellular space. Therefore, gadolinium-enhanced MRA should be performed during the early

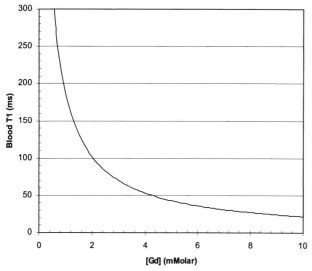

Figure 29–1. Plot of blood T1 versus gadolinium concentration in the blood, with relaxivity of gadolinium (R_1) = 4.5 mMolar^{-1}s^{-1}. Species with short T1s appear bright in gadolinium-enhanced angiographic images. Standard dosages (blood concentration) of gadolinium lower the blood T1 from 1200 ms to less than 280 ms.

arterial phase of the contrast agent. The timing of the bolus must be very precise such that imaging coincides with the arrival of the contrast agent bolus in the vascular territory of interest.

Assuming adequate mixing of the contrast agent and blood, the arterial blood gadolinium concentration is related to the injection rate and cardiac output as follows[3, 18]:

$$[Gd] = \frac{(\text{rate of Gd injection in mol/s})}{(\text{cardiac output in liter/s})} \quad [2]$$

The blood T1 for a given cardiac output and injection rate can be calculated using equations [1] and [2].

Bolus Timing

Proper delay between the start of the contrast agent injection and the start of the CMR scan is crucial. For maximum arterial enhancement, the contrast injection has to be timed such that the acquisition of the central lines of k-space coincides with the peak arterial gadolinium concentration. If inadvertently the central lines of k-space are acquired before the peak arterial enhancement, a coarse ringing artifact combined with widening of the vessel margins occurs.[19] When the central lines of k-space are acquired too late, that is, after the peak arterial enhancement, there is reduced arterial and increased venous signal intensity, resulting in a sometimes confusing overlap between arterial and venous structures. This problem reduces the accuracy of the imaging modality. There are different types of CMR pulses sequences with the central lines of k-space acquired in the very beginning of

the scan, after 25 percent of the scan, and after 50 percent of the scan.

To time the gadolinium injection bolus correctly, the exact transit time of the contrast agent from the injection site (typically in the arm) to the blood vessels of interest has to be determined. The transit time of the contrast agent depends on several factors including stroke volume, heart rate, valvular function, injection rate, and site of the IV catheter. Currently there are several different methods to determine this transit time.

The *test bolus technique* involves the injection of a small dose (1–2 ml) of gadolinium, at the same rate as the actual contrast injection, followed by a 10- to 15-ml saline flush.[20, 21] A thick two-dimensional (2-D) gradient-echo section is acquired every 1 to 2 seconds through the vascular territory of interest for approximately 1 minute. The contrast transit time can be determined visually or by drawing a region of interest and measuring the signal intensity on sequential images (Figure 29–2).

Another way of timing the arrival of the contrast agent bolus is through the use of *automatic triggering*.[22, 23] This technique involves automatic synchronization of central k-space image data with the arterial phase of contrast material bolus infusion (for example, MR Smartprep, GE Medical Systems, Milwaukee, WI). A spin-echo sequence with orthogonal 90° and 180° is used to monitor the signal

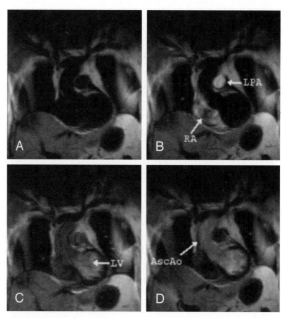

Figure 29–2. Bolus timing: A small-volume contrast injection together with a two-dimensional inversion recovery segmented k-space gradient-echo sequence can be used to calculate the transit time individually for each subject, prior to gadolinium-enhanced imaging. In this example one slice was acquired every 2 seconds. One milliliter of gadolinium-DTPA followed by a 10-ml saline flush was injected at a rate of 1.6 ml/s. The slice is oriented in the coronal plane, and the images displayed correspond to time points *(A)* 8, *(B)* 14, *(C)* 16, and *(D)* 22 seconds after contrast injection. Labeled are the right atrium (RA), left pulmonary artery (LPA), left ventricle (LV), and ascending aorta (AscAo).

intensity in a single large, 4 × 4 × 12-cm voxel, which is placed in the vascular territory of interest. The arrival of the gadolinium bolus corresponds with an increase of signal intensity in the large voxel. Either a technician or a computer program senses the arrival of the gadolinium and triggers the 3-D gradient-echo sequence.

The *time-resolved contrast-enhanced 3-D MRA* is a technique that is virtually independent of the exact contrast transit time.[24] It involves the rapid collection of successive 3-D data sets beginning immediately after contrast injection so that at least one of the 3-D data sets aligns with the arterial phase of the contrast injection. With this technique, acquisition times for acquiring each 3-D data set are on the order of 2 to 10 seconds. The initiation of the injection and initiation of the first image acquisition are simultaneous, and multiple vascular phases are collected (i.e., arterial, tissue perfusion, venous) as the contrast agent flows through the vascular bed of interest (Figure 29–3). In addition to simplifying bolus timing, this technique also allows evaluation of the temporal enhancement of structures, such as the kidneys, that are caught within the 3-D volume. Because organ function depends on its level of perfusion, regional blood flow information might aid in the assessment of hemodynamically significant stenoses.

3-D CONTRAST-ENHANCED MRA COMPARED WITH 3-D CONTRAST-ENHANCED CT ANGIOGRAPHY

Both MRA and computed tomography angiography (CTA) are in some respects quite similar because each acquires image data in a tomographic manner. CE 3-D MRA does have several advantages over contrast-enhanced helical CTA. Both methods use an intravenous injection of a contrast agent and make images during a short period of time, when the agent is in the vessels of interest. The major difference is how the image is made and the side effects of the injected contrast agent. MRA and CTA images are made using a magnetic field and x-ray beams, respectively. The MRA data set may be acquired in any orientation whereas the CTA is acquired in the transverse plane.

The paramagnetic agents, used for CE MRA, have proven to be much safer than the iodinated agents. The incidence of anaphylaxis with paramagnetic agents is rare.[5, 6] The paramagnetic contrast agents are not nephrotoxic and can be administered safely to patients with impaired renal function. As is discussed later in relation to renal artery stenosis, contrast-enhanced MRA allows angiographic investigation, without grave concern for inducing renal failure in this important patient population.[15–17]

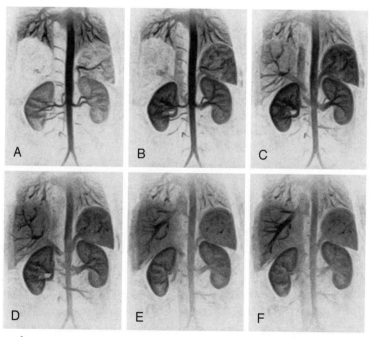

Figure 29–3. Inverted gray scale maximum-intensity projections of six sequentially obtained gadolinium-enhanced volumes in a healthy subject. Each 3-D volume is acquired with a temporal resolution of 5.3 seconds. Volumes *A* through *F* show the different phases of enhancement, i.e., arterial, parenchymal, and venous, that can be distinguished after gadolinium injection. The first image volume *(A)* shows enhancement of the aorta, splenic artery, renal arteries and branches, and iliac arteries. In the second volume *(B)*, there is renal and splenic parenchymal enhancement. On volume *C*, the portal vein and branches are enhanced, and there is initial hepatic enhancement. The renal and splenic veins can be distinguished. Volumes *E* and *F* show strong enhancement of the hepatic vein and a reduction in the signal intensity of the aorta as well as renal and splenic vasculature.

CTA data are acquired exclusively in the transverse plane, whereas MRA allows imaging to be performed in the sagittal, coronal and oblique imaging planes, providing greater volume coverage. Conventional magnetic resonance machines allow maximal FOVs from 40 to 50 cm. This feature is particularly important for the evaluation of the extent of aortic disease, both proximally into the thorax and distally into the iliac arteries. MRA can depict vessels in their full extent at high resolution with fewer slices than can CTA, which requires many thin sections for adequate distinction of the vascular anatomy. This distinction illustrates another disadvantage of CTA, i.e., the relatively large radiation dose to which the patient is exposed when obtaining these multiple thin sections.

In contrast to CTA, MRA can also provide functional information, such as flow velocity and flow volume. These measurements are not made using CE techniques. Rather, they can be calculated using PC sequences. With the gated cine sequences, the blood flow can be displayed dynamically during the cardiac cycle.

MRA data can be easily reconstructed and displayed using the multiplanar reconstruction (MPR) and maximum intensity projection (MIP) algorithms (Figure 29–4). Postprocessing CE CTA images is considerably more complicated. It involves segmenting out structures based on their density, such as bone, that obscure the vascular anatomy from each individual slice. This procedure can be tedious and lengthy.

One major clinical advantage of CTA over MRA is that the easier set-up in a nonmagnetic environment allows clinically unstable patients to be more comfortably studied with CTA than with MRA.

MAGNETIC RESONANCE ANGIOGRAPHY OF THE BODY

Aorta

MRA of the aorta is among the most common studies performed. Various radiological imaging modalities have been employed for the assessment of disease of the thoracic and abdominal aorta. For the visualization of the thoracic aorta, plain film radiography, CT, ultrasound, and conventional x-ray aortography are all established techniques. For imaging of the abdominal aorta, conventional angiography, ultrasound, and CT are widely used. Of these three techniques, CT is currently the most widely used because of its widespread availability, low cost, and diagnostic accuracy. More recently, MRA has emerged as a robust and reliable technique for imaging both the thoracic and abdominal aorta.[25–28] Indications for evaluating the aorta include occlusive disease in patients with generalized atherosclerosis,[33, 34] suspected aneurysm,[35] dissection,[36] follow-up surgical repair, and congenital malformations such as coarctation.

Both contrast enhancement and non–contrast enhancement are used to image the aorta with CMR. The non-CE techniques include the black-blood spin-echo (SE) sequences and the bright-blood gradient-recalled-echo (GRE) sequences. The development of turbo spin-echo (TSE) or fast spin-echo (FSE) techniques has significantly reduced the scan time and is quickly replacing spin-echo imaging. For TSE and GRE imaging of the thoracic aorta, ECG-gating is necessary to reduce artifacts associated with the motion of the heart during the cardiac cycle.[28] In contrast, imaging of the abdominal aorta

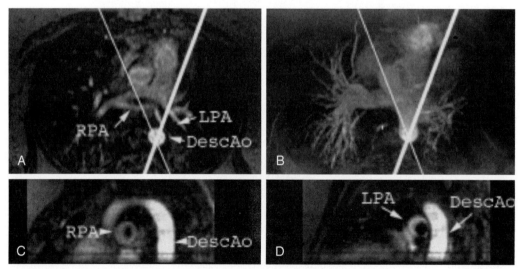

Figure 29–4. To display blood vessels, which span many images, two postprocessing techniques are commonly used. The maximum-intensity-projection (MIP) algorithm uses only the largest single intensity at each position in a projection showing the blood vessels with the brightest signal intensities. The multiplanar-reformation (MPR) algorithm reslices a stack of images such that the anatomy can be viewed retrospectively in any anatomical orientation. *A,* Original axial image from a 3-D gadolinium-enhanced acquisition of a patient with a saddle embolism. RPA = right pulmonary artery; LPA = left pulmonary artery; DescAo = descending aorta. *B,* MIP image showing the pulmonary vasculature. *C,* MPR along the thin line shown in *A* and *B* showing the embolism in the RPA. *D,* MPR along the thick line shown in *A* and *B* showing the embolism in the LPA.

is not directly affected by cardiac motion and ECG-gating is usually not required. However, studies of the abdominal aorta may be subject to pulsation artifacts, which can be reduced by the use of spatial saturation bands for SE techniques and flow compensation algorithms for GRE techniques. The black-blood sequences are used to provide anatomical information (aneurysm size, anatomical location), and to evaluate any diseases involving the aortic wall, such as atherosclerosis and aortitis. The bright-blood sequences consist of TOF techniques. Especially useful for evaluation of the thoracic aorta are the ECG-gated bright-blood cine-MR sequences, which provide functional information displaying the blood flow through the aorta dynamically during the cardiac cycle. The PC techniques can be used to calculate the flow velocity and flow volume. These parameters may be particularly useful for the hemodynamic evaluation of focal stenoses, such as in coarctation of the aorta.[29]

The black-blood (SE) and bright-blood (GRE) sequences are excellent techniques for the imaging of vessels in healthy subjects with steady laminar flow, which is perpendicular to the imaging plane. Due to vessel wall irregularities such as stenoses, many patients with vascular disease have turbulent flow, which is neither laminar nor perpendicular to the imaging plane. As a result, image quality is seriously degraded by local artifacts. Decreased flow velocities, caused by impaired cardiac function or aneurysmatic changes of the vessel wall, provide less contrast between the vessel and the surrounding background tissue. The differentiation of slow flow from thrombosis and from artifacts can be quite difficult. This difficulty is another limitation that makes these techniques that are strongly dependent on flow less accurate for clinical use in many patients. The shortcomings of these methods illustrate the need for techniques not dependent on blood flow, such as the CE 3-D MRA techniques.

With the CE 3-D MRA techniques, many of these shortcomings are overcome. CE 3-D MRA is not dependent on blood flow and provides excellent vessel-to-background contrast. The image quality is not affected by slow flow and, consequently, differentiation of thrombus from slow flow is less likely to be a problem. Gadolinium-enhanced 3-D MRA gradient-echo sequences are fast techniques. For the thoracic and abdominal aorta, cardiac triggering (ECG) is generally *not* required.[25, 30, 31] Acquisition times are approximately 20 seconds, with a sagittal or oblique sagittal volume, allowing a complete study of the thoracic aorta and branches to be performed in a single breath-hold.

For imaging of both the thoracic and the abdominal aorta, a body-phased-array surface coil should be used to optimize the signal-to-noise ratio of the image. In large patients, the body-phased-array coil may not provide enough coverage of the volume of interest or does not allow the patient to be comfortably moved into the machine's bore. In these patients, the body coil is often the best choice. For studies of the thoracic aorta, special attention

should be given to the positioning of the body-phased-array coil to ensure that the arch vessels are included in the FOV of the coil. Patients are routinely imaged in the supine position.

Our current imaging protocol for the thoracic aorta consists of a localizer sequence, a contrast bolus timing sequence, a T1-weighted pre-contrast sequence, the 3-D pre- and post-CE MRA sequence, and a T1-weighted postcontrast sequence. As a localizer sequence, an ECG-gated 2-D GRE or fast SE sequence is appropriate. It is advisable to acquire the localizer during breath-holding in the same breath-holding position, i.e., end-inspiration or end-expiration, as is planned for the CE 3-D MRA. The anatomy during breath-holding is shown on the localizer, and positioning the 3-D imaging slab on this localizer ensures that this anatomy will be included in the important contrast acquisition. A single-slice segmented k-space gradient-echo sequence can be performed in many orientations (axial, sagittal, coronal, or parasagittal plane) for proper timing of the contrast bolus. For evaluation of the thoracic aorta, the gradient-echo 3-D contrast-enhanced MRA sequence may be obtained either in the sagittal, coronal or parasagittal plane. The 3-D volumes, which are acquired in the coronal plane, do provide more anatomical coverage (inclusion of a greater portion of the arch vessels) but may require a longer acquisition time than feasible in a breath-holding period for many patients (this difficulty may be a problem in older scanners not equipped with high-powered gradients). For this reason, the sagittal or parasagittal plane may be preferred for imaging of the thoracic aorta. This approach provides less volume coverage but has equal resolution and slice thickness as the coronal approach. The abdominal aorta is typically imaged in the coronal plane, ensuring the smallest number of slices and the shortest acquisition time. The renal and iliac vessels should be included in the 3-D volume. For the abdominal aorta, the imaging FOV varies typically between 30 and 34 cm. Imaging of the thoracic aorta typically requires a larger FOV, ranging from 36 to 40 cm. These dimensions depend on the extent of the aortic disease and the patient's body habitus. The slice thicknesses for imaging of both the thoracic and abdominal aorta vary between 1.5 and 3.5 mm. First a test 3-D data set is acquired precontrast to ensure accurate positioning and to search for possible disturbing aliasing artifacts. This precontrast 3-D volume can also be used as baseline data for subtraction to enhance vessel visualization. Typically two postcontrast data sets are obtained, with an interval of approximately 10 seconds between the two. During this time, the patient can breathe freely. The probability of obtaining a diagnostic imaging study is increased by acquiring two postcontrast data sets. The second data set predominantly shows the enhancement of veins because it is acquired later. Subtraction of the latter from the first CE 3-D volume ideally results in a data set showing the arteries only.

Another advantage of obtaining two postcontrast 3-D data sets is that structures with slow flow such as large aneurysms may require additional time to fully enhance and are therefore only visualized optimally on the second 3-D volume. Because CE imaging predominantly depicts the vascular lumen, every aorta protocol should also include ECG-gated T1-weighted SE postcontrast scans to visualize vessel walls. These sequences are acquired in the transverse/axial plane, typically with two to three signal averages for the reduction of respiratory motion artifacts. In addition, depending on the nature of the vessel disease, GRE sequences, showing the pulsatile blood flow, and also PC sequences, quantifying flow dynamics, may be desirable.[32]

The majority of thoracic and abdominal aortic aneurysms occur secondary to atherosclerosis. Other causes include infection, inflammation, trauma, syphilis, cystic medial necrosis, and valvular disease (thoracic aorta). There seems to be a strong association of aneurysmatic changes of the aortic wall with smoking[37] and hypertension. Aneurysms can further be subdivided by their gross appearance into fusiform (circumferential enlargement of the involved segment) and saccular shaped (asymmetrical or focal outpouching of the involved segment). Aneurysms can be categorized as either being true, with all three aortic wall layers involved in the aneurysm, or false (pseudoaneurysms). In the latter case, there is a break in the intima and media such that the wall of the aneurysm is formed by the adventitia only. Most abdominal aortic aneurysms are localized infrarenally and are fusiform in shape. Typically the patient is asymptomatic and a pulsatile mass is palpated on physical examination. If these aneurysms are untreated, life-threatening complications such as rupture, thrombosis, and embolism frequently occur.

Mycotic aneurysms are a rare entity resulting from weakening of the vessel wall by a bacterial infection. This infection can be either primary or a secondary infection in a preexisting aneurysm. Mycotic aneurysms are usually saccular and are most often found in a suprarenal location. They may demonstrate paravertebral fluid or soft tissue mass, paraaortic gas, or osteomyelitic involvement of an adjacent vertebral body. Mycotic aneurysms are more common in women and may present either insidiously or with sepsis and rupture. Although a high percentage of mycotic aneurysms yield negative cultures, the most commonly associated organisms are *staphylococcus* and *salmonella*.

MRA can be used to visualize both intraluminal and extraluminal anatomy. The intraluminal anatomy is shown by the CE 3-D GRE technique, enabling detailed visualization of the aneurysm morphology providing cross-sectional, and, if desired, reformatted images in any orientation. Advantageously, the 3-D technique clearly shows the relationship of the aneurysm to the origin of the brachiocephalic (Figure 29–5) or renal arteries. It is chiefly this property that has led to an increased use

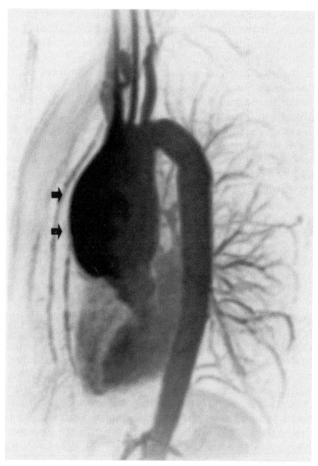

Figure 29–5. 49-year-old patient with a saccular aneurysm of the ascending aorta (maximal diameter of 6 cm): Oblique MIP inverted gray scale image of a gadolinium-enhanced 3-D data set shows the involvement of the arch vessels in the aneurysm.

of this technique for imaging of aortic aneurysms. The precise dimensions of the aneurysm can be evaluated by subvolume MIP and MPR postprocessing techniques. The extraluminal anatomy of the aneurysm is visualized on the T1-weighted SE sequences. These sequences should be performed *after* the 3-D gadolinium-enhanced scans but while there is still a noticeable concentration of the contrast agent present in the circulation. This presence of contrast aids in the differentiation of flowing blood from mural thrombus. Finally, delayed enhancement of the vessel wall, indicating inflammation as is seen in aortitis and mycotic aneurysms, when using these T1-weighted sequences after contrast. T2-weighted SE images should be obtained for mycotic aneurysms. The T2 images show perivascular edema or fluid collections, and GRE cine sequences can also be used to demonstrate flow within the aneurysm.[38]

Several clinical studies have reported high accuracy of CE 3-D MRA for the detection of aneurysms involving the thoracic and abdominal aorta.[3, 25, 27] A sensitivity of 100 percent and specificity of 100 percent have been reported for the morphological analysis of aneurysms. Analysis involves assessment of maximum aneurysm diameter, luminal patency, and the proximal and distal aneurysm extent.[26, 39] CE MRA has an 88 to 100 percent sensitivity and a 97 to 100 percent specificity for the detection of stenoses or occlusions of the aorta.[3, 40]

Aortic dissection occurs when blood dissects through the endothelial lining and into the media of the aortic wall through an intimal tear. In most instances, it is related to degeneration of an aging aorta and may be accelerated by hypertension. In younger patients, there is often an underlying process such as a bicuspid or unicuspid aortic valve, coarctation, pregnancy, connective tissue disorders such as Marfan or Ehlers-Danlos syndrome, relapsing polychondritis, Turner or Noonan syndrome, thoracic cage deformities, and, rarely, systemic lupus erythematosus or giant cell arteritis. Dissection may also occur at sites of iatrogenic trauma such as at the location of a prior aortic incision, cross clamping, or traumatic catheterization.

The two most common sites for the initiation of aortic dissection are the proximal ascending aorta and the descending aorta just distal to the left subclavian artery. These locations reflect sites of maximal mechanical strain on the aorta caused by flexion. The dissection channel generally spirals. In the ascending aorta, the false lumen is usually ante-

rior and to the right. In the arch, the false lumen is superior and slightly posterior. In the descending aorta, it is posterior and to the left. The false channel may compress the true lumen. Branch arteries may be perfused by either the false lumen, the true lumen, or both. Arch and iliac vessels are supplied by the false lumen in approximately one half of patients. Renal arteries are fed by the false lumen in approximately 25 percent of cases.

Therapeutic management depends on the proximal extent of the dissection. This is reflected in the Stanford classification system, which describes all aneurysms involving the ascending aorta as type A (60%) and all other dissections as type B. Type A dissections represent a surgical emergency with surgical resection of the dissected segment and replacement with a synthetic graft. If the aortic valve is also compromised, the valve is also replaced. Patients with type B dissection are usually treated medically, with surgical treatment reserved for cases in which visceral or lower extremity circulation is threatened or medical therapy has failed.

CE 3-D MRA is an excellent technique for the diagnosis and follow-up of chronic aortic dissections. The intimal flaps are typically clearly visualized on both the source images and MPRs (Figures 29–6 and 29–7). On the projection images, the intimal flaps might be obscured and not visible. There-

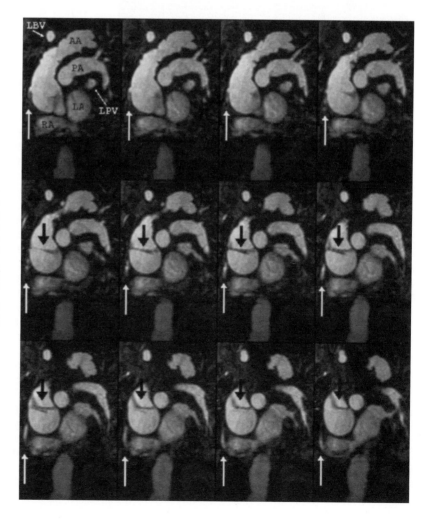

Figure 29–6. Twelve consecutive left-anterior-oblique images from a 3-D gadolinium-enhanced data set of the thoracic aorta. In the ascending aorta there is a dissection flap (type A dissection, black arrow). A coronary artery bypass graft is also seen. This graft extends from the proximal part of the ascending aorta. The posterior descending artery at the base of the heart is the site of the graft's anastomosis (white arrow). Labeled are the aortic arch (AA), pulmonary artery (PA), left atrium (LA), right atrium (RA), left pulmonary vein (LPV), and left brachiocephalic vein (LBV).

fore, the source images remain the most important tools in the diagnosis of aortic dissection. It is not advisable to make a diagnosis based solely on the projection images. The involvement of branch vessels, if any, can be determined using reformations of the original 3-D images. It can be evaluated whether the branch vessels are perfused by the true lumen, the false lumen, or both. The extension of the dissecting membrane into branch vessels should also be evaluated. The axial reformations are particularly helpful for the assessment of the communications between the true and false lumen, i.e., entry and reentry tears.

Studies from 1996 and 1997 report high sensitivity (92–96%) and high specificity (100%) of the CE 3-D MRA technique for the diagnosis of thoracic aortic dissection.[25, 27] The abdominal arteries and abdominal aortic atherosclerotic narrowing are also well seen using contrast-enhanced MRA (Figures 29–8 and 29–9). The contrast-enhanced technique yields high-quality images in preference to the 2-D TOF technique (Figure 29–10). Accessory arteries from the aorta are well visualized (Figure 29–11).

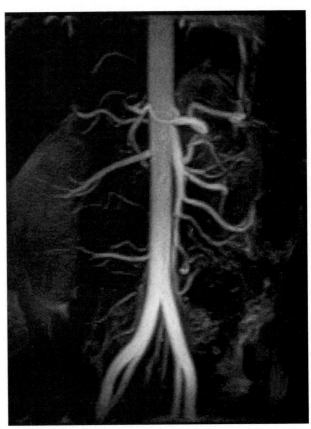

Figure 29–8. Coronal MIP image of a healthy subject, which demonstrates the current ability of gadolinium-enhanced MR angiography (MRA) to depict the abdominal vasculature.

Renal Arteries

Two to five percent of all hypertension cases are directly caused by stenosis of the renal artery and/or its branches.[41] Renovascular hypertension, defined as high blood pressure caused by renal artery stenosis, is the most common cause of secondary hypertension. There are two main lesion groups that cause renal artery stenosis. The first main group comprises atherosclerotic lesions, causing 90 to 95 percent of all cases of renovascular disease.[42] These lesions have twice the prevalence in men than in women. Atherosclerotic lesions are more typical for older age groups, with an average age of 55 years. The stenosis can be caused either by a plaque in the renal artery itself or an aortic atherosclerotic plaque impinging on the origin of the renal artery and narrowing its proximal portion. The second main group of lesions causing renal artery stenosis is a heterogeneous group of intrinsic structural abnormalities of the arterial renal artery wall termed *fibromuscular dysplasia*. This type of stenosis is most frequently seen in women and typically manifests at a younger age (average 35 years) than atherosclerotic lesions. These lesions consist of fibrous and fibromuscular dysplasia of the renal artery wall and typically have no atherosclerotic components.[42]

Renal artery stenosis is a progressive disease that

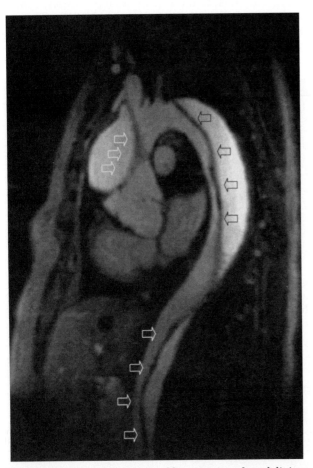

Figure 29–7. Single left-anterior-oblique section of a gadolinium-enhanced 3-D gradient-echo acquisition showing a type A dissection (ascending aorta) and a type B dissection (descending and abdominal aorta). As a result of slower blood flow, the gadolinium concentration is higher in the false lumen than in the true lumen. Therefore, the signal intensity in both false lumina is greater than in the true lumen.

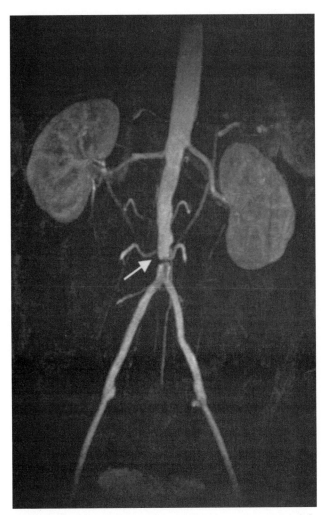

Figure 29–9. 49-year-old patient with a severe stenosis of the abdominal aorta. Coronal MIP image from a 3-D gadolinium-enhanced data set is shown. There is a severe atherosclerotic narrowing of the distal abdominal aorta just above the iliac bifurcation (arrow). Atherosclerotic changes of the aortic wall are also shown. (Courtesy of M.B.J.M. Korst, University Hospital Nijmegen, Nijmegen, Netherlands.)

may result in gradual and silent (almost one half of patients are not hypertensive)[43] loss of functional renal tissue. The incidence increases with age.[44] It has been estimated that renal artery stenosis accounts for 5 to 15 percent of all patients developing end-stage renal disease each year.[43]

For patients with hypertension due to renal artery stenosis, interventional therapy is far superior to pharmacological therapy. Renovascular hypertension tends to be resistant to medical therapy, and regulation of the blood pressure does not reduce the ischemic injury to the kidney. Interventional therapy with surgery or angioplasty restores blood flow to the ischemic kidney and also reduces the need or complexity for lifelong antihypertensive therapy, thereby reducing the number of patients who require dialysis and decreasing the morbidity and mortality associated with long-term hypertension.[45]

The challenge to the clinician is to identify pa-

tients with renal artery stenosis in a safe and cost-effective manner. The diagnosis should be made sufficiently early in the natural history, such that the patient may benefit maximally from intervention (prior to development of end-stage renal dysfunction). The tests that have been used to diagnose renal artery stenosis can be subdivided into tests that detect physiological abnormalities (functional) and ones that detect an anatomical lesion of the renal artery (morphological). No single test can provide both types of information.[46] Tests that detect physiological abnormalities associated with renal artery stenosis have included the rapid sequence intravenous pyelogram (no longer used), measurement of peripheral and renal vein renin activity,[47] the captopril test with sampling of plasma renin, and captopril renography. Tests that directly detect an anatomical lesion of the renal artery include intravenous digital subtraction angiography, CTA,[48] and renal MRA. With duplex sonography, anatomical lesions can be detected indirectly by virtue of changes in flow patterns. Although each of these tests has shown promise, none has been universally accepted.

The captopril test is a relatively safe and inexpensive test. It is based on the increase in plasma renin activity after captopril administration. However, its accuracy is affected by age,[49] race,[50] renal function,[51] intake of antihypertensive medications,[51–54] renal arteriolar changes,[51] food intake before the test,[55] and dilution of plasma. Consequently there is a great variability in reported sensitivies (38–100%) and specificities (72–100%) among different studies. The administration of captopril can worsen renal function in the setting of renal artery stenosis, although the effect is transient.

Radionuclide renography is accurate in demonstrating differential glomerular filtration rates between both kidneys.[56, 57] When captopril is used in association with 99mTc-DTPA or Mag3 renography, the difference is magnified by resultant decrease in the glomerular filtration rate of the kidney affected by renal artery stenosis. Captopril acts by reducing the angiotensin-II–mediated constriction of the efferent arteriole, thus lowering glomerular pressure.[58] In contrast, the glomerular filtration rate in patients with primary hypertension does not change after captopril administration.[59] Consequently, the use of these drugs enhances the sensitivity and specificity of the routine renal scintigram with reported sensitivities ranging from 91 to 94 percent and specificities ranging from 93 to 97 percent.[60–62]

The current gold standard for determining the anatomical presence of renal artery stenosis is conventional x-ray angiography. X-ray angiography provides excellent resolution of the presence and extent of a stenosis, usually identifying the specific cause.[63] Digital subtraction technology has reduced the amount of iodinated contrast volume needed and the risk of contrast nephropathy.[64] The use of smaller catheters has decreased morbidity associated with the arteriotomy, allowing renal angiogra-

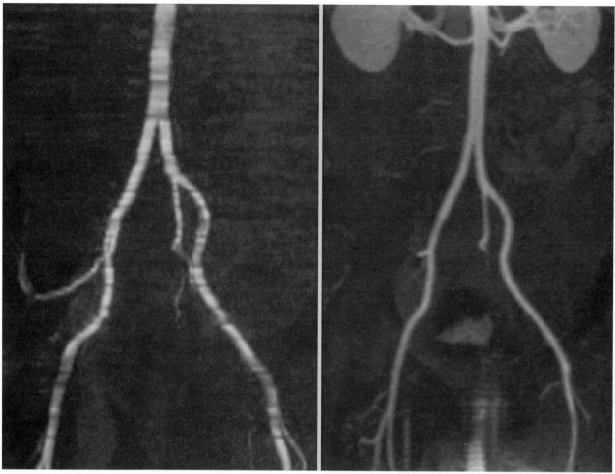

Figure 29–10. 2-D electrocardiogram (ECG)-gated time-of-flight *(left)* and 3-D gadolinium-enhanced *(right)* techniques can be used for imaging of the abdominal arteries. The 2-D technique often suffers from strip artifacts due to irregular flow and slice misregistration, which might obscure underlying pathology.

phy to be performed on an outpatient basis.[65] Risks associated with angiography include allergic reaction, reduction in renal function after administration of contrast medium, and arteriotomy and catheter manipulation complications. The presence of stenosis at angiography does not necessarily imply that the hypertension is related (or will reverse after treatment) to the lesion. Recently, several studies have reported high sensitivity (88–96%) and high specificity (87–100%) of helical CT angiography for the grading of renal artery stenosis greater than 50 percent of the luminal diameter.[66–69] This technique involves capturing the arterial phase of a properly timed bolus of intravenous contrast agent over a region of interest. Currently, CTA has lower spatial and temporal resolution than conventional angiography or digital subtraction angiography. Image postprocessing is more time-consuming than for MRA. The most serious drawback of CTA is the large load of contrast material, which can be contraindicated in patients with impaired renal function.

Doppler ultrasonography enables the indirect detection of an anatomical lesion by measurement of changes in renal artery blood flow patterns. It has also been used for the detection of renal artery ste-

nosis, and either the main renal artery or intrarenal branch vessels can be interrogated. Unfortunately, technically inadequate studies have been reported in up to 40 percent of cases with ultrasonographic examination of the main renal arteries. The failures are predominantly caused by obesity, bowel gas, aortic aneurysm, and/or recent surgery.[70] Consequently, several studies have demonstrated high sensitivity (84–93%) and high specificity (95–98%) for detecting renal artery stenosis,[71–73] whereas other studies have reported a sensitivity as low as 0 percent[74, 75] and a specificity of 37 percent.[74] Several Doppler ultrasound techniques have been employed for the detection of renal artery stenosis. The renal aortic ratio involves measuring the peak velocity in the renal artery to that from the adjacent abdominal aorta. Handa and associates[76] demonstrated that the acceleration index in segmental arteries may be used to quantify the severity of the damping of the Doppler waveform and predict significant renal artery stenosis. They reported a sensitivity of 100 percent and specificity of 93 percent for renal artery stenosis greater than 50 percent. The examination was diagnostic in 98 percent of the 149 vessels. Patriquin and coworkers[77] showed that stenosis

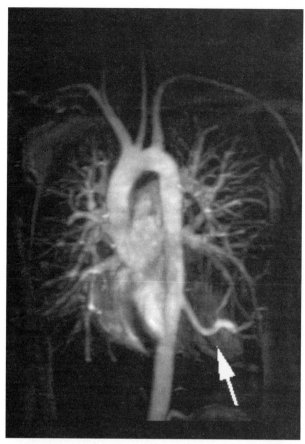

Figure 29–11. Pulmonary sequestration: In this 25-year-old patient, the lower pulmonary lobe receives its blood supply from an artery arising from the descending aorta (arrow). This MIP image, from a gadolinium-enhanced 3-D data set obtained in the coronal orientation, shows the major vessels in the thorax. (Courtesy of D. Janssen, University Hospital Nijmegen, Nijmegen, Netherlands.)

greater than 75 percent was detected with a sensitivity and specificity of 100 percent, using the acceleration index in segmental and interlobar arteries of children. Stavros and colleagues[78] also evaluated segmental renal artery branches. They reported a sensitivity of 89 percent and a specificity of 83 percent for detecting stenosis of more than 60 percent. Currently, the use of Doppler sonography as a screening tool for renal artery stenosis is controversial. The diagnostic value of the examination is highly dependent on the expertise of the operator, the accessory vessels are often poorly visualized,[74, 79, 80] and, as mentioned previously, although several studies report high sensitivities and specificities, it is not clear that high technical success is widely reproducible.

The aforementioned shortcomings of conventional angiography, digital subtraction angiography, and other techniques, added to the recently available cost-effective treatment options for renal artery stenosis, illustrate the need for a novel noninvasive screening modality for this disease. Many investigators have regarded renal MRA as this promising technique for noninvasive screening of patients with suspected renal artery disease. Thus, various MRA imaging strategies for visualizing the renal arteries have been proposed, including TOF and PC techniques with 2-D[81, 82] or 3-D acquisitions.[83–86] Another method that has been applied to image the renal arteries is signal targeting with alternating radiofrequency (STAR). This method involves tagging blood in a feeding vessel (suprarenal abdominal aorta) and then imaging it after it has moved into a target vessel (renal artery).[87, 88]

Though many studies initially reported promising results, the widespread use of these methods has been limited by an inability of these techniques to fully and reliably visualize the renal arteries due to motion artifacts caused by respiration, vessel tortuosity, in-plane flow saturation effects, and limited spatial resolution. The renal arteries are typically 4 to 5 mm in diameter, similar to the left main coronary artery. Added to these problems, the accessory renal arteries are often smaller and typically not well depicted.

The use of gadolinium-enhanced 3-D MRA has overcome many of these limitations, being a fast technique that enables imaging of a large 3-D volume, providing accurate and reliable visualization of the major renal arteries and accessory renal arteries in their full length (Figure 29–12). A typical

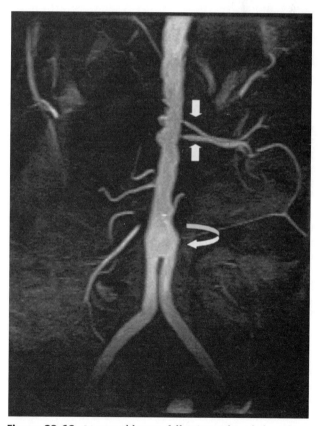

Figure 29–12. 64-year-old man, following right-sided nephrectomy, with renal artery stenosis. Coronal MIP image shows the abdominal aorta and branches. There are two renal arteries that both supply the left kidney (arrows). Just above the iliac bifurcation, a focal aneurysm (curved arrow) of the abdominal aorta is seen.

imaging protocol for gadolinium-enhanced MRA of the renal arteries is quite similar to investigations of the aorta. It consists of a localizer sequence, a contrast bolus timing sequence, and the 3-D pre- and post-CE MRA sequence. A body-phased-array coil should be used for optimal signal. As a localizer sequence, a 2-D gradient-refocused sequence or FSE sequence is appropriate, prescribed in the sagittal and/or coronal plane. It should be acquired during breath-holding in the same position, i.e., end-expiration or end-inspiration, as that in which the 3-D CE sequence will be performed. The 3-D CE MRA volume for evaluating the renal arteries is prescribed in a similar fashion as in imaging of the aorta, though for most studies, the FOV can typically be reduced to 32 cm or less. The imaging volume should be positioned in the coronal plane, preferably slightly tilted parallel to the abdominal aorta (number of slices 28 to 40; slice thickness of 2.5 to 3 mm). These parameters provide high-resolution images within a breath-holding period of less than 30 seconds.

The 3-D PC techniques depict blood flow through the renal arteries and depict turbulence and slow flow, both of which are indicative of stenosis, as signal voids.[86] Therefore, although the gadolinium-enhanced 3-D sequences provide anatomical images of the renal arteries, PC sequences provide additional functional information. Consequently some studies suggest the use of 3-D PC techniques as an adjunct to the gadolinium-enhanced renal MRA.[89, 90] In addition, postcontrast T1-weighted images should be obtained to look for the renal excretion of gadolinium-DTPA, which typically occurs within minutes after the contrast injection. This last study is especially important for the evaluation of transplanted kidneys.

For the detection of renal artery stenosis greater than 50 percent with the 3-D gadolinium-enhanced techniques (Figure 29–13), reported sensitivities and specificities vary from 93 to 100 percent and 71 to 92 percent, respectively.[40, 89, 91–93] These techniques are also superior to noncontrast MRA techniques in the detection of accessory renal arteries,[89, 92] and have found use in the preoperative detection of accessory renal arteries in living renal allograft donors.[94] The 3-D gadolinium-enhanced technique has also been successfully applied in the postoperative evaluation of transplanted renal arteries.[95] In these patients, it is recommended to add an axial T2-weighted sequence with fat saturation, prior to giving gadolinium contrast, to assess for perinephric fluid collections and hydronephrosis. Due to the pelvic location of transplanted kidneys, these studies are generally less susceptible to respiratory motion and can be carried out without breath-holding.

Mesenteric Arteries

Chronic intestinal ischemia is caused in the majority of cases by atherosclerotic narrowing or ob-

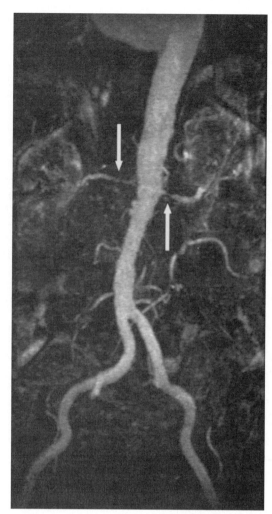

Figure 29–13. 71-year-old patient with renal artery stenosis. On the coronal MIP image of a gadolinium-enhanced 3-D data set, the abdominal aorta with branching renal arteries is shown. The proximal part of the right renal artery is fragile (arrow from above), whereas the distal part seems unaffected. There is a proximal stenosis in the left renal artery (arrow from below).

struction of the major splanchnic arteries. These narrowings or obstructions compromise the blood flow to the intestine, resulting in postprandial abdominal pain, called intestinal (abdominal) ischemia. This pain typically starts within a half hour of eating and persists for hours. The significant weight loss in these patients is primarily due to a decreased intake of food. Mucosal damage caused by the chronic ischemia also adds to the weight loss.

Although patients with generalized atherosclerotic disease often have involvement of the splanchnic circulation, few of these patients have the symptoms of mesenteric ischemia. This finding is due to the extensive collateral circulation between the celiac artery (CA) and superior mesenteric arteries (SMAs), and to a lesser extent the inferior mesenteric arteries (IMAs). Therefore, chronic stenosis, or even occlusion of all three major mesenteric arteries, frequently occurs without abdominal symptoms.[96, 97] The high therapeutic success rates of various surgical techniques for reestablishing blood flow have illustrated the need for a reliable imaging technique to diagnose stenosis and preoperatively map the mesenteric arteries.[98–102]

Although conventional invasive x-ray angiography remains the gold standard, it does have many disadvantages, including invasiveness, cost, and nephrotoxicity. More recently, Doppler sonography has been suggested for imaging of the mesenteric vessels.[103–106] However, sonography has several severe limitations including a 25 percent rate of nonvisualization of both vessels and failure to obtain adequate signal due to overlying bowel gas, excess adipose tissue, or vessel wall calcification. Possibly, it also overestimates the prevalence of double-vessel disease.[106]

Noncontrast MRA techniques such as PC MRA have failed to reliably depict the mesenteric arteries. The disappointing results achieved with these techniques (even when systolically gated[107]) are likely to be due to the triphasic flow pattern in these arteries. In addition, there are several other shortcomings that preclude 3-D PC MRA from being used as a screening tool for mesenteric artery stenosis. These limitations include phase ghosting, motion artifact related to long scan times, and uncertainty regarding choice of appropriate velocity encoding gradient value.

Recently, the 3-D gadolinium-enhanced techniques have been successfully applied to image the mesenteric circulation.[108–110] Two main factors have added to this success:

1. Most stenoses occur within 1 cm from the orifice, where the diameter of the SMA and CA is the greatest and well within the resolution capacity of the 3-D techniques.
2. The fact that these techniques are flow independent. With the 3-D gadolinium-enhanced technique, the IMA, which has a significantly smaller caliber, is less reliably visualized and is therefore prone to be more frequently graded incorrectly.[109]

A 3-D MRA study of the mesenteric vessels is carried out in the following manner. Prior to the examination, the patients can be administered glucagon intravenously to reduce bowel motion and they may be given a high caloric meal, which results in an increased blood flow to the intestine, allowing visualization of the smaller branch vessels. The patients should be positioned supine, preferentially with their arms placed above their heads to minimize aliasing artifacts, and for optimal signal, the body-phased-array coil should be used. As with all 3-D gadolinium-enhanced imaging, the protocol is composed of a localizer sequence, contrast bolus sequence, and 3-D gradient-echo sequence precontrast and postcontrast. The 3-D slab is positioned in the coronal plane and should also include the aorta and the portal vein, in addition to the splanchnic vessels. If the primary clinical question is to determine disease involving the proximal mesenteric arteries, the slab should be positioned in the sagittal

plane. The number of slices should be 30 to 32, with slice thickness of 2 to 3 mm. Overall, these parameters should be tailored for the individual patient so that the entire 3-D slab can be imaged within a single breath-holding period. As discussed previously for other vascular territories, a 3-D volume should be acquired before injecting the contrast agent to check for correct positioning. This precontrast image data set can later be subtracted from the postcontrast images to increase vessel conspicuity. A delayed 3-D data set should be acquired to image the venous phase of the contrast bolus injection, which shows the portal vein and mesenteric veins.

In addition, the severity of the mesenteric ischemia can be functionally assessed by using other noncontrast CMR techniques, which provide quantitative information such as blood flow and blood oxygen saturation. Postprandial PC cine CMR of the SMA, superior mesenteric vein (SMV), and portal vein showed that in patients with mesenteric ischemia, the percentage change in postprandial blood flow in both the SMA and SMV is significantly less than that seen in healthy subjects.[111, 112] CMR evaluation of blood oxygen saturation[113] in the superior mesenteric vein can be used as a measure of the degree of acute flow reduction in the superior mesenteric artery, which suggests mesenteric ischemia in a canine model.[114] These functional CMR techniques are not suitable for screening purposes. However, when added to a 3-D contrast MRA protocol, they may aid in the diagnosis of mesenteric ischemia.

Peripheral Vessels

The leading cause of occlusive arterial disease of the extremities in patients older than 40 years of age is atherosclerosis. The highest incidence of peripheral vascular disease (PVD) occurs in the sixth and seventh decades of life and is slightly more common among men than women.[115] The very same risk factors that have been identified for other types of cardiovascular disease, i.e., coronary artery disease and cerebrovascular disease, also have been found to correlate with PVD. These risk factors include hypertension, low levels of high-density lipoprotein cholesterol, and high levels of triglycerides. Risk factors that show a particularly strong correlation with PVD are cigarette smoking and diabetes mellitus/impaired glucose intolerance.[116] Atherosclerotic lesions have a predilection for sites of increased turbulence or complex flow, arterial branching points, and areas with increased mechanical wall stress.

Several noninvasive techniques have been developed for the detection of PVD and for the evaluation of the severity of the stenosis. These tests include Doppler ultrasonography, pulse volume recording, segmental blood pressure measurement, postocclusive reactive hyperemia testing, transcutaneous oximetry, and color-assisted ultrasound imaging.[117]

These noninvasive tests are safe, can be performed on an outpatient basis, and are readily repeatable. It should be noted, however, that the ultrasound techniques are operator dependent and that the accuracy of these tests does depend on the skill and experience of the examiner.[118] There are several drawbacks of imaging of the peripheral arteries with ultrasound. Obese extremities are difficult to evaluate, and high-quality ultrasound images distal to the midpopliteal level are difficult to obtain consistently in many patients. The estimation of the degree of luminal narrowing by peak velocity measurements can be inaccurate. These noninvasive tests are used to evaluate and determine the extent of vascular stenosis and/or occlusion. Once it has been determined that the patient's symptoms are due to hemodynamically significant vascular disease, it should be determined whether the patient is eligible for a revascularization procedure. Due to the risks and toxicities associated with this technique, only patients who are considered for an intervention should undergo conventional angiography. These complications are well described[119, 120] and have decreased somewhat with the widespread use of digital subtraction technology and smaller catheter size.[121] The systemic complications associated with conventional angiography include nausea, vasovagal attack, angina, allergic reaction, renal dysfunction, and death. Local complications include hematoma, dissection, thrombosis, embolus, and pseudoaneurysm, sometimes requiring additional therapy such as blood transfusion or surgery. The estimated prevalence of local complications varies from 4.1 percent to 23.2 percent.[122] Conventional peripheral x-ray angiography can be uncomfortable, inasmuch as large volumes of iodinated contrast are required, with subsequent filling of many muscular arterial branches.[123]

Interpretation of contrast angiograms may be inaccurate due to poor filling (slow flow) into distal vessels. Conventional angiography can fail to demonstrate distal vessels suitable for reconstructive surgery in up to 70 percent of patients with severe disease.[124, 125] In addition, background structures such as cortical bone can make interpretation of crossing arteries difficult. Asymmetrical stenoses can be underestimated by conventional angiography, and film magnification can affect the interpretation of lesions. The dose of contrast medium and timing of its administration affect the quality of the study. Contrast angiography is expensive. The examination often requires several hours to perform, with a minimum of 4 hours required for recovery of the patient after uncomplicated procedures.

Because of its several advantages over conventional angiography, many investigators have attempted to develop an MRA technique to reliably image the peripheral arteries. Peripheral MRA using the 2-D TOF technique has been evaluated and compared to conventional angiography with mixed results.[126–129] Currently, it is widely believed that the 2-D TOF MRA techniques do not have the diagnostic accuracy to replace conventional angiography for several reasons. In the peripheral arteries, the nature of the flow is highly pulsatile, with possible retrograde flow during diastole, yielding a decreased blood signal on the TOF images. This pulsatile flow also causes ghost artifacts, which can be reduced by synchronizing the data acquisition to a single phase of the cardiac cycle, though at the expense of increased imaging time.[130, 131] Vessels with horizontal flow, such as collateral vessels, tend to have less signal intensity on 2-D TOF images and may appear stenotic, and retrograde and collateral flow is often missed. Retrograde flow can also complicate the use of presaturation slabs. Turbulent flow causes de-

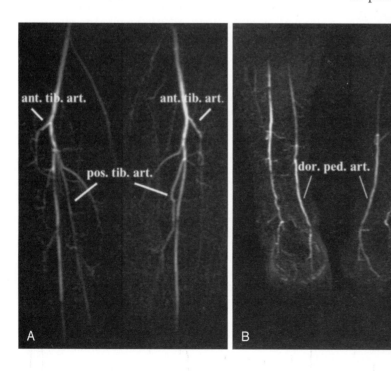

Figure 29–14. 62-year-old patient with peripheral vascular disease. Coronal MIP images of the calf vessels *(A)* and vessels of the ankle *(B)*. There is complete occlusion of bilateral anterior tibials (*A*, ant. tib. art.). Bilateral posterior tibials are severely stenosed (*A*, pos. tib. art.). There is reconstitution distally in the ankle and foot *(B)* dor. ped. art = dorsal pedal artery.

phasing, which results in signal loss in areas of stenosis and/or aneurysms, causing an overestimation of the degree of luminal narrowing. Motion of the patient during the acquisition of 2-D data results in an apparent offset of the vessel that can simulate a vascular lesion. Background suppression can be imperfect, particularly if there is abundant fat.[132] In addition, 2-D TOF should be prescribed in a plane perpendicular to the blood flow direction. In the peripheral arteries, this plane is the transaxial plane. Thus, multiple 2-D slabs have to be acquired in the transaxial direction to cover the total length of the vessels in the extremities. The blood flow in the peripheral arteries is typically slow and is visualized best when using thin slice acquisitions (1.5–3 mm). It can therefore take up to 1.5 to 2 hours to complete a study visualizing the peripheral arteries in their total length, i.e., from the iliac bifurcation of the aorta to the dorsalis pedis artery. These long imaging times have also contributed to the limited use of the 2-D TOF techniques in routine clinical practice. Noncontrast 3-D TOF studies have similar limitations, as well as a long examination time.

Despite the aforementioned limitations of the noncontrast 2-D TOF techniques, Owen and colleagues[133] demonstrated that these MRA techniques can detect runoff vessels distal to occlusions in patients with peripheral arterial occlusive disease with greater sensitivity than conventional angiography. This finding may have consequences for the surgical management and alter the treatment plan between revascularization and amputation.

The introduction of the 3-D CE peripheral MRA techniques has altered the imaging strategies for the peripheral arteries. The 3-D contrast techniques and conventional CE angiography have similar approaches to imaging vessels, inasmuch as both follow the contrast agent as it travels through the vascular region of interest. This approach makes the 3-D CE techniques relatively independent of flow, in contrast to the 2-D noncontrast TOF techniques. With the CE 3-D techniques, the imaging data can be acquired in the coronal plane so that the vessels in the extremities can be imaged in their total length in under 4 minutes. The adult legs are long and the imaging volume encompassing the peripheral arteries is too large to be covered in a single acquisition. Consequently, either the imaging range should be restricted or multiple acquisitions should be performed. There are two main approaches for acquiring multiple acquisitions with the 3-D contrast techniques.

One approach is to give multiple low doses of gadolinium chelate, while covering a different part of the vasculature each time, as was done in Figure 29–14. With this technique, higher signal-to-noise ratio and higher resolution can be achieved. Imaging can be completed in 2 to 4 minutes, a period short enough that the legs can easily be held still. Subtraction techniques can correct for background tissue

enhancement and venous signal resulting from previous injections.[134]

The other approach is quite similar to the technique applied in conventional angiography, i.e., the bolus chase technique. With this MRA technique, a single-contrast medium bolus is tracked by fast multiple 3-D slabs as it flows down through the leg vasculature. To follow the contrast agent, the table or the patient needs to be moved, typically two to three times between sets of 3-D volumes, to ensure total coverage of the leg vessels.[135–137] With this technique, the peripheral arteries can be visualized in their full length in 4 minutes as well.[138] Care should be taken that the patient is positioned such that the arteries remain within the imaging volume over the full range of table motion. Despite the long delays, especially for imaging the more distal arteries, there is no significant venous enhancement. This finding is most likely due to substantial gadolinium extraction during its first pass through the capillary bed, which lowers the venous concentration relative to the arterial concentration during the injection. Some MR vendors have now semiautomated the table motion process.

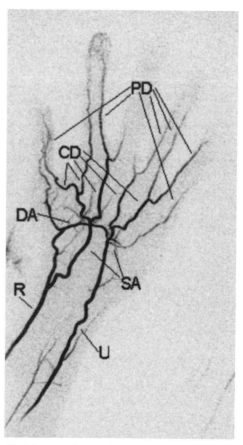

Figure 29–15. Normal vascular anatomy of the hand as shown by 3-D gadolinium-enhanced MRA. Several blood vessels have been labeled on this inverted gray scale image: radial artery (R), ulnar artery (U), superficial palmar arch (SA), deep palmar arch (DA), common palmar digital arteries (CD), and proper palmar digital arteries (PD). The superficial palmar arch is incomplete as in approximately 60% of the cases.

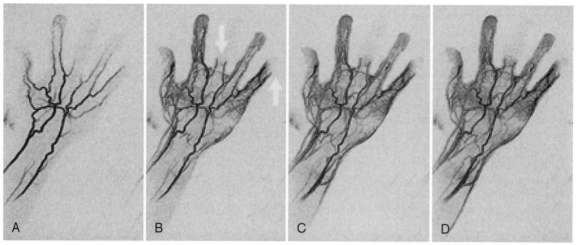

Figure 29-16. Inverted gray scale MR projection angiogram from 3-D volumes acquired *(A)* 30, *(B)* 50, *(C)* 70, and *(D)* 90 seconds after injection of the gadolinium contrast agent. In this normal volunteer, a tourniquet was briefly applied to the middle finger. Relative decrease in perfusion distal to the tourniquet (arrow down) can be seen. Note the loss of signal due to mispositioning of the coil in the fifth digit (arrow up). Arteries of the hand are labeled in Figure 29–15. Venous contamination is seen in *B* but not in *A*.

An imaging study of the peripheral arteries using the 3-D CE techniques is done in the following manner. First, the timing of the contrast bolus is determined. To ensure correct positioning of the vessels within the imaging volume, a TOF sequence can be done. Then multiple 3-D slabs are imaged *without* contrast, in the same position using the exact same parameters as with which the CE 3-D slabs will be done. The FOV is typically 38 to 50 × 41 to 50 cm and the 3-D slabs contain 32 sections, each 3 mm thick (which can be interpolated to 64 sections, each 1.5 mm thick). As mentioned previously, these volumes are almost always acquired in the coronal plane. The nonenhanced slabs ensure correct positioning and can later be used as mask images for the subtraction techniques. The CE 3-D volumes are obtained either by infusing one large (0.3 mmol/kg) bolus and tracking the bolus into the legs or by infusing smaller (0.1 mmol/kg) multiple doses of gadopentetate dimeglumine per slab.

With the 3-D CE technique, the reported sensitivities and specificities for detecting and grading hemodynamically significant stenoses of the peripheral arteries of the lower extremities are 89 to 97 percent and a specificity of 95 to 98 percent,[134, 135, 138] respectively.

The 3-D contrast MRA techniques are furthermore also useful for the assessment of bypass graft patency and for the detection of arteriovenous fistulas of the peripheral arteries.[12]

Historically MRA of the hand and wrist has received considerably less attention in the literature than has MRA of the lower extremities. Although the 2-D TOF and PC techniques have been employed for imaging of the hand by some groups,[139–141] the use of these techniques is complicated by the necessity of imaging perpendicular to the flow. This limitation requires separate acquisitions for the superficial and deep palmar arches and digital arteries.

Because of the relatively lengthy acquisition times, the TOF and PC images of the hand and wrist are often degraded by patient motion.

Although there has been considerable interest in the adaptation of gadolinium-enhanced MRA for imaging of the lower extremities, optimization of techniques for gadolinium-enhanced MR imaging of the hand has received relatively less attention to date (Figures 29–15 and 29–16). Rofsky and associates[142] described the use of a 3-D gradient-echo sequence with a gadolinium chelate to image the hand and wrist. They administered a double dose of gadolinium-DTPA at an even rate throughout the entire 2- to 3.5-minute acquisition and imaged a single 3-D slab. In their experience, the gadolinium-enhanced 3-D sequences are very suitable for obtaining detailed anatomy of the hand and wrist.

CONCLUSION

Though noncontrast TOF and PC MRA is often suitable for normal/healthy subjects, 3-D CE gradient-echo techniques are currently the most reliable and diagnostically accurate techniques for evaluation of many vascular territories throughout the body with MRA. These techniques are safe, cost efficient, and fast. Consequently gadolinium-enhanced angiography already rivals conventional x-ray angiography in many cases as an initial angiographic exam. Further refinement, with improvements in hardware, contrast agents, and pulse sequences, will undoubtedly broaden the clinical utility of gadolinium-enhanced MRA even more in the near future.

References

1. Wehrli FW: Time-of-flight effects in magnetic resonance imaging of flow. Magn Reson Med 14:187, 1990.

2. Dumoulin CL: Phase contrast magnetic resonance angiography techniques. Magn Reson Imaging Clin N Am 3:399, 1995.

3. Prince MR: Gadolinium-enhanced magnetic resonance aortography. Radiology 191:155, 1994.

4. Prince MR: Body magnetic resonance angiography with gadolinium contrast agents. Magn Reson Imaging Clin N Am 4:11, 1996.

5. Goldstein HA, Kashanian FK, Blumetti RF, et al: Safety assessment of gadopentetate dimeglumine in U.S. clinical trials. Radiology 174:17, 1990.

6. Niendorf HP, Haustein J, Cornelius I, et al: Safety of gadolinium-DTPA: Extended clinical experience. Magn Reson Med 22:222, 1991.

7. Axel L: Blood flow effects in magnetic resonance imaging. AJR Am J Roentgenol 143:1157, 1984.

8. Bradley-WG J, Waluch V: Blood flow: Magnetic resonance imaging. Radiology 154:443, 1985.

9. Dumoulin CL, Hart HR-J: Magnetic resonance angiography. Radiology 161:717, 1986.

10. Dumoulin CL, Souza SP, Walker MF, Wagle W: Three-dimensional phase contrast angiography. Magn Reson Med 9:139, 1989.

11. Wedeen VJ, Rosen BR, Chesler D, Brady TJ: Magnetic resonance velocity imaging by phase display. J Comput Assist Tomogr 9:530, 1985.

12. Prince MR, Grist TM, Debatin JF (eds): 3D Contrast Magnetic Resonance Angiography. 2nd ed. Berlin, Springer-Verlag, 1999.

13. Weinmann HJ, Brasch RC, Press WR, Wesbey GE: Characteristics of gadolinium-DTPA complex: A potential NMR contrast agent. AJR Am J Roentgenol 142:619, 1984.

14. Niendorf HP, Haustein J, Louton T, et al: Safety and tolerance after intravenous administration of 0.3 mmol/kg Gd-DTPA. Results of a randomized, controlled clinical trial. Invest Radiol 26(Suppl 1):S221, 1991.

15. Haustein J, Niendorf HP, Krestin G, et al: Renal tolerance of gadolinium-DTPA/dimeglumine in patients with chronic renal failure. Invest Radiol 27:153, 1992.

16. Prince MR, Arnoldus C, Frisoli JK: Nephrotoxicity of high-dose gadolinium compared with iodinated contrast. J Magn Reson Imaging 6:162, 1996.

17. Rofsky NM, Weinreb JC, Bosniak MA, et al: Renal lesion characterization with gadolinium-enhanced magnetic resonance imaging: Efficacy and safety in patients with renal insufficiency. Radiology 180:85, 1991.

18. Prince MR: Contrast-enhanced magnetic resonance angiography: Theory and optimization. Magn Reson Imaging Clin N Am 6:257, 1998.

19. Maki JH, Prince MR, Londy FJ, Chenevert TL: The effects of time varying intravascular signal intensity and k-space acquisition order on three-dimensional magnetic resonance angiography image quality. J Magn Reson Imaging 6:642, 1996.

20. Earls JP, Rofsky NM, DeCorato DR, et al: Breath-hold single-dose gadolinium-enhanced three-dimensional magnetic resonance aortography: Usefulness of a timing examination and magnetic resonance power injector. Radiology 201:705, 1996.

21. Krinsky G, Rofsky N, Flyer M, et al: Gadolinium-enhanced three-dimensional magnetic resonance angiography of acquired arch vessel disease. AJR Am J Roentgenol 167:981, 1996.

22. Foo TK, Saranathan M, Prince MR, Chenevert TL: Automated detection of bolus arrival and initiation of data acquisition in fast, three-dimensional, gadolinium-enhanced magnetic resonance angiography. Radiology 203:275, 1997.

23. Prince MR, Chenevert TL, Foo TK, et al: Contrast-enhanced abdominal magnetic resonance angiography: Optimization of imaging delay time by automating the detection of contrast material arrival in the aorta. Radiology 203:109, 1997.

24. Korosec FR, Frayne R, Grist TM, Mistretta CA: Time-resolved contrast-enhanced 3D magnetic resonance angiography. Magn Reson Med 36:345, 1996.

25. Krinsky GA, Rofsky NM, DeCorato DR, et al: Thoracic aorta:

comparison of gadolinium-enhanced three-dimensional magnetic resonance angiography with conventional magnetic resonance imaging. Radiology 202:183, 1997.

26. Prince MR, Narasimham DL, Stanley JC, et al: Gadolinium-enhanced magnetic resonance angiography of abdominal aortic aneurysms. J Vasc Surg 21:656, 1995.

27. Prince MR, Narasimham DL, Jacoby WT, et al: Three-dimensional gadolinium-enhanced magnetic resonance angiography of the thoracic aorta. AJR Am J Roentgenol 166:1387, 1996.

28. Flamm SD, VanDyke CW, White RD: Magnetic resonance imaging of the thoracic aorta. Magn Reson Imaging Clin N Am 4:217, 1996.

29. Steffens JC, Bourne MW, Sakuma H, et al: Quantification of collateral blood flow in coarctation of the aorta by velocity encoded cine magnetic resonance imaging. Circulation 90:937, 1994.

30. Krinsky G, Reuss PM: Magnetic resonance angiography of the thoracic aorta. Magn Reson Imaging Clin N Am 6:293, 1998.

31. Revel D, Loubeyre P, Delignette A, et al: Contrast-enhanced magnetic resonance tomoangiography: A new imaging technique for studying thoracic great vessels. Magn Reson Imaging 11:1101, 1993.

32. Ho VB, Prince MR: Thoracic magnetic resonance aortography: Imaging techniques and strategies. Radiographics 18:287, 1998.

33. Lerich R, Morel A: The syndrome of thrombosis of the aortic bifurcation. Ann Surg 127:193, 1948.

34. Watt JK: Pattern of aorto-iliac occlusion. Br Med J 2:979, 1966.

35. Thompson JE, Garrett WV, Patman RD, et al: Surgery for abdominal aortic aneurysms. In Bergan J, Yao J (eds): Aneurysms: Diagnosis and Treatment. New York, Grune and Stratton, 1982, p 287.

36. Daily PO, Trueblood HW, Stinson EB, et al: Management of acute aortic dissections. Ann Thorac Surg 10:237, 1970.

37. Dapunt OE, Galla JD, Sadeghi AM, et al: The natural history of thoracic aortic aneurysms. J Thorac Cardiovasc Surg 107:1323, 1994.

38. Moriarty JA, Edelman RR, Tumeh SS: CT and MRI of mycotic aneurysms of the abdominal aorta. J Comput Assist Tomogr 16:941, 1992.

39. Arlart IP, Gerlach A, Kolb M, et al: Magnetic resonance angiography using Gd-DTPA in staging of abdominal aortic aneurysm: A correlation with DSA and CT. Rofo Fortschr Geb Rontgenstr Neuen Bildgeb Verfahr 167:257, 1997.

40. Snidow JJ, Johnson MS, Harris VJ, et al: Three-dimensional gadolinium-enhanced magnetic resonance angiography for aortoiliac inflow assessment plus renal artery screening in a single breath hold. Radiology 198:725, 1996.

41. Badr KF, Brenner BM: Vascular injury to the kidney. Disorders of the kidney and urinary tract. In Fauci AS, Braunwald E, Isselbacher KJ, et al (eds): Harrison's Principles of Internal Medicine. 14th ed. New York, McGraw-Hill, 1998, p 1558.

42. Spargo BH, Haas M: The kidney; vascular diseases, renovascular hypertension. In Rubin E, Farber JL (eds): Pathology. 2nd ed. Philadelphia, JB Lippincott, 1994, p 805.

43. Rimmer JM, Gennari FJ: Atherosclerotic renovascular disease and progressive renal failure. Ann Intern Med 118:712, 1993.

44. Hunt JC, Strong CG: Renovascular hypertension. Mechanisms, natural history and treatment. Am J Cardiol 32:562, 1973.

45. Hunt JC, Sheps SG, Harrison-EG J, et al: Renal and renovascular hypertension. A reasoned approach to diagnosis and management. Arch Intern Med 133:988, 1974.

46. Nally-JV J, Olin JW, Lammert GK: Advances in noninvasive screening for renovascular disease. Cleve Clin J Med 61:328, 1994.

47. Martin LG, Cork RD, Wells JO: Renal vein renin analysis: Limitations of its use in predicting benefit from percutaneous angioplasty. Cardiovasc Intervent Radiol 16:76, 1993.

48. Rubin GD, Walker PJ, Dake MD, et al: Three-dimensional

spiral computed tomographic angiography: An alternative imaging modality for the abdominal aorta and its branches. J Vasc Surg 18:656, 1993.

49. Idrissi A, Fournier A, Renaud H, et al: The captopril challenge test as a screening test for renovascular hypertension. Kidney Int Suppl 25:S138, 1988.

50. Svetkey LP, Kadir S, Dunnick NR, et al: Similar prevalence of renovascular hypertension in selected blacks and whites. Hypertension 17:678, 1991.

51. Muller FB, Sealey JE, Case DB, et al: The captopril test for identifying renovascular disease in hypertensive patients. Am J Med 80:633, 1986.

52. Frederickson ED, Wilcox CS, Bucci M, et al: A prospective evaluation of a simplified captopril test for the detection of renovascular hypertension. Arch Intern Med 150:569, 1990.

53. Gosse P, Dupas JY, Reynaud P, et al: Captopril test in the detection of renovascular hypertension in a population with low prevalence of the disease. A prospective study. Am J Hypertens 2(3 Pt 1):191, 1989.

54. Svetkey LP, Himmelstein SI, Dunnick NR, et al: Prospective analysis of strategies for diagnosing renovascular hypertension. Hypertension 14:247, 1989.

55. Postma CT, van-der SP, Hoefnagels WH, et al: The captopril test in the detection of renovascular disease in hypertensive patients. Arch Intern Med 150:625, 1990.

56. Ginjaume M, Casey M, Barker F, Duffy G: A comparison between four simple methods for measuring glomerular filtration rate using technetium-99m DTPA. Clin Nucl Med 11:647, 1986.

57. Ziessman HA, Balseiro J, Fahey FH, et al: ^{99m}Tc-glucoheptonate for quantitation of differential renal function. AJR Am J Roentgenol 148:889, 1987.

58. Textor SC, Tarazi RC, Novick AC, et al: Regulation of renal hemodynamics and glomerular filtration in patients with renovascular hypertension during converting enzyme inhibition with captopril. Am J Med 76:29, 1984.

59. Wenting GJ, Tan TH, Derkx FH, et al: Splint renal function after captopril in unilateral renal artery stenosis. Br Med J Clin Res Ed 288:886, 1984.

60. Chen CC, Hoffer PB, Vahjen G, et al: Patients at high risk for renal artery stenosis: A simple method of renal scintigraphic analysis with Tc-99m DTPA and captopril. Radiology 176:365, 1990.

61. Dondi M, Franchi R, Levorato M, et al: Evaluation of hypertensive patients by means of captopril enhanced renal scintigraphy with technetium-99m DTPA. J Nucl Med 30:615, 1989.

62. Mann SJ, Pickering TG, Sos TA, et al: Captopril renography in the diagnosis of renal artery stenosis: Accuracy and limitations. Am J Med 90:30, 1991.

63. Scott JA, Rabe FE, Becker GJ, et al: Angiographic assessment of renal artery pathology: How reliable? AJR Am J Roentgenol 141:1299, 1983.

64. Taliercio CP, Vlietstra RE, Fisher LD, Burnett JC: Risks for renal dysfunction with cardiac angiography. Ann Intern Med 104:501, 1986.

65. Saint GG, Aube M: Safety of outpatient angiography: A prospective study. AJR Am J Roentgenol 144:235, 1985.

66. Beregi JP, Elkohen M, Deklunder G, et al: Helical CT angiography compared with arteriography in the detection of renal artery stenosis. AJR Am J Roentgenol 167:495, 1996.

67. Galanski M, Prokop M, Chavan A, et al: Accuracy of CT angiography in the diagnosis of renal artery stenosis. Rofo Fortschr Geb Rontgenstr Neuen Bildgeb Verfahr 161:519, 1994.

68. Johnson PT, Halpern EJ, Kuszyk BS, et al: Renal artery stenosis: CT angiography-comparison of real-time volume-rendering and maximum intensity projection algorithms. Radiology 211:337, 1999.

69. Kaatee R, Beek FJ, de LE, et al: Renal artery stenosis: Detection and quantification with spiral CT angiography versus optimized digital subtraction angiography. Radiology 205:121, 1997.

70. Lewis BD, James EM: Current applications of duplex and color Doppler ultrasound imaging: Abdomen. Mayo Clin Proc 64:1158, 1989.

71. Hansen KJ, Tribble RW, Reavis SW, et al: Renal duplex sonography: Evaluation of clinical utility. J Vasc Surg 12:227, 1990.

72. Kohler TR, Zierler RE, Martin RL, et al: Noninvasive diagnosis of renal artery stenosis by ultrasonic duplex scanning. J Vasc Surg 4:450, 1986.

73. Taylor DC, Kettler MD, Moneta GL, et al: Duplex ultrasound scanning in the diagnosis of renal artery stenosis: A prospective evaluation. J Vasc Surg 7:363, 1988.

74. Berland LL, Koslin DB, Routh WD, Keller FS: Renal artery stenosis: Prospective evaluation of diagnosis with color duplex ultrasound compared with angiography. Work in progress. Radiology 174:421, 1990.

75. Desberg AL, Paushter DM, Lammert GK, et al: Renal artery stenosis: Evaluation with color Doppler flow imaging. Radiology 177:749, 1990.

76. Handa N, Fukanaga R, Ogawa S, et al: A new accurate and non-invasive screening method for renovascular hypertension: The renal artery Doppler technique. J Hypertens Suppl 6:S458, 1988.

77. Patriquin HB, Lafortune M, Jequier JC, et al: Stenosis of the renal artery: Assessment of slowed systole in the downstream circulation with Doppler sonography. Radiology 184:479, 1992.

78. Stavros AT, Parker SH, Yakes WF, et al: Segmental stenosis of the renal artery: Pattern recognition of tardus and parvus abnormalities with duplex sonography. Radiology 184:487, 1992.

79. Mollo M, Pelet V, Mouawad J, et al: Evaluation of colour duplex ultrasound scanning in diagnosis of renal artery stenosis, compared to angiography: A prospective study on 53 patients. Eur J Vasc Endovasc Surg 14:305, 1997.

80. Strotzer M, Fellner CM, Geissler A, et al: Noninvasive assessment of renal artery stenosis. A comparison of magnetic resonance angiography, color Doppler sonography, and intraarterial angiography. Acta Radiol 36:243, 1995.

81. Kent KC, Edelman RR, Kim D, et al: Magnetic resonance imaging: A reliable test for the evaluation of proximal atherosclerotic renal arterial stenosis. J Vasc Surg 13:311, 1991.

82. Debatin JF, Spritzer CE, Grist TM, et al: Imaging of the renal arteries: Value of magnetic resonance angiography. AJR Am J Roentgenol 157:981, 1991.

83. Bass JC, Prince MR, Londy FJ, Chenevert TL: Effect of gadolinium on phase-contrast magnetic resonance angiography of the renal arteries. AJR Am J Roentgenol 168:261, 1997.

84. Borrello JA, Li D, Vesely TM, et al: Renal arteries: clinical comparison of three-dimensional time-of-flight magnetic resonance angiographic sequences and radiographic angiography. Radiology 197:793, 1995.

85. Loubeyre P, Trolliet P, Cahen R, et al: Magnetic resonance angiography of renal artery stenosis: Value of the combination of three-dimensional time-of-flight and three-dimensional phase-contrast magnetic resonance angiography sequences. AJR Am J Roentgenol 167:489, 1996.

86. Wasser MN, Westenberg J, van der Hulst VP, et al: Hemodynamic significance of renal artery stenosis: Digital subtraction angiography versus systolically gated three-dimensional phase-contrast magnetic resonance angiography. Radiology 202:333, 1997.

87. Edelman RR, Siewert B, Adamis M, et al: Signal targeting with alternating radiofrequency (STAR) sequences: Application to magnetic resonance angiography. Magn Reson Med 31:233, 1994.

88. Wielopolski PA, Adamis M, Prasad P, et al: Breath-hold 3D STAR magnetic resonance angiography of the renal arteries using segmented echo planar imaging. Magn Reson Med 33:432, 1995.

89. De Cobelli F, Vanzulli A, Sironi S, et al: Renal artery stenosis: Evaluation with breath-hold, three-dimensional, dynamic, gadolinium-enhanced versus three-dimensional, phase-contrast magnetic resonance angiography. Radiology 205:689, 1997.

90. Prince MR, Schoenberg SO, Ward JS, et al: Hemodynamically significant atherosclerotic renal artery stenosis: Magnetic resonance angiographic features. Radiology 205:128, 1997.

91. Bakker J, Beek FJ, Beutler JJ, et al: Renal artery stenosis and accessory renal arteries: accuracy of detection and visualization with gadolinium-enhanced breath-hold magnetic resonance angiography. Radiology 207:497, 1998.

92. Hany TF, Debatin JF, Leung DA, Pfammatter T: Evaluation of the aortoiliac and renal arteries: Comparison of breath-hold, contrast-enhanced, three-dimensional magnetic resonance angiography with conventional catheter angiography. Radiology 204:357, 1997.

93. Rieumont MJ, Kaufman JA, Geller SC, et al: Evaluation of renal artery stenosis with dynamic gadolinium-enhanced magnetic resonance angiography. AJR Am J Roentgenol 169:39, 1997.

94. Buzzas GR, Shield CF III, Pay NT, et al: Use of gadolinium-enhanced, ultrafast, three-dimensional, spoiled gradient-echo magnetic resonance angiography in the preoperative evaluation of living renal allograft donors. Transplantation 64:1734, 1997.

95. Johnson DB, Lerner CA, Prince MR, et al: Gadolinium-enhanced magnetic resonance angiography of renal transplants. Magn Reson Imaging 15:13, 1997.

96. Cunningham CG, Reilly LM, Stoney R: Chronic visceral ischemia. Surg Clin North Am 72:231, 1992.

97. Kurland B, Brandt LJ, Delany HM: Diagnostic tests for intestinal ischemia. Surg Clin North Am 72:85, 1992.

98. Calderon M, Reul GJ, Gregoric ID, et al: Long-term results of the surgical management of symptomatic chronic intestinal ischemia. J Cardiovasc Surg Torino 33:723, 1992.

99. Geelkerken RH, van Bockel JH, de Roos WK, et al: Chronic mesenteric vascular syndrome. Results of reconstructive surgery. Arch Surg 126:1101, 1991.

100. Rheudasil JM, Stewart MT, Schellack JV, et al: Surgical treatment of chronic mesenteric arterial insufficiency. J Vasc Surg 8:495, 1988.

101. Roberts L, Wertman-DA J, Mills SR, et al: Transluminal angioplasty of the superior mesenteric artery: An alternative to surgical revascularization. AJR Am J Roentgenol 141:1039, 1983.

102. Sniderman K: Transluminal angioplasty in the management of chronic intestinal ischaemia. In Strandness DE, Van Breda A (eds): Vascular Diseases: Surgical and Interventional Therapy. New York, Churchill Livingstone, 1994, p 803.

103. Jager KA, Fortner GS, Thiele BL, Strandness DE: Noninvasive diagnosis of intestinal angina. J Clin Ultrasound 12:588, 1984.

104. Koslin DB, Mulligan SA, Berland LL: Duplex assessment of the splanchnic vasculature. Semin Ultrasound CT MR 13:34, 1992.

105. Moneta GL, Yeager RA, Dalman R, et al: Duplex ultrasound criteria for diagnosis of splanchnic artery stenosis or occlusion. J Vasc Surg 14:511, 1991.

106. Roobottom CA, Dubbins PA: Significant disease of the celiac and superior mesenteric arteries in asymptomatic patients: Predictive value of Doppler sonography. AJR Am J Roentgenol 161:985, 1993.

107. Wasser MN, Geelkerken RH, Kouwenhoven M, et al: Systolically gated 3D phase contrast MRA of mesenteric arteries in suspected mesenteric ischemia. J Comput Assist Tomogr 20:262, 1996.

108. Hany TF, Schmidt M, Schoenenberger AW, Debatin JF: Contrast-enhanced three-dimensional magnetic resonance angiography of the splanchnic vasculature before and after caloric stimulation. Original investigation. Invest Radiol 33:682, 1998.

109. Meaney JF, Prince MR, Nostrant TT, Stanley JC: Gadolinium-enhanced magnetic resonance angiography of visceral arteries in patients with suspected chronic mesenteric ischemia. J Magn Reson Imaging 7:171, 1997.

110. Shirkhoda A, Konez O, Shetty AN, et al: Mesenteric circulation: Three-dimensional magnetic resonance angiography with a gadolinium-enhanced multiecho gradient-echo technique. Radiology 202:257, 1997.

111. Burkart DJ, Johnson CD, Reading CC, Ehman RL: MR measurements of mesenteric venous flow: Prospective evaluation in healthy volunteers and patients with suspected chronic mesenteric ischemia. Radiology 194:801, 1995.

112. Li KC, Whitney WS, McDonnell CH, et al: Chronic mesenteric ischemia: Evaluation with phase-contrast c cine magnetic resonance imaging. Radiology 190:175, 1994.

113. Wright GA, Hu BS, Macovski A: 1991 I.I. Rabi Award. Estimating oxygen saturation of blood in vivo with magnetic resonance imaging at 1.5 T. J Magn Reson Imaging 1:275, 1991.

114. Li KC, Pelc LR, Puvvala S, Wright GA: Mesenteric ischemia due to hemorrhagic shock: Magnetic resonance imaging diagnosis and monitoring in a canine model. Radiology 206:219, 1998.

115. Creager MA, Dzau VJ: Vascular diseases of the extremities. In Isselbacher KJ, Braunwald E, Wilson JD, et al (eds): Harrison's Principles of Internal Medicine. 13th ed. New York, McGraw-Hill, 1994, p 1135.

116. Criqui MH, Denenberg JO, Langer RD, Fronek A: The epidemiology of peripheral arterial disease: Importance of identifying the population at risk. Vasc Med 2:221, 1997.

117. Creager MA: Clinical assessment of the patient with claudication: The role of the vascular laboratory. Vasc Med 2:231, 1997.

118. O'Keeffe ST, Persson AV: Use of noninvasive vascular laboratory in diagnosis of venous and arterial disease. Cardiol Clin 9:429, 1991.

119. Hessel SJ, Adams DF, Abrams HL: Complications of angiography. Radiology 138:273, 1981.

120. Shehadi WH, Toniolo G: Adverse reactions to contrast media: A report from the Committee on Safety of Contrast Media of the International Society of Radiology. Radiology 137:299, 1980.

121. Waugh JR, Sacharias N: Arteriographic complications in the DSA era. Radiology 182:243, 1992.

122. Olivecrona H: Complications of cerebral angiography. Neuroradiology 14:175, 1977.

123. Katzen BT: Peripheral, abdominal, and interventional applications of DSA. Radiol Clin North Am 23:227, 1985.

124. Patel KR, Semel L, Clauss RH: Extended reconstruction rate for limb salvage with intraoperative prereconstruction angiography. J Vasc Surg 7:531, 1988.

125. Flanigan DP, Williams LR, Keifer T, et al: Prebypass operative arteriography. Surgery 92:627, 1982.

126. Baum RA, Rutter CM, Sunshine JH, et al: Multicenter trial to evaluate vascular magnetic resonance angiography of the lower extremity. American College of Radiology Rapid Technology Assessment Group. JAMA 274:875, 1995.

127. Quinn SF, Demlow TA, Hallin RW, et al: Femoral magnetic resonance angiography versus conventional angiography: Preliminary results. Radiology 189:181, 1993.

128. Mulligan SA, Matsuda T, Lanzer P, et al: Peripheral arterial occlusive disease: Prospective comparison of magnetic resonance angiography and color duplex ultrasound with conventional angiography. Radiology 178:695, 1991.

129. Yucel EK, Kaufman JA, Geller SC, Waltman AC: Atherosclerotic occlusive disease of the lower extremity: Prospective evaluation with two-dimensional time-of-flight magnetic resonance angiography. Radiology 187:637, 1993.

130. Ho KY, de HM, Oei TK, et al: Magnetic resonance angiography of the iliac and upper femoral arteries using four different inflow techniques. AJR Am J Roentgenol 169:45, 1997.

131. Selby K, Saloner D, Anderson CM, et al: Magnetic resonance angiography with a cardiac-phase-specific acquisition window. J Magn Reson Imaging 2:637, 1992.

132. Edelman RR: Magnetic resonance angiography: Present and future. AJR Am J Roentgenol 161:1, 1993.

133. Owen RS, Carpenter JP, Baum RA, et al: Magnetic resonance imaging of angiographically occult runoff vessels in peripheral arterial occlusive disease. N Engl J Med 326:1577, 1992.

134. Rofsky NM, Johnson G, Adelman MA, et al: Peripheral vascular disease evaluated with reduced-dose gadolinium-enhanced magnetic resonance angiography. Radiology 205:163, 1997.

135. Meaney JF, Ridgway JP, Chakraverty S, et al: Stepping-table gadolinium-enhanced digital subtraction magnetic reso-

nance angiography of the aorta and lower extremity arteries: Preliminary experience. Radiology 211:59, 1999.

136. Wang Y, Lee HM, Avakian R, et al: Timing algorithm for bolus chase magnetic resonance digital subtraction angiography. Magn Reson Med 39:691, 1998.

137. Wang Y, Lee HM, Khilnani NM, et al: Bolus-chase magnetic resonance digital subtraction angiography in the lower extremity. Radiology 207:263, 1998.

138. Ho KY, Leiner T, de HM, et al: Peripheral vascular tree stenoses: Evaluation with moving-bed infusion-tracking magnetic resonance angiography. Radiology 206:683, 1998.

139. Disa JJ, Chung KC, Gellad FE, et al: Efficacy of magnetic resonance angiography in the evaluation of vascular malformations of the hand. Plast Reconstr Surg 99:136, 1997.

140. Dobson MJ, Hartley RW, Ashleigh R, et al: Magnetic resonance angiography and magnetic resonance imaging of symptomatic vascular malformations. Clin Radiol 52:595, 1997.

141. Holder LE, Merine DS, Yang A: Nuclear medicine, contrast angiography, and magnetic resonance imaging for evaluating vascular problems in the hand. Hand Clin 9:85, 1993.

142. Rofsky NM: Magnetic resonance angiography of the hand and wrist. Magn Reson Imaging Clin N Am 3:345, 1995.

SECTION V

FUNCTIONAL CARDIOVASCULAR DISEASE

CHAPTER **30**

Valvular Heart Disease

Raad H. Mohiaddin and Philip J. Kilner

Diagnostic imaging methods are important in the evaluation of valvular heart disease and are usually requested (1) to define valvular morphology including leaflets, annulus, and supporting apparatus; (2) to grade the severity of valvular stenosis or regurgitation; and (3) to assess other cardiovascular structures affected by valvular dysfunction (atrial and ventricular volumes, myocardial mass, atrial thrombi and poststenotic dilatation, etc.). This information is useful to delineate etiology, to assess a patient's prognosis, and to define the role of surgical intervention. Echocardiography, x-ray angiography, and cardiac catheterization are usually used to provide such information. An alternative method is cardiovascular magnetic resonance (CMR), which can provide much of the required information noninvasively and without the need for catheters, a contrast agent, or x-ray irradiation.

In valvular heart disease, quantitative information provided by CMR complements and goes beyond that obtainable by echocardiography. CMR can, in most cases, directly measure regurgitation volume, and appropriate velocity mapping allows measurement of peak transstenotic jet velocities. Cine imaging provides wide, unrestricted views for visualization of chamber dimensions, movements and relations of chambers, and visualization of stenotic and regurgitant jets. Cine imaging in multiple contiguous slices in the short-axis plane allows accurate, reproducible measurements of ventricular volume and myocardial mass, which is important in relation to the remodeling associated with chronic valve disease that contributes to the optimization of medical, interventional, or surgical management. CMR may also assess atrial and ventricular chamber size and function, detection of associated intracardiac thrombus, detection of paravalvular abscess and fistulas, and assessment of prosthetic valves. Although conventional CMR is less suitable than echocardiography for direct visualization of valve leaflets and suspensory apparatus or infective vegetations, fast imaging methods are now available and have been shown to improve the capability of CMR in these areas.[1]

When compared with cardiac catheterization, CMR has both limitations and advantages. Like Doppler echocardiography, it is unable to measure pressure directly, but its noninvasiveness, its ability to measure flow and to map velocities through ob-structive lesions, and its ability to obtain tomographic views of complex three-dimensional anatomy, give it distinct advantages. In cases where shunting or stenoses of valves and large vessels are concerned, especially when these are combined with anatomical deformation, CMR with velocity mapping can offer the surgeon more comprehensive information, more safely, and more economically than the catheterization laboratory.

CARDIOVASCULAR MAGNETIC RESONANCE TECHNIQUES

Although individual spin-echo images may show valve structures (Figure 30–1), they cannot be relied upon to show adequate detail. More recently, other black-blood techniques have been successful in this regard.[1a] Multislice spin-echo images can, however, be useful for visualization of relations and sizes of chambers and vessels that may contribute to evaluation of heart valve disease, and multislice spin-echo acquisitions provide scout images for initial placement of gradient-echo slices, which then serve as scouts for further cine and velocity acquisitions.

Assessment of heart valves by CMR depends mainly on cine gradient-echo imaging, with phase-shift velocity mapping for measurements of flow. Cine imaging shows flowing blood as white, and turbulence caused by high-velocity jets in either

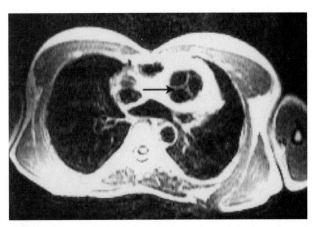

Figure 30–1. A transverse spin-echo image of thickened aortic leaflets (arrow) in a patient with transposition of the great arteries.

regurgitation or stenosis causes signal loss, much like that seen using color Doppler echocardiographic techniques. Thus, cine imaging is used to detect valve dysfunction and to localize jets. However, the choice of imaging sequence is important, because the longer the echo time, the more signal loss that is seen.[1b] This situation has consequences when faster imaging (shorter echo time) is used. Conventional cine imaging is time consuming, but whereas reduction of acquisition time avoids unwanted effects of respiratory motion and saves time, one or more of the following sources of information relevant to valve disease may be compromised.

- Effects of flow (e.g., turbulence) on CMR signal recovered
- Delineation of blood-muscle and blood-valve boundaries
- Measurements of flow velocity and flow volume by velocity mapping

These problems have to be borne in mind when using any of the rapid CMR methods that have been developed. The main techniques are segmented k-space gradient-echo[2] and echo-planar imaging (EPI), including spiral imaging.[3–6] Although the segmented k-space gradient-echo technique is slower than the EPI methods, it is still more commonly used. This is mainly because the segmented k-space technique is less demanding on the hardware and software of the system than the EPI methods and can usually be implemented on unmodified machines. These approaches can be used during breath-holding, which significantly reduces breathing-related artifacts. Volume imaging of the heart to measure mass and stroke volumes of the left and right ventricle can now be achieved in under 15 minutes using the breath-hold techniques (see Chapter 9).

Velocity mapping is an extension of gradient-echo cine imaging but requires longer acquisition times and therefore is usually not performed during a

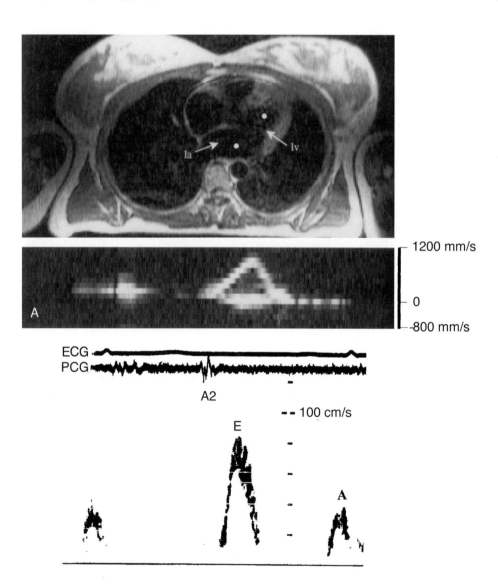

Figure 30–2. *A*, A transverse spin-echo image at midventricular level (upper panel). The two markers on the left atrium (la) and left ventricle (lv) defined the position and direction of the Fourier velocity (FV) imaging for the mitral valve. A selected frame from the complete cine acquisition of FV across the mitral valve acquired at peak ventricular filling (lower panel) with velocity through the valve plotted vertically and position along the cylinder plotted horizontally. *B*, The corresponding Doppler velocity trace with velocity (cm/s) through the valve plotted vertically and time (R-R interval, ms) plotted horizontally. ECG = electrocardiogram; PCG = phonocardiogram. (From Mohiaddin RH, Gatehouse PD, Henien M, Firmin DN: Cine magnetic resonance Fourier velocimetry of blood flow through cardiac valves: Comparison with Doppler echocardiography. J Magn Reson Imaging 7:657, 1997. Reprinted by permission of Wiley-Liss, Inc., a subsidiary of John Wiley & Sons, Inc.)

single breath-hold. The development of velocity mapping as an extension of CMR can be compared with Doppler techniques in relation to echocardiography, except that, in many respects, CMR has capabilities beyond those of Doppler.[7] CMR velocity mapping is able to measure accurate velocities in pixels throughout the plane of acquisition. It can acquire data in any orientation, unrestricted by windows of access, and it allows choice of the direction in which velocities are measured with respect to the imaging plane. The other major advantage of CMR velocity mapping is that volume flow can be calculated with high accuracy because of the simultaneous acquisition of mean through-plane velocity and area of the vessel. Doppler echocardiography is accurate for measuring velocity, but poor at assessing volume flow, especially when the flow pattern is complex. In addition, CMR velocity mapping is the only imaging technique with the potential to acquire comprehensive information (three spatial dimensions, three velocity components, and time) and is well suited for studying spatial and temporal patterns of flow in the human cardiovascular system.[8] This capacity is important because flow in heart cavities and in vessels that are curved or branching can be complex, with components of velocity in various directions.[9] It should be borne in mind again that just as for imaging of turbulent flow to show valve disease, the echo time of velocity mapping is important in evaluating valve disease and needs to be carefully chosen (see later discussion). Other methods such as CMR Fourier velocity mapping have been developed to assess flow through cardiac valves (Figure 30–2). In this method, the spatial phase encoding gradient pulses are replaced by velocity phase encoding bipolar gradient waveforms. Thus, the resulting image data set has only a single spatial dimension (the readout dimension) and one velocity dimension, which is sensitive only to the component of velocity parallel to the velocity encoding gradient pulse.[10]

ASSESSMENT OF STENOTIC VALVULAR DISEASE

Jet Flow and Signal Loss

Turbulent flow causes signal loss on gradient-echo images, which is important for depiction of regurgitant or stenotic jets.[11] Relationships between signal and flow are, however, complex. On the whole, coherent through-plane flow is associated with brightening of signal whereas turbulence causes signal loss, but loss of signal can also be caused by shear, acceleration, and higher orders of motion, not necessarily in the presence of turbulence. Turbulence itself is not homogeneous, but has an inner dynamic structure of rapidly interacting eddies and countereddies with certain orientations in relation to a jet.

Jet flow through an orifice is generally associated with a region of acceleration (convergence of streamlines) into the orifice, a relatively stable jet core extending beyond the orifice, and, lateral and distal to the core, a parajet region of intense turbulence. This turbulent zone dissipates energy, becoming less intensely turbulent as it is swept downstream. Each of these zones, whose relative sizes and shapes depend on orifice size, shape, and flow, may be identifiable on gradient-echo images or velocity maps. Their appearances on cine images depend not only on properties of flow but also on the nature of the acquisition—slice thickness, pixel dimensions, sequence design and, to an important degree, on echo time of the sequence used.[1b] In general, longer echo times are associated with more signal loss due to turbulence, shear, or acceleration. The shorter the echo time, the higher the level of turbulence intensity required to cause signal loss (Figure 30–3),[12] which means that echo time is an important consideration when choosing a sequence for visualization of jet flow.

Although the size of the signal loss in the turbulent jet distal to the valvular stenosis is related to the pressure gradient across the valve, it is difficult to use this as a measure of severity of stenosis because other factors can lead to turbulence and signal loss without stenosis. For example, it is not uncommon to see turbulence distal to a nonstenotic bicuspid aortic valve because of the disturbed flow across it.

The Modified Bernoulli Equation

A stenotic valve may be assessed by measuring the flow velocity in the jet of blood passing through a stenosis. In the case of a given flow, an increasingly narrow stenosis leads to an increase in the velocity of flow through the orifice. The relationship between the velocity of the jet and the difference in pressure on either side of the stenosis can be approximated by the modified Bernoulli equation, which in its simplest form is

$$P_1 - P_2 = 4(V_2^2 - V_1^2) \qquad [1]$$

where V_2 = velocity distal to the obstruction, V_1 = velocity proximal to obstruction, P_1 = pressure proximal to obstruction, and P_2 = pressure distal to obstruction.

When velocity is expressed in m/s, the pressure difference $(P_1 - P_2)$ is in mm Hg. This simplified version of the equation makes the assumption that true velocity has been recorded and that the blood velocity before the stenosis was negligible. In most clinical situations the proximal velocity (V_1) is rarely significantly greater than 1 m/s and is usually disregarded. The fully modified equation is

$$\Delta P = 4V_2^2 \qquad [2]$$

The Bernoulli equation is a complex physical equation relating to fluid flow in cylinders, and it is

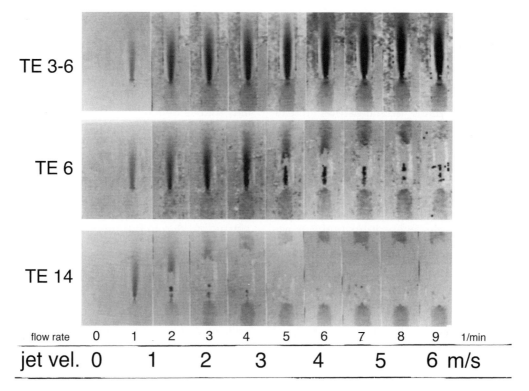

Figure 30–3. In vitro, poststenotic jet velocity mapping using a gradient-echo sequence. Water with copper sulfate was propelled continuously through 22-mm diameter tubing containing a 6-mm concentric stenosis at flow rates up to 9 liter/min, giving jet velocities of up to 6 m/s. Signal loss is almost total with the 14-ms echo time sequence compared with retention of signal within the jet at velocities up to 6 m/s using a short echo time of 3.6 ms. TE = echo time in ms. (From Kilner PJ, Firmin DN, Rees RSO, et al: Valve and great vessel stenosis: Assessment with CMR jet velocity mapping. Radiology 178:229, 1991.)

important to remember that there are several errors or limitations that occur when using the maximum jet velocity to predict the pressure drop across an obstruction. For example, the viscous resistance increases as vessel or valve size decreases, and the pressure drop associated with flow acceleration can be significant in certain clinical situations (e.g., when estimating the pressure drop across a severely stenotic aortic valve).

The velocity across a cardiac valve is increased with obstruction or with increased flow across the valve. However, the rate of decline of the velocity from the early peak velocity is altered by obstruction rather than flow. The slope of the decline in the maximum velocity has been used to estimate the pressure half-time (the time it takes until the initial pressure drop is halved), for example, in mitral stenosis.

Jet Velocity Mapping for Assessment of Valvular Stenosis

CMR jet velocity mapping can be valuable for assessment of stenoses where ultrasonic access is limited, for example, where there is calcification and deformation of an aortic valve, especially where it has proved impossible to cross the valve with a catheter, or in surgically placed ventriculopulmonary conduits.[13, 14] Combined cine imaging and ve-

locity mapping give information on the location, nature, and severity of obstruction. Mapping of high-velocity poststenotic jets is possible only if sequences with very short echo times are used. Accurate location of a sufficiently thin slice (6–8 mm) in relation to the jet core is also essential.

Accurate location depends on recognition of the jet core on preliminary cine images. A series of orthogonal cuts can home in with increasing accuracy (as long as the patient remains still) on the area of interest. It is usually possible to select a cine sequence that, when correctly aligned, will show signal from the bright jet core outlined by signal void in the parajet region. It is preferable to use a slightly longer echo time for preliminary cine imaging than that chosen for final velocity mapping.

For measurement of jet velocity, the echo time must be short enough to recover signal from the jet core. To give approximate guidance:

- For low-velocity jets, up to 2 m/s, which may be significant at atrioventricular level, use a longer echo time, TE = 6 to 8 ms.
- For velocities up to 3 or 4 m/s (mild to moderate obstruction of outflow valves and conduits), use TE = 4 to 6 ms.
- For higher velocities (severe stenosis of outflow valves or conduits), use the shortest available echo time, e.g., TE ≤ 3 ms.

Because shortening of echo time tends to reduce overall signal-to-noise ratio and velocity sensitivity, and may be associated with disturbing levels of acoustic noise, use of very short echo times should be reserved for high-velocity jets. Occasionally, jets of regurgitation and severe aortic stenosis can be so narrow, turbulent, and fragmented that they are not suitable for assessment by CMR jet velocity mapping, with excessive signal loss and partial volume effects preventing accurate jet velocity measurement.

For evaluation of jet flow through stenoses, there are certain advantages in mapping velocities in a plane aligned with the direction of flow, with velocity encoded in the read gradient direction.[15] This arrangement allows the location of stenosis and jet to be seen in relation to upstream and downstream regions (Figure 30–4). Through-plane velocity mapping may also be used. The plane is aligned so as to transect the jet immediately distal to the orifice. Velocity is encoded in the slice-select gradient direction (Figures 30–5 and 30–6). This set-up has different advantages: It allows visualization and measurement of jet cross-sectional area, and possibly measurement of volume flow through the orifice.[16]

ASSESSMENT OF VALVULAR REGURGITATION

In regurgitant heart valve disease, the optimal time for surgical intervention depends on a balanced assessment of when valve dysfunction is sufficiently severe to warrant the short- and long-term risk of surgery, but before the heart muscle sustains irreversible damage. The need for intervention is also partly determined by the severity of the symptoms, but other measurements such as the severity of regurgitation, ventricular volumes, and ventricular function are also important. Measurement of valvular regurgitation is important but is difficult to quantify with current echocardiographic and angiographic techniques. There are several CMR methods available to assess the severity of valvular regurgitation.

Cine Imaging of Regurgitant Jets

The gradient-echo sequence with relatively long echo time (e.g., 14 ms), as described earlier, identifies areas of turbulent blood flow as areas of signal loss within the high blood signal of the receiving chamber. The size of the signal loss has been used as a semiquantitative measure of the severity of valvular regurgitation.[17] This technique is similar to color Doppler and suffers from the same problems, namely that technical factors such as gain, adjustment, and filter setting are important. Both methods also suffer from other less technical problems, which include the difficulty in separating the turbulent volumes when dual valve disease exists (such as mitral stenosis and aortic regurgitation), and the fact that many other factors, other than the severity of the regurgitation, affect the size of the regurgitant jet including the shape and size of the regurgitant orifice and the size of the receiving chamber. Thus, though useful for detecting the presence of valvular regurgitation, turbulent signal loss is not an accurate guide to severity.

On the whole, regurgitant jets have little or no coherent jet core, and imaging depends on visualization of the shape and extent of signal loss from turbulence. In addition, a small spot of signal loss may be present upstream of the orifice (just within the compartment from which flow is arising). The area of signal void is caused by acceleration in the

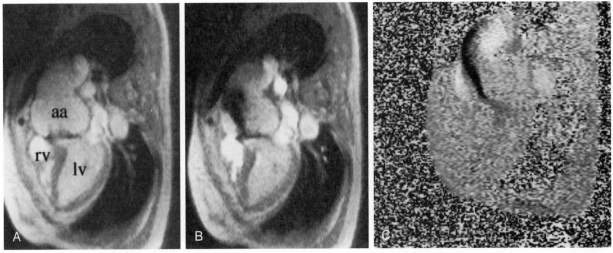

Figure 30–4. *A* and *B*, Diastolic and systolic gradient-echo (echo time 14 ms) images in an oblique cut aligned along the inflow-and-outflow-long-axis plane of left ventricle in a patient with xenograft aortic valve replacement. The aortic root is grossly enlarged, and there is a jet of systolic signal loss distal to the aortic valve indicating severe aortic valve stenosis. *C*, A corresponding systolic velocity map image (echo time 3 ms). The signal loss distal to the stenotic aortic valve is recaptured using a short echo time, and the velocity map shows a narrow stenotic jet with peak velocity of 3.9 m/s in the center of the jet. aa = aortic root, lv = left ventricle, rv = right ventricle.

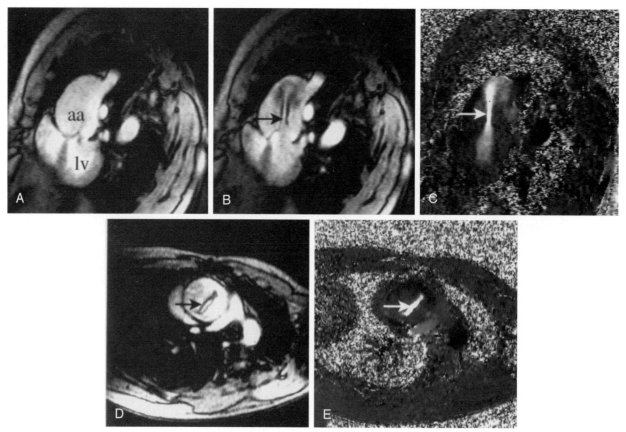

Figure 30–5. *(A)* Diastolic and *(B)* systolic gradient-echo images (TE 6 ms) in a tilted coronal plane through the aortic valve of a patient with bicuspid aortic valve stenosis. Note the jet with a bright white core in *B*. *C*, The corresponding systolic velocity map. The ascending aorta is dilated and the narrow jet of aortic valve stenosis (arrow) can be depicted on the gradient-echo image *(D, arrow)*, whereas maximum velocity along the jet can be measured from the velocity map image *(E, arrow)*. *D*, Gradient-echo (TE 6 ms) image and the corresponding systolic velocity map *(E)* acquired above the aortic valve of the same patient. The aortic valve is bicuspid, and the cross-sectional area of the narrow jet can be measured.

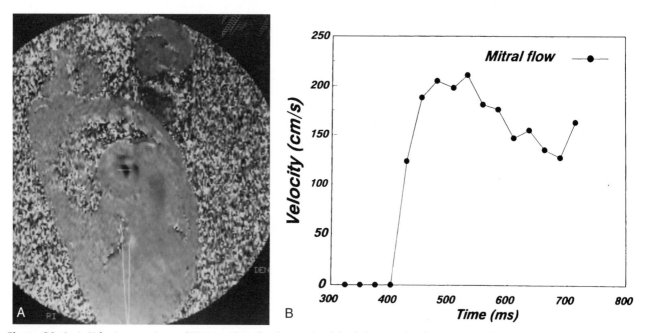

Figure 30–6. *A*, Velocity map image (TE 3.6 ms) in the short axis of the left ventricle of a patient with mitral valve stenosis. Velocity is encoded through the plane from base to apex. Ventricular filling with a narrow velocity jet is seen in black with a velocity profile (peak velocity 2 m/s) displayed in the center of the stenotic jet. *B*, Mitral valve blood velocity throughout left ventricular filling measured from the complete through-plane cine velocity map acquisition, from which the pressure half-time can be calculated, as in echocardiography.

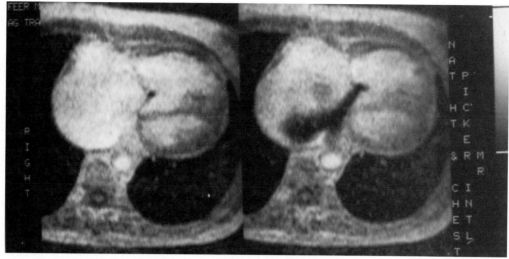

Figure 30–7. Diastolic *(left)* and systolic *(right)* gradient-echo images (TE 14 ms) in a transverse plane in a patient with tricuspid regurgitation and severe right ventricular enlargement and systolic dysfunction due to cardiac sarcoidosis. The right atrium is enlarged, and the signal loss caused by the eccentric tricuspid regurgitation jet is clearly seen on the systolic image.

region of flow convergence, and its size may give an approximate guide to the severity of regurgitation.[18] This proximal isovolumic surface area (PISA) technique, however, also suffers from a number of assumptions that cause significant limitations in clinical practice.

For adequate visualization of a regurgitant jet, it is necessary to align the imaging plane with the orifice and the principal axis of jet flow. Because regurgitant jets can be eccentric, flattened, and obliquely directed, sometimes attaching to a cusp or leaflet, there can be no fixed rules for alignment of planes. More than one cine acquisition is likely to be needed to discover the shape and extent of a regurgitant jet.

The choice of the initial cine imaging plane depends on the valve or valves affected: For the *tricuspid valve*, a transaxial cut, tilted slightly towards the ventricular apex (aligned from a coronal spinecho image) is suitable (Figure 30–7). For the *pulmonary valve*, select a sagittal cut, rotated slightly to align with the pulmonary trunk (from transaxial images) (see later discussion). For the *mitral* and *aortic valves*, a useful oblique cut can be aligned from coronal images—an inflow-and-outflow longaxis plane (see Figure 30–4). It lies diagonally with respect to coronal images, aligned with the left ventricular outflow tract (LVOT) and aortic valve. More posteriorly, it generally passes through the mitral orifice, conveniently depicting movements of both mitral leaflets. The horizontal and vertical long-axis planes of the left ventricle are also suitable planes for visualization of mitral valve regurgitation (Figure 30–8). For the *aortic valve*, a coronal cut aligned with the LVOT and aortic valve, from transaxials, is often useful (Figure 30–9). Whichever slice is tried first, a further orthogonal cut (or cuts) should be performed, aligned with the principal line of regurgitant flow, as deduced from appearances of the first acquisition.

Allowing for the reservations stated above, the severity of regurgitation may be estimated semiquantitatively from the extent of the jet with respect to the cavity, together with observations of jet shape, its time course, cavity dimensions, and movements. Severe regurgitation is generally associated with a jet extending right across the receiving cavity, for example, in the case of mitral regurgitation, impinging on the posterior atrial wall, opposite the valve. Mild regurgitation, on the other hand, is associated with a small, localized jet extending only a short distance back from the valve. But, as with Doppler ultrasound, this type of assessment is subjective, and can be misleading if slice placement is inaccurate, or where an asymmetrical jet attaches to a wall or cusp. CMR at least has the advantage of free and unrestricted orientation of oblique planes, allowing optimal views to be sought in a series of orthogonal cuts.

The operator should always be alert during cine imaging for unexplained findings, for example, additional jets that might be related to flow through a ventricular septal defect (VSD), fistula (Figure 30–10),[19] or perforated valve leaflet.

Quantification of Regurgitation by Ventricular Volume Measurements

Regurgitation of any single heart valve, if isolated and uncomplicated by shunting, may be calculated by comparison of the left and right ventricular stroke volumes measured by planimetry from a standard multislice short-axis volume acquistion (Figure 30–11). The ventricle with the leaking valve will have excess stroke volume that represents the volume of leakage. The accuracy of this approach is dependent on the image quality and accuracy of planimetry, which currently needs to be done manu-

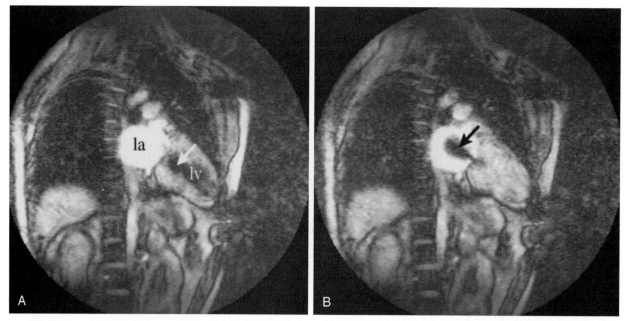

Figure 30–8. Diastolic *(A)* and systolic *(B)* gradient-echo (TE 14 ms) images in the vertical long-axis plane of the left ventricle acquired in a patient with combined mitral valve stenosis and regurgitation. In diastole, the stenotic jet is seen extending from the mitral valve into the body of the left ventricle (white arrow), whereas in systole, the regurgitant jet is seen extending from the mitral valve into the left atrium (black arrow). la = left atrium; lv = left ventricle.

ally. But, if appropriately performed, CMR measurements of right and left ventricular volumes should have greater accuracy and reproducibility than those attempted by any other technique (see Chapter 9).[20–23]

Where more than one valve is regurgitant, it is possible to combine ventricular volume measure-

ment with aortic and pulmonary flow measurements (see later discussion) to calculate regurgitant volumes of all four valves independently. This method depends on accuracy of techniques used, which may need to be validated for a particular CMR and image processing system, but the principle works, and although accuracy is likely to be lower in com-

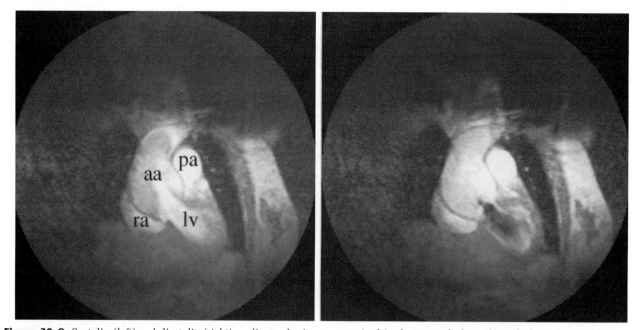

Figure 30–9. Systolic *(left)* and diastolic *(right)* gradient-echo images acquired in the coronal plane through the ascending aorta and the aortic valve of a patient with Marfan syndrome. The aortic root is grossly dilated. There is a jet of diastolic signal loss extending from the aortic valve into the body of the left ventricle indicating severe aortic valve regurgitation. aa = ascending aorta; pa = pulmonary trunk; lv = left ventricle; ra = right atrium.

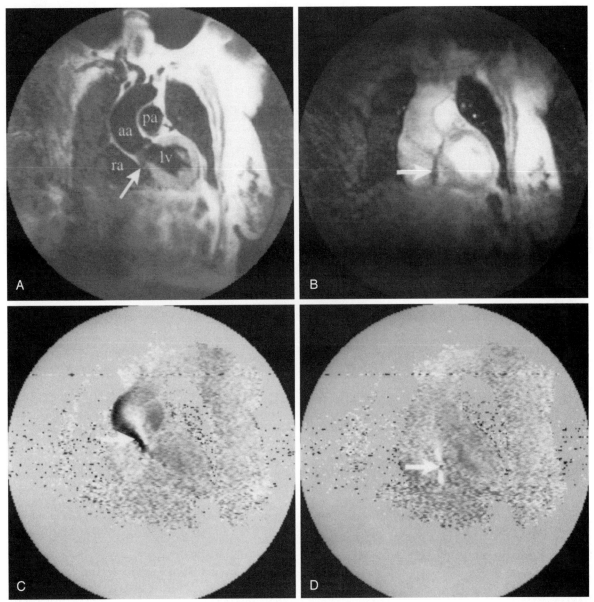

Figure 30–10. *A,* Spin-echo image in a coronal plane showing a fistula between the aortic root and the right atrium in a patient with aortic valve disease. *B,* Diastolic image selected from cine gradient-echo images acquired in a plane similar to *A.* A turbulent jet extends from the aortic root into the body of the right atrium (arrow). *C* and *D,* Corresponding systolic and diastolic velocity maps. Cranial flow velocity in the ascending aorta is seen in black, and caudal velocity through the fistula is seen in white (arrow). aa = ascending aorta; ra = right atrium; lv = left ventricle, pa = pulmonary trunk. (From Allum C, Knight C, Mohiaddin RH, Poole-Wilson PA: Images in cardiovascular medicine. Use of magnetic resonance imaging to demonstrate a fistula from the aorta to the right atrium. Circulation 97:1024, 1998.)

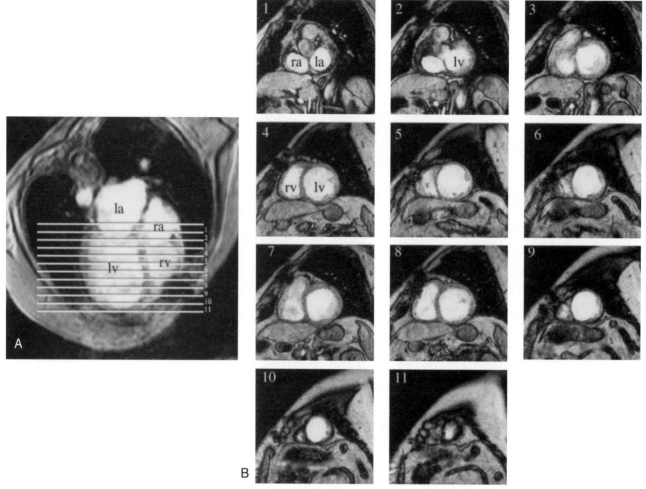

Figure 30–11. *A*, Diastolic frame selected from breath-hold cine imaging acquired in the horizontal long axis of the left ventricle in patients with ischemic cardiomyopathy and isolated mitral valve regurgitation. This view is necessary to pilot the subsequent multiple breath-hold cine acquisitions in the short axis of the left ventricle. *B*, The left and right ventricles are divided into multiple short-axis slices of 1-cm thickness, from the mitral valve to the apex. For each slice, the endomyocardial border is marked and the area measured at end diastole and end systole. Ventricular end-diastolic and end-systolic volumes are calculated by summing the areas in the contiguous images. Mitral valve regurgitant volume in this example equals left ventricular stroke volume minus right ventricular stroke volume. Regurgitant volume may be expressed as a percentage of left ventricular stroke volume (regurgitant fraction). The left ventricular mass can be similarly calculated for each slice by subtracting the endocardial area from the epicardial area. The sum of these areas is then multiplied by the specific gravity of myocardium (1.05 kg/liter) to give a measurement in grams. la = left atrium; ra = right atrium; lv = left ventricle; rv = right ventricle.

plex situations, this is also true of all other assessment techniques as well.

Quantification of Regurgitation by Cardiovascular Magnetic Resonance Velocity Mapping

Magnetic resonance phase-shift velocity mapping, if correctly implemented, is accurate and versatile. It has important roles in relation to quantification of regurgitant valve lesions. Recovery of signal from flowing blood is a prerequisite, so gradient-echo sequences, with the refinement of even-echo rephasing, are used. Phase velocity mapping techniques are based on the frequency changes experienced by nuclei that move relative to applied magnetic gradients.[24–26] Inasmuch as the direction, steepness, and timing of applied gradients can be chosen, a range of flow velocities can be measured, ranging from those of slow diastolic flow in the aorta associated with regurgitation to those of high-velocity poststenotic jets. But for successful clinical application, the operator should understand not only valve pathology but also technical velocity mapping choices. It is necessary to select a plane, echo time, velocity encoding direction, and sensitivity appropriate for a particular investigation.

Velocity can be encoded in directions that lie either in or through an image plane, and each pixel across a velocity map can carry quantitative velocity information. In these respects, CMR velocity mapping is more versatile and comprehensive than Doppler ultrasound. Cine velocity mapping records multiple phases through the cardiac cycle, for example from 16 to 30 phases, or more, depending on the temporal resolution and extent of heart cycle coverage required.

The severity of atrioventricular valvular regurgitation may be quantified by the subtraction of flow in a great vessel measured by velocity mapping from the stroke volume of the associated ventricle, usually measured from the multislice short-axis technique (for example, mitral regurgitant volume equals the left ventricular stroke volume minus aortic flow). This method allows true isolation of the left and right sides of the heart for assessment of regurgitant fraction and regurgitant volumes. The technique, however, cannot directly separate regurgitant flow from two regurgitant valves on one side of the heart. In these cases, aortic or pulmonary valve regurgitation may be quantified from the backflow of blood in the proximal great vessels as assessed by velocity mapping (see later discussion). Such a measurement would allow accurate regurgitant volumes for each cardiac valve to be individually calculated in the presence of any mixture of valve lesions.

Measurement of Ventricular Mass and Volumes

It has been shown that measurements of left ventricular mass in patients with hypertension and dilated hearts are more reproducible when performed by CMR than by echocardiography.[27, 28] This reproducibility makes detection of progression of increased left ventricular mass possible in follow-up investigation of, for example, chronic aortic stenosis. It should also be borne in mind, however, that volume overload states also induce ventricular hypertrophy, and this measurement therefore is also relevant to chronic aortic regurgitation, for example. Measurement of right ventricular mass is also possible but less straightforward because it is a very thin-walled chamber, but with marked trabeculations. However, it can be usefully followed where there is hypertrophy due to obstruction of a ventriculopulmonary conduit, or where the right ventricle is systemic.[29] Volume measurements of the left and right ventricles are also valuable in the assessment of ventricular response to valve disease, and again these are more reproducible using a three-dimensional approach such as CMR, rather than two-dimensional or M-mode echocardiography, particularly if ventricular shape is distorted through remodeling.[28] The long-term follow-up of changes in chronic disease is aided by using reproducible CMR techniques.[30]

Aortic and Pulmonary Regurgitation

CMR velocity mapping can measure aortic or pulmonary regurgitant fraction, something that cannot yet be achieved accurately or reproducibly by any other modality (Figures 30–12 and 30–13).[31, 32] It does this by mapping velocities of flow *through* a plane transecting the ascending aorta or pulmonary trunk. Velocity is encoded in the direction of the slice-select gradient, which allows measurement of systolic forward flow and any diastolic reversed flow, after closure of the valve. The cross-sectional area of the lumen is outlined manually and measured in each frame through the heart cycle, and the mean axially directed velocity is measured within that area. A flow curve is then plotted, and systolic forward flow and diastolic reversed flow are computed by integration. Regurgitant fraction can be calculated as diastolic reversed flow as a percentage of systolic forward flow. For the aortic valve, a regurgitant fraction above approximately 40 percent is severe, above 25 percent is moderate, and below 25 percent is mild, although these figures require further investigation. On the pulmonary side, free regurgitation seems to result in a regurgitant fraction of about 40 percent, although there may be a higher fraction if complicated by pulmonary artery branch stenosis or otherwise elevated pulmonary vascular resistance. CMR methods also facilitate calculation of regurgitant volume index, the ratio of regurgitant volume to end-diastolic left ventricular volume.[32a]

For aortic flow measurement, there has been in vitro and limited in vivo investigation of the suitability of planes for velocity mapping.[33] It has been suggested that a plane transecting the aortic root, immediately distal to aortic valve cusps, is most suitable, but further work is needed to confirm whether this level or a slightly more distal slice is preferable. The issue revolves around the inaccuracies introduced by compliance of the aortic root, and movement of the aortic valve during the cardiac cycle, when a fixed imaging plane relative to space outside the human body is used to make the flow measurement. Work on using a technique that tracks the aortic valve throughout the cardiac cycle as it descends in systole and ascends in diastole has also been performed using tagging, and this technique may provide the optimal solution in the future.[33a] Until this method is more widely validated and used, however, it is reasonable to use a plane close to the aortic valve and understand that in patients with very dilated aortic root, the measurement may be less accurate.

For pulmonary flow measurement, a suitable plane should cut the pulmonary trunk (or ventriculopulmonary conduit) before its bifurcation. It should be aligned across an oblique sagittal cine showing flow from right ventricle to pulmonary trunk. An echo time of about 6 ms is probably adequate for forward and regurgitant flow measurement. The sensitivity (velocity window) needs to be just sufficient to measure peak systolic velocity as well as the much lower diastolic velocities.

There are technical issues associated with the analysis of velocity mapping acquisitions that must be considered if the technique is to be used correctly

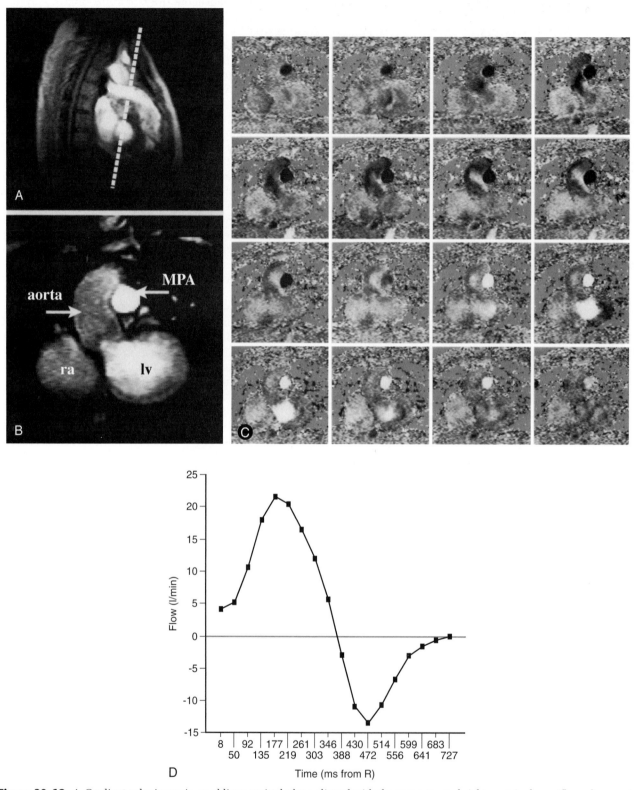

Figure 30–12. *A*, Gradient-echo image in an oblique sagittal plane aligned with the reconstructed right ventricular outflow after repair of tetralogy of Fallot. Cine images showed no evidence of an effective pulmonary valve. *B*, A magnitude image and *(C)* detail from all 16 frames of a cine velocity map (TE 6 ms) acquisition. Velocity encoded through the plane transacting the reconstructed pulmonary trunk, as indicated in *A*. Systolic forward flow appears black and diastolic reverse flow, white. The frames run from left to right in line ordered from above downward. *D*, Graph showing systolic forward flow (upward curve) and diastolic reversed flow (downward curve) measured by multiplying mean velocity by cross-sectional area of the pulmonary trunk for each frame through the cycle. Integration of areas under the curves allows measurement of forward and reversed flow, and of regurgitant fraction, which was 40% in this case of free pulmonary regurgitation. lv = left ventricle; ra = right atrium, MPA = main pulmonary artery.

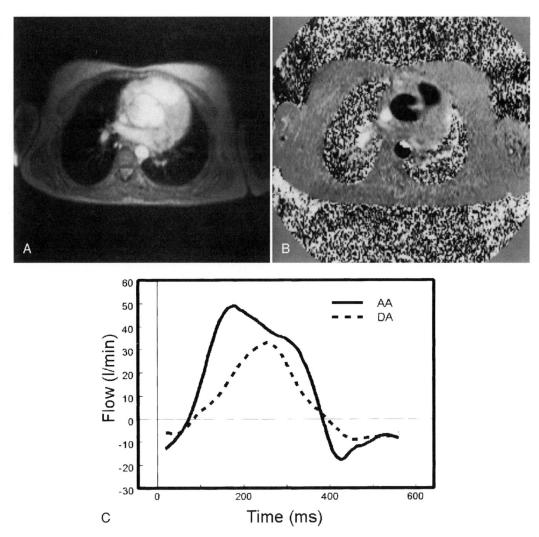

Figure 30–13. A systolic gradient-echo *(A)* and a corresponding velocity image *(B)* acquired in a plane perpendicular to the proximal ascending aorta of a patient with Marfan syndrome and aortic valve regurgitation. *C*, The flow volume curve in the ascending aorta (AA) and descending thoracic aorta (DA) measured from the complete cine CMR velocity mapping show a large retrograde net flow during diastole. Note the aliasing artifact in the descending aorta. (Flow to the feet is depicted in white, but the descending aorta is largely black due to the setting of a velocity sensitivity that was too low for this portion of the cardiac cycle. The aliasing was corrected using postprocessing techniques in this case to derive the flow curves.) (From Mohiaddin RH, Longmore DB: Functional aspects of cardiovascular nuclear magnetic resonance imaging: Techniques and application. Circulation 88:264, 1993.)

clinically. These issues revolve around ensuring that the data are correctly analyzed. First, the region of interest around the great vessel must be accurate and properly positioned for each phase of the cardiac cycle because the vessels move. If present, aliasing must be corrected for in each image using appropriate postprocessing software (preferably, aliasing should be prevented by correct acquisition but sometimes peak velocities are higher than anticipated). Edge pixels and other artifacts of imaging should be excluded from the analysis. Finally, and very importantly, it is crucial to ensure that the background mean velocity is set to zero, which usually means placing a series of static markers in the chest wall around the image and forcing these values to zero velocity to ensure prevention of any general velocity offset or indeed any gradient in

velocity offset across the images in any direction. This method ensures that the diastolic reverse flows, which are usually small in comparison with the systolic flows, are not unduly compromised in accuracy in proportional terms, inasmuch as the ratio of the two is clearly very sensitive to the zero baseline level. One way in which the diastolic flows can be more accurately assessed is by using a sequence with dual velocity sensitivity, with higher velocity sensitivity in systole and lower sensitivity in diastole. This capability is not widely available on most scanners at present. Finally, it is also important that the cardiac cycle be treated as a whole. If the final portion of diastole is curtailed, as is commonly the case with CMR acquisitions, the way in which this is treated by the analysis software must be ascertained. A simple approach is a linear extrapolation of the final diastolic flow to zero by the end of the RR interval.

VENTRICULAR FILLING AND PULMONARY VENOUS FLOW

Normal flow through the mitral and tricuspid valves takes place in two phases that can clearly be recognized by CMR velocity mapping (Figure 30–14).[34] The initial passive flow in early diastole (E wave) is produced because the ventricle relaxes and allows flow from the slightly higher pressure in the atrium. This flow normally takes place very quickly and there is a mid-diastolic reduction or cessation of flow before a second phase of flow caused by atrial contraction (A wave). CMR map-

ping of the E wave agrees well with those obtained by Doppler echocardiography, but CMR may underestimate peak A wave velocity. This underestimation is due to beat length variability that occurs in the T-P time interval.[35]

In mitral stenosis, for example, the initial diastolic flow persists at high velocity because the narrow orifice of the valve cannot relieve the pressure difference between the atrium and the ventricle (see Figure 30–6B). The pressure half-time in mitral stenosis can be measured by CMR in a manner similar to that by echocardiography,[15] but this measurement is not commonly done in clinical practice because few patients fail to be adequately assessed for this valve lesion by Doppler.

Pulmonary venous flow velocity is a reflection of the pressure gradient between the pulmonary veins

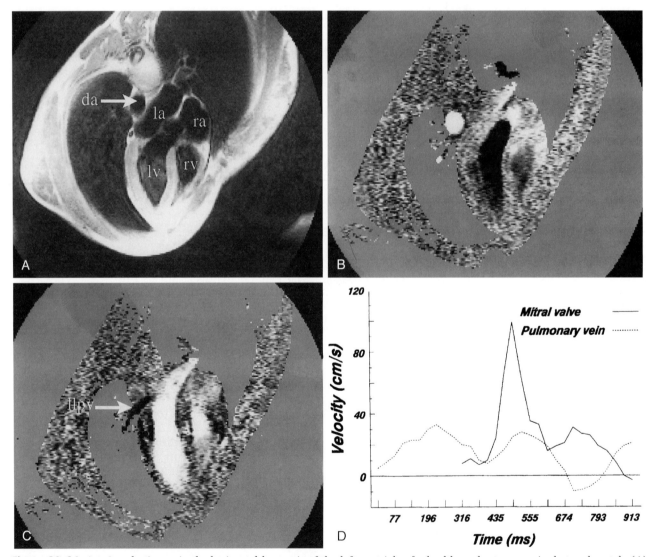

Figure 30–14. A spin-echo image in the horizontal long axis of the left ventricle of a healthy volunteer acquired at end systole *(A)*, with velocity maps encoded from base to apex in midventricular systole *(B)* and early ventricular diastole *(C)* with vertical velocity encoding. In systole, ventricular emptying is seen in black, and in diastole filling towards the apex is seen in white and signal intensity is related to velocity. *D,* Mitral valve and pulmonary vein blood velocity throughout the cardiac cycle in a healthy volunteer measured from the in-plane cine velocity map acquisition. lv = left ventricle; rv = right ventricle; la = left atrium; ra = right atrium; da = desending aorta; llpv = left lower pulmonary vein. (From Mohiaddin RH, Amanuma M, Kilner PJ, et al: Magnetic resonance phase-shift velocity mapping of mitral and pulmonary venous flow. J Comput Assist Tomogr 15:237, 1991.)

and the left atrium,[36] and changes of this gradient under pathological conditions affect the flow pattern in the pulmonary veins.[37] Normal pulmonary venous flow measured by CMR velocity mapping shows two peaks of forward flow, one during ventricular systole and the other in diastole.[38] A small backflow during atrial systole occurs. A similar reverse flow has been demonstrated in the pulmonary veins by transesophageal Doppler echocardiography during atrial systole and transmitral A flow peak.[39, 40] A noncompliant left ventricle produces high left atrial pressure during atrial systole, causing the retrograde flow in the pulmonary veins to become relatively larger than the flow through the mitral valve. An attenuated pulmonary vein systolic peak has been demonstrated in patients with mitral valve regurgitation, and the degree of this attenuation correlates well with the severity of regurgitation.[41]

ATRIAL VOLUMES AND ATRIAL THROMBI

Changes in atrial volumes throughout the cardiac cycle can be measured noninvasively by CMR.[42] During normal ventricular ejection, the atrioventricular valves are actively pulled down by the contracting ventricular myocardium toward the cardiac apex, leading to an increase in atrial size, a drop in atrial pressure, and a promotion of atrial filling from the great veins. Maximum atrial size is therefore achieved during end-ventricular systole. In early ventricular diastole, there is a rapid phase of atrial emptying in which the atria act as a passive conduit between the central veins and the ventricles. This phase is followed by active atrial contraction, reducing the volumes of the atria to their minimum size. In patients with atrioventricular valve stenosis and regurgitation, atrial size is increased with severity, and CMR of the left atrium has been used for assessment of mitral valve following mitral valvuloplasty.[43]

It is important to identify the presence of thrombus in the atrium, particularly in the presence of mitral stenosis or valvular disease with atrial fibrillation. Conventional spin-echo imaging can identify atrial thrombus, but care must be taken to distinguish between slow-moving blood, which may appear as increased signal on these images.[44] Gradient-echo imaging and velocity mapping should thus be used to confirm the presence and size of any apparent mass. Thrombus can also be distinguished from other atrial masses using gadolinium-based contrast agents,[45] and newer black-blood sequences may also help in the future in increasing confidence in thrombus detection. For now, transesophageal echocardiography should be considered the preferred technique, particularly in patients with atrial fibrillation.

SAFETY OF CARDIOVASCULAR MAGNETIC RESONANCE IN PATIENTS WITH MECHANICAL HEART VALVES

Prosthetic valves with many different designs are now in widespread use. Evaluation of prosthetic valve function by catheter can be more difficult than with native valves and there is an increased potential morbidity for invasive investigation in these patients. The metal portion of valve prostheses is not visible by CMR because it contains no mobile hydrogen atoms. There is distortion of the applied magnetic field by alteration of the local magnetic field between the prosthesis and biological tissue and the eddy current induced in the valve, which leads to loss of signal from tissues for a variable distance around the prosthesis. This distance is small for spin-echo images, and neighboring structures are seen normally, but the defect in the image is much larger in gradient-echo images, making it difficult to assess turbulent jets in the region of the valve (Figure 30–15). Valves with more metal components cause larger artifacts, e.g., the Starr-Edwards valve. Metal valves are not ferromagnetic, however, and at currently available magnetic field strengths, there is no effect on the field upon the working of the valve.[46–49] The presence of mechanical heart valves, therefore, does not preclude CMR or general MR of another organ system. Torque on the ring caused in the magnetic field is very small compared with the strength of anchorage necessary for the valve to stay in place and the stresses induced by cardiac motion.

EVALUATION OF MECHANICAL HEART VALVES

The multiple orifices of mechanical heart valve replacements generate complex flows that may not be accurately interrogated by Doppler ultrasound techniques. CMR velocity mapping can contribute to evaluation of replacement heart valves in vitro and in vivo,[49a, 49b] although metal rings cause local signal void, dependent on the type of alloy and on echo time of the sequence used. Shortening of echo time minimizes the signal void. Patterns of flow downstream of a valve and effects of upstream geometry can be studied.[50] Volume flow, regurgitant fraction, and ventricular function may be studied as for a native valve, although it may be necessary to ensure that planes chosen for flow measurement lie outside the magnetic field distortion caused by metal of the ring.

CONCLUSION

For investigation of valvular heart disease, the most important strengths of CMR lie in the un-

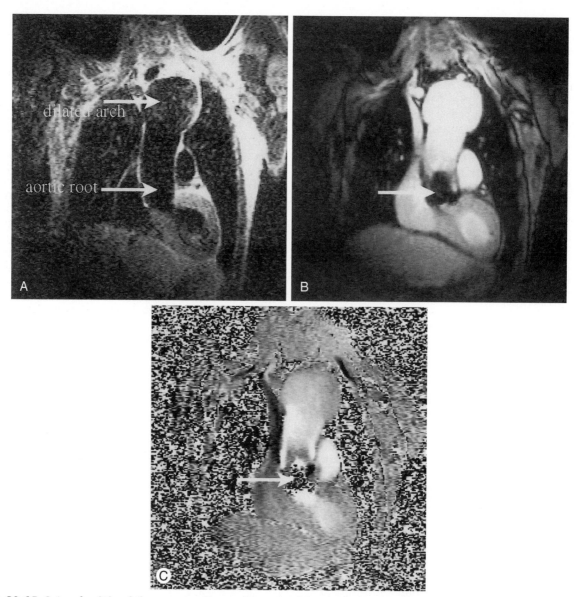

Figure 30–15. Spin-echo *(A)* and the corresponding gradient-echo *(B)* and velocity *(C)* images acquired in a coronal plane through the aortic valve and the ascending aorta of a patient with Marfan syndrome and aortic valve replacement (Starr-Edwards). The ascending aorta has been previously repaired but the aortic arch is dilated. Note the signal loss around the aortic valve (arrows) seen on the gradient-echo and the velocity mapping images.

restricted visualization and quantification of regurgitation, and of ventricular volumes, function, and mass. Freedom of access and free orientation of planes, together with ability to measure poststenotic jet velocities, also give CMR a role in cases such as calcified aortic stenosis, obstruction of ventriculopulmonary conduits, and in patients with chest deformity or lung abnormality where echocardiographic access may be limited.

ACKNOWLEDGMENTS

We wish to thank our colleagues at the magnetic resonance unit of the Royal Brompton Hospital for their help and support. We also wish to acknowledge CORDA—the Heart Charity, the Wellcome Trust, and The British Heart Foundation for support and grants.

References

1. Davis CP, McKinnon GC, Debatin JF, et al: Single-shot versus interleaved echo-planar CMR imaging: Application to visualization of cardiac valve leaflets. J Cardiovasc Magn Reson 5:107, 1995.

1a. Arai AE, Epstein FH, Bove KE, Wolff SD: Visualization of aortic valve leaflets using black blood MRI. J Magn Reson Imaging 10:771, 1999.

1b. Suzuki J, Caputo GR, Kondo C, Higgins CB: Cine MR imaging of valvular heart disease: Display and imaging parameters affect the size of the signal void caused by valvular regurgitation. AJR Am J Roentgenol 155:723, 1990.

2. Edelman RR, Wallner B, Singer A, et al: Segmented Tur-

boFLASH: Method for breath-hold CMR imaging of the liver with flexible contrast. Radiology 177:515, 1990.

3. Mansfield P: Multiplanar image formation using NCMR spin echoes. J Phys C 10:L55, 1997.

4. Meyer CH, Hu BS, Nishimura DG, Macovski A: Fast spiral coronary artery imaging. Magn Reson Med 28:202, 1992.

5. Gatehouse PD, Firmin DN, Collins S, Longmore DB: Real time blood flow imaging by spiral scan phase velocity mapping. Magn Reson Med 31:504, 1994.

6. Pike GB, Meyer CH, Brosnan TJ, Pelc NJ: Magnetic resonance velocity imaging using a fast spiral phase contrast sequence. Magn Reson Med 32:476, 1994.

7. Mohiaddin RH, Longmore DB: Functional aspects of cardiovascular nuclear magnetic resonance imaging: Techniques and application. Circulation 88:264, 1993.

8. Mohiaddin RH, Yang GZ, Kilner PJ: Visualization of flow by vector analysis of multidirectional cine magnetic resonance velocity mapping. J Comput Assist Tomogr 18:383, 1994.

9. Mohiaddin RH: Flow patterns in the dilated ischaemic left ventricle studied by magnetic resonance imaging with velocity vector mapping. J Magn Reson Imaging 5:493, 1995.

10. Mohiaddin RH, Gatehouse PD, Henien M, Firmin DN: Cine magnetic resonance Fourier velocimetry of blood flow through cardiac valves: Comparison with Doppler echocardiography. J Magn Reson Imaging 7:657, 1997.

11. Evans AJ, Blinder RA, Herfkens RJ, et al: Effects of turbulence on signal intensity in gradient echo images. Invest Radiol 23:512, 1988.

12. Kilner PJ, Firmin DN, Rees RSO, et al: Valve and great vessel stenosis: Assessment with CMR jet velocity mapping. Radiology 178:229, 1991.

13. Heidenreich PA, Steffens JC, Fujita N, et al: The evaluation of mitral stenosis with velocity-encoded cine CMR. Am J Cardiol 75:365, 1995.

14. Martinez JE, Mohiaddin RH, Kilner PJ, et al: Obstruction in extracardiac ventriculopulmonary conduits: Value of nuclear magnetic resonance imaging with velocity mapping and Doppler echocardiography. J Am Coll Cardiol 20:338, 1992.

15. Kilner PJ, Manzara CC, Mohiaddin RH, et al: Magnetic resonance jet velocity mapping in mitral and aortic valve stenosis. Circulation 87:1239, 1993.

16. Sondergard L, Stahlberg F, Thomsen C, et al: Accuracy and precision of CMR velocity mapping in measurement of stenotic cross sectional area, flow rate and pressure gradient. J Magn Reson Imaging 3:433, 1993.

17. Sechtem U, Pflugfelder PW, Cassidy MM, et al: Mitral and aortic regurgitation: Quantification of regurgitant volumes with cine CMR imaging. Radiology 167:425, 1988.

18. Wagner S, Auffermann W, Buser P, et al: Diagnostic accuracy and estimation of the severity of valvular regurgitation from the signal void on cine magnetic resonance imaging. Am Heart J 118:760, 1989.

19. Allum C, Knight C, Mohiaddin RH, Poole-Wilson PA: Images in cardiovascular medicine. Use of magnetic resonance imaging to demonstrate a fistula from the aorta to the right atrium. Circulation 97:1024, 1998.

20. Hundley WG, Li HF, Willard JE, et al: Magnetic resonance imaging assessment of the severity of mitral regurgitation. Circulation 92:1151, 1995.

21. Dulce MC, Mostbeck GH, Friese KK, et al: Quantification of the left ventricular volumes and function with cine CMR imaging: Comparison of geometric models with three-dimensional data. Radiology 188:371, 1993.

22. Aurigemma G, Reichek N, Schiebler M, Axel L: Evaluation of aortic regurgitation by cardiac cine CMR: Planar analysis and comparison to Doppler echocardiography. Cardiology 78:340, 1991.

23. Sakuma H, Fujita N, Foo TK, et al: Evaluation of left ventricular volume and mass with breath-hold cine CMR imaging. Radiology 188:377, 1993.

24. Moran PR: A flow velocity zeugmatographic interlace for NCMR imaging in humans. Magn Reson Imaging 1:197, 1982.

25. Bryant DJ, Payne JA, Firmin DN, Longmore DB: Measurement of flow with NCMR imaging using a gradient pulse and phase difference technique. J Comput Assist Tomogr 8:588, 1984.

26. van Dijk P: Direct cardiac NCMR imaging of heart wall and blood flow velocity. J Comput Assist Tomogr 8:429, 1984.

27. Bottini PB, Carr AA, Prisant M, et al: Magnetic resonance imaging compared to echocardiography to assess left ventricular mass in the hypertensive patient. Am J Hypertens 8:221, 1995.

28. Bellenger NG, Davies LC, Francis JM, et al: Sample size reduction in studies of remodelling in heart failure using cardiovascular magnetic resonance. J Cardiovasc Magn Reson 2:271, 2000.

29. Lorenz CH, Walker ES, Graham TP, Powers TA: Right ventricular performance and mass by use of cine MRI late after atrial repair of transposition of the great arteries. Circulation 92(suppl II):233, 1995.

30. Bellenger NG, Francis JM, Davies CL, et al: Establishment and performance of a magnetic resonance cardiac function clinic. J Cardiovasc Magn Reson 2:15, 2000.

31. Dulce MC, Mostbeck GH, O'Sullivan RN, et al: Severity of aortic regurgitation: Interstudy reproducibility of measurements with velocity-encoded cine CMR imaging. Radiology 185:235, 1992.

32. Rebergen SA, Chin JGL, Ottenkamp J, et al: Pulmonary regurgitation in the late post-operative follow-up of tetralogy of Fallot: Volumetric quantification by nuclear magnetic resonance velocity mapping. Circulation 88:2257, 1993.

32a. Newton MN, Kissinger KV, Chuong ML, et al: Regurgitant volume index: Use of cardiac MRI for the assessment of aortic regurgitation [abstract]. J Am Coll Cardiol 33:527A, 1998.

33. Chatzimavroudis GP, Walker PG, Oshinski JN, et al: Slice location dependence of aortic regurgitation measurements with CMR phase velocity mapping. Magn Reson Med 37:545, 1997.

33a. Kozerke S, Scheidegger MB, Pedersen EM, Boesiger P: Heart motion adapted cine phase-contrast flow measurements through the aortic valve. Magn Reson Med 42:970, 1999.

34. Mohiaddin RH, Amanuma M, Kilner PJ, et al: Magnetic resonance phase-shift velocity mapping of mitral and pulmonary venous flow. J Comput Assist Tomogr 15:237, 1991.

35. Karwatowski SP, Brecker SJD, Yang GZ, et al: Mitral valve flow measured with cine CMR velocity mapping in patients with ischaemic heart disease: Comparison with Doppler echocardiography. J Magn Reson Med 5:89, 1995.

36. Appleton CP, Hatle LK, Popp RL: Cardiac tamponade and pericardial effusion: Respiratory variation in transvalvular flow velocities studied by Doppler echocardiography. J Am Coll Cardiol 11:1020, 1988.

37. Schiavone WA, Calafiore PA, Salcedo EE: Transesophageal Doppler echocardiographic demonstration of pulmonary venous flow velocity in restrictive cardiomyopathy and constrictive pericarditis. Am J Cardiol 63:1286, 1989.

38. Mohiaddin RH, Amanuma M, Longmore DB: Magnetic resonance measurement of pulmonary venous flow and distensibility. Am J Noninvasive Cardiol 6:13, 1992.

39. Keren G, Sherez J, MeGidish R, et al: Pulmonary venous flow pattern: Its relationship to cardiac dynamics: A pulsed Doppler echocardiographic study. Circulation 71:1105, 1985.

40. Nishimura RA, Abel MD, Hatle LK, Tajik JA: Relation of pulmonary vein to mitral flow velocities by transoesophageal Doppler echocardiography: Effect of different loading conditions. Circulation 81:1488, 1990.

41. Van Rossum A, Sprenger KH, Peels FC, et al: In vivo validation of quantitative flow imaging in arteries and veins using magnetic resonance phase shift techniques. Eur Heart J 12:117, 1991.

42. Mohiaddin RH, Hasegawa M: Measurement of atrial volumes by magnetic resonance imaging in healthy volunteers and in patients with myocardial infarction. Eur Heart J 16:106, 1995.

43. Park JH, Han MC, Im JI, et al: Mitral stenosis: Evaluation with CMR imaging after percutaneous balloon valvuloplasty. Radiology 177:533, 1990.

44. Dooms G, Higgins CB: CMR imaging of cardiac thrombi. J Comput Assist Tomogr 10:415, 1986.

45. Weinmann HJ, Lanaido M, Mutzel W: Pharmacokinetics of Gd-DTPA/dimeglumine after IV injection. Physiol Chem Phys Med NCMR 16:167, 1984.

46. Soulen R, Higgins CB, Budinger TF: Magnetic resonance imaging of prosthetic cardiac valves. Radiology 54:705, 1985.

47. Randall PA, Kohman LJ, Scalzetti EM, et al: Magnetic resonance imaging of prosthetic cardiac valves in vitro and in vivo. Am J Cardiol 62:973, 1988.

48. Shellock FG: CMR imaging of metallic implants and materials: A compilation of the literature. AJR Am J Roentgenol 151:811, 1988.

49. Shellock FG, Crues JV: High-field-strength CMR imaging and metallic biomedical implants: An ex vivo evaluation of deflection forces. AJR Am J Roentgenol 151:389, 1988.

49a. Hasenkam JM, Ringgard S, Houlind K, et al: Prosthetic heart valve evaluation by magnetic resonance imaging. Eur J Cardiothorac Surg 16:300, 1999.

49b. Botnar R, Nagel E, Scheidegger MB, et al: Assessment of prosthetic aortic valve performance by magnetic resonance velocity imaging. Magma 10:18, 2000.

50. Kon M, McCormak K, Kilner P, et al: Magnetic resonance phase velocity mapping of flow patterns associated with mechanical valve prostheses: Development of in vitro model with curved aortic arch and pulsatile flow (Abstract 2147). Proceedings of the International Society of Magnetic Resonance in Medicine, Sydney, Australia, 1998.

31

Cardiovascular Magnetic Resonance in Cardiomyopathies

Matthias G. Friedrich

Cardiomyopathies are chronic, progressive myocardial diseases with distinct patterns of morphological, functional, and electrophysiological changes. On clinical, morphological, and histological grounds they have been classified into four categories: dilated cardiomyopathy (DCM), hypertrophic cardiomyopathy (HCM) with and without intracavitary obstruction, restrictive cardiomyopathy, and arrhythmogenic right ventricular cardiomyopathy (ARVC).[1] Although originally understood as being "primary" or "idiopathic," several etiologic factors leading to the phenotype have now been identified. A genetic predisposition with a possible additional effect of inflammatory or toxic injuries to the myocardium may contribute to the development of DCM,[2] as well as ARVC.[3–6] Genetic defects occur in HCM.[7] Restrictive cardiomyopathy may still appear idiopathic, but can also result from infiltrative (e.g., amyloid) systemic diseases. The diagnosis of the cardiomyopathies is established by exclusion of other cardiovascular etiologies and an accurate characterization of the phenotype. Therapy is guided by the individual stage and hemodynamic relevance of the disease, and long-term follow-up is needed. Thus, imaging techniques are important for both the diagnosis and therapy of the cardiomyopathies. The modalities frequently used in these diseases are echocardiography, x-ray angiography, radionuclide ventriculography, and cardiovascular magnetic resonance (CMR).

THE ROLE OF DIAGNOSTIC MODALITIES IN CARDIOMYOPATHIES—GENERAL ASPECTS

Echocardiography

In routine clinical practice, transthoracic echocardiography (TTE) serves as the standard technique to assess left ventricular parameters, including M-mode, two-dimensional (2-D), and Doppler techniques. It is widely available, noninvasive, fast, portable, and straightforward to perform in most patients. Ejection fraction is estimated from the end-diastolic and end-systolic diameters obtained in the parasternal view using the Teichholz formula or—especially in case of regional wall motion abnormalities—by ventricular area/length measurements from apical views using Simpson's rule.

However, standard M-mode and 2-D results suffer from substantial interstudy and interobserver variability.[8–11] Thus, the reliability of serial comparisons in a patient with cardiomyopathy may be limited, especially when different observers are involved.[12] Other problems include the poor ultrasound transmission of adjacent tissues such as lung and sternal or costal bones that may restrict the echocardiographic fields of view and transducer positions. Consequently, M-mode and 2-D echocardiography are susceptible to angular errors of the imaging plane, and diameters as well as the shortening fraction can be overestimated. Control for plane localization and parallel shift of the ultrasonic plane out of the targeted center of the ventricle may be problematic. This problem may result in reduced accuracy (especially of systolic) intraventricular dimensions and subsequently the ejection fraction,[13] even with the use of transesophageal echocardiography (TEE),[14] or newer techniques such as acoustic quantification,[15] automated border detection,[16] or three-dimensional (3-D) postprocessing.[13] Doppler echocardiography allows assessment of intracavitary flow, including intracavitary gradients and abnormalities of valve function. Mitral inflow characteristics, for example, show distinct patterns in different stages of progressive diastolic dysfunction. However, the waveforms go through phases of pseudonormalization[17] and thus may have a lack of sensitivity, although recent developments have attempted to meet this problem.[18, 19]

The endocardial border often is difficult to detect, especially in the apex, although there may be improved visualization with the use of intravenous preparations of fluorocarbon-filled microbubbles.[20–22] However, their routine use is limited by cost and the fact that microbubbles are destroyed by conventional ultrasound frequencies[23, 24] and thus have a rather short half-life of only a few minutes.[25, 26] Harmonic imaging includes the signal from the second ultrasound echo into analysis and has also been found to reduce the proportion of unacceptable images.[27] Recent studies indicate that

3-D postprocessed echocardiography may be as accurate as CMR,[28] but its value in clinical cardiology is not yet clarified. Still, the reliability of volumetric results relies on the quality of the raw image data set, which may be nondiagnostic in about 15 percent of the patients. TEE may overcome some of the limitations of the transthoracic approach, but it is semi-invasive, uncomfortable, and not free of risks for cardiovascular patients.[29]

One of the most important limitations of echocardiography is the lack of techniques to characterize tissue pathology itself. Despite the initial hope to identify specific pathology-related changes of echogenicity,[30] results to date have been disappointing.

X-Ray Left Ventricular Cineangiography

Left ventricular (LV) cineangiography following injection of iodinated contrast media gives a projection of the contracting left ventricle. Biplane data acquisition is possible with a high temporal and high in-plane spatial resolution, which provides information on cardiac volume and function of the ventricles.[31] Intraventricular thrombi or tumors may be detected, and the investigation of coronary artery integrity in the same session is important to exclude coronary artery disease. Limitations of the method include the risks to the patient of vessel injury, plaque disruption, volume overload, and arrhythmia, the radiation exposure to patient and operators, and an overall mortality of 0.14 percent.[32] The luminographic nature of the procedure precludes a complete visualization of myocardial anatomy, and the hemodynamic results are affected by adrenergic stimulation of the patient's cardiovascular system.[33] In patients with dilated cardiomyopathy, the main task of x-ray angiography is the exclusion of significant coronary artery disease. In the setting of hypertrophic cardiomyopathy, it is used to confirm the pressure gradient within the left ventricular outflow tract (LVOT). Right ventricular (RV) angiography visualizes regional or global disturbances of systolic RV function in ARVC, but is less commonly performed.[34]

Radionuclide Left Ventricular Angiography

LV angiography using technetium-99m–labeled red cells can be used to assess ventricular function[35–37] but has largely been superseded by echocardiography, which avoids the radiation exposure. It is still used, however, in patients undergoing anthracycline cardiotoxicity because of its good interstudy reproducibility.[38] However, this reproducibility does not exceed that of 3-D echocardiography[39] or CMR.[40]

Electron Beam Tomography

Electron beam tomography (EBT) has also been applied to assess LV function[41] and mass,[42] and clinical applications are reported in congestive heart failure,[43] HCM,[44] and ARVC.[45] In combination with an iodinated contrast agent, EBT has been shown to noninvasively visualize coronary anatomy[46] and thus may be useful for the exclusion of an ischemic etiology. Another important potential contribution of EBT may be the visualization of fibrosis.[47] However, this technique is limited by radiation, the use of contrast media, and relative lack of availability.

CARDIOVASCULAR MAGNETIC RESONANCE IN CARDIOMYOPATHIES—GENERAL ASPECTS

CMR noninvasively visualizes left and right ventricular morphology and function with high accuracy and reproducibility.[48, 49] It is superior to 2-D echocardiography in determination of ventricular mass[8, 50] and volumes.[51] CMR is being increasingly accepted as the in vivo gold standard for identifying the phenotype of cardiomyopathies in the diagnosis and follow-up of these patients. The power of CMR to obtain visual information on the pathological processes of the myocardium and perform tissue analysis in cardiomyopathies has not yet been fully exploited. This technique will continue to be developed because of faster gradient systems and a wider spectrum of sequences. It is probable that these innovations will overcome the limitations of density projections such as x-ray or analysis of reflected ultrasound. Because proton relaxivity depends on its chemical environment, pathological processes with distinct local chemistry may allow a specific identification of diseased tissue, a principle recognized some years ago,[52, 53] but not fully exploited.

System Requirements

CMR requirements include a high spatial and temporal resolution with short acquisition times. Thus, state-of-the art, 1.5 T magnetic resonance (MR) systems are desirable. Technical MR system design has advanced and gradient systems today are far more powerful than 5 years ago. 1.5 T scanners equipped with gradient systems of up to 40 mT/m are state-of-the-art tools for CMR studies. Phased-array surface coils dedicated for cardiac studies improve image quality substantially,[54–57] although signal analysis may be compromised by a signal intensity gradient and inhomogeneities. However, the focus of CMR studies in cardiomyopathies is primarily simple volumes, function, and mass, which can be assessed by lower grade systems.[51, 58] Tissue enhancement displayed by contrast-enhanced CMR can also be achieved at lower field,[59] as can the assessment of anatomy, function, and blood flow,[60–64] at the cost of longer acquisition times and lower spatial resolution.[65]

Limitations of Cardiovascular Magnetic Resonance

CMR is contraindicated when metallic or electronic objects may harm the patient during exposure to the magnetic field. This problem is to be considered in the case of implanted pacemakers, implanted cardiodefibrillators, and other electronic implants, cerebrovascular aneurysm clips, or projectiles in the magnet room. Surgical implants, valve prostheses, and coronary stents are generally harmless. Claustrophobia is present in about 5 percent of the patients, but can be managed by careful sedation before moving the patient into the bore. Newer CMR system designs with shorter magnets will also reduce the proportion of prematurely terminated studies. Other limitations are the lack of wide availability of scanners, costs, and limited knowledge and personal experience of this technique.

The time needed for a study has become shorter and most of the studies can be run by breath-hold sequences. Furthermore, real-time acquisition studies will further shorten scan time.[65a] Thus, a complete, comprehensive CMR study should be performed within 30 to 45 minutes. However, there is a huge amount of data generated within this time, and the limiting factor will be postprocessing and evaluation. The relevance of this problem will depend on the software developments in this field.

CMR has not yet reached its zenith of temporal and spatial resolution. Echo times (TE) will probably be below 1 ms, and the "natural" limit for the pixel size is 20 μm, which is the diffusion distance of protons within the minimal scan time. Thus, there is still space for development, which will also include visualization of coronary arteries, which is possible,[66–69] even on a low-field system,[70] but not yet robust enough to become clinical routine. However, it is accepted as suitable to exclude anomalous coronary artery anatomy,[71, 72] and preliminary data for multivessel disease is also promising.[72a] Blood pool agents improve contrast-to-noise ratio markedly[73] and may increase the sensitivity of coronary magnetic resonance angiography (MRA) in patients with DCM. Other techniques such as spiral coronary MRA are also promising.[72b] Thus, CMR may have the diagnostic power to do the whole diagnostic work-up of a patient with cardiomyopathy.

CARDIOVASCULAR MAGNETIC RESONANCE APPROACH TO THE PATIENT WITH CARDIOMYOPATHY

Morphology and Function

Cardiomyopathies are characterized by specific alterations of ventricular and myocardial geometry and/or function. To assess volumes and mass, white-blood gradient-echo sequences generally are applied with 10 to 30 phases per heartbeat. Breath-hold techniques with acquisition times of 15 to 20 seconds reduce blurring of endocardium-blood border and should be preferred, although a small shift of the heart's position between serial breath-hold studies may occur. Developments of breath-hold multislice techniques covering the entire ventricles are under way. To cover the whole diastolic phase, techniques have been developed with continuous data acquisition and retrospective gating.[74] The inclusion of the end-diastolic phase may be particularly important for the analysis of time-volume curves with respect to late diastole. Whereas under routine clinical circumstances a biplanar approach (long-axis and short-axis views) may be sufficient (Figure 31–1),[75, 76] the entire coverage of the left ventricle with short-axis views from the mitral plane to the apex is preferable for accurate measure-

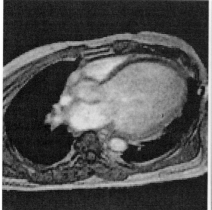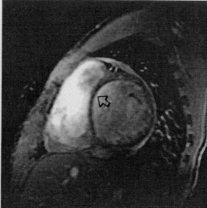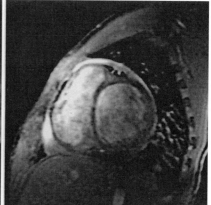

Figure 31–1. Dilated cardiomyopathy. ECG-gated gradient-echo images (echo time [TE] 4.6 ms) in a long-axis view *(left)* showing extensive global dilatation of the left ventricle. The end-diastolic diameter is 88 mm. Diastolic *(middle)* and systolic *(right)* gradient-echo image in a short-axis view revealing global hypokinesia with septal (arrow) akinesia.

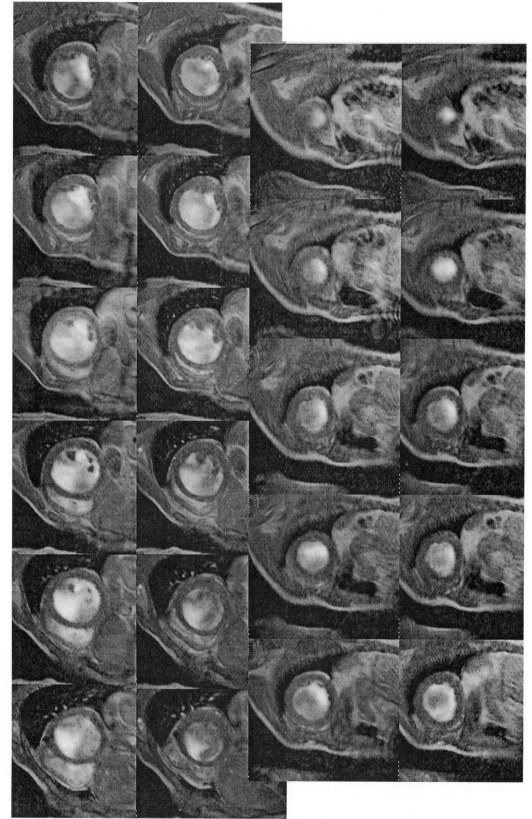

Figure 31–2. Dilated cardiomyopathy. Gradient-echo image set in the short-axis orientation covering the entire left ventricle (upper two rows end-diastolic; lower two rows systolic phase).

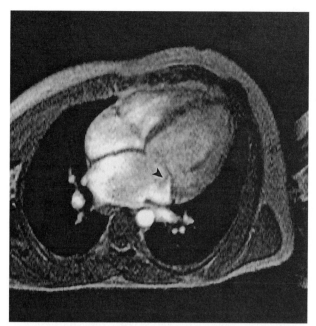

Figure 31–3. Mitral regurgitation in a patient with a dilated cardiomyopathy. Systolic gradient-echo image (TE 2.6 ms) in the 4-chamber view. The small mitral regurgitant jet (arrowhead) originates between the tips of the mitral leaflets.

ments of volume and mass (Figure 31–2),[76a, 77] especially if serial evaluations after therapeutic interventions are planned. Reliable angulation of the images by a series of at least three angulated scouts is crucial because the anatomical axis of the heart is not perpendicular to any of the orthogonal planes of the magnetic field (see Chapter 9). The slice thickness should be less than or equal to 10 mm; in case of circumscribed or subtle global changes, it should be reduced adequately. It is important to notice that there is a substantial shortening of the ventricular long axis[78] leading to a smaller number of slices covering the heart in systole compared with diastole.[78, 79]

Small doses of gadolinium diethylenetriamine penta-acetic acid (Gd-DTPA) contrast may enhance image quality in patients with arrhythmia or other causes of a low contrast-to-noise ratio.[80] Automated edge detection may facilitate the evaluation process in clinical routine.[81, 82] Steady state free precession (SSFP) methods, with enhanced endocardial border definition, are being used more frequently.[82a]

A frequent finding in patients with cardiomyopathies is mild to moderate mitral regurgitation. Promoting factors are dilatation of the mitral annulus (DCM, infiltrative cardiomyopathies) and papillary muscle dysfunction due to infiltration (sarcoidosis, amyloidosis, hemochromatosis, or tumor). Because mitral valve competence is of prognostic value, it should be assessed using gradient-echo sequences (with adjusted echo time [TE]) or phase contrast techniques (Figure 31–3). If quantification of mitral regurgitation is required for therapeutic decision-making, an established technique using flow analysis should be performed.[83, 84] Eyeball quantification

of the regurgitant jet, as in echocardiography,[85] can be misleading and should be used with caution, especially with short echo times.

Tissue Characterization

For delineation of cardiac anatomy, black-blood T1-weighted spin-echo techniques are preferable because there is an excellent contrast between the myocardium and adjacent structures such as epicardial fat and intracavitary blood. Slice orientation depends on the clinical situation and should include views orthogonal to the anatomical axis of the heart, especially the transaxial plane. Gd-DTPA administration followed by a repeat T1 study may be helpful in infiltrative and inflammatory myocardial disease. T2-weighted image quality has been markedly improved by short T1 inversion recovery (STIR) techniques, and fluid accumulation such as edema and effusion in inflammatory or malignant diseases may easily be visualized (Figure 31–4). Cardiomyopathic tissue transformation can be identified, including granulomatous infiltrates in sarcoidosis or iron deposits in hemochromatosis. However, CMR still lacks reliable techniques to detect intramyocardial fibrosis as a frequent type of pathology with prognostic relevance. Preliminary studies with contrast-enhanced CMR suggest that the increase of interstitial space in fibrotic tissue may be reflected by gadolinium accumulation,[86] but clinical and controlled data are not available.

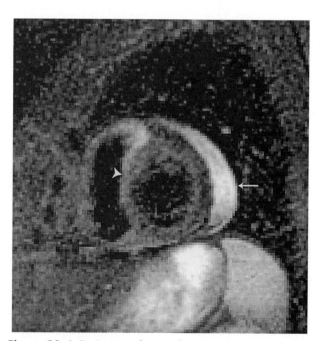

Figure 31–4. Perimyocarditis with pericardial effusion. T2-weighted short T1 inversion recovery (STIR; TE 68 ms) image in a short-axis view showing signal increases due to myocardial edema in the septum (arrowhead) and a posterior pericardial effusion (arrow).

Metabolism

CMR techniques include spectroscopy of several nuclei, of which ^{1}H and ^{31}P have been applied in several studies on cardiomyopathies. Changes of high-energy phosphates as studied by ^{31}P CMR in cardiomyopathy have been reported for DCM[87-89] and HCM.[90] However, CMR spectroscopy is still an experimental approach. Whereas ^{1}H spectroscopy suffers from a strong signal from water-bound protons and difficulties of spectral interpretation, ^{31}P spectroscopy is limited by the weakness of the phosphorus signal. Thus, voxels have to be too large to cover circumscribed myocardial regions and spectra are often altered by blood or adjacent tissue (e.g., skeletal muscle). Newer techniques deal with irregularly shaped voxels and a significantly lower degree of spectral contamination.[91] This development may allow reproducible acquisition of reliable and informative myocardial spectra, even of local pathologies. However, CMR spectroscopic techniques require extensive experience of the investigator, strong physicist support, and sophisticated hardware and software. Thus, the number of centers with access to this promising tool is currently limited.

DILATED CARDIOMYOPATHY

DCM is characterized by progressive dilatation of the heart with loss of contractile function. Its etiology is unclear in about half of cases,[92] but the typical pattern of DCM may be the end-stage of a disease process initiated by myocardial inflammation, toxic agents such as alcohol or anthracyclines, or a genetic disorder.[2] The histological hallmark of DCM is a progressive interstitial fibrosis with a numerical decrease of contractile myocytes. In advanced stages, it is also associated with relative wall thinning. Endomyocardial biopsy may be of prognostic value,[93] and be part of a diagnostic approach. However, disappointing results of sensitivity and specificity have been reported,[94] especially in myocarditis.[95] Echocardiography is a standard tool and provides information on LV size and function, although the variability of 2-D results may limit its value. 3-D techniques may improve the accuracy[28] with which EBT may be able to detect myocardial fibrosis,[47] but data in DCM are not available. The main targets of CMR studies in DCM are LV morphology and function, and gradient-echo sequences are suitable (see Figures 31–1 and 31–2). CMR has a low interobserver and intraobserver variability of left ventricular mass and volume measurements,[48, 96] and good correlations to results obtained with radionuclide ventriculography.[97] It is superior to echocardiography.[50, 98] CMR has been used to analyze wall thickening in DCM,[99] visualize impaired fiber shortening,[100] and calculate end-systolic wall stress, which may be a very sensitive parameter for changes in LV function.[101] The right ventricle is also frequently affected in DCM, and its morphology and

function is accurately assessed by CMR.[102, 103] Using CMR, atrial volumes and function have been assessed.[104] In patients with DCM, enlargement and reduction of the atrial ejection fraction were found.[105] CMR may be the method of choice for a longitudinal follow-up in patients with DCM after pharmacological intervention,[106] or after cardiomyoplasty.[107] CMR-derived parameters also serve as reliable end points in clinical studies of DCM.[58] In an analysis of study patients with DCM, the estimated sample size needed to detect LV parameter changes in a clinical trial on pharmacological interventions was lower when CMR was chosen instead of 2-D echocardiography.[51, 107a] Thus, costs could be reduced markedly and time could be saved in clinical research.

But the application of CMR as a method with a higher sensitivity to subtle changes should not be confined to scientific considerations. The high accuracy allows the physician to fine tune therapy that may well prove beneficial for the patient's quality of life and prognosis. Moreover, the frequency of repeating sometimes inconsistent studies using less precise methods during follow-up may be reduced. Both improvement of therapy adjustment and reduction of admissions for repeat studies are likely to overcome the additional costs of a CMR study. However, an analysis of cost effectiveness of CMR when replacing other modalities is awaited.

CMR spectroscopy studies have shown high-energy phosphate metabolism is altered in dilated cardiomyopathy.[108] Moreover, a low ratio of phosphocreatine (PCr) to adenosine triphosphate (ATP) was shown to be of prognostic value in DCM.[89] A study with a similar technique related these changes to a reduction of creatine kinase activity.[109] Future studies will shed more light on these exciting fields of research and further clinical studies are warranted.

Another important aspect of DCM is the understanding of its onset; in addition, early detection of etiological events might lead to effective prevention strategies. The incidence of inflammation-induced forms of DCM is unclear,[2] but a substantial proportion of DCM patients may in fact suffer from viral myocarditis.[110] In a series of endomyocardial biopsies in patients with clinically suspected HCM and DCM, the proportion of inflammatory changes as detected in biopsy material was as high as 25 percent.[111] Thus, inflammation as well as autoimmunological mechanisms and persistence of active viruses may serve as a trigger to initiate myocardial tissue transformation. In a recent study, contrast enhanced T1-weighted CMR visualized reversible myocardial signal changes in acute myocarditis.[59] Presumably caused by a combination of increased inflow (inflammatory hyperemia), slow interstitial wash-in/washout kinetics, (capillary leakage and edema) as well as diffusion into cells (necrosis), myocardial Gd-DTPA accumulation was significantly higher in patients with myocarditis than in healthy volunteers. Long-term follow-up revealed persistent changes in patients with clinical and

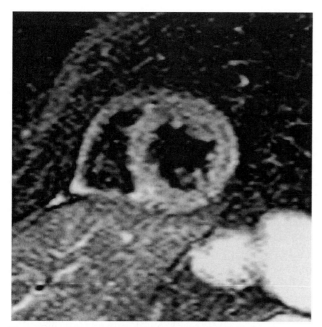

Figure 31–5. Acute viral myocarditis. T2-weighted STIR image in a short-axis view showing a diffuse bright myocardial signal. The signal intensity was 2.4 times higher than that of skeletal muscle in the same slice (normal ratio < 2.0).

functional evidence for ongoing inflammation.[112] Signal enhancement was also found to be increased in patients with Chagas myocarditis.[113] There is also evidence for similar changes in chronic DCM.[114] Associated edema during acute inflammation may be detected by conventional and breath-hold T2-weighted CMR.[115, 116] (Figure 31–5). A small study by Bellotti and associates suggests that contrast-enhanced CMR could also increase the sensitivity of endomyocardial biopsy by visualization of inflammatory areas to define of the site of biopsy.[117] Regional changes of contrast media may also occur when the myocardium is involved in systemic vasculitis (Figure 31–6). Thus, CMR may be a very helpful tool for the diagnosis and noninvasive follow-up of patients with inflammatory myocardial disease presenting as DCM. However, further studies are needed to enhance the specificity of contrast-enhanced CMR by additional or improved imaging techniques. Moreover, similar approaches are warranted in toxic injuries to the myocardium.

Progressive fibrotic replacement of the myocardium characterizes advanced stages of the disease. In a pilot study, the attempt was made to visualize myocardial fibrosis by contrast-enhanced CMR,[8, 86] but further results are awaited.

HYPERTROPHIC CARDIOMYOPATHY

HCM causes myocardial hypertrophy with impaired diastolic function and, in its obstructive form, systolic narrowing of the LVOT. It is a common cause of sudden death in the young.[118] Histo-

logically, areas of hypertrophy show a pattern of myofibrillar disarray and patchy areas of necrotic tissue. Although endocardial biopsy might therefore be used clinically for diagnosis, it is mainly performed for scientific reasons such as analysis of gene expression.[119] This use is because the sensitivity[120] and specificity of this invasive procedure in restrictive cardiomyopathy[121] and pressure-induced hypertrophy[122] are not good compared with the procedural risk. Echocardiographic findings include wall thickening, and in obstructive HCM, accelerated flow in the LVOT. For clinical follow-up, the pressure gradient can be estimated from measurements of velocity and calculation by the modified Bernoulli formula, and the correlation to invasive data is reasonable.[123] However, there is a high variation of results within a patient over time, probably due to the susceptibility of flow velocities to the hemodynamic status.[124] Moreover, the pressure gradient may be overestimated.[125] Thus, although frequently used, echocardiography may provide a limited basis to characterize morphology and hemodynamic relevance of HCM in the individual.

CMR studies have been used to study morphology, mass, function, tissue characterization, and hemodynamic relevance of obstruction. Due to its high sensitivity of detecting regional morphological changes and its noninvasiveness, CMR may be of value in screening families of index patients. CMR reliably quantifies LV mass and is superior to 2-D echocardiography,[50] which seems to be susceptible to overestimation.[126] Also, CMR is more accurate than 2-D echocardiography to assess regional hypertrophy patterns[127, 128] and is very suitable to determine different phenotypes of this disease, e.g., apical forms,[129] or cases associated with other diseases.[130] Postsurgical changes after myectomy or

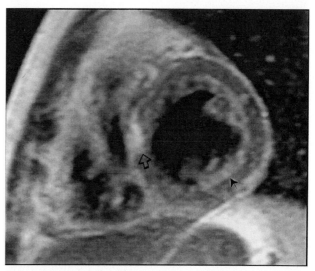

Figure 31–6. Myocardial involvement in systemic lupus erythematosus in a patient with known disease and new onset of angina, tachycardia, ST segment changes, arrhythmia, and small pericardial effusion. T1-weighted spin-echo image after application of Gd-DTPA. Subendocardial contrast media accumulation (black arrow) and a suspected septal focus (open arrow).

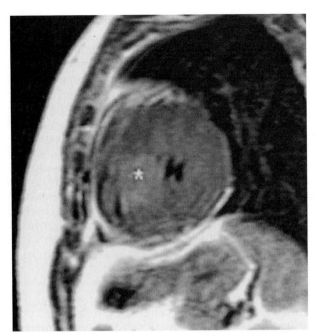

Figure 31–7. Symmetric hypertrophic cardiomyopathy in a patient with exertional angina. Midventricular systolic T1-weighted short-axis view after administration of Gd-DTPA (TE 25 ms, repetition time [TR] 950 ms). The small lumen reflects the decreased end-systolic volume. The wedge-shaped area marked by the asterisk shows an increased enhancement due to regional myocardial infarction caused by a septal artery ablation.

selective septal embolization can be reliably monitored and quantified.[131, 131a] Standard gradient-echo sequences are suitable for functional studies, visualization of turbulent flow in LVOT obstruction, and

mass quantification (Figures 31–7 and 31–8). In severe forms, the contracting ventricular walls may become close to each other and the end-systolic volume may remain under 10 ml. In these circumstances, careful contour definition is necessary and a contiguous set of slices is needed to prevent underestimation of end-systolic volume.

The turbulent jet during systolic LVOT obstruction is easily detected but is best visualized with short echo times (about 4 ms for typical blood flow velocities; see Figure 31–8). The systolic anterior motion of the anterior mitral valve leaflet may contribute significantly to the LVOT obstruction and is a typical feature of obstructive HCM. It is detectable by CMR,[132] and best visualized in the four-chamber view, an LVOT view, or a short-axis view through the valvular plane perpendicular to the outflow tract.

Mitral valve regurgitation is frequent in HCM and probably due to a pathological change of leaflet geometry, and assessment of this problem should be included in a CMR work-up (Figure 31–9). The distribution of hypertrophy should be assessed by T1-weighted spin-echo techniques, which should include long-axis views to detect regional hypertrophy such as apical or midventricular forms of HCM. CMR may be helpful in the exclusion or verification of hypertrophy due to extracardiac causes such as amyloidosis.[133] Preliminary studies indicate that hypertrophic tissue may reveal native signal heterogeneities.[134] The tissue signal of HCM-associated hypertrophy may also differ from normal myocardium in contrast-enhanced CMR.[135] CMR techniques have been established to investigate phosphate metabo-

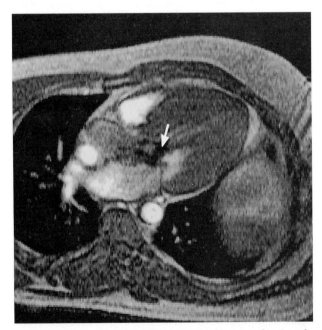

Figure 31–8. Hypertrophic obstructive cardiomyopathy. Systolic gradient echo in a long-axis view including the left ventricular outflow tract (LVOT; TE 4 ms). The increase of flow velocity and subsequent signal loss (arrow) due to LVOT obstruction is easily detected.

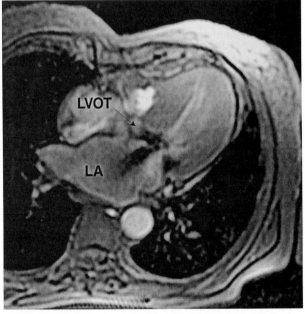

Figure 31–9. Hypertrophic obstructive cardiomyopathy with mitral regurgitation. Systolic gradient echo (TE 4 ms) in a long-axis view with a signal void indicative of an LVOT obstruction (arrow). There is a signal void in the left atrium (LA) due to a mitral regurgitation jet.

lism in HCM. Myocardial PCr/ATP ratio and the signal of phosphomonoesters were found to be changed in patients with HCM.[90, 136] Spindler and colleagues were able to correlate diastolic dysfunction to a decrease of myocardial energy reserve related to high-energy phosphate metabolism.[137] Phosphorus metabolism was also found to be altered in skeletal muscles of HCM.[138] Blood flow analysis in the coronary sinus using CMR is feasible, and a preliminary study has suggested alterations of coronary flow reserve in patients with HCM.[139]

A promising approach to assess the hemodynamic relevance of LVOT obstruction may be the noninvasive measurement of the effective outflow tract area by CMR planimetry of transplanar flow in the LVOT during systole (Figure 31–10).[131a] This method may overcome the problem of pressure recovery with subsequent overestimation of the pressure gradient,[125] and the high interstudy variability of pressure gradient measurements by echocardiography,[124] which is presumably present in invasive studies as well. Diastolic function (see Chapter 5), as a powerful clinical and prognostic factor in hypertrophy, is not in the routine measurements of CMR. Preliminary clinical results suggest that the analysis of the early untwisting motion of the myocardium may be a helpful tool to assess early diastolic function in hypertrophic heart diseases.[141] Other functional changes detected by the use of myocardial tagging are a reduction of posterior rotation, a reduced radial displacement of the inferior septal myocardium,[142] heterogeneity of regional function,[143] and reduced 3-D myocardial shortening.[144] These findings may be more sensitive in the detection and quantification of functional impairment than conventional parameters such as mitral valve inflow patterns in echocardiography. Future studies will have to show the feasibility and practicability of these approaches in clinical routine. CMR may also be very important in the follow-up of patients after surgical[132] or pharmacological interventions. Long-term follow-up is sensitive to morphological changes during the natural course of the disease.[145] The acute and chronic morphological and functional changes due to interventional ablation of a septal artery in obstructive HCM are easily detected by T1-weighted images (Figure 31–11). Of course, pacemaker implantation as a frequent therapeutic option in HCM precludes patients from a CMR follow-up.

ARRHYTHMOGENIC RIGHT VENTRICULAR CARDIOMYOPATHY

Arrhythmogenic right ventricular cardiomyopathy (ARVC) is characterized by a progressive degeneration of the right and, to a lesser extent, the left ventricular myocardium, with localized disturbance of myocardial function. Morphological features include fibrous and/or fatty replacement of myocardial tissue, extensive wall thinning, and atypical arrangement of trabecular muscles. In 1994, a set of criteria was defined to establish the diagnosis.[146] There are several genetic defects described leading to an ARVC phenotype,[147–151] but the pathogenesis is still under discussion.[5, 6] The associated morphological spectrum ranges from subtle changes to extensive fibrofatty dysplasia of the right ventricle,[152, 153] leading to right-sided congestive heart failure in rare cases. However, the clinical course is

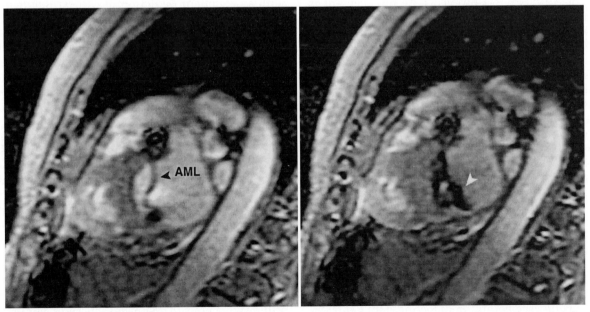

Figure 31–10. Hypertrophic obstructive cardiomyopathy. Diastolic and systolic gradient echo in a short-axis view perpendicular to the LVOT (TE 4 ms). The diastolic image *(left)* shows a signal loss at the base of the mitral leaflet. AML = anterior mitral leaflet. In systole *(right),* there is a turbulent jet crossing the plane filling the whole LVOT area. In addition, there is a jet originating from a small area of mitral regurgitation (arrowhead).

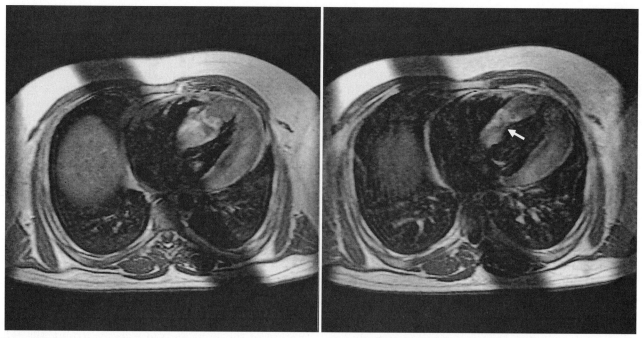

Figure 31–11. Effect of septal artery ablation in a severely symptomatic patient with hypertrophic cardiomyopathy. T1-weighted spin-echo images (TE 25 ms, non–breath-hold, acquisition time ≈ 6 min) in axial orientation after administration of Gd-DTPA. *Left:* 3 days after the intervention, there is contrast media accumulation in the thickened septum. *Right:* 6 months later, the septum shows a significant involution (arrow) with subsequent reduction of LVOT obstruction (images not shown).

generally determined by the occurrence of severe ventricular arrhythmias with a substantial risk of sudden cardiac death.[154] The morphological substrate to ventricular arrhythmias probably is fibromuscular bundles isolated from each other by fatty tissue leading to reentry phenomena.[155]

Endomyocardial biopsy of affected myocardium shows fibrous and/or fatty replacement of myocytes and is accepted as strongly suggestive for the diagnosis of ARVC.[156, 157] However, fatty replacement is also found in other cardiac diseases such as dilated cardiomyopathy, myocarditis, or alcoholic myocardial injury,[158, 159] and its specificity has been questioned.[6] All in all, the correlation of morphology to electrophysiological findings in ARVC is not well understood. Because the septum is involved in only 20 percent of the cases,[160] the specimen should be obtained from the free wall of the right ventricle. However, this procedure may pose a substantial risk to the patient,[161] especially since the free wall may be as thin as 1 mm. Furthermore, the sensitivity of biopsy may be impaired by the segmental pattern of the changes.[146] Finally, the hereditary occurrence in 30 percent of cases[146] requires the assessment of relatives. Thus, a noninvasive approach to establish the diagnosis of ARVC based on history, electrocardiogram (ECG), and cardiac imaging techniques would be appropriate.

Echocardiography is able to show regional or global changes of myocardial contractility[162, 163] and thus may be very helpful in detecting right ventricular systolic dysfunction during a routine study. However, visibility of the apex and the right ventricular outflow tract (RVOT) is limited, areas of wall thinning may be very difficult to detect, and fat signal is not unique to differentiate it from surrounding tissue or fluid. Studies have shown a limited accuracy compared to electrocardiographic and angiographic criteria.[164] So, echocardiography is a useful tool to detect findings leading to a further diagnostic work-up but lacks the power to rule out or confirm the diagnosis of ARVC.

CMR visualizes ventricular cavities and walls with an excellent depiction of myocardial anatomy. To obtain the high spatial resolution needed in this entity, we recommend a prone position of the subject and the use of a surface or phased-array coil. A slice thickness of less than 6 mm is obligatory. T1-weighted spin-echo studies visualize fatty infiltration and wall thinning as well as dysplastic trabecular structures (Figure 31–12). Orthogonal image planes (axial and sagittal) and additional short-axis views reveal the best results. Additional studies with fat saturation may help to differentiate fat from fibrous tissue. The addition of a saturation band over the atria in T1-weighted images may reduce slow-flow artifacts of inflowing blood and thereby enhance the image quality at the endocardial border of the right ventricular apex.[165]

Standard gradient-echo sequences detect regional wall motion changes such as global or local hypokinesia, localized early diastolic bulging, or circumscribed saccular outpouchings (Figure 31–13).[165–170] In a series with 33 ARVC patients, Auffermann and coworkers found increased right ventricular volumes and a lower right ventricular ejection fraction in patients with inducible ventricular tachycardia, compared to those without.[169] Additional find-

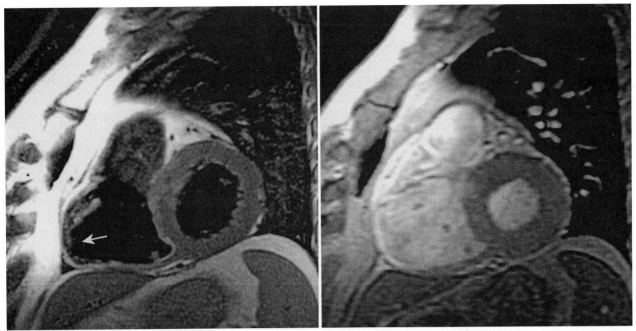

Figure 31–12. Arrhythmogenic right ventricular cardiomyopathy (ARVC). *Left:* Spin echo in the short-axis view (TE 40 ms, TR 1882 ms, slice thickness 5 mm). There is a marked RV dilatation (arrow), wall thinning, and possible myocardial fatty infiltration. *Right:* Systolic gradient-echo image in the same slice orientation showing extensive right ventricular dyskinesia.

ings included wall thinning and regional functional abnormalities. However, the sensitivity of CMR to detect intramyocardial fat was lower compared with that of endomyocardial biopsy, and fibrosis is not visualized.[170] Recently, the specificity of singular findings typical for ARVC has been questioned.[171] On the other hand, the sensitivity and specificity are good when findings such as wall thinning and regional loss of contractility are combined.[171]

Morphological and functional abnormalities were also detected by CMR in up to 76 percent of patients with idiopathic RVOT tachycardia.[172–175] The results concerning the incidence of regional wall motion abnormalities are conflicting but was found to be up to 97 percent.[175] Morphological features included RVOT enlargement,[174] wall thinning,[173, 175] and fatty replacement (Figure 31–14).[173, 175] Whereas in other studies there was no clear correlation of these findings to the electrophysiological phenotype,[174, 175] Globits and colleagues were able to demonstrate regional abnormalities at sites with successful ablation in 6 of 8 patients.[173] The etiological and pathophysiological relation of idiopathic RVOT tachycardia with its favorable prognosis to ARVC is under discussion. CMR will play an important role as the tool of choice for follow-up studies.

RESTRICTIVE CARDIOMYOPATHY

Primary infiltration of the myocardium by fibrosis or other tissues leads to the rare entity of restrictive cardiomyopathy. It is characterized by severe diastolic dysfunction, biatrial dilatation, normal LV size, and systolic function. Atrial thrombi are common. The main differential diagnostic consideration is constrictive pericarditis, which needs to be excluded in patients with suspected restrictive cardiomyopathy. M-mode echocardiography allows the measurement of atrial enlargement, preserved LV size, and sometimes myocardial wall thicken-

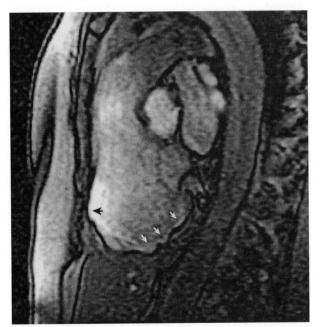

Figure 31–13. Arrhythmogenic right ventricular cardiomyopathy. Systolic gradient-echo image in a sagittal view. There are several small dyskinetic areas (arrows) of the diaphragmal right ventricular wall.

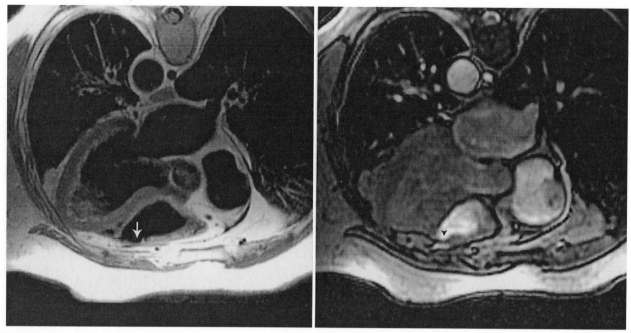

Figure 31–14. Findings in idiopathic right ventricular outflow tract (RVOT) tachycardia. *Left:* Spin echo in an axial plane (TE 40 ms, TR 1552 ms, slice thickness 5 mm) with localized wall thinning (arrow). *Right:* Gradient echo in the same orientation showing hypokinesia in the same region (arrowhead).

ing.[176, 177] Doppler studies have also attempted to differentiate restrictive cardiomyopathy from other cardiomyopathies[178] and constrictive pericarditis.[179, 180] However, pericardial thickening as an important diagnostic clue in constrictive pericarditis may not be adequately visualized.[176]

In a pilot study, EBT visualized endomyocardial fibrosis[47] as an important clue to this disease. CMR studies in restrictive cardiomyopathy, however, focus on myocardial morphology and function as well as on the exclusion of constrictive pericardial disease. LV size and wall thickness are quantified in gradient-echo image sets, but long-axis views are also needed. Biatrial dilatation is easily visualized with the four-chamber view. CMR volumetry of the enlarged atria may be recommended.[181, 182] Although in theory CMR might be as powerful as echocardiography to assess atrial thrombi,[183] slow-flowing blood in the atria may lead to false-positive results when spin-echo techniques are used.[184] The choice of a longer TE, saturation of inflowing blood, and combination with gradient-echo sequences may be helpful. Concomitant mitral regurgitation should be visualized. The exclusion of relevant pericardial thickening rules out constrictive pericarditis and may prevent unnecessary operative exploration of the pericardium.[185] This exclusion is possible by T1-weighted spin-echo techniques (Figure 31–15).[186, 187]

INFILTRATIVE SECONDARY CARDIOMYOPATHIES AND ENDOMYOCARDIAL DISEASES

A group of diseases affecting ventricular function, known as *infiltrative cardiomyopathies,* include sar-

coidosis, amyloidosis, and hemochromatosis. The myocardium may be infiltrated in these systemic diseases leading to impairment of function and/or conduction abnormalities. Because infiltration of the tissue is accompanied by changes of myocardial signal properties, CMR may become a powerful diagnostic tool, although the specificity may be limited.

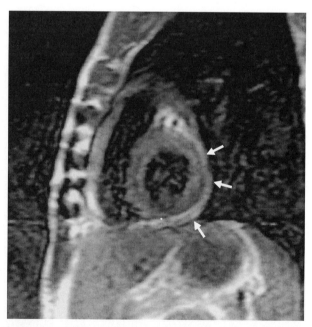

Figure 31–15. Constrictive pericarditis in a patient with a history of tuberculous pericarditis and a restrictive left ventricular pressure pattern. T1-weighted spin echo (TE 25 ms, TR 726 ms) showing a thickened pericardium (arrows) surrounding the left ventricle.

Sarcoidosis

The incidence of myocardial involvement in systemic sarcoidosis is about 20 to 30 percent,[188] although the heart was involved in about half of the cases in a large Japanese autopsy series.[189] Up to 50 percent of deaths in sarcoidosis may be related to cardiac involvement[190] leading to sudden death and congestive heart failure. Histological findings are granulomas, which often are transmural. Probably because of patchy involvement, however, the sensitivity of endomyocardial biopsy is below 30 percent.[191] Echocardiography may detect regional abnormalities of systolic function,[192, 193] or impairment of diastolic function.[194] Sarcoid lesions may lead to different signal intensities, possibly because of different stages of disease activity. Muscular sarcoidosis was reported as high-signal intensity areas in T2-weighted CMR.[195] In another study of skeletal muscle sarcoidosis, the granulomatous nodules exhibited a central region with low-signal intensity in T1- and T2-weighted imaging, but were surrounded by a high-signal ring.[196] Gd-DTPA seems to accumulate in sarcoid lesions of the brain.[197] This behavior may be compatible with fibrotic, not active granulomatous nodules with inflammatory response of the surrounding tissue. Similar findings were reported in several cases of sarcoid infiltration of the heart.[198–202] Thus, T2-weighted followed by T1-weighted spin-echo techniques in short and long axis before and after a 0.1 mmol/kg dose of Gd-DTPA may be useful to detect or exclude suspected granulomas (Figure 31–16). Occasionally, the CMR findings could be used to guide endomyocardial biopsy.[199] Because the follow-up is very important to guide therapy, CMR may be very helpful in these patients. However, sufficient data and standard protocols are still to be generated.

Amyloidosis

Infiltration of the heart by amyloid deposits is found in almost all cases of primary amyloidosis and in about one fourth of familial amyloidosis.[203] Such infiltration leads to a loss of atrial[204] or LV function and left-sided congestive heart failure.[205] Because the infiltration often is diffuse, endomyocardial biopsy is very sensitive in the detection of amyloid infiltrates.[206] Echocardiography reveals increased wall thickness, a marked increase of myocardial echogenicity, the combination of small ventricles with large atria, and diastolic dysfunction.[207–209] There are only a few reports of CMR in cardiac amyloidosis.[133, 210] Similar to sarcoidosis, the CMR approach is directed towards the detection of signal changes after Gd-DTPA administration. At present, it is uncertain how to differentiate between different infiltrative diseases. Because of the rather low incidence of these diseases, the difficulties to enroll a sufficient number of patients are obvious. Furthermore, the underlying mechanisms of a change in magnetization and relaxation have to be clarified and standard protocols have to be developed.

Hemochromatosis

Hemochromatosis of the myocardium is characterized by sometimes extensive iron deposits lead-

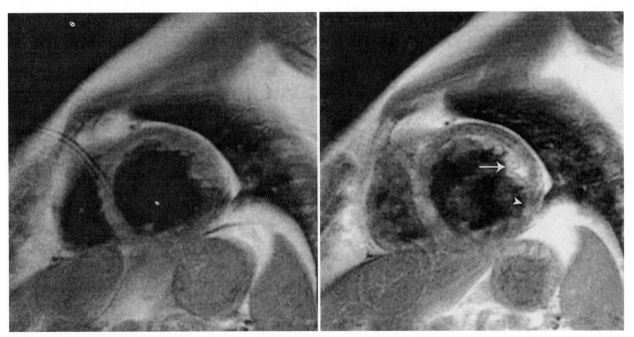

Figure 31–16. Cardiac involvement in sarcoidosis in a patient with clinical signs of heart failure, nonspecific electrocardiogram (ECG) changes, and invasive exclusion of coronary artery disease. Diastolic T1-weighted short-axis view before *(left)* and after *(right)* administration of Gd-DTPA (TE 23 ms, TR 635 ms). There is a circumscribed area of increased Gd-DTPA uptake in the lateral segment including the papillary muscle (arrow). The arrowhead marks the central region of an extensive posterolateral hypokinesia.

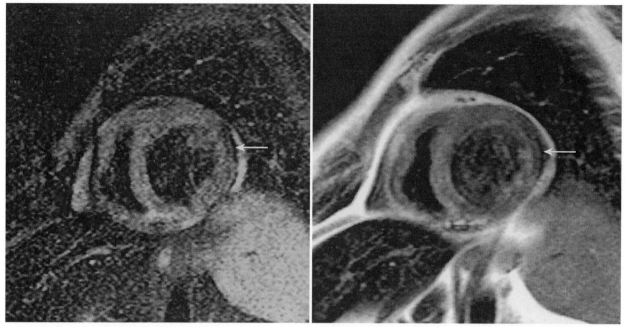

Figure 31–17. Cardiac involvement in hemochromatosis. *Left:* T2-weighted STIR (TE 68 ms, TR 1552 ms). *Right:* T1-weighted fast spin echo (TE 23 ms, TR 768 ms). Both techniques show a low intensity appearance of the liver and an area with signal reduction in the anterolateral myocardium (arrows) consistent with iron deposition.

ing to wall thickening, ventricular dilatation, progressive loss of function with congestive heart failure, and subsequent death. Because of a predominantly subepicardial deposition of iron, endomyocardial biopsy may fail to confirm the diagnosis.[211] Whereas echocardiography is useful to detect impairment of LV function and preserved wall thickness,[212] iron deposition does not lead to a significant change of myocardial echogenicity, and differentiation from DCM is not possible. The CMR approach is directed towards the detection of iron deposits as the specific marker for the disease. Iron has very strong paramagnetic properties, and deposits result in signal loss in T1-weighted as well as T2- and T2*-weighted imaging of different regions of the body.[213–218] Similar findings have also been reported for cardiac hemochromatosis (Figure 31–17).[219] The pattern of focal signal loss in a dysfunctional myocardium combined with a dark liver may be sufficient to confirm the diagnosis of cardiac hemochromatosis by CMR alone. LV function should be carefully assessed. A preliminary report of 106 patients with thalassemia major found indices of left ventricular systolic function correlated with myocardial T2* in the setting of severe iron deposition.[219a] Serum ferritin and liver iron deposition (T2*) are not reliable predictors of myocardial iron deposition.

Because intensified chelation therapy may improve LV function,[220] CMR may be an ideal tool to follow-up infiltration and LV parameters in these patients.

ENDOMYOCARDIAL DISEASES

Endomyocardial fibrosis occurs in two forms, one occurring in tropics and the other in temperate cli-

mate, termed Löffler endocarditis. Both lead to primarily posterobasal concentric wall thickening, followed by extensive subendocardial fibrosis and frequent apical thrombus formation. Both ventricles may be affected. The course is determined by progressive diastolic dysfunction and a decrease of stroke volume. Histological features, as detected by endomyocardial biopsy, may be myocyte hypertrophy and/or fibrosis, although the specificity is limited by similarity of findings in HCM,[221] and sometimes in DCM. Echocardiography shows apical obliteration of the ventricles with a high sensitivity,[222] but it is difficult to distinguish between fibrosis, thrombus, and myocardium. EBT may be of special interest because of its capability to visualize fibrosis and calcifications that may also be present.[47]

The morphological and functional features can be visualized and quantified by CMR (Figure 31–18).[223, 224] Fibrosis or calcification may be visible as a dark rim in bright-blood–prepared gradient-echo sequences but may also reveal an intermediate signal intensity.[225] Thus, there is a certain lack of sensitivity and specificity to detect fibrosis and/or calcification. Differentiation from apical infarction is easy by visualization of the preserved (or increased) wall thickness and the V-shaped outer form of the apex in the long-axis views.

CONCLUSION

The evolving role of CMR for the understanding and treatment of cardiomyopathies appears to be substantial. Establishing the diagnosis is generally possible by a single noninvasive CMR study, and the follow-up of function is sensitive to even small

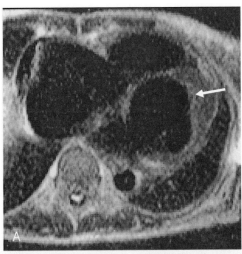

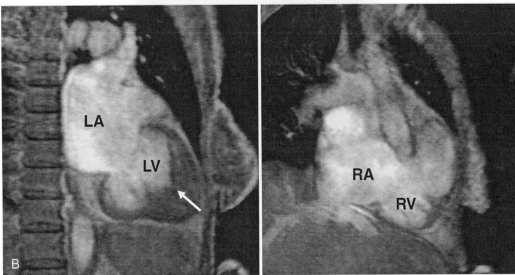

Figure 31–18. Endomyocardial fibrosis in a patient with a variant of Churg-Strauss vasculitis. *A,* T1-weighted spin echo in an axial orientation (TE 25 ms, TR 821 ms) crossing the apex. There is an apical wall thickening of intermediate signal intensity (arrow) with a subsequent reduction of left ventricular stroke volume. *B, Left:* Gradient-echo images in a long-axis view of the left ventricle (LV) and the left atrium (LA) with similar apical thickening (arrow) as in *A. Right;* Right ventricle (RV) and right atrium (RA). Ventricles are small and both atria are dilated due to the restrictive ventricular physiology.

changes. CMR data on ventricular morphology, volumes, and function are very reliable and the use of CMR-derived endpoints in clinical studies may lead to a substantial decrease of sample size. In addition to established approaches, CMR analysis of myocardial tissue analysis should be a focus of future studies. If coronary MRA becomes available in a routine clinical setting, a single CMR study could provide a complete and comprehensive diagnostic procedure in patients with suspected cardiomyopathies.

References

1. Richardson P, McKenna W, Bristow M, et al: Report of the 1995 World Health Organization/International Society and Federation of Cardiology Task Force on the Definition and Classification of Cardiomyopathies. Circulation 93:841, 1996.
2. Bender JR: Idiopathic dilated cardiomyopathy. An immuno-logic, genetic, or infectious disease, or all of the above? Circulation 83:704, 1991.
3. Sabel KG, Blomstrom-Lundqvist C, Olsson SB, Enestrom S: Arrhythmogenic right ventricular dysplasia in brother and sister: Is it related to myocarditis? Pediatr Cardiol 11:113, 1990.
4. Fontaine G, Fontaliran F, Andrade FR, et al: The arrhythmogenic right ventricle. Dysplasia versus cardiomyopathy. Heart Vessels 10:227, 1995.
5. Basso C, Thiene G, Corrado D, et al: Arrhythmogenic right ventricular cardiomyopathy. Dysplasia, dystrophy, or myocarditis? Circulation 94:983, 1996.
6. Burke AP, Farb A, Tashko G, Virmani R: Arrhythmogenic right ventricular cardiomyopathy and fatty replacement of the right ventricular myocardium: Are they different diseases? Circulation 97:1571, 1998.
7. Spirito P, Seidman CE, McKenna WJ, Maron BJ: The management of hypertrophic cardiomyopathy. N Engl J Med 336:775, 1997.
8. Germain P, Roul G, Kastler B, et al: Inter study variability in left ventricular mass measurement. Comparison between M mode echography and MRI. Eur Heart J 13:1011, 1992.
9. Gottdiener JS, Livengood SV, Meyer PS, Chase GA: Should echocardiography be performed to assess effects of antihy-

pertensive therapy? Test retest reliability of echocardiography for measurement of left ventricular mass and function. J Am Coll Cardiol 25:424, 1995.

10. van Royen N, Jaffe CC, Krumholz HM, et al: Comparison and reproducibility of visual echocardiographic and quantitative radionuclide left ventricular ejection fractions. Am J Cardiol 77:843, 1996.

11. Jensen Urstad K, Bouvier F, Hojer J, et al: Comparison of different echocardiographic methods with radionuclide imaging for measuring left ventricular ejection fraction during acute myocardial infarction treated by thrombolytic therapy. Am J Cardiol 81:538, 1998.

12. Otterstad JE, Froeland G, St. John Sutton M, Holme I: Accuracy and reproducibility of biplane two dimensional echocardiographic measurements of left ventricular dimensions and function. Eur Heart J 18:507, 1997.

13. Kupferwasser I, Mohr Kahaly S, Stahr P, et al: Transthoracic three dimensional echocardiographic volumetry of distorted left ventricles using rotational scanning. J Am Soc Echocardiogr 10:840, 1997.

14. Nessly ML, Bashein G, Detmer PR, et al: Left ventricular ejection fraction: Single plane and multiplanar transesophageal echocardiography versus equilibrium gated pool scintigraphy. J Cardiothorac Vasc Anesth 5:40, 1991.

15. Chandra S, Bahl VK, Reddy SC, et al: Comparison of echocardiographic acoustic quantification system and radionuclide ventriculography for estimating left ventricular ejection fraction: Validation in patients without regional wall motion abnormalities. Am Heart J 133:359, 1997.

16. Lucariello RJ, Sun Y, Doganay G, Chiaramida SA: Sensitivity and specificity of left ventricular ejection fraction by echocardiographic automated border detection: Comparison with radionuclide ventriculography. Clin Cardiol 20:943, 1997.

17. Rakowski H, Appleton C, Chan KL, et al: Canadian consensus recommendations for the measurement and reporting of diastolic dysfunction by echocardiography from the Investigators of Consensus on Diastolic Dysfunction by Echocardiography. J Am Soc Echocardiogr 9:736, 1996.

18. Takatsuji H, Mikami T, Urasawa K, et al: A new approach for evaluation of left ventricular diastolic function: Spatial and temporal analysis of left ventricular filling flow propagation by color M mode Doppler echocardiography. J Am Coll Cardiol 27:365, 1996.

19. Sohn DW, Chai IH, Lee DJ, et al: Assessment of mitral annulus velocity by Doppler tissue imaging in the evaluation of left ventricular diastolic function. J Am Coll Cardiol 30:474, 1997.

20. Cohen JL, Cheirif J, Segar DS, et al: Improved left ventricular endocardial border delineation and opacification with OPTISON (FS069), a new echocardiographic contrast agent. Results of a phase III Multicenter Trial. J Am Coll Cardiol 32:746, 1998.

21. Grayburn PA, Weiss JL, Hack TC, et al: Phase III multicenter trial comparing the efficacy of 2% dodecafluoropentane emulsion (EchoGen) and sonicated 5% human albumin (Albunex) as ultrasound contrast agents in patients with suboptimal echocardiograms. J Am Coll Cardiol 32:230, 1998.

22. Hundley WG, Kizilbash AM, Afridi I, et al: Administration of an intravenous perfluorocarbon contrast agent improves echocardiographic determination of left ventricular volumes and ejection fraction: Comparison with cine magnetic resonance imaging. J Am Coll Cardiol 32:1426, 1998.

23. Villarraga HR, Foley DA, Aeschbacher BC, et al: Destruction of contrast microbubbles during ultrasound imaging at conventional power output. J Am Soc Echocardiogr 10:783, 1997.

24. Wei K, Skyba DM, Firschke C, et al: Interactions between microbubbles and ultrasound: In vitro and in vivo observations. J Am Coll Cardiol 29:1081, 1997.

25. Skyba DM, Camarano G, Goodman NC, et al: Hemodynamic characteristics, myocardial kinetics and microvascular rheology of FS 069, a second generation echocardiographic contrast agent capable of producing myocardial opacification from a venous injection. J Am Coll Cardiol 28:1292, 1996.

26. Main ML, Escobar JF, Hall SA, Grayburn PA: Safety and efficacy of QW7437, a new fluorocarbon based echocardiographic contrast agent. J Am Soc Echocardiogr 10:798, 1997.

27. Caidahl K, Kazzam E, Lidberg J, et al: New concept in echocardiography: Harmonic imaging of tissue without use of contrast agent. Lancet 352:1264, 1998.

28. Chuang ML, Hibberg MG, Beaudin RA, et al: Importance of imaging method over imaging modality in non-invasive determination of left ventricular volumes and ejection fraction: Assessment by two-dimensional and three-dimensional echocardiography and magnetic resonance imaging. J Am Coll Cardiol 35:477, 2000.

29. Tam JW, Burwash IG, Ascah KJ, et al: Feasibility and complications of single plane and biplane versus multiplane transesophageal imaging: A review of 2947 consecutive studies. Can J Cardiol 13:81, 1997.

30. Skorton DJ, Collins SM: Clinical potential of ultrasound tissue characterization in cardiomyopathies. J Am Soc Echocardiogr 1:69, 1988.

31. Rogers WJ, Smith LR, Hood WP Jr, et al: Effect of filming projection and inter-observer variability on angiographic biplane left ventricular volume determination. Circulation 59:96, 1979.

32. American College of Cardiology/American Heart Association Ad Hoc Task Force on Cardiac Catheterization: ACC/AHA guidelines for cardiac catheterization and cardiac catheterization laboratories. J Am Coll Cardiol 18:1149, 1991.

33. Takenaka A, Iwase M, Sobue T, Yokota M: The discrepancy between echocardiography, cineventriculography and thermodilution. Evaluation of left ventricular volume and ejection fraction. Int J Card Imaging 11:255, 1995.

34. Daubert C, Descaves C, Foulgoc JL, et al: Critical analysis of cineangiographic criteria for diagnosis of arrhythmogenic right ventricular dysplasia. Am Heart J 115:448, 1988.

35. Schelbert HR, Verba JW, Johnson AD, et al: Nontraumatic determination of left ventricular ejection fraction by radionuclide angiocardiography. Circulation 51:902, 1975.

36. McKillop JH, Bristow MR, Goris ML, et al: Sensitivity and specificity of radionuclide ejection fractions in doxorubicin cardiotoxicity. Am Heart J 106:1048, 1983.

37. Suzuki J, Yanagisawa A, Shigeyama T, et al: Early detection of anthracycline induced cardiotoxicity by radionuclide angiocardiography. Angiology 50:37, 1999.

38. Lopez M, Vici P, Di Lauro K, et al: Randomized prospective clinical trial of high dose epirubicin and dexrazoxane in patients with advanced breast cancer and soft tissue sarcomas. J Clin Oncol 16:86, 1998.

39. Nosir YF, Salustri A, Kasprzak JD, et al: Left ventricular ejection fraction in patients with normal and distorted left ventricular shape by three dimensional echocardiographic methods: A comparison with radionuclide angiography. J Am Soc Echocardiogr 11:620, 1998.

40. Anagnostopoulos C, Gunning MG, Davies G, et al: Simultaneous biplane first pass radionuclide ventriculography using 99Tc-m tetrofosmin: A comparison with magnetic resonance imaging. Nucl Med Commun 19:435, 1998.

41. Lipton MJ, Higgins CB, Boyd DP: Computed tomography of the heart: Evaluation of anatomy and function. J Am Coll Cardiol 5:55, 1985.

42. Mousseaux E, Beygui F, Fornes P, et al: Determination of left ventricular mass with electron beam computed tomography in deformed, hypertrophic human hearts. Eur Heart J 15:832, 1994.

43. Schmermund A, Rensing BJ, Sheedy PF, Rumberger JA: Reproducibility of right and left ventricular volume measurements by electron beam CT in patients with congestive heart failure. Int J Card Imaging 14:201, 1998.

44. Yoshida M, Takamoto T: Left ventricular hypertrophic patterns and wall motion dynamics in hypertrophic cardiomyopathy: An electron beam computed tomographic study. Intern Med 36:263, 1997.

45. Tada H, Shimizu W, Ohe T, et al: Usefulness of electron beam computed tomography in arrhythmogenic right ventricular dysplasia. Relationship to electrophysiological ab-

normalities and left ventricular involvement. Circulation 94:437, 1996.

46. Achenbach S, Moshage W, Ropers D, et al: Value of electron beam computed tomography for the noninvasive detection of high grade coronary artery stenoses and occlusions. N Engl J Med 339:1964, 1998.

47. Mousseaux E, Hernigou A, Azencot M, et al: Endomyocardial fibrosis: Electron beam CT features. Radiology 198:755, 1996.

48. Benjelloun H, Cranney GB, Kirk KA, et al: Interstudy reproducibility of biplane cine nuclear magnetic resonance measurements of left ventricular function. Am J Cardiol 67:1413, 1991.

49. Pattynama PM, Lamb HJ, Van der Velde EA, et al: Reproducibility of MRI derived measurements of right ventricular volumes and myocardial mass. Magn Reson Imaging 13:53, 1995.

50. Bottini PB, Carr AA, Prisant LM, et al: Magnetic resonance imaging compared to echocardiography to assess left ventricular mass in the hypertensive patient. Am J Hypertens 8:221, 1995.

51. Strohm O, Schulz-Menger J, Pilz B, et al: Measurement of left ventricular dimensions and function in patients with dilated cardiomyopathy. J Magn Reson Imaging 13:367:2001.

52. Higgins CB, Lanzer P, Stark D, et al: Assessment of cardiac anatomy using nuclear magnetic resonance imaging. J Am Coll Cardiol 5:77, 1985.

53. Lund G, Morin RL, Olivari MT, Ring WS: Serial myocardial T2 relaxation time measurements in normal subjects and heart transplant recipients. J Heart Transplant 7:274, 1988.

54. Hardy CJ, Bottomley PA, Rohling KW, Roemer PB: An NMR phased array for human cardiac ^{31}P spectroscopy. Magn Reson Med 28:54, 1992.

55. Fayad ZA, Connick TJ, Axel L: An improved quadrature or phased array coil for MR cardiac imaging. Magn Reson Med 34:186, 1995.

56. Constantinides CD, Westgate CR, O'Dell WG, et al: A phased array coil for human cardiac imaging. Magn Reson Med 34:92, 1995.

57. Bottomley PA, Lugo Olivieri CH, Giaquinto R: What is the optimum phased array coil design for cardiac and torso magnetic resonance? Magn Reson Med 37:591, 1997.

58. Osterziel KJ, Strohm O, Schuler J, et al: Randomised, double blind, placebo controlled trial of human recombinant growth hormone in patients with chronic heart failure due to dilated cardiomyopathy. Lancet 351:1233, 1998.

59. Friedrich MG, Strohm O, Schulz-Menger J, et al: Contrast media enhanced magnetic resonance imaging visualizes myocardial changes in the course of viral myocarditis. Circulation 97:1802, 1998.

60. Seelos KC, von Smekal A, Vahlensieck M, et al: Cardiac abnormalities: Assessment with T2 weighted turbo spin echo MR imaging with electrocardiogram gating at 0.5 T. Radiology 189:517, 1993.

61. Börnert P, Jensen D: Coronary artery imaging at 0.5 T using segmented 3D echo planar imaging. Magn Reson Med 34:779, 1995.

62. Mohiaddin RH, Gatehouse PD, Henien M, Firmin DN: Cine MR Fourier velocimetry of blood flow through cardiac valves: Comparison with Doppler echocardiography. J Magn Reson Imaging 7:657, 1997.

63. Henk CB, Schlechta B, Grampp S, et al: Pulmonary and aortic blood flow measurements in normal subjects and patients after single lung transplantation at 0.5 T using velocity encoded cine MRI. Chest 114:771, 1998.

64. Wiesmann F, Gatehouse PD, Panting JR, et al: Comparison of fast spiral, echo planar, and fast low angle shot MRI for cardiac volumetry at 0.5T. J Magn Reson Imaging 8:1033, 1998.

65. Marti Bonmati L, Kormano M: MR equipment acquisition strategies: Low field or high field scanners. Eur Radiol 5:263, 1997.

65a. Nayak KS, Pauly JM, Nishimura DG, Hu BS: Rapid ventricular assessment using real-time interactive multislice MRI. Magn Reson Med 45:371, 2001.

66. Pennell DJ, Bogren HG, Keegan J, et al: Assessment of coronary artery stenosis by magnetic resonance imaging. Heart 75:127, 1996.

67. Post JC, van Rossum AC, Hofman MB, et al: Three dimensional respiratory gated MR angiography of coronary arteries: Comparison with conventional coronary angiography. AJR Am J Roentgenol 166:1399, 1996.

68. Post JC, van Rossum AC, Hofman MB, et al: Clinical utility of two dimensional magnetic resonance angiography in detecting coronary artery disease. Eur Heart J 18:426, 1997.

69. Achenbach S, Kessler W, Moshage WE, et al: Visualization of the coronary arteries in three dimensional reconstructions using respiratory gated magnetic resonance imaging. Coron Artery Dis 8:441, 1997.

70. Keegan J, Gatehouse PD, Taylor AM, et al: Coronary artery imaging in a 0.5 Tesla scanner: Implementation of real time, navigator echo controlled segmented k space FLASH and interleaved spiral sequences. Magn Reson Med 41:392, 1999.

71. Post JC, van Rossum AC, Bronzwaer JG, et al: Magnetic resonance angiography of anomalous coronary arteries. A new gold standard for delineating the proximal course? Circulation 92:3163, 1995.

72. Taylor AM, Thorne SA, Rubens MB, et al: Coronary artery imaging in grown up congenital heart disease: Complementary role of magnetic resonance and x-ray coronary angiography. Circulation 101:1670, 2000.

72a. Danias PG, Kim WY, Stuber M, et al: Coronary magnetic resonance angiography—A prospective international multicenter study (abstract). European Society of Cardiology, 2001 (in press).

72b. Börnert P, Stuber M, Botnar RM, et al: Direct comparison of 3D spiral vs. cartesian gradient echo coronary magnetic resonance angiography. Magn Reson Med, 2001 (in press).

73. Hofman MB, Henson RE, Kovacs SJ, et al: Blood pool agentvb strongly improves 3D magnetic resonance coronary angiography using an inversion pre pulse. Magn Reson Med 41:360, 1999.

74. Feinstein JA, Epstein FH, Arai AE, et al: Using cardiac phase to order reconstruction (CAPTOR): A method to improve diastolic images. J Magn Reson Imaging 7:794, 1997.

75. Dell'Italia LJ, Blackwell GG, Pearce DJ, et al: Assessment of ventricular volumes using cine magnetic resonance in the intact dog. A comparison of measurement methods. Invest Radiol 29:162, 1994.

76. Lawson MA, Blackwell GG, Davis ND, et al: Accuracy of biplane long axis left ventricular volume determined by cine magnetic resonance imaging in patients with regional and global dysfunction. Am J Cardiol 77:1098, 1996.

76a. Friedrich MG, Schulz-Menger J, Strohm O, et al: The diagnostic impact of 2D versus 3D left ventricular volumetry by MRI in patients with suspected heart failure. MAGMA 11:16, 2000.

77. Sinha S, Mather R, Sinha U, et al: Estimation of the left ventricular ejection fraction using a novel multiphase, dark-blood, breath-hold MR imaging technique. AJR Am J Roentgenol 169:101, 1997.

78. Rogers WJ Jr, Shapiro EP, Weiss JL, et al: Quantification of and correction for left ventricular systolic long-axis shortening by magnetic resonance tissue tagging and slice isolation. Circulation 84:721, 1991.

79. Marcus JT, Gotte MJW, DeWaal LK, et al: The influence of through plane motion on left ventricular volumes measured by magnetic resonance imaging: Implications for image acquisition and analysis. J Cardiovasc Magn Res 1:1, 1998.

80. Pennell DJ, Underwood SR, Longmore DB: Improved cine MR imaging of left ventricular wall motion with gadopentetate dimeglumine. J Magn Reson Imaging 3:13, 1993.

81. Singleton HR, Pohost GM: Automatic cardiac MR image segmentation using edge detection by tissue classification in pixel neighborhoods. Magn Reson Med 37:418, 1997.

82. Furber A, Balzer P, Cavaro-Menard C, et al: Experimental validation of an automated edge detection method for a simultaneous determination of the endocardial and epicardial borders in short axis cardiac MR images: Application in normal volunteers. J Magn Reson Imaging 8:1006, 1998.

82a. Carr J, Simonetti O, Kroeker R et al: Segmented true FISP—An improved technique for cine MR angiography (abstract). Proceedings of International Society for Magnetic Resonance in Medicine, I-199, 2000.

83. Sechtem U, Pflugfelder PW, Cassidy MM, et al: Mitral or aortic regurgitation: Quantification of regurgitant volumes with cine MR imaging. Radiology 167:425, 1988.

84. Fujita N, Chazouilleres AF, Hartiala JJ, et al: Quantification of mitral regurgitation by velocity encoded cine nuclear magnetic resonance imaging. J Am Coll Cardiol 23:951, 1994.

85. McCully RB, Enriquez-Sarano M, Tajik AJ, Seward JB: Overestimation of severity of ischemic/functional mitral regurgitation by color Doppler jet area. Am J Cardiol 74:790, 1994.

86. Aso H, Takeda K, Ito T, et al: Assessment of myocardial fibrosis in cardiomyopathic hamsters with gadolinium DTPA–enhanced magnetic resonance imaging. Invest Radiol 33:22, 1998.

87. Neubauer S, Krahe T, Schindler R, et al: [31]P magnetic resonance spectroscopy in dilated cardiomyopathy and coronary artery disease. Altered cardiac high energy phosphate metabolism in heart failure. Circulation 86:1810, 1992.

88. Neubauer S, Horn M, Pabst T, et al: Contributions of ^{31}P magnetic resonance spectroscopy to the understanding of dilated heart muscle disease. Eur Heart J 16(suppl O):115, 1995.

89. Neubauer S, Horn M, Cramer M, et al: Myocardial phosphocreatine to ATP ratio is a predictor of mortality in patients with dilated cardiomyopathy. Circulation 96:2190, 1997.

90. Jung WI, Sieverding L, Breuer J, et al: ^{31}P NMR spectroscopy detects metabolic abnormalities in asymptomatic patients with hypertrophic cardiomyopathy. Circulation 97:2536, 1998.

91. Loffler R, Sauter R, Kolem H, et al: Localized spectroscopy from anatomically matched compartments: Improved sensitivity and localization for cardiac ^{31}P MRS in humans. J Magn Reson 134:287, 1998.

92. Kasper EK, Agema WR, Hutchins GM, et al: The causes of dilated cardiomyopathy: A clinicopathologic review of 673 consecutive patients. J Am Coll Cardiol 23:586, 1994.

93. Pelliccia F, d'Amati G, Cianfrocca C, et al: Histomorphometric features predict 1 year outcome of patients with idiopathic dilated cardiomyopathy considered to be at low priority for cardiac transplantation. Am Heart J 128:316, 1994.

94. Yonesaka S, Becker AE: Dilated cardiomyopathy: Diagnostic accuracy of endomyocardial biopsy. Br Heart J 58:156, 1987.

95. O'Connell JB, Henkin RE, Robinson JA, et al: Gallium 67 imaging in patients with dilated cardiomyopathy and biopsy proven myocarditis. Circulation 70:58, 1984.

96. Semelka RC, Tomei E, Wagner S, et al: Interstudy reproducibility of dimensional and functional measurements between cine magnetic resonance studies in the morphologically abnormal left ventricle. Am Heart J 119:1367, 1990.

97. Gaudio C, Tanzilli G, Mazzarotto P, et al: Comparison of left ventricular ejection fraction by magnetic resonance imaging and radionuclide ventriculography in idiopathic dilated cardiomyopathy. Am J Cardiol 67:411, 1991.

98. Friedrich MG, Strohm O, Osterziel KJ, Dietz R: Growth hormone therapy in dilated cardiomyopathy monitored with MRI. MAGMA 6:152, 1998.

99. Buser PT, Auffermann W, Holt WW, et al: Noninvasive evaluation of global left ventricular function with use of cine nuclear magnetic resonance. J Am Coll Cardiol 13:1294, 1989.

100. MacGowan GA, Shapiro EP, Azhari H, et al: Noninvasive measurement of shortening in the fiber and cross fiber directions in the normal human left ventricle and in idiopathic dilated cardiomyopathy. Circulation 96:535, 1997.

101. Fujita N, Duerinckx AJ, Higgins CB: Variation in left ventricular regional wall stress with cine magnetic resonance imaging: Normal subjects versus dilated cardiomyopathy. Am Heart J 125:1337, 1993.

102. Sechtem U, Pflugfelder PW, Gould RG, et al: Measurement of right and left ventricular volumes in healthy individuals with cine MR imaging. Radiology 163:697, 1987.

103. Globits S, Pacher R, Frank H, et al: Comparative assessment of right ventricular volumes and ejection fraction by thermodilution and magnetic resonance imaging in dilated cardiomyopathy. Cardiology 86:67, 1995.

104. Jarvinen VM, Kupari MM, Hekali PE, Poutanen VP: Right atrial MR imaging studies of cadaveric atrial casts and comparison with right and left atrial volumes and function in healthy subjects. Radiology 191:137, 1994.

105. Jarvinen VM, Kupari MM, Poutanen VP, Hekali PE: Right and left atrial phasic volumetric function in mildly symptomatic dilated and hypertrophic cardiomyopathy: Cine MR imaging assessment. Radiology 198:487, 1996.

106. Doherty NE, Seelos KC, Suzuki J, et al: Application of cine nuclear magnetic resonance imaging for sequential evaluation of response to angiotensin converting enzyme inhibitor therapy in dilated cardiomyopathy. J Am Coll Cardiol 19:1294, 1992.

107. Kalil Filho R, Bocchi E, Weiss RG, et al: Magnetic resonance imaging evaluation of chronic changes in latissimus dorsi cardiomyoplasty. Circulation 90:102, 1994.

107a. Bellenger NG, Davies LC, Francis JM, et al: Reduction in sample size for studies of remodeling in heart failure by the use of cardiovascular magnetic resonance. J Cardiovasc Magn Reson 2:271, 2000.

108. Hardy CJ, Weiss RG, Bottomley PA, Gerstenblith G: Altered myocardial high energy phosphate metabolites in patients with dilated cardiomyopathy. Am Heart J 122:795, 1991.

109. Liao R, Nascimben L, Friedrich J, et al: Decreased energy reserve in an animal model of dilated cardiomyopathy. Relationship to contractile performance. Circ Res 78:893, 1996.

110. Kawai C: From myocarditis to cardiomyopathy: Mechanisms of inflammation and cell death: Learning from the past for the future. Circulation 99:1091, 1999.

111. Leatherbury L, Chandra RS, Shapiro SR, Perry LW: Value of endomyocardial biopsy in infants, children and adolescents with dilated or hypertrophic cardiomyopathy and myocarditis. J Am Coll Cardiol 12:1547, 1988.

112. Friedrich MG, Strohm O, Schulz-Menger JE, et al: Noninvasive diagnosis of acute myocarditis by contrast enhanced magnetic resonance imaging. Proceedings of the ISMRM 6th Annual Meeting, Sydney, 1998, p 912.

113. Kalil R, Bocchi EA, Ferreira BM, et al: Magnetic resonance imaging in chronic Chagas cardiopathy. Correlation with endomyocardial biopsy findings. Arq Bras Cardiol 65:413, 1995.

114. Koito H, Suzuki J, Ohkubo N, et al: Gadolinium diethylenetriamine pentaacetic acid enhanced magnetic resonance imaging of dilated cardiomyopathy: Clinical significance of abnormally high signal intensity of left ventricular myocardium. J Cardiol 28:41, 1996.

115. Gagliardi MG, Polletta B, Di Renzi P: MRI for the diagnosis and follow up of myocarditis. Circulation 99:458, 1999.

116. Friedrich MG, Strohm O, Schulz-Menger J, et al: Response to the author. Circulation 99:458, 1999.

117. Bellotti G, Bocchi EA, de Moraes AV, et al: In vivo detection of Trypanosoma cruzi antigens in hearts of patients with chronic Chagas' heart disease. Am Heart J 131:301, 1996.

118. Drory Y, Turetz Y, Hiss Y, et al: Sudden unexpected death in persons less than 40 years of age. Am J Cardiol 68:1388, 1991.

119. Kai H, Muraishi A, Sugiu Y, et al: Expression of proto oncogenes and gene mutation of sarcomeric proteins in patients with hypertrophic cardiomyopathy. Circ Res 1998.83:594.

120. Kawai C, Sakurai T, Fujiwara H, et al: Hypertrophic obstructive and non obstructive cardiomyopathy in Japan. Diagnosis of the disease with special reference to endomyocardial catheter biopsy. Eur Heart J 4:121, 1983.

121. Angelini A, Calzolari V, Thiene G, et al: Morphologic spectrum of primary restrictive cardiomyopathy. Am J Cardiol 80:1046, 1997.

122. Nunoda S, Genda A, Sekiguchi M, Takeda R: Left ventricular endomyocardial biopsy findings in patients with essential hypertension and hypertrophic cardiomyopathy with

special reference to the incidence of bizarre myocardial hypertrophy with disorganization and biopsy score. Heart Vessels 1:170, 1985.

123. Sasson Z, Yock PG, Hatle LK, et al: Doppler echocardiographic determination of the pressure gradient in hypertrophic cardiomyopathy. J Am Coll Cardiol 11:752, 1988.

124. Kizilbash AM, Heinle SK, Grayburn PA: Spontaneous variability of left ventricular outflow tract gradient in hypertrophic obstructive cardiomyopathy. Circulation 97:461, 1998.

125. Levine RA, Jimoh A, Cape EG, et al: Pressure recovery distal to a stenosis: Potential cause of gradient "overestimation" by Doppler echocardiography. J Am Coll Cardiol 13:706, 1989.

126. Missouris CG, Forbat SM, Singer DR, et al: Echocardiography overestimates left ventricular mass: A comparative study with magnetic resonance imaging in patients with hypertension. J Hypertens 14:1005, 1996.

127. Posma JL, Blanksma PK, van der Wall EE, et al: Assessment of quantitative hypertrophy scores in hypertrophic cardiomyopathy: Magnetic resonance imaging versus echocardiography. Am Heart J 132:1020, 1996.

128. Pons Llado G, Carreras F, Borras X, et al: Comparison of morphologic assessment of hypertrophic cardiomyopathy by magnetic resonance versus echocardiographic imaging. Am J Cardiol 79:1651, 1997.

129. Soler R, Rodriguez E, Rodriguez JA, et al: Magnetic resonance imaging of apical hypertrophic cardiomyopathy. J Thorac Imaging 12:221, 1997.

130. Pongratz G, Friedrich M, Unverdorben M, et al: Hypertrophic obstructive cardiomyopathy as a manifestation of a cardiocutaneous syndrome (Noonan syndrome). Klin Wochenschr 69:932, 1991.

131. Franke A, Schondube FA, Kuhl HP, et al: Quantitative assessment of the operative results after extended myectomy and surgical reconstruction of the subvalvular mitral apparatus in hypertrophic obstructive cardiomyopathy using dynamic three dimensional transesophageal echocardiography. J Am Coll Cardiol 31:1641, 1998.

131a. Schulz-Menger J, Strohm O, Waigand J, et al: The value of magnetic resonance imaging of the left ventricular outflow tract in patients with hypertrophic obstructive cardiomyopathy after septal artery embolization. Circulation 101:1764, 2000.

132. White RD, Obuchowski NA, Gunawardena S, et al: Left ventricular outflow tract obstruction in hypertrophic cardiomyopathy: Presurgical and postsurgical evaluation by computed tomography magnetic resonance imaging. Am J Card Imaging 10:1, 1996.

133. Fattori R, Rocchi G, Celletti F, et al: Contribution of magnetic resonance imaging in the differential diagnosis of cardiac amyloidosis and symmetric hypertrophic cardiomyopathy. Am Heart J 136:824, 1998.

134. Zahler R, Chelmow D, Gore J, et al: Heterogeneous signal intensity in magnetic resonance images of hypertrophied left ventricular myocardium. Magn Reson Imaging 7:517, 1989.

135. Tsukihashi H, Ishibashi Y, Shimada T, et al: Changes in gadolinium DTPA enhanced magnetic resonance signal intensity ratio in hypertrophic cardiomyopathy. J Cardiol 24:185, 1994.

136. de Roos A, Doornbos J, Luyten PR, et al: Cardiac metabolism in patients with dilated and hypertrophic cardiomyopathy: Assessment with proton decoupled P 31 MR spectroscopy. J Magn Reson Imaging 2:711, 1992.

137. Spindler M, Saupe KW, Christe ME, et al: Diastolic dysfunction and altered energetics in the alphaMHC403/ + mouse model of familial hypertrophic cardiomyopathy. J Clin Invest 101:1775, 1998.

138. Thompson CH, Kemp GJ, Taylor DJ, et al: Abnormal skeletal muscle bioenergetics in familial hypertrophic cardiomyopathy. Heart 78:177, 1997.

139. Kawada N, Sakuma H, Yamakado T, et al: Hypertrophic cardiomyopathy: MR measurement of coronary blood flow and vasodilator flow reserve in patients and healthy subjects. Radiology 211:129, 1999.

141. Stuber M, Scheidegger M, Fischer S, et al: Alterations in the local myocardial motion pattern in patients suffering from pressure overload due to aortic stenosis. Circulation 100:361, 1999.

142. Maier SE, Fischer SE, McKinnon GC, et al: Evaluation of left ventricular segmental wall motion in hypertrophic cardiomyopathy with myocardial tagging. Circulation 86:1919, 1992.

143. Kramer CM, Reichek N, Ferrari VA, et al: Regional heterogeneity of function in hypertrophic cardiomyopathy. Circulation 90:186, 1994.

144. Young AA, Kramer CM, Ferrari VA, et al: Three dimensional left ventricular deformation in hypertrophic cardiomyopathy. Circulation 90:854, 1994.

145. Suzuki J, Shimamoto R, Nishikawa J, et al: Morphological onset and early diagnosis in apical hypertrophic cardiomyopathy: A long term analysis with nuclear magnetic resonance imaging. J Am Coll Cardiol 33:146, 1999.

146. McKenna WJ, Thiene G, Nava A, et al: Diagnosis of arrhythmogenic right ventricular dysplasia/cardiomyopathy. Task Force of the Working Group Myocardial and Pericardial Disease of the European Society of Cardiology and of the Scientific Council on Cardiomyopathies of the International Society and Federation of Cardiology. Br Heart J 71:215, 1994.

147. Rampazzo A, Nava A, Danieli GA, et al: The gene for arrhythmogenic right ventricular cardiomyopathy maps to chromosome 14q23 q24. Hum Mol Genet 3:959, 1994.

148. Rampazzo A, Nava A, Erne P, et al: A new locus for arrhythmogenic right ventricular cardiomyopathy (ARVD2) maps to chromosome 1q42 q43. Hum Mol Genet 4:2151, 1995.

149. Severini GM, Krajinovic M, Pinamonti B, et al: A new locus for arrhythmogenic right ventricular dysplasia on the long arm of chromosome 14. Genomics 31:193, 1996.

150. Rampazzo A, Nava A, Miorin M, et al: ARVD4, a new locus for arrhythmogenic right ventricular cardiomyopathy, maps to chromosome 2 long arm. Genomics 45:259, 1997.

151. Coonar AS, Protonotarios N, Tsatsopoulou A, et al: Gene for arrhythmogenic right ventricular cardiomyopathy with diffuse non-epidermolytic palmoplantar keratoderma and woolly hair (Naxos disease) maps to 17q21. Circulation 97:2049, 1998.

152. Corrado D, Basso C, Thiene G, et al: Spectrum of clinicopathologic manifestations of arrhythmogenic right ventricular cardiomyopathy/dysplasia: A multicenter study. J Am Coll Cardiol 30:1512, 1997.

153. Fontaine G, Fontaliran F, Frank R: Arrhythmogenic right ventricular cardiomyopathies: Clinical forms and main differential diagnoses. Circulation 97:1532, 1998.

154. Committees UI US: Survivors of out of hospital cardiac arrest with apparently normal heart. Need for definition and standardized clinical evaluation. Consensus Statement of the Joint Steering Committees of the Unexplained Cardiac Arrest Registry of Europe and of the Idiopathic Ventricular Fibrillation Registry of the United States. Circulation 95:265, 1997.

155. Fontaliran F, Arkwright S, Vilde F, Fontaine G: Arrhythmogenic right ventricular dysplasia and cardiomyopathy. Clinical and anatomic pathologic aspects, nosologic approach. Arch Anat Cytol Pathol 46:171, 1998.

156. Nava A, Thiene G, Canciani B, et al: Clinical profile of concealed form of arrhythmogenic right ventricular cardiomyopathy presenting with apparently idiopathic ventricular arrhythmias. Int J Cardiol 35:195, 1992.

157. Angelini A, Basso C, Nava A, Thiene G: Endomyocardial biopsy in arrhythmogenic right ventricular cardiomyopathy. Am Heart J 132:203, 1996.

158. Vikhert AM, Tsiplenkova VG, Cherpachenko NM: Alcoholic cardiomyopathy and sudden cardiac death. J Am Coll Cardiol 8:3a, 1986.

159. Hasumi M, Sekiguchi M, Hiroe M, et al: Endomyocardial biopsy approach to patients with ventricular tachycardia with special reference to arrhythmogenic right ventricular dysplasia. Jpn Circ J 51:242, 1987.

160. Widmaier S, Jung WI, Sieverding L, et al: Detection of

monoester signals in human myocardium by ^{31}P MRS. MAGMA 6:161, 1998.

161. Deckers JW, Hare JM, Baughman KL: Complications of transvenous right ventricular endomyocardial biopsy in adult patients with cardiomyopathy: A seven year survey of 546 consecutive diagnostic procedures in a tertiary referral center. J Am Coll Cardiol 19:43, 1992.

162. Robertson JH, Bardy GH, German LD, et al: Comparison of two dimensional echocardiographic and angiographic findings in arrhythmogenic right ventricular dysplasia. Am J Cardiol 55:1506, 1985.

163. Blomstrom Lundqvist C, Beckman Suurkula M, Wallentin I, et al: Ventricular dimensions and wall motion assessed by echocardiography in patients with arrhythmogenic right ventricular dysplasia. Eur Heart J 9:1291, 1988.

164. Kisslo J: Two dimensional echocardiography in arrhythmogenic right ventricular dysplasia. Eur Heart J 10:22, 1989.

165. Blake LM, Scheinman MM, Higgins CB: MR features of arrhythmogenic right ventricular dysplasia. AJR Am J Roentgenol 162:809, 1994.

166. Casolo GC, Poggesi L, Boddi M, et al: ECG gated magnetic resonance imaging in right ventricular dysplasia. Am Heart J 113:1245, 1987.

167. Frank H, Rimpfl T, Weber H, et al: Diagnosis of arrhythmogenic right ventricular disease using magnetic resonance tomography. Z Kardiol 80:569, 1991.

168. Ricci C, Longo R, Pagnan L, et al: Magnetic resonance imaging in right ventricular dysplasia. Am J Cardiol 70:1589, 1992.

169. Auffermann W, Wichter T, Breithardt G, et al: Arrhythmogenic right ventricular disease: MR imaging vs angiography. AJR Am J Roentgenol 161:549, 1993.

170. Menghetti L, Basso C, Nava A, et al: Spin echo nuclear magnetic resonance for tissue characterisation in arrhythmogenic right ventricular cardiomyopathy. Heart 76:46, 1996.

171. Friedrich MG, Strohm O, Schulz-Menger J, et al: Evidence for a low specificity of right ventricular morphological changes related to arrhythmogenic right ventricular cardiomyopathy as assessed by magnetic resonance imaging (abstract). Circulation 98:857, 1998.

172. Carlson MD, White RD, Trohman RG, et al: Right ventricular outflow tract ventricular tachycardia: Detection of previously unrecognized anatomic abnormalities using cine magnetic resonance imaging. J Am Coll Cardiol 24:720, 1994.

173. Globits S, Kreiner G, Frank H, et al: Significance of morphological abnormalities detected by MRI in patients undergoing successful ablation of right ventricular outflow tract tachycardia. Circulation 96:2633, 1997.

174. Proclemer A, Basadonna PT, Slavich GA, et al: Cardiac magnetic resonance imaging findings in patients with right ventricular outflow tract premature contractions. Eur Heart J 18:2002, 1997.

175. White RD, Trohman RG, Flamm SD, et al: Right ventricular arrhythmia in the absence of arrhythmogenic dysplasia: MR imaging of myocardial abnormalities. Radiology 207:743, 1998.

176. Benotti JR, Grossman W, Cohn PF: Clinical profile of restrictive cardiomyopathy. Circulation 61:1206, 1980.

177. Keren A, Billingham ME, Popp RL: Features of mildly dilated congestive cardiomyopathy compared with idiopathic restrictive cardiomyopathy and typical dilated cardiomyopathy. J Am Soc Echocardiogr 1:78, 1988.

178. Appleton CP, Hatle LK, Popp RL: Relation of transmitral flow velocity patterns to left ventricular diastolic function: New insights from a combined hemodynamic and Doppler echocardiographic study. J Am Coll Cardiol 12:426, 1988.

179. Hatle LK, Appleton CP, Popp RL: Differentiation of constrictive pericarditis and restrictive cardiomyopathy by Doppler echocardiography. Circulation 79:357, 1989.

180. Morgan JM, Raposo L, Clague JC, et al: Restrictive cardiomyopathy and constrictive pericarditis: Non invasive distinction by digitised M mode echocardiography. Br Heart J 61:29, 1989.

181. Mohiaddin RH, Hasegawa M: Measurement of atrial volumes by magnetic resonance imaging in healthy volunteers and in patients with myocardial infarction. Eur Heart J 16:106, 1995.

182. Jarvinen VM, Kupari MM, Poutanen VP, Hekali PE: A simplified method for the determination of left atrial size and function using cine magnetic resonance imaging. Magn Reson Imaging 14:215, 1996.

183. Johnson DE, Vacek J, Gollub SB, et al: Comparison of gated cardiac magnetic resonance imaging and two dimensional echocardiography for the evaluation of right ventricular thrombi: A case report with autopsy correlation. Cathet Cardiovasc Diagn 14:266, 1988.

184. Raggi P, Daniels M, Shanoudy H, Jarmukli NF: MRI misinterpretation of spontaneous echo contrast as a large left atrial thrombus. Int J Card Imaging 12:85, 1996.

185. Restrictive cardiomyopathy or constrictive pericarditis? Lancet 2:372, 1987.

186. Sechtem U, Higgins CB, Sommerhoff BA, et al: Magnetic resonance imaging of restrictive cardiomyopathy. Am J Cardiol 59:480, 1987.

187. Masui T, Finck S, Higgins CB: Constrictive pericarditis and restrictive cardiomyopathy: Evaluation with MR imaging. Radiology 182:369, 1992.

188. Flora GS, Sharma OP: Myocardial sarcoidosis: A review. Sarcoidosis 6:97, 1989.

189. Sugie T, Hashimoto N, Iwai K: Clinical and autopsy studies on prognosis of sarcoidosis. Nippon Rinsho 52:1567, 1994.

190. Perry A, Vuitch F: Causes of death in patients with sarcoidosis. A morphologic study of 38 autopsies with clinicopathologic correlations. Arch Pathol Lab Med 119:167, 1995.

191. Sekiguchi M, Yazaki Y, Isobe M, Hiroe M: Cardiac sarcoidosis: Diagnostic, prognostic, and therapeutic considerations. Cardiovasc Drugs Ther 10:495, 1996.

192. Valantine H, McKenna WJ, Nihoyannopoulos P, et al: Sarcoidosis: A pattern of clinical and morphological presentation. Br Heart J 57:256, 1987.

193. Burstow DJ, Tajik AJ, Bailey KR, et al: Two dimensional echocardiographic findings in systemic sarcoidosis. Am J Cardiol 63:478, 1989.

194. Angomachalelis N, Hourzamanis A, Vamvalis C, Gavrielides A: Doppler echocardiographic evaluation of left ventricular diastolic function in patients with systemic sarcoidosis. Postgrad Med J 68(suppl 1):S52, 1992.

195. Kurashima K, Shimizu H, Ogawa H, et al: MR and CT in the evaluation of sarcoid myopathy. J Comput Assist Tomogr 15:1004, 1991.

196. Otake S, Banno T, Ohba S, et al: Muscular sarcoidosis: Findings at MR imaging. Radiology 176:145, 1990.

197. Seltzer S, Mark AS, Atlas SW: CNS sarcoidosis: Evaluation with contrast enhanced MR imaging. AJNR 12:1227, 1991.

198. Riedy K, Fisher MR, Belic N, Koenigsberg DI: MR imaging of myocardial sarcoidosis. AJR Am J Roentgenol 151:915, 1988.

199. Dupuis JM, Victor J, Furber A, et al: Value of magnetic resonance imaging in cardiac sarcoidosis. Apropos of a case. Arch Mal Coeur Vaiss 87:105, 1994.

200. Eliasch H, Juhlin-Dannfelt A, Sjogren I, Terent A: Magnetic resonance imaging as an aid to the diagnosis and treatment evaluation of suspected myocardial sarcoidosis in a fighter pilot. Aviat Space Environ Med 66:1010, 1995.

201. Chandra M, Silverman ME, Oshinski J, Pettigrew R: Diagnosis of cardiac sarcoidosis aided by MRI. Chest 110:562, 1996.

202. Doherty MJ, Kumar SK, Nicholson AA, McGivern DV: Cardiac sarcoidosis: The value of magnetic resonance imaging in diagnosis and assessment of response to treatment. Respir Med 92:697, 1998.

203. Gertz MA, Kyle RA, Thibodeau SN: Familial amyloidosis: A study of 52 North American born patients examined during a 30 year period. Mayo Clin Proc 67:428, 1992.

204. Plehn JF, Southworth J, Cornwell GG: Brief report: Atrial systolic failure in primary amyloidosis. N Engl J Med 327:1570, 1992.

205. Kyle RA, Spittell PC, Gertz MA, et al: The premortem recog-

nition of systemic senile amyloidosis with cardiac involvement. Am J Med 101:395, 1996.

206. Pellikka PA, Holmes DR Jr, Edwards WD, et al: Endomyocardial biopsy in 30 patients with primary amyloidosis and suspected cardiac involvement. Arch Intern Med 148:662, 1988.

207. Klein AL, Oh JK, Miller FA, et al: Two dimensional and Doppler echocardiographic assessment of infiltrative cardiomyopathy. J Am Soc Echocardiogr 1:48, 1988.

208. Falk RH, Plehn JF, Deering T, et al: Sensitivity and specificity of the echocardiographic features of cardiac amyloidosis. Am J Cardiol 59:418, 1987.

209. Klein AL, Hatle LK, Burstow DJ, et al: Doppler characterization of left ventricular diastolic function in cardiac amyloidosis. J Am Coll Cardiol 13:1017, 1989.

210. Matsuoka H, Hamada M, Honda T, et al: Precise assessment of myocardial damage associated with secondary cardiomyopathies by use of Gd DTPA enhanced magnetic resonance imaging. Angiology 44:945, 1993.

211. Olson LJ, Edwards WD, McCall JT, et al: Cardiac iron deposition in idiopathic hemochromatosis: Histologic and analytic assessment of 14 hearts from autopsy. J Am Coll Cardiol 10:1239, 1987.

212. Olson LJ, Baldus WP, Tajik AJ: Echocardiographic features of idiopathic hemochromatosis. Am J Cardiol 60:885, 1987.

213. Siegelman ES, Mitchell DG, Rubin R, et al: Parenchymal versus reticuloendothelial iron overload in the liver: Distinction with MR imaging. Radiology 179:361, 1991.

214. Gandon Y, Guyader D, Heautot JF, et al: Hemochromatosis: diagnosis and quantification of liver iron with gradient echo MR imaging. Radiology 193:533, 1994.

215. Siegelman ES, Mitchell DG, Semelka RC: Abdominal iron deposition: Metabolism, MR findings, and clinical importance. Radiology 199:13, 1996.

216. Ernst O, Sergent G, Bonvarlet P, et al: Hepatic iron overload: Diagnosis and quantification with MR imaging. AJR Am J Roentgenol 168:1205, 1997.

217. Jager HJ, Mehring U, Gotz GF, et al: Radiological features of the visceral and skeletal involvement of hemochromatosis. Eur Radiol 7:1199, 1997.

218. Sparacia G, Banco A, Midiri M, Iaia A: MR imaging technique for the diagnosis of pituitary iron overload in patients with transfusion dependent beta thalassemia major. AJNR 19:1905, 1998.

219. Waxman S, Eustace S, Hartnell GG: Myocardial involvement in primary hemochromatosis demonstrated by magnetic resonance imaging. Am Heart J 128:1047, 1994.

219a. Anderson LJ, Holden S, Davies BA, et al: A novel method of cardiac iron measurement using magnetic resonance T2* imaging vs thalassemia: Validation and clinical application (abstract). Circulation 102:II-403, 2000.

220. Rahko PS, Salerni R, Uretsky BF: Successful reversal by chelation therapy of congestive cardiomyopathy due to iron overload. J Am Coll Cardiol 8:436, 1986.

221. Katritsis D, Wilmshurst PT, Wendon JA, et al: Primary restrictive cardiomyopathy: Clinical and pathologic characteristics. J Am Coll Cardiol 18:1230, 1991.

222. Schneider U, Jenni R, Turina J, et al: Long term follow up of patients with endomyocardial fibrosis: Effects of surgery. Heart 79:362, 1998.

223. D'Silva SA, Kohli A, Dalvi BV, Kale PA: MRI in right ventricular endomyocardial fibrosis. Am Heart J 123:1390, 1992.

224. Celletti F, Fattori R, Napoli G, et al: Assessment of restrictive cardiomyopathy of amyloid or idiopathic etiology by magnetic resonance imaging. Am J Cardiol 83:798, 1999.

225. Huong DL, Wechsler B, Papo T, et al: Endomyocardial fibrosis in Behçet's disease. Ann Rheum Dis 56:205, 1997.

CHAPTER 32

Cardiac Transplantation

William T. Evanochko and Gerald M. Pohost

The methods for clinical detection of cardiac allograft rejection include endomyocardial biopsy, studies of left ventricular (LV) function, and examination of peripheral T-cell to B-cell ratios and the metabolic products of lymphocyte breakdown. These methods are suboptimal. Abnormal LV function is a relatively late sign of rejection, and blood studies suffer from limitations in specificity and sensitivity. Endomyocardial biopsy is expensive and inconvenient to apply frequently and, over a period of time, samples from regions upon which biopsies have been previously performed confound histological interpretation. The biopsy method also has a small but inherent procedural risk, and because distribution of early rejection is focal, multiple samples are required. In addition, with the progressive lowering of recipient age for cardiac transplantation, even infants and neonates are now being considered for this therapy. The difficulty and potential dangers of routine endomyocardial biopsy techniques in very young patients have underscored the need for truly noninvasive methods that are sensitive and specific for detecting cardiac rejection. Cardiovascular magnetic resonance (CMR) methods provide a potential solution to detection of cardiac allograft rejection.

This review discusses the role of CMR, including angiography (MRA) and spectroscopy (MRS) for detecting cardiac rejection in both animal models and patient studies. Although all of these CMR techniques have demonstrated the potential for diagnostic applicability, none are used in routine clinical practice at present. Therefore this review highlights the more relevant studies using CMR. Most CMR studies have focused on imaging signal changes associated with alterations in the relaxation parameters T1 and T2. There are only a few reports describing the application of MRA to cardiac transplantation, and the primary focus of MRS has been to evaluate the bioenergetic profile of the tissue by studying high-energy phosphate (^{31}P) compounds.

IMAGING

Animal Models

One of the earliest CMR reports in heterotopic cardiac transplantation appeared in 1986. Aherne and associates[1] studied 15 dogs, with 6 untreated controls and 9 receiving immunosuppression. The T1 and T2 values were evaluated at 4 time periods ranging from 2 days to 29 days after transplantation. They reported that only increases in T2 correlated with biopsy data. It is likely that such T2 increases were a reflection of the correlation between T2 values and water content and the occurrence of edema with rejection.

In 1987, Nishimura and coworkers[2] studied mongrel dogs that underwent heterotopic heart transplantation. CMR of the donor hearts showed high signal intensity in the rejecting hearts predominantly in the right ventricular free wall and the intraventricular septum with less signal in the left ventricular posterolateral wall. Once again, prolonged T1 and T2 values were the major result reported, but interestingly they observed that relaxation property changes in the right ventricle were more sensitive than those in the left ventricle.

In 1988, Konstam and colleagues[3] evaluated contrast agent–enhanced CMR to detect cardiac rejection in a rat model. Employing Gd-DTPA (gadolinium diethylenetriamine pentaacetic acid), they used T1-weighted images in hearts 2 to 12 days after transplantation. Pathological scores were graded I, II, or III to indicate the severity of the rejection. They reported that nearly all cases with grade II or III rejection manifested more than one area of intense myocardial signal enhancement. Their data suggested that contrast-enhanced CMR might be useful clinically.

To summarize, these early studies predominantly measured relaxation times T1 and T2 and gadolinium enhancement, with T2 elevations providing a useful approach for detecting cardiac transplant rejection. However, probably owing to poor specificity, T2-weighted imaging is not being used routinely for clinical diagnosis.

Patient Studies

The application of CMR to study the morphology of the transplanted heart sheds little light on the viability of the organ. Figure 32–1 shows spin-echo magnetic resonance (MR) images from transplanted human hearts to illustrate this point. Image *A* displays the heart from a 54-year-old man 6 months after transplantation. The patient had biopsy evi-

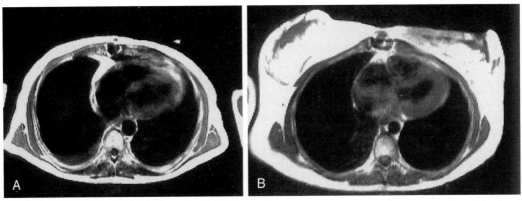

Figure 32–1. Multislice spin-echo MR images from transplanted human hearts. *A* displays the heart from a 54-year-old white man 6 months after transplantation (rejecting by biopsy). Note the morphology is nearly identical to a heart from a 24-year-old white woman 5 years after transplantation *(B)*, which was not rejecting by biopsy.

dence of rejection (stage ≥ 3 using the International Society of Heart Transplantation classification system). However, the morphology is nearly identical to a heart from a 24-year-old woman 5 years after transplantation, which is shown in Image *B* and in which the biopsy demonstrated no evidence of rejection. Because there is much more information in MR images than simply morphology, others have addressed a broader range of parameters as they apply to the transplanted human heart, namely T1, T2, left ventricular (LV), and right ventricular (RV) function.

Among the first to use CMR for the diagnosis of cardiac allograft rejection in patients were Wisenberg and coworkers[4] when, in 1987, they examined patients with cardiac allografts and biopsy at 0.15 Tesla. Nineteen patients were examined within 24 days after transplantation. The authors noted that only one patient had graft rejection by biopsy. They indicated, however, that even nonrejecting grafts uniformly had increased T1 and T2 values. Interestingly, these nonrejecting patients when studied more than 25 days after transplantation displayed normal T1s and T2s. Only patients with allograft rejection had elevated T1 and T2 values more than 25 days postoperatively.

Lund and associates[5] in 1988 performed T2 measurements in cardiac transplant patients on the same day as the biopsy specimen was taken. Out of the 35 measurements (9 patients), only 3 measurements had a rejection episode as determined by pathology and these 3 had significantly elevated T2 values. Clearly, the numbers were limited but the results were consistent with previous work.

In 1996, Lauerma and colleagues[6] used CMR to assess right and left atrial function in patients with transplanted hearts. The purpose of this study was to determine both the volumes and cyclic volume changes of the cardiac atria after heart transplantation in otherwise healthy recipients. Ten cardiac transplant patients underwent conventional gradient-echo cine CMR. It was noted that the minimum volumes of both atria were significantly larger, and fractional emptying was smaller in transplanted

hearts as compared with controls. Both the reservoir and stroke volumes were smaller and conduit volumes were larger in the patient population as compared with controls. Finally, compared with controls, the atrial filling and emptying rates were lower in transplanted hearts. This group concluded that the surgical technique used in heart transplantation results in large atria with decreased emptying and filling rates even in physically healthy transplant recipients, and that CMR is a reliable method for assessing atrial volumes and cyclic atrial function in this patient population. No relation between rejection and atrial function was examined.

A study by Walpoth and coworkers[7] was conducted to assess the effectiveness of immunosuppression for cardiac rejection using CMR. They tested the hypothesis that orthotopic heart transplantation with bicaval and pulmonary venous anastomoses preserves atrial contractility. They argued that the standard biatrial anastomotic technique of orthotopic cardiac allograft causes impaired function and enlargement of the atria, which should be detected by cine CMR. Sixteen patients underwent standard gated spin-echo and gradient-echo cine sequence examinations. They found that right atrial emptying fraction was significantly higher in the bicaval than in the biatrial group. In addition, left atrial emptying was significantly higher in the bicaval than in the biatrial, but was significantly lower in both transplant groups than the control group. They also noted that the left atrium was larger in the biatrial group than in the controls. Comparing the biatrial technique with the bicaval technique, they concluded that left and right atrial emptying fractions are significantly depressed in the former and markedly improved in the latter. They suggested that the beneficial effects of the bicaval technique on atrial function might improve allograft exercise performance.

More recent studies have examined the use of other CMR parameters. In 1997, Globits and associates[8] studied the cine CMR assessment of early LV remodeling in orthotopic heart transplant recipients. Using short-axis cine to study 11 patients 2 months after transplant, they measured LV volumes, mass, and end-systolic wall stress. They reported that there were no significant differences in ventricular volumes and ejection fractions between heart transplant recipients and control subjects. Heart transplant recipients, however, were noted to have a significantly higher LV mass. Such myocardial hypertrophy leads to a significant reduction in end-systolic wall stress compared with controls. Of particular interest was that these patients demonstrated a significantly reduced end-systolic wall stress/volume ratio, thereby indicating reduced LV performance as early as 2 months after transplantation. They reported a significant correlation between LV mass and average cyclosporine levels, but no other correlations could be made. As a result, they concluded that cyclosporine may be contributing to early LV remodeling after transplantation with reduced myocardial contractility.

ANGIOGRAPHY

Patient Studies

Mohiaddin and colleagues[9] have used MRA to study the coronary arteries in heart transplant recipients. Using fast low-angle shot MRA, they observed the proximal coronary arteries and assessed the sensitivity and specificity of coronary MRA in 16 heart transplant recipients to detect coronary stenosis. MRA and x-ray angiography studies were performed within 4 months of each other, and the average time after transplant for the first MRA was 6 years. The authors concluded that although this method is feasible, its sensitivity and specificity are limited. The best sensitivity and specificity were observed in the right coronary artery, 100 percent and 75 percent, respectively. The left anterior descending (LAD) and the left circumflex artery (LCx) had a specificity of 86 percent, but the sensitivity was poor. The combined overall sensitivity was 56 percent and specificity 82 percent for the right coronary artery (RCA), left main stem (LMS), and LAD. They attributed this finding to the need for improved acquisition techniques and the occasional interference from metal clips. In addition, they indicated that respiratory gating methods should improve the accuracy of coronary MRA in the heart transplant patient population.

Davis and associates[10] also were among the first to report on the feasibility of MRA to detect flow-limiting focal stenoses and quantify the altered coronary artery orientation. They studied 15 male adult transplant patients and identified the left main coronary artery in all transplant recipients with normal coronary anatomy. In addition, they found that MRA demonstrated a $+25°$ anterior (clockwise) right coronary artery ostial rotation with a corresponding realignment of the left main coronary artery ostium. Compared with routine coronary angiography, which found 9 discrete stenoses with 50 percent or

greater compromise of the luminal diameter in 5 transplant recipients, MRA successfully identified 7 of these stenoses. The authors concluded that using a breath-hold electrocardiogram (ECG)-gated segmented k-space technique, MRA can identify focal stenoses in cardiac allograft recipients. In addition, they suggested that the quantification of the anterior rotation of the coronary ostia might be used to guide coronary engagement for subsequent interventions.

SPECTROSCOPY

Laboratory Studies

In 1984, the first ^{31}P MRS studies of cardiac allograft rejection were described by Walpoth and co-workers[11] using excised myocardial tissue. Although changes in high-energy phosphate metabolites were noted, these ex-vivo samples can undergo loss of their phosphocreatine (PCr) leading to uninterpretable change in their phosphoenergetic profile. Accordingly, interpretation of such results may be misleading. In 1986, Hall and associates[12] used in-vivo ^{31}P MRS to study acute cardiac rejection in a dog model. In addition, they compared data obtained from two-dimensional (2-D) echocardiography, radiolabeled antimyosin antibody, and endomyocardial biopsy. The model involved transplanting a heart into the neck of a dog (heterotopic) and generating serial ^{31}P profiles by surface coil spectroscopy. They reported that PCr decreased during mild to moderate rejection and these data correlated with radiolabeled antimyosin antibody uptake. They also noted that this change usually preceded echocardiographic evidence of ventricular dysfunction.

In 1987, our laboratory[13] used a rat model to evaluate serial changes in ^{31}P MRS. Brown Norway rat hearts were transplanted subcutaneously into the anterior region of the neck of Lewis rats to form the allograft group, and the control isograft group used hearts from Lewis donors rats transplanted into Lewis recipients. Among the ^{31}P profiles studied were PCr to inorganic phosphate (PCr/Pi), PCr to beta adenosine triphosphate (PCr/ATP), and the intracellular pH of the transplanted hearts. Surface coil ^{31}P MRS was acquired at 4.7 Tesla daily for 7 days. Considering the recovery required due to surgical procedure, measurements obtained on day 2 (rather than day 1) were taken as baseline. Figure 32–2A demonstrates serial spectra obtained from controls of the same transplanted heart. Note the lack of serial changes in the ^{31}P spectra from this isograft. Figure 32–2B demonstrates the ^{31}P spectra obtained from the heterograft model. Note the decrease in the PCr signal as the model progresses to complete rejection by day 5. Figure 32–3 illustrates the different ratios studied. PCr/Pi *(A)* was unchanged or increased in the isografts, whereas allografts demonstrated a continual decrease in PCr/Pi with the difference becoming significant by day 4 when compared with day 2 of the allograft group

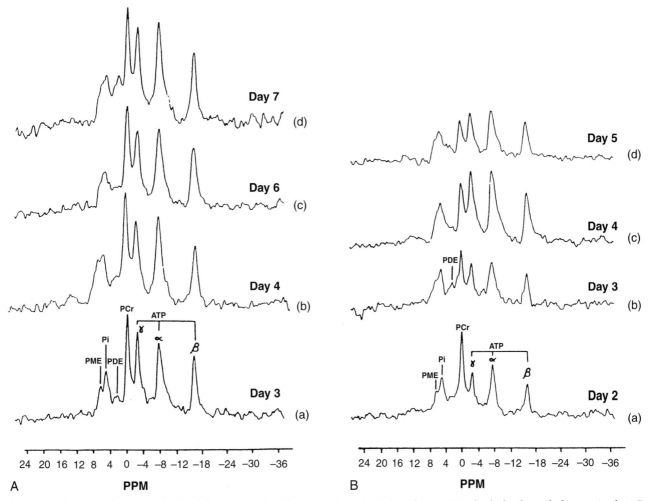

Figure 32–2. *A,* Serial spectra obtained from controls of the same transplanted rat heart. Note the lack of serial changes in the ³¹P spectra from this isograft. *B* demonstrates the ³¹P spectra obtained from our allograft model. Note the decrease in the PCr signal as the model progresses to complete rejection by day 5. (From Canby RC, Evanochko WT, Kirklin JK, et al: Monitoring the bioenergetics of cardiac allograft rejection using in vivo P-31 NMR spectroscopy. J Am Coll Cardiol 9:1067, 1987. Reprinted with permission from the American College of Cardiology.)

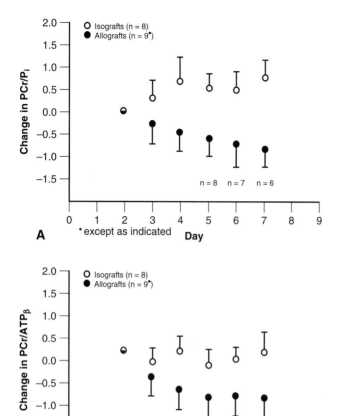

Figure 32–3. Plots of the various ratios obtained from Figure 32–2 spectra. A, PCr/Pi was unchanged or increased in the isografts, whereas allografts demonstrated a continual decrease in PCr/Pi with the difference becoming significantly different by day 4 when compared with day 2 allografts ($p < .005$) and by day 3 when compared with the isograft group ($p < .05$). B, The PCr/ATP ratio from the isograft group did not change throughout the study; however, allografts demonstrated a significant decrease as early as day 3 ($p < .01$), although a significant difference between isografts and allografts did not manifest until day 4 ($p < .005$). (From Canby RC, Evanochko WT, Kirklin JK, et al: Monitoring the bioenergetics of cardiac allograft rejection using in vivo P-31 NMR spectroscopy. J Am Coll Cardiol 9:1067, 1987. Reprinted with permission from the American College of Cardiology.)

($p < .005$) and by day 3 when compared with the isograft group ($p < .05$). The PCr/ATP ratio (B) from the isograft group did not change throughout the study. However, allografts demonstrated a significant decrease by day 3 ($p < .01$), although a significant difference between isografts and allografts did not manifest until day 4 ($p < .005$). It should be noted that PCr/Pi is more sensitive than PCr/ATP; however, at normal clinical field strengths (1.5 T), the PCr/Pi cannot be measured due to the confounding effects of blood 2,3-diphosphoglycerate (2,3-DPG) obscuring myocardial Pi. Accordingly, most studies have relied on the less sensitive PCr/ATP ratio. In addition, the isograft group showed no

change in intracellular pH; however, the allograft group demonstrated an early alkaline shift followed by acidosis.

Using a rat model of cardiac rejection, Suzuki and colleagues[14] demonstrated that ^{31}P MRS was useful for evaluating the therapeutic response of a new immunosuppressant, 15-deoxyspergualin (DSG). This study demonstrated that DSG inhibited rejection. In addition, ^{31}P MRS indicated that cyclosporine did not fully inhibit rejection by itself, and that 14 mg/kg of cyclosporine, a tolerogenic dose, had a cardiotoxic effect. In 1987, Haug and associates[15] evaluated a variety of high-energy phosphate ratios in a rat model. Although the heart was transplanted heterotopically into the groin with the possibility of contamination from adjacent skeletal muscle, these investigators noted significantly higher ^{31}P MRS ratios in the immunosuppressed group in contrast to those animals that were not immunosuppressed. It was concluded that the ^{31}P profile could have clinical utility for the diagnosis of cardiac allograft rejection.

In 1988, Fraser and coworkers[16] performed notable experiments to determine if the metabolic indices observed by ^{31}P MRS preceded functional and morphological changes of cardiac allograft rejection. Using beagles and a model with hearts transplanted heterotopically in the neck, they concluded that abnormal ^{31}P MRS–derived indices preceded either functional or histological changes. Thus, this study stressed the importance of ^{31}P MRS as an early indicator of cardiac rejection. This group continued to study this phenomenon using an animal model and made an important finding—that the biochemical changes, as observed by ^{31}P MRS, in early rejection are reversible if appropriate augmented immunotherapy is applied.[17] An important point in this regard is that the rejection process can be monitored at a metabolic level, and reversal of these metabolic events can be detected noninvasively, possibly before any significant myocyte loss, makes this technology preferred to the current situation in which there is obligate cell death even prior to the institution of augmented immunotherapy in many instances.

Work from Duke University also evaluated ^{31}P MRS of rejection, and their data concur with previous animal modeling studies. Using a canine model, D'Amico and associates[18] detected a 28 percent mean decrease in PCr/Pi when animals displayed moderate to severe rejection based on the Billingham criteria. They also used positron emission tomography (PET) and concluded that PET did not detect cardiac allograft rejection. The continued work from the Johns Hopkins group, headed by Baumgartner,[19] has shed new insight into coronary blood flow and ^{31}P indices of rejection. Sixteen beagles received cervical cardiac allografts from mongrel dogs and were immunosuppressed with cyclosporine and prednisone. Measurement of mean coronary flow showed no decrease during rejection, although the peak coronary blood flow was noted to decrease. They observed no increase in blood lactate; however, they demonstrated a significant decrease in PCr/Pi. The reduced peak coronary blood flow returned to normal with augmented immunotherapy. The changes in peak coronary blood flow suggested an impairment of coronary reserve. Despite the fact that this model was not a working heart, the authors concluded that ischemia was not a significant component of early rejection. Naturally, the working heart model would be more relevant to clinical rejection.

An interesting paper by van Dobbenburgh and associates[20] stated that high-energy phosphate metabolism is altered irrespective of cardiac allograft rejection. Van Dobbenburgh and associates using an animal model obtained donor hearts from either Lewis rats (L) or Brown Norway rats (BN). These hearts were transplanted into the neck of Lewis rats resulting in a nonrejecting group (L-L) and a rejecting group (L-BN), similar to the study of Canby and colleagues.[13] Both groups were serially studied using ^{31}P MRS for 1 to 8 days. They reported a decrease in PCr/ATP from day 1 to 3 in both groups after surgery, with stabilization by day 3. PCr/ATP remained depressed throughout the 8-day study interval. Histological evaluation showed an increase in the severity of rejection in L-BN, which coincided with a further decrease in the PCr/ATP ratio from the L-BN after day 4, but this did not occur in the control group through day 8. These authors postulated that since high-energy phosphate metabolism was affected in the unloaded heterotopically transplanted heart, irrespective of the presence of rejection, the PCr/ATP ratio was not a specific marker of acute rejection in this model. However, they did state that the PCr appears to be a specific and sensitive marker of acute rejection, but only in a late, severe stage. One must question whether the authors considered the ischemic insult that occurs in the excised heart and that this insult with anticipated decrease in PCr/ATP would tend to show changes in PCr/ATP unrelated to rejection.

Therefore, in summary, the animal studies from our laboratory and others have overwhelmingly supported the use of ^{31}P MRS as a means to detect rejection. These results encouraged us to perform preliminary studies in patients to determine if image-guided volume-localized spectroscopy can be used to determine alterations in the bioenergetics of the transplanted heart in man.

SPECTROSCOPY

Patient Studies

The first MRS study of cardiac transplant rejection in humans was that of Herfkens and coworkers.[21] They reported 6 patients that tended to demonstrate a decrease in PCr/ATP with rejection using ^{31}P MRS. However, Bottomley and colleagues[22] observed anomalies in cardiac high-energy phosphate metabolism in rejecting and control heart. Although PCr/ATP was abnormal in some patients with moderate rejection, the data were inconsistent. Although they showed a statistically significant difference between rejecting myocardium and normal volunteers, they were unable to differentiate between mild and moderate rejection. Unfortunately, these two studies examined a mixture of patients including some patients with early acute rejection, some with late acute rejection, and some with chronic rejection. A study of Wolfe and associates[23] demonstrated a trend towards a lower PCr/ATP with rejection as compared with no rejection; thus they were unable to distinguish between mild and moderate rejection.

In 1996, van Dobbenburgh and colleagues[24] measured the hypothermic donor heart to determine if its metabolic condition is related to functional recovery after transplantation. Their premise was that although strict selection criteria are being used for the acceptance of human donor hearts for transplantation, problems with respect to functional recovery on reperfusion sometimes still occur. Therefore, evaluation of the viability of a human donor heart before implantation may be of value. They used MRS to assess excised human donor hearts arrested with St Thomas' Hospital No. 2 cardioplegic solution before implantation, which was correlated with myocardial function measured with thermodilution in heart transplant patients. They reported that no significant correlation was found early after transplantation between the cardiac index of heart transplant and the PCr/ATP ratio or pH at the time of reperfusion. Interestingly, they stated that 1 week after transplantation, a significant correlation was observed between the cardiac index and the ATP ratio. They concluded that functional recovery after human heart transplantation is related to the metabolic condition of the hypothermic donor heart.

Preliminary studies with ^{31}P MRS have used either the single voxel (ISIS) or the multiple voxel approach (1DCSI with ISIS and gradient phase encoding).[25] Figure 32–4A demonstrates ISIS-derived spectra from a cardiac transplant patient that was not rejecting at the time of biopsy; note the high level of PCr.[25] Figure 32–4B–D are from another patient. The first spectrum demonstrated an abnormal bioenergetic profile (Figure 32–4B) with decreased PCr signal. It is also of interest to note the similarities between a patient with allograft rejection in Figure 32–4B and the rat model with rejection in Figure 32–2B (day 4). The spectra shown in Figure 32–4C and D depict the response to augmented immunotherapy at 4 and 6 weeks later. Thus, ^{31}P MRS can depict alterations in the high-energy phosphate profiles in transplanted hearts; it was abnormal before detection by biopsy and was improved when the patient received augmented immunotherapy. Figure 32–5 demonstrates spectra obtained from a normal volunteer with 1DCSI. The time for the entire study was approximately 75 minutes, and spectra from each slice (obtained in approximately 20 minutes)

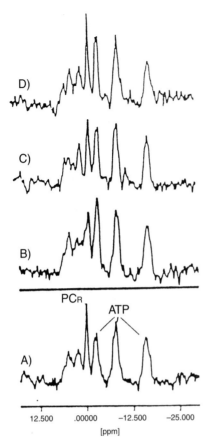

Figure 32–4. [31]P spectra acquired with the single-voxel (ISIS) technique from cardiac transplant patients. Spectrum A is from a patient with mild rejection. Spectra B, C, and D are serial studies from another patient experiencing moderate rejection (B), and mild rejection (C and D), the latter denoting the response to augmented immunotherapy. Note the similarity between A and D.

displayed adequate signal-to-noise ratio for diagnostic purposes. Thirty-two slice profiles 1 cm thick were acquired, and both the column selection and phase encoding selection were angulated to maximize the overlap of the selected slices and the left ventricle of the heart (see Figure 32–5B). Broad signals derived primarily from skeletal muscle and fat are observed in slice #29. Delineation from skeletal muscle to cardiac muscle occurs in slice #28, which results in a combination of skeletal and cardiac muscle. Slices #25, #26, and #27 are from myocardium alone. By employing such strategies, regional variations not uncommon in allograft rejection in the myocardial high-energy phosphates could be studied. Figure 32–6 is an identical study to Figure 32–5 but depicts spectra from each voxel in a patient in whom PCr/ATP ratios were depressed in several such voxels. Note the decrease in PCr in slices #23 to #26. Although there was a significant decrease in the PCr/ATP ratio in all transplants, the decreases were not predictive of the level of rejection determined by endomyocardial biopsy. This PCr/ATP decrease is similar to that seen by Bottomley and associates[22] in patients with mildly rejecting transplants. Decreased PCr/ATP ratios have

also been observed in patients with cardiomyopathies, coronary artery disease,[26–28] and hypertrophied myocardium.[29]

One possible explanation for the observed change in the PCr/ATP ratio is the impact of changes in the vasculature with early rejection.[30] The vasculature shows diffuse capillary endothelial cell swelling and interstitial hemorrhage and edema, which lead to ischemia. While patients may be asymptomatic, ^{31}P MRS demonstrates a decrease in PCr/ATP ratio. An alternate possibility is that the equilibrium constant of the myocardial creatine kinase has changed as a result of rejection. To test this new hypothesis, one could mildly stress the patient. If ischemia was present, one would expect a further decrease in the PCr/ATP ratio.[26, 31] Alternatively, if ischemia was not the cause, the PCr/ATP ratio would not change. We have started testing transplant patients using mild isometric handgrip exercise to evaluate the possibility that transient changes can occur in the PCr/ATP ratio.[32] Spectra are obtained at rest, with handgrip exercise and following stress. PCr/ATP is corrected for both T1 differences and blood contamination. A spectrum from one transplant subject is presented (Figure 32–7). In this case, the PCr/ATP ratio clearly decreases with exercise and recovers afterwards. In all subjects studied, the handgrip exercise generated a small increase in the rate-pressure product (RPP) (average + 12%, Figure 32–8). Although there was no significant difference in RPP in the study, the normal subjects usually showed an increase in heart rate, whereas the transplants, being denervated, showed an increase in blood pressure during exercise. As expected, the normal subjects showed no significant change in PCr/ATP during exercise (average + 2%), whereas 8 of the 27 transplant patients showed a significant decrease in PCr/ATP (> 2 SD, Figure 32–8) yet no significant change in RPP. The remaining transplant patients showed no significant change in PCr/ATP during exercise. We are currently following these patients to determine the relationship between change in PCr/ATP ratios and long-term outcome. The decrease in PCr/ATP in transplant patients may be the first sign of microvascular rejection.

CONCLUSION

The ultimate role that CMR will play in the management of the cardiac transplant patient is still unclear. Progress has been made in both imaging and MRS applications. With other possible nuclei available for clinical application (e.g., [23]Na), additional information could be acquired. Some key points that should be reiterated for MRS methods are as follows. The PCr/Pi ratio is probably more sensitive than PCr/ATP. However for widely available clinical systems, the PCr/Pi is not attainable due to the contamination from blood phosphate. Thus, clinical studies must rely on the less sensitive PCr/ATP ratio. With the advent of high-field (≥ 3 T)

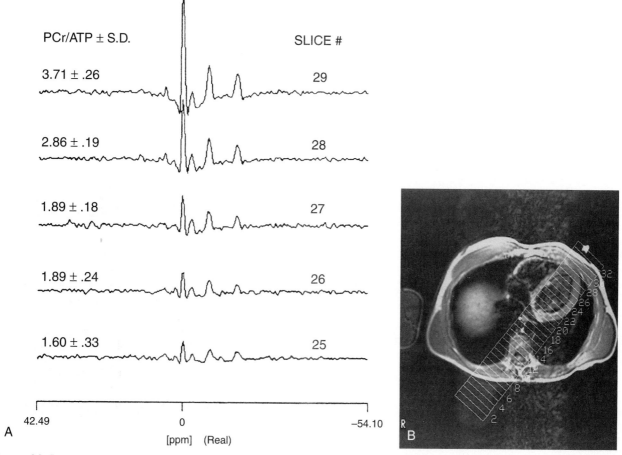

Figure 32–5. Representative one-dimensional chemical shift imaging (1DCSI) data from a normal volunteer. *A*, The ^{31}P spectra obtained through the chest wall. *B*, Slice numbers correspond to the ^{1}H image from the same patient. Each spectrum should have a good signal-to-noise (S/N) ratio for both analysis and diagnostic purposes.

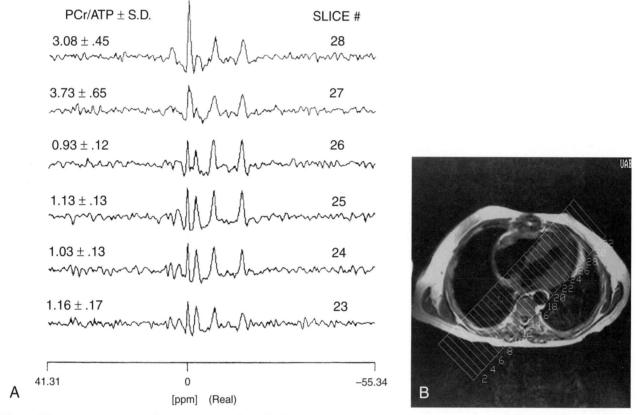

Figure 32–6. *A*, Demonstrates the 1DCSI approach applied to a cardiac transplant patient who was experiencing moderate rejection. Note the excellent S/N for each slice and the depressed PCr signal and the associated PCr/ATP ratios. *B*, The image displays the orientation and slice numbers corresponding to the spectra shown in *A*.

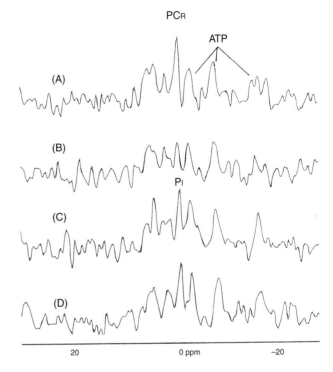

Figure 32–7. Representative ^{31}P magnetic resonance spectroscopy (MRS) stress test from a 45-year-old black man 3 years after transplant. Spectrum A is baseline and spectrum B is during exercise. Spectra C and D are serial recovery spectra. Note the significant drop in the PCr resonance in spectrum B. Spectrum C manifests both a return of PCr and a concomitant increase in Pi. Recovery spectrum C shows the decrease in the Pi resonance. All four spectra demonstrate the dynamics of the ^{31}P MRS stress test.

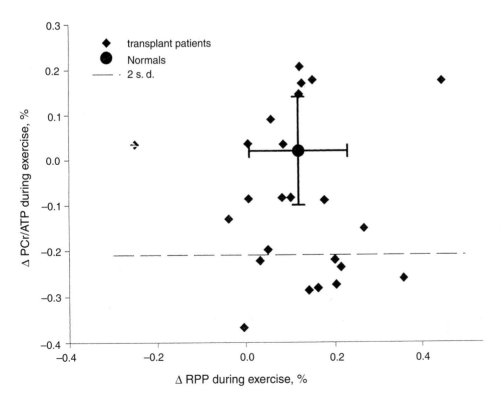

Figure 32–8. Graph of the percent change in the PCr/ATP ratio compared with the change in rate pressure product (RPP). The diamond symbols are the transplant patients and the single solid circle is the sum of the controls. The dashed line is the 2-SD distance from controls. Note that eight separate patients had significant decreases in the PCr/ATP ratio compared with normal controls at the 2-SD mark. Also note that no diamonds even approached the 2 SD in the positive direction. The RPP shows no significant difference between the two groups.

clinical systems, PCr/Pi can be measured. Likewise, at these higher fields, intracellular pH is easily identified and may play an important role in assessing the rejection process. The fact remains clear, however, that ^{31}P MRS methods can assess alterations in the high-energy phosphates in transplanted hearts that can be detected before biopsy evidence. Furthermore, an improvement in PCr/ATP shows the response to the augmented immunotherapy. Magnetic resonance methods have great potential to become the most important diagnostic modality for allograft rejection.

References

1. Aherne T, Tscholakoff D, Finkbeiner W, et al: Magnetic resonance imaging of cardiac transplants: The evaluation of rejection of cardiac allografts with and without immunosuppression. Circulation 74:145, 1986.
2. Nishimura T, Sada M, Sasaki H, et al: Identification of cardiac rejection with magnetic resonance imaging in heterotopic heart transplantation model. Heart Vessels 3:135, 1987.
3. Konstam MA, Aronovitz MJ, Runge VM, et al: Magnetic resonance imaging with gadolinium-DTPA for detecting cardiac transplant rejection in rats. Circulation 78(suppl III):87, 1988.
4. Wisenberg G, Pflugfelder PW, Kostuk WJ, et al: Diagnostic applicability of magnetic resonance imaging in assessing human cardiac allograft rejection. Am J Cardiol 60:130, 1987.
5. Lund G, Morin RL, Olivari MT, Ring WS: Serial myocardial T2 relaxation time measurements in normal subjects and heart transplant recipients. J Heart Transplant 7:274, 1988.
6. Lauerma K, Harjula A, Jarvinen V, et al: Assessment of right and left atrial function in patients with transplanted hearts with the use of magnetic resonance imaging. J Heart Lung Transplant 15:360, 1996.
7. Walpoth BH, Lazeyras F, Tschopp A, et al: Assessment of cardiac rejection and immunosuppression by magnetic resonance imaging and spectroscopy. Transplant Proc 27:2088, 1995.
8. Globits S, De Marco T, Schwitter J, et al: Assessment of early left ventricular remodeling in orthotopic heart transplant recipients with cine magnetic resonance imaging: Potential mechanisms. J Heart Lung Transplant 16:504, 1997.
9. Mohiaddin RH, Bogren HG, Lazim F, et al: Magnetic resonance coronary angiography in heart transplant recipients. Coron Artery Dis 7:591, 1996.
10. Davis SF, Kannam JP, Wielopolski P, et al: Magnetic resonance coronary angiography in heart transplant recipients. J Heart Lung Transplant 15:580, 1996.
11. Walpoth BH, McGregor CG, Aziz S, et al: Assessment of myocardial rejection by NMR (P-31) (abstract). Circulation 70(suppl II):165, 1984.
12. Hall TS, Baumgartner WA, Borkon AM, et al: Diagnosis of acute cardiac rejection with antimyosin monoclonal antibody, phosphorus nuclear magnetic resonance imaging, two dimensional echocardiography and endocardial biopsy. J Heart Transplant 5:419, 1986.
13. Canby RC, Evanochko WT, Kirklin JK, et al: Monitoring the bioenergetics of cardiac allograft rejection using in vivo P-31 NMR spectroscopy. J Am Coll Cardiol 9:1067, 1987.
14. Suzuki S, Kanashiro M, Amemiya H: Immunosuppressive effect of a new drug, 15-deoxyspergualin, in heterotopic rat

15. heart transplantation: In vivo energy metabolic studies by 31P-NMR spectroscopy. Trans Proc 19:3982, 1987.
15. Haug CE, Shapiro II, Shan L, Weil R: P-31 NMR spectroscopic evaluation of heterotopic cardiac allograft rejection in the rat. Transplantation 44:175, 1987.
16. Fraser CD, Chacko VP, Jacobus WE, et al: Metabolic changes preceding functional and morphologic indices of rejection in heterotopic cardiac allografts. Transplantation 46:346, 1988.
17. Fraser CD, Chacko VP, Jacobus WE, et al: Evidence from 31P NMR studies of cardiac allografts that early rejection is characterized by reversible biochemical changes. Transplantation 48:1068, 1989.
18. D'Amico TA, Buchanan SA, Gall SA, et al: Diagnosis of cardiac allograft rejection using PET and MRS (abstract). Circulation 82(suppl III):613, 1990.
19. Bando K, Fraser CD, Chacko VP, et al: Coronary blood flow does not decrease during allograft rejection in heterotopic heart transplant. J Heart Lung Transplant. 10:251, 1991.
20. van Dobbenburgh JO, Kasbergen C, Slootweg PJ, et al: Heterotopic heart transplantation alters high-energy phosphate metabolism irrespective of cardiac allograft rejection. Mol Cell Biochem 163–164:247, 1996.
21. Herfkens RJ, Charles HC, Negro-Vilar R, Van Trigt P: In vivo P-31 NMRS of human heart transplants (abstract). Proceedings of Society of Magnetic Resonance in Medicine, 1988, p 827.
22. Bottomley PA, Weiss RG, Hardy CJ, Baumgartner WA: Myocardial high-energy phosphate metabolism and allograft rejection in patients with heart transplants. Radiology 181:67, 1991.
23. Wolfe CL, Caputo G, Chew W, et al: Detection of cardiac transplant rejection by magnetic resonance imaging and spectroscopy (abstract). Proceedings of the Society of Magnetic Resonance in Medicine, 1991, p 574.
24. van Dobbenburgh JO, Lahpor JR, Woolley SR, et al: Functional recovery after human heart transplantation is related to the metabolic condition of the hypothermic donor heart. Circulation 94:2831, 1996.
25. Evanochko WT, Bouchard A, Kirklin J, et al: Detection of cardiac transplant rejection in patients by 31P NMR spectroscopy (abstract). Proceedings of the Society of Magnetic Resonance in Medicine, New York, 1990, p 246.
26. Weiss RG, Bottomley PA, Hardy CJ, Gerstenblith G: Regional myocardial metabolism of high-energy phosphates during isometric exercise in patients with coronary artery disease. N Engl J Med 323:1593, 1990.
27. Hardy CJ, Weiss RG, Bottomley PA, Gerstenblith G: Altered cardiac energy status in human cardiomyopathy: Correlation with etiology and ejection fraction (abstract). Proceedings of the Society of Magnetic Resonance in Medicine, 1990, p 931.
28. Neubauer SN, Krahe T, Schindler R, et al: 31P Magnetic resonance spectroscopy in dilated cardiomyopathy and coronary artery disease. Circulation 86:1810, 1992.
29. Conway MA, Allis J, Ouwerkerk R, et al: Detection of low phosphocreatine to ATP ratio in failing hypertrophied human myocardium by 31P magnetic resonance spectroscopy. Lancet 338:973, 1991.
30. Lones MA, Czer LSC, Trento A, et al: Clinical-pathologic features of humoral rejection in cardiac allografts: A study in 81 consecutive patients. J Heart Lung Transplant 14:151, 1995.
31. Yabe T, Mitsunami K, Okada M, et al: Detection of myocardial ischemia by 31P magnetic resonance spectroscopy during handgrip exercise. Circulation 89:1709, 1994.
32. Evanochko WT, Buchthal SB, den Hollander JA, et al: Cardiac transplant patients assessed by the P-31 MRS stress test (abstract). J Cardiovasc Magn Reson 1:96, 1999.

CHAPTER 33

Clinical Cardiac Magnetic Resonance Spectroscopy

Robert G. Weiss, Roberto Kalil-Filho, and Paul A. Bottomley

Magnetic resonance imaging (MRI) displays the distribution of hydrogen (^{1}H) nuclei primarily in water and fat to define cardiac and vascular anatomy, function, and perfusion. The same physics principles are used by magnetic resonance spectroscopy (MRS) to noninvasively quantify the regional chemistry and metabolism of tissues under physiological conditions. MRS can be conducted on a number of nuclei, including phosphorus (^{31}P), hydrogen (^{1}H), and carbon (^{13}C).[1, 2] ^{31}P MRS is unique in its ability to noninvasively quantify high-energy phosphate compounds such as adenosine triphosphate (ATP) and creatine phosphate (PCr) that fuel ongoing myocardial contractile function and that are necessary for viability. ^{1}H MRS can be used to identify lipids and quantify creatine, a marker of myocyte viability. ^{13}C MRS can be used to study carbon substrate utilization and intermediary metabolism. In this chapter we provide an overview of common cardiac spectroscopic approaches and review prominent clinical studies, focusing on those investigating ischemic heart disease, heart failure, and transplant rejection.

PRINCIPLES OF MAGNETIC RESONANCE SPECTROSCOPY

Cardiac Magnetic Resonance Spectra

A magnetic resonance (MR) spectrum is a plot of the MR signal intensity as a function of MR frequency, typically measured in units of parts per million (ppm) relative to the resonant frequency of a reference compound, such as PCr in ^{31}P studies in vivo, or tetramethyl silane (TMS) for ^{1}H or carbon (^{13}C) MRS.[3] The MR spectrum is generated from the Fourier transformation (FT) of a transient, time-dependent MR signal recorded in the absence of any spatial localization or imaging magnetic field gradients. The chemical bonds and surrounding chemical environment determine the magnetic shielding of a nucleus and hence affect its resonant frequency. Thus different chemical moieties resonate at slightly different frequencies and appear at different positions in the MR spectrum. The integrated signal area of each peak is proportional to the number of such nuclei or to the concentration of the given compound.

The most prominent peak in the normal human cardiac ^{1}H spectrum is the water resonance at about 4.7 ppm. This peak can be suppressed by several MRS methods to permit observation of less abundant moieties, and -CH_2- and CH_3 peaks from lipids at 1.3 and 0.9 ppm,[4] like that in pericardial fat. The total cardiac creatine pool can be observed and quantified in ^{1}H spectra at 3.0 ppm.[5, 6] Resonances associated with deoxy- and oxymyoglobin at 75 ppm and −2.8 ppm that correlate with intracellular pO_2 have been reported in rat and dog hearts,[7, 8] but this work has not yet been extended to the human heart.

The normal human ^{31}P cardiac spectrum (Figure 33–1) typically exhibits a single sharp PCr peak at 0 ppm; three distinct ATP-related phosphate peaks at about −2.7 ppm (γ-ATP), −7.8 ppm (α-ATP), and −16.3 ppm (β-ATP). In addition, variable amounts of phosphodiester (PD) at 2 to 3 ppm; inorganic phosphate (Pi), whose chemical shift varies from 3.9 to 5.1 ppm depending on the intracellular pH; two blood 2,3-diphosphoglycerate (DPG) peaks at 5.4 and 6.3 ppm, often as a contaminant from the ventricle; and phosphomonoester (PM) resonances at 6.3 to 6.8 ppm[9, 10] may be seen. Calibration curves permit intracellular pH measurements from the Pi chemical shift.[1, 12] Figure 33–1 shows examples of in vivo regional ^{31}P MRS spectra from a normal human heart.

The natural abundance ^{13}C spectrum from the human heart is dominated by fat resonances: -CH_2- at about 30 ppm, glycerol and carboxyl (-CO) resonances at 170 ppm. Sometimes a glycogen resonance at about 101 ppm is also detectable.[13] Because ^{13}C is a 1 percent naturally abundant, stable, nonradioactive isotope, ^{13}C-enriched substrates can provide the basis for tracer studies of glycolytic and tricarboxylic acid cycle metabolism. Although many ^{13}C MRS studies have been performed in animals,[14–18] human cardiac applications with ^{13}C-labeled substrates remain to be reported.

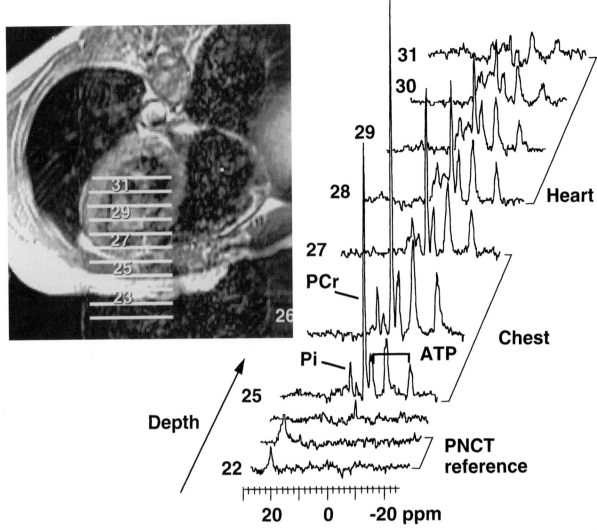

Figure 33–1. ¹H MR image of a normal human heart, and spatially localized surface coil ³¹P MR spectra obtained on a 1.5 T NMR imaging system. *Upper left:* Spin-echo ¹H MR axial image of a normal human chest with white lines demarcating the nominal locations of the respective ³¹P MR spectra shown at right. *Right:* Cardiac-gated ³¹P MR spectra acquired from 1-cm thick coronal slices using a one-dimensional (1-D) chemical shift imaging (CSI) sequence from outside the chest (*bottom*, slices 22–23), through the skeletal muscle in the chest (slices 25–27) to the heart (slices 28–31). These spectra were all acquired in about 10 minutes. (Modified from Bottomley PA, Atalar E, Weiss RG: Human cardiac high-energy phosphate metabolite concentrations by 1D-resolved NMR spectroscopy. Magn Reson Med 35:664, 1996. Reprinted by permission of Wiley-Liss, Inc., a subsidiary of John Wiley & Sons, Inc.)

Acquiring Cardiac Magnetic Resonance Spectra

The typical human cardiac MRS examination uses surface ³¹P receiver coils. In order to optimize MR sensitivity, the coils are positioned on the chest wall in closest proximity to the regional myocardium of interest.[19, 20] A separate, larger MRS transmitter coil is used to provide a substantially uniform excitation field. For ³¹P, this transmitter coil is typically a larger surface coil or body coil. Conventional ¹H coils are available for MRI. Once the coil is properly positioned on the patient, MR images are acquired to identify the region of interest and verify the desired coil placement relative to the heart. Optimization of

magnetic field homogeneity (shimming), which can be done automatically, is usually quite helpful.

¹H nuclei have intrinsically higher MR sensitivity than ³¹P nuclei, thereby allowing ¹H detection of several metabolites to be obtained from nearly all regions of the heart. In contrast, the depth from which ³¹P nuclear magnetic resonance (NMR) spectra can be acquired is relatively limited, and usually confined to the anterior wall and septum. This restriction is due to the intrinsically lower sensitivity of the ³¹P nucleus, the low concentration of ATP and PCr (mM range), and to the dramatic decrease in surface coil sensitivity with depth. Three technical advances promise improvement in the depth of ³¹P detection in the heart and include nuclear Overhauser enhancement (nOe),[12, 21] phased arrays of sur-

face detection coils,[22] and higher magnetic field strengths. Nuclear Overhauser enhancement involves irradiating the sample at the ^1H resonant frequency during the course of the ^{31}P experiment, resulting in a transfer of ^1H nuclear magnetization to the ^{31}P nuclei from nearby protons. The enhancement at the heart with nOe can be 60 percent (1.6-fold) for PCr and 30 to 40 percent for ATP in humans.[12, 21] Secondly, phased arrays of ^{31}P surface coils can provide an additional increase of up to about 40 percent compared with the best positioned single surface coil of the same diameter.[22] Moreover, the extended fields of view enable more myocardium to be optimally positioned relative to any coil; therefore, sensitivity enhancements of much more than about 40 percent can result for cardiac regions that are not optimally positioned relative to a single coil.[22] Finally, although most clinical MR units operate at 1.5 to 2 T, whole body research machines operating at 3 T and 4 T offer potential increases in signal-to-noise ratio (SNR) or reductions in the size of the spectroscopy volume element (voxel) that are nearly linear with the increase in field strength, assuming sample dominant coil-noise conditions.[23–26]

The details of MRS pulse sequences and spatial-localization schemes for cardiac application have been described in other reviews[1, 27] and are beyond the scope of this chapter. This review focuses on the observations from such techniques.

Metabolite Quantification from Cardiac Magnetic Resonance Spectroscopy Spectra

Metabolite quantification is usually done by best-fitting curves to the peaks in the spectrum and measuring the areas beneath the curves. The resulting ratios of metabolite concentration can be converted to absolute concentrations if the signal from a known concentration standard is also measured. Though theoretically straightforward, the signals must first be corrected for sample volume, inhomogeneities in the sensitivity of the MRS coils, and other distortions due to MR relaxation effects.[28]

For normal human myocardium, the PCr/ATP ratio by ^{31}P MRS varies from 0.9 ± 0.3[29] to 2.1 ± 0.4.[30] The variability in reported PCr/ATP ratio is probably due to systematic errors from relaxation effects; contamination from blood, especially when voxels intersect the ventricular cavities; and contamination from chest wall skeletal muscle. Blood contains ATP but no PCr and hence may reduce the observed PCr/ATP ratio. The amount of blood ATP contaminant can be estimated and subtracted from the spectrum by measuring the blood DPG peak.[31]

Metabolites in the ^1H MRS spectrum can also be quantified by comparing their peak areas to those of other ^1H-containing substances of known concentrations. The quantification of muscle and heart creatine has been reported using the H_2O peak as a reference.[6]

Normal Cardiac High-Energy Phosphate and Creatine Concentrations

The consensus for a normal myocardial PCr/ATP ratio averaged from a number of published studies that correct for T1 distortion and account for blood contamination is 1.83 ± 0.12.[1, 25, 31–35] This figure is probably the best current estimate of the true human cardiac PCr/ATP ratio under physiological conditions because the metabolites appear to be 100 percent MRS-visible,[36] and biochemical assays of biopsy specimens taken at surgery[37] can result in PCr loss. Concentration measurements obtained using a reference signal and correcting for tissue volumes are about 10 mol/g wet weight for PCr and 6 µmol/g wet weight for ATP.[28, 38] Concentration measurements yield about 26 µmol/g wet weight for total cardiac creatine.[6] The pseudo-first-order forward creatine kinase reaction rate for the human heart estimated at 4 T using the saturation transfer method is about 0.5 s^{-1}, corresponding to a PCr flux of about 5 µmol/g wet wt.s)$^{-1}$.[23]

CLINICAL STUDIES

Ischemic Heart Disease

Stress-Induced Transient Ischemia

Myocardial ischemia is typically defined as an imbalance between oxygen supply and demand. Myocardial ischemia rapidly inhibits oxidative metabolism, which results in increases in inorganic phosphate from the rapid depletion of creatine phosphate (PCr) and a subsequent, slower decline of ATP. The metabolic consequences of ischemia are initially a decline in the PCr/ATP and PCr/Pi ratios.[39] During prolonged ischemia, there is depletion of both PCr and ATP and the development of severe acidosis. Myocardial stress has been induced inside MR scanners by the administration of pharmacological agents[40] or by physiological exercise (Figure 33–2).[41–44]

Transient, exercise-induced ischemia was evaluated by ^{31}P MRS in our laboratory in 16 patients with severe (≥ 70% diameter) coronary stenoses of the left anterior descending or left main coronary arteries.[41] Subjects were studied before, during, and after continuous isometric handgrip exercise at 30 percent of their maximum. The mean cardiac PCr/ATP ratio of the anterior left ventricular wall at rest was 1.45 ± 0.31, slightly lower than that in normal subjects at rest ($p = 0.05$). During physiological stress, the mean anterior left ventricular PCr/ATP ratio fell to 0.91 ± 0.24 ($p < 0.001$) while PCr/ATP from the more superficial skeletal muscle remained

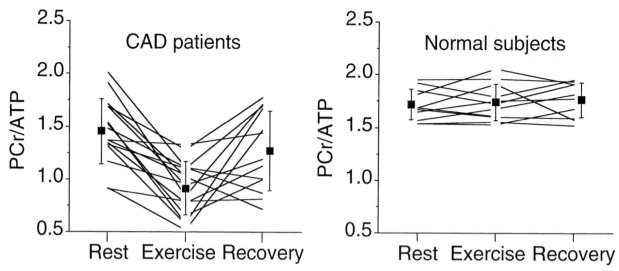

Figure 33–2. Summary myocardial PCr/ATP ratios obtained by ³¹P MRS (using 1D CSI in about 6–10 minutes) at rest, during isometric handgrip exercise, and during subsequent recovery in subjects with coronary artery disease (CAD, *left*) and in normal subjects *(right).* Significant stress-induced declines in the cardiac PCr/ATP ratio, indicative of an imbalance of oxygen supply-demand at the cellular level, are observed only in subjects with critical coronary disease and not in normal subjects. (Modified from Weiss RG, Bottomley PA, Hardy CJ, Gerstenblith G: Regional myocardial metabolism of high-energy phosphates during isometric exercise in patients with coronary artery disease. N Engl J Med 323:1593, 1990, with permission. Copyright © 1990 Massachusetts Medical Society. All rights reserved.)

unchanged. The myocardial PCr/ATP ratio recovered toward baseline in acquisitions commenced a minute after exercise. Although age-matched healthy subjects achieved a similar heart rate–blood pressure product, no stress-induced declines in the cardiac PCr/ATP ratio were observed in them or in 9 patients with valvular or myopathic (nonischemic) disease. These data suggest that the stress-induced significant decline in the PCr/ATP ratio may be specific for ischemic disease. A small group of patients (n = 5) were studied after successful revascularization with balloon angioplasty or bypass surgery. The previously noted stress-induced declines in cardiac PCr/ATP were not seen after revascularization.[41] This study demonstrated that MRS can detect transient changes in myocardial high-energy phosphates, characteristic of reversible myocellular ischemia, in humans. These observations were subsequently confirmed and extended by others.[42] The PCr/ATP ratio in 11 normal/healthy subjects did not change during isometric handgrip stress from that at rest, but a decline in the PCr/ATP ratio from 1.60 ± .19 at rest to 0.96 ± .28 with exercise ($p < 0.001$) was observed in 15 subjects with critical coronary disease and a reversible ²⁰¹Tl defect.[42] The PCr/ATP ratio in 12 subjects with a fixed ²⁰¹Tl defect was low at rest (1.24 ± 0.30) and did not change significantly during exercise stress. The findings support the notion that ³¹P MRS with isometric stress may be a sensitive and specific method for detecting reversible cardiac ischemia in *viable* myocardium.

³¹P MRS stress testing during isometric handgrip exercise is now being used to determine whether ischemic metabolic changes occur in subjects with microvascular disease, manifesting as ischemic symptoms and findings on routine stress testing but

no significant coronary stenoses at coronary angiography. A report indicated a stress-induced decline in the PCr/ATP ratio in a minority of patients,[45] an increase in others, but overall no significant difference in the mean change in PCr/ATP between such subjects and normal volunteers.

Prolonged Ischemia, Infarction, and Loss of Viability

Severe prolonged ischemic injury can lead to infarction with loss of high-energy phosphate and total creatine stores. Reduced myocardial PCr/Pi ratios at rest, in the absence of any detectable myocardial PCr/ATP changes, were first observed in patients with recent anterior myocardial infarctions using spatially localized ³¹P MRS.[10] Significant reductions in the myocardial levels of PCr (50% reduction) and ATP (65% reduction) in subjects with Q wave infarctions[46] have more recently been reported and represent an important step forward from ratio measurements alone in this setting. In addition, ²⁰¹Tl scintigraphy extent score correlated negatively with [ATP] in some subjects.[46] Thus the available clinical evidence is consistent with prior basic findings that high-energy phosphates are absent in infarcted, irreversibly injured myocardium.

Because spatial resolution and depth penetration limit the ability of ³¹P NMR spectroscopy to evaluate metabolite depletion as a marker of infarcted, nonviable cardiac tissues, the alternative use of ¹H MRS with its intrinsically higher sensitivity to detect metabolite depletion is appealing. The quantification of the methyl resonances of creatine at 3.0 ppm with ¹H MRS was initially described in brain and skeletal muscle[5] and recently extended to the heart (Figure 33–3).[6] The detection of total creatine (CR) by ¹H

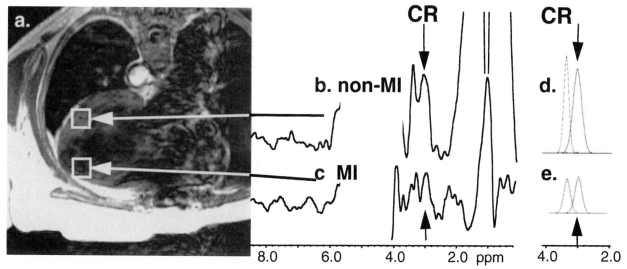

Figure 33–3. Spin-echo axial ¹H MR image of a patient with a large transmural myocardial infarction (MI; a) and corresponding water-suppressed ¹H MR cardiac spectra from a posterior noninfarcted region (b) and an anterior infarcted region (c). The total creatine peak, CR, is shown at about 3 ppm, and the corresponding computer fits to the respective metabolite peaks are shown to the right (d and e). The nominal size of the spectral volumes of interest is about 6 ml, and the spectra were acquired in about 4 to 6 minutes each. Note that there is a 50 to 70% reduction in CR in the infarcted region. The spectra are normalized by the intensity of the unsuppressed water peak from their respective voxels. (Modified from Bottomley PA, Weiss RG: Creatine depletion in non-viable, infarcted myocardium measured by noninvasive MRS. Lancet 351:714, 1998. © The Lancet Ltd.)

MRS promises nearly a 20-fold improvement in sensitivity over that of PCr detection by ³¹P MRS due to the higher sensitivity of the ¹H nucleus over that of ³¹P, the larger number of nuclei per molecule (-CH₃ vs. a single ³¹P in PCr), and the higher concentration of total CR over PCr. In practice, this higher sensitivity has enabled detection of metabolites from the posterior wall of the heart with ¹H MRS, thus far not possible with ³¹P MRS at 1.5 T. ¹H MRS quantification of creatine also promises smaller voxels (4–6 ml) and potentially shorter acquisition times (4–6 min) than conventional ³¹P MRS (~20 ml and 6–20 min, respectively).[6] The approach of quantifying total (phosphorylated + nonphosphorylated) creatine in cardiac muscle was validated in human skeletal muscle and in canine myocardium.[5, 6] Human myocardial creatine concentrations, determined with ¹H MRS, is 28 ± 6 μmol/g wet weight (mean ± SD) in normal human myocardial tissue and 26 ± 11 μmol/g wet weight in noninfarcted regions of subjects with remote myocardial infarction (p = NS).[6] However, total creatine is significantly depleted to 10 ± 8 μmol/g wet weight in nonviable, infarcted regions ($p < 0.001$). This technical advance and new approach of quantifying creatine with ¹H MRS allows metabolite quantification by MRS for the first time throughout the entire human heart (including the remote posterior wall not accessible by conventional ³¹P MRS), with smaller voxels (4–6 ml) for detecting infarcted, nonviable myocardium.

It should be emphasized that the scientific appeal of characterizing reversibly injured myocardium with ³¹P and ¹H MRS quantification of preserved high-energy phosphates and creatine over the more established approach of detecting uptake of an administered radioactive tracer with nuclear or PET technology,[47] is that the former quantifies endogenous metabolites essential for myocellular viability and is not confounded by questions of tracer availability under partial and no-flow conditions, problematic with the latter technologies.

Stunned Viable Myocardium

Transient, self-limited contractile dysfunction following brief episodes of severe ischemia and in the absence of necrosis was first described in 1975[48] and later termed *stunned* myocardium.[49, 50] The advent of thrombolytic therapy and primary balloon angioplasty for acute myocardial infarction has made postischemic dysfunction in viable tissues a more common clinical scenario. However, it is often difficult to distinguish contractile dysfunction due to stunning in viable tissues from regional dysfunction due to myocardial necrosis. Both typically manifest similar systolic dysfunction by conventional imaging with two-dimensional (2-D) echocardiography, nuclear ventriculography, and CMR. For all of these reasons, it is increasingly important to understand the mechanisms contributing to stunning in human myocardium and to develop a method for identifying and distinguishing stunned/viable from necrotic myocardium.

Although myocardial metabolism could theoretically remain inhibited for prolonged periods following ischemia, all animal models of stunned viable myocardium consistently exhibit normal high-energy phosphate levels (normal PCr/ATP ratios).[51–53] The first study of myocardial high-energy phosphates in humans was reported in 29 subjects experiencing their first anterior myocardial infarction who underwent successful reperfusion therapy with balloon angioplasty or thrombolytic therapy.[54] Regional contractile function, assessed by conventional CMR, was depressed in the first 4 days after the event, but improved over the succeeding 30 days, demonstrating that the interrogated regions were indeed stunned. ^{31}P MRS demonstrated that at 4 days following infarction, the regional cardiac PCr/ATP ratios were similar in stunned myocardium and in normal subjects, and to those later recorded in patients at 30 days. Thus, consistent with prior animal studies, altered relative high-energy phosphates are not present in stunned myocardium and therefore cannot contribute to local contractile dysfunction. In addition, the data suggest that ^{31}P MRS may provide a method for distinguishing dysfunctional myocardium due to stunning in viable tissues with normal PCr/ATP from that of infarcted, nonviable tissues with depressed [PCr], [ATP], and [CR] by ^{1}H MRS.

Cardiomyopathy and Congestive Heart Failure

Human heart failure is a progressive, complex syndrome with substantial morbidity and mortality in which symptoms are often manifest as exertional dyspnea or fatigue. Many clinicians define heart failure as an inability of the heart to pump blood commensurate with the needs of the body at rest and/or with activity, or to do so only at a the expense of an increased diastolic pressure. Biochemical energy is required to sustain myocardial function at rest and to increase it during physical activity. The creatine kinase reaction serves as a major energy reservoir in the heart. The hypothesis that abnormalities in myocardial energy production or energy reserve are present and actually underlie human heart failure has been suggested for several years[55-57] and is gaining support. Growing experimental evidence in many animal models indicate that the contractile reserve of normal hearts can be impaired by metabolic interventions that reduce the metabolic reserve of the creatine kinase reaction and that many animal models of heart failure manifest reduced creatine kinase (CK) metabolite concentrations and maximal CK activity.[55-57] The role of reduced energy reserve in human heart failure has been widely studied by ^{31}P MRS.

In the first report from a patient with left ventricular hypertrophy and heart failure studied with localized spectroscopy, the myocardial PCr/ATP was reduced as compared with those of normal individuals.[58] The first significant reductions in the cardiac PCr/ATP ratio in patients with dilated cardiomyopathy and heart failure were roughly 20 to 30 percent at rest below those of normal subjects.[31] This finding has been confirmed by other investigators,[34, 35] whereas some investigators have found nonsignificant or no reductions[29, 59] in subjects with cardiomyopathy and varying degrees of heart failure. Heterogeneity of heart failure etiology and severity may be an important factor contributing to the discrepant findings inasmuch as subjects with more severe

heart failure exhibit the most severe metabolic abnormalities.[31, 34, 60] In a study of 19 subjects with dilated cardiomyopathy, the resting PCr/ATP ratio was reduced in those with severe heart failure (1.4 ± 0.5 vs. 2.0 ± 0.5) and correlated negatively with the New York Heart Association (NYHA) class of failure severity.[34] Following clinical improvement with medical therapy for heart failure, normalization of cardiac PCr/ATP ratios has been reported.[34]

In another study, abnormalities in myocardial CK energetics were shown to be an independent predictor of mortality in patients with dilated cardiomyopathy and heart failure (Figure 33–4).[35] Patients with heart failure (mean ejection fraction [EF] = 30 ± 2%, no angiographic coronary disease, NYHA Class I–III) were initially studied with ^{31}P NMR spectroscopy after a brief period of medical therapy, and then followed over 2.5 years for total and cardiovascular mortality. Two groups were defined: those with cardiac PCr/ATP ratios less than 1.6 and those with PCr/ATP ratios of 1.6 or greater. At the beginning of the study, the ages, cause of cardiomyopathy, medications, and symptoms were similar in the two groups, although the group with PCr/ATP below 1.6 had somewhat lower ejection fractions (27 ± 2%) and, by study design, lower mean PCr/ATP ratios (1.3 ± 0.05) than those of the higher PCr/ATP group (EF = 33 ± 2% and PCr/ATP ratio = 1.98 ± 0.07, which was similar to that of normal subjects [1.94 ± 0.11]). At a mean follow-up of 2.5 years, total mortality in the low PCr/ATP group was 40 percent while that in normal PCr/ATP group was only 11 percent. Kaplan-Meier analysis showed significantly reduced total ($p = 0.036$) and cardiovascular ($p = 0.016$) mortality for patients with normal PCr/ATP ratios, as compared with those with reduced (< 1.6) ratios. In addition, Cox multivariate analysis demonstrated that both the cardiac PCr/ATP ratio and NYHA class independently predicted cardiovascular mortality. In fact, both the significance and the magnitude of the difference in PCr/ATP ratio exceeded that of NYHA class and EF.[35] This finding is the first demonstration that the energetic status of the heart is an independent predictor of prognosis and mortality in heart failure. It is consistent with the premise that metabolic abnormalities in the CK energy reserve of the heart, quantified noninvasively by ^{31}P MRS, are closely linked to the progression of heart failure in humans.

All of these ^{31}P MRS studies increasingly suggest, but do not unequivocally prove, that reduced energy reserve contributes to reduced contractile reserve in clinical heart failure.

Left Ventricular Hypertrophy

Left ventricular hypertrophy (LVH) is associated with increased mortality, is an independent risk factor for the development of ischemic heart disease, and can eventually evolve into a dilated cardiomyopathy. The increased distance from the epicardium to the endocardium in hypertrophy and the in-

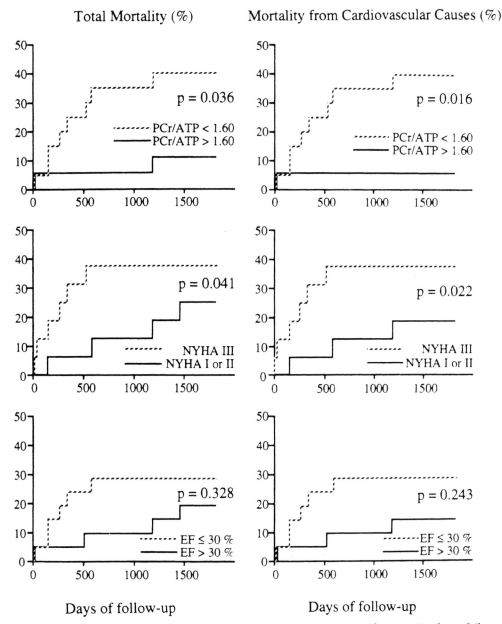

Figure 33–4. Predictors of total *(left)* and cardiovascular *(right)* mortality in patients with congestive heart failure. Note that a low resting cardiac PCr/ATP ratio (< 1.60, *top panel*) is a better predictor of total and especially cardiovascular mortality over the next 3 to 5 years in subjects with heart failure than the usual clinical endpoints of New York Heart Association (NYHA) class symptoms *(middle panels)* or left ventricular ejection fraction (EF, *bottom panels*). (From Neubauer S, Horn M, Cramer M, et al: Myocardial phosphocreatine-to-ATP ratio is a predictor of mortality in patients with dilated cardiomyopathy. Circulation 96:2190, 1997, with permission.)

creased demand in settings of hypertensive hypertrophy contribute to an imbalance in supply/demand. Normal myocardial PCr/ATP ratios may be observed in asymptomatic subjects with LVH.[29, 30] In other subjects with hypertrophic cardiomyopathy, however, reduced cardiac PCr/ATP ratios also have been reported in at least three different studies.[30, 32, 61] In a subset of these patients in whom Pi was detectable, intracellular pH was significantly lower than in controls.[32] In patients with LVH due to aortic valvular disease, subjects with heart failure symptoms had lower resting PCr/ATP ratios than those without heart failure,[60] which suggests that heart failure may be a factor contributing to the variability

in PCr/ATP findings. Future studies with serial measures of high-energy phosphates, especially in the presence and absence of interventions that cause regression of LVH, may enhance our understanding of this pathophysiology.

Detecting Heart Transplant Rejection

Cardiac transplant rejection is an important cause of morbidity and mortality in transplanted patients. The presence and severity of clinical transplant rejection is currently assessed by endomyocardial biopsies obtained using a transvenous right-heart bioptome. To suppress rejection episodes, transplant recipients receive chronic immunosuppressive medication. Many heart transplants are often histologically classified as non- or mildly rejecting. A decision to augment immunosuppressive therapy for moderate or severe rejection is typically based on the histological appearance of myocyte necrosis. Clinical management decisions therefore are dependent on the ability to quantify the severity of rejection.

^{31}P MRS studies of animals with nonworking, heterotopically implanted hearts show significant decreases in myocardial PCr/Pi and PCr/ATP ratios in the first days or weeks following transplantation. Importantly, these changes appeared to *precede* the histological evidence for allograft rejection,[62–65] thus raising the possibility that ^{31}P MRS indices may be clinically useful for managing the growing transplant population.

^{31}P MRS studies of transplant patients have generally confirmed that myocardial PCr/ATP or PCr/Pi is reduced in transplanted human hearts. Although some preliminary reports suggested the possibility that ^{31}P MRS may predict histological rejection,[66, 67] subsequent larger studies indicate that ^{31}P MRS can not reliably distinguish milder from more severe rejection with myocyte necrosis.[33, 68, 69] Thus, although myocardial PCr/ATP is typically significantly lower in transplanted hearts up to 5.5 years following transplantation (1.6 ± 0.5 vs. 1.9 ± 0.2, $p < 0.01$), individual ^{31}P MRS measures poorly correlate with histological indices for rejection.[33] PCr/ATP reductions may be more common early after transplantation, perhaps due to ischemia and reperfusion injury during transplantation, and/or subsequent edema.[69]

The differences between metabolic and histological findings probably reflect fundamental differences in the properties measured. Dead cells are depleted of high-energy phosphates, so a PCr/ATP ratio cannot index myocyte necrosis. A lower PCr/ATP ratio could warn of impending or ongoing damage, if it were manifested by a sufficiently large myocyte population so as to be observable by ^{31}P MRS.[33] But if this lower PCr/ATP does not progress to cell death or resolves in response to the chronic immunosuppressive medication, then the ^{31}P indices will register as false-positive predictions of rejection when compared to the histological standard.

THE FUTURE OF CARDIAC NUCLEAR MAGNETIC RESONANCE SPECTROSCOPY

There have been a number of recent technical advances and important clinical studies exploiting the strengths of cardiac MRS to probe myocardial metabolism. Despite the unique scientific insights into cardiac metabolism that NMR spectroscopy can provide, it is not currently used in the general clinical evaluation of heart disease. The ultimate role of cardiac NMR spectroscopy in clinical cardiology will likely depend on the success in overcoming the technical problems of low sensitivity and implementation. The expanded availability of high-field (≥ 3 T) human MR systems will lead to improved SNR, but practical clinical protocols as well as the impact on prognosis or diagnosis in large patient populations remain to be defined. We envision that recent advances to improve the spatial resolution of MRS-generated metabolite data, to display spectroscopic data in graphical modes (metabolite maps), and to generate metabolic data in conjunction with anatomical and functional information from conventional CMR will greatly enhance the clinical utility of this very powerful scientific tool.

References

1. Bottomley PA: MR Spectroscopy of the human heart: The status and the challenges. Radiology 191:593, 1994.
2. Ingwall JS, Weiss RG: ^{31}P NMR spectroscopy: The noninvasive tool for the study of the biochemistry of the cardiovascular system. Trends Cardiovasc Med 3:29, 1993.
3. Gadian DG: Nuclear Magnetic Resonance and Its Applications to Living Systems. Oxford, Oxford University Press, 1982, p 30.
4. Den Hollander JA, Evanochko WT, Pohost GM: Observation of cardiac lipids in humans by localized ^{1}H magnetic resonance spectroscopic imaging. Magn Reson Med 32:175, 1994.
5. Bottomley PA, Lee Y, Weiss RG: Total creatine in muscle: Imaging and quantification with proton MR spectroscopy. Radiology 204:403, 1997.
6. Bottomley PA, Weiss RG: Creatine depletion in non-viable, infarcted myocardium measured by noninvasive MRS. Lancet 351:714, 1998.
7. Kreutzer U, Wang DS, Jue T: Observing the ^{1}H NMR signal of the myoglobin Val-E11 in myocardium: An index of cellular oxygenation. Proc Natl Acad Sci U S A 89:4731, 1992.
8. Chen W, Zhang J, Eljgelshoven MH, et al: Determination of deoxymyoglobin changes during graded myocardial ischemia: An in vivo ^{1}H NMR spectroscopy study. Magn Reson Med 38:193, 1997.
9. Bottomley PA: Noninvasive study of high-energy phosphate metabolism in human heart by depth-resolved ^{31}P NMR spectroscopy. Science 229:769, 1985.
10. Bottomley PA, Herfkens RJ, Smith LS, Bashore TM: Altered phosphate metabolism in myocardial infarction: P-31 MR spectroscopy. Radiology 165:703, 1987.
11. Flaherty JF, Weisfeldt ML, Bulkley BH, et al: Mechanisms of myocardial cell damage assessed by phosphorus-31 nuclear magnetic resonance. Circulation 65:561, 1982.
12. Kolem H, Sauter R, Friedrich M, et al: Nuclear Overhauser enhancement and proton decoupling in phosphorus chemical shift imaging of the human heart. In Pohost GM (ed): Cardiovascular Applications of Magnetic Resonance. Mt. Kisco, NY, Futura, 1993, p 417.
13. Bottomley PA, Hardy CJ, Roemer PB, Mueller OM: Proton-decoupled, Overhauser-enhanced, spatially localized carbon-13 spectroscopy in humans. Magn Reson Med 12:348, 1989.
14. Weiss RG, Chacko VP, Glickson JD, Gerstenblith G: Comparative ^{13}C and ^{31}P NMR assessment of altered metabolism during graded reductions in coronary flow in intact rat hearts. Proc Natl Acad Sci U S A 86:6426, 1989.
15. Lewandowski ED: Nuclear magnetic resonance evaluation of metabolic and respiratory support of work load in intact rabbit hearts. Circ Res 70:576, 1992.
16. Laughlin MR, Petit WAJ, Dizon JM, et al: NMR measurements of in vivo myocardial glycogen metabolism. J Biol Chem 263:2285, 1988.

17. Malloy CR, Sherry AD, Jeffrey FMH: Carbon flux through citric acid cycle pathways in perfused heart by ^{13}C NMR spectroscopy. FEBS Lett 212:58, 1987.

18. Weiss RG, de Albuquerque CP, Vandegaer KM, et al: Attenuated glycogenolysis reduces glycolytic catabolite accumulation during ischemia in preconditioned rat hearts. Circ Res 79:435, 1996.

19. Bottomley PA, Hardy CJ, Weiss RG: Correcting human heart ^{31}P NMR spectra for partial saturation. Evidence that saturation factors for PCr/ATP are homogeneous in normal and diseased states. J Magn Reson 95:341, 1991.

20. Bottomley PA, Hardy CJ, Roemer PB, Weiss RG: Problems and expediencies in human ^{31}P spectroscopy: The definition of localized volumes, dealing with saturation and the technique-dependence of quantification. NMR Biomed 20:284, 1989.

21. Bottomley PA, Hardy CJ: Proton Overhauser enhancements in human cardiac phosphorus NMR spectroscopy at 1.5 T. Magn Reson Med 24:384, 1992.

22. Hardy CJ, Bottomley PA, Rohling KW, Roemer PB: An NMR phased array for human cardiac ^{31}P spectroscopy. Magn Reson Med 28:54, 1992.

23. Bottomley PA, Hardy CJ: Mapping creatine kinase reaction rates in human brain and heart with 4 Tesla saturation transfer ^{31}P NMR. J Magn Reson 99:443, 1992.

24. Hardy CJ, Bottomley PA, Roemer PB, Redington RW: Rapid ^{31}P spectroscopy on a 4 T whole-body system. Magn Reson Med 8:104, 1988.

25. Menon RS, Hendrich K, Hu X, Ugurbil K: ^{31}P NMR spectroscopy of the human heart at 4T: Detection of substantially uncontaminated cardiac spectra and differentiation of subepicardium and subendocardium. Magn Reson Med 26:368, 1992.

26. Hetherington HP, Luney DJ, Vaughan JT, et al: 3D ^{31}P spectroscopic imaging of the human heart at 4.1T. Magn Reson Med 33:427, 1995.

27. Weiss RG, Bottomley PA: Cardiac magnetic resonance spectroscopy: Principles and applications. *In* Skorton DJ, Schelbert HR, Wolf GL, Brundage BH (eds): Marcus Cardiac Imaging: A Companion to Braunwald's Heart Disease. Philadelphia, WB Saunders, 1996, p 784.

28. Bottomley PA, Hardy CJ, Roemer PB: Phosphate metabolite imaging and concentration measurements in human heart by nuclear magnetic resonance. Magn Reson Med 14:425, 1990.

29. Schaefer S, Gober JR, Schwartz GG, et al: In vivo phosphorus-31 spectroscopic imaging in patients with global myocardial disease. Am J Cardiol 65:1154, 1990.

30. Masuda Y, Tateno Y, Ikehira H, et al: High-energy phosphate metabolism of the myocardium in normal subjects and patients with various cardiomyopathies: The study using ECG gated MR spectroscopy with a localization technique. Jpn Circ J 56:620, 1992.

31. Hardy CJ, Weiss RG, Bottomley PA, Gerstenblith G: Altered myocardial high-energy phosphate metabolites in patients with dilated cardiomyopathy. Am Heart J 122:795, 1991.

32. deRoos A, Doornbos J, Luyten PR, et al: Cardiac metabolism in patients with dilated and hypertrophic cardiomyopathy: Assessment with proton-decoupled P-31 MR spectroscopy. J Magn Reson Imaging 2:711, 1992.

33. Bottomley PA, Weiss RG, Hardy CJ, Baumgartner WA: Myocardial high-energy phosphate metabolism and allograft rejection in patients with heart transplants. Radiology 181:67, 1991.

34. Neubauer S, Krahe T, Schindler R, et al: [31]P Magnetic resonance spectroscopy in dilated cardiomyopathy and coronary artery disease: Altered cardiac high-energy phosphate metabolism in heart failure. Circulation 86:1810, 1992.

35. Neubauer S, Horn M, Cramer M, et al: Myocardial phosphocreatine-to-ATP ratio is a predictor of mortality in patients with dilated cardiomyopathy. Circulation 96:2190, 1997.

36. Humphrey SM, Garlick PB: NMR-invisible ATP and Pi in normoxic and reperfused rat hearts: a quantitative study. Am J Physiol 260:H6, 1991.

37. Swain JL, Sabina RL, Peyton RB, et al: Derangements in myocardial purine and pyrimidine nucleotide metabolism in patients with coronary artery disease and left ventricular hypertrophy. Proc Natl Acad Sci U S A 79:655, 1982.

38. Yabe T, Mitsunami K, Inubushi T, Kinoshita M: Quantitative measurements of cardiac phosphorus metabolites in coronary artery disease by ^{31}P magnetic resonance spectroscopy. Circulation 92:15, 1995.

39. Jacobus WE, Taylor GJ, Hollis DP, Nunnally RL: Phosphorus nuclear magnetic resonance of perfused working rat hearts. Nature 265:756, 1977.

40. Schaefer S, Schwartz GG, Steinman SK, et al: Metabolic response of the human heart to inotropic stimulation: In vivo phosphorus-31 studies of normal and cardiomyopathic myocardium. Magn Reson Med 25:260, 1992.

41. Weiss RG, Bottomley PA, Hardy CJ, Gerstenblith G: Regional myocardial metabolism of high-energy phosphates during isometric exercise in patients with coronary artery disease. N Engl J Med 323:1593, 1990.

42. Yabe T, Mitsunami K, Okada M, et al: Detection of myocardial ischemia by ^{31}P magnetic resonance spectroscopy during handgrip exercise. Circulation 89:1709, 1994.

43. Conway MA, Bristow JD, Blackledge MJ, et al: Cardiac metabolism during exercise measured by magnetic resonance spectroscopy. Lancet 861:692, 1988.

44. Conway MA, Bristow JD, Blackledge MJ, et al: Cardiac metabolism during exercise in healthy volunteers measured by ^{31}P magnetic resonance spectroscopy. Br Heart J 65:25, 1991.

45. Buchthal SD, den Hollander JA, Merz CN, et al: Abnormal myocardial phosphorus-31 nuclear magnetic resonance spectroscopy in women with chest pain but normal coronary angiograms. N Engl J Med 342:829, 2000.

46. Mitsunami K, Okada M, Inoue T, et al: In vivo ^{31}P nuclear magnetic resonance spectroscopy in patients with old myocardial infarction. Jpn Circ J 56:614, 1992.

47. Tillisch JH, Brunken R, Marshall RC, et al: Reversibility of cardiac wall-motion abnormalities predicted by positron tomography. N Engl J Med 314:884, 1986.

48. Hendrickx GR, Millard RW, McRitchie RJ, et al: Regional myocardial functional and electrophysiological alterations after brief coronary occlusion in conscious dog. J Clin Invest 57:978, 1975.

49. Kloner RA, Deboer LWV, Darsee JR, et al: Prolonged abnormalities of myocardium salvaged by reperfusion. Am J Physiol 241:H591, 1981.

50. Braunwald E, Kloner RA: The stunned myocardium: Prolonged, postischemic ventricular dysfunction. Circulation 66:1146, 1982.

51. Laster SB, Becker LC, Ambrosio G, Jacobus WE: Reduced aerobic metabolic efficiency in globally "stunned" myocardium. J Mol Cell Cardiol 21:419, 1989.

52. Zimmer SD, Ugurbil K, Michurski SP, et al: Alterations in oxidative function and respiratory regulation in the postischemic myocardium. J Biol Chem 264,21:12402, 1989.

53. Weiss RG, Gerstenblith G, Lakatta EG: Calcium oscillations index the extent of calcium loading and predict functional recovery during reperfusion in rat myocardium. J Clin Invest 85:757, 1990.

54. Kalil-Filho R, de Albuquerque CP, Weiss RG, et al: Normal high energy phosphate ratios in "stunned" human myocardium. J Am Coll Cardiol 30:1228, 1997.

55. Ingwall JS: Is cardiac failure a consequence of decreased energy reserve? Circulation 87:VII-58, 1993.

56. Ingwall JS, Nascimben L, Gwathmey JK: Heart failure: Is the pathology due to calcium overload or to mismatch of energy supply and demand? *In* Gwathmey JK, Briggs M, Allen PD (eds): Inotropic Drugs: Basic Research and Clinical Practice. New York, Marcel Dekker, 1994.

57. Krause SM: Metabolism in the failing heart. Heart Failure 1:267, 1988.

58. Rajagopalan B, Blackledge MJ, McKenna WJ: Measurement of phosphocreatine to ATP ratio in normal and diseased human heart by ^{31}P magnetic resonance spectroscopy using the rotating frame-depth selection technique. Ann N Y Acad Sci 508:321, 1987.

59. Auffermann W, Chew WM, Wolfe CL, et al: Normal and diffusely abnormal myocardium in humans: Functional and

metabolic characterization with P-31 MR spectroscopy and cine MR imaging. Radiology 179:253, 1991.

60. Conway MA, Allis J, Ouwerkerk R, et al: Detection of low phosphocreatine to ATP ratio in failing hypertrophied human myocardium by ^{31}P magnetic resonance spectroscopy. Lancet 338:973, 1991.

61. Sakuma H, Takeda K, Tagami T, et al: ^{31}P MR spectroscopy in hypertrophic cardiomyopathy: Comparison with ^{201}Tl myocardial perfusion imaging. Am Heart J 125:1323, 1993.

62. Canby RC, Evanochko WT, Barrett LV, et al: Monitoring the bioenergetics of cardiac allograft rejection using in vivo P-31 nuclear magnetic resonance spectroscopy. J Am Coll Cardiol 9:1067, 1997.

63. Haug CE, Shapiro JL, Chan L, Weil R: P-31 nuclear magnetic resonance spectroscopic evaluation of heterotopic cardiac allograft rejection in the rat. Transplantation 44:175, 1987.

64. Press WH, Flannery BP, Teukolsky SA, Vetterling WT: Numerical recipes: The art of scientific computing. Cambridge, Cambridge University Press, 1988, p 366.

65. Fraser CD, Chacko VP, Jacobus WE, Baumgartner WA: Early phosphorus 31 nuclear magnetic bioenergetic changes potentially predict rejection in heterotopic cardiac allografts. J Heart Transplant 9:197, 1990.

66. Herfkens RJ, Charles HC, Negro-Vilar R, VanTrigt P: In vivo phosphorus-31 NMR spectroscopy of human heart transplants (abstract). Proc Soc Magn Reson Med II:287, 1988.

67. Evanochko WT, Bouchard A, Kirklin JK, et al: Detection of cardiac transplant rejection in patients by ^{31}P NMR spectroscopy (abstract). Proc Soc Magn Reson Med I:246, 1990.

68. Evanochko WT, Den Hollander JA, Luney D, et al: ^{31}P MRS in human heart transplants: A clinical update (abstract). Proc Soc Magn Reson Med 3:1092, 1993.

69. van Dobbenburgh JO, Lahpor JR, Woolley SR, et al: Functional recovery after human heart transplantation is related to the metabolic condition of the hypothermic donor heart. Circulation 94:2831, 1996.

70. Bottomley PA, Atalar E, Weiss RG: Human cardiac high-energy phosphate metabolite concentrations by 1D-resolved NMR spectroscopy. Magn Reson Med 35:664, 1996.

CHAPTER 34

Magnetic Resonance Assessment of Myocardial Oxygenation

Debiao Li, William F. Oellerich, and Robert J. Gropler

Under physiological conditions, myocardial blood flow, oxygen consumption (MVO_2), and myocardial mechanical function are intimately related. Thus, it is not surprising that the most common disease processes involving the heart manifest themselves as imbalances between myocardial oxygen supply and demand. As a consequence, the noninvasive assessment of imbalances in myocardial oxygen supply and demand, particularly on a regional basis, is of critical importance in both cardiovascular research and clinical cardiology. The noninvasive quantification of MVO_2 was not possible until it was shown that positron emission tomography (PET), using ^{11}C-acetate, permits accurate quantification of MVO_2.[1–4] Using this approach, numerous investigators have demonstrated the salutary effects of restoring nutritive perfusion on MVO_2 and cardiac function and the importance of preserving MVO_2 as a descriptor and probable determinant of myocardial viability in both experimental animals and humans with acute or chronic ischemia.[5–8] However, PET studies are limited by relatively poor spatial resolution, limited availability, and use of potentially harmful ionizing radiation.

Magnetic resonance imaging (MRI) has become the method of choice for many biomedical applications. It is noninvasive, does not require iodinated contrast media or ionizing radiation, and is widely available. It can, potentially, provide functional and anatomical information in the same sitting. Cardiovascular magnetic resonance (CMR) applications have already been developed, including anatomical imaging of the heart and great vessels, coronary artery imaging, coronary artery flow quantification, myocardial wall motion assessment, and myocardial perfusion measurement.

More recently, inroads have been made in the use of CMR to determine regional myocardial venous blood deoxyhemoglobin concentration.[9–15] The concentration of deoxyhemoglobin in myocardial venous blood reflects the combined effects of myocardial blood flow and oxygen extraction (which together reflect MVO_2). Consequently, we believe that the change in myocardial venous blood oxygenation secondary to imbalances between oxygen supply and demand would be useful in assessing disease processes that involve such imbalances such as coronary artery disease or impaired coronary vascular reserve. Noninvasive assessment of myocardial venous blood oxygenation may permit the measurement of oxygen extraction. When coupled with flow, these data would allow for measurement of MVO_2. Anatomical, functional, and metabolic information can then be obtained in a single CMR sitting, thereby providing a comprehensive examination for the diagnosis of coronary artery disease and the evaluation of therapies designed to improve the balance between myocardial blood flow and oxygen demand.

For the myocardial system, assuming the fraction of oxygenated hemoglobin in the arterial blood is 100 percent while that in the venous blood is Y, then the oxygen consumption of the myocardium (MVO_2) can be given by

$$MVO_2 = F \times Hct \times (1\text{-}Y) \qquad [1]$$

where F (in ml/g/min) is the blood flow to myocardium and Hct (%) is the hematocrit of blood. By assessing changes of F and Y, one can evaluate the changes in MVO_2.

Another potential application of myocardial venous blood oxygenation assessment is for the noninvasive assessment of myocardial perfusion reserve. Under pharmacological stress induced by dipyridamole or adenosine, normal coronary blood flow will increase by severalfold, whereas the oxygen consumption of the heart will have minimal change. As a result, the myocardial venous blood oxygen saturation will be increased. On the other hand, if the coronary artery is partially or completely occluded, the increase of blood flow (and thus the venous blood oxygen saturation) in the myocardial region subtended by the artery will be significantly less than that in a normally supplied region. By evaluating regional differences in myocardial venous blood oxygenation, one may be able to assess the functional significance of coronary artery disease. Extensive research has been conducted to evaluate myocardial perfusion using CMR by the administration of extraneous contrast media.[16–19] The potential advantage of myocardial perfusion evaluation through venous blood oxygenation as an endogenous contrast medium is that multiple examinations can be performed consecutively without the need to wait for the clearance of extraneous contrast materials between studies.

In this chapter, we first review the basic concepts of CMR myocardial venous blood oxygenation assessment. We then discuss the methods used to evaluate myocardial oxygenation and present initial results in animal models and human subjects.

BLOOD OXYGEN LEVEL DEPENDENT EFFECT

Hemoglobin is highly concentrated in red blood cells. In the circulating form, hemoglobin alternates between the oxy- and the deoxy- forms as oxygen is exchanged in lungs and capillaries. Deoxygenated hemoglobin has unpaired electrons and is paramagnetic (with a positive magnetic susceptibility). When oxygen is attached to hemoglobin, the molecule of oxyhemoglobin contains no unpaired electrons and is diamagnetic (with a slightly negative magnetic susceptibility).[20] Because water and proteins are diamagnetic, a local magnetic field inhomogeneity will occur surrounding deoxyhemoglobin, causing signal loss to blood and the tissue containing blood in T2- and T2*-weighted imaging. This effect is referred to as the blood oxygen level dependent (BOLD) effect.

In a T2*-weighted CMR experiment, the image signal intensity is a function of the echo time (TE) and the T2* of the tissue:

$$S = K \exp(-TE/T2^*) = K \exp(-TE \cdot R2^*) \quad [2]$$

where $R2^* = 1/T2^*$, T2* is the apparent transverse relaxation rate of the tissue, and K is a factor related to the spin density and tissue T1 relaxation time. Throughout this chapter, either T2* or R2* is used for convenience of the discussion. In blood containing deoxyhemoglobin,[21, 22]

$$R2^* \sim \Delta\chi \sim [\text{deoxyHb}] \sim Hct (1-Y) \quad [3]$$

where $\Delta\chi$ is the bulk magnetic susceptibility difference between deoxygenated hemoglobin and water, [deoxyHb] is the concentration of the deoxyhemoglobin in blood, and Y is the fraction of hemoglobin molecules that are deoxygenated in blood. Higher-order dependence of blood R2* on (1-Y) may also exist due to the diffusion effect, but it is believed that the linear term is dominant. By measuring the change in blood R2*, one can evaluate the changes in blood oxygenation.[23, 24] As the blood oxygen saturation level decreases (or the fraction of deoxygenated hemoglobin increases), blood R2* increases. Figure 34–1 demonstrates this relationship using data obtained from in vivo pig studies. T2 measurements have also been used to evaluate blood oxygen saturation.[25, 26]

Similarly, for an image voxel containing both tissue and blood, the voxel R2* can be given by

$$R2^* \sim V_f Hct(1-Y) \quad [4]$$

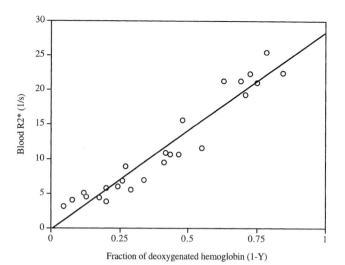

Figure 34–1. Blood R2* as a function of the fraction of deoxygenated hemoglobin measured in four pigs. The change of the fraction of deoxygenated hemoglobin was induced by changing the ventilation rate of the pigs. The solid line is a linear fitting to the data points. Y is the fraction of oxygenated hemoglobin in blood.

where V_f is the blood volume fraction in the tissue.

The BOLD effect has been widely used for brain functional imaging.[27–30] Recently, attempts have been made to use the BOLD effect to evaluate myocardial venous blood oxygenation.[9–15] Compared to brain, the heart has a larger blood volume fraction (~10% vs. ~4%). In addition, the venous blood oxygen saturation in the heart is ~30 percent compared with ~60 percent in the brain.[9] This will allow for a wider range of signal change with stimulated flow in the myocardium. In contrast to BOLD brain imaging, the challenge for BOLD cardiac imaging has been to eliminate image artifacts caused by heart motion due to both cardiac and respiratory cycles, by pulsatile blood flow in the cardiac chambers and the aorta, and by susceptibility variations between tissue and air in the thoracic space.

MYOCARDIAL OXYGENATION EVALUATION USING CARDIOVASCULAR MAGNETIC RESONANCE

A number of in vivo animal and isolated heart studies have demonstrated a change in the myocardial signal during interventions designed to alter the blood oxygenation using T2- or T2*-weighted imaging.[9–13] Significant signal loss was observed in both the left ventricular chamber and myocardium during apnea in rats.[9] In an isolated rabbit heart model, a substantial correlation was found between the gradient-echo image intensity of the myocardium and deoxyhemoglobin concentration levels.[13] In a more physiologically relevant setting, myocardial signal *increased* significantly after the infusion of dipyridamole,[11] presumably because of an increase in myocardial blood flow in the absence of a corresponding increase in oxygen demand, resulting in a decrease in myocardial venous blood deoxyhemoglobin concentration. In contrast, occlusion of the left anterior descending coronary artery resulted in signal reduction,[11, 12] likely reflecting an increase in myocardial venous blood deoxyhemoglobin concentration as oxygen extraction increased due to inadequate blood flow. It should be noted that most of these studies were conducted at high field strengths (2–4.7 T).

Similar effects have been observed in human hearts on conventional 1.5 T clinical imaging systems.[14, 15] Using a T2*-weighted echo-planar imaging (EPI) technique, Niemi and associates[15] demonstrated a close correlation between changes in myocardial signal intensity and those of blood flow velocity measured in the left anterior descending coronary artery during dipyridamole-induced coronary vasodilation in seven healthy volunteers. In a preliminary study from the same group in patients with coronary artery disease, myocardial signal changes were compared with measurements of myocardial blood flow obtained by PET.[31] A lack of increase in myocardial signal intensity with dipyridamole was observed in myocardial regions having a blunted perfusion response to the drug as observed by PET.

Human Studies Correlating Myocardial R2* and Venous Blood Oxygen Saturation

To demonstrate that myocardial R2* is a function of venous blood oxygen saturation and not a direct function of coronary blood flow, we imaged healthy volunteers with the infusion of two different pharmacological stress agents: dipyridamole and dobutamine.[14, 32] Both agents induce an increase in myocardial blood flow but with differing effects on myocardial venous blood oxygenation. Dipyridamole is a potent coronary vasodilator that typically induces a three- to fourfold increase in myocardial perfusion, but induces minimal effects on myocar-

dial oxygen consumption.[33] As a consequence, myocardial venous blood oxygen saturation increases as oxygen supply (blood flow) exceeds demand (oxygen consumption). In contrast, dobutamine is a potent beta-agonist whose primary pharmacological effect is to increase cardiac work.[33] This increase results in an increase in myocardial oxygen consumption, which induces the increase in myocardial perfusion. Thus, oxygen supply and demand remain largely balanced and there is little change in myocardial venous blood oxygen saturation.[34]

To verify the hypothesis that MR BOLD imaging can detect the different effects of the two agents on myocardial venous blood oxygenation, we developed a segmented, multiecho, gradient-echo sequence.[14, 32] To eliminate the image artifacts superimposed on the myocardium caused by the pulsatile ventricular blood flow at relatively long echo times, a double-inversion magnetization preparation[35] was applied immediately after the electrocardiogram (ECG) trigger. This magnetization preparation has no effect on myocardial signal.[36] However, it inverts the magnetization of blood outside the imaging slice. During the trigger delay time (also the inversion recovery time), the blood moves into the imaging slice while its magnetization recovers towards an equilibrium magnetization with a time constant of T1, going through zero along the way. Data were collected when blood magnetization was close to zero, resulting in dramatic suppression of the blood signal.

Data were collected during diastole when there is the least cardiac motion. Three echoes (TE: 6, 14, and 21 ms) were collected after the same radiofrequency (RF) pulse (Figure 34–2). Five to seven phase encoding lines of each echo were collected during each cardiac cycle with a line repetition time of 25.8 ms. The in-plane resolution was 2 to 3 mm, and the image thickness was 6 mm. The imaging time for each scan was 15 to 20 seconds, allowing for each sequence to be acquired during a single breath-hold to minimize respiratory-related artifacts.

Ten human volunteers (age: 25 ± 7 years, 4 men

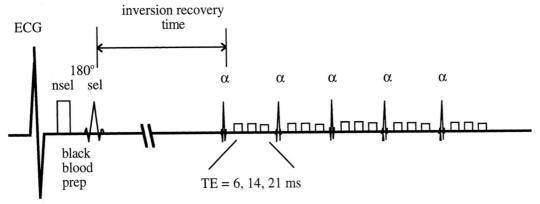

Figure 34–2. Schematic diagram of the imaging sequence to measure myocardial T2*. Five lines are collected during each heartbeat for each of the three echoes. nsel = nonselective; sel = selective; prep = preparation; ECG = electrocardiogram; TE = echo time.

and 6 women) with a low likelihood for coronary artery disease were studied. Each subject was imaged at baseline for more than six consecutive times, and continuously during and after the infusion of dipyridamole (n = 5) and dobutamine (n = 5) at an interval of 1 to 2 minutes. Dipyridamole was infused at a rate of 0.14 mg/kg/min over 4 minutes. Dobutamine was infused at a rate of 15 μg/kg/min.

Myocardial signals were measured at different echo times in the same region of interest outlined manually in the ventricular septum and anterior wall of the left ventricle. Myocardial T2* was estimated using a least squares fitting scheme.[37]

The time courses of the myocardial T2* response to the administration of dipyridamole and dobutamine are shown in Figure 34–3. Note that myocardial T2* increased significantly (20–36%) over baseline values during a period of approximately 10 to 20 minutes after the initiation of dipyridamole infusion, and then recovered to baseline. On the other hand, no significant change in myocardial T2* was observed following the administration of dobutamine, although both dipyridamole and dobutamine increased blood flow in the coronary artery significantly.[14] These results demonstrate that the response of myocardial T2* to the administration of the two stress agents is consistent with the expected venous blood oxygen saturation.

Direct in Vivo Correlation Between Myocardial T2* and Global Coronary Venous Blood Oxygen Saturation

To validate the direct in vivo correlation between myocardial R2* and venous blood oxygen saturation, a well-controlled dog model was used. A wide range of global myocardial venous blood oxygen saturation levels was created in five normal dogs. Hyperemic conditions were induced by the intravenous administration of dipyridamole and dobutamine. To induce hypoxemia, the oxygen content of the inspired gas was reduced by ventilating dogs with a mixture of 10 percent oxygen and 90 percent nitrogen, which reduced the oxygen saturation in both arteries and veins. To correlate myocardial R2* with global venous blood oxygenation, venous blood oxygen saturation levels were measured directly by coronary sinus sampling. Myocardial perfusion was quantified by the administration of radiolabeled microspheres.

A similar CMR technique as in human studies (illustrated in Figure 34–2) was used to measure myocardial T2* of the dogs. The number of echoes collected after each RF pulse was increased to eight (TE: 3.7, 6.7, 9.8, 12.8, 15.8, 18.9, 21.9, 25.0 ms). This increase allows for improved reliability of the myocardial R2* measurement. The repetition time (TR) of the RF pulses was 30 ms. Five phase encoding lines were collected during the diastolic period of each heartbeat. The field of view of the images was 188 × 250 mm, the data acquisition matrix was 120 × 256, and the spatial resolution was 1.6 × 0.98 × 5 mm. A retrospective respiratory gating method was used to eliminate the respiratory motion artifacts.[38] Assuming the RR interval of the dogs was approximately 600 ms, the imaging time of each scan was approximately 1.2 minutes. Typical images obtained using the technique and the region of interest for myocardial T2* estimation are shown in Figure 34–4. Examples of least squares fitting to estimate myocardial T2* are shown in Figure 34–5 at two different levels of venous blood oxygen saturation.

After animal preparation and scout scans to locate a short-axis view of the heart, measurements of myocardial T2* were obtained at baseline, during

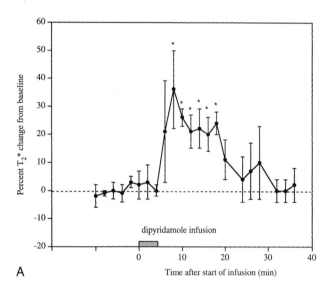

A

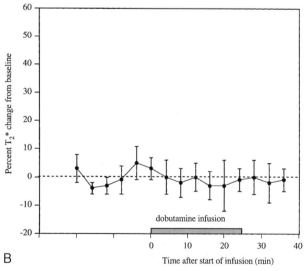

B

Figure 34–3. *A,* Myocardial T2* response to infusion of dipyridamole. The peak T2* increase occurred between 10 and 20 minutes after the start of the infusion, coinciding with the expected peak pharmacological effect. * indicates significant increase from baseline ($p < 0.05$). *B,* Myocardial T2* response to infusion of dobutamine. No significant change of T2* was observed.

TE = 3.7 ms   6.7 ms   9.8 ms   12.8 ms

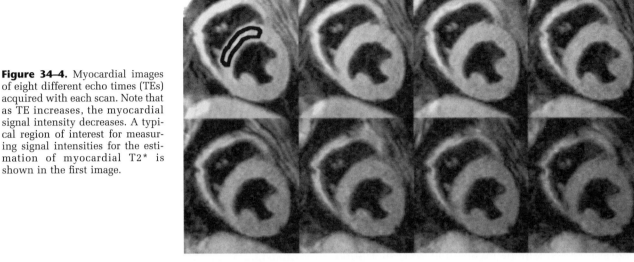

Figure 34–4. Myocardial images of eight different echo times (TEs) acquired with each scan. Note that as TE increases, the myocardial signal intensity decreases. A typical region of interest for measuring signal intensities for the estimation of myocardial T2* is shown in the first image.

TE = 15.8 ms   18.9 ms   21.9 ms   25.0 ms

and after infusion of dipyridamole, dobutamine, and when the dogs were ventilated with hypoxic air. The imaging protocol is shown in Table 34–1. Paired arterial and coronary sinus blood samples were withdrawn at the six different stages of the study. Blood oxygen saturation levels were measured using a blood gas analyzer interfaced with an oximeter.

Coronary sinus blood oxygen saturation levels ranged from 9 to 80 percent with experimental interventions with dipyridamole, dobutamine, or hypoxic air. Myocardial perfusion values at baseline and various interventions are shown in Figure 34–6.

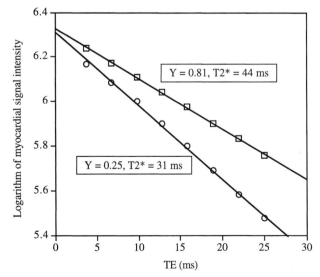

Figure 34–5. Linear least squares fitting of the logarithm of the myocardial signal intensity as a function of the echo time (TE) for the estimation of myocardial T2*. The negative inverse of the slope of each line is the corresponding T2* estimate. Y is the fraction of oxygenated hemoglobin in coronary sinus.

Administration of dipyridamole and dobutamine, and ventilation of hypoxic air all increased myocardial perfusion significantly, but significant myocardial T2* increase occurred only after dipyridamole infusion, which indicates that myocardial T2* is not a direct function of myocardial perfusion. The relationship between the changes of myocardial R2* from baseline and the percent deoxyhemoglobin (%deoxyHb) in the coronary sinus in five dogs is demonstrated in Figure 34–7. A linear regression correlation coefficient of 0.84 (r) was found from the data, indicating strong correlation between myocardial R2* and %deoxyHb in the coronary sinus.

Note that in the hypoxic condition (%deoxyHb > 80%), the increase of R2* as a function of increased %deoxyHb is greater than that predicted from the regression line, which may indicate the effect of blood volume change and can be explained by Equation 4. Both dipyridamole and hypoxic air increase myocardial blood volume fraction by more than 50 percent. However, their effect on R2* manifests itself differently. With the administration of dipyridamole, blood oxygen saturation in coronary sinus increases, resulting in a decrease in R2*. However, the blood volume fraction in the myocardium also increases, which increases the deoxyhemoglobin content of a voxel and tends to cause increase in myocardial R2*, the opposite effect of increased oxygen saturation. Because a decrease in myocardial R2* was observed in our studies, increased oxygen saturation clearly has the dominant effect over increased blood volume, but the apparent R2* change as a function of %deoxyHb is reduced because of the accompanied blood volume effect (left side of Figure 34–7). In contrast, during hypoxia, both %deoxyHb in coronary sinus and the blood volume fraction increase, and their effects enhance each

Table 34–1. Protocol of Dog Studies

baseline	dipyridamole (0.14 mg/kg/min over 4 min)	baseline	dobutamine (15 µg/kg/min)	baseline	ventilation with 10% oxygen and 90% nitrogen

Note: The order of interventions was switched in different studies to prevent any potential systematic errors.

other. As a result, the apparent R2* change as a function of %deoxyHb is greater than that if blood volume fraction remains the same (right side of Figure 34–7). Our studies showed that by measuring myocardial blood volume fraction changes using technetium-99m–labeled red blood cells at each intervention and correcting their effects on myocardial R2*, a more linear relationship was found between R2* and the deoxyhemoglobin concentration. Thus, accurate assessment of myocardial oxygen saturation using CMR will likely require a correction for blood volume. Future studies are needed to clarify this important issue.

In a separate study conducted by Koelling and coworkers,[39] various doses of adenosine were infused to swine, and myocardial R2* was measured using a gradient-echo EPI sequence at baseline and during each dose of adenosine. A strong correlation ($r^2 = 0.8$) between myocardial R2* and coronary venous oxygen saturation was also observed. Like dipyridamole, adenosine has little effect on myocardial oxygen consumption. Thus, change of coronary venous oxygen saturation will reflect the change in myocardial perfusion. In this situation, T2* evaluation during adenosine infusion may become a useful tool for assessing myocardial flow.

Using BOLD Imaging to Detect Regional Myocardial Oxygenation Differences

The ultimate test and utility of MR BOLD imaging is to detect regional differences in myocardial oxygenation caused by focal coronary artery disease. With the administration of a coronary vasodilator such as dipyridamole or adenosine, increased myocardial blood flow occurs without much change in myocardial oxygen consumption. However, myocardial blood flow to territories distal to the coronary artery stenosis may be limited and it could be anticipated that myocardial oxygenation in the affected region may be increased compared with the resting state, but the increase may be smaller than that in normal myocardium.

Preliminary studies in animals and patients have demonstrated the feasibility of using CMR T2* measurements to detect ischemic regions.[12, 31] We studied a dog with coronary artery occlusion created by placing a thrombogenic copper coil through the right carotid artery into the left anterior descending artery (LAD).[40] A calculated T2* image of the heart after the dipyridamole infusion is shown in Figure 34–8A. As predicted, the myocardial T2* in the

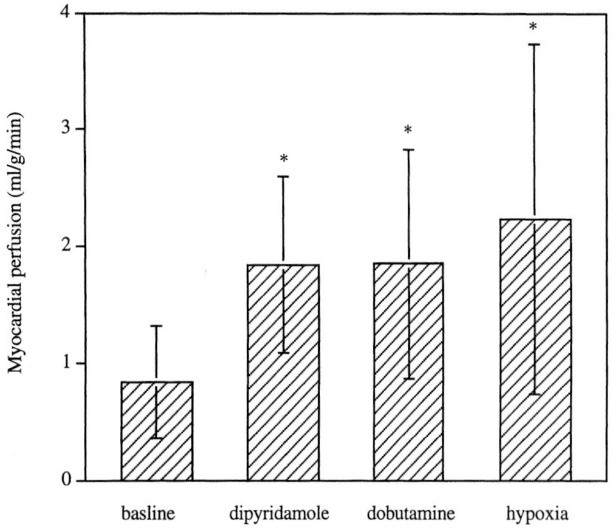

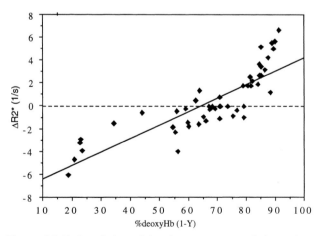

Figure 34–6. Myocardial perfusion for each experimental intervention presented as mean ± standard deviation. * indicates significant difference from baseline ($p < 0.05$).

Figure 34–7. Correlation of the change in myocardial R2* from baseline (ΔR2*) with coronary venous %deoxyHb measured directly from blood samples using a co-oximeter. The solid line is the linear regression of the data. The correlation coefficient (r) is 0.84.

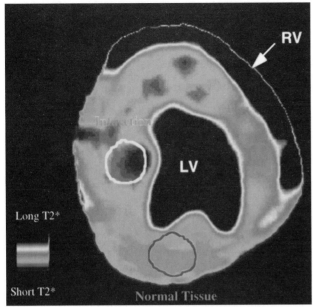

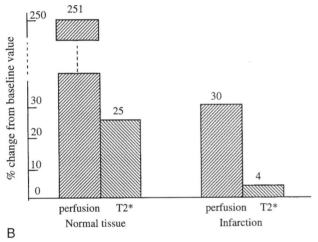

Figure 34–8. *A*, A calculated T2* image of a dog with an occluded left anterior descending artery (LAD) obtained after dipyridamole administration (see also color plates). Note the shorter T2* in the regions of the anterior wall and anterior papillary muscle that are supplied by the LAD than that in remote normal myocardial regions of the left ventricular wall. LV = left ventricle; RV = right ventricle. *B*, Correlation between regional myocardial perfusion and T2* changes after the administration of dipyridamole.

resulting ischemic anterior wall and anterior papillary muscle was much shorter than that of remote normally perfused regions. The relationship between regional myocardial T2* and perfusion measured by using radiolabeled microspheres is shown in Figure 34–8*B*. Clinical results on regional myocardial oxygenation evaluation using BOLD CMR are very limited. Further studies remain to be performed to define the sensitivity and the meaning of myocardial T2* changes. Such studies will help to define the clinical role of BOLD CMR in the assessment of patients with known or suspected ischemic heart disease.

ACKNOWLEDGMENTS

The authors wish to thank E. Mark Haacke, PhD, for discussions on MR BOLD imaging, Gabriele M. Beck, MS, for help in MR technical development, and Terry L. Sharp, RT, for assistance in animal preparations.

References

1. Brown MA, Marshall DR, Sobel BE: Delineation of myocardial oxygen utilization with carbon-11-labeled acetate. Circulation 76:687, 1987.
2. Brown MA, Myears DW, Bergmann SR: Noninvasive assessment of canine myocardial oxidative metabolism with carbon-11 acetate and positron emission tomography. J Am Coll Cardiol 12:1054, 1988.
3. Brown MA, Myears DW, Bergmann SR: Validity of estimates of myocardial oxidative metabolism with carbon-11 acetate in positron emission tomography despite altered patterns of substrate utilization. J Nucl Med 30:187, 1989.
4. Buxton DB, Nienaber CA, Luxen A, et al: Noninvasive quantitation of regional myocardial oxygen consumption in vivo with [1-11C] acetate in dynamic positron emission tomography. Circulation 79:134, 1989.
5. Henes CG, Bergmann SR, Walsh MN: Assessment of myocardial oxidative metabolic reserve with positron emission tomography and carbon-11 acetate. J Nucl Med 30:1489, 1989.
6. Gropler RJ, Siegel BA, Sampathkumaran KS: Dependence of recovery of contractile function on maintenance of oxidative metabolism after myocardial infarction. J Am Coll Cardiol 19:989, 1992.
7. Gropler RJ, Geltman EM, Sampathkumaran KS: Functional recovery after revascularization for chronic coronary artery disease is dependent on maintenance of oxidative metabolism. J Am Coll Cardiol 20:569, 1992.
8. Gropler RJ, Geltman EM, Sampathkumaran KS: Comparison of C-11 acetate with F-18 fluorodeoxyglucose for delineating viable myocardium by positron emission tomography. J Am Coll Cardiol 22:1587, 1993.
9. Wendland MF, Saeed M, Lauerma K, et al: Endogenous susceptibility contrast in myocardium during apnea measured using gradient recalled echo planar imaging. Magn Reson Med 29:273, 1993.
10. Atalay MK, Forder JR, Chacko VP, et al: Oxygenation in the rabbit myocardium: Assessment with susceptibility-dependent MR imaging. Radiology 189:759, 1993.
11. Balaban RS, Taylor JF, Turner R: Effect of cardiac flow on gradient recalled echo images of the canine heart. NMR Biomed 7:89, 1994.
12. Stillman AE, Wilke N, Jerosch-Herold M, et al: BOLD contrast of the heart during occlusion and reperfusion. Works in Progress Supplement, SMR 1st Meeting, 1994, Dallas, p S24.
13. Atalay M, Reeder SB, Zerhouni E, Forder JR: Blood oxygenation dependence of T1 and T2 in the isolated, perfused rabbit heart at 4.7T. Magn Reson Med 34:623, 1995.
14. Li D, Dhawale P, Rubin PJ, et al: Myocardial signal response to dipyridamole and dobutamine: Demonstration of the BOLD effect using a double-echo gradient-echo sequence. Magn Reson Med 36:16, 1996.
15. Niemi P, Poncelet BP, Kwong K, et al: Myocardial intensity changes associated with flow stimulation in blood oxygenation sensitive magnetic resonance imaging. Magn Reson Med 36:78, 1996.
16. Wilke N, Simm C, Zhang J, et al: Contrast-enhanced first pass myocardial perfusion imaging: Correlation between myocardial blood flow in dogs at rest and during hyperemia. Magn Reson Med 29:485, 1993.

17. Wendland MF, Saeed M, Mausi T, et al: First pass of an MR susceptibility contrast agent through normal and ischemic heart: Gradient-recalled echo-planar imaging. J Magn Reson Imaging 3:755, 1993.

18. Edelman RR, Li W: Contrast-enhanced echo-planar MR imaging of myocardial perfusion: preliminary study in humans. Radiology 190:771, 1994.

19. Wilke N, Jerosch-Herold M, Wang Y, et al: Myocardial perfusion reserve: Assessment with multisection, quantitative, first-pass MR imaging. Radiology 204:373, 1997.

20. Pauling L, Coryell CD: The magnetic properties and structure of hemoglobin, oxyhemoglobin and carbonmonoxyhemoglobin. Proc Natl Acad Sci U S A 22:210, 1936.

21. Kennan RP, Scanley BE, Gore JC: Physiologic basis for BOLD MR signal changes due to hypoxia/hyperoxia: Separation of blood volume and magnetic susceptibility effects. Magn Reson Med 37:953, 1997.

22. Kim SG, Ugurbil K: Comparison of blood oxygenation and cerebral blood flow effects in fMRI: Estimation of relative oxygen consumption change. Magn Reson Med 38:59, 1997.

23. Chien D, Levin DL, Anderson CM: MR gradient echo imaging of intravascular blood oxygenation: T_2^* determination in the presence of flow. Magn Reson Med 32:540, 1994.

24. Li D, Wang Y, Waight DJ: Blood oxygen saturation assessment *in vivo* using T2* estimation. Magn Resort Med 39:685, 1998.

25. Wright GA, Hu BS, Macovski A: Estimating oxygen saturation of blood in vivo with MR imaging at 1.5T. J Magn Reson Imaging 1:275, 1991.

26. Li KCP, Dalma RL, Ch'en IY, et al: Chronic mesenteric ischemia: Use of in vivo MR imaging measurements of blood oxygen saturation in the superior mesenteric vein for diagnosis. Radiology 204:71, 1997.

27. Belliveau JW, Kennedy DN, Mckinstry RC, et al: Functional mapping of human visual cortex by magnetic resonance imaging. Science 254:716, 1991.

28. Kwong KK, Belliveau JW, Chesler DA, et al: Dynamic magnetic resonance imaging of human brain activity during primary sensory stimulation. Proc Natl Acad Sci U S A 89:5675, 1992.

29. Ogawa S, Menon RS, Tank DW, et al: Functional brain mapping by blood oxygenation level-dependent contrast magnetic resonance imaging: A comparison of signal characteristics with a biophysical model. Biophys J 64:803, 1993.

30. Lai S, Hopkins AL, Haacke EM, et al: Identification of vascular structures as a major source of signal contrast in high resolution 2D and 3D functional activiation imaging of the motor cortex as 1.5T: Preliminary results. Magn Reson Med 30:387, 1993.

31. Poncelet B, Weisskoff RM, Zervos G, et al: EPI detection of changes in coronary flow velocity and myocardial tissue perfusion during hyperemia in patients with coronary artery disease. Proceedings of the Society of Magnetic Resonance: Third Scientific Meeting and Exhibition, 1995, Nice, France, p 20.

32. Li D, Oellerich WF, Beck G, Gropler RJ: Assessment of myocardial response to pharmacologic interventions using an improved MR imaging technique to estimate T2* values. AJR Am J Roentgenol 172:141, 1999.

33. McGuinness ME, Talbert RL: Pharmacologic stress testing: Experience with dipyridamole, adenosine, and dobutamine. Am J Hosp Pharm 51:328, 1994.

34. Massie BM, Schwartz GG, Garcia J, et al: Myocardial metabolism during increased work states in the porcine left ventricle in vivo. Circ Res 74:64, 1994.

35. Edelman RR, Chien D, Kim D: Fast selective black blood MR imaging. Radiology 181:655, 1991.

36. Simonetti OP, Finn JP, White RD, et al: "Black blood" T2-weighted inversion-recovery MR imaging of the heart. Radiology 199:49, 1996.

37. Bevington PR: Data reduction and error analysis for the physical sciences. New York, McGraw-Hill, 1969.

38. Li D, Kaushikkar S, Haacke EM, et al: Coronary arteries: Three-dimensional MR imaging with retrospective respiratory gating. Radiology 201:857, 1996.

39. Koelling TM, Poncelet BP, Schmidt CJ, et al: Gradient-echo EPI BOLD T2* contrast changes with varying doses of adenosine and correlation with coronary venous oxygen saturation in swine. International Society for Magnetic Resonance in Medicine, 5th Scientific Meeting, 1997, Vancouver, BC, Canada, p 475.

40. Bergmann SR, Lerch RA, Fox KAA, et al: Temporal dependence of beneficial effects of coronary thrombolysis characterized by positron tomography. Am J Med 73:573, 1982.

CHAPTER 35

Interventional Cardiovascular Magnetic Resonance

Paul R. Hilfiker, Simon Wildermuth, Jörg F. Debatin, and Gustav K. von Schulthess

Cardiovascular magnetic resonance (CMR) seems well suited for monitoring vascular interventions. CMR causes no radiation exposure, is capable of combining high temporal and spatial resolution, and provides cross-sectional images in any desired plane. Intravenously administered paramagnetic contrast induces a T1-shortening of blood,[1] translating into a selective signal increase in the arterial system. Three-dimensional (3-D) magnetic resonance angiography (MRA) uses this effect. With contrast-enhanced 3-D MRA, arterial signal is no longer dependent on flow effects (in contrast to time-of-flight and phase contrast MRA). Reflecting its reliability, robustness, and accuracy, the technique has already been implemented into clinical practice in centers around the world (see Chapter 29). The 3-D data sets can be postprocessed to provide a comprehensive exoscopic and endoscopic appreciation of the vascular morphology[2] and potentially serve as roadmaps.

Patient access remains a significant problem for any scenario involving CMR guidance and control of intravascular interventions. The ideal interventional MRA scanner has yet to be designed, because at this time, a combination of open access, high field strength, and fast gradient systems has not yet been achieved. Recent hardware developments have provided more patient access by shortening the bore while maintaining field strength and gradient performance. Real-time interactive CMR is also now a clinical reality.[3] For the concept of interventional MRA to evolve from a hypothetical concept to a practical possibility, further progress will need to be made in this respect.

Fundamental to the success and safety of intravascular procedures such as percutaneous transluminal angioplasty (PTA) is the visualization of the catheter and guidewire relative to the area of treatment. To date visualization is achieved with x-ray fluoroscopy. Exposure to potentially harmful ionizing radiation, limited soft tissue contrast, and the inability to image in cross section have motivated the exploration of alternate imaging strategies. There are two fundamental approaches to device localization with

CMR: electrically passive techniques based on visualization of the susceptibility-induced signal void caused by the instrument,[4, 5] and electrically active techniques, first suggested by Ackerman and colleagues in 1986.[6]

ACTIVE VISUALIZATION OF DEVICES: MAGNETIC RESONANCE TRACKING

CMR tracking was developed by Dumoulin and associates[7] for active monitoring of devices under real-time conditions. A small receive-only coil is incorporated into the tip of the device. Following nonselective radiofrequency (RF) excitation of a volume defined in size by the dimensions of the field of view, a gradient-recalled echo (GRE) is generated by the miniature receive coil, which is effectively the point of signal (Figure 35–1). Because only a point in 3-D space has to be identified, only four echoes are needed for localization. The real-time position of the coil within the catheter or guidewire is simply superimposed on previously acquired MR images (Figure 35–2). There are several unique characteristics of the CMR-based active tracking technique. The technique is virtually independent of device orientation and provides maximum flexibility with regard to the characteristics of the underlying CMR image on which MR-tracking is based. A position update of 18 updates per second with a display delay of under 10 ms ensures real-time tracking of the instrument.[5] In vitro phantom experiments demonstrated the CMR tracking technique to be highly accurate with regard to positioning of the tracking coil.[5] With biplanar display implementation, it is possible to track any instrument in real time simultaneously in any two planes.[5] Considerable challenges, however, still need to be overcome. Visualization of the guidewire and possibly also the catheter must extend beyond the mere tip.

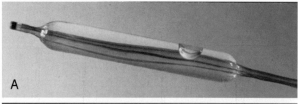
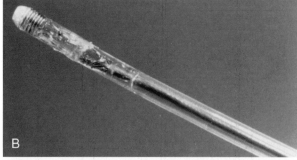
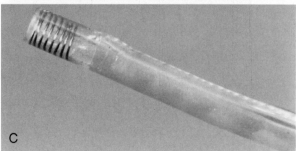

Figure 35–1. Examples of prototype interventional CMR receiver coils. The coil is connected to a coaxial cable running through the instruments. Shown are the distal ends of a balloon catheter (*A*), a guidewire (*B*), and a straight standard 5F catheter (*C*). At the end of each catheter and guidewire is a plug connecting the tracking coil to the CMR system. (From Baert AL, Guenther RW, von Schulthess, et al (eds): Interventional Magnetic Resonance Imaging. New York, Springer-Verlag, 1997. © Springer-Verlag.)

CARDIOVASCULAR MAGNETIC RESONANCE–GUIDED INTRAVASCULAR PROCEDURES—SELECTED EXAMPLES

Cardiovascular Magnetic Resonance Tracking of Guidewire-Catheter Combination

A CMR tracking guidewire compatible with a standard 5F catheter was evaluated in a glass phantom simulating the aorta and its branches.[8] Based upon time-of-flight, maximum intensity projection (TOF MIP) images, (roadmap images depicting the morphology of the phantom), various branches of the phantom were targeted. These experiments demonstrated the ability to actively track two devices simultaneously, accomplished by attaching the coils within each device to separate receivers.[8] The guidewire-catheter composition was also assessed in an in vivo experiment, conducted on a fully anesthetized swine in the usual manner. The combined

guidewire-catheter composition was maneuvered into the superior mesenteric artery, both renal arteries, and the carotid arteries, as well as the right coronary artery, with ease. Similar to angiographic roadmap techniques, the tracking process on roadmap images is sensitive to patient motion. The development of ultrafast imaging techniques[9] will reduce the time required to update such roadmaps. In addition, cyclical motion such as breathing could be compensated for by the use of navigator sequences.[10, 11] Until the implementation of these techniques into an interventional MRA environment, the motion-induced need for time-consuming roadmap updates will remain a significant hurdle.

Embolization

A 5F CMR tracking catheter inserted through the carotid introducer and tracked into the right renal artery based on previously acquired TOF MIP roadmap images. CMR contrast (Gd-DTPA, Schering AG, Berlin, FRG) was injected through the catheter lumen. Prior to, as well as in 10-s intervals following contrast administration, coronal sections were acquired through the kidneys using a fast, multiplanar spoiled gradient echo sequence (FMP-SPGR). These dynamically acquired images revealed isolated early enhancement of the right kidney with subsequent enhancement of the contralateral kidney 30 seconds after contrast administration. Subsequently the right renal artery was embolized by injecting Ethibloc (Ethicon, Germany) embolization material through the lumen of the catheter with its tip lodged in the right renal artery. Dynamic FMP-SPGR imaging was repeated. The images now demonstrated isolated enhancement of the contralateral kidney commencing 10 seconds after contrast application due to backflow out of the occluded right renal arterial system.[12]

Balloon Occlusion and Percutaneous Transluminal Angioplasty

The functionality of balloon inflation was demonstrated by occlusion of a sacral artery in a swine.[13] The vessel was occluded over the length of the balloon. Collaterals filled the vessel distal to the occlusion. The experiment demonstrates the ability to place the balloon in a predefined vascular segment.[13]

The function of the CMR tracking PTA catheter was assessed first in vitro using a flow phantom, as well as in vivo in fully anesthesized swine. For the in vitro experiment, a harvested 12-cm segment of human common femoral artery was used. The stenosis was crossed under CMR tracking guidance. The balloon was inflated, and tracking remained possible. Inflating the balloon reduced the stenosis, as documented by the postdilation MR angiogram.[13]

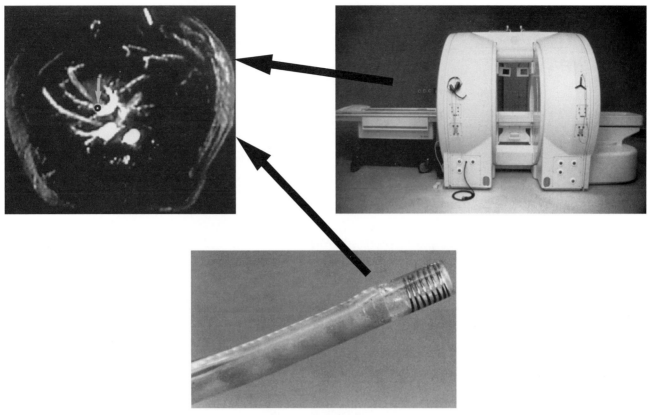

Figure 35–2. Set-up used for instrument tracking combining a CMR road map with a tip localization point (arrow), the position of which can be updated in real time. The tip of the catheter has been placed in the portal vein. The radiofrequency (RF) coil is connected to a coaxial cable, which is interfaced to a workstation. (From Baert AL, Guenther RW, von Schulthess, et al (eds): Interventional Magnetic Resonance Imaging. New York, Springer-Verlag, 1997. © Springer-Verlag.)

To date, five patients have undergone dilation of the common iliac artery without complications. Good results were documented by the postdilation MR angiogram and with conventional angiography (Pfammatter, unpublished data).

Transjugular Intrahepatic Puncture of the Portal System

The ability to track catheters is not limited to the confines of vessels, as shown by the transhepatic puncture of the portal system. The transhepatic puncture of the portal system from the hepatic vein represents the most critical step in the placement of a transjugular intrahepatic portosystemic shunt.[14, 15] With MRA, the perihepatic vascular anatomy is easily displayed in any desired plane. Based on these images, the puncture could be accurately targeted and monitored in real time with CMR tracking. The real-time monitoring ability in two planes allowed for the interactive correction of the needle's course.[12] The ability to guide the transhepatic puncture in real time in two planes simultaneously does emphasize a potential advantage of CMR guidance over conventional techniques: Motion of the instrument can be tracked outside the vascular confines on images displaying the vascular anatomy without the need to repeatedly administer contrast.

Intravascular Magnetic Resonance

In in vitro experiments, intravascular CMR images can depict the vessel wall almost free of artifacts, with high spatial resolution. One of the greatest challenges of intravascular CMR imaging lies in the suppression of motion artifacts due to flow. A receive-only coil, consisting of a single-loop wire, is mounted onto the surface of a balloon and covered with a second balloon. The inflatable balloon allows adjustment of the coil to different sizes, and flow artifacts are reduced due to transient occlusion of the vessel.[16] In vivo experiments in rabbits showed good correlation of the intravascular CMR to the histological specimen.[17] In combination with other recent hardware and software developments, intravascular CMR imaging promises to become an integral element in the overall concept of interventional CMR (Figure 35–3). It is conceivable that the morphological data could be used to guide treatment decisions with regard to PTA and/or stenting.

Outlook for Cardiac Interventions Using Magnetic Resonance

To date, no cardiac interventions in humans are possible under CMR control. There are several prob-

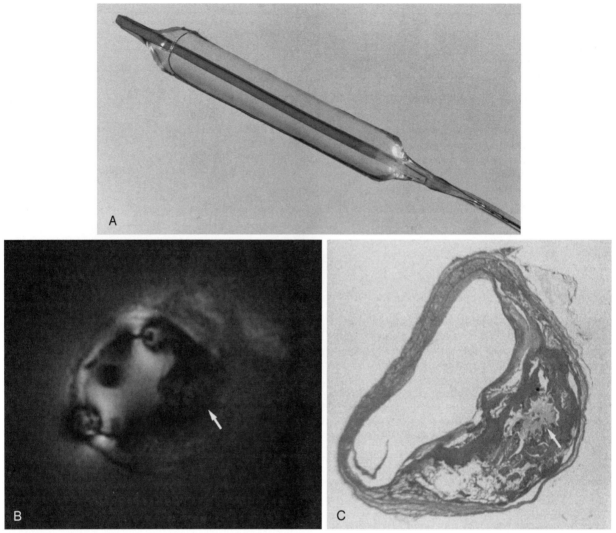

Figure 35–3. *(A),* Balloon-mounted endovascular single-loop prototype coil (6-mm diameter). Ex vivo plaque demonstrating calcifications (arrow) on a T2-weighted image *(B)* correlating to calcifications in the histological section *(C)*. (From Zimmermann GG, Erhart P, Schneider J, et al: Intravascular MR imaging of atherosclerotic plaque: "Ex vivo" analysis of human femoral arteries with histologic correlation. Radiology 204:769, 1997.)

lems to resolve: inadequate depiction of the coronary arteries, motion of the heart, instrument visualization, and access to the patient. Access to the patient can, for example, be addressed with the development of a short-bore CMR system with a high magnetic field. Such an advancement, together with the innovative developments mentioned above and more CMR-compatible instrumentation, will make the heart the next target of interventional CMR. Recent animal work in atheroma imaging, angioplasty, coronary catheterization, and radiofrequency ablation is encouraging.[18–21]

References

1. Prince MR, Yucel EK, Kaufman JA, et al: Dynamic gadolinium-enhanced three-dimensional abdominal MR arteriography. J Magn Reson Imaging 3:877, 1993.
2. Davis CP, Ladd ME, Romanowski BJ, et al: Human aorta: Preliminary results with virtual endoscopy based on three-dimensional MR imaging data sets. Radiology 199:37, 1996.
3. Yang PC, Kerr AB, Liu AC, et al: New real-time interactive cardiac magnetic resonance imaging system complements echocardiography. J Am Coll Cardiol 32:2049, 1998.
4. Dumoulin CL, Souza SP, Darrow RD, et al: Simultaneous acquisition of phase-contrast angiograms and stationary-tissue images with Hadamard encoding of flow-induced phase shifts. J Magn Reson Imaging 1:399, 1991.
5. Leung DA, Debatin JF, Wildermuth S, et al: Intravascular MR tracking catheter: Preliminary experimental evaluation. AJR Am J Roentgenol 164:1265, 1995.
6. Ackerman JL, Offut MC, Buxton RB, Brady TJ: Rapid 3D tracking of small RF coils. Book of Abstracts: Society of Magnetic Resonance in Medicine. Berkeley, CA, 1986, p 1131.
7. Dumoulin CL, Souza SP, Darrow RD: Real-time position monitoring of invasive devices using magnetic resonance. Magn Reson Med 29:411, 1993.
8. Hilfiker PR, Zimmermann GG, Ladd ME, et al: Evaluation of a catheter guidewire concept in an open 0.5 T MR scanner: Comparison of active realtime MR visualization with conventional angiography in an aorta phantom. Cardiovasc Intervent Radiol 20:S101, 1997.
9. Wetter DR, McKinnon GC, Debatin JF, von Schulthess GK:

Cardiac echo-planar MR imaging: Comparison of single- and multiple-shot techniques. Radiology 194:765, 1995.

10. Pelc NJ, Bernstein MA, Shimakawa A, Glover GH: Encoding strategies for three-direction phase-contrast MR imaging of flow. J Magn Reson Imaging 1:405, 1991.

11. Hausmann R, Lewin JS, Laub G: Phase-contrast MR angiography with reduced acquisition time: New concepts in sequence design. J Magn Reson Imaging 1:415, 1991.

12. Wildermuth S, Debatin JF, Leung DA, et al: MR imaging-guided intravascular procedures: Initial demonstration in a pig model. Radiology 202:578, 1997.

13. Wildermuth S, Dumoulin CL, Pfammatter T, et al: MR guided percutaneous angioplasty: Assessment of tracking safety, catheter handling and functionality. Cardiovasc Intervent Radiol 21:404, 1998.

14. Skeens J, Semba C, Dake M: Transjugular intrahepatic portosystemic shunts. Ann Rev Med 46:95, 1995.

15. Roizental M, Kane RA, Takahashi J, et al: Portal vein: US-guided localization prior to transjugular intrahepatic portosystemic shunt placement. Radiology 196:868, 1995.

16. Zimmermann GG, Erhart P, Schneider J, et al: Intravascular MR imaging of atherosclerotic plaque: "Ex vivo" analysis of human femoral arteries with histologic correlation. Radiology 204:769, 1997.

17. Zimmermann GG, Quick HH, Hilfiker PR, et al: Intravascular MR imaging: First in vivo experience with an arteriosclerotic animal model. Fortschr Röntgenstr 162:180, 1997.

18. Yang X, Bolster BD, Kraitchman DL, Atalar E: Intravascular MR-monitored balloon angioplasty: An in vivo feasibility study. J Vasc Interv Radiol 9:953, 1998.

19. Correia LCL, Atalar E, Kelemen MD, et al: Intravascular magnetic resonance imaging of aortic atherosclerotic plaque composition. Arterioscler Thromb Vasc Biol 17:3926, 1997.

20. Lardo AC, McVeigh ER, Jumrussirikul P, et al: Visualization and temporal/spatial characterization of cardiac radiofrequency ablation lesions using magnetic resonance imaging. Circulation 102:698, 2000.

21. Serfaty JM, Yang X, Aksit P, et al: Toward MRI-guided coronary catheterisation: Visualization of guiding catheters, guidewires, and anatomy in real-time. J Magn Reson Imaging 12:590, 2000.

Common Abbreviations Used in the Text

1-D	one-dimensional
2-D	two-dimensional
3-D	three-dimensional
AA	ascending thoracic aorta
ACE	angiotensin converting enzyme
ASD	atrial septal defect
ATP	adenosine triphosphate
AV	atrioventricular
BOLD	blood oxygen level dependent
C	Celsius
ceMRA	contrast-enhanced magnetic resonance angiography
CK	creatine kinase
CMR	cardiovascular magnetic resonance
CNR	contrast-to-noise ratio
CO	cardiac output
CSPAMM	complementary spatial modulation of magnetization
CT	computed tomography
CTA	computed tomography angiography
CXR	chest x-ray
DA	descending thoracic aorta
EBCT	electron beam computed tomography
ECG	electrocardiogram
ED	end-diastolic
EF	ejection fraction
EPI	echo-planar imaging
ES	end-systolic
FARM	fast acquisition relaxation mapping
FDG	fluorodeoxyglucose
FID	free induction decay
FISP	fast imaging with steady state precession
FOV	field of view
FSE	fast spin echo
Gd-DTPA	gadolinium diethylenetriamine pentaacetic acid
GRE	gradient recalled echo
^{1}H	hydrogen/proton
Hct	hematocrit
IR	inversion recovery
K	kelvin
LA	left atrium
LAD	left anterior descending coronary artery
LCX	left circumflex coronary artery
LM	left main coronary artery
LV	left ventricle
LVEDV	left ventricular end-diastolic volume
LVEDVI	left ventricular end-diastolic volume index

LVEF	left ventricular ejection fraction
LVESV	left ventricular end-systolic volume
LVH	left ventricular hypertrophy
LVOT	left ventricular outflow tract
MR	magnetic resonance
MRA	magnetic resonance angiography
MRI	magnetic resonance imaging
MRS	magnetic resonance spectroscopy
ms	milliseconds
MTC	magnetization transfer contrast
MTT	mean transit time
MVO_2	maximum oxygen consumption
NS	not significant
NSA	number of signal averages
NYHA	New York Heart Association
^{31}P	phosphorus-31
PA	main pulmonary artery
PC	phase contrast
PCr	phosphocreatine
PET	positron emission tomography
Pi	inorganic phosphate
ppm	parts per million
Qp	pulmonic flow
Qs	systemic flow
RA	right atrium
RCA	right coronary artery
RD (+)	redistribution (+)
RD (−)	redistribution (−)
RF	radiofrequency
RHD	right hemidiaphragm
ROI	region of interest
RV	right ventricle
s	seconds
SE	spin echo
SI	signal intensity
SNR	signal-to-noise ratio
SPAMM	spatial modulation of magnetization
SPECT	single-photon emission computed tomography
SSFP	steady state free precession
SV	stroke volume
T	Tesla
TE	echo time
TEE	transesophageal echocardiogram
TGA	transposition of the great arteries
^{201}Tl	thallium-201
TOF	time of flight
TR	repetition time
TSE	turbo spin echo
TTC	2,3,5-triphenyltetrazolium chloride
VSD	ventricular septal defect

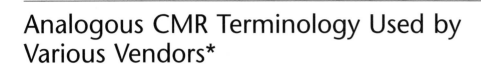

Analogous CMR Terminology Used by Various Vendors*

Generic	General Electric	Marconi	Philips	Siemens	Toshiba
Spin Echo	SE	SE	SE	SE	SE
Fast (Turbo) Spin Echo	FSE	FSE	TSE	TSE	
Gradient Echo	GRASS, GRE, MPGR	FAST-II	FFE	GRE	FE
Fast (Turbo) Gradient Echo	RAPID SPGR	T1-FAST	TFE	TurboFLASH	FFE
Steady State Free Precession	Fiesta		Balanced FFE	TrueFISP	
Time-of-flight Angio	2D, 3D TOF	TOF	2D, 3D Inflow	TOF	2D, 3D TOF
Phase Contrast	2D, 3D PCA	PC	2D, 3D PCA	PC	2D, 3D PSI
Angiography Options					
MOTSA	MOTSA	Slab Tracking	Multichunk	Multislab or MOTSA	Multi-coverage
Dynamic Contrast Tracking	SmartPrep		BolusTrak	Care Bolus	VisualPrep
Suppression Techniques					
Inversion Recovery	STIR, fseSTIR, fastFlair	IR, STIR, FLAIR	STIR, TSE-STIR, TSE-Flair	IR, STIR, IR-TSE	fastSTIR, fastFLAIR, fastIR
Fat Suppression	FatSAT	FatSAT	SPIR	FatSAT, STIR	WFS, WFOP, MSOFT, DIET
Other					
Auto Shim	Auto Shim		Dynamic FOV Shimming	Auto Shim	AAS
Half-Fourier	Fractional NEX	Phase Conjugate Syn.	Half-Scan	Half-Fourier	HFI, AFI
Asymmetric Echo	Fractional Echo	Read Conjugate Syn.	Partial Echo	Optimized Bandwidth	Matched BW
Bandwidth	RBW	Read Conjugate Syn.	WFS	OPT. BW	BW, LCS
Surface Coil Internal Correction	Image Intensity Correction		Homogeneity Correction	Normalize Filter	Surface coil internal correction
Reduced Acquisition	Asymmetric FOV	RFOV	RFOV	RFOV	RFOV
Phase Oversampling	NPW	Oversampling	Foldover Suppression	Oversampling	PNW
Navigator Echoes	Navigator Echoes		Navigator Echoes	Navigator Echoes	

MOTSA, Multiple Overlapping Thin Slab Acquisition.
*With appreciation to Dr. Christine Lorenz.

Index

ISBN 0-443-07519-0

DATE DUE

FEB '04